Biology

EIGHTH EDITION

Volume 2

Eldra P. Solomon
University of South Florida

Linda R. Berg
St. Petersburg College

Diana W. Martin
Rutgers University

THOMSON
™
BROOKS/COLE

Australia • Canada • Mexico • Singapore • Spain • United Kingdom • United States

THOMSON
BROOKS/COLE

Biology, Eighth Edition, Volume 2
Eldra P. Solomon, Linda R. Berg, and Diana W. Martin

Publisher: PETER ADAMS

Development Editor: SUZANNAH ALEXANDER

Assistant Editor: LAUREN OLIVEIRA

Editorial Assistant: KATE FRANCO

Technology Project Manager: KELI AMANN

Marketing Manager: KARA KINDSTROM

Marketing Assistant: CATIE RONQUILLO

Marketing Communications Manager: BRYAN VANN

Project Manager, Editorial Production: CHERYLL LINTHICUM,
    JENNIFER RISDEN

Creative Director: ROB HUGEL

Art Director: LEE FRIEDMAN

Print Buyer: JUDY INOUYE

Permissions Editor: BOB KAUSER

Production Service: JAMIE ARMSTRONG, NEWGEN–AUSTIN

Text Designer: JOHN WALKER

Photo Researcher: KATHLEEN OLSON

Copy Editor: CYNTHIA LINDLOF

Illustrator: PRECISION GRAPHICS, NEWGEN–AUSTIN

Cover Designer: ROBIN TERRA

Cover Image: GAIL SHUMWAY/TAXI/GETTY IMAGES

Cover Printer: QUEBECOR WORLD/VERSAILLES

Compositor: NEWGEN–AUSTIN

Printer: QUEBECOR WORLD/VERSAILLES

Thomson Higher Education
10 Davis Drive
Belmont, CA 94002-3098
USA

Library of Congress Control Number: 2006924995

Volume 2: ISBN 13: 978-0-495-30979-6

Volume 2: ISBN 10: 0-495-30979-6

DEDICATION

To our families, friends, and colleagues who gave freely of their love, support, knowledge, and time as we prepared this eighth edition of *Biology*

ESPECIALLY TO
Rabbi Theodore and Freda Brod
Alan and Jennifer
Chuck and Margaret

ABOUT THE AUTHORS

ELDRA P. SOLOMON has written several leading college-level textbooks in biology and in human anatomy and physiology. Her books have been translated into more than 10 languages. Dr. Solomon earned an M.S. from the University of Florida and an M.A. and Ph.D. from the University of South Florida. Dr. Solomon taught biology and nursing students for more than 20 years. She is adjunct professor and member of the Graduate Faculty of the University of South Florida.

In addition to being a biologist and science author, Dr. Solomon is a bio-psychologist with a special interest in the neurophysiology of traumatic experience. Her research has focused on the relationships among stress, emotions, and health and on post-traumatic stress disorder.

Dr. Solomon has presented her work in plenary sessions and scientific meetings at many national and international conferences. She has been profiled more than 20 times in leading publications, including *Who's Who in America*, *Who's Who in Science and Engineering*, *Who's Who in Medicine and Healthcare*, *Who's Who in American Education*, *Who's Who of American Women*, and *Who's Who in the World*.

LINDA R. BERG is an award-winning teacher and textbook author. She received a B.S. in science education, an M.S. in botany, and a Ph.D. in plant physiology from the University of Maryland. Her research focused on the evolutionary implications of steroid biosynthetic pathways in various organisms.

Dr. Berg taught at the University of Maryland at College Park for 17 years and at St. Petersburg College in Florida for 8 years. During her career, she taught introductory courses in biology, botany, and environmental science to thousands of students. At the University of Maryland, she received numerous teaching and service awards. Dr. Berg is also the recipient of many national and regional awards, including the National Science Teachers Association Award for Innovations in College Science Teaching, the Nation's Capital Area Disabled Student Services Award, and the Washington Academy of Sciences Award in University Science Teaching.

During her career as a professional science writer, Dr. Berg has authored or co-authored several leading college science textbooks. Her writing reflects her teaching style and love of science.

DIANA W. MARTIN is the Director of General Biology, Division of Life Sciences, at Rutgers University, New Brunswick Campus. She received an M.S. at Florida State University, where she studied the chromosomes of related plant species to understand their evolutionary relationships. She earned a Ph.D. at the University of Texas at Austin, where she studied the genetics of the fruit fly, *Drosophila melanogaster,* and then conducted postdoctoral research at Princeton University. She has taught general biology and other courses at Rutgers for more than 20 years and has been involved in writing textbooks since 1988. She is immensely grateful that her decision to study biology in college has led to a career that allows her many ways to share her excitement about all aspects of biology.

Brief Contents

Contents

Part 3   THE CONTINUITY OF LIFE: GENETICS   211

10   CHROMOSOMES, MITOSIS, AND MEIOSIS   211

11   THE BASIC PRINCIPLES OF HEREDITY   234

12   DNA: THE CARRIER OF GENETIC INFORMATION   260

Contents   **xvii**

Preface

This eighth edition of Solomon, Berg, Martin *Biology* continues to convey our vision of the dynamic science of biology and how it affects every aspect of our lives, from our health and behavior to the challenging environmental issues that confront us. Recent discoveries in the biological sciences have increased our understanding of both the unity and diversity of life's processes and adaptations. With this understanding, we have become more aware of our interdependence with the vast diversity of organisms with which we share planet Earth.

BIOLOGY IS A BOOK FOR STUDENTS

The study of biology can be an exciting journey of discovery for beginning biology students. To that end, we seek to help students better appreciate Earth's diverse organisms, their remarkable adaptations to the environment, and their evolutionary and ecological relationships. We also want students to appreciate the workings of science and the contributions of scientists whose discoveries not only expand our knowledge of biology but also help shape and protect the future of our planet. *Biology* provides insight into what science is, how scientists work, what the roles are of the many scientists who have contributed to our current understanding of biology, and how scientific knowledge affects daily life.

Since the first edition of *Biology*, we have tried to present the principles of biology in an integrated way that is accurate, interesting, and conceptually accessible to students. In this eighth edition of *Biology*, we continue this tradition. We also continue to present biology in an inquiry-based framework. Many professors interpret inquiry as a learning method that takes place in the laboratory as students perform experiments. Laboratory research is certainly an integral part of inquiry-based learning. But inquiry is also a way of learning in which the student actively pursues some line of knowledge outside the laboratory. In *Biology*, we have always presented the history of scientific advances, including scientific debates, to help students understand that science is a process (that is, a field of investigative inquiry) as well as a body of knowledge (the product of inquiry). **NEW** In *Biology*, Eighth Edition, we make a concerted effort to further integrate inquiry-based learning into the textbook with the inclusion of two new features: *Key Experiment* figures and *Analyzing Data* questions. *Key Experiment* figures encourage students to evaluate investigative approaches that real scientists have taken.

Analyzing Data **questions** require students to actively interpret experimental data presented in the text.

Throughout the text, we stimulate interest by relating concepts to experiences within the student's frame of reference. By helping students make such connections, we facilitate their mastery of general concepts. We hope the combined effect of an engaging writing style and interesting features will fascinate students and encourage them to continue their study of biology.

THE SOLOMON/BERG/MARTIN LEARNING SYSTEM

In the eighth edition we have refined our highly successful Learning System, which concentrates on learning objectives and outcomes. Mastering biology is challenging, particularly because the subject of biology is filled with so many facts that must be integrated into the framework of general biological principles. To help students, we provide Learning Objectives both for the course and for each major section of every chapter. At the end of each section, we provide Review questions based on the learning objectives so students can assess their mastery of the material presented in the section.

Throughout the book, students are directed to Thomson-NOW, a powerful online diagnostic tool that helps students assess their study needs and master the chapter objectives. After taking a pretest on ThomsonNOW, students receive feedback based on their answers as well as a Personalized Study plan with links to animations and other resources keyed to their specific learning needs. Selected illustrations in the text are also keyed to *Animated* figures in ThomsonNOW.

Course Learning Objectives

At the end of a successful study of introductory biology, the student can demonstrate mastery of the subject by responding accurately to the following Course Learning Objectives, or Outcomes:

- Design an experiment to test a given hypothesis, using the procedure and terminology of the scientific method.

- Cite the cell theory, and relate structure to function in both prokaryotic and eukaryotic cells.

- Describe the theory of evolution, explain why it is the principal unifying concept in biology, and discuss natural selection as the primary agent of evolutionary change.

- Explain the role of genetic information in all species, and discuss applications of genetics that affect society.

- Describe several mechanisms by which cells and organisms transfer information, including the use of nucleic acids, chemical signals (such as hormones and pheromones), electrical signals (for example, neural transmission), signal transduction, sounds, and visual displays.

- Argue for or against the classification of organisms in three domains and six kingdoms, characterizing each of these clades; based on your knowledge of genetics and evolution, give specific examples of the unity and diversity of these organisms.

- Compare the structural adaptations, life processes, and life cycles of a prokaryote, protist, fungus, plant, and animal.

- Define *homeostasis,* and give examples of regulatory mechanisms, including feedback systems.

- Trace the flow of matter and energy through a photosynthetic cell and a nonphotosynthetic cell, and through the biosphere, comparing the roles of producers, consumers, and decomposers.

- Describe the study of ecology at the levels of an individual organism, a population, a community, and an ecosystem.

Pedagogical Features

We use numerous learning strategies to increase the student's success:

- **NEW** A list of *Key Concepts* at the beginning of each chapter provides a chapter overview.

- *Learning Objectives* at the beginning of each major section in the chapter indicate, in behavioral terms, what the student must do to demonstrate mastery of the material in that section.

- Each major section of the chapter is followed by a series of *Review* questions that assess comprehension by asking the student to describe, explain, compare, contrast, or illustrate important concepts. The Review questions are based on the section Learning Objectives.

- *Concept Statement Subheads* introduce sections, previewing and summarizing the key idea or ideas to be discussed in that section.

- *Sequence Summaries* within the text simplify and summarize information presented in paragraph form. For example, paragraphs describing blood circulation through the body or the steps by which cells take in certain materials are followed by a Sequence Summary listing the structures or steps.

- *Focus On* boxes explore issues of special relevance to students, such as the effects of smoking or alcohol abuse. These boxes also provide a forum for discussing certain topics of current interest in more detail, such as the smallest ancient humans (*Homo floresiensis*), worldwide declining amphibian populations, and Alzheimer's disease.

- Numerous *Tables,* many illustrated, help the student organize and summarize material presented in the text.

- A *Summary with Key Terms* at the end of each chapter is organized around the chapter Learning Objectives. This summary provides a review of the material, and because selected key terms are boldface in the summary, students are provided the opportunity to study vocabulary words within the context of related concepts.

- End-of-chapter questions provide students with the opportunity to evaluate their understanding of the material in the chapter. *Test Your Understanding* consists of multiple-choice questions, some of which are based on the recall of important terms, whereas others challenge students to integrate their knowledge. Answers to the Test Your Understanding questions are provided in Appendix E. A series of thought-provoking *Critical Thinking* questions encourages the student to apply the concepts just learned to new situations or to make connections among important concepts. **NEW** Every chapter has one or more *Evolution Link* questions in the Critical Thinking section. **NEW** Many chapters contain one or more *Analyzing Data* questions based on data presented in the chapter.

- The *Glossary* at the end of the book, the most comprehensive glossary found in any biology text, provides precise definitions of terms. The Glossary is especially useful because it is extensively cross-referenced and includes pronunciations. The vertical blue bar along the margin facilitates rapid access to the Glossary. The companion website also includes glossary flash cards with audio pronunciations.

- **NEW** An updated and expanded *art program* brings to life, reinforces, and expands concepts discussed in the text. This edition includes *Key Experiment* figures, which emphasize the scientific process in both classic and modern research; examples include Figures 9-7, 17-2, 36-7, and 50-8. Also new to this edition are *Key Points* stated in process diagrams of complex topics; examples include Figures 28-19 and 54-2. Many of these figures have numbered parts that show sequences of events in biological processes or life cycles. Numerous photographs, both alone and combined with line art, help students grasp concepts by connecting the "real" to the "ideal." The line art uses devices such as *orientation icons* to help the student put the detailed figures into the broad context. We use symbols and colors consistently throughout the book to help students connect concepts. For example, the same four colors and shapes are used throughout the book to identify guanine, cytosine, adenine, and thymine.

WHAT'S NEW: AN OVERVIEW OF *BIOLOGY*, EIGHTH EDITION

Three themes are interwoven throughout *Biology*: the evolution of life, the transmission of biological information, and the flow of energy through living systems. As we introduce the concepts of modern biology, we explain how these themes are connected and how life depends on them.

Educators present the major topics of an introductory biology course in a variety of orders. For this reason, we carefully designed the eight parts of this book so that they do not depend heavily on preceding chapters and parts. This flexible organization means that an instructor can present the 56 chapters in any number of sequences with pedagogical success. Chapter 1, which introduces the student to the major principles of biology, provides a good springboard for future discussions, whether the professor prefers a "top-down" or "bottom-up" approach.

In this edition, as in previous editions, we examined every line of every chapter for accuracy and currency, and we made a serious attempt to update every topic and verify all new material. The following brief survey provides a general overview of the organization of *Biology* and some changes made to the eighth edition.

Part 1: The Organization of Life

The six chapters that make up Part 1 provide basic background knowledge. We begin Chapter 1 with a new discussion of research on heart repair and then introduce the main themes of the book—evolution, energy transfer, and information transfer. Chapter 1 then examines several fundamental concepts in biology and the nature of the scientific process, including a discussion of systems biology. Chapters 2 and 3, which focus on the molecular level of organization, establish the foundations in chemistry necessary for understanding biological processes. Chapters 4 and 5 focus on the cell level of organization. Chapter 4 contains additional information on microtubules, microfilaments, and motor proteins; Chapter 5 has expanded coverage of transport proteins and a new section on uniporters, symporters, and antiporters.

NEW In the eighth edition we have added a new chapter, Chapter 6, on cell communication, because recent studies of cell signaling, including receptor function and signal transduction, are providing new understanding of many life processes, particularly at the cell level.

Part 2: Energy Transfer through Living Systems

Because all living cells need energy for life processes, the flow of energy through living systems—that is, capturing energy and converting it to usable forms—is a basic theme of *Biology*. Chapter 7 examines how cells capture, transfer, store, and use energy. Chapters 8 and 9 discuss the metabolic adaptations by which organisms obtain and use energy through cellular respiration and photosynthesis. New to Chapter 9 is a summary section on the importance of photosynthesis to plants and other organisms.

Part 3: The Continuity of Life: Genetics

We have updated and expanded the eight chapters of Part 3 for the eighth edition. We begin this unit by discussing mitosis and meiosis in Chapter 10, which includes new sections on binary fission and prometaphase in mitosis. Chapter 11, which consid-

ers Mendelian genetics and related patterns of inheritance, has a new section on recognition of Mendel's work, including discussion of the chromosome theory of inheritance. We then turn our attention to the structure and replication of DNA in Chapter 12, including a new section on proofreading and repair of errors in DNA. Chapter 13 has a discussion of RNA and protein synthesis, including a new section and summary table on the different kinds of eukaryotic RNA. Gene regulation is discussed in Chapter 14, which includes new paragraphs on genomic imprinting and epigenetic inheritance. In Chapter 15, we focus on DNA technology and genomics, including new material on DNA analysis, genomics, expressed sequence tags, RNAi, and the importance of GM crops to U.S. agriculture. These chapters build the necessary foundation for exploring the human genome in Chapter 16, which has expanded discussion of abnormalities in chromosome structure and on X-linked genes affecting intelligence. In Chapter 17, we introduce the role of genes in development, emphasizing studies on specific model organisms that have led to spectacular advances in this field; changes include new material on RNAi as a powerful tool in developmental genetics and discussion of cancer as a stem cell disease.

NEW The art program in the genetics section includes many new pieces, such as Figures 10-8 (cohesins), 10-11 (binary fission), 12-14 (DNA repair), 13-11 (linkage of amino acid to its specific RNA), 13-18 (photo of plants with altered miRNA), 14-1 (lean pig with *IGF2* mutation), 14-13 (protein degradation by ubiquitin-proteasomes), 15-9 (Southern blotting technique), 15-15 (DNA fingerprinting), 16-5 (common abnormalities in chromosome structure), 16-10 (gene therapy in mouse bone marrow cells), and 17-6 (six model organisms).

Part 4: The Continuity of Life: Evolution

Although we explore evolution as the cornerstone of biology throughout the book, Part 4 delves into the subject in depth. We provide the history behind the discovery of the theory of evolution, the mechanisms by which it occurs, and the methods by which it is studied and tested. Chapter 18 introduces the Darwinian concept of evolution and presents several kinds of evidence that support the theory of evolution. In Chapter 19, we examine evolution at the population level. Chapter 20 describes the evolution of new species and discusses aspects of macroevolution. Chapter 21 summarizes the evolutionary history of life on Earth. In Chapter 22, we recount the evolution of the primates, including humans. Many topics and examples have been added to the eighth edition, such as new material on developmental biology as evidence for evolution in Galápagos finches, the founder effect and genetic drift with respect to Finns and Icelanders, ways of defining a species, the Archaeon and Proterozoic eons, and a new *Focus On* box on *Homo floresiensis*.

Part 5: The Diversity of Life

In this edition of *Biology,* we continue to emphasize the cladistic approach. We use an evolutionary framework to discuss each group of organisms, presenting current hypotheses of how groups of organisms are related. Chapter 23 discusses *why* organisms are classified and provides insight into the scientific process of deciding *how* they are classified. New advances have enabled us to further clarify the connection between evolutionary history and systematics in the eighth edition. Chapter 24, which focuses on the viruses and prokaryotes, contains a major revision of prokaryote classification and new sections on evolution in bacteria and on biofilms. Chapter 25 reflects the developing consensus on protist diversity and summarizes evolutionary relationships among the eight major eukaryote groups. Chapter 26 describes the fungi and includes an updated phylogeny with a discussion of members of phylum Glomeromycota and their role as mycorrhizal fungi. Chapters 27 and 28 present the members of the plant kingdom; Chapter 28 includes new material that updates the evolution of the angiosperms. In Chapters 29 through 31, which cover the diversity of animals, we place more emphasis on molecular systematics. We have also added new sections on animal origins and the evolution of development.

Part 6: Structure and Life Processes in Plants

Part 6 introduces students to the fascinating plant world. It stresses relationships between structure and function in plant cells, tissues, organs, and individual organisms. In Chapter 32, we introduce plant structure, growth, and differentiation. Chapters 33 through 35 discuss the structural and physiological adaptations of leaves, stems, and roots. Chapter 36 describes reproduction in flowering plants, including asexual reproduction, flowers, fruits, and seeds. Chapter 37 focuses on growth responses and regulation of growth. In the eighth edition, we present the latest findings generated by the continuing explosion of knowledge in plant biology, particularly at the molecular level. New topics include the molecular basis of mycorrhizal and rhizobial symbioses, molecular aspects of floral initiation at apical meristems, self-incompatibility, and the way that auxin acts by signal transduction. Chapter 36 also discusses a new experiment figure on the evolutionary implications of a single mutation in a gene for flower color that resulted in a shift in animal pollinators.

Part 7: Structure and Life Processes in Animals

In Part 7, we provide a strong emphasis on comparative animal physiology, showing the structural, functional, and behavioral adaptations that help animals meet environmental challenges. We use a comparative approach to examine how various animal groups have solved similar and diverse problems. In Chapter 38, we discuss the basic tissues and organ systems of the animal body, homeostasis, and the ways that animals regulate their body temperature. Chapter 39 focuses on body coverings, skeletons, and muscles. In Chapters 40 through 42, we discuss neural signaling, neural regulation, and sensory reception. In Chapters 43 through 50, we compare how different animal groups carry on specific life processes, such as internal transport, internal defense, gas exchange, digestion, reproduction, and development. Each chapter in this part considers the human adaptations for the life processes being discussed. Part 7 ends with a discussion of

behavioral adaptations in Chapter 51. Reflecting recent research findings, we have added new material on homeostasis and the process of contraction in whole muscle. We have also updated and added new material on neurotransmitters, information processing, thermoreceptors, electroreceptors and magnetoreception, endocrine regulation of blood pressure, Toll-like receptors, cytokines, cancer treatment, HIV and autoimmune diseases, new nutrition guidelines, obesity, maintenance of fluid and electrolyte balance, steroid action that allows rapid signaling, melanocyte-stimulating hormones (MSH), diabetes, anabolic steroids, menopause, the estrous cycle, contraception, and STDs. In Chapter 51, we have added new information on cognition, communication, interspecific and intraspecific selection, parental care, and helping behavior.

Part 8: The Interactions of Life: Ecology

Part 8 focuses on the dynamics of populations, communities, and ecosystems and on the application of ecological principles to disciplines such as conservation biology. Chapters 52 through 55 give the student an understanding of the ecology of populations, communities, ecosystems, and the biosphere, whereas Chapter 56 focuses on global environmental issues. Among the many changes in this unit are new material on secondary productivity, top-down and bottom-up processes, characteristics of endangered species, deforestation, and a greatly expanded section on conservation biology.

A COMPREHENSIVE PACKAGE FOR LEARNING AND TEACHING

A carefully designed supplement package is available to further facilitate learning. In addition to the usual print resources, we are pleased to present student multimedia tools that have been developed in conjunction with the text.

Resources for Students

***Study Guide to Accompany* Biology, Eighth Edition,** by Ronald S. Daniel of California State Polytechnic University, Pomona; Sharon C. Daniel of Orange Coast College; and Ronald L. Taylor. Extensively updated for this edition, the study guide provides the student with many opportunities to review chapter concepts. Multiple-choice study questions, coloring-book exercises, vocabulary-building exercises, and many other types of active-learning tools are provided to suit different cognitive learning styles.

A Problem-Based Guide to Basic Genetics by Donald Cronkite of Hope College. This brief guide provides students with a systematic approach to solving genetics problems, along with numerous solved problems and practice problems.

Spanish Glossary. **NEW** This Spanish glossary of biology terms is available to Spanish-speaking students.

Website. The content-rich companion website that accompanies *Biology*, Eighth Edition, gives students access to high-quality resources, including focused quizzing, a Glossary complete with pronunciations, *InfoTrac® College Edition* readings and exercises, Internet activities, and annotated web links. For these and other resources, visit www.thomsonedu.com/biology/solomon.

ThomsonNOW. NEW This updated and expanded online learning tool helps assess students' personal study needs and focus their time. By taking a pretest, they are provided with a Personalized Study plan that directs them to text sections and narrated animations—many of which are new to this edition—that they need to review. If they need to brush up on basic skills, the **How Do I Prepare?** feature walks them through tutorials on basic math, chemistry, study skills, and word roots.

Audio. NEW This edition of *Biology* is accompanied by a range of "mobile content" resources, including downloadable study skill tips and concept reviews in MP3 format for use on a portable MP3 player.

Virtual Biology Laboratory 3.0 by Beneski and Waber. **NEW** These 14 online laboratory experiments, designed within a simulation format, allow students to "do" science by acquiring data, performing experiments, and using data to explain biological concepts. Assigned activities automatically flow to the instructor's grade book. New self-designed activities ask students to plan their procedures around an experimental question and write up their results.

Additional Resources for Instructors

The instructors' Examination Copy for this edition lists a comprehensive package of print and multimedia supplements, including online resources, available to qualified adopters. Please ask your local sales representative for details.

ACKNOWLEDGMENTS

The development and production of the eighth edition of *Biology* required extensive interaction and cooperation among the authors and many individuals in our family, social, and professional environments. We thank our editors, colleagues, students, family, and friends for their help and support. Preparing a book of this complexity is challenging and requires a cohesive, talented, and hardworking professional team. We appreciate the contributions of everyone on the editorial and production staff at Brooks/Cole/Thomson Learning who worked on this eighth edition of *Biology*. We thank Michelle Julet, Vice President and Editor-in-Chief, and Peter Adams, Executive Editor, for their commitment to this book and for their support in making the eighth edition happen. We appreciate Stacy Best and Kara Kindstrom, our Marketing Managers, whose expertise ensured that you would know about our new edition.

We appreciate the hard work of our dedicated Developmental Editor, Suzannah Alexander, who provided us with valuable input as she guided the eighth edition through its many phases. We especially thank Suzannah for sharing her artistic talent and for her great ideas for visual presentations. We appreciate the help of Senior Content Project Manager Cheryll Linthicum, Content Project Manager Jennifer Risden, and Project Editor Jamie Armstrong, who expertly shepherded the project. We thank Editorial Assistants Kristin Marrs and Kate Franco for quickly providing us with resources whenever we needed them. We appreciate the efforts of photo editor Don Murie. We thank Creative Director Rob Hugel, Art Director Lee Friedman, Text Designer John Walker, and Cover Designer Robin Terra. We also thank Joy Westberg for developing the Instructor's Preface.

We are grateful to Keli Amann, Technology Project Manager, who coordinated the many high-tech components of the computerized aspects of our Learning System. We thank Lauren Oliveira, Assistant Editor, for coordinating the print supplements. These dedicated professionals and many others at Brooks/Cole provided the skill, attention, and good humor needed to produce *Biology*, Eighth Edition. We thank them for their help and support throughout this project.

A **Biology Advisory Board** greatly enhanced our preparation of the eighth edition of *Biology*. We thank these professionals for their insight and suggestions:

Susan R. Barnum, Miami University, Oxford, OH, Molecular biology of cyanobacteria

Virginia McDonough, Hope College, Molecular biology of lipid metabolism

David K. Bruck, San Jose State University, Plant cell biology

Lee F. Johnson, The Ohio State University, Molecular genetics, biochemistry

Karen A. Curto, University of Pittsburgh, Cell signaling

Robert J. Kosinski, Clemson University, Introductory biology laboratory development

Tim Schuh, St. Cloud State University, Developmental biology

Robert W. Yost, Indiana University–Purdue University Indianapolis, Physiology, animal form and function

We are grateful to Bruce Mohn of Rutgers University for his expert advice on dinosaur and bird evolution and to Bill Norris of Western New Mexico University for his help with the book's cladograms. We thank Dr. Susan Pross, University of South Florida, College of Medicine, for her helpful suggestions for updating the immunology chapter. We greatly appreciate the expert assistance of Mary Kay Hartung of Florida Gulf Coast University, who came to our rescue whenever we had difficulty finding needed research studies from the Internet. We thank doctoral student Lois Ball of the University of South Florida, Department of Biology, who reviewed several chapters and offered helpful suggestions. We thank obstetrician, gynecologist Dr. Amy Solomon for her input regarding pregnancy, childbirth, conception, and sexually transmitted diseases.

We thank our families and friends for their understanding, support, and encouragement as we struggled through many revisions and deadlines. We especially thank Mical Solomon, Dr. Amy Solomon, Dr. Kathleen M. Heide, Alan Berg, Jennifer Berg, Dr. Charles Martin, and Margaret Martin for their support and input. We greatly appreciate the many hours Alan Berg devoted to preparing the manuscript.

Our colleagues and students who have used our book have provided valuable input by sharing their responses to past editions of *Biology.* We thank them and ask again for their comments and suggestions as they use this new edition. We can be reached through the Internet at our website www.thomsonedu.com/ biology/solomon or through our editors at Brooks/Cole, a division of Thomson Learning.

We express our thanks to the many biologists who have read the manuscript during various stages of its development and provided us with valuable suggestions for improving it. Eighth edition reviewers include the following:

Joseph J. Arruda, Pittsburg State University

Amir M. Assadi-Rad, San Joaquin Delta College

Douglas J. Birks, Wilmington College

William L. Bischoff, University of Toledo

Catherine S. Black, Idaho State University

Andrew R. Blaustein, Oregon State University

Scott Bowling, Auburn University

W. Randy Brooks, Florida Atlantic University

Mark Browning, Purdue University

Arthur L. Buikema Jr., Virginia Tech

Anne Bullerjahn, Owens Community College

Carolyn J. W. Bunde, Idaho State University

Scott Burt, Truman State University

David Byres, Florida Community College–Jacksonville

Jeff Carmichael, University of North Dakota

Domenic Castignetti, Loyola University of Chicago

Geoffrey A. Church, Fairfield University

Barbara Collins, California Lutheran University

Linda W. Crow, Montgomery College

Karen J. Dalton, Community College of Baltimore County

Mark Decker, University of Minnesota

Jonathan J. Dennis, University of Alberta

Philippa M. Drennan, Loyola Marymount University

David W. Eldridge, Baylor University

H. W. Elmore, Marshall University

Cheryld L. Emmons, Alfred University

Robert C. Evans, Rutgers University, Camden

John Geiser, Western Michigan University

William J. Higgins, University of Maryland

Jeffrey P. Hill, Idaho State University

Walter S. Judd, University of Florida

Mary Jane Keleher, Salt Lake Community College

Scott L. Kight, Montclair State University

Joanne Kivela Tillotson, Purchase College SUNY

Will Kleinelp, Middlesex County College

Kenneth M. Klemow, Wilkes University

Jonathan Lyon, Merrimack College

Blasé Maffia, University of Miami

Kathleen R. Malueg, University of Colorado, Colorado Springs

Patricia Matthews, Grand Valley State University

David Morgan, Western Washington University

Darrel L. Murray, University of Illinois at Chicago

William R. Norris, Western New Mexico University

James G. Patton, Vanderbilt University

Mitch Price, Penn State

Susan Pross, University of South Florida

Jerry Purcell, San Antonio College

Kenneth R. Robinson, Purdue University

Darrin Rubino, Hanover College

Julie C. Rutherford, Concordia College

Andrew M. Scala, Dutchess Community College

Pramila Sen, Houston Community College

Mark A. Sheridan, North Dakota State University

Marcia Shofner, University of Maryland

Phillip Snider, Gadsden State Community College

David Stanton, Saginaw Valley State University

William Terzaghi, Wilkes University

Keti Venovski, Lake Sumter Community College

Steven D. Wilt, Bellarmine University

James R. Yount, Brevard Community College

We would also like to thank the hundreds of previous edition reviewers, both professors and students, who are too numerous to mention. Without their contributions, *Biology,* Eighth Edition, would not have been the same. They asked thoughtful questions, provided new perspectives, offered alternative wordings to clarify difficult passages, and informed us of possible errors. We are truly indebted to their excellent feedback.

To the Student

Biology is a challenging subject. The thousands of students we have taught have differed in their life goals and learning styles. Some have had excellent backgrounds in science; others, poor ones. Regardless of their backgrounds, it is common for students taking their first college biology course to find that they must work harder than they expected. You can make the task easier by using approaches to learning that have been successful for a broad range of our students over the years. Be sure to use the Learning System we use in this book. It is described in the Preface.

Make a Study Schedule

Many college professors suggest that students study 3 hours for every hour spent in class. This major investment in study time is one of the main differences between high school and college. To succeed academically, college students must learn to manage their time effectively. The actual number of hours you spend studying biology will vary depending on how quickly you learn the material, as well as on your course load and personal responsibilities, such as work schedules and family commitments.

The most successful students are often those who are best organized. At the beginning of the semester, make a detailed daily calendar. Mark off the hours you are in each class, along with travel time to and from class if you are a commuter. After you get your course syllabi, add to your calendar the dates of all exams, quizzes, papers, and reports. As a reminder, it also helps to add an entry for each major exam or assignment 1 week before the test or due date. Now add your work schedule and other personal commitments to your calendar. Using a calendar helps you find convenient study times.

Many of our successful biology students set aside 2 hours a day to study biology rather than depend on a weekly marathon session for 8 or 10 hours during the weekend (when that kind of session rarely happens). Put your study hours into your daily calendar, and stick to your schedule.

Determine Whether the Professor Emphasizes Text Material or Lecture Notes

Some professors test almost exclusively on material covered in lecture. Others rely on their students' learning most, or even all, of the content in assigned chapters. Find out what your pro-fessor's requirements are, because the way you study will vary accordingly.

How to study when professors test lecture material

If lectures are the main source of examination questions, make your lecture notes as complete and organized as possible. Before going to class, skim over the chapter, identifying key terms and examining the main figures, so that you can take effective lecture notes. Spend no more than 1 hour on this.

Within 24 hours after class, rewrite (or type) your notes. Before rewriting, however, read the notes and make marginal notes about anything that is not clear. Then read the corresponding material in your text. Highlight or underline any sections that clarify questions you had in your notes. Read the entire chapter, including parts that are not covered in lecture. This extra information will give you breadth of understanding and will help you grasp key concepts.

After reading the text, you are ready to rewrite your notes, incorporating relevant material from the text. It also helps to use the Glossary to find definitions for unfamiliar terms. Many students develop a set of flash cards of key terms and concepts as a way to study. Flash cards are a useful tool to help you learn scientific terminology. They are portable and can be used at times when other studying is not possible, for example, when riding a bus.

Flash cards are not effective when the student tries to second-guess the professor. ("She won't ask this, so I won't make a flash card of it.") Flash cards are also a hindrance when students rely on them exclusively. Studying flash cards instead of reading the text is a bit like reading the first page of each chapter in a mystery novel: It's hard to fill in the missing parts, because you are learning the facts in a disconnected way.

How to study when professors test material in the book

If the assigned readings in the text are going to be tested, you must use your text intensively. After reading the chapter introduction, read the list of Learning Objectives for the first section. These objectives are written in behavioral terms; that is, they ask you to "do" something in order to demonstrate mastery. The objectives give you a concrete set of goals for each section of the chapter. At the end of each section, you will find Review questions keyed to the Learning Objectives. Test yourself, going back over the material to check your responses.

Read each chapter section actively. Many students read and study passively. An active learner always has questions in mind and is constantly making connections. For example, there are many processes that must be understood in biology. Don't try to blindly memorize these; instead, think about causes and effects so that every process becomes a story. Eventually, you'll see that many processes are connected by common elements.

You will probably have to read each chapter two or three times before mastering the material. The second and third times through will be much easier than the first, because you'll be reinforcing concepts that you have already partially learned.

After reading the chapter, write a four- to six-page chapter outline by using the subheads as the body of the outline (first-level heads are boldface, in color, and all caps; second-level heads are in color and not all caps). Flesh out your outline by adding important concepts and boldface terms with definitions. Use this outline when preparing for the exam.

Now it is time to test yourself. Answer the Test Your Understanding questions, and check your answers. Write answers to each of the Critical Thinking questions. Finally, review the Learning Objectives in the Chapter Summary, and try to answer them before reading the summary provided. If your professor has told you that some or all of the exam will be short-answer or essay format, write out the answer for each Learning Objective. Remember that this is a self-test. If you do not know an answer to a question, find it in the text. If you can't find the answer, use the Index.

Learn the Vocabulary

One stumbling block for many students is learning the many terms that make up the language of biology. In fact, it would be much more difficult to learn and communicate if we did not have this terminology, because words are really tools for thinking. Learning terminology generally becomes easier if you realize that most biological terms are modular. They consist of mostly Latin and Greek roots; once you learn many of these, you will have a good idea of the meaning of a new word even before it is defined. For this reason, we have included an Appendix on Understanding Biological Terms. To be sure you understand the precise definition of a term, use the Index and Glossary. The more you use biological terms in speech and writing, the more comfortable you will be.

Form a Study Group

Active learning is facilitated if you do some of your studying in a small group. In a study group, the roles of teacher and learner can be interchanged: a good way to learn material is to teach. A study group lets you meet challenges in a nonthreatening environment and can provide some emotional support. Study groups are effective learning tools when combined with individual study of text and lecture notes. If, however, you and other members of your study group have not prepared for your meetings by studying individually in advance, the study session can be a waste of time.

Prepare for the Exam

Your calendar tells you it is now 1 week before your first biology exam. If you have been following these suggestions, you are well prepared and will need only some last-minute reviewing. No all-nighters will be required.

During the week prior to the exam, spend 2 hours each day actively studying your lecture notes or chapter outlines. It helps many students to read these notes out loud (most people listen to what they say!). Begin with the first lecture/chapter covered on the exam, and continue in the order on the lecture syllabus. Stop when you have reached the end of your 2-hour study period. The following day, begin where you stopped the previous day. When you reach the end of your notes, start at the beginning and study them a second time. The material should be very familiar to you by the second or third time around. At this stage, use your textbook only to answer questions or clarify important points.

The night before the exam, do a little light studying, eat a nutritious dinner, and get a full night's sleep. That way, you'll arrive in class on exam day with a well-rested body (and brain) and the self-confidence that goes with being well prepared.

Eldra P. Solomon
Linda R. Berg
Diana W. Martin

Understanding Diversity: Systematics

Kipunji (*Rungwecebus kipunji*), a recently discovered species of monkey. Kipunjis inhabit highland forests in Tanzania. This monkey has a long, stiff, upright crest of hair on its head and thick, brown fur, an adaptation to the cold temperatures in its mountain habitat. Adult kipunjis emit a distinctive loud, low-pitched honk-bark.

Dr. Tim Davenport / WCS

KEY CONCEPTS

Traditionally, biologists have named organisms using a binomial system (in which each species has a genus name followed by a specific epithet) and have classified organisms in taxonomic categories arranged in a hierarchy from most inclusive (domain) to least inclusive (species).

Based on current knowledge, the tree of life consists of three domains: Bacteria, Archaea, and Eukarya.

Phylogeny (evolutionary history) is based on common ancestry inferred from shared characters, including structural, developmental, behavioral, and molecular similarities, as well as from fossil evidence.

Phylogenetic trees are diagrams of hypothetical evolutionary relationships; cladograms are phylogenetic trees constructed by analyzing shared derived characters.

The total number of living species is estimated to be between 4 million and 100 million, but only about 1.7 million species have been described. Biologists estimate that to date they have identified less than 10% of bacteria, about 5% of fungi species, only about 2% of nematode (roundworm) species, and less than 20% of insect species. Thousands of new species are discovered each year, and even new primates are found. For example, in 2003 and 2004 two independent research teams discovered a monkey, now known as kipunji (*Rungwecebus kipunji*), at two sites in southern Tanzania (see photo). Whether or not the monkey remains classified in this genus will depend on DNA analysis.

The variety of living organisms and the variety of ecosystems they form are referred to as **biological diversity,** or **biodiversity.** The totality of life on Earth represents our biological heritage, and the quality of life for all organisms depends on the health and balance of this worldwide web of life-forms. For example, we depend on organisms to maintain the life-sustaining composition of gases in the atmosphere, form soil, break down wastes, recycle nutrients, and provide food for one another. We humans exploit many species for economic benefit. One very anthropocentric (human-centered) example of the practical importance of biodiversity is that more than 40% of the prescriptions that pharmacists dispense in the United States derive from living organisms. Investigators are just learning how to effectively screen organisms for potential drugs. Unfortunately, human activity is seriously reducing biodiversity, and species are becoming extinct faster than researchers can study them (see the discussion on declining biodiversity in Chapter 56).

Alarmed by these extinctions, biologists are mobilizing to more rapidly identify new species and to conserve biodiversity. Many biologists agree that describing and classifying all the surviving species of the world should be a major scientific goal of the 21st century. An important step toward that goal is the Catalogue of Life, the work of a collaborative international research team that manages a comprehensive global database made available on the Internet. The Catalogue of Life categorizes and lists all of the world's known species.

In this chapter, we explore some of the approaches and methods used by biologists to classify organisms and to infer their evolutionary history and relationships. Information comes from many sources, including fossils, biogeography, patterns of development, structural and behavioral characteristics, and molecular similarities and differences. Researchers organize this information into large databases and manipulate it using computer algorithms to generate hypotheses regarding relationships. From these data, biologists are constructing the "tree of life." However, as more sophisticated molecular methods and discovery of new fossils provide new data, old data must be reinterpreted. As we gain greater understanding of how organisms are related, we revise the way we classify organisms, and the structure of the tree of life changes. ∎

CLASSIFYING ORGANISMS

Learning Objectives

1 State at least two justifications for the use of scientific names and classifications of organisms.
2 Describe the binomial system of naming organisms, and arrange the Linnaean categories in hierarchical fashion, from most inclusive to least inclusive.

The scientific study of the diversity of organisms and their evolutionary relationships is called **systematics.** An important aspect of systematics is **taxonomy,** the science of naming, describing, and classifying organisms. The term **classification** means arranging organisms into groups based on similarities that reflect evolutionary relationships among lineages.

Organisms are named using a binomial system

Because there are millions of kinds of organisms, scientists need a system for accurately identifying them. Imagine that you were going to develop a classification system. How would you use what you already know about living things to assign them to categories? Would you place insects, bats, and birds in one category because they all have wings and fly? And would you, perhaps, place squid, whales, fish, penguins, and Olympic backstroke champions in another category because they all swim? Or would you classify organisms according to a culinary scheme, placing lobsters and tuna in the same part of the menu, perhaps identifying them as "seafood"? Any of these schemes might be valid, depending on your purpose. Similar methods have been used throughout history. St. Augustine in the fourth century classified animals as useful, harmful, or superfluous—to humans. During the Renaissance, scholars began to develop categories based on the characteristics of the organisms themselves.

Of the many classification systems that were developed, the one designed by Carolus Linnaeus in the mid-18th century (described briefly in Chapter 1) has survived with some modification to the present day. Linnaeus grouped organisms according to their similarities, mainly structural similarities.

Before the mid-18th century, each species had a lengthy descriptive name, sometimes consisting of 10 or more Latin words! Linnaeus simplified scientific classification, developing a **binomial system of nomenclature** in which each species is assigned a unique two-part name. The first part of a binomial scientific name designates the **genus** (pl., *genera*), and the second part is called the **specific epithet.**

The genus name is always capitalized, whereas the specific epithet is usually not. Both names are underlined or italicized. The genus, or generic, name can be used alone to designate all species in the genus (for example, the genus *Quercus* includes all oak species). Note that the specific epithet alone is *not* the species' name. In fact, the same specific epithet can be used as the second name of species in different genera. For example, *Quercus alba* is the scientific species name for the white oak, and *Salix alba* is the species name for the white willow (*alba* comes from a Latin word meaning "white"). Thus, both parts of the name must be used to accurately identify the species. However, the specific epithet is never used alone; it must always follow the full or abbreviated genus name, such as *Quercus alba* or *Q. alba.*

Scientific names are generally derived from Greek or Latin roots or from latinized versions of the names of persons, places, or characteristics. For example, the generic name for the bacterium *Escherichia coli* is based on the name of the scientist, Theodor Escherich, who first described it. The specific epithet *coli* reminds us that *E. coli* lives in the colon (large intestine).

Most areas of biology depend on scientific names and classification. To study the effects of pollution on an aquatic community, biologists must be able to accurately record the relative numbers of each type of organism present and changes in their populations over time. This requires that all investigators identify each species accurately by name.

Scientific names permit biology to be a truly international science. Even though the common names of an organism may vary in different locations and languages, an organism can be universally identified by its scientific name. A researcher in Puerto Rico knows exactly which organisms were used in a study published by a Russian scientist and therefore can repeat or extend the Russian's experiments using the same species.

Each taxonomic level is more general than the one below it

Linnaeus devised a system for assigning species to a hierarchy of increasingly broader groups. As you move up the hierarchy, each group is more inclusive, that is, it includes the groups below it. When he set up his system, Linnaeus did not have a theory of evolution in mind. Nor did he have any idea of the vast number of extant (living) and extinct organisms that would later be discovered. Yet his system has proved remarkably flexible and adaptable to new biological knowledge and theory. Very few other 18th-century inventions survive today in a form that would still be recognizable to their originators.

The range of taxonomic categories from species to domain forms a hierarchy (❙ Fig. 23-1 and ❙ Table 23-1). Closely related species are assigned to the same genus, and closely related genera are grouped in a single **family.** Families are grouped into **orders,** orders into **classes,** classes into **phyla,** phyla into **kingdoms,** and kingdoms into **domains.** A species is considered a true biological entity, but taxonomic categories above the species level are artificial constructs used for cataloguing the diverse life-forms on Earth.

A **taxon** (pl., *taxa*) is a formal grouping of organisms at any given level, such as species, genus, or phylum. For example, class Mammalia is a taxon that includes many different orders. A taxon can be separated into subgroupings, such as subphyla or superclasses. Subphylum Vertebrata, a subgroup of phylum Chordata, is a taxon that contains several classes, including Amphibia and Mammalia.

Biologists are moving away from Linnaean categories

Biologists are finding it difficult to name new species as quickly as they are being discovered. They are also working hard to reclassify organisms as they adjust hypotheses about relationships among organisms. Many systematists are moving away from the hierarchical Linnaean categories because these categories are limiting and do not fit well with recent findings.

One group of biologists has proposed a very different approach to classification known as **PhyloCode,** which is based on evolutionary relationships. In this system, taxonomic ranks are not used. A species is defined as a segment of a population lineage. Organisms are grouped into clades; a **clade** is defined as a group of organisms with a common ancestor. Advocates of PhyloCode argue that this system avoids the problem of changing names when discoveries of new species require shifts in ranks. The system also makes it easier to name clades one at a time as they are discovered.

A new approach to identifying organisms is based on recently developed molecular methods. In 2003, Paul Hebert, of the University of Guelph in Canada, and his colleagues proposed in the British *Proceedings of the Royal Society* that we identify all living things by DNA rather than by their physical structure. Hebert holds that each organism has a DNA **bar code** (much like grocery store products). Biologists can use bar codes to identify the species of an organism in any stage of life, even as an egg or seed. Another advantage of bar coding is that this method can be used to distinguish among species that look alike.

Hebert has used a segment of mitochondrial DNA consisting of only 648 base pairs. This short sequence of DNA, part of the gene that codes for the enzyme cytochrome *c* oxidase, can be used as a label for the genome of a species. When applied to bird species, results from bar coding were strikingly consistent with relationships inferred from traditional taxonomic methods. For example, differences between members of closely related species were about 18 times as great as differences among birds within a single species. Information gained suggested that four bird species should each be divided into two new species.

Hebert's bar code does not work with all groups. Some groups, such as bacteria, will require different bar code markers. A component of ribosomes is being used as a bar code to identify bacteria. The sequence used is a small subunit ribosomal RNA (SSU rRNA), specifically the 16S rRNA nucleotide sequence.[1] The gene that codes SSU rRNA is highly conserved; it is present in all known organisms, and its function has remained the same through time.

Review

❙ What are the key features of the binomial system of nomenclature?

❙ Why is it important to use scientific, rather than common, names for organisms?

❙ Classify the domestic cat, using taxa from domain to species.

DETERMINING THE MAJOR BRANCHES IN THE TREE OF LIFE

Learning Objectives

3 Describe the six kingdoms and three domains of organisms, and give the rationales for and against this system of classification.

4 Given its distinguishing characters, classify an organism in the appropriate domain and kingdom.

The history of taxonomy at the kingdom level is a good example of the process of science. From the time of Aristotle to the mid-19th century, biologists divided organisms into two kingdoms: **Plantae** and **Animalia.** After the development of microscopes,

---

[1]The number *16S* refers to the sedimentation coefficient (a measure of relative size) when centrifuged.

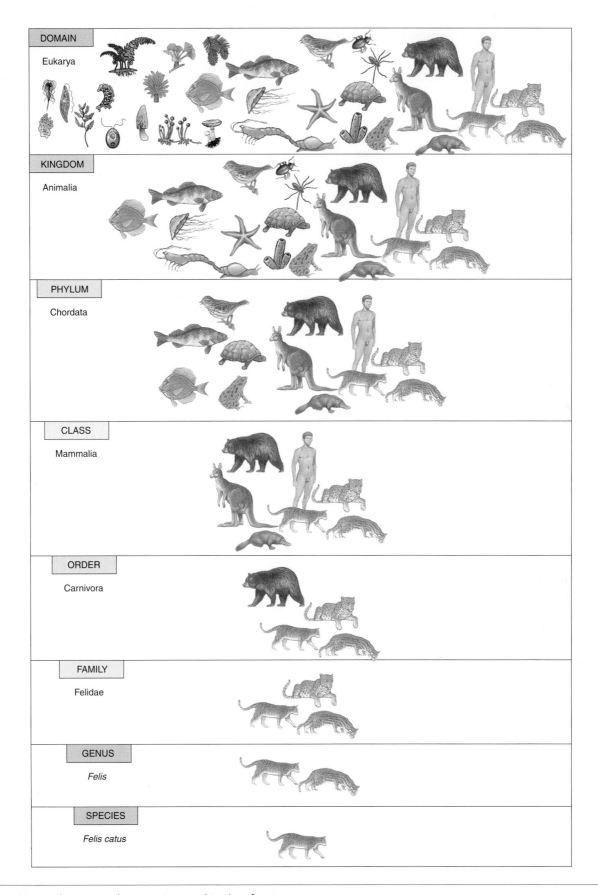

Figure 23-1 The principal categories used in classification

The domestic cat (*Felis catus*) is classified here to illustrate the hierarchical organization of our taxonomic system. Each level is more inclusive than the one below it, meaning that it includes more groups of organisms.

TABLE 23-1

Classification of Corn

Domain	**Eukarya**
	Organisms that have nuclei and other membrane-enclosed organelles
Kingdom	**Plantae**
	Terrestrial, multicellular, photosynthetic organisms
Phylum	**Anthophyta**
	Vascular plants with flowers, fruits, and seeds
Class	**Monocotyledones**
	Monocots: Flowering plants with one seed leaf (cotyledon) and flower parts in threes
Order	**Commelinales**
	Monocots with reduced flower parts, elongated leaves, and dry 1-seeded fruits
Family	**Poaceae**
	Grasses with hollow stems; fruit is a grain; and abundant endosperm in seed
Genus	*Zea*
	Tall annual grass with separate female and male flowers
Species	*Zea mays*
	Only one species in genus—corn

it became increasingly obvious that many organisms could not be easily assigned to either the plant or the animal kingdom. For example, the unicellular organism *Euglena,* which has been classified at various times in the plant kingdom and in the animal kingdom, carries on photosynthesis in the light but in the dark uses its flagellum to move about in search of food (see Fig. 25-6). In 1866, a German biologist, Ernst Haeckel, proposed that a third kingdom, **Protista,** be established to accommodate bacteria and other microorganisms, such as *Euglena,* that did not seem to fit into the plant or animal kingdoms. Today, many biologists place algae (including multicellular forms), protozoa, water molds, and slime molds in kingdom Protista. Thus, protists are a diverse group of mainly unicellular, mainly aquatic eukaryotic organisms.

In 1937, the French marine biologist Edouard Chatton suggested the term *procariotique* ("before nucleus") to describe bacteria, and the term *eucariotique* ("true nucleus") to describe all other cells. This dichotomy between prokaryotes and eukaryotes is now universally accepted by biologists as a fundamental evolutionary divergence. In the 1960s, advances in electron microscopy and biochemical techniques revealed further cell differences that inspired many new proposals for classifying organisms.

In 1969, R. H. Whittaker proposed a five-kingdom classification based mainly on cell structure and the way that organisms derive nutrition from their environment. Whittaker suggested that the fungi (which include the mushrooms, molds, and yeasts) be removed from the plant kingdom and classified in their own kingdom, **Fungi.** After all, fungi are not photosynthetic and must

absorb nutrients produced by other organisms. Fungi also differ from plants in the composition of their cell walls, in their body structures, and in their modes of reproduction. Kingdom **Prokaryotae** was established to accommodate the bacteria, which are fundamentally different from all other organisms in that they do not have distinct nuclei and other membranous organelles and do not undergo mitotic division (see Fig. 4-6).

In the late 1970s, Carl Woese (pronounced "woes"), of the University of Illinois, and his colleagues began to study the evolutionary relationships among organisms by analyzing their 16S rRNA. Using sequence analysis, Woese used variations in this universal molecule to challenge the long-held view that all prokaryotes were closely related and very similar to one another. He showed that there are two fundamentally different groups of prokaryotes, archaea and bacteria. He proposed that the prokaryotes account for two of the three major branches of organisms.

Woese's hypothesis gained support in 1996 when Carol J. Bult of the Institute for Genomic Research in Rockville, Maryland, reported in the journal *Science* that she and her colleagues had sequenced the complete genome of a methane-producing archaea, *Methanococcus jannaschii.* When these researchers compared gene sequences with those of two previously sequenced bacteria, they found that fewer than half of the genes matched. Given this molecular evidence, most biologists now divide the prokaryotes into kingdom **Bacteria** and kingdom **Archaea.**

In this eighth edition of *Biology,* we classify organisms into six kingdoms: Bacteria, Archaea, Protista, Fungi, Plantae, and Animalia (▌Table 23-2). As we discuss in Chapter 25, the Protista are quite diverse, and biologists will likely divide this group into several kingdoms.

Systematists construct **phylogenetic trees** to graphically represent the evolutionary history of a group of species. We can construct a type of phylogenetic tree called a **cladogram** to illustrate the evolutionary relationships among groups of organisms (▌Fig. 23-2).

Most systematists use a level of classification above the kingdom, called a **domain,** based on fundamental molecular differences among the bacteria, archaea, and eukaryotes. They classify organisms in three domains: **Archaea** (which corresponds to kingdom Archaea), **Bacteria** (which corresponds to kingdom Bacteria), and **Eukarya** (eukaryotes) (▌Fig. 23-3). As we will discuss in Chapter 24, the viruses are a special case and are not classified in any of the three domains. Biologists have inferred that the three domains are the three main branches of the tree of life.

An important character that distinguishes the archaea from the bacteria is the absence of the compound peptidoglycan in the cell walls of the archaea (discussed in Chapter 24). Neither do the archaea have the simple RNA polymerase characteristic of bacteria. Gene sequencing indicates that the archaea have a combination of bacteria-like and eukaryote-like genes.

As biologists continue to investigate the origin and relationships of the domains, they have discovered that evolution is not always linear. During the course of evolution, genes not only were passed down "vertically" from one generation to the next but were also exchanged laterally. Such gene swapping between organisms in one taxon and related organisms in another taxon is called **horizontal gene transfer,** or *lateral gene transfer.* This pro-

TABLE 23-2

Domains and Kingdoms

Domain	Kingdom	Characteristics	Ecological Role and Comments
Bacteria	Bacteria	Prokaryotes (lack distinct nuclei and other membranous organelles); unicellular; microscopic; cell walls generally composed of peptidoglycan.	Most are decomposers; some parasitic (and pathogenic); some chemosynthetic autotrophs; some photosynthetic; important in recycling nitrogen and other elements; some used in industrial processes.
Archaea	Archaea	Prokaryotes; unicellular; microscopic; peptidoglycan absent in cell walls; differ biochemically from bacteria.	Methanogens are anaerobes that inhabit sewage, swamps, and animal digestive tracts; extreme halophiles inhabit salty environments; extreme thermophiles inhabit hot, sometimes acidic environments.
Eukarya	Protista	Eukaryotes; mainly unicellular or simple multicellular. Three informal groups (not taxa) include protozoa, algae, and slime molds and water molds.	Protozoa are an important part of zooplankton. Algae are important producers, especially in marine and freshwater ecosystems; important oxygen source.
	Fungi	Eukaryotes; heterotrophic; absorb nutrients; do not photosynthesize; body composed of threadlike hyphae that form tangled masses that infiltrate food or habitat; cell walls of chitin.	Decomposers; some parasitic (and pathogenic); some form important symbiotic relationships with plant roots (mycorrhizae) or algae (lichens); some used as food; yeast used in making bread and alcoholic beverages; some used to make industrial chemicals or antibiotics; responsible for much spoilage and crop loss.
	Plantae	Eukaryotes; multicellular; photosynthetic; possess multicellular reproductive organs; alternation of generations; cell walls of cellulose.	Terrestrial biosphere depends on plants in their role as primary producers; important source of oxygen in Earth's atmosphere.
	Animalia	Eukaryotes; multicellular heterotrophs; many exhibit tissue differentiation and complex organ systems; most able to move about by muscular contraction; nervous tissue coordinates responses to stimuli.	Consumers; some specialized as herbivores, carnivores, or detritus feeders.

cess is apparently common and can occur in several ways, such as by exchange of DNA among different populations or species of bacteria or by interbreeding between closely related groups. Indeed, the common ancestor of all living things may have been a *community* of species that traded their genes. Horizontal gene transfer appears to be a continuous process between members of the three domains.

Eukaryotic cells most likely evolved from prokaryotic cells that lived symbiotically within one another. Recall that mitochondria and chloroplasts are thought to have originated from prokaryotes that lived as endosymbionts inside other cells (see Chapter 21). Based on what is now known about horizontal gene transfer, some systematists argue that there is no simple tree of life and have suggested alternative hypotheses. One group hypothesizes a complex bush with many connecting branches (Fig. 23-4a). James A. Lake and Maria C. Rivera, of the University of California at Los Angeles, have recently proposed another approach—that the tree of life is actually a ring of life (Fig. 23-4b). These systematists hypothesize, based on their analysis of hundreds of genes, that horizontal gene transfer between bacteria and archaea gave rise to the eukaryotes.

The evolution of systematics reflects the creative and dynamic process of science. Systematists are very responsive to new data, and consequently, classification of organisms at all levels is a continuously changing process. Although many biologists are moving away from a hierarchical system, some suggest adding new groups so that the classical approach keeps up with new findings. Whatever approach is used, the goal of systematics is to base classification on evolutionary change and relationships.

Review

- What are some of the defining characteristics of each of the six kingdoms and three domains?
- What are the advantages of a "six-kingdom" system over a "two-kingdom" system?
- In which domain and kingdom would you classify each of the following: *Salix alba*, amoeba, *Escherichia coli*, tapeworm, and black bread mold?
- Why do representations of the structure of the tree of life constantly change?

RECONSTRUCTING PHYLOGENY

Learning Objectives

5 Critically review the difficulties encountered in choosing taxonomic criteria.
6 Apply the concept of shared derived characters to the classification of organisms.
7 Describe how molecular homologies contribute to the science of systematics.
8 Contrast monophyletic, paraphyletic, and polyphyletic taxa.

Many biologists classify organisms in categories called kingdoms.

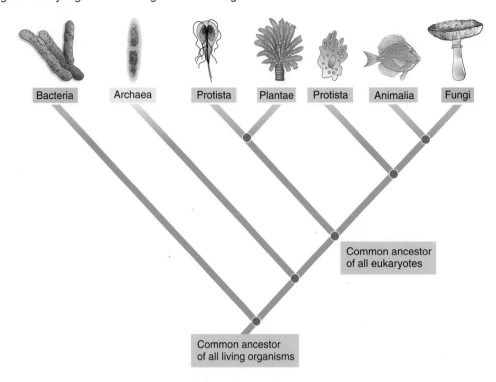

Figure 23-2 *Animated* The six-kingdom system of classification

This cladogram illustrates the evolutionary relationships among organisms in the six kingdoms. The branches portray ancestral populations of each group through time. Each node (*circle*) represents the point when two groups diverged from each other.

Modern biologists link classification with evolution. Their goal is to reconstruct **phylogeny** (literally, "production of phyla"), the evolutionary history of a group of organisms from a common ancestor. As they determine evolutionary relationships among and between species and higher taxa, systematists build classifications based on common ancestry. Consequently, systematics reflects the integration of all evolutionary processes and evolutionary evidence and thus is at the center of how we understand and explain the nonrandom occurrence and relationships of life-forms on Earth.

Once phylogenies are established, they help answer other questions in biology. For example, phylogenies help us understand evolutionary patterns that might provide clues to the origin and spread of HIV and other pathogens. Phylogenies also help biologists identify new species and predict their characteristics. Systematists may classify a new species based on specific characters that it shares with other organisms in a particular taxon. They may then infer that the new species shares other characters with organisms in that group.

As you read the following sections, remember that phylogenies are testable hypotheses. They are supported or falsified by the available data. Systematics proceeds by constantly re-evaluating data, hypotheses, and theoretical constructs. As new data are discovered and old data are reinterpreted, the ideas of system-

Many biologists classify organisms in three major categories called domains.

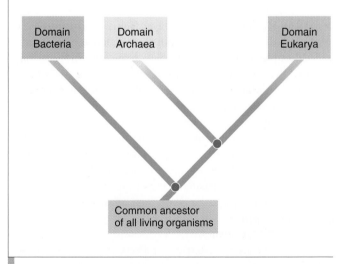

Figure 23-3 *Animated* The three domains

This cladogram illustrates the evolutionary relationships among organisms in the three domains. Kingdoms Protista, Plantae, Animalia, and Fungi are assigned to domain Eukarya.

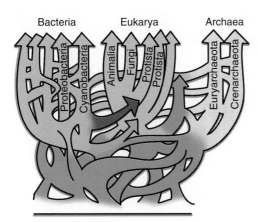

a The three-domain approach drawn to show horizontal gene transfer as a continuous process between domains and also among groups within each domain. Only some of the branches are labeled. The diagonal brown arrow represents mitochondrial endosymbiosis; the diagonal green arrow represents chloroplast endosymbiosis.

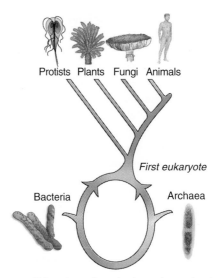

b The ring of life is based on the hypothesis that lateral gene transfer between bacteria and archaea gave rise to the eukaryotes.

Figure 23-4   Other ways to represent the tree of life

[(a) based on W. F. Doolittle, *Science*, Vol. 284, pp. 2124–2128, 1999; (b) based on M. C. Rivera and J. A. Lake, *Nature*, Vol. 431, pp. 152–155, Sept. 9, 2004.]

atists change. As a result, our understanding of how organisms are related and the way we classify organisms are continuously being revised. Systematics is a dynamic science that changes as biologists discover new species and use ever more sophisticated techniques to investigate the evolutionary relationships among organisms.

You learned in Chapter 18 that evolution is the accumulation of inherited changes within populations over time. Recall that a **population** is made up of all the individuals of the same species that live in a particular area. A population has a dimension in space—its geographic range—and also a dimension in time. Each population extends backward in time. Somewhat like branches of a tree, a population may diverge from other populations enough to become a new species (which may be depicted by a new tree branch; see Chapters 19 and 20). Species have various degrees of evolutionary relationships with one another, depending on the degree of genetic divergence since their populations branched from a common ancestor.

Homologous structures are important in determining evolutionary relationships

Just how to determine phylogeny and how to group species into higher taxa—genera, families, orders, classes, or phyla—can be difficult decisions. Biologists base their judgments about the degree of relationship on the extent of similarity among species. They examine structural, physiological, developmental, behavioral, and molecular traits, as well as fossil evidence. When examining these traits, they look for homologies among different organisms. Recall from Chapter 18 that **homology** refers to the

presence, in two or more species, of a structure derived from a recent common ancestor. For example, the bones in the wings of a bird and a bat are homologous. As we will discuss, identifying homologous traits is very important in inferring phylogeny.

It is often difficult to determine whether traits are similar as a result of homology. Although the wings of a butterfly are adapted for flight, their structure is different from bird wings, and these animals do not share a common winged ancestor. Similar structures sometimes evolve when unrelated or distantly related species become adapted to similar environmental conditions. Thus, wings may occur in two or more species not derived from a recent common ancestor. Recall from Chapter 18 that independent evolution of similar structures in distantly related organisms is known as **convergent evolution.** Sharks and dolphins have similar, but independently derived, body forms because they have become adapted to similar environments (aquatic) and lifestyles (predatory).

Another challenge in deciding on homology is **reversal,** in which a trait reverts to its ancestral state. A reversal removes a similarity that had evolved. A characteristic that superficially appears homologous but is actually independently acquired by convergent evolution or reversal is described as exhibiting **homoplasy.**

Shared derived characters provide clues about phylogeny

Organisms sharing many homologous structures are considered closely related, whereas organisms sharing few homologous characters are presumed less closely related. However, distinguishing between homology and homoplasy is not always straightforward. Determining traits that indicate evolutionary relationships is extremely important.

How does a systematist interpret the significance of these similarities? In making decisions about taxonomic relationships, the systematist first examines the characteristics in the largest group (such as phylum or class) of organisms being studied and interprets them as indicating the most *remote* common ancestry. These **shared ancestral characters,** or **plesiomorphies,** are features that were present in an ancestral species and remain present in all groups descended from that ancestor. For example, the vertebral column, present in all vertebrates, is an ancestral character for study of classes within the subphylum Vertebrata. Studying the presence or absence of the vertebral column does not help us discriminate among various classes of vertebrates: we could not distinguish between amphibians and mammals, because all individuals in these classes have a vertebral column.

When two populations become separated and begin to evolve independently, some of their homologous traits change as a result of mutation, natural selection, and genetic drift. The novel traits that evolve are referred to as **shared derived characters,** or **synapomorphies.** Note that shared derived characters originate in a *recent* common ancestor and are present in its descendants. Species that share derived characters form a clade. Systematists use shared derived characters to identify points where groups diverged from one another. A trait viewed as a *derived character* in a more inclusive (broader) taxon may also be considered an *ancestral character* in a less inclusive (narrower) taxon.

More recent common ancestry is indicated by classification into less and less inclusive taxonomic groups with more and more specific shared derived characters. For example, the three small bones in the middle ear are useful in identifying a branch point between reptiles and mammals. The evolution of this derived character was a unique event, and only mammals have these bones. However, if we compare mammals with one another, the three ear bones are a shared ancestral character, because all mammals have them. Consequently, they have no value in distinguishing among mammalian taxa. Other characters must be used to establish branch points among the mammals.

If we compare dogs, goats, and dolphins (all of which are mammals), we find that dogs and goats have abundant hair whereas dolphins do not. Hair is an ancestral trait in mammals and therefore cannot be used as evidence that dogs and goats share a more recent common ancestor. In contrast, the virtual absence of hair in mature dolphins is a derived character within mammals. When we compare dogs, dolphins, and whales, we find that dolphins and whales share this derived character, providing evidence that these animals evolved from a common ancestor not shared by dogs.

Biologists carefully choose taxonomic criteria

Both fishes and dolphins have streamlined body forms, but this characteristic is homoplastic and does not indicate close evolutionary relationships. The dolphin shares important homologous derived characters (synapomorphies) with mammals such as humans: mammary glands, which produce milk for the young;

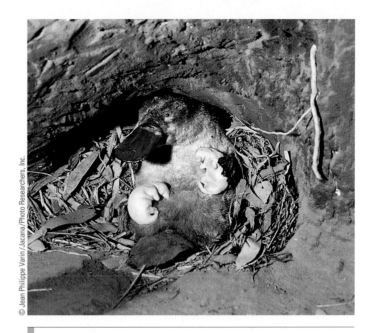

Figure 23-5 Is this animal a bird?

A few mammals, such as the duck-billed platypus, lay eggs, have beaks, and lack teeth. However, the platypus does not have feathers, and it nourishes its young with milk secreted from mammary glands.

three small bones in the middle ear; and a muscular diaphragm that helps move air into and out of the lungs. Thus, the dolphin is classified as a mammal.

Although dolphins have more shared derived characters in common with humans than with fishes, some characters are shared by all three. Among these shared ancestral characters are a dorsal tubular nerve cord, and during embryonic development, a notochord (skeletal rod) and rudimentary gill slits. These shared ancestral characters (plesiomorphies) indicate a common ancestry and serve as a basis for classification. The ancestry is more remote between the dolphin and the fish than between the dolphin and the human. Therefore, although fishes, humans, and dolphins are grouped in a more inclusive taxon, the phylum Chordata, humans and dolphins are also classified in class Mammalia, a less inclusive taxon within phylum Chordata, indicating that they are more closely related.

Determining which traits best illustrate evolutionary relationships can be challenging. What, for example, are the most important taxonomic characteristics of a bird? We might list feathers, beak, wings, absence of teeth, egg laying, and endothermy (an endothermic animal uses metabolic heat to maintain a constant body temperature despite variations in surrounding temperature). The duck-billed platypus and a few other mammals (members of a group called monotremes) have many of these same characteristics: beaks, endothermy, absence of teeth, and egg laying. Yet we do not classify them as birds (■ Fig. 23-5). No mammal, however, has feathers. Is this trait absolutely diagnostic of birds? According to the conventional taxonomic wisdom, the presence or absence of feathers determines what is and is not a bird. This criterion was not always valid, however. A number of dinosaur fossils bear impressions of feathers!

Organisms are typically classified on the basis of a combination of traits rather than on any single trait. The significance of these combinations is determined inductively, that is, by integrating and interpreting the data. Such induction is a necessary part of the process of science. Systematists may hypothesize that all birds should have beaks, feathers, no teeth, and so forth. Then they re-examine the living world and observe whether any organisms might reasonably be called birds that do not fit the current definition of "birdness." If not, the definition stands. If too many exceptions emerge, the definition may be modified or abandoned. Sometimes the systematist determines that an apparent exception—the bat, for instance—resembles a bird only superficially and should not be considered one. The bat has all the basic characteristics of a mammal, such as hair and mammary glands that produce milk for the young.

Molecular homologies help clarify phylogeny

When a new species evolves, it does not always exhibit obvious phenotypic differences when compared to closely related species. For example, two distinct species of fruit flies may appear identical. Some of their DNA, proteins, and other molecules, however, are different. Such variations in the structure of specific macromolecules among species, just like differences in anatomical structure, result from mutations. Macromolecules that are functionally similar in two different types of organisms are considered homologous if their subunit sequence is similar. Advances in molecular biology have provided the tools for biologists to compare the macromolecules of various organisms.

The science of **molecular systematics** focuses on molecular structure to clarify evolutionary relationships. DNA, RNA, and amino acid sequencing are used to compare the macromolecules of organisms being studied. Such comparisons provide systematists with valuable information about the degree of relatedness among organisms. The more that subunit sequences of two species correspond, the more closely related the species are considered to be. The number of differences in certain DNA or RNA nucleotide sequences or in amino acid sequences in two groups of organisms may reflect how much time has passed since the groups branched from a common ancestor. (This can be true only if the rates of change have remained constant.) Thus, specific macromolecules can be used as **molecular clocks** (see Chapter 18).

Many systematists look to ribosomal RNA structure to help determine phylogenies. All known organisms have ribosomes that function in protein synthesis, and certain ribosomal RNA nucleotide sequences have been highly conserved in evolution. Recall that the division of organisms into three domains was based, in large part, on the comparison of ribosomal RNA by Carl Woese and his research team. All prokaryotic ribosomes contain three types of RNA, named in order of increasing size: 5S, 16S, and 23S. The 5S and 16S RNAs have been extensively used to determine evolutionary relationships among bacteria because the number of base pairs is manageable and because these molecules are transcribed from highly conserved regions of DNA.

The very precise tools provided by molecular biology have put systematics on the cutting edge of biological research. Biologists are no longer limited by subjective decisions regarding anatomical similarities among organisms. For example, until recently the African elephant and the Indian elephant were the only two species of living elephants recognized. Using molecular methods, biologists have demonstrated that there are at least two distinct species of African elephants. Each species has its own distinctive DNA bar code.

Kerstin Lindblad-Toh, codirector of Broad's Genome Sequencing and Analysis program at MIT and Harvard, and her colleagues recently provided another example of applied molecular systematics (❙ Fig. 23-6). They sequenced the genome of the domestic dog and compared coding regions of more than 13,000

Key Experiment

QUESTION: How is the dog related to other canids?

HYPOTHESIS: Analysis of molecular data from a variety of canids will clarify the evolutionary relationships of canids.

EXPERIMENT: Researcher Kerstin Lindblad-Toh and her colleagues sequenced the genome of the domestic dog. They compared coding regions of more than 13,000 dog genes with corresponding human and mouse genes. Then, they selected 12 exons and 4 introns and sequenced them in 30 out of 34 living canids.

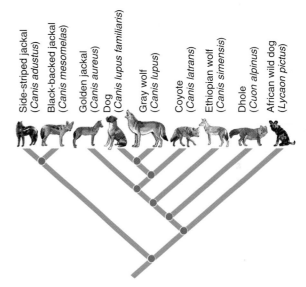

RESULTS AND CONCLUSION: Analysis of the comparisons of exon and intron nucleotide sequences allowed these researchers to build a cladogram of Canidae species. The data indicated that the dog is most closely related to the gray wolf. (Sequence divergence of nuclear exon and intron sequences is 0.04% and 0.21, respectively.) The two African jackals are the most basal members of this clade (branch at left), suggesting an African origin for the canids.

Figure 23-6   Molecular phylogeny

Molecular similarities indicate that the dog is closely related to the gray wolf. Note that some biologists list the scientific name for the domestic dog as *Canis familiaris* and others, including the *Smithsonian Institution* and the *American Society of Mammalogists*, consider the dog a subspecies of the gray wolf and list its name as *Canis lupus familiaris*.

Groups of organisms can be monophyletic, paraphyletic, or polyphyletic. Most systematists recognize only monophyletic groups.

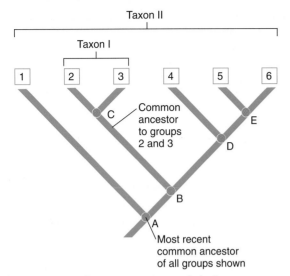

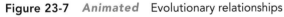

(a) Taxon I and taxon II are monophyletic groups, or clades. Each includes a common ancestor and all its descendants.

Figure 23-7 *Animated* Evolutionary relationships

In all three figures, node A represents the branch point from the common ancestor of all groups shown. Node B represents the branch point from the common ancestor of groups 2 through 6. Node C represents the branch point from the common ancestor of groups 2 and 3. Node D represents the branch point from the common ancestor of groups 4, 5, and 6. Node E represents the branch point from the common ancestor of groups 5 and 6.

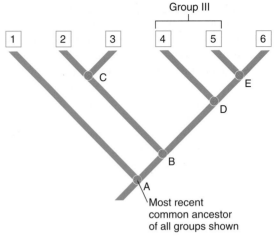

(b) Group III is paraphyletic. It includes some, but not all of the descendants of a common ancestor.

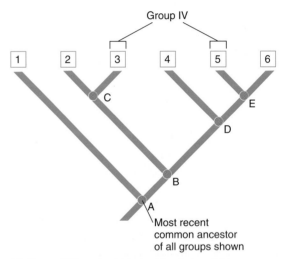

(c) Group IV is polyphyletic. Members of this group do not share the same recent common ancestor.

dog genes with human and mouse genes. Then, they selected exons and introns and compared their nucleotide sequences in 30 living canids (members of the family Canidae to which the dog belongs). The data they gathered enabled these investigators to construct a cladogram of the phylogenetic relationships of canids. They confirmed earlier research showing that the dog's closest relative is the gray wolf. Lindblad-Toh is using her nucleotide data to identify genes responsible for specific diseases and will apply her findings to the human genome in an effort to better understand the many diseases that canids have in common with humans.

Comparison of ribosomal RNA sequences has also been used to challenge the once widely accepted idea that fungi are closely related to plants. According to ribosomal RNA analysis, fungi are more closely related to animals than to plants; that is, animals and fungi share a more recent common ancestor, perhaps a flagellate protist.

Taxa are grouped based on their evolutionary relationships

Systematists recognize three kinds of taxonomic groupings based on structural similarities, molecular data, and other criteria: monophyletic, paraphyletic, and polyphyletic. A **monophy-**

letic taxon includes an ancestral species and all its descendants (Fig. 23-7a). Mammals, for example, form a monophyletic taxon because all mammals are thought to have evolved from a common ancestral mammal and all descendants of this ancestor are mammals. Monophyletic taxa are natural groupings, because they represent true evolutionary relationships and include all close relatives. **Sister taxa,** or sister groups, share a more recent common ancestor with one another than either taxon does with any other group shown on a cladogram. In Figure 23-7a, taxa 2 and 3 are sister taxa; taxa 5 and 6 are also sister taxa.

A **paraphyletic taxon** is a group that contains a common ancestor and some, but not all, of its descendants (Fig. 23-7b). As discussed later in this chapter, the class Reptilia is paraphyletic because it does not include all descendants of the most recent common ancestor of reptiles. Birds are thought to share a recent common ancestor with reptiles.

A **polyphyletic group** consists of several evolutionary lines that do not share the same recent common ancestor (Fig. 23-7c). Biologists might have mistakenly classified the members of such a group together because these organisms share similar (homoplastic) features arising from convergent evolution (see Chapter 18). Systematists avoid constructing polyphyletic taxa, because they are unnatural and misrepresent evolutionary relationships. Polyphyletic taxa are sometimes accepted temporarily until research provides additional data.

Review

- How are shared ancestral characters and shared derived characters different?
- Why don't shared ancestral characters provide evidence for relationships between organisms within a taxon that has those traits? Give an example.
- How is molecular biology contributing to the science of systematics?
- Why do systematists prefer monophyletic taxa to polyphyletic taxa?

CONSTRUCTING PHYLOGENETIC TREES

Learning Objectives

9 Contrast the classification of reptiles and birds by an evolutionary systematist and a cladist.

10 Describe the construction of a cladogram by using outgroup analysis.

In determining the relationships among organisms, systematists use a variety of data and methods. How data are analyzed and interpreted depends on the systematist's approach. In the **phenetic approach,** relationships are based on the number of shared characteristics. Phylogenetic relationships are defined by statistics, such as genetic distances, which summarize overall similarities among taxa based on collected data. For example, a researcher can compare the base sequences of a given gene or coding region to arrive at an average percentage of difference between the taxa being analyzed. A computer program then compares these values and generates trees that cluster similar taxa and place divergent taxa on more distant branches.

Phenetic techniques are currently used with molecular data. For example, if each amino acid in a protein is considered a character, amino acid sequences of various animals can be determined in the laboratory and compared by computer. The information about differences in amino acid sequences can be used to construct phylogenetic diagrams. Species are placed at relative distances from one another, reflecting the extent of difference in amino acid sequence.

Cladistics, also known as **phylogenetic systematics,** is an approach in which shared derived characters are analyzed and used to infer evolutionary relationships. Cladists emphasize common

ancestry rather than phenotypic similarity as the basis for classification. They base their assessment on shared derived characters that can be structural, behavioral, physiological, or molecular. The characters must be homologous. According to cladistics, dolphins are classified with mammals rather than with fishes, because dolphins and mammals share derived characters not present in fishes, indicating that dolphins and mammals share a more recent common ancestor.

Cladists use shared derived characters to reconstruct evolutionary relationships of organisms and express them in *cladograms.* In a cladogram, the **branches** depict the evolutionary history of each group. Each branch point, referred to as a **node,** represents the divergence, or splitting, of two or more new groups from a common ancestor. The most recent common ancestor of each monophyletic group would be found at the node. Thus, a cladogram uses the positions of branch points to illustrate the evolutionary relationships among taxa. Shared derived characters can be indicated by labels or by bars across the branches. Cladograms may be rooted. The **root** indicates the most recent common ancestor of the clade depicted.

Consider the evolutionary grouping of mammals, lizards, snakes, crocodiles, dinosaurs, and birds (Fig. 23-8a). Birds, along with dinosaurs, are thought to share a common ancestor with modern crocodiles and alligators (node D in Fig. 23-8b). Crocodiles, dinosaurs, and birds, then, constitute a monophyletic group, or clade. Similarly, snakes and lizards form a clade that is the closest group to birds, dinosaurs, and crocodiles. Mammals form an additional clade.

Evolutionary systematists use phenotypic similarity and a combination of shared ancestral characters and shared derived characters to establish evolutionary relationships and build classifications. Although evolutionary systematists recognize paraphyletic taxa, groups that include some, but not all, subgroups of organisms that share the same most recent common ancestor, most of the taxa that they currently recognize are monophyletic. These biologists recognize class Reptilia as a valid group containing snakes, lizards, crocodiles, dinosaurs, and turtles, even though this class is paraphyletic; it does not include all subgroups, such as birds, that evolved from the ancestral reptile (see Fig. 23-8b). Evolutionary systematists assign birds to a separate class because birds have diverged markedly from the reptiles. As we have discussed, cladists do not recognize reptiles as a natural grouping because reptiles do not form a monophyletic group.

Outgroup analysis is used in constructing and interpreting cladograms

A crucial step in most cladistic analyses is **outgroup analysis,** a method for estimating which attributes are shared derived characters in a given group of organisms. An **outgroup** is a taxon that is considered to have diverged earlier than the taxa under investigation, the **ingroups.** Therefore, the outgroup represents an approximation of the ancestral condition. An ideal outgroup is the closest relative of the group being studied, its sister taxon, and has not been highly modified since its origin. Recall that sister taxa evolved from the same recent common ancestor. Systematists argue that an outgroup is likely to retain the ancestral state

The cladistic approach, shown in (a), is preferred because it recognizes reptiles and birds as a monophyletic group.

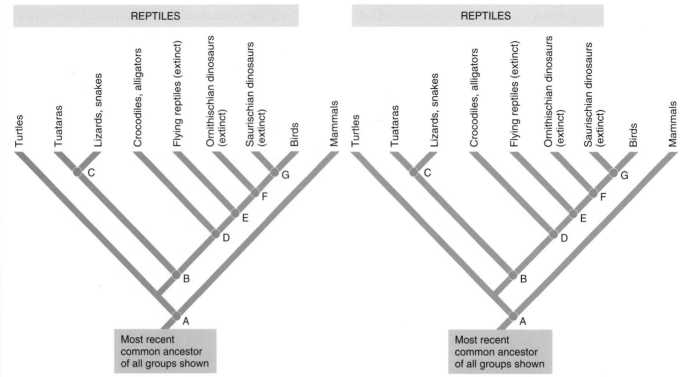

(a) Cladists classify birds and reptiles together because they have a recent common ancestor and are a monophyletic group. The cladogram shows the branching points in the evolution of the major groups of reptiles. Node F represents the branching of two clades of dinosaurs and of birds from a common ancestor, and the subsequent branching of birds from the dinosaurs.

(b) The evolutionary systematics approach considers both common ancestry and extent of divergence that has occurred since two taxa split. Reptiles are a paraphyletic group. Lizards, snakes, and crocodiles are phenotypically most similar, but crocodiles, dinosaurs, and birds are most closely related because they evolved most recently from a common ancestor.

Figure 23-8 Two approaches to the classification of reptiles, birds, and mammals

for characters being used in the analysis, allowing the biologists to identify the evolutionary changes leading to derived characters. To help you understand outgroup analysis, we now consider a specific example.

The first step in constructing a cladogram is to select the taxa, which may consist of individuals, species, genera, or other taxonomic levels. Here we use a representative group of eight chordates: the lancelet, lamprey, sunfish, frog, lizard, bear, chimpanzee, and human (▌Fig. 23-9). The next step is to select the homologous characters to be analyzed. In our example, we use seven characters. For each character, we must define all the different conditions, or states, as they exist in our taxa. For simplicity, we consider our characters to have only two different states: *present* or *absent*. Keep in mind that many characters used in cladistics have more than two states. For example, black, brown, yellow, and red may be only a few of the many possible states for the character of hair color.

The last, and often the most difficult, step in preparing the data is to organize the character states into their correct evolutionary order. For this step, we use outgroup analysis. In our

example, the lancelet, a small marine chordate with a fishlike appearance, is the chosen outgroup. It belongs to a taxon that is considered to have diverged earlier than any of the other taxa under investigation but to be closely related to vertebrates; thus, the lancelet represents an approximation of the ancestral condition for vertebrates. Therefore, the character state "absent" for a particular character state such as vertebrae is the ancestral (plesiomorphic) condition, and the character state "present" is the derived condition for the characters in Figure 23-9.

A cladogram is constructed by considering shared derived characters

Our objective is to construct a cladogram that requires the fewest number of evolutionary changes in the characters. Taxa are grouped by the presence of shared derived characters. To form a valid monophyletic group, all members must share at least one derived character. Membership in a clade cannot be established by shared ancestral characters (plesiomorphies).

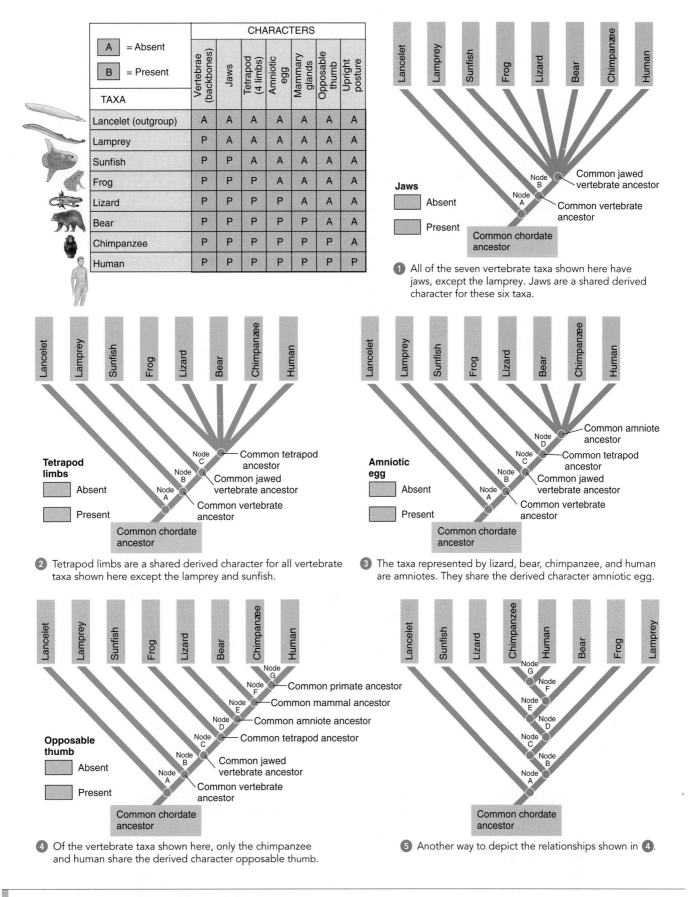

Figure 23-9 *Animated* Building a cladogram

In this example, the lancelet is the chosen outgroup that represents an approximation of the ancestral condition. Refer to the table as you follow the steps in ❶ through ❹ to build a cladogram. In ❺ we illustrate another way to build a cladogram showing the same relationships as in ❹.

Biologists use the principle of parsimony to choose between alternative hypotheses of evolutionary relationships.

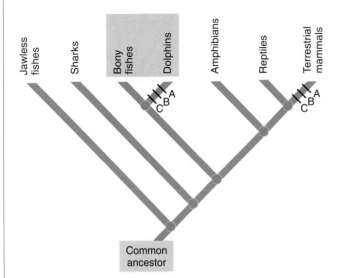

(a) Hypothesis 1: Dolphins and bony fishes are close relatives.

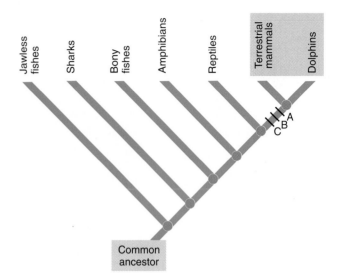

(b) Hypothesis 2: Dolphins and terrestrial mammals are close relatives.

Figure 23-10 Applying parsimony to cladogram construction

Are dolphins more closely related to bony fishes or to terrestrial mammals? Three characters are examined: character A = hair; B = mammary glands; C = middle-ear bones. In hypothesis 1, all three characters examined had to evolve independently twice. Hypothesis 2 is chosen because it requires that each character evolve only once.

In our example, notice that all taxa, except the outgroup lancelet, have vertebrae. We can therefore conclude that these seven vertebrate taxa form a valid clade. Next, among the seven vertebrate taxa, note that jaws are present in all groups except lampreys (a group of jawless vertebrates). Using these data, we construct a preliminary cladogram (Fig. 23-9 ①).

The base, or root, of the cladogram represents the common ancestor for all taxa being analyzed. In Figure 23-9 ①, node A represents the common chordate ancestor from which the outgroup (lancelet) and the seven vertebrate taxa evolved. Similarly, node B represents the common ancestor of the vertebrates; and node C, the common ancestor of the jawed vertebrates. Continuing with this procedure, notice that among the six-jawed taxa, all but sunfish are tetrapods (animals with four limbs) (Fig. 23-9 ②). Among the five tetrapods, all but frogs have amniotic eggs (Fig. 23-9 ③). (In an amniotic egg, the embryo is surrounded by a fluid-filled sac, called an *amnion*.) The branching process is continued, using the data in the table in the upper left-hand corner of the figure, until all clades are established (Fig. 23-9 ④).

In a cladogram each branch point represents a major evolutionary step

Notice that humans and chimpanzees share a recent common ancestor at node G in Figure 23-9 ④. This hypothesis is supported by the derived characters they share (synapomorphies). In the same way, bears are more closely related to the human–chimpanzee clade than to any other clade considered in our example, as indicated by the common ancestor at node F.

Note that shared derived characters are nested. As you trace the tree from its root to its tips, each branch reflects the addition of one or more shared derived characters. When we compare the nodes, the order of divergence (branching) is indicated by distance from the base of the diagram. The farther a node is located up the cladogram, the more recently the group diverged. In our example, node G represents the most recent divergence, and node A represents the most ancient divergence. Thus, in our example, humans are closely related to chimpanzees (through node G) but more distantly related to bears (through node F). Therefore, humans and chimpanzees are collectively assigned to a less inclusive taxon (order Primates), whereas humans, chimpanzees, and bears are assigned to a broader, more inclusive taxon (class Mammalia). In addition, the cladogram reveals that lizards are more closely related to the mammal clade than to frog, sunfish, or any other clade. Can you explain why?

Two important concepts guide interpretation of cladograms. First, the relationships among taxa are determined only by tracing along the branches back to the most recent common ancestor and not by the relative placement of the branches along the horizontal axis. It is possible to represent the same relationships with many different types of branching diagrams. For example, the cladogram in Figure 23-9 ⑤ is equivalent to the one in Figure 23-9 ④. (Verify this by comparing the numbered nodes and by checking the relationships described earlier.)

A second important concept is that the cladogram tells us which taxa shared a common ancestor and how recently they shared a common ancestor. The ancestor itself remains unspecified. The cladogram does not establish direct ancestor–descendant relationships among taxa. In other words, a cladogram does not suggest that a taxon gave rise to any other taxon.

A cladogram illustrates the pattern of evolutionary relationships but does not indicate time. The lengths of the branches in a cladogram do not give us information. A **phylogram** provides information about the relative number of mutations that have occurred within each lineage. This information is depicted by the length of the branches. An **ultrametric tree** is rooted, and the tips are all equally distant from the root. The branch points of an ultrametric tree indicate when in geologic time critical events took place.

Systematists use the principle of parsimony to make decisions

When systematists consider the evolutionary relationships of organisms, they must choose between multiple, competing cladograms. How do they choose the most accurate branching pattern? The most common criterion is the principle of **parsimony**—they choose the simplest explanation to interpret the data. Parsimony, a guiding principle in many areas of research, is based on the experience that the simplest explanation is probably the correct one. Applied to choice of cladograms, parsimony requires that the cladogram with the fewest changes in characters (the one with the fewest homoplasies) be accepted as most probable (▮ Fig. 23-10). In actual practice it is often possible to generate several cladograms that are equally parsimonious. The excluded choices are those with more homoplasies, because shared derived characters are more likely than homoplasies.

Systematists also use **maximum likelihood,** especially when analyzing molecular data. This method depends on the probability that nucleotide sequences in DNA and RNA change at a constant rate over time. Complex computer programs analyze large data sets and report the probability of a particular tree.

Many evolutionary questions remain. New hypotheses about how organisms are related suggest new taxonomic schemes. Systematics offers tools to help us test these hypotheses. In Chapters 24 through 31, we explore many evolutionary puzzles and discuss current hypotheses about relationships and classifications of organisms.

Review

▮ In what way are shared derived characters important to cladists?

▮ What does a node represent in a cladogram?

▮ What is outgroup analysis?

▮ How do systematists apply the principle of parsimony?

SUMMARY WITH KEY TERMS

Learning Objectives

1 State at least two justifications for the use of scientific names and classifications of organisms (page 483).

▮ **Systematics** is the scientific study of the diversity of organisms and their evolutionary relationships. **Taxonomy** is the branch of systematics devoted to naming, describing, and classifying organisms. The process of assigning organisms into groups based on their similarities or relationships is **classification.**

▮ Scientific names are important because they allow biologists from different countries with different languages to communicate about organisms. Biologists in distant locations must know with certainty that they are studying the same (or different) organisms.

▮ Classifications help biologists organize their knowledge.

2 Describe the binomial system of naming organisms, and arrange the Linnaean categories in hierarchical fashion, from most inclusive to least inclusive (page 483).

▮ Biologists name organisms using the **binomial system of nomenclature** developed by Linnaeus in the mid-18th century. In this system the basic unit of classification is the **species.**

▮ The name of each species has two parts: the **genus** name followed by the **specific epithet.**

▮ The hierarchical system of classification used in this book includes **domain, kingdom, phylum, class, order, family, genus,** and **species.** Each formal grouping at any given level is a **taxon.**

3 Describe the six kingdoms and three domains of organisms, and give the rationales for and against this system of classification (page 484).

▮ The six-kingdom classification recognizes the kingdoms **Archaea, Bacteria, Protista, Fungi, Plantae,** and **Animalia.** Members of Archaea and Bacteria are prokaryotes. Kingdom Protista consists of a diverse group of mainly unicellular, mainly aquatic eukaryotic organisms. The fungi, which include molds, yeasts, and mushrooms, absorb nutrients produced by other organisms. Kingdoms Plantae and Animalia both consist of multicellular eukaryotes. For additional characteristics of the six kingdoms, refer to Table 23-2.

▮ The three-domain classification system assigns organisms to domains **Archaea, Bacteria,** or **Eukarya.** Domain Bacteria is distinguished by the presence of the compound peptidoglycan in cell walls. Peptidoglycan is absent in the cell walls of the Archaea.

▮ Domain Eukarya includes the protists, fungi, plants, and animals.

ThomsonNOW™ **Explore the tree of life by clicking on the figures in ThomsonNOW.**

4 Given its distinguishing characters, classify an organism in the appropriate domain and kingdom (page 484).

▮ To classify an organism in a domain, first determine whether it is a prokaryote or eukaryote. If it is a prokaryote, decide whether it should be assigned to domain Archaea or domain Bacteria by learning about its characters, such as presence or absence of peptidoglycan in its cell wall. If it is not a prokaryote, it should be assigned to domain Eukarya. Check the characteristics in Table 23-2 to determine the appropriate kingdom.

5 Critically review the difficulties encountered in choosing taxonomic criteria (page 487).
- The goal of systematics is to determine evolutionary relationships, or **phylogeny,** based on shared characteristics. Historically, scientists have depended mainly on structural similarities to make decisions about phylogeny. **Homology,** the presence in two or more species of a trait derived from a recent common ancestor, implies evolution from a common ancestor.
- It can be challenging to decide if a character is homologous. Some seemingly homologous characters are acquired independently by **convergent evolution,** independent evolution of similar structures in distantly related organisms, or by **reversal,** reversion of a trait to its ancestral state. The term **homoplasy** refers to such superficially similar characters that are not homologous.

6 Apply the concept of shared derived characters to the classification of organisms (page 487).
- **Shared ancestral characters (plesiomorphies)** suggest a distant common ancestor.
- **Shared derived characters (synapomorphies)** indicate a more recent common ancestor. Systematists carefully consider shared derived characters in choosing taxonomic criteria.

7 Describe how molecular homologies contribute to the science of systematics (page 487).
- **Molecular systematics** depends on molecular structure to clarify phylogeny. Comparisons of nucleotide sequences in DNA and RNA, and of amino acid sequences in proteins, provide important information about how closely organisms are related. For example, comparison of nucleotide sequences in ribosomal RNA has led to important taxonomic decisions regarding domains, kingdoms, and species.

8 Contrast monophyletic, paraphyletic, and polyphyletic taxa (page 487).

- A **monophyletic taxon** includes all the descendants of the most recent common ancestor. A **paraphyletic taxon** consists of a common ancestor and some, but not all, of its descendants. The organisms in a **polyphyletic group** evolved from different recent ancestors.

9 Contrast the classification of reptiles and birds by an evolutionary systematist and a cladist (page 493).
- Scientists applying the principles of **cladistics** insist that taxa be monophyletic. Each monophyletic taxon, or **clade,** consists of a common ancestor and all its descendants. Cladists use shared derived characters to determine these relationships. Cladists classify reptiles and birds in a single clade.
- Scientists using an **evolutionary systematics** approach classify reptiles and birds in separate classes even though they are paraphyletic groups. Evolutionary systematics is based on shared ancestral characters as well as shared derived characters.

10 Describe the construction of a cladogram by using outgroup analysis (page 493).
- Cladists use shared derived characters to reconstruct evolutionary relationships and diagram these relationships in phylogenetic trees called **cladograms.** The cladogram indicates which taxa shared a common ancestor and how recently they shared that ancestor.
- Each branch point, referred to as a **node,** represents the divergence, or splitting, of two or more new groups from a common ancestor. The most recent common ancestor of each monophyletic group would be found at the node. In interpreting cladograms, cladists determine the relationships among taxa by tracing along the branches back to the nodes.
- Cladists use **outgroup analysis** to determine which characters in a given group of taxa are ancestral and which are derived. An **outgroup** is a taxon that represents the ancestral condition because it diverged earlier than any of the other taxa being investigated.
- Cladists use the principle of **parsimony:** They choose the simplest explanation to interpret the data.

ThomsonNOW **Learn more about interpreting and constructing cladograms by clicking on the figures in ThomsonNOW.**

TEST YOUR UNDERSTANDING

1. The science of describing, naming, and classifying organisms is (a) systematics (b) taxonomy (c) cladistics (phylogenetic systematics) (d) phenetics (e) evolutionary systematics

2. In the binomial system of nomenclature, the scientific name of each species consists of which two parts? (a) class, specific epithet (b) family, genus (c) genus, specific epithet (d) family, species (e) genus, species

3. The mold that produces penicillin is *Penicillium notatum.* *Penicillium* is the name of its (a) genus (b) order (c) family (d) species (e) specific epithet

4. Closely related genera may be grouped together in a single (a) phylum (b) domain (c) species (d) family (e) kingdom

5. Related classes are grouped in the same (a) genus (b) phylum (c) order (d) paraphyletic taxon (e) family

6. In the six-kingdom system, the kingdom that includes the protozoa is (a) Plantae (b) Protista (c) Archaea (d) Eukarya (e) Fungi

7. Decomposers such as molds and mushrooms belong to which domain and kingdom? (a) Eukarya, Plantae (b) Bacteria, Protista (c) Archaea, Fungi (d) Eukarya, Protista (e) Eukarya, Fungi

8. A group of organisms that contains a recent common ancestor and all its descendants is (a) polyphyletic (b) paraphyletic (c) monophyletic (d) an example of horizontal gene transfer (e) plesiomorphic

9. The presence of homologous structures in two different groups of organisms suggests that (a) the organisms evolved from a common ancestor (b) convergent evolution has oc-

curred (c) they belong to a polyphyletic group (d) homoplasy has occurred (e) independently acquired characters may evolve when organisms inhabit similar environments

10. The dolphin and the human both have the ability to nurse their young, whereas the less closely related fish does not. The ability to nurse their young is a (a) shared derived character of mammals (b) shared ancestral character of all vertebrates (c) plesiomorphy (d) homologous behavior (e) homoplasy

11. Relative constancy in the rates of DNA and protein evolution permits biologists to use these macromolecules as (a) molecular clocks (b) polymerase chains (c) clades (d) paraphyletic clues (e) outgroups

12. Using DNA as a molecular bar code (a) is effective only with bacteria (b) is possible because of similarities in sugars (c) could cause major mistakes in distinguishing among closely related species (d) is an effective way to identify vertical gene transfer (e) could help taxonomists identify and describe new species

13. The conclusion that fungi are more closely related to animals than to plants was based in large part on comparing (a) molecular clocks (b) polymerase chains (c) nucleotides in DNA (d) ribosomal RNA sequences (e) amino acid sequences

14. Phenetics (a) is a systematic approach based on phenotypic similarities (b) emphasizes common ancestry (c) emphasizes polyphyletic groups (d) focuses on ancestral characters (e) strives to differentiate between homology and homoplasy

15. Some systematists classify crocodiles and birds in the same taxon because they are contained within a monophyletic group. These systematists follow which approach? (a) phyletic (b) cladistic (c) evolutionary systematic (d) polyphyletic (e) both cladistic and evolutionary systematic

16. In cladistic analysis (a) ancestral characters are used to reconstruct phylogenies (b) characters must be homoplastic (c) polyphyletic groups are preferred (d) the method called maximum likelihood is rarely used (e) outgroup analysis is used to determine a taxon that diverged earlier than the other taxa being studied

17. When cladists use the principle of parsimony, they (a) choose the simplest explanation to interpret the data (b) select multiple hypotheses to explain each relationship (c) do not use outgroups (d) typically use a polyphyletic approach (e) hypothesize that the most complex explanation is most probably the correct one

CRITICAL THINKING

1. What are the advantages of PhyloCode? Of bar-coding organisms?

2. **Evolution Link.** Are members of a genus similar because they share a common ancestor, or do they belong to the same genus because they are similar? How might your answer vary depending on which approach to systematics you are following?

3. **Evolution Link.** The TATA-binding protein (TBP) is thought to be necessary for transcription in all eukaryotic cell nuclei. Studies show that archaea, but not bacteria, have a protein structurally and functionally similar to TBP. What does this similarity suggest regarding the evolution of archaea and eukaryotes? How might knowledge of this similarity affect how systematists classify these organisms?

4. **Analyzing Data.** Construct a cladogram based on the following data. Mosses are plants with no vascular tissue. Horsetails, ferns, gymnosperms (pines and other plants with naked seeds), and angiosperms (flowering plants) are all vascular plants. Seeds are absent in all but the gymnosperms and angiosperms.

5. **Analyzing Data.** In the illustration, what kind of grouping is represented by the bracketed area? What type of group is formed by 2 and 3? What type of group is formed by 2, 4, and 6?

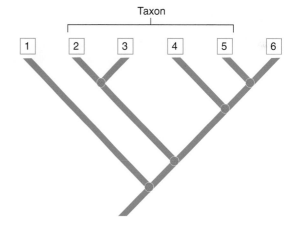

Additional questions are available in ThomsonNOW at www.thomsonedu.com/login

24

Viruses and Prokaryotes

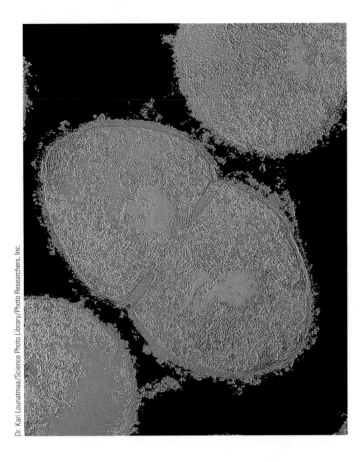

Dr. Kari Lounatmaa/Science Photo Library/Photo Researchers, Inc.

Color-enhanced TEM of a bacterium (*Streptococcus pyogenes*) dividing by binary fission. This bacterium, a pathogen that inhabits the human nose and throat, can cause scarlet fever and inflammation of the heart tissue. The strain shown here is resistant to antibiotics, and infection can be fatal.

KEY CONCEPTS

A virus is a small particle consisting of a DNA or RNA genome surrounded by a protein coat.

Viroids and prions are smaller than viruses; a prion consists only of proteins.

In contrast to eukaryotic cells, prokaryotic cells do not have membrane-enclosed organelles such as nuclei and mitochondria.

Prokaryotes make up two of the three domains: Bacteria and Archaea.

Evolution occurs rapidly in prokaryotes; natural selection acts on the genetic variation provided by mutations and genetic recombination and is facilitated by rapid reproduction.

Great diversity has evolved in the mode of nutrition, the metabolism, and the ecological roles of prokaryotes.

Prokaryotes (bacteria and archaea) have inhabited our planet for more than 3.5 billion years—much longer than eukaryotes, which evolved about 1.5 billion to 1.6 billion years ago. Although prokaryotes are microscopic, they are so numerous that they, together with the fungi, account for approximately half of Earth's biomass, the mass of living material. In contrast, plants account for about 35% and animals for about 15% of the biomass of our planet.

The Dutch microscopist Anton van Leeuwenhoek discovered bacteria and other microorganisms in 1674 when he looked at a drop of lake water through a glass lens. During the late 1800s, many microorganisms, including bacteria, fungi, and protozoa, were identified as **pathogens,** agents that cause disease (see photograph). Although bacteria cause many diseases, for example, respiratory infections and food poisoning in humans, only a small minority of bacterial species are pathogens. In fact, bacteria play an essential role in the biosphere as decomposers, breaking down organic molecules into their components. Along with fungi, prokaryotes are nature's chief recyclers. Without these microorganisms, elements such as carbon, nitrogen, phosphorus, and sulfur would remain locked up in the wastes and dead bodies of plants, animals, and other organisms and would be unavailable for the synthesis of new cells and organisms.

Some prokaryotes are producers that carry on photosynthesis. Others convert atmospheric nitrogen to ammonia and then to nitrates, forms that are used by plants (see Fig. 54-8). This conversion enables plants and animals (because they eat plants) to manufacture essential nitrogen-containing compounds such as proteins and nucleic acids.

In contrast to prokaryotes, viruses are acellular. Biologists long considered them nonliving particles, but some biologists now view them as life-forms. Viruses contain the nucleic acids necessary to make copies of themselves, and they reproduce by invading living cells and commandeering their metabolic machinery. These tiny, but potent, particles infect cells and produce a wide variety of diseases in plants and animals. Viruses influence the biodiversity of prokaryotes and algae, as well as many ecological processes, including the recycling of nutrients.

This chapter examines the diversity and characteristics of viruses, prokaryotes, and the smaller viroids and prions. They are not a natural group of closely related organisms, and we discuss them in a single chapter only for convenience, not to reflect shared ancestry. ■

VIRUSES

Learning Objectives

1 Describe the structure of a virus, and contrast a virus with a living cell.
2 Trace the evolutionary origin of viruses according to current hypotheses.
3 Characterize bacteriophages, and contrast a lytic cycle with a lysogenic cycle.
4 Compare viral infection of animals and plants, and identify specific diseases caused by animal viruses.
5 Describe the reproductive cycle of a retrovirus, such as human immunodeficiency virus (HIV).

During the late 1800s, botanists searched for the cause of tobacco mosaic disease, which stunts the growth of tobacco plants and gives the infected tobacco leaves a spotted, mosaic appearance. They found they could transmit the disease to healthy plants by daubing their leaves with the sap of diseased plants. In 1892, Dmitri Ivanowsky, a Russian botanist, showed that the sap was still infective after it had been passed through porcelain filters designed to filter out all known bacteria. A few years later, in 1898, Martinus Beijerinck, a Dutch microbiologist, provided evidence that the agent that caused tobacco mosaic disease had many characteristics of a living organism. He hypothesized that the infective agent could reproduce only within a living cell and named it *virus* (from the Latin word *virus*, which means "poison").

Early in the 20th century, scientists discovered infective agents, like those responsible for tobacco mosaic disease, that could cause disease in animals or kill bacteria. These pathogens were so small that they could not be seen with the light microscope. They also passed through filters that removed all known bacteria. Curiously, they could not be grown in laboratory cultures unless living cells were present.

Further studies of viruses have confirmed that they are indeed small. The poliovirus is about 30 nm in diameter (about the size of a ribosome). Almost a million of them lined up end to end would span only an inch. A larger virus, such as the poxvirus that causes smallpox, can measure up to 300 nm long and 200 nm wide. The largest known virus is the Mimivirus, with a diameter of at least 400 nm, the size of a small bacterium (mycoplasma). Mimivirus infects amoebae (which are protists).

A virus consists of nucleic acid surrounded by a protein coat

Sir Peter Medawar, 1960 Nobel laureate in Medicine, defined a virus as "a piece of bad news wrapped in protein." We now know that most known viruses are not pathogens, that is, they do not cause disease. A **virus,** or **virion,** is a tiny, infectious particle consisting of a nucleic acid core (its genetic material) surrounded by a protein coat called a **capsid.**

Viruses are not made of cells and cannot independently perform metabolic activities. They do not have the components necessary to carry on cellular respiration or to synthesize proteins and other molecules. Viruses reproduce, but only within the complex environment of the living **host cells** they infect. Viruses use their genetic information to force the host cell to replicate the viral nucleic acid. They take over the transcriptional and translational mechanisms of the host cell. Viruses are **obligate intracellular parasites,** which means that they survive only by using the resources of a host cell.

A virus may contain either DNA or RNA

Until recently, biologists thought that whereas other living organisms contain both deoxyribonucleic acid (DNA) and ribonucleic acid (RNA), a virus contains *either* DNA or RNA, not both. However, investigators recently sequenced the genome of Mimivirus and reported that this large DNA virus also has six transfer RNAs. Viruses typically have single-stranded (ss) DNA, double-stranded (ds) DNA, ssRNA, or dsRNA. As we will discuss, the type of nucleic acid is important in classifying viruses.

The capsid is a protective protein coat

After viruses take over the metabolic machinery of a cell, the host synthesizes capsids for the viruses. Some viruses are surrounded by an outer membranous envelope, and the host cell also produces these envelopes.

Capsids consist of protein subunits, called *capsomeres.* The capsomeres determine the shape of a virus. Viral capsids generally are either helical or polyhedral, or a combination of both shapes. Helical viruses, such as the tobacco mosaic virus (TMV), appear as long rods or threads (▌Fig. 24-1a). The capsid is a hollow cylinder made up of proteins that form a groove into which the RNA fits.

Polyhedral viruses, such as the adenoviruses (which cause a number of human illnesses, including some respiratory infections), appear somewhat spherical (▌Fig. 24-1b). The capsomeres are organized in equilateral triangles. The capsid of a large virus can consist of several hundred capsomeres. The most common polyhedral structure is an *icosahedron,* a structure with 20 identical surface faces (each face is a triangle).

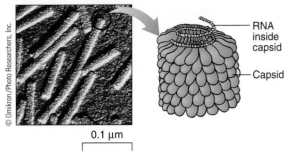

(a) TEM of tobacco mosaic virus, a rod-shaped virus with a helical arrangement of capsid proteins.

0.1 µm

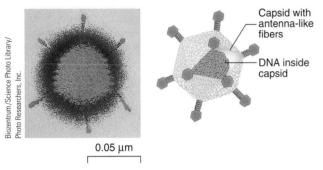

(b) Color-enhanced TEM of an adenovirus, which has a capsid composed of 252 subunits (visible as tiny ovals) arranged into a 20-sided polyhedron. Twelve of the subunits have projecting protein spikes that permit the virus to recognize host cells.

0.05 µm

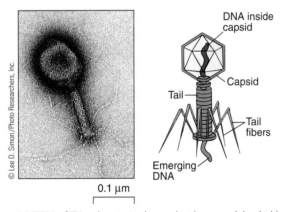

(c) TEM of T4, a bacteriophage that has a polyhedral head and a helical tail.

0.1 µm

Figure 24-1 *Animated* Virus structure

Some viruses have both helical and polyhedral components. Viruses that infect bacteria are called **bacteriophages** ("bacteria eaters"), or **phages** (▮ Fig. 24-1c). The T4 phage that infects the bacterium *Escherichia coli* consists of a polyhedral "head" attached to a helical "tail," a shape commonly found in bacteriophages. Many phages have tail fibers that attach to the host cell.

Viruses surrounded by an outer membranous envelope are referred to as *enveloped viruses.* Typically, the virus acquires the envelope, which is a part of the host cell's plasma membrane, as it leaves the host cell. Interestingly, while in the host cell, the virus synthesizes certain proteins and inserts them into the host's plasma membrane. Thus, the viral envelope consists of phos-

pholipids and proteins of the host's plasma membrane, as well as distinctive proteins produced by the virus itself. Some viruses produce envelope proteins that extend out from the envelope as spikes. As we will discuss, these spikes can be very important in the virus's interaction with the host cell.

Viruses may have evolved from cells

Where did viruses come from? The most widely held hypothesis is that viruses derive from bits of nucleic acid that "escaped" from cellular organisms. Viruses may have originated as mobile genetic elements such as *transposons* (see Chapter 13) or *plasmids* (small, circular DNA fragments discussed in Chapter 15 and later in this chapter). Such fragments could have moved from one cell and entered another through damaged cell membranes.

According to this *escaped gene hypothesis,* some viruses may trace their origin to animal cells, others to plant cells or to bacterial cells. Their multiple origins may explain why many viruses are species specific; perhaps they infect only those species that are closely related to the organisms from which they originated. This hypothesis is supported by the genetic similarity between some viruses and their host cells—a closer similarity than exists between one type of virus and another.

A very different hypothesis suggests that viruses arose early in the history of life, even before the three domains diverged. Evidence for this hypothesis comes from similarities found in the protein structures of some viral capsids and in genetic similarities between some viruses that infect archaea and some that infect bacteria. Molecular biologists studying this hypothesis consider it improbable that these similarities evolved independently. This evidence suggests that viruses diverged very early. However, viruses are parasites, dependent on their hosts. How could they have existed before their hosts evolved?

Several investigators have suggested that DNA viruses have a common origin and that they evolved before the evolution of the three domains. According to this hypothesis, ancestral DNA viruses were important in the evolution of eukaryotes. The investigators who sequenced the Mimivirus genome reported that its genome is larger than the genomes of 20 different bacteria and archaea. Also unlike other viruses, Mimivirus has genes that encode components for protein translation. These researchers suggest that this giant virus evolved from a more complex ancestor and gradually lost some of the genes necessary for protein synthesis. This hypothesis explains how viruses could have existed before their hosts evolved.

No matter how viruses came to be, they are here in large numbers and are important players in evolution. They mutate constantly, producing many new alleles. They replicate rapidly, and some of their genes are incorporated into the genomes of host cells. In fact, viruses may be a major source of new genes for their host cells.

The International Committee on Taxonomy of Viruses classifies viruses

Viruses present a taxonomic challenge to biologists because they do not show the characteristics that define living organisms (see Chapter 1). They are acellular, do not carry on metabolic activities, and reproduce only by taking over the reproductive machinery of other cells. Viruses do not manufacture ribosomes so do not have distinctive rRNA. For these reasons, viruses cannot be classified in any of the three domains.

Viruses can be classified based on their **host range,** the types of organisms they infect. Host ranges include bacteria, algae, protozoa, fungi, plants, invertebrates, and vertebrates. Viruses are further classified based on their characteristics. Two major defining traits are what type of nucleic acid the virus contains and whether the nucleic acid is single or double stranded. Other traits considered are the presence of an envelope, the size and shape of the virus, and the method by which it is transmitted from host to host.

The International Committee on Taxonomy of Viruses (ICTV), a group of virologists, decides on the criteria for classifying and naming viruses. The ICTV has classified viruses into approximately 56 families and 233 genera based on host range and characteristics. Note that this classification system is not a traditional Linnaean system and does not assign viruses to domains, kingdoms, or phyla.

Bacteriophages are viruses that attack bacteria

Much of our knowledge of viruses has come from studying bacteriophages, because they can be cultured easily within living bacteria in the laboratory. Researchers have identified more than 2000 phages. Bacteriophages are among the most complex viruses (see Fig. 24-1c). Their most common structure consists of a long nucleic acid molecule (usually dsDNA) coiled within a polyhedral head.

Before the age of sulfa drugs and antibiotics, phages were used clinically to treat infection. In the 1940s, they were abandoned (at least in Western countries) in favor of antibiotics, which were more dependable and easier to use. Now, with the widespread and growing problem of bacterial resistance to antibiotics, these bacteria-killing viruses are once again the focus of research. With new knowledge of phages and sophisticated technology, several research groups are investigating phages to determine which ones kill which species of bacteria. Scientists are genetically engineering phages so that bacteria will be slower to evolve resistance to them.

Phages can also be used to improve food safety. For example, certain phages can kill deadly strains of *E. coli* in cattle. These bacteria do not appear to make cattle ill but can cause illness and death in people who eat contaminated, undercooked hamburgers.

Viruses reproduce only inside host cells

The reproductive cycle of viruses begins with a virus coming into contact with a host cell. The virus typically attaches to the surface of the host cell. The viral nucleic acid must enter the host cell and synthesize the components it needs to reproduce itself. Then, viral components are assembled and viruses are released from the cell—ready to invade other cells. Two types of reproduction among viruses are lytic and lysogenic cycles.

Lytic reproductive cycles destroy host cells

In a **lytic cycle,** the virus lyses (destroys) the host cell. When the virus infects a susceptible host cell, it forces the host to use its metabolic machinery to replicate viral particles. Viruses that have only a lytic cycle are described as **virulent,** which means that they cause disease, and often death.

Five steps are typical in lytic viral reproduction (❚ Fig. 24-2):

① **Attachment (or absorption).** The virus attaches to receptors on the host cell. This process ensures that the virus infects only its specific host.

② **Penetration.** The virus penetrates the host plasma membrane and moves into the cytoplasm. Many viruses that infect animal cells enter the host cell intact. Some phages inject only their nucleic acid into the cytoplasm of the host cell; the capsid remains on the outside.

③ **Replication and synthesis.** The viral genome contains all the information necessary to produce new viruses. Once inside, the virus degrades the host cell nucleic acid and uses the molecular machinery of the host cell to replicate its own nucleic acid. Many antiviral drugs interfere with replication of viral nucleic acid. After replication, capsid proteins and other needed molecules are synthesized.

④ **Assembly.** The newly synthesized viral components are assembled into new viruses.

⑤ **Release.** Assembled viruses are released from the cell. Generally, lytic enzymes, produced by the phage late in the replication process, destroy the host plasma membrane. Phage release typically occurs all at once and results in rapid cell lysis. Animal viruses are often released slowly or bud off from the plasma membrane.

Once released, the viruses infect other cells, and the process begins anew. The time required for viral reproduction, from attachment to the bacterium to the release of new viruses, varies from less than 20 minutes to more than an hour.

How do bacteria protect themselves from phage infection? You may recall from Chapter 15 that bacteria produce **restriction enzymes,** enzymes that cut up the foreign DNA of the phage. The bacterial cell protects its own DNA by slightly modifying it after replication so that the restriction enzyme does not recognize the sites it would cut.

Temperate viruses integrate their DNA into the host DNA

Temperate viruses do not always destroy their hosts. In a **lysogenic cycle,** the viral genome usually becomes integrated into the host bacterial DNA and is then referred to as a **provirus,** or **prophage.** When the bacterial DNA replicates, the provirus also replicates (❚ Fig. 24-3). The viral genes that code for viral structural proteins may be repressed indefinitely. Bacterial cells carrying proviruses are called *lysogenic cells.* Certain external conditions (such as ultraviolet light and X-rays) cause temperate viruses to revert to a lytic cycle and then destroy their host. Sometimes temperate viruses become lytic spontaneously.

Viruses reproduce by seizing control of the metabolic machinery of a host cell.
The host cell is destroyed in a lytic infection.

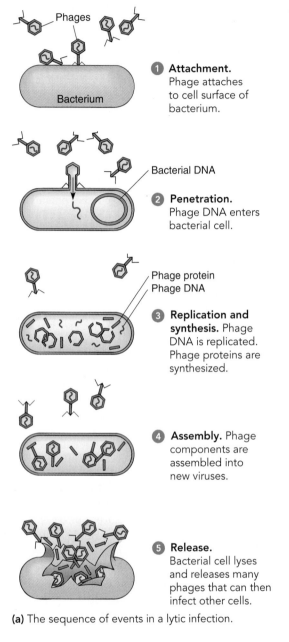

Phages

1 Attachment. Phage attaches to cell surface of bacterium.

Bacterium

Bacterial DNA

2 Penetration. Phage DNA enters bacterial cell.

Phage protein
Phage DNA

3 Replication and synthesis. Phage DNA is replicated. Phage proteins are synthesized.

4 Assembly. Phage components are assembled into new viruses.

5 Release. Bacterial cell lyses and releases many phages that can then infect other cells.

(a) The sequence of events in a lytic infection.

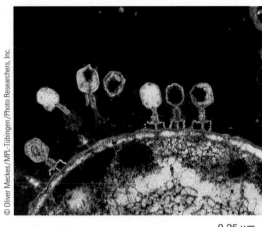

0.25 µm

(b) Color-enhanced TEM of phages infecting a bacterium, *Escherichia coli*.

Figure 24-2 *Animated* Lytic cycle

Bacterial cells containing certain temperate viruses may exhibit new properties. This change is called **lysogenic conversion.** An interesting example involves the bacterium *Corynebacterium diphtheriae*, which causes diphtheria. Two strains of this species exist, one that produces a toxin (and causes diphtheria) and one that does not. The only difference between these two strains is that the toxin-producing bacteria are infected by a specific temperate phage. The phage DNA codes for the powerful toxin that causes the symptoms of diphtheria. Similarly, the bacterium

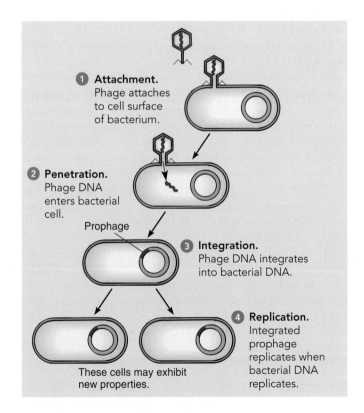

1 Attachment. Phage attaches to cell surface of bacterium.

2 Penetration. Phage DNA enters bacterial cell.

Prophage

3 Integration. Phage DNA integrates into bacterial DNA.

4 Replication. Integrated prophage replicates when bacterial DNA replicates.

These cells may exhibit new properties.

Figure 24-3 *Animated* Lysogenic cycle

Temperate phages integrate their nucleic acid into the host cell DNA, making it a lysogenic cell.

TABLE 24-1

Some Viruses That Infect Vertebrates

Group	Diseases Caused	Characteristics
DNA Viruses with Envelope		
Poxviruses	Smallpox, cowpox,* monkeypox, and economically important diseases of domestic fowl	dsDNA; large, complex viruses; replicate in the cytoplasm of the host cell
Herpesviruses	Cold sores (herpes simplex virus type 1); genital herpes, a sexually transmitted disease (herpes simplex virus type 2); chickenpox and shingles (herpes varicella–zoster virus); infectious mononucleosis and Burkitt's lymphoma (Epstein–Barr virus)	dsDNA; medium to large, enveloped viruses; replicate in the host nucleus†
DNA Viruses with No Envelope		
Adenoviruses	Respiratory tract disorders (e.g., sore throat, tonsillitis), conjunctivitis, and gastrointestinal disorders are caused by more than 40 types of adenoviruses in humans; other varieties infect other animals	dsDNA; replicate in the host nucleus
Papovaviruses‡	Human warts and some degenerative brain diseases; some cancers, including cervical cancer	dsDNA
Parvoviruses	Infections in dogs, swine, arthropods, rodents; gastroenteritis in humans (transmitted by consumption of infected shellfish)	ssDNA; some require a helper virus in order to multiply
RNA Viruses with Envelope		
Togaviruses	Rubella (German measles)	ssRNA that can serve as mRNA; large diverse group of medium-sized, enveloped viruses; many transmitted by arthropods
Orthomyxoviruses	Influenza (flu) in humans and other animals	ssRNA that serves as template for mRNA synthesis; medium-sized viruses that often exhibit projecting spikes
Paramyxoviruses	Rubeola (measles) and mumps in humans; distemper in dogs	ssRNA; resemble orthomyxoviruses but somewhat larger
Rhabdoviruses	Rabies	ssRNA
Coronaviruses	Upper respiratory infections; SARS	ssRNA
Flaviviruses	Yellow fever; West Nile virus; hepatitis C (the most common reason for liver transplants in the United States)	ssRNA
Filoviruses	Hemorrhagic fever, including that caused by the Ebola virus	ssRNA
Bunyaviruses	St. Louis encephalitis; hantavirus pulmonary syndrome (caused by Sin Nombre virus, a hantavirus)	ssRNA
Retroviruses	AIDS; some types of cancer	ssRNA viruses that contain reverse transcriptase for transcribing the RNA genome into DNA; two identical molecules of ssRNA
RNA Viruses with No Envelope		
Picornaviruses	Polio (poliovirus); hepatitis A (hepatitis A virus); intestinal disorders (enteroviruses); common cold (rhinoviruses); aseptic meningitis (coxsackievirus, echovirus)	ssRNA that can serve as mRNA; diverse group of small viruses
Reoviruses	Vomiting and diarrhea; encephalitis	dsRNA

*The vaccinia (cowpox) virus is used to produce genetically engineered vaccines.

†These viruses frequently cause latent infections; some cause tumors.

‡The virus SV40 has been used as a vector to transport genes into cells.

Clostridium botulinum, which causes botulism, a serious form of food poisoning, is harmless unless it contains certain provirus DNA that induces synthesis of the toxin.

Many viruses infect vertebrates

Hundreds of different viruses infect humans and other vertebrates. Animal viruses cause hog cholera, foot-and-mouth disease, canine distemper, influenza, and certain types of cancer (such as feline leukemia and cervical cancer). Viruses cause chickenpox, herpes simplex (one type causes genital herpes), mumps, rubella (German measles), rubeola (measles), rabies, warts, infectious mononucleosis, influenza, viral hepatitis, and AIDS (▮ Table 24-1 and *Focus On: Influenza and Other Emerging and Re-emerging Diseases*). Most humans suffer from two to six viral infections each year, including common colds.

INFLUENZA AND OTHER EMERGING AND RE-EMERGING DISEASES

The development of antibiotics, vaccines, and improved health care has made it possible for public health officials to effectively control some diseases. However, many familiar diseases, such as influenza, malaria, tuberculosis, and bacterial pneumonias, continue to infect large numbers of people and sometimes reappear in forms that are resistant to drug treatments. **Re-emerging diseases** are those that have been almost eradicated and then strike unpredictably, causing an epidemic, sometimes in a new geographic area. **Emerging diseases,** those new to the human population—such as AIDS, severe acute respiratory syndrome (SARS), Ebola, eastern equine encephalitis, and West Nile virus—often appear suddenly. According to the U.S. Centers for Disease Control and Prevention, more than 200 new, continual, or re-emerging pathogens have the potential to strike globally.

Historically, new viral strains have claimed many human lives. For example, in 1918 an influenza (flu) pandemic killed more than 20 million people throughout the world. Flu pandemics also occurred in 1957 and 1968. The outbreak of influenza A (H5N1) in 2005 was a major warning to the World Health Organization (WHO) and other public health agencies to mobilize resources and prepare for a new pandemic. Some epidemiologists warn that an influenza epidemic today could kill huge numbers of people, perhaps as many as one third of the human population. Pathogens can strike quickly and spread rapidly.

Typically, new strains of the avian influenza virus (orthomyxovirus) at first infect mainly birds and are not able to spread among mammals. However, a virus of the avian strain can mutate and become virulent. The avian virus can also exchange genes with a virus that has the genes necessary to spread among pigs and other mammals. Pigs can be susceptible to both avian and human strains of the virus. Inside the pig's body, the viruses can exchange genes and thereby produce a virus that can infect humans. Gene exchange can also occur inside a human host and produce a viral strain that can spread from person to person. If the new combination of viral genes is unfamiliar to the human immune system and can spread easily, a pandemic can occur (see figure).

Each year new strains of influenza virus evolve and become infectious, so the challenge to scientists and public health officials is ongoing. In addition, the virus has a very brief incubation period (about 2 days) and so spreads rapidly. Minimizing the effects of a pandemic will require careful surveillance and willingness for countries to rapidly share information. Effective vaccines and treatments must be developed, and containment measures (such as quarantines) must be planned.

As new technology is developed, researchers can more quickly characterize emerging disease organisms and explore treatments and vaccines. Hepatitis E, a viral disease that may be re-emerging, was documented as an epidemic in India in 1955 but was not isolated as a separate virus until the development of modern immunologic research tools in the early 1980s. In contrast, SARS, first recognized as a global threat in 2003, was successfully contained in less than 4 months. Researchers isolated and identified the SARS virus as a coronavirus in a matter of a few weeks. This information coupled with public health response helped speed efforts to diagnose and treat the disease and to reduce transmission, thus preventing a global epidemic of this deadly disease. However, WHO cautions that the SARS virus could hide in some animal or environmental reservoir and resurface when conditions again become favorable for spread to its new human host. Epidemiologists are modeling the SARS epidemic in an effort to answer basic questions, such as whether we can move from control of local outbreaks to global eradication of this disease. The SARS virus does not become infectious for about 10 days, which allows more time to track its spread and contain it than for a flu virus.

Even at the level of our current knowledge about viruses and epidemiology, just how prepared are we to contain a particularly virulent virus? How well are we containing HIV (the virus that causes AIDS), which infects 5 million and kills 3 million people worldwide every year?

Since the September 11, 2001, terrorist attacks in the United States, bioterrorism has become a critical concern worldwide. Bioterrorism is the intentional use of microorganisms or toxins derived from living organisms to cause death or disease in humans, animals, or plants on which humans depend. For example, in 2001 *Bacillus anthracis*, the causative agent of anthrax, was intentionally released and disseminated through the U.S. postal system. Terrorists could conceivably initiate epidemics of anthrax, smallpox, plague, and other potentially fatal diseases. The quest for effective treatments and vaccines for these diseases has taken on new urgency.

Whether disease outbreaks occur naturally or intentionally, countries must be prepared to protect their citizens. How well are public health agencies prepared to quickly detect and investigate new outbreaks? How well are they prepared to help the health-care community contain the spread of disease? Are we training physicians, nurses, and other health-care providers to recognize, report, and treat infectious diseases? Do we have sufficient supplies of medications and vaccines that would be necessary to contain an epidemic? Are systems of communication effective in reporting and sharing information among health-care agencies, providers, and the public?

Human activity, including social factors such as urbanization, global travel, and war, contributes to epidemics of infectious disease. For example, as human populations concentrate in cities, large numbers

---

Many vertebrate viruses have lytic reproductive cycles

Most viruses cannot survive very long outside a living host cell, so their survival depends on their being transmitted from animal to animal. However, their host range may be quite limited because attachment to a host cell is very specific. The type of attachment proteins on the surface of a virus determines what type of cell it can infect. Receptors typically vary with each species and sometimes with each type of tissue. Thus, many human viruses can infect only humans because the viral attachment proteins combine only with receptor sites found on human cell surfaces. The measles virus and poxviruses infect many types of human tissue

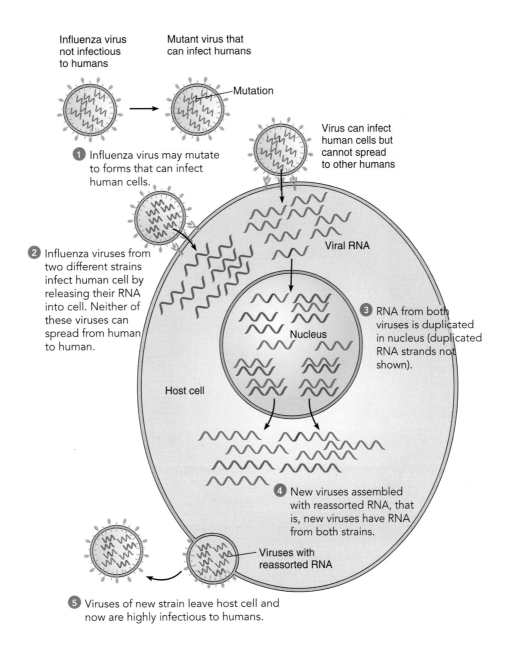

Influenza virus not infectious to humans

Mutant virus that can infect humans

Mutation

1 Influenza virus may mutate to forms that can infect human cells.

Virus can infect human cells but cannot spread to other humans

2 Influenza viruses from two different strains infect human cell by releasing their RNA into cell. Neither of these viruses can spread from human to human.

Viral RNA

3 RNA from both viruses is duplicated in nucleus (duplicated RNA strands not shown).

Nucleus

Host cell

4 New viruses assembled with reassorted RNA, that is, new viruses have RNA from both strains.

Viruses with reassorted RNA

5 Viruses of new strain leave host cell and now are highly infectious to humans.

of people come into close contact, permitting the rapid spread of viruses. Living conditions, including sanitation, nutrition, physical stress, level of health care, and sexual practices, are important factors in the spread of disease. In the United States and other highly developed countries, infectious disease accounts for about 4% to 8% of deaths compared with death rates of 30% to 50% in developing regions. Lessons learned from dealing with emerging viruses and the resurgence of old ones will help us contain future epidemics, but much more research is needed.

because their attachment proteins combine with receptor sites on a variety of cells. In contrast, polioviruses attach to specific types of human cells, such as those that line the digestive tract and motor neurons of the brain and spinal cord.

Some viruses, such as the adenoviruses, have fibers that project from the capsid and adhere to complementary receptors on the host cell. Other viruses, such as those that cause herpes and rabies, are surrounded by a lipoprotein envelope with projecting glycoprotein spikes that attach to a host cell. The influenza virus also has rodlike spikes that project from its envelope. The glycoprotein spikes bind with specific receptors present only on cells lining the respiratory tract of certain vertebrates, including wild

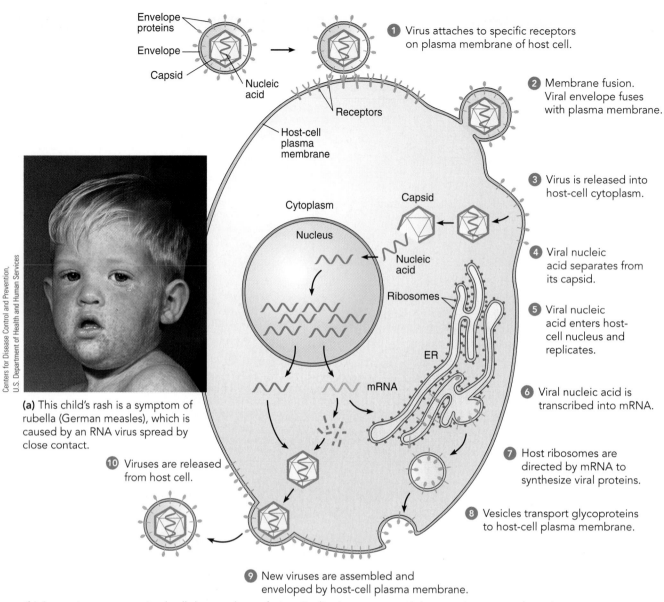

Envelope proteins

Envelope

Capsid

Nucleic acid

1 Virus attaches to specific receptors on plasma membrane of host cell.

2 Membrane fusion. Viral envelope fuses with plasma membrane.

Receptors

Host-cell plasma membrane

3 Virus is released into host-cell cytoplasm.

Cytoplasm

Capsid

Nucleus

Nucleic acid

4 Viral nucleic acid separates from its capsid.

Ribosomes

5 Viral nucleic acid enters host-cell nucleus and replicates.

ER

6 Viral nucleic acid is transcribed into mRNA.

mRNA

7 Host ribosomes are directed by mRNA to synthesize viral proteins.

10 Viruses are released from host cell.

8 Vesicles transport glycoproteins to host-cell plasma membrane.

9 New viruses are assembled and enveloped by host-cell plasma membrane.

Centers for Disease Control and Prevention, U.S. Department of Health and Human Services

(a) This child's rash is a symptom of rubella (German measles), which is caused by an RNA virus spread by close contact.

(b) Some viruses enter animal cells by membrane fusion. Replication occurs, and the new viruses are released.

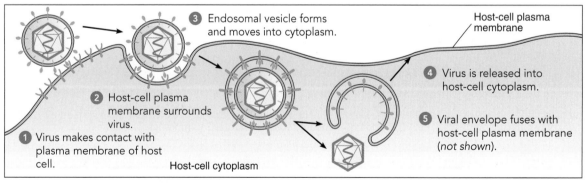

3 Endosomal vesicle forms and moves into cytoplasm.

Host-cell plasma membrane

4 Virus is released into host-cell cytoplasm.

2 Host-cell plasma membrane surrounds virus.

5 Viral envelope fuses with host-cell plasma membrane (*not shown*).

1 Virus makes contact with plasma membrane of host cell.

Host-cell cytoplasm

(c) Some viruses enter the host cell by endocytosis.

Figure 24-4 Viruses cause diseases in animals

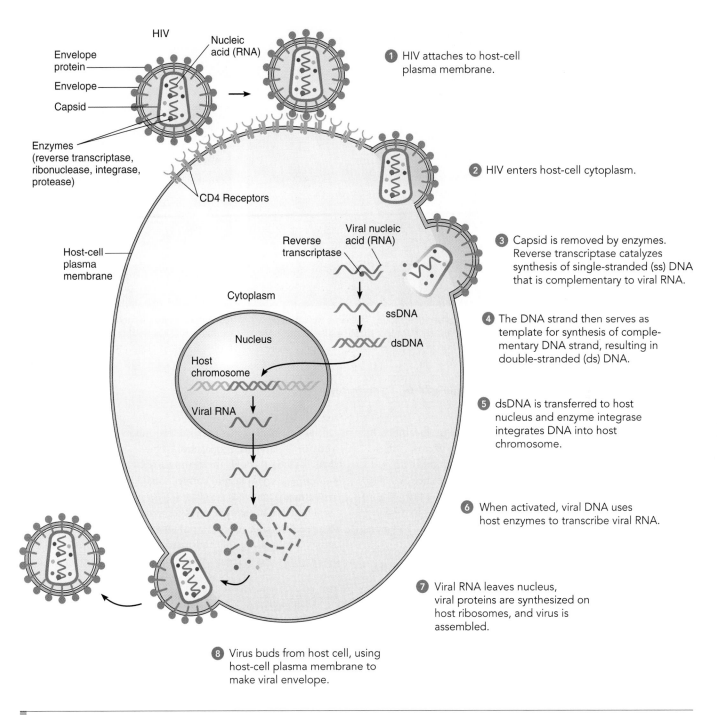

Figure 24-5 *Animated* Life cycle of HIV, the retrovirus that causes AIDS

HIV infects T helper cells, specialized cells of the host's immune system. The virus attaches to protein receptors, known as CD4, on the plasma membrane of the T helper cell. HIV has two identical single-stranded RNA molecules.

and domestic birds, horses, pigs, and humans. Influenza vaccine prevents attachment by stimulating the host's antibodies to cover the spikes projecting from the virus.

Viruses have several ways to penetrate animal cells (▐ Fig. 24-4). After attachment to a host-cell receptor, some enveloped viruses fuse with the animal cell's plasma membrane. The viral capsid and nucleic acid are both released into the animal cell. Other viruses enter the host cell by *endocytosis.* In this process, the plasma membrane of the animal cell invaginates to form a membrane-bound vesicle that contains the virus. Endocytosis is

advantageous to the virus because the endocytotic vesicle delivers the virus deep into the cytosol.

In DNA animal viruses, replication of viral DNA and protein synthesis are similar to the processes by which the host cell would normally carry out its own DNA replication and protein synthesis. In most RNA viruses, RNA synthesis takes place with the help of an RNA-dependent RNA polymerase.

Retroviruses are RNA viruses that have a DNA polymerase called **reverse transcriptase,** which is used to transcribe the RNA genome into a DNA intermediate (▐ Fig. 24-5). This DNA becomes integrated into the host DNA by an enzyme also carried by the virus. Copies of the viral RNA are synthesized as the incorporated DNA is transcribed by host RNA polymerases. The **human immunodeficiency virus (HIV)** that causes acquired immunodeficiency syndrome (AIDS) is a retrovirus. Certain cancer-causing viruses are also retroviruses.

After viral genes are transcribed, the viral structural proteins are synthesized. The capsid is produced, and then assembly of new virus particles takes place. Finally, release occurs. Viruses that do not have an outer envelope exit by cell lysis. The plasma membrane ruptures, releasing many new viral particles. Enveloped viruses obtain their lipoprotein envelopes by picking up a fragment of the host plasma membrane as they leave the infected cell (see Figs. 24-4b and 24-5).

Viral proteins damage the host cell in several ways. These proteins may alter the permeability of the plasma membrane or may inhibit synthesis of host nucleic acids or proteins. Viruses sometimes damage or kill their host cells by their sheer numbers. A poliovirus can produce 100,000 new viruses within a single host cell!

Antiviral drugs are being developed

Antibiotics are specific for fighting bacteria, so they do not kill viruses. However, researchers have developed several antiviral drugs. Most of these drugs inhibit the replication of many RNA and DNA viruses. Amantadine has been used mainly to treat patients with influenza. The drug inhibits penetration or uncoating of viral nucleic acids. However, recent strains of influenza virus have become resistant to Amantadine, perhaps as a result of using the drug in poultry feed in certain areas of Asia. A newer group of antiviral drugs, including Tamiflu, inhibit neuraminidase, a viral enzyme necessary for the virus to leave the host cell. Other antiviral drugs are in clinical trials. Some inhibit viral attachment to host cells, and others interfere with replication of viral nucleic acid.

Some viruses infect plant cells

Viruses cause many important plant diseases and are responsible for huge agricultural losses and lower crop quality. Infected crops almost always produce lower yields. The genome of most plant viruses consists of single-stranded RNA. Most plant viruses do not have envelopes. Symptoms of viral infection include reduced plant size and spots, streaks, or mottled patterns on leaves, flowers, or fruits (▌Fig. 24-6).

As they feed on plant tissues, insects such as aphids and leafhoppers spread viral diseases among plants. Because of plants' thick cell walls, viruses cannot penetrate plant cells unless the

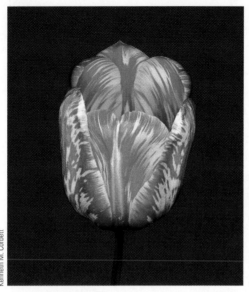

Kenneth M. Corbett

© Jack M. Bostrack / Visuals Unlimited

(a) Virus-streaked tulip. The virus that causes this relatively harmless disease affects pigment formation in the petals.

(b) Pepper leaves infected with tobacco mosaic virus. The leaf is characteristically mottled with light green areas.

Figure 24-6 Plant viruses

cells are damaged. Plant viruses are inherited through infected seeds or by asexual propagation. Once a plant is infected, the virus spreads through the plant body by passing through plasmodesmata (cytoplasmic connections) that penetrate the walls between adjacent cells (see Fig. 5-27).

There are no known cures for most viral diseases of plants, so infected plants are commonly burned. Some agricultural scientists are focusing their efforts on preventing viral disease by developing virus-resistant strains of important crop plants.

Review

▌ What characteristics of life are absent in a virus?

▌ What are the steps in a lytic cycle? Draw diagrams to illustrate your answer.

▌ How is a lysogenic cycle different from a lytic cycle?

▌ How does a virus enter a vertebrate cell? How does the virus damage the cell?

VIROIDS AND PRIONS

Learning Objective

6 Compare and contrast viroids and prions.

The ICTV recognizes four types of subviral agents. They are smaller and simpler than viruses. Here we will discuss two of them: viroids and prions.

Viroids are the smallest known pathogens

In 1961, scientists discovered an infective agent in potatoes that had a short RNA strand but no protein coat. The agent, named a **viroid,** is much smaller than a virus. Each viroid consists of a very short strand of naked RNA (only 250 to 400 nucleotides) that serves as a template copied by host RNA polymerases. The virus has no protective protein coat and no associated proteins to assist in duplication. Viroids are extremely hardy and resist heat and ultraviolet radiation because of the condensed folding of their RNA.

Viroids cause diseases in many plants. These infective agents, transmitted by pollen or seeds, are generally found within the host cell nucleus and may interfere with gene regulation. They cause stunted or distorted growth and sometimes kill the plant.

Prions are protein particles

Stanley Prusiner, professor of neurology and biochemistry at the University of California School of Medicine, San Francisco, began his studies of prions in the early 1970s, motivated by the death of a patient from Creutzfeldt–Jakob disease (CJD), a degenerative brain disease. Prusiner discovered that the infective agent was not affected by radiation (which typically mutates nucleic acids), and he could not find DNA or RNA in the particles. In 1982, he named the infective agent **prion,** for "proteinaceous infectious particle."

Prusiner's hypothesis that a pathogen could exist and transfer information without nucleic acids went against all accepted biological dogma. However, Prusiner and other researchers continued to study prions, and their findings supported Prusiner's hypothesis. In 1997, Prusiner was awarded the Nobel Prize in Physiology or Medicine for his discovery of prions—a new biological principle of infection.

Prusiner and others have shown that animals have a gene that encodes a normally harmless protein known as PrP. This protein, which consists of 208 amino acids, may help process copper. Normal forms of the protein are found on the surfaces of brain cells and many other types of cells. Sometimes, the PrP protein folds into a different shape, an insoluble form that can cause disease. This misfolded protein is the prion. Mutations in the gene that encodes the PrP protein increase the risk that the protein will misfold and become a prion. The prion then somehow induces other PrP molecules to misfold into the pathogenic form (❚ Fig. 24-7). Prions apparently can aggregate and accumulate in the brain and in certain other tissues and cause serious damage.

Prions are found in the brains of patients with **transmissible spongiform encephalopathies (TSEs).** This group of fatal degenerative brain diseases, which have been identified in birds and mammals, are called TSEs because when infected, the brain appears to develop holes and becomes somewhat spongelike. The TSEs are in some ways like viral illnesses. However, there is a long latency period (about 5 years in cows and up to 10 years in humans) between contracting the infection and developing symptoms of the illness. Also, the TSEs have proven very difficult to treat. Genetically engineered mice that lack the prion protein gene are immune to TSE infection.

The oldest known prion disease is *scrapie* in sheep and goats. When infected, animals lose coordination, become irritable, and itch so severely that they scrape off their wool or hair. Bovine spongiform encephalopathy (BSE) is a related prion disease popularly referred to as "mad cow disease" because some diseased cattle become aggressive. BSE became epidemic in cattle in the United Kingdom in the 1990s.

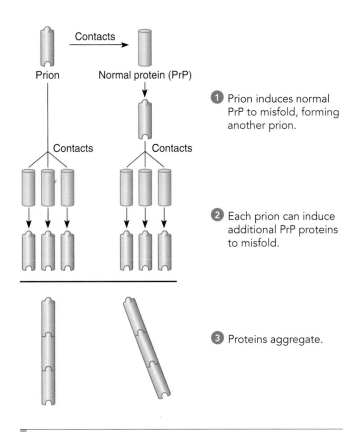

Figure 24-7 A model for how the prion population expands

Prions contact normal PrP proteins and induce them to misfold and become prions. Each new prion can then contact additional PrP molecules and induce them to misfold, thus expanding the prion population. Prions can form clumps by aggregating.

More than 150 people have died from a human variety of BSE, providing evidence that the disease is transmissible from cow to human. The human disease is called vCJD because it is a variant of CJD, which is caused by the transformation of PrP proteins into prions. The infectious agent of vCJD has been recovered from infected human neural tissue and appears similar to the prion that causes scrapie in sheep. New regulations have been instituted in several countries to safeguard the food supply. Human-to-human transmission of vCJD has been associated with tissue and organ transplants and transfusion with contaminated blood.

Chronic wasting disease, an illness related to mad cow disease, has spread among deer and elk populations in North America. Studies are under way to determine whether this disease can infect livestock or humans.

Like viruses, prions exhibit variability. Investigators now understand that there are different strains of prions. Strains may differ from one another in the conformation of their proteins, in their properties, and in the diseases they cause. Prions are different from viruses in that they can arise spontaneously, most likely as a result of mutation. Researchers are studying ways to block prions from forming. They are also looking for ways to stimulate cells to destroy prions. Early diagnosis is critical, and the search is on for a blood test that would detect prion disease.

Review

▮ How do viroids differ from prions?
▮ How do viroids and prions differ from viruses?

PROKARYOTES

Learning Objectives

7 Describe the structure and common shapes of prokaryotic cells.

8 Describe asexual reproduction in prokaryotes, and summarize three mechanisms (transformation, transduction, and conjugation) that may lead to genetic recombination.

9 Describe the modes of nutrition and metabolic adaptations of prokaryotes.

In contrast to viruses, viroids, and prions, which consist only of nucleic acid and/or protein, **prokaryotes** are cellular organisms. Recall from Chapters 4 and 23 that the cell structure of prokaryotes is fundamentally different from that of the eukaryotic cells of other living organisms. Microbiologists assign prokaryotes to two domains: **Archaea** and **Bacteria.**

Most prokaryotic cells are very small. Typically, their diameter ranges from 0.5 to 1.0 μm, and their length, from 1.0 to 5.0 μm. Their cell volume is only about one thousandth that of small eukaryotic cells, and their length is only about one tenth. Most prokaryotes are unicellular, but some form colonies or filaments containing specialized cells.

Prokaryotes have several common shapes

Prokaryotes have two basic shapes: spherical and rod shaped. Spherical prokaryotes, known as **cocci** (sing., *coccus*), occur singly in some species and in groups of independent cells in others

(▮ Fig. 24-8a). Cells may be grouped in twos (*diplococci*), in long chains (*streptococci*), or in irregular clumps that look like bunches of grapes (*staphylococci*). Rod-shaped prokaryotes, called **bacilli** (sing., *bacillus*), may occur as single rods or as long chains of rods (▮ Fig. 24-8b). Some prokaryotes form spirals. If the spiral-shaped cell is flexible, it is a **spirochete;** if rigid, it is a **spirillum** (pl., *spirilla*) (▮ Fig. 24-8c). A spirillum shaped like a comma is called a **vibrio.**

Prokaryotic cells lack membrane-enclosed organelles

Prokaryotic cells do not have membrane-enclosed organelles typical of eukaryotic cells. Thus, these cells do not have nuclei, mitochondria, chloroplasts, endoplasmic reticula, Golgi complexes, or lysosomes (▮ Fig. 24-9). Although the prokaryotic cell does not have a membrane-enclosed nucleus, it does have a **nuclear area,** also referred to as the **nucleoid,** which contains DNA.

The dense cytoplasm of the prokaryotic cell contains ribosomes (smaller than those found in eukaryotic cells) and storage granules that hold glycogen, lipid, or phosphate compounds. Enzymes needed for metabolic activities may be located in the cytoplasm. Although the membranous organelles of eukaryotic cells are absent, in some prokaryotic cells the plasma membrane is extensively folded inward. Enzymes needed for cellular respiration and photosynthesis may be associated with the plasma membrane or its folds.

A cell wall typically covers the cell surface

Most prokaryotic cells have a **cell wall** surrounding the plasma membrane. The cell wall provides a rigid framework that sup-

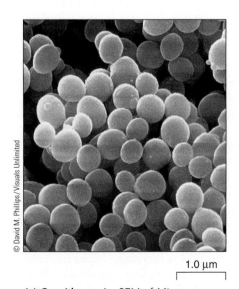

(a) Cocci bacteria. SEM of *Micrococcus.*

1.0 μm

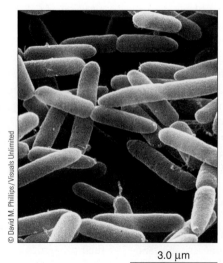

(b) Bacilli bacteria. SEM of *Salmonella.*

3.0 μm

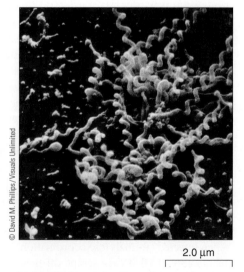

(c) Spirilla bacteria. SEM of *Spiroplasma.*

2.0 μm

▮ **Figure 24-8** Common shapes of prokaryotes

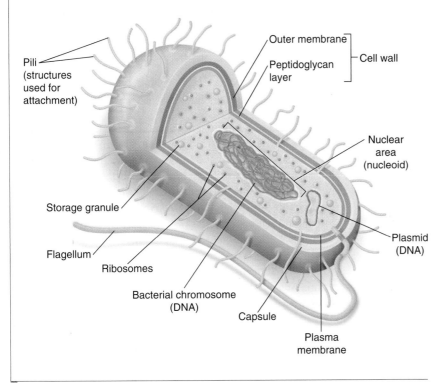

Figure 24-9 *Animated* Structure of a prokaryotic cell

This bacillus is a gram-negative bacterium (discussed in text). Note the absence of a nuclear envelope surrounding the bacterial DNA.

plasma membrane but contains polysaccharides bonded to lipids (Fig. 24-10).

Distinguishing between gram-positive and gram-negative bacteria is important in treating certain diseases. For example, the antibiotic penicillin interferes with peptidoglycan synthesis, ultimately resulting in a fragile cell wall that cannot protect the cell (see Chapter 7 discussion of drugs that are enzyme inhibitors). Predictably, penicillin works most effectively against gram-positive bacteria.

Some species of bacteria produce a **capsule** or slime layer that surrounds the cell wall. In free-living species, the capsule may provide the cell with added protection against phagocytosis (engulfment; see Chapter 5) by other microorganisms. In disease-causing bacteria, the capsule may protect against phagocytosis by the host's white blood cells. For example, the ability of *Streptococcus pneumoniae* to cause bacterial pneumonia depends on its capsule. A strain of *S. pneumoniae* that lacks a capsule does not cause the disease. Bacteria also use their capsules to attach to surfaces such as rocks, plant roots, or human teeth (where they cause dental plaque).

Some bacteria have hundreds of hairlike appendages known as **pili** (sing., *pilus*). These protein structures help bacteria adhere to one another or attach to certain surfaces, such as the cells they infect. Some elongated pili, called *sex pili,* are important in transmitting DNA between bacteria.

ports the cell, maintains its shape, and keeps it from bursting under hypotonic conditions (see Chapter 5). Most bacteria are adapted to hypotonic surroundings. When wall-less forms of bacteria are produced experimentally, they must be maintained in isotonic solutions to keep them from bursting. However, cell walls are of little help when a bacterium is in a hypertonic environment, as in food preserved by a high sugar or salt content. For this reason, most bacteria grow poorly in jellies, jams, salted fish, and other foods preserved in these ways.

The bacterial cell wall includes **peptidoglycan,** a complex polymer that consists of two unusual types of sugars (amino sugars) linked with short polypeptides. The sugars and polypeptides are linked to form a single macromolecule that surrounds the entire plasma membrane.

Differences in bacterial cell wall composition are of great interest to microbiologists and are important clinically. In 1888, the Danish physician Christian Gram developed the Gram staining procedure. Bacteria that absorb and retain crystal violet stain in the laboratory are referred to as **gram-positive,** whereas those that do not retain the stain when rinsed with alcohol are **gram-negative.** The cell walls of gram-positive bacteria are very thick and consist primarily of peptidoglycan. The cell walls of gram-negative bacteria have two layers: a thin peptidoglycan layer and a thick outer membrane. The outer membrane resembles the

Many types of prokaryotes are motile

Can you imagine trying to swim through molasses? Water has the same relative viscosity to prokaryotes that molasses has to humans. Most motile prokaryotes manage to move by means of rotating **flagella.** The number and location of flagella are important in classifying some bacterial species.

Unlike eukaryotic flagella, prokaryotic flagella do not consist of microtubules (see Chapter 4). A bacterial flagellum consists of three parts: a basal body, a hook, and a single filament (Fig. 24-11). The basal body is a complex structure that anchors the flagellum into the cell wall by disc-shaped plates. The curved hook connects the basal body to the long, hollow filament that extends into the outside environment. The basal body is a motor. The bacterium uses energy from ATP to pump protons out of the cell. Diffusion of these protons back into the cell powers the motor that spins the flagellum. The rotary motion produced pushes the cell much like a propeller pushes a ship through the water. The flagellum rotates counterclockwise, pushing the cell forward.

Many prokaryotes exhibit **chemotaxis,** movement in response to chemicals in the environment. For example, bacteria move toward food. Certain bacteria also move toward one another; however, bacteria move away from certain harmful chemicals.

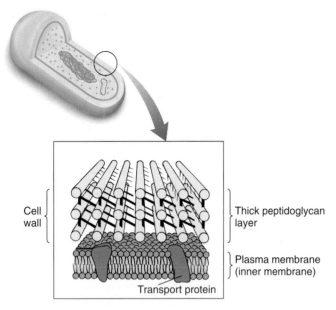

(a) **Gram-positive cell wall.** A thick layer of peptidoglycan molecules is held together by amino acids.

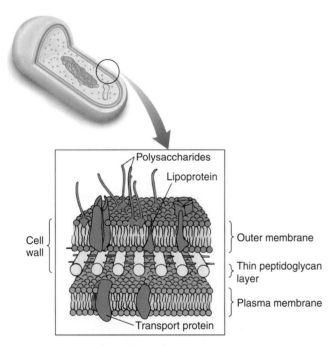

(b) **Gram-negative cell wall.** A thin layer of peptidoglycan is covered by an outer membrane.

Figure 24-10 *Animated* Bacterial cell walls

Prokaryotes have a circular DNA molecule

The genetic material of a prokaryote lies in the nuclear area but is not surrounded by a nuclear envelope. In most species, the genetic material is contained in a single, circular DNA molecule. If stretched out to its full length, this molecule would be about 1000 times as long as the cell itself. Unlike eukaryotic chromosomes, prokaryote DNA has little protein associated with it.

In addition to their genomic DNA, most bacteria have a small amount of genetic information present as one or more **plasmids,** smaller circular fragments of DNA. Plasmids replicate independently of the genomic DNA (see Chapter 15) or become integrated into it. Bacterial plasmids often have genes that code for catabolic enzymes, for genetic exchange, or for resistance to antibiotics.

Most prokaryotes reproduce by binary fission

Prokaryotes are extremely successful organisms in terms of their numbers and distribution. Their success is in large part due to their remarkable ability to reproduce rapidly. Prokaryotes reproduce asexually, generally by **binary fission,** a process in which one cell divides into two similar cells, as in the chapter opening photograph (also see Fig. 10-11). First, the circular DNA replicates; and then an ingrowth of both the plasma membrane and the cell wall forms a transverse wall.

Binary fission occurs with remarkable speed. Under ideal conditions, some bacterial species divide in less than 20 minutes. At this rate, if nothing interfered, one bacterium could give rise to more than 1 billion bacteria within 10 hours! However, bacteria cannot reproduce at this rate for very long before lack of food or the accumulation of waste products slows their population expansion.

A less common form of asexual reproduction among bacteria is **budding.** In budding, a cell develops a bulge, or bud, that enlarges, matures, and eventually separates from the mother cell. A few species of bacteria (actinomycetes) divide by **fragmentation.** Walls develop within the cell, which then separates into several new cells.

Bacteria transfer genetic information

Although sexual reproduction involving the fusion of gametes does not occur in prokaryotes, genetic material can be exchanged among bacteria and among archaea. Such exchange of genes, or **gene transfer,** results in genetic recombination. **Horizontal gene transfer** occurs when an organism transfers genetic material to another organism that is not its offspring. (Transfer of genetic material from parent to offspring is called *vertical gene transfer.*) Gene transfer among bacteria takes place by three different mechanisms: transformation, transduction, and conjugation.

1. When bacteria die, they release DNA that can be taken in by certain other bacteria. In **transformation,** a bacterial cell takes up fragments of foreign DNA (or RNA) released by another bacterium. The DNA must bind to DNA-binding proteins on the surface of the bacterium. Once it enters the bacterial cell, some of the DNA is incorporated into the host's own genome by reciprocal recombination (the DNA is exchanged) between the new DNA and the host chromosome (❚ Fig. 24-12). Recall from Chapter 12 that Oswald T. Avery and his colleagues identified DNA as the agent that transformed bacterial cells experimentally and showed that DNA is the chemical basis of heredity. Foreign DNA can also be taken up as plasmids. When this occurs, DNA does not

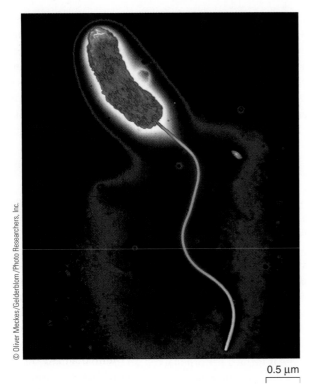

(a) Color-enhanced TEM of *Vibrio cholerae*, the flagellate bacterium that causes cholera in humans.

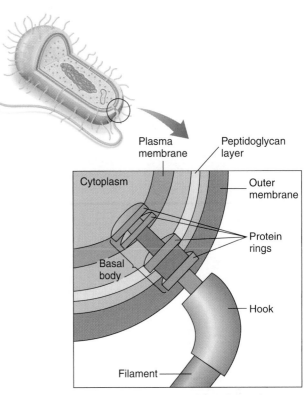

(b) Structure of a bacterial flagellum. The motor is the basal body, which consists of a series of disc-shaped plates that anchor the flagellum to the cell wall and plasma membrane and spin the hook and filament of the flagellum.

Figure 24-11 Bacterial flagella

undergo recombination; it remains as a plasmid separate from the bacterial chromosome.

2. In a different process of gene transfer, **transduction,** a phage carries bacterial genes from one bacterial cell into another (❙ Fig. 24-13). Normally, a phage contains only its own DNA. However, sometimes a phage incorporates some of the bacterial DNA of its host. Then, when the phage infects another bacterium, it transfers that DNA to its new host.

3. In **conjugation,** two cells of different mating types come together, and genetic material is transferred from one to the other (❙ Fig. 24-14). In contrast to transformation and transduction, conjugation involves contact between two cells.

Conjugation has been most extensively studied in the bacterium *E. coli.* In the *E. coli* population there are **donor cells,** or **F⁺ cells,** that have DNA that can be transmitted to **recipient cells,** or **F⁻ cells.** F⁺ cells have a DNA sequence known as the **F factor** (F stands for fertility) that is necessary for a bacterium to serve as a donor during conjugation. The F factor, which consists of about 20 genes, can be in the form of a plasmid or it can be part of the DNA in the bacterial chromosome.

F genes encode enzymes essential for transferring DNA. Certain F genes encode **sex pili,** long, hairlike extensions that project from the cell surface. The sex pilus recognizes and binds to the surface of an F⁻ cell, forming a cytoplasmic conjugation bridge between the two cells. The F plasmid replicates itself, and DNA is transferred from donor to recipient bacterium through the conjugation bridge. F plasmids may also have other types of genes, including those that determine resistance to antibiotics.

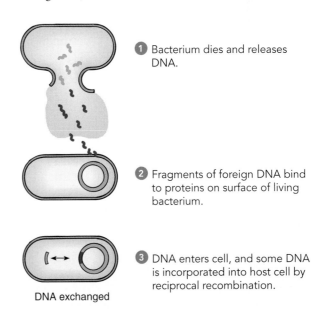

1 Bacterium dies and releases DNA.

2 Fragments of foreign DNA bind to proteins on surface of living bacterium.

3 DNA enters cell, and some DNA is incorporated into host cell by reciprocal recombination.

DNA exchanged

Figure 24-12 Transformation

In this process, foreign DNA from a bacterium that has died enters a host bacterium. DNA is exchanged by recombination.

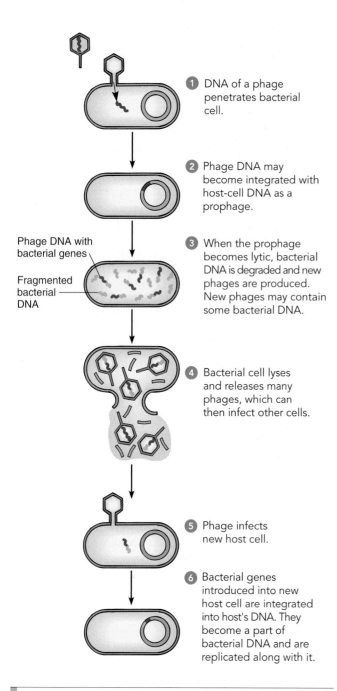

① DNA of a phage penetrates bacterial cell.

② Phage DNA may become integrated with host-cell DNA as a prophage.

Phage DNA with bacterial genes

Fragmented bacterial DNA

③ When the prophage becomes lytic, bacterial DNA is degraded and new phages are produced. New phages may contain some bacterial DNA.

④ Bacterial cell lyses and releases many phages, which can then infect other cells.

⑤ Phage infects new host cell.

⑥ Bacterial genes introduced into new host cell are integrated into host's DNA. They become a part of bacterial DNA and are replicated along with it.

Figure 24-13 Transduction

In transduction, a phage transfers bacterial DNA from one bacterium to another. Transduction is an important means of gene transfer.

Evolution proceeds rapidly in bacterial populations

Bacteria reproduce rapidly by binary fission, so mutations are quickly passed on to new generations, and the effects of natural selection are quickly evident. Bacteria that have some advantage rapidly pass the genes that facilitate their survival to future generations.

Horizontal gene transfer greatly contributes to the rapid evolution that takes place in prokaryotes. Acquisition of new DNA and genetic recombination are important sources for diversification and adaptation. New DNA introduced into a bacterium's genome represents new raw material for evolution. New genes are subject to mutation and are acted on by the forces of natural selection. The changes in the genetic material are passed on to succeeding generations by binary fission. Changes that result in adaptation can quickly spread through future bacterial populations. Recent studies provide evidence that genetic recombination is also an important mechanism in the evolution of archaea.

Some bacteria form endospores

When the environment becomes unfavorable, for example, when nutrients are limited or the environment becomes very dry, some types of bacteria become dormant. The cell loses water, shrinks slightly, and remains inactive until water is again available. Certain bacteria form dormant, extremely durable cells called **endospores.** After the endospore forms, the cell wall of the original cell lyses, releasing the endospore. Formation of endospores is not a type of reproduction in bacteria; endospores are not comparable to the reproductive spores of fungi and plants. Only one endospore is formed per original cell, so the total number of individuals does not increase. Archaea do not form endospores; these prokaryotes produce unique enzymes on the cell surface that protect them from cold, heat, and desiccation.

Endospores survive in very dry, hot, or frozen environments or at times when food is scarce (❚ Fig. 24-15). Some endospores are so resistant that they can survive an hour or more of boiling or centuries of freezing. When environmental conditions are again suitable for growth, the endospore germinates, forming an active, growing bacterial cell.

Several types of bacteria that form endospores cause disease. The endospore of *Bacillus anthracis,* the bacterium that causes anthrax, is so hardy that this pathogen has become a concern as an agent of biological warfare. The bacteria (*Clostridium tetani*) that cause tetanus and the bacteria (*C. perfringens*) that cause gas gangrene typically enter the body with soil when a person suffers a deep cut or puncture wound. Patients may also be exposed to these serious diseases when surgical instruments are not effectively sterilized, which results in survival of endospores.

Many bacteria form biofilms

Many types of bacteria that inhabit watery environments form dense films called **biofilms** that attach to solid surfaces. The bacteria in a biofilm form layers up to 200 μm thick. The bacteria secrete a slimy, gluelike substance rich in polysaccharides and become embedded in this matrix. Most biofilms consist of many species of bacteria and may include other types of organisms, such as fungi and protozoa. Scientists have found evidence of ancient biofilms in sediments from coastal marine environments 3.5 billion years old.

The dental plaque that forms on teeth is a familiar example of a biofilm. These films also commonly form on the surfaces of contact lenses and catheters. They sometimes develop on surgical implants such as pacemakers and joint replacements.

Metabolic diversity has evolved among prokaryotes

Based on the two main ways of obtaining nutrition, prokaryotes are classified as heterotrophs or autotrophs (*-troph* comes from a Greek word that means "nutrition"). Most prokaryotes are **heterotrophs** that obtain carbon atoms from the organic compounds of other organisms. Some bacteria are **autotrophs** that are able to use carbon dioxide as a source of carbon for manufacturing their organic molecules.

Based on two principal ways of capturing energy, prokaryotes are classified as chemotrophs and phototrophs. **Chemotrophs** obtain their energy from chemical compounds. **Phototrophs** capture energy from light. When we consider the source of carbon *and* the source of energy, we can classify prokaryotes into four main groups:

1. The majority of prokaryotes are **chemoheterotrophs.** They depend on organic molecules for both their carbon and energy. Many bacterial chemoheterotrophs are free-living decomposers that obtain their carbon and energy from dead organic matter. These bacteria are sometimes called *saprotrophs.* Some chemoheterotrophs are pathogens, which obtain their nourishment from the organisms they infect. They harm their hosts by causing diseases. Other heterotrophic bacteria benefit their hosts, for example, by producing needed vitamins.

2. **Photoheterotrophs,** like the purple nonsulfur bacteria, obtain their carbon from other organisms but use chlorophyll and other photosynthetic pigments to trap energy from sunlight.

3. **Photoautotrophs,** like the cyanobacteria, use the energy from sunlight to synthesize organic compounds from carbon dioxide and other inorganic compounds.

4. **Chemoautotrophs** also use carbon dioxide as a carbon source, but they do not use sunlight as their energy source. Instead, they obtain energy by oxidizing inorganic chemical substances such as ammonia (NH_3) and hydrogen sulfide (H_2S).

Most prokaryotes require oxygen

Whether they are heterotrophs or autotrophs, most bacterial cells are **aerobic** and require oxygen for cellular respiration. Many are **facultative anaerobes** that use oxygen for cellular respiration if it is available but can carry on metabolism anaerobically when necessary. Other bacteria are **obligate anaerobes** that carry out anaerobic respiration; they respire with terminal electron acceptors other than oxygen, for example, sulfate (SO_4^{2-}), nitrate (NO_3^-), or iron (F^{2+}). Some

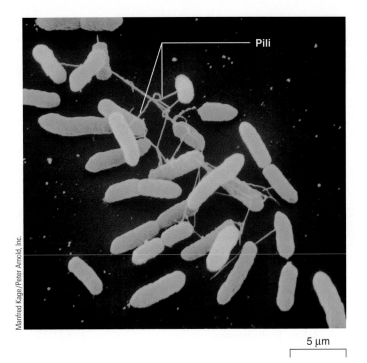

Manfred Kage/Peter Arnold, Inc.

5 µm

(a) Color-enhanced SEM of *Serratia marcescens* bacteria, which are connected by pili. Stimulated by the contact, the cells pull close together and form conjugation bridges between donor and recipient cells (*not shown*).

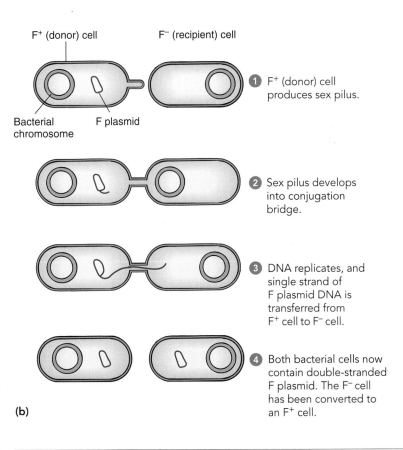

(b)

① F^+ (donor) cell produces sex pilus.

② Sex pilus develops into conjugation bridge.

③ DNA replicates, and single strand of F plasmid DNA is transferred from F^+ cell to F^- cell.

④ Both bacterial cells now contain double-stranded F plasmid. The F^- cell has been converted to an F^+ cell.

Figure 24-14 *Animated* Conjugation

In conjugation, a donor bacterium transfers plasmid DNA to a recipient bacterium.

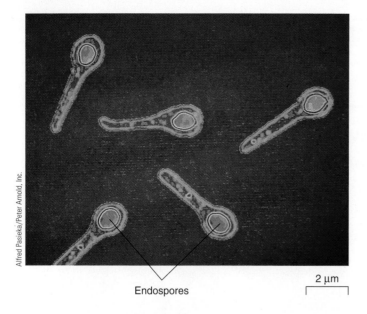

Endospores

2 μm

Figure 24-15 Endospores

Color-enhanced TEM of *Clostridium tetani*, the bacterium that causes tetanus. Each bacterial cell (*blue*) contains one endospore (*orange*), a resistant, dehydrated cell that develops within the original cell.

obligate anaerobes, including certain archaea, are actually killed by even low concentrations of oxygen.

Review

- In what ways do prokaryotic cells differ from eukaryotic cells?

- How do prokaryotes reproduce? What mechanisms result in gene transfer?

- How do facultative anaerobes differ from obligate anaerobes? How do they differ from aerobes?

THE TWO PROKARYOTE DOMAINS

Learning Objectives

10 Compare the three domains: Archaea, Bacteria, and Eukarya.

11 Distinguish among the main groups of archaea based on their ecology, identify the archaean phyla (see Table 24-3), and describe the main groups of bacteria (see Table 24-4).

Under a microscope, most prokaryotes appear rather similar in size and form. However, using sequence analysis of small subunit 16S ribosomal RNA (SSU rRNA), Carl Woese and his coworkers demonstrated that there are two fundamentally different groups of prokaryotes (see Chapter 23). Each group has **signature sequences,** regions of SSU rRNA that have unique nucleotide sequences. The reason is that after they diverged, prokaryote populations diversified and mutations occurred that affect RNA sequences. Based on such analyses, Woese hypothesized that

ancient prokaryotes split into two lineages early in the history of life.

Based on Woese's work, microbiologists now classify the modern descendants of these two ancient lines in two domains: Archaea and Bacteria (█ Fig. 24-16). Domain **Archaea,** which corresponds to kingdom Archaea, includes a group of prokaryotes that produce methane gas from simple carbon sources and two groups that live in extreme environments. Domain **Bacteria,** which corresponds to kingdom Bacteria, includes all other prokaryotes. In this system, all eukaryotes are classified in domain **Eukarya.**

Several key characteristics distinguish archaea from bacteria. In contrast to bacteria, archaea do not have peptidoglycan in their cell walls. Although their plasma membranes are structurally similar, they are chemically unique. In the plasma membranes of bacteria and eukaryotes, straight-chain fatty acids are linked to glycerol molecules by *ester linkages.* In contrast, fatty acid components are not found in archaea. Instead, branched-chain hydrocarbons (synthesized from isoprene units) are bonded to glycerol by *ether linkages* (█ Table 24-2).

$$R-\overset{\overset{\textstyle O}{\|}}{C}-O-R \qquad R-\overset{\overset{\textstyle H}{|}}{\underset{\underset{\textstyle H}{|}}{C}}-O-R$$

Ester linkage   **Ether linkage**

The absence of a second electronegative oxygen atom makes ether linkages stronger than ester linkages. This unique membrane structure may contribute to the ability of the archaea to survive and thrive in harsh environments.

In some ways, archaea are more like eukaryotes than like bacteria. For example, archaea do not have the simple RNA polymerase found in bacteria. Like eukaryotes, their translation process begins with methionine, whereas in bacteria, translation begins with formylmethionine (see Chapter 13). In addition, several antibiotics that affect bacteria do not affect archaea or eukaryotes.

Prokaryote taxonomy is now based largely on RNA sequencing. Groups that branched off earlier had more time to accumulate mutations in their SSU rRNA. Their nucleotide sequences are less similar than those of groups that diverged more recently. However, some microbiologists are developing phylogenetic trees based on entire genomes. They argue that there may be 1000 genes that code proteins for every 1 gene that codes an rRNA. These researchers prefer to consider the proportions of genes (or proteins) that genomes of various groups have in common.

Although prokaryotic taxonomy is controversial and continuously changing, more than 1220 genera and 6000 species of prokaryotes have been classified. The editors of *Bergey's Manual of Systematic Bacteriology,* considered the definitive reference text by microbiologists, have divided prokaryotes into more than 25 phyla based on 16S rRNA analysis.[1]

---

[1] *Bergey's Manual of Systematic Bacteriology,* 2nd ed., Springer, New York, 2001, 2005; the first two of five volumes of the second edition have now been published.

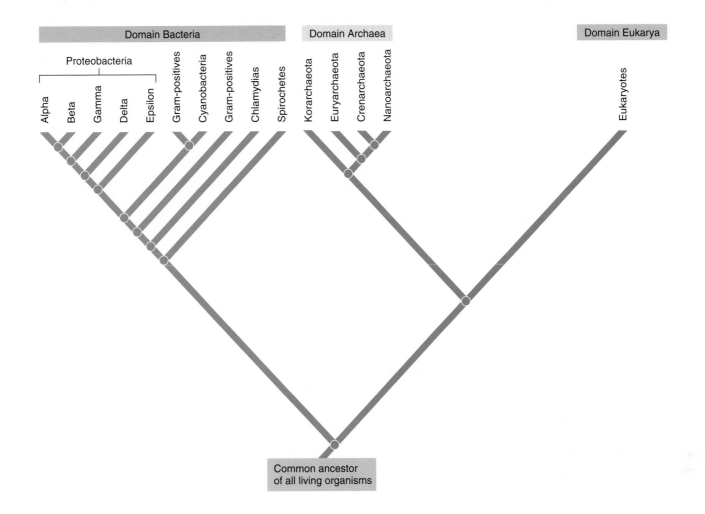

This highly simplified diagram depicts some representative taxa of domain Bacteria and domain Archaea. The relationships are based on analyses of nucleotide sequences of small subunit rRNA. As taxonomists consider additional data, these relationships will be modified.

Note that because biologists cannot extract rRNA from fossils, no extinct species are shown in this phylogenetic tree. Thus, this tree does not show evolutionary lineages.

Some archaea survive in harsh environments

Many members of domain Archaea inhabit environments thought to be similar to conditions on early Earth. We can identify three main types of archaea based on their metabolism and ecology: methanogens, extreme halophiles, and extreme thermophiles.

The **methanogens** (methane producers) are a large, diverse group that inhabit oxygen-free environments in sewage and swamps and are common in the digestive tracts of humans and other animals. They are obligate anaerobes that produce methane gas from simple carbon compounds. The methanogens are important in recycling components of organic products of organisms that inhabit swamps. Methanogens that inhabit the digestive tracts of cows and other grazing animals produce methane, which is belched out by the animals. Methanogens produce more than 80% of the methane (more than 2 billion tons each year) in Earth's atmosphere.

Extreme halophiles are heterotrophs that live in saturated brine solutions such as salt ponds, the Dead Sea, and Great Salt Lake (❚ Fig. 24-17). The extreme halophiles use aerobic respiration to make ATP. However, they also carry out a form of photosynthesis in which they capture the energy of sunlight using a purple pigment (*bacteriorhodopsin*). (This pigment is very similar to the pigment rhodopsin involved in animal vision.)

Extreme thermophiles require a very high temperature for growth and normally inhabit hot (45°C to 110°C), sometimes acidic, environments. One species is found in the hot sulfur springs of Yellowstone Park at temperatures near 60°C and pH values of 1 to 2 (the pH of concentrated sulfuric acid). Others inhabit volcanic areas under the sea. One species, found in hot deep-sea vents on the seafloor, lives at temperatures ranging from 80°C to 110°C.

Archaea also inhabit less extreme conditions. For example, nonextreme archaea are abundant in the soil and in the cold ocean surface waters near Antarctica. Microbiologists have sug-

TABLE 24-2

Comparison of the Three Domains

Characteristic	Bacteria	Archaea	Eukarya
Nuclear envelope	Absent	Absent	Present
Membrane-enclosed organelles	Absent	Absent	Present
Circular chromosome	Present (linear in some species)	Present	Absent
Number of chromosomes	Typically one (may also have plasmid)	Typically one (may also have plasmid)	Typically many
Histones associated with DNA	Absent	Present	Present
Peptidoglycan in cell wall	Present	Absent	Absent
Structure of lipids in plasma membrane	Straight-chain fatty acids bonded to glycerol by ester linkages	Branched-chain hydrocarbons linked to glycerol by ether linkages	Straight-chain fatty acids bonded to glycerol by ester linkages
Size of ribosomes	70S*	70S	80S except in mitochondria and chloroplasts
RNA polymerase	One relatively simple RNA polymerase	Several relatively complex RNA polymerases	Several relatively complex RNA polymerases
Translation	Begins with formylmethionine	Begins with methionine	Begins with methionine

*The numbers 70S and 80S refer to the sedimentation coefficient (a measure of relative size) when centrifuged.

Figure 24-17 Seawater evaporating ponds

These salt ponds near San Francisco Bay are colored pink, orange, and yellow from the large number of extreme halophiles that inhabit them. The colors are a result of pigments in the cell membranes. These bacteria are harmless, and the ponds are used to produce salt commercially.

gested that the archaea may be important in biogeochemical cycles and in marine food chains. No pathogenic archaea have been identified.

Bergey's Manual divides Archaea into four phyla: Crenarchaeota, Euryarchaeota, Korarchaeota, and Nanoarchaeota. The archaean phyla are summarized in ▌ Table 24-3.

Bacteria are the most familiar prokaryotes

The bacteria are widely distributed in the environment and are better known to microbiologists than are the archaea. *Bergey's Manual* divides the bacteria into more than 20 phyla. Major groups of bacteria include the gram-negative proteobacteria, cyanobacteria, gram-positive bacteria, chlamydias, and spirochetes. These groups are summarized in ▌ Table 24-4.

Review

▌ What are three main types of archaea, based on their ecology?

▌ How do these bacteria differ (listed in Table 24-4): (1) proteobacteria, (2) cyanobacteria, and (3) chlamydias?

▌ IMPACT OF PROKARYOTES

Learning Objective

12 Discuss the ecological roles of prokaryotes, their importance as pathogens, and their commercial importance.

Prokaryotes are vital members of the biosphere. They interact with other organisms in many ways. An intimate relationship between members of two or more species is called **symbiosis.** The partners in a symbiotic relationship are called **symbionts.** Symbiotic relationships arise by coevolution. Three forms of symbiosis are mutualism, commensalism, and parasitism.

Mutualism is a symbiotic relationship in which both partners benefit. Cows and other ruminants (cud-chewing animals) have mutualistic relationships with bacteria that inhabit their digestive tracts. Ruminants lack enzymes for digesting cellulose, so they provide the bacteria with a nutrient-rich home and the bacteria digest the cellulose for them.

TABLE 24-3

Three Groups of Archaea*

Crenarchaeota

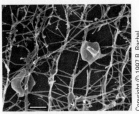

Copyright © 1997 R. Rachel

10 μm

Some extremely thermophilic (thrive at 70°C, or some thrive at temperatures greater than 100°C); others psychrophilic (can live at temperatures below 15°C). Many are chemoautotrophs. Important part of plankton in cold, oxygen-rich seas.

SEM of *Pyrodictium*
Pyrodictium is found in hydrothermal vents under the ocean. Their optimum growth temperature is 105°C. These archaea have irregular, disk-shaped cells connected by extensive networks of protein fibers that help them attach to sulfur granules.

Euryarchaeota

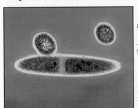

Dr. Kari Lounatmaa/Science Photo Library/Photo Researchers, Inc.

5 μm

Includes methanogens, halophiles (not all extreme), and acidophiles. Some extreme thermophiles included in this phylum.

TEM of a methanogen (*Methanospirillum hungatii*) undergoing cell division
Two other archaea are visible in cross section. When not dividing, these archaea are spiral shaped.

Nanoarchaeota

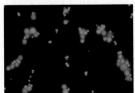

Courtesy K. O. Stetter, University of Regensburg, Germany

5 μm

Nanoarchaeum equitans, the first-described member of this group, is a very small (400 nm), anaerobic, extreme thermophile. Symbiotic with another archaean. As of 2005, *Nanoarchaeum* was the smallest known organism.

Fluorescence LM of *Nanoarchaeum equitans*
Nanoarchaeum equitans (tiny cells) attached to its archaean host *Ignicoccus spec* (big cells).

*Members of a fourth group, Korarchaeota, have been discovered in hot springs in the ocean floor and on land. The Korarchaeota may be the oldest group of archaeans.

Some investigators estimate that up to 100 trillion bacteria inhabit the nutrient-rich human intestine. Many of these are mutualistic bacteria. For example, in exchange for nutrients and a place to live, *Bacteroides thetaiotaomicron* break down indigestible complex carbohydrates into sugars that their human host can absorb. They produce certain vitamins that the host absorbs and uses. These bacteria also promote proliferation of blood vessels that improve intestinal function. Investigators have reported that *B. thetaiotaomicron* can turn on specific genes in cells of the host's intestine; apparently, these bacteria can induce synthesis of a compound that may kill competing bacteria.

In **commensalism,** one partner benefits and the other is neither harmed nor helped. Many bacteria that inhabit the human intestine are commensals that live on unused food. In **parasitism,** one partner lives on or in the other. The **parasite** benefits and the **host** is harmed. Pathogenic bacteria are parasites.

Bacteria, especially actinomycetes and myxobacteria, are the most numerous inhabitants of soil. As described in the chapter introduction, prokaryotes are essential decomposers, chemoheterotrophs that break down dead organic matter and wastes. They use the products of decomposition as an energy source. When prokaryotes break down organic compounds, many of their components, including nitrogen, oxygen, carbon, phosphorus, sulfur, and certain trace elements, are recycled. (The roles of bacteria in biogeochemical cycles, particularly the nitrogen cycle, are discussed in Chapter 54.)

Nitrogen is constantly removed from the soil by plants and other natural processes, as well as by human activities, such as agriculture. Plant growth depends on the availability of usable nitrogen, so it must be continually added to the soil. Several types of bacteria, including cyanobacteria, transform atmospheric nitrogen to forms that can be used by plants (Fig. 24-18).

TABLE 24-4

Major Groups of Bacteria

Proteobacteria (Gram-negative)

David Scharf/Peter Arnold, Inc.

10 μm

SEM of *Escherichia coli* colony

A large, very diverse group of gram-negative bacteria. Based on rRNA sequences, the group is divided into five subgroups designated as alpha, beta, gamma, delta, and epsilon.

Alpha Proteobacteria
Includes many symbionts of plants and animals, and some pathogens. **Rhizobium** species live symbiotically in root nodules of legumes (e.g., beans) and convert atmospheric nitrogen to a form usable by plants (nitrogen fixation). **Rickettsias** are very small, rod-shaped bacteria. A few species are pathogenic to humans and other animals; transmitted by arthropods through bites or through contact with their excretions. Among these are typhus (transmitted by fleas and lice) and Rocky Mountain spotted fever (transmitted by ticks).

Beta Proteobacteria
Several diverse groups, including *Nitrosomonas,* which oxidizes ammonia. Pathogenic bacteria in this group include *Neisseria gonorrhoeae,* which causes gonorrhea.

Gamma Proteobacteria
Includes the **enterobacteria,** decomposers that live on decaying plant matter, pathogens, and a variety of bacteria that inhabit humans. Although *Escherichia coli* is a normal inhabitant of the animal intestinal tract, certain strains can cause moderate to severe diarrhea. One species of *Salmonella* infects food and produces a toxin that causes a form of food poisoning; another species causes typhoid fever.

Vibrios are mainly marine; some are bioluminescent. *Vibrio cholerae* causes cholera.

Pseudomonads are heterotrophs that produce nonphotosynthetic pigments; cause disease in plants and animals, including humans.

Purple sulfur bacteria are photoautotrophs that do not produce oxygen.

Delta Proteobacteria
Includes the **myxobacteria** (slime bacteria), which secrete slime and glide or creep along. When nutrients are exhausted, these bacteria aggregate into stalked, multicellular reproductive structures called fruiting bodies. Bacterial cells within the fruiting body enter a resting stage. When conditions are favorable, resting cells become active.

Epsilon Proteobacteria
A small group of bacteria that inhabit the animal digestive tract. *Helicobacter* can cause peptic ulcers.

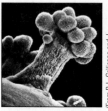

(From P. L. Grilicone and J. Pangborn, *Journal of Bacteriology* 124:1558, 1975.)

10 μm

SEM of fruiting body of the myxobacterium *Stigmatella aurantiaca*
Protective resting cells within the fruiting bodies are very resistant to heat and drying.

Cyanobacteria (Gram-negative)

Gram-negative; photosynthetic. Inhabit ponds, lakes, swimming pools, moist soil, dead logs, tree bark. Contain chlorophyll *a* and use a photosynthetic process similar to that of plants and algae. Many species also fix nitrogen.

Gram-positive Bacteria

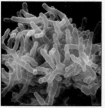

© David M. Phillips / Visuals Unlimited

5 μm

SEM of *Actinomycetes naeslundi,* a soil-dwelling bacterium that forms filamentous colonies

An extremely diverse group. A few examples follow.

Actinomycetes superficially resemble fungi. However, they have peptidoglycan in their cell walls, lack nuclear envelopes, and have other prokaryotic characteristics. Most actinomycetes are saprotrophs that decompose organic materials in soil. Some are anaerobic. Several species of the genus *Streptomyces* produce antibiotics such as streptomycin, erythromycin, chloramphenicol, and the tetracyclines. Some actinomycetes cause serious lung disease or generalized infections in humans and other animals.

Lactic acid bacteria ferment sugar, producing lactic acid as the main end product. Inhabit decomposing plant material, milk, and other dairy products; responsible for the characteristic taste of yogurt, pickles, sauerkraut, and green olives. Among the normal inhabitants of the human mouth and vagina.

Mycobacteria are slender, irregular rods; contain a waxy substance in their cell walls. One species causes tuberculosis; another causes leprosy.

Streptococci inhabit the mouth and digestive tract of humans and some other animals. Among the harmful species are those that cause "strep throat," dental caries, a form of pneumonia, scarlet fever, and rheumatic fever.

Staphylococci normally live in nose and on skin. Opportunistic pathogens that cause disease when the immunity of the host is lowered. *Staphylococcus aureus* causes boils and skin infections (some extremely serious); may infect wounds. Certain strains of *S. aureus* cause a form of food poisoning; other strains cause toxic shock syndrome.

Clostridia are anaerobic. One species causes tetanus; another causes gas gangrene. *Clostridium botulinum* can cause botulism, an often fatal type of food poisoning.

The **mycoplasmas** are a group of bacteria that lack cell walls. They may have evolved from bacteria with gram-positive cell walls. They inhabit soil and sewage; some are parasitic on plants or animals. Some inhabit human mucous membranes but do not generally cause disease; one species causes a mild type of bacterial pneumonia in humans.

© David M. Phillips / Visuals Unlimited

5 μm

SEM of *Mycoplasma* on fibroblast cells

continues

TABLE 24-4

Continued

Chlamydias (Gram-negative)

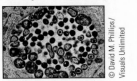

5 μm

© David M. Phillips / Visuals Unlimited

TEM of *Chlamydia trachomatis* in human oviduct cell

Chlamydias lack peptidoglycan in their cell walls. They are energy parasites, completely dependent on their hosts for ATP. Infect almost every species of bird and mammal. A strain of *Chlamydia* causes trachoma, the leading cause of blindness in the world. Sexually transmitted chlamydias are the major cause of pelvic inflammatory disease in women.

Spirochetes (Gram-negative)

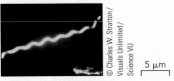

5 μm

© Charles W. Stratton / Visuals Unlimited / Science VU

LM of *Treponema pallidum*, the spirochete that causes syphilis

Spirochetes are spiral-shaped bacteria with flexible cell walls; move by means of unique internal flagella called *axial filaments*. Some species are free-living, whereas others form symbiotic associations; a few are parasitic. The spirochete of greatest medical importance is *Treponema pallidum*, which causes syphilis. Lyme disease, a tick-borne disease of humans and some other animals, is caused by a spirochete belonging to genus *Borrelia*.

Rhizobial bacteria, a type of motile bacteria that inhabit the soil, form mutualistic relationships with the roots of legumes, a large family of plants that includes important crops such as beans, peas, and peanuts. The infected plant cells form tumorlike nodules in which the bacteria reside and fix nitrogen (see Fig. 54-9). The bacteria supply the plant with all the nitrogen it requires, and the plant provides the bacteria with organic compounds, including sugar needed for cellular respiration. Because many soils are deficient in nitrogen, legumes that have formed mutualistic associations with rhizobial bacteria have a decided advantage over other plants. When the legumes die and are decomposed, the fixed nitrogen is released into and enriches the soil.

Many prokaryotes, such as the cyanobacteria, carry out photosynthesis, using water as the electron source and generating oxygen. During this process, they fix huge amounts of carbon dioxide into organic molecules.

Microbiologists are only beginning to unravel the mysteries of prokaryote ecology. For example, proteobacteria in the SAR11 clade are among the most successful organisms on Earth. Although they are one of the most abundant organisms in the Atlantic Ocean, they were not successfully cultured until 2002, and very little is known about their ecological role. There is some evidence that they are important in cycling carbon, nitrogen, and sulfur in the ocean.

Some types of bacteria appear to influence the course of evolution of other species. The proteobacteria *Wolbachia* infect many invertebrates, including insects, spiders, crustaceans, and nematodes (roundworms). *Wolbachia* are transmitted from generation to generation in the eggs of their hosts, so male hosts are not useful to them. Consequently, these parasites limit or eradicate males from the population. In some insect species, infected males can reproduce only when they mate with infected females carrying the same *Wolbachia* strain. In other species, these parasites convert male insects into females. Some infected females reproduce parthenogenetically (that is, their eggs develop without fertilization). Because they affect reproduction in their hosts, *Wolbachia* and other reproductive parasites may influence evolutionary divergence and even extinction in some species.

Heterocysts

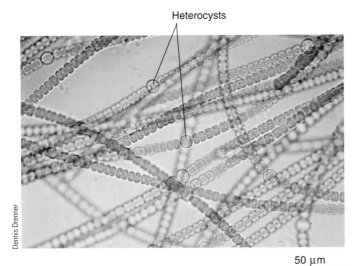

50 μm

Dennis Drenner

Figure 24-18 LM of *Anabaena*, a filamentous cyanobacterium that fixes nitrogen

Nitrogen fixation takes place in the rounded cells, called *heterocysts*.

Some prokaryotes cause disease

Some prokaryote species have coevolved with eukaryotes and are interdependent with them. All plants and animals harbor a popu-

TABLE 24-5

Important Bacterial Diseases and Their Causative Agents

Disease	Pathogen	Epidemiology/Comments
Anthrax	*Bacillus anthracis*	Most commonly occurs in domestic animals such as cattle. Can be transmitted to humans from infected animals or animal products. Endospores can live in soil for many years. Anthrax is not spread from person to person. Infection can occur in three forms: cutaneous, by inhalation, and gastrointestinal.
Botulism	*Clostridium botulinum*	Contracted by eating foods that contain the endotoxin or from infected wound. Infant botulism is caused by ingesting endospores. Causes muscle paralysis and can cause death from respiratory failure.
Chlamydia	*Chlamydia trachomatis*	One of the most frequently reported sexually transmitted disease in the U.S. About 75% of infected women and 50% of infected men have no symptoms. If untreated, the infection spreads and damages reproductive organs; can lead to infertility. This pathogen can also infect eyes and causes millions of cases of blindness worldwide each year.
Cholera	*Vibrio cholerae*	Contracted by eating food or drinking water contaminated with the bacterium. Common in areas with inadequate sewage treatment and impure water. Infects intestine and can cause severe diarrhea. Rapid fluid loss can lead to dehydration and death.
Diphtheria	*Corynebacterium diphtheriae*	Transmitted from person to person by intimate respiratory and physical contact. Endemic in developing countries. Not common in U.S. since vaccine became available in 1920s. Infects the heart muscle and respiratory passageways.
Epidemic typhus	*Rickettsia prowazekii*	Transmitted by infected body lice. After an 8- to 12-day incubation period, symptoms include fever, severe headache, muscle aches, and chills. Several days later a rash appears. About 40% of untreated patients die.
Gonorrhea	*Neisseria gonorrhoeae*	Common sexually transmitted disease.
Hansen disease (leprosy)	*Mycobacterium leprae*	Thought to be spread from person to person in nasal secretions. Worldwide, this disease has disabled up to 2 million people. Typically affects the skin, nerves, and mucous membranes.
Lyme disease	*Borrelia burgdoferi*	Transmitted to humans by bite of infected blacklegged ticks. Symptoms include skin rash, headache, fever, and fatigue. If untreated, infection can spread to joints, heart, and nervous system.
Peptic ulcer disease	*Helicobacter pylori*	Causes peptic ulcer, a lesion in the lining of the stomach or duodenum (upper part of small intestine).
Pertussis (whooping cough)	*Bordetella pertussis*	Highly communicable from person to person. Causes spasms of severe coughing. Vaccination available.
Plague	*Yersinia pestis*	Transmitted from wild rodents, squirrels, and cats to people by infected fleas. If untreated, can cause death. Killed millions of people in Europe during the Middle Ages.
Pneumonia	*Streptococcus pneumoniae*	Transmitted from person to person. Strains of *S. pneumoniae* are resistant to one or more antibiotics commonly used to treat this infection. Incidence has decreased since introduction of a vaccine.
Salmonella (salmonellosis)	*Salmonella* sp.	Transmitted to people in contaminated chicken, eggs, or other food. Also transmitted from feces of infected animals, e.g., lizards, turtles, chicks, birds, even dogs and cats. Symptoms include fever, diarrhea, and stomach pain.
Tetanus	*Clostridium tetani*	Bacteria enter the body through a wound in the skin. Affects the nervous system; causes lockjaw, stiffness in neck, difficulty swallowing, fever, and muscle spasms. Vaccination available.
Travelers' diarrhea	Enterotoxigenic *Escherichia coli* is most common cause	Ingested in contaminated food and water. From 30% to 50% of travelers to high-risk areas (Central and South America, Africa, Middle East, and most of Asia) develop travelers' diarrhea.
Syphilis	*Treponema pallidum*	Sexually transmitted disease passed through direct contact with a syphilis sore. If untreated, eventually damages brain, liver, bone, and spleen and can cause death.
Tuberculosis	*Mycobacterium tuberculosis*	Transmitted from person to person through inhalation of air containing the pathogen. Symptoms include fatigue, cough, fever, weight loss. Not common in U.S. However, multidrug-resistant form is a growing threat.
Typhoid fever	*Salmonella typhi*	Transmitted from person to person in food or water contaminated with feces. Risk greatest for travelers in developing countries, but vaccine available. Symptoms include high fever, headache, and loss of appetite. Can cause death if not treated.

lation of microorganisms that are considered normal **microbiota** (also called microflora)—harmless symbiotic prokaryotes. In fact, the number of bacteria that normally inhabit the human body (approximately 700 trillion cells) exceeds the number of the body's human cells (about 70 trillion cells). The presence of certain bacterial populations prevents harmful microorganisms (including other bacteria) from flourishing.

A small percentage of bacterial species are important pathogens of plants and animals. Some of the normal bacterial (and viral) inhabitants are opportunistic pathogens that cause disease only under certain conditions. For example, when the immune system is compromised, opportunistic bacteria increase in number and cause disease. Some important bacterial diseases and the pathogens that cause them are listed in ▌Table 24-5.

Many scientists have contributed to our understanding of infectious disease

The idea that some unknown agent causes disease was debated long before Leeuwenhoek discovered microorganisms with his microscope in the late 1600s. However, not until much later did scientists develop the tools or the methods needed to accurately understand the relationships between bacteria and disease. During the late 19th century several physicians, microbiologists, and chemists working independently laid the foundations for the science of microbiology. French chemist Louis Pasteur disproved the prevailing views of spontaneous generation by demonstrating that sterilization of a sugar and protein culture prevented bacterial growth. Pasteur also developed a rabies vaccine, showing that people can be stimulated to develop immunity to disease.

The German physician Robert Koch was the first to clearly demonstrate that bacteria cause infectious disease. In 1876, he showed that *Bacillus anthracis* causes anthrax. Using a microscope, Koch observed the bacteria in the blood and spleens of dead sheep. When he inoculated mice with the infected sheep blood, he was able to identify *B. anthracis* in the blood of the mice. He also cultured *B. anthracis* and showed that when he injected the bacteria into healthy mice, they developed anthrax.

Koch proposed a set of guidelines, now known as **Koch's postulates,** that are still used to demonstrate that a specific pathogen causes specific disease symptoms: (1) the pathogen must be present in every individual with the disease, (2) a sample of the microorganism taken from the diseased host can be grown in pure culture, (3) a sample of the pure culture causes the same disease when injected into a healthy host, and (4) the microorganism can be recovered from the experimentally infected host.

Many adaptations contribute to pathogen success

Pathogenic microorganisms can enter the body in food, dust, or droplets or through wounds. Many diseases are transmitted by insect or animal bites. To cause disease, a pathogen must adhere to a specific cell type, multiply, and produce toxic substances. Adherence and multiplication occur only when the pathogen competes successfully with the normal microbiota and counteracts the host's defenses against invasion.

Helicobacter pylori, the most common cause of peptic ulcers (ulcers of the stomach and duodenum), is an extremely successful pathogen (❙ Fig. 24-19). It is also associated with chronic gastritis (stomach inflammation) and with stomach cancer (the second most common type of cancer in the world). *Helicobacter pylori* inhabits the intestinal tracts of an estimated 40% of adults in developed countries and 80% of adults in developing countries. Among its many adaptations is its ability to produce an alkaline shield around itself that protects it from stomach acid. Also contributing to its success are several powerful flagella used to propel the pathogen through the thick mucus lining the stomach.

Pathogens produce a variety of substances that increase their success. Some bacteria produce **exotoxins,** strong poisons that are either secreted from the cell or leak out when the bacterial cell is destroyed. The toxin, not the presence of the bacteria themselves, is responsible for the disease. As mentioned previously, diphtheria is caused by a gram-positive bacillus (*Corynebacterium diph-*

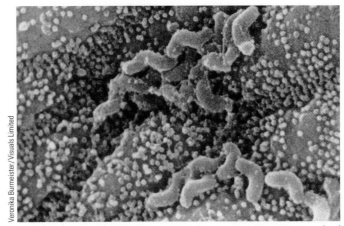

(a) SEM of *Helicobacter pylori* attached to the epithelial lining of the stomach.

Veronika Burmeister/Visuals Limited

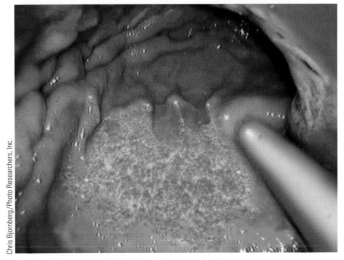

(b) An ulcer caused by *Helicobacter pylori.*

Chris Bjornberg/Photo Researchers, Inc.

Figure 24-19 *Helicobacter pylori,* the bacteria that cause peptic ulcers

theriae) that produces a toxin only when lysogenized by a phage. The diphtheria toxin kills cells and causes inflammation.

Botulism, a type of food poisoning that can lead to paralysis and sometimes death, results from eating improperly canned food. Botulism is caused by an exotoxin released by the gram-positive, endospore-forming *Clostridium botulinum.* During the canning process, food must be heated sufficiently to kill any highly heat-resistant endospores that may be present. If not, the endospores can germinate. The resulting bacterial population grows and releases an exotoxin so powerful that 1 g could kill a million humans! Like many exotoxins, the one that causes botulism is inactivated by heating. (Food must be heated to 80°C for 10 minutes, or boiled for 3 to 4 minutes.) The botulism exotoxin, marketed under the trade name Botox, is used in extremely tiny amounts to treat several medical conditions involving spasms (involuntary muscle contractions). Because Botox is a neurotoxin that works by paralyzing muscles, it can also relax facial wrinkles caused by contraction of the underlying muscles. However, its effects last for only 3 to 8 months.

Endotoxins are not secreted by pathogens but are components of the cell walls of most gram-negative bacteria. These compounds affect the host only when they are released from dead bacteria. Endotoxins bind to the host's macrophages and stimulate them to release substances that cause fever and other symptoms of infection. Unlike exotoxins, which cause specific symptoms, endotoxins appear to affect the entire body. Endotoxins are not destroyed by heating.

Many bacteria have become resistant to antibiotics

Bacteria mutate frequently and reproduce rapidly. Many antibiotics target protein synthesis. For example, streptomycin and related antibiotics block the initiation of protein synthesis. The tetracyclines block aminoacyl tRNA from binding to the A site on the ribosome.

Drug resistance may result from an accumulation of mutations in plasmid or chromosomal DNA. Plasmids that have genes for antibiotic resistance are called **R factors.** They have genes for resistance to a specific drug and for transferring the resistance to other bacteria. Some R factors have several genes for drug resistance, each encoding resistance to a different drug.

Overuse of antibiotics is the principal contributing factor to drug resistance. In any bacterial population, there are likely at least a few bacteria that are genetically resistant to a particular antibiotic. The bacteria that are *not* resistant are killed by the antibiotic, leaving the resistant bacteria to multiply and produce a resistant population. Note that this is a common, present-day example of natural selection.

Another type of drug resistance has recently been described, explaining why some infections, for example, urinary tract infections, are difficult to cure. When *E. coli* infect the bladder, the immune system launches a powerful defense. The bacteria subvert the attack by forming *biofilms* (discussed previously). Each biofilm is surrounded by a matrix rich in polysaccharides and by a protective shell. The bacteria in the biofilm are resistant to antibiotics, as well as to host defenses. By one estimate, bacteria growing in a biofilm are up to 1000 times as resistant to antibiotics as are the same type of bacteria that have not formed a biofilm. Within biofilms, differential gene expression and mutation lead to diverse bacterial colonies. The biofilm strategy also explains some other types of chronic or recurrent infections. A high percentage of infections acquired in hospitals involve biofilms.

Prokaryotes are used in many commercial processes

Some microorganisms produce *antibiotics.* These compounds limit competition for nutrients by inhibiting or destroying other microorganisms. By the 1950s, antibiotics had become important clinical tools that transformed the treatment of infectious disease. Today about 100 clinically useful antibiotics are available, and literally tons of antibiotics are produced annually. Pharmaceutical companies obtain most antibiotics from three groups of microorganisms: a large group of gram-positive soil bacteria, the

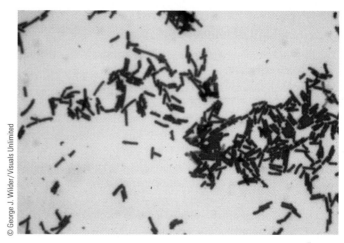

5 μm

Figure 24-20 LM of lactic acid bacteria (*Lactobacillus acidophilus*)

actinomycetes; gram-positive bacteria of the genus *Bacillus;* and molds (eukaryotes belonging to kingdom Fungi).

Because of their prolific reproduction rates, bacteria are ideal "factories" for the production of biomolecules. Researchers have genetically engineered bacteria (see Chapter 15) to produce certain vaccines, human growth hormone, insulin, and many other clinically important compounds. Most of the insulin used to treat diabetics is now derived from genetically engineered bacteria. Researchers are developing genetically engineered bacteria for production of many other medically and agriculturally useful products.

Microbial fermentation helps produce many foods and beverages. Lactic acid bacteria are used in producing acidophilus milk, yogurt, pickles, olives, and sauerkraut (■ Fig. 24-20). Several types of bacteria are used in the production of cheese. Bacteria are involved in making fermented meats such as salami and in producing vinegar, soy sauce, chocolate, certain B vitamins (B_{12} and riboflavin), and citric acid, a compound added to candy and to most soft drinks.

Bacteria are used in sewage treatment and to break down solid wastes in landfills. They are also used in **bioremediation,** a process in which a contaminated site is exposed to microorganisms that break down the toxins, leaving behind harmless metabolic by-products such as carbon dioxide and chlorides. More than 1000 different species of bacteria and fungi have been used to clean up various forms of pollution, and microbiologists are searching for others.

Archaea are also economically important. For example, archaeal enzymes have been added to laundry and industrial detergents to increase their performance at higher temperatures and pH levels. Another archaeal enzyme has been useful in the food industry to convert corn starch to dextrins.

Review

■ What important roles do prokaryotes play in ecosystems?

■ Why is each of Koch's postulates essential?

■ What is bioremediation?

Learning Objectives

1 Describe the structure of a virus, and contrast a virus with a living cell (page 501).

■ A **virus,** or **virion,** is a subcellular particle consisting of a DNA or RNA genome surrounded by a protein coat, called a **capsid.**

■ In contrast to a living cell, viruses cannot metabolize on their own. Viruses contain the nucleic acids necessary to make copies of themselves, but in order to reproduce, they must invade living cells and commandeer their metabolic machinery.

ThomsonNOW **Learn more about virus structure by clicking on the figure in ThomsonNOW.**

2 Trace the evolutionary origin of viruses according to current hypotheses (page 501).

■ Viruses may be bits of nucleic acid that originally "escaped" from animal, plant, or bacterial cells.

■ Some biologists hypothesize that viruses must have evolved before the three domains diverged; they think it unlikely that the similarities between viruses that infect archaea and those that infect bacteria evolved twice.

3 Characterize bacteriophages, and contrast a lytic cycle with a lysogenic cycle (page 501).

■ **Bacteriophages,** or **phages,** are viruses that infect bacteria.

■ A viral reproductive cycle can be lytic or lysogenic. In a **lytic cycle,** the virus destroys the host cell. There are five steps in a lytic cycle: **attachment** to the host cell; **penetration** of viral nucleic acid into the host cell; **replication** of the viral nucleic acid; **assembly** of newly synthesized components into new viruses; and **release** from the host cell.

■ **Temperate viruses** do not always destroy their hosts. In a **lysogenic cycle,** the viral genome is replicated along with the host DNA. The nucleic acid of some phages becomes integrated into the bacterial DNA and is then called a **prophage.** Bacterial cells that carry prophages are lysogenic cells. In **lysogenic conversion,** bacterial cells containing certain temperate viruses exhibit new properties.

ThomsonNOW **Watch the lytic and lysogenic cycles by clicking on the figure in ThomsonNOW.**

4 Compare viral infection of animals and plants, and identify specific diseases caused by animal viruses (page 501).

■ Viruses enter animal cells by membrane fusion or by endocytosis. Viral nucleic acid is replicated within the host cell, proteins are synthesized, and new viruses are assembled and released from the cell. Among the diseases caused by DNA viruses are smallpox, herpes, respiratory infections, and gastrointestinal disorders. RNA viruses cause influenza, upper respiratory infections, AIDS, and some types of cancer.

■ Most plant viruses are RNA viruses. Plant viruses can be spread among plants by insect vectors. Viruses spread through the plant via plasmodesmata.

5 Describe the reproductive cycle of a retrovirus, such as human immunodeficiency virus (HIV) (page 501).

■ **Retroviruses,** such as HIV, use **reverse transcriptase** to transcribe their RNA genome into a DNA intermediate that becomes integrated into the host DNA. Copies of the viral RNA are then synthesized.

ThomsonNOW **Watch the HIV life cycle by clicking on the figure in ThomsonNOW.**

6 Compare and contrast viroids and prions (page 510).

■ **Viroids** and **prions** are smaller than viruses. A viroid consists of a short strand of RNA with no protein coat. A prion consists only of protein. Prions cause **transmissible spongiform encephalopathies (TSEs).**

7 Describe the structure and common shapes of prokaryotic cells (page 512).

■ Prokaryotic cells do not have membrane-enclosed organelles such as nuclei and mitochondria.

■ Bacterial cells have several common shapes: spherical (**cocci**), rod shaped (**bacilli**), and spiral. Spiral bacteria include the **spirillum,** which is a rigid helix, and **spirochete,** which is a flexible helix. **Vibrios** are comma shaped.

■ Most bacteria have **cell walls** composed of **peptidoglycan.** The walls of **gram-positive** bacteria are very thick and consist mainly of peptidoglycan. The cell walls of **gram-negative** bacteria consist of a thin peptidoglycan layer and an outer membrane resembling the plasma membrane. Some species of bacteria produce a **capsule** that surrounds the cell wall.

■ Some bacteria have **pili,** protein structures that extend from the cell and help bacteria adhere to one another or to certain other surfaces.

■ Bacterial **flagella** are structurally different from eukaryotic flagella; each flagellum consists of a basal body, hook, and filament. They produce a rotary motion.

ThomsonNOW **Learn more about the structure of prokaryotes and their cell walls by clicking on the figures in ThomsonNOW.**

8 Describe asexual reproduction in prokaryotes, and summarize three mechanisms (transformation, transduction, and conjugation) that may lead to genetic recombination (page 512).

■ The genetic material of a prokaryote typically consists of a circular DNA molecule and one or more **plasmids,** smaller circular fragments of DNA.

■ Prokaryotes reproduce asexually by **binary fission** (the cell divides, forming two cells), **budding** (a bud forms and separates from the mother cell), or **fragmentation** (walls form inside the cell, which then separates into several cells.)

■ In bacteria, genetic material may be exchanged by transformation, transduction, or conjugation. In **transformation,** a bacterial cell takes in DNA fragments released by another cell. In **transduction,** a phage carries bacterial DNA from one bacterial cell into another. In **conjugation,** two cells of different mating types exchange genetic material.

ThomsonNOW **Watch conjugation by clicking on the figure in ThomsonNOW.**

9 Describe the modes of nutrition and metabolic adaptations of prokaryotes (page 512).

■ Most prokaryotes are **heterotrophs** that obtain energy and carbon from other organisms; some are **autotrophs** that make their own organic molecules from simple raw materials.

■ The majority of bacteria are **chemoheterotrophs.** Most are free-living **decomposers** that obtain both carbon and energy from dead organic matter. **Photoheterotrophs** obtain carbon from other organisms but use chlorophyll and other photosynthetic pigments to trap energy from sunlight.

- Autotrophs may be **photoautotrophs,** which obtain energy from sunlight, or **chemoautotrophs,** which obtain energy by oxidizing inorganic chemicals such as ammonia.

- Most bacteria are **aerobic,** that is, they require oxygen for cellular respiration. Some prokaryotes are **facultative anaerobes** that metabolize anaerobically when necessary; others are **obligate anaerobes** that can carry on metabolism *only* anaerobically.

10 Compare the three domains: Archaea, Bacteria, and Eukarya (page 518).

- Prokaryotes are assigned to domain **Archaea** and domain **Bacteria.**

- Unlike those of bacteria, the cell walls of archaea do not have peptidoglycan. The translational mechanisms of archaea more closely resemble eukaryotic mechanisms than those of other prokaryotes. Domain **Eukarya** includes the four kingdoms of eukaryotes.

11 Distinguish among the main groups of archaea based on their ecology, identify the archaean phyla (see Table 24-3), and describe the main groups of bacteria (see Table 24-4) (page 518).

- Three main groups of archaea are methanogens, extreme halophiles, and extreme thermophiles. **Methanogens** produce methane gas from simple carbon compounds. They inhabit anaerobic environments such as marshes, marine sediments, and the digestive tracts of animals. **Extreme halophiles** inhabit saturated salt solutions. **Extreme thermophiles** can inhabit environments at temperatures higher than 100°C. Archaeans are currently assigned to four phyla (see Table 24-3).

- Major groups of bacteria include the gram-negative proteobacteria, cyanobacteria, gram-positive bacteria, chlamydias, and spirochetes (see Table 24-4).

12 Discuss the ecological roles of prokaryotes, their importance as pathogens, and their commercial importance (page 520).

- Prokaryotes play essential ecological roles as decomposers and are important in recycling nutrients. Some prokaryotes carry out photosynthesis.

- Many bacteria are symbiotic with other organisms. **Mutualism** is a symbiotic relationship in which both partners benefit. In **commensalism,** one partner benefits and the other is neither harmed nor helped. In **parasitism,** the **parasite** benefits and the **host** is harmed. Some parasitic bacteria are important **pathogens** of plants and animals.

- Anton van Leeuwenhoek, Louis Pasteur, and Robert Koch were important pioneers in microbiology. **Koch's postulates** are a set of guidelines used to demonstrate that a specific pathogen causes specific disease symptoms.

- Some pathogenic bacteria release strong poisons called **exotoxins;** others produce **endotoxins,** poisonous components of their cell walls that are released when bacteria die. See Table 24-5 for a description of some pathogenic bacteria.

- Many bacteria have become resistant to antibiotics. Plasmids that have genes for antibiotic resistance are called **R factors.**

- Bacteria are important in many commercial processes. For example, some bacteria produce antibiotics. Several types of bacteria are used to produce cheese. Lactic acid bacteria are used in the production of yogurt, pickles, olives, and sauerkraut.

TEST YOUR UNDERSTANDING

1. The genome of a virus consists of (a) DNA (b) RNA (c) prions (d) DNA and RNA (e) DNA or RNA

2. The capsid of a virus consists of (a) protein (b) nucleic acid (c) helical proteins (d) a simple carbohydrate (e) RNA and lipid

3. The types of organisms a particular virus can infect are referred to as its (a) assembly (b) commensal cohort (c) host range (d) prophage range (e) restriction cycle

4. Viruses that kill host cells are (a) lysogenic (b) lytic (c) viroids (d) prophages (e) temperate

5. Arrange the following list into the correct sequence for viral reproduction.

   1. penetration 2. assembly 3. replication 4. attachment 5. release

   (a) 1, 2, 3, 4, 5 (b) 5, 2, 3, 4, 1 (c) 4, 1, 3, 2, 5 (d) 4, 1, 2, 3, 5 (e) 3, 1, 2, 4, 5

6. In lysogenic conversion (a) bacterial cells may exhibit new properties (b) the host cell dies (c) prions sometimes convert to viroids (d) reverse transcriptase transcribes DNA into RNA (e) prion proteins convert to infective agents

7. Peptidoglycan is a chemical compound found in the cell walls of (a) most viroids (b) most Archaea (c) all prokaryotes (d) most Bacteria (e) most Eukarya

8. Retroviruses (a) are DNA viruses (b) are RNA viruses (c) use reverse transcriptase to transcribe DNA into RNA (d) use prion protein to transcribe RNA (e) are DNA viruses that cause AIDS

9. Bacterial flagella (a) are homologous with eukaryotic flagella (b) exhibit a rotary motion (c) consist of a basal body and nine pairs of microtubules (d) are important in transduction (e) are characteristic of gram-positive bacteria

10. In conjugation (a) two bacterial cells of different mating types come together, and genetic material is transferred from one to another (b) a bacterial cell develops a bulge that enlarges and eventually separates from the mother cell (c) fragments of DNA released by a broken cell are taken in by another bacterial cell (d) a phage carries bacterial genes from one bacterial cell into another (e) walls develop in the cell, which then divides into six new cells

11. Endospores (a) are formed by some viruses (b) are extremely durable cells (c) are comparable to the reproductive

spores of fungi and plants (d) cause fever and other symptoms in the host (e) are extremely vulnerable to infection by phages

12. The majority of heterotrophic bacteria are (a) free-living chemoheterotrophs (b) photoautotrophs (c) chemoautotrophs (d) facultative anaerobes (e) obligate anaerobes

13. Which of the following do *not* belong to domain Archaea? (a) bacteria that produce methane from carbon dioxide and hydrogen (b) thermophiles (c) halophiles (d) bacteriophages (e) prokaryotes with cell walls that lack peptidoglycan

14. Which of the following scientists proposed a set of guidelines to demonstrate that a specific pathogen causes specific disease symptoms? (a) Leeuwenhoek (b) Pasteur (c) Koch (d) Prusiner (e) Woese

15. Bacteria that thrive in puncture wounds are likely to be (a) chemoheterotrophs (b) photoautotrophs (c) chemoautotrophs (d) endospores (e) obligate anaerobes

16. Prions (a) consist of a short strand of RNA with no protein coat (b) are deadly viruses that have caused several hundred deaths in Uganda and other parts of Africa (c) appear to consist only of protein; no nucleic acid component has been found (d) are responsible for causing hepatitis C (e) are not transmissible from human to human

17. Bacteria that are autotrophs (a) do not require atmospheric oxygen for cellular respiration (b) manufacture their own organic molecules from simple raw materials (c) must obtain organic compounds from other organisms (d) get their nourishment from dead organisms (e) produce endospores when oxygen levels are too low for active growth

18. Pathogens (a) are agents that cause disease (b) include most bacteria (c) include most myxobacteria (d) include many commensals (e) are cyanobacteria that cause disease

19. Lactic acid bacteria (a) have gram-negative cell walls (b) fix nitrogen under anaerobic conditions (c) were formerly known as blue-green algae (d) contribute to the unique taste of yogurt, pickles, and sauerkraut (e) secrete slime on which they glide

20. Which group of bacteria contains the gram-positive, anaerobic bacterium that causes botulism? (a) clostridia (b) actinomycetes (c) enterobacteria (d) spirochetes (e) streptococci

CRITICAL THINKING

1. Imagine that you discover a new microorganism. After careful study you determine that it should be classified in domain Archaea. What characteristics might lead you to this classification?

2. How would life on Earth be affected if all prokaryotes were suddenly annihilated?

3. **Evolution Link.** Historically, biologists thought that viruses, because of their simple structure, evolved before cellular organisms. Later, the hypothesis that viruses escaped from cells gained favor. Recently, microbiologists have again suggested that viruses evolved early in the history of life. Based on what you have learned about viruses, present an argument for one of these hypotheses.

Additional questions are available in ThomsonNOW at www.thomsonedu.com/login

25

Protists

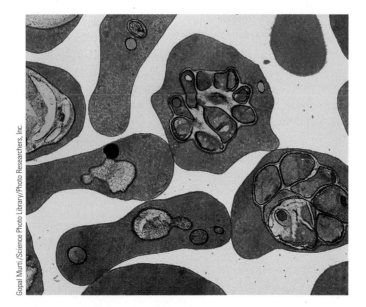

Gopal Murti /Science Photo Library/Photo Researchers, Inc.

Human red blood cells filled with malarial parasites. In this colorized TEM, the parasites multiply inside human red blood cells (*red*), safe from the body's immune system. Malaria is a tropical disease caused by the protist *Plasmodium* and transmitted to humans by the mosquito *Anopheles gambiae* as it feeds on human blood.

KEY CONCEPTS

Protists are a diverse group of eukaryotic organisms, most of which are microscopic.

Protists vary in body plan (unicellular, colonial, coenocytic, multicellular), method of motility (pseudopodia, cilia, flagella), nutrition type (autotrophic, heterotrophic), and mode of reproduction (asexual, sexual).

Protists are descendants of early eukaryotes.

Animals, fungi, and plants evolved from protist ancestors.

Biologists are making progress in understanding the evolutionary relationships among various protist taxa.

In Chapter 23, you learned that the kingdom Protista was first proposed in 1866, to include bacteria and other microorganisms that differ from plants and animals. When Robert Whittaker recommended the five-kingdom system of classification in 1969, only unicellular eukaryotic organisms were placed in kingdom Protista. Since then the boundaries of this kingdom have expanded to include any eukaryote that is not a fungus, animal, or land plant. More recently, some biologists have proposed eliminating the kingdom Protista by splitting it into several kingdoms based largely on molecular data that have clarified certain evolutionary relationships. Many relationships among protists remain uncertain, however.

Kingdom Protista currently consists of a vast assortment of primarily aquatic eukaryotic organisms whose diverse body forms, types of reproduction, modes of nutrition, and lifestyles make them difficult to characterize. **Protists** are unicellular, colonial, or simple multicellular organisms that have a eukaryotic cell organization. The word *protist,* from the Greek for "the very first," reflects the idea that protists were the first eukaryotes to evolve.

Because of their huge numbers, members of kingdom Protista are crucial to the natural balance of the living world. Protists are an important source of food for other organisms, and photosynthetic protists supply oxygen to aquatic and terrestrial ecosystems. Certain protists are economically important, and others cause devastating diseases (see photograph).

Biologists currently recognize as many as 60 major taxa of protists. Consideration of all protist groups is beyond the scope of this text, but we discuss a number of representative taxa. ∎

INTRODUCTION TO THE PROTISTS

Learning Objectives

1   Characterize the features common to the members of kingdom Protista.
2   Discuss in general terms the diversity inherent in the protist kingdom, including means of locomotion, modes of nutrition, interactions with other organisms, habitats, and modes of reproduction.

Protists are members of Eukarya, the third domain on the tree of life. Eukaryotic cells are characteristic of protists as well as of complex multicellular organisms belonging to the kingdoms Fungi, Animalia, and Plantae. However, having a eukaryotic cell structure clearly differentiates protists from members of the prokaryotic domains Bacteria and Archaea. Recall from Chapter 4 that unlike prokaryotic cells, eukaryotic cells have nuclei and other membrane-enclosed organelles such as mitochondria and plastids, 9 + 2 flagella, and multiple chromosomes in which DNA and proteins form a complex called *chromatin.* Sexual reproduction, meiosis, and mitosis are also characteristic of eukaryotes.

Size varies considerably within the protist kingdom, from microscopic protists to giant kelps, which are multicellular brown algae that reach 75 m (about 250 ft) in length. Most protists are **unicellular,** with each cell forming a complete organism capable of performing all the functions characteristic of life. Some protists form **colonies,** loosely connected groups of cells; some are **coenocytic,** consisting of a multinucleate mass of cytoplasm; and some are **multicellular,** composed of many cells. Unlike animals, plants, and many fungi, most multicellular protists have relatively simple body forms without specialized tissues.

Size and structural complexity are not the only variable features of protists. During the course of evolutionary history, organisms in kingdom Protista have evolved diversity in their (1) means of locomotion, (2) ways of obtaining nutrients, (3) in-teractions with other organisms, (4) habitats, and (5) modes of reproduction.

Protists, most of which are motile at some point in their life cycle, have various means of locomotion. They may move by pushing out cytoplasmic extensions (*pseudopodia*) along the leading edge and retracting the cytoplasm that trails behind, as an amoeba does; by flexing individual cells; by gliding over surfaces; by waving *cilia,* short, hairlike organelles; or by lashing *flagella,* long, whiplike organelles (▌Fig. 25-1; also see Fig. 4-25). Some protists have two or more means of locomotion—for example, both flagella and pseudopodia.

Methods of obtaining nutrients differ widely in kingdom Protista. Most algae are autotrophic and photosynthesize as plants do. Some heterotrophic protists obtain their nutrients by absorption, as fungi do, whereas others resemble animals in that they ingest food. Some protists switch their modes of nutrition and are autotrophic at certain times and heterotrophic at others.

Although many protists are free-living, others form stable symbiotic associations with unrelated organisms. These intimate associations range from **mutualism,** a more or less equal partnership where both partners benefit; to **commensalism,** where one partner benefits and the other is unaffected; to **parasitism,** where one partner (the parasite) lives on or in another (the host) and metabolically depends on it (see Chapter 53). Some parasitic protists are important *pathogens* (disease-causing agents) of plants or animals. Throughout this chapter, we describe specific examples of symbiotic associations involving protists.

Most protists are aquatic and live in the ocean or in freshwater ponds, lakes, and streams. They make up most of the **plankton,** the floating, often microscopic organisms that inhabit surface waters and are the base of the food web in aquatic ecosystems. Other aquatic protists attach to rocks or other surfaces in the water. Even parasitic protists are aquatic, because they live in the watery environments of other organisms' body fluids. Terrestrial protists are restricted to damp places such as soil, cracks in bark, and leaf litter.

Reproduction is varied in kingdom Protista. Almost all protists reproduce asexually, and many also reproduce sexually. However, most protists do not develop multicellular reproductive organs, nor do they form embryos the way more complex organisms do.

Review

▌   What are the features of a "typical" protist?
▌   How do protists vary in their means of obtaining nutrients?
▌   What are some of the ways protists interact with other organisms?

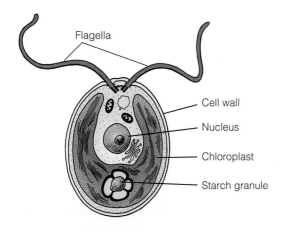

Flagella
Cell wall
Nucleus
Chloroplast
Starch granule

▌ **Figure 25-1**   *Chlamydomonas,* a unicellular protist

Chlamydomonas is a motile organism with two flagella and a cup-shaped chloroplast.

EVOLUTION OF THE EUKARYOTES

Learning Objectives

3   Discuss the hypothesis of serial endosymbiosis, and briefly explain some of the evidence that supports it.
4   Describe the kinds of data biologists use to classify eukaryotes.

Biologists hypothesize that protists were the first eukaryotic cells to evolve from ancestral prokaryotes. They may have appeared in the fossil record as early as 2.2 billion years ago. The first eukaryotes were probably *zooflagellates,* heterotrophic unicellular protists that have flagella. However, other than a few protists with hard shells, such as diatoms and foraminiferans, most ancient protists did not leave many fossils because their bodies were too soft to leave permanent traces. Evolutionary studies of protists focus primarily on molecular and structural comparisons of present-day organisms, which contain many clues about their evolutionary history.

Mitochondria and chloroplasts probably originated from endosymbionts

According to the hypothesis of **serial endosymbiosis,** certain eukaryotic organelles, particularly mitochondria and chloroplasts, arose from symbiotic relationships between larger cells and smaller prokaryotes that were incorporated and lived within them (see Fig. 21-7). Cell biologists hypothesize that mitochondria originated from aerobic bacteria. Studies of mitochondrial DNA suggest it is a remnant from the mitochondrion's past, when it was an independent organism. Ribosomal RNA (rRNA) sequences from mitochondria closely match rRNAs found in purple bacteria, suggesting that ancient purple bacteria were the ancestors of mitochondria.

Chloroplast evolution is more complex, as there were probably several endosymbiotic events. Molecular evidence supports the view that incorporation of an ancient cyanobacterium within a host cell, known as the *primary endosymbiosis,* resulted in the chloroplasts in today's red algae, green algae, and plants. Biologists hypothesize that these chloroplasts, which are enclosed by two external membranes, later provided other nonphotosynthetic eukaryotes with their chloroplasts during *secondary endosymbiosis* (❚ Fig. 25-2).

Secondary endosymbiosis occurred frequently in eukaryote evolution, as evidenced by the presence of additional chloroplast membranes. For example, *three* membranes envelop the chloroplasts of euglenoids and dinoflagellates, and *four* membranes surround the chloroplasts of diatoms, golden algae, and brown algae. Understanding how these membranes originated is an essential aspect of serial endosymbiosis, and many researchers are studying the origin of chloroplasts in different organisms.

Even nonphotosynthetic protists may contain chloroplast relics from secondary endosymbiotic events. Apicomplexans—protists such as *Plasmodium,* which causes malaria—have a nonfunctional chloroplast surrounded by four membranes. Some researchers hypothesize that drugs that inhibit chloroplasts may also kill *Plasmodium.* This nonfunctional chloroplast, which is probably derived from either a red alga or a green alga, therefore has great medical potential in treating malaria.

A consensus is emerging in eukaryote classification

Scientists re-evaluate evolutionary relationships among the eukaryotes as additional evidence becomes available. Two types of

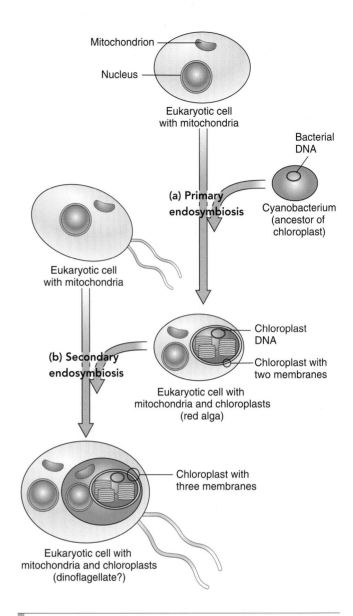

Figure 25-2 Chloroplast evolution by primary and secondary endosymbiosis

(a) In primary endosymbiosis, an ancient eukaryotic cell engulfed a cyanobacterium, which survived and evolved into a chloroplast. The ancient cell is depicted as a eukaryotic cell because mitochondria almost certainly evolved before chloroplasts. **(b)** In a secondary endosymbiotic event, a heterotrophic eukaryotic cell (with mitochondria) engulfed a eukaryotic cell with chloroplasts (a red alga is depicted). The red alga survived and evolved into a chloroplast surrounded by three membranes (a dinoflagellate chloroplast is depicted). Other secondary endosymbiotic events resulted in more complex chloroplast membrane structures.

modern research, molecular and ultrastructure, have contributed substantially to scientific understanding of the phylogenetic relationships among protists. Molecular data were initially obtained for the gene that codes for small subunit ribosomal RNA in different eukaryotes (SSU rRNA; see Chapter 23). More recently, biologists compared other nuclear genes, many of which code for proteins, in different protist taxa.

Unraveling phylogenetic relationships among the protists has been difficult, but progress is occurring. At present, some relationships remain unclear.

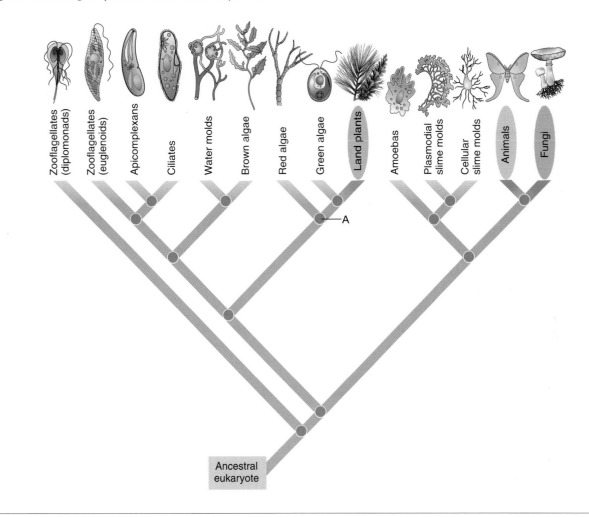

Figure 25-3 Evolutionary relationships among eukaryotes

Relationships among eukaryotes are diverse and debated; this clado-gram presents just one interpretation. It is based on comparisons of four protein sequences among many eukaryotic phyla. Foraminifer-ans, actinopods, dinoflagellates, diatoms, and golden algae are not included. Groups within ovals represent separate eukaryotic kingdoms in the six-kingdom system used in this text. All other eukaryotes (with-out the ovals) are in kingdom Protista, a paraphyletic group. Node A indicates the common ancestor of red algae, green algae, and plants. (Based on S. L. Baldauf, A. J. Roger, I. Wenk-Siefert, and W. F. Doolittle, "A Kingdom-Level Phylogeny of Eukaryotes Based on Combined Pro-tein Data," *Science*, Vol. 290, Nov. 3, 2000.)

Ultrastructure is the fine details of cell structure revealed by electron microscopy. In many cases, ultrastructure data comple-ment molecular data. Electron microscopy reveals similar struc-tural patterns among those protist taxa that comparative molecu-lar evidence suggests are **monophyletic**—that is, evolved from a common ancestor (see Chapter 23). For example, molecular and ultrastructure data suggest that water molds, diatoms, golden algae, and brown algae—protist taxa that at first glance seem to share few characteristics—are a monophyletic group.

Given the diversity in protist structures and molecules, most biologists regard the protist kingdom as a **paraphyletic** group—

that is, the protist kingdom contains some, but not all, of the descendants of a common eukaryote ancestor. Molecular and ul-trastructural analyses continue to help biologists clarify relation-ships among the various protist phyla and among protists and the other eukaryotic kingdoms (Fig. 25-3).

Biologists also use these data to develop various proposals that split organisms in kingdom Protista into several kingdoms. One proposal, for example, removes red algae and green algae from the protists and classifies them within the plant kingdom. Figure 25-3 supports such a classification change, because it sug-gests they share a common ancestor (see node A).

■ How does serial endosymbiosis explain the origin of mitochondria and chloroplasts?

■ What kinds of evidence support the hypothesis that the protist kingdom is a paraphyletic group?

REPRESENTATIVE PROTISTS

Learning Objectives

5 Explain why zooflagellates are no longer classified in a single
phylum, and distinguish among diplomonads, euglenoids,
and choanoflagellates.

6 Briefly describe and compare the following alveolates:
ciliates, dinoflagellates, and apicomplexans.

7 Briefly describe and compare the following heterokonts:
water molds, diatoms, golden algae, and brown algae.

8 Describe the foraminiferans and actinopods, and explain
why many biologists classify them in the monophyletic group
cercozoa.

9 Support the hypothesis that red algae and green algae
should be included in a monophyletic group with land plants.

10 Briefly describe and compare the following amoebozoa:
amoebas, plasmodial slime molds, and cellular slime molds.

The protists include those organisms traditionally known as protozoa, algae, and "lower fungi." Biologists continue to resolve evolutionary relationships among the protists and between the protists and other eukaryotes (plants, animals, and fungi). As a result
of ongoing research, many biologists are developing different ways
to classify eukaryotes, including new names for major taxa. In the

classification scheme we are currently using in this text, almost
all known eukaryotes are assigned to one of eight major clades:
excavates, discicristates, alveolates, heterokonts, plants, cercozoa, amoebozoa, and opisthokonts (∎ Table 25-1). ∎ Figure 25-4
shows these eight major clades superimposed over Figure 25-3.

As additional data accumulate, these groups may be elevated
to the status of kingdoms, resulting in a 10-kingdom system that
replaces the 6-kingdom system currently in use. (Of the 10 kingdoms, 8 are eukaryotes; the remaining 2 are prokaryotes—the
bacteria and the archaea.) However, it is also possible that new
data will prompt biologists to abandon or reorganize at least
some of the eight clades.

Excavates are anaerobic zooflagellates

Zooflagellates are mostly heterotrophic, unicellular (a few are
colonial) organisms with spherical or elongated bodies. They
move rapidly by lashing one to many long, whiplike **flagella** that
are usually located at the anterior (front) end. Some zooflagellates ingest food by means of a definite "mouth," or *oral groove*.
Biologists traditionally classified zooflagellates in a single phylum. However, evidence indicates that zooflagellates are polyphyletic, and most biologists have separated them into several
groups, such as the excavates and discicristates, based on close
evolutionary relationships.

Most **excavates** are endosymbionts and live in anoxic (without oxygen) environments. Unlike virtually all other protists, excavates lack mitochondria, or if they are present, the mitochondria are atypical. Excavates do not carry out aerobic respiration;
they obtain energy by the anaerobic pathway of glycolysis (pre-

TABLE 25-1

Major Monophyletic Eukaryote Clades

Monophyletic Group	Representative Members	Key Characters
Excavates	Diplomonads and other zooflagellates	Mitochondria absent or atypical; cells specialized for glycolysis instead of aerobic respiration; deep (excavated) oral groove
Discicristates	Euglenoids and other zooflagellates	Disc-shaped cristae in mitochondria
Alveolates	Ciliates Dinoflagellates Apicomplexans	Alveoli, flattened vesicles just inside the plasma membrane; tubular cristae in mitochondria
Heterokonts	Water molds Diatoms Golden algae Brown algae	No flagella or two flagella, one plain and one with hairs
Plants	Red algae Green algae Land plants	Chloroplasts surrounded by two membranes (inner and outer membranes)
Cercozoa	Foraminiferans Actinopods	Tests (hard shells) through which cytoplasmic projections extend
Amoebozoa	Amoebas Plasmodial slime molds Cellular slime molds	Naked amoebas (no tests) with pseudopods
Opisthokonts	Fungi Choanoflagellates Animals	No flagella or single posterior flagellum on motile cells; flattened cristae in mitochondria

Source: S. L. Baldauf, "The Deep Roots of Eukaryotes," *Science*, Vol. 300, June 13, 2003.

Many biologists currently classify eukaryotes into eight major groups. However, classification of protists is ongoing.

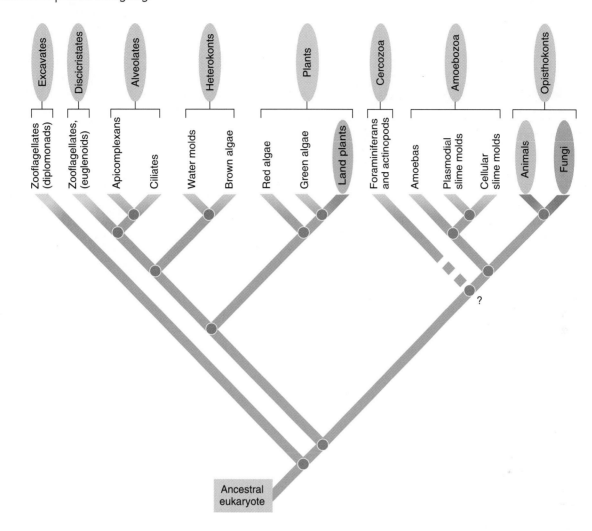

Figure 25-4 The eight monophyletic clades of eukaryotes

The major groups of eukaryotes are superimposed over Figure 25-3. Because classification of the protists is in a state of flux, biologists will modify this diagram as more data are collected and evaluated. (Based on S. L. Baldauf, "The Deep Roots of Eukaryotes," *Science*, Vol. 300, June 13, 2003.)

sumably by fermentation). These protists are so named because most have a deep, or *excavated,* oral groove.

From an evolutionary perspective, one of the most interesting groups of modern-day excavates is the **diplomonads,** which retain some characteristics thought to have been present in ancient protists. (Note in Figure 25-4 that excavates are depicted as branching off closest to the ancestral eukaryote; this denotes that excavates are the *outgroup* in the cladogram and therefore represent an approximation of the ancestral condition for eukaryotes.) Diplomonads have one or two nuclei, no functional mitochondria, no Golgi complex, and up to eight flagella.

Giardia is a parasitic diplomonad. Comparative ribosomal RNA sequencing suggests that diplomonads such as *Giardia* may be more closely related to prokaryotes than are any other protists

(Fig. 25-5a). Interestingly, *Giardia*'s cell structure—it has two haploid nuclei—suggests a partial explanation of how diploid eukaryotes may have arisen from haploid prokaryotes. Some biologists hypothesize that the first eukaryotes had a single haploid nucleus. Most eukaryotes today have a single diploid nucleus formed at some stage in the life cycle when two haploid nuclei fuse. Perhaps the ancestors of *Giardia* represent an intermediate stage in eukaryotic evolution when cells each had two haploid nuclei but fusion had not yet occurred:

Haploid prokaryote ⟶ eukaryote with single haploid nucleus ⟶ eukaryote with two haploid nuclei ⟶ eukaryote with single diploid nucleus

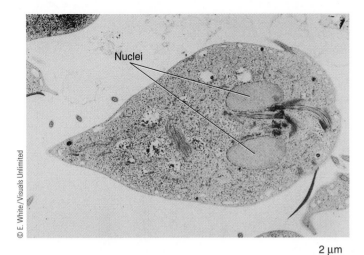

(a) This colorized TEM of *Giardia intestinalis*, a parasitic diplomonad, reveals two nuclei, suggesting that *Giardia* retains an ancestral condition that was an intermediate stage in the evolution of diploidy in eukaryotic life cycles.

2 µm

Figure 25-5 The excavates

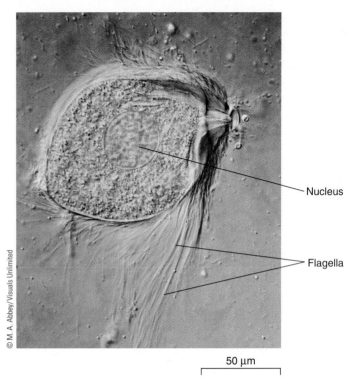

(b) LM of *Trichonympha*, an excavate that lives in the gut of wood-eating termites and cockroaches. *Trichonympha* has hundreds of flagella.

50 µm

Giardia also lacks mitochondria, although it contains certain genes that code for proteins associated with mitochondria in other organisms. *Giardia* also possesses reduced structures that somewhat resemble mitochondria. This information suggests to some biologists that an early eukaryotic ancestor of *Giardia* may have possessed mitochondria, which were somehow lost or reduced at a later time during its evolutionary history.

Giardia intestinalis causes backpackers' diarrhea, a common infection among campers and hikers, particularly in the mountains of the western United States. *Giardia* is eliminated as a resistant cyst in the feces of many vertebrate animals. These cysts are a common contaminant in mountain streams. Campers and hikers become infected when they drink or rinse dishes in the "clean" mountain water. In a heavy infection, much of the wall of the small intestine is coated with these zooflagellates, which interfere with the absorption of digested nutrients and cause weight loss, abdominal cramps, and diarrhea.

Trichonymphs are specialized excavates with hundreds of flagella that live in the guts of termites and wood-eating cockroaches (Fig. 25-5b). These anaerobic zooflagellates lack mitochondria but have Golgi complexes. Trichonymphs ingest wood chips from the wood that termites or roaches eat. The trichonymphs rely on endosymbiotic bacteria to digest cellulose in the wood. The insects, trichonymphs, and bacteria all obtain their nutrients from this source. This is an excellent example of mutualism.

Discicristates include euglenoids and trypanosomes

Discicristates are zooflagellates named for their *disc*-shaped mitochondrial *cristae.* Discicristates include euglenoids, which are somewhat familiar to most students. Most **euglenoids** are unicellular flagellates, and about one third of them are photosynthetic (Fig. 25-6a and b). They generally have two flagella: one long and whiplike and one that is often so short that it does not extend outside the cell. Some euglenoids, such as *Euglena*, change shape continually as they move through the water, because their **pellicle,** or outer covering, is flexible. Euglenoids reproduce asexually by mitosis; none has been observed to reproduce sexually.

At various times scientists classified euglenoids in the plant kingdom (with the algae) and in the animal kingdom (when protozoa were considered animals). Autotrophic euglenoids have chlorophylls *a* and *b* and yellow and orange **carotenoids,** the same photosynthetic pigments that green algae and plants have. Although the euglenoids have the same pigments as the green algae and plants, they are not closely related to either group, as shown in Figure 25-3. They store energy reserves as *paramylon,* a polysaccharide. (Different protist groups produce a variety of storage compounds.) Some photosynthetic euglenoids lose their chlorophyll when grown in the dark, and they obtain their nutrients heterotrophically, by ingesting organic matter. Other species of euglenoids, such as *Peranema,* are always colorless and heterotrophic. Some heterotrophic species absorb organic compounds from the surrounding water, whereas others engulf bacteria and protists by **phagocytosis;** they digest the prey within food vacuoles.

Euglenoids inhabit freshwater ponds and puddles, particularly those with large concentrations of organic material. For this reason, researchers use them as indicator species of organic pollution. Some euglenoids also live in marine waters and mudflats.

Trypanosomes are colorless discicristates, many of which are parasitic and cause disease. In vertebrates, including humans, trypanosomes live in the blood. For example, *Trypanosoma brucei* is a human parasite that causes African sleeping sickness (Fig. 25-6c). It is transmitted by the bite of infected tsetse flies. Early symptoms include recurring attacks of fever. Later, when the trypanosomes have invaded the central nervous system, infected people are lethargic and have difficulty speaking or walking. An estimated 300,000 new cases occur each year, and if untreated, African sleeping sickness can cause death. Trypanosomes, like euglenoids, reproduce asexually by mitosis; none has been observed to reproduce sexually.

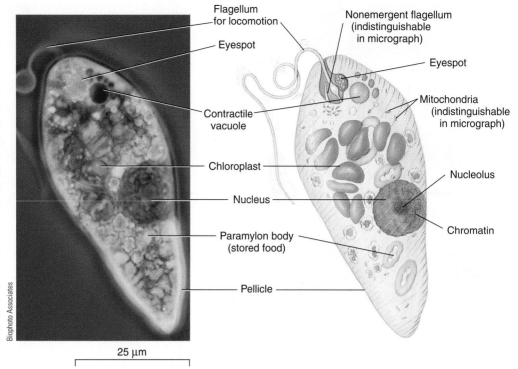

(a) Note the complex cell structure shown in this LM of *Euglena gracilis,* a unicellular, flagellate euglenoid.

(b) *Euglena*'s pellicle is flexible and changes shape easily. The eyespot may shield a light detector at the base of the long flagellum, thereby helping *Euglena* move to light of an appropriate intensity.

Alveolates have flattened vesicles under the plasma membrane

The unifying features of protists classified as **alveolates** are similar ribosomal DNA sequences and **alveoli** (sing., *alveolus*), flattened vesicles located just inside the plasma membrane. In some alveolates, the vesicles contain plates of cellulose. Alveolates include the ciliates, dinoflagellates, and apicomplexans.

Ciliates use cilia for locomotion

Ciliates are among the most complex of eukaryotic cells. These unicellular alveolates have a pellicle that gives them a definite but changeable shape. In *Paramecium,* the surface of the cell is covered with several thousand fine, short, hairlike **cilia** that extend through pores in the pellicle to facilitate movement (Fig. 25-7a and b; also see Fig. 1-1a). The cilia beat with such precise coordination that the organism can back up and turn around as well as move forward.

Not all ciliates are motile. Some sessile forms have stalks, and others, although capable of some swimming, are more likely to remain attached to a rock or other surface at one spot (Fig. 25-7c). Their cilia set up water currents that draw food toward them.

One group of ciliates, the **hypotrichs,** have greatly modified cilia and move with a creeping–darting motion. Hypotrichs lack cilia over much of the body except on the ventral (lowermost) surface, where they occur in stiff tufts called **cirri** (sing., *cirrus,* from the Latin for "a curl of hair") (Fig. 25-7d). Hypotrichs crawl about by means of these cirri, which beat in a coordinated manner.

Ciliates have a wide range of feeding habits and diets; most ingest bacteria or other tiny protists. Their cilia draw the food

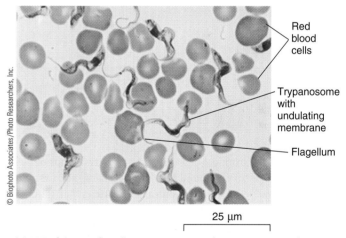

(c) LM of the zooflagellate *Trypanosoma brucei* among red blood cells in a human blood smear. *Trypanosoma brucei* causes sleeping sickness.

Figure 25-6 The discicristates

into a simple opening in some species and into a funnel-like oral groove in others. A vacuole forms around the food at the end of the opening, and the food is digested. Special organelles called **contractile vacuoles** control water regulation in freshwater ciliates. Being hypertonic to their environment, freshwater ciliates continually take in water by osmosis; the contractile vacuole continually expels excess water.

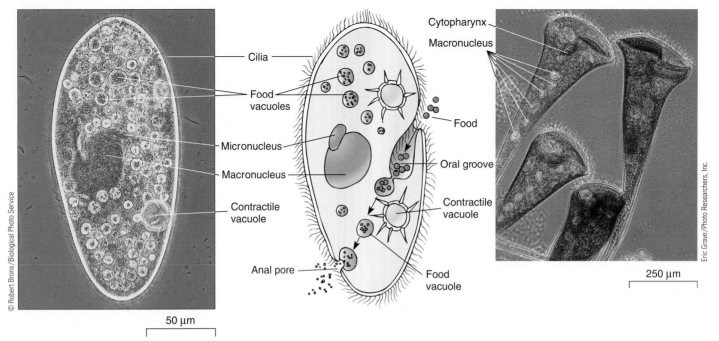

(a) Note the complex cell structure in this phase contrast LM of *Paramecium*, a freshwater ciliate. Like many ciliates, *Paramecium* has multiple nuclei: a macronucleus and one or more smaller micronuclei.

(b) Food particles are swept into *Paramecium*'s ciliated oral groove and incorporated into food vacuoles. Lysosomes fuse with the food vacuoles, and the food is digested and absorbed; undigested wastes are eliminated through the anal pore.

(c) LM of *Stentor*, a sessile ciliate. Note the numerous cilia that direct food particles into its funnel-like cytopharynx ("throat"). The elongated macronucleus resembles a string of beads.

Figure 25-7 Ciliates

[(d) by H. Machemer in K. G. Grell, *Protozoology*, © 1973 Springer-Verlag.]

(d) The hypotrich *Euplotes* has stiff ciliary tufts called cirri.

Ciliates differ from other protists in having two kinds of nuclei: one or more small, diploid **micronuclei** that function in reproduction; and a larger, polyploid **macronucleus** that controls cell metabolism and growth. Most ciliates are capable of a sexual process called **conjugation,** in which two individuals come together and exchange genetic material. Biologists divide each ciliate species into several different mating types. Individuals with different mating types are identical in appearance but genetically different in terms of sexual compatibility. Because there are no known physical differences between the mating types, it is not appropriate to refer to them as "male" and "female."

During conjugation in *Paramecium,* two individuals of compatible mating types press their oral surfaces together (Fig. 25-8). Within each individual the macronucleus disintegrates and the micronucleus undergoes meiosis, forming four haploid nuclei.[1] Three of these degenerate, leaving one, which divides mitotically to form two identical haploid nuclei. One of these remains within the cell, and the other nucleus crosses

through the oral region into the other individual. The migrant haploid micronucleus fuses with the stationary haploid nucleus in that cell, and the two ciliates separate. The new diploid micronucleus in each cell divides a varying number of times, with one micronucleus developing into a new macronucleus.

A single act of conjugation yields two cross-fertilizations as each cell fertilizes the other. Conjugation results in two "new" cells that are genetically identical to each other but different from what they were before conjugation. Mitosis and cell division need not follow immediately after conjugation. Ciliates usually divide perpendicularly to their longitudinal axis.

Most dinoflagellates are a part of marine plankton

Most **dinoflagellates** are unicellular, although a few are colonial. Their alveoli contain interlocking cellulose plates impregnated with silicates (Fig. 25-9a). The typical dinoflagellate has two flagella. One flagellum wraps around a transverse groove in the center of the cell like a belt, and the other lies in a longitudinal groove (perpendicular to the transverse groove), projecting be-

---

[1]The details of conjugation vary somewhat among species.

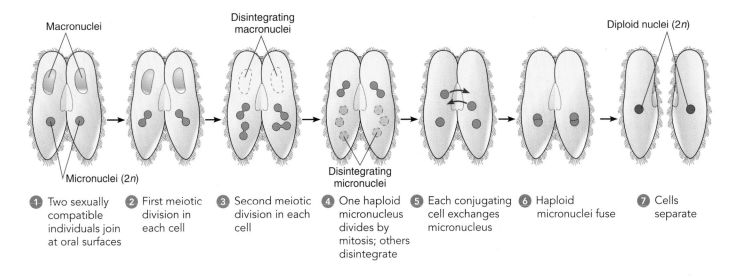

① **Two sexually compatible individuals join at oral surfaces**

② **First meiotic division in each cell**

③ **Second meiotic division in each cell**

④ **One haploid micronucleus divides by mitosis; others disintegrate**

⑤ **Each conjugating cell exchanges micronucleus**

⑥ **Haploid micronuclei fuse**

⑦ **Cells separate**

Labels: Macronuclei · Disintegrating macronuclei · Diploid nuclei (2n) · Micronuclei (2n) · Disintegrating micronuclei

Figure 25-8 *Animated*   Conjugation in *Paramecium caudatum*

hind the cell. The undulation of these flagella propels the dinoflagellate through the water like a spinning top. Indeed, the name is derived from the Greek *dinos,* meaning "whirling." Many marine dinoflagellates are bioluminescent.

Most dinoflagellates are photosynthetic and possess chlorophylls *a* and *c* and carotenoids, including **fucoxanthin,** a yellow-brown carotenoid (Fig. 25-9b). However, some are colorless and ingest other microorganisms for food. Dinoflagellates usually store energy reserves as oils or polysaccharides.

Many dinoflagellates are endosymbionts that live in the bodies of marine invertebrates such as mollusks, jellyfish, and corals (see Fig. 53-11). These symbiotic dinoflagellates, called **zooxanthellae,** lack cellulose plates and flagella. Zooxanthellae photosynthesize and provide carbohydrates for their invertebrate partners. Zooxanthellae contribute substantially to the productivity of coral reefs. Other dinoflagellates that are endosymbionts lack pigments and are parasites that live off their hosts.

Reproduction in the dinoflagellates is primarily asexual, by mitosis, although a few species reproduce sexually. The dinoflagellate nucleus is distinctive because the chromosomes lack histones, are permanently condensed, and always evident. Meiosis and mitosis are unusual because the nuclear envelope remains intact throughout cell division and the spindle lies *outside* the nucleus. The chromosomes attach to the nuclear envelope, and the spindle separates the new nuclei from each other.

Ecologically, dinoflagellates are important producers in marine ecosystems. A few dinoflagellates are known to have occasional population explosions, or blooms. These blooms, which frequently color coastal waters orange, red, or brown, are known as **red tides.** The environmental conditions that initiate dinoflagellate blooms are not known, but many experts think that human-produced coastal pollution is a factor. Some dinoflagel-

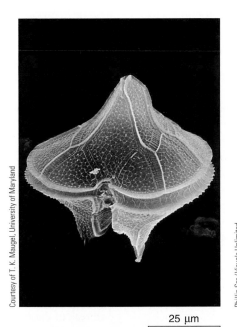

Courtesy of T. K. Maugel, University of Maryland

25 µm

(a) SEM of *Protoperidinium.* Note the cellulose plates that encase the unicellular body and the sutures, or junctions, between adjacent plates. The two flagella (*not visible*) are located in grooves.

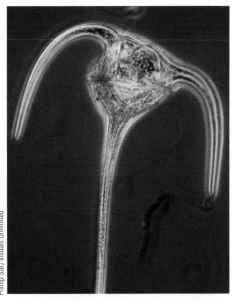

Philip Sze/Visuals Unlimited

(b) LM of *Ceratium,* which is photosynthetic. Because of the pigments, the cellulose plates are difficult to see in living cells.

Figure 25-9 Dinoflagellates

late species that form red tides produce a toxin that attacks the nervous systems of fishes, leading to fish kills. Birds sometimes die after eating contaminated fish. Research has also linked red tides to manatee and dolphin deaths in Florida.

Humans sometimes get paralytic shellfish poisoning by eating filter-feeding mollusks, such as oysters, mussels, or clams, that have fed on certain dinoflagellates. The toxin may be produced by a bacterial endosymbiont living within the dinoflagellate. The poison is a neurotoxin that impairs breathing in humans who eat the contaminated shellfish; death from respiratory failure occasionally occurs. The dinoflagellates do not seem to harm the shellfish.

Apicomplexans are spore-forming parasites of animals

Apicomplexans are a large group of parasitic, spore-forming alveolates, some of which cause serious diseases in humans. They contain a vestigial plastid and may have evolved from parasitic dinoflagellates living in the intestines of marine invertebrates. Apicomplexans lack specific structures for locomotion (cilia, flagella, or pseudopodia); instead, they move by flexing. Apicomplexans have an *apical complex* of microtubules that attaches the parasite to its host cell; the apical complex is visible only by using electron microscopy. At some stage in their life cycle, apicomplexans produce **sporozoites,** small infective agents transmitted to the next host; for this reason, some biologists call these organisms **sporozoa.** Many apicomplexans spend part of their complex life cycle in one host species and part in a different host species.

Malaria is caused by an apicomplexan. According to the World Health Organization, approximately 500 million people currently have malaria, and more than 1 million people, mostly children in developing countries, die from the disease each year. Although for centuries Chinese, Greek, Arabic, and Roman writings described the disease, its causative agent and mode of transmission were not identified until the end of the 19th century. British scientist Ronald Ross received the Nobel Prize in 1902 for his role in elucidating the life cycle. Four species of *Plasmodium* cause malaria in humans. *Plasmodium* sporozoites enter human blood through the bite of an infected female *Anopheles* mosquito (▌Fig. 25-10). *Plasmodium* first enters liver cells, where it multiplies, and then red blood cells, where it continues to proliferate (see chapter opening micrograph). When each infected red blood cell bursts, many new parasites are released. The released parasites infect new red blood cells, and the process is repeated. The simultaneous bursting of millions of red cells causes the symptoms of malaria: a chill, followed by high fever as a result of toxic substances that are released and affect various organs of the body.

Today malaria is recurring in many countries where it was under control for decades. Formerly effective control methods—antimalarial drugs and pesticides—have lost much of their effectiveness. Chloroquine and several other antimalarial drugs are taken prophylactically to prevent malaria, but *Plasmodium* has evolved resistance to several of these, necessitating the use of a combination of drugs. Moreover, pesticides are used to control *Plasmodium*'s vector (carrier), the mosquito, but mosquitoes have evolved resistance to many pesticides. Researchers are currently testing new antimalarial drugs and several possible vaccines against malaria. In addition, the recent sequencing of the genomes of *P. falciparum* and the *Anopheles* mosquito has the potential to provide new diagnostics, drugs, and vaccines.

Motile cells of heterokonts are biflagellate

The **heterokonts** include water molds, diatoms, golden algae, and brown algae. At first glance, heterokonts appear too diverse to classify together. The name *heterokont,* derived from the Greek words for "different flagella," provides a clue. Most heterokonts have motile cells with two flagella, one of which has tiny hairlike projections off the shaft.

Water molds produce biflagellate reproductive cells

Water molds were once classified as fungi because of their superficial resemblance. Both water molds and fungi have a body, called a **mycelium,** that grows over organic material, digesting it and then absorbing the predigested nutrients (▌Fig. 25-11a). The threadlike **hyphae** that make up the mycelium in water molds are *coenocytic,* meaning that there are no cross walls, and the body of these heterokonts consists of a single multinucleate cell (see Fig. 26-1e). The cell walls of water molds are composed of cellulose (as in plants), chitin (as in fungi), or both.

When food is plentiful and environmental conditions are favorable, water molds reproduce asexually (▌Fig. 25-11b). A hyphal tip swells and a cross wall is formed, separating the hyphal tip from the rest of the mycelium. Within this structure, called a **zoosporangium,** tiny biflagellate zoospores form, each of which swims about, lands and encysts, and eventually develops into a new mycelium. When environmental conditions worsen, water molds initiate sexual reproduction. After fusion of male and female nuclei, thick-walled **oospores** develop from the oospheres (female gametes). Water molds often spend the winter as oospores.

Some water molds have played infamous roles in human history. For example, the Irish potato famine of the 19th century was precipitated by the water mold *Phytophthora infestans,* which causes late blight of potatoes. (The genus *Phytophthora* is named from Greek words meaning "plant destruction.") During several rainy, cool summers in Ireland in the 1840s, the water mold multiplied unchecked, causing potato tubers to rot in the fields. Because potatoes were the staple of the Irish peasants' diet, many people starved. Estimates of the number of deaths from the outbreak of the potato famine range from 250,000 to more than 1 million. The famine prompted a mass migration out of Ireland to the United States and other countries.

Late blight is still a problem today. New strains of *P. infestans* have appeared in the United States, Canada, northern Europe, Russia, South America, Japan, South Korea, the Middle East, and Africa. These strains resist the chemicals usually used to control the disease. Because today most people eat a varied diet, late

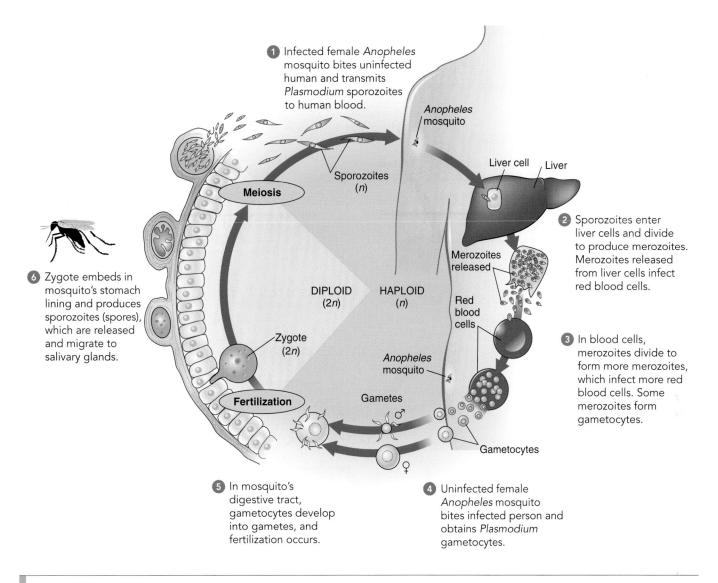

① Infected female *Anopheles* mosquito bites uninfected human and transmits *Plasmodium* sporozoites to human blood.

Sporozoites (*n*)

Meiosis

Anopheles mosquito

Liver cell Liver

② Sporozoites enter liver cells and divide to produce merozoites. Merozoites released from liver cells infect red blood cells.

Merozoites released

⑥ Zygote embeds in mosquito's stomach lining and produces sporozoites (spores), which are released and migrate to salivary glands.

DIPLOID (2*n*) HAPLOID (*n*)

Red blood cells

Zygote (2*n*)

Anopheles mosquito

③ In blood cells, merozoites divide to form more merozoites, which infect more red blood cells. Some merozoites form gametocytes.

Fertilization

Gametes

Gametocytes

♂

♀

⑤ In mosquito's digestive tract, gametocytes develop into gametes, and fertilization occurs.

④ Uninfected female *Anopheles* mosquito bites infected person and obtains *Plasmodium* gametocytes.

Figure 25-10 *Animated* The life cycle of *Plasmodium*, the causative agent of malaria

blight does not cause famine, but it costs potato growers millions of dollars annually in lost crops.

A close relative of the late blight water mold, *P. ramorum*, causes sudden oak death, which is killing oak forests in California and Oregon. Plant pathologists are concerned that the disease may spread to Midwestern and Eastern forests. This particular water mold also attacks redwoods, Douglas firs, bay trees, maples, and several other plant species, but most have only twig and leaf infections, not the rapid death observed in oaks.

Diatoms have shells composed of two parts

Most **diatoms** are unicellular algae, although a few exist as colonies. The cell wall of each diatom consists of two shells that overlap where they fit together, much like a petri dish. Silica is deposited in the shell, and this glasslike material is laid down in intricate

patterns (█ Fig. 25-12a). There are two basic groups of diatoms: those with radial symmetry (wheel shaped) and those with bilateral symmetry (boat shaped or needle shaped). Although some diatoms are part of the floating plankton, others live on rocks and sediments, where they move by gliding. This gliding movement is facilitated by the secretion of a slimy material from a small groove along the shell.

Diatoms contain the photosynthetic pigments chlorophylls *a* and *c* and carotenoids, including fucoxanthin; their pigment composition gives them a yellow or brown color. Energy reserves are stored as oils or the water-soluble carbohydrate *chrysolaminarin*, which is similar to the laminarin stored in brown algae.

Diatoms most often reproduce asexually by mitosis. When a diatom divides, the two halves of its shell separate, and each becomes the larger half of a new diatom shell (█ Fig. 25-12b). Because the glass shell cannot grow, some diatom cells get progres-

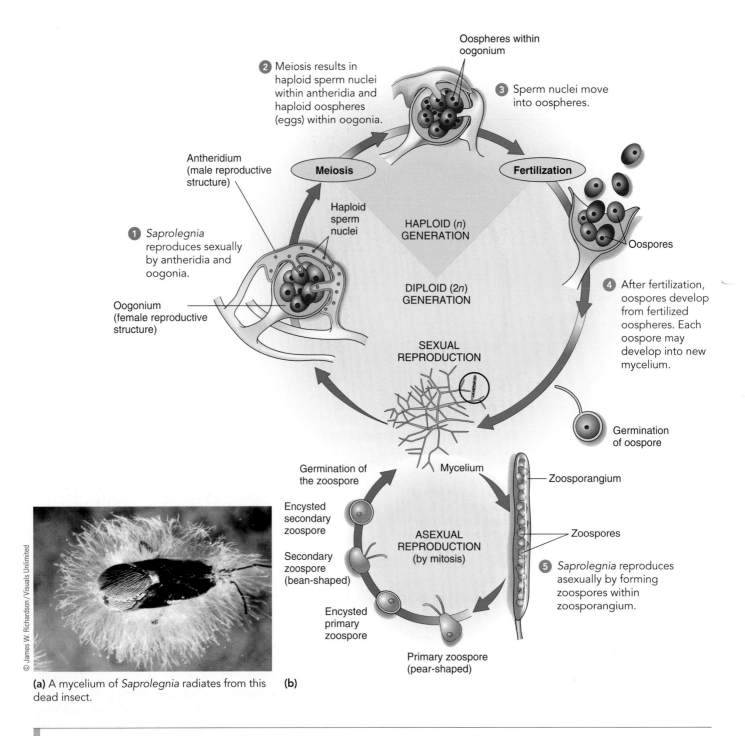

① *Saprolegnia* reproduces sexually by antheridia and oogonia.

② Meiosis results in haploid sperm nuclei within antheridia and haploid oospheres (eggs) within oogonia.

Oospheres within oogonium

③ Sperm nuclei move into oospheres.

Antheridium (male reproductive structure)

Meiosis

Haploid sperm nuclei

Fertilization

HAPLOID (*n*) GENERATION

Oospores

DIPLOID (*2n*) GENERATION

Oogonium (female reproductive structure)

④ After fertilization, oospores develop from fertilized oospheres. Each oospore may develop into new mycelium.

SEXUAL REPRODUCTION

Germination of oospore

Germination of the zoospore

Mycelium

Zoosporangium

Encysted secondary zoospore

ASEXUAL REPRODUCTION (by mitosis)

Zoospores

Secondary zoospore (bean-shaped)

⑤ *Saprolegnia* reproduces asexually by forming zoospores within zoosporangium.

Encysted primary zoospore

Primary zoospore (pear-shaped)

(a) A mycelium of *Saprolegnia* radiates from this dead insect.

(b)

© James W. Richardson / Visuals Unlimited

Figure 25-11   The life cycle of *Saprolegnia*, a water mold

---

sively smaller with each succeeding generation. When a diatom reaches a fraction of its original size, sexual reproduction occurs, with the production of shell-less gametes. Sexual reproduction restores the diatom to its original size because the resulting *zygote,* a 2*n* cell that results from the fusion of *n* gametes, grows substantially before producing a new shell.

Diatoms are common in fresh water, but they are especially abundant in relatively cool ocean water. They are major producers in aquatic ecosystems because of their extremely large numbers. At least one species is toxic and linked to shellfish poisonings,

marine mammal strandings, and the deaths of sea lions along the central California coast.

When diatoms die, their shells trickle to the ocean floor and accumulate in layers that eventually become sedimentary rock. After millions of years, geologic upheaval has exposed some of these deposits on land. Called *diatomaceous earth,* these deposits are mined and used as filtering, insulating, and soundproofing materials. As a filtering agent, diatomaceous earth is used to refine raw sugar and to process vegetable oils. Because of its abrasive properties, diatomaceous earth is a common ingredient in

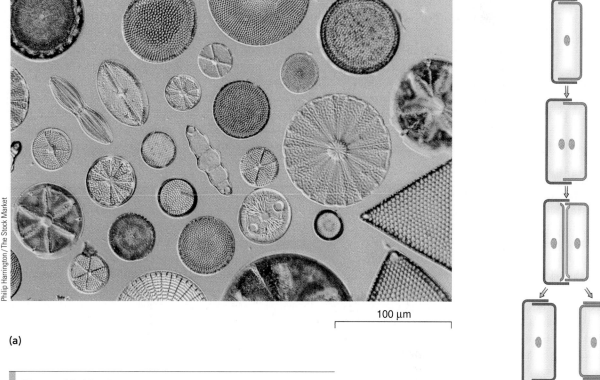

(a)

Figure 25-12 Diatoms

(a) LM of diatoms, unicellular heterokonts with strikingly beautiful symmetrical shells that contain silica. **(b)** Asexual reproduction in diatoms. After cell division, each new cell retains half of the original shell. The newly synthesized half of the shell always fits *inside* the original half. As a result, one of the new cells is slightly smaller.

(b)

scouring powders and metal polishes; it is no longer added to most toothpastes because it is too abrasive for tooth enamel. The intricately detailed diatom shells are often used to test microscope resolution down to 1 μm.

in diatoms, energy reserves are stored as oils or carbohydrates. A few species ingest bacteria and other particles of food. Ecologically, golden algae are important producers in marine environ-

Most golden algae are unicellular biflagellates

Golden algae are found in both freshwater and marine environments. Most species are biflagellate, unicellular organisms, although some are colonial (Fig. 25-13a). A few of these heterokonts lack flagella and are similar to amoebas in appearance except that golden algae contain chloroplasts. Tiny scales of either silica or calcium carbonate may cover the cells. Reproduction in golden algae is primarily asexual and involves the production of biflagellate, motile spores called **zoospores.**

Most golden algae are photosynthetic and produce the same pigments as diatoms: chlorophylls *a* and *c* and carotenoids, including fucoxanthin. The pigment composition of golden algae gives them a golden or golden brown color. As

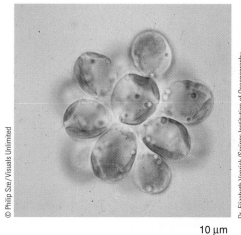

(a) LM of a colonial golden alga (*Synura*) found in freshwater lakes and ponds.

(b) SEM of a coccolithophorid (*Emiliania huxleyi*). Note the overlapping scales of calcium carbonate.

Figure 25-13 Golden algae

ments. They compose a significant portion of the ocean's **nanoplankton,** extremely minute algae (2 to 10 μm) that are major producers because of their great abundance.

Classification of golden algae is controversial. Some biologists lump diatoms and golden algae in a single phylum, whereas others classify them as brown algae. At the other extreme, some biologists divide the golden algae into two phyla by placing many of the marine species, such as **coccolithophorids** (Fig. 25-13b), in a separate phylum.

Brown algae are multicellular heterokonts

Brown algae include the giants of the protist kingdom. They are the largest and most complex of all algae commonly called seaweeds. All brown algae are multicellular and range in size from a few centimeters (about an inch) to 75 m (about 250 ft). Their body forms are branched filaments; tufts; fleshy "ropes"; or thick, flattened branches. The largest brown algae, called *kelps,* are tough and leathery in appearance.

Many kelps have leaflike **blades** in which most photosynthesis occurs, stemlike **stipes,** and rootlike anchoring **holdfasts** (Fig. 25-14a). They often have gas-filled *bladders* that provide buoyancy. (The blades, stipes, and holdfasts of brown algae are not homologous to the leaves, stems, and roots of plants. Brown

algae and plants arose from different unicellular ancestors, as shown in Figure 25-3.)

Brown algae are photosynthetic and have chlorophylls *a* and *c* and carotenoids, including fucoxanthin, in their chloroplasts. The main energy storage reserve in brown algae is a carbohydrate called *laminarin.*

Reproduction is varied and complex in the brown algae. Their reproductive cells, both asexual zoospores and sexual gametes, are usually biflagellate. Most have a life cycle that exhibits **alternation of generations,** in which they spend part of their life as multicellular haploid organisms and part as multicellular diploid organisms (see Fig. 10-17c).

Brown algae are commercially important for several reasons. Their cell walls contain a polysaccharide called *algin* that is harvested from kelps such as *Macrocystis* and used as a thickening and stabilizing agent in ice cream, toothpaste, shaving cream, hair spray, and hand lotion. Brown algae are an important human food, particularly in eastern Asia, and they are rich sources of certain vitamins and minerals such as iodine.

Brown algae are common in cooler marine waters, especially along rocky coastlines, where they live mainly in the intertidal zone or relatively shallow offshore waters. Kelps form extensive underwater "forests," or kelp beds (Fig. 25-14b). They are essential in that ecosystem as important food producers, and they

(a) *Laminaria* is widely distributed on rocky coastlines of temperate and polar seas. It grows to 2 m (6.5 ft).

(b) A kelp (*Macrocystis pyrifera*) bed is ecologically important to aquatic organisms, including the sea lion shown here. Photographed off the coast of California.

Figure 25-14 Brown algae

provide habitat for many marine invertebrates, fishes, and mammals. The diversity of life supported by kelp beds rivals that found in coral reefs. There is also an extensive population of floating brown algae in a central area of the North Atlantic Ocean called the Sargasso Sea, named for the brown alga *Sargassum.* The Sargasso Sea is not greatly affected by the surface ocean currents rotating around the margins of the North Atlantic, so the floating *Sargassum* remains there.

(a) *Polysiphonia*, which is widely distributed throughout the world, has a highly branched body of interwoven filaments. It grows to 30 cm (11.7 in).

(b) *Bossiella* is a coralline red alga encrusted with calcium carbonate. It lives in the Pacific Ocean, where it grows to 12 cm (4.7 in).

Figure 25-15 *Animated*   Red algae

Red algae, green algae, and land plants are collectively classified as plants

In the classification scheme adopted in this text, the monophyletic group of **plants** includes red algae and green algae, which are currently classified in kingdom Protista, and land plants, which are in a separate kingdom (see Chapters 27 and 28). Biologists classify these groups together based on molecular data and on the presence of chloroplasts bounded by outer and inner membranes (see Fig. 9-4).

Red algae do not produce motile cells

The vast majority of **red algae** are multicellular organisms, although there are a few unicellular species. The multicellular body form of red algae commonly consists of complex, interwoven filaments that are delicate and feathery (Fig. 25-15a); a few red algae are flattened sheets of cells. Most multicellular red algae attach to rocks or other substrates by a basal holdfast. Reproduction in the red algae is remarkably complex, with an alternation of sexual and asexual stages. Although sexual reproduction is common, no flagellate cells develop during the life cycle.

The chloroplasts of red algae contain **phycoerythrin,** a red pigment, and **phycocyanin,** a blue pigment, in addition to chlorophylls *a* and *d* and carotenoids. The red algae store energy reserves as *floridean starch,* a polysaccharide similar to glycogen. They have the same pigment composition as the cyanobacteria, a group of photosynthetic bacteria (see Chapter 24).

The cell walls of red algae often contain thick, sticky polysaccharides that have commercial value. For example, *agar* is a polysaccharide extracted from certain red algae and used as a food thickener and culture medium, a substrate on which to grow microorganisms and propagate some plants, such as orchids. Another polysaccharide extracted from red algae, *carrageenan*, is a food additive used to stabilize chocolate milk and to provide a thick, creamy texture to ice cream and other soft processed foods. Carrageenan is also used to stabilize paints and cosmetics. Red algae are a source of vitamins (particularly A and C) and minerals, especially in Japan and other eastern Asian countries where people eat red algae fresh, dried, or toasted in such traditional foods as sushi and nori.

Red algae primarily live in warm tropical ocean waters, although a few species occur in fresh water and in soil. Some red algae, known as *coralline algae,* incorporate calcium carbonate in their cell walls from the ocean water (Fig. 25-15b). The hard calcium carbonate may protect coralline algae from the rigors of wave action. These coralline red algae are extremely important in building "coral" reefs—perhaps more crucial than coral animals in this process.

Green algae share many similarities with land plants

Green algae have pigments, energy reserve products, and cell walls that are chemically identical to those of land plants. Green algae are photosynthetic, with chlorophylls *a* and *b* and carotenoids present in chloroplasts of a wide variety of shapes. They store their main energy reserves as starch. Most green algae have cell walls with cellulose, although some lack walls. Because of these and other similarities, biologists generally accept that land plants arose from ancestral green algae (see Fig. 25-3). Using recent molecular and ultrastructure data, some biologists want to reclassify this extremely diverse group as members of the plant kingdom.

Green algae exhibit a variety of body types, from single cells to colonial forms, to coenocytic algae (multinucleate), to multicellular filaments and sheets (Fig. 25-16). The multicellular forms do not have cells differentiated into tissues, a characteristic that separates them from land plants. Most green algae have, or produce, flagellate cells during their life cycle, although a few are totally nonmotile.

Reproduction in the green algae is as varied as their body forms, with both sexual and asexual reproduction. Many green algae have life cycles with an alternation of multicellular haploid and diploid generations. Asexual reproduction is by mitosis and cell division in single cells or by fragmentation in multicellular forms. Many green algae produce spores asexually by mitosis; if these spores have flagella and are motile, they are called *zoospores* (Fig. 25-17).

Sexual reproduction in the green algae involves gamete formation in unicellular **gametangia** (sing., *gametangium*), reproductive structures in which gametes are produced. Green algae have three types of sexual reproduction—isogamous, anisogamous, and oogamous. If the two flagellate gametes that fuse are identical in size and appearance, sexual reproduction is **isogamous** (see Fig. 25-17). **Anisogamous** sexual reproduction involves the fusion of two flagellate gametes of different sizes (Fig. 25-18). Some green algae are **oogamous** and produce a nonmotile egg and a flagellate male gamete. In addition to sexual reproduction by the fusion of gametes, some green algae exchange genetic information by a form of conjugation in which the genetic material of one cell passes into a recipient cell (Fig. 25-19).

Green algae are found in both aquatic and terrestrial environments. Aquatic green algae primarily inhabit fresh water, although there are many marine species. Terrestrial green algae are restricted to damp soil, cracks in tree bark, and other moist places. Many green algae are symbionts with other organisms; some live as endosymbionts in body cells of invertebrates, and a few grow together with fungi as "dual organisms" called *lichens* (discussed in Chapter 26). Regardless of where they live, green algae are ecologically important as producers.

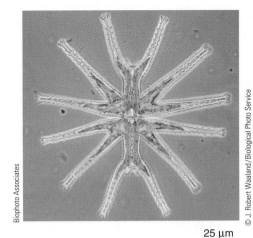

25 μm

(a) LM of a widely distributed desmid (*Micrasterias*), a unicellular green alga with mirror-image halves.

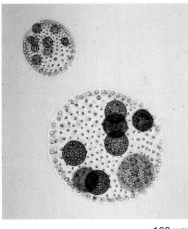

100 μm

(b) LM of two *Volvox* colonies, each composed of up to 50,000 cells. New colonies are inside the parental colonies, which eventually break apart.

(c) *Ulva*'s thin, sheetlike form suggests its common name, sea lettuce. Ulva typically grows to 30 cm (11.7 in).

(d) *Chara*, a green alga commonly called a stonewort, is closely related to land plants. *Chara* is widely distributed in fresh water, where it grows to about 30 cm (11.7 in).

Figure 25-16 Green algae

Cercozoa are amoeboid cells enclosed in shells

The **cercozoa** are amoeboid cells that often have hard outer shells, called **tests,** through which cytoplasmic projections extend. The cytoplasmic projections suggest the name *cercozoa*, from the Greek *cerco*, meaning "rod." Foraminiferans and actinopods are cercozoa.

Foraminiferans extend cytoplasmic projections that form a threadlike, interconnected net

Almost all **foraminiferans** are marine cercozoa that produce elaborate tests (Fig. 25-20a). The ocean contains enormous numbers of foraminiferans, which secrete chalky, many-chambered tests with pores through which cytoplasmic projections are

extended. (*Foraminifera* is derived from the Latin for "bearing openings.") The cytoplasmic projections form a sticky, interconnected net that entangles prey. Many foraminiferans contain unicellular algal endosymbionts (green algae, red algae, or diatoms) that provide food by photosynthesis. Many foraminiferan species live on the ocean floor, but others are part of the plankton.

Dead foraminiferans settle on the bottom of the ocean, where their tests form a gray mud that is gradually transformed into chalk. With geologic uplifting, these chalk formations become part of the land, for example, the White Cliffs of Dover in England (Fig. 25-20b). (The White Cliffs of Dover are the remains of a variety of carbonate organisms, not only foraminiferans.) Because foraminiferan tests often appear in rock layers covering

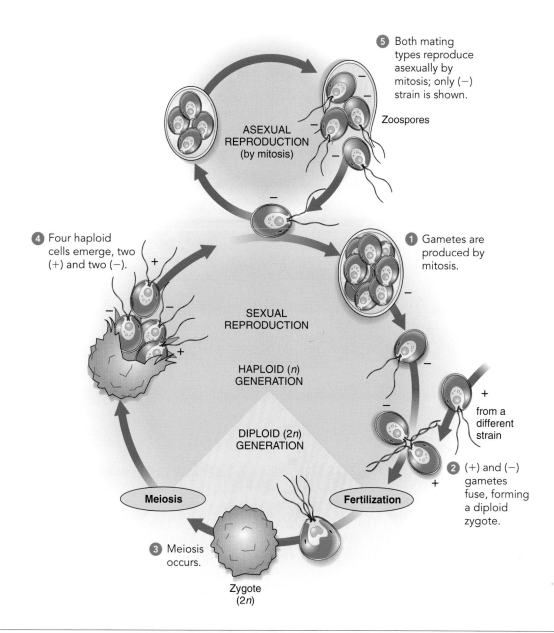

Figure 25-17 *Animated* The life cycle of *Chlamydomonas*

Chlamydomonas, a unicellular haploid green alga with two mating types, (+) and (−), is an example of isogamous sexual reproduction.

Labels in figure:

5 Both mating types reproduce asexually by mitosis; only (−) strain is shown.

Zoospores

ASEXUAL REPRODUCTION (by mitosis)

4 Four haploid cells emerge, two (+) and two (−).

1 Gametes are produced by mitosis.

SEXUAL REPRODUCTION

HAPLOID (*n*) GENERATION

DIPLOID (2*n*) GENERATION

+ from a different strain

2 (+) and (−) gametes fuse, forming a diploid zygote.

Meiosis

Fertilization

3 Meiosis occurs.

Zygote (2*n*)

---

oil deposits, geologists exploring for oil look for foraminiferan tests in rock strata. Foraminiferans are well preserved in the fossil record, and biologists use some as **index fossils,** markers to help identify sedimentary rock layers (see Chapter 18).

Actinopods project slender axopods

Actinopods are mostly marine plankton cercozoa with long, filamentous cytoplasmic projections called **axopods** that protrude through pores in their shells (‖ Fig. 25-21). A cluster of microtubules strengthens each axopod. Unicellular algae and other prey become entangled in these axopods and are engulfed outside the main body of the actinopod; cytoplasmic streaming carries the prey inside the body. Many actinopods contain algal endosymbionts that provide them with the products of photosynthesis.

Some actinopods, called **radiolarians,** secrete elaborate, beautiful glassy shells made of silica. Radiolarians are an important constituent of marine plankton. When radiolarians and other actinopods die, their shells settle and become an ooze (sediment) that may be several meters thick on the ocean floor.

Amoebozoa have lobose pseudopodia

Most **amoebozoa** do not have tests and produce temporary cytoplasmic projections called **pseudopodia** (sing., *pseudopodium,* meaning "false foot") at some point in their life cycle. The

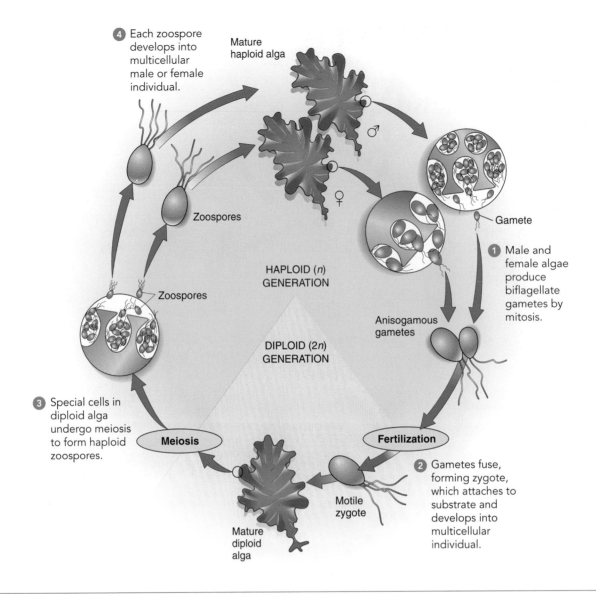

Figure 25-18 The life cycle of *Ulva*

The green alga *Ulva* alternates between haploid and diploid multicellular generations, which are identical in overall appearance. Gametes are isogamous or anisogamous (*shown*), depending on the species.

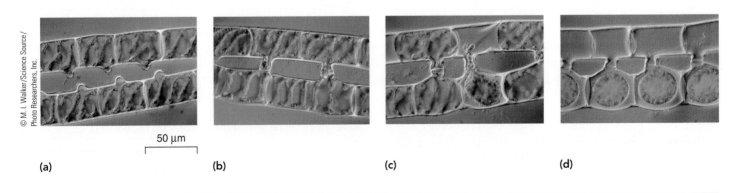

50 μm

(a)   (b)   (c)   (d)

Figure 25-19 Conjugation in *Spirogyra*

(a, b) Filaments of two different mating types of the green alga *Spirogyra* align, and conjugation tubes grow between cells of the two haploid filaments. (c) The contents of one cell passes into the other

through the conjugation tube. (d) The two cells fuse, forming a diploid zygote. Following a period of dormancy, the rounded zygote undergoes meiosis, restoring the haploid condition.

500 µm

Biophoto Associates

(a)

© Lynn McLaren/Photo Researchers, Inc.

(b)

Figure 25-20 Foraminiferans

(a) SEM of a foraminiferan test. Note the pores through which cytoplasm extrudes. Most foraminiferans have multichambered shells like this one. (b) The White Cliffs of Dover, England, consist largely of the tests of foraminiferans.

pseudopodia of amoebozoa are *lobose*—that is, rounded and wide—as opposed to the slender cytoplasmic projections characteristic of the cercozoa. Biologists currently classify amoebas, plasmodial slime molds, and cellular slime molds as amoebozoa.

Amoebas move by forming pseudopodia

Amoebas are unicellular amoebozoa found in soil, fresh water, the ocean, and other organisms (as parasites). Because of the extreme flexibility of their outer plasma membrane, many members of this group have an asymmetrical body form and continu-

ally change shape as they move. (The word *amoeba* derives from a Greek word meaning "change.") An amoeba moves by pushing out lobose pseudopodia from the surface of the cell. More cytoplasm flows into the pseudopodia, enlarging them until all the cytoplasm has entered and the organism as a whole has moved. Pseudopodia also capture and engulf food by surrounding and forming a vacuole around it (Fig. 25-22). Food particles are digested when the food vacuole fuses with a lysosome containing digestive enzymes. Digested materials are absorbed from the food vacuole, which gradually shrinks as it empties. Amoebas

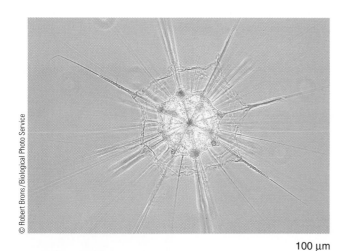

© Robert Brons / Biological Photo Service

100 µm

Figure 25-21 Actinopods

LM of an unidentified living actinopod from the Red Sea. Note the many slender axopods that project from the cell. The shell is not visible, because cytoplasm covers it on all sides (the shell is an endoskeleton).

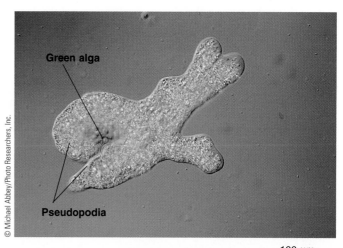

Green alga

Pseudopodia

© Michael Abbey/Photo Researchers, Inc.

100 µm

Figure 25-22 *Animated* Amoeba

LM of a giant amoeba (*Chaos carolinense*). This unicellular protist, which moves and feeds by pseudopodia, is surrounding and ingesting a colonial green alga. *Chaos* amoebas are generally scavengers that feed on debris in freshwater habitats, but they ingest living organisms when the opportunity arises.

reproduce asexually, splitting into two equal parts after mitotic division of the nucleus; sexual reproduction has not been observed.

Parasitic amoebas include *Entamoeba histolytica*, which causes *amoebic dysentery*, a serious human intestinal disease characterized by severe diarrhea, bloody stools, and ulcers in the intestinal wall. In especially severe cases, the organism spreads from the large intestine and causes abscesses in the liver, lungs, or brain. *Entamoeba histolytica* is transmitted as cysts in contaminated drinking water. A **cyst** is a thick-walled, resistant, resting stage in the life cycle of some protists. Other amoebas, such as *Acanthamoeba*, are usually free-living but produce opportunistic infections such as eye infections in contact lens wearers.

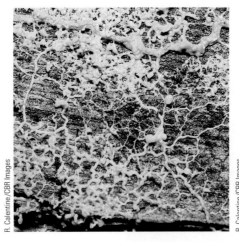

250 μm

(a) The brightly pigmented plasmodium, shown on a dead log, feeds on bacteria and other microorganisms.

(b) The reproductive structures are sporangia on stalks.

Figure 25-23 The plasmodial slime mold *Physarum polycephalum*

Plasmodial slime molds feed as multinucleate plasmodia

The feeding stage of a **plasmodial slime mold** is a **plasmodium,** a multinucleate mass of cytoplasm that can grow up to 30 cm (1 ft) in diameter (Fig. 25-23a). The slimy plasmodium streams over damp decaying logs and leaf litter, often forming a network of channels that covers a large surface area. As it creeps along, it ingests bacteria, yeasts, spores, and decaying organic matter.

When the food supply dwindles or there is insufficient moisture, the plasmodium crawls to an exposed surface and starts reproducing. Stalked structures of intricate complexity and beauty usually form from the drying plasmodium (Fig. 25-23b). Within these structures, called **sporangia,** meiosis produces haploid spores that are extremely resistant to adverse environmental conditions.

When conditions become favorable, the spores germinate, and a haploid reproductive cell emerges from each. This haploid cell is either a biflagellate *swarm cell* or an amoeboid *myxamoeba,* depending on available moisture; flagellate cells form in wet conditions. Swarm cells and myxamoebas act as gametes, which fuse to form a zygote with a diploid nucleus. The resultant diploid nucleus divides many times by mitosis, but the cytoplasm does not divide, so the result is a multinucleate plasmodium.

The plasmodial slime mold *Physarum polycephalum* is a model organism that researchers use to study many fundamental biological processes, such as growth, cytoplasmic streaming, and the function of the cytoskeleton.

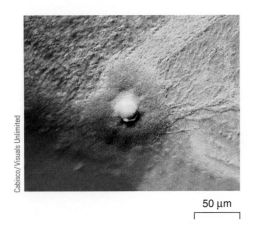

50 μm

50 μm

100 μm

(a)

(b)

(c)

Figure 25-24 *Animated* The cellular slime mold *Dictyostelium discoideum*

(a) Hundreds of amoeboid cells stream together and **(b)** form a migrating sluglike aggregate. When it stops migrating, the aggregate forms a fruiting body on a stalk. **(c)** The fruiting body releases spores, each of which opens in a favorable environment to liberate an amoeboid cell.

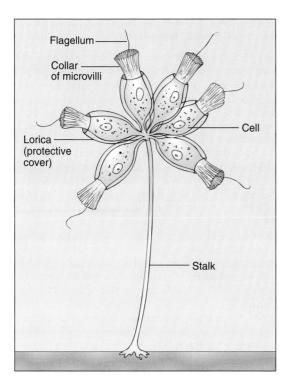

Figure 25-25 Choanoflagellate

Choanoflagellates are free-living zooflagellates that obtain food by waving their flagella, causing water currents to carry bacteria and other small particles of food into the collar of microvilli. A colonial form is shown. Each cell is 5 to 10 μm long, not including the flagellum.

Cellular slime molds feed as individual amoeboid cells

The **cellular slime molds** are amoebozoa with close affinities to amoebas and plasmodial slime molds. During its feeding stage, each cellular slime mold is an individual amoeboid cell that behaves as a separate, solitary organism. Each cell creeps over rotting logs and soil or swims in fresh water, ingesting bacteria and other particles of food as it goes. Each amoeboid cell has a haploid nucleus and reproduces by mitosis, as a true amoeba does.

When moisture or food becomes inadequate, certain cells send out a chemical signal, cyclic adenosine monophosphate (cAMP), that causes them to aggregate by the hundreds or thousands (Fig. 25-24a). During this stage the cells creep about for short distances as a single multicellular unit, called a **pseudoplasmodium,** or slug (Fig. 25-24b). Each cell of the slug retains

its plasma membrane and individual identity. Eventually the slug settles and reorganizes, forming a stalked fruiting body containing spores (Fig. 25-24c). After being released, each spore opens and a single haploid amoeboid cell—the feeding stage—emerges. The spore-forming reproductive cycle is asexual, although sexual reproduction is observed occasionally. The life cycles of most cellular slime molds lack a flagellate stage.

The cellular slime mold *Dictyostelium discoideum* is a model organism for the study of cell differentiation, cell communication, and cell motility and adhesion. Its biology has been studied intensively, particularly as it relates to **cell signaling** molecules, such as cAMP, which are found in many organisms in addition to the cellular slime molds.

Opisthokonts include choanoflagellates, fungi, and animals

The **opisthokonts** are a monophyletic group that includes members of three kingdoms: choanoflagellates (kingdom Protista), fungi (kingdom Fungi), and animals (kingdom Animalia). Opisthokonts are characterized by a single posterior flagellum in flagellate cells such as sperm and motile spores. We consider fungi and animals in later chapters (see Chapters 26, 29, 30, and 31).

Choanoflagellates are collared zooflagellates. These small, inconspicuous marine and freshwater opisthokonts include both free-swimming and **sessile** species that are permanently attached by a thin stalk to bacteria-rich debris. Their single flagellum is surrounded at the base by a delicate collar of microvilli that trap food (Fig. 25-25). Choanoflagellates are of special interest because of their striking resemblance to collar cells in sponges (see Fig. 29-9). Other animal phyla, such as flatworms and rotifers, also contain choanoflagellate-like cells. Given structural similarities and comparative ribosomal RNA and DNA sequence data, many biologists hypothesize that choanoflagellates are related to the ancestor of animals—that is, living choanoflagellates and animals may share a common choanoflagellate-like ancestor.

Review

- What are some unique features of diplomonads?
- What characteristics do ciliates, dinoflagellates, and apicomplexans share?
- What are heterokonts?
- How do cercozoa and amoebozoa differ?
- Why do many biologists classify red algae and green algae with land plants?

SUMMARY WITH KEY TERMS

Learning Objectives

1 Characterize the features common to the members of kingdom Protista (page 531).
- **Protists** are mostly unicellular eukaryotic organisms that live in aquatic environments.
- Protists range in size from microscopic **unicellular** organisms to **colonies** (loosely connected groups of cells) to **coenocytes** (multinucleate masses of cytoplasm) to **multicellular** organisms (composed of many cells).

2 Discuss in general terms the diversity inherent in the protist kingdom, including means of locomotion, modes of nutrition, interactions with other organisms, habitats, and modes of reproduction (page 531).
- Protists have various means of locomotion, including pseudopodia, flagella, and cilia. A few are nonmotile.
- Protists obtain their nutrients autotrophically or heterotrophically.

- Protists are free-living or symbiotic, with symbiotic relationships ranging from **mutualism** to **parasitism.**
- Most protists live in the ocean or in freshwater ponds, lakes, and streams. Parasitic protists live in the body fluids of their hosts.
- Many protists reproduce both sexually and asexually; others reproduce only asexually.

3 Discuss the hypothesis of serial endosymbiosis, and briefly explain some of the evidence that supports it (page 531).
- According to the hypothesis of **serial endosymbiosis,** mitochondria and chloroplasts arose from symbiotic relationships between larger cells and the smaller prokaryotes that were incorporated and lived within them.
- Mitochondria probably originated from aerobic bacteria. Ribosomal RNA data suggest that ancient purple bacteria were the ancestors of mitochondria.
- Chloroplasts of red algae, green algae, and plants probably arose in a single primary endosymbiotic event in which a cyanobacterium was incorporated into a cell. Multiple secondary endosymbioses led to chloroplasts in euglenoids, dinoflagellates, diatoms, golden algae, and brown algae and to the nonfunctional chloroplasts in apicomplexans.

4 Describe the kinds of data biologists use to classify eukaryotes (page 531).
- Relationships among protists are determined largely by **ultrastructure,** which is the fine details of cell structure revealed by electron microscopy, and comparative molecular data. Most biologists consider the protist kingdom as a **paraphyletic** group that contains some, but not all, descendants of a common eukaryotic ancestor.

5 Explain why zooflagellates are no longer classified in a single phylum, and distinguish among diplomonads, euglenoids, and choanoflagellates (page 534).
- **Zooflagellates** are mostly unicellular heterotrophs that move by means of whiplike **flagella.** Evidence indicates that zooflagellates are polyphyletic, and most biologists have separated them into several monophyletic groups.
- **Diplomonads** are **excavates,** organisms with a deep, or excavated, oral groove. Diplomonads that have one or two nuclei, no mitochondria, no Golgi complex, and up to eight flagella.
- **Euglenoids** are **discicristates,** organisms with disclike cristae in their mitochondria. Euglenoids are unicellular and flagellate. Some euglenoids are photosynthetic. The discicristate *Trypanosoma* causes African sleeping sickness.
- **Choanoflagellates** are **opisthokonts** (organisms with a single posterior flagellum in flagellate cells) and are related to fungi and animals. A collar of microvilli surrounds their single flagellum at the base.

6 Briefly describe and compare the following alveolates: ciliates, dinoflagellates, and apicomplexans (page 534).
- **Ciliates** are **alveolates** that move by hairlike **cilia,** have **micronuclei** (for sexual reproduction) and **macronuclei** (for controlling cell metabolism and growth), and undergo a complex sexual reproduction called **conjugation.**

ThomsonNOW™ **Watch conjugation by clicking on the figure in ThomsonNOW.**

- **Dinoflagellates** are mostly unicellular, biflagellate, photosynthetic alveolates of great ecological importance as major producers in marine ecosystems. Their **alveoli,** flattened vesicles under the plasma membrane, often contain cellulose plates impregnated with silicates. Some dinoflagellates produce toxic blooms known as **red tides.**
- **Apicomplexans** are parasites that produce **sporozoites** and are nonmotile. An apical complex of microtubules attaches the apicomplexan to its host cell. The apicomplexan *Plasmodium* causes malaria.

ThomsonNOW™ **Watch the life cycle of the malaria parasite by clicking on the figure in ThomsonNOW.**

7 Briefly describe and compare the following heterokonts: water molds, diatoms, golden algae, and brown algae (page 534).
- **Water molds** are **heterokonts,** organisms that have two different kinds of flagella. Water molds have a coenocytic **mycelium.** They reproduce asexually by forming biflagellate **zoospores** and sexually by forming **oospores.** The water mold *Phytophthora* causes serious plant diseases, such as late blight of potato and sudden oak death.
- **Diatoms,** which are major producers in aquatic ecosystems, are mostly unicellular heterokonts, with shells containing silica. Some diatoms are part of floating **plankton,** and others live on rocks and sediments where they move by gliding.
- **Golden algae** are mostly unicellular, biflagellate freshwater and marine heterokonts that are of ecological importance as a major component of the ocean's extremely minute **nanoplankton. Coccolithophorids** are golden algae covered by tiny, overlapping scales of calcium carbonate.
- **Brown algae** are multicellular heterokonts that are ecologically important in cooler ocean waters. The largest brown algae (kelps) possess leaflike **blades,** stemlike **stipes,** anchoring **holdfasts,** and gas-filled bladders for buoyancy.

8 Describe the foraminiferans and actinopods, and explain why many biologists classify them in the monophyletic group cercozoa (page 534).
- The **cercozoa** are amoeboid cells that often have hard outer shells, called **tests,** through which cytoplasmic projections extend.
- **Foraminiferans** secrete many-chambered tests with pores through which cytoplasmic projections extend to move and obtain food.
- **Actinopods** are mostly marine plankton that obtain food by means of **axopods,** slender cytoplasmic projections that extend through pores in their shells. **Radiolarians** are actinopods with glassy shells.

9 Support the hypothesis that red algae and green algae should be included in a monophyletic group with land plants (page 534).
- Red algae, green algae, and land plants, collectively called **plants,** are considered a monophyletic group based on molecular data and on the presence of chloroplasts bounded by outer and inner membranes.
- **Red algae,** which are mostly multicellular seaweeds, are ecologically important in warm tropical ocean waters. Red algae that incorporate calcium carbonate in their cell walls are important in reef building.
- **Green algae** exhibit a wide diversity in size, structural complexity, and reproduction. Botanists hypothesize that ancestral green algae gave rise to land plants.

10 Briefly describe and compare the following amoebozoa: amoebas, plasmodial slime molds, and cellular slime molds (page 534).
- **Amoebas** move and obtain food by **phagocytosis,** using cytoplasmic extensions called **pseudopodia.** The parasitic amoeba *Entamoeba histolytica* causes amoebic dysentery.
- The feeding stage of **plasmodial slime molds** is a multinucleate **plasmodium.** Reproduction is by haploid spores produced within **sporangia.**
- **Cellular slime molds** feed as individual amoeboid cells. They reproduce by aggregating into a **pseudoplasmodium** (slug), then forming asexual spores.

ThomsonNOW™ **Watch the life cycles of a red alga, green alga, and cellular slime mold by clicking on the figures in ThomsonNOW.**

1. Which of the following is *not* true of the protists? (a) they are unicellular, colonial, coenocytic, or simple multicellular organisms (b) their cilia and flagella have a 9 + 2 arrangement of microtubules (c) they are prokaryotic, as bacteria and archaea are (d) some are free-living, and some are endosymbionts (e) most are aquatic and live in the ocean or in freshwater ponds

2. Amoebas move and obtain food by means of (a) pseudopodia (b) flagella (c) cilia (d) gametangia (e) endosymbiosis

3. Foraminiferans (a) are endosymbionts in many marine invertebrates (b) were responsible for the Irish potato famine in the 19th century (c) secrete many-chambered tests with pores through which cytoplasmic extensions project (d) have numerous axopods that may aid in trapping and holding prey (e) contain phycocyanin and phycoerythrin, pigments found in no other protist group

4. Unicellular protists that are free-living or parasitic, move by means of flagella, and do not photosynthesize are informally called (a) euglenoids (b) dinoflagellates (c) myxamoebas (d) zooflagellates (e) apicomplexans

5. *Paramecium* and other ciliates often display a sexual phenomenon called (a) oogamy (b) conjugation (c) anisogamy (d) red tide (e) alternation of generations

6. Parasitic alveolates that form spores at some stage in their life belong to which group? (a) actinopods (b) ciliates (c) coccolithophorids (d) apicomplexans (e) dinoflagellates

7. Malaria (a) is transmitted by the bite of a female tsetse fly (b) is caused by a parasitic zooflagellate, *Giardia intestinalis* (c) is a serious form of amoebic dysentery caused by *Entamoeba histolytica* (d) is caused by an apicomplexan that spends part of its life cycle in the *Anopheles* mosquito and part in humans (e) is transmitted when people drink water tainted by a red tide

8. Alveolates characterized by two flagella, one wrapped around the center of the cell like a belt and the other projecting behind the cell, are (a) actinopods (b) ciliates (c) coccolithophorids (d) apicomplexans (e) dinoflagellates

9. Photosynthetic protists with shells composed of two halves that fit together like a petri dish are (a) golden algae (b) diatoms (c) euglenoids (d) brown algae (e) foraminiferans

10. Chlorophyll *a,* chlorophyll *b,* and carotenoids are found in (a) green algae, red algae, and land plants (b) green algae, euglenoids, and land plants (c) brown algae, green algae, and golden algae (d) brown algae, diatoms, and golden algae (e) green algae, euglenoids, and diatoms

11. Which protists have photosynthetic pigments similar to those of the cyanobacteria? (a) golden algae (b) diatoms (c) euglenoids (d) brown algae (e) red algae

12. The multicellular bodies of _____ are differentiated into blades, stipes, holdfasts, and gas-filled floats.

(a) golden algae (b) diatoms (c) euglenoids (d) kelps (e) green algae

13. The feeding stage of plasmodial slime molds is a multinucleate (a) plasmodium (b) pseudoplasmodium (c) pseudopodium (d) gametangium (e) mycelium

14. Cellular slime molds (a) include *Physarum* and *Phytophthora* (b) are more closely related to prokaryotes than are any other protists (c) are responsible for late blight of potatoes, which led to starvation in Ireland in the 1840s (d) have isogamous sexual reproduction (e) form a pseudoplasmodium when cells aggregate in response to cyclic AMP

15. Water molds reproduce asexually by forming _____ and sexually by forming _____. (a) oospores; holdfasts (b) zoospores; zooxanthellae (c) zoospores; oospores (d) holdfasts; isogametes (e) oospores; isogametes

16. Label the diagrams. Use Figures 25-7b and 25-6b to check your answers.

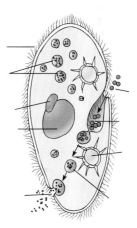

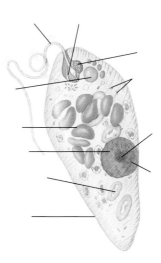

1. **Evolution Link.** Why is the protist kingdom considered paraphyletic? Use Figure 25-3 to help explain your answer.

2. **Analyzing Data.** Where on Figure 25-3 would you place the common ancestor of brown algae and water molds? The common ancestor of animals and fungi? The common ancestor of animals and plants?

3. **Analyzing Data.** Given what you have learned about dinoflagellates, where would you place them in the Figure 25-3 cladogram? Justify your decision.

4. **Analyzing Data.** In this chapter we discussed the hypothesis that choanoflagellates and animals share a common choanoflagellate-like ancestor. If this hypothesis is true, where would you place the choanoflagellate branch on Figure 25-3? Explain.

Additional questions are available in ThomsonNOW at www.thomsonedu.com/login

26

Kingdom Fungi

Jeff Lepore/Photo Researchers, Inc.

Fungal spores. The rounded earthstar (*Geastrum saccatum*) releases a puff of microscopic spores after the sac, which is about 1.3 cm (0.5 in) wide, is hit by a raindrop. This fungus is common in leaf litter under trees throughout North America.

KEY CONCEPTS

Fungi are eukaryotic heterotrophs that absorb nutrients from their surroundings.

A fungus may be a unicellular yeast or a filamentous, multicellular mold consisting of long, branched hyphae that form a mycelium.

Most fungi reproduce both asexually and sexually by means of spores.

According to current hypotheses, fungi evolved from a unicellular, flagellate protist and diverged into five main groups.

Fungi are of major ecological, economic, biological, and medical importance.

Mushrooms, morels, and truffles, delights of the gourmet, share a recent common ancestry with baker's yeast, the black mold that forms on stale bread, and the mildew that collects on damp shower curtains. All of these life-forms belong to kingdom Fungi, a diverse group of about 100,000 known species, most of which are terrestrial. Some biologists estimate that there may be as many as 1.5 million species of fungi.

Fungi grow best in moist habitats, but they are found universally wherever organic material is available. They require moisture to grow, and they obtain water from a humid atmosphere as well as from the medium on which they live. When the environment becomes dry, fungi survive by going into a resting stage or by producing spores (see photograph) that resist desiccation (drying out).

Some fungi grow to enormous size. In Washington State, a fungal clone (*Armillaria ostoyae*) has been identified that covers more than 1500 acres. This giant fungus, which is mainly underground, developed from a single spore that germinated more than a thousand years ago. The fungus has fragmented and is no longer one continuous body.

Like prokaryotes, most fungi are decomposers that obtain nutrients from dead organic matter. They are vital members of ecosystems because they break down the organic compounds found in dead organisms, leaves, garbage, sewage, and other waste. When they break down organic compounds, carbon and other elements are released into the environment, where they are recycled.

Many fungi form vital symbiotic relationships. For example, most terrestrial plants have fungal partners that live in close association with their roots. The fungi help the plants obtain phosphate ions and other needed minerals from the soil. In exchange, the plants provide the fungi with organic nutrients. Some fungi live symbiotically with algae and cyanobacteria as lichens. Others are parasites and pathogens that cause disease in animals or plants. ■

CHARACTERISTICS OF FUNGI

Learning Objectives

1   Describe the distinguishing characteristics of kingdom Fungi.
2   Describe the body plan of a fungus.
3   Describe the life cycle of a typical fungus, including sexual and asexual reproduction.

Although they vary strikingly in size and shape, all **fungi** (sing., *fungus*) are eukaryotes; their cells contain membrane-enclosed nuclei, mitochondria, and other membranous organelles. Fungi share certain key characters, including their way of obtaining nutrition.

The optimum pH for most species is about 5.6, but various fungi can tolerate and grow in environments where the pH ranges from 2 to 9. Many fungi are less sensitive to high osmotic pressures than are bacteria. As a result, they can grow in concentrated salt solutions or in sugar solutions such as jelly, which discourage or prevent bacterial growth. Fungi also thrive over a wide temperature range. Even refrigerated food may be invaded by fungi.

Fungi absorb food from the environment

Unlike plants, fungi cannot produce their own organic materials from a simple carbon source (carbon dioxide). Like animals, fungi are heterotrophs. They depend on preformed carbon molecules produced by other organisms. However, fungi do not ingest food and then digest it in the body as animals do. Instead, they infiltrate a food source and secrete digestive enzymes onto it. Digestion takes place outside the body. When complex molecules are broken down into smaller compounds, the fungus absorbs the predigested food into its body.

The fungus is very efficient at absorbing nutrients and growing. It rapidly converts nutrients into new cell material. If excessive amounts of nutrients are available, fungi store them, usually as lipid droplets or glycogen.

Fungi have cell walls that contain chitin

Like the cells of bacteria, certain protists, and plants, fungal cells are enclosed by cell walls during at least some stage in their life cycle. Fungal cell walls, however, have a different chemical composition from cell walls of other organisms. In most fungi, the cell wall consists of complex carbohydrates, including **chitin**, a polymer that consists of subunits of a nitrogen-containing sugar (see Fig. 3-11). Chitin is resistant to breakdown by most microorganisms. Chitin is also a component of the external skeletons of insects and other arthropods.

Most fungi have a filamentous body plan

Mycologists, biologists who study fungi, identify two main types of fungi, based on body plan: yeasts and molds. The simplest fungi are the **yeasts,** which are unicellular, with a round or oval shape. Yeasts are widely distributed in the soil; on leaves, fruits, and cured meats; and on and in our bodies. Yeasts are essential in making bread and fermenting alcoholic beverages. Some yeasts have been very important in medicine and as model organisms in modern biology (discussed later in this chapter).

Most fungi are **molds.** The vegetative (nonreproductive) body plan of molds consists of long, branched, threadlike filaments called **hyphae** (sing., *hypha*) (■ Fig. 26-1a and b). Hyphae are an adaptation to the fungal mode of nutrition. Growth occurs at the tips of the hyphae; as the hyphae elongate, the fungus grows into and infiltrates food sources. The fungus absorbs nutrients through its very large surface area.

A hypha first develops from a unicellular spore. As hyphae grow, they form a tangled mass or tissuelike aggregation known as a **mycelium** (pl., *mycelia*). The cobweblike mold sometimes seen on bread is the mycelium of a fungus. What is not seen is the extensive mycelium that grows into the bread. Depending on environmental conditions, some fungi can alternate between a yeast phase and a phase in which they produce hyphae.

In most fungi, hyphae are divided by cross walls, called **septa** (sing., *septum*), into individual cells containing one or more nuclei (■ Fig. 26-1c and d). As we will discuss, the presence of septa is an important character in the two largest fungal phyla (which include the most complex fungi). The septa of many fungi are perforated by a pore that may be large enough to permit organelles to flow from cell to cell. Some fungi, called **coenocytic** fungi, lack septa. In these species, nuclear division is not followed by cytoplasmic division. As a result, a coenocytic fungus is one elongated, multinucleated, giant cell (■ Fig. 26-1e).

Fungi reproduce by spores

Most fungi reproduce by means of microscopic **spores,** reproductive cells that can develop into new organisms (■ Fig. 26-2). In most groups, spores are nonmotile. They are dispersed by wind, water, or animals. The air we breathe is filled with hundreds of thousands of fungal spores.

Fungi produce spores either sexually or asexually. With asexual reproduction, new fungi are produced quickly, but there is little genetic variability. Sexual reproduction involves meiosis and generates new genotypes.

Spores are usually produced on specialized aerial hyphae or in fruiting structures. Positioned up above the ground, spores can

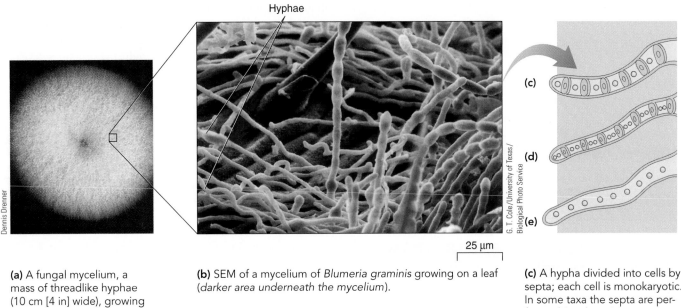

Hyphae

(a) A fungal mycelium, a mass of threadlike hyphae (10 cm [4 in] wide), growing on agar in a culture dish. In nature, fungal mycelia are rarely so symmetric.

(b) SEM of a mycelium of *Blumeria graminis* growing on a leaf (*darker area underneath the mycelium*).

25 μm

Dennis Drenner

G. T. Cole/University of Texas/ Biological Photo Service

(c) A hypha divided into cells by septa; each cell is monokaryotic. In some taxa the septa are perforated (*as shown*).

(d) A septate hypha in which each cell is dikaryotic (has two nuclei).

(e) A coenocytic hypha.

Figure 26-1 Fungus body plan

be easily dispersed. Structures in which spores are produced are called **sporangia** (sing., *sporangium*). The aerial hyphae of some fungi produce spores in large, complex reproductive structures, referred to as **fruiting bodies.** The familiar part of a mushroom is a large fruiting body. We do not normally see the bulk of the fungus, a nearly invisible mycelium buried out of sight in the rotting material or soil on which it grows.

Many fungi reproduce asexually

Yeasts reproduce asexually, primarily by forming buds that pinch off from the parent cell (▌Fig. 26-3). Many species of molds also reproduce asexually. Spores are produced by mitosis and then released into the air or water. **Conidiophores** (from the Greek, meaning "dust-bearers") are specialized hyphae that produce asexual spores called **conidia** (sing., *conidium*). The arrangement of conidia on conidiophores varies from species to species.

Most fungi reproduce sexually

Many mold species reproduce sexually when they come into contact with other mating types. In contrast to the majority of animal and plant cells, most fungal cells contain haploid nuclei. In sexual reproduction, the hyphae of two genetically compatible mating types come together, and their cytoplasm fuses, a process known as **plasmogamy.** The resulting cell has two haploid nuclei, one from each fungus. This cell gives rise by mitosis to other cells

with two nuclei. At some point the two haploid nuclei fuse. This process, called **karyogamy,** results in a cell containing a diploid nucleus known as a **zygote nucleus.** In some groups, the zygote nucleus is the only diploid nucleus.

In the two largest fungal phyla, the ascomycetes and basidiomycetes (discussed later in this chapter), plasmogamy occurs (the hyphae fuse), but karyogamy (the fusion of the two different nuclei) does not follow immediately. For a time, the nuclei re-

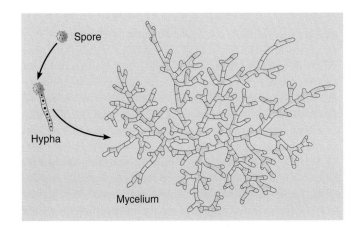

Spore

Hypha

Mycelium

Figure 26-2 *Animated* Germination of a spore to form a mycelium

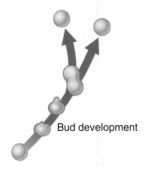

(a) Yeasts can reproduce asexually by budding.

Bud development

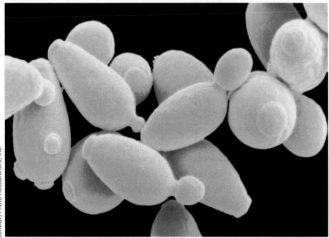

(b) SEM (color enhanced) of *Saccharomyces cerevisiae*. Also known as baker's or brewer's yeast, this fungus can ferment sugar, producing carbon dioxide and alcohol in the process. It is used in brewing beer, in wine production, and in baking bread. Note that several yeast cells are budding, a form of sexual reproduction.

Figure 26-3 Yeasts are unicellular fungi

main separate within the fungal cytoplasm. Hyphae that contain two genetically distinct, sexually compatible nuclei within each cell are described as **dikaryotic** (see Fig. 26-1d). This condition is referred to as $n + n$ rather than $2n$, because there are two separate haploid nuclei. Hyphae that contain only one nucleus per cell are described as **monokaryotic.** The presence of a dikaryotic stage is an important defining character of the ascomycetes and basidiomycetes.

Fungi are a diverse group with many variations in their life cycles. Most fungi (but not all) reproduce both asexually and sexually. ▌Figure 26-4 illustrates a generalized life cycle. When a fungal spore comes into contact with an appropriate food source, the spore germinates and begins to grow. A threadlike hypha emerges from the tiny spore and grows at its tip, branching frequently. In many fungi, cells from individuals of two different mating types fuse (plasmogamy), resulting in a dikaryotic mycelium. In most

fungi, karyogamy (fusion of the two nuclei) occurs, producing a zygote nucleus that undergoes meiosis. Each haploid spore produced can germinate and develop into a new mycelium by mitosis. In asexual reproduction, spore-producing structures develop and spores are produced by mitosis.

Fungi communicate chemically by secreting signaling molecules called **pheromones.** At least one pheromone has been identified in each fungal phylum. The pheromone binds with a compatible receptor on a different mating type. For example, in the phylum Zygomycetes, a pheromone induces the formation of specialized aerial hyphae. Another pheromone causes the tips of aerial hyphae of opposite mating types to grow toward each other and fuse prior to sexual reproduction.

Review

▌ What characteristics distinguish fungi from other organisms?

▌ How does the body of a yeast differ from that of a mold?

▌ Draw a generalized life cycle of a fungus.

FUNGAL DIVERSITY

Learning Objectives

4 Give arguments to support the hypothesis that fungi are opisthokonts, more closely related to animals than to plants.

5 Give arguments to support the hypothesis that chytrids may have been the earliest fungal group to evolve from the most recent common ancestor of fungi.

6 List distinguishing characteristics, describe a typical life cycle, and give examples of each of the following fungal groups: chytridiomycetes, zygomycetes, glomeromycetes, ascomycetes, and basidiomycetes.

For centuries, biologists classified fungi in the plant kingdom. However, some systematists disagreed with this classification, arguing that there are important structural differences between fungi and plants. Like plants, fungi have cell walls; but unlike plants, fungal cell walls do not contain cellulose. Rather, they contain chitin, a polysaccharide found in insect skeletons. The fungal mode of nutrition is also very different from that of plants Whereas plants are photosynthetic autotrophs, fungi, like animals, are heterotrophs. In 1969, R. H. Whittaker proposed that fungi be assigned to a separate kingdom—Fungi.

As with systematics for other kingdoms, fungal systematics is a challenging and continuously changing process. For example, slime molds and water molds were formerly classified as fungi but have now been assigned to the protist kingdom (see Chapter 25).

Fungi are assigned to the opisthokont clade

As discussed in Chapter 25, fungi, animals, and a few protists, including the choanoflagellates, form a monophyletic group, the **opisthokonts.** Both genetic and structural similarities sup-

Most fungi can reproduce both asexually (rapid proliferation) and sexually (new genotypes).

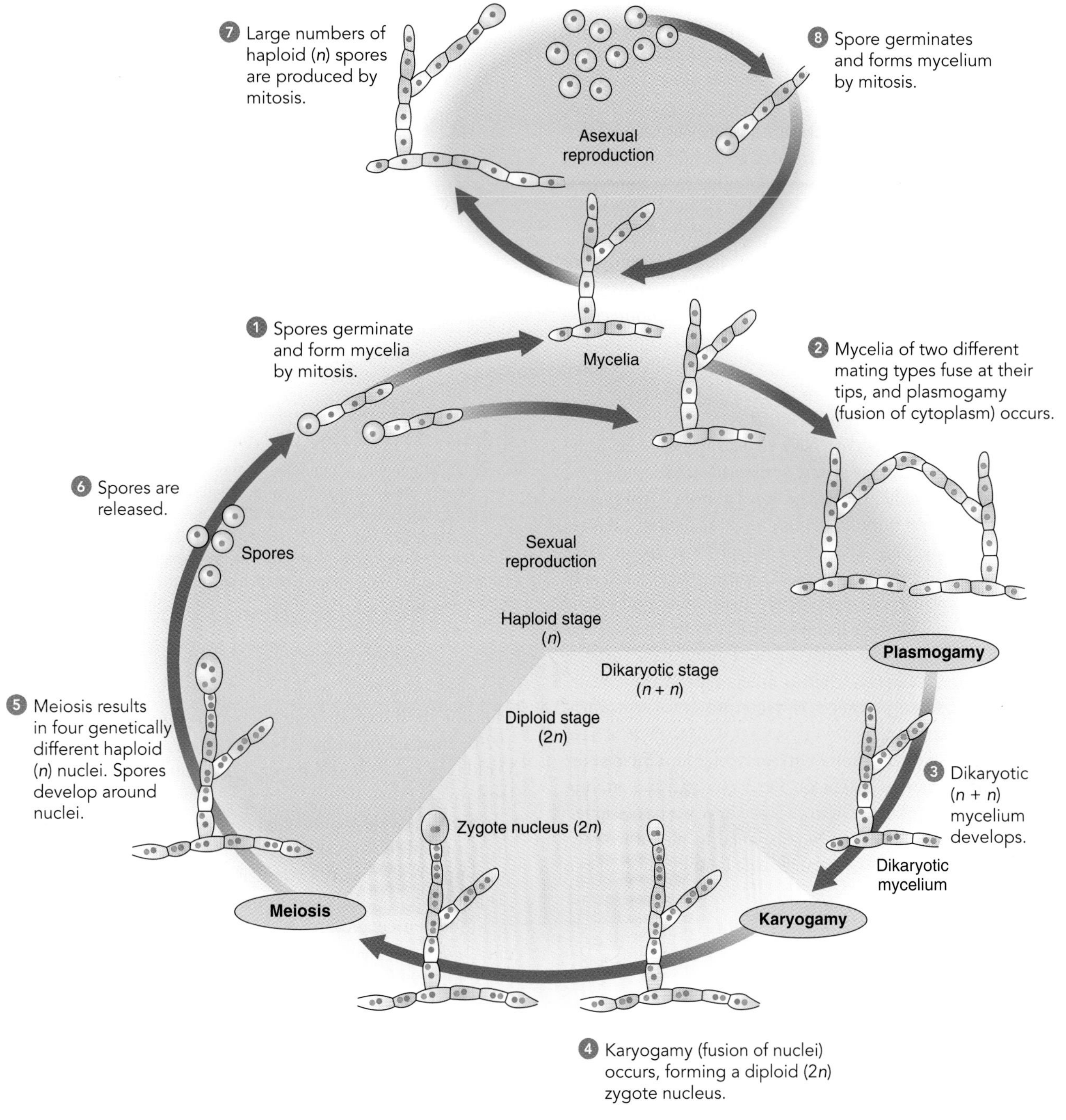

7 Large numbers of haploid (*n*) spores are produced by mitosis.

8 Spore germinates and forms mycelium by mitosis.

Asexual reproduction

1 Spores germinate and form mycelia by mitosis.

Mycelia

2 Mycelia of two different mating types fuse at their tips, and plasmogamy (fusion of cytoplasm) occurs.

6 Spores are released.

Spores

Sexual reproduction

Haploid stage (*n*)

Dikaryotic stage (*n* + *n*)

Diploid stage (2*n*)

Plasmogamy

5 Meiosis results in four genetically different haploid (*n*) nuclei. Spores develop around nuclei.

3 Dikaryotic (*n* + *n*) mycelium develops.

Dikaryotic mycelium

Zygote nucleus (2*n*)

Meiosis

Karyogamy

4 Karyogamy (fusion of nuclei) occurs, forming a diploid (2*n*) zygote nucleus.

Figure 26-4 The basic sequence of events in most fungal life cycles

port this grouping. Like animals, fungi have platelike cristae in their mitochondria. Another character shared by this clade is that flagellate cells propel themselves with a single posterior flagellum. In other eukaryote groups, flagellate cells move by means of one or more anterior flagella. Based on structural characters and on molecular data, the fungi, once considered part of the plant kingdom, are now viewed as more closely related to animals.

Diverse groups of fungi have evolved

Fossil evidence has not been very helpful to systematists studying evolutionary relationships among different fungal groups. Most fossilized fungi recovered to date have been microscopic. For example, fossilized fungal spores have been found in amber more than 225 million years old, and fossils of hyphae associated with cyanobacteria or algae have been dated as more than 550 million years old. Few large fungal fossils, such as mushrooms, have been found.

Historically, fungi have been classified mainly on the characteristics of their sexual spores and fruiting bodies. More recently, molecular data, such as comparative DNA and RNA sequences, have clarified relationships among fungal groups. Currently, mycologists assign fungi to five main phyla: Chytridiomycota, Zygomycota, Glomeromycota, Ascomycota, and Basidiomycota (‖ Fig. 26-5 and ‖ Table 26-1). Microsporidia, a group of intracellular parasites, are classified in this text with the zygomycetes. About 95% of all named fungi have been assigned to phyla Ascomycota and Basidiomycota. These phyla are considered sister taxa because they share a more recent common ancestor with each other than either does with any other group. Fungi of phyla Ascomycota and Basidiomycota have septate hyphae and a dikaryotic stage during the sexual part of their life cycle.

Until recently, mycologists had grouped about 25,000 species of fungi that did not fit into the major groups as deuteromycetes (phylum Deuteromycota). This was a polyphyletic group (they do not share a recent common ancestor) lumped together simply as a matter of convenience. Mycologists classified fungi as deuteromycetes if no sexual stage had been observed for them at any point during their life cycle. Some of these fungi have lost the ability to reproduce sexually, whereas others reproduce sexually only rarely. Most fungi classified as deuteromycetes reproduce only by means of conidia (as ascomycetes do). Mycologists have identified relationships between deuteromycetes and their sexually reproducing relatives, based on DNA comparisons among various species. Most of the deuteromycetes have been reassigned to phylum Ascomycota, and a few reassigned to phylum Basidiomycota. (By convention, the asexual stage is still identified as a deuteromycete.)

Chytrids have flagellate spores

At one time, biologists considered the **chytrids,** also known as **chytridiomycetes** (phylum Chytridiomycota), to be funguslike protists, similar in many respects to the water molds. However, both structural and molecular characters indicate that the approximately 1000 species of chytridiomycetes are members of kingdom Fungi. Like fungi, their cell walls contain chitin. Molecular comparisons, particularly of DNA and RNA sequences, have provided compelling evidence that chytrids are indeed fungi.

Chytrids are small, relatively simple fungi that inhabit ponds and damp soil. A few species have been found in salt water. Most chytrids are decomposers that degrade organic matter. However, a few species cause disease in plants and animals. Although not

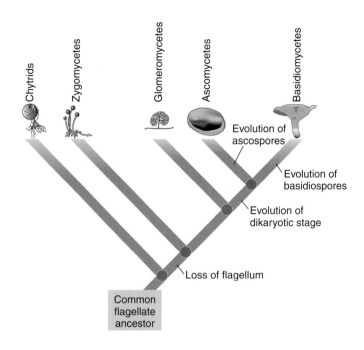

Figure 26-5 Fungal evolution: cladogram of major phyla of fungi

This cladogram shows phylogenetic relationships among living fungi, based on comparisons of ribosomal and nuclear gene sequence data for many species. The chytrids were the lineage that branched off first during fungal evolution. Note that ascomycetes and basidiomycetes are sister clades. Remember that the phylogeny of fungi is a work in process.

identified until 1998, a parasitic chytrid has been partly responsible for declining amphibian populations dating back to the 1970s. Infected frogs have been identified in many parts of the world (see *Focus On: Declining Amphibian Populations,* in Chapter 56).

Most chytrids are unicellular or composed of a few cells that form a simple body, called a **thallus.** (The term *thallus* describes the simple body plan of certain algae, fungi, or plants.) The thallus may have slender extensions, called **rhizoids,** that anchor it to a food source and absorb food (‖ Fig. 26-6). Chytrids are the only fungi that have flagellate cells. Their spores bear a single, posterior flagellum. Sexual reproduction has not been identified in most chytrids. Species that reproduce sexually have flagellate gametes.

Some chytrids produce branched, coenocytic mycelia. *Allomyces,* a large, common chytrid, has an unusual life cycle compared to that of most fungi. It undergoes an **alternation of generations** (common in plants, but rare in fungi), spending part of its life as a multicellular haploid (n) thallus and part as a multicellular diploid ($2n$) thallus (‖ Fig. 26-7). The haploid and diploid thalli are similar in appearance. At the tips of its branches, the haploid thallus bears two types of sporangia, structures in which gametes form by mitosis. Each sporangium produces a different type of flagellate gamete.

TABLE 26-1

Characteristics of Currently Recognized Phyla of Kingdom Fungi

Phylum and Common Types	Asexual Reproduction	Sexual Reproduction	Other Key Characters
Chytridiomycota (chytrids or chytridiomycetes) *Allomyces*	Flagellate, diploid zoospores produced by mitosis in zoosporangia	Flagellate, haploid gametes in some species	Haploid zoospores produced in resting sporangia; form haploid thallus
Zygomycota (zygomycetes) Black bread mold. Microsporidia are classified with the zygomycetes	Haploid spores produced in sporangia	Zygospores develop in zygosporangia	Important decomposers; some are insect parasites. Microsporidia are opportunistic pathogens that infect animals.
Glomeromycota (glomeromycetes)	Large, multinucleate blastospores	Has not been observed	Form arbuscular mycorrhizae with plant roots
Ascomycota (ascomycetes) Yeasts, powdery mildews, molds, morels, truffles	Conidia pinch off from conidiophores	Ascospores develop in asci	Have a dikaryotic stage; form important symbiotic relationships as lichens and mycorrhizae
Basidiomycota (basidiomycetes or club fungi) Mushrooms, bracket fungi, puffballs, rusts, smuts	Uncommon	Basidiospores develop on club-shaped basidia	Have a dikaryotic stage; many form mycorrhizae with tree roots

Each type of gamete secretes a pheromone that attracts the other type. The two types of gametes fuse; plasmogamy and karyogamy occur, forming a motile zygote. Each zygote can develop into a diploid thallus. The thallus bears two kinds of spore cases, zoosporangia and resting sporangia. Zoosporangia produce flagellate diploid **zoospores** that develop into new diploid thalli. Meiosis occurs within resting sporangia, producing recombinant haploid zoospores. Each zoospore has the potential to develop into a haploid thallus.

Molecular evidence suggests that chytrids were probably the earliest fungal group to evolve. Scientists hypothesize that the common ancestor of all plants, fungi, and animals was a flagellated ancient protist. Chytrids, which produce flagellate cells at some stage in their life history, have retained this feature of their protist ancestor. No other fungal group has flagellate cells. At some point in their evolutionary history, other fungal groups apparently lost the ability to produce motile cells, perhaps during the transition from aquatic to terrestrial habitats.

Zygomycetes reproduce sexually by forming zygospores

Mycologists have named more than 1100 species of **zygomycetes** (phylum Zygomycota). Of all the fungi, members of this taxon appear most closely related to the chytrids. However, zygomycetes are not a monophyletic group, and as mycologists learn more about fungal relationships, they may divide this phylum into several taxa or reassign some of its members to other existing phyla.

Most zygomycetes are decomposers that live in the soil on decaying plant or animal matter (❚ Fig. 26-8). Some zygomycetes form mycorrhizal relationships with plant roots. A few species cause disease in plants and animals, including humans.

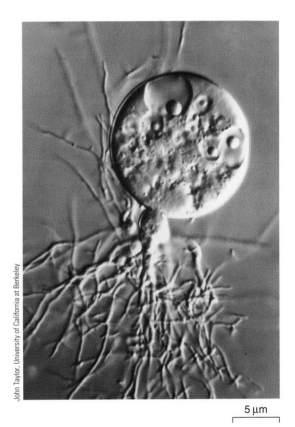

John Taylor, University of California at Berkeley

5 µm

Figure 26-6 Chytrid

Nomarski differential interference micrograph of a common chytrid (*Chytridium convervae*). Many chytrids have a microscopic body form consisting of a rounded, coenocytic thallus and branched rhizoids that superficially resemble roots. The rhizoids may anchor the chytrid thallus and absorb predigested food.

Chytrids have flagellate reproductive cells.

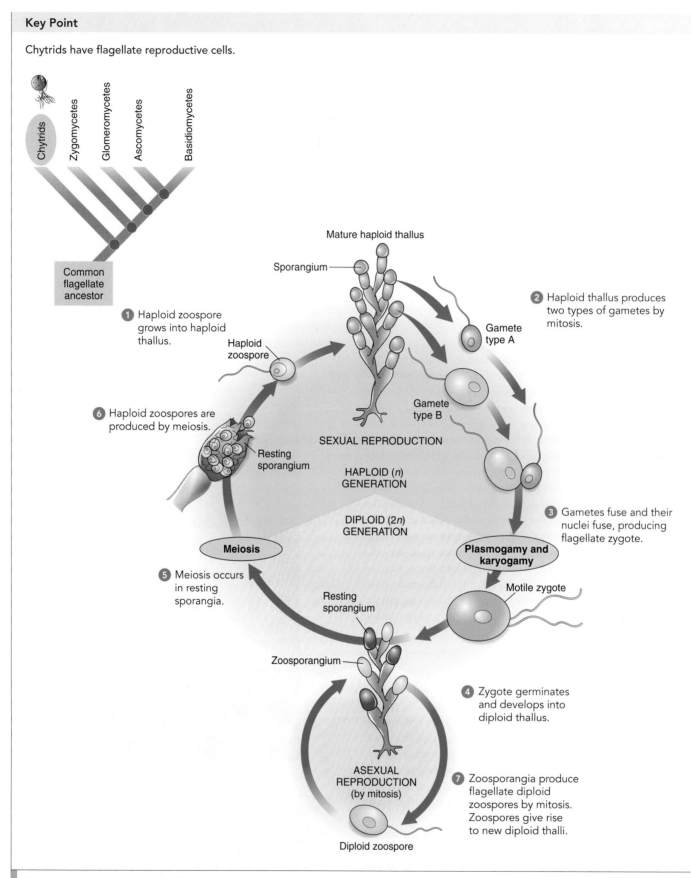

Figure 26-7 Life cycle of *Allomyces arbuscula*, a chytrid

Allomyces alternates between haploid and diploid stages, which are similar in appearance.

Figure 26-8 *Pilobolus,* a zygomycete that grows in animal dung

Stalked sporangia of *Pilobolus* protruding from a pile of dung, which contains an extensive mycelium of the fungus. The stalked sporangia, which are 5 to 10 mm tall, act like shotguns and forcefully discharge sporangia (the black tips) away from the dung onto nearby grass. When animals such as cattle or horses eat the grass, the spores pass unharmed through the animal's digestive tract and are deposited in a fresh pile of dung.

During sexual reproduction, zygomycetes produce sexual spores, called **zygospores.** The zygospores are typically produced in spore sacs called sporangia. The hyphae in zygomycetes are coenocytic; that is, they lack regularly spaced septa. However, septa do form to separate the hyphae from reproductive structures.

Perhaps the most familiar zygomycete is the black bread mold, *Rhizopus stolonifer,* a decomposer that breaks down bread and other foods. If preservatives are not added, bread left at room temperature often becomes covered with a black, fuzzy growth in a few days. Bread becomes moldy when a spore falls on it and then germinates and grows into a mycelium (█ Fig. 26-9). Hyphae penetrate the bread and absorb nutrients. Eventually, certain hyphae grow upward and develop sporangia at their tips. Clusters of more than 50,000 black asexual spores can develop within each sporangium. The spores are released when the delicate sporangium ruptures. The spores give the black bread mold its characteristic color.

Sexual reproduction in the black bread mold occurs when the hyphae of two different mating types, designated plus (+) and minus (−), grow into contact with one another. The bread mold is **heterothallic,** meaning that an individual fungal hypha is self-sterile and mates only with a hypha of a different mating type. That is, sexual reproduction occurs only between a member of a (+) strain and one of a (−) strain, not between members of two (+) strains or members of two (−) strains. Because there are no physical differences between the two mating types, it is not appropriate to refer to them as "male" and "female."

When hyphae of opposite mating types grow in close proximity, they signal one another with pheromones. In response to these chemical signals, the tips of the hyphae come together and form **gametangia,** which serve as gametes. Plasmogamy occurs

as the gametangia fuse. Then karyogamy occurs as the (+) and (−) nuclei fuse to form the diploid zygote nucleus. The zygote develops into a zygospore. Zygospores are surrounded by a thick protective covering called a **zygosporangium,** which contains many diploid nuclei.

The zygospore may lie dormant for several months and can survive desiccation and extreme temperatures. Meiosis probably occurs at or just before germination of the zygospore.

When the zygospore germinates, an aerial hypha develops with a sporangium at the tip. Mitosis within the sporangium produces haploid spores. These spores may be all (+) spores, all (−) spores, or a mixture of (+) and (−) spores. When released, the spores germinate to form new hyphae. Only the zygote and zygospore of a black bread mold are diploid; all the hyphae and the asexual spores are haploid.

Microsporidia have been a taxonomic mystery

Microsporidia are small, unicellular parasites that infect eukaryotic cells. At present, these common pathogens are classified with the zygomycetes. They are opportunistic pathogens that infect animals. For example, microsporidia infect people with compromised immune systems, such as those with AIDS. Microsporidia cause a variety of diseases involving many organ systems, and some species cause lethal infections. Some species of microsporidia appear to be host specific.

Microsporidia have two developmental stages inside their host: a feeding stage and a reproductive stage. Some microsporidian species divide by binary fission, and others divide into several cells. Some species undergo nuclear fusion and meiosis before producing spores. The spores, which have thick protective walls, can pass from cell to cell inside the host or can be excreted in urine or through the skin. The spores, the only stage with distinct characters, are used to identify groups.

Each spore is equipped with a unique structure, a long, threadlike **polar tube.** When the spore enters the gut of a new host, it discharges its polar tube and penetrates the lining of the gut. Acting as a hypodermic needle, the polar tube injects the contents of the spore into the host cell (█ Fig. 26-10).

Microbiologists estimate that there may be more than a million species of microsporidia, but only about 1500 species have been named. Microsporidia were originally classified with yeasts and bacteria. In 1976, they were assigned to the protozoa. In the 1980s, biologists viewed microsporidia as the most primitive example of a eukaryote. They were the smallest and simplest known eukaryotes, and the genomes of some species are smaller than most bacterial genomes. Microsporidia lack mitochondria, flagella, and Golgi complexes. Their ribosomes resemble those of prokaryotes.

In 1998, British biologist Thomas Cavalier-Smith reassigned microsporidia to kingdom Fungi. Molecular studies show that microsporidia have gene sequences indicating that they originally had mitochondria. Microbiologists now generally agree that these organisms have become simpler as they adapted to their parasitic way of life. Other molecular studies have provided additional evidence for their taxonomic relationship with fungi.

Like most fungi, most zygomycetes reproduce both asexually and sexually.

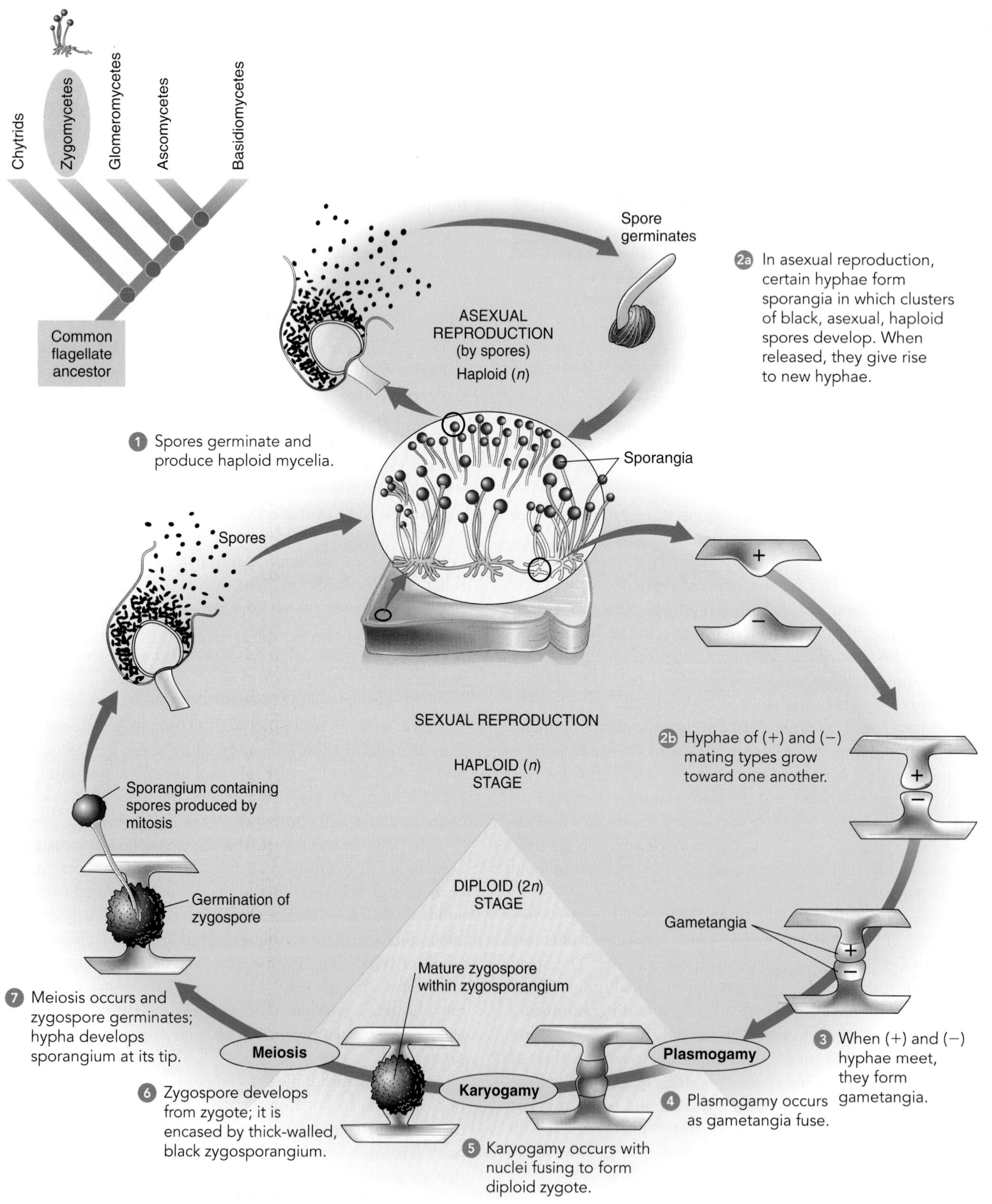

ASEXUAL REPRODUCTION (by spores) Haploid (n)

Spore germinates

2a In asexual reproduction, certain hyphae form sporangia in which clusters of black, asexual, haploid spores develop. When released, they give rise to new hyphae.

1 Spores germinate and produce haploid mycelia.

Sporangia

Spores

SEXUAL REPRODUCTION

HAPLOID (n) STAGE

2b Hyphae of (+) and (−) mating types grow toward one another.

DIPLOID (2n) STAGE

Sporangium containing spores produced by mitosis

Germination of zygospore

Gametangia

Mature zygospore within zygosporangium

7 Meiosis occurs and zygospore germinates; hypha develops sporangium at its tip.

Meiosis

Karyogamy

Plasmogamy

3 When (+) and (−) hyphae meet, they form gametangia.

6 Zygospore develops from zygote; it is encased by thick-walled, black zygosporangium.

5 Karyogamy occurs with nuclei fusing to form diploid zygote.

4 Plasmogamy occurs as gametangia fuse.

Chytrids
Zygomycetes
Glomeromycetes
Ascomycetes
Basidiomycetes

Common flagellate ancestor

Figure 26-9 *Animated* Life cycle of the black bread mold (*Rhizopus stolonifer*), a zygomycete

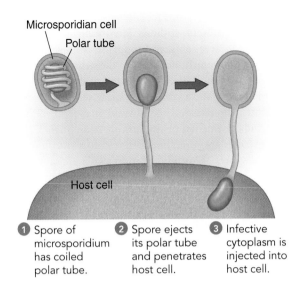

① Spore of microsporidium has coiled polar tube.

② Spore ejects its polar tube and penetrates host cell.

③ Infective cytoplasm is injected into host cell.

Figure 26-10 Infection by microsporidium

Establishing that they are closely related to fungi is important in developing medications that will be effective in treating microsporidian infections.

Glomeromycetes are symbionts with plant roots

Glomeromycetes (phylum Glomeromycota) are symbionts that form intracellular associations within the roots of most trees and herbaceous plants. (Recall that a symbiotic association is an intimate relationship between organisms of different species.) Glomeromycetes were previously considered zygomycetes, but in 2002 taxonomists concluded that they form a separate monophyletic group. This phylogeny is based on comparison of ssRNA genes.

Glomeromycetes have coenocytic (no septa) hyphae. They reproduce asexually with large, multinucleate spores called *blastospores*. Sexual reproduction has not been documented.

The symbiotic relationships between fungi and the roots of plants are called **mycorrhizae** (from Greek words meaning "fungus roots") (Fig. 26-11; also see Fig. 35-10). The roots supply the fungus with sugars, amino acids, and other organic substances. The mycorrhizal fungus decomposes organic material in the soil and also benefits the plant by extending the reach of its roots. The slender mycelia are far thinner than roots and can extend into narrow spaces, absorbing nutrients that the plant could not capture on its own. Thus, with the help of the mycorrhizal fungus, the plant can take in more water and nutrient minerals such as phosphorus and nitrogen.

What we have just described is a mutualistic symbiotic relationship: both partners benefit. Studies show that if a plant grows in phosphate-deficient soil or if it has a limited root system, its growth is enhanced by having a fungal partner. However, a plant in a phosphate-rich soil with a well-developed root system may not need fungal partners. For these plants, the fungus may be a parasite.

Much remains to be learned about mycorrhizal relationships. Other, yet unknown, benefits of the symbiosis may define the relationship as mutualistic. For example, some fungi release alkaloids that protect the plants from herbivores and pathogens. Plants also exchange nutrients with one another through fungi that connect them.

Glomeromycetes extend their hyphae through the cell walls of root cells but do not penetrate the plasma membrane. As the hyphae push forward, the root plasma membrane surrounds them. Thus, the hyphae can be thought of as fingers pushing into a glove formed by the plasma membrane. Because they appear intracellular, these fungi are referred to as **endomycorrhizal fungi.** The most widespread endomycorrhizae are called **arbuscular mycorrhizae** because the hyphae inside the root cells form branched, tree-shaped structures known as arbuscules (see Fig. 26-11). The arbuscules are the sites of nutrient exchange between the plant and the fungus. Arbuscular mycorrhizae live entirely underground.

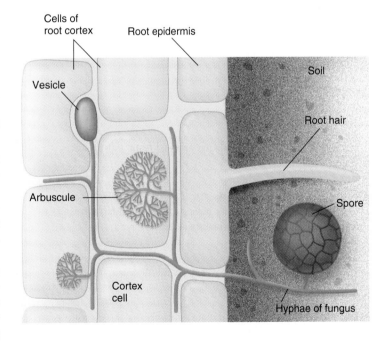

Figure 26-11 Arbuscular mycorrhizae

This mycelium has grown into the root. Its hyphae branch between the cells of the root. Hyphae have penetrated through the cell walls of two root cells and have branched extensively to form arbuscules. The tip of one hypha between root cells has enlarged and serves as a vesicle that stores food. The tip of a hypha in the soil has enlarged, forming a spore. The spaces between the root cells have been magnified.

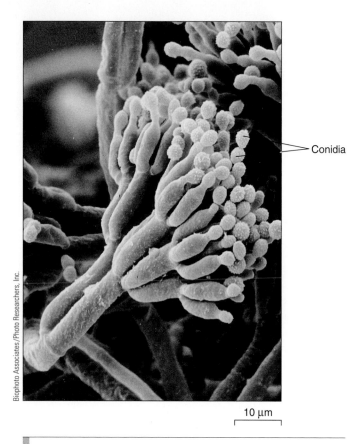

Conidia

Figure 26-12 Conidia

Conidia are asexual reproductive cells produced by ascomycetes and some basidiomycetes. Biologists use the arrangement of conidia on conidiophores to identify species of these fungi. Shown is a SEM of *Penicillium* conidiophores, which resemble paintbrushes. Note the conidia pinching off the tips of the "brushes."

Biophoto Associates/Photo Researchers, Inc.

10 μm

Scientists have discovered mycorrhizal fungi within ancient plant fossils in rocks that are about 400 million years old. These findings suggest that when plants moved onto the land, their fungal partners moved with them. In fact, botanists have suggested that fungal hyphae may have provided plants with water and minerals before their own root systems evolved.

Ascomycetes reproduce sexually by forming ascospores

Ascomycetes (phylum Ascomycota) comprise a large group of fungi consisting of more than 32,000 described species. The diverse ascomycetes include most yeasts; the powdery mildews; most of the blue-green, pink, and brown molds that cause food to spoil; decomposer cup fungi; and the edible morels and truffles.

The ascomycetes, more than any other group of fungi, impact humans. As we will discuss in a later section, ascomycetes are used to flavor cheeses, to bake bread (yeast), and to ferment alcohol. Some are enjoyed as foods (morels and truffles). Asco-

mycetes are used to produce antibiotics. They have also served as valuable model organisms for biologists studying cellular processes, including protein synthesis. Many fungi in this group form mycorrhizae with tree roots, and about 40% join with green algae or cyanobacteria to form lichens. On the negative side, ascomycetes cause most fungal diseases of plants and animals, including humans. For example, ascomycetes cause serious plant diseases such as Dutch elm disease, ergot disease on rye, powdery mildew on fruits and ornamental plants, and chestnut blight.

Ascomycetes are sometimes referred to as *sac fungi* because their sexual spores are produced in microscopic sacs called **asci** (sing., *ascus*). Their hyphae usually have septa, but these cross walls have pores so that cytoplasm is continuous from one cell compartment to another.

In most ascomycetes, asexual reproduction involves production of spores called conidia, which form at the tips of certain specialized hyphae known as conidiophores (Fig. 26-12). Production of these spores is a means of rapidly propagating new mycelia when environmental conditions are favorable. Conidia occur in various shapes, sizes, and colors in different species. The color of the conidia produces the characteristic brown, blue-green, pink, or other tints of many of these molds.

Some species of ascomycetes are heterothallic and have different mating strains. Others are **homothallic,** which means they are self-fertile and have the ability to mate with themselves. In both heterothallic and homothallic ascomycetes, sexual reproduction takes place after two gametangia come together and their cytoplasm mingles.

Let us examine the life cycle of a typical ascomycete (Fig. 26-13). In our example, plasmogamy takes place as hyphae of two different mating types come together and fuse. Within this fused structure, pairs of haploid nuclei, one from each parent hypha, associate but do not fuse. New hyphae, with dikaryotic cells, develop from the fused structure. The hyphae branch repeatedly until the hyphal tips reach the site where asci will be produced. As the many sac-shaped asci develop, each containing two dissimilar nuclei (one from each parent), they are surrounded by intertwining haploid (monokaryotic) hyphae. These hyphae help make a fruiting body known as an **ascocarp** (Fig. 26-14a).

Karyogamy occurs in each ascus. The two nuclei fuse and form a diploid zygote nucleus. The zygote nucleus then undergoes meiosis to form four haploid nuclei with different genotypes. One mitotic division of each of the four nuclei usually follows, resulting in eight haploid nuclei. Each haploid nucleus becomes incorporated into a thick-walled **ascospore;** thus, there are typically eight haploid ascospores within the ascus (Fig. 26-14b). The ascospores are usually released through a pore, slit, or hinged lid at the tip of the ascus. Air currents carry individual ascospores, often for long distances. If one lands in a suitable location, it germinates and forms a new mycelium. The fungus can reproduce asexually by producing conidia that can develop into new mycelia.

Phylum Ascomycota includes more than 300 species of unicellular yeasts. Asexual reproduction of yeasts is mainly by **budding;** in this process a small protuberance (bud) grows and

Ascomycetes produce asexual spores called conidia and sexual spores called ascospores.

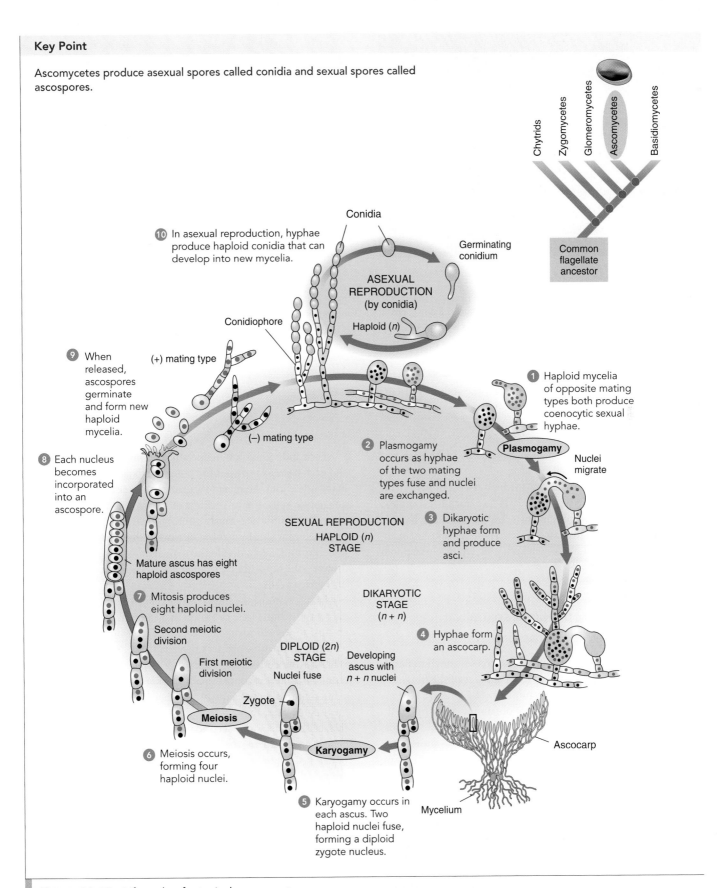

Figure 26-13   Life cycle of a typical ascomycete

Sexual reproduction requires haploid mycelia of different mating types. Note the dikaryotic stage and the separation of plasmogamy and karyogamy. Steps 5–8 take place within an ascus in the ascocarp.

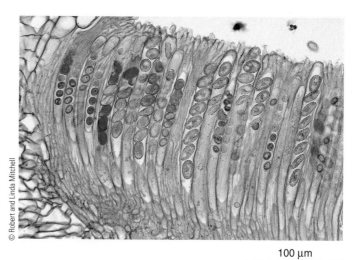

100 μm

(a) The ascocarp (fruiting body) of the common brown cup (*Peziza badio-confusa*) is shaped like a saucer or bowl and is 3 to 10 cm (1 to 4 in) wide. It is found on damp soil in woods throughout North America. Photographed in Muskegon, Michigan.

(b) Asci, each containing eight ascospores, line the inner portion of the ascocarp.

Figure 26-14 *Animated* Sexual reproduction in the ascomycetes

eventually separates from the parent cell (see Fig. 26-3). Each bud can grow into a new yeast cell.

Yeasts reproduce sexually by forming ascospores. During sexual reproduction, two haploid yeasts fuse, forming a diploid zygote. The zygote undergoes meiosis, and the resulting haploid nuclei are incorporated into ascospores. These spores remain enclosed for a time within the original cell wall, which corresponds to an ascus.

Basidiomycetes reproduce sexually by forming basidiospores

The more than 30,000 species of **basidiomycetes** (phylum Basidiomycota) include the largest and most familiar of the fungi: the mushrooms, bracket fungi, and puffballs (Fig. 26-15). Many basidiomycetes are decomposers that obtain nutrients by breaking down organic matter. Some species cause great economic loss because they cause dry rot in buildings. Certain basidiomycetes form mycorrhizae. Others, such as wheat rust and corn smut, infect important crops. A few basidiomycetes cause human disease.

Sometimes called *club fungi,* basidiomycetes derive their name from their microscopic club-shaped **basidia** (sing., *basidium*). Basidia are comparable in function to the asci of ascomycetes. Each basidium is an enlarged hyphal cell that undergoes meiosis to form four **basidiospores** (Fig. 26-16). Note that basidiospores develop on the *outside* of a basidium, whereas ascospores develop *within* an ascus.

Each individual fungus produces millions of basidiospores, and each basidiospore has the potential to give rise to a new **primary mycelium.** Hyphae of a primary mycelium consist of monokaryotic cells. The mycelium of a basidiomycete, such as the commonly cultivated mushroom *Agaricus brunnescens,* consists of a mass of white, branching, threadlike hyphae that live mostly underground. Septa divide the hyphae into cells, but as in ascomycetes, the septa are perforated and allow cytoplasmic streaming between cells.

Let us examine the life cycle of a typical basidiomycete. Asexual reproduction is less common in basidiomycetes than in other groups, so we will focus here on sexual reproduction. We begin with two compatible primary mycelia (Fig. 26-17 on page 571). When in the course of its growth a hypha of a primary mycelium encounters a compatible monokaryotic hypha, typically of a different mating type, the two hyphae fuse (plasmogamy). As in the ascomycetes, the two haploid nuclei remain separate within each cell. In this way a **secondary mycelium** with dikaryotic hyphae is produced, in which each cell contains two haploid nuclei. The $n + n$ hyphae of the secondary mycelium grow rapidly and extensively.

When environmental conditions are favorable, the hyphae form compact masses, called *buttons,* along the mycelium. Each button grows into a fruiting body that we know as a mushroom. A mushroom, which consists of a stalk and a cap, is more formally referred to as a **basidiocarp.** Each basidiocarp consists of intertwined, matted hyphae. The lower surface of the cap usually consists of many thin, perpendicular plates called **gills** that radiate from the stalk to the edge of the cap.

Karyogamy takes place within the young basidia on the gills of the mushroom. The haploid nuclei fuse in the dikaryotic cells, forming diploid zygote nuclei. These are the only diploid cells that form during a basidiomycete's life history. Meiosis then takes place, forming four haploid nuclei with different genotypes. These nuclei move to the outer edge of the basidium. Fingerlike extensions of the basidium develop, into which the nuclei and some cytoplasm move; each of these extensions becomes a basidiospore. A septum forms that separates the basidiospore from the rest of the basidium by a delicate stalk that breaks when the basidiospore is forcibly discharged. Each basidiospore can germinate and give rise to a primary mycelium.

Many basiodmycetes produce "fairy rings" in lawns and forests (█ Fig. 26-18). A fairy ring may first appear as a dark green ring surrounding an inner brown circle. The size of the ring ranges from a few centimeters to more than 15 m (about 51 ft) in diameter. The green ring consists of grass, well nourished by the nutrients released as the fungi decompose organic material. Grass dies, producing the inner brown circle, because the mass of mycelia decreases the movement of water into the area. As the fungi grow outward, the circle widens. The rings grow at a rate of a few centimeters to more than a meter per year. After rainfall or irrigation, a ring of mushrooms may appear just outside the green circle. The name "fairy ring" comes from a legend that a ring of mushrooms appeared where fairies had danced in a circle the night before.

Review

▌ What evidence supports the hypothesis that chytrids were the earliest fungal group to evolve from the common ancestor of fungi?

▌ What are the distinguishing characteristics of each of the following fungal groups: (1) zygomycetes, (2) glomeromycetes, (3) ascomycetes, and (4) basidiomycetes?

▌ How does the life cycle of a typical basidiomycete differ from that of a typical ascomycete? (Draw diagrams to support your answer.)

▌ Distinguish among (1) ascocarp, ascus, and ascospore and among (2) basidiocarp, basidium, and basidiospore.

(a) Basidia line the gills of the Jack-o'-lantern mushroom (*Omphalotus olearius*), a poisonous species whose gills produce a greenish glow in the dark. Each cap is about 15 cm (6 in) wide. Photographed at the base of an oak tree in Maryland, the Jack-o'-lantern occurs throughout eastern North America and California.

(b) The elegant stinkhorn (*Phallus ravenelii*) has a foul smell that attracts flies. The flies help disperse the slimy mass of basidiospores. Fruiting bodies of elegant stinkhorns grow to 18 cm (7 in) tall. Photographed in Pennsylvania.

(c) Turkey-tail (*Trametes versicolor*) is a common bracket fungus. Bracket fungi grow on both dead and living trees and produce shelflike fruiting bodies. Basidiospores are produced in pores located underneath each shelf.

Figure 26-15 Basidiomycete fruiting bodies

ECOLOGICAL IMPORTANCE OF FUNGI

Learning Objectives

7   Explain the ecological significance of fungi as decomposers.
8   Describe the important ecological role of mycorrhizae.
9   Characterize the unique nature of a lichen.

Fungi make vital contributions to the ecological balance of our planet. Like bacteria, most fungi are free-living decomposers, chemoheterotrophs that absorb nutrients from organic wastes and dead organisms. For example, many fungal decomposers degrade cellulose and lignin, the main components of plant cell walls. When fungi degrade wastes and dead organisms, they release water, carbon (as CO_2), and mineral components of organic compounds, and these elements are recycled (see biogeochemical cycles in Chapter 54). Without this continuous decomposition, essential nutrients would remain locked up in huge mounds of animal carcasses, feces, branches, logs, and leaves. The nutrients would be unavailable for use by new generations of organisms, and life would eventually cease.

Fungi form important symbiotic relationships with animals, plants, bacteria, and protists. These symbioses have major effects on ecosystems.

Fungi form symbiotic relationships with some animals

Because animals do not have the enzymes necessary to digest cellulose and lignin, cattle and other grazing animals cannot, by themselves, obtain needed nutrients from the plant material they eat. Their survival depends on fungi that inhabit their guts because fungi, like many other microorganisms, do have the enzymes that break down these organic compounds. The fungi benefit by living in a nutrient-rich environment.

Fungi also form symbiotic associations with ants and termites. More than 200 species of ants farm fungi. Leaf-cutting ants bring leaves to their fungi and protect them from competitors and predators. The ants also disperse the fungi to new locations. In exchange, the fungi digest the leaves, providing nutrients for the ants. This symbiosis can involve other organisms. The farmed fungi can be infested by fungal parasites. In response, the ants culture bacteria (actinomycetes) that produce antibiotics to control these parasites. These symbiotic relationships, the most complex known, are the product of 50 million years of coevolution.

Mycorrhizae are symbiotic relationships between fungi and plant roots

Mycorrhizae occur in about 80% of plants (and more than 90% of all plant families). As discussed in the section on glomeromycetes, mycorrhizal fungi decompose organic material in the soil

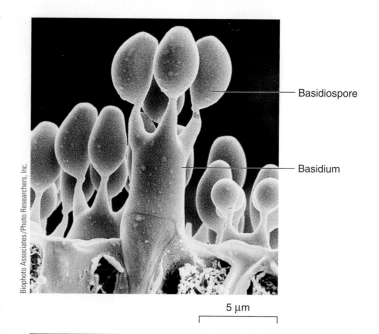

Basidiospore

Basidium

Biophoto Associates/Photo Researchers, Inc.

5 μm

Figure 26-16   SEM of a basidium
Each basidium produces four basidiospores.

and increase the surface area of a plant's roots so that the plant can absorb more water and nutrient minerals. In exchange, the roots supply the fungus with organic nutrients.

The importance of mycorrhizae first became evident when horticulturalists observed that orchids do not grow unless an appropriate fungus lives with them. Similarly, many forest trees, such as pines, decline and eventually die from mineral deficiencies when transplanted to mineral-rich grassland soils that lack the appropriate mycorrhizal fungi. When forest soil containing the appropriate fungi or their spores is added to the soil around these trees, they quickly resume normal growth. Studies performed with various types of plants, including cedar, have confirmed the role of mycorrhizae in plant growth (Fig. 26-19).

As we have discussed, glomeromycetes form *endomycorrhizal* connections; they infiltrate the cells of plant roots. At least 5000 species of ascomycetes and basidiomycetes also form mycorrhizal connections, but their hyphae coat the plant root rather than penetrate its cells. They are referred to as **ectomycorrhizae.** Interestingly, researchers have shown that some mycorrhizal fungi harbor bacteria in their cytoplasm. Although the role of the bacteria is not yet clear, their presence suggests that they may be members of a three-domain partnership: fungus, plant, and bacteria.

Mycorrhizal fungi connect plants, allowing nutrient transfer among them. Scientists have measured the movement of organic materials from one tree species to another through shared mycorrhizal connections. Mycorrhizal fungi also release chemicals that protect the plant against herbivores and pathogens.

Mycorrhizae improve the soil by decreasing water loss and erosion. Ecologists are studying the role of mycorrhizal fungi

Basidiomycetes produce sexual basidiospores on the gills of basidiocarps (fruiting bodies).

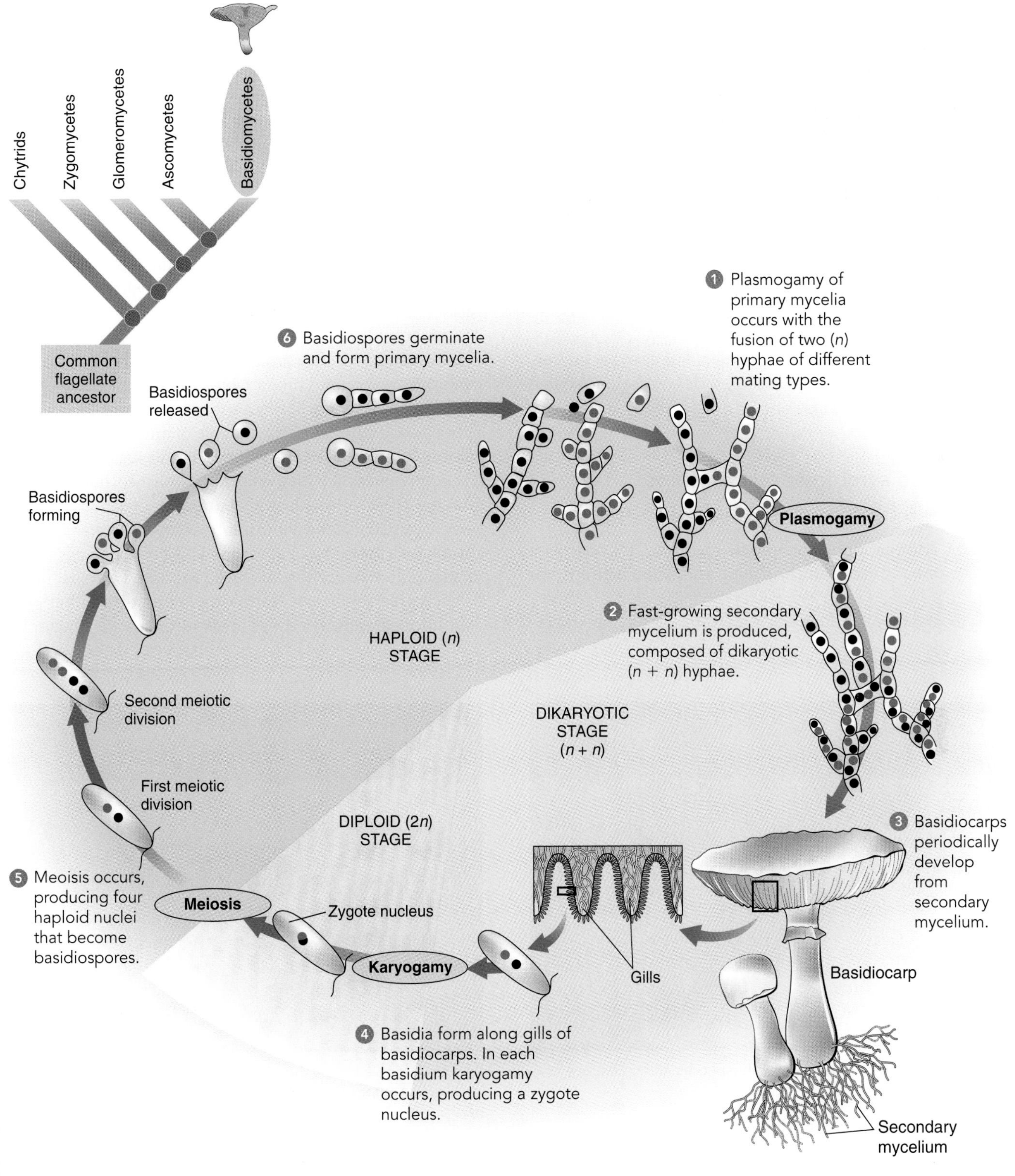

① Plasmogamy of primary mycelia occurs with the fusion of two (*n*) hyphae of different mating types.

⑥ Basidiospores germinate and form primary mycelia.

Basidiospores released

Basidiospores forming

Common flagellate ancestor

Chytrids

Zygomycetes

Glomeromycetes

Ascomycetes

Basidiomycetes

HAPLOID (*n*) STAGE

Plasmogamy

② Fast-growing secondary mycelium is produced, composed of dikaryotic (*n* + *n*) hyphae.

DIKARYOTIC STAGE (*n* + *n*)

Second meiotic division

First meiotic division

③ Basidiocarps periodically develop from secondary mycelium.

DIPLOID (2*n*) STAGE

⑤ Meoisis occurs, producing four haploid nuclei that become basidiospores.

Meiosis

Zygote nucleus

Karyogamy

Gills

Basidiocarp

Secondary mycelium

④ Basidia form along gills of basidiocarps. In each basidium karyogamy occurs, producing a zygote nucleus.

Figure 26-17 *Animated* Life cycle of a typical basidiomycete

Note the dikaryotic stage and the separation of plasmogamy and karyogamy. Steps ④–⑤ take place within the basidia of the basidiocarp. Asexual reproduction is uncommon in this group.

Figure 26-18 A fairy ring

in reclaiming soils damaged by pollution. For example, mycorrhizae can modify toxic heavy metals, such as cadmium, so that plants cannot absorb them.

Lichens are symbiotic relationships between a fungus and a photoautotroph

Although a **lichen** looks like a single organism, it is actually a dual organism. A lichen is a symbiotic association between two organisms: a fungus and a *photoautotroph* (❙ Fig. 26-20a). Almost one fifth of all known fungal species form these symbiotic re-

lationships; about 14,000 kinds of lichens have been described. Fossils suggest that fungi developed symbiotic partnerships with photoautotrophs before the evolution of vascular plants.

The photoautotrophic component of a lichen is either a green alga, a cyanobacterium, or both. The fungus is most often an ascomycete, although in some tropical lichens the fungal partner is a basidiomycete. Most photoautotrophic organisms found in lichens also occur as free-living species in nature, but the fungal components are generally found only as a part of the lichen. Typically, the fungus forms most of the lichen thallus (body). The fungus surrounds hundreds of photosynthetic partners and holds them in place. Lichens are named for the fungal component.

In the laboratory, researchers can isolate the fungal and photoautotrophic components of some lichens and grow them separately in appropriate culture media. The photoautotroph grows more rapidly when separated, whereas the fungus grows more slowly and requires many complex carbohydrates. Neither organism resembles a lichen in appearance when grown separately. The photoautotroph and fungus can be reassembled as a lichen thallus, but only if they are placed in a culture medium under conditions that cannot support either of them independently.

What is the nature of this partnership? The lichen was originally considered a definitive example of mutualism. The photoautotroph carries on photosynthesis, producing energy-rich carbon compounds for both members of the lichen, but it is unclear how the photoautotroph benefits from the relationship. It has been suggested that the photoautotroph obtains water and nutrient minerals from the fungus as well as protection against desiccation. More recently, some biologists have suggested that the lichen partnership is not really a case of mutualism but one of controlled parasitism of the photoautotroph by the fungus.

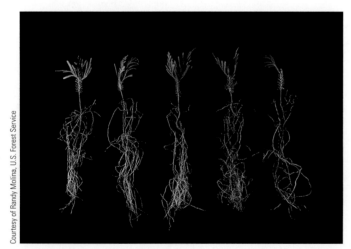

(a) Control seedlings grown in low phosphorus in the absence of the fungus

(b) These seedlings are the same age as the control plants and were grown under conditions identical to those of the control, except that their roots have formed mycorrhizal associations.

Figure 26-19 The effect of mycorrhizae on western red cedar (*Thuja plicata*) seedlings

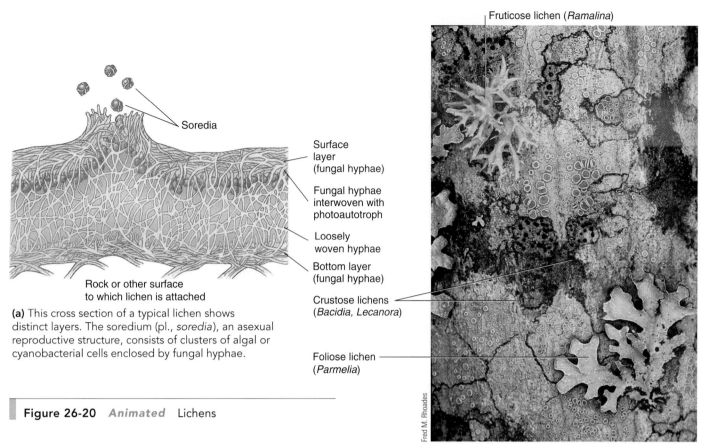

Soredia

Surface
layer
(fungal hyphae)

Fungal hyphae
interwoven with
photoautotroph

Loosely
woven hyphae

Bottom layer
(fungal hyphae)

Rock or other surface
to which lichen is attached

(a) This cross section of a typical lichen shows distinct layers. The soredium (pl., *soredia*), an asexual reproductive structure, consists of clusters of algal or cyanobacterial cells enclosed by fungal hyphae.

Fruticose lichen (*Ramalina*)

Crustose lichens
(*Bacidia, Lecanora*)

Foliose lichen
(*Parmelia*)

Fred M. Rhoades

(b) Lichens vary in color, shape, and overall appearance. Three growth forms—crustose, foliose, and fruticose—are shown on a maple branch in Washington State.

Figure 26-20 *Animated* Lichens

Lichens typically exhibit one of three different growth forms (Fig. 26-20b). *Crustose lichens* are flat and grow tightly against their substrate (the surface they are growing on); *foliose lichens* are also flat, but they have leaflike lobes and are not so tightly pressed to the substrate; and *fruticose lichens* grow erect and have many branches.

Able to tolerate extremes of temperature and moisture, lichens grow in almost all terrestrial environments except polluted cities. They exist farther north than any plants of the arctic region and are equally at home in the steaming equatorial rain forest. They grow on tree bark, leaves, and exposed rock surfaces, from solidified lava to tombstones. In fact, lichens are often the first organisms to inhabit rocky areas, and their growth in these areas is important in forming soil from rock. They secrete acid that gradually etches tiny cracks in the rock, releasing minerals. This process sets the stage for further disintegration of the rock by wind and rain. When the lichens themselves die and are decomposed, they become part of the soil.

Reindeer mosses of the arctic region, which serve as the main source of food for migrating herds of caribou, are actually lichens, not mosses. Some lichens produce colored pigments. One of them, orchil, is used to dye woolens, and another, litmus, is widely used in chemistry laboratories as an acid–base (pH) indicator.

Lichens vary greatly in size. Some are almost invisible, whereas others, such as the reindeer mosses, may cover many square kilometers of land with an ankle-deep growth. Growth proceeds slowly; the radius of a lichen may increase by less than 1 mm each year. Some mature lichens are thought to be thousands of years old.

Lichens absorb minerals from the air, rainwater, and the surface on which they grow. They cannot excrete the elements they absorb, and perhaps for this reason they are extremely sensitive to toxic compounds. This sensitivity was first reported in 1866 by a Finnish biologist who observed that lichens growing on tree trunks in Paris were poorly developed or sterile. He deduced that lichens could be used to measure air purity. Today, reduction in lichen growth is used as a sensitive indicator of air pollution, particularly from sulfur dioxide. Investigators demonstrated a relationship between lung cancer and air pollution by comparing the locations of low lichen biodiversity (and therefore of air pollution) with the locations of lung cancer deaths in young males. The return of lichens to an area indicates an improvement in air quality.

Lichens reproduce mainly by asexual means, usually by fragmentation, a process in which special dispersal units of the lichen, called **soredia,** break off and, if they land on a suitable surface, establish themselves as new lichens. Soredia contain cells of both partners. In some lichens, the fungus produces ascospores, which may be dispersed by wind and find an appropriate algal partner only by chance.

Review

- What is the ecological importance of fungal decomposers?
- What is the importance of mycorrhizae?
- What is a lichen?

ECONOMIC, BIOLOGICAL, AND MEDICAL IMPACT OF FUNGI

Learning Objectives

10 Summarize some of the ways that fungi impact humans economically.

11 Summarize the importance of fungi to biology and medicine; describe how fungi infect plants and humans, identifying at least three fungal plant diseases and three fungal animal diseases.

The same powerful digestive enzymes that fungi use to decompose wastes and dead organisms can also be used with great efficiency to reduce wood, fiber, and food to their basic components. Many species of basidiomycetes have enzymes that break down the lignin in wood. (Lignin is the second most abundant organic compound on Earth, second only to cellulose.) From the human perspective, various fungi cause incalculable damage to stored goods and building materials each year. Bracket fungi, for example, cause enormous losses by decaying wood, both in living trees and in stored lumber.

Fungi are more destructive to plants than any other disease-causing organism. Their activities cost billions of dollars in agricultural damage yearly. Some fungi cause diseases in humans and other animals. Yet fungi also contribute to our quality of life. They are responsible for economic gains as well as losses. People eat them and grow them to make various chemicals, such as citric acid and other industrial chemicals.

Fungi provide beverages and food

Humans exploit the capability of yeasts to produce ethyl alcohol and carbon dioxide from sugars such as glucose by fermentation (see Chapter 8) to make wine, beer, and other fermented beverages, as well as bread. Wine is produced when yeasts ferment fruit sugars, and beer results when yeasts ferment sugars derived from starch in grains (usually barley). During the process of making bread, carbon dioxide produced by yeast becomes trapped in dough as bubbles, causing the dough to rise; this gives leavened bread its light texture. Both the carbon dioxide and the alcohol produced by the yeast escape during baking.

The unique flavor of cheeses such as Roquefort, Brie, Gorgonzola, and Camembert is produced by species of *Penicillium*. For example, *P. roquefortii,* found in caves near the French village of Roquefort, is used to make Roquefort cheese. By French law, only cheeses produced in this area can be called Roquefort cheese. (The blue spots in Roquefort and certain other cheeses are masses of conidia.)

Aspergillus tamarii and certain other fungi are used to produce soy sauce by fermenting soybeans with the fungi for at least 3 months. (In the United States soy sauce is often made by adding flavoring to salt water rather than by soaking fermented soybeans.) Soy sauce enriches other foods with more than just its special flavor. It also adds vital amino acids from both the soybeans and the fungi themselves, which, in some parts of the world, supplement a low-protein rice diet.

Among the basidiomycetes, there are some 200 kinds of edible mushrooms and about 70 species of poisonous ones. Some edible mushrooms are cultivated commercially. The mushroom *Agaricus brunnescens* is the principal fungal species grown extensively for food. About 30 other mushroom species, such as oyster, shiitake, portobello, and straw mushrooms, are available in supermarkets. Morels, which superficially resemble mushrooms, and truffles, which produce underground fruiting bodies, are ascomycetes (Fig. 26-21). These gourmet delights are now being cultivated as mycorrhizal fungi on the roots of tree seedlings.

Edible and poisonous mushrooms can look very much alike and may even belong to the same genus. There is no simple way to tell them apart; an expert must identify them. Some of the most poisonous mushrooms belong to the genus *Amanita* (Fig. 26-22). Toxic species of this genus have been appropriately called such names as "destroying angel" (*A. virosa*) and "death cap" (*A. phalloides*). Eating a single mushroom of either species can be fatal.

Certain species of mushrooms cause intoxication and hallucinations. The sacred mushrooms of the Aztecs—*Conocybe* and *Psilocybe*—are still used in religious ceremonies by native peoples of Central America for their hallucinogenic properties. The chemical ingredient *psilocybin* is responsible for the trances and visions experienced by those who eat these mushrooms. Psilocybin is related to lysergic acid diethylamide (LSD). Ingestion of psychoactive mushrooms is dangerous because negative reactions vary considerably, from mild indigestion, sweating, and heart palpitations, to death. In addition, the possession and use of such mushrooms are illegal in the United States and some other countries.

Fungi are important to modern biology and medicine

The yeast *Saccharomyces cerevisiae* (baker's yeast), an ascomycete, has served as a model eukaryotic cell. It was the first eukaryote whose genome was sequenced, and with its 6000 genes, it has the smallest genome of any eukaryotic model organism. Molecular biologists are in the process of determining the functions of the proteins encoded by its genes. Biologists have used *S. cerevisiae* to study molecular genetics, including how genes regulate cell division. Researchers continue to use this yeast to study such problems as genetic recombination, and the correlation between cell

Richard Shiell/Dembinsky Photo Associates

(a) The yellow morel (*Morchella esculenta*), which grows 6 to 10 cm (2.5 to 4 in) tall, is found throughout North America. Photographed in Michigan.

Figure 26-21 Edible ascomycetes

John D. Cunningham/Visuals Unlimited

(b) The Oregon white truffle (*Tuber gibbosum*), which is found underground near Douglas firs and possibly oak trees in British Columbia and Northern California, is 1 to 5 cm (0.4 to 2 in) wide. People find these subterranean ascocarps with the help of trained dogs or pigs. Here, truffles are shown whole and sectioned to show the conspicuous white, marbled tissue.

age and cancer. *S. cerevisiae* is also being used to study the mechanism of action of antifungal drugs and resistance to these drugs.

Biologists have used the ascomycete *Aspergillus nidulans,* an opportunistic pathogen of humans, to study mitosis and other cell processes. This fungus has provided valuable knowledge about the genetics of microtubules. Biologists are using recombinant DNA techniques to manipulate yeasts and certain filamentous fungi to produce important biological molecules, such as hormones. Among the many genes that have been cloned in yeast are those for insulin, human growth hormone, and molecules important in immune function. These procedures allow researchers to produce unlimited amounts of these compounds for study and eventual medical use.

Researchers are investigating fungi—for example, certain species of microsporidia—for the biological control of pathogens and insect pests. Some of these species are already being used to parasitize insect pests. In some cases, they interfere with reproduction in their insect host. Other microsporidia completely control the metabolism and reproduction of the host. Researchers are studying the use of microsporidia in controlling the spread of malaria. A recent study showed that when female *Anopheles* mosquitoes are infected with microsporidia, their blood feeding is decreased. Fungal infection of the mosquito also interferes with development of *Plasmodium,* the protist that causes malaria.

Fungi produce useful drugs and chemicals. Discovered in 1928 by the British bacteriologist Alexander Fleming, penicillin, produced by the mold *Penicillium notatum,* is still among the most widely used and effective antibiotics (see Chapter 1). Other drugs derived from fungi include the cephalosporin antibiotics (produced by *Cephalosporium*), statins (used to lower blood cholesterol levels), and cyclosporine (used to suppress immune responses in patients who receive organ transplants). Fumagillin, a chemical produced by the ascomycete *Aspergillus fumigatus,* inhibits the formation of new blood vessels. Because solid tumors need a rich blood supply, fumagillin shows promise as an anti-

cancer agent. Fumagillin is also used to treat diseases caused by microsporidia.

The ascomycete *Claviceps purpurea* infects the flowers of rye plants and other cereals. It produces a structure called an *ergot* where a seed would normally form in the grain head. When live-

James W. Richardson/CBR Images

Figure 26-22 Poisonous mushrooms

The destroying angel (*Amanita virosa*) is an extremely poisonous mushroom that is distinguished, as are other amanitas, by the ring of tissue around its stalk and by the underground cup from which the stalk protrudes. About 50 g (2 oz) of this mushroom can kill an adult man. The destroying angel, which is 7.5 to 20 cm (3 to 8 in) tall, is found in grass or near trees throughout North America.

stock eat this grain or when humans eat bread made from ergot-contaminated rye flour, they may be poisoned by the extremely toxic substances in the ergot. However, some ergot compounds are now used clinically in small quantities as drugs to induce labor, to stop uterine bleeding, to treat high blood pressure, and to relieve one type of migraine headache.

Some fungi cause animal diseases

Some fungi cause superficial infections in which only the skin, hair, or nails are infected. Ringworm, athlete's foot, and jock itch are examples of superficial fungal infections. Because these fungi infect dead layers of skin that are not fed by capillaries, the immune system cannot launch an effective response. Ascomycetes cause these types of infections.

Many pathogenic fungi are opportunists that cause infections only when the body's immune system is compromised, for example, in patients with HIV. Cancer patients and organ transplant recipients who are given medication to suppress their immune systems are also at risk. *Candida* is an ascomycete that inhabits the human mouth and vagina. The immune system and the normal bacteria of these regions normally prevent this yeast from causing infection. However, when the immune system is compromised, *Candida* multiplies, causing thrush, a painful yeast infection of the mouth, throat, and vagina.

The ascomycete *Aspergillus fumigatus* is usually harmless but causes aspergillosis in people with lowered immune function. During the course of aspergillosis, the fungus can invade the lungs, heart, brain, kidneys, and other vital organs and cause death.

Other fungi infect internal tissues and organs and may spread through many regions of the body. Histoplasmosis, for example, is a serious infection of the lungs caused by inhaling spores of a fungus common in soil contaminated with bird feces. Most people in the Eastern and Midwestern parts of the United States have been exposed to this fungus at some time, and an estimated 40 million Americans have had mild infections. Fortunately, the infection is usually confined to the lungs and is of short duration, but if the infection spreads through the blood to the heart, brain, or other parts of the body, it can be serious and sometimes fatal.

Some fungi produce poisonous compounds collectively called **mycotoxins.** A few species of *Aspergillus,* for example, produce potent mycotoxins called **aflatoxins** that harm the liver and are known carcinogens. Foods on which aflatoxin-producing fungi commonly grow include peanuts, pecans, corn, and other grains. Other foods that may contain traces of aflatoxins include animal products such as milk, eggs, and meat (from animals that consumed feed contaminated by aflatoxin). Avoiding aflatoxin in the diet is impossible, but exposure should be minimized as much as possible. Any human food or animal forage product that has become moldy should be suspected of aflatoxin contamination and be discarded.

Fungi contribute to sick-building syndrome, a situation in which occupants of a building experience acute adverse health effects linked to the time they spend in a given building. Mold-related insurance claims amount to hundreds of millions of dol-

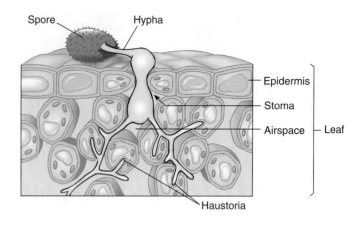

Figure 26-23 How a fungus parasitizes a plant

In this example, the hypha enters the leaf through a stoma. The hypha grows, branching extensively through the internal air spaces, and penetrates plant cells with specialized hyphal extensions called haustoria.

lars each year. When conditions are moist, molds can grow on carpets, leather, cloth, wood, insulation, and food. Mold spores, fragments, and mold products make their way into the air, and people are exposed through inhalation as well as by skin contact. Exposure to molds and their toxins has been linked to depressed immune function, irritation of the throat and respiratory passageways, infection, and toxicity. The most common responses to mold exposure are allergic reactions that range from mild to severe illnesses, including hay fever, sinusitis, asthma, and dermatitis.

Fungi cause many important plant diseases

Fungi are responsible for about 70% of all major crop diseases. They cause serious epidemic diseases that spread rapidly and often result in complete crop failure. All plants are apparently susceptible to some fungal infection. Damage may be localized in certain tissues or structures of the plant, or the disease may be systemic and spread throughout the entire plant. Fungal infections may cause stunting of plant parts or of the plant; they may cause wartlike growths or kill the plant.

A plant often becomes infected after hyphae enter through stomata (pores) in the leaf or stem or through wounds in the plant body (Fig. 26-23). Alternatively, the fungus may produce *cutinase,* an enzyme that dissolves the waxy cuticle that covers the surface of leaves and stems. After dissolving the cuticle, the fungus easily invades the plant tissues. As the mycelium grows, it may stay mainly between the plant cells or it may penetrate the cells. Parasitic fungi often produce special hyphal branches called **haustoria** (sing., *haustorium*) that penetrate the host cells and obtain nourishment from the cytoplasm.

Ascomycetes cause important plant diseases, including powdery mildew, chestnut blight, Dutch elm disease, apple scab, wilt on potatoes, and brown rot, which attacks cherries, peaches, plums, and apricots (Fig. 26-24a). Early in the 20th century the chestnut blight fungus (*Cryphonectria parasitica*) was accidentally imported on diseased Asian chestnuts and quickly at-

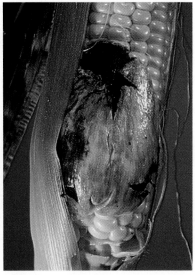

(a) Brown rot of peaches is caused by *Monilinia fruticola*, an ascomycete. Photographed in Oregon.

(b) Corn smut on an ear of sweet corn is caused by *Ustilago maydis*, a basidiomycete.

Figure 26-24 Fungi that cause plant diseases

tacked native American chestnut trees, which had no resistance to the fungus. By the late 1940s, the blight had killed or damaged several billion mature trees—almost every North American chestnut tree throughout its entire natural range. Plant biologists are working to save the American chestnut tree from extinction. One approach has been to develop a disease-resistant variety by breeding the American chestnut with the Chinese chestnut, a related species that is resistant to chestnut blight. Other biologists are genetically engineering a virus that infects the chestnut blight fungus and reduces its virulence.

Basidiomycetes cause smuts and rusts that attack corn, wheat, oats, and other grains (▌Fig. 26-24b). Some fungal parasites, such as the stem rust of wheat, have complex life cycles that involve two or more different host plants and the production of several kinds of spores. For example, wheat rust must infect a barberry plant at one stage in its life cycle. As a result of this discovery, the eradication of barberry plants in wheat-growing regions has reduced infection with wheat rust. Eradication of the barberry has not eliminated wheat rust, however, because the fungus overwinters on wheat at the southern end of the Grain Belt and forms asexual spores. During the spring, wind blows these spores for hundreds of kilometers, reinfecting northern areas of the United States and Canada.

Review

▌ Some dictionaries erroneously define a morel as a type of mushroom. Why is this definition incorrect?

▌ How are fungi important in modern biology and medicine?

▌ What are three important fungal diseases of humans?

▌ What are three important fungal diseases of plants?

SUMMARY WITH KEY TERMS

Learning Objectives

1 Describe the distinguishing characteristics of kingdom Fungi (page 556).
 ▌ **Fungi** are eukaryotic heterotrophs that secrete digestive enzymes onto their food source and then absorb the predigested food.
 ▌ Fungi are characterized by cell walls that contain **chitin.**
2 Describe the body plan of a fungus (page 556).
 ▌ A fungus may be a unicellular **yeast** or a filamentous, multicellular **mold.** The body of most multicellular fungi consists of long, threadlike filaments called **hyphae** that branch and form a tangled mass called a **mycelium.**

ThomsonNOW™ **Learn more about mycelium by clicking on the figure in ThomsonNOW.**

 ▌ In most fungi, perforated **septa,** or cross walls, divide the hyphae into individual cells. In some fungi (zygomycetes and glomeromycetes), the hyphae are **coenocytic,** that is, they form an elongated, multinuclear cell.
3 Describe the life cycle of a typical fungus, including sexual and asexual reproduction (page 556).
 ▌ Most fungi reproduce both sexually and asexually by means of **spores.** Spores are produced on aerial hyphae.

When a fungal spore lands in a suitable spot, the spore germinates.

- When fungi of two different mating types meet, their hyphae fuse, a process called **plasmogamy.** The cytoplasm fuses, but the nuclei remain separate. The fungi enter a **dikaryotic** ($n + n$) stage in which each new cell formed has one nucleus of each type.

- **Karyogamy,** fusion of the nuclei, takes place in the hyphal tip and results in a diploid ($2n$) **zygote nucleus.**

- Meiosis takes place, producing four genetically different haploid (n) nuclei. Each nucleus becomes part of a spore. When the spores germinate, they form new mycelia by mitosis.

- Genetically similar asexual spores can be produced by mitosis. When these spores germinate, they also develop into mycelia.

4 Give arguments to support the hypothesis that fungi are opisthokonts, more closely related to animals than to plants (page 558).

- Like animals, some fungi have flagellate cells—for example, chytrid gametes and spores; the flagellate cells propel themselves with a single posterior flagellum.

- Like animal cells, fungal cells have platelike cristae in their mitochondria.

- Based on chemical and structural characters, fungi are classified, along with animals and choanoflagellates, as **opisthokonts.**

5 Give arguments to support the hypothesis that chytrids may have been the earliest fungal group to evolve from the most recent common ancestor of fungi (page 558).

- **Chytrids,** or **chytridiomycetes,** produce flagellate cells at some stage in their life cycle. No other fungi have flagella. Thus, chytrids probably were the earliest fungi to evolve from a flagellate protist, the common ancestor of all fungi.

6 List distinguishing characteristics, describe a typical life cycle, and give examples of each of the following fungal groups: chytridiomycetes, zygomycetes, glomeromycetes, ascomycetes, and basidiomycetes (page 558).

- Chytrids reproduce both asexually and sexually. Their gametes and zoospores are flagellate. *Allomyces,* a common chytrid, spends part of its life as a multicellular haploid **thallus** and part as a multicellular diploid thallus. The haploid thallus produces two types of flagellate gametes that fuse. Both plasmogamy and karyogamy occur, producing a flagellate zygote. The diploid thallus bears zoosporangia that produce diploid **zoospores** and resting sporangia in which haploid zoospores form by meiosis. The haploid zoospores form new haploid thalli.

- **Zygomycetes,** such as the black bread mold, *Rhizopus,* form a haploid thallus that produces both asexual spores and sexual spores. Asexual spores germinate and form new thalli. In sexual reproduction, hyphae of two different haploid mating types form gametangia. Plasmogamy occurs as the gametangia fuse. Karyogamy occurs, and a diploid zygote is formed, from which a **zygospore** develops. Meiosis produces recombinant haploid zygospores. When zygospores germinate, each hypha develops a sporangium at its tip. Spores are released and develop into new hyphae. **Microsporidia,** now considered zygomycetes, are opportunistic pathogens that penetrate and infect animal cells with their long, threadlike **polar tubes.**

- **Glomeromycetes** (phylum Glomeromycota) are symbionts that form intracellular associations called **mycorrhizae** with the roots of many plants. Because they extend their hyphae into root cells, glomeromycetes are **endomycorrhizal fungi.** The most common endomycorrhizae are called **arbuscular mycorrhizae** because the hyphae inside the root cells form branched, tree-shaped structures known as **arbuscules.** Glomeromycetes have coenocytic hyphae. They reproduce asexually with large, multinucleate spores called blastospores.

- **Ascomycetes** produce asexual spores called **conidia;** sexual spores called **ascospores** are produced in **asci.** The asci line a fruiting body called an **ascocarp.** Haploid mycelia of opposite mating types produce septate hyphae. Plasmogamy occurs, and nuclei are exchanged. A dikaryotic $n + n$ stage occurs in which hyphae form and produce asci and an ascocarp. Karyogamy occurs, followed by meiosis. The recombinant nuclei divide by mitosis, producing eight haploid nuclei that develop into ascospores. When the ascospores germinate, they can form new mycelia. Ascomycetes include yeasts, cup fungi, morels, truffles, and pink, brown, and blue-green molds. Some ascomycetes form mycorrhizae; others form lichens.

- **Basidiomycetes** produce sexual spores called **basidiospores** on the outside of a **basidium.** Basidia develop on the surface of **gills** in mushrooms, a type of **basidiocarp** (a fruiting body). Hyphae in this phylum have septa. Plasmogamy occurs with the fusion of two hyphae of different mating types. A dikaryotic secondary mycelium forms. Then a basidiocarp develops, and basidia form. Karyogamy occurs, producing a diploid zygote nucleus. Meiosis produces four haploid nuclei that become basidiospores. When basidiospores germinate, they form haploid primary mycelia. Basidiomycetes include mushrooms, puffballs, bracket fungi, rusts, and smuts.

ThomsonNOW™ **Explore fungus life cycles by clicking on the figures in ThomsonNOW.**

7 Explain the ecological significance of fungi as decomposers (page 570).

- Most fungi are decomposers that break down organic compounds in dead organisms, leaves, garbage, and wastes into simpler nutrients that can be recycled.

8 Describe the important ecological role of mycorrhizae (page 570).

- Mycorrhizae are mutualistic relationships between fungi and the roots of plants. The fungus supplies water and nutrient minerals to the plant, and the plant secretes organic compounds needed by the fungus. Glomeromycetes form endomycorrhizal associations with roots. Ascomycetes and basidiomycetes form **ectomycorrhizae** with tree roots; they do not penetrate the root cells.

9 Characterize the unique nature of a lichen (page 570).

- A **lichen** is a symbiotic combination of a fungus and a photoautotroph (an alga or cyanobacterium). The photoautotroph provides the fungus with organic compounds, shelter, water, and minerals. Lichens have three main growth forms: crustose, foliose, and fruticose.

ThomsonNOW™ **Learn more about lichens by clicking on the figure in ThomsonNOW.**

10 Summarize some of the ways that fungi impact humans economically (page 574).

- Mushrooms, morels, and truffles are foods; yeasts are vital in the production of beer, wine, and bread; certain fungi are used to produce cheeses and soy sauce. Fungi are also used to make citric acid and other industrial chemicals.

11 Summarize the importance of fungi to biology and medicine; describe how fungi infect plants and humans, identifying at least three fungal plant diseases and three fungal animal diseases (page 574).

■ Modern biologists use the yeast *Saccharomyces cerevisiae* and other fungi as model organisms for research in molecular biology and genetics. Fungi are also being investigated for the biological control of insects, such as the mosquitoes that transmit malaria.

■ Fungi are used to make many medications, including penicillin and other antibiotics.

■ Fungi are opportunistic pathogens in humans. They cause human diseases, including ringworm, athlete's foot, candidiasis, and histoplasmosis.

■ Some fungi produce **mycotoxins,** such as **aflatoxins,** which can cause liver damage and cancer.

■ Fungal hyphae infect plants through stomata. Hyphal branches called **haustoria** penetrate plant cells and obtain nourishment from the cytoplasm. Fungi cause many important plant diseases, including wheat rust, Dutch elm disease, and chestnut blight.

TEST YOUR UNDERSTANDING

1. Fungi produce cell walls containing the polymer _____. (a) chitin (b) cyclosporine (c) cellulose (d) peptidoglycan (e) lysergic acid

2. Which of the following fungi does *not* have a mycelium? (a) black bread mold (b) yeast (c) decomposer cup fungus (d) cultivated mushroom (e) *Penicillium*

3. The condition described as *n + n* is (a) monokaryotic (b) diploid (c) a primary mycelium (d) coenocytic (e) dikaryotic

4. With the exception of chytridiomycetes, fungi are generally disseminated by (a) water currents (b) fragmentation of hyphae (c) soredia (d) airborne spores (e) flagellate zoospores

5. Which statement is *not* true of the chytrids? (a) they are simple aquatic fungi (b) they produce motile cells with single, posterior flagella (c) they undergo both sexual and asexual reproduction (d) the black bread mold is a representative of this group (e) they were the earliest fungi to evolve

6. Which statement is *not* true of the zygomycetes? (a) many members of this group form endomycorrhizae with tree roots (b) their sexual spores are called zygospores (c) they undergo both sexual and asexual reproduction (d) plasmogamy and karyogamy occur (e) they have coenocytic hyphae

7. Which statement is *not* true of the ascomycetes? (a) their sexual spores are produced in asci (b) their asexual spores are produced in basidia (c) some species in this group are yeasts (d) their asexual spores are called conidia (e) some species cause serious plant diseases

8. The ascomycete life cycle typically includes (a) mainly diploid thalli (b) the formation of a thick zygosporangium (c) the production of eight haploid ascospores within an ascus (d) production of microsporidia (e) the production of ascospores, zoospores, and conidia at different stages

9. Which statement is *not* true of the basidiomycetes? (a) they have a diploid thallus that produces zoospores (b) their sexual spores are called basidiospores (c) they produce a secondary mycelium with *n + n* hyphae (d) mushrooms and bracket fungi are examples of this group (e) this group includes both edible and poisonous species

10. The familiar portion of a mushroom is actually a large fruiting body called a(an) (a) ascocarp (b) basidium (c) basidiocarp (d) gametangium (e) ascus

11. Glomeromycetes (a) reproduce mainly by sexual spores called glomerospores (b) are characterized by unique structures called polar tubes (c) include some species that associate with cyanobacteria to form lichens (d) include many opportunistic pathogens that cause human disease (e) form arbuscular endomycorrhizae with tree roots

12. A symbiotic association between a photoautotroph and a fungus is called a(an) _____ (a) arbuscular endomycorrhiza (b) ectomycorrhiza (c) lichen (d) pathogenic agent (e) lignin decomposing agent

13. Fungi are (a) eukaryotes and opisthokonts (b) prokaryotes and opisthokonts (c) flagellate and dikaryotic (d) prokaryotic pathogens (e) heterotrophs with cellulose cell walls

14. Mutualistic relationships between fungi and the roots of plants are called (a) lichens (b) mycorrhizae (c) pathogenic associations (d) parasitic haustoria (e) mycotoxic symbioses

15. *Amanita virosa* (a) is the mushroom commonly cultivated for food (b) is the yeast that ferments wine and beer (c) produces the unique flavor of many cheeses (d) is a highly toxic mushroom (e) has been ingested for its hallucinogenic properties

16. Which of the following describes how a fungus infects a plant? (a) it infiltrates leaves with lichens (b) it forms relationships by attaching mycorrhizae to stems (c) it secretes powerful digestive juices onto the leaves (d) it uses haustoria to penetrate stem tissue (e) its hyphae enter leaves through a stoma

17. Mycotoxins (a) are released by most lichens (b) typically cause mild allergic reactions (c) can harm the liver (d) reduce the effects of sick-building syndrome (e) are produced mainly by chytrids

18. *Saccharomyces cerevisiae* (a) is used by biologists as a model eukaryotic cell (b) was the first prokaryote whose genome was sequenced (c) has an extremely complex genome with more than 6 million genes (d) is a mold used to produce antibiotics (e) is a common chytrid

1. Explain the statement "Mushrooms are like the tips of ice-bergs." If you do not see mushrooms in your lawn, can you conclude that no fungi live there? Why or why not?

2. Biologists have discovered that many mycorrhizal fungi are sensitive to a low pH. What human-caused environmental problem may prove catastrophic for these fungi? How may this problem affect their plant partners?

3. **Evolution Link.** How are the life cycles of the marine alga *Ulva* (see Fig. 25-18) and *Allomyces* similar? What hypoth-eses do these similarities suggest about their evolutionary relationship?

4. **Evolution Link.** Justify (a) classifying fungi as opisthokonts; (b) classifying microsporidia as fungi; (c) grouping ascomycetes and basidiomycetes as sister clades.

Additional questions are available in ThomsonNOW at www.thomsonedu.com/login

The Plant Kingdom: Seedless Plants

Darrell Gulin/Getty Images

Moss-covered rocks.
Shown is Elowa Falls, Columbia River Gorge, Oregon.

KEY CONCEPTS

Biologists infer that plants evolved from aquatic green algal ancestors known as a charophytes.

Adaptations to life on land that have evolved in plants include a waxy cuticle to prevent water loss; multicellular gametangia; stomata; and for most plants, vascular tissues containing lignin.

Plants undergo an alternation of generations between multicellular gametophyte and sporophyte generations.

Mosses and other bryophytes lack vascular tissues and do not form true roots, stems, or leaves.

In club mosses and ferns, lignin-hardened vascular tissues that transport water and dissolved substances throughout the plant body have evolved.

About 445 million years ago (mya), the ocean was filled with vast numbers of fishes, mollusks, and crustaceans as well as countless microscopic algae, and the water along rocky coastlines was home to large seaweeds. Occasionally, perhaps, an animal would crawl out of the water onto land, but it never stayed there permanently because there was little to eat on land—not a single blade of grass, no fruit, and no seeds.

During the next 30 million years, a time corresponding roughly to the Silurian period (see Table 21-1), plants appeared in abundance and colonized the land. Where did they come from? Although plants living today exhibit great diversity in size, form, and habitat (see photograph), biologists hypothesize that they all evolved from a common ancestor that was an ancient green alga. Recall from Chapter 25 that red algae, green algae, and land plants are collectively classified as "plants." Biologists infer this common ancestor because modern green algae share many biochemical and metabolic traits with modern plants. Both green algae and plants contain the same photosynthetic pigments: chlorophylls *a* and *b* and carotenoids, including xanthophylls (yellow pigments) and carotenes (orange pigments). Both store excess carbohydrates as starch and have cellulose as a major component of their cell walls. In addition, plants and some green algae share certain details of cell division, including formation of a cell plate during cytokinesis (see Chapter 10).

Land plants are complex multicellular organisms that range in size from minute, almost microscopic duckweeds to massive giant sequoias, some of the largest organisms that have ever lived. The plant kingdom consists of hundreds of thousands of species that live in varied habitats, from frozen arctic tundra to lush tropical rain forests to harsh deserts to moist stream banks. ■

ADAPTATIONS OF PLANTS

Learning Objectives

1 Discuss some environmental challenges of living on land, and describe how several plant adaptations meet these challenges.

2 Name the green algal group from which plants are hypothesized to have descended, and describe supporting evidence.

What are some features of plants that have let them colonize so many types of environments? One important difference between plants and algae is that a waxy **cuticle** covers the aerial portion of a plant. Essential for existence on land, the cuticle helps prevent desiccation, or drying out, of plant tissues by evaporation. Plants that are adapted to moister habitats may have a very thin layer of wax, whereas those adapted to drier environments often have a thick, crusty cuticle. (Many desert plants also have a reduced surface area, particularly of leaves, which minimizes water loss.)

Plants obtain the carbon they need for photosynthesis from the atmosphere as carbon dioxide (CO_2). To be fixed into organic molecules such as sugar, CO_2 must first diffuse into the chloroplasts that are inside green plant cells. Because a waxy cuticle covers the external surfaces of leaves and stems, however, gas exchange through the cuticle between the atmosphere and the interior of cells is negligible. Tiny pores called **stomata** (sing., *stoma*), which dot the surfaces of leaves and stems of almost all plants, facilitate gas exchange; algae lack stomata.

The sex organs, or **gametangia** (sing., *gametangium*), of most plants are multicellular, whereas the gametangia of algae are unicellular (Fig. 27-1). Each plant gametangium has a layer of sterile (nonreproductive) cells that surrounds and protects the delicate gametes (eggs and sperm cells). In plants, the fertilized egg develops into a multicellular **embryo** (young plant) within the female gametangium. Thus, the embryo is protected during its development. In algae, the fertilized egg develops away from its gametangium; in some algae, the gametes are released before fertilization, whereas in others the fertilized egg is released.

The plant life cycle alternates haploid and diploid generations

Plants have a clearly defined **alternation of generations** in which they spend part of their lives in a multicellular haploid stage and part in a multicellular diploid stage (Fig. 27-2).[1] The haploid portion of the life cycle is called the **gametophyte generation** because it gives rise to haploid gametes by mitosis. When two gam-

---

[1]For convenience we limit our discussion to plants that are not polyploid, although polyploidy is very common in the plant kingdom. We therefore use the terms *diploid* and *2n* (and *haploid* and *n*) interchangeably, although these terms are not actually synonymous.

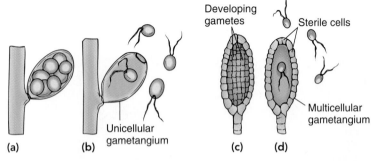

Figure 27-1 Generalized reproductive structures of algae and plants

(a, b) In algae, gametangia are generally unicellular. When the gametes are released, only the wall of the original cell remains. **(c, d)** In plants, the gametangia are multicellular, but only the inner cells become gametes. The gametes are surrounded by a protective layer of sterile (nonreproductive) cells.

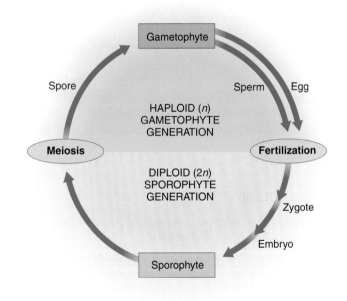

Figure 27-2 The basic plant life cycle

Plants have an alternation of generations, spending part of the cycle in a haploid gametophyte stage and part in a diploid sporophyte stage. Depending on the plant group, the haploid or the diploid stage may be greatly reduced.

etes fuse, the diploid portion of the life cycle, called the **sporophyte generation,** begins. The sporophyte generation produces haploid spores by the process of meiosis; these spores represent the first stage in the gametophyte generation.

Let us examine alternation of generations more closely. The haploid gametophytes produce male gametangia, known as **antheridia** (sing., *antheridium*), in which sperm cells form, and/or female gametangia, known as **archegonia** (sing., *archegonium*), each bearing a single egg (Fig. 27-3). Sperm cells reach the female gametangium in a variety of ways, and one sperm cell fertilizes the egg to form a **zygote,** or fertilized egg.

The diploid zygote is the first stage in the sporophyte generation. The zygote divides by mitosis and develops into a multicellular embryo, the young sporophyte plant. Embryo development takes place within the archegonium; thus, the embryo is protected as it develops. Eventually the embryo grows into a mature sporophyte plant. The mature sporophyte has special cells called *sporogenous cells* (spore-producing cells, also called *spore mother cells*) that divide by meiosis to form haploid **spores.**

Fertilization of egg by sperm cell ⟶ zygote ⟶ embryo ⟶ mature sporophyte plant ⟶ sporogenous cells ⟶ meiosis ⟶ spores

All plants produce spores by meiosis, in contrast with algae and fungi, which may produce spores by meiosis or mitosis. The spores represent the first stage in the gametophyte genera-

tion. Each spore divides by mitosis to produce a multicellular gametophyte, and the cycle continues. Plants therefore alternate between a haploid gametophyte generation and a diploid sporophyte generation.

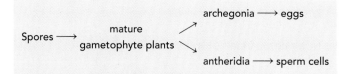

Spores ⟶ mature gametophyte plants ⟶ archegonia ⟶ eggs / antheridia ⟶ sperm cells

Four major groups of plants evolved

Recent ultrastructural and molecular data indicate that land plants probably descended from a group of green algae called **charophytes** or **stoneworts** (see Fig. 25-16d). Molecular comparisons, particularly of DNA and RNA sequences, provide compelling evidence that charophytes are closely allied to plants. These comparisons among plants and various green algae include sequences of chloroplast DNA, certain nuclear genes, and ribosomal RNA. In each case, the closest match occurs between charophytes and plants, indicating that modern charophytes and plants probably share a recent common ancestor.

The plant kingdom consists of four major groups: bryophytes, seedless vascular plants, and two groups of seeded vascular plants: gymnosperms and angiosperms (flowering plants) (Fig. 27-4; see also Table 27-1, which is an overview of plant phyla). The mosses and other bryophytes are small nonvascular plants that lack a specialized vascular, or conducting, system to transport nutrients, water, and essential minerals (inorganic nutrients) throughout the plant body. In the absence of such a system, bryophytes rely on diffusion and osmosis to obtain needed

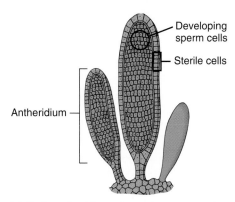

(a) Each antheridium, the male gametangium, produces numerous sperm cells.

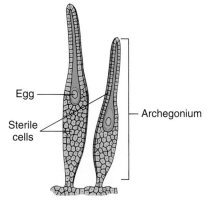

(b) Each archegonium, the female gametangium, produces a single egg.

Figure 27-3   Plant gametangia

Shown are generalized moss gametangia.

TABLE 27-1
The Plant Kingdom
Nonvascular plants with a dominant gametophyte generation (bryophytes)
Phylum Bryophyta (mosses)
Phylum Hepatophyta (liverworts)
Phylum Anthocerophyta (hornworts)
Vascular plants with a dominant sporophyte generation
Seedless plants
Phylum Lycopodiophyta (club mosses)
Phylum Pteridophyta (ferns and their allies, the whisk ferns and horsetails)
Seed plants
Plants with naked seeds (gymnosperms)
Phylum Coniferophyta (conifers)
Phylum Cycadophyta (cycads)
Phylum Ginkgophyta (ginkgoes)
Phylum Gnetophyta (gnetophytes)
Seeds enclosed within a fruit
Phylum Anthophyta (angiosperms or flowering plants)
Class Eudicotyledones (eudicots)
Class Monocotyledones (monocots)

Key Point

This cladogram shows hypothetical evolutionary relationships among living plants, based on current evidence. The four major groups of plants are bryophytes, seedless vascular plants, and two groups of seed plants—gymnosperms and angiosperms.

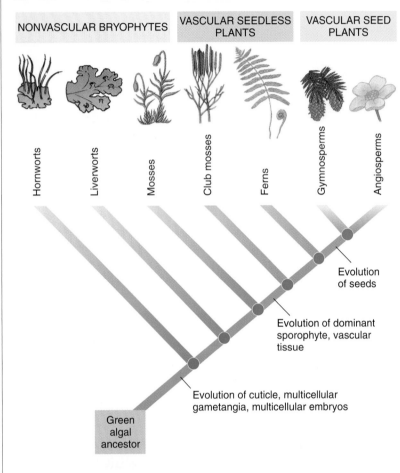

Figure 27-4 *Animated* Plant evolution

Cladograms such as this one represent an emerging consensus that is open to change as new discoveries are made. Although the arrangement of nonvascular, seedless vascular, and seed plant groupings is widely recognized, the order in which the hornworts, liverworts, and mosses evolved is not yet resolved.

plant cell wall chemistry, including lignin). The stiffening property of lignin enabled plants to grow tall (which let them maximize light interception). The successful occupation of the land by plants, in turn, made the evolution of terrestrial animals possible by providing them with both habitat and food.

Club mosses and ferns (which include whisk ferns and horsetails) are seedless vascular plants that, like the bryophytes, reproduce and disperse primarily via spores. Seedless vascular plants arose and diversified during the Silurian and Devonian periods of the Paleozoic era, between 444 mya and 359 mya. Club mosses and ferns extend back more than 420 million years and were of considerable importance as Earth's dominant plants in past ages. Fossil evidence indicates that many species of these plants were the size of immense trees. Most living representatives of club mosses and ferns are small.

The gymnosperms are vascular plants that reproduce by forming seeds. Gymnosperms produce seeds borne exposed (unprotected) on a stem or in a cone. Plants with seeds as their primary means of reproduction and dispersal first appeared about 359 mya, at the end of the Devonian period. These early seed plants diversified into many varied species of gymnosperms.

The most recent plant group to appear is the flowering plants, or angiosperms, which arose during the early Cretaceous period of the Mesozoic era, about 130 mya. Like gymnosperms, flowering plants reproduce by forming seeds. Flowering plants, however, produce seeds enclosed within a fruit.

Review

- What are the most important environmental challenges that plants face living on land?
- What adaptations do plants have to meet these environmental challenges?
- From which group of green algae are plants hypothesized to have descended?
- What is alternation of generations in plants?

materials. This reliance means that bryophytes are restricted in size; if they were much larger, some of their cells could not obtain enough necessary materials. Bryophytes do not form seeds, the reproductive structures discussed in Chapter 28. Bryophytes reproduce and disperse primarily via haploid spores. Recent molecular and fossil evidence, discussed later in the chapter, suggests that bryophytes may have been the earliest plants to colonize land.

The other three groups of plants—seedless vascular plants, gymnosperms, and flowering plants—have vascular tissues and are thus known as vascular plants. The two vascular tissues are **xylem,** for conducting water and dissolved minerals, and **phloem,** for conducting dissolved organic molecules such as sugar. A key step in the evolution of vascular plants was the ability to produce **lignin,** a strengthening polymer in the walls of cells that function for support and conduction (see Chapter 32 for a discussion of

BRYOPHYTES

Learning Objectives

3   Summarize the features that distinguish bryophytes from other plants.
4   Name and briefly describe the three phyla of bryophytes.
5   Describe the life cycle of mosses, and compare their gametophyte and sporophyte generations.

The **bryophytes** (from the Greek words meaning "moss plant") consist of about 16,000 species of mosses, liverworts, and hornworts; bryophytes are the only living nonvascular plants (▮ Table 27-2). Because they have no means for extensive internal transport of water, sugar, and essential minerals, bryophytes are typically quite small. They generally require a moist environment for

TABLE 27-2

A Comparison of Major Groups of Seedless Plants

Plant Group	Dominant Stage of Life Cycle	Representative Genera
Nonvascular; reproduce by spores (bryophytes)		
Mosses (phylum Bryophyta)	Gametophyte: leafy plant	*Polytrichum, Sphagnum, Physcomitrella*
Liverworts (phylum Hepatophyta)	Gametophyte: thalloid or leafy plant	*Marchantia*
Hornworts (phylum Anthocerophyta)	Gametophyte: thalloid plant	*Anthoceros*
Vascular; reproduce by spores		
Club mosses (phylum Lycopodiophyta)	Sporophyte: roots, rhizomes, erect stems, and leaves (microphylls)	*Lycopodium, Selaginella*
Ferns (phylum Pteridophyta)	Sporophyte: roots, rhizomes, and leaves (megaphylls)	*Pteridium, Polystichum, Azolla, Platycerium*
Whisk ferns (phylum Pteridophyta)	Sporophyte: rhizomes and erect stems; no true roots or leaves	*Psilotum*
Horsetails (phylum Pteridophyta)	Sporophyte: roots, rhizomes, erect stems, and leaves (reduced megaphylls)	*Equisetum*

active growth and reproduction, but some bryophytes tolerate dry areas.

The bryophytes are divided into three distinct phyla: mosses (phylum Bryophyta), liverworts (phylum Hepatophyta), and hornworts (phylum Anthocerophyta) (❚ Fig. 27-5). These three groups differ in many ways and may or may not be closely related. They are usually studied together because they lack vascular tissues and have similar life cycles.

Moss gametophytes are differentiated into "leaves" and "stems"

Mosses (phylum Bryophyta), which include about 9900 species, usually live in dense colonies or beds (see Fig. 27-5a). Each individual plant has tiny, hairlike absorptive structures called *rhizoids* and an upright, stemlike structure that bears leaflike blades, each normally consisting of a single layer of undifferentiated cells except at the midrib. Because mosses lack vascular tissues, they do not have true roots, stems, or leaves; the moss structures are not homologous to roots, stems, or leaves in vascular plants. Some moss species have water-conducting cells and sugar-conducting

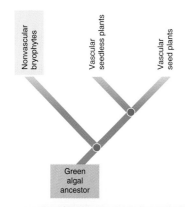

(a) A close-up of haircap moss (*Polytrichum commune*) gametophytes. The haircap moss is a popular ground cover in rock gardens, particularly in Japan.

(b) Flattened, ribbonlike lobes characterize the gametophyte of the common liverwort (*Marchantia polymorpha*). *Marchantia* grows on moist soil, from the tropics to arctic regions.

(c) The gametophyte with mature sporophytes (the "horns" projecting out of the gametophyte) of the common hornwort (*Anthoceros natans*).

Figure 27-5 Bryophytes

cells, although these cells are not lignified and as specialized or as effective as the conducting cells of vascular plants.

Alternation of generations is clear in the life cycle of mosses (❚ Fig. 27-6). As shown in ❶, the green moss gametophyte often bears its gametangia at the top of the plant. Many moss species have separate sexes: male plants that bear antheridia and female plants that bear archegonia. Other mosses produce antheridia and archegonia on the same plant.

Fertilization occurs when one of the sperm cells fuses with the egg within the archegonium ❷. Sperm cells, which have flagella, are transported from antheridium to archegonium by flowing water, such as splashing rain droplets. A raindrop lands on the top of a male gametophyte, and sperm cells are released into it from the antheridia. Another raindrop landing on the male plant may splash the sperm-laden droplet into the air and onto the top of a nearby female plant. Alternatively, insects may touch the sperm-laden fluid and inadvertently carry it for considerable distances. Once in a film of water on the female moss, a sperm cell swims into the archegonium, which secretes chemicals to attract and guide the sperm cells, and fuses with the egg.

The diploid zygote, formed by fertilization ❸, grows by mitosis into a multicellular embryo that develops into a mature moss sporophyte ❹. This sporophyte grows out of the top of the female gametophyte and remains attached and nutritionally dependent on the gametophyte throughout its existence (❚ Fig. 27-7). Initially green and photosynthetic, the sporophyte becomes golden brown at maturity. It consists of three main parts: a *foot*, which anchors the sporophyte to the gametophyte and absorbs minerals and nutrients from it; a *seta*, or stalk; and a *capsule*, which contains sporogenous cells (spore mother cells). The capsule of some species is covered by a caplike structure, the *calyptra*, which is derived from the archegonium.

The sporogenous cells undergo meiosis to form haploid spores (see Fig. 27-6 ❺). When the spores are mature, the capsule opens and releases the spores, which are then transported by wind or rain. If a moss spore lands in a suitable spot, it germinates

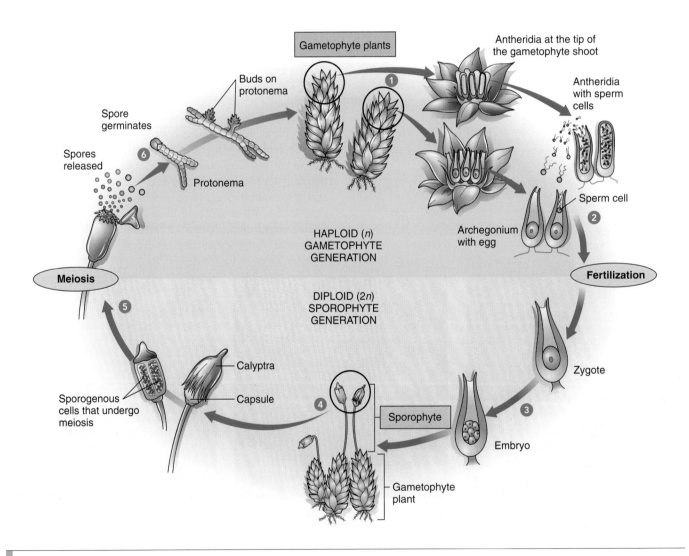

Figure 27-6 *Animated* The life cycle of mosses

The gametophyte generation is dominant in the moss life cycle. After sexual reproduction, the sporophyte grows out of the gametophyte. See text for a detailed description.

David Cavagnaro

Figure 27-7 Moss sporophytes

Each consisting of a foot, seta, and capsule, the sporophytes grow out of the top of the gametophytes. Spores are produced within the capsule. Shown is the haircap moss (*Polytrichum commune*).

and grows into a filament of cells called a **protonema** ❻. The protonema, which superficially resembles a filamentous green alga, forms buds, each of which grows into a green gametophyte, and the life cycle continues.

Biologists consider the haploid gametophyte generation the dominant generation in mosses because it lives independently of the diploid sporophyte. In contrast, the moss sporophyte is attached to and nutritionally dependent on the gametophyte.

Mosses make up an inconspicuous but significant part of their environment. They play an important role in forming soil. Because they grow tightly packed in dense colonies, mosses hold soil in place and help prevent erosion.

Commercially, the most important mosses are the peat mosses in the genus *Sphagnum*. One of the distinctive features of *Sphagnum* "leaves" is the presence of many large dead cells that absorb and hold water. This feature makes peat moss particularly beneficial as a soil conditioner. Added to sandy soils, for example, peat moss helps absorb and retain moisture. In some countries, such as Ireland and Scotland, people cut out layers of dead peat moss that have accumulated for hundreds of years in peat bogs, dry them, and burn them for fuel.

The name "moss" is often misused to refer to plants that are not truly mosses. For example, reindeer "moss" is a lichen that is a dominant form of vegetation in the arctic tundra, Spanish "moss" is a flowering plant, and club "moss" (discussed later in the chapter) is a relative of ferns.

Liverwort gametophytes are either thalloid or leafy

Liverworts (phylum Hepatophyta) consist of about 6000 species of nonvascular plants with a dominant gametophyte generation, but the gametophytes of some liverworts are quite different from those of mosses. Their body form is often a flattened, lobed structure called a **thallus** (pl., *thalli*) that is not differentiated into leaves, stems, or roots. The common liverwort, *Marchantia polymorpha*, is thalloid (see Fig. 27-5b). Liverworts are so named because the lobes of their thalli superficially resemble the lobes of the human liver; *wort* is derived from the Old English word *wyrt*, meaning "plant." On the underside of the liverwort thallus are hairlike rhizoids that anchor the plant to the soil. Other liverworts, known as *leafy liverworts,* superficially resemble mosses, with leaflike blades, "stems," and rhizoids rather than a lobed thallus. As in the mosses, leafy liverwort "leaves" consist of a single layer of undifferentiated cells. Like other bryophytes, liverworts are small, generally inconspicuous plants that are largely restricted to damp environments. Unlike mosses, hornworts, and other plants, liverworts lack stomata (although some liverworts have surface pores thought to be analogous to stomata).

Liverworts reproduce both sexually and asexually (❙Fig. 27-8). As shown in ❶, their sexual reproduction involves production of archegonia and antheridia on the haploid gametophyte. In some liverworts, these gametangia are borne on stalked structures called *archegoniophores,* which bear archegonia, and *antheridiophores,* which bear antheridia. Their life cycle is basically the same as that of mosses, although some of the structures look quite different. Splashing raindrops transport sperm cells to the archegonia, where fertilization takes place ❷. The resulting zygote develops into a multicellular embryo that becomes a mature sporophyte ❸. The liverwort sporophyte, which is usually somewhat spherical, is attached to the gametophyte, as in mosses. As shown in ❹, sporogenous cells in the capsule of the sporophyte undergo meiosis, producing haploid spores. Each spore has the potential to develop into a green gametophyte ❺, and the cycle continues.

Some liverworts reproduce asexually by forming tiny balls of tissue called **gemmae** (sing., *gemma*), which are borne in a saucer-shaped structure, the gemmae cup, directly on the liverwort thallus. Splashing raindrops and small animals help disperse gemmae. When a gemma lands in a suitable place, it grows into a new liverwort thallus. Liverworts may also reproduce asexually by thallus branching and growth. The individual thallus lobes elongate, and each becomes a separate plant when the older part of the thallus that originally connected the individual lobes dies.

Hornwort gametophytes are inconspicuous thalloid plants

Hornworts (phylum Anthocerophyta) are a small group of about 100 species of bryophytes whose gametophytes superficially resemble those of the thalloid liverworts. Hornworts live in disturbed habitats such as fallow fields and roadsides.

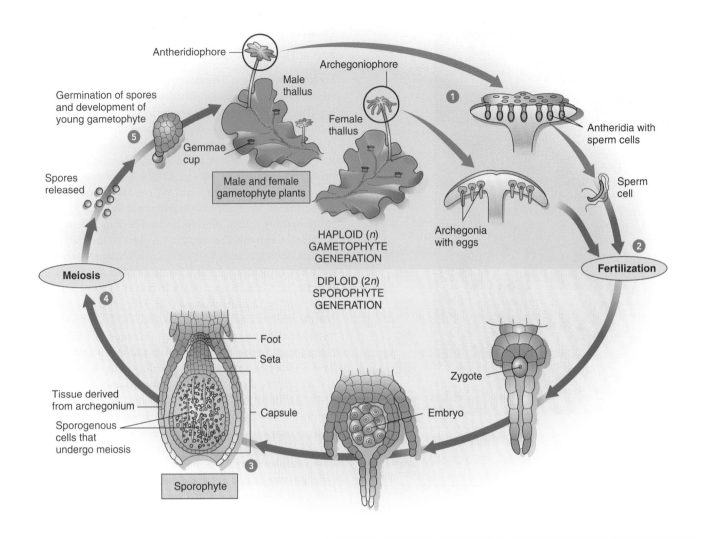

Figure 27-8 *Animated* The life cycle of the common liverwort (*Marchantia polymorpha*)

The dominant generation is the gametophyte, represented by separate male and female thalli. The stalked, umbrella-shaped structures are the antheridiophores, with antheridia that produce sperm cells, and the archegoniophores, with archegonia that each bear an egg cell. See text for a detailed description.

Hornworts may or may not be closely related to other bryophytes. For example, their cell structure, particularly the presence of a single large chloroplast in each cell, resembles certain algal cells more than plant cells. In contrast, mosses, liverworts, and other plants have many disc-shaped chloroplasts per cell.

In the common hornwort (*Anthoceros natans*), archegonia and antheridia are embedded in the gametophyte thallus rather than on archegoniophores and antheridiophores. After fertilization and development, the needlelike sporophyte projects out of the gametophyte thallus, forming a spike or "horn"—hence the name *hornwort*. A single gametophyte often produces multiple sporophytes (see Fig. 27-5c). Meiosis occurs, forming spores within each **sporangium** (pl., *sporangia*), or spore case. The sporangium splits open from the top to release the spores; each spore can give rise to a new gametophyte thallus. A unique feature of hornworts is that the sporophytes, unlike those of mosses and liverworts, continue to grow from their bases for the remainder of the gametophyte's life.

Bryophytes are used for experimental studies

Botanists use certain bryophytes as experimental models to study many fundamental aspects of plant biology, including genetics, growth and development, plant ecology, plant hormones, and *photoperiodism*, which is plant responses to varying periods of night and day length. The moss *Physcomitrella patens* is a particularly important research organism for studying plant evolution because its features and genome can be compared to those of flowering plants. In this regard, *Physcomitrella* is the plant equivalent of the fruit fly *Drosophila*, which is an important model organism for studies of animal inheritance, development, and evolution. As

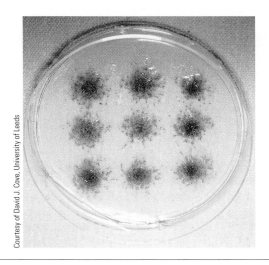

Courtesy of David J. Cove, University of Leeds

Figure 27-9 Mosses as research organisms

Researchers inoculated this petri dish with offspring from a genetic cross in the moss *Physcomitrella patens*. They will test the cultures for their phenotypes with respect to vitamin requirements. (The culture medium in which the mosses are currently growing is supplemented with all the vitamins required by the parental strains.)

experimental organisms, *Physcomitrella* and other bryophytes are easy to grow on artificial media and do not require much space because they are so small (Fig. 27-9).

Details of bryophyte evolution are based on fossils and on structural and molecular evidence

Plants are a **monophyletic group**—that is, all plants probably evolved from a common ancestral green alga. Fossil evidence indicates that bryophytes are ancient plants, probably the first group of plants to arise from the common plant ancestor. The fossil record of ancient bryophytes is incomplete, consisting mostly of spores and small tissue fragments, and can be interpreted in different ways. As a result, it does not provide a definite answer on bryophyte evolution.

The oldest known recognizable plant fossils are dated at about 425 million years old. These fossils resemble modern liverworts in many respects, but the spores are virtually identical to those in 470-million-year-old rocks. In 2003, scientists reported fossil fragments of tiny liverwort-like plants associated with ancient spores that were discovered in Oman. This evidence suggests that liverwort-like plants may have been the earliest plants to colonize land.

Structural and molecular evidence, however, supports the hypothesis that *hornworts* may be the most ancient group of plants alive today. A 1998 cladistic analysis using structural data was the first to strongly support this hypothesis, which was reinforced in 2000 by DNA evidence. The study of nuclear, chloroplast, and mitochondrial genes from numerous plant species, from bryophytes to flowering plants, suggests that an early plant ancestor gave rise to two lineages, one from which the hornworts descended and the other from which all other plants descended.

- Which of the following are parts of the gametophyte generation in mosses: antheridia, zygote, embryo, capsule, archegonia, sperm cells, egg cell, spores, and protonema?
- How are mosses, liverworts, and hornworts similar? How is each group distinctive?

SEEDLESS VASCULAR PLANTS

Learning Objectives

6 Discuss the features that distinguish seedless vascular plants from algae and bryophytes.
7 Name and briefly describe the two phyla of seedless vascular plants.
8 Describe the life cycle of ferns, and compare their sporophyte and gametophyte generations.
9 Compare the generalized life cycles of homosporous and heterosporous plants.

The most important adaptation found in seedless vascular plants, though absent in algae and bryophytes, is specialized vascular tissues—xylem and phloem—for support and conduction. This system of conduction lets vascular plants grow larger than the bryophytes because water, minerals, and sugar are transported to all parts of the plant. Although seedless vascular plants in temperate environments are relatively small, tree ferns in the tropics may grow to heights of 18 m (60 ft). All seedless vascular plants have true stems with vascular tissues, and most also have true roots and leaves.

Botanists have extensively studied the evolution of the leaf as the main organ of photosynthesis. The two basic types of true leaves—microphylls and megaphylls—evolved independently of one another (Fig. 27-10). The **microphyll,** which is usually small and has a single vascular strand, probably evolved from small, projecting extensions of stem tissue (*enations*). Only one group of living plants, the club mosses, has microphylls. In contrast, **megaphylls** probably evolved from stem branches that gradually filled in with additional tissue (*webbing*) to form most leaves as we know them today. Megaphylls have more than one vascular strand, as we would expect if they evolved from branch systems. Ferns (with the exception of whisk ferns, discussed later in the chapter), gymnosperms, and flowering plants have megaphylls.

Recent evidence suggests megaphylls evolved over a 40-million-year period in the Late Paleozoic era in response to a gradual decline in the level of atmospheric CO_2. As CO_2 declined, plants developed a flattened blade with more stomata for gas exchange. (More stomata allowed cells inside the leaf to get enough CO_2.)

There are two main clades of seedless vascular plants: the club mosses and the ferns. Biologists originally considered horsetails and whisk ferns distinct enough to be classified in separate phyla. However, many kinds of evidence, such as DNA comparisons and similarities in sperm structure, have resulted in their being reclassified as ferns. As shown in Figure 27-4, ferns, including horsetails and whisk ferns, are a monophyletic group and the closest living relatives of seed plants.

ANCIENT PLANTS AND COAL FORMATION

Industrial society depends on energy from fossil fuels that formed from the remains of ancient organisms. One of the most important fossil fuels is coal, which people burn to produce electricity and to manufacture items of steel and iron. Although mined, coal is not an inorganic mineral like gold or aluminum but an organic material formed from the remains of ancient vascular plants, particularly those of the Carboniferous period, approximately 320 mya. Five main groups of plants contributed to coal formation. Three were seedless vascular plants: the club mosses, horsetails, and ferns. The other two were seed plants: seed ferns (now extinct) and early gymnosperms.

It is hard to imagine that relatives of the small, relatively inconspicuous club mosses, ferns, and horsetails of today were so significant in forming vast beds of coal. However, many members of these groups that existed during the Carboniferous

period were giants compared with their modern counterparts and formed immense forests (see Fig. 21-11).

The climate during the Carboniferous period was warm, moist, and mild. Plants in most locations could grow year-round because of the favorable conditions. Forests of these plants often grew in low-lying, swampy areas that periodically flooded when the sea level rose. As the sea level receded, these plants would re-establish.

When these large plants died or were blown over in storms, they decomposed incompletely because they were covered by swamp water. (The anaerobic conditions of the water prevented wood-rotting fungi from decomposing the plants, and anaerobic bacteria do not decompose wood rapidly.) Thus, over time the partially decomposed plant material accumulated and consolidated.

Layers of sediment formed over the plant material each time the water level rose and flooded the low-lying swamps. With time, heat and pressure built up in these accumulated layers and converted the plant material to coal and the sediment layers to sedimentary rock. Much later, geologic upheavals raised the layers of coal and sedimentary rock. Coal is usually found in seams, underground layers that vary in thickness from 2.5 cm to more than 30 m (100 ft).

The various grades of coal (lignite, sub-bituminous, bituminous, and anthracite) formed as a result of the different temperatures and pressures to which the layers were exposed. Coal exposed to high heat and pressure during its formation is drier, is more compact (and therefore harder), and has a higher heating value (that is, a higher energy content).

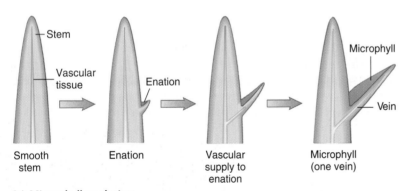

(a) Microphyll evolution

Smooth stem → Enation → Vascular supply to enation → Microphyll (one vein)

Labels: Stem, Vascular tissue, Enation, Microphyll, Vein

Figure 27-10
Evolution of microphylls and megaphylls

Dichotomous branching (in **b**) is branching into two equal halves. Webbing (in **b**) is the evolutionary process in which the spaces between close branches become filled with chlorophyll-containing cells.

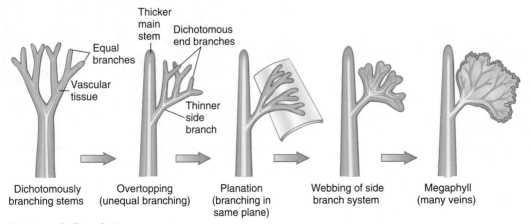

(b) Megaphyll evolution

Dichotomously branching stems → Overtopping (unequal branching) → Planation (branching in same plane) → Webbing of side branch system → Megaphyll (many veins)

Labels: Equal branches, Vascular tissue, Thicker main stem, Dichotomous end branches, Thinner side branch

Club mosses are small plants with rhizomes and short, erect branches

Club mosses (phylum Lycopodiophyta) were important plants millions of years ago, when species that are now extinct often reached great size (▮Fig. 27-11a). These large, treelike plants were major contributors to our present-day coal deposits (see *Focus On: Ancient Plants and Coal Formation*).

The 1200 or so species of club mosses living today, such as *Lycopodium* (▮Fig. 27-11b), are small (less than 25 cm, or 10 in, tall), attractive plants common in temperate woodlands. They possess true roots; both rhizomes and erect aerial stems; and small, scale-like leaves (microphylls). Sporangia are borne on reproductive leaves that are either clustered in conelike strobili at the tips of stems or scattered in reproductive areas along the stem. Club mosses are evergreen and often fashioned into Christmas wreaths and other decorations. In some areas they are endangered by overharvesting.

That common names are sometimes misleading in biology is vividly evident in this group of plants. The most common names for the phylum Lycopodiophyta are "club mosses" and "ground pines," yet these plants are neither mosses, which are non-vascular, nor pines, which are seed plants.

Ferns are a diverse group of spore-forming vascular plants

Most of the 11,000 species of **ferns** (phylum Pteridophyta) are terrestrial, although a few have adapted to aquatic habitats (▮Fig. 27-12a and b). Ferns range from the tropics to the Arctic Circle, with most species living in tropical rain forests, where they perch high in the branches of trees (▮Fig. 27-12c). In temperate regions, ferns commonly inhabit swamps, marshes, moist woodlands, and stream banks. Some species grow in fields, rocky crevices on cliffs or mountains, or even deserts.

The life cycle of ferns involves a clearly defined alternation of generations. The ferns grown as house-plants (such as Boston fern, maidenhair fern, and staghorn fern) represent

the larger, more conspicuous sporophyte generation. As shown in ▮Figure 27-13 ❶, the fern sporophyte consists of a horizontal underground stem, or *rhizome*, that bears leaves, called *fronds*, and true roots. As each young frond first emerges from the ground, it is tightly coiled and resembles the top of a violin, hence the name *fiddlehead*. As fiddleheads grow, they unroll and expand to form fronds. Fern fronds are usually compound (the blade is divided into several leaflets), with the leaflets forming beautifully complex leaves. Fronds, roots, and rhizomes all contain vascular tissues.

Spore production usually occurs in certain areas on the fronds, which develop sporangia. Many species bear the sporangia in clusters, called **sori** (sing., *sorus*) ❷. Within sporangia, sporogenous cells (spore mother cells) undergo meiosis to form haploid spores. The sporangia burst open and discharge spores that may germinate and grow by mitosis into gametophytes ❸.

The mature fern gametophyte, which bears no resemblance to the sporophyte, is a tiny (less than half the size of one of your fingernails), green, often heart-shaped structure that grows flat against the ground ❹. Called a **prothallus** (pl., *prothalli*), the fern gametophyte lacks vascular tissues and has tiny, hairlike absorptive rhizoids to anchor it. The prothallus usually produces both archegonia and antheridia on its underside ❺. Each archegonium contains a

(a) Reconstruction of *Lepidodendron*. This ancient club moss was the size of a large tree—to 40 m (about 130 ft). Numerous fossils of *Lepidodendron* were preserved in coal deposits, particularly in Great Britain and the central United States.

(b) The sporophyte of *Lycopodium*, a club moss, has small, scalelike leaves (microphylls) that are evergreen. Spores are produced in sporangia on reproductive leaves clustered in a conelike strobilus (*as shown*) or, in other species, scattered along the stem.

Figure 27-11 Club mosses

(**a,** Redrawn from M. Hirmer, *Handbuch der Paläobotanik*, R. Olderbourg, Munich, 1927.)

single egg, whereas numerous sperm cells are produced in each antheridium.

Ferns use water as a transport medium. The flagellate sperm cells swim, usually from a nearby prothallus, to the neck of an archegonium through a thin film of water on the ground underneath the prothallus. As shown in ⑥, after one of the sperm cells fertilizes the egg, a diploid zygote grows by mitosis into a multicellular embryo (an immature sporophyte). At this stage, the sporophyte embryo is attached to and dependent on the gametophyte; but as the embryo matures, the prothallus withers and dies, and the sporophyte becomes free-living.

The fern life cycle alternates between the dominant, diploid sporophyte with its rhizome, roots, and fronds, and the haploid gametophyte (prothallus). The sporophyte generation is dominant not only because it is larger than the gametophyte but also because it persists for an extended period (most fern sporophytes are perennials), whereas the gametophyte dies soon after reproducing.

Whisk ferns are classified as reduced ferns

Only about 12 species of **whisk ferns** (phylum Pteridophyta) exist today, and the fossil record contains several extinct species. Whisk ferns, which live mainly in the tropics and subtropics, are relatively simple in structure and lack true roots and leaves but have vascularized stems. *Psilotum nudum,* a representative whisk fern, has both a horizontal underground rhizome and vertical aerial stems (Fig. 27-14). Whenever the stem forks, or branches, it always divides into two equal halves. Botanists call this forking **dichotomous branching**. In contrast, when most plant stems branch, one stem is more vigorous and becomes the main trunk.

The upright stems of *Psilotum* are green and are the main organs of photosynthesis. Tiny, round sporangia, borne directly on the erect, aerial stems, contain sporogenous cells that undergo

(a) The Christmas fern (*Polystichum acrostichoides*) is green at Christmas, making it a popular decoration. This fern, photographed in the Great Smoky Mountains in Tennessee, has fronds that grow to 0.6 m (2 ft) in length.

(b) The tiny mosquito fern (*Azolla caroliniana*) is a free-floating aquatic fern that does not resemble "typical" ferns. Although each individual plant grows to only about 1.3 cm (0.5 in) in length, *Azolla* populations sometimes grow so densely across ponds that they reportedly smother mosquito larvae. Photographed in Massachusetts.

(c) The staghorn fern (*Platycerium bifurcatum*) is native to Australian rain forests and is widely cultivated elsewhere. In nature the staghorn fern is an epiphyte, a plant that grows attached to another organism (in this case, a tree trunk) but derives no nourishment from it. The individual leaves grow to 0.9 m (3 ft) in length.

Figure 27-12 Ferns

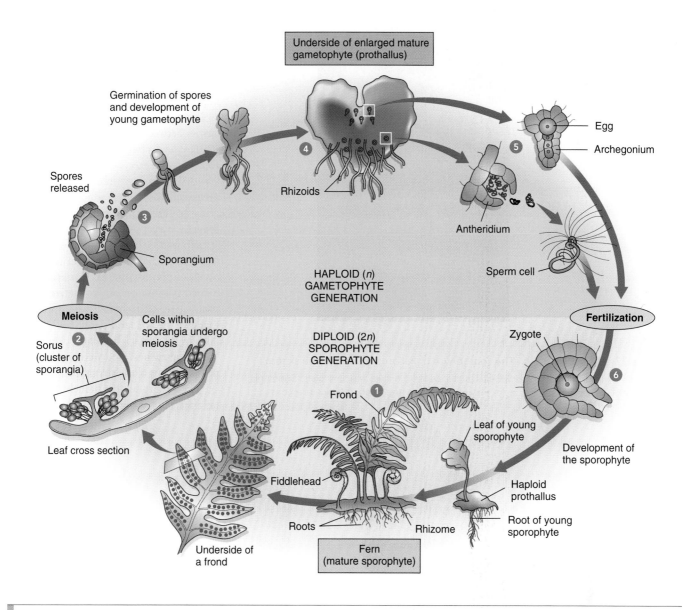

Figure 27-13 *Animated* The life cycle of ferns

Note the clearly defined alternation of generations between the gametophyte (prothallus) and sporophyte (leafy plant) generations. See text for a detailed description.

meiosis to form haploid spores. After dispersal, the spores germinate to form haploid prothalli. Because they grow underground, the prothalli of whisk ferns are difficult to study. They are non-photosynthetic as a result of their subterranean location, and they apparently have a symbiotic relationship with mycorrhizal fungi that provides them with sugar and essential minerals (see Chapter 26).

Botanists have carefully studied whisk ferns in recent years. Molecular data, including comparisons of nucleotide sequences of ribosomal RNA, chloroplast DNA, and mitochondrial DNA in living species, support the hypothesis that the whisk ferns should be classified as reduced ferns rather than as surviving representa-

tives of extinct vascular plants (see discussion of rhyniophytes later in the chapter).

Horsetails are a second evolutionary line of ferns

About 300 mya the **horsetails** (phylum Pteridophyta) were among the dominant plants and grew as large as modern trees (❙ Fig 27-15a). Because they contributed to Earth's vast coal deposits, these ancient horsetails, like ancient club mosses, are still significant today.

The few surviving horsetails, about 15 species in the genus *Equisetum*, grow mostly in wet, marshy habitats and are less than

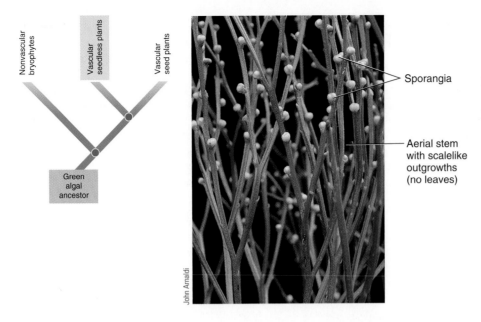

Sporangia

Aerial stem
with scalelike
outgrowths
(no leaves)

John Arnaldi

Figure 27-14 *Animated* The sporophyte of *Psilotum nudum*, a whisk fern

The stem is the main organ of photosynthesis in this rootless, leafless, vascular plant. Sporangia, which are initially green but turn yellow as they mature, are borne on short lateral branches directly on the stems.

1.3 m (4 ft) tall, but extremely distinctive (▮ Fig. 27-15b). They are widely distributed on every continent except Australia. Traditionally classified in their own phylum, horsetails are now grouped with ferns. This reclassification is based on molecular similarities between horsetails and other ferns.

Horsetails have true roots, stems (both rhizomes and erect aerial stems), and small leaves. The hollow, jointed stems are impregnated with silica, which gives them a gritty texture. Small leaves, interpreted as reduced megaphylls, are fused in whorls at each node (the area on the stem where leaves attach). The green stem is the main organ of photosynthesis. Horsetails are so named because certain vegetative (nonreproductive) stems have whorls of branches that give the appearance of a bushy horse's tail. In the past, horsetails were called "scouring rushes" and were used to scrub pots and pans along stream banks.

Each reproductive branch of a horsetail bears a terminal conelike **strobilus**

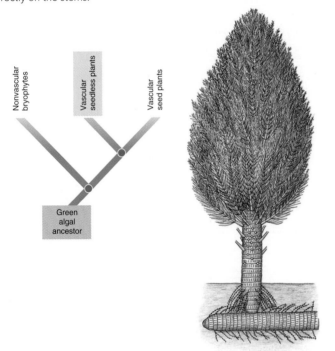

(a) Reconstruction of *Calamites*, an ancient horsetail that grew as tall as many modern trees—to 20 m (about 65 ft). *Calamites* had an underground rhizome where roots and aerial shoots originated.

J. Robert Waaland/Biological Photo Service

Strobilus

Vegetative
shoots

Reproductive
shoots

(b) *Equisetum telematia*, a horsetail with a wide distribution in Eurasia, Africa, and North America, has unbranched reproductive shoots bearing conelike strobili and separate, highly branched vegetative (nonreproductive) shoots. In some horsetail species, both reproductive and vegetative shoots are unbranched.

Figure 27-15 *Animated* Horsetails

(a, Redrawn from L. Emberger, *Les Plantes Fossiles,* Masson et Cie, Paris, 1968.)

(pl., *strobili*). The strobilus consists of several stalked, umbrella-like structures, each of which bears 5 to 10 sporangia in a circle around a common axis.

The horsetail life cycle is similar in many respects to the fern life cycle. In horsetails, as in ferns, the sporophyte is the conspicuous plant, whereas the gametophyte is a minute, lobed thallus ranging in width from the size of a pinhead to about 1 cm across. The sporophyte and gametophyte are both photosynthetic and nutritionally independent at maturity. Like ferns, horsetails require water as a medium for flagellate sperm cells to swim to the egg.

Some ferns and club mosses are heterosporous

In the life cycles examined thus far, plants produce only one type of spore as a result of meiosis. This condition, known as **homospory,** is characteristic of bryophytes, horsetails, whisk ferns, and most ferns and club mosses. However, certain ferns and club mosses exhibit **heterospory,** in which they produce two types of spores: microspores and megaspores. ▌Figure 27-16 illustrates the generalized life cycle of a heterosporous plant.

Spike moss (*Selaginella*), a club moss, is an example of a heterosporous plant (▌Fig. 27-17). As shown in ❶, the sporophyte plant produces sporangia within a conelike strobilus. Each strobilus usually bears two kinds of sporangia: microsporangia and megasporangia. *Microsporangia* are sporangia that produce mi-

crosporocytes (also called *microspore mother cells*), which undergo meiosis to form microscopic, haploid **microspores** ❷. Each microspore develops into a male gametophyte that produces sperm cells within antheridia ❸.

As shown in ❹, *megasporangia* in the *Selaginella* strobilus produce *megasporocytes* (also called *megaspore mother cells*). When megasporocytes undergo meiosis, they form haploid **megaspores,** each of which develops into a female gametophyte that produces eggs in archegonia ❺.

In *Selaginella,* the development of male gametophytes from microspores and of female gametophytes from megaspores occurs within their respective spore walls, using stored food provided by the sporophyte. As a result, the male and female gametophytes are not truly free-living, unlike the gametophytes of other seedless vascular plants. Fertilization is followed by the development of a new sporophyte ❻.

Heterospory evolved several times during the history of land plants. It was a significant development in plant evolution because it was the forerunner of the evolution of seeds. Heterospory characterizes the two most successful groups of plants existing today, the gymnosperms and the flowering plants, both of which produce seeds.

Seedless vascular plants are used for experimental studies

Botanists use many seedless vascular plants as experimental models to study certain aspects of plant biology, such as physiology, growth, and development. Ferns and other seedless vascular plants are useful in studying how apical meristems give rise to plant tissues. An **apical meristem** is the area at the tip (apex) of a root or shoot where growth—cell division, elongation, and differentiation—occurs. Ferns and other seedless vascular plants have a single large *apical cell* located at the center tip of the apical meristem. This apical cell is the source, by mitosis, of all the cells that eventually make up the root or shoot. The apical cell divides in an orderly fashion, and the smaller daughter cells produced by the apical cell, in turn, divide and give rise to different parts of the root or shoot. It is possible to trace mature cells in the root or shoot back to their origin from the single apical cell.

Ferns are interesting research plants for studies in genetics because they are **polyploids** and have multiple sets of chromosomes. (Many ferns have hundreds of chromosomes.) However, gene expression in ferns is exactly what one would expect of a *diploid* plant. Apparently, genes in the extra sets of chromosomes are gradually silenced and therefore not expressed.

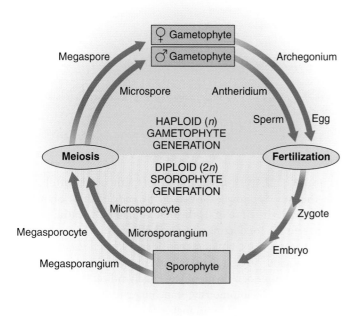

▌ **Figure 27-16** The basic life cycle of heterosporous plants

Two types of spores, microspores and megaspores, are produced during the life cycle of heterosporous plants.

Seedless vascular plants arose more than 420 mya

Currently, the oldest known megafossils of early vascular plants are from Silurian (420 mya) deposits in Europe. (Plant *megafossils* are fossilized roots, stems, leaves, and reproductive structures.) Megafossils of several kinds of small, seedless vascular

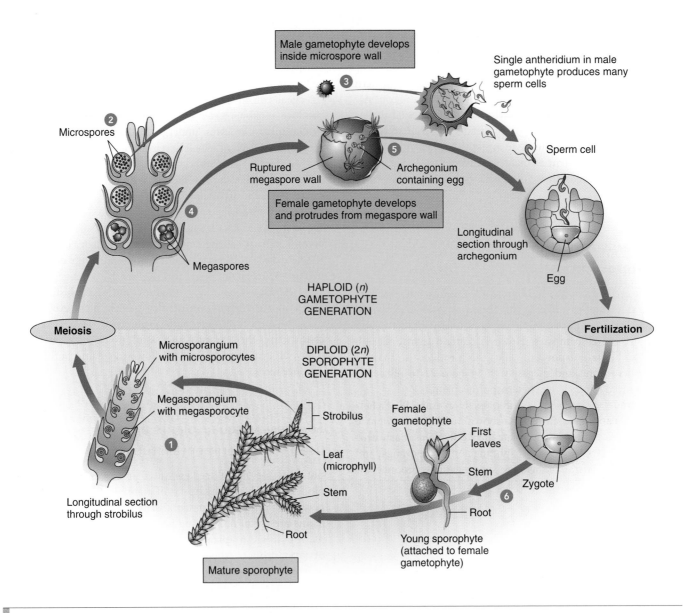

Figure 27-17 The life cycle of spike moss (*Selaginella*)

Spike moss is heterosporous, producing two types of spores in one strobilus. The megaspores develop into female gametophytes, and the microspores become male gametophytes. See text for a detailed description.

plants were also discovered in Silurian deposits in Bolivia, Australia, and northwestern China. Microscopic spores of early vascular plants appear in the fossil record earlier than megafossils, suggesting that even older megafossils of simple vascular plants may be discovered.

Botanists assign the oldest vascular plants to phylum Rhyniophyta, which, according to the fossil record, arose some 420 mya and became extinct about 380 mya. The rhyniophytes are so named because many fossils of these extinct plants were found in fossil beds near Rhynie, Scotland. *Rhynia gwynne-vaughanii* is an example of an early vascular plant that superficially resembled whisk ferns in that it consisted of leafless upright stems that branched dichotomously from an underground rhizome (Fig. 27-18). *Rhynia* lacked roots, although it had absorptive rhizoids. Sporangia formed at the ends of short branches. The internal structure of its rhizome contained a central core of xylem cells for conducting water and minerals.

For many years, botanists considered *Rhynia major*, a plant that grew about 50 cm (20 in) tall and probably lived in marshes, a classic example of a rhyniophyte. Fossils indicate that this plant had rhizoids, dichotomously branching rhizomes, and upright stems that terminated in sporangia. However, recent microscopic studies of fossil rhizomes indicate that the central core of tissue lacked the xylem cells characteristic of vascular plants. For that reason, *R. major* was reclassified into a new genus, *Aglaophyton*, and is no longer considered a rhyniophyte (Fig. 27-19). Science

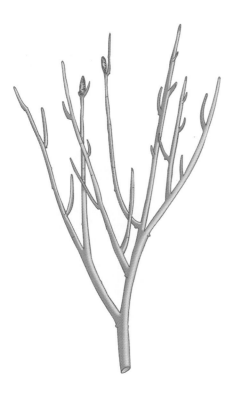

Figure 27-18 Reconstruction of *Rhynia gwynne-vaughanii*

This leafless plant, one of Earth's earliest vascular plants, is now extinct. It grew about 18 cm (7 in) tall. (Redrawn from D. Edwards, "Evidence for the Sporophytic Status of the Lower Devonian Plant *Rhynia gwynne-vaughanii*," *Review of Palaeobotany and Palynology*, Vol. 29, 1980.)

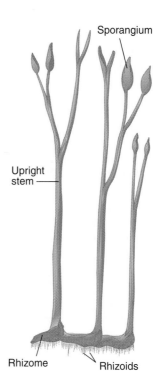

Figure 27-19 Reconstruction of *Aglaophyton major*

Recent evidence indicates that this plant, although superficially similar to other early vascular plants, lacked conducting tissues that are characteristic of vascular plants. For this reason, it has been reclassified into a new genus and is no longer considered a rhyniophyte. (Redrawn from J. D. Mauseth, *Botany: An Introduction to Plant Biology*, 2nd ed., Saunders College Publishing, Philadelphia, 1995.)

is an ongoing enterprise, and over time, existing knowledge is re-evaluated in light of newly discovered evidence. *Aglaophyton major* is an excellent example of the self-correcting nature of science, which is in a perpetually dynamic state and changes in response to newly available techniques and data.

Review

▪ What adaptations do ferns have that both algae and bryophytes lack?

▪ How does one distinguish between megaphylls and microphylls?

▪ Which of the following are parts of the sporophyte generation in ferns: frond, sperm cells, egg cell, roots, sorus, sporangium, spores, prothallus, rhizome, antheridium, archegonium, and zygote?

▪ Why are whisk ferns and horsetails now classified as ferns?

▪ How does heterospory modify the life cycle?

SUMMARY WITH KEY TERMS

Learning Objectives

1 Discuss some environmental challenges of living on land, and describe how several plant adaptations meet these challenges (page 582).

▪ The colonization of land by plants required the evolution of many anatomical, physiological, and reproductive adaptations. Plants have a waxy **cuticle** to protect against water loss and **stomata** for gas exchange needed for photosynthesis.

▪ Plant life cycles have an **alternation of generations** in which they spend part of their life cycle in a haploid **game-**tophyte generation and part in a diploid **sporophyte generation.** The gametophyte plant produces gametes by mitosis. During fertilization these gametes fuse to form a **zygote,** the first stage of the sporophyte generation. The zygote develops into a multicellular **embryo** that the gametophyte protects and nourishes. The mature sporophyte plant develops from the embryo and produces sporogenous cells (spore mother cells). These undergo meiosis to form **spores,** the first stage in the gametophyte generation.

- Most plants have multicellular **gametangia** with a protective jacket of sterile cells surrounding the gametes. **Antheridia** are gametangia that produce sperm cells, and **archegonia** are gametangia that produce eggs.
- Ferns and other vascular plants have **xylem** to conduct water and dissolved minerals and **phloem** to conduct dissolved sugar.

2 Name the green algal group from which plants are hypothesized to have descended, and describe supporting evidence (page 582).
- Plants probably arose from a group of green algae called **charophytes.** This conclusion is based in part on molecular comparisons of DNA and RNA sequences, which show the closest match between charophytes and plants.

ThomsonNOW **Explore plant evolution by clicking on the figure in ThomsonNOW.**

3 Summarize the features that distinguish bryophytes from other plants (page 584).
- Unlike other land plants, bryophytes are nonvascular and lack xylem and phloem. Bryophytes are the only plants with a dominant gametophyte generation. Their sporophytes remain permanently attached and nutritionally dependent on the gametophytes.

4 Name and briefly describe the three phyla of bryophytes (page 584).
- Mosses (phylum Bryophyta) have gametophytes that are green plants that grow from a filamentous **protonema.**
- Many **liverworts** (phylum Hepatophyta) have gametophytes that are flattened, lobelike **thalli;** others are leafy.
- Hornworts (phylum Anthocerophyta) have thalloid gametophytes.

5 Describe the life cycle of mosses, and compare their gametophyte and sporophyte generations (page 584).
- The green moss gametophyte bears archegonia and/or antheridia at the top of the plant. During fertilization, a sperm cell fuses with an egg cell in the archegonium. The zygote grows into an embryo that develops into a moss sporophyte, which is attached to the gametophyte. Meiosis occurs within the capsule of the sporophyte to produce spores. When a spore germinates, it grows into a protonema that forms buds that develop into gametophytes.

ThomsonNOW **Watch the life cycles of the mosses and liverworts by clicking on the figures in ThomsonNOW.**

6 Discuss the features that distinguish seedless vascular plants from algae and bryophytes (page 589).
- Seedless vascular plants have several adaptations that algae and bryophytes lack, including vascular tissues and a dominant sporophyte generation. As in bryophytes, reproduction in seedless vascular plants depends on water as a transport medium for motile sperm cells.

7 Name and briefly describe the two phyla of seedless vascular plants (page 589).
- Sporophytes of **club mosses** (phylum Lycopodiophyta) consist of roots, rhizomes, erect branches, and leaves that are **microphylls.**
- Ferns (phylum Pteridophyta) are the largest and most diverse group of seedless vascular plants. The fern sporophyte consists of a rhizome that bears fronds and true roots. Phylum Pteridophyta also includes whisk ferns and horsetails. Sporophytes of **whisk ferns** have **dichotomously branching** rhizomes and erect stems; they lack true roots and leaves. **Horsetail** sporophytes have roots, rhizomes, aerial stems that are hollow and jointed, and leaves that are reduced **megaphylls.**

8 Describe the life cycle of ferns, and compare their sporophyte and gametophyte generations (page 589).
- Fern sporophytes have roots, rhizomes, and leaves that are megaphylls. Their leaves, or fronds, bear sporangia in clusters called **sori.** Meiosis in sporangia produces haploid spores. The fern gametophyte, called a **prothallus,** develops from a haploid spore and bears both archegonia and antheridia.

ThomsonNOW **Watch the life cycle of the ferns by clicking on the figure in ThomsonNOW.**

9 Compare the generalized life cycles of homosporous and heterosporous plants (page 589).
- **Homospory,** the production of one kind of spore, is characteristic of bryophytes, most club mosses, and most ferns, including whisk ferns and horsetails. In homospory, spores give rise to gametophyte plants that produce both egg cells and sperm cells.
- **Heterospory,** the production of two kinds of spores (microspores and megaspores), occurs in certain club mosses, certain ferns, and all seed plants. **Microspores** give rise to male gametophytes that produce sperm cells. **Megaspores** give rise to female gametophytes that produce eggs. The evolution of heterospory was an essential step in the evolution of seeds.

TEST YOUR UNDERSTANDING

1. The bryophytes (a) include mosses, liverworts, and hornworts (b) include whisk ferns, horsetails, and club mosses (c) are small plants that lack a vascular system (d) a and c (e) b and c

2. The waxy layer that covers aerial parts of plants is the (a) cuticle (b) archegonium (c) protonema (d) stoma (e) thallus

3. A strengthening compound found in cell walls of vascular plants is (a) xanthophyll (b) lignin (c) cutin (d) cellulose (e) carotenoid

4. Stomata (a) help prevent desiccation of plant tissues (b) transport water and minerals through plant tissues (c) allow gas exchange for photosynthesis (d) strengthen cell walls (e) produce male gametes

5. The female gametangium, or _____, produces an egg; the male gametangium, or _____, produces sperm cells. (a) antheridium; archegonium (b) archegonium; megaphyll (c) megasporangium; antheridium (d) archegonium; antheridium (e) megasporangium; megaphyll

6. Liverworts and hornworts share life-cycle similarities with (a) ferns (b) mosses (c) horsetails (d) club mosses (e) whisk ferns

7. The green, gametangia-bearing moss plant (a) is the haploid gametophyte generation (b) is the diploid sporophyte generation (c) is called a protonema (d) contains cells with single large chloroplasts (e) b and c

8. Seedless vascular plants have _____ to conduct water and minerals and _____ to conduct sugar. (a) cuticle; xylem (b) phloem; stomata (c) phloem; xylem (d) stomata; cuticle (e) xylem; phloem

9. A(an) _____ is a leaf that arose from a branch system. (a) antheridium (b) microphyll (c) megaphyll (d) sorus (e) microspore

10. These plants have vascularized stems but lack true roots and leaves. (a) mosses (b) club mosses (c) horsetails (d) whisk ferns (e) hornworts

11. These plants have hollow, jointed stems that are impregnated with silica. (a) mosses (b) club mosses (c) horsetails (d) whisk ferns (e) hornworts

12. Which of the following statements about ferns is *not* true? (a) ferns have motile sperm cells that swim through water to the egg-containing archegonium (b) ferns are vascular plants (c) ferns are the most economically important group of bryophytes (d) the fern sporophyte consists of a rhizome, roots, and fronds (e) the diversity of ferns is greatest in the tropics

13. Plants probably descended from a group of green algae called (a) rhyniophytes (b) *Calamites* (c) epiphytes (d) charophytes (e) club mosses

14. Which of the following is *not* a characteristic of plants? (a) cuticle (b) unicellular gametangia (c) stomata (d) multicellular embryo (e) alternation of generations

15. In plant life cycles (a) the first products of meiosis are gametes (b) spores are part of the diploid sporophyte generation (c) the embryo gives rise to a zygote (d) the first stage in the diploid sporophyte generation is the zygote (e) the first stage in the haploid gametophyte generation is the prothallus

CRITICAL THINKING

1. Which group probably colonized the land first, plants or animals? Explain.

2. What adaptations do bryophytes have that algae lack?

3. **Evolution Link.** How may the following trends in plant evolution be adaptive to living on land?

   a. dependence on water for fertilization ⟶ no need for water as a transport medium

   b. homospory ⟶ heterospory

4. **Analyzing Data.** According to the cladogram in Figure 27-4, which plants evolved first: nonvascular bryophytes, seedless vascular plants, or seed plants? Which clade would be considered the outgroup: hornworts, liverworts, or mosses?

 Thomson **NOW!**

Additional questions are available in ThomsonNOW at www.thomsonedu.com/login

The Plant Kingdom: Seed Plants

M. F. Merlet/Science Photo Library/Photo Researchers, Inc.

Magnolia fruit with protruding seeds. Magnolia (*Magnolia grandiflora*) is a flowering plant, and its seeds are enclosed within a fruit.

KEY CONCEPTS

Seed plants include gymnosperms and angiosperms.

Gymnosperms produce exposed seeds, usually in cones borne on the sporophytes.

Conifers are the most diverse and numerous of the four living gymnosperm phyla.

Angiosperms produce ovules enclosed within carpels; following fertilization, seeds develop from the ovules, and the ovaries of carpels become fruits.

Angiosperms, which compose a single phylum, dominate the land and exhibit great diversity in both vegetative and reproductive structures.

Gymnosperms and angiosperms evolved from ancestral seedless vascular plants.

Chapter 27 focused on plants that reproduce by means of *spores,* haploid reproductive units that give rise to gametophytes. Although the most successful and widespread plants also produce spores, their primary means of reproduction and dispersal is by **seeds,** which represent an important adaptation for life on land (see photograph). Each seed consists of an embryonic sporophyte, nutritive tissue, and a protective coat. Seeds develop from the fertilized egg cell, the female gametophyte, and its associated tissues. The two groups of seed plants, gymnosperms and angiosperms (flowering plants), exhibit the greatest evolutionary complexity in the plant kingdom and are the dominant plants in most terrestrial environments.

Seeds are reproductively superior to spores for three main reasons. First, a seed contains a multicellular young plant with embryonic root, stem, and one or more leaves already formed, whereas a spore is a single cell. Second, a seed contains an abundant food supply. After germination, food stored in the seed nourishes the plant embryo until it becomes self-sufficient. Because a spore is a single cell, few food reserves exist for the plant that develops from a spore. Third, a seed is protected by a multicellular **seed coat** that is very thick and hard in some plants, as, for example, in lima beans. Like spores, seeds live for extended periods at reduced rates of metabolism and germinate when conditions become favorable.

Seeds and seed plants are intimately connected with the development of human civilization. From prehistoric times, early humans collected and used seeds for food. The food stored in seeds is a concentrated source of proteins, oils, carbohydrates, and vitamins, which are nourishing for humans as well as for germinating plants. Seeds are easy to store (if kept dry), so humans can collect them during times of plenty to save for times of need. Few other foods are stored as conveniently or for as long. Although flowering plants produce most seeds that humans consume, the seeds of certain gymnosperms—the piñon pine, for example—are edible. ▪

AN INTRODUCTION TO SEED PLANTS

Learning Objective

1   Compare the features of gymnosperms and angiosperms.

In Chapter 27, you learned that some seedless vascular plants are heterosporous. However, *all* seed plants are heterosporous and produce two types of spores: microspores and megaspores. In fact, heterospory is a requirement of seed production.

Following fertilization in seed plants, an **ovule,** which is a *megasporangium* and its enclosed structures, develops into a seed. Seed plants also have **integuments,** layers of sporophyte tissue that surround and enclose the megasporangium. After fertilization takes place, the seed coat develops from the integuments.

Botanists divide seed plants into two groups based on whether or not an ovary wall surrounds their ovules (an *ovary* is a structure that contains one or more ovules). The two groups of seed plants are the **gymnosperms** and the **angiosperms** (▪ Table 28-1). The word *gymnosperm* is derived from the Greek for "naked seed." Gymnosperms produce seeds that are totally exposed or borne on the scales of cones. In other words, an ovary wall does not surround the ovules of gymnosperms. Pine, spruce, fir, hemlock, and ginkgo are examples of gymnosperms.

The term *angiosperm* is derived from the Greek expression that means "seed enclosed in a vessel or case." Angiosperms are flowering plants that produce their seeds within a *fruit* (a mature ovary). Thus, the ovules of angiosperms are protected. Flowering plants, which are extremely diverse, include corn, oaks, water lilies, cacti, apples, grasses, palms, and buttercups.

Both gymnosperms and flowering plants have vascular tissues: **xylem,** for conducting water and dissolved minerals (inorganic nutrients); and **phloem,** for conducting dissolved sugar. Both have life cycles with an **alternation of generations,** that is, they spend part of their lives in the multicellular diploid sporophyte stage and part in the multicellular haploid gametophyte stage. The sporophyte generation is the dominant stage in each group, and the gametophyte generation is significantly reduced in size and entirely dependent on the sporophyte generation. Unlike the plants we have considered so far (bryophytes and ferns; see Chapter 27), gymnosperms and flowering plants do not have free-living gametophytes. Instead, the female gametophyte is attached to and nutritionally dependent on the sporophyte generation.

Review

▪ What is an ovule?
▪ How do gymnosperm and angiosperm seeds differ?

GYMNOSPERMS

Learning Objectives

2   Trace the steps in the life cycle of a pine, and compare its sporophyte and gametophyte generations.
3   Summarize the features that distinguish gymnosperms from bryophytes and ferns.
4   Name and briefly describe the four phyla of gymnosperms.

The gymnosperms include some of the most interesting members of the plant kingdom. For example, a giant sequoia (*Sequoiadendron giganteum*) known as the General Sherman Tree, in Sequoia National Park in California, is one of the world's most massive organisms. It is 82 m (267 ft) tall and has a girth of 23.7 m (77 ft) measured 1.5 m (5 ft) above ground level. Another gymnosperm, a coast redwood (*Sequoia sempervirens*) known as the Mendocino

---

TABLE 28-1

A Comparison of Gymnosperms and Angiosperms

Characteristic	Gymnosperms	Angiosperms
Growth habit	Woody trees and shrubs	Woody or herbaceous
Conducting cells in xylem	Tracheids	Vessel elements and tracheids
Reproductive structures	Cones (usually)	Flowers
Pollen grain transfer	Wind (usually)	Animals or wind
Fertilization	Egg and sperm $\longrightarrow$ zygote; double fertilization in gnetophytes	Double fertilization: egg and sperm $\longrightarrow$ zygote; two polar nuclei and sperm $\longrightarrow$ endosperm
Seeds	Exposed or borne on scales of cones	Enclosed within fruit derived from ovary
Number of species	About 840	More than 300,000
Geographic distribution	Worldwide	Worldwide

Figure 28-1 Bristlecone pine at sunrise

Bristlecone pines (*Pinus aristata*), found in mountainous parts of California, Nevada, and Utah, are the world's longest-lived trees.

tree, is among the world's tallest trees, measuring 112 m (364 ft) in the year 2000. Botanists using tree-ring analysis determined that one of the oldest living trees, a bristlecone pine (*Pinus aristata*) in the White Mountains of California, is 4900 years old (▌ Fig. 28-1).

Gymnosperms are usually classified into four phyla, which represent four different evolutionary lines (▌ Fig. 28-2). Numbering 630 species, the largest phylum of gymnosperms is Coniferophyta, commonly called *conifers.* Two phyla of gymnosperms, Ginkgophyta (the ginkgoes) and Cycadophyta (the cycads), are evolutionary remnants of groups that were more significant in the past. The fourth phylum, Gnetophyta (gnetophytes), is a collection of some unusual plants that share certain traits not found in other gymnosperms.

Conifers are woody plants that produce seeds in cones

The **conifers** (phylum Coniferophyta), which include pines, spruces, hemlocks, and firs, are the most familiar group of gymnosperms (▌ Fig. 28-3a). These woody trees or shrubs produce annual additions of secondary tissues (wood and bark; see Chapter 34); there are no herbaceous (nonwoody) conifers. The wood (*secondary xylem*) consists of **tracheids,** which are long, tapering cells with pits through which water and dissolved minerals move from one cell to another.

Many conifers produce **resin,** a viscous, clear or translucent substance consisting of several organic compounds that may protect the plant from attack by fungi or insects. The resin collects in resin ducts, tubelike cavities that extend throughout the roots, stems, and leaves. Cells lining the resin ducts produce and secrete resin.

Most conifers have leaves called **needles** that are commonly long and narrow, tough, and leathery (▌ Fig. 28-3b). Pines bear clusters of two to five needles, depending on the species. In a few conifers, such as American arborvitae, the leaves are scalelike and

Key Point

Seed plants include four gymnosperm phyla and one phylum of flowering plants (angiosperms).

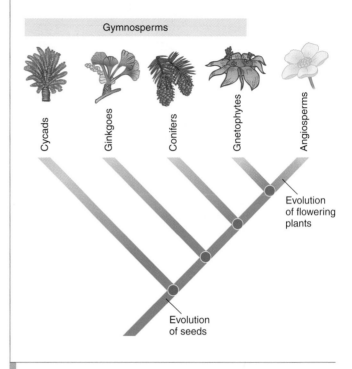

Figure 28-2 Gymnosperm and angiosperm evolution

This cladogram shows one hypothesis of phylogenetic relationships among living seed plants, based on structural evidence and molecular comparisons. The arrangement of the phyla shown here may change as future analyses help clarify relationships.

cover the stem (▌ Fig. 28-3c). Most conifers are evergreen and bear their leaves throughout the year. Only a few, such as the dawn redwood, larch, and bald cypress, are deciduous and shed their needles at the end of each growing season.

Most conifers are **monoecious:** they have separate male and female reproductive parts in different locations on the same plant. These reproductive parts are generally borne in *strobili* (commonly called *cones*), hence the name *conifer,* which means "cone-bearing."

Conifers occupy extensive areas, ranging from the Arctic to the tropics, and are the dominant vegetation in the forested regions of Alaska, Canada, northern Europe, and Siberia. In addition, they are important in the Southern Hemisphere, particularly in wet, mountainous areas of temperate and tropical regions in South America, Australia, New Zealand, and Malaysia. Southwestern China, with more than 60 species of conifers, has the greatest regional diversity of conifer species in the world. California, New Caledonia (an island east of Australia), southeastern China, and Japan also have considerable diversity of conifer species.

Ecologically, conifers contribute food and shelter to animals and other organisms, and their roots hold the soil in place and help prevent soil erosion. Humans use conifers for their wood (for building materials as well as paper products), medicinal

Figure 28-3 Conifers

(a) Colorado blue spruce (*Picea pungens*) is a coniferous evergreen tree native to Colorado, Wyoming, Utah, and New Mexico. Spruces are important ornamental plants, as shown in this New Jersey garden.

(b, c) Leaf variation in conifers. (b) Needles of white pine (*Pinus strobus*). (c) Small, scalelike leaves of American arborvitae (*Thuja occidentalis*).

value (such as the anticancer drug Taxol from the Pacific yew), turpentine, and resins. Because of their attractive appearance, conifers such as firs, spruces, pines, and cedars are grown for landscape design and decorative holiday trees and wreaths.

Pines represent a typical conifer life cycle

The genus *Pinus,* by far the largest genus in the conifers, consists of about 100 species. As shown in ▌Figure 28-4 ❶, a pine tree is a mature sporophyte. Pine is heterosporous and therefore produces microspores and megaspores in separate cones.[1] Male cones, usually 1 cm or less in length, are smaller than female cones and are generally produced on the lower branches each spring (▌Fig. 28-5). The more familiar, woody female cones, which are on the tree year-round, are usually found on the upper branches of the tree and bear seeds after reproduction. Female cones vary considerably in size. The sugar pine (*P. lambertiana*) that grows in California produces the world's longest female cones, which reach lengths of 60 cm (2 ft).

Each male cone, also called a *pollen cone,* consists of **sporophylls,** leaflike scales that bear sporangia on the underside. At the base of each sporophyll are two **microsporangia,** which contain numerous **microsporocytes,** also called *microspore mother cells.*

In Figure 28-4 ❷, each microsporocyte undergoes meiosis to form four haploid microspores. **Microspores** then develop into extremely reduced male gametophytes. Each immature male gametophyte, also called a **pollen grain,** consists of four cells, two of which—a *generative cell* and a *tube cell*—are involved in reproduction. The other two cells soon degenerate. Two large air sacs on each pollen grain provide buoyancy for wind dissemination. In ❸, male cones shed pollen grains in great numbers, and wind currents carry some to the immature female cones.

Many botanists think that the female cones (also called *seed cones*) are modified branch systems. As shown in ❹, each cone scale bears two ovules, or **megasporangia,** on its upper surface. Within each megasporangium, meiosis of a **megasporocyte,** or *megaspore mother cell,* produces four haploid **megaspores.** One of these divides mitotically, developing into the female gametophyte, which produces an egg within each of several archegonia. The other three megaspores are nonfunctional and soon degenerate.

When the ovule is ready to receive pollen, it produces a sticky droplet at the opening where the pollen grains land. **Pollination,** the transfer of pollen to the female cones, occurs in the spring for a week or 10 days, after which the pollen cones wither and drop off the tree. One of the many pollen grains that adhere to the sticky female cone grows a **pollen tube,** an outgrowth that digests its way through the megasporangium to the egg within the archegonium. The germinated pollen grain with its pollen tube is the mature male gametophyte. The tube cell, which is involved

---

[1]It may be helpful to review alternation of generations in Chapter 27, including Figure 27-16, which depicts a heterosporous life cycle, before studying the pine life cycle.

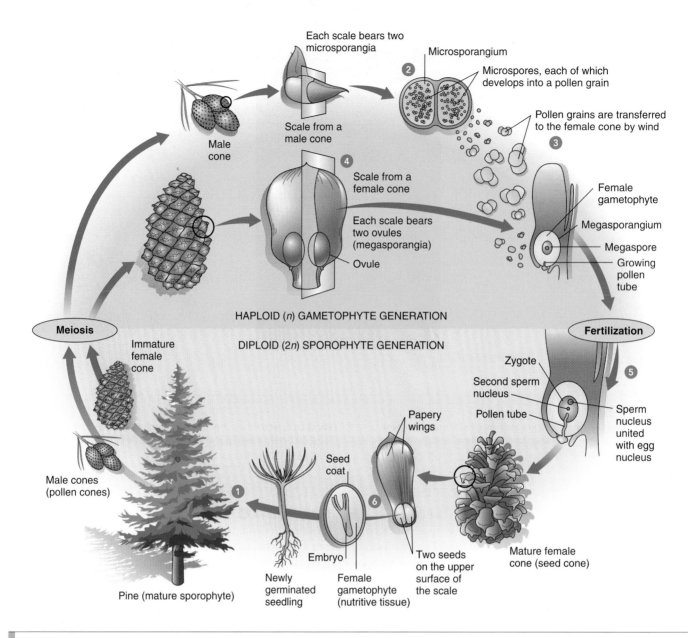

Each scale bears two
microsporangia

Microsporangium

2

Microspores, each of which
develops into a pollen grain

Scale from a
male cone

Pollen grains are transferred
to the female cone by wind

3

Male
cone

4

Scale from a
female cone

Female
gametophyte

Megasporangium

Each scale bears
two ovules
(megasporangia)

Megaspore

Growing
pollen
tube

Ovule

HAPLOID (*n*) GAMETOPHYTE GENERATION

Meiosis

DIPLOID (2*n*) SPOROPHYTE GENERATION

Fertilization

Immature
female
cone

Zygote

Second sperm
nucleus

Papery
wings

Pollen tube

5

Sperm
nucleus
united
with egg
nucleus

Seed
coat

1

6

Male cones
(pollen cones)

Embryo

Two seeds
on the upper
surface of
the scale

Mature female
cone (seed cone)

Pine (mature sporophyte)

Newly
germinated
seedling

Female
gametophyte
(nutritive tissue)

Figure 28-4 *Animated* Life cycle of pine

One major advantage of gymnosperms over the seedless vascular plants is the production of
wind-borne pollen grains. See text for a detailed description.

in the growth of the pollen tube, and the generative cell enter the
pollen tube.

The generative cell divides and forms a *stalk cell* and a *body
cell;* the body cell later divides and forms two nonmotile (non-
flagellate) sperm cells. When it reaches the female gametophyte,
the pollen tube discharges the two sperm cells near the egg. In ⑤,
one of these sperm cells fuses with the egg, in the process of **fertilization,** to form a zygote, or fertilized egg, which subsequently
grows into a young pine embryo in the seed. The other sperm cell
degenerates.

In ⑥, the developing embryo, which consists of an embryonic
root and an embryonic shoot with several cotyledons (embryonic
leaves), is embedded in haploid female gametophyte tissue that
becomes the nutritive tissue in the mature pine seed. A tough,
protective seed coat derived from the integuments surrounds the

embryo and nutritive tissue. The seed coat forms a thin, papery
wing at one end that enables dispersal by air currents.

A long time elapses between the appearance of female cones
on a tree and the maturation of seeds in those cones. When pollination occurs during the first spring, the female cone is still immature, and meiosis of the megasporocytes (megaspore mother
cells) has not yet occurred. After the megaspore has formed, it
takes more than a year for eggs to form within archegonia. Meanwhile, the pollen tube grows slowly through the megasporangium
to the archegonia. About 15 months after pollination, fertilization
occurs and the embryo begins to develop. Seed maturation takes
several additional months, although some seeds remain within
the female cones for several years before being shed.

In the pine life cycle, the sporophyte generation is dominant,
and the gametophyte generation is restricted in size to micro-

Figure 28-5 *Animated* Male and female cones in lodgepole pine (*Pinus contorta*)

(*Top*) Mature female cones have opened to shed their seeds. (*Bottom*) Clusters of male cones produce copious amounts of pollen grains in the spring.

scopic structures in the cones. Although the female gametophyte produces archegonia, the male gametophyte is so reduced that it does not produce antheridia. The gametophyte generation in pines, as in all seed plants, depends totally on the sporophyte generation for nourishment.

A major adaptation in the pine life cycle is elimination of the need for external water as a sperm transport medium. Instead, air currents carry pine pollen grains to female cones, and non-flagellate sperm cells accomplish fertilization by moving through a pollen tube to the egg. Pine and other conifers are plants whose reproduction is totally adapted for life on land.

Cycads have seed cones and compound leaves

Cycads (phylum Cycadophyta) were very important during the Triassic period, which began about 251 million years ago (mya) and is sometimes referred to as the "Age of Cycads." Most species are now extinct, and the few surviving cycads, about 140 species, are tropical and subtropical plants with stout, trunklike stems and compound leaves that resemble those of palms or tree ferns (Fig. 28-6). Many cycads are endangered, primarily because they are popular as ornamentals and are gathered from the wild and sold to collectors.

Cycad reproduction is similar to that in pines except that cycads are **dioecious** and therefore have seed cones on female plants and pollen cones on male plants. Their seed structure is most like that of the earliest seeds found in the fossil record. Cycads have also retained motile sperm cells, each of which has many hair-like flagella. Motile sperm cells are a vestige retained from the ancestors of cycads, in which sperm cells swam from antheridia to archegonia. In cycads, air or insects carry pollen grains to the female plants and their cones; there the pollen grain germinates and grows a pollen tube. The sperm cells are released at the top of this tube and swim to the egg.

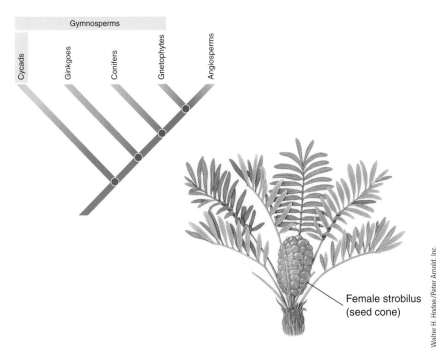

(a) A female Florida arrowroot (*Zamia integrifolia*) produces seed cones. This plant is the only cycad native to the United States. Like *Zamia*, most cycads are short plants, less than 2 m (6.5 ft) tall.

(b) This female cycad (*Encephalartos transvenosus*) in South Africa has a trunk that reaches a height of about 9 m (30 ft) and resembles a palm. Note its immense seed cones, to 0.8 m (30 in) in length.

Figure 28-6 Cycads

Ginkgo biloba is the only living species in its phylum

Ginkgo (phylum Ginkgophyta) is represented by a single living species, the maidenhair tree, *Ginkgo biloba* (▌Fig. 28-7). This native of eastern China grew in the wild in only two locations, although people had cultivated it for its edible seeds in China and Japan for centuries. *Ginkgo* is the oldest genus (and species) of living trees. Fossil ginkgoes 200 million years old have been discovered that are similar to the modern ginkgo.

People often plant ginkgo in North America and Europe today, particularly in parks and along city streets, because it is hardy and somewhat resistant to air pollution. Its leaves are deciduous and turn a beautiful yellow before being shed in the fall.

Like cycads, ginkgo is dioecious, with separate male and female trees. It has flagellate sperm cells—an evolutionary vestige that is not required, because ginkgo produces airborne pollen grains. Ginkgo seeds are completely exposed rather than contained within cones. Male trees are usually planted, because the female trees bear seeds whose fleshy outer coverings give off a foul odor that smells like rancid butter. In China and Japan, where people eat the seeds, the female trees are more common.

Ginkgo has been an important medicinal plant for centuries. Extracts from the leaves may enhance neurological functioning by increasing blood flow to the brain. Several studies are underway to determine if ginkgo improves memory in elderly people.

Gnetophytes include three unusual genera

The **gnetophytes** (phylum Gnetophyta) consist of about 70 species in three diverse genera (*Gnetum*, *Ephedra*, and *Welwitschia*). Gnetophytes share certain features that make them unique among the gymnosperms. For example, gnetophytes have more efficient water-conducting cells, called *vessel elements,* in their xylem (see Chapter 32). Flowering plants also have vessel elements in their xylem, but of the gymnosperms, only the gnetophytes do. In addition, the cone clusters that some gnetophytes produce resemble flower clusters, and certain details in their life cycles resemble those of flowering plants.

The genus *Gnetum* contains tropical vines, shrubs, and trees with broad leaves that resemble those of flowering plants (▌Fig. 28-8a). Species in the genus *Ephedra* include many shrubs and vines that grow in deserts and other dry temperate and tropical regions. Some *Ephedra* species resemble horsetails in that they have jointed green stems with tiny leaves (▌Fig. 28-8b). Commonly called *joint fir, Ephedra* has been used medicinally for centuries. An Asiatic *Ephedra* is the source of ephedrine, which stimulates the heart and raises blood pressure. Ephedrine is sold

(a) Close-up of a branch from a female ginkgo, showing the exposed seeds and the distinctive, fan-shaped leaves.

(b) Ginkgo (*Ginkgo biloba*) tree in a formal garden in northern England.

Figure 28-7 Ginkgo, or maidenhair tree

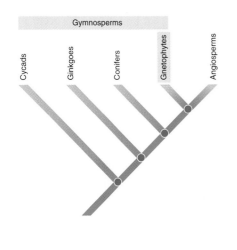

Gymnosperms

Cycads · Ginkgoes · Conifers · Gnetophytes · Angiosperms

(a) The leaves of *Gnetum gnemon* resemble those of flowering plants. Note the exposed seeds.

John D. Cunningham / Visuals Unlimited

(b) A male joint fir (*Ephedra*) has pollen cones clustered at the nodes. In the 19th century, European pioneers used species native to the American Southwest to make a beverage, Mormon tea.

David Cavagnaro

Figure 28-8 Gnetophytes

over the counter in weight-control medications and herbal energy boosters; several deaths have been reported from chronic use or overdose of products containing ephedrine.

The third gnetophyte genus, *Welwitschia*, contains a single species found in deserts of southwestern Africa (▌ Fig. 28-8c). Most of *Welwitschia*'s body—a long taproot—grows underground. Its short, wide stem forms a shallow disc, up to 0.9 m (3 ft) in diameter, from which two ribbonlike leaves extend. These two leaves continue to grow from the stem throughout the plant's life, but their ends are usually broken and torn by the wind, giving the appearance of numerous leaves. Each leaf grows to about 2 m (6.5 ft) in length. When *Welwitschia* reproduces, cones form around the edge of its disclike stem.

Review

▌ What is the dominant generation in the pine life cycle? How does pollination occur in gymnosperms?

▌ What features distinguish gymnosperms from other plants?

▌ What are the four groups of gymnosperms?

▌ What features distinguish cycads from ginkgo? From gnetophytes?

FLOWERING PLANTS

Learning Objectives

5   Summarize the features that distinguish flowering plants from other plants.

6   Briefly explain the life cycle of a flowering plant, and describe double fertilization.

7   Contrast eudicots and monocots, the two largest classes of flowering plants.

8   Discuss the evolutionary adaptations of flowering plants.

Robert and Linda Mitchell

(c) *Welwitschia mirabilis* is native to deserts in southwestern Africa. It survives on moisture-laden fogs that drift inland from the ocean. Photographed in the Namib Desert, Namibia.

Flowering plants, or angiosperms (phylum Anthophyta), are the most successful plants today, surpassing even the gymnosperms in importance. They have adapted to almost every habitat and, with at least 300,000 species, are Earth's dominant plants. Flowering plants come in a wide variety of sizes and forms, from herbaceous violets to massive eucalyptus trees. Some flowering plants—tulips and roses, for example—have large, conspicuous flowers; others, such as grasses and oaks, produce small, inconspicuous flowers.

Flowering plants are vascular plants that reproduce sexually by forming flowers and, after a unique double fertilization process (discussed later), seeds within fruits. The fruit protects the developing seeds and often aids in their dispersal (see Chapter 36). Flowering plants have efficient water-conducting cells called **ves-**

sel elements in their xylem, and efficient sugar-conducting cells called **sieve tube elements** in their phloem (see Chapter 32).

Flowering plants are extremely important to humans because our survival as a species literally depends on them. All our major food crops are flowering plants, including cereal crops such as rice, wheat, corn, and barley. Woody flowering plants such as oak, cherry, and walnut provide valuable lumber. Flowering plants give us fibers such as cotton and linen and medicines such as digitalis and codeine. Products as diverse as rubber, tobacco, coffee, chocolate, and aromatic oils for perfumes come from flowering plants. **Economic botany** is the subdiscipline of botany that deals with plants of economic importance.

Monocots and eudicots are the two largest classes of flowering plants

Phylum Anthophyta is divided into several classes with only a few members each and two very large classes: the monocots (class Monocotyledones) and the eudicots (class Eudicotyledones) (▌Fig. 28-9). The smaller classes will be discussed later in the chapter in the context of their evolutionary significance; for now, we restrict our discussion of flowering plants to the monocots and eudicots, which collectively represent about 97% of all flowering plant species. Eudicots are more diverse and include many more species (at least 200,000) than the monocots (at least 90,000). ▌Table 28-2 provides a comparison of some of the general features of the two classes.

Monocots include palms, grasses, orchids, irises, onions, and lilies. Monocots are mostly herbaceous plants with long, narrow leaves that have parallel veins (the main leaf veins run parallel to one another). The parts of monocot flowers usually occur in threes. For example, a flower may have three sepals, three petals, six stamens, and a compound pistil consisting of three fused carpels (these flower parts are discussed shortly). Monocot seeds have a single **cotyledon,** or embryonic seed leaf; **endosperm,** a nutritive tissue, is usually present in the mature seed.

Eudicots include oaks, roses, mustards, cacti, blueberries, and sunflowers. Eudicots are either herbaceous (such as a tomato plant) or woody (such as a hickory tree). Their leaves vary in shape but usually are broader than monocot leaves, with netted veins (branched veins resembling a net). Flower parts usually occur in fours or fives or multiples thereof. Two cotyledons are present in eudicot seeds, and endosperm is usually absent in the mature seed, having been absorbed by the two cotyledons during seed development.

Flowers are involved in sexual reproduction

Flowers are reproductive shoots usually composed of four parts —sepals, petals, stamens, and carpels—arranged in whorls (circles) on the end of a flower stalk, or **peduncle** (▌Fig. 28-10). The peduncle may terminate in a single flower or a cluster of flowers known as an **inflorescence.** The tip of the flower stalk that bears the flower parts is known as the **receptacle.**

All four floral parts are important in the reproductive process, but only the stamens (the "male" organs) and carpels (the "female" organs) produce gametes. A flower that has all four parts is **complete,** whereas an **incomplete** flower lacks one or more of these four parts. A flower with both stamens and carpels is **perfect,** whereas an **imperfect** flower has stamens or carpels, but not both.

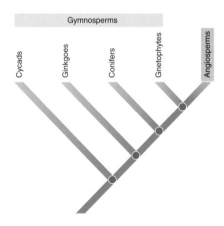

Figure 28-9 Flowering plants

John Gerlach / Tom Stack & Associates

(a) Monocot. *Trillium erectum,* like most monocots, has floral parts in threes. Note the three green sepals, three red petals, six stamens, and three stigmas (the compound pistil consists of three fused carpels).

Richard H. Gross

(b) Eudicot. Most eudicots, such as this *Tacitus,* have floral parts in fours or fives. Note the five petals, 10 stamens, and five separate pistils. Five sepals are also present but barely visible against the background.

Darwin Dale/Photo Researchers, Inc.

(a)

Figure 28-10 *Animated* Floral structure

(a) An *Arabidopsis thaliana* flower. **(b)** This cutaway view of an *Arabidopsis* flower shows the details of basic floral structure. Each flower has four sepals (*two are shown*), four petals (*two are shown*), six stamens, and one long pistil. Four of the stamens are long, and two are short (*two long and two short are shown*). Pollen grains develop within sacs in the anthers. In *Arabidopsis*, the compound pistil consists of two carpels that each contain numerous ovules. This flower is both a complete and a perfect flower.

Female floral parts

Stigma
Style
PISTIL (consisting of one or more carpels)
Ovary
Ovules (each producing one egg cell)

Male floral parts

Pollen grain (each will produce two sperm cells)
Anther
Filament
STAMEN
Petal
Sepal
Receptacle
Peduncle

(b)

TABLE 28-2

Distinguishing Features of Monocots and Eudicots

Feature	Monocots	Eudicots
Flower parts	Usually in threes	Usually in fours or fives
Pollen grains	One furrow or pore	Three furrows or pores
Leaf venation	Usually parallel	Usually netted
Vascular bundles in stem cross section	Usually scattered or more complex arrangement	Arranged in a circle (ring)
Roots	Fibrous root system	Taproot system
Seeds	Embryo with one cotyledon	Embryo with two cotyledons
Secondary growth (wood and bark)	Absent	Often present

Petals   Sepals

(a) The leaflike sepals of a rosebud (*Rosa*) enclose and protect the inner flower parts.

James Mauseth /University of Texas

Figure 28-11   Parts of a flower

Sepals, which make up the lowermost and outermost whorl on a floral shoot, are leaflike in appearance and often green (❘ Fig. 28-11a). Sepals cover and protect the other flower parts when the flower is a bud. As the blossom opens, the sepals fold back to reveal the more conspicuous petals. The collective term for all the sepals of a flower is the **calyx.**

The whorl just above the sepals consists of **petals,** which are broad, flat, and thin (like sepals and leaves) but vary in shape and are frequently brightly colored. Petals play an important role in attracting animal pollinators to the flower (see Chapter 36). Sometimes petals are fused to form a tube (as in trumpet honeysuckle flowers) or other floral shape (as in snapdragons, whose petals form two lips). The petals of a flower are referred to collectively as the **corolla.**

Just inside the petals is a whorl of **stamens** (❘ Fig. 28-11b). Each stamen is composed of a thin stalk, called a **filament,** and a saclike **anther,** where meiosis occurs to form microspores that develop into pollen grains. Each pollen grain produces two cells surrounded by a tough outer wall. One cell eventually divides to form two male gametes, or sperm cells, and the other produces a pollen tube through which the sperm cells travel to reach the ovule.

In the center of most flowers is one or more closed **carpels,** the "female" reproductive organs. Carpels bear ovules, which, as you may recall, are structures with the potential to develop into seeds. The carpels of a flower are separate or fused into a single structure. The female part of the flower is also called a **pistil** (see Fig. 28-11b). A pistil may consist of a single carpel (a simple pistil) or a group of fused carpels (a compound pistil) (❘ Fig. 28-12). Each pistil generally has three sections: a **stigma,** on which the pollen grain lands; a **style,** a necklike structure through which the pollen tube grows; and an **ovary,** an enlarged structure that contains one or more ovules. Each young ovule contains a female gametophyte that forms one female gamete (an egg), two *polar nuclei,* and several other haploid cells. After fertilization, the ovule develops into a seed; and the ovary, into a fruit.

Marion Lobstein

Stamen

Pistil

(b) A twinleaf (*Jeffersonia diphylla*) flower has eight yellow stamens. Note the simple pistil with its green ovary in the center of the flower.

The life cycle of flowering plants includes double fertilization

Flowering plants undergo alternation of generations in which the sporophyte generation is larger and nutritionally independent. The gametophyte generation in flowering plants is microscopic and nutritionally dependent on the sporophyte. Flowering plants, like gymnosperms and certain other vascular plants, are heterosporous and produce two kinds of spores: microspores and megaspores. As shown in ❘ Figure 28-13 ❶, sexual reproduction occurs in the flower.

In ❷, each young ovule within an ovary contains a megasporocyte (megaspore mother cell) that undergoes meiosis to produce four haploid megaspores. Three of these usually disintegrate, and one divides mitotically and ❸ develops into a mature female gametophyte, also called an **embryo sac.** The most widely studied type of embryo sac contains seven cells with eight haploid nuclei. Six of these cells, including the egg cell, contain a single nucleus each, and a central cell has two nuclei, called **polar nuclei.** The egg and the central cell with two polar nuclei are directly involved in fertilization; the other five cells in the embryo sac apparently have no direct role in the fertilization process and disintegrate. As the *synergids* (the two cells closely associated with the egg) disintegrate, however, they release chemicals that may affect the direction of pollen tube growth.

Each pollen sac, or microsporangium, of the anther contains numerous microsporocytes (microspore mother cells) ❹, each of which undergoes meiosis to form four haploid microspores. Every microspore develops into ❺ an immature male gameto-

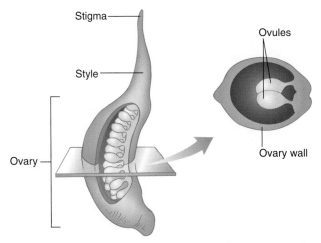

(a) Simple pistil. This simple pistil consists of a single carpel.

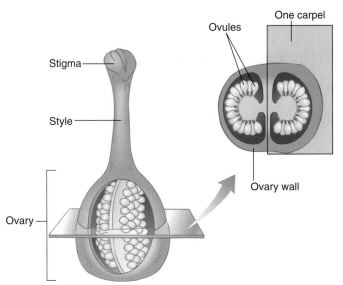

(b) Compound pistil. This compound pistil has two united carpels. In most flowers with single pistils, the pistils are compound, consisting of two or more fused carpels.

Figure 28-12 Simple and compound pistils

phyte, also called a *pollen grain.* Pollen grains are extremely small; each consists of two cells: the *tube cell* and the *generative cell.*

The anthers split open and begin to shed pollen. A variety of agents, including wind, water, insects, and other animal pollinators (see Chapter 36), transfer pollen grains to the stigma. As shown in ⑥, if compatible with the stigma, the pollen grain germinates; that is, the tube cell forms a pollen tube that grows down the style and into the ovary. The germinated pollen grain with its pollen tube is the mature male gametophyte. Next, the generative cell divides to form two nonflagellate sperm cells. The sperm cells move down the pollen tube and are discharged into the embryo sac. Both sperm cells are involved in fertilization.

Something happens during sexual reproduction in flowering plants that does not occur anywhere else in the living world. As shown in ⑦, when the two sperm cells enter the embryo sac, *both*

participate in fertilization. One sperm cell fuses with the egg, forming a zygote that grows by mitosis and develops into a multicellular embryo in the seed. The second sperm cell fuses with the two haploid polar nuclei of the central cell to form a triploid ($3n$) cell that grows by mitosis and develops into **endosperm,** a nutrient tissue rich in lipids, proteins, and carbohydrates that nourishes the growing embryo. This fertilization process, which involves two separate nuclear fusions, is called **double fertilization** and is, with two exceptions, unique to flowering plants. (Double fertilization has been reported in the gymnosperms *Ephedra nevadensis* and *Gnetum gnemon.* This process differs from double fertilization in flowering plants in that an additional zygote, rather than endosperm, is produced. The second zygote later disintegrates.)

Seeds and fruits develop after fertilization

As a result of double fertilization and subsequent growth and development, each seed contains a young plant embryo and nutritive tissue (the endosperm), both of which are surrounded by a protective seed coat. In monocots the endosperm persists and is the main source of food in the mature seed. In most eudicots the endosperm nourishes the developing embryo, which subsequently stores food in its cotyledons.

As a seed develops from an ovule following fertilization (Fig. 28-13 ⑧), the ovary wall surrounding it enlarges dramatically and develops into a **fruit.** In some instances, other tissues associated with the ovary also enlarge to form the fruit. Fruits serve two purposes: to protect the developing seeds from desiccation as they grow and mature and to aid in the dispersal of seeds (see Chapter 36). For example, dandelion fruits have feathery plumes that are lifted and carried by air currents. Animals often assist in dispersing seeds found in edible fruits (❙ Fig. 28-14). Once a seed lands in a suitable place, it may germinate and develop into a mature sporophyte that produces flowers, and the life cycle continues as described.

Flowering plants have many adaptations that account for their success

The evolutionary adaptations of flowering plants account for their success in terms of their ecological dominance and their great number of species. Seed production as the primary means of reproduction and dispersal, an adaptation shared with the gymnosperms, is clearly significant and provides a definite advantage over seedless vascular plants. Closed carpels, which give rise to fruits surrounding the seeds, and the process of double fertilization increase the likelihood of reproductive success. The evolution of a variety of interdependencies with many types of insects, birds, and bats, which disperse pollen from one flower to another of the same species, is another reason for angiosperm success. Pollen transfer results in cross-fertilization, which mixes the genetic material and promotes genetic variation among the offspring.

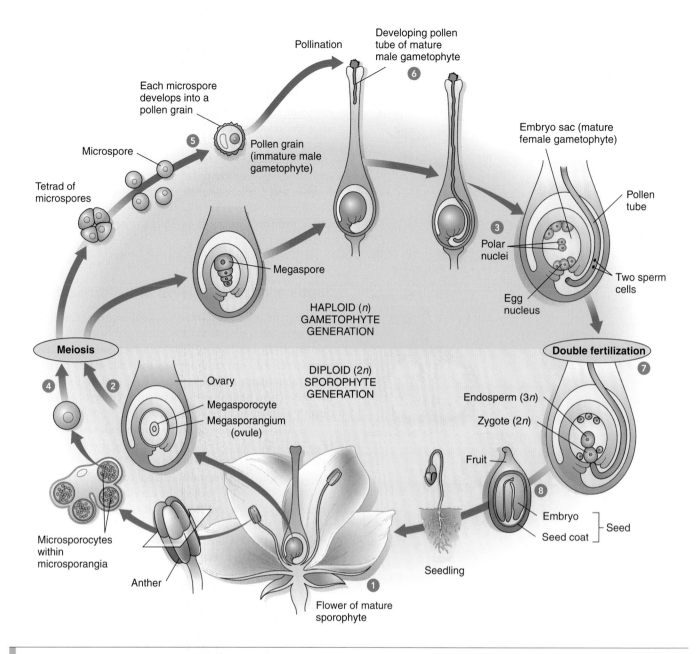

Pollination

Developing pollen
tube of mature
male gametophyte

6

Each microspore
develops into a
pollen grain

5

Microspore

Pollen grain
(immature male
gametophyte)

Tetrad of
microspores

Embryo sac (mature
female gametophyte)

Pollen
tube

Megaspore

HAPLOID (*n*)
GAMETOPHYTE
GENERATION

Polar
nuclei

3

Two sperm
cells

Egg
nucleus

Meiosis

DIPLOID (2*n*)
SPOROPHYTE
GENERATION

Double fertilization

4

2

Ovary

Megasporocyte

Megasporangium
(ovule)

Endosperm (3*n*)

Zygote (2*n*)

7

Fruit

Microsporocytes
within
microsporangia

Embryo

Seed coat

Seed

8

Anther

Seedling

1

Flower of mature
sporophyte

Figure 28-13 *Animated* Life cycle of flowering plants

A significant feature of the flowering plant life cycle is double fertilization, in which one sperm cell unites with the egg, forming a zygote, and the other sperm cell unites with the two polar nuclei, forming a triploid cell that gives rise to endosperm. See text for a detailed description.

Several distinctive features have contributed to the success of flowering plants in addition to their highly successful reproduction involving flowers, fruits, and seeds. Recall that most flowering plants have very efficient water-conducting vessel elements in their xylem, as well as tracheids. In contrast, the xylem of almost all seedless vascular plants and gymnosperms consists exclusively of tracheids. Most flowering plants also have efficient carbohydrate-conducting sieve tube elements in their phloem. Vascular plants other than flowering plants and gnetophytes lack vessel elements and sieve tube elements.

The leaves of flowering plants, with their broad, expanded blades, are very efficient at absorbing light for photosynthesis.

Abscission (shedding) of these leaves during cold or dry periods reduces water loss and has enabled some flowering plants to expand into habitats that would otherwise be too harsh for survival. The stems and roots of flowering plants are often modified for food or water storage, another feature that helps flowering plants survive in severe environments.

Probably most crucial to the evolutionary success of flowering plants, however, is the overall adaptability of the sporophyte generation. As a group, flowering plants readily adapt to new habitats and changing environments. This adaptability is evident in the great diversity exhibited by the various species of flowering plants. For example, the cactus is remarkably well adapted to des-

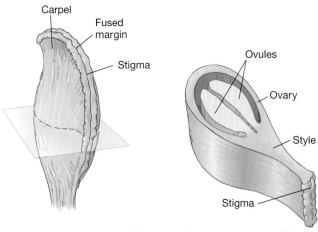

(a) The carpel resembles a folded leaf in which the ovules borne on its upper surface are enclosed.

(b) A cross section of the carpel, cut along the dashed line in **(a)**.

Figure 28-15   Carpel of *Drimys piperita*

Sepals are the most leaflike of the four floral organs, and botanists generally agree that sepals are specialized leaves. Although petals of many flowering plant species are leaflike in appearance, botanists generally view petals as modified stamens that later became sterile and leaflike. Cultivated roses and camellias provide evidence supporting this hypothesis; in some varieties the stamens have been transformed into petals, forming showy flowers with large numbers of petals.

The remarkably leaflike stamens and carpels of certain tropical trees and other species support the origin of stamens and carpels from leaves or leaflike organs. Consider, for example, the carpel of *Drimys*, a genus of flowering trees and shrubs native to Southeast Asia, Australia, and South America. This carpel resembles a leaf that is folded inward along the midrib, thereby enclosing the ovules, and joined along the entire length of the leaf's margin (Fig. 28-15).

The fundamental question is whether these leaflike stamens and carpels are early organs that were conserved (retained) during the course of evolution or are highly specialized organs that do not resemble early stamens and carpels. Many botanists who have studied this question have concluded that stamens and carpels are probably derived from leaves. Not all botanists accept the origin of stamens and carpels from highly modified leaves, however. As we noted in Chapter 1, uncertainty and debate are part of the scientific process, and scientists can never claim to know a final answer.

During the course of more than 130 million years of angiosperm evolution, flower structure diversified as floral organs fused or became reduced in size or number. These changes led to greater complexity in floral structure in some species and to greater simplicity in other species. Interpreting the floral structures of so many different angiosperm species is sometimes difficult, but it is important because correct interpretations are essential to devising a phylogenetic classification scheme.

Figure 28-14   Guava fruit and seeds

Animals eat fleshy fruits such as guava. The seeds are frequently swallowed whole and pass unharmed through the animals' digestive tracts.

ert environments. Its stem stores water; its leaves (spines) have a reduced surface area available for transpiration (loss of water vapor; see Chapter 33) and may also protect against thirsty herbivorous animals; and its thick, waxy cuticle reduces water loss. In contrast, the water lily is well adapted for wet environments, in part because it has air channels that provide adequate oxygen to stems and roots living in oxygen-deficient water and mud.

Studying how flowers evolved provides insights into the evolutionary process

In evolution, new structures or organs often originate by modification of previously existing structures or organs (see Chapter 20). Much evidence supports the classical interpretation that the four organs of a flower—sepals, petals, stamens, and carpels—arose from highly modified leaves. This evidence includes comparisons of the arrangement of vascular tissues in both flowers and leafy stems and of the developmental stages of floral parts and leaves.

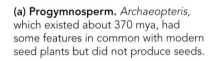

(a) Progymnosperm. *Archaeopteris,* which existed about 370 mya, had some features in common with modern seed plants but did not produce seeds.

(b) Seed fern. *Emplectopteris* produced seeds on fernlike leaves. Seed ferns existed from about 360 mya to 250 mya.

Figure 28-16   Evolution of seed plants

(**a**, Redrawn from C. B. Beck, "Reconstructions of *Archaeopteris* and Further Consideration of Its Phylogenetic Position," *American Journal of Botany,* Vol. 49, 1962. **b,** Redrawn from H. N. Andrews, *Ancient Plants and the World They Lived In,* Comstock, New York, 1947.)

Review

- How do nonreproductive adaptations of flowering plants differ from those of gymnosperms?
- How does the flowering plant life cycle differ from that of the gymnosperms?
- What are the two classes of flowering plants, and how can one distinguish between them?
- How does pollination occur in flowering plants?
- How does fertilization differ in gymnosperms and flowering plants?

THE EVOLUTION OF SEED PLANTS

Learning Objective

9   Summarize the evolution of gymnosperms from seedless vascular plants, and trace the evolution of flowering plants from gymnosperms.

One group that descended from ancestral seedless vascular plants was the **progymnosperms,** all of which are now extinct. Progymnosperms had two derived features absent in their immediate ancestors: leaves with branching veins (*megaphylls*) and woody tissue (*secondary xylem*) similar to that of modern gymnosperms. Pro-

gymnosperms, however, reproduced by spores, not seeds. *Archaeopteris,* a progymnosperm that lived about 370 mya, is the earliest known tree with "modern" woody tissue (❚ Fig. 28-16a).

Fossils of several progymnosperms with reproductive structures intermediate between those of spore plants and seed plants have been discovered. For example, the evolution of microspores into pollen grains and of megasporangia into ovules (seed-producing structures) can be traced in fossil progymnosperms. Plants producing seeds appeared during the late Devonian period, more than 359 mya. The fossil record indicates that different groups of seed plants apparently arose independently several times.

As mentioned previously, fossilized remains of ginkgo are found in 200-million-year-old rocks, and other groups of gymnosperms were well established by 160 mya to 100 mya. Although the gymnosperms are an ancient group, some questions persist about the exact pathways of gymnosperm evolution. The fossil record indicates that progymnosperms probably gave rise to conifers and to another group of extinct plants called **seed ferns,** which were seed-bearing woody plants with fernlike leaves (❚ Fig. 28-16b). The seed ferns, in turn, probably gave rise to cycads and possibly ginkgo, as well as to several gymnosperm groups now extinct. The origin of gnetophytes remains unclear, although molecular data indicate that they are closely related to conifers.

Our understanding of the evolution of flowering plants has made great progress in recent years

Flowering plants are the most recent group of plants to evolve. The fossil record, although incomplete, suggests that flowering plants descended from gymnosperms. By the middle of the Jurassic period, about 180 mya, several gymnosperm lines existed with some features resembling those of flowering plants. Among other traits, these derived gymnosperms possessed leaves with broad, expanded blades and the first modified seed-bearing leaves, which nearly enclose the ovules. Beetles were evidently visiting these plants, and biologists have suggested that perhaps this relationship was the beginning of **coevolution,** a mutual adaptation between plants and their animal pollinators (see Chapter 36).

One important task facing paleobotanists (biologists who study fossil plants) is determining which of the ancient gymnosperms are in the direct line of evolution leading to the flowering plants. Given the structural data, most botanists hypothesize that flowering plants arose only once—that is, there is only one line of evolution from the gymnosperms to the flowering plants. The gnetophytes are the gymnosperm group considered by some

Carpel

Ovule

5 mm

Figure 28-17 The oldest known fossil angiosperm

This fossil of *Archaefructus* shows a carpel-bearing stem. Discovered in northeastern China, it is about 125 million years old.

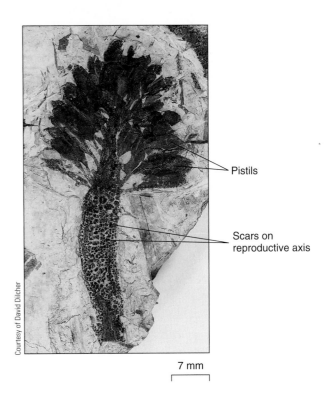

Pistils

Scars on reproductive axis

7 mm

Figure 28-18 Fossil flower

The fossilized flower of the extinct plant *Archaeanthus linnenbergeri*, which lived about 100 mya. The scars on the reproductive axis (receptacle) may show where stamens, petals, and sepals were originally attached but abscised (fell off). Many spirally arranged pistils were still attached at the time this flower was fossilized.

botanists to be the closest living relatives of flowering plants; both structural similarities and certain comparative molecular data support this conclusion. Like angiosperms, gnetophytes have vessels, lack archegonia, have flowerlike compound strobili, and undergo double fertilization. However, some molecular data refute a close link between the gnetophytes and flowering plants. It is hoped additional studies will clarify the relationships among the various gymnosperm phyla and the flowering plants.

The oldest definitive trace of flowering plants in the fossil record consists of ovules enclosed in tiny podlike fruits interpreted as carpels in Jurassic and Lower Cretaceous rocks some 125 million to 145 million years old (Fig. 28-17). The oldest fossilized flowers are about 118 million to 120 million years old. Based on the fossil record as well as structural and molecular data of living angiosperms, the first flowering plants were probably small, weedy shrubs or herbaceous plants adapted to disturbed habitats. They may have been fragile plants that were not easily preserved. If so, the reason there are few fossils of early angiosperms may be that the environment in which they evolved was repeatedly disturbed and not favorable for their preservation as fossils.

Botanists hypothesize that the rapid diversification of angiosperms did not occur until early flowering plants had invaded lowland regions. By 90 mya, during the Cretaceous period, flowering plants had diversified and had begun to replace gymnosperms as Earth's dominant plants. Fossils of flowering plant leaves, stems, flowers, fruits, and seeds are numerous and diverse. They outnumber fossils of gymnosperms and ferns in Late Cre-

taceous deposits, indicating the rapid success of flowering plants once they appeared (Fig. 28-18). Many angiosperm species apparently arose from changes in chromosome number (see discussion of allopolyploidy and sympatric evolution in Chapter 20).

The basal angiosperms comprise three clades

Recent cladistic analyses of structural features and molecular comparisons have helped clarify relationships among the angiosperm classes. The evidence indicates that three clades of basal angiosperms evolved before the divergence of core angiosperms (Fig. 28-19a). **Basal angiosperms** consist of about 170 species thought to be ancestral to all other flowering plants.

The oldest surviving clade of basal angiosperms is represented by a single living species, *Amborella trichopoda* (Fig. 28-19b). A shrub native to New Caledonia, an island in the South Pacific, *Amborella* may be the nearest living relative to the ancestor of all flowering plants. The water lilies and related families compose the second clade of basal angiosperms (Fig. 28-19c). This clade, which contains about 70 species of aquatic or wetland herbs, may be the second-oldest surviving lineage. Star anise and relatives—the third clade of basal angiosperms—consist of about 100 species of vines, trees, and shrubs found mostly in warmer climates. Star anise is important economically because it is a source of spice and anise oil (Fig. 28-19d). Star anise is also used to make Tamiflu, one of the treatments for influenza.

There are six clades of angiosperms, three of basal angiosperms and three of core angiosperms. The monocot and eudicot clades have the greatest number of species.

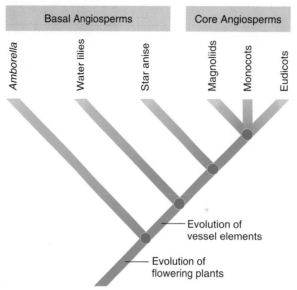

(a) One hypothesis of relationships among the flowering plants, based on fossil and molecular evidence. *Amborella*, water lilies, and star anise are living plants whose ancestors apparently branched off the angiosperm family tree early. These early clades were followed by the magnoliids, the monocot branch, and the eudicots.

(b) *Amborella trichopoda,* a basal angiosperm.

(c) Water lily (*Nymphaea*), a basal angiosperm.

(d) Star anise (*Illicium verum*), a basal angiosperm.

(e) *Magnolia grandiflora,* a core angiosperm.

Figure 28-19 Evolution of flowering plants

The core angiosperms comprise magnoliids, monocots, and eudicots

Most angiosperm species belong to a clade of **core angiosperms,** which is divided into three subclades: magnoliids, monocots, and eudicots. **Magnoliids** include species in the magnolia, laurel, and black pepper families, as well as several related families (∥ Fig. 28-19e). Although magnoliids have been traditionally classified with the eudicots as "dicots," molecular evidence such as DNA sequence comparisons indicates that the magnoliids are neither eudicots nor monocots. Native to tropical or warm temperate regions, magnoliids include several economically important plants, such as avocado, black pepper, nutmeg, and bay laurel.

Note in Figure 28-19a that the magnoliids, monocots, and eudicots are all depicted as arising from a single branch point. Exact relationships among these three subclades, each of which is probably monophyletic, remain to be clarified. In other words, at this time we cannot say with certainty which of the three groups was the first to branch off from the other two.

Review

▪ What features distinguish progymnosperms from seed ferns?

▪ Describe the significant features of the oldest known fossil angiosperm.

▪ Are monocots considered basal or core angiosperms? Explain your answer.

Learning Objectives

1 Compare the features of gymnosperms and angiosperms (page 601).

- The two groups of seed plants are the **gymnosperms** and the **angiosperms.** Gymnosperms produce seeds that are totally exposed or borne on the scales of cones; an ovary wall does not surround the ovules of gymnosperms. Angiosperms are flowering plants that produce their seeds within a fruit (a mature ovary).

2 Trace the steps in the life cycle of a pine, and compare its sporophyte and gametophyte generations (page 601).

- A pine tree is a mature sporophyte; pine gametophytes are extremely small and nutritionally dependent on the sporophyte generation. Pine is heterosporous and produces microspores and megaspores in separate cones.

- Male cones produce **microspores** that develop into **pollen grains** (immature male gametophytes) that are carried by air currents to female cones.

- Female cones produce **megaspores.** One of each four megaspores produced by meiosis develops into a female gametophyte within an **ovule (megasporangium).**

- After **pollination,** the transfer of pollen to the female cones, a **pollen tube** grows through the megasporangium to the egg within the archegonium. After **fertilization,** the zygote develops into an embryo encased inside a seed adapted for wind dispersal.

3 Summarize the features that distinguish gymnosperms from bryophytes and ferns (page 601).

- Unlike bryophytes, **gymnosperms** are vascular plants. Unlike bryophytes and ferns, gymnosperms produce seeds. Gymnosperms also produce wind-borne pollen grains, a feature that ferns and other seedless vascular plants lack.

4 Name and briefly describe the four phyla of gymnosperms (page 601).

- **Conifers** (phylum Coniferophyta), the largest phylum of gymnosperms, are woody plants that bear **needles** (leaves that are usually evergreen) and produce seeds in cones. Most conifers are **monoecious** and have male and female reproductive parts in separate cones on the same plant.

- **Cycads** (phylum Cycadophyta) are palmlike or fernlike in appearance. They are **dioecious**—they have male and female reproductive structures on separate plants—but reproduce with pollen and seeds in conelike structures.

- *Ginkgo biloba,* the only surviving species in phylum Ginkgophyta, is a deciduous, dioecious tree. The female **ginkgo** produces fleshy seeds directly on branches.

- **Gnetophytes** (phylum Gnetophyta) share a number of traits with angiosperms.

5 Summarize the features that distinguish flowering plants from other plants (page 607).

- **Flowering plants,** or angiosperms (phylum Anthophyta), constitute the phylum of vascular plants that produce flowers and seeds enclosed within a **fruit.** They are the most diverse and most successful group of plants.

- The flower, which may contain **sepals, petals, stamens,** and **carpels,** functions in sexual reproduction. Unlike those of gymnosperms, the ovules of flowering plants are enclosed within an **ovary.** After fertilization, the ovules become seeds, and the ovary develops into a fruit.

6 Briefly explain the life cycle of a flowering plant, and describe double fertilization (page 607).

- The sporophyte generation is dominant in flowering plants; gametophytes are extremely reduced in size and nutritionally dependent on the sporophyte generation. Flowering plants are heterosporous and produce microspores and megaspores within the flower.

- Each microspore develops into a pollen grain (immature male gametophyte). One of each four megaspores produced by meiosis develops into an **embryo sac** (female gametophyte). The embryo sac contains seven cells with eight nuclei; the egg cell and the central cell with two **polar nuclei** participate in fertilization.

- **Double fertilization,** which results in the formation of a diploid zygote and triploid **endosperm,** is characteristic of flowering plants.

ThomsonNOW™ **Explore plant life cycles by clicking on the figures in ThomsonNOW.**

7 Contrast eudicots and monocots, the two largest classes of flowering plants (page 607).

- Most **monocots** (class Monocotyledones) have floral parts in threes, and their seeds each contain one **cotyledon.** The nutritive tissue in their mature seeds is endosperm.

- **Eudicots** (class Eudicotyledones) usually have floral parts in fours or fives or multiples thereof, and their seeds each contain two cotyledons. The nutritive organs in their mature seeds are usually the cotyledons, which have absorbed the nutrients in the endosperm.

8 Discuss the evolutionary adaptations of flowering plants (page 607).

- Flowering plants reproduce sexually by forming flowers. After double fertilization, seeds form within fruits. Flowering plants have efficient water-conducting **vessel elements** in their xylem and efficient carbohydrate-conducting **sieve tube elements** in their phloem. Wind, water, insects, or other animals transfer pollen grains in various flowering plants.

ThomsonNOW™ **Learn more about flower structure by clicking on the figure in ThomsonNOW.**

9 Summarize the evolution of gymnosperms from seedless vascular plants, and trace the evolution of flowering plants from gymnosperms (page 614).

- Seed plants arose from seedless vascular plants. **Progymnosperms** were seedless vascular plants that had megaphylls and "modern" woody tissue. Progymnosperms probably gave rise to conifers as well as to **seed ferns,** which in turn likely gave rise to cycads and possibly ginkgo.

- The evolution of the gnetophytes, particularly their relationship to flowering plants, is unclear.

- Flowering plants probably descended from ancient gymnosperms that had specialized features, such as leaves with broad, expanded blades and closed carpels. Flowering plants likely arose only once.

1. Seed plants *lack* which of the following structure(s)? (a) ovules surrounded by integuments (b) microspores and megaspores (c) vascular tissues (d) a large, nutritionally independent sporophyte (e) a large, nutritionally independent gametophyte

2. Conifers, cycads, ginkgo, and gnetophytes are collectively called (a) club mosses (b) gymnosperms (c) angiosperms (d) eudicots (e) seedless vascular plants

3. Most conifers are _____, having male and female reproductive parts at different locations on the same plant. (a) incomplete (b) imperfect (c) monoecious (d) dioecious (e) perfect

4. The immature male gametophytes of pine are called (a) ovules (b) stamens (c) seed cones (d) pollen grains (e) polar nuclei

5. The transfer of pollen grains from the male to the female reproductive structure is known as (a) pollination (b) fertilization (c) embryo sac development (d) seed development (e) fruit development

6. Motile sperm cells are found as vestiges in these two gymnosperm groups: (a) monocots, eudicots (b) gnetophytes, conifers (c) gnetophytes, flowering plants (d) cycads, conifers (e) cycads, ginkgo

7. There are at least _____ species of flowering plants. (a) 235 (b) 3000 (c) 30,000 (d) 300,000 (e) 3,000,000

8. This class of flowering plants includes the palms, grasses, and orchids. (a) eudicots (b) gnetophytes (c) cycads (d) monocots (e) conifers

9. The pistil has three sections: (a) stigma, style, and anther (b) anther, filament, and ovule (c) stigma, style, and ovary (d) ovary, ovule, and sepal (e) corolla, stamen, and sepal

10. A simple pistil consists of a single (a) calyx (b) carpel (c) ovule (d) filament (e) petal

11. A flower that lacks stamens is both _____ and _____. (a) complete; imperfect (b) incomplete; perfect (c) complete; perfect (d) incomplete; imperfect

12. After fertilization, the _____ develop(s) into a fruit and the _____ develop(s) into a seed. (a) ovary; ovule (b) polar nuclei; ovule (c) ovary; endosperm (d) ovule; ovary (e) ovule; polar nuclei

13. The female gametophyte in flowering plants is also called the (a) polar nuclei (b) anther (c) embryo sac (d) endosperm (e) sporophyll

14. This flowering plant may be the nearest living relative to the ancestor of all flowering plants. (a) *Amborella* (b) *Archaeopteris* (c) *Gnetum* (d) water lily (e) *Archaeanthus*

15.  This cross section through an ovary reveals that the pistil is (a) simple, with one carpel (b) compound, with two fused carpels (c) compound, with three fused carpels (d) compound, with six separate carpels (e) compound, with six fused carpels

CRITICAL THINKING

1. How are cones and flowers alike? How are they different? (*Hint:* Your answer should consider microspores/megaspores and seeds.)

2. How do the life cycles of seedless plants (see Chapter 27) and seed plants differ? In what fundamental way are they alike?

3. **Evolution Link.** Most flowers contain both male and female reproductive structures, in contrast to the cones of gymnosperms, which are either male or female. Explain how bisexual flowers might be an advantageous evolutionary adaptation to the flowering plants that possess them.

4. **Evolution Link.** Contrast the algae, mosses, ferns, gymnosperms, and angiosperms with respect to their dependence on water as a transport medium for reproductive cells. Suggest a hypothesis to explain how the differences might be adaptive to living on land.

5. **Analyzing Data.** According to the cladogram in Figure 28-19, which plant(s) is/are the outgroup? Which feature does the outgroup lack that all other angiosperms possess?

Thomson™
NOW!

Additional questions are available in ThomsonNOW at www.thomsonedu.com/login

The Animal Kingdom:
An Introduction to Animal Diversity

Charles V. Angelo/Photo Researchers, Inc.

The tube sponge
(*Callyspongia vaginalis*).
This animal, which ranges in
color from purple to blue to
gray, is common on coral reefs
in the Caribbean, from Florida
to Mexico.

Animal phylogeny has become an exciting and rapidly changing field of study. Biologists have described and named more
than 1.5 million species of animals, and 15,000 to 20,000 new species are named each year. Millions more probably remain to be
discovered and classified. Interestingly, an estimated 99% of all
animal species that ever inhabited our planet are extinct. Biologists
have assigned the extant (living) members of **kingdom Animalia** to
about 35 phyla, but the relative positions of phyla and classes are
a work in progress. As they consider new data, systematists redraw
the tree of animal life.

Although most animal species are readily recognizable as animals, the identity of some others is less obvious. Early naturalists
thought sponges were plants, because they did not move from
place to place. Some people still mistake certain marine animals,
such as sponges and corals, for plants (see photograph). Locomotion is not a requirement for being classified as an animal.

In this chapter we discuss some of the criteria biologists use to
determine evolutionary relationships and to classify animals. Then
we describe three phyla that diverged early in the evolutionary history of this kingdom. In Chapters 30 and 31, we discuss two major
clades of animals: the protostomes and the deuterostomes. Many
hypotheses are presented in these chapters, and it is important
to remember that biologists continually respond to new data by
redrawing branches of the animal phylogenetic tree. ∎

KEY CONCEPTS

Animals are multicellular, eukaryotic heterotrophs.

Biologists classify animals based on their body plan and
features of their early development.

Molecular data indicate that bilateral animals split into three
major clades: two protostome groups—Lophotrochozoa
(such as flatworms, mollusks, and annelids) and Ecdysozoa
(such as nematodes and arthropods)—and deuterostomes
(echinoderms and chordates).

Sponges (phylum Porifera) are characterized by collar cells
and by loosely associated cells that do not form true tissues.

Members of phylum Cnidaria (hydras, jellyfish, sea anemones) are characterized by radial symmetry, two tissue layers,
and cnidocytes, cells that contain stinging organelles.

Members of phylum Ctenophora (comb jellies) have biradial
symmetry, two tissue layers, eight rows of cilia, and tentacles with adhesive glue cells.

ANIMAL CHARACTERS

Learning Objective

1   Describe several characters common to most animals.

Animals are so diverse that for almost any definition, we can find exceptions. However, the following characters describe most animals:

1. Animals are multicellular eukaryotes.

2. Animals are **heterotrophs.** As consumers, they depend on producers for their raw materials and energy. In contrast to the fungi, most animals ingest their food first and then digest it inside the body, usually within a digestive system.

3. Cells that make up the animal body are specialized to perform specific functions. In all but the simplest animals, cells are organized to form tissues, and tissues are organized to form organs. In small animals with simple body plans, life processes such as gas exchange, circulation of materials, and waste disposal can take place by diffusion of gases and other substances directly to and from the environment. In large, complex animals, specialized organ systems have evolved that perform these life processes.

4. Animals have diverse body plans. The term **body plan** refers to the basic structure and functional design of the body. An animal's body plan and lifestyle are adapted to its methods of obtaining food and reproducing.

5. Most animals are capable of locomotion at some time during their life cycle. Some animals (such as sponges and corals) move about as larvae (immature forms) but are **sessile** (firmly attached to the ground or some other surface) as adults (see chapter opening photograph).

6. Most animals have nervous systems and muscle systems that enable them to respond rapidly to stimuli in their environment.

7. Most animals are diploid organisms that reproduce sexually, with large, nonmotile eggs and small, flagellate sperm. A haploid sperm unites with a haploid egg, forming a **zygote** (fertilized egg).

8. Animals go through a period of embryonic development. The zygote undergoes **cleavage,** a series of mitotic cell divisions. During cleavage the zygote develops into a hollow ball of cells called a **blastula.** Cells of the blastula undergo **gastrulation,** a process of forming and segregating specific layers of tissue, called *germ layers.* Although some animals develop directly into adults, the majority first develop into a **larva,** a sexually immature form that may look very different from the adult (Fig. 29-1). The larva differs from the adult in many ways, including where it lives (its habitat), how it moves, and what it eats. Larvae typically go through **metamorphosis,** a developmental process that converts the immature animal into a juvenile form that can then grow into an adult.

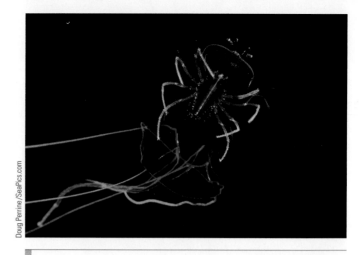

Doug Perrine/SeaPics.com

Figure 29-1   Larva

Spiny lobster larva (*Panulirus* sp.) hitching a ride on a jellyfish. Photographed in open ocean at night, Hawaii.

Review

■   For centuries, scientists classified sponges as plants, but now they are classified as animals. What criteria do biologists use to classify them as animals? (*Hint:* What is their mode of nutrition? Also, consider their larval stage.)

ADAPTATIONS TO HABITATS

Learning Objective

2   Compare the advantages and disadvantages of life in the ocean, in fresh water, and on land.

Fossil evidence suggests that animals evolved in shallow, marine environments during the Proterozoic eon, at least 600 million years ago (mya; see Chapter 21). Although animals are now distributed in virtually every environment, members of most animal phyla still inhabit marine environments.

Marine habitats offer many advantages

The buoyancy of sea water provides support, and the large volume keeps the water temperature relatively stable. The body fluids of most invertebrates have about the same osmotic concentration as sea water, so fluid and salt balance are more easily maintained than in fresh water. **Plankton,** which consists of the mainly microscopic animals and protists that are suspended in water and float with its movement, provides a ready source of food for many aquatic animals.

Life in the ocean also presents some challenges. Although the continuous motion of water brings nutrients to animals and washes their wastes away, the animals must be able to cope with the water's movements and the currents that could sweep them away. Squids, fishes, and marine mammals have evolved as strong swimmers, usually able to direct their movements and maintain their location. However, most invertebrates and young vertebrates cannot swim strongly, and they have adapted in various ways to the tides and currents. Some sessile animals attach permanently to a stable structure such as a rock. Others burrow in the sand and silt that cover the sea bottom. Many invertebrates have adapted by maintaining a small body size and becoming part of the plankton. As they are tossed about, their food supply continues to surround them.

Some animals are adapted to freshwater habitats

Far fewer kinds of animals make their homes in fresh water than in the ocean, because living in this habitat is more difficult. Fresh water is hypotonic to the tissue fluids of animals, so water tends to move into the animal by osmosis. To survive in this habitat, freshwater species must have mechanisms for removing excess water while retaining salts. This *osmoregulation* requires an expenditure of energy.

Fresh water offers a much less constant environment than sea water. Animals that inhabit fresh water must have adaptations for surviving variations in oxygen content, temperature, turbidity (because of sediments suspended in the water), and even water volume. In addition, fresh water generally contains less food than the sea.

Terrestrial living requires major adaptations

Living on land is even more difficult than living in fresh water, and the evolution of terrestrial animals involved major adaptations. Analyzing the fossil record, many biologists hypothesize that the first air-breathing terrestrial animals were scorpion-like arthropods that came ashore in the Silurian period about 444 mya. The first vertebrates to inhabit terrestrial environments, the amphibians, did not appear until the Devonian period, about 30 million years later.

The chief problem facing all terrestrial organisms is desiccation (drying out). Water is constantly lost by evaporation and is often difficult to replace. A body covering adapted to minimize fluid loss helps solve this problem in many terrestrial animals (❙ Fig. 29-2). Location of the respiratory surface deep within the animal also helps prevent fluid loss. Thus, the gills of aquatic animals are typically located externally, but lungs and tracheal tubes of terrestrial animals are internal.

Reproduction on land also poses challenges to protect gametes and the developing offspring from desiccation. Aquatic animals typically shed their gametes in the water, where fertilization occurs. The surrounding water also serves as an effective shock absorber that protects the delicate embryos as they develop. Some land animals, including most amphibians, return to the water for reproduction, and their larval forms develop in the water.

The evolution of internal fertilization has permitted many terrestrial animals, including earthworms, land snails, insects, reptiles, birds, and mammals, to meet the desiccation challenge. Because these terrestrial animals transfer sperm from the body of the male directly into the body of the female by copulation,

E. R. Degginger/Photo Researchers, Inc.

Figure 29-2 Adaptations to terrestrial life

The tough, horny skin of the green iguana (*Iguana iguana*) has scales and is water resistant. Leathery eggs protect the embryos from drying out.

a watery medium continuously surrounds the sperm. Another important adaptation to reproduction on land is the tough, protective shell that surrounds the eggs of many species. Secreted by the female, this shell protects the developing embryo from drying out. An alternative adaptation for terrestrial reproduction is development of the embryo within the moist body of the mother.

The temperature extremes of terrestrial habitats also present challenges. In later chapters we discuss behavioral and physiological adaptations for maintaining body temperature.

Review

❙ What are some advantages of marine environments over freshwater and terrestrial habitats?

❙ What are some animal adaptations to the terrestrial environment?

ANIMAL ORIGINS

Learning Objective

3 Use current hypotheses to trace the early evolution of animals.

Biologists generally agree that animals share a common ancestor with a group of protists known as *choanoflagellates* (see Fig. 25-25). The cells of these colonial flagellates became specialized to perform a specific function, such as movement, feeding, or reproduction. As this division of labor evolved, a colony of flagellates reached the level of cooperation and coordination that qualified it to be considered a single organism—the first animal. The choanoflagellates, fungi, and animals are a monophyletic group known as **opisthokonts**. (Recall from Chapter 25 that opisthokonts are characterized by a posterior flagellum on motile cells.)

Historically, biologists depended on fossils, on similarities in body plan, and on patterns of development to determine evolutionary relationships among various groups of animals. Now, molecular methods are providing additional data that help answer questions about phylogeny. **Molecular systematics,** the science that focuses on molecular structure to clarify evolutionary relationships, is contributing greatly to the process of reconstructing animal phylogeny.

Molecular analysis indicates that the structure of genes that control development, RNA molecules, and many other molecules are very similar among all animal groups that have been studied. According to the *principle of parsimony* (see Chapter 23), such complex molecules are unlikely to have evolved multiple times. Thus, these data support the hypothesis that animals evolved only once. They are a monophyletic group.

Molecular systematics helps biologists interpret the fossil record

The evolutionary history of animals has been vigorously debated, because early animals were soft-bodied forms that left few fossils. The scarcity of fossils has made it difficult to determine the age, rate of divergence, and number of branches of animal groups. The earliest known animal fossils are **Ediacaran fossils** from the Ediacaran period (600 mya to 542 mya). These fossils of small, simple animals suggest that sponges, jellyfish, and comb jellies were present at that time (see Fig. 21-8).

Paleontologists have discovered many large, complex animal fossils in Chengjiang, an Early Cambrian (525 mya to 520 mya) fossil site in China, and in the Burgess Shale in British Columbia, a Middle Cambrian fossil site (520 mya to 515 mya). Fossils of most extant phyla (and also many extinct animals) have been found at these sites. The rapid appearance of an amazing variety of body plans during this time is known as the **Cambrian Radiation,** or less formally, **Cambrian explosion** (see Fig. 21-9). According to the Cambrian explosion hypothesis, which is based on the fossil record, the many major modifications in body plan that occurred during this time account for many branches in the animal tree.

Studies of large molecular data sets suggest that most animal clades actually diverged over a very long period during the Proterozoic eon (2.5 bya to 542 mya). Thus, the animal phyla that first left fossils during the Cambrian Radiation may have evolved several hundred million years *before* they appear in the fossil record. Biologists estimate that certain groups are about twice as old as the oldest fossils found to date. According to this view, the Cambrian Radiation was a rapid evolution of new animal body plans among clades that already existed. Perhaps fossils of these early animals remain to be discovered in Proterozoic rocks.

Biologists develop hypotheses about the evolution of development

Changes in animal body plan are linked to changes in patterns of embryonic development. Biologists have long used similarities and differences in embryonic development to hypothesize how animal groups are related. Traditionally, biologists depended mainly on structural changes to compare the process of development in various groups.

During the past 20 years, researchers have begun to study the molecular basis of developmental processes. They have identified the genes that direct the early development of the body plan and have discovered that many of these genes have been conserved during animal evolution. The same basic set of genes controls early development in all animal groups. Furthermore, the same genes are used in similar fashion to regulate development.

Evolutionary developmental biology, sometimes referred to as **Evo Devo,** has become an important approach to studying animal relationships. Biologists compare molecular events, such as gene regulation during development, in various animal groups. Similarities in molecular development among different animal groups suggest that similarities existed in the common ancestor of these groups.

Recall from Chapter 17 that *Hox* genes are a group of regulatory genes that help control early development. All major animal groups, except sponges, have *Hox* genes. Investigators suggest that all the *Hox* gene groups had evolved by the beginning of the Cambrian period. Mutations in *Hox* genes could have resulted in rapid changes in animal body plans. For example, regulation by *Hox* gene groups has been linked with the development of wings or legs.

Review

▮ What was the Cambrian explosion?

▮ According to the current hypothesis, when did most major groups of animals evolve?

RECONSTRUCTING ANIMAL PHYLOGENY

Learning Objectives

4 Describe how biologists use structural characters (including variations in body symmetry, number of tissue layers, and type of body cavity) and patterns of early development to infer relationships among animal phyla.

5 Describe three major contributions to animal phylogeny made by molecular systematists. Include the identification of three major clades of bilateral animals.

The basic animal body plan has been highly conserved throughout our evolutionary history. Biologists use similarities and differences in structure and early development to infer evolutionary relationships among animal groups. Variations in major characters of the body plan provide clues to animal relationships. Biologists compare variations in body symmetry, number of tissue layers, type of body cavity, and pattern of development. In addition, biologists now have molecular tools to enhance our understanding of animal phylogeny.

Animals exhibit two main types of body symmetry

Symmetry refers to the arrangement of body structures in relation to the body axis. Most sponges are not symmetrical, so when a sponge is cut in half, the two halves are not similar to each

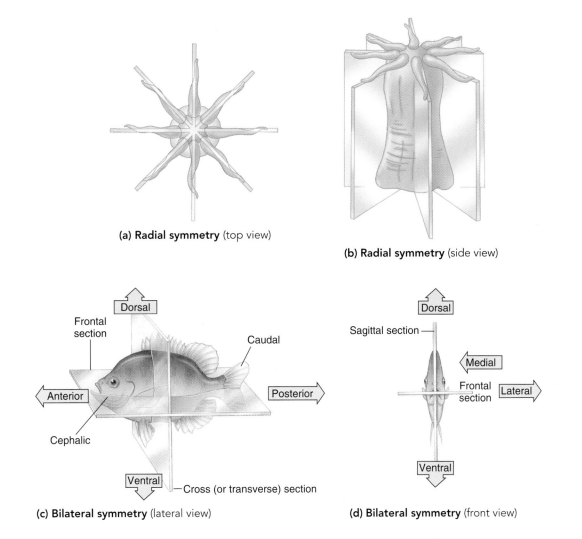

(a) Radial symmetry (top view)

(b) Radial symmetry (side view)

(c) Bilateral symmetry (lateral view)

(d) Bilateral symmetry (front view)

Figure 29-3 *Animated* Radial and bilateral symmetry

(a, b) In radial symmetry, multiple planes can be drawn through the central axis; each divides the animal into two mirror images. **(c, d)** In bilateral symmetry, the head end of the animal is its anterior end, and the opposite end is its posterior end. The back of the animal is its dorsal surface, and the belly is its ventral surface. The diagrams also illustrate various ways the body can be sectioned (cut) to study its internal structure. A sagittal section (lengthwise vertical cut) divides the animal into right and left parts. A frontal, or longitudinal, cut (lengthwise horizontal) divides the body into dorsal and ventral parts. Sections are used in illustrations throughout this book to show the structure and arrangement of tissues and organs.

other. Most other animals exhibit either radial or bilateral body symmetry.

Cnidarians (jellyfish, sea anemones, and their relatives), ctenophores (comb jellies), and adult echinoderms (sea stars and their relatives) have **radial symmetry.** In radial symmetry, the body has the general form of a wheel or cylinder, and similar structures are regularly arranged as spokes from a central axis (❙ Fig. 29-3a and b). Multiple planes can be drawn through the central axis, each dividing the organism into two mirror images. An animal with radial symmetry receives stimuli equally from all directions in the environment.

Many radially symmetrical animals have modified radial symmetry. For example, sea anemones and ctenophores (comb jellies) actually have **biradial symmetry,** in which parts of the body have become specialized so that only two planes can divide the body into similar halves.

Most animals exhibit **bilateral symmetry,** at least in their larval stages. A bilaterally symmetrical animal can be divided through only one plane (which passes through the midline of the body) to produce roughly equivalent right and left halves that are mirror images (❙ Fig. 29-3c and d).

As bilateral symmetry evolved, a corresponding evolutionary trend led toward **cephalization,** the development of a head where sensory structures are concentrated. In many animal groups, concentrations of nerve cells in the head form a brain, and a nerve cord extends from the brain toward the rear end of the animal. Biologists consider bilateral symmetry and cephalization as adaptations to locomotion. The head end of the animal meets its environment first and is best equipped to capture food or respond to danger.

Some definitions of basic terms and directions will help in locating body structures in bilaterally symmetrical animals. The

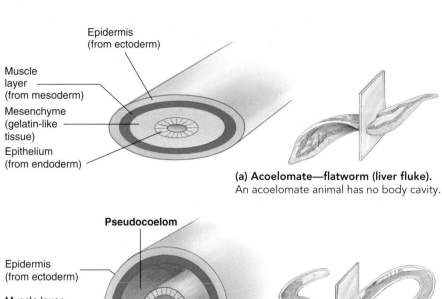

Epidermis
(from ectoderm)

Muscle
layer
(from mesoderm)

Mesenchyme
(gelatin-like
tissue)

Epithelium
(from endoderm)

(a) Acoelomate—flatworm (liver fluke).
An acoelomate animal has no body cavity.

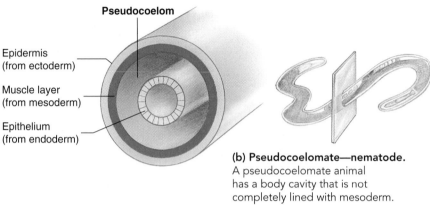

Pseudocoelom

Epidermis
(from ectoderm)

Muscle layer
(from mesoderm)

Epithelium
(from endoderm)

(b) Pseudocoelomate—nematode.
A pseudocoelomate animal
has a body cavity that is not
completely lined with mesoderm.

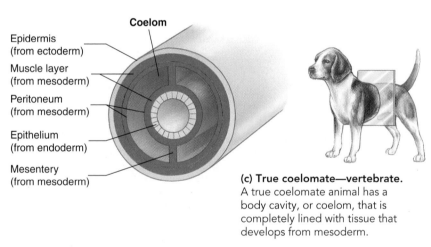

Coelom

Epidermis
(from ectoderm)

Muscle layer
(from mesoderm)

Peritoneum
(from mesoderm)

Epithelium
(from endoderm)

Mesentery
(from mesoderm)

(c) True coelomate—vertebrate.
A true coelomate animal has a
body cavity, or coelom, that is
completely lined with tissue that
develops from mesoderm.

Figure 29-4 *Animated* Three basic body plans in triploblastic animals

The germ layer from which each tissue was derived is indicated in parentheses. Ectoderm
is shown in blue, mesoderm in red, and endoderm in yellow.

A bilaterally symmetrical animal has three axes, each at right angles to the other two: an anterior-posterior axis extending from head to tail, a dorsal-ventral axis extending from back to belly, and a left-right axis extending from side to side. We can distinguish three planes or sections that divide the body into specific parts. A **sagittal plane** divides the body into right and left parts. A sagittal plane passes from anterior to posterior and from dorsal to ventral. A **frontal plane** divides a bilateral body into dorsal and ventral parts. A **transverse section,** or **cross section,** cuts at right angles to the body axis and separates anterior and posterior parts.

Animal body plans are linked to the level of tissue development

Sponges have several types of cells, but they are not organized into **tissues,** groups of closely associated, similar cells that work together to carry out specific functions. In the early development of all animals except sponges, cells form layers, called **germ layers.** The outer germ layer, or **ectoderm,** gives rise to the tissues that form the outer covering of the body and to nervous tissue. The inner layer, or **endoderm,** forms the lining of the digestive tube and other digestive structures. These layers develop into specific types of tissues.

Biologists describe cnidarians and ctenophores as **diploblastic** because they have two tissue layers. Other animals are **triploblastic.** They have a third germ layer, the **mesoderm,** that gives rise to most other body structures, including muscles, skeletal structures, and circulatory system (when present).

Biologists group animals according to type of body cavity

Biologists can classify triploblastic animals based on the presence and type of body cavity, or **coelom** (pronounced "see′-lum"), a fluid-filled space between the body wall and the digestive tube (❙ Fig. 29-4). The flatworms and ribbon worms are triploblastic but have a solid body; that is, they have no body cavity. They are called **acoelomates** (*a-,* "without"; and *coelom,* "cavity").

In most animals, the body cavity is completely lined with mesoderm. Such a body cavity is a *true coelom.* Animals with true coeloms are **coelomates.** The coelom is a space that separates the body wall from the digestive tube, or gut, producing a **tube-**

back surface of an animal is its **dorsal** surface; the underside (belly) is its **ventral** surface. **Anterior** (or **cephalic**) means toward the head end of the animal; **posterior,** or **caudal,** means toward the tail end. A structure is said to be **medial** if it is located toward the midline of the body and **lateral** if it is toward one side of the body; for example, the human ear is lateral to the nose. In human anatomy, the term **superior** refers to a structure located above some point of reference, or toward the head end of the body. The term **inferior** is used in human anatomy to mean located below some point of reference, or toward the feet.

within-a-tube body plan. The body wall, which forms the outer tube, is covered with tissue that develops from ectoderm. Tissue derived from endoderm lines the inner digestive tube, which has an opening at each end: the mouth and the anus.

In some (typically, small) animals, the body cavity is not completely lined with mesoderm; this type of body cavity is called a **pseudocoelom** ("false coelom"). Animals with a pseudocoelom, such as nematodes (roundworms) and rotifers, are **pseudocoelomates,** and biologists formerly classified them as a separate group. Recent evidence indicates that pseudocoelomates are not a monophyletic group and probably evolved through a process of simplification from more than one group of animals with a true coelom.

Bilateral animals form two main groups based on differences in development

Basic differences in their pattern of early development distinguish two main evolutionary lines of bilateral animals: Protostomia and Deuterostomia. **Protostomia** includes flatworms, mollusks, annelids, arthropods, and several other phyla. **Deuterostomia** includes the echinoderms (such as sea stars and sea urchins) and chordates (members of the phylum that includes the vertebrates).

One important difference in the development of protostomes and deuterostomes is the pattern of cleavage, the first several cell divisions of the embryo. In many protostomes, the early cell divisions are diagonal to the polar axis (the long axis of the egg), resulting in a somewhat spiral arrangement of cells; any one cell lies between the two cells above or below it (▌Fig. 29-5a). This pattern of division is known as **spiral cleavage.** In **radial cleavage,** characteristic of the deuterostomes, the early divisions are either parallel or at right angles to the polar axis. The resulting cells lie directly above or below one another (▌Fig. 29-5b).

In the protostomes, the developmental fate of each embryonic cell is typically fixed very early. For example, if the first four cells of an annelid embryo are separated, each cell develops into only a fixed quarter of the larva; this pattern of cleavage is called **determinate cleavage.** In contrast, deuterostomes typically undergo **indeterminate cleavage.** For example, if the first four cells of a sea star embryo are separated, each cell can form a complete, though small, larva. If a few cells are removed from a blastula undergoing indeterminate cleavage, other cells compensate, and the embryo develops normally. In contrast, if a few cells are removed from the blastula of an embryo undergoing determinate cleavage, some structure, such as a limb, does not develop.

During gastrulation a group of cells moves inward, forming a sac that becomes the embryonic gut. The opening to the outside is called the **blastopore.** In most protostomes, the blastopore develops into the mouth. The word *protostome* comes from Greek words meaning "first" and "the mouth." In deuterostomes the blastopore does not give rise to the mouth but generally develops into the anus. A second opening that forms later in development gives rise to the mouth. The word *deuterostome* derives from words meaning "second" and "the mouth."

Another, though less reliable, difference between protostome and deuterostome development is the manner in which the coelom forms. In most protostomes, the mesoderm splits, and the split widens into a cavity that becomes the coelom (▌Fig. 29-6). This method of coelom formation is known as **schizocoely.** In deuterostomes, the mesoderm forms as "outpocketings" of the developing gut, a process called **enterocoely.** These outpocketings eventually pinch off and form pouches; the cavity within the pouches becomes the coelom.

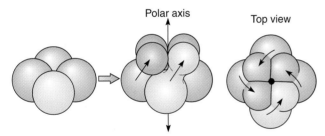

(a) Spiral cleavage is characteristic of protostomes. Note the spiral arrangement, with the upper cells centered between the lower cells.

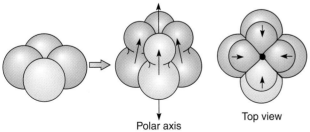

(b) Radial cleavage is characteristic of deuterostomes. The early divisions are either parallel to the polar axis or at right angles to it. The cells are stacked, with the upper cells centered directly above the lower cells.

▌ **Figure 29-5** Spiral and radial cleavage

The pattern of cleavage can be appreciated by comparing the positions of the purple cells in **(a)** and **(b)**.

Biologists have identified major animal groups based on structure

Biologists have inferred evolutionary relationships among animals, based on the structural variations and pattern of development we have just discussed. Their hypotheses are illustrated in ▌Figure 29-7. Biologists assign sponges to a basal group, **Parazoa** (*para*, "alongside"; and *zoa*, "animals"), based on their asymmetry and their simple body plan. The cells of sponges are loosely associated and do not form true tissues. Other animals have true tissues and are classified as **Eumetazoa** (*eu*, "true"; *meta*, "later"; and *zoa*, "animals"). Thus, biologists divide the animal kingdom into two major animal clades: Parazoa and Eumetazoa. Parazoa is considered a sister group of the Eumetazoa.

The eumetazoan branch of the animal tree splits into the **Radiata,** animals with radial symmetry, and the **Bilateria,** animals with bilateral symmetry. Bilateral animals are characterized by

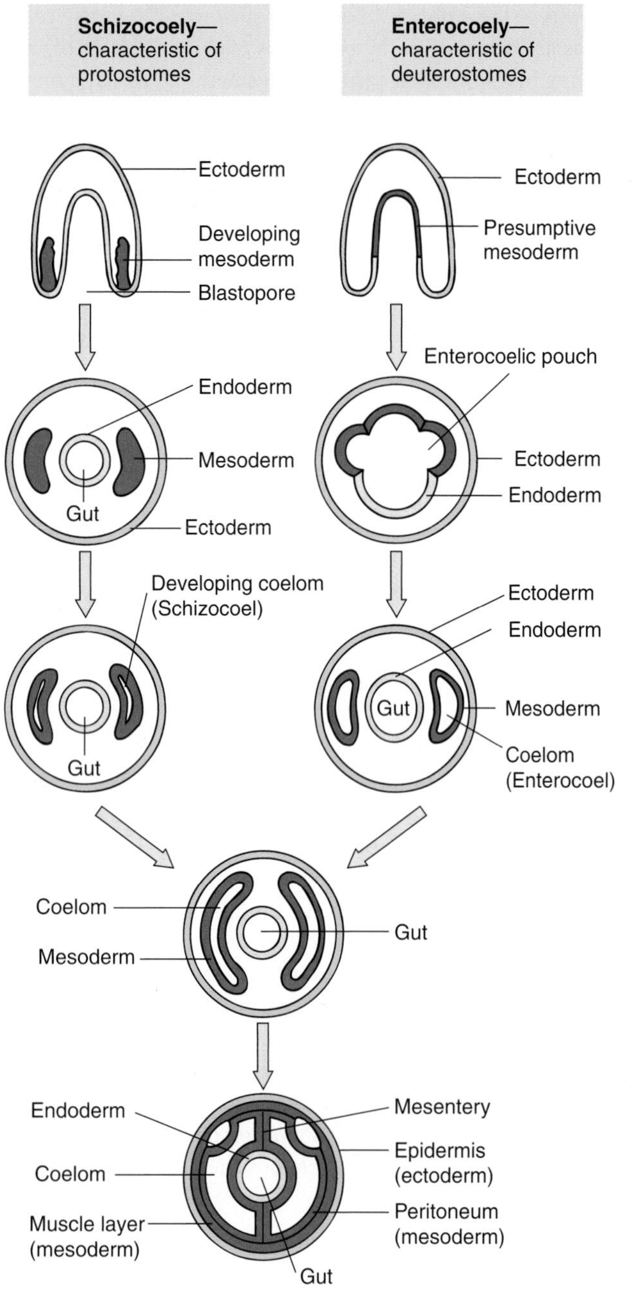

Schizocoely—characteristic of protostomes

Enterocoely—characteristic of deuterostomes

Ectoderm
Developing mesoderm
Blastopore

Ectoderm
Presumptive mesoderm

Endoderm
Mesoderm
Ectoderm
Gut

Enterocoelic pouch
Ectoderm
Endoderm

Developing coelom (Schizocoel)
Gut

Ectoderm
Endoderm
Gut
Mesoderm
Coelom (Enterocoel)

Coelom
Mesoderm
Gut

Endoderm
Coelom
Muscle layer (mesoderm)
Gut
Mesentery
Epidermis (ectoderm)
Peritoneum (mesoderm)

Figure 29-6 Two types of coelom formation

The coelom originates in the embryo from blocks of mesoderm that split off from each side of the embryonic gut. In protostomes, the coelom typically forms by the process of schizocoely, in which the mesoderm (*red*) splits. The split widens, forming a cavity that becomes the coelom. In enterocoely, characteristic of deuterostomes, the mesoderm outpockets from the gut, forming pouches. The cavity within these pouches becomes the coelom. Ectoderm is shown in blue, endoderm in yellow. The diagrams in the top row are longitudinal sections of developing embryos; other diagrams are cross sections.

three distinct tissue layers: ectoderm, endoderm, and mesoderm. Historically, biologists considered acoelomates to be the simplest bilateral animals and hypothesized that the pseudocoelomates evolved next, and finally the coelomates. The coelomate animals

split into two groups: protostomes and deuterostomes. As we will discuss, some of these hypotheses have been challenged by molecular data.

Molecular data contribute to our understanding of animal relationships

Molecular systematists are using molecular tools to clarify the evolutionary relationships among animal groups. So far, these systematists have validated much of the phylogeny that biologists have determined based on structure and development. However, molecular data do challenge some traditional conclusions. For example, *when* various animal phyla diverged has been the topic of much debate.

Molecular systematics has confirmed that animal body plans usually evolved from simple to complex. However, there are important exceptions. For example, biologists have traditionally viewed flatworms and ribbon worms as simple groups that appeared early in animal evolution, based on their lack of a body cavity. However, because analysis of molecular data suggests that these worms evolved from more complex animals and then became structurally simpler over time, systematists have repositioned these groups.

As explained in the last section, biologists classify bilateral animals into protostome and deuterostome groups. Molecular data show that the protostomes split into two major clades: **Lophotrochozoa** (pronounced "loh-foh-troh-koh-zoh′-ah") and **Ecdysozoa** (pronounced "ek-die-so-zoh′-ah") (▌Fig. 29-8 and ▌Table 29-1). The name *Lophotrochozoa* is derived from the names of two major animal groups assigned to this clade: the lophophorate phyla and the Trochozoa. The lophophorate phyla are characterized by a lophophore, a ciliated ring of tentacles surrounding the mouth. The *trochozoa* part of the name refers to the trochophore larva, a type of larva characteristic of two major Lophotrochozoa phyla—the mollusks and annelids. The name *Ecdysozoa* is based on one of the characteristics of this group: the animals in this group molt (a process called *ecdysis*).

The Lophotrochozoa include the flatworms, ribbon worms, mollusks, annelids, the lophophorate phyla, and rotifers. The Ecdysozoa include the nematodes and arthropods. Note that this phylogeny assigns bilateral animals to three major clades: Lophotrochozoa, Ecdysozoa, and Deuterostomia.

Over millions of years, evolutionary forces acting on the basic animal body plan have produced changes resulting in a remarkable diversity of body forms. One very important innovation has been **segmentation,** a body plan in which certain structures are repeated, producing a series of body compartments. Each compartment can be regulated somewhat independently of the others, which means that various parts of the body can become specialized to perform various functions. Perhaps the most obvious example of segmentation can be found in the earthworm. However, as will be discussed in the following chapters, arthropods and vertebrates also have segmented body plans. Thus, segmented animals are found within each of the three major clades of bilateral animals. Molecular data suggest that segmentation evolved independently three times. This view is reflected in the cladogram in Figure 29-8 in which annelids and arthropods are

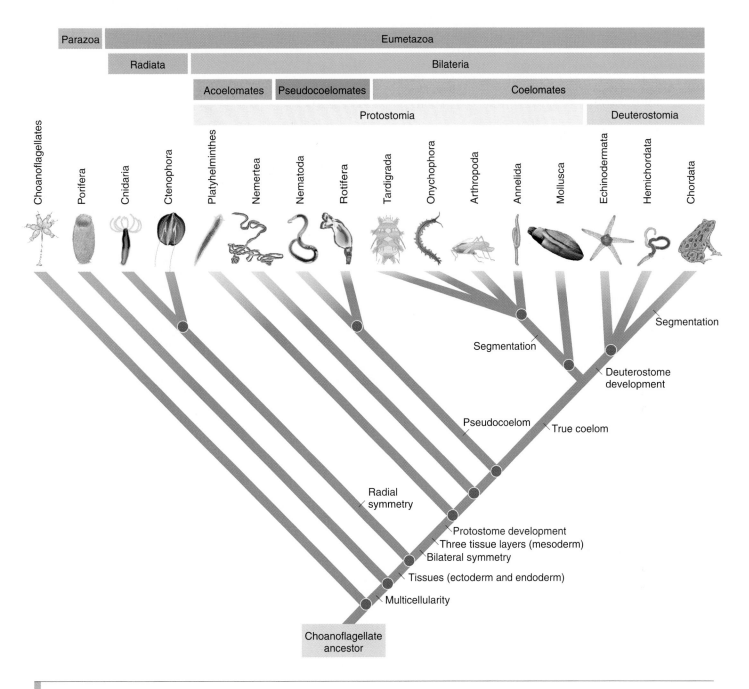

Figure 29-7 Evolutionary relationships of major animal phyla, based on structure

Note that bilateral animals are grouped by several criteria, including type of body cavity: acoelomate, pseudocoelomate, or coelomate. Using the criteria indicated by the colored bars above the cladogram, identify which animal phyla belong in each group. Labels indicate when certain key traits originated.

shown in different clades. Contrast this view with the traditional approach depicted in Figure 29-7. Annelids and arthropods are shown as closely related; segmentation evolved twice. In each independent origin of segmentation, apparently natural selection acted on many of the same genes (for example, *Hox* genes; see Fig. 17-13).

Now that we have briefly discussed animal body plans and some of the criteria for determining phylogenetic relationships, we survey representative animal phyla. In the remainder of this chapter, we discuss two groups that appeared early in animal evolution: the Parazoa and the Radiata. The cladogram shown in Figure 29-8 depicts one current view of the relationships among the major phyla of animals.

Review

▌ How are animals classified based on type of symmetry?

▌ What are some differences between protostomes and deuterostomes?

▌ What are the three main clades of bilateral animals, and how do they differ from one another?

TABLE 29-1

Overview of the Animal Kingdom

Major Groups and Phyla	Body Plan	Key Characteristics
Parazoa		
Porifera (sponges)	Asymmetrical; cells loosely arranged; body is sac with pores, central cavity, and osculum	Choanocytes (collar cells); aquatic, mainly marine
Radiata		
Cnidaria (hydras, jellyfish, corals)	Radial symmetry; diploblastic; gastrovascular cavity with one opening	Tentacles with cnidocytes (stinging cells) that discharge nematocysts; two body forms: polyp and medusa; some form colonies; mainly marine
Ctenophora (comb jellies)	Biradial symmetry; diploblastic; gastrovascular cavity with mouth and anal pores	Eight rows of cilia that resemble combs; tentacles with adhesive glue cells; marine predators
Protostomes: Lophotrochozoan Branch		
Platyhelminthes (flatworms: planarians, tapeworms, flukes)	Bilateral symmetry; triploblastic; simple organ systems; gastrovascular cavity with one opening	No body cavity; some cephalization; some are free-living carnivores; many are parasites
Nemertea (proboscis worms; also known as ribbon worms)	Bilateral symmetry; triploblastic; organ systems; complete digestive tube*	Proboscis (long, muscular tube that can be everted to capture prey); only coelomate space is rhynchocoel surrounding proboscis; mainly marine carnivores
Mollusca (clams, snails, squids)	Bilateral symmetry; triploblastic; organ systems; complete digestive tube	Soft body usually covered by dorsal shell; flat, muscular foot; mantle covers visceral mass; most have a radula (belt of teeth)
Annelida (some marine worms, earthworms, leeches)	Bilateral symmetry; triploblastic; organ systems; complete digestive tube	Segmented body; most have setae, bristles that provide traction during crawling
Lophophorates (brachiopods, phoronids, bryozoans)	Bilateral symmetry; triploblastic; organ systems; complete digestive tube	Lophophore (ciliated ring of tentacles) surrounds mouth, captures suspended particles in water; mainly sessile, marine animals
Rotifera (wheel animals)	Bilateral symmetry; triploblastic; organ systems; complete digestive tube	Crown of cilia at anterior end; cell number constant; microscopic, aquatic animals
Protostomes: Ecdysozoan Branch		
Nematoda (roundworms, e.g., *Ascaris*, hookworms, trichina worms)	Bilateral symmetry; triploblastic; organ systems; complete digestive tube	Cylindrical, threadlike body; pseudocoelom; widely distributed in soil and aquatic sediments; important as decomposers; many are predators; some are parasites
Arthropoda (centipedes, spiders, crabs, lobsters, insects)	Bilateral symmetry; triploblastic; organ systems; complete digestive tube	Segmented body; exoskeleton; paired, jointed appendages; insects and many crustaceans have compound eyes; insects have tracheal tubes for air exchange
Deuterostomes		
Echinodermata (sea stars, sea urchins; sand dollars)	Larva bilateral, ciliated; adult pentaradial; triploblastic; organ systems; complete digestive tube	Endoskeleton with spines; water vascular system functions in locomotion, feeding, and gas exchange; tube feet; marine
Hemichordata (acorn worms)	Bilateral symmetry; triploblastic; organ systems; complete digestive tube	Ring of cilia surrounds mouth; three-part body: proboscis, collar, and trunk; wormlike marine animals
Chordata (tunicates, lancelets, vertebrates)	Bilateral symmetry; triploblastic; organ systems; complete digestive tube	Notochord; dorsal, tubular nerve cord; pharyngeal slits during some time in life cycle; postanal tail; segmented body

*A complete digestive tube has a mouth for food intake and an opening (anus) for elimination of wastes.

Based on molecular data, three major clades of bilateral animals are shown: Lophotrochozoa, Ecdysozoa, and Deuterostomia.

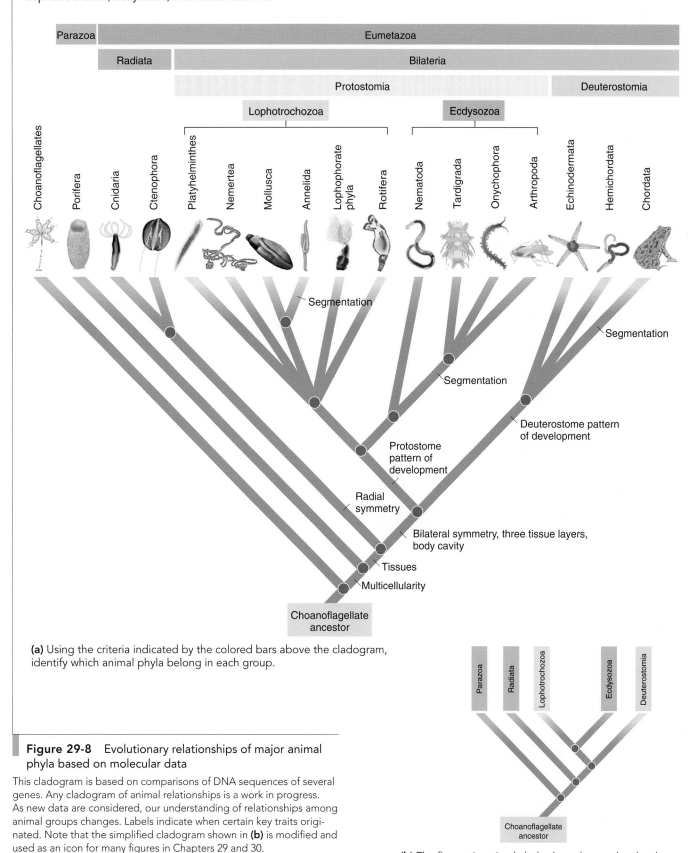

(a) Using the criteria indicated by the colored bars above the cladogram, identify which animal phyla belong in each group.

Figure 29-8 Evolutionary relationships of major animal phyla based on molecular data

This cladogram is based on comparisons of DNA sequences of several genes. Any cladogram of animal relationships is a work in progress. As new data are considered, our understanding of relationships among animal groups changes. Labels indicate when certain key traits originated. Note that the simplified cladogram shown in **(b)** is modified and used as an icon for many figures in Chapters 29 and 30.

(b) The five main animal clades based on molecular data.

THE PARAZOA: SPONGES

Learning Objective

6 Identify distinguishing characteristics of phylum Porifera.

Sponges, the only members of the Parazoa, diverged early from the lineage leading to other animals. The earliest known animal fossils are sponges found in China that are about 600 million years old. Because they have been evolving independently for so long, they are very different from other animals.

Collar cells characterize sponges

Biologists have identified about 10,000 species of sponges and assigned them to phylum **Porifera.** The name *Porifera,* meaning "to have pores," aptly describes the sponges, whose bodies are perfo-rated by tiny holes. Sponges are aquatic, mainly marine animals that are most abundant in warm waters. They range in size from 1 to 200 cm (0.4 to 79 in) in height. Many sponges are asymmetrical, but they vary in shape from flat, encrusting growths to balls, cups, fans, or vases. Living sponges may be brightly colored—green, orange, red, yellow, blue, or purple—or they may be white or drab (❚ Fig. 29-9). Some species are inhabited by symbiotic bacteria or algae that give them color.

Although they are multicellular and can be large, sponges function much like choanoflagellates—colonial, unicellular protists. Recall that according to a current hypothesis, animals (including sponges) and choanoflagellates share a common choanoflagellate ancestor. Choanoflagellates are characterized by a single flagellum surrounded by a collar of microvilli (see Fig. 25-25).

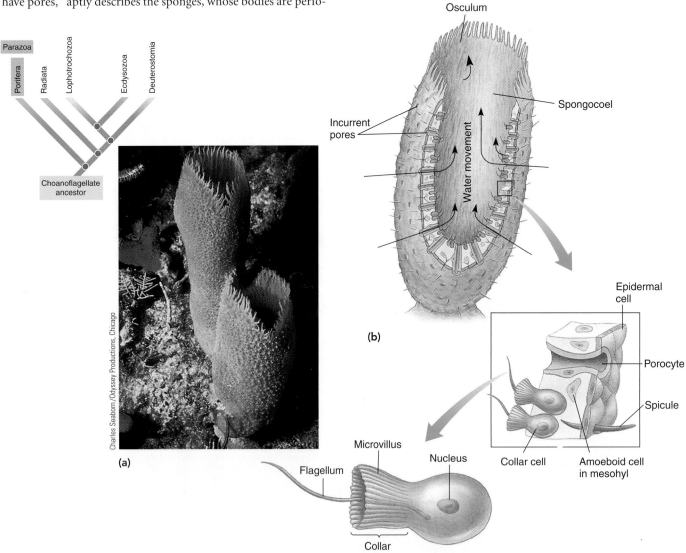

Figure 29-9 *Animated* Sponge structure

(a) Tube sponges (*Spinosella plicifera*) from the Caribbean, attached to the coral reef substrate. **(b)** Diagram of a simple sponge cut to expose its organization. Collar cells (choanocytes) beat their flagella, producing a current of water that enters through the pores. The water passes through the spongocoel and exits through the osculum. Food particles in the stream of water are trapped by the collars.

Sponges have **collar cells,** or **choanocytes,** flagellate cells that are strikingly similar to the choanoflagellates.

Sponge larvae have flagella and can swim about. Adult sponges attach to some solid object and have long been described as sessile. However, biologists have observed adults of several species moving slowly (about 4 mm per day), perhaps by the cumulative movement of cells along the sponge's lower surface.

In a simple sponge, water enters through hundreds of tiny pores (*ostia*); passes into the central cavity, or **spongocoel** (not a digestive cavity); and flows out through the sponge's open end, the **osculum.** In most types of sponges, the body wall is extensively folded, and complex systems of canals provide increased surface area for food capture. Although sponges are multicellular, their cells are loosely associated and do not form true tissues. However, a division of labor exists among the several types of cells that make up the sponge, with certain cells specializing in nutrition, support, contraction, or reproduction. Epidermal cells form the outer layer of the sponge and line the canals. Specialized tubelike cells, called *porocytes,* form the pores of a simple sponge. These cells regulate the diameter of the pores by contracting. Many sponge cells are extremely versatile and can change form and function.

Collar cells make up the inner layer of certain sponges. Each cell is equipped with a tiny collar surrounding the base of the flagellum. The collar is an extension of the plasma membrane and consists of microvilli. Collar cells create the water current that brings food and oxygen to the cells and carries away carbon dioxide and other wastes. Collar cells also trap and phagocytose food particles. Together, the collar cells of some sponges can pump a volume of water equal to the volume of the sponge each minute!

Between the outer and inner cell layers of the sponge body is a gelatin-like layer, the *mesohyl,* supported by slender skeletal spikes, or **spicules.** Amoeboid cells, which wander about in the mesohyl, secrete the spicules, which consist of calcium carbonate, silica, or a fibrous protein material known as *spongin.* Other amoeboid cells in the mesohyl are important in digestion and food transport.

Sponges are *suspension feeders,* adapted for trapping and eating whatever food the water brings to them. As water circulates through the body, food is trapped along the sticky collars of the choanocytes. Food particles are either digested within the collar cell or transferred to an amoeboid cell for digestion. The amoeboid cell transports nutrients to epidermal cells. Undigested food passes out through the osculum and is simply eliminated into the water.

Gas exchange and excretion of wastes depend on diffusion into and out of individual cells. Although cells of the sponge can react to stimuli, sponges do not have specialized nerve cells and so cannot react as a whole. Behavior appears limited to basic metabolic necessities such as capturing food and regulating the flow of water through the body.

Sponges reproduce both asexually and sexually. In asexual reproduction, a small fragment or bud may break free from the parent sponge and give rise to a new sponge. Such fragments may attach to the parent sponge and thus form a colony. Most sponges are **hermaphrodites,** meaning the same individual can produce both eggs and sperm. Some of the amoeboid cells develop into sperm cells; others, into egg cells. However, hermaphroditic sponges usually produce eggs and sperm at different times, and they cross-fertilize with other sponges. They release mature sperm into the water, which are taken in by other sponges of the same species.

Fertilization and early development take place within the jelly-like mesohyl. Zygotes develop into flagellate larvae that leave the parent along with the stream of outflowing water. After swimming for a while, a larva finds a solid object, attaches to it, and settles down to a sessile life.

Sponges have a remarkable ability to repair themselves when injured and to regenerate lost parts. When the cells of a sponge are separated experimentally, they recognize one another and their place in the whole and aggregate to re-form a complete sponge.

Review

▌ Why are choanocytes significant? What is their function in sponges?

▌ In what ways do the cells of sponges specialize?

THE RADIATA: ANIMALS WITH RADIAL SYMMETRY AND TWO CELL LAYERS

Learning Objectives

7 Identify distinguishing characteristics of phylum Cnidaria, describe four classes of this phylum, and give examples of animals that belong to each class.

8 Identify distinguishing characteristics of phylum Ctenophora.

Recall that all animals except the sponges are classified as Eumetazoa. The Radiata, animals with radial symmetry and two tissue layers, may have been the first eumetazoans to evolve. Biologists currently classify two phyla as Radiata: Cnidaria and Ctenophora. However, the relationships of these groups to each other, and the relationship of the Radiata to other animals, is still a matter of debate.

Cnidarians have unique stinging cells

Most of the 10,000 or so species of phylum **Cnidaria** (pronounced "ni-dah´-ree-ah") are marine. The radially symmetrical cnidarian body is organized as a hollow sac with the mouth and surrounding tentacles located at one end. We consider four classes (▌ Table 29-2). Class **Hydrozoa** includes hydras and *hydroids,* such as *Obelia* and the Portuguese man-of-war; class **Scyphozoa** is made up of jellyfish; class **Cubozoa** includes the "box jellyfish"; and class **Anthozoa** includes sea anemones and corals. Some cnidarians live a solitary existence, whereas many others, such as corals, form colonies.

TABLE 29-2

Major Classes of Phylum Cnidaria

Class and Representative Animals	Characteristics
Hydrozoa *Hydra, Obelia,* Portuguese man-of-war	Mainly marine, but some freshwater species; alternation of polyp and medusa stages in most species (polyp form only in *Hydra*); some form colonies
Scyphozoa Jellyfish	Mainly marine; typically inhabit coastal water, free-swimming medusa most prominent form; polyp stage often reduced
Cubozoa "Box jellyfish"	Inhabit tropical and semitropical waters; have polyp stage, but medusa form most prominent; square shape when viewed from above; actively hunt prey; complex eyes that form blurred images
Anthozoa Sea anemones, corals, sea fans	Marine; solitary or colonial polyps; in most no medusa stage; gastrovascular cavity divided by partitions into chambers, increasing area for digestion; sessile

Cnidarians have two body shapes: the polyp and the medusa (❚ Fig. 29-10). The **polyp** form, represented by *Hydra,* typically has a dorsal mouth surrounded by tentacles. In the **medusa** (pl., *medusae*), or jellyfish form, the mouth is located in the lower concave, or *oral,* surface; the convex upper surface is the *aboral* surface. Some cnidarians have the polyp shape during one stage of their life cycle and the medusa form during another stage. The Portuguese man-of-war and some other cnidarians consist of colonies of many individuals, some of which are polyps and others medusae.

Cnidarians get their name from specialized cells, called **cnidocytes** (from a Greek word meaning "sea nettles"), that contain stinging organelles. Cnidocytes are located mainly in the epidermis, especially on the tentacles. The cnidocytes contain stinging "thread capsules," or **nematocysts** (❚ Fig. 29-11). When stimulated, a nematocyst releases a coiled, hollow thread. Some types of nematocyst threads are sticky. Others are long and coil around prey. A third type bears barbs or spines that can inject a protein toxin that paralyzes prey animals, such as small crustaceans. Each cnidocyte has a small, projecting trigger (*cnidocil*) on its outer surface. Stimuli such as touch or chemicals dissolved in the water stimulate the nematocyst to fire its thread.

Cnidarians use their tentacles to capture prey and push it into the mouth. The mouth leads into the **gastrovascular cavity,** where digestion takes place. The mouth is the only opening into the gastrovascular cavity and so must serve for both ingestion of food and expulsion of wastes. Gas exchange and excretion occur by diffusion. The body wall is thin enough that no cell is far from the surface.

More highly organized than sponges, cnidarians are diploblastic; that is, they have two definite tissue layers. The ectoderm gives rise to the outer **epidermis,** a protective layer covering the body. The endoderm gives rise to the inner **gastrodermis,** which lines the gastrovascular cavity and functions in digestion. These thin layers are separated by a gelatinous, mainly acellular **mesoglea.**

Cnidarians have nerve cells that form **nerve nets** connecting sensory cells in the body wall to contractile and gland cells. Sense organs—for example, photoreceptors that detect light—are positioned around the edge of the body. An impulse set up by one sensory cell passes in all directions more or less equally. Nerve cells are not organized to form a brain or nerve cord.

Both the epidermis and gastrodermis have cells specialized to contract (however, they are not true muscle cells). Contractile fibers in the epidermal cells are arranged lengthwise, and those in the gastrodermis are arranged in a circular pattern. These two sets of contractile cells act on the water-filled gastrovascular cavity, which forms a **hydrostatic skeleton.** This skeleton supports the body and allows movement. By contracting one set of contractile cells or the other, the hydra can shorten, lengthen, or bend its body (see Fig. 39-2).

Class Hydrozoa includes solitary and colonial forms

Although not really typical, the solitary *Hydra* is the cnidarian that beginning biology students most often study. To the naked eye, *Hydra* looks like a bit of frayed string. This tiny animal is found in freshwater ponds (❚ Fig. 29-12). Because it has a remarkable ability to regenerate, biologists named *Hydra* after the multiheaded monster of Greek mythology that could grow two new heads for each head cut off. When *Hydra* is cut into several pieces, each piece can regenerate all the missing parts and become a whole animal.

Hydra lives in fresh water and typically attaches to a rock, aquatic plant, or detritus by a disc of cells at its base. Hydras reproduce asexually by budding during periods when environmental conditions are optimal. However, they differentiate as males and females and reproduce sexually in the fall or when pond water becomes stagnant. The zygote may become covered with a shell that protects it through the winter or until conditions become more favorable.

Many hydrozoans form colonies consisting of hundreds or thousands of individuals. A colony begins with a single polyp that reproduces asexually by budding. However, instead of separating from the parent, the bud remains attached and eventually forms additional buds. Several types of individuals may arise in the same colony, some specialized for feeding, some for reproduction, and others for defense.

Some marine cnidarians are remarkable for their alternation of sexual and asexual stages. Their life cycle differs from the alternation of generations in plants in that both sexual and asexual forms are diploid; only sperm and eggs are haploid. The life cycle of the colonial marine hydrozoan *Obelia* illustrates alternation of sexual and asexual stages (❚ Fig. 29-13).

Class Scyphozoa includes the "true" jellyfish

Among the jellyfish, the medusa is the dominant body form. Scyphozoan medusae are generally larger than hydrozoan medusae, and they have a thick, viscous mesoglea that gives firmness to the body. In scyphozoans, the polyp stage is small and inconspicuous or may even be absent. The largest jellyfish, *Cyanea,* may be more than 2 m (6.5 ft) in diameter and have tentacles 30 m (98 ft) long. These orange and blue "monsters," among the largest invertebrates, are dangerous to swimmers in the North Atlantic Ocean.

Class Cubozoa includes the "box jellyfish"

When viewed from above, cubozoans have a square shape. They have four tentacles, or groups of tentacles. These fast-swimming jellyfish have complex eyes that form blurred images, and they actively hunt for prey. Found in waters off the northern Australian coast, the sea wasp (*Chironex fleckeri*) has long tentacles that can extend more than 3 m (10 ft). The sea wasp produces a toxin that can cause respiratory failure and cardiac arrest in humans. Fatalities are mainly associated with severe stings caused by contact with tentacles over a large area of the body.

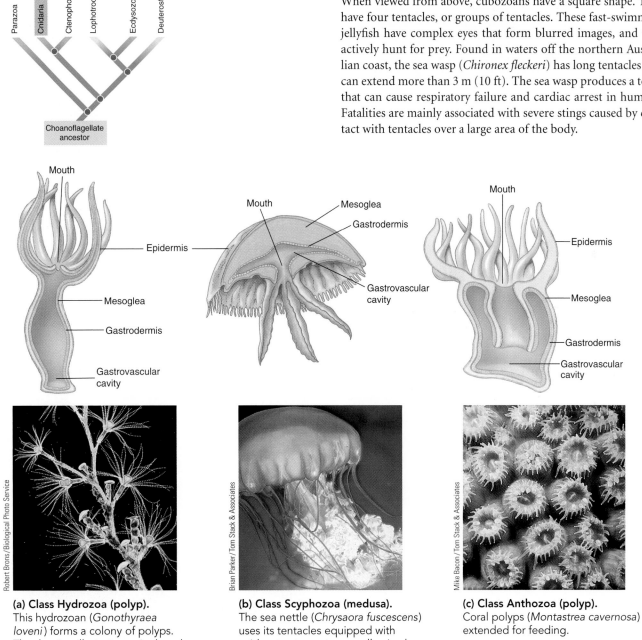

(a) Class Hydrozoa (polyp).
This hydrozoan (*Gonothyraea loveni*) forms a colony of polyps. The drawing illustrates a single polyp.

(b) Class Scyphozoa (medusa).
The sea nettle (*Chrysaora fuscescens*) uses its tentacles equipped with cnidocytes to capture small animals (zooplankton) suspended in the water.

(c) Class Anthozoa (polyp).
Coral polyps (*Montastrea cavernosa*) extended for feeding.

Figure 29-10 *Animated* Polyp and medusa body forms of cnidarians

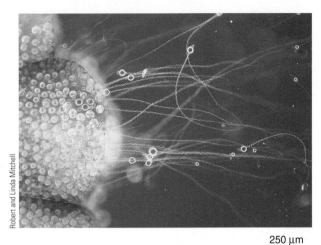

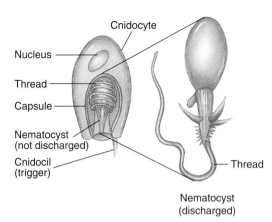

250 μm

(a) LM of discharged nematocysts of a Portuguese man-of-war (*Physalia physalis*). Photographed in the Gulf of Mexico.

(b) Undischarged and discharged nematocyst. The cnidocil, or trigger, is a mechanoreceptor that discharges the nematocyst when it senses contact with an object.

Figure 29-11 *Animated* Nematocysts

When cnidarian stinging cells (cnidocytes) are stimulated, the nematocyst discharges, ejecting a thread that may entangle or penetrate the prey. Some nematocysts secrete a toxic substance that immobilizes the prey.

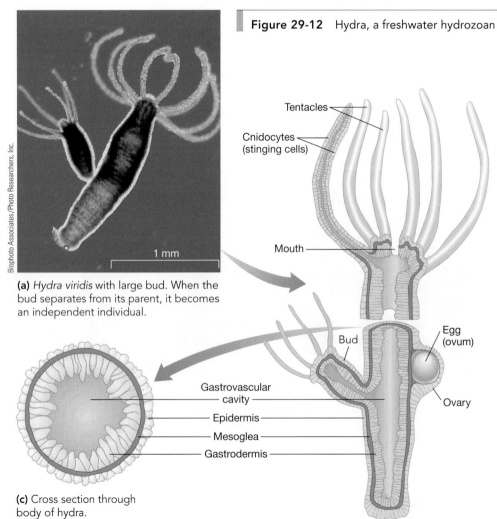

Figure 29-12 Hydra, a freshwater hydrozoan

(a) *Hydra viridis* with large bud. When the bud separates from its parent, it becomes an independent individual.

(c) Cross section through body of hydra.

(b) Hydra cut longitudinally to show internal structure. Asexual reproduction by budding is represented on left; sexual reproduction is represented by ovary on right. Male hydras develop testes that produce sperm.

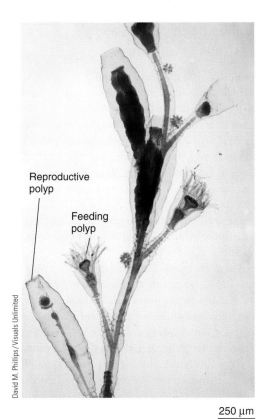

Reproductive polyp

Feeding polyp

David M. Phillips / Visuals Unlimited

250 μm

(a) LM of *Obelia*. Some polyps have tentacles and are specialized for feeding, whereas others are specialized for reproduction.

Class Anthozoa includes only polyps

Sea anemones and corals, members of class Anthozoa, have either individual or colonial polyps but no free-swimming medusa stage. The polyp produces eggs and sperm, and the fertilized egg develops into a small, ciliated larva called a **planula.** This larval form may swim to a new location before attaching to develop into a polyp.

Anthozoans differ from hydrozoans in that a series of vertical partitions partially divides the gastrovascular cavity into connected chambers. The partitions increase the surface area for digestion, enabling an anemone to digest an animal as large as a crab. Although corals can capture prey, many tropical species depend for nutrition on photosynthetic algae (*zooxanthellae*) that live within cells lining the coral's digestive cavity (see Chapter 25 and Fig. 53-11). The relationship between coral and zooxanthellae is symbiotic and mutually beneficial. The algae provide the coral with oxygen and with carbon and nitrogen compounds. In exchange, the coral supplies the algae with waste products such as ammonia, from which the algae make nitrogenous compounds for both partners.

In warm, shallow seas, much of the bottom is covered with coral or anemones, most of them brightly colored. Coral reefs consist of colonies of millions of corals and of certain algae (mainly coralline red algae). Living colonies occur only in the uppermost regions of such reefs, adding their own skeletons to the forming rock. Coral reefs are among the most productive of all ecosystems, rivaling tropical rain forests in species diversity (see

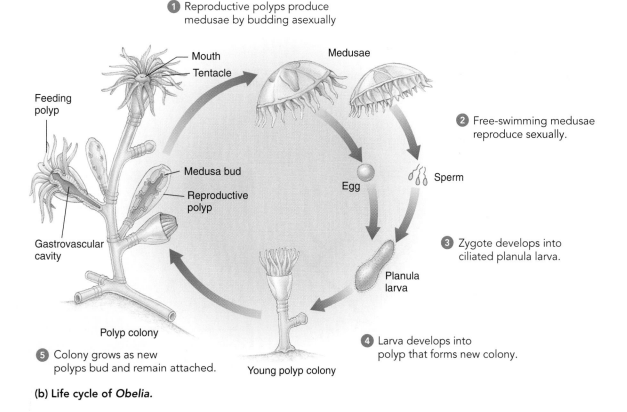

① Reproductive polyps produce medusae by budding asexually

② Free-swimming medusae reproduce sexually.

③ Zygote develops into ciliated planula larva.

④ Larva develops into polyp that forms new colony.

⑤ Colony grows as new polyps bud and remain attached.

Mouth
Tentacle
Feeding polyp
Medusa bud
Reproductive polyp
Gastrovascular cavity
Polyp colony
Medusae
Egg
Sperm
Planula larva
Young polyp colony

(b) Life cycle of *Obelia*.

Figure 29-13 *Animated* *Obelia*, a marine colonial hydrozoan

Figure 29-14 Bleached staghorn coral

This staghorn coral (*Acropora*) was photographed off Heron Island on the Great Barrier Reef off the coast of Australia.

Comb jellies have adhesive glue cells that trap prey

The approximately 100 species of phylum **Ctenophora,** the comb jellies, are fragile, luminescent marine animals. Some species are as small as a pea, whereas others are larger than a tomato. The outer surface of a ctenophore bears eight rows of cilia that resemble combs (Fig. 29-15). The coordinated beating of the cilia in these combs moves the animal through the water. A sense organ functions in balance and helps the animal orient itself. Some ctenophores have two tentacles. They do not have the stinging nematocysts characteristic of the cnidarians. However, their tentacles are equipped with adhesive glue cells. When prey comes in contact with the tentacle, the glue cells burst open, releasing sticky threads that trap the prey.

Ctenophores are biradially symmetrical, meaning you could obtain equal halves by cutting through the body axis in two different ways. Because ctenophores have radial symmetry and feeding tentacles, their body plan is somewhat similar to that of a cnidarian medusa. Ctenophores are also like cnidarians in that they consist of two cell layers separated by a thick, jellylike mesoglea. Because of these similarities, biologists classify ctenophores near the cnidarians. However, their development is different, and unlike cnidarians, their digestive system has a mouth for food intake at one end and two anal pores for the egestion of water and wastes at the other end. The similarities between ctenophores and cnidarians may be a result of convergent evolution from living in a similar environment, the ocean. Where, then, do the ctenophores belong on our cladogram (see Fig. 29-8)? Systematists will decide as they gather new data.

Review

- In what ways do cnidarians differ from sponges?
- Sketch the major events of the life cycle of *Obelia*.
- How is a ctenophore like a cnidarian? In what ways is it different?

Fig. 55-22). A single reef can serve as home for more than 3000 species of fishes and other marine organisms, and an estimated one fourth of all marine species depend on coral reefs.

During the past several years many coral reefs, especially those in coastal waters, have suffered serious damage from human activities. These include overfishing, mining reefs for building materials, polluting coastal waters with industrial chemicals, and smothering coral with the silt washed downstream from clear-cut forests. **Coral bleaching** is the stress-induced loss of the colorful symbiotic algae that inhabit coral cells (Fig. 29-14). Without their algae, coral become malnourished and die. Environmental factors suspected to contribute to coral bleaching include unusually low or high (associated with global warming) temperatures, pollution, changes in salinity, disease, and high doses of ultraviolet radiation (associated with the destruction of the ozone layer).

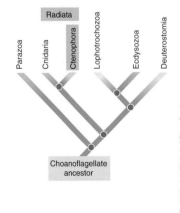

Figure 29-15 Ctenophore (comb jelly)

Ctenophores are free-swimming, bioluminescent hermaphrodites capable of self-fertilization. The sea gooseberry (*Pleurobrachia*) has two long tentacles with adhesive cells used to capture prey.

Learning Objectives

1 Describe several characters common to most animals (page 620).

- Members of **kingdom Animalia** are eukaryotic, multicellular, heterotrophic organisms with cells specialized to perform specific functions.

- Animals have diverse body plans. The **body plan** is the basic structure and functional design of the body.

- Most animals are capable of locomotion at some time during their life cycle, can respond adaptively to external stimuli, and can reproduce sexually.

- In sexual reproduction, sperm and egg unite to form a **zygote.** The zygote undergoes **cleavage,** a series of cell divisions that produce a hollow ball of cells called a **blastula.** The blastula undergoes **gastrulation,** forming embryonic tissues.

ThomsonNOW™ **Explore the characteristics of animals by clicking on the figures in ThomsonNOW.**

2 Compare the advantages and disadvantages of life in the ocean, in fresh water, and on land (page 620).

- Marine environments have relatively stable temperatures, provide buoyancy, and provide readily available food. Fluid and salt balance are more easily maintained than in fresh water. Currents and other water movements are a disadvantage.

- Fresh water offers a less constant environment and less food. Because fresh water is hypotonic to tissue fluid, animals must osmoregulate.

- Terrestrial animals must have adaptations that protect them from drying out and from temperature changes. They also require adaptations to protect their gametes and embryos.

3 Use current hypotheses to trace the early evolution of animals (page 621).

- Biologists hypothesize, based on molecular data, that most animal clades actually diverged over a long period during the Proterozoic eon.

- During the **Cambrian Radiation** new animal body plans rapidly evolved among clades already existing. Fossils of these animals are first found in Cambrian fossil sites.

- The *Hox* gene group controls early development in animal groups. Many of the *Hox* genes had evolved by the beginning of the Cambrian period, and mutations in these genes could have resulted in rapid changes in animal body plans.

4 Describe how biologists use structural characters (including variations in body symmetry, number of tissue layers, and type of body cavity) and patterns of early development to infer relationships among animal phyla (page 622).

- Biologists hypothesize that cnidarians and ctenophores are closely related because they share **radial symmetry;** most other animals exhibit **bilateral symmetry,** at least in their larval stages. **Cephalization,** the development of a head, evolved along with bilateral symmetry.

- Biologists have also inferred relationships based on level of tissue development and type of body cavity. Embryonic tissues, called **germ layers,** include the outer layer, **ectoderm,** which gives rise to the body covering and the nervous system; the inner layer, **endoderm,** which lines the gut and other digestive organs; and a middle layer, **mesoderm,** which gives rise to most other body structures.

- Bilateral animals have traditionally been classified as **acoelomate** (no body cavity); **pseudocoelomate** (body cavity not completely lined with mesoderm); or **coelomate,** an animal with a true **coelom** (a body cavity completely lined with mesoderm).

- Two major evolutionary branches of bilateral animals are **Protostomia** (mollusks, annelids, and arthropods) and **Deuterostomia** (echinoderms and chordates). In protostomes the **blastopore,** the opening from the embryonic gut to the outside, develops into the mouth; in deuterostomes the blastopore typically becomes the anus.

- Protostomes undergo **spiral cleavage,** in which early cell divisions are diagonal to the polar axis. Deuterostomes undergo **radial cleavage,** in which the early cell divisions are either parallel or at right angles to the polar axis, so the cells lie directly above or below one another.

- Protostomes undergo **determinate cleavage,** in which the fate of each embryonic cell is fixed very early. Deuterostomes undergo **indeterminate cleavage,** in which the fate of each cell in early development is more flexible.

5 Describe three major contributions to animal phylogeny made by molecular systematics. Include the identification of three major clades of bilateral animals (page 622).

- Molecular systematics has confirmed much of animal phylogeny based on structural characters, including the axiom that animal body plans usually evolved from simple to complex.

- Molecular systematics has provided evidence for exceptions to the "simple-to-complex" rule. For example, molecular data indicate that flatworms and ribbon worms evolved from more complex animals and became simpler over time. Molecular data further suggest that pseudocoelomate animals do not form a natural group and probably evolved from coelomate ancestors.

- Many biologists now subdivide the protostomes into two clades based on molecular data: **Lophotrochozoa** and **Ecdysozoa.** The Lophotrochozoa include the flatworms, ribbon worms, mollusks, annelids, the lophophorate phyla (groups that have a ciliated ring of tentacles surrounding the mouth), and rotifers. The Ecdysozoa, animals that molt, include the nematodes and arthropods. Thus, the three major clades of bilateral animals are Lophotrochozoa, Ecdysozoa, and Deuterostomia.

6 Identify distinguishing characteristics of phylum Porifera (page 630).

- Phylum **Porifera** consists of the sponges, animals characterized by flagellate **collar cells (choanocytes).** Sponges are the only members of the **Parazoa,** the sister group of the Eumetazoa.

- The sponge body is a sac with tiny openings through which water enters; a central cavity, or **spongocoel;** and an open end, or **osculum,** through which water exits. The cells of sponges are loosely associated; they do not form true tissues.

ThomsonNOW™ **Learn more about sponge structure by clicking on the figure in ThomsonNOW.**

7 Identify distinguishing characteristics of phylum Cnidaria, describe four classes of this phylum, and give examples of animals that belong to each class (page 631).

■ Phylum **Cnidaria** is characterized by radial symmetry, two tissue layers, and **cnidocytes,** cells that contain stinging organelles called **nematocysts.** The **gastrovascular cavity** has a single opening that serves as both mouth and anus. Nerve cells form irregular, nondirectional **nerve nets** that connect sensory cells with contractile and gland cells.

■ The life cycle of many cnidarians includes a sessile **polyp** stage (a form with a dorsal mouth surrounded by tentacles) and a free-swimming **medusa** (jellyfish) stage.

■ Phylum Cnidaria includes four classes. Class **Hydrozoa** (hydras, hydroids, and the Portuguese man-of-war) are typically polyps and may be solitary or colonial. Class **Scyphozoa** (the jellyfish) are generally medusae. Class

Cubozoa, the "box jellyfish," have complex eyes that form blurred images. Class **Anthozoa** (sea anemones and corals) are polyps and may be solitary or colonial; anthozoans differ from hydrozoans in the organization of the gastrovascular cavity.

ThomsonNOW™ **Learn more about cnidarian body forms, nematocysts, and life cycles by clicking on the figures in ThomsonNOW.**

8 Identify distinguishing characteristics of phylum Ctenophora (page 631).

■ Phylum **Ctenophora** consists of the comb jellies, fragile, luminescent marine predators with biradial symmetry. Ctenophores have eight rows of cilia that resemble combs. They are diploblastic and have tentacles with adhesive glue cells.

TEST YOUR UNDERSTANDING

1. Which of the following is *not* a defining characteristic of animals? (a) heterotrophic (b) multicellular (c) eukaryotic (d) presence of a coelom (e) formation of a zygote that undergoes cleavage

2. Which of the following is *not* an adaptation to terrestrial living? (a) internal fertilization (b) shell surrounding egg (c) adaptations for maintaining body temperature (d) surface for gas exchange deep in body (e) ability to maintain location

3. The Parazoa (a) are coelomates (b) include cnidarians (c) include all animals with radial symmetry (d) do not form true tissues (e) have a pseudocoelom

4. Cephalization (a) evolved along with bilateral symmetry (b) is the development of a digestive system (c) is characteristic of cnidarians (d) involves a concentration of excretory organs (e) first evolved in deuterostomes

5. The germ layer that gives rise to the outer covering of the body and the nervous system is the (a) gastrodermis (b) ectoderm (c) contractile layer (d) endoderm (e) mesoderm

6. Radial symmetry is characteristic of (a) protostomes (b) acoelomates (c) deuterostomes (d) cnidarians (e) poriferans

7. A true coelom is completely lined with (a) flagella (b) ectoderm (c) a contractile layer (d) endoderm (e) mesoderm

8. Protostomes are characterized by (a) spiral cleavage (b) indeterminate cleavage (c) enterocoely (d) radial symmetry (e) a distinctive body plan that includes a pseudocoelom

9. The evolution of animals (a) followed an orderly progression from simple to complex (b) will be better understood as biologists continue to collect molecular and other types of data (c) has been determined by studying classification (d) began with the ctenophores (e) began with their common ancestor, a parazoan

10. During cleavage, a zygote (a) undergoes metamorphosis (b) becomes a larva (c) undergoes a series of mitotic divisions and becomes a blastula (d) becomes diploid (e) reproduces sexually

11. Collar cells (choanocytes) are characteristic of (a) phylum Porifera (b) Cnidaria (c) most coelomates (d) Lophotrochozoa (e) Ecdysozoa

12. Which of the following is an example of a deuterostome? (a) a lophotrochozoan (b) a coral (c) a chordate (d) a planarian (e) a pseudocoelomate

13. Cnidocytes (a) are characteristic of sponges (b) are found among cells lining the gastrovascular cavity (c) contain stinging organelles (d) are lined with mesoderm (e) have two main shapes

14. The marine hydrozoan *Obelia* (a) is seen mainly in the medusa form (b) lacks cnidocytes (c) has reproductive polyps that produce eggs and sperm (d) is one of the largest jellyfish known (e) alternates sexual and asexual stages during its life cycle

15. Corals (a) are coelomates (b) form symbiotic relationships with certain algae (c) lack a polyp stage (d) are remarkably resistant to environmental stressors (e) have bilateral symmetry

16. Ctenophores (a) tend to be sessile (b) have a digestive system with only one opening (c) exhibit bilateral symmetry (d) are characterized by adhesive glue cells on their tentacles (e) are free-swimming polyps

17. Label the branches of the diagram. Use Figure 29-8b to check your answers.

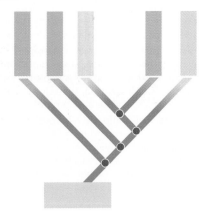

1. Imagine that you discover a new animal in a rain forest. How would you decide to which phylum it belongs? What are some characteristics that might contribute to your decision?

2. Several international monitoring projects are gathering data to help us understand coral reef destruction. Why is it important to take action to protect coral reefs?

3. **Evolution Link.** Biologists traditionally used the type of body cavity to infer relationships among animal groups. In what ways has molecular systematics challenged some of those hypotheses?

4. **Evolution Link.** Animal body plans have generally evolved from simple to more complex. How can we explain the exceptions?

5. **Analyzing Data.** Imagine that a biologist discovers that ctenophores actually branched off from some ancestor that gave rise to the deuterostomes. Where would you place the ctenophore branch on Figure 29-8a?

Additional questions are available in ThomsonNOW at www.thomsonedu.com/login

The Animal Kingdom: The Protostomes

Marty Snyderman / Visuals Unlimited

A bearded fireworm (*Hermodice carunculata*). The fireworm is a segmented worm (an annelid) up to 30 cm (11.8 in) long. It lives in shallow marine waters—in coral reefs, in beds of turtle grass, or under rocks. Fireworm bristles, which are filled with venom, can break off in the skin and cause irritation.

In Chapter 29 we introduced the animal kingdom and discussed Parazoa (sponges) and the radially symmetrical cnidarians and ctenophores. We explored animal phylogeny, examining differences in a purely structural approach and then a structural approach informed by molecular data. Both approaches hypothesize that bilateral animals form two major branches: protostomes and deuterostomes. Molecular data show that the protostomes evolved along two branches: the Lophotrochozoa and the Ecdysozoa. These three clades—Lophotrochozoa, Ecdysozoa, and Deuterostomia—are sometimes referred to as *superphyla*.

In this chapter we begin our discussion of bilateral animals, the most familiar and by far the most numerous animals on Earth. More than 99% of animal species belong to the Bilateria. These diverse forms have adapted to life in almost every imaginable habitat in the ocean and in freshwater and terrestrial environments. The bearded fireworm shown in the photograph is an example of the diverse animals found among the Bilateria.

The remarkable success of the bilateral animals can be attributed to the evolution of adaptations that facilitated food capture, escape from predators, and reproduction. Among their key adaptations are cephalization, a central nervous system, muscles, a coelom, a tube-within-a-tube body plan, and compartmentalization of the body. Large animals developed body systems for gas exchange, waste disposal, and internal transport of nutrients, respiratory gases, wastes, and other materials. In this chapter we focus on the evolution, innovations in body plans, phylogenetic relationships, and life histories of several phyla of the Lophotrochozoa and Ecdysozoa. ∎

KEY CONCEPTS

The evolution of the coelom has been associated with important innovations in body plan, including cephalization, the tube-within-a-tube body plan, compartmentalization, and segmentation.

Protostomes are a monophyletic group that gave rise to two major clades: Lophotrochozoa and Ecdysozoa.

The most biologically successful protostomes in terms of diversity and numbers are the mollusks and the arthropods.

The remarkable biological success of the insects can be attributed to the evolution of complex body plans and life cycles, for example, their exoskeleton, segmentation, specialized jointed appendages, ability to fly, and metamorphosis.

IMPORTANCE OF THE COELOM

Learning Objective

1 Cite specific examples of the evolutionary significance of the coelom.

The true **coelom,** one of the most important early animal adaptations, is a fluid-filled body cavity completely lined by mesoderm that lies between the digestive tube and the outer body wall (see Chapter 29). During the Proterozoic eon, animals were small, simple forms. Evolution of the coelom was an important step in the evolution of larger, more complex animals.

With the evolution of the coelom came a new body design, the **tube-within-a-tube body plan.** The body wall is the outer tube. The inner tube is the digestive tube, which is attached to the body wall at its ends. Typically, the digestive tube has a mouth at one end for taking in food and an anus at the other end for eliminating wastes. Because the coelom separates the muscles of the body wall from those in the wall of the digestive tract, the digestive tube can move food along independently of body movements.

Recall that a **pseudocoelom** is a body cavity that is not lined with mesoderm. Animals with a coelom, and to a lesser extent those with a pseudocoelom, have another major advantage over acoelomate animals. Because it is an enclosed compartment (or series of compartments) of fluid under pressure, the coelom can serve as a **hydrostatic skeleton** in which contracting muscles push against a tube of fluid. In many coelomates, such as the fireworm (shown in the chapter opening photograph), a hydrostatic skeleton permits a greater range of movement than is possible for acoelomate animals. The hydrostatic skeleton also shapes the body of soft animals. The evolution of various shapes and divisions of the coelom provided the opportunity for animals to become specialized in swimming, crawling, or walking. Animals that can move quickly to capture food or avoid predators are more likely to survive.

Evolution of the coelom provided a space where internal organs could develop. Most coelomate animals have well-developed circulatory, excretory, and nervous systems. Many internal organs are suspended within folds of the tissue lining the coelom and can move independently of the outer body wall. For example, the pumping action of the heart is possible because of the surrounding space the coelom provides. The fluid-filled coelom also protects internal organs by cushioning them. In addition, the coelom provides space for the gonads to develop. During the breeding season of many animals, such as birds, the gonads enlarge within the coelom as they fill with ripe gametes.

In some animals, fluid within the coelom helps transport materials such as food, oxygen, and wastes. Cells bathed by the coelomic fluid can exchange materials with it. The cells receive nutrients and oxygen from the coelomic fluid and excrete wastes into it. Some coelomates have excretory structures that remove wastes directly from the coelomic fluid.

Review

■ What are four major modifications in body plan that evolved along with the coelom?

THE LOPHOTROCHOZOA

Learning Objectives

2 Characterize the protostomes, describe their two main evolutionary branches, and give examples of animals assigned to each branch.

3 Identify distinguishing characteristics of phylum Nemertea and phylum Platyhelminthes; describe the main classes of phylum Platyhelminthes, giving examples of animals that belong to each class.

4 Describe the adaptive advantages of cephalization.

5 Describe the distinguishing characteristics of phylum Mollusca and the four molluscan classes discussed, giving examples.

6 Describe the distinguishing characteristics of phylum Annelida and the three annelid classes discussed, giving examples.

7 Describe the distinguishing characteristics of the lophophorate phyla.

The **Lophotrochozoa** include the platyhelminths, nemerteans (ribbon worms), mollusks, annelids, the lophophorate phyla, and the rotifers (see Fig. 29-8). Recall from Chapter 29 that the name *Lophotrochozoa* comes from the names of two major animal groups included: the Lophophorata and the Trochozoa. The *Lophophorata* include the lophophorate phyla, which are characterized by a lophophore, a ciliated ring of tentacles surrounding the mouth. The name *Trochozoa* refers to the trochophore larva that characterizes its two major groups—the mollusks and annelids.

Flatworms are bilateral acoelomates

Members of phylum **Platyhelminthes,** the **flatworms,** are the simplest bilaterally symmetrical, triploblastic (having three embryonic tissue layers) animals. Their flat, elongated bodies are solid. Thus, they are **acoelomate,** meaning that they have no body cavity. These soft-bodied animals left few fossils to provide clues to their evolutionary history.

Biologists historically assigned the 20,000 species to four classes. Class **Turbellaria** consists of the free-living flatworms, including planarians and their relatives. Classes **Trematoda** and **Monogenea** include the flukes, which are either internal or external parasites. Class **Cestoda** includes the tapeworms, which as adults are intestinal parasites of vertebrates (■ Table 30-1).

Historically, the flatworms were thought to be a basal group. However, molecular data, specifically 18S ribosomal RNA sequences, indicate that phylum Platyhelminthes, as traditionally classified, is not a monophyletic group. Systematists have reclassified two orders (from class Turbellaria), based on this molecular data, that differ significantly from other flatworms. These two orders have been given their own phylum (Acoelomorpha). This group may be closely related to the common ancestor that linked radially symmetrical animals with bilateral animals.

The remaining flatworms appear more closely related to coelomates. If so, they originally had a body cavity, but their body plan became simplified later in their evolutionary history. Many biologists currently classify them as lophotrochozoans.

TABLE 30-1

Classes of Phylum Platyhelminthes

Class and Representative Animals	Characteristics
Turbellaria Planarians	Mainly free-living; mainly marine; body covered by ciliated epidermis; typically carnivorous; prey on tiny invertebrates
Trematoda and Monogenea Flukes	All parasites with a wide range of vertebrate and invertebrate hosts; may require intermediate hosts; adults have suckers for attachment to host
Cestoda Tapeworms	Parasites of vertebrates; complex life cycle usually with one or two intermediate hosts; larval host may be invertebrate; typically have suckers and sometimes hooks for attachment to host; eggs produced within proglottids, which are shed; no digestive or nervous systems

This phylogeny is controversial. As biologists consider new data, their assessment of the phylogenetic relationships of flatworms may change.

Flatworms have definite anterior and posterior ends. They exhibit **cephalization,** the evolution of a head with a concentration of sense organs at the anterior end. A simple brain and paired sense organs are concentrated in the "head" region. An animal with a front end generally moves forward. With a concentration of sense organs in the part of the body that first meets its environment, the animal can find food, shelter, and mates or detect an enemy quickly.

Flatworms are triploblastic, that is, they have three definite tissue layers. In addition to its outer epidermis derived from ectoderm, and an inner endodermis derived from endoderm, the flatworm has a middle tissue layer that develops from mesoderm.

Flatworms have no organs for circulation or gas exchange. These functions depend largely on diffusion through the body wall. The activities of some organs are coordinated and form simple organ systems, for example, the digestive and nervous systems. As in the cnidarians, the digestive system is a **gastrovascular cavity** with only one opening, a mouth. The gastrovascular cavity is often extensively branched.

Flatworms typically have a simple **nervous system.** The brain consists of two masses of nervous tissue, called **ganglia,** in the head region. In many species, the ganglia connect to two nerve cords that extend the length of the body. This nervous system is sometimes referred to as a "ladder-type nervous system," because a series of nerves connects the cords like the rungs of a ladder.

Parasitic flatworms—flukes and tapeworms—are highly adapted to and modified for their parasitic lifestyle. They have

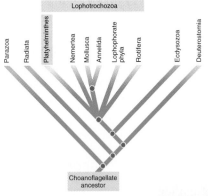

(a) Internal structure.

(b) LM of a living planarian (*Dugesia dorotocephala*). Note the prominent auricles, used to locate food.

1 mm

T. E. Adams/Peter Arnold, Inc.

(c) Cross section through a planarian.

Figure 30-1 *Animated* The common planarian, *Dugesia*

suckers or hooks for holding on to their hosts. The bodies of those that live in digestive tracts resist the digestive enzymes secreted by their hosts. Many have complicated life cycles and produce large numbers of eggs. Other adaptations include the loss of certain structures such as sense organs. Tapeworms have also lost the digestive system.

Turbellarians are free-living flatworms

Most members of class Turbellaria are free-living, marine flatworms. However, many inhabit freshwater habitats, and some are terrestrial. Turbellarian flatworms typically have a muscular pharynx that takes in food; a simple brain, eyespots, and other sensory organs in the head; **protonephridia,** structures that function in *osmoregulation* (fluid balance) and metabolic waste disposal (excretion); and reproductive organs.

Planarians are turbellarian flatworms found in ponds and quiet streams throughout the world. The common American planarian *Dugesia* is about 15 mm (0.6 in) long, with what appear to be crossed eyes and flapping "ears" called **auricles** (▌ Fig. 30-1). The auricles actually serve as organs of chemoreception, important in locating food. Planarians are capable of learning. Memory is not localized within the ganglia but appears to be retained throughout the nervous system.

Planarians are carnivorous; they trap small animals in a mucous secretion. The digestive system consists of a single opening (the mouth); a tubelike, muscular **pharynx** (the first portion of the digestive tube); and a branched gastrovascular cavity. A planarian projects its pharynx outward through its mouth, using it to suck in prey. The long, highly branched gastrovascular cavity distributes food to all parts of the body so each cell can receive nutrients by diffusion.

Although excretion also takes place mainly by diffusion, some metabolic wastes are excreted by the protonephridia. Protonephridia are blind tubules that end in **flame bulbs,** collecting cells equipped with cilia. The beating of the cilia channels waste through the system of tubules and eventually out of the body through excretory pores. Osmoregulation and protonephridia are discussed further in Chapter 47 (see Fig. 47-2).

Planarians reproduce either asexually or sexually. In asexual reproduction, an individual constricts in the middle and divides into two planarians. Each regenerates its missing parts. Sexually, these animals are **hermaphrodites.** During the warm months of the year, each is equipped with a complete set of male and female organs. Two planarians come together in copulation and exchange sperm cells so that their eggs are cross-fertilized.

Flukes parasitize other animals

Although their body plan resembles that of the free-living flatworms, the **flukes,** members of classes Trematoda and Monogenea, have structures, such as hooks and suckers, for attachment to their host. Flukes also have extremely prolific reproductive organs.

Flukes that are parasitic in humans include blood flukes, widespread in tropical areas of the world, and liver flukes, common in Asia, particularly in areas where humans use their own feces for fertilizing crops. Blood flukes of the genus *Schistosoma* infect about 200 million people who live in tropical areas. Both blood flukes and liver flukes go through complicated life cycles involving several different forms. There is an alternation of sexual and asexual stages, and parasitism on one or more intermediate hosts, such as snails and fishes (▌ Fig. 30-2). The aquatic snails that serve as intermediate hosts thrive in ponds, rice paddies, and the marshy areas that develop when dams are built.

Tapeworms inhabit the intestine of vertebrates

Adult members of the more than 5000 species of class Cestoda live as parasites in the intestine of probably every kind of vertebrate, including humans. Tapeworms are long, flat, ribbonlike animals strikingly specialized for their parasitic mode of life. Among their many adaptations are suckers and sometimes hooks on the "head," or **scolex,** that enable the parasite to attach to the host's intestine (▌ Fig. 30-3).

The reproductive adaptations and abilities of tapeworms are extraordinary. The body of the tapeworm consists of a long chain of segments called **proglottids.** Each proglottid is an entire reproductive machine equipped with both male and female reproductive organs and contains up to 100,000 eggs. Because an adult tapeworm may have as many as 2000 segments, its reproductive potential is staggering. A single tapeworm can produce 600 million eggs per year! Proglottids farthest from the tapeworm's head contain the ripest eggs; these segments are shed from the host's body along with the feces.

The tapeworm has no mouth or digestive system. Digested food from the host is absorbed across the worm's body wall. The tapeworm also lacks well-developed sense organs. Some tapeworms have complex life cycles, spending their larval stage within the body of an intermediate host and their adult life within the body of a different, final host. ▌ Figure 30-4 illustrates the life cycle of the beef tapeworm, which can infect humans when they eat undercooked beef containing the larvae.

Phylum Nemertea is characterized by the proboscis

Phylum **Nemertea,** the ribbon worms or proboscis worms, is a relatively small group (about 1200 species) of free-living animals (▌ Fig. 30-5 on page 646). Most burrow in marine sediments, but a few species inhabit deep sea water, fresh water, or damp soil. Many nemerteans are carnivorous, feeding on crustaceans and annelids. Nemerteans have long, narrow bodies, either cylindrical or flattened, ranging in length from 1 mm to more than 30 m (100 ft). Some are a vivid orange, red, or green, with black or colored stripes.

Their most remarkable organ, the **proboscis,** from which one of their common names, *proboscis worms,* derives, is a long, hollow, muscular tube that can be rapidly everted (turned inside out) from the anterior end of the body. Used to capture prey, the sticky proboscis can be wrapped around a small animal. In some species the proboscis is sharp, and in various species it secretes

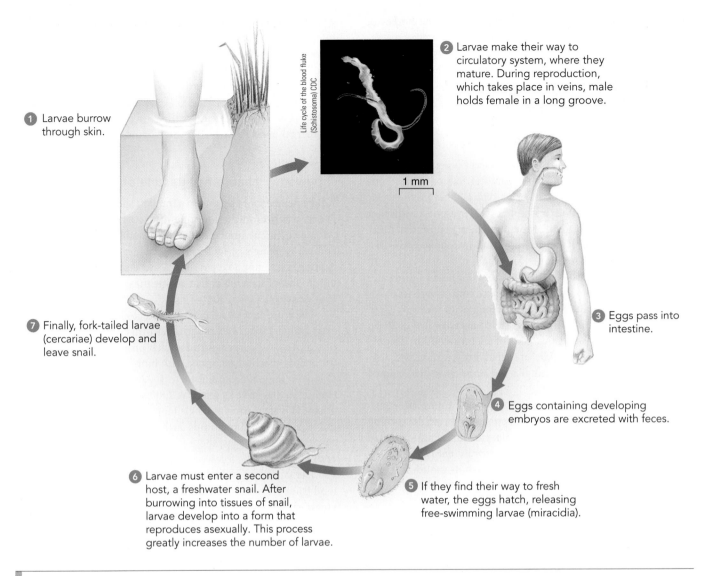

1 Larvae burrow through skin.

2 Larvae make their way to circulatory system, where they mature. During reproduction, which takes place in veins, male holds female in a long groove.

Life cycle of the blood fluke (*Schistosoma*) CDC

1 mm

3 Eggs pass into intestine.

4 Eggs containing developing embryos are excreted with feces.

5 If they find their way to fresh water, the eggs hatch, releasing free-swimming larvae (miracidia).

6 Larvae must enter a second host, a freshwater snail. After burrowing into tissues of snail, larvae develop into a form that reproduces asexually. This process greatly increases the number of larvae.

7 Finally, fork-tailed larvae (cercariae) develop and leave snail.

Figure 30-2 *Animated* Life cycle of the blood fluke (*Schistosoma*)

The LM shows an adult male enfolding a smaller female.

toxic fluid that immobilizes prey. The proboscis is a *derived character* that distinguishes nemerteans from all other invertebrate groups.

Historically, the nemerteans were of special evolutionary interest because biologists thought they were the earliest animals to have a circulatory system and a complete digestive tract. Nemerteans have no heart; blood is circulated by contractions of muscular blood vessels and by movements of the body. Although the nemerteans are functionally acoelomate, the system of blood vessels may be homologous to the coelom of other coelomate animals.

The chamber surrounding the proboscis is a true coelomic space, known as a **rhynchocoel.** It develops in the embryo as the coelom does in coelomate protostomes (see Chapter 29). Biologists now view the rhynchocoel as a remnant of the coelom of a coelomate ancestor. Because of its presence and recent molecular evidence, biologists now classify nemerteans with the lophotrochozoans.

Mollusks have a muscular foot, visceral mass, and mantle

Phylum **Mollusca** includes clams, oysters, snails, slugs, octopods, and the largest of all the invertebrates, the giant squid, which averages 9 to 16 m (about 30 to 53 ft) in length, including its tentacles. More than 50,000 living species and 35,000 fossil species (second only to the arthropods in number) have been described. In this chapter we discuss four of the eight recognized classes. We illustrate representatives of these classes in ▌Figure 30-6 and describe them in ▌Table 30-2.

Although most mollusks are marine, many snails and clams live in fresh water, and some species of snails and slugs inhabit the land. Mollusks probably evolved early in the history of the protostome clade, soon after the evolution of the coelom but before the origin of the segmented body that is characteristic of annelids. Although mollusks vary widely in outward appearance, most share six basic characteristics:

Cath Ellis/Science Photo Library/Photo Researchers, Inc.

250 μm

Figure 30-3 Scolex of a tapeworm

The small tapeworm (*Acanthrocirrus retrisrostris*) reaches maturity in the intestine of wading birds that eat barnacles. The color-enhanced SEM shows the pistonlike cluster of hooks that can be withdrawn into the head or thrust out and buried in the host's tissue. You can see two of the four powerful suckers beneath the hooks.

1. A soft body, usually covered by a dorsal shell composed mainly of calcium carbonate.

2. A broad, flat, muscular **foot,** located ventrally, which is used for locomotion.

3. The body organs (viscera) concentrated as a **visceral mass** located above the foot.

4. A **mantle,** a thin sheet of tissue that covers the visceral mass and usually contains glands that secrete a shell. The mantle generally overhangs the visceral mass, forming a mantle cavity that contains gills and other structures.

5. A rasplike structure called a **radula,** which is a belt of teeth in the mouth region. (The radula is not present in clams or their relatives, which are suspension feeders.)

6. A coelom, generally reduced to small compartments around certain organs, including the heart and excretory organs (*metanephridia*). The main body cavity is typically a **hemocoel,** a space containing blood (see following discussion of open circulatory system). The hemocoel is not a coelom.

Mollusks have all of the organ systems typical of complex animals. The digestive system is a tube, often coiled, consisting of a mouth, buccal cavity (mouth cavity), esophagus, stomach, intestine, and anus. The mollusk radula, located within the buccal cavity, projects out of the mouth and is used to scrape particles of food from the surface of rocks or the ocean floor. Sometimes the radula is used to drill a hole in the shell of a prey animal or to tear off pieces of a plant.

Most mollusks have an **open circulatory system** in which the blood, called **hemolymph,** bathes the tissues directly. The heart pumps blood into a single blood vessel, the aorta, which may branch into other vessels. Eventually, blood flows into a network of large spaces called *sinuses,* bringing the blood into direct contact with the tissues. This network makes up the hemocoel, or blood cavity. From the sinuses, blood drains into vessels that conduct it to the gills, where it is recharged with oxygen. After passing through the gills, the blood returns to the heart. Thus, blood flow in a mollusk follows this pattern:

Heart ⟶ aorta ⟶ smaller blood vessels ⟶ blood sinuses (hemocoel) ⟶ blood vessels to gills ⟶ heart

In open circulatory systems, blood pressure tends to below, and tissues are not very efficiently oxygenated. Because

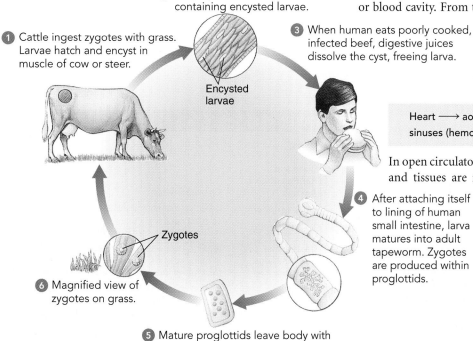

❶ Cattle ingest zygotes with grass. Larvae hatch and encyst in muscle of cow or steer.

❷ Magnified view of muscle containing encysted larvae.

Encysted larvae

❸ When human eats poorly cooked, infected beef, digestive juices dissolve the cyst, freeing larva.

❹ After attaching itself to lining of human small intestine, larva matures into adult tapeworm. Zygotes are produced within proglottids.

❺ Mature proglottids leave body with feces and burst open, releasing zygotes.

❻ Magnified view of zygotes on grass.

Zygotes

Figure 30-4 *Animated* Life cycle of the beef tapeworm

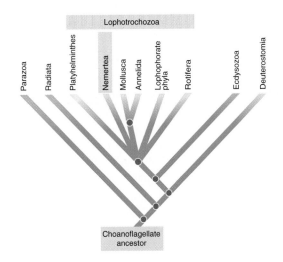

Figure 30-5 Nemerteans (proboscis worms)

(a) Proboscis worm (*Lineus*) from the Pacific coast of Panama.

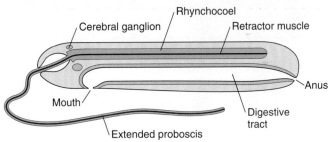

(b) Lateral view of a typical nemertean. Note the complete digestive tract that extends from mouth to anus, giving this animal a tube-within-a-tube body plan.

most mollusks are slow-moving animals with low metabolic rates, this type of circulatory system is adequate. The active cephalopods (the class that includes the squids and octopods), have a **closed circulatory system** in which blood flows through a complete circuit of blood vessels.

Most marine mollusks pass through one or more larval stages. The first stage is typically a **trochophore larva,** a free-swimming, top-shaped larva with two bands of cilia around its middle (Fig. 30-7). In many mollusks (gastropods such as snails and bivalves such as clams), the trochophore larva develops into a **veliger larva,** which has a shell, foot, and mantle. The veliger larva is unique to the mollusks.

Chitons may be similar to ancestral mollusks

Class **Polyplacophora** (meaning "many plates") consists of the **chitons,** marine animals with flattened bodies (see Fig. 30-6a). Their most distinctive feature is a shell composed of eight separate but overlapping dorsal plates. The head is reduced, and there are no eyes or tentacles.

The chiton inhabits rocky intertidal zones, using its broad, flat foot to move and to hold on firmly to rocks. By pressing its mantle against the substratum and lifting the inner edge of the mantle, the chiton can produce a partial vacuum. The resulting suction lets the animal adhere powerfully to its perch. Using its radula for grazing, the chiton scrapes algae and other small organisms off rocks and shells.

Gastropods are the largest group of mollusks

Class **Gastropoda,** which includes snails, slugs, conchs, sea slugs, and their relatives, is the largest and most diverse group of mollusks (see Fig. 30-6b). In fact, with more than 70,000 extant species, gastropods constitute the second-largest class in the animal kingdom, second only to insects. Most gastropods inhabit marine waters, but others make their homes in brackish or fresh water or on land. We think of snails as having a single, spirally coiled shell into which they can withdraw the body, and many do. However,

other gastropods, such as limpets, have shells like flattened dunce caps. Still others, such as garden slugs and the beautiful marine slugs known as **nudibranchs,** have no shell at all (Fig. 30-8).

Many gastropods have a well-developed head with tentacles. Two simple eyes may be located on stalks that extend from the head. The gastropod uses its broad, flat foot for creeping. Most land snails do not have gills. Instead, the mantle is highly vascularized and functions as a lung. Biologists describe these garden snails and slugs as **pulmonate** ("having a lung").

TABLE 30-2

Major Classes of Phylum Mollusca

Class and Representative Animals	Characteristics
Polyplacophora Chitons	Primitive marine animals with shell consisting of eight separate transverse plates; head reduced; broad foot used for locomotion
Gastropoda Snails, slugs, nudibranchs	Marine, freshwater, or terrestrial; coiled shell in many species; torsion of visceral mass; well-developed head with tentacles and eyes
Bivalvia Clams, oysters, mussels	Marine or freshwater; body laterally compressed; two-part shell hinged dorsally; hatchet-shaped foot; suspension feeders
Cephalopoda Squids, octopods	Marine; predatory; foot modified into tentacles, usually bearing suckers; well-developed eyes; closed circulatory system

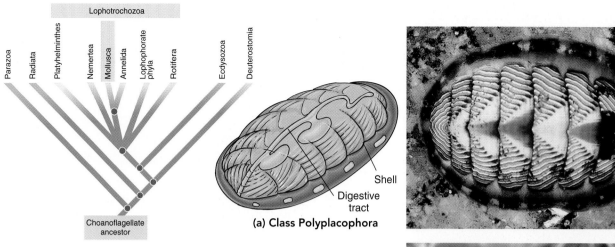

(a) Class Polyplacophora

Shell
Digestive tract

Kjell B. Sandved

Figure 30-6 *Animated*
The mollusk body plan

(a) Chitons are sluggish marine animals with shells composed of eight overlapping plates. This sea cradle chiton (*Tonicella lineata*), which reaches about 5 cm (2 in) in length, inhabits rocks covered with corraline algae in coastal waters off the U.S. Pacific Northwest. **(b)** The broad, flat foot of the gastropod is an adaptation to its mobile lifestyle. The mystery snail (*Pomacea bridgesi*) inhabits fresh water and can be found burrowing in the mud. **(c)** The compressed body of the horseneck clam (*Tresus capax*) is adapted for burrowing in the North Pacific mud. Its shell grows to about 20 cm (8 in) in length. **(d)** The squid body is streamlined for swimming. To avoid being seen by potential predators, the squid can change color to blend with its background. This northern shortfin squid (*Illex illecebrosus*), which grows to a mantle length of up to about 31 cm (about 1 ft), inhabits the North Atlantic ocean.

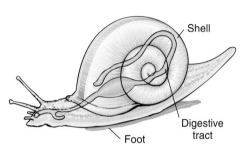

Shell
Digestive tract
Foot

(b) Class Gastropoda

E. R. Degginger/Dembinsky Photo Associates

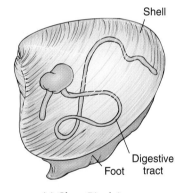

Shell
Digestive tract
Foot

(c) Class Bivalvia

Tom McHugh/Photo Researchers, Inc.

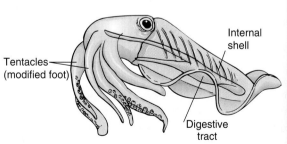

Tentacles (modified foot)
Internal shell
Digestive tract

(d) Class Cephalopoda

Jeffrey Rotman/Peter Arnold, Inc.

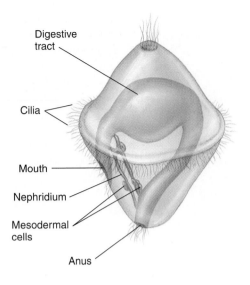

Figure 30-7 A trochophore larva

The first larval stage of a marine mollusk, the trochophore larva, is also characteristic of annelids. Just above the mouth a band of ciliated cells functions as a swimming organ and, in some species, collects suspended food particles.

Figure 30-8 Nudibranch

A nudibranch (*Chromodoris porterae*) feeds on a lightbulb tunicate (*Clavelina huntsmani*).

Torsion, a twisting of the visceral mass, is a unique feature of gastropods. This twisting is unrelated to the coiling of the shell. As the bilateral larva develops, one side of the visceral mass grows more rapidly than the other side. This uneven growth rotates the visceral mass. The visceral mass and mantle twist permanently up to 180 degrees relative to the head. As a result, the digestive tract becomes somewhat U shaped, and the anus comes to lie above the head and gill (Fig. 30-9). Subsequent growth is dorsal and usually in a spiral coil. Torsion limits space in the body, and typically the gill, metanephridium (excretory organ), and gonad

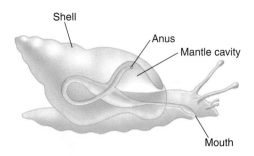

Figure 30-9 *Animated* Torsion in a gastropod

As the bilateral larva develops, the visceral mass twists 180 degrees relative to the head. The digestive tube coils, and the anus becomes relocated near the mouth.

are absent on one side. Some biologists hypothesize that torsion is an adaptation that protects the head by allowing it to enter the shell first during withdrawal from potential predators. Without torsion, the foot would be withdrawn first.

Bivalves typically burrow in the mud

Class **Bivalvia** includes the clams, oysters, mussels, scallops, and their relatives (see Fig. 30-6c). Typically, the soft body of members of class Bivalvia is laterally compressed and completely enclosed by a two-part shell that hinges dorsally and opens ventrally (Fig. 30-10). This arrangement allows the hatchet-shaped foot to protrude ventrally for locomotion and for burrowing in mud. The two parts, or **valves,** of the shell are connected by an elastic ligament. Stretching of the ligament opens the shell. Large, strong adductor muscles attached to the shell permit the animal to close its shell.

The inner, pearly layer of the bivalve shell is made of calcium carbonate secreted in thin sheets by the epithelial cells of the mantle. Known as *mother-of-pearl*, this material is valued for making jewelry and buttons. Should a bit of foreign matter lodge between the shell and the mantle, the epithelial cells covering the mantle secrete concentric layers of calcium carbonate around the intruding particle. Many species of oysters and clams form pearls in this way.

Some bivalves, such as oysters, attach permanently to the substrate. Others burrow slowly through rock or wood, seeking protected dwellings. The shipworm, *Teredo,* a clam that damages dock pilings and other marine installations, is just looking for a home. A few bivalves, such as scallops, swim rapidly by clapping their two shells together with the contraction of a large adductor muscle (the part of the scallop that humans eat).

Clams and oysters are suspension feeders that trap food particles in sea water. They take water in through an extension of the mantle called the **incurrent siphon.** Water leaves by way of an **excurrent siphon.** As the water passes over the gills, mucus secreted by the gills traps food particles in the stream of water. Cilia move the food to the mouth. An oyster can filter more than 30 L (about

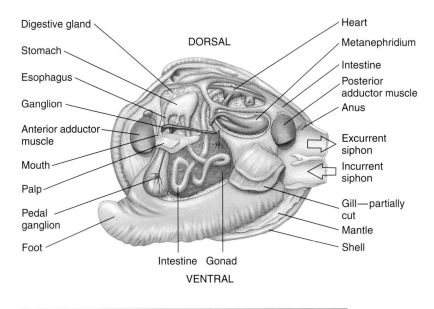

Digestive gland
Stomach
Esophagus
Ganglion
Anterior adductor muscle
Mouth
Palp
Pedal ganglion
Foot

DORSAL

Heart
Metanephridium
Intestine
Posterior adductor muscle
Anus
Excurrent siphon
Incurrent siphon
Gill—partially cut
Mantle
Shell

Intestine   Gonad
VENTRAL

Figure 30-10 *Animated*   Internal anatomy of a clam

The two shells of a bivalve hinge dorsally and open ventrally.

32 qt) of water per hour! As suspension feeders, bivalves have no need for a radula, and indeed they are the only group of mollusks that lack this structure.

Cephalopods are active, predatory animals

In contrast with most other mollusks, members of class **Cephalopoda** (meaning "head-foot") are fast-swimming predatory animals (see Fig. 30-6d). The cephalopod mouth is surrounded by tentacles, or arms: 8 in octopods, 10 in squids, and as many as 90 in the chambered nautilus. The large head has well-developed eyes that form images. Although they develop differently, their complex eyes are structurally similar to vertebrate eyes and function in much the same way. The octopus has no shell; and the shell of the squid, located inside the body, is greatly reduced.

Nautilus has a coiled shell consisting of many chambers built up over time. Each year the animal lives in the newest and largest chamber of the series. *Nautilus* secretes a mixture of gases similar to air into the other chambers. By regulating the amount of gas in the chambers, the animal controls its depth in the ocean.

The tentacles of squids, octopods, and cuttlefish are covered with suckers for seizing and holding prey. In addition to a radula, the mouth has two strong, horny beaks used to kill prey and tear it to bits. The thick, muscular mantle is fitted with a funnel-like **siphon.** By filling the cavity with water and ejecting it through the siphon, the cephalopod achieves forceful jet propulsion. Squids and cuttlefish have streamlined bodies adapted for efficient swimming.

In addition to speed, the cephalopod has two other adaptations that facilitate escape from its predators, which include certain whales and moray eels. It can confuse the enemy by rapidly changing colors. By expanding and contracting *chromatophores,* cells in the skin that contain pigment granules, the cephalopod

can display an impressive variety of mottled colors. Another defense mechanism is its **ink sac,** which produces a thick, black liquid the animal releases in a dark cloud when alarmed. While its enemy pauses, temporarily blinded and confused, the cephalopod escapes. The ink inactivates the chemical receptors of some predators, making them incapable of detecting their prey.

The octopus feeds on crabs and other arthropods, catching and killing them with a poisonous secretion of its salivary glands. During the day the octopus usually hides among the rocks; in the evening it emerges to hunt for food. Its motion is incredibly fluid, giving little hint of the considerable strength in its eight arms.

Small octopods survive well in aquariums and have been studied extensively. Because octopods are relatively intelligent and can make associations among stimuli, researchers have used them as models for studying learning and memory. Their highly adaptable behavior more closely resembles that of the vertebrates than the more stereotypic patterns of behavior seen in other invertebrates (see Chapter 51).

Annelids are segmented worms

Members of phylum **Annelida,** the segmented worms, have bilateral symmetry and a tubular body that may be partitioned into more than 100 ringlike segments. This phylum, composed of about 15,000 species, includes three main classes: the polychaetes, a group of marine and freshwater worms; the earthworms; and the leeches (⬛ Table 30-3 and ⬛ Fig. 30-11).

The term *Annelida* (meaning "ringed") refers to the series of rings, or segments, that make up the annelid body. Both the body wall and many of the internal organs are segmented. Some structures, such as the digestive tract and certain nerves, extend the length of the body, passing through successive segments. Other structures, such as excretory organs, are repeated in each segment. In polychaetes and earthworms, segments are separated from one another internally by transverse partitions called **septa.**

TABLE 30-3

Major Classes of Phylum Annelida

Class and Representative Animals	Characteristics
Polychaeta Sandworms, tubeworms	Mainly marine; each segment bears a pair of parapodia with many setae; well-developed head; separate sexes; trochophore larva
Oligochaeta Earthworms	Terrestrial and freshwater worms; few setae per segment; lack well-developed head; hermaphroditic
Hirudinea Leeches	Most are blood-sucking parasites that inhabit fresh water; appendages and setae absent; prominent muscular suckers

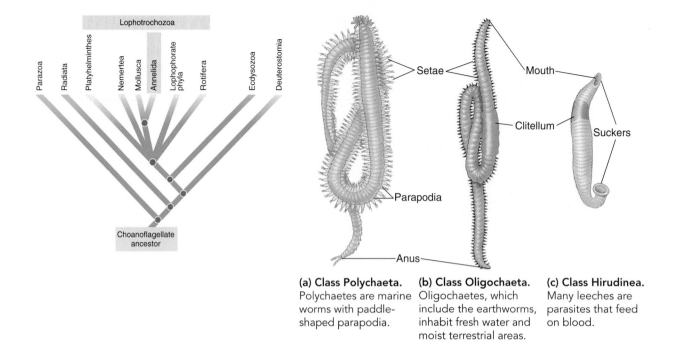

(a) Class Polychaeta. Polychaetes are marine worms with paddle-shaped parapodia.

(b) Class Oligochaeta. Oligochaetes, which include the earthworms, inhabit fresh water and moist terrestrial areas.

(c) Class Hirudinea. Many leeches are parasites that feed on blood.

Figure 30-11 Three major classes of annelids

An important advantage of **segmentation** is that it facilitates locomotion. The coelom is divided into segments, and each segment has its own muscles. This arrangement allows the animal to elongate one part of its body while shortening another part. The annelid's hydrostatic skeleton is important in movement. In polychaetes and earthworms, bristlelike structures called **setae** (sing., *seta*), located on each segment, provide traction as the worm moves along by alternating contraction of its longitudinal and circular muscles.

The annelid nervous system generally is a ventral nerve cord and a simple brain consisting of a pair of ganglia. A pair of ganglia and lateral nerves is repeated in each segment. Annelids have a large, well-developed coelom, a closed circulatory system, and a complete digestive tract extending from mouth to anus. Respiration is *cutaneous,* that is, through the skin, or by gills. Typically, a pair of excretory tubules called **metanephridia** is located in each segment (described in Chapter 47).

Most polychaetes are marine worms with parapodia

Class **Polychaeta** includes marine worms that swim freely in the sea, burrow in the mud near the shore, or live in tubes they secrete or make by cementing bits of shell and sand together with mucus. Each body segment typically bears a pair of paddle-shaped appendages called **parapodia** (sing., *parapodium*) that function in locomotion and in gas exchange. These fleshy structures bear many stiff setae (the term *polychaete* means "many bristles"; see the fireworm in the chapter opening photograph). Most polychaetes have a well-developed head with eyes and antennae. The head may also be equipped with tentacles and palps (feelers).

Polychaetes develop from free-swimming trochophore larvae similar to those of mollusks.

Behavioral patterns that ensure fertilization have evolved in many polychaete species. By responding to certain rhythmic variations, or cycles, in the environment, nearly all the females and males of a given species release their gametes into the water at the same time. For example, more than 90% of reef-dwelling *Palolo* worms of the South Pacific shed their eggs and sperm within a 2-hour period on one night of the year. The posterior portion of the *Palolo* worm, loaded with eggs or sperm, breaks off from the rest of the body. It comes to the surface and bursts, releasing its gametes. This mass reproductive event occurs in October or November at the beginning of the last lunar quarter. Local islanders gather great numbers of the swarming polychaetes, which they bake or eat raw.

Earthworms help maintain fertile soil

The approximately 3000 species of class **Oligochaeta** live almost exclusively in fresh water and in moist terrestrial habitats. These worms lack parapodia, have few bristles per segment (the term *oligochaete* means "few bristles"), and lack a well-developed head. All oligochaetes are hermaphroditic.

Lumbricus terrestris, the common earthworm, is about 20 cm (8 in) long. Its body has more than 100 segments, separated externally by grooves that indicate the internal position of the septa (Fig. 30-12). The earthworm's body is somewhat protected from drying by a thin, transparent **cuticle** secreted by the cells of the epidermis. Mucus secreted by gland cells of the epidermis forms an additional protective layer over the body surface. The body

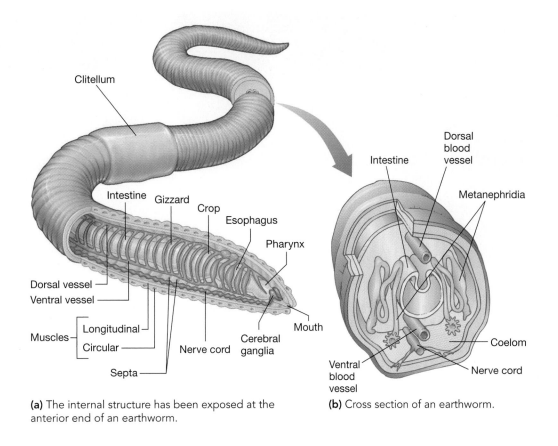

(a) The internal structure has been exposed at the anterior end of an earthworm.

(b) Cross section of an earthworm.

Figure 30-12 *Animated* Body plan of an earthworm

wall has an outer layer of circular muscles and an inner layer of longitudinal muscles.

An earthworm literally eats its way through the soil. It ingests its own weight in soil and nutritious decaying vegetation every 24 hours! Earthworms enhance the formation and maintenance of fertile soil. As they move about, earthworms turn and aerate the soil and enrich it with nitrogenous wastes.

The earthworm processes its meal of soil in a complex digestive system. Food is swallowed through the muscular **pharynx** and passes through the **esophagus,** which is modified to form a thin-walled **crop** where food is stored and a thick-walled, muscular **gizzard.** In the gizzard, food is ground to bits by sand grains consumed along with the meal. The rest of the digestive system is a long, straight **intestine** that chemically digests food and absorbs nutrients. Wastes pass out of the intestine to the exterior through the **anus.**

The efficient closed circulatory system consists of two main blood vessels that extend longitudinally. The dorsal blood vessel, just above the digestive tract, collects blood from vessels in the segments, then contracts to pump the blood anteriorly. In the region of the esophagus, five pairs of blood vessels propel blood from the dorsal to the ventral blood vessel. Located just below the digestive tract, the ventral blood vessel carries blood posteriorly. Small vessels branch from it and deliver blood to the various structures in each segment, as well as to the body wall. Within

these structures blood flows through tiny capillaries before returning to the dorsal blood vessel.

Gas exchange takes place through the moist skin. Oxygen is transported by hemoglobin, the respiratory pigment in the blood plasma (the liquid part of the blood). The excretory system consists of paired metanephridia, repeated in almost every segment of the body.

The nervous system consists of a pair of **cerebral ganglia** (a simple brain) just above the pharynx and a subpharyngeal ganglion just below the pharynx. A ring of nerve fibers connects these ganglia. From the lower ganglion a ventral nerve cord extends beneath the digestive tract to the posterior end of the body. A pair of fused ganglia is present in each segment along the nerve cord. Nerves extend laterally from the ganglia to the muscles and other structures of that segment. The ganglia coordinate muscle contractions of the body wall, allowing the worm to creep along.

Like other oligochaetes, earthworms are hermaphroditic. During copulation, two worms, headed in opposite directions, press their ventral surfaces together. These surfaces become glued together by the thick mucous secretions of each worm's *clitellum,* a thickened ring of epidermis. Sperm are then exchanged and stored in the *seminal receptacles* (small sacs) of the other worm. A few days later each clitellum secretes a membranous cocoon containing a sticky fluid. As the cocoon is slipped forward, eggs are laid in it. Sperm are added as the cocoon passes over the

T. E. Adams/Visuals Unlimited

500 μm

St. Bartholomew's Hospital/Science Photo Library/Photo Researchers, Inc.

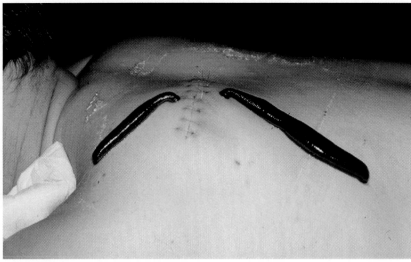

(a) LM of a blood-sucking leech (*Helobdella stagnalis*) that feeds on mammals. The digestive tract (*dark area*) of its swollen body is filled with ingested blood.

(b) Medicinal leeches (*Hirudo medicinalis*) are used here to treat hematoma, an accumulation of blood within tissues that results from injury or disease.

Figure 30-13 Leeches

openings of the seminal receptacles. As the cocoon slips free over the worm's head, its openings constrict so that a spindle-shaped capsule is formed. The fertilized eggs develop into tiny worms within this capsule. This reproductive pattern is an adaptation to terrestrial life; the cocoon protects the delicate gametes and young worms from drying out.

Many leeches are blood-sucking parasites

Most leeches, members of class **Hirudinea,** inhabit fresh water, but some live in the sea or in moist areas on land. About 75% of the known species of leeches are blood-sucking parasites. Some leeches are nonparasitic predators that capture small invertebrates such as earthworms and snails. Leeches differ from other annelids in having neither setae nor parapodia.

Leeches have muscular suckers at both the anterior and posterior ends of their body. Most parasitic leeches attach themselves to a vertebrate host, bite through the skin, and suck out a quantity of blood, which is stored in pouches in the digestive tract. *Hirudin,* an anticoagulant secreted by glands in the crop, ensures leeches a full meal of blood. In 30 minutes, a leech can suck out as much as 10 times its own weight in blood! Some leeches feed only about twice each year because they digest their food slowly over several months.

Leeches have been used since ancient times for drawing blood from areas swollen by poisonous stings and bites. During the 19th century they were widely used to remove "bad blood," thought to be the cause of many diseases. Leeches are used in modern medicine to remove excess fluid and blood that accumulate within body tissues as a result of injury, disease, or surgery

(Fig. 30-13). The leech attaches its sucker near the site of injury, makes an incision, and secretes hirudin. The hirudin prevents the blood from clotting and dissolves clots already existing.

The lophophorate phyla are distinguished by a ciliated ring of tentacles

With a few exceptions, lophophorates are marine animals adapted for life on the ocean floor. These coelomates are distinguished by their **lophophore,** a ring of ciliated tentacles that surrounds the mouth. The lophophore is an adaptation for capturing suspended particles in the water. Lophophorates do not have a distinct head.

Biologists are still debating the evolutionary position of the **lophophorate phyla.** They were traditionally considered a monophyletic group and a sister group to the deuterostomes, based on certain structural similarities. However, brachiopods and bryozoans have bristles composed of chitin, suggesting a relationship with the annelids and other lophotrochozoans. Although their phylogeny is still being debated, based on molecular data (18S rDNA nucleotide sequences), the lophophorate phyla are now classified as Lophotrochozoa.

The lophophorate phyla include Brachiopoda, Phoronida, and Bryozoa. **Brachiopods,** or lampshells, are solitary marine animals that inhabit cold water. Until the mid-19th century, they were thought to be mollusks because they are suspension feeders that superficially resemble clams and other bivalve mollusks. Their body is enclosed between two shells and has a mantle and mantle cavity (Fig. 30-14a). However, they differ from bivalve

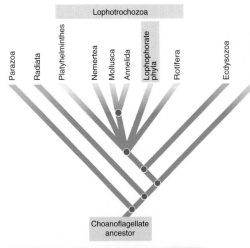

Lophotrochozoa

Parazoa · Radiata · Platyhelminthes · Nemertea · Mollusca · Annelida · Lophophorate phyla · Rotifera · Ecdysozoa · Deuterostomia

Choanoflagellate ancestor

(a) Phylum Brachiopoda. Like other brachiopods, these northern lamp shells (*Terebratulina septentrionalis*) superficially resemble clams.

Andrew J. Martinez/Photo Researchers, Inc.

Stan Elems/Visuals Unlimited

(b) Phylum Phoronida. A single lophophore of the phoronid (*Phoronopsis viridis*). This animal is about 20 cm (8 in) long, but only 3 mm (1/8 in) in diameter.

M. I. Walker/Photo Researchers, Inc.

(c) Phylum Bryozoa (ectoprocts). Like many bryozoans, this freshwater species (*Cristatella mucedo*) forms colonies by asexual budding.

Figure 30-14 Representative lophophorates

mollusks in that the shells are dorsal and ventral rather than lateral; each shell is symmetrical about the midline, and the two shell valves are typically of unequal size.

Brachiopods attach to the substrate by a long stalk. Cilia on the lophophore beat, bringing water laden with food into the slightly opened shell. Although there are now only about 350 living species, brachiopods were abundant and diversified during the Paleozoic era. Paleobiologists have described about 30,000 fossil species.

Only about 20 extant species of **phoronids** are known. They are wormlike, sessile animals found in coastal marine sediments. Adult phoronids secrete chitinous tubes in which they live (Fig. 30-14b). They extend their lophophores from their tubes for feeding. Phoronids vary in length from 2 cm to more than 20 cm (less than 1 in to about 8 in), but they are only 1 to 3 mm in diameter.

Bryozoans, also known as ectoprocts or "moss animals," are microscopic aquatic animals. They form sessile colonies by asexual budding (Fig. 30-14c). Each colony can consist of millions of individuals and can extend in length up to about 30 cm (about 1 ft). The colonies typically appear plantlike, but some have the appearance of coral. Approximately 5000 extant species are known.

Rotifers have a crown of cilia

Among the less familiar invertebrates are the "wheel animals" of phylum **Rotifera.** Although no larger than many protozoa, these aquatic, microscopic animals are multicellular. About 2000 species have been described. Most inhabit fresh water, but some live in marine environments or damp soil. Marine rotifers are very important in cycling nutrients in marine ecosystems.

Rotifers were formerly classified as pseudocoelomates, but many biologists now think that rotifers evolved from animals with a coelom (see Figs. 29-7 and 29-8). Because rotifers have a cuticle, some systematists classify them as ecdysozoans related to nematodes and arthropods. However, other systematists classify rotifers with the lophotrochozoans.

Rotifers have a characteristic crown of cilia on their anterior end. The cilia beat rapidly during swimming and feeding, giving the appearance of a spinning wheel (Fig. 30-15). Rotifers feed on tiny organisms suspended in the stream of water drawn into the mouth by the cilia. A muscular organ posterior to the mouth grinds the food. The complete digestive tract ends in an anus through which wastes are eliminated.

Rotifers have a nervous system with a "brain" and sense organs, including eyespots. Protonephridia with flame cells remove excess water from the body and may also excrete metabolic wastes.

Biologists have discovered that rotifers, like some other animals with a pseudocoelom, are "cell constant." Each member of a given species has exactly the same number of cells. Indeed, each part of the body has a precisely fixed number of cells arranged in a characteristic pattern. Cell division does not take place after embryonic development, and mitosis cannot be induced. Growth results from an increase in cell size. One of the challenging problems of biological research is discovering the difference between such nondividing cells and the dividing cells of other animals. Do you think rotifers could develop cancer?

Another interesting characteristic of rotifers is their ability to survive very dry conditions and temperature extremes. These interesting animals can remain in a dormant state for months or years. When conditions become favorable, they recover and become active once again.

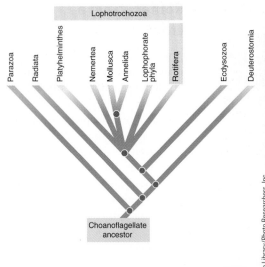

Figure 30-15 *Animated* Rotifers (wheel animals)

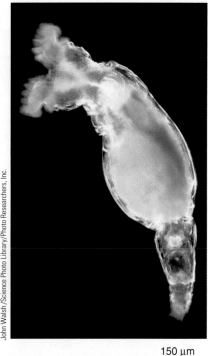

150 μm

(a) LM of an Antarctic rotifer (*Philodina gregaria*) that survives the winter by forming a cyst. It reproduces in great numbers, sometimes coloring the water red.

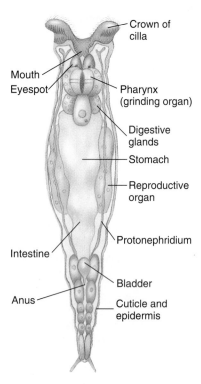

(b) Longitudinal section showing rotifer anatomy. The motion of its cilia draws particles of food such as algae into the mouth.

Review

- What are some advantages of bilateral symmetry and cephalization?
- On what basis have biologists now classified nemerteans as lophotrochozoans?
- How does the lifestyle of a gastropod differ from that of a cephalopod? Identify adaptations that have evolved in each for its particular lifestyle.
- What are two distinguishing characteristics for each of the following: (1) mollusks, (2) annelids, and (3) nematodes?

THE ECDYSOZOA

Learning Objectives

8 List the distinguishing characteristics of phylum Nematoda.

9 Describe the distinguishing characteristics of phylum Arthropoda, and distinguish among the subphyla and classes of this phylum; give examples of animals that belong to each group.

10 Identify adaptations that have contributed to the biological success of insects.

The **Ecdysozoa** include the nematodes (roundworms) and arthropods, animals characterized by a cuticle, a noncellular body covering secreted by the epidermis. The name *Ecdysozoa* refers to the process of **ecdysis**, or **molting**, characteristic of animals in this group. During ecdysis, an animal sheds its cuticle. The taxonomic validity of the Ecdysozoa has been supported by many types of data, most notably by a 2001 study that analyzed more than 300 ribosomal RNA genes and 138 structural characters.

Roundworms are of great ecological importance

Members of phylum **Nematoda,** the **roundworms,** play key ecological roles as decomposers and predators of smaller organisms. Many soil nematodes eat bacteria. Nematodes are numerous and widely distributed in soil and in marine and freshwater sediments. More than 20,000 species have been described, and many more remain to be discovered. The nematodes rival the insects in number of individuals. A spadeful of soil may contain more than a million of these mainly microscopic worms, which thrash around, coiling and uncoiling. *Caenorhabditis elegans*, a free-living soil nematode, is an important model research organism for biologists studying the genetic control of development (see Chapter 17).

The elongated, cylindrical, threadlike nematode body is pointed at both ends and covered with a tough, flexible cuticle (Fig. 30-16). Secreted by the underlying epidermis, the thick cuticle gives the nematode body shape and offers some protection. The epidermis is unusual in that it does not consist of distinct cells.

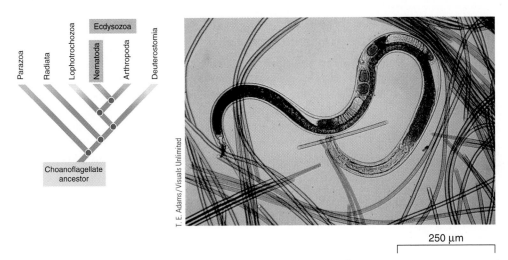

Parazoa
Radiata
Lophotrochozoa
Ecdysozoa
Nematoda
Arthropoda
Deuterostomia

Choanoflagellate
ancestor

T. E. Adams / Visuals Unlimited

250 μm

Figure 30-16 LM of a free-living nematode

This unidentified aquatic nematode is shown among the cyanobacterium *Oscillatoria*, which it eats.

Nematodes have a fluid-filled pseudocoelom that serves as a hydrostatic skeleton. It transmits the force of muscle contraction to the enclosed fluid. Movement of fluid in the pseudocoelom is also important in nutrient transport and distribution. Like ribbon worms, nematodes exhibit bilateral symmetry, a complete digestive tract, three definite tissue layers, and definite organ systems; however, they lack specific circulatory structures. The sexes are usually separate, and the male is generally smaller than the female.

Traditionally, biologists classified nematodes as a single phylum based on the pseudocoelom, wormlike body structure, and other characters such as absence of a well-defined head (see Fig. 29-7). However, molecular evidence suggests that nematodes are not a monophyletic group and that they evolved from more complex animals by a simplification in body plan. Systematists now classify nematodes with the Ecdysozoa (see Fig. 29-8). Like arthropods and other ecdysozoans, the nematode sheds its cuticle periodically.

Although most nematodes are free-living, others are important parasites in plants and animals. More than 50 species of roundworms are human parasites, including *Ascaris*, hookworms, pinworms, and the trichina worm. The common intestinal parasite *Ascaris* spends its adult life in the human intestine, where it ingests partly digested food (❚ Fig. 30-17). *Ascaris* is a white worm about 25 cm (10 in) long. Like most parasites, it has remarkable reproductive adaptations that ensure survival of its species. A mature female may produce as many as 200,000 eggs per day!

Hookworms attach to the lining of the intestine and suck blood, potentially causing serious tissue damage and blood loss. **Pinworms** are the most common worms found in children. The tiny pinworm eggs are often ingested by eating with hands contaminated with them. The **trichina worm** lives inside a variety of animals, including pigs, rats, and bears. Humans typically become infected by eating undercooked, infected meat. Larvae encyst in skeletal muscle.

Arthropods are characterized by jointed appendages and an exoskeleton of chitin

Phylum **Arthropoda,** the most biologically successful group of animals, includes more than 1 million species, and biologists predict millions more will be identified. More than 80% of all known animals are arthropods! They are more diverse and live in a greater range of habitats than do the members of any other animal phylum. The following adaptations have greatly contributed to their success:

1. The arthropod body, like that of the annelid, is segmented. *Segmentation* is important from an evolutionary perspective because it provides the opportunity for specialization of body regions. In arthropods, groups of segments are specialized, by virtue of their shape, muscles, or the appendages they bear, to perform particular functions.

2. A hard **exoskeleton,** composed of chitin and protein, covers the entire body and appendages. The exoskeleton serves as a coat of armor that protects against predators and helps prevent excessive loss of moisture. It also supports the underlying soft tissues. Arthropods move effectively because distinct muscles attach to the inner surface of the exoskeleton and operate the joints of the body and appendages.

   A disadvantage of the exoskeleton is that it is nonliving, and the arthropod periodically outgrows it. *Molting* is the process of shedding an old exoskeleton and growing a larger one. The shed exoskeleton represents a net metabolic loss, and molting also leaves the arthropod temporarily vulnerable to predators.

3. **Paired, jointed appendages,** from which this phylum gets its name (*arthropod* means "jointed foot"), are modified for many functions. They serve as swimming paddles, walking legs, mouthparts for capturing and manipulating food, sensory structures, or organs for transferring sperm.

4. The nervous system, which resembles that of the annelids, consists of a "brain" (cerebral ganglia) and ventral nerve cord with ganglia. In some arthropods, successive ganglia may fuse. Arthropods have a variety of very effective sense organs. Many have organs of hearing and **antennae** that sense taste and touch. Most insects and many crustaceans have **compound eyes** composed of many light-sensitive units called **ommatidia** (see Fig. 42-17). The compound eye can form an image and is especially adapted for detecting movement.

Arthropods have an *open circulatory system*. A dorsal, tubular heart pumps hemolymph into a dorsal artery, which may branch

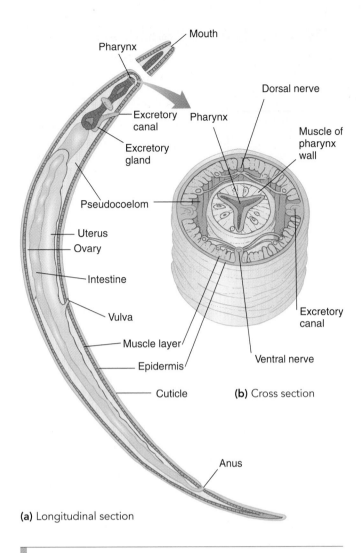

Mouth

Pharynx

Excretory
canal

Excretory
gland

Pseudocoelom

Uterus

Ovary

Intestine

Vulva

Muscle layer

Epidermis

Cuticle

Anus

(a) Longitudinal section

Dorsal nerve

Pharynx

Muscle of
pharynx
wall

Excretory
canal

Ventral nerve

(b) Cross section

| **Figure 30-17** *Animated* The roundworm *Ascaris*

(a) Note the complete digestive tract that extends from mouth to anus. **(b)** This cross section through *Ascaris* shows the tube-within-a-tube body plan. The protective cuticle that covers the body helps the animal resist drying.

into smaller arteries. From the arteries, hemolymph flows into large spaces that collectively make up the hemocoel. Eventually hemolymph re-enters the heart through openings, called *ostia*, in its walls.

The exoskeleton presents a barrier to diffusion of oxygen and carbon dioxide through the body wall, necessitating the evolution of specialized respiratory systems for gas exchange. Most aquatic arthropods have gills that function in gas exchange, whereas many terrestrial forms have a system of internal branching air tubes called **tracheae,** or **tracheal tubes.** Other terrestrial arthropods have platelike **book lungs.**

The coelom is small and filled chiefly by the organs of the reproductive system. The arthropod digestive system is a tube similar to that of the earthworm. Excretory structures vary somewhat from class to class.

Arthropod evolution and classification are controversial

Two groups of animals thought to be closely related to arthropods are the onychophorans and the tardigrades (❚ Fig. 30-18). Like arthropods, onychophorans and tardigrades have a thick cuticle and must molt in order to grow. The body is segmented in all three groups, and all share similarities in the circulatory system.

The wormlike members of phylum **Onychophora,** or velvet worms, most likely branched early from the arthropod line. They have paired appendages, but they are not jointed. As in arthropods, their jaws are derived from appendages. Some biologists place the tardigrades, or water bears, as the sister taxon of the arthropods (see Fig. 2-12). Assigned to phylum **Tardigrada,** these tiny animals have unbranched, clawed legs.

Arthropod fossils have been identified that date back to the Proterozoic eon. Early in their evolutionary history, more than 540 million years ago (mya), arthropods diverged into several groups. Arthropods evolved rapidly during the Cambrian Radiation, as evidenced by many diverse fossils in the Burgess Shale and other ancient sites.

We discuss five main arthropod groups, based on emerging molecular and other data: the extinct trilobites, and the extant Myriapoda, Chelicerata, Crustacea, and Hexapoda (❚ Table 30-4). Some systematists consider each of these groups a phylum; others view them as subphyla or classes. In this edition, we classify these groups as subphyla. However, as you study the following sections, bear in mind that arthropod systematics is a work in progress. Both the relationship of arthropods to other protostomes and the relationships among arthropods continue to be topics of lively debate.

Trilobites were early arthropods

Among the earliest arthropods to evolve, **trilobites** inhabited shallow Paleozoic seas more than 500 mya. These arthropods, which have been extinct for about 250 million years, lived on the

(a) Onychophoran. Note the soft segmented body and series of non-jointed legs. Extant onychophorans live in humid forests, although many extinct species inhabited the ocean.

(b) Tardigrades. Commonly known as water bears, tardigrades normally live in moist habitats, such as thin films of water on mosses. Most are less than 0.5 mm long.

| **Figure 30-18** Onychophorans and tardigrades

These animals are not arthropods but are thought to be close relatives.

TABLE 30-4

Extant Arthropod Subphyla

Subphylum and Selected Classes	Body Divisions	Appendages: Antennae, Mouthparts	Appendages: Legs	Gas Exchange	Development	Main Habitat
Myriapoda Class Chilopoda (centipedes) Class Diplopoda (millipedes)	Head with segmented body	Uniramous. Antennae: one pair Mouthparts: mandibles, maxillae	Chilopods: one pair/segment Diplopods: usually two pairs/segment	Tracheae	Direct	Terrestrial
Chelicerata Class Merostoma (horseshoe crabs) Class Arachnida (spiders, scorpions, ticks, mites)	Cephalothorax and abdomen	Uniramous. Antennae: none Mouthparts: chelicerae, pedipalps	Merostomes: five pairs of walking legs Arachnids: four pairs on cephalothorax	Merostomes: gills Arachnids: book lungs or tracheae	Direct, except mites and ticks	Merostomes: marine Arachnids: mainly terrestrial
Crustacea Class Malacostraca (lobsters, crabs, shrimp, isopods) Class Cirripedia (barnacles) Class Copepoda (copepods)	Head, thorax, and abdomen	Biramous. Antennae: two pairs Mouthparts: mandibles, two pairs of maxillae (for handling food)	Typically one pair/segment	Gills	Usually larval stages (nauplius larva)	Marine or freshwater; a few are terrestrial
Hexapoda Class Insecta (bees, ants, grasshoppers, roaches, flies, beetles)	Head, thorax, and abdomen	Uniramous. Antennae: one pair Mouthparts: mandibles, maxillae	Three pairs on thorax	Tracheae	Typically, larval stages; most with complete metamorphosis	Mainly terrestrial

sea bottom and filtered mud to obtain food. Most ranged from 3 to 10 cm (about 1 to 4 in), but a few reached almost 1 m (39 in) in length.

Covered by a hard, segmented exoskeleton, the trilobite body was a flattened oval divided into three parts: an anterior head bearing a pair of antennae and a pair of compound eyes; a thorax; and a posterior abdomen (▌Fig. 30-19; also see introduction to Chapter 21). At right angles to these divisions, two dorsal grooves extended the length of the animal, dividing the body into a median lobe and two lateral lobes. (The name *trilobite* derives from this division of the body into three longitudinal parts.) Each segment had a pair of segmented **biramous appendages,** appendages with two jointed branches extending from their base. In trilobites, each appendage consisted of an inner walking leg and an outer branch with gills.

Subphylum Myriapoda includes centipedes and millipedes

Members of subphylum **Myriapoda** are characterized by unbranched appendages called **uniramous appendages,** jawlike **mandibles,** and a single pair of antennae. Centipedes (class **Chilopoda**) and millipedes (class **Diplopoda**) are terrestrial and are typically found beneath stones or wood in the soil in both temperate and tropical regions. Both of these animals have a head and an elongated trunk with many segments, each bearing uniramous legs (▌Fig. 30-20).

Centipedes ("hundred-legged") have one pair of legs on each segment behind the head. Most centipedes do not have enough legs to merit their name; typically, they have approximately 30, although a few species have 100 or more. Because centipedes have long legs, they are able to run rapidly. Centipedes are carnivorous and feed on other animals, mostly insects. Larger centipedes eat snakes, mice, and frogs. Using their poison claws located just behind the head on the first trunk segment, centipedes capture and kill their prey.

Millipedes ("thousand-legged") have two pairs of legs on most body segments. Diplopods are not as agile as chilopods, and most species can crawl only slowly over the ground. However, they can powerfully force their way through earth and rotting wood. Millipedes are generally herbivorous and feed on both living and dead vegetation.

Chelicerates have no antennae

Subphylum **Chelicerata** includes the merostomes (horseshoe crabs) and the arachnids. The chelicerate body consists of a *cephalothorax* (fused head and thorax) and an abdomen. Chelicerates are the only arthropods without antennae. They have no chewing mandibles. Instead, the first appendages, located immediately anterior to the mouth, are a pair of **chelicerae** (sing., *chelicera*), fanglike feeding appendages. The second appendages are the **pedipalps.** The chelicerae and pedipalps are modified to perform different functions in various groups, including manipulation of

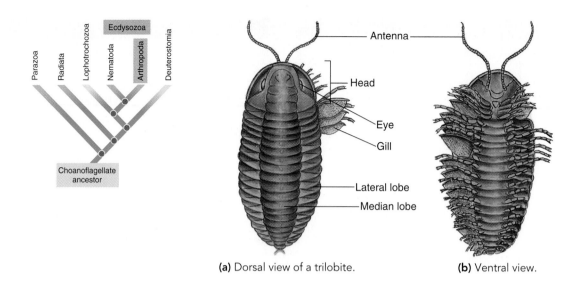

(a) Dorsal view of a trilobite.

(b) Ventral view.

Figure 30-19 Trilobites

Biologists consider these extinct marine arthropods the most primitive members of the phylum.

(a) Centipede (*Lithobius*), a member of class Chilopoda. Centipedes have one pair of uniramous appendages per segment.

(b) Millipede (*Diplopoda pachydesmus*), a member of class Diplopoda. Millipedes have two pairs of uniramous appendages per segment.

Figure 30-20 Myriapods

food, locomotion, defense, and copulation. The four pairs of legs on the cephalothorax are specialized for walking.

Almost all the **merostomes** are extinct. Only the horseshoe crabs have survived, essentially unchanged for more than 350 million years. *Limulus polyphemus,* the species common along the shore of North America, is horseshoe shaped (Fig. 30-21a). Its long, spikelike tail is used in locomotion, not for defense or offense. Horseshoe crabs feed on mollusks, worms, and other invertebrates that they find on the sandy ocean floor.

Arachnids include spiders, scorpions, ticks, harvestmen (daddy longlegs), and mites (Fig. 30-21b and c). Most of the approximately 65,000 named species are carnivorous and prey on insects and other small arthropods. Arachnids have six pairs of jointed appendages. In spiders, the first appendages, the chelicerae, are used to penetrate prey; some species use the chelicerae to inject poison into their prey. Spiders use the second pair of appendages, the pedipalps, to hold and chew food. Many arachnid species have pedipalps modified as sense organs for tasting food or for reproduction (for sperm transfer and courtship displays). Scorpions use their very large pedipalps as pincers for capturing prey. Chelicerates use the remaining four pairs of appendages for walking.

Typically, spiders have eight eyes arranged in two rows of four each along the anterior dorsal edge of the cephalothorax. The eyes detect movement, locate objects, and in some species form a relatively sharp image.

Gas exchange in arachnids takes place by either tracheal tubes, book lungs, or both. A *book lung* consists of 15 to 20 plates (like pages of a book) that contain tiny blood vessels. Air enters the body through abdominal slits and circulates between the plates. As air passes over the blood vessels in the plates, oxygen diffuses into the blood and carbon dioxide diffuses out of the blood into the air. As many as four pairs of book lungs provide an extensive surface area for gas exchange.

(a) The only living merostomes are a few closely related species of horseshoe crabs (*Limulus polyphemus*). Seasonally, horseshoe crabs return to beaches for mating. In this photograph, males are competing for a female.

(b) A female black widow spider (*Latrodectus mactans*) rests on her web. The venom of the black widow spider is a neurotoxin.

Figure 30-21 Chelicerates

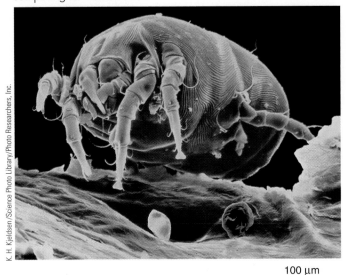

100 μm

(c) Color-enhanced SEM of a house dust mite (*Dermatophagoides*), a common inhabitant of homes. This mite is associated with house dust allergies.

The spider has unique glands in its abdomen that secrete silk, an elastic protein that is spun into fibers by organs called *spinnerets*. The silk is liquid as it emerges from the spinnerets but hardens after it leaves the body. Spiders use silk to build nests, to encase their eggs in a cocoon, and in some species, to trap prey in a web. Many spiders lay down a silken dragline that serves both as a safety line and a means of communication between members of a species. From the type of dragline, another spider can determine the sex and maturity level of the spinner.

Although all spiders (and scorpions) have poison glands useful in capturing prey, only a few produce poison toxic to humans. The most widely distributed poisonous spider in the United States is the black widow (see Fig. 30-21b). Its venom is a neurotoxin that interferes with transmission of messages from nerves to muscles. The shiny black female is about 3.8 cm (1.5 in) long with a red or orange hourglass-shaped marking on the underside of its abdomen. Adult male black widow spiders are smaller and

are harmless. After mating, the female sometimes kills and eats the male.

The brown recluse spider (*Loxosceles reclusa*) is more slender than the black widow and has a violin-shaped dorsal stripe on its cephalothorax. Its venom destroys the tissues surrounding the bite. Although painful, spider bites cause fewer than five fatalities per year in the United States.

Although some mites contribute to soil fertility by breaking down organic matter, many are serious nuisances. They eat crops, infest livestock and pets, and inhabit our own bodies. Certain mites cause mange in dogs and other domestic animals. Chiggers (red bugs), the larval form of red mites, attach themselves to the skin and secrete an irritating digestive fluid that may cause itchy red welts.

Larger than mites, ticks are parasites on dogs, deer, and many other animals. They transmit diseases such as Rocky Mountain spotted fever, Texas cattle fever, relapsing fever, and Lyme disease (see Fig. 53-1).

Crustaceans are vital members of marine food webs

Subphylum **Crustacea** includes lobsters, crabs, shrimp, barnacles, and their relatives (Fig. 30-22). Paleobiologists have dated crustacean fossils back to the Cambrian period (more than 505 mya). At present, most biologists agree that crustaceans form a clade with members of subphylum Hexapoda (insects); these two subphyla are sister taxa (see Chapter 23).

Many crustaceans are primary consumers of algae and detritus. Countless billions of microscopic crustaceans are part of marine zooplankton, the free-floating, mainly microscopic organisms in the upper layers of the ocean. These crustaceans are food for many fishes and other marine animals, such as certain baleen whales.

A distinctive feature of crustaceans is the *nauplius larva*, which is often the first stage after hatching. This larva has only the most anterior three pairs of appendages. Crustaceans are

Tim Dyes

(a) Goose barnacles (*Lepas fascicularis*) hanging from driftwood. These stalked barnacles occur in large numbers on intertidal rocks along the West Coast of the United States.

Fred Bavendam/Peter Arnold, Inc.

(b) This sponge crab (*Criptodromia octodenta*), photographed in Australia, wears a living sponge on its back as camouflage.

Figure 30-22 Crustaceans

Isopods are mainly tiny (5 to 15 mm in length) marine crustaceans that inhabit the ocean floor. However, this group includes some familiar terrestrial animals: pill bugs and sow bugs. The typically microscopic *copepods* are the most abundant crustaceans. Marine copepods are the most numerous component of zooplankton.

The largest and most familiar order of crustaceans, Decapoda, contains more than 10,000 species of lobsters, crayfish, crabs, and shrimp. Most decapods are marine, but a few, such as the crayfish, certain shrimp, and a few crabs, live in fresh water. The crustaceans in general and the decapods

also characterized by mandibles, biramous appendages, and two pairs of antennae. Their antennae serve as sensory organs for touch and taste. Most adult crustaceans have compound eyes, and many crustaceans have **statocysts,** sense organs that detect gravity (see Fig. 42-6).

The hard **mandibles,** used for biting and grinding food, are the third pair of appendages; they lie on each side of the ventral mouth. Posterior to the mandibles are two pairs of appendages, the first and second **maxillae,** used for manipulating and holding food. Several other pairs of appendages are present. Some are modified for walking. Others may be specialized for swimming, sperm transmission, carrying eggs and young, or sensation.

Crustaceans are the only group of arthropods that are primarily aquatic; not surprisingly, they typically have gills for gas exchange. Two large glands located in the head excrete metabolic wastes and regulate salt balance. Most crustaceans have separate sexes. During copulation, the male uses specialized appendages to transfer sperm into the female. Newly hatched animals may resemble adults, or they may undergo successive molts as they pass through a series of larval stages before they develop an adult body.

Barnacles, the only sessile crustaceans, differ markedly in external anatomy from other members of the subphylum. They are marine suspension feeders that secrete limestone cups within which they live (see Fig. 30-22a). The larvae of barnacles are free-swimming forms that go through several molts. They eventually become sessile and develop into the adult form. The 19th-century naturalist Louis Agassiz described the barnacle as "nothing more than a little shrimplike animal standing on its head in a limestone house and kicking food into its mouth." The bane of marine boaters, barnacles can proliferate on ship bottoms in great numbers. Their presence can reduce the speed of a ship by more than 30%.

in particular show striking specialization and differentiation of parts in the various regions of the animal. In the lobster, the appendages in the different parts of the body differ markedly in form and function (Fig. 30-23).

The five segments of the lobster's head and the eight segments of its thorax are fused into a cephalothorax, which is covered on the top and sides by a shield, the **carapace,** composed of chitin impregnated with calcium salts. The two pairs of antennae serve as chemoreceptors and tactile sense organs. The mandibles are short and heavy, with opposing surfaces used in grinding and biting food. Behind the mandibles are two pairs of accessory feeding appendages: the first and second maxillae.

The appendages of the first three segments of the thorax are the maxillipeds, which aid in chopping up food and passing it to the mouth. The fourth segment of the thorax has a pair of large **chelipeds,** or pinching claws. The last four thoracic segments bear walking legs.

The appendages of the first abdominal segment are part of the reproductive system and function in the male as sperm-transferring structures. The following four abdominal segments bear paired **swimmerets,** small paddlelike structures used by some decapods for swimming and by the females of all species for holding eggs. Each branch of the sixth abdominal appendages consists of a large flattened structure. Together with the flattened posterior end of the abdomen, they form a tail fan used for swimming backward.

Insects belong to subphylum Hexapoda

With more than 1 million described species (and perhaps millions more not yet identified), class **Insecta** of subphylum **Hexapoda** is the most successful group of animals on our planet in terms of diversity, geographic distribution, number of species, and num-

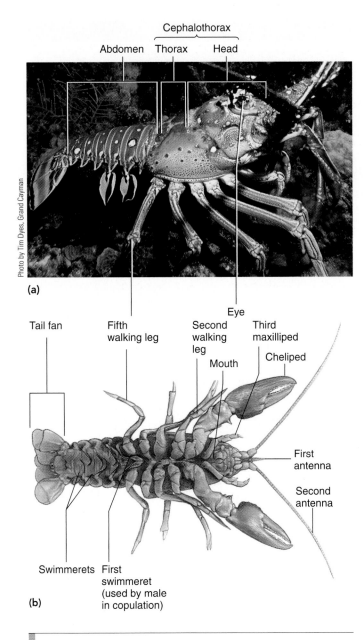

Figure 30-23   Anatomy of the lobster

(a) Like other decapods, the spiny lobster (*Panulirus argus*) has five pairs of walking legs. The first pair of walking legs is modified as chelipeds (*large claws*). Photographed in Grand Cayman. **(b)** Ventral view of a lobster. Note the variety of specialized appendages.

Key Point

Insects, the most successful of all animals, are articulated, tracheated hexapods.

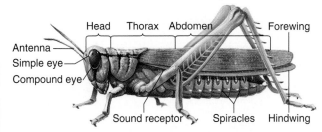

(a) External structure. Note the three pairs of segmented legs.

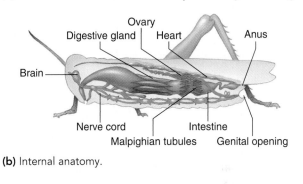

(b) Internal anatomy.

Figure 30-24   *Animated*   Anatomy of the grasshopper

ber of individuals (■ Table 30-5).What they lack in size, insects make up in sheer numbers. If we could weigh all the insects in the world, their weight would exceed that of all the remaining terrestrial animals. Although primarily terrestrial, some insect species live in fresh water, a few are truly marine, and others inhabit the shore between the tides.

The earliest fossil insects—primitive, wingless species—date to the Devonian period more than 360 mya. Insect fossils from the Carboniferous period more than 300 mya include both wingless and primitive winged species. Cockroaches, mayflies,

and cicadas are among the insects that have survived relatively unchanged from the Carboniferous period to the present day. Until recently, biologists thought that insects made up a monophyletic taxon and were closely related to myriapods (centipedes and millipedes). Recent molecular and developmental studies suggest that insects are not a monophyletic group and are more closely related to crustaceans. Molecular data also suggest that a group of wingless soil organisms known as *springtails* (order Collembola) may have evolved independently of the insects. Some evidence suggests that springtails branched earlier than crustaceans off the line leading to insects. Unlike insects, springtails have internal mouthparts.

We describe an insect as an **articulated** (jointed), **tracheated** (having tracheal tubes for gas exchange) **hexapod** (having six feet). The insect body consists of three distinct parts: head, thorax, and abdomen (■ Fig. 30-24). An insect's uniramous (unbranched) appendages include three pairs of legs that extend from the adult thorax and, in many orders, one or two pairs of wings. One pair of antennae protrudes from the head, and the sense organs include both simple and compound eyes. The complex mouthparts, which include mandibles and maxillae, are adapted for piercing, chewing, sucking, or lapping.

The tracheal system of insects has made an important contribution to their diversity. Air enters the tracheal tubes, or tracheae, through **spiracles,** tiny openings in the body wall. Oxygen passes directly to the internal organs. This effective oxygen delivery system permits insects to have the high metabolic rate neces-

TABLE 30-5

Some Orders of Insects

Order, Number of Species, Some Characteristics, and Representative Adult

Insects with Incomplete Metamorphosis
(egg ⟶ larva ⟶ adult)

Homoptera (43,000)

Aphids, leafhoppers, cicadas, scale insects
Two pairs of membranous wings; piercing–sucking mouthparts form beak; parasites of plants; vectors of plant diseases

Buffalo treehopper
Stictocephala bubalus

Odonata (6000)

Dragonflies, damselflies
Two pairs of long membranous wings; chewing mouthparts; large, compound eyes; active predators; larvae very different from adult

Damselfly
Ischnura

Orthoptera (20,000)

Grasshoppers, crickets
Forewings leathery, hindwings membranous; chewing mouthparts; most herbivorous, some cause crop damage; some predatory

Fork-tailed bush katydid
Scudderia furcata

Blattaria (3700)

Cockroaches
When wings present, forewings leathery, hindwings membranous; chewing mouthparts; legs adapted for running

American cockroach
Periplaneta americana

Isoptera (2000)

Termites
Two pairs of wings, or none; wings shed by sexual forms after mating; chewing mouthparts; social insects, form large colonies; eat wood

Eastern subterranean termite
Reticulitermes flavipes

Phthiraptera (3000)

Sucking lice
No wings; piercing–sucking mouthparts; ectoparasites of mammals; head and body louse and crab louse are human parasites; vectors of typhus fever

Human head and body louse
Pediculus humanus

Hemiptera (true bugs) (68,000)

Chinch bugs, bedbugs, water striders
Hindwings membranous; forewings smaller; piercing–sucking mouthparts form beak; most herbivorous; some parasitic

Chinch bug
Blissus leucopterus

Insects with Complete Metamorphosis
(egg ⟶ larva ⟶ pupa ⟶ adult)

Lepidoptera (140,000)

Moths, butterflies
Usually two pairs of membranous, colorful, scaled wings; sucking mouthparts; larvae are wormlike caterpillars that eat plants; adults suck flower nectar; important pollinators

Luna moth
Aetias luna

Diptera (150,000)

Houseflies, mosquitoes
Only forewings functional in flying; hindwings small, knoblike balancing organs (halteres); mouthparts usually adapted for sucking (and piercing in some); larvae are maggots that damage domestic animals or food; adults transmit diseases such as sleeping sickness, yellow fever, and malaria

Deerfly
Chrysops vittatus

Siphonaptera (1750)

Fleas
No wings; piercing–sucking mouthparts; legs adapted for clinging and jumping; parasites on birds and mammals; vectors of bubonic plague and typhus

Dog flea
Ctenocephalides canis

Coleoptera (350,000)

Beetles, weevils
Forewings modified as protective coverings for membranous hindwings (which are sometimes absent); chewing mouthparts; largest order of insects; most herbivorous; some aquatic

Colorado potato beetle
Leptinotarsa decemlineata

Hymenoptera (190,000)

Ants, bees, wasps
Usually two pairs of membranous wings; mouthparts may be modified for sucking or lapping nectar; many are social insects; some sting; important pollinators

Bald-faced hornet
Vespula maculata

Insects with No Metamorphosis (egg ⟶ immature form ⟶ adult)

Thysanura (320)

Silverfish
No wings; biting–chewing mouthparts; two to three "tails" extend from posterior tip of abdomen; inhabit dead leaves; eat starch in books

Silverfish
Lepisma saccharina

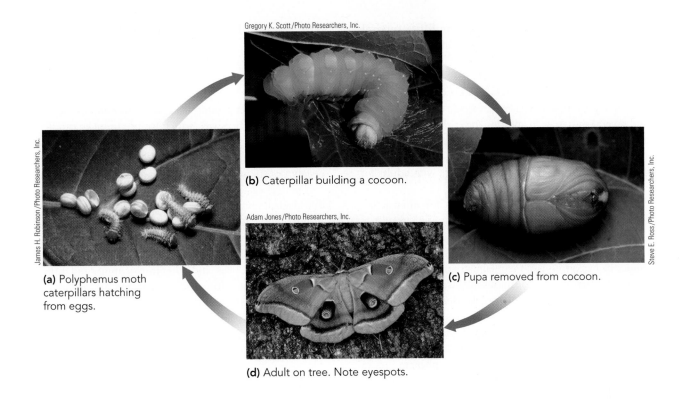

Gregory K. Scott/Photo Researchers, Inc.

(b) Caterpillar building a cocoon.

Adam Jones/Photo Researchers, Inc.

James H. Robinson/Photo Researchers, Inc.

(a) Polyphemus moth caterpillars hatching from eggs.

Steve E. Ross/Photo Researchers, Inc.

(c) Pupa removed from cocoon.

(d) Adult on tree. Note eyespots.

Figure 30-25 *Animated* Complete metamorphosis in a Polyphemus moth

This moth (*Antheraea polyphemus*) belongs to the family of giant silk moths.

sary for activities such as flight, even though they have an open circulatory system.

Excretion is accomplished by two or more **Malpighian tubules,** which receive metabolic wastes from the blood, concentrate the wastes, and discharge them into the intestine. Unique to terrestrial arthropods, Malpighian tubules also perform the extremely important function of conserving water (see Fig. 47-4).

The sexes are separate, and fertilization takes place internally. Several molts occur during development. The immature stages between molts are called **instars.** Primitive insects with no wings, such as silverfish, have direct, or simple, development: the young hatch as juveniles that resemble the adult form. Other insects, such as grasshoppers and cockroaches, undergo **incomplete metamorphosis** in which the egg gives rise to a larva that resembles the adult in many ways but lacks functional wings and reproductive structures. The larva goes through a series of molts during which it becomes more like the adult.

Most insects, including bees, butterflies, and fleas, undergo **complete metamorphosis** with four distinct stages in the life cycle: **egg, larva, pupa,** and **adult** (❙ Fig. 30-25). The wormlike larva does not look at all like the adult. For example, caterpillars have a body form entirely different from that of butterflies. Typically, an insect spends most of its life as a larva. Eventually the larva stops feeding, molts, and enters a pupal stage, usually within a protective cocoon or underground burrow. The pupa does not feed and typically cannot defend itself. Energy reserves stored during the larval stage are spent remodeling its body. When it emerges as an adult, it is equipped with functional wings and reproductive organs.

Certain species of bees, ants, and termites exist as colonies or societies made up of several different types of individuals, each adapted for some particular function. The members of some insect societies communicate with one another by "dances" and by means of **pheromones,** substances secreted to the external environment. Social insects and their communication are discussed in Chapter 51.

Adaptations that contribute to the biological success of insects. What are the secrets of insect success? The tough exoskeleton protects insects from predators and helps prevent water loss. Segmentation allows mobility and flexibility and permits regional specialization of the body. For example, cephalization and highly developed sense organs are important in both offense and defense.

The insect body plan has been modified and specialized in so many ways that insects are adapted to a remarkable number of lifestyles. The jointed appendages are specialized for various types of feeding, sensory functions, and different types of locomotion (walking, jumping, swimming). For example, grasshopper mouthparts are adapted for biting and chewing leaves. Moth and butterfly mouthparts are adapted for sucking nectar from flowers, and mosquito mouthparts are adapted for sucking blood. Aphids and leafhoppers have mouthparts specialized to pierce plants and feed on plant juices.

Another adaptation that contributes to insect success is the ability to fly. Unlike other terrestrial invertebrates, which creep slowly along on or under the ground, many insects fly rapidly through the air. Their wings and small size facilitate their wide distribution and, in many instances, their immediate survival. For example, when a pond dries up, adult aquatic insects can fly to another habitat.

The reproductive capacity of insects is remarkable. Under ideal conditions, the fruit fly *Drosophila* can produce 25 generations in a single year! Insect eggs are protected by a thick membrane, and several eggs may, in addition, be enclosed in a protective egg case. By dividing the insect life cycle into different stages, metamorphosis reduces intraspecific competition; larval forms have different lifestyles, so they do not compete with adults for food or habitats.

Insects have many interesting adaptations for offense and defense (see Fig. 7-8). We all are familiar with the stingers of bees and wasps, which are specialized egg-laying structures (ovipositors). Many insects have **cryptic coloration,** which allows them to blend into the background of their habitat; some look like dead twigs or leaves. Others mimic poisonous insects, deriving protection from this resemblance. Still others "play dead" by remaining motionless.

Insects communicate by tactile, auditory, visual, or chemical signals. For example, certain ant species use pheromones to mark trails and to warn of danger. Other insects use pheromones to attract a mate.

Impact of insects on humans. Not all insects compete with humans for food or cause us to scratch, swell up, or recoil from their presence. Bees, wasps, beetles, and many other insects pollinate flowers of crops and fruit trees. Some insects destroy other insects that are harmful to humans. For example, dragonflies eat mosquitoes. And some organic farmers and home gardeners buy ladybird beetles, which are adept at ridding plants of aphids and other insect pests. Insects are important members of many food

webs. Many birds, mammals, amphibians, reptiles, and some fishes depend on insects for food. Some beetles and the larvae (maggots) of flies are detritus feeders—they break down dead plants and animals and their wastes, permitting nutrients to be recycled.

Various insect products are useful to humans. Bees produce honey as well as beeswax, which we use to make candles, lubricants, and other products. Shellac is made from lac, a substance given off by certain scale insects that feed on the sap of trees. And the labor of silkworms provides us with beautiful fabric.

On the negative side, insects destroy billions of dollars worth of crops each year. Termites destroy buildings, and moths damage clothing. Fire ants not only inflict painful stings but cause farmers serious economic loss because of their large mounds, which damage mowers and other farm equipment. Blood-sucking flies, screwworms, lice, fleas, and other insects annoy and transmit disease in humans and domestic animals. Mosquitoes are vectors of yellow fever and malaria (see Fig. 25-10). Body lice transmit the typhus rickettsia, and houseflies sometimes transmit typhoid fever and dysentery. Tsetse flies transmit African sleeping sickness, and fleas may be vectors of bubonic plague.

Review

- What functions does the pseudocoelom have in nematodes?
- What are four key arthropod characteristics, and how does each contribute to arthropod success?
- How do myriapods differ from chelicerates?
- What are the functions of each of these appendages in a lobster: maxillae, chelipeds, and swimmerets?
- Define an insect. What are four adaptations that have contributed to insect success?

SUMMARY WITH KEY TERMS

Learning Objectives

1 Cite specific examples of the evolutionary significance of the coelom (page 641).
   - The true **coelom** is a fluid-filled body cavity completely lined by mesoderm that lies between the digestive tube and the outer body wall. The coelom allows the **tube-within-a-tube body plan.** The body wall is the outer tube. The inner tube is the digestive tube.
   - Because the coelom is an enclosed compartment (or series of compartments) of fluid under pressure, it can serve as a **hydrostatic skeleton** in which contracting muscles push against a tube of fluid.
   - The coelom is a space in which internal organs, including gonads, can develop. The coelom helps transport materials and protects internal organs.

2 Characterize the protostomes, describe their two main evolutionary branches, and give examples of animals assigned to each branch (page 641).
   - The **protostomes** are characterized by spiral cleavage, determinate cleavage, and the development of the mouth from the blastopore.
   - The protostomes include two main branches: Lophotrochozoa and Ecdysozoa. The **Lophotrochozoa** include

the platyhelminths, nemerteans, mollusks, annelids, the lophophorate phyla, and the rotifers. The **Ecdysozoa** include the nematodes (roundworms) and arthropods.

3 Identify distinguishing characteristics of phylum Nemertea and phylum Platyhelminthes; describe the main classes of phylum Platyhelminthes, giving examples of animals that belong to each class (page 641).
   - Phylum **Nemertea** (ribbon worms) is characterized by the **proboscis,** a muscular tube used in capturing food and in defense. The coelom is reduced; it consists of the **rhynchocoel,** a space surrounding the proboscis. Nemerteans have a tube-within-a-tube body plan, a complete digestive tract with mouth and anus, and a circulatory system.
   - Members of phylum **Platyhelminthes,** the **flatworms,** are **acoelomate** animals with bilateral symmetry, **cephalization,** three definite tissue layers, and well-developed organs. Many flatworms are **hermaphrodites;** a single animal produces both sperm and eggs. Flatworms have a ladder-type nervous system, typically consisting of sense organs and a simple brain composed of two **ganglia** connected to two nerve cords that extend the length of the body. They have **protonephridia,** organs that function in osmoregulation and disposal of metabolic wastes.

- Phylum Platyhelminthes includes four classes: Class **Turbellaria** comprises free-living flatworms, including **planarians**. Classes **Trematoda** and **Monogenea** include the parasitic **flukes**; the parasitic tapeworms compose class **Cestoda**. The parasitic flukes and tapeworms typically have suckers or hooks for holding on to their hosts; they have complicated life cycles with intermediate hosts and produce large numbers of eggs.

ThomsonNOW **Watch the fluke and tapeworm life cycles by clicking on the figures in ThomsonNOW.**

4 Describe the adaptive advantages of cephalization (page 641).
- Flatworms show the beginnings of cephalization, the evolution of a head with the concentration of sense organs and nerve cells (a simple brain) at the anterior end. Cephalization increases the effectiveness of a bilateral animal to actively find food, shelter, and mates and to detect enemies.

5 Describe the distinguishing characteristics of phylum Mollusca and the four molluscan classes discussed, giving examples (page 641).
- Members of phylum **Mollusca** are soft-bodied animals usually covered by a shell. They have a ventral **foot** for locomotion and a **mantle** that covers the **visceral mass**, a concentration of body organs.
- Mollusks have an **open circulatory system** with the exception of cephalopods, which have a **closed circulatory system**. A rasplike **radula** functions as a scraper in feeding in all groups except the bivalves, which are suspension feeders. Typically, marine mollusks have a free-swimming, ciliated **trochophore larva**.
- Class **Polyplacophora** includes the marine **chitons**, which have shells that consist of eight overlapping plates. Class **Gastropoda**, the largest group of mollusks, includes the snails, slugs, and their relatives. In gastropods, the body undergoes **torsion**, a twisting of the visceral mass. Class **Bivalvia** includes the aquatic clams, scallops, and oysters; a two-part shell, hinged dorsally, encloses the bodies of these suspension feeders. Class **Cephalopoda** includes the squids, octopods, and *Nautilus*. Cephalopods are active, predatory swimmers. Tentacles surround the mouth, which is located in the large head.

6 Describe the distinguishing characteristics of phylum Annelida and the three annelid classes discussed, giving examples (page 641).
- Phylum **Annelida**, the segmented worms, includes many aquatic worms, earthworms, and leeches. Annelids have conspicuously long bodies with **segmentation** both internally and externally; their large, compartmentalized coelom serves as a hydrostatic skeleton.
- Class **Polychaeta** consists of marine worms characterized by **parapodia**, appendages used for locomotion and gas exchange. The parapodia bear many bristlelike structures called **setae**. Polychaetes also differ from other annelids in having a well-defined head with sense organs. Class **Oligochaeta**, which comprises the earthworms, is characterized by a few short setae per segment. The body is divided into more than 100 segments separated internally by **septa**. Class **Hirudinea**, the leeches, is characterized by the absence of setae and appendages. Parasitic leeches are equipped with suckers for holding on to their host.

7 Describe the distinguishing characteristics of the lophophorate phyla (page 641).
- The **lophophorate phyla**, marine animals that have a lophophore, include the **brachiopods, phoronids,** and bryozoans. The **lophophore,** a ciliated ring of tentacles surrounding the mouth, is specialized for capturing suspended particles in the water.

8 List the distinguishing characteristics of phylum Nematoda (page 654).
- Phylum **Nematoda,** the **roundworms,** consists of highly successful ecdysozoans. Nematodes have a pseudocoelom. The body is covered by a tough **cuticle** that helps prevent desiccation. Parasitic nematodes in humans include *Ascaris,* **hookworms, trichina worms,** and **pinworms.**

9 Describe the distinguishing characteristics of phylum Arthropoda, and distinguish among the subphyla and classes of this phylum; give examples of animals that belong to each group (page 654).
- Phylum **Arthropoda** is composed of segmented animals with **paired, jointed appendages** and an armorlike **exoskeleton** of chitin. **Molting** is necessary for the arthropod to grow. Arthropods have an open circulatory system with a dorsal heart that pumps **hemolymph.** Aquatic forms have gills for gas exchange; terrestrial forms have either **tracheae** or **book lungs.**
- The **trilobites** are extinct marine arthropods covered by a hard, segmented shell. Each segment had a pair of **biramous appendages,** appendages with two jointed branches, an inner walking leg and an outer gill branch.
- Subphylum **Myriapoda** includes class **Chilopoda**, the centipedes, and class **Diplopoda,** the millipedes. Members of this subphylum have **uniramous** (unbranched) **appendages** and a single pair of antennae.
- Subphylum **Chelicerata** includes the **merostomes** (horseshoe crabs) and the **arachnids** (spiders, mites, and their relatives). The chelicerate body consists of a cephalothorax and abdomen; there are six pairs of uniramous, jointed appendages, of which four pairs serve as legs. The first appendages are **chelicerae;** the second are **pedipalps.** These appendages are adapted for manipulation of food, locomotion, defense, or copulation. Chelicerates have no antennae and no mandibles.
- Subphylum **Crustacea** includes lobsters, crabs, shrimp, pill bugs, and barnacles. The body consists of a cephalothorax and abdomen. Crustaceans typically have five pairs of walking legs. Appendages are biramous. Crustaceans have two pairs of **antennae** that sense taste and touch. The third appendages are **mandibles** used for chewing. Two pairs of **maxillae,** posterior to the mandibles, manipulate and hold food.
- Subphylum **Hexapoda** includes class **Insecta.** An insect is an **articulated, tracheated hexapod**; its body consists of head, thorax, and abdomen. Insects have uniramous appendages and a single pair of antennae. These arthropods have tracheae for gas exchange and **Malpighian tubules** for excretion.

10 Identify adaptations that have contributed to the biological success of insects (page 654).
- The biological success of the insects results from many adaptations, including a versatile exoskeleton, segmentation, specialized jointed appendages, highly developed sense organs, and ability to fly. **Metamorphosis,** transition from one developmental form to another, reduces intraspecific competition. Insects have developed effective reproductive strategies, effective mechanisms for defense and offense, and the ability to communicate.

ThomsonNOW **Explore the body plans of the protostomes by clicking on the figures in ThomsonNOW.**

1. Which of the following is associated with the evolution of the coelom? (a) radial symmetry (b) tube-within-a-tube body plan (c) incomplete metamorphosis (d) bilateral symmetry (e) development of ganglia

2. Cephalization (a) evolved along with bilateral symmetry (b) refers to the development of a digestive system (c) is associated with radial symmetry (d) involves a concentration of excretory organs (e) first evolved in arthropods

3. Rudimentary cephalization, protonephridia, and auricles characterize some members of phylum (a) Arthropoda (b) Cnidaria (c) Platyhelminthes (d) Mollusca (e) Crustacea

4. Trochophore larvae are characteristic of (a) Arthropoda (b) Cnidaria (c) Platyhelminthes (d) Mollusca (e) Crustacea

5. Tapeworms are classified in phylum (a) Porifera (b) Cnidaria (c) Platyhelminthes (d) Ctenophora (e) Coelomata

6. Which of the following is *not* an adaptation to parasitic life? (a) production of a few well-protected eggs (b) hooks (c) suckers (d) reduced digestive system (e) intermediate host

7. The hydrostatic skeleton (a) results in torsion (b) permits contracting muscles to recover quickly (c) permits a greater range of movement than possible in acoelomate animals (d) is a unique characteristic of annelids (e) permits ecdysis

8. Which of the following is *not* characteristic of nemerteans (proboscis worms)? (a) large coelom (b) rhynchocoel (c) complete digestive tube (d) muscular tube for capturing food (e) circulatory system

9. Which of the following characteristics is associated with phylum Mollusca? (a) mandibles (b) mantle (c) pedipalps (d) chelipeds (e) setae

10. Which of the following belong to phylum Mollusca? (a) gastropods and crustaceans (b) oligochaetes and polychaetes (c) chelicerates and bryozoans (d) crustaceans and nemerteans (e) gastropods and cephalopods

11. Which of the following are classified as Ecdysozoa? (a) mollusks (b) annelids (c) nematodes (d) nemerteans (e) lophophorates

12. Which of the following is *not* a nematode? (a) hookworm (b) trichina worm (c) *Ascaris* (d) leech (e) pinworm

13. Which of the following is *not* true of the lophophorates? (a) this group includes the bryozoans (b) some have shells and look somewhat like mollusks (c) this group is characterized by a ciliated ring of tentacles at anterior end (d) this group consists mainly of marine animals (e) some of its members are articulated, tracheated hexapods

14. Torsion in mollusks (a) is characteristic of bivalves (b) is a twisting of the visceral mass (c) involves coiling of the molluscan shell (d) begins in the adult stage (e) depends on action of parapodia

15. Trilobites (a) were early mollusks (b) are members of phylum Onychophora (c) are characterized by parapodia and setae (d) were early arthropods (e) are an evolutionary link between annelids and arthropods

16. Which of the following belong to subphylum Hexapoda? (a) insects (b) horseshoe crabs (c) crustaceans (d) centipedes (e) spiders

17. The correct sequence in insect complete metamorphosis is (a) egg $\longrightarrow$ immature form $\longrightarrow$ adult (b) egg $\longrightarrow$ trochophore larva $\longrightarrow$ veliger larva $\longrightarrow$ adult (c) egg $\longrightarrow$ pupa $\longrightarrow$ larva $\longrightarrow$ adult (d) egg $\longrightarrow$ larva $\longrightarrow$ pupa $\longrightarrow$ adult (e) adult $\longrightarrow$ larva $\longrightarrow$ egg $\longrightarrow$ pupa

18. Which of the following is characteristic of insects? (a) biramous appendages (b) mandibles (c) chelicerae (d) eight legs (e) two pairs of antennae

19. Which of the following is *not* characteristic of arthropods? (a) exoskeleton (b) pseudocoelom (c) paired, jointed appendages (d) chitin (e) segmentation

20. Spiders are characterized by (a) mandibles and maxillae (b) six pairs of legs on the abdomen (c) one pair of antennae (d) biramous appendages (e) chelicerae and pedipalps

1. In what ways are mollusks similar to platyhelminths? How are they different?

2. Hypothesize why oysters secrete calcium carbonate layers around foreign particles.

3. In what ways are nematodes similar to arthropods? How are they different?

4. **Evolution Link.** Discuss important adaptations made possible by the development of the coelom.

5. **Evolution Link.** Discuss the idea that every evolutionary adaptation has both advantages and disadvantages, using each of the following characters as an example: (a) cephalization, (b) the arthropod exoskeleton, and (c) segmentation with specialization.

6. **Evolution Link.** Insects that undergo complete metamorphosis outnumber those that do not by more than 10 to 1. Hypothesize an explanation.

Additional questions are available in ThomsonNOW at www.thomsonedu.com/login

The Animal Kingdom: The Deuterostomes

© Ed Robinson / Tom Stack & Associates

Representative deuterostomes. This marine habitat photographed in Hawaii includes an echinoderm known as a pencil urchin (*Heterocentrotus mammilatus*); and chordates, represented by fishes, the clown wrasse (*Coris gaimard*) with its yellow tail and the Moorish idol (*Zanchus cornutus*).

KEY CONCEPTS

The echinoderms and the chordates are the two most successful deuterostome lineages in terms of diversity, number of species, and number of individuals.

Echinoderms are characterized by radial symmetry in adults, a water vascular system, tube feet, and spiny skin.

At some time in its life, a chordate has a notochord; dorsal, tubular nerve cord; pharyngeal slits; and a muscular post-anal tail.

Shared derived characters of vertebrates include a vertebral column, cranium, neural crest cells, and an endoskeleton of cartilage or bone.

Jaws and fins were key adaptations that contributed to the success of jawed fishes.

Limbs, a body covering that retards water loss, and the amniotic egg, with its shell and amnion, were key adaptations that contributed to the success of terrestrial vertebrates.

What does a sea star have in common with a fish, lizard, hawk, or human? You may think it strange to group the echinoderms—the sea stars, sea urchins, and sand dollars—with the chordates, the phylum to which we humans and other animals with a backbone belong. However, even though these animals look and behave very differently, structural and molecular data suggest that chordates and echinoderms share a common ancestor and are closely related.

In Chapter 30 we discussed the two major protostome branches of the animal kingdom, the lophotrochozoans and the ecdysozoans. In this chapter we focus on the third major branch of the animal kingdom, the deuterostomes. Some biologists speculate that the common ancestor of the deuterostomes was an animal that obtained food by filtering ocean water. Today, the two major phyla assigned to the deuterostomes are the echinoderms and chordates. The largest chordate subphylum is Vertebrata, which includes the animals with which we are most familiar—fishes, amphibians, reptiles, birds, and mammals. Echinoderms are represented in the photograph by the pencil urchin, and chordates are represented by the fishes. ■

WHAT ARE DEUTEROSTOMES?

Learning Objective

1   Identify shared derived characters of deuterostomes.

Deuterostomia, the third major branch of the animal kingdom, includes the **echinoderms**—the sea stars, sea urchins, and sand dollars—and the **chordates,** the phylum to which we humans and other **vertebrates** (animals with a backbone) belong. Biologists also include the **hemichordates,** a small group of wormlike marine animals, with the deuterostomes. A three-part body including a proboscis, collar, and trunk is a shared character of hemichordates. The most familiar of the hemichordates are the **acorn worms,** animals that live buried in mud or sand. The hemichordates have a characteristic ring of cilia surrounding the mouth.

Deuterostomes evolved from a common ancestor during the Proterozoic eon more than 550 million years ago (mya). Deuterostomes are characterized by several **shared derived characters (synapomorphies),** evolutionary novelties absent in their common ancestor. Their defining characters are similarities in their patterns of development (see Chapter 29). For example, deuterostomes are characterized by radial, rather than spiral, cleavage. Their cleavage is indeterminate, which means the fate of their cells is fixed later in development than is the case in protostomes. In deuterostomes, the mouth does not develop from the blastopore as in protostomes. The blastopore of deuterostomes becomes the anus (or is located near the future site of the anus), and the mouth develops from a second opening at the anterior end of the embryo—thus the name *deuterostome,* which is derived from the Greek words for "second mouth." Pharyngeal slits are characteristic of deuterostomes, but the extant echinoderms have lost this character. Basal deuterostomes have a type of larva with a loop-shaped band of cilia used for locomotion.

Review

■   What are three shared derived characters of deuterosomes?

ECHINODERMS

Learning Objective

2   Identify three shared derived characters of echinoderms, and describe the main classes of echinoderms.

The **echinoderms** (phylum **Echinodermata**) have one of the most highly derived body plans in the animal kingdom. Echinoderm larvae are bilaterally symmetrical, ciliated, and free-swimming. However, during development the body reorganizes, and the adult exhibits **pentaradial symmetry,** in which the body is arranged in five parts around a central axis.

The most unique derived character of echinoderms is the **water vascular system,** a network of fluid-filled canals and chambers. Sea water flows into and out of the water vascular system through an opening in the body wall. Cilia lining the canals of the system move the water along. The water vascular system functions in feeding and gas exchange and serves as a hydrostatic skeleton important in locomotion.

Branches of the water vascular system lead to numerous tiny **tube feet** that extend when filled with fluid. Each tube foot receives fluid from the main system of canals. A rounded muscular sac, or **ampulla,** at the base of the foot, stores fluid and is used to operate the tube foot. A valve separates each tube foot from other parts of the system. When the valve shuts, the ampulla contracts, forcing fluid into the tube foot so that it extends. At the bottom of the foot is a suction-type structure that presses against and adheres to whatever surface the tube foot is on.

Another unique echinoderm character is the **endoskeleton,** an internal skeleton, that is covered by a thin, ciliated epidermis. The endoskeleton consists of calcium carbonate ($CaCO_3$) plates and spines. The name *Echinodermata,* derived from words meaning "spiny skinned," was inspired by the spines that project outward from the endoskeleton. Some groups have pincerlike, modified spines called **pedicellariae** on the body surface. These structures, found only among the echinoderms, keep the surface of the animal free of debris.

Echinoderms have a well-developed coelom, and the coelomic fluid transports materials. Although its structure varies in different groups, the complete digestive system is the most prominent body system. A variety of respiratory structures are found in the various classes. No excretory organs are present. The nervous system is simple, generally consisting of a nerve ring with nerves that extend out from it. Echinoderms have no brain. The sexes are usually separate, and eggs and sperm are generally released into the water, where fertilization takes place.

The echinoderms evolved from bilaterally symmetrical ancestors, probably during the Early Cambrian period. They achieved maximum diversity by the middle of the Paleozoic era, about 400 mya. By the beginning of the Mesozoic era 251 mya, they had declined, leaving five main groups that have survived to the present day (■ Fig. 31-1): class Crinoidea, sea lilies and feather stars; class Asteroidea, sea stars; class Ophiuroidea, basket stars and brittle stars; class Echinoidea, sea urchins and sand dollars; and class Holothuroidea, sea cucumbers. All members of phylum Echinodermata inhabit marine environments. They are found in the ocean at all depths. Biologists have identified about 7000 living and more than 13,000 extinct species.

Members of class Crinoidea are suspension feeders

Class **Crinoidea,** the oldest class of living echinoderms, includes the feather stars and the sea lilies (see Fig. 31-1a). Although many extinct crinoids are known, there are relatively few living species. The feather stars are motile crinoids, although they often remain in the same location for long periods. Sea lilies are sessile and remain attached to the ocean floor by a stalk.

Crinoids remove suspended food from the water. In all other echinoderms, the mouth is located on the underside of the disc toward the substratum, but in crinoids the **oral surface** (mouth surface) is on the upper side of the disc. Several branched, feathery arms also extend upward. Numerous tube feet shaped like small tentacles are located along the feathery arms. These tube feet are coated with mucus that traps microscopic organisms.

Many members of class Asteroidea capture prey

Sea stars, or starfish, are members of class **Asteroidea** (see Fig. 31-1b). Their bodies consist of a central disc from which extend 5 to more than 20 arms, or rays (Fig. 31-2). The undersurface of each arm has hundreds of pairs of tube feet. The mouth lies in the center of the underside of the disc. The endoskeleton consists of a series of calcareous plates that permit some movement in the arms. Delicate dermal gills, small extensions of the body wall, carry on gas exchange.

Most sea stars are carnivorous predators and scavengers that feed on crustaceans, mollusks, annelids, and even other echinoderms. Occasionally they catch small fish. The sea star's water vascular system does not permit rapid movement, so its prey usually consists of slow-moving or stationary animals, such as clams.

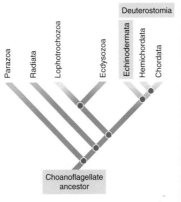

(a) **Crinoidea.** Feather stars use their slender, jointed appendages to cling to the surface of a rock or coral reef. They can creep and often move away to escape predators.

Figure 31-1 Echinoderms

(b) **Asteroidea.** Orange and red sea star (*Fromia monilis*) on bubble coral.

(c) **Ophiuroidea.** Daisy brittle star (*Ophiopholis aculeata*), photographed in Muscongus Bay, Maine.

(d) **Echinoidea.** With its flattened, circular body, the sand dollar (*Dendraster excentricus*) is adapted for burrowing on the ocean floor.

(e) **Holothuroidea.** A sea cucumber (*Thelonota*) raises its body to spawn.

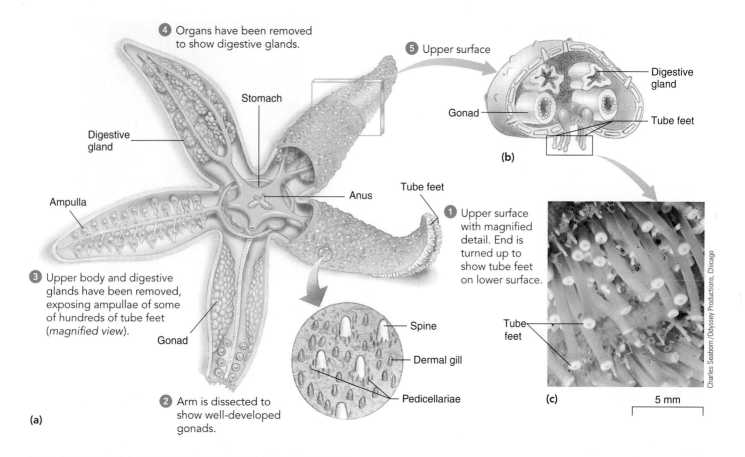

(a)

④ Organs have been removed to show digestive glands.

Digestive gland

Stomach

Ampulla

③ Upper body and digestive glands have been removed, exposing ampullae of some of hundreds of tube feet (*magnified view*).

Gonad

② Arm is dissected to show well-developed gonads.

Anus

Tube feet

② Arm is dissected to show well-developed gonads.

Spine

Dermal gill

Pedicellariae

⑤ Upper surface

Digestive gland

Gonad

Tube feet

(b)

① Upper surface with magnified detail. End is turned up to show tube feet on lower surface.

Tube feet

(c)

5 mm

Charles Seaborn /Odyssey Productions, Chicago

Figure 31-2 *Animated* Body plan of a sea star

(a) A sea star viewed from above, with its arms in various stages of dissection. Similar structures are present in each arm. The two-part stomach is in the central disc with the anus on the aboral (*upper*) surface and the mouth beneath on the oral surface. **(b)** Cross section through arm and tube feet. **(c)** LM of tube feet of a sea star.

To attack a clam or other bivalve mollusk, the sea star mounts it and assumes a humped position as it straddles the edge opposite the hinge. Then, holding itself in position with its tube feet, the sea star slides its thin, flexible stomach out through its mouth and between the closed, or slightly gaping, valves (shell parts) of the clam. The sea star secretes enzymes that digest the soft parts of the clam to the consistency of a thick soup while the clam is still in its own shell. The partly digested meal passes into the sea star body, where it is further digested by enzymes secreted from glands located in each arm.

The circulatory system in sea stars is poorly developed and probably of little help in circulating materials. Instead, the coelomic fluid, which fills the large coelom and bathes the internal tissues, assumes this function. Metabolic wastes pass to the outside by diffusion across the tube feet and dermal gills. The nervous system consists of a ring of nervous tissue encircling the mouth and a nerve extending from this ring into each arm.

Sea daisies are a curious group of echinoderms characterized by small, disc-shaped, flat bodies (less than 1 cm in diameter) with no arms or mouth. They inhabit bacteria-rich wood sunk in deep water and apparently absorb bacteria through their body surface. When sea daisies were discovered in 1986, biologists established a class (**Concentricycloidea**) to accommodate them. However, in 2005 they were reclassified as an order in the class Asteroidea.

Class Ophiuroidea is the largest class of echinoderms

Basket stars and brittle stars (serpent stars), members of class **Ophiuroidea,** are the largest group of echinoderms, both in number of species and of individuals (see Fig. 31-1c). These animals resemble sea stars in that their bodies consist of a central disc with arms. However, the arms are long and slender and more sharply set off from the central disc. Ophiuroids move more rapidly than sea stars by using their arms to perform rowing or even swimming movements. Their tube feet lack suckers and are not used in locomotion. Instead, they are used to collect and handle food and may also serve a sensory function, perhaps that of taste.

Members of class Echinoidea have movable spines

Sea urchins and sand dollars, the animals of class **Echinoidea,** have no arms (see Fig. 31-1d and chapter opening photograph). Their skeletal plates are flattened and fused to form a solid shell called a **test.** The flattened body of the sand dollar is adapted for burrowing in the sand, where it feeds on tiny organic particles. Sand dollars have smaller spines than do sea urchins.

The sea urchin body is covered with spines that in some species can penetrate flesh and are difficult to remove. So threatening are these spines that swimmers on tropical beaches are often cautioned to wear shoes when venturing offshore, where these living pincushions may live in abundance. Sea urchins use their tube feet for locomotion. They also push themselves along with their movable spines. Many sea urchins graze on algae, scraping the seafloor with their calcareous teeth.

Members of class Holothuroidea are elongated, sluggish animals

Sea cucumbers, members of class **Holothuroidea,** are appropriately named, for some species are about the size and shape of a cucumber. The elongated body is a flexible, muscular sac (see Fig. 31-1e). The mouth is usually surrounded by a circle of tentacles that are modified tube feet. The endoskeleton consists of microscopic plates embedded in the body wall. More highly developed than that of other echinoderms, the circulatory system functions to transport oxygen and perhaps nutrients.

Sea cucumbers are sluggish animals that usually live on the bottom of the sea, sometimes burrowing in the mud. Some graze with their tentacles, whereas others stretch their branched tentacles out in the water and wait for dinner to float by. Algae and other morsels are trapped in mucus along the tentacles.

An interesting habit of some sea cucumbers is evisceration, in which the digestive tract, respiratory structures, and gonads are ejected from the body, usually when environmental conditions are unfavorable. When conditions improve, the lost parts are regenerated. Even more curious is the fact that when certain sea cucumbers are irritated or attacked, they direct their rear end toward the enemy and shoot red tubules out of their anus! These unusual weapons are sticky, and the attacking animal may become hopelessly entangled. Some of the tubules release a toxic substance.

Review

■ What are three derived characters of echinoderms? Describe each.

■ How do sea stars differ from crinoids? From sea urchins?

CHORDATE CHARACTERS

Learning Objective

3 Describe five shared derived characters of chordates.

Biologists currently divide phylum **Chordata** into three subphyla: Urochordata, marine animals called *tunicates;* Cephalochordata, marine animals called *lancelets;* and Vertebrata, animals with backbones. Of these extant chordate groups, the urochordates were the earliest to evolve (■ Fig. 31-3).

Chordates are deuterostome coelomates with bilateral symmetry, a tube-within-a-tube body plan, and three well-developed germ layers (■ Fig. 31-4). Typically, chordates have an endoskeleton and a closed circulatory system with a ventral heart. Chordates have segmented bodies, but specialization is so pronounced that the basic segmentation of the body plan may not be apparent. For example, in vertebrates, although segmentation of the body is not obvious, muscles and nerves are segmentally organized.

Five shared derived characters distinguish the chordates. These characters, which evolved in connection with their evolv-

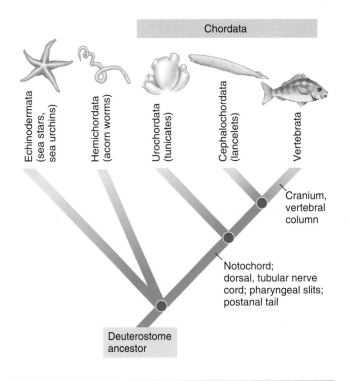

Figure 31-3 Evolutionary relationships of the chordates

This cladogram shows possible phylogenetic relationships among deuterostomes, including chordates (*tan region*), based on structural and DNA data.

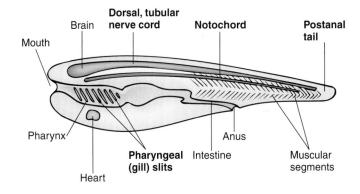

Figure 31-4 Generalized chordate body plan

Note the notochord; dorsal, tubular nerve cord; pharyngeal (gill) slits; and postanal tail.

ing methods of locomotion and obtaining food, include the notochord; the dorsal, tubular nerve cord; pharyngeal slits; a postanal tail; and an endostyle or thyroid gland.

1. All chordates have a **notochord** during some time in their life cycle. The notochord is a dorsal, longitudinal rod composed of cartilage; it is firm but flexible and supports the body.

2. At some time in their life cycle, chordates have a **dorsal, tubular nerve cord.** The chordate nerve cord differs from the nerve cord of most other animals in that it is located dorsally rather than ventrally, is hollow rather than solid, and is single rather than double.

3. Chordates have **pharyngeal slits** (also called *pharyngeal gill slits*) or pharyngeal pouches during some time in their life cycle. In the embryo, a series of alternating branchial (gill) arches and grooves develop in the body wall in the pharyngeal (throat) region. Pharyngeal pouches extend laterally from the anterior portion of the digestive tract toward the grooves.

Early chordates, like many living chordates, were probably suspension feeders. The arrangement of pharyngeal pouches and slits permitted them to take water in through the mouth, concentrate small particles of food in the gut, and allow the water to exit from the body through the slits. In fishes and some other aquatic vertebrates, the pharyngeal slits and the arches supporting them became modified for gas exchange. Early in vertebrate evolution, anterior pharyngeal arches evolved into jaws.

4. Chordates have a larva or embryo with a muscular **postanal tail,** an appendage that extends posterior to the anus.

5. Another derived chordate character is the **endostyle** or its derivative, the thyroid gland. The endostyle is a groove in the floor of the pharynx. It secretes mucus and functions as a net that traps food particles in the sea water passing through the pharynx.

Review

- What is a notochord?
- What is the significance of pharyngeal slits?

INVERTEBRATE CHORDATES

Learning Objectives

4 Describe the invertebrate chordate subphyla.
5 Discuss the evolution of chordates.

Some chordates are not vertebrates. The invertebrate chordates include tunicates and lancelets. Biologists assign the tunicates to subphylum Urochordata and the lancelets to subphylum Cephalochordata.

Tunicates are common marine animals

The **tunicates,** which make up subphylum **Urochordata,** include the sea squirts and their relatives. Larval tunicates have typical chordate characteristics and superficially resemble tadpoles. The expanded body has a pharynx with slits, and the long muscular tail contains a notochord and a dorsal, tubular nerve cord. Some tunicates (*appendicularians*) retain their chordate features and ability to swim. These animals are common members of the zooplankton.

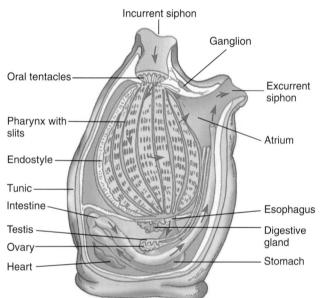

Figure 31-5 *Animated*
Tunicate body plan

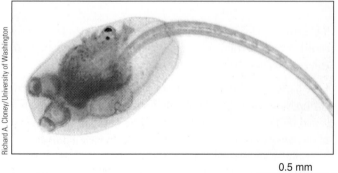

(a) Incurrent (*top*) and excurrent (*side*) siphons of a sea peach (*Halocynthia aurantium*), a solitary tunicate.

(b) Lateral view of an adult tunicate. The blue arrows represent the flow of water, and the red arrows represent the path of food.

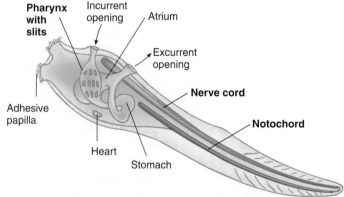

(c) Swimming larval stage of a colonial species, *Distaplia occidentalis*.

(d) Internal structure of a larval tunicate (*lateral view*).

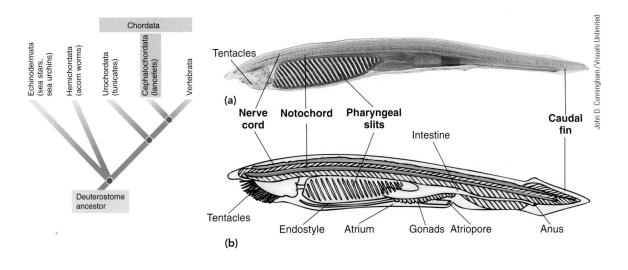

Figure 31-6 *Animated* Cephalochordate body plan

(a) Photograph of a lancelet, *Branchiostoma* (amphioxus). Note the prominent pharyngeal gill slits. (b) Longitudinal section showing internal structure.

Most tunicates are **ascidians,** commonly known as sea squirts (class Ascidiacea). A sea squirt larva swims for a time, then attaches itself to the sea bottom and loses its tail, notochord, and much of its nervous system. Adult sea squirts are barrel-shaped, sessile marine animals unlike other chordates. Indeed, they are often mistaken for sponges or cnidarians (▌Fig. 31-5). Only the pharyngeal slits, endostyle, and the structure of its larva indicate that the sea squirt is a chordate.

Adult tunicates develop a protective covering, or **tunic,** that may be soft and transparent or quite leathery. Curiously, the tunic consists of a carbohydrate much like cellulose. The tunic has two openings: the incurrent siphon, through which water and food enter; and the excurrent siphon, through which water, waste products, and gametes pass to the outside. Sea squirts get their name from their practice of forcefully expelling a stream of water from the excurrent siphon when irritated.

Tunicates are suspension feeders that remove plankton from the stream of water passing through the pharynx. Food particles are trapped in mucus secreted by cells of the *endostyle,* a groove that extends the length of the pharynx. Ciliated cells of the pharynx move the stream of food-laden mucus into the esophagus. Much of the water entering the pharynx passes out through the pharyngeal slits into an **atrium** (chamber) and is discharged through the excurrent siphon.

Some species of tunicates form large colonies in which members share a common tunic and excurrent siphon. Colonial forms often reproduce asexually by budding. Sexual forms are usually hermaphroditic.

Lancelets may be closely related to vertebrates

Most species of subphylum **Cephalochordata** belong to the genus *Branchiostoma,* which consists of animals commonly known as **lancelets,** or **amphioxus.** Lancelets are translucent, fish-shaped

animals, 5 to 10 cm (2 to 4 in) long and pointed at both ends. They are widely distributed in shallow seas, either swimming freely or burrowing in the sand near the low-tide line. In some parts of the world, lancelets are an important source of food. One Chinese fishery reports an annual catch of 35 tons (about 1 billion lancelets)!

Chordate characteristics are highly developed in lancelets. The notochord extends from the anterior tip ("head"; hence the name *Cephalochordata*) to the posterior tip. Many pairs of pharyngeal slits are evident in the large pharyngeal region, and a dorsal, tubular nerve cord extends the entire length of the animal (▌Fig. 31-6). Although superficially similar to fishes, lancelets have a far simpler body plan. They do not have paired fins, jaws, sense organs, a heart, or a well-defined head or brain.

Like the tunicates, lancelets use their cilia to draw a current of water into the mouth and then strain out microscopic organisms. Food particles are trapped in mucus in the pharynx and are then carried back to the intestine.

Water passes through the pharyngeal slits into the atrium, a chamber with a ventral opening (the *atriopore*) just anterior to the anus. Metabolic wastes are excreted by segmentally arranged, ciliated *protonephridia* that open into the atrium. Unlike circulation in other invertebrates, the blood flows anteriorly in the ventral vessel and posteriorly in the dorsal vessel in lancelets. This circulatory pattern is similar to that of fishes.

Recent molecular and structural findings suggest that lancelets are the vertebrates' closest living relative. Important shared derived characters include the presence of the notochord, nerve cord, and postanal tail in the adults. Another is the presence of **somites,** a series of paired blocks of mesoderm that develop on each side of the notochord. Somites are temporary structures that define the segmentation of the embryo. In vertebrates, they give rise to the vertebrae, ribs, and certain skeletal muscles. Another similarity between cephalochordates and vertebrates is the lancelet nerve cord, which has specialized regions that correspond to the vertebrate forebrain and midbrain.

Systematists are making progress in understanding chordate phylogeny

Both structural and molecular data indicate that subphyla Cephalochordata and Vertebrata are **sister taxa,** groups that diverged from a recent common ancestor. That ancestor may have resembled a tunicate larva.

Researchers have sequenced the genome of the sea squirt *Ciona intestinalis,* the most-studied ascidian tunicate. *Ciona* is an excellent model organism for studying chordate development because its genome consists of only about 16,000 genes, about half the number found in vertebrates. Interestingly, about 80% of *Ciona*'s genes are found in vertebrates. In fact, *Ciona* has most vertebrate gene families, but in simplified form. For example, *Ciona* has only a single copy of each gene family involved in cell signaling and regulation of development, whereas vertebrates have two or more copies of these gene families. (Some of these "extra" copies probably evolved to code for new structures or functions.) *Ciona* also has genes for some vertebrate structures and processes, even though it does not express these genes. Biologists hypothesize, based on these molecular data, that the sea squirt genome may correspond to that of the chordate ancestor.

Fossils of early chordates also contribute to our understanding of chordate phylogeny. A lancelet-like animal, which biologists named *Pikaia,* was discovered in the Burgess Shale of British Columbia, Canada. This animal, which dates to the Cambrian period, had a primitive notochord with muscles attached to it that allowed for locomotion. About 4 cm (1.5 in) in length, *Pikaia* had a tail fin and probably filtered food from the water. *Pikaia* is considered an early chordate.

Many well-preserved fossils occur in **Chengjiang,** an Early Cambrian fossil site in China. At this site, fine-grained rocks about 530 million years old have preserved soft-bodied animals, such as *Yunnanozoon, Haikouella,* and *Haikouichthys,* in great detail. Some biologists have interpreted *Yunnanozoon,* which somewhat resembles *Pikaia,* as an early chordate, but other biologists think it was a hemichordate.

Haikouella was about 2 to 3 cm (approximately 1 in) long and had a nerve cord, notochord, gills, muscle segments, and a brain. Most biologists view *Haikouella* as an early chordate. *Haikouichthys* was a fishlike animal about 2.5 cm long. It had several characters found in vertebrates, and some researchers think it was an early vertebrate.

Conodonts were simple fishlike chordates with fins, large eyes, and complex toothlike hooks that were probably used in capturing prey. These animals had gill arches and muscular segments. Conodonts were abundant during the Precambrian to Late Triassic; they have been identified in several fossil beds around the world. Recent cladistic analyses suggest that conodonts were early vertebrates.

Review

■ How are the main derived chordate characters evident in a tunicate larva, in an adult tunicate, and in a lancelet?

■ How does sequencing animal genomes help clarify relationships among chordate groups?

INTRODUCING THE VERTEBRATES

Learning Objective

6 Describe four shared derived characters of vertebrates.

Consisting of about 48,000 species, the **vertebrates,** members of subphylum **Vertebrata,** are less diverse and much less numerous than the insects but rival them in their adaptations to an enormous variety of lifestyles (■ Fig. 31-7). In addition to the basic chordate characteristics, vertebrates have a number of shared derived characters not found in other groups.

The vertebral column is a key vertebrate character

The vertebrates are distinguished from other chordates in having a backbone, or **vertebral column,** that forms the skeletal axis of the body. This flexible support develops around the notochord, and in most species it largely replaces the notochord during embryonic development. The vertebral column consists of cartilaginous or bony segments called **vertebrae.** Dorsal projections of the vertebrae enclose the nerve cord along its length. Anterior to the vertebral column, a cartilaginous or bony **cranium,** or braincase, encloses and protects the brain, the enlarged anterior end of the nerve cord.

The cranium and vertebral column are part of the **endoskeleton.** In contrast with the nonliving exoskeleton of many invertebrates, the vertebrate endoskeleton is a living tissue that grows with the animal. Some vertebrates (jawless and cartilaginous fishes) have skeletons made of cartilage. However, in most vertebrates, the skeleton is mainly bone, a tissue that contains fibers made of the protein collagen. The hard matrix of bone consists of the compound hydroxyapatite, composed mainly of calcium phosphate.

Many characters common to vertebrates have been derived from a group of cells called **neural crest cells.** These cells, found only in vertebrates, appear early in development and migrate to various parts of the embryo. Neural crest cells give rise to or influence the development of many structures, including nerves, head muscles, cranium, and jaws.

Recall that invertebrates show an evolutionary trend toward **cephalization,** the concentration of nerve cells and sense organs in a definite head. Vertebrate evolution is characterized by *pronounced* cephalization. The brain has become larger and more elaborate, and its various regions have specialized to perform different functions. Either 10 or 12 pairs of **cranial nerves** emerge from the brain and extend to various organs of the body. Vertebrates have well-developed sense organs concentrated in the head: eyes; ears that serve as organs of balance and, in some vertebrates, for hearing as well; and organs of smell and taste.

Most vertebrates have two pairs of appendages. The fins of fishes are appendages that stabilize the fish in the water. Paired pectoral and pelvic fins are also used in steering. Biologists hypothesize that jointed appendages that facilitated locomotion on land evolved from the lobed (divided) fins of lungfishes.

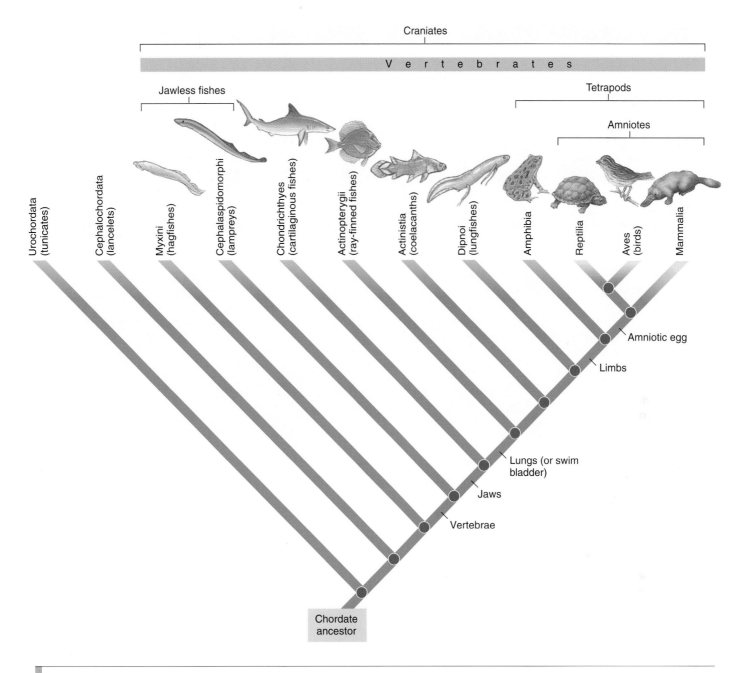

Figure 31-7 *Animated* Evolutionary relationships of vertebrates

This cladogram represents one phylogenetic interpretation of vertebrate phylogeny. Biologists have not yet reached a consensus as to whether hagfishes are true vertebrates. The relationships of the amniotes in this cladogram are highly simplified. Figure 31-20 is a more detailed cladogram for these groups. The evolution of certain key traits is indicated. This cladogram provides a framework for understanding the evolution of vertebrate diversity and adaptations to aquatic and terrestrial environments. As new data are collected and considered, details of the cladogram will likely change.

Vertebrates have a closed circulatory system with a ventral heart and blood containing hemoglobin. The complete digestive tract has specialized regions and large digestive glands (the liver and pancreas). Several **endocrine glands,** which are ductless glands, secrete hormones. Paired kidneys regulate fluid balance. The sexes are typically separate.

Early vertebrates were probably marine animals. As we have discussed, tunicates and cephalochordates use cilia to beat a stream of water into the mouth. They filter particles of food from the water. In contrast, vertebrates use muscles for feeding. Some use muscles to draw in a current of water from which both food and oxygen can be extracted. Muscles in the wall of the digestive tube are more powerful than cilia and probably contributed to an increase in both size and activity of early vertebrates. A muscular pharynx may have permitted animals to crush small prey after capture. The evolution of more effective sensory systems, a more complex brain, and organs that could support increased activity was important in the shift from filter-feeding to a more active lifestyle.

TABLE 31-1

Extant Vertebrate Classes

Class		Examples	Characteristics
Myxini		Hagfishes	Jawless, marine fishes that lack paired appendages; the notochord is their only axial support; lack vertebrae, so some biologists classify as craniates, but not as vertebrates
Cephalaspidomorphi		Lampreys	Jawless, freshwater and marine fishes with skeleton of cartilage; complete cranium and rudimentary vertebrae; hatch as small larvae
Chondrichthyes		Sharks, rays, skates, chimaeras	Jawed marine and freshwater fishes with skeleton of cartilage; notochord replaced by vertebrae in adult; gills; placoid scales; two pairs of fins; oviparous, ovoviparous, or viviparous (a few species); well-developed sense organs (including lateral line system)
Actinopterygii (ray-finned fishes)		Perch, salmon, tuna, trout	Bony, marine and freshwater fishes; gills; swim bladder; generally oviparous
Actinistia (lobe-finned fishes)		Coelacanths	Bony fishes; marine, nocturnal predators on fish; lobed fins
Dipnoi		Lungfishes	Bony freshwater fishes; four similarly sized limbs, which are similar in structure and position to those of tetrapods
Amphibia		Salamanders, frogs and toads, caecilians	Tetrapods; aquatic larva typically undergoes metamorphosis into terrestrial adult; gas exchange through lungs and/or moist skin; heart consists of two atria and single ventricle; systemic and pulmonary circulation
Reptilia		Turtles, lizards, snakes, alligators	Amniotes with horny scales; adapted for reproduction on land (internal fertilization, leathery shell, amnion); lungs; ventricles of heart partly divided*
Aves		Robins, pelicans, eagles, ducks, penguins, ostriches	Amniotes with feathers; anterior limbs modified as wings; compact, streamlined body; four-chambered heart; endothermic; vocal calls and complex songs
Mammalia		Monotremes (Holotheria), marsupials (Metatheria), placental mammals (Eutheria)	Amniotes with hair; females nourish young with mammary glands; differentiation of teeth; three middle-ear bones; diaphragm; four-chambered heart; endothermic; highly developed nervous system

*A cladistic classification of amniotes is shown in Figure 31-20.

Vertebrate taxonomy is a work in progress

The study of evolutionary relationships of vertebrates is an important focus of research. Biologists consider the vertebrates and lancelets as sister groups that diverged from a common ancestor. The earliest vertebrates to evolve were **agnathans** (*a*, "without"; and *gnathos*, "jaw"): the hagfishes, followed by the lampreys. Biologists assign hagfishes to class **Myxini** and lampreys to class **Cephalaspidomorphi.**

Some systematists have argued that the hagfishes do not quite qualify as vertebrates. Although they have many vertebrate characters, including a cranium, hagfishes have no trace of vertebrae. The notochord is their only axial support. Some biologists use the term **Craniata,** based on the presence of a cranium, to designate a clade that includes the vertebrates plus the hagfishes. They view the Myxini as the sister group of all other craniates. Recent molecular analyses, however, support classifying hagfishes as vertebrates.

If we include the Myxini, the extant vertebrates are currently assigned to 10 classes: 6 classes of fishes and 4 of tetrapods (four-limbed vertebrates) (▌Table 31-1 and see Fig. 31-7). Some biologists give fishes and tetrapods superclass status: superclass **Pisces** (fishes) and superclass **Tetrapoda.** In this classification scheme, the fishes include the hagfishes, lampreys, and four classes of jawed fishes: Chondrichthyes, Actinopterygii, Actinistia, and Dipnoi.

The four-limbed land vertebrates, or **tetrapods,** include class Amphibia, the frogs, toads, and salamanders; class Reptilia, the lizards, snakes, turtles, and alligators; class Aves, the birds; and class Mammalia, the mammals. As we discuss later in the chapter, many cladists classify birds with reptiles rather than assign them to a separate class. The reptiles, birds, and mammals form a clade known as *amniotes,* animals fully adapted to life on land.

Review

▌ What derived characters distinguish the vertebrates from the rest of the chordates?

JAWLESS FISHES

Learning Objective

7 Distinguish among the major groups of jawless fishes.

Some of the earliest known vertebrates, collectively referred to as **ostracoderms,** consisted of several groups of small, armored, jawless fishes that lived on the bottom and strained their food from the water (see Fig. 21-10). Thick bony plates protected their heads from predators, and thick scales covered their trunks and tails. Most ostracoderms lacked fins. Fragments of ostracoderm scales have been found in rocks from the Cambrian period, but most ostracoderm fossils are from the Ordovician and Silurian periods. They became extinct by the end of the Devonian period.

Like ostracoderms, present-day hagfishes and lampreys have neither jaws nor paired fins. They are eel-shaped animals, up to 1 m (about 3 ft) long. Their smooth skin lacks scales, and they are supported by a cartilaginous skeleton and well-developed notochord.

Hagfishes, assigned to class Myxini, are marine scavengers (▐ Fig. 31-8). They burrow for worms and other invertebrates or prey on dead and disabled fishes. Hagfishes use toothlike projections from their tongue to pull off flesh from prey. For leverage, a hagfish can tie itself into a knot. As a defense mechanism, hag-

Tom McHugh / Photo Researchers, Inc.

▌ **Figure 31-8** *Animated* Pacific hagfish (*Eptatretus stoutii*)

The hagfish exudes slime when alarmed.

fishes secrete large amounts of fluid that forms sticky slime in sea water. The hagfish becomes so slippery that predators cannot grasp it.

Lampreys are jawless vertebrates assigned to class Cephalaspidomorphi. Some spend their adult lives in the ocean and return to fresh water to reproduce. Many species of adult lampreys are parasites on other fishes (▐ Fig. 31-9). Adult parasitic lampreys have a circular sucking disc around the mouth, which lies on the ventral side of the anterior end of the body. Using this disc to attach to a fish, the lamprey bores through the skin of its host with horny (made of keratin rather than bone) teeth on the disc and tongue. Then the lamprey injects an anticoagulant into its host and sucks out blood and soft tissues.

As in hagfish, the notochord persists throughout life and is not replaced by a vertebral column. However, lampreys have ru-

Tom Stack / Tom Stack & Associates

(a) Three lampreys are attached to a carp by their suction-cup mouths. Note the absence of jaws and paired fins.

Courtesy of Dr. Kiyoko Uehara

(b) Suction-cup mouth of adult lamprey (*Entosphenus japonicus*). Note the rasplike teeth.

▌ **Figure 31-9** *Animated* Lampreys

diments of vertebrae, cartilaginous segments called *neural arches* that extend dorsally around the spinal cord.

Review

■ How do lampreys and hagfishes resemble ostracoderms?
■ How do hagfishes differ from other fishes?

EVOLUTION OF JAWS AND LIMBS: JAWED FISHES AND AMPHIBIANS

Learning Objective

8 Trace the evolution of jawed fishes and early tetrapods, and identify major taxa of jawed fishes and amphibians.

Imagine trying to hunt for food and eat it without jaws or limbs. This was the challenge of early vertebrates and helps explain why the jawless fishes are mainly scavengers and parasites. The evolution of jaws from a portion of the gill arch skeleton and the development of fins allowed fishes to become active predators. With jaws an animal can grasp and hold on to live prey while eating it. The evolution of fins allowed fishes to swim faster and with more control. Fishes with jaws and fins had many new opportunities for capturing food.

Fossil evidence suggests that jaws and paired fins evolved during the Late Silurian and Devonian periods. Two early groups of jawed fishes, now extinct, were the **acanthodians,** armored fishes with paired spines and pectoral and pelvic fins, and **placoderms,** armored fishes with paired fins (∎ Fig. 31-10). The success of the jawed vertebrates probably contributed to the extinction of the ostracoderms.

When the fins of certain fishes evolved into limbs about 370 mya, a different type of locomotion was possible. These fishes were able to move about in shallow waters and wetlands in search

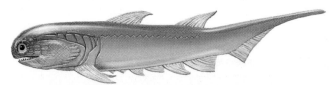

(a) *Climatius* was a spiny-skinned acanthodian with large fin spines and five pairs of accessory fins between the pectoral and pelvic pairs. *Climatius* was a small fish that reached a length of 8 cm (3 in).

(b) *Dinichthys* was a giant placoderm that grew to a length of 8 m (26 ft). (Most placoderms were only about 20 cm, or 8 in, long.) Its head and thorax were covered by bony armor, but the rest of the body and tail were naked.

∎ **Figure 31-10** *Animated* Early jawed fishes

Acanthodians and placoderms flourished in the Devonian period.

(a) Blue-spotted stingray (*Taeniura lymma*). Stingrays typically feed on shellfish and bottom-dwelling fishes.

(b) The great white shark (*Carcharodon carcharias*), photographed in Australia, is considered the most dangerous shark to humans. This shark is actually white only on its ventral apsect; the rest of the body is brownish gray or bluish gray.

∎ **Figure 31-11** *Animated* Cartilaginous fishes

of food. When the first tetrapods—the amphibians—moved onto the land, new opportunities for habitats and food became available.

Members of class Chondrichthyes are cartilaginous fishes

Members of class **Chondrichthyes,** the cartilaginous fishes, evolved as successful marine forms in the Devonian period. Class Chondrichthyes, which is considered monophyletic, includes the sharks, rays, and skates (∎ Figs. 31-11 and 31-12a). Most species are ocean dwellers, but a few have invaded fresh water. With the exception of whales, the sharks are the largest living vertebrates. Some whale sharks (*Rhincodon*) exceed 15 m (49 ft) in length, making them the largest fish.

Most rays and skates are flattened creatures that live partly buried in the sand. Their enormous pectoral fins propel them along the bottom, where they feed on mussels and clams. The stingray has a whiplike tail with a barbed spine at its base that can inflict a painful wound. The electric ray has electric organs on either side of the head. These modified muscles can discharge

Figure 31-12 Anatomy of a shark

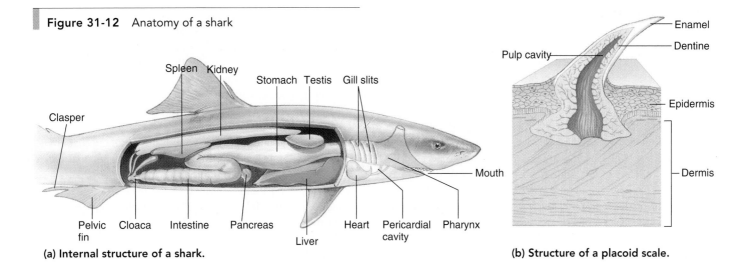

(a) Internal structure of a shark.

(b) Structure of a placoid scale.

enough electric current (up to 200 volts) to stun fairly large fishes, as well as human swimmers.

The chondrichthyes retain their cartilaginous embryonic skeleton. Although this skeleton is not replaced by bone, a deposit of calcium salts may strengthen it. All chondrichthyes have paired jaws and two pairs of fins. The skin contains **placoid scales.** Each scale is a toothlike structure consisting of an outer layer of enamel and an inner layer of dentine (Fig. 31-12b). The lining of the mouth contains larger, but essentially similar, scales that serve as teeth. The teeth of other vertebrates are homologous with these scales. Shark teeth are embedded in the flesh and not attached to the jawbones; new teeth develop continuously in rows behind the functional teeth and migrate forward to replace any that are lost.

The shark body is adapted for swimming. Lift is provided by body shape and fins and by the action of swimming swiftly. The shark stores a great deal of oil in its large liver (which may account for up to 30% of the body weight). Fats and oils decrease the overall density of fishes and contribute to buoyancy. Even so, the shark body is denser than water, so sharks tend to sink unless they are actively swimming.

Most sharks are streamlined predators that swim actively and catch and eat other fishes as well as crustaceans and mollusks. The largest sharks and rays, like the largest whales, are suspension feeders that strain plankton from the water. They gulp water through the mouth. As the water passes through the pharynx and out the gill slits, food particles are trapped in a sievelike structure.

Predatory sharks are attracted to blood, so a wounded swimmer or a skin diver towing speared fish is a target. However, despite the common portrayal of sharks in books and films as monstrous enemies, most do not go out of their way to attack humans. In fact, of the approximately 350 known shark species, fewer than 30 are known to attack humans.

The shark has a complex brain and a spinal cord that is protected by vertebrae. The well-developed sense organs very effectively locate prey in the water. Sharks may detect other animals electrically before sensing them by sight or smell. **Electroreceptors** on the shark's head sense weak electric currents generated by the muscular activity of animals. The **lateral line organ,** found in all fishes, is a groove along each side of the body with many tiny openings to the outside. Sensory cells in the lateral line organ are sensitive to waves and other motion in the water, alerting the shark to the presence of predator or prey (see Fig. 42-7).

Cartilaginous fishes have no lungs. Gas exchange takes place through their five to seven pairs of gills. A current of water enters the mouth and passes over the gills and out the gill slits, constantly providing the fish with a fresh supply of dissolved oxygen. Sharks that actively swim depend on their motion to enhance gas exchange. Rays, skates, and sharks that spend time on the ocean floor use muscles of the jaw and pharynx to pump water over their gills.

The digestive tract of sharks consists of the mouth cavity; a long pharynx leading to the stomach; a short, straight intestine; and a **cloaca,** which opens on the underside of the body and is characteristic of many vertebrates (see Fig. 31-12a). The liver and pancreas discharge digestive juices into the intestine. The cloaca receives digestive wastes, as well as metabolic wastes from the urinary system. In females, the cloaca also serves as a reproductive organ.

The sexes are separate, and fertilization is internal. In the mature male, each pelvic fin has a slender, grooved section, known as a **clasper,** used to transfer sperm into the female's cloaca. The eggs are fertilized in the upper part of the female's oviducts. Part of the oviduct is modified as a shell gland, which secretes a protective coat around the egg.

Skates and some species of sharks are **oviparous;** that is, they lay eggs. Many species of sharks, however, are **ovoviviparous,** meaning their young are enclosed in eggs and incubated within the mother's body. During development, the young depend on stored yolk for their nourishment rather than on transfer of materials from the mother. The young are born after hatching from the eggs.

A few species of sharks are **viviparous.** Not only do the embryos develop within the uterus, but much of their nourishment is delivered to them by the mother's blood. Nutrients are transferred between the blood vessels in the lining of the uterus and the yolk sac surrounding each embryo.

The ray-finned fishes gave rise to modern bony fishes

Although bony fishes appear earlier in the fossil record than cartilaginous fishes, both groups may have evolved about the same time, during the Devonian period. The two groups share many characteristics (such as continuous tooth replacement), but they also differ in important ways.

Most bony fishes are characterized by a bony skeleton with many vertebrae. Bone has advantages over cartilage because it provides excellent support and effectively stores calcium. Most species have flexible median and paired fins, supported by long rays made of cartilage or bone. Overlapping, bony dermal scales cover the body. A lateral bony flap, the **operculum,** extends posteriorly from the head and protects the gills.

Unlike most sharks, bony fishes are oviparous. They lay an impressive number of eggs and fertilize them externally. The ocean sunfish, for example, lays more than 300 million eggs! Of course, most of the eggs and young become food for other animals. The probability of survival is increased by certain behavioral adaptations. For example, many species of fishes build nests for their eggs and protect them.

During the Devonian period, the bony fishes diverged into two major groups: the **Sarcopterygii** and the **ray-finned fishes,** class **Actinopterygii.** Fossils of the earliest sarcopterygians date back to the Devonian period, about 400 mya. Lungs and fleshy, **lobed fins** characterized these fishes. The flexible fins were not supported by rays of bone except at their tips. Instead, a group of bones with joints between them supported the fins. These fins had muscles that could move them and support the body. Lobed fins may have evolved as an adaptation to life on the bottom of the sea. Fish could use the fins to push against the bottom.

Early sarcopterygians evolved along two separate lines: the **lungfishes** (class **Dipnoi**) and **lobe-finned fishes** (class **Actinistia**). Both lineages have lobed fins. Three genera of lungfishes survive today in the rivers of tropical Africa, Australia, and South America.

The ray-finned fishes underwent two important adaptive radiations. The first gave rise during the Late Paleozoic era to a group of fishes that are now mostly extinct. The second radiation began during the Early Mesozoic era and gave rise to the modern bony fishes.

The common ancestor of the modern bony fishes had primitive lungs that could exchange gases in air. Lungs for gas exchange were retained by the lobe-finned fishes and lungfishes. In the ray-finned fishes, the lungs became modified as a **swim bladder,** an air sac that helps regulate buoyancy (Fig. 31-13). Bones and muscles are heavier than water, and without the swim bladder the fish would sink. By regulating gas exchange between its blood and swim bladder, a fish can control the amount of gas in the swim bladder and thus change the overall density of its body. This ability allows a bony fish, in contrast to a shark, to hover at a given depth of water without much muscular effort.

There are more species of bony fishes than of any other group of vertebrates. Biologists have identified about 49,000 living species of freshwater and saltwater bony fishes, of many shapes and colors (Fig. 31-14). Bony fishes range in size from that of the recently discovered stout infantfish (genus *Paedocypris*), which averages less than about 8 mm (about 0.3 in) long, to that of the ocean sunfish (or *Mola;* see Fig. 20-14c), which may reach 4 m (13 ft) and weigh about 1500 kg (about 3300 lb).

The diversity of bony fishes may have resulted from the action of *Hox* genes. Recall that *Hox* genes are important in determining pattern development in embryos (see Chapter 17). Specifically, *Hox* genes determine the fate of cells in the anterior-posterior axis. These genes occur in clusters on specific chromosomes. All invertebrates and early chordates that have been studied have one *Hox* gene cluster, but tetrapods have four clusters. The zebrafish has seven *Hox* clusters. During the radiation of ray-finned fishes, entire chromosomes duplicated, resulting in one or more additional clusters of *Hox* genes. These genes could have provided the genetic material for the evolution of the diverse species of ray-finned fishes.

Descendants of the lungfishes moved onto the land

Biologists thought the lobe-finned fishes were extinct by the end of the Paleozoic era, so in 1938 the scientific community was very excited when a commercial fisherman caught one off the coast of South Africa. Since that time more than 200 specimens of these giant "living fossils" have been found in the deep waters off the southeastern coast of Africa, the Comores Islands, Madagascar, and Indonesia (Fig. 31-15). These fishes, known as **coelacanths,**

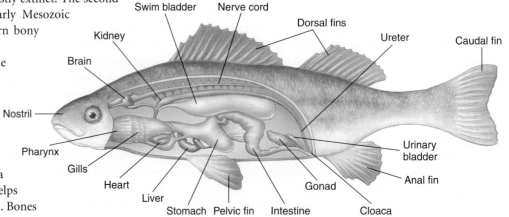

Figure 31-13 *Animated* Perch, a representative bony fish

The swim bladder is a hydrostatic organ that enables the fish to change the density of its body and remain stationary at a given depth. Pectoral fins (*not shown*) and pelvic fins are paired.

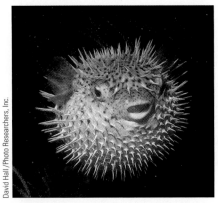

(a) The porcupinefish (*Diodon hystrix*) swallows air or water to inflate its body, a strategy that discourages potential predators. Photographed in the Virgin Islands.

David Hall/Photo Researchers, Inc.

(b) The parrotfish (*Scarus gibbus*) feeds on coral, grinds it in its digestive tract, and extracts the coralline algae. The fish eliminates a fine white sand. These fishes contribute to white sand beaches in many parts of the world. Parrotfish begin life as females and later become males.

Jeffrey L. Rotman/Peter Arnold, Inc.

Figure 31-14 Modern bony fishes

Peter Arnold, Inc.

(c) The leaflike extensions of the body wall of this Australian leafy sea dragon (*Phycodurus eques*) camouflage it in surrounding kelp.

Mark Erdmann

Figure 31-15 Diver swimming with coelacanth

For many years, biologists thought that ancestors of this lobe-finned fish (*Latimeria chalumnae*) gave rise to the amphibians.

measure nearly 2 m (about 6 ft) in length. They are nocturnal predators on other fishes.

Until the 1980s, the prevailing hypothesis was that coelacanths were the living fossil representatives of the group that gave rise to the land vertebrates. Then, a few investigators challenged this hypothesis with structural data. The external nasal openings of fossil and living lungfishes suggested that they, rather than the coelacanths, are the ancestors of the land vertebrates. Other similarities between lungfishes and land vertebrates are tooth enamel and four limbs that have the same structure and position (Fig. 31-16). (Recall that the fins of lungfishes are also lobed.)

During the 1990s, the lungfish hypothesis gained support from comparisons of mitochondrial DNA. More recently, underwater studies of coelacanths have shown that these fish use their muscular fins to slowly swim rather than to "walk" on the ocean bottom. In contrast, living lungfishes do use their fins to "walk"

on the river bottom. Biologists now generally agree that ancestors of lungfishes, rather than coelacanths, gave rise to the tetrapods.

Molecular data support the structural evidence that the limbs of tetrapods evolved from the lobed fins of fishes. Proteins encoded by certain regulatory genes, including *Hox* genes, have been found at the same developmental times and in the same locations in both limbs and fins. These findings indicate that limb and fin development is governed by the same genes.

What drove the evolution of tetrapod limbs? One hypothesis explains that during the frequent seasonal droughts in the Devonian period, swamps became stagnant or dried up completely. Fishes with lobed fins had a tremendous advantage for survival under those conditions. They were strikingly preadapted for moving onto the land. They had lungs for breathing air, and their sturdy, fleshy fins allowed them to "walk" along in shallow water. These fins could support the fish's weight, so it could emerge onto dry land and make its way to another pond or stream.

Figure 31-16 South American lungfish (*Lepidosiren paradoxus*)

Lungfishes were common in Devonian and Carboniferous times (400 mya to 300 mya). South American lungfish have paired lungs on either side of the throat and can survive for long periods if the river they inhabit dries up.

Figure 31-17 *Animated* Tiktaalik

This fish, which grew to 2.75 m (9 ft), had limblike fins and other tetrapod features.

A more recent hypothesis holds that limbs evolved in a fully aquatic environment. Animals with more developed limbs could move along in shallow water or creep through dense aquatic vegetation more efficiently than their lobe-finned ancestors. *Tiktaalik* was a transitional form between fishes and tetrapods (Fig. 31-17). Biologists consider *Tiktaalik,* which lived about 375 mya (during the Devonian period), a fish because it had scales and fins. However, *Tiktaalik* also had tetrapod features such as a movable neck and ribs that supported lungs.

Acanthostega was one of the earliest known tetrapods. Fossils from the Devonian period show that this animal had four legs with well-formed digits. However, its limbs were not sufficiently strong for walking effectively on land. *Acanthostega* was an aquatic animal, not an amphibian. *Ichthyostega* is the earliest amphibian fossil found to date. This animal had strong legs supported by well-developed shoulder and hip bones and muscles. Natural selection favored individuals such as *Ichthyostega* that could explore shallow wetlands and make their way onto dry land. The ability to move about on land, however awkwardly, provided access to new food sources. Terrestrial plants were already established, and terrestrial insects and arachnids were rapidly evolving. A vertebrate that could survive on land had less competition for food. However, success on land required the evolution of several major adaptations in addition to legs.

As already discussed, the lungfishes had lungs for gas exchange. This was important because gills cannot function in air. Life on land also required the evolution of vertebrae and muscles strong enough to support the weight of the body in air. Body coverings and other mechanisms were needed to protect animals against the drying effect of air. Evolution of ears that could hear sound transmitted through air and olfactory mechanisms for detecting airborne odors contributed to the ability to find food and mates and to avoid predators. At first, animals returned to the water to reproduce, but eventually adaptations evolved that allowed animals to reproduce in terrestrial environments. The early vertebrate experience on land was so successful that this lineage gave rise to the amphibians and, later, the reptiles, birds, and mammals.

Amphibians were the first successful land vertebrates

Biologists classify modern amphibians (class **Amphibia**) in three orders. Order **Urodela** ("visible tail") includes salamanders, mud puppies, and newts, all animals with long tails; order **Anura** ("no tail") is made up of frogs and toads, with legs adapted for hopping; and order **Apoda** ("no feet") contains the wormlike caecilians (Fig. 31-18). Although some adult amphibians are quite successful as land animals and live in dry environments, most return to the water to reproduce. Eggs and sperm are typically released in the water.

Amphibians undergo **metamorphosis,** a transition from larva to adult. The embryos of most frogs and toads develop into larvae called **tadpoles.** These larvae have tails and gills, and most feed on aquatic plants. After a time, the tadpole undergoes metamorphosis, a process regulated by hormones secreted by the *thyroid gland.* During metamorphosis, gills and gill slits disappear, the tail is resorbed, and limbs emerge. The digestive tract shortens, and food preference shifts from plant material to a carnivorous diet; the mouth widens; a tongue develops; the tympanic membrane (eardrum) and eyelids appear; and the eye lens changes shape. Many biochemical changes accompany the transformation from a completely aquatic life to an amphibious one.

Several salamanders, such as the mud puppy *Necturus,* do not undergo complete metamorphosis; they retain many larval characteristics even when sexually mature adults. Recall from Chapter 20 that this is an example of **paedomorphosis** (see Fig. 20-15). This type of development permits these salamanders to remain aquatic rather than having to compete on land.

Gerald and Buff Corsi / Tom Stack & Associates

(a) This red dart frog (*Dendrobates pumilio*) is a poison arrow frog.

Lynda Richardson / Corbis

(b) The spotted salamander (*Ambystoma maculatum*) stays underground most of the time. In early spring, these salamanders congregate in woodland ponds to reproduce.

Figure 31-18 *Animated* Modern amphibians

The coloration of amphibians may conceal them in their habitat or may be very bright and striking. Many of the brightly colored species are poisonous (see Fig. 31-18a). Their distinctive colors warn predators that they are not encountering an ordinary amphibian. Some frogs camouflage themselves by changing color.

Adult amphibians do not depend solely on their primitive lungs for the exchange of respiratory gases. Their moist, glandular skin, which lacks scales and is plentifully supplied with blood vessels, also serves as a respiratory surface. The numerous mucous glands within the skin help keep the body surface moist, which is important in gas exchange. The mucus also makes the animal slippery, facilitating its escape from predators. Some amphibians have glands in their skin that secrete poisonous substances harmful to predators.

The amphibian heart is divided into three chambers: Two **atria** receive blood, and a single **ventricle** pumps it into the arteries. A double circuit of blood vessels keeps oxygen-rich and oxygen-poor blood partially separate. Blood passes through the **systemic circulation** to the various tissues and organs of the body. Then, after returning to the heart, it is directed through the **pulmonary circulation** to the lungs and skin, where it is recharged with oxygen. The oxygen-rich blood returns to the heart to be pumped out into the systemic circulation again. We discuss the comparative anatomy of the heart and circulation of various vertebrate classes in Chapter 43.

Review

- From an evolutionary perspective, what is the significance of each of the following: (1) placoderms, (2) lungfishes, (3) ray-finned fishes, (4) *Ichthyostega*, and (5) *Tiktaalik*?
- What changes allow amphibians to move from aquatic life as a tadpole to terrestrial life as an adult?

AMNIOTES

Learning Objectives

9 Describe three vertebrate adaptations to terrestrial life.
10 Describe the reptiles and birds, and give an argument for including the birds in the reptile clade.
11 Contrast monotremes, marsupials, and placental mammals; and give examples of animals that belong to each group.

The evolution of reptiles from ancestral amphibians required many adaptations that allowed them to be completely terrestrial. Evolution of the **amniotic egg** was an extremely important event because it allowed terrestrial vertebrates to complete their life cycles on land. This egg contains an **amnion,** a membrane that forms a fluid-filled sac around the embryo. The amnion provides the embryo with its own private "pond," permitting independence from a watery external environment. In addition to keeping the embryo moist, the amniotic fluid serves as a shock absorber to cushion the developing embryo.

The evolution of the amniotic egg is so important to the success of terrestrial vertebrates—reptiles, birds, and mammals—that biologists refer to these animals as **amniotes.** Amniotes are a monophyletic group, because they have a common ancestor that was itself an amniote.

In addition to the amnion, the amniotic egg has three other extraembryonic (not part of the developing body itself) membranes: yolk sac, chorion, and allantois (Fig. 31-19). These membranes protect the developing embryo, store nutrients (*yolk*

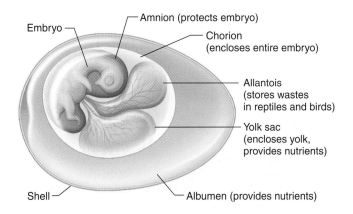

Figure 31-19 *Animated* An amniotic egg
The amnion is a protective membrane surrounding the embryo. Other extraembryonic membranes are the yolk sac, allantois, and chorion.

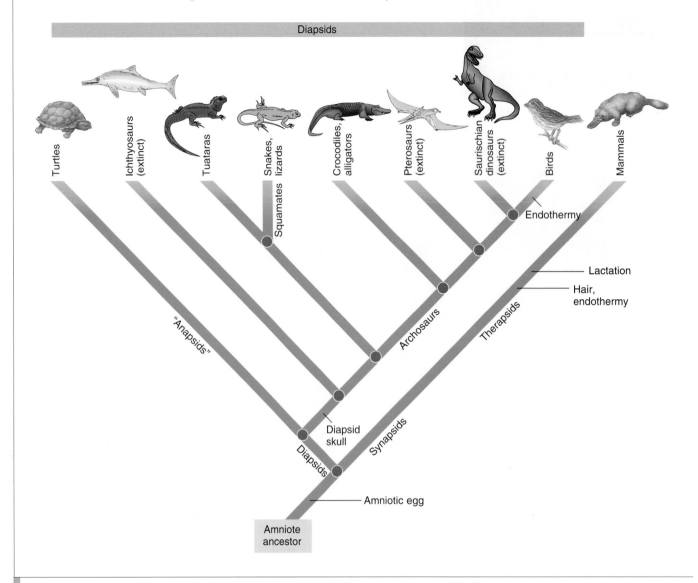

Figure 31-20   Evolutionary relationships of some amniotes

This cladogram shows proposed evolutionary relationships among reptiles, birds, and mammals. Many of these relationships are currently being studied and may be reinterpreted.

sac), carry on gas exchange (*chorion* and *allantois*), and store wastes (*allantois*). We discuss development of the extraembryonic membranes in Chapter 50. Although most extant mammals do not lay eggs, their embryos have an amnion and other extraembryonic membranes.

Another important adaptation to terrestrial life is a body covering that minimizes water loss. Such a covering presents another challenge because it severely decreases gas exchange across the body surface. This challenge has been met by the evolution of efficient lungs and circulatory systems for exchange of oxygen and carbon dioxide. Amniotes also have physiological mechanisms for conserving water. For example, much of the fluid filtered from the blood by the kidneys is reabsorbed in the kidney tubules and urinary bladder to avoid water loss during excretion of metabolic wastes.

Our understanding of amniote phylogeny is changing

Biology instructors once taught that the three classes of amniotes were Reptilia, Aves, and Mammalia. However, cladistic analysis indicates that class Reptilia is not a monophyletic group. Because it includes some, but not all, of its descendants, it is paraphyletic (see Fig. 23-7). Our discussion of amniotes reflects the cladistic view, although we describe modern birds separately for convenience.

TABLE 31-2

Some Major Amniote Groups

Subclass (Superorder/Order)	Important Characters
Subclass Diapsida	Skull with two pairs of temporal openings
Order Testudines (Chelonia) Turtles, terrapins, tortoises Turtles formerly classified as anapsids	Enclosed in bony shell; beak of keratin instead of teeth; some can withdraw head and legs into shell
Superorder Lepidosauria	Overlapping scales; most successful of modern reptiles
Order Squamata Snakes, lizards, amphisbaenians (worm lizards)	Flexible armor of overlapping, horny scales, which are shed
Lizards	Slender body, typically with four limbs; movable eyelids
Snakes	Elongated body with no limbs; flexible, loosely jointed jaw; forked tongue protrusible
Amphisbaenaea (worm lizards)	Elongated body with no limbs (except one genus); eyes hidden under skin
Order Rhynchocephalia Tuataras	Third eye under skin of forehead; only two extant species
Superorder Ichthyosauria (Extinct) Marine reptiles of Mesozoic	Body somewhat similar to that of modern fishes
Superorder Archosauria	Mainly terrestrial; some specialized for flight
Order Crocodilia Crocodiles, alligators, caimans, gavials	Four-chambered heart; closest extant relative to birds
Pterosauria (Extinct) Flying reptiles of Mesozoic	Membranous wings
Saurischia (Extinct) Mesozoic dinosaurs (*Tyrannosaurus, Diplodocus*) Birds descended from this lineage	Some were two-legged carnivores; others were four-legged herbivores
Ornithischia (Extinct) Mesozoic dinosaurs (*Triceratops*)	Bipedal and quadrupedal herbivores
Subclass Synapsida	Skull with one pair of temporal openings
Therapsida (Extinct) Mammals descended from this lineage	Many mammal-like characters; became dominant land animals during Middle Permian

Biologists hypothesize that the earliest amniotes resembled lizards. By the Late Carboniferous period, about 290 mya, amniotes had undergone an impressive adaptive radiation and diverged into two main branches: **diapsids** and **synapsids** (▐ Fig. 31-20 and ▐ Table 31-2). The term *diapsid* refers to the two pairs of openings in the temporal bones that characterize the skulls of these animals. Synapsid skulls have one pair of temporal openings. The diapsids comprise all reptilian groups, including most extinct reptiles. Biologists now classify the birds as diapsids, based on cladistic analysis. The synapsids include the extinct therapsid reptiles and the mammals.

Until very recently, systematists included a third branch, the **anapsids.** These animals have no temporal opening in the skull. The turtles were considered the only extant anapsids. However, molecular data suggest that the turtles are actually diapsids.

A second great adaptive radiation of amniotes occurred during the Mesozoic era, which ended about 66 mya. During that time reptiles were the dominant terrestrial animals (see Chapter 21). In fact, the Mesozoic era is known as the "Age of Reptiles." These animals had radiated into an impressive variety of ecological lifestyles (see Figs. 21-12, 21-13, and 21-14). Some could fly, others became marine, and many filled terrestrial habitats. Among the major groups were the **pterosaurs,** the flying reptiles; the **saurischian dinosaurs,** a group that included *Tyrannosaurus* and *Diplodocus;* and the **ornithischian dinosaurs,** including *Triceratops.*

Some of the dinosaurs were among the largest land animals to have ever walked on Earth. Some dinosaurs apparently traveled in social groups and took care of their young. Fossil evidence supports the hypothesis that at least some dinosaurs were **endotherms,** meaning they used metabolic energy to maintain a constant body temperature despite changes in the temperature of the environment. An advantage of endothermy is that it allows animals to be more active. The natural history and evolution of dinosaurs and other early reptiles are discussed in more detail in Chapter 21.

The reptiles were the dominant land animals for almost 200 million years. Then, toward the end of the Mesozoic era, many, including all the dinosaurs and pterosaurs, disappeared from the fossil record. In fact, more than half of all animal species became extinct at that time (see Chapter 21). As a result of this mass extinction, there are many more extinct reptiles than living species.

Reptiles have many terrestrial adaptations

Many reptilian characters are adaptations to terrestrial life. The female reptile secretes a protective leathery shell around the egg, which helps prevent the developing embryo from drying out. However, sperm cannot penetrate this shell. Fertilization must take place within the body of the female before the shell is added. In this process, the male uses a copulatory organ (penis) to transfer sperm into the female reproductive tract. An amnion surrounds the embryo as it develops within the protective shell.

The hard, dry, horny scales that retard drying are another adaptation to life on land. This scaly protective armor, which also protects the reptile from predators, is shed periodically. The dry reptilian skin does not allow effective gas exchange. Reptilian lungs are better developed than the saclike lungs of amphibians. Divided into many chambers, the reptilian lung provides a greatly increased surface area for gas exchange. Most reptiles have a three-chambered heart that is more efficient than the amphibian heart. The single ventricle has a partition, though incomplete, that separates oxygen-rich and oxygen-poor blood, facilitating oxygenation of body tissues. In crocodiles the partition is complete, and the heart has four chambers.

Like fishes and amphibians, extant reptiles generally lack metabolic mechanisms for regulating body temperature. They are **ectothermic,** meaning that their body temperature fluctuates with the temperature of the surrounding environment. Some reptiles have behavioral adaptations that let them maintain a body temperature higher than that of their environment. For example, you may have observed a lizard basking in the sun, which raises its body temperature and so increases its metabolic rate. This permits the lizard to hunt actively for food. When the body of a reptile is cold, its metabolic rate is low and the animal tends to be sluggish. Ectothermy may explain why reptiles are more successful in warm than in cold climates.

Many reptiles are carnivorous (meat eating). Their paired limbs, usually with five toes, are well adapted for running and climbing in search of prey. In addition, their well-developed sense organs enable them to locate prey.

We can assign extant reptiles to four groups

Biologists assign the extant reptiles to four orders (we ignore birds for the moment), though some systematists rank them as classes. Order Testudines includes the turtles, terrapins, and tortoises. Order Squamata includes lizards, snakes, and amphisbaenians (worm lizards). Order Rhynchocephalia includes the **tuataras,** lizardlike animals that live in burrows. Order Crocodilia contains crocodiles, alligators, caimans, and gavials (Fig. 31-21).

Turtles have protective shells

Members of order **Testudines** are enclosed in a protective shell made of bony plates overlaid by horny scales. Some terrestrial species can withdraw their heads and legs completely into their shells. Turtles do not have teeth. Their horny beak covers the jaws. The size of adult turtles ranges in length from about 8 cm (3 in) to more than 2 m (6.5 ft) in the great marine species, which may weigh 450 kg (almost 1000 lb). The forelimbs of marine turtles are modified into flippers. Biologists usually refer to the aquatic forms as *turtles* and to the land species as *tortoises*. Freshwater types are sometimes called *terrapins*.

Lizards and snakes are the most common modern reptiles

Lizards, snakes, and amphisbaenians (worm lizards) are assigned to order **Squamata.** These animals have rows of scales that overlap like shingles on a roof, forming a continuous, flexible armor that is shed periodically. Lizards range in size from certain geckos, which weigh as little as 1 g (less than 0.1 oz), to the Komodo dragon of Indonesia, which may weigh 100 kg (220 lb). Their body sizes and shapes vary greatly. Some lizards, such as the glass snake (which is really a lizard), are legless.

Snakes are characterized by a flexible, loosely jointed jaw structure that lets them swallow animals larger than the diameter of their own jaws. Snakes have elongated bodies with no legs, although pythons have vestigial hindlimb bones (see Fig. 18-14). (Remember that although not all the tetrapods have four limbs, all evolved from four-limbed ancestors.) Their eyes, covered by a transparent scale, do not have movable eyelids. Also absent are an external ear opening, a tympanic membrane (eardrum), and a middle-ear cavity.

The snake uses its forked tongue as an accessory sensory organ for touch and smell. Chemicals from the ground or air adhere to the tongue. The snake rubs the tip of its tongue across the opening of a sense organ in the roof of the mouth that detects odors. Pit vipers and some boas also have a prominent **pit organ** on each side of the head that detects heat from endothermic prey (see Fig. 42-3). These snakes use their pit organs to locate and capture small nocturnal mammals.

Some snakes, such as king snakes, pythons, and boa constrictors, kill their prey by rapidly wrapping themselves around the animal and squeezing so it cannot breathe. Others have fangs, which are hollow teeth connected to venom glands. When the snake bites, it pumps venom through the fangs into the prey. Some snake venoms cause the breakdown of red blood cells; oth-

Carlyn Iverson

(a) The paddlelike appendages of this green turtle (*Chelonia mydas*) are adapted for swimming.

Pete Oxford / Minden Pictures

(b) The emerald tree boa (*Corallus canina*) inhabits tropical South American forests. It rarely leaves the trees it inhabits to come to the ground. Emerald boas use their heat-sensitive pit organs to locate prey.

Betty and Nathan Cohen / Visuals Unlimited

(c) Tuataras (*Sphenodon punctatus*) are nocturnal predators that reach a length of 60 cm (more than 2 ft) or longer. There are two species of tuataras. Both are classified as endangered.

Frans Lanting / Minden Pictures

Figure 31-21 *Animated*
Representative extant reptiles

(d) A Nile crocodile (*Crocodilus niloticus*) emerges from its leathery egg.

Tuataras superficially resemble lizards

The two extant species of tuataras inhabit small islands off the coast of New Zealand. They originally inhabited the two main islands, but when humans brought rats to New Zealand, the rats decimated the population of tuataras by eating their eggs. Tuataras look somewhat like iguanas but have certain distinct characters. For example, they are the only amniotes with vertebrae that are concave at both ends; such vertebrae are characteristic of fish and some amphibians.

Crocodilians have an elongated skull

The extant members of order **Crocodilia** are the only surviving reptiles (except birds) of the archosaur lineage (see Fig. 20-18). This is the lineage that gave rise to most of the dinosaurs that dominated the Mesozoic era. Modern crocodilians include three groups: (1) the crocodiles of Africa, Asia, and America; (2) the alligators of the southern United States and China, plus the caimans of Central America; and (3) the gavials of South Asia. Most species live in swamps, in rivers, or along seacoasts, feeding on various kinds of animals.

Crocodiles are the largest living reptiles; some exceed 7 m (23 ft) in length. The cranial skeleton is adapted for aquatic life. The crocodile can be distinguished from the alligator or caiman by its long, slender snout and by the large fourth tooth on the bottom jaw that is visible when the mouth is closed.

Are birds really dinosaurs?

Although the bones of birds are fragile and disintegrate quickly, paleobiologists have found a few fossils of early birds. The first birds looked very much like dinosaurs. They had teeth (which modern birds lack), a long tail, and bones with thick walls.

Cladistic analyses indicate that birds evolved from the lineage of saurischian dinosaurs, specifically from the **theropods,** a group of bipedal, mainly carnivorous animals (see Fig. 21-15). In fact, many biologists classify birds as theropods. Many extinct theropods were long-tailed animals that moved about on two feet and had forelimbs with three clawed fingers. Extinct theropods

ers, such as that of the coral snake, are neurotoxins, which interfere with nerve function. Venomous snakes of the United States include rattlesnakes, copperheads, cottonmouths, and coral snakes. All these except the coral snakes are pit vipers.

Amphisbaenians, or worm lizards, are well adapted for their burrowing lifestyle. They have elongated, legless bodies, and their eyes are hidden under their skin. Most are less than 15 cm (6 in) long. Only one species is known in the United States.

From a painting by Rudolph Freund, courtesy of Carnegie Museum of Natural History

(a) This reconstruction represents the view that *Archaeopteryx* was a climbing animal that had at least some ability to use its wings and feathers for gliding.

Figure 31-22 *Archaeopteryx* and *Caudipteryx*

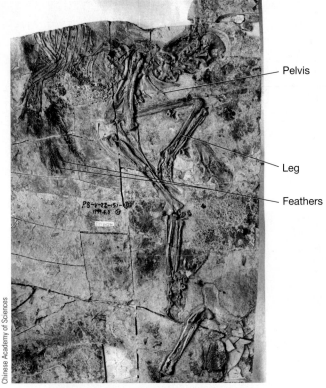

Chinese Academy of Sciences

Pelvis

Leg

Feathers

(b) *Caudipteryx* exhibits both dinosaur and bird characteristics. This fossil lacks a head, but the rest of its skeleton is well preserved. The bones in the foot and the shape and orientation of the pelvis are similar to those of dinosaurs, but the fossil also has birdlike characters, including impressions of feathers.

and modern birds have feet with three digits and thin-walled, hollow bones. Both have a **furcula,** or wishbone, which is formed by the two clavicles (collarbones) fusing in the midline. Fossil evidence indicates that some extinct theropods had feathers. Using structural and molecular data, many biologists now view birds as living dinosaurs!

The earliest known bird, *Archaeopteryx* (meaning "ancient wing"), was about the size of a pigeon. Ten specimens of this genus have been found in Bavaria in Jurassic limestone, which was laid down about 150 mya. Unlike those of extant birds, the jawbones of *Archaeopteryx* were armed with teeth, and its long reptilian tail was covered with feathers (Fig. 31-22a). Each of its short, broad wings had three claw-bearing, functional digits. Like modern birds, *Archaeopteryx* had wings, feathers, and a furcula. Its feathers were very similar to those of modern birds. Biologists do not know whether *Archaeopteryx* shared a common ancestor with other birds or whether *Archaeopteryx* itself gave rise to modern birds.

Recently, researchers used computed tomography and computerized three-dimensional reconstruction to see inside the *Archaeopteryx* skull. They found that *Archaeopteryx* had a birdlike brain with comparatively larger cerebral lobes than those of other reptiles. It had enlarged visual centers and highly developed inner ear canals similar to those of modern birds. These structures would support the coordination and agility needed for flight.

Cretaceous rocks have yielded fossils of other early birds. *Hesperornis,* which lived in North America, was a toothed, aquatic diving bird with small wings and broad, lobed feet for swimming. *Ichthyornis* was a toothed, flying bird about the size of a small tern. By the end of the Cretaceous period, fully modern birds had evolved, including early members of several modern bird families.

Some dinosaurs had feathers

Feathers are an amazing example of biological engineering. Until recently, feathers were considered a key derived character of birds. Indeed, birds are the only extant animals with feathers. However, recent molecular and fossil evidence indicates that feathers also evolved in dinosaurs.

Biologists have suggested several hypotheses to explain the early functions of feathers. They may have been important in courtship rituals or in camouflage. They may have been an adaptation that conserved body heat. Insulating feathers could have contributed to the evolution of endothermy (the ability to maintain a constant body temperature), permitting animals to be more active.

Chinese, American, and Canadian paleontologists have discovered a remarkable array of fossils in the Yixian Formation in China. These fossils date back to the Early Cretaceous period (128 mya to 124 mya). These researchers have identified many theropod dinosaur fossils that demonstrate the evolution of feathers from very simple, tubular structures to complex modern forms. One feathered dinosaur discovered there is *Caudipteryx*, which was about the size of a turkey. It had only a few teeth in the front of its upper jaw, and its bones show many birdlike characteristics (▌Fig. 31-22b). *Caudipteryx* has well-preserved imprints of complex feathers on its tail and forelimbs. Analyzing fossil evidence and recent molecular data, some biologists hypothesize that feather evolution took place in terrestrial, bipedal dinosaurs before the evolution of birds or flight.

In 2003, paleontologists reported a new species of feathered dinosaur, *Microraptor gui*. *Microraptor* had asymmetrical feathers on its forelimbs and hind limbs. Modern birds use asymmetrical feathers for flight. These small dinosaurs, which lived about 126 mya, coexisted with early birds. Their small size and body structure support the hypothesis that flying evolved from gliding; that is, bird ancestors climbed trees and used their feathered limbs for gliding. (The evolution of feathers is discussed in *Focus On: The Origin of Flight in Birds*, in Chapter 21).

Modern birds are adapted for flight

Biologists have described about 9000 species of extant birds and classified them into about 23 monophyletic groups. Birds inhabit a variety of habitats and can be found on all continents, most islands, and even the open sea. Birds are a diverse group (▌Fig. 31-23). The largest living birds are the ostriches of Africa, which may be up to 2 m (6.5 ft) tall and weigh 136 kg (300 lb), and the great condors of the Americas, with wingspans of up to 3 m (10 ft). The smallest known bird is Helena's hummingbird of Cuba, with a length of less than 6 cm (a little over 2 in) and a weight of less than 4 g (0.14 oz).

Like their reptilian ancestors, modern birds lay eggs and have reptilian-type scales on their legs. Birds have evolved remarkable specializations for flight. Their feathers are very light, yet flexible and strong, and present a flat surface to the air. Feathers also protect the body and decrease water and heat loss. The anterior limbs of birds are wings, usually adapted for flight. The posterior limbs are modified for walking, swimming, or perching.

Birds also have other adaptations for flight. Their bodies are compact and streamlined, and the fusion of many bones gives them the rigidity needed for flying. Their bones are strong but very light. Many are hollow, containing large air spaces. The avian jaw is light, and instead of teeth, birds have a light, horny beak. The breastbone is broad and keeled for the attachment of the large flight muscles.

Birds have efficient lungs with **air sacs,** thin-walled extensions that occupy spaces between the internal organs and within certain bones. They have a unique "one-way" flow of air through their respiratory system (discussed in Chapter 45). Birds have a four-chambered heart and a double circuit of blood flow. Blood delivers oxygen to the tissues and then recharges with oxygen in the lungs before the heart pumps it back into the systemic circulation. The very effective respiratory and circulatory systems provide the cells with enough oxygen to permit a high metabolic rate, which is necessary for the tremendous muscular activity that flying requires. Some of the heat a bird generates by metabolic activities helps it maintain a constant body temperature. Because they are endotherms, birds can remain active in cold climates.

Birds excrete nitrogenous wastes mainly as semisolid uric acid. Because birds typically do not have a urinary bladder, these solid wastes are delivered into the cloaca. They leave the body

(a) A male peacock (*Pavo cristatus*) impressively displaying his feathers.

(b) This Atlantic puffin (*Fratercula arctica*) has caught several fish.

(c) Two fledgling wood storks (*Mycteria americana*), about 8 to 9 weeks old, preen their feathers atop their nest in South Carolina's White Hall rookery.

▌ **Figure 31-23** Modern birds

Although they are diverse, most birds are highly adapted for flight, and their basic structure is similar.

with the feces, which are dropped frequently. This adaptive mechanism helps maintain a light body weight.

Birds have become adapted to a variety of environments, and various species have different types of beaks, feet, wings, tails, and behavioral patterns. Not all birds fly. Some, such as penguins, have small, flipperlike wings used in swimming. **Ratites** are a group of mainly large, flightless birds, including the ostriches, cassowaries, and kiwis. These birds have only vestigial wings, but they have well-developed legs used for running.

Bills are specifically adapted for the type of food the bird eats (see Fig. 1-11). Birds must eat frequently because they have a high metabolic rate and typically do not store much fat. Although the choice of food varies widely among species, most birds eat energy-rich foods such as seeds, fruits, worms, mollusks, or arthropods. Warblers and some other species eat mainly insects. Owls and hawks eat rodents, rabbits, and other small mammals. Some hawks catch snakes and lizards. Vultures feed on dead animals. Pelicans, gulls, terns, and kingfishers catch fish.

An interesting feature of the bird digestive system is the **crop,** an expanded, saclike portion of the digestive tract below the esophagus, in which food is temporarily stored. The stomach is divided into a **proventriculus,** which secretes gastric juices; and a thick, muscular **gizzard,** which grinds food. The bird swallows small bits of gravel that act as "teeth" in the gizzard to mechanically break down food.

Birds have a well-developed nervous system, with a brain that is proportionately larger than that of reptiles. Birds rely heavily on vision, and their eyes are relatively larger than those of other vertebrates. Hearing is also well developed.

In striking contrast with the relatively silent reptiles, birds are very vocal. Most have short, simple *calls* that signal danger or influence feeding, flocking, or interaction between parent and young. *Songs* are usually more complex than calls and are performed mainly by males. Birds sing songs to attract and keep a mate and to claim and defend a territory.

One of the most fascinating aspects of bird behavior is the annual migration made by many species. Some birds, such as the golden plover and Arctic tern, fly from Alaska to Patagonia, South America, and back each year, covering perhaps 40,250 km (25,000 mi) en route. Migration and navigation are discussed in Chapter 51.

Many birds have beautiful, striking colors. The color is due partly to pigments deposited during the development of the feathers and partly to reflection and refraction of light of certain wavelengths. Many birds, especially females, are protectively colored by their plumage. During the breeding season the male often assumes brighter colors to help him attract a mate.

Mammals are characterized by hair and mammary glands

Derived characters that distinguish mammals (class Mammalia) include **hair,** which insulates and protects the body; **mammary glands,** which produce milk for the young; **differentiation of**

Figure 31-24 Therapsid (*Lycaenops*)

The therapsids were synapsid reptiles. *Lycaenops*, a carnivore about the size of a coyote, lived in South Africa during the Late Permian period.

Painting by John C. Germann, Department of Library Services, American Museum of Natural History

teeth into incisors, canines, premolars, and molars; and three **middle-ear bones** (malleus, incus, and stapes) that conduct vibrations. The evolution of the **cochlea,** an organ of hearing in the inner ear, gives mammals an excellent sense of hearing.

A muscular **diaphragm** helps move air into and out of the lungs. Like birds, mammals are endotherms, but biologists think that mammals evolved endothermy independently of birds. Maintaining a constant body temperature is enhanced by the covering of insulating hair, by the four-chambered heart, and by separate pulmonary and systemic circulations. Red blood cells without nuclei serve as efficient oxygen transporters.

Contributing significantly to the success of the mammals, the nervous system is more highly developed than in any other group of animals. The cerebrum is especially large and complex, with an outer gray region called the **cerebral cortex.** In mammals, the highly specialized cerebral cortex, called the **neocortex,** has six layers of neurons. Specific regions in the neocortex are specialized for functions such as vision, hearing, touch, movement, emotional response, and higher cognitive functions.

Fertilization in mammals is internal, and with the exception of the primitive monotremes that lay eggs, mammals are viviparous. Most mammals develop a **placenta,** an organ of exchange between developing embryo and mother. As the mother's blood passes through blood vessels in the placenta, it delivers nourishment and oxygen to the embryo and carries off wastes. By carrying their developing young internally, mammals avoid the hazards of having their eggs consumed by predators. By nourishing the young and caring for them, the parents offer both protection and an "education" on how to obtain food and avoid being eaten.

The limbs of mammals are variously adapted for walking, running, climbing, swimming, burrowing, or flying. In four-legged mammals, the limbs are more directly under the body than they are in extant reptiles, which contributes to speed and agility.

New fossil discoveries are changing our understanding of the early evolution of mammals

Mammals descended from **therapsids,** a group of synapsid reptiles, during the Triassic period more than 200 mya. The therapsids were dog-like carnivores with differentiated teeth and legs adapted for running (Fig. 31-24). Until recently, early mammals were thought to be small, about the size of a mouse or shrew. However, recent fossil evidence indicates that early mammals were much more diverse than previously imagined. Some early mammals were fairly large, with limbs adapted for swimming or burrowing. Some had fur, and some may have been endothermic.

How did mammals manage to coexist with reptiles during the approximately 160 million years that reptiles ruled Earth? Many adaptations permitted early mammals to compete for a place on this planet. Perhaps one of the most important was their skill at being inconspicuous. Many were **arboreal** (tree dwelling) and **nocturnal** (active at night), searching for food (mainly insects and plant material, and perhaps reptile eggs) at night while the reptiles were inactive. Although their eyes may have been small, early mammals had well-developed sense organs for hearing and smell.

As many reptiles became extinct, mammals adapted to the niches (lifestyles) that reptiles abandoned. During this time, the flowering plants, including many trees, underwent an adaptive radiation, providing new habitats, sources of food, and protection from predators. Numerous new varieties of mammals evolved. During the Early Cenozoic era (more than 55 mya), the mammals underwent an adaptive radiation; they became widely distributed and adapted to an impressive variety of ecological lifestyles.

Modern mammals are assigned to three subclasses

By the end of the Cretaceous period, three main groups of mammals had evolved (Fig. 31-25). Today, mammals inhabit virtually every corner of Earth; they live on land, in fresh and salt water, and in the air. They range in size from the tiny pigmy shrew, weighing about 2.5 g (less than 0.1 oz), to the blue whale, which may weigh up to 136,000 kg (150 tons) and which is probably one of the largest animals that has ever lived.

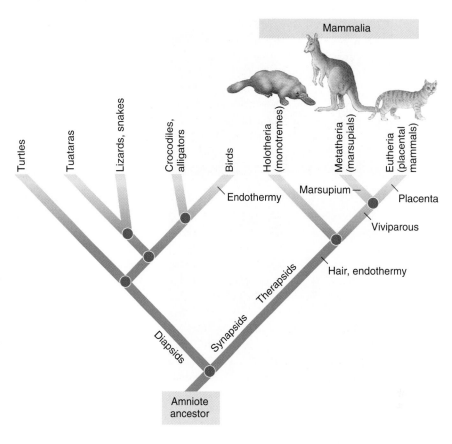

Figure 31-25 Evolution of mammals

This cladogram shows the current interpretation of the evolution of mammals.

Modern mammals are classified in three subclasses: **Holotheria** includes the egg-laying mammals, or monotremes; **Metatheria** includes the marsupials, or pouched mammals; and **Eutheria** includes the placental mammals.

The **monotremes** are the only living order of subclass Holotheria. One genus includes the duck-billed platypus and a second, the spiny anteaters or echidnas (Fig. 31-26; see also

Figure 31-26 Spiny anteater

The spiny anteater (*Tachyglossus aculeatus*) is an egg-laying monotreme.

(a) Eastern gray kangaroo (*Macropus giganteus*) with young, known as a joey. The kangaroo is native to Australia.

John Cancalosi/Peter Arnold, Inc.

(b) A kangaroo soon after birth. Marsupials are born in the embryonic state and continue to develop in the safety of the mother's marsupium (pouch).

Robert Anderson, reprinted with permission of Hubbard Scientific Company

Figure 31-27 Marsupials

Placentals   **Marsupials**

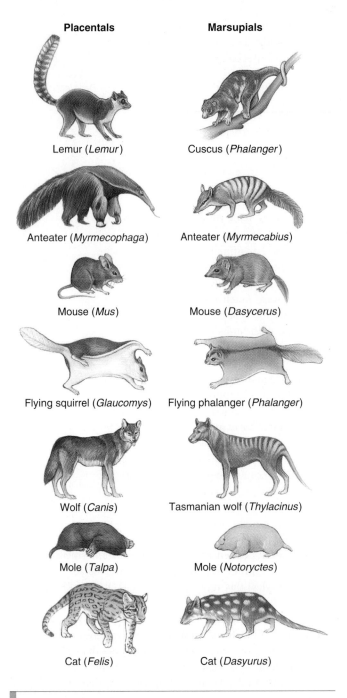

Lemur (*Lemur*)   Cuscus (*Phalanger*)

Anteater (*Myrmecophaga*)   Anteater (*Myrmecabius*)

Mouse (*Mus*)   Mouse (*Dasycerus*)

Flying squirrel (*Glaucomys*)   Flying phalanger (*Phalanger*)

Wolf (*Canis*)   Tasmanian wolf (*Thylacinus*)

Mole (*Talpa*)   Mole (*Notoryctes*)

Cat (*Felis*)   Cat (*Dasyurus*)

Figure 31-28 Convergent evolution in placental and marsupial mammals

For each mammal in one group, there is a counterpart in the other group. Their similarities include both lifestyles and structural features.

Figs. 1-16 and 23-5). These animals live in Australia and Tasmania; two of the three spiny anteater species also live in New Guinea. The females lay eggs that they carry in a pouch on the abdomen or keep warm in a nest. When the young hatch, they lap up milk secreted by their mother's mammary glands. Unlike other mammals, monotremes do not have nipples. With their long beaks and long, sticky tongues, spiny anteaters capture ants and termites. The duck-billed platypus, which lives in burrows along river banks, preys on freshwater invertebrates. It has webbed feet and a flat, beaverlike tail, which help it swim.

Marsupials include pouched mammals such as kangaroos and opossums (**Fig. 31-27a**). Embryos begin their development

(a) Wildebeest (*Connochaetes taurinus*), photographed in Tanzania. The dominant plains antelope in many areas of eastern and southern Africa, wildebeest migrate long distances during the dry season in search of food and water.

(c) Polar bears (*Ursus maritimus*), the largest living land carnivores, feed mainly on seals.

(b) Humpback whales (*Megaptera novaeangliae*) exhibiting a rare double breach.

▌**Figure 31-29** Placental mammals

in the mother's uterus, where they are nourished by fluid and yolk. The young are born after a few weeks in an undeveloped stage. In many species, the young crawl to the **marsupium** (pouch), where they complete their development. The young marsupial attaches its mouth to a mammary gland nipple and is nourished by its mother's milk (▌Fig. 31-27b).

At one time, marsupials probably inhabited much of the world, but placental mammals largely outcompeted them. Now marsupials live mainly in Australia and in Central and South America. The only marsupial species that ranges into the United States is the Virginia opossum. Australia became geographically isolated from the rest of the world before placental mammals migrated there, and the marsupials remained the dominant mammals (see discussion of continental drift in Chapter 18). Australian marsupials underwent adaptive radiation, paralleling the evolution of placental mammals elsewhere. Thus, marsupials that live in Australia and adjacent islands correspond to North American placental wolves, bears, rats, moles, flying squirrels, and even cats (▌Fig. 31-28).

Most familiar to us are **placental mammals,** characterized by development of a placenta (▌Fig. 31-29). The placenta forms from both embryonic membranes and the maternal uterine wall. In it the blood vessels of the embryo come very close to the blood vessels of the mother, so materials can be exchanged by diffusion. (The two circulations do not normally mix.) The placenta allows the young to remain within the mother's body until embryonic development is complete.

Placental mammals are born at a more mature stage than marsupials. Indeed, among some species the young can walk and begin to interact with other members of the group within a few minutes after birth. Biologists assign extant placental mammals to about 19 orders. ▌Table 31-3 gives a brief summary of some of these orders. Remember that aquatic mammals, such as dolphins, whales, and seals, evolved from terrestrial ancestors (see Fig. 18-8).

Review

▌ What vertebrate adaptations are important for terrestrial life?

▌ What are some arguments for classifying birds and reptiles in a single clade?

▌ Which amniotes are endothermic? What are some advantages of endothermy?

▌ What are the three main clades of mammals? How do they differ?

TABLE 31-3

Some Orders of Living Placental Mammals

Order, Representative Members, and Some Characteristics

Insectivora (moles, hedgehogs, shrews)

Nocturnal; eat insects; considered most primitive placental mammals. Shrew is smallest living mammal; some weigh less than 5 g.

African hedgehog
Atelerix albiventris

Chiroptera (bats)

Adapted for flying; fold of skin extends from elongated fingers to body and legs, forming wing. Guided in flight by type of biological sonar; emit high-frequency squeaks and are guided by echoes from obstructions. Eat insects and fruit or suck blood of other animals.

Bat
Eptesicus fuscus

Carnivora (cats, dogs, wolves, foxes, bears, otters, mink, weasels, skunks)

Carnivores with sharp, pointed canine teeth and molars for shearing; keen sense of smell; complex social interactions; among fastest, strongest, and smartest animals. Seals, sea lions, and walruses are large marine carnivores with limbs adapted as flippers for swimming; these mammals, known as pinnipeds, are now classified in order Carnivora.

Wolf
Canis lupus

Xenarthra (sloths, anteaters, armadillos)

Teeth reduced or no teeth. Sloths are sluggish animals that hang upside down from branches; often protectively colored by green algae that grow on their hair. Armadillos are protected by bony plates; eat insects and small invertebrates.

Nine-banded armadillo
Dasypus bellus

Rodentia (squirrels, beavers, rats, mice, hamsters, porcupines, guinea pigs)

Gnawing animals with chisel-like incisors. As they gnaw, teeth are worn down, and so must grow continually.

Flying squirrel
Glaucomys volans

Lagomorpha (rabbits, hares, pikas)

Like rodents, have chisel-like incisors; typically have long hind legs adapted for jumping; many have long ears.

Pika
Ochontona sp.

Primates (lemurs, monkeys, apes, humans)

Highly developed brains and eyes; nails instead of claws; opposable thumb; eyes directed forward; omnivores; most species arboreal. (Primate evolution discussed in Chapter 22.)

Ring-tailed lemur
Lemur catta

Perissodactyla (horses, zebras, tapirs, rhinoceroses)

Herbivores; hoofed with odd number of digits per foot; one or three toes; teeth adapted for chewing; usually large animals with long legs.

Tapir
Tapirus sp.

Artiodactyla (cattle, sheep, pigs, deer, giraffes)

Hoofed with even numbers of digits per foot; most have two toes, some have four. Most have antlers or horns. Herbivores; most are ruminants, chew a cud and have series of stomachs inhabited by bacteria that digest cellulose.

American elk
Cervus elaphus

Proboscidea (elephants)

Largest land animals; weigh up to 7 tons; large head; broad ears; long, muscular, flexible trunk (proboscis); thick loose skin is characteristic; two upper incisors are elongated as tusks. Includes extinct mastodons and woolly mammoths.

African elephant
Loxodonta africana

Sirenia (sea cows, manatees)

Herbivorous, aquatic mammals with fin-like forelimbs and no hind limbs. They are probably the basis for most tales about mermaids.

Manatee
Trichechus manatus

Cetacea (whales, dolphins, porpoises)

Adapted for aquatic life with fish-shaped body and broad, paddlelike forelimbs (flippers); posterior limbs absent; many have thick layer of blubber under skin; some are suspension feeders; mate and bear young in the water; very intelligent.

Humpback whale
Megaptera noveangliae

Learning Objectives

1 Identify shared derived characters of deuterostomes (page 668).

- **Deuterostomes** include echinoderms, hemichordates, and chordates. **Hemichordates** (acorn worms) are marine deuterostomes with a three-part body, including proboscis, collar, and trunk. Shared derived characters include radial, indeterminate cleavage; the blastopore becomes (or is near the future site of) the anus; basal deuterostomes have a larva with a loop-shaped ciliated band used for locomotion.

2 Identify three shared derived characters of echinoderms, and describe the main classes of echinoderms (page 668).

- Phylum **Echinodermata** includes marine animals with a spiny "skin," **water vascular system, tube feet,** and **endoskeleton.** The larvae exhibit bilateral symmetry; most of the adults exhibit **pentaradial symmetry.**
- Class **Crinoidea** includes sea lilies and feather stars. The **oral surface** of crinoids is turned upward; some crinoids are sessile.
- Class **Asteroidea** consists of the sea stars. They have a central disc with five or more arms. They use tube feet for locomotion.
- Class **Ophiuroidea** includes the brittle stars, which resemble sea stars but have longer, more slender arms that are set off more distinctly from the central disc. They use their arms for locomotion. Their tube feet lack suckers and are not used in locomotion.
- Class **Echinoidea** includes the sea urchins and sand dollars. Echinoids lack arms; they have a solid shell and are covered with spines.
- Class **Holothuroidea** consists of sea cucumbers, animals with elongated flexible bodies. The mouth is surrounded by a circle of modified tube feet that serve as tentacles.

3 Describe five shared derived characters of chordates (page 671).

- Phylum **Chordata** is divided into three subphyla: Urochordata, Cephalochordata, and Vertebrata. At some time in its life cycle, a chordate has a flexible, supporting **notochord;** a dorsal, **tubular nerve cord; pharyngeal (gill) slits;** a muscular **postanal tail;** and an **endostyle,** or thyroid gland.

4 Describe the invertebrate chordate subphyla (page 672).

- The **tunicates,** which belong to subphylum **Urochordata,** are suspension-feeding, marine animals with **tunics.** Larvae have typical chordate characteristics and are free-swimming. Adults of most groups are sessile suspension feeders.
- Subphylum **Cephalochordata** consists of the **lancelets,** small, segmented, fishlike animals that exhibit chordate characteristics.

ThomsonNOW **Learn more about the body plans of the deuterostomes by clicking on the figures in ThomsonNOW.**

5 Discuss the evolution of chordates (page 672).

- The urochordates (tunicates) were probably the first of the extant chordates to evolve.
- Subphyla Cephalochordata and **Vertebrata** are viewed as **sister taxa,** groups that diverged from a recent common ancestor.

ThomsonNOW **Explore the evolutionary relationships of vertebrates by clicking on the figure in ThomsonNOW.**

6 Describe four shared derived characters of vertebrates (page 674).

- The vertebrates have a **vertebral column** that forms the chief skeletal axis of the body and a braincase called a **cranium. Neural crest cells** give rise to or influence the development of many structures, including the cranium and jaws. Vertebrates have pronounced cephalization, a complex brain, and muscles attached to the **endoskeleton** for movement.

7 Distinguish among the major groups of jawless fishes (page 677).

- Some of the earliest known vertebrates, called **ostracoderms,** were jawless fishes that are now extinct.
- The extant jawless fishes, or **agnathans,** are the hagfishes, assigned to class **Myxini,** and the lampreys, to class **Cephalaspidomorphi.** Because hagfishes have no trace of vertebrae, some systematists have argued that they should not be classified as vertebrates. They classify the vertebrates plus the hagfishes as craniates (Craniata). However, molecular data support classifying hagfishes as vertebrates. Jaws and paired fins are absent in both hagfishes and lampreys. Hagfishes are marine scavengers that secrete slime as a defense mechanism. Many lampreys are parasites on other fishes.

8 Trace the evolution of jawed fishes and early tetrapods, and identify major taxa of jawed fishes and amphibians (page 678).

- Class **Chondrichthyes,** the cartilaginous fishes, includes the sharks, rays, and skates. These fishes have jaws, two pairs of fins, and **placoid scales.** Skates and some species of sharks are **oviparous** and lay eggs. Many species of sharks are **ovoviviparous;** their young are enclosed by eggs that are incubated in the mother's body. A few shark species are **viviparous;** the young develop in the mother's uterus and are nourished by transfer of nutrients from the mother's blood.
- The bony fishes are assigned to three classes: **Actinopterygii,** ray-finned fishes; **Actinistia,** coelacanths; and **Dipnoi,** lungfishes. During the Devonian, bony fishes gave rise to two evolutionary lines: the Actinopterygii, or **ray-finned fishes,** and the **Sarcopterygii,** or **lobe-finned fishes.**
- The ray-finned fishes gave rise to the modern bony fishes. In these fishes, the lungs have been modified as a **swim bladder,** an air sac for regulating buoyancy.
- The sarcopterygii probably gave rise to the **lungfishes** (class Dipnoi) and **coelacanths** (class Actinistia). Biologists think the lungfishes gave rise to the **tetrapods,** the land vertebrates. *Tiktaalik* was a transitional form between fishes and tetrapods.
- Early amphibians were mainly aquatic animals that ventured onto the land to find food or escape predators. These early tetrapods had limbs strong enough to support the weight of their bodies on land.
- Class **Amphibia** includes salamanders, frogs and toads, and caecilians. Most amphibians return to the water to reproduce. Frog embryos develop into **tadpoles,** which undergo **metamorphosis** to become adults. Amphibians use their moist skin as well as lungs for gas exchange.

They have a three-chambered heart and **systemic** and **pulmonary circulations.**

ThomsonNOW Learn more about jawless, early jawed, cartilaginous, and bony fishes by clicking on the figures in ThomsonNOW.

9 Describe three vertebrate adaptations to terrestrial life (page 683).

- Terrestrial vertebrates, or **amniotes,** include reptiles, birds, and mammals. The evolution of the **amniotic egg** with its shell and amnion was an important adaptation for life on land. The **amnion** is a membrane that forms a fluid-filled sac around the embryo. Amniotes have a body covering that retards water loss. They also have physiological mechanisms that conserve water.

ThomsonNOW Learn more about the amniotic egg by clicking on the figure in ThomsonNOW.

10 Describe the reptiles and birds, and give an argument for including the birds in the reptile clade (page 683).

- Class **Reptilia,** as traditionally defined, is paraphyletic. It includes dinosaurs, turtles, lizards, snakes, and alligators. Biologists classify amniotes in two main groups: diapsids and synapsids. **Diapsids** include the turtles, ichthyosaurs, tuataras, squamates (snakes and lizards), crocodiles, pterosaurs, saurischian dinosaurs, and birds. The **synapsids** gave rise to the **therapsids,** which gave rise to the mammals.

- Many biologists consider the birds as feathered dinosaurs; they classify the birds along with most of the reptiles as diapsids.

- Extant reptiles (excluding birds) are classified in four groups: turtles, terrapins, and tortoises; lizards, snakes,

and amphisbaenians; tuataras; and crocodiles, alligators, caimans, and gavials.

- In reptiles, fertilization is internal, and most reptiles secrete a leathery protective shell around the egg. The embryo develops protective membranes, including an amnion that keeps it moist. Reptiles have dry skin with horny scales, lungs with many chambers, and a three-chambered heart with some separation of oxygen-rich and oxygen-poor blood.

- Birds have many adaptations for powered flight, including feathers; wings; and light, hollow bones containing air spaces. Birds have a four-chambered heart and very efficient lungs, and they excrete solid metabolic wastes (uric acid). They are **endotherms,** that is, they maintain a constant body temperature. Birds have a well-developed nervous system and excellent vision and hearing.

11 Contrast monotremes, marsupials, and placental mammals; and give examples of animals that belong to each group (page 683).

- **Mammals** have **hair, mammary glands, differentiated teeth,** and three **middle-ear bones.** They are endotherms and have a highly developed nervous system and a muscular **diaphragm.**

- **Monotremes** (subclass **Holotheria**) include the duck-billed platypus and spiny anteaters. Monotremes lay eggs.

- **Marsupials** (subclass **Metatheria**) include pouched mammals, such as kangaroos and opossums. The young are born in an embryonic stage. They complete their development in their mother's **marsupium,** where they are nourished with milk from the mammary glands.

- **Placental mammals** (subclass **Eutheria**) are characterized by the **placenta,** an organ of exchange that develops between the embryo and the mother.

TEST YOUR UNDERSTANDING

1. Which of the following is *not* a deuterostome? (a) echinoderm (b) chordate (c) hemichordate (d) arthropod (e) hagfish

2. Which of the following belongs to subphylum Vertebrata? (a) lophophorate (b) lamprey (c) lancelet (d) tunicate (e) crinoid

3. Which of the following is/are found in tunicates? (a) notochord (b) tube feet (c) anal gill slits (d) two pairs of appendages (e) vertebral column

4. Which of the following was an early jawed fish? (a) placoderm (b) crinoid (c) therapsid (d) saurischian (e) ostracoderm

5. Arrange the following animals to reflect an accepted sequence of evolution.

   1. lungfish 2. dinosaur 3. amphibian 4. bird

   (a) 3, 1, 2, 4 (b) 1, 3, 2, 4 (c) 2, 1, 3, 4 (d) 1, 2, 3, 4

6. Which of the following characteristics is *not* associated with sea stars? (a) tube feet (b) water vascular system (c) central disc with five or more arms (d) spiny skeleton (e) notochord

7. Which of the following characteristics is associated with amphibians? (a) mantle (b) placoid scales (c) three-chambered heart (d) swim bladder (e) amnion

8. Which of the following is *not* characteristic of birds? (a) hollow bones (b) ectothermy (c) amnion (d) high metabolic rate (e) reptilian-like scales on legs

9. Which of the following is *not* true of the duck-billed platypus? (a) is classified as a marsupial (b) lays eggs (c) has a notochord at embryo stage (d) has pharyngeal slits at embryo stage (e) is a predator

10. Which of the following is *not* a placental mammal? (a) shrew (b) human (c) dolphin (d) spiny anteater (e) bat

11. Which of the following is a shared derived character of mammals? (a) placenta (b) amnion (c) three middle-ear bones (d) endostyle (e) notochord

12. The vertebral column (a) is a shared derived character of chordates (b) forms the skeletal axis of the chordate body (c) may consist of cartilaginous or bony segments (d) is well developed in lancelets (e) three of the preceding are correct

13. A shark is characterized by (a) placoid scales (b) bony skeleton (c) water vascular system (d) amnion (e) muscular diaphragm

14. Reptiles (a) have dry, scaly skin (b) are amniotes (c) gave rise to the birds and mammals (d) a, b, and c (e) a and c

15. Amniotes (a) include amphibians, reptiles, birds, and mammals (b) are especially well adapted for fresh water (c) have physiological mechanisms for conserving water (d) typically excrete ammonia (e) gave rise to the amphibians

16. Which of the following is *not* true? (a) feathers evolved in dinosaurs (b) birds are endothermic (c) birds excrete uric acid (d) birds have an amniotic egg (e) *Archaeopteryx* was the ancestor of feathered dinosaurs

CRITICAL THINKING

1. Sea urchins have radial symmetry. Explain why they are not classified as Radiata along with the cnidarians.

2. Which are more specialized, birds or mammals? Argue for each case.

3. Imagine that you discover an interesting new animal. You find it has a dorsal, tubular nerve cord; a cranium; moist skin; and a heart with two atria and a single ventricle. How would you classify the animal? Explain each step in your decision.

4. **Evolution Link.** Discuss the relationships among the echinoderms and chordates, describing shared derived characters that support grouping them as deuterostomes.

5. **Evolution Link.** Support the hypothesis that lungfishes gave rise to tetrapods.

6. **Evolution Link.** Many biologists now consider birds as living dinosaurs. Justify this position.

7. **Evolution Link.** Some biologists consider monotremes to be therapsid reptiles rather than mammals. Give arguments for and against this position.

Additional questions are available in ThomsonNOW at www.thomsonedu.com/login

Plant Structure, Growth, and Differentiation

Runk/Schoenberger from Grant Heilman

Oak seedling. A massive oak tree may grow from this young basket oak (*Quercus prinus*) seedling.

KEY CONCEPTS

Plants exhibit a variety of patterns of growth, reproduction, and longevity.

The structure of plant cells, tissues, and organs matches their functions.

The vascular plant body is differentiated into three tissue systems: the ground tissue system, the vascular tissue system, and the dermal tissue system.

Primary growth lengthens roots and shoots throughout the life span of most plants.

Woody plants have both primary growth and secondary growth, in which they increase in thickness.

About 90% of the approximately 330,000 species of plants are *flowering plants*, vascular plants characterized by flowers, double fertilization, endosperm, and seeds enclosed within fruits (see Chapter 28). Because flowering plants are the largest, most successful group of plants, they are the focus of much of this chapter and of Chapters 33 to 37.

Let us begin our study of the structure and functions of plants by examining a familiar species, the oak, and its fruit, the acorn. A new oak often begins its life after a squirrel has buried an acorn in the soil. The seed within first absorbs water from the surrounding soil. *Germination* occurs as a root emerges and works its way into the soil, absorbing additional water and anchoring the young plant. A shoot grows upward and breaks through the soil. At this point the young oak (see photograph) is a *seedling* that still depends on the food supply within the seed.

The stem continues to elongate, and small leaves develop, expand, and begin to photosynthesize. The young tree grows taller by growth at the tips of its branches. The roots likewise elongate at their tips. As it ages, the tree also forms wood and bark, increasing in girth (thickness) along the sides of the more mature stems and roots. Growth and expansion of both the root and shoot systems continue throughout the oak's life.

After several years, the oak tree becomes reproductively mature. Oaks reproduce by forming separate male and female flowers. The minute male flowers are found together on slender, drooping, caterpillar-shaped *catkins* (clusters); in contrast, the tiny female flowers are often solitary. After wind carries pollen from a male flower to a female flower, a sperm cell fertilizes the egg and an acorn develops, partially enclosed by a scaly cup. Within the acorn is a seed containing an embryo of a miniature oak, along with a supply of stored food. Thus, the cycle of life repeats itself.

In this chapter we examine the external structure of the flowering plant body; the organization of its cells, tissues, and tissue systems; and its basic growth patterns. ∎

PLANT STRUCTURE AND LIFE SPAN

Learning Objectives

1 Distinguish between herbaceous and woody plants.
2 Discuss the differences among annuals, biennials, and perennials, and give an example of each.
3 Contrast two different life history strategies in plants.

Plants range in size from the minute floating water-meal (*Wolffia microscopica*), the smallest flowering plant known, to Australian gum trees (*Eucalyptus*), some of Earth's tallest trees (▌Fig. 32-1). The more than 300,000 species of flowering plants that live in and are adapted to the many environments on Earth represent remarkable variety. Yet all of these—from desert cacti with enormously swollen stems, to cattails partly submerged in marshes, to orchids growing in the uppermost tree branches of lush tropical rain forests—are recognizable as plants. Almost all plants have the same basic body plan, which consists of roots, stems, and leaves.

Plants are either herbaceous or woody. *Herbaceous* plants are nonwoody. In temperate climates, the aerial parts (stems and leaves) of herbaceous plants die back to the ground at the end of the growing season. In contrast, the aerial parts of *woody* plants (trees and shrubs) persist. Botanically speaking, woody plants produce hard, lignified secondary tissues (the cell walls of secondary tissues contain *lignin*), and herbaceous plants do not. Lig-

nin and the production of secondary tissues are discussed later in the chapter.

Annuals are herbaceous plants (such as corn, geranium, and marigold) that grow, reproduce, and die in 1 year or less. Other herbaceous plants (such as carrot, Queen Anne's lace, cabbage, and foxglove) are **biennials;** they take 2 years to complete their life cycles before dying. During their first season, biennials produce extra carbohydrates, which they store and use during their second year, when they typically form flowers and reproduce. **Perennials** are herbaceous and woody plants that have the potential to live for more than 2 years. In temperate climates, the aerial (aboveground) stems of herbaceous perennials such as iris, rhubarb, onion, and asparagus die back each winter. Their underground parts (roots and underground stems) become dormant during the winter and send out new growth each spring. (In **dormancy,** an organism reduces its metabolic state to a minimum level to survive unfavorable conditions.) Similarly, in certain tropical climates with pronounced wet and dry seasons, the aerial parts of herbaceous perennials die back and the underground parts become dormant during the unfavorable dry season. Other tropical plants, such as orchids, are herbaceous perennials that grow year-round.

All woody plants are perennials, and some of them live for hundreds or even thousands of years. In temperate climates, the

(a) Duckweed (*Spirodela*), the larger green plants in the photograph, are tiny floating aquatic plants about 1 cm (0.4 in) across. If you look closely at the frog's body, you will see minute green dots. Each is a water-meal (*Wolffia*) plant. These tiny floating herbs, about 1.5 mm (0.06 in) wide, are the smallest known flowering plants.

(b) This red tingle tree (*Eucalyptus jacksonii*) was photographed in Walpole Noralup National Park, Australia. Red tingles, which are rare and endemic to Western Australia, have massive bases, with girths up to 26 m (85 ft), and heights of 75 m (244 ft).

▌**Figure 32-1** Size variation in plants

aboveground stems of woody plants become dormant during the winter. Many temperate woody perennials are **deciduous;** that is, they shed their leaves before winter and produce new stems with new leaves the following spring. Other woody perennials are **evergreen** and shed their leaves over a long period, so some leaves are always present. Because they have permanent woody stems that are the starting points for new growth the following season, many trees attain massive sizes.

Plants have different life history strategies

Woody perennials often live for hundreds of years, whereas some herbaceous annuals may live for only a few weeks or months. Such characteristic features of an organism's life cycle, particularly as they relate to its survival and successful reproduction, are known as **life history strategies.** Biologists try to understand the relative advantages of each life history strategy, including any trade-offs involved in the allocation of various resources. It appears that in some environments a longer life span is advantageous, whereas in other environments a shorter life span increases a species' chances for reproductive success—that is, for long-term survival.

When an environment is relatively favorable, it is filled with plants competing for available space. Because such an environment is so crowded, it has few open spots in which new plants can become established. When a plant dies, the empty area is quickly filled by another plant, but not necessarily by the same species as before. Thus, an adult perennial survives well, but young plants, whether perennials or annuals, do not. A plant with a long life span thrives in this type of environment because it can occupy a piece of soil and continue to produce seeds for many years. In a tropical rain forest, for example, competition prevents most young plants from becoming established, and woody perennials predominate. These perennials expend considerable energy in making woody tissues, but this energy investment allows them to thrive and reproduce for a long period.

In a relatively unfavorable environment, many possible sites are usually available. This type of environment is not crowded, and young plants usually do not have to compete against large, fully established plants. Here, smaller, short-lived plants have the reproductive advantage. These plants are opportunists; that is, they grow and mature quickly during the brief periods when environmental conditions are most favorable. As a result, all their resources are directed into producing as many seeds as possible before dying. In deserts following a rainy period, for example, annuals are more prevalent than woody perennials.

Thus, each species has a characteristic life history strategy, with some plants adapted to variable environments and others adapted to stable environments. The longer life span characteristic of woody perennials is just one of several successful life history plans. We return to life history strategies in our discussion of population ecology (see Chapter 52).

Review

▪ Consider a common plant such as a marigold. Is it herbaceous or woody? An annual, biennial, or perennial? Explain.

▪ Consider a common plant such as a carrot. Is it herbaceous or woody? An annual, biennial, or perennial? Explain.

▪ Consider a common plant such as an oak. Is it herbaceous or woody? An annual, biennial, or perennial? Explain.

THE PLANT BODY

Learning Objectives

4   Discuss the functions of various parts of the vascular plant body, including the nutrient- and water-absorbing root system and the photosynthesizing shoot system.

5   Describe the structure and functions of the ground tissue system (parenchyma tissue, collenchyma tissue, and sclerenchyma tissue).

6   Describe the structure and functions of the vascular tissue system (xylem and phloem).

7   Describe the structure and functions of the dermal tissue system (epidermis and periderm).

The plant body of flowering plants (and other vascular plants) is usually organized into a root system and a shoot system (▪ Fig. 32-2). The **root system** is generally underground. The aboveground portion, the **shoot system,** usually consists of a vertical stem that bears leaves and, in flowering plants, flowers and fruits that contain seeds.

Each plant typically grows in two different environments: the dark, moist soil and the illuminated, relatively dry air. Plants usually have both roots and shoots because they require resources from both environments. Thus, roots branch extensively through the soil, forming a network that anchors the plant firmly in place and absorbs water and dissolved minerals (inorganic nutrients) from the soil. Leaves, the flattened organs of photosynthesis, are attached more or less regularly on the stem, where they absorb the sun's light and atmospheric carbon dioxide (CO_2) used in photosynthesis.

The plant body consists of cells and tissues

As in other organisms, the basic structural and functional unit of plants is the cell. Plants have evolved a variety of cell types, each specialized for particular functions.

Like animal cells, plant cells are organized into tissues. A **tissue** is a group of cells that forms a structural and functional unit. *Simple tissues* are composed of only one kind of cell, whereas *complex tissues* have two or more kinds of cells.

In vascular plants, tissues are organized into three tissue systems, each of which extends throughout the plant body (▪ Fig. 32-3). Each tissue system contains two or more kinds of tissues (▪ Table 32-1 on page 703). Most of the plant body is composed of the **ground tissue system,** which has a variety of functions, including photosynthesis, storage, and support. The **vascular tissue system,** an intricate conducting system that extends throughout the plant body, is responsible for conduction of various substances, including water, dissolved minerals, and

During the evolution of plants, the root and shoot systems specialized to obtain resources from the soil and air, respectively.

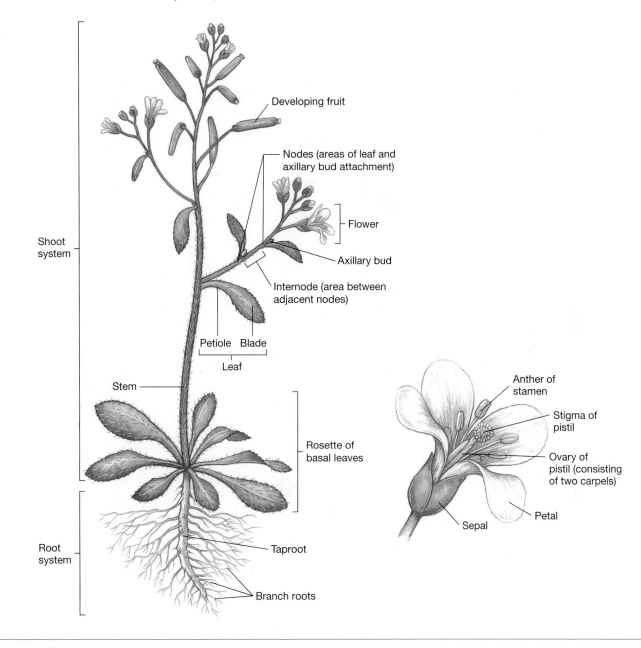

Figure 32-2 The plant body

A plant body consists of a root system, usually underground, and a shoot system, usually aboveground. Shown is mouse-ear cress (*Arabidopsis thaliana*), a small plant in the mustard family that is a model plant for biological research. *Arabidopsis* is native to North Africa and Eurasia and has been naturalized (was introduced and is now growing in the wild) in California and the eastern half of the United States.

food (dissolved sugar). The vascular tissue system also functions in strengthening and supporting the plant. The **dermal tissue system** provides a covering for the plant body.

Roots, stems, leaves, flower parts, and fruits are **organs,** because each is composed of all three tissue systems. The tissue systems of different plant organs form an interconnected network throughout the plant. For example, the vascular tissue system of a leaf is continuous with the vascular tissue system of the stem to which it is attached, and the vascular tissue system of the stem is continuous with the vascular tissue system of the root.

The tissue systems are continuous throughout the plant. For example, the vascular tissue system in a leaf is continuous with the vascular tissue system in the stem to which it is attached.

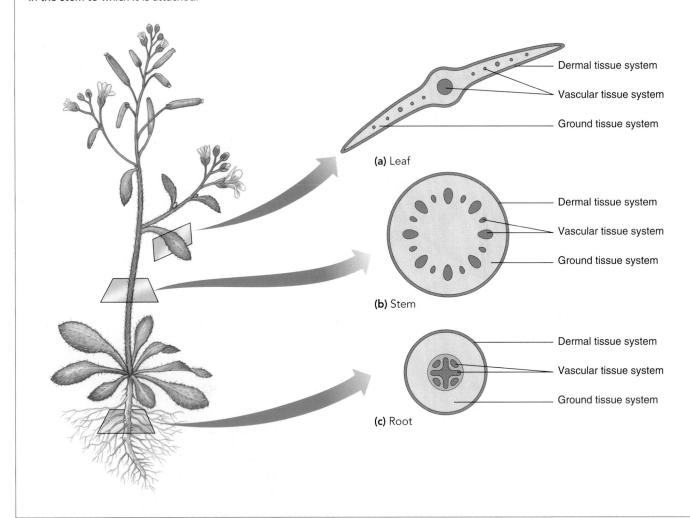

(a) Leaf
- Dermal tissue system
- Vascular tissue system
- Ground tissue system

(b) Stem
- Dermal tissue system
- Vascular tissue system
- Ground tissue system

(c) Root
- Dermal tissue system
- Vascular tissue system
- Ground tissue system

Figure 32-3 *Animated* The three tissue systems in the plant body

This figure shows the distribution of the ground tissue system, vascular tissue system, and dermal tissue system in the **(a)** leaves, **(b)** stems, and **(c)** roots of a herbaceous eudicot such as *Arabidopsis*.

The ground tissue system is composed of three simple tissues

The bulk of a herbaceous plant is its ground tissue system, which is composed of three tissues: parenchyma, collenchyma, and sclerenchyma (▮ Table 32-2). These tissues can be distinguished by their cell wall structures. Recall that plant cells are surrounded by a cell wall that provides structural support (see Chapter 4). A growing plant cell secretes a thin **primary cell wall,** which stretches and expands as the cell increases in size. After the cell stops growing, it sometimes secretes a thick, strong **secondary cell wall,** which is deposited *inside* the primary cell wall—that is, between the primary cell wall and the plasma membrane (see Fig. 4-29).

Plant cell walls have several important roles. They are involved in growth; the ability of primary cell walls to expand allows cells to increase in size. Scientists are increasingly aware that **signal transduction** takes place in plant cell walls, because many carbohydrate and protein molecules in plant cell walls communicate with other molecules both inside and outside the cell. The cell wall is also the first line of defense against disease-causing organisms.

TABLE 32-1

Tissue Systems, Tissues, and Cell Types of Flowering Plants

Tissue System	Tissue	Cell Types	Main Functions of Tissue
Ground tissue system	Parenchyma tissue Collenchyma tissue Sclerenchyma tissue	Parenchyma cells Collenchyma cells Sclerenchyma cells (sclereids or fibers)	Storage, secretion, photosynthesis Support Support, strength
Vascular tissue system	Xylem	Tracheids Vessel elements Xylem parenchyma cells Fibers (sclerenchyma cells)	Conduction of water and nutrient minerals, support Conduction of water and nutrient minerals, support Storage Support, strength
	Phloem	Sieve tube elements Companion cells Phloem parenchyma cells Fibers (sclerenchyma cells)	Conduction of sugar in solution, support May control functioning of sieve tube elements, loading sugar into sieve tube elements Storage Support, strength
Dermal tissue system	Epidermis	Epidermal cells Guard cells Trichomes	Protective covering over surface of plant body Regulate stomata Variable functions
	Periderm	Cork cells Cork cambium cells Cork parenchyma cells	Protective covering over surface of plant body Meristematic (form new cells) Storage

Parenchyma cells have thin primary walls

Parenchyma tissue, a simple tissue composed of parenchyma cells, is found throughout the plant body and is the most common type of cell and tissue (❙ Fig. 32-4a). The soft parts of a plant, such as the edible part of an apple or a potato, consist largely of parenchyma.

Parenchyma cells perform several important functions, such as photosynthesis, storage, and secretion. Parenchyma cells that function in photosynthesis contain green chloroplasts, whereas nonphotosynthetic parenchyma cells lack chloroplasts and are often colorless. Materials stored in parenchyma cells include starch grains, oil droplets, water, and salts, which are sometimes visible as crystals. Resins, tannins, hormones, enzymes, and sugary nectar are examples of substances that may be secreted by parenchyma cells. The various functions of parenchyma require that parenchyma cells be alive and metabolically active.

Parenchyma cells have the ability to *differentiate* into other kinds of cells, particularly when a plant has been injured. If xylem (water-conducting) cells are severed, for example, adjacent parenchyma cells may divide and differentiate into new xylem cells within a few days. (Recall from Chapter 17 that it is possible to induce certain plant cells to become the equivalent of embryonic cells that can then differentiate into specialized cells.)

Collenchyma cells have unevenly thickened primary walls

Collenchyma tissue, a simple plant tissue composed of collenchyma cells, is an extremely flexible structural tissue that provides much of the support in soft, nonwoody plant organs (❙ Fig. 32-4b). Support is crucial for plants, in part because it lets them grow upward to compete with other plants for available sunlight in a plant-crowded area. Plants lack the bony skeletal system typical of many animals; instead, individual cells, including collenchyma cells, support the plant body.

Collenchyma cells, which are usually elongated, are alive at maturity. Their primary cell walls are unevenly thickened and are especially thick in the corners. Collenchyma is not found uniformly throughout the plant and often occurs as long strands near stem surfaces and along leaf veins. The "strings" in a celery stalk, for example, consist of collenchyma tissue.

Sclerenchyma cells have both primary walls and thick secondary walls

Another simple plant tissue specialized for structural support is **sclerenchyma** tissue, whose cells have both primary and secondary cell walls. The root of the word *sclerenchyma* is derived from the Greek word *sclero*, meaning "hard." The secondary cell walls of sclerenchyma cells become strong and hard because of extreme thickening. Thus, sclerenchyma cells are unable to stretch or elongate. At functional maturity, when sclerenchyma tissue is providing support for the plant body, its cells are often dead.

Sclerenchyma tissue may be located in several areas of the plant body. Two types of sclerenchyma cells are sclereids and fibers. **Sclereids** are cells of variable shape common in the shells of nuts and in the stones of fruits such as cherries and peaches. The slightly gritty texture of pears results from the presence of clusters of sclereids. **Fibers,** which are long, tapered cells that often occur in patches or clumps, are particularly abundant in

TABLE 32-2

Cell Types in the Ground Tissue System

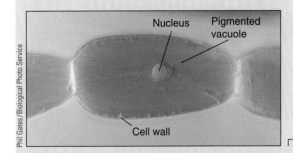

Phil Gates/Biological Photo Service

Nucleus

Pigmented vacuole

Cell wall

100 µm

Parenchyma cell

Description
Living, actively metabolizing; thin primary cell walls

Functions
Storage; secretion, and/or photosynthesis

Location and comments
Throughout the plant body; shown is an LM of a stamen hair of a spiderwort (*Tradescantia virginiana*) flower; note the large pigmented vacuole; the nucleus is not inside the vacuole but is lying on *top* of the vacuole

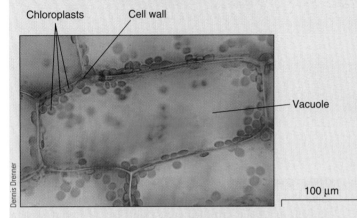

Dennis Drenner

Chloroplasts

Cell wall

Vacuole

100 µm

Parenchyma cell

Description
Living, actively metabolizing; thin primary cell walls

Functions
Storage; secretion; photosynthesis

Location and comments
Throughout the plant body; shown is an LM of leaf cells from an aquatic plant (*Elodea*); note the many chloroplasts in the thin layer of cytoplasm surrounding the large, transparent vacuole

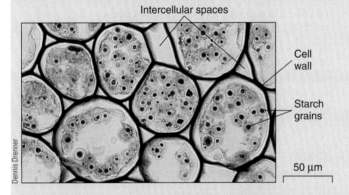

Dennis Drenner

Intercellular spaces

Cell wall

Starch grains

50 µm

Parenchyma cell

Description
Living, actively metabolizing; thin primary cell walls

Functions
Storage; secretion; photosynthesis

Location and comments
Throughout the plant body; shown is an LM of a cross section of part of a buttercup (*Ranunculus*) root; note the starch grains filling the cells

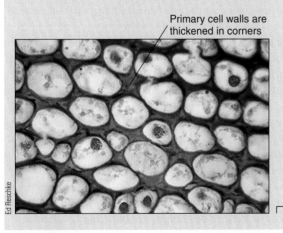

Ed Reschke

Primary cell walls are thickened in corners

50 µm

Collenchyma cell

Description
Living; unevenly thickened primary cell walls

Function
Elastic support

Location and comments
Just under stem epidermis; shown is an LM of a cross section of an elderberry (*Sambucus*) stem; note the unevenly thickened cell walls that are especially thick in the corners, making the cell contents assume a rounded shape in cross section

TABLE 32-2

Continued

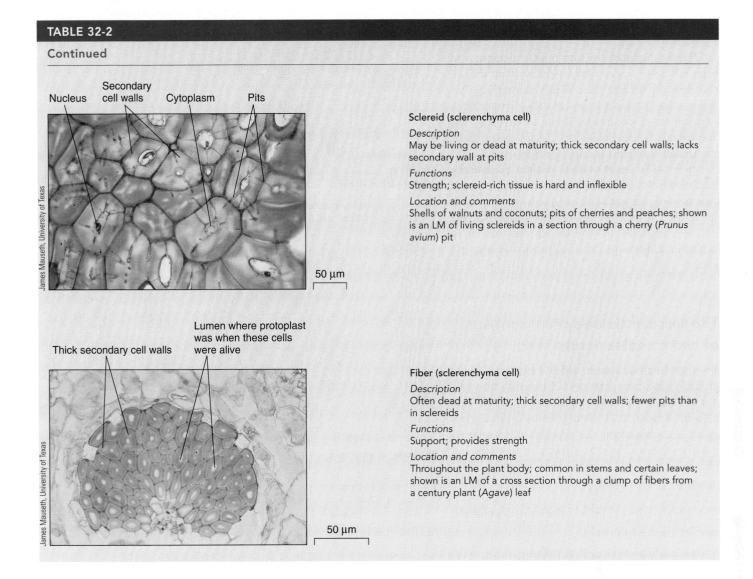

Nucleus Secondary cell walls Cytoplasm Pits

Lumen where protoplast was when these cells were alive

Thick secondary cell walls

50 µm

James Mauseth, University of Texas

Sclereid (sclerenchyma cell)

Description
May be living or dead at maturity; thick secondary cell walls; lacks secondary wall at pits

Functions
Strength; sclereid-rich tissue is hard and inflexible

Location and comments
Shells of walnuts and coconuts; pits of cherries and peaches; shown is an LM of living sclereids in a section through a cherry (*Prunus avium*) pit

Fiber (sclerenchyma cell)

Description
Often dead at maturity; thick secondary cell walls; fewer pits than in sclereids

Functions
Support; provides strength

Location and comments
Throughout the plant body; common in stems and certain leaves; shown is an LM of a cross section through a clump of fibers from a century plant (*Agave*) leaf

the wood, inner bark, and leaf ribs (veins) of flowering plants (❙ Fig. 32-4c).

Cells of the three simple tissues vary in their cell wall chemistry

Parenchyma, collenchyma, and sclerenchyma cells can be distinguished by the chemistry of their cell walls. Cell walls may contain cellulose, hemicelluloses, pectin, and lignin. **Cellulose,** the most abundant polymer in the world, accounts for about 40% to 60% of the dry weight of plant cell walls. As discussed in Chapter 3, cellulose is a **polysaccharide** composed of glucose units joined by β-1,4 bonds. Each cellulose molecule consists of thousands of glucose subunits joined to form a flat, ribbonlike chain. From 40 to 70 of these chains lie parallel to one another and connect by hydrogen bonding to form a **cellulose microfibril,** a strong, tiny strand visible under the electron microscope (see Fig. 3-10a).

Cellulose microfibrils are cemented together by a matrix of hemicelluloses and pectins. **Hemicelluloses** are a group of polysaccharides that are more soluble than cellulose. Hemicelluloses vary in their chemical composition from one species to another. Some hemicelluloses are composed of *xyloglucan,* which consists of a backbone of β-1,4-glucose molecules to which side chains of *xylose,* a five-carbon sugar, are attached. Despite their name, the chemical structure of the hemicelluloses is significantly different from that of cellulose. **Pectin,** another cementing polysaccharide, is less variable in its monomer composition than are the hemicelluloses. Pectin monomer units are *α-galacturonic acid,* a six-carbon molecule that is a derivative of glucose.

An important component of secondary plant cell walls, particularly those of wood, is **lignin.** Composing as much as 35% of the dry weight of the secondary cell wall, lignin is an extremely complex strengthening polymer made up of monomers derived from certain amino acids. Scientists have not yet determined the exact chemical structure of lignin because it is hard to remove from cellulose and other cell wall materials to which it is covalently bound.

Having discussed the four main components in plant cell walls, we can now generalize about the cell chemistry of parenchyma, collenchyma, and sclerenchyma cells. The thin primary cell walls of parenchyma cells contain predominantly cellulose, although they also contain hemicelluloses and pectin. Both parenchyma and collenchyma cells have primary cell walls, but their walls are chemically distinct because the thickened areas of collenchyma walls contain large quantities of pectin in addition to cellulose and hemicelluloses. The thick secondary walls of sclerenchyma cells are chemically different because they are rich in lignin, in addition to cellulose, hemicelluloses, and pectin.

The vascular tissue system consists of two complex tissues

The vascular tissue system, which is embedded in the ground tissue, transports needed materials throughout the plant via two complex tissues: xylem and phloem (Table 32-3). Both xylem and phloem are continuous throughout the plant body. (Chapter 34 discusses the mechanisms of transport in xylem and phloem.)

The conducting cells in xylem are tracheids and vessel elements

Xylem conducts water and dissolved minerals from the roots to the stems and leaves and provides structural support. In flowering plants, xylem is a complex tissue composed of four cell types: tracheids, vessel elements, parenchyma cells, and fibers. Two of the four cell types found in xylem, the **tracheids** and **vessel elements,** conduct water and dissolved minerals. Xylem also contains fibers that provide support, and parenchyma cells, known as *xylem parenchyma,* that perform storage functions.

Tracheids and vessel elements are highly specialized for conduction. As they develop, both types of cells undergo apoptosis, or programmed cell death. As a result, mature tracheids and vessel elements are dead and therefore hollow; only their cell walls remain. Tracheids, the chief water-conducting cells in gymnosperms (such as pine) and seedless vascular plants (such as ferns), are long, tapering cells located in patches or clumps (Fig. 32-5a). Water is conducted upward, from roots to shoots, passing from one tracheid into another through *pits,* thin areas in the tracheids' cell walls where a secondary wall did not form.

In addition to tracheids, flowering plants possess extremely efficient vessel elements (Fig. 32-5b). The cell diameters of vessel elements are usually greater than those of tracheids. Vessel elements are hollow, but unlike tracheids, the end walls have holes, known as *perforations,* or are absent. Vessel elements are stacked

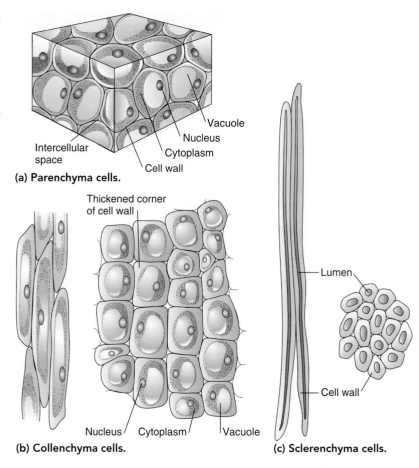

(a) Parenchyma cells.

Vacuole
Nucleus
Cytoplasm
Cell wall
Intercellular space

Thickened corner of cell wall

Lumen

Cell wall

(b) Collenchyma cells.

Nucleus Cytoplasm Vacuole

(c) Sclerenchyma cells.

Figure 32-4 *Animated* Cell types: parenchyma, collenchyma, and sclerenchyma

(a) A 3-D view of parenchyma cells. Parenchyma cells vary in size and structure, depending on their various functions within the plant body. **(b)** Collenchyma cells in longitudinal section (*left*) and cross section. Note the elongated cells, evident in longitudinal section, and the unevenly thickened cell walls, evident in cross section. **(c)** Sclerenchyma cells (fibers) in longitudinal section (*left*) and cross section. Mature fibers are often dead at functional maturity and therefore lack nuclei and cytoplasm; the lumen is the space formerly occupied by the living cell.

one on top of the other, and water is conducted readily from one vessel element into the next. A stack of vessel elements, called a **vessel,** resembles a miniature water pipe. Vessel elements also have pits in their side walls that permit the lateral transport (sideways movement) of water from one vessel to another.

Sieve tube elements are the conducting cells of phloem

Phloem conducts food materials, that is, carbohydrates formed in photosynthesis, throughout the plant and provides structural support. In flowering plants, phloem is a complex tissue composed of four cell types: sieve tube elements, companion cells, fibers, and phloem parenchyma cells (Fig. 32-5c and d). Fibers, which are usually extensive in the phloem of herbaceous plants, provide additional structural support for the plant body.

Food materials are conducted in solution—that is, dissolved in water—through the **sieve tube elements,** which are among

TABLE 32-3

Selected Cell Types in the Vascular Tissue System

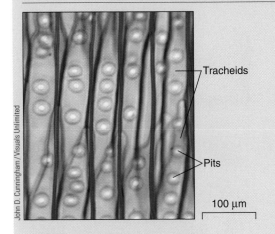

John D. Cunningham/Visuals Unlimited

— Tracheids

— Pits

100 μm

Tracheid

Description
Dead at maturity; lacks secondary wall at pits

Functions
Conduction of water and nutrient minerals; support

Location and comments
Occurs in clumps in xylem throughout plant body; shown is an LM of a longitudinal section of tracheids from white pine (*Pinus strobus*) wood

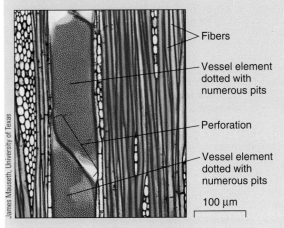

James Mauseth, University of Texas

— Fibers

— Vessel element dotted with numerous pits

— Perforation

— Vessel element dotted with numerous pits

100 μm

Vessel element

Description
Dead at maturity; end walls have perforations; lacks secondary wall at pits

Functions
Conduction of water and nutrient minerals; support

Location and comments
Xylem throughout plant body; vessel elements are more efficient than tracheids in conduction; shown is an LM of a longitudinal section of two vessel elements from an unknown woody eudicot

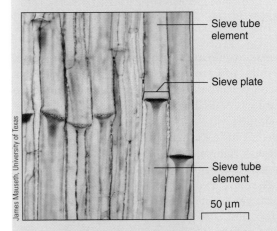

James Mauseth, University of Texas

— Sieve tube element

— Sieve plate

— Sieve tube element

50 μm

Sieve tube element

Description
Living but lacks nucleus and other organelles at maturity; end walls are sieve plates

Function
Conduction of sugar in solution

Location and comments
Phloem throughout plant body; shown is an LM of a longitudinal section through a clump of sieve tube elements in a squash (*Cucurbita*) petiole (leaf stalk)

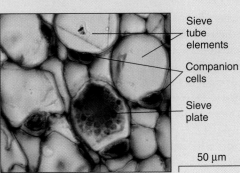

J. Robert Waaland/Biological Photo Service

— Sieve tube elements

— Companion cells

— Sieve plate

50 μm

Companion cell

Description
Living; has cytoplasmic connections with sieve tube element

Function
Assists in moving sugars into and out of sieve tube element

Location and comments
Phloem throughout plant body; shown is an LM of phloem from a squash (*Cucurbita*) petiole (leaf stalk) in cross section

the most specialized plant cells. Sieve tube elements are joined end to end to form long **sieve tubes.** The cells' end walls, called *sieve plates,* have a series of holes through which cytoplasm extends from one sieve tube element into the next. Sieve tube elements are living at maturity, but many of their organelles, including the nucleus, vacuole, mitochondria, and ribosomes, disintegrate or shrink as they mature.

Sieve tube elements are among the few eukaryotic cells that can function without nuclei. These cells typically live for less than a year. There are, however, notable exceptions. Certain palms have sieve tube elements that remain alive for approximately 100 years!

Adjacent to each sieve tube element is a **companion cell** that assists in the functioning of the sieve tube element. The companion cell is a living cell, complete with a nucleus. This nucleus is thought to direct the activities of both the companion cell and sieve tube element. Numerous **plasmodesmata**—cytoplasmic connections through which cytoplasm extends from one cell to another (see Chapter 5)—occur between a companion cell and its adjoining sieve tube element. The companion cells play an essential role in moving sugar into the sieve tube elements for transport to other parts of the plant.

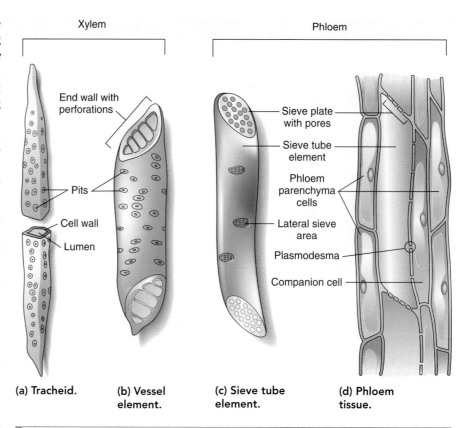

(a) Tracheid. (b) Vessel element. (c) Sieve tube element. (d) Phloem tissue.

Figure 32-5 *Animated* Longitudinal views of cell types in xylem and phloem tissues

(a) A 3-D view of a tracheid (xylem cell). The tracheid is opened to reveal the cell's appearance in cross section. At maturity, tracheids are usually dead. **(b)** A 3-D view of a vessel element (xylem cell). The end walls of vessel elements have a variety of openings, or perforations, depending on the species. Vessel elements are joined end to end to produce minute "water pipes" that run the length of the plant, from roots to leaves and other shoot parts. **(c)** A 3-D view of a sieve tube element, showing the sieve plate. **(d)** Longitudinal section of phloem tissue, showing sieve tube elements, companion cells, and phloem parenchyma cells. Fiber cells are not shown.

The dermal tissue system consists of two complex tissues

The dermal tissue system, the epidermis and periderm, provides a protective covering over plant parts (❚ Table 32-4). In herbaceous plants, the dermal tissue system is a single layer of cells called the epidermis. Woody plants initially produce an epidermis, but it splits apart as a result of the production of additional woody tissues inside the epidermis as the plant increases in girth. Periderm, a tissue several to many cell layers thick, forms under the epidermis to provide a new protective covering as the epidermis is destroyed. Periderm, which replaces the epidermis in the stems and roots of older woody plants, composes the outer bark.

Epidermis is the outermost layer of a herbaceous plant

The **epidermis** is a complex tissue composed primarily of relatively unspecialized living cells. Dispersed among these cells are more specialized guard cells and outgrowths called trichomes (discussed later). In most plants, the epidermis consists of a single layer of flattened cells (❚ Fig. 32-6). Epidermal cells generally contain no chloroplasts and are therefore transparent, so light can penetrate the interior tissues of stems and leaves. In both stems and leaves, photosynthetic tissues lie *beneath* the epidermis.

An important requirement of the aerial parts (stems, leaves, flowers, and fruits) of a plant is the ability to control water loss. Epidermal cells of aerial parts secrete a waxy **cuticle** over the surface of their exterior walls; this waxy layer greatly restricts water loss from plant surfaces.

Although the cuticle is extremely efficient at preventing most water loss through epidermal cells, it also slows the diffusion of CO_2, needed for photosynthesis, from the atmosphere into the leaf or stem. The diffusion of CO_2 is facilitated by **stomata** (sing., *stoma*). Stomata are minute pores in the epidermis that are surrounded by two cells called **guard cells** (see Figure 33-4). Many gases, including CO_2, oxygen, and water vapor, pass through the stomata by diffusion. Stomata are generally open during the day when photosynthesis is occurring, and the water loss that also takes place when stomata are open provides some evaporative cooling. At night, stomata usually close. In drought conditions, the need to conserve water overrides the need to cool the leaves and exchange gases. Thus, during a drought, the stomata close in the daytime. Stomata are discussed in greater detail in Chapter 33.

The epidermis also contains special outgrowths, or hairs, called **trichomes,** which occur in many sizes and shapes and

TABLE 32-4

Selected Cell Types in the Dermal Tissue System

James Mauseth, University of Texas

Labels: Cuticle, Epidermis, Collenchyma
Scale: 100 μm

Epidermal cell

Description
Relatively unspecialized cell with thin primary wall; outer wall often thicker and covered by a noncellular waxy layer (cuticle)

Functions
Protective covering over surface of plant body; helps reduce water loss

Location and comments
Epidermis is usually one cell thick; shown is an LM of a cross section through epidermis of an ivy (*Hedera helix*) stem

Dwight R. Kuhn

Labels: Guard cell, Stomatal pore, Epidermal cell, Guard cell
Scale: 10 μm

Guard cell

Description
Chloroplast-containing cell that occurs in pairs; pair changes shape to open and close stomatal pore

Function
Opens and closes stomatal pore

Location and comments
Epidermis of stems and leaves; shown is an LM of the epidermis of a spiderwort (*Tradescantia virginiana*) leaf

Biophoto Associates

Labels: Trichomes, Epidermis
Scale: 100 μm

Trichome

Description
Hair or other epidermal outgrowth; may be unicellular or multicellular; occurs in variety of sizes and shapes

Functions
Varied; absorption; secretion; excretion; protection; reduction of water loss

Location and comments
Epidermis; shown is an SEM of a nettle (*Solanum carolinense*) leaf, which has trichomes that break off inside the skin of animals and inject irritating substances that cause a stinging sensation

Dennis Drenner

Labels: Remnants of epidermis, Cork cells, Cork cambium, Cork parenchyma, Periderm
Scale: 50 μm

Cork cell

Description
Dead at maturity; cell walls impregnated with waterproof material (suberin)

Functions
Reduces water loss and prevents disease-causing organisms from penetrating

Location and comments
Produced in large numbers; cork often forms just under the epidermis; replaces epidermis in older stems and roots; shown is an LM of a cross section through periderm of a geranium (*Pelargonium*) stem

---

have a variety of functions. Plants that tolerate salty environments such as the seashore often have specialized trichomes on their leaves to remove excess salt that accumulates in the plant. The presence of trichomes on the aerial parts of desert plants may increase the reflection of light off the plants, thereby cooling the internal tissues and decreasing water loss. Other trichomes have a protective function. For example, the trichomes on stinging nettle leaves and stems contain irritating substances that may discourage herbivorous animals from eating the plant. **Root hairs** are simple, unbranched trichomes that increase the surface area

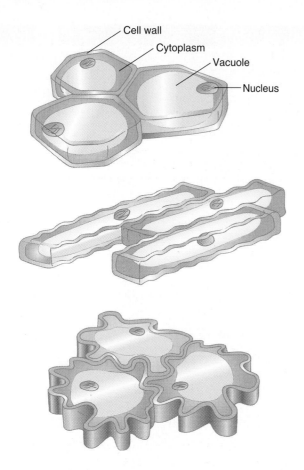

Figure 32-6 begins here with labels: Cell wall, Cytoplasm, Vacuole, Nucleus

Figure 32-6 Cell types: epidermal cells

Diagrams of 3-D views of epidermal cells from three different plants. Note that regardless of the cell shape, epidermal cells fit tightly together.

of the root epidermis (which comes into contact with the soil) for more effective water and mineral absorption.

Periderm replaces epidermis in woody plants

As a woody plant begins to increase in girth, its epidermis is sloughed off and replaced by **periderm.** Periderm forms the outer bark of older stems and roots. It is a complex tissue composed mainly of cork cells and cork parenchyma cells. **Cork cells** are dead at maturity, and their walls are heavily coated with a waterproof substance called *suberin,* which helps reduce water loss. **Cork parenchyma** cells function primarily in storage.

Review

- What are some of the functions of roots? Of shoots?
- What are the three tissue systems in plants? Describe the functions of each.
- How do parenchyma, collenchyma, and sclerenchyma tissues differ in cell structure and function?
- What are the functions of xylem and phloem? Describe the conducting cells that occur in each of these complex tissues?
- How are epidermis and periderm alike? How are they different?

PLANT MERISTEMS

Learning Objectives

8 Discuss what is meant by growth in plants, and state how it differs from growth in animals.

9 Distinguish between primary and secondary growth.

10 Distinguish between apical meristems and lateral meristems.

Plant growth involves three processes: cell division, cell elongation (the lengthening of a cell), and cell differentiation. Cell division is an essential part of plant growth that results in an increase in the number of cells. These new cells elongate as the cytoplasm grows and the vacuole fills with water, which exerts pressure on the cell wall and causes it to expand. In an onion root cell, the vacuole increases in size by some 30 to 150 times during elongation. Plant cells also differentiate, or specialize, into the various cell types just discussed. These cell types constitute the mature plant body and perform the various functions required in a multicellular organism. Although differentiation does not contribute to an increase in size, it is an important aspect of growth and development because it is essential for tissue formation.

One difference between plants and animals is the *location* of growth. When a young animal is growing, all parts of its body

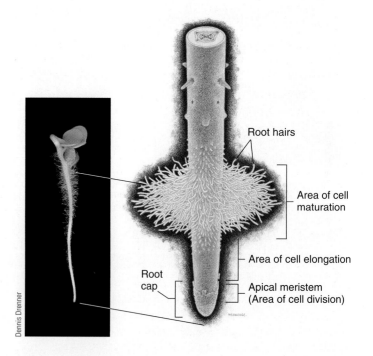

Labels: Root hairs, Area of cell maturation, Area of cell elongation, Apical meristem (Area of cell division), Root cap

Dennis Drenner

Figure 32-7 *Animated* A root tip

The root apical meristem (where cells divide and thus increase in number) is protected by a root cap. Farther from the tip is an area of cell elongation where cells enlarge and begin to differentiate. The area of cell maturation has fully mature, differentiated cells. Note the young root hairs in this area and on the root of a young radish (*Raphanus sativus*) seedling, which is approximately 5 cm (2 in) long (*left*).

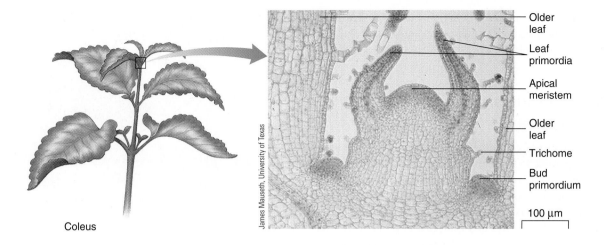

Labels on figure:
- Older leaf
- Leaf primordia
- Apical meristem
- Older leaf
- Trichome
- Bud primordium
- 100 µm

James Mauseth, University of Texas

Coleus

Figure 32-8 *Animated* A shoot apex

LM of a longitudinal section through a shoot apex of *Coleus*, showing the shoot apical meristem, leaf primordia, and bud primordia.

grow, although not necessarily at the same rate. However, when plants grow, their cells divide only in specific areas, called **meristems,** which are composed of cells whose primary function is to form new cells by mitotic division. Meristematic cells do not differentiate. Instead, they retain the ability to divide by mitosis, a trait that many differentiated cells lose. The persistence of mitotically active meristems means that plants, unlike most animals, can grow throughout their entire life span.

This ability of roots and stems to grow throughout a plant's life is known as **indeterminate growth.** In contrast, many leaves and flowers have **determinate growth;** that is, they stop growing after reaching a certain size. The size of leaves and flowers with determinate growth varies from species to species and from individual to individual, depending on the plant's genetic programming and on environmental conditions, such as availability of sunlight, water, and essential minerals.

Two kinds of meristematic growth may occur in plants. **Primary growth** is an increase in stem and root length. All plants have primary growth, which produces the entire plant body in herbaceous plants and the young, soft shoots and roots in woody trees and shrubs. **Secondary growth** is an increase in the girth of a plant. For the most part, only gymnosperms and woody eudicots have secondary growth.[1] Tissues produced by secondary growth comprise the wood and bark, which make up most of the bulk of trees and shrubs. A few annuals—geranium and sunflower, for example—undergo limited secondary growth despite the fact that they lack obvious wood and bark tissues.

[1] Recall from Chapter 28 that on the basis of structural features and molecular data, flowering plants are divided into two main groups, informally called *eudicots* and *monocots*. Oak, sycamore, ash, cherry, apple, and maple are examples of woody eudicots; whereas bean, daisy, and snapdragon are herbaceous eudicots. Palm, corn, bluegrass, lily, and tulip are monocots.

Primary growth takes place at apical meristems

Primary growth occurs as a result of the activity of **apical meristems,** areas located at the tips of roots and shoots, including within the buds of stems. (**Buds** are dormant embryonic shoots that eventually develop into branches.)

Primary growth is evident when a root tip is examined (Fig. 32-7). A protective layer of cells called a **root cap** covers the root tip. Directly behind the root cap is the root apical meristem, which consists of meristematic cells. Meristematic cells, which are cube shaped, remain small because they are continually dividing. (In meristems, as daughter cells begin to enlarge, one or both will divide again.)

Farther from the root tip, just behind the area of cell division, is an area of cell elongation where the cells have been displaced from the meristem. Here the cells are no longer dividing but instead are growing longer, pushing the root tip ahead of them, deeper into the soil. Some differentiation also begins to occur in the area of cell elongation, and immature tissues such as differentiating xylem and phloem become evident. The tissue systems continue to develop and differentiate into primary tissues (epidermis, xylem, phloem, and ground tissues) of the adult plant. Farther from the tip, behind the area of cell elongation, most cells have completely differentiated and are fully mature. Root hairs are evident in this area.

A shoot apex (such as the terminal bud) is quite different in appearance from a root tip (Fig. 32-8). A dome of minute, regularly arranged meristematic cells—the shoot apical meristem—is located in each shoot apex. **Leaf primordia** (developing leaves) and **bud primordia** (developing buds) arise from the shoot apical meristem. The tiny leaf primordia cover and protect the shoot apical meristem. As the cells formed by the shoot apical meristem elongate, the shoot apical meristem is pushed upward. Subsequent cell divisions produce additional stem tissue and cause new leaf and bud primordia to appear. Farther from the stem tip, the immature cells differentiate into the three tissue systems of the mature plant body.

Secondary growth takes place at lateral meristems

Trees and shrubs undergo both primary and secondary growth. These plants increase in length by primary growth and increase in girth by secondary growth. The increase in girth, which occurs in areas that are no longer elongating, is due to cell divisions that take place in **lateral meristems,** areas extending along the entire length of the stems and roots except at the tips. Two lateral meristems, the vascular cambium and the cork cambium, are responsible for secondary growth, which is the formation of secondary tissues: secondary xylem, secondary phloem, and periderm (▮ Fig. 32-9).

The **vascular cambium** is a layer of meristematic cells that forms a long, thin, continuous cylinder within the stem and root. It is located between the wood and bark of a woody plant. Division of cells of the vascular cambium adds more cells to the wood (secondary xylem) and inner bark (secondary phloem).

The **cork cambium,** located in the outer bark, is composed of a thin cylinder or irregular arrangement of meristematic cells. Cells of the cork cambium divide and form **cork cells** toward the outside and one or more underlying layers of **cork parenchyma** cells that function in storage. Collectively, cork cells, cork cambium, and cork parenchyma make up the periderm.

We are now ready to give a more precise definition of bark. **Bark,** the outermost covering over woody stems and roots, consists of all plant tissues located outside the vascular cambium. Bark has two regions, a living inner bark composed of secondary

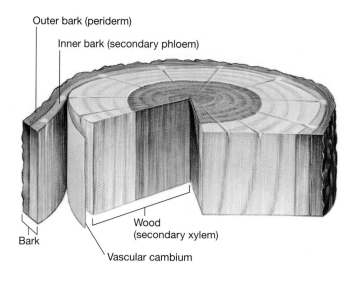

Outer bark (periderm)
Inner bark (secondary phloem)
Bark
Wood (secondary xylem)
Vascular cambium

▮ **Figure 32-9** *Animated* Secondary growth
The vascular cambium, a thin layer of cells sandwiched between the wood and bark, produces the secondary vascular tissues: the wood, which is secondary xylem, and the inner bark, which is secondary phloem. The cork cambium produces the periderm, the outer bark tissue that replaces the epidermis in a plant with secondary growth.

phloem and a mostly dead outer bark composed of periderm. Chapters 34 and 35 present a more comprehensive discussion of secondary growth.

Review

▮ What is the role of plant meristems? Does animal growth involve meristems?

▮ What is primary growth? Secondary growth?

▮ What are apical meristems? Lateral meristems?

SUMMARY WITH KEY TERMS

Learning Objectives

1 Distinguish between herbaceous and woody plants (page 699).
   ▮ Plants are either herbaceous (nonwoody) or woody. In temperate climates, the aerial parts of herbaceous plants die back, whereas the aerial parts of woody plants persist.

2 Discuss the differences among annuals, biennials, and perennials, and give an example of each (page 699).
   ▮ **Annuals,** such as corn, geranium, and marigold, are herbaceous plants that grow, reproduce, and die in 1 year or less. **Biennials,** such as carrot and Queen Anne's lace, take 2 years to complete their life cycles before dying. **Perennials,** such as asparagus and oak trees, are herbaceous and woody plants that have the potential to live for more than 2 years.

3 Contrast two different life history strategies in plants (page 699).
   ▮ Plants have a variety of **life history strategies** to ensure their successful reproduction and survival.

   ▮ Long-lived trees thrive in a tropical rain forest, where competition prevents most small, short-lived plants from becoming established. Small, short-lived plants thrive in a relatively unfavorable environment, such as a desert following a rainy period.

4 Discuss the functions of various parts of the vascular plant body, including the nutrient- and water-absorbing root system and the photosynthesizing shoot system (page 700).
   ▮ The vascular plant body typically consists of a root system and a shoot system. The **root system** is generally underground and obtains water and dissolved minerals for the plant. Roots also anchor the plant firmly in place.

   ▮ The **shoot system** is generally aerial and obtains sunlight and exchanges gases such as CO_2, oxygen, and water vapor. The shoot system consists of a vertical stem that bears leaves (the main organs of photosynthesis) and reproductive structures (flowers and fruits in flowering plants). **Buds,** undeveloped embryonic shoots, develop on stems.

- Although separate **organs** (roots, stems, leaves, flower parts, and fruits) exist in the plant, tissue systems are integrated throughout the plant body, providing continuity from organ to organ. The plant body is composed of three tissue systems: ground, vascular, and dermal.

ThomsonNOW Learn more about the plant body by clicking on the figure in ThomsonNOW.

5 Describe the structure and functions of the ground tissue system (parenchyma tissue, collenchyma tissue, and sclerenchyma tissue) (page 700).
- The **ground tissue system** consists of three tissues with a variety of functions. **Parenchyma** tissue is composed of living parenchyma cells that have thin **primary cell walls**. Functions of parenchyma tissue include photosynthesis, storage, and secretion.
- **Collenchyma** tissue consists of collenchyma cells with unevenly thickened primary cell walls. This tissue provides flexible structural support.
- **Sclerenchyma** tissue is composed of sclerenchyma cells—**sclereids** or **fibers**—that have both primary cell walls and **secondary cell walls**. Sclerenchyma cells are often dead at maturity but provide structural support.

ThomsonNOW Learn more about parenchyma, collenchyma, and sclerenchyma cells by clicking on the figure in ThomsonNOW.

6 Describe the structure and functions of the vascular tissue system (xylem and phloem) (page 700).
- The **vascular tissue system** conducts materials throughout the plant body and provides strength and support.
- **Xylem** is a complex tissue that conducts water and dissolved minerals. The actual conducting cells of xylem are **tracheids** and **vessel elements.**
- **Phloem** is a complex tissue that conducts sugar in solution. **Sieve tube elements** are the conducting cells of phloem; they are assisted by **companion cells.**

ThomsonNOW Learn more about xylem and phloem by clicking on the figure in ThomsonNOW.

7 Describe the structure and functions of the dermal tissue system (epidermis and periderm) (page 700).
- The **dermal tissue system** is the outer protective covering of the plant body.
- The **epidermis** is a complex tissue that covers the herbaceous plant body. The epidermis that covers aerial parts secretes a waxy **cuticle** that reduces water loss. **Stomata** permit gas exchange between the interior of the shoot system and the surrounding atmosphere. **Trichomes** are outgrowths, or hairs, that occur in many sizes and shapes and have a variety of functions.
- The **periderm** is a complex tissue that covers the woody parts of the plant body in woody plants.

8 Discuss what is meant by growth in plants, and state how it differs from growth in animals (page 710).
- Growth in plants, unlike that in animals, is localized in specific regions called **meristems**. Growth involves three processes: cell division, cell elongation, and cell differentiation.

9 Distinguish between primary and secondary growth (page 710).
- **Primary growth** is an increase in stem or root length. Primary growth occurs in all plants.

ThomsonNOW See root and shoot development by clicking on the figures in ThomsonNOW.

- **Secondary growth** is an increase in stem or root girth (thickness). Secondary growth typically occurs in long cylinders of meristematic cells throughout the length of older stems and roots.

ThomsonNOW Learn more about secondary growth by clicking on the figures in ThomsonNOW.

10 Distinguish between apical meristems and lateral meristems (page 710).
- Primary growth results from the activity of **apical meristems** that are localized at the tips of roots and shoots and within the buds of stems.
- The two **lateral meristems** responsible for secondary growth are the **vascular cambium** and the **cork cambium.**

TEST YOUR UNDERSTANDING

1. Plants that complete their life cycles in 1 year are called _____; those that complete them in 2 years are _____; and those that live year after year are _____. (a) annuals; perennials; biennials (b) biennials; annuals; perennials (c) annuals; biennials; perennials (d) perennials; annuals; biennials (e) perennials; biennials; annuals

2. Which of the following plant life history strategies would be successful in a relatively favorable environment such as a tropical rain forest? (a) long life span with flowers and seeds produced each year (b) long life span with flowers and seeds produced only when the plant is very young (c) short life span with flowers and seeds produced each year (d) short life span with flowers and seeds produced only when the plant is very young (e) very short life span with flowering at end of life

3. Most of the plant body consists of the _____ tissue system. (a) ground (b) vascular (c) periderm (d) dermal (e) cortex

4. The cell walls of parenchyma cells (a) contain large quantities of pectin in the thickened corners (b) are rich in lignin but do not contain hemicelluloses and pectin (c) are predominantly cellulose, although they also contain hemicelluloses and pectin (d) contain cellulose, hemicelluloses, and lignin in approximately equal amounts (e) contain hemicelluloses, pectin, and lignin but no cellulose

5. Which tissue system provides a covering for the plant body? (a) ground (b) vascular (c) periderm (d) dermal (e) cortex

6. Storage, secretion, and photosynthesis are the functions of (a) collenchyma (b) vessel elements (c) lateral meristems (d) sclerenchyma (e) parenchyma

7. The two simple tissues that are specialized for support are (a) parenchyma and collenchyma (b) collenchyma and sclerenchyma (c) sclerenchyma and parenchyma (d) parenchyma and xylem (e) xylem and phloem

8. Sclereids and fibers are examples of which plant tissue? (a) parenchyma (b) collenchyma (c) sclerenchyma (d) xylem (e) epidermis

9. Which of the following statements about the vascular tissue system is *not* true? (a) xylem and phloem are continuous throughout the plant body (b) xylem not only conducts water and dissolved minerals but also provides support (c) four cell types occur in phloem: sieve tube elements, companion cells, tracheids, and vessel elements (d) sieve tube elements lack nuclei (e) vessel elements are hollow, and their end walls have perforations or are entirely dissolved away

10. Conduction of water and minerals in xylem occurs in vessel elements and (a) sieve tube elements (b) tracheids (c) collenchyma (d) cork cells (e) phloem

11. Conduction of sugar in solution in the sieve tube elements is aided by (a) cork cells (b) sclerenchyma (c) parenchyma (d) guard cells (e) companion cells

12. The outer tissue that covers plants with primary growth is _____, whereas _____ covers plants with secondary growth. (a) cuticle; cork parenchyma (b) periderm; phloem (c) epidermis; periderm (d) epidermis; collenchyma (e) cellulose; lignin

13. The noncellular waxy layer secreted by the epidermis over its aerial surface is called (a) lignin (b) cuticle (c) periderm (d) cellulose (e) trichome

14. Minute pores known as _____ dot the surface of the epidermis of leaves and stems; each pore is bordered by two _____. (a) stomata; guard cells (b) stomata; fibers (c) sieve tube elements; companion cells (d) sclereids; guard cells (e) cuticle; guard cells

15. Localized areas within the plant body where cell divisions occur are known as (a) organs (b) fibers (c) meristems (d) cork parenchyma (e) stomata

16. Primary growth, an increase in the length of a plant, occurs at the (a) cork cambium (b) apical meristem (c) vascular cambium (d) lateral meristem (e) periderm

17. The two lateral meristems responsible for secondary growth are the (a) cork cambium and apical meristem (b) apical meristem and cork parenchyma (c) vascular cambium and apical meristem (d) vascular cambium and cork cambium (e) cork cambium and cork parenchyma

CRITICAL THINKING

1. Grasses have a special meristem situated at the base of the leaves. Relate this information to what you know about the growth of grass after you mow the lawn.

2. A couple carved a heart with their initials into a tree trunk 4 ft above ground level; the tree was 25 ft tall at the time. Twenty years later the tree was 50 ft tall. How far above the ground were the initials? Explain your answer.

3. Sclerenchyma in plants is the functional equivalent of bone in humans (both sclerenchyma and bone provide support). However, sclerenchyma is dead, whereas bone is living tissue. What are some of the advantages of a plant having dead support cells? Can you think of any disadvantages?

4. **Evolution Link.** Flowering plants have both tracheids and vessel elements in their xylem. Conifers, which evolved earlier than flowering plants, have tracheids only. How do these differences in xylem structure help explain the difference in the success of these two plant groups?

Additional questions are available in ThomsonNOW at www.thomsonedu.com/login

Leaf Structure and Function

David Sieren / Visuals Unlimited

Red maple leaves. Note how the leaves of red maple (*Acer rubrum*) are arranged to efficiently capture light.

KEY CONCEPTS

Leaf structure reflects its primary function of photosynthesis.

Opening and closing of stomata affect carbon dioxide availability in the daily cycle of photosynthesis.

Transpiration keeps leaves from getting overheated and promotes water transport in the plant.

Leaf abscission is the seasonal removal of deciduous leaves.

Some leaves are modified for special functions in addition to photosynthesis and transpiration.

Imagine that you are taking a course in engineering and are asked to design an efficient solar collector that can convert the radiant energy it collects into chemical energy. Where would you start? It might be helpful to check in the library to see how solar collectors have been designed in the past. In this instance, it would also be wise to ask a biology student if anything comparable exists in nature. The answer, of course, is yes. Plants have organs that are effective solar collectors and energy converters: *leaves.*

Plants allocate many resources to the production of leaves. Each year, a large maple tree (see photograph) may produce 47 m^2 (500 ft^2) of leaves, which may weigh more than 113 kg (250 lb). The metabolic cost of producing so many leaves is high, but leaves are essential to the tree's survival. Leaves gather the sunlight necessary for **photosynthesis,** the biological process that converts radiant energy into the chemical energy of carbohydrate molecules. Plants use these molecules as starting materials to synthesize all other organic compounds and as fuel to provide energy for metabolism. During a single summer, the leaves of a maple tree will fix about 454 kg (1000 lb) of carbon dioxide (CO_2) into organic compounds.

The structure of a leaf is superbly adapted for its primary function of photosynthesis. Most leaves are thin and flat, a shape that allows optimal absorption of light energy and the efficient internal diffusion of gases such as CO_2 and O_2. As a result of their ordered arrangement on the stem, leaves efficiently catch the sun's rays. The leaves form an intricate green mosaic, bathed in sunlight and atmospheric gases.

To control water loss, a thin, transparent layer of wax covers the leaf surface. Such structural adaptations are compromises between competing needs, and some features that optimize photosynthesis *promote* water loss. For example, plants have minute pores that allow gas exchange for photosynthesis, but these openings also let water vapor escape into the atmosphere. Thus, leaf structure represents a trade-off between photosynthesis and water conservation. ■

LEAF FORM AND STRUCTURE

Learning Objectives

1 Discuss variation in leaf form, including simple versus compound leaves, leaf arrangement on the stem, and venation patterns.
2 Describe the major tissues of the leaf (epidermis, photosynthetic ground tissue, xylem, and phloem), and label them on a diagram of a leaf cross section.
3 Compare leaf anatomy in eudicots and monocots.
4 Relate leaf structure to its function of photosynthesis.

Foliage leaves are the most variable of plant organs, so much so that plant biologists developed specific terminology to describe their shapes, margins (edges), vein patterns, and the way they attach to stems. Because each leaf is characteristic of the species on which it grows, many plants can be identified by their leaves alone. Leaves may be round, needlelike, scalelike, cylindrical, heart shaped, fan shaped, or thin and narrow. They vary in size from those of the raffia palm (*Raphia ruffia*), whose leaves often grow more than 20 m (65 ft) long, to those of water-meal (*Wolffia*), whose leaves are so small that 16 of them laid end to end measure only 2.5 cm (1 in) (see Fig. 32-1a).

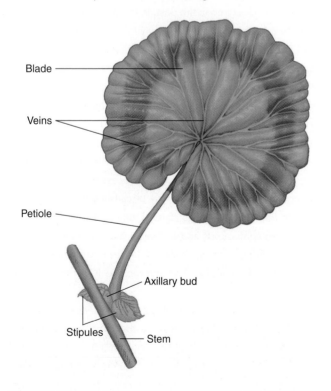

Figure 33-1 Parts of a leaf

A geranium leaf consists of a blade, a petiole, and two stipules at the base of the leaf. Note the axillary bud in the leaf axil.

The broad, flat portion of a leaf is the **blade;** the stalk that attaches the blade to the stem is the **petiole.** Some leaves also have **stipules,** which are leaflike outgrowths usually present in pairs at the base of the petiole (▌Fig. 33-1). Some leaves do not have petioles or stipules.

Leaves may be *simple* (having a single blade) or *compound* (having a blade divided into two or more leaflets) (▌Fig. 33-2a). Sometimes it is difficult to tell whether a plant has formed one compound leaf or a small stem bearing several simple leaves. One easy way to determine if a plant has simple or compound leaves is to look for axillary buds, so called because each develops in a leaf *axil* (the angle between the stem and petiole). Axillary buds form at the base of a leaf, whether it is simple or compound. However, axillary buds never develop at the base of leaflets. Also, the leaflets of a compound leaf lie in a single plane (you can lay a compound leaf flat on a table), whereas simple leaves usually are not arranged in one plane on a stem.

Leaves are arranged on a stem in one of three possible ways (▌Fig. 33-2b). Plants such as beeches and walnuts have an *alternate leaf arrangement*, with one leaf at each **node,** the area of the stem where one or more leaves are attached. In an *opposite leaf arrangement*, as occurs in maples and ashes, two leaves grow at each node. In a *whorled leaf arrangement*, as in catalpa trees, three or more leaves grow at each node.

Leaf blades may possess *parallel venation*, in which the primary **veins**—strands of vascular tissue—run approximately parallel to one another (generally characteristic of monocots), or *netted venation*, in which veins are branched in such a way that they resemble a net (generally characteristic of eudicots; ▌Fig. 33-2c).[1] Netted veins can be *pinnately netted*, with major veins branching off in succession along the entire length of the **midvein** (main or central vein of a leaf), or *palmately netted*, with several major veins radiating out from one point.

Leaf structure consists of an epidermis, photosynthetic ground tissue, and vascular tissue

The leaf is a complex organ composed of several tissues organized to optimize photosynthesis (▌Fig. 33-3). The leaf blade has upper and lower surfaces consisting of an epidermal layer. The **upper**

---

[1] Recall that flowering plants, the focus of this chapter, are divided into two main groups, informally called *eudicots* and *monocots* (see Chapter 28). Examples of eudicots include beans, petunias, oaks, cherry trees, roses, and snapdragons; monocots include corn, lilies, grasses, palms, tulips, orchids, and bananas.

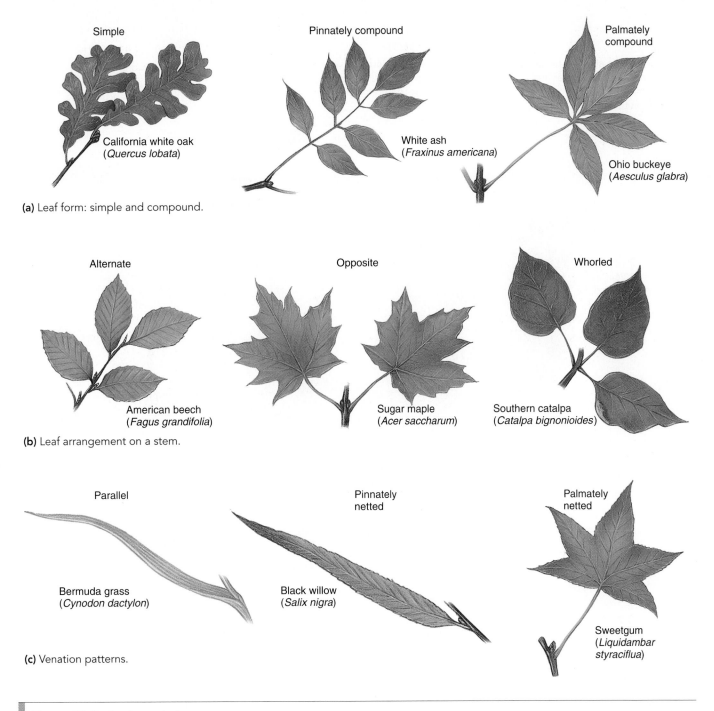

(a) Leaf form: simple and compound.

Simple

California white oak
(*Quercus lobata*)

Pinnately compound

White ash
(*Fraxinus americana*)

Palmately compound

Ohio buckeye
(*Aesculus glabra*)

(b) Leaf arrangement on a stem.

Alternate

American beech
(*Fagus grandifolia*)

Opposite

Sugar maple
(*Acer saccharum*)

Whorled

Southern catalpa
(*Catalpa bignonioides*)

(c) Venation patterns.

Parallel

Bermuda grass
(*Cynodon dactylon*)

Pinnately netted

Black willow
(*Salix nigra*)

Palmately netted

Sweetgum
(*Liquidambar styraciflua*)

Figure 33-2 Leaf morphology

All leaves shown are woody eudicot trees from North America, except Bermuda grass, which is a herbaceous monocot native to Europe and Asia.

epidermis covers the upper surface, and the **lower epidermis** covers the lower surface. Most cells in these layers lack chloroplasts and are relatively transparent. One interesting feature of leaf epidermal cells is that the cell wall facing toward the outside environment is somewhat thicker than the cell wall facing inward. This extra thickness may provide the plant with additional protection against injury or water loss.

Because leaves have such a large surface area exposed to the atmosphere, water loss by evaporation from the leaf's surface is unavoidable. However, epidermal cells secrete a waxy layer, the **cuticle,** that reduces water loss from their exterior walls (see Table 32-4). The cuticle, which consists primarily of a waxy substance called **cutin,** varies in thickness in different plants, in part as a result of environmental conditions. As one might expect, the leaves of plants adapted to hot, dry climates have extremely thick cuticles. Furthermore, a leaf's exposed (and warmer) upper epidermis generally has a thicker cuticle than its shaded (and cooler) lower epidermis.

Leaves contain all three tissue systems found in plants: the dermal tissue system, represented by the upper and lower epidermis; the ground tissue system, represented by the mesophyll; and the vascular tissue system, represented by the xylem and phloem in the veins.

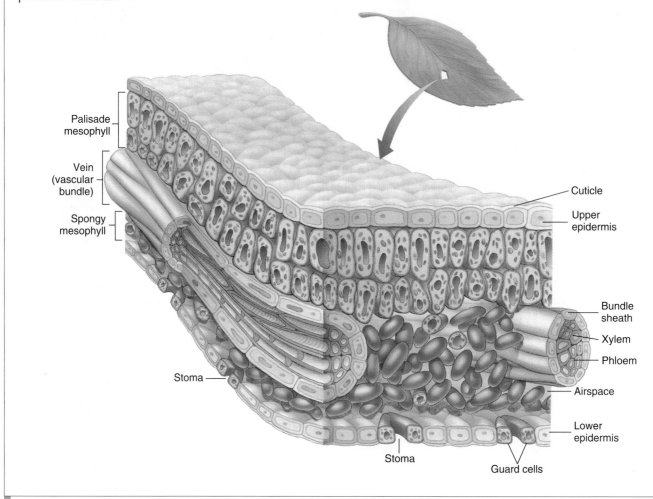

Figure 33-3 *Animated* Tissues in a typical leaf blade

An upper epidermis and lower epidermis cover the blade. The photosynthetic ground tissue, called *mesophyll*, is often arranged into palisade and spongy layers. Veins branch throughout the mesophyll.

The epidermis of many leaves is covered with various hair-like structures called **trichomes**, which have several functions (see Table 32-4). Trichomes of some plants help reduce water loss from the leaf surface by retaining a layer of moist air next to the leaf and by reflecting sunlight, thereby protecting the plant from overheating. Some trichomes secrete stinging irritants for deterring *herbivores*, animals that feed on plants. In addition, a leaf covered with trichomes is difficult for an insect to walk over or eat. Other trichomes excrete excess salts absorbed from a salty soil.

The leaf epidermis contains minute openings, or **stomata** (sing., *stoma*), for gas exchange between leaf cells and the environment; the stomata are evenly spaced to optimize this gas exchange. Each stoma is flanked by two specialized epidermal **guard cells** (Fig. 33-4) responsible for opening and closing the stoma. Guard cells are usually the only epidermal cells with chloroplasts.

Stomata are especially numerous on the lower epidermis of horizontally oriented leaves (an average of about 100 stomata per square millimeter) and in many species are located *only* on the lower surface. The lower epidermis of apple (*Malus sylvestris*) leaves, for example, has almost 400 stomata per square millimeter, whereas the upper epidermis has none. This adaptation reduces water loss because stomata on the lower epidermis are shielded from direct sunlight and are therefore cooler than those on the upper epidermis. In contrast, floating leaves of aquatic plants, such as water lilies, have stomata only on the upper epidermis.

Guard cells are associated with special epidermal cells called **subsidiary cells** that are often structurally different from other epidermal cells. Subsidiary cells provide a reservoir of water and ions that move into and out of the guard cells as they change shape during stomatal opening and closing (discussed later in the chapter).

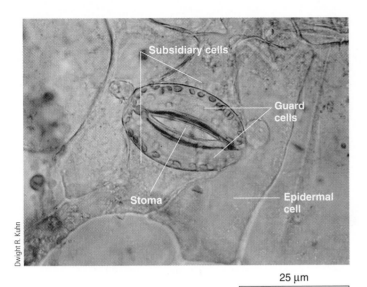

Figure 33-4 **Stoma**

LM of an open stoma from the leaf epidermis of inch plant (*Zebrina pendula*). Note the chloroplasts present in the guard cells and the thicker inner wall of each guard cell.

The photosynthetic ground tissue of the leaf, called the **mesophyll** (from the Greek *meso*, "the middle of"; and *phyll*, "leaf"), is sandwiched between the upper epidermis and the lower epidermis. Mesophyll cells, which are parenchyma cells (see Chapter 32) packed with chloroplasts, are loosely arranged, with many air spaces between them that facilitate gas exchange. These intercellular air spaces account for as much as 70% of the leaf's volume.

In many plants, the mesophyll is divided into two sublayers. Toward the upper epidermis, the columnar cells are stacked closely together in a layer called **palisade mesophyll.** In the lower portion, the cells are more loosely and irregularly arranged in a layer called **spongy mesophyll.** The two layers have different functions. Palisade mesophyll is the main site of photosynthesis in the leaf. Photosynthesis also occurs in the spongy mesophyll, but the primary function of the spongy mesophyll is to allow diffusion of gases, particularly CO_2, within the leaf.

Palisade mesophyll may be further organized into one, two, three, or even more layers of cells. The presence of additional layers is at least partly an adaptation to environmental conditions. Leaves exposed to direct sunlight contain more layers of palisade mesophyll than do shaded leaves on the same plant. In direct sunlight, the light is strong enough to effectively penetrate multiple layers of palisade mesophyll, allowing all layers to photosynthesize efficiently.

The veins, or **vascular bundles,** of a leaf extend through the mesophyll. Branching is extensive, and no mesophyll cell is more than two or three cells away from a vein. Therefore, the slow process of diffusion does not limit the movement of needed resources between mesophyll cells and veins. Each vein contains two types of vascular tissue: xylem and phloem. **Xylem,** which conducts water and dissolved minerals (inorganic nutrients), is usually located in the upper part of a vein, toward the upper epidermis. **Phloem,** which conducts dissolved sugars, is usually confined to the lower part of a vein.

One or more layers of nonvascular cells surround the larger veins and make up the **bundle sheath.** Bundle sheaths are composed of parenchyma or sclerenchyma cells (see Chapter 32). Frequently the bundle sheath has support columns, called **bundle sheath extensions,** that extend through the mesophyll from the upper epidermis to the lower epidermis (Fig. 33-5). Bundle

Figure 33-5 **Bundle sheath extension**

LM of a wheat (*Triticum aestivum*) midvein in cross section. Note the bundle sheath extensions to both the upper epidermis and lower epidermis. Photographed using fluorescence microscopy.

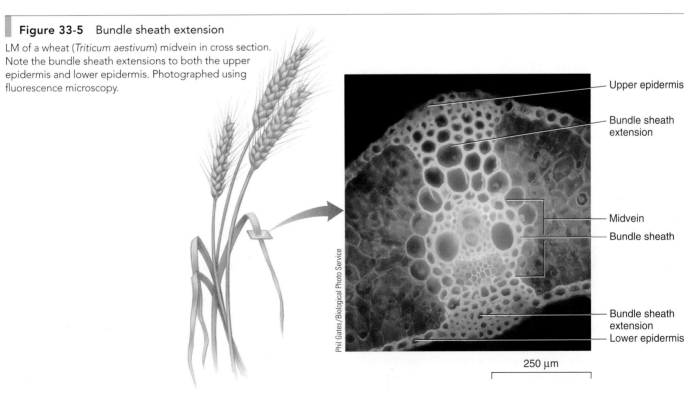

sheath extensions may be composed of parenchyma, collenchyma, or sclerenchyma cells.

Leaf structure differs in eudicots and monocots

Each eudicot leaf is usually composed of a broad, flattened blade and a petiole. As mentioned previously, eudicot leaves typically have netted venation. In contrast, many monocot leaves lack a petiole; they are narrow, and the base of the leaf often wraps around the stem to form a sheath. Parallel venation is characteristic of monocot leaves.

Eudicots and certain monocots also differ in internal leaf anatomy (▌Fig. 33-6). Although most eudicots and monocots have both palisade and spongy layers, some monocots (corn and other grasses) do not have mesophyll differentiated into distinct palisade and spongy layers. Because eudicots have netted veins, a cross section of a eudicot blade often shows veins in both cross-sectional and lengthwise views. In a cross section of a monocot leaf, in contrast, the parallel venation pattern produces evenly spaced veins, all of which appear in cross section.

Differences between the guard cells in eudicot and certain monocot leaves also occur (▌Fig. 33-7). The guard cells of eudicots and many monocots are shaped like kidney beans. Other monocot leaves (those of grasses, reeds, and sedges) have guard cells shaped like dumbbells. These structural differences affect how the cells swell or shrink to open or close the stoma.

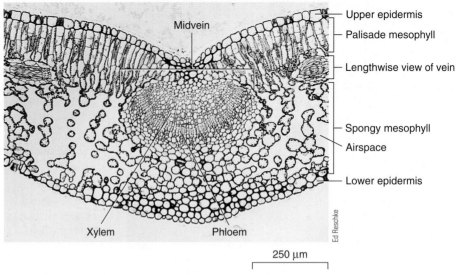

(a) Eudicot leaf. LM of part of a leaf cross section of privet (*Ligustrum vulgare*). The mesophyll has distinct palisade and spongy sections.

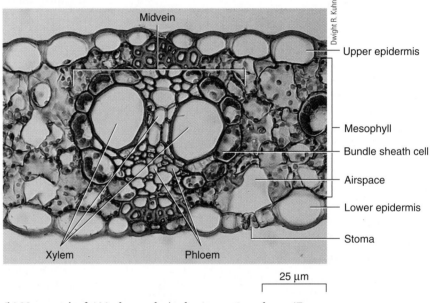

(b) Monocot leaf. LM of part of a leaf cross section of corn (*Zea mays*). Corn leaves lack distinct regions of palisade and spongy mesophyll.

Figure 33-6 Eudicot and monocot leaf cross sections

Leaf structure is related to function

How is leaf structure related to its primary function of photosynthesis? The epidermis of a leaf is relatively transparent and allows light to penetrate to the interior of the leaf where the photosynthetic ground tissue, the mesophyll, is located. Stomata, which dot the leaf surfaces, permit the exchange of gases between the atmosphere and the leaf's internal tissues. CO_2, a raw material of photosynthesis, diffuses into the leaf through stomata, and the oxygen produced during photosynthesis diffuses rapidly out of the leaf through stomata. Stomata also permit other gases, including air pollutants, to enter the leaf (see *Focus On: Air Pollution and Leaves*).

Water required for photosynthesis is obtained from the soil and transported in the xylem to the leaf, where it diffuses into the mesophyll and moistens the surfaces of mesophyll cells. The loose arrangement of the mesophyll cells, with air spaces between cells, allows for rapid diffusion of CO_2 to the mesophyll cell surfaces; there it dissolves in a film of water before diffusing into the cells.

The veins not only supply the photosynthetic ground tissue with water and minerals (from the roots, by way of the xylem) but also carry (in the phloem) dissolved sugar produced during photosynthesis to all parts of the plant. Bundle sheaths and bundle sheath extensions associated with the veins provide additional support to prevent the leaf, which is structurally weak because of the large amount of air space in the mesophyll, from collapsing under its own weight.

AIR POLLUTION AND LEAVES

The air we breathe is often dirty and contaminated with many pollutants, particularly in urban areas. Air pollution consists of gases, liquids, or solids present in the atmosphere in levels high enough to harm humans and other organisms as well as nonliving materials. Although air pollutants can come from natural sources, for example, from a lightning-caused fire or a volcanic eruption, human activities are a major cause of global air pollution. Motor vehicles and industry are the two main human sources of air pollution.

Air pollution can damage all parts of a plant, but leaves are particularly susceptible because of their structure and function. The thin blade provides a large surface area that comes into contact with the surrounding air. The thousands of stomatal pores that dot the epidermis and allow gas exchange with the atmosphere also permit pollutants to diffuse into the leaf. Just as lungs, the organs of gas exchange in humans and other terrestrial vertebrates, are affected by air pollution, so too the leaves of a plant are affected.

Many studies have shown that high levels of most forms of air pollution reduce the overall productivity of crop plants. The worst pollutant in terms of yield loss is ozone, a toxic gas produced when sunlight catalyzes a reaction between pollutants emitted by motor vehicles and industries. Ozone has been observed to reduce soybean yields by as much as 35%, and the average yield reduction from exposure to ozone is 15%. Ozone inhibits photosynthesis because it damages the mesophyll cells, probably by altering the permeability of their cell membranes. Exposure to low levels of air pollution often causes a decline in photosynthesis without other symptoms of injury. Lesions on leaves and other obvious symptoms appear at much higher levels of air pollution.

Air pollution combined with other environmental stressors (such as low winter temperatures; prolonged droughts; insects; and bacterial, fungal, and viral diseases) causes plants to decline and die. More than half of the red spruce trees in the mountains of the northeastern United States have died since the mid-1970s, and sugar maples in eastern Canada and the United States are also dying. Many still-living trees are exhibiting symptoms of **forest decline**, characterized by gradual deterioration and eventual death of trees.

The general symptoms of forest decline are reduced vigor and growth, but some plants exhibit specific symptoms, such as yellowing of needles in conifers. Forest decline is more pronounced at higher elevations, possibly because most trees growing at high elevations are at the limit of their normal range and are therefore more susceptible to wind and low temperatures.

Many factors interact to decrease the health of trees, and no single factor accounts for the recent instances of forest decline. Several air pollutants have been implicated, including acid precipitation, ozone, and toxic heavy metals such as lead, cadmium, and copper. Power plants, ore smelters, refineries, and motor vehicles produce these pollutants. Insects and weather factors such as drought and severe winters may also be important. To complicate matters further, the actual causes of forest decline may vary from one tree species to another and from one location to another. Thus, forest decline appears to result from the combination of multiple stressors. When one or more stressors weaken a tree, then an additional stressor may be enough to cause its death.

Leaves are adapted to help a plant survive in its environment

Leaf structure reflects the environment to which a particular plant is adapted. Although both aquatic plants and those adapted to dry conditions perform photosynthesis and have the same basic leaf anatomy, their leaves are modified to enable them to survive different environmental conditions. The leaves of water lilies have petioles long enough to allow the blade to float on the water's surface. Large air spaces in the mesophyll make the floating blade buoyant. The petioles and other submerged parts have an internal system of air ducts; oxygen moves through these ducts from the floating leaves to the underwater roots and stems, which live in a poorly aerated environment.

Conifers are an important group of woody trees and shrubs that includes pine, spruce, fir, redwood, and cedar.[2] Most conifers are evergreen, which means they lose leaves throughout the year rather than during certain seasons. Conifers dominate a large portion of Earth's land area, particularly in northern forests and mountains. The leaves of most conifers are waxy needles. Their needles have structural adaptations that help them survive winter, the driest part of the year. (Winter is arid even in areas of heavy snows because roots cannot absorb water from soil when the soil temperature is very low.) Indeed, many of the structural features of needles are also found in many desert plants.

❚ Figure 33-8 shows a cross section of a pine needle. Note that the needle is somewhat thickened rather than thin and blade-like. The needle's relative thickness, which results in less surface area exposed to the air, reduces water loss. Other features that help conserve water include the thick, waxy cuticle and sunken stomata; these permit gas exchange while minimizing water loss. Thus, needles help conifers tolerate the dry winds that occur during winter (*dry* refers here to low relative humidity). With the warming of spring, soil water again becomes available, and the needles quickly resume photosynthesis.

Review

❚ How are leaves adapted to conserve water?

❚ What is the photosynthetic ground tissue of a leaf called? What are its two sublayers?

❚ What are the two types of vascular tissue in a vascular bundle? Which vascular tissue is usually located on the upper part of the vascular bundle?

❚ How is the leaf organized to deliver the raw materials and remove the products of photosynthesis?

---

[2]As discussed in Chapter 28, conifers are gymnosperms, one of the two groups of seed plants (the other group is the flowering plants, or angiosperms). Unlike flowering plants, whose seeds are enclosed in fruits, conifers bear "naked" seeds on the scales of female cones.

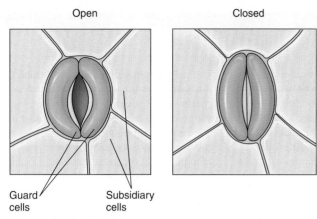

(a) Guard cells of eudicots and many monocots are bean shaped.

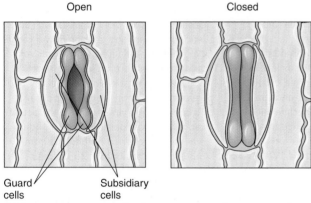

(b) Some monocot guard cells are narrow in the center and thicker at each end.

Figure 33-7 Variation in guard cells

Guard cells are associated with epidermal cells called *subsidiary cells*.

STOMATAL OPENING AND CLOSING

Learning Objectives

5   Explain the role of blue light in the opening of stomata.
6   Outline the physiological changes that accompany stomatal opening and closing.

Stomata are adjustable pores that are usually open during the day when CO_2 is required for photosynthesis and closed at night when photosynthesis is shut down (see the section on CAM photosynthesis in Chapter 9 for an interesting exception). The opening and closing of stomata are controlled by changes in the shape of the two guard cells that surround each pore. The guard cells' shape is determined by their rigidity. When water moves into guard cells from surrounding nonguard cells, the guard cells become turgid (swollen) and bend, producing a pore. When water leaves the guard cells, they become flaccid (limp) and collapse against one another, closing the pore. What causes water to move into and out of guard cells?

Blue light triggers stomatal opening

Data from numerous experiments and observations are beginning to explain the complexities of stomatal movements. Let us begin with stomatal opening, which occurs when the plant detects light from the rising sun. You already know that light is a form of energy; plants absorb light and convert it to chemical energy in the process of photosynthesis. However, light is also an important *environmental signal* for plants—that is, light provides plants with information about their environment that they use to modify various activities at the molecular and cellular levels.

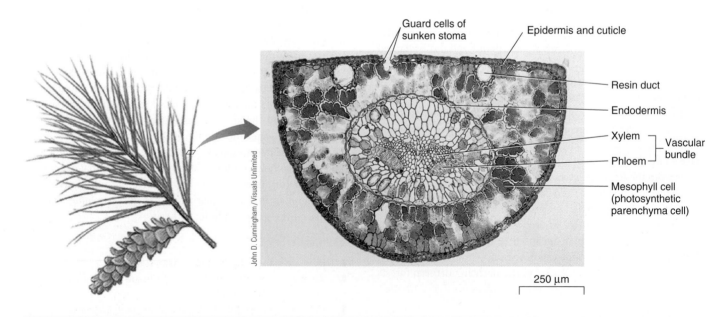

John D. Cunningham / Visuals Unlimited

250 μm

Figure 33-8 LM of a pine needle in cross section

The thick, waxy cuticle and sunken stomata are two structural adaptations that enable pine (*Pinus*) to retain its needles throughout the winter.

www.thomsonedu.com/biology/solomon

The accumulation of osmotically active ions (K⁺ and Cl⁻) in the guard cells is driven by a proton (H⁺) gradient.

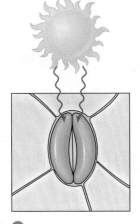

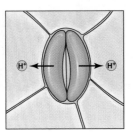

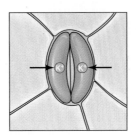

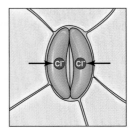

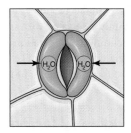

1 Blue light activates proton pumps.

2 Protons are pumped out of guard cells, forming electro-chemical gradient.

3 Potassium ions enter guard cells through voltage-activated ion channels.

4 Chloride ions also enter guard cells through ion channels.

5 Water enters guard cells by osmosis, and stoma opens.

Figure 33-9 *Animated* Mechanism of stomatal opening

In stomatal opening and several other plant responses, **blue light,** which has wavelengths from 400 to 500 nm, is an environmental signal. Any plant response to light must involve a **pigment,** a molecule that absorbs the light before the induction of a particular biological response. Data such as the responses of stomata to different colors of light suggest that the pigment involved in stomatal opening and closing is yellow (yellow pigments strongly absorb blue light). The yellow pigment is thought to be located in the guard cells, probably in their plasma membranes.

In step **1** of ▮ Figure 33-9, blue light, which is a component of sunlight, triggers the activation of proton pumps, located in the guard cell plasma membrane. Blue light also triggers the synthesis of malic acid and the **hydrolysis** (splitting) of starch (discussed later).

In step **2**, the proton pumps use ATP energy to actively transport protons (H⁺) out of the guard cells. The H⁺ that are pumped are formed when malic acid produced in the guard cells ionizes to form H⁺ and negatively charged malate ions. As the proton pumps in the plasma membranes of guard cells transport protons out of the guard cells, an **electrochemical gradient**—that is, a charge and concentration difference—forms on the two sides of the guard cell plasma membrane.

In step **3**, the resulting electrochemical gradient of H⁺ drives the facilitated diffusion of large numbers of potassium ions *into* guard cells. This movement occurs through **voltage-activated potassium channels,** which open when a certain voltage (difference in charge between the two sides of the guard cell plasma membrane) is attained. This movement of potassium ions has been experimentally measured by the **patch clamp technique,** in which researchers seal the tip of a micropipette to a tiny patch of

membrane that contains a single ion channel. They measure the flow of ions through that channel between the cytoplasm and the solution in the micropipette. (Interestingly, in animals, voltage-activated ion channels are found in the plasma membranes of nerve cells and are involved in transmitting neural impulses.)

As shown in step **4**, chloride ions are also taken into the guard cells through ion channels in the guard cell plasma membrane. The negatively charged chloride and malate ions help to electrically balance the positively charged potassium ions.

The potassium, chloride, and malate ions accumulate in the vacuoles of the guard cells, thus increasing the solute concentration in the vacuoles. You may recall from the discussion of osmosis in Chapter 5 that when a cell has a solute concentration greater than that of surrounding cells, water flows *into* the cell. Thus, in step **5**, water enters the guard cells from surrounding epidermal cells by osmosis. The increased turgidity of the guard cells changes their shape, because the thickened inner cell walls do not expand as much as the outer walls, so the stoma opens.

Blue light activates proton pumps ⟶ proton pump moves H⁺ out of guard cells ⟶ K⁺ and Cl⁻ diffuse into guard cells through voltage-activated ion channels ⟶ water diffuses by osmosis into guard cells ⟶ guard cells change shape and stoma opens

Stomata close in the late afternoon or early evening, but not by an exact reversal of the opening process. Recent studies have demonstrated that the concentration of potassium ions in guard cells slowly decreases during the day. However, the concentration of sucrose, another osmotically active substance, increases during the day—maintaining the open pore—and then slowly decreases as evening approaches. This sucrose comes from the hydrolysis of the polysaccharide starch, which is stored in the guard cell chloroplasts. As evening approaches, the sucrose concentration in the guard cells declines as sucrose is converted back to starch (which is osmotically inactive), water leaves by osmosis, the guard cells lose their turgidity, and the pore closes.

To summarize, different mechanisms appear to regulate the opening and closing of stomata. The uptake of potassium and chloride ions is mainly associated with stomatal opening, and the declining concentration of sucrose is mainly associated with stomatal closing. In Chapter 37, we discuss other blue-light responses in plants.

Additional factors affect stomatal opening and closing

Although light and darkness trigger the opening and closing of stomata, other environmental factors are also involved, including CO_2 concentration. A low concentration of CO_2 in the leaf induces stomata to open, even in the dark. The effects of light and CO_2 concentration on stomatal opening are interrelated. Photosynthesis, which occurs in the presence of light, reduces the internal concentration of CO_2 in the leaf, triggering stomatal opening. Another environmental factor that affects stomatal opening and closing is dehydration (water stress). During a prolonged drought, stomata remain closed, even during the day. Stomatal opening and closing are under hormonal control, particularly by the plant hormone *abscisic acid* (see Table 37-1).

The opening and closing of stomata are also regulated by an internal biological clock that in some way measures time. For example, after plants are placed in continual darkness, their stomata continue to open and close at more or less the same time each day. Such biological rhythms that follow an approximate 24-hour cycle are known as **circadian rhythms.** Other examples of circadian rhythms are provided in Chapter 37.

Review

- How does blue light trigger stomatal opening?
- What physiological changes occur in guard cells during stomatal opening? During stomatal closing?

TRANSPIRATION AND GUTTATION

Learning Objectives

7 Discuss transpiration and its effects on plants.
8 Distinguish between transpiration and guttation.

Despite leaf adaptations such as the cuticle, approximately 99% of the water that a plant absorbs from the soil is lost by evap-

(a) In the late afternoon of a hot day, the leaves have wilted because of water loss. Note that wilting helps reduce the surface area from which transpiration occurs.

(b) The next morning, water in the leaves has been replenished. Transpiration is negligible during the night, and the plants recover by absorbing water from the soil.

Figure 33-10   Temporary wilting in squash (*Cucurbita pepo*) leaves

oration from the leaves and, to a lesser extent, the stems. Loss of water vapor by evaporation from aerial plant parts is called **transpiration.**

The cuticle is extremely effective in reducing water loss by transpiration. It is estimated that only 1% to 3% of the water lost from a plant passes directly through the cuticle. Most transpiration occurs through open stomata. The numerous stomatal pores that are so effective in gas exchange for photosynthesis also provide openings through which water vapor escapes. In addition, the loose arrangement of the spongy mesophyll cells provides a large surface area within the leaf from which water can evaporate.

Several environmental factors influence the rate of transpiration. More water is lost from plant surfaces at higher temperatures. Light increases the transpiration rate, in part because it triggers stomatal opening and in part because it increases the leaf's temperature. Wind and dry air increase transpiration, but humid air *decreases* transpiration because the air is already saturated, or nearly so, with water vapor.

Although transpiration may seem wasteful, particularly to farmers in arid lands, it is an essential process that has adaptive value. Transpiration is responsible for water movement in plants, and without it water would not reach the leaves from the soil (see discussion of the tension–cohesion model in Chapter 34). The large amount of water that plants lose by transpiration may provide some additional benefits. Transpiration, like sweating in humans, cools the leaves and stems. When water passes from a liquid state to a vapor, it absorbs a great deal of heat. When the water molecules leave the plant as water vapor, they carry this heat with them. Thus, the cooling effect of transpiration may prevent leaves from overheating, particularly in direct sunlight.

Another benefit of transpiration is that it distributes essential minerals throughout the plant. The water that a plant transpires is initially absorbed from the soil, where it is present not as pure water but as a dilute solution of dissolved mineral salts. The water and dissolved minerals are then transported in the xylem throughout the plant, including its leaves. Water moves from the plant to the atmosphere during transpiration, but minerals remain in plant tissues. Many of these minerals are required for the plant's growth. It has been suggested that transpiration enables a plant to take in sufficient water to provide enough essential minerals and that plants cannot satisfy their mineral requirements if the transpiration rate is not high enough.

There is no doubt, however, that under certain circumstances excessive transpiration can be harmful to a plant. On hot summer days, plants frequently lose more water by transpiration than they can take in from the soil. Their cells experience a loss of turgor, and the plants wilt (❙ Fig. 33-10). If a plant is able to recover overnight, because of the combination of negligible transpiration (recall that stomata are closed) and absorption of water from the soil, the plant is said to have experienced *temporary wilting*. Most plants recover from temporary wilting with no ill effects. In cases of prolonged drought, however, the soil may not contain sufficient moisture to permit recovery from wilting. A plant that cannot recover is said to be *permanently wilted* and will die.

Transpiration is an important part of the **hydrologic cycle,** in which water cycles from the ocean and land to the atmosphere and then back to the ocean and land (see Fig. 54-11). As a result

of transpiration, water evaporates from leaves and stems to form clouds in the atmosphere. Thus, transpiration eventually results in precipitation. As you may expect, forest trees release substantial amounts of moisture into the air by transpiration. Researchers have determined that at least half the rain that falls in the Amazon rainforest basin is recycled again and again by transpiration and precipitation.

Some plants exude liquid water

Many leaves have special structures through which liquid water is literally forced out. This loss of liquid water, known as **guttation,** occurs when transpiration is negligible and available soil moisture is high. Guttation typically occurs at night because the stomata are closed, but water continues to move into the roots by osmosis. People sometimes mistake the early-morning water droplets formed on leaf margins by guttation for dew (❙ Fig. 33-11). Unlike dew, which condenses from cool night air, guttation droplets come from within the plant. (The mechanism for guttation is discussed in Chapter 34.)

Review

❙ What is transpiration? How is leaf structure related to transpiration?

❙ How do environmental factors (sunlight, temperature, humidity, and wind) influence the rate of transpiration?

❙ How does guttation differ from transpiration?

LEAF ABSCISSION

Learning Objective

9 Define leaf *abscission*, explain why it occurs, and describe the physiological and anatomical changes that precede it.

All trees shed leaves, a process known as **abscission.** Many conifers shed their needles in small numbers year-round. The leaves of deciduous plants turn color and abscise, or fall off, once a year— as winter approaches in temperate climates or at the beginning of the dry period in tropical climates with pronounced wet and dry seasons. In temperate forests, most woody plants with broad leaves shed their leaves to survive the low temperatures of winter. During winter, the plant's metabolism, including its photosynthetic machinery, slows down or halts temporarily.

Another reason for abscission is related to a plant's water requirements, which become critical during the physiological drought of winter. As mentioned previously, as the ground chills, absorption of water by the roots is inhibited. When the ground freezes, *no* absorption occurs. If the broad leaves were to stay on the plant during the winter, the plant would continue to lose water by transpiration but would be unable to replace it with water absorbed from the soil.

Leaf abscission is a complex process that involves many physiological changes, all initiated and orchestrated by changing levels of plant hormones, particularly *ethylene*. Briefly, the process is this: As autumn approaches, sugars, amino acids, and many es-

Ed Reschke/Peter Arnold, Inc.

❙ **Figure 33-11** Guttation

Shown is a compound leaf of strawberry (*Fragaria*). Many people mistake guttation for early-morning dew.

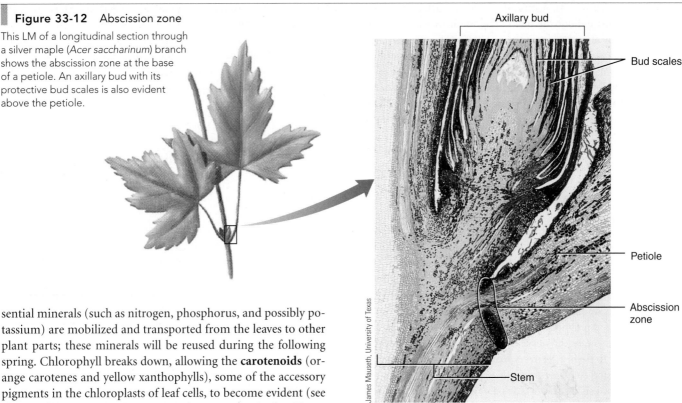

Figure 33-12 Abscission zone

This LM of a longitudinal section through a silver maple (*Acer saccharinum*) branch shows the abscission zone at the base of a petiole. An axillary bud with its protective bud scales is also evident above the petiole.

James Mauseth, University of Texas

Axillary bud

Bud scales

Petiole

Abscission zone

Stem

0.5 mm

sential minerals (such as nitrogen, phosphorus, and possibly potassium) are mobilized and transported from the leaves to other plant parts; these minerals will be reused during the following spring. Chlorophyll breaks down, allowing the **carotenoids** (orange carotenes and yellow xanthophylls), some of the accessory pigments in the chloroplasts of leaf cells, to become evident (see Fig. 3-14b). Accessory pigments are always present in the chloroplasts but are masked by the green of the chlorophyll. In addition, red water-soluble pigments called **anthocyanins** may accumulate in the vacuoles of epidermal leaf cells in some species; anthocyanins may protect leaves against damage by ultraviolet radiation. The various combinations of carotenoids and anthocyanins are responsible for the brilliant colors found in autumn landscapes in temperate climates.

In many leaves, abscission occurs at an abscission zone near the base of the petiole

The area where a petiole detaches from the stem is a structurally distinct area called the **abscission zone.** Composed primarily of thin-walled parenchyma cells, it is anatomically weak because it contains few fibers (▌ Fig. 33-12). A protective layer of cork cells develops on the stem side of the abscission zone. These cells have a waxy, waterproof material impregnated in their walls. Enzymes then dissolve the **middle lamella** (the "cement" that holds the primary cell walls of adjacent cells together) in the abscission zone (see Fig. 4-29). Once this process is completed, nothing holds the leaf to the stem but a few xylem cells. A sudden breeze is enough to make the final break, and the leaf detaches. The protective layer of cork remains, sealing off the area and forming a leaf scar.

Although the structure of abscission zones and the physiological changes associated with abscission are well known, only recently have botanists investigated the genes responsible for the development of an abscission zone and the physiological changes of leaf abscission. Using **DNA microarrays,** for example, bota-

nists have identified about 35 genes that are expressed in leaves only as winter approaches. (As shown in Figure 15-13, a DNA microarray contains hundreds of different DNA molecules placed on a glass slide or chip). Some of these genes code for degradative enzymes involved in breaking down proteins and other molecules; others have functions not yet identified.

Review

▌ Why do many woody plants living in temperate zones lose their leaves in autumn?

▌ What physiological changes occur during leaf abscission?

▌ What is the abscission zone?

MODIFIED LEAVES

Learning Objective

10 List at least four examples of modified leaves, and give the function of each.

Although photosynthesis is the main function of leaves, certain leaves have special modifications for other functions. Some plants have leaves specialized for deterring herbivores. **Spines,** modified leaves that are hard and pointed, are found on many desert plants, such as cacti (▌ Fig. 33-13a). In the cactus, the main organ of photosynthesis is the stem rather than the leaf. Spines discourage animals from eating the succulent stem tissue.

(a) The leaves of this pincushion cactus (*Mammilaria*) are spines for deterring herbivores.

(b) Tendrils of bur cucumber (*Echinocystis lobata*) are modified leaves that wind around objects and aid in climbing.

(c) A terminal bud and two axillary buds of a maple (*Acer*) twig have overlapping bud scales to protect buds.

(d) The leaves of bulbs such as the onion (*Allium cepa*) are fleshy for storage of food materials and water.

(e) The succulent leaves of the string-of-beads plant (*Senecio rowleyanus*) are spherical to minimize surface area, thereby conserving water.

Figure 33-13 Leaf modifications

Vines are climbing plants whose stems cannot support their own weight, so they often possess **tendrils** that help keep the vine attached to the structure on which it is growing. The tendrils of many vines, such as peas, cucumbers, and squash, are specialized leaves (❙ Fig. 33-13b). However, some tendrils, such as those of ivy, Virginia creeper, and grape, are specialized stems.

The winter buds of a dormant woody plant are covered by **bud scales,** modified leaves that protect the delicate meristematic tissue of the bud from injury, freezing, or drying out (❙ Fig. 33-13c).

Leaves may also be modified for storage of water or food. For example, a **bulb** is a short, underground stem to which large, fleshy leaves are attached (❙ Fig. 33-13d). Onions and tulips form bulbs. Many plants adapted to arid conditions, such as jade plant, medicinal aloe, and string-of-beads, have succulent leaves for water storage (❙ Fig. 33-13e). These leaves are usually green and also function in photosynthesis.

Some leaves are modified for sexual reproduction (see discussion of flower evolution in Chapter 28, including Fig. 28-15) or asexual reproduction (see Fig. 36-22).

Modified leaves of carnivorous plants capture insects

Carnivorous plants are plants that capture insects. Most carnivorous plants grow in poor soil that is deficient in certain essential minerals, particularly nitrogen. These plants meet some of their mineral requirements by digesting insects and other small animals. The leaves of carnivorous plants are adapted to attract, capture, and digest their animal prey.

Some carnivorous plants have passive traps. The leaves of a pitcher plant, for example, are shaped so that rainwater collects and forms a reservoir that also contains acid secreted by the plant (❚ Fig. 33-14). Some pitchers are quite large; in the tropics, pitcher plants may be large enough to hold 1 L (approximately 1 qt) or more of liquid. An insect attracted by the odor or nectar of the pitcher may lean over the edge and fall in. Although it may make repeated attempts to escape, the insect is prevented from crawling out by the slippery sides and the rows of stiff hairs that point downward around the lip of the pitcher. The insect eventually drowns, and part of its body disintegrates and is absorbed.

Most insects are killed in pitcher plants. However, the larvae of several insects (certain flies, midges, and mosquitoes) and a large community of microorganisms live inside the pitchers. These insect species obtain their food from the insect carcasses, and the pitcher plant digests what remains. It is not known how these insects survive the acidic environment inside the pitcher.

The Venus flytrap is a carnivorous plant with active traps. Its leaf blades resemble tiny bear traps (see Fig. 1-3). Each side of the leaf blade contains three small, stiff hairs. If an insect alights and brushes against two of the hairs, or against the same hair twice in quick succession, the trap springs shut with amazing rapidity—about 100 milliseconds. The spines along the margins of the blades fit closely together to prevent the insect from escaping. After the leaf initially traps the insect, the leaf continues to slowly close for the next several hours. Digestive glands on the surface of the trap secrete enzymes in response to the insect pressing against them. Days later, after the insect has died and been digested, the trap reopens and the indigestible remains fall out.

Bill Lea/Dembinsky Photo Associates

Figure 33-14 A common pitcher plant

This species (*Sarracenia purpurea*), whose pitchers grow to 30.5 cm (12 in), is widely distributed in acidic bogs and marshes in eastern North America. Young pitchers are green but turn red as they age. Note the dead beetle in the "pitcher."

Review

❚ What are the primary functions of each of the following modified leaves: spines, tendrils, and bud scales?
❚ What are the functions of bulbs? Of succulent leaves?
❚ What are some of the specialized features of the leaves of carnivorous plants?

SUMMARY WITH KEY TERMS

Learning Objectives

1 Discuss variation in leaf form, including simple versus compound leaves, leaf arrangement on the stem, and venation patterns (page 716).
  ❚ Leaves typically consist of a broad, flat **blade** and a stalk-like **petiole.** Some leaves also have small, leaflike outgrowths from the base called **stipules.**
  ❚ Leaves may be simple (having a single blade) or compound (having a blade divided into two or more leaflets).
  ❚ Leaf arrangement on a stem may be alternate (one leaf at each **node**), opposite (two leaves at each node), or whorled (three or more leaves at each node).
  ❚ Leaves may have parallel or netted venation. Netted venation may be palmately netted, with several major **veins** radiating from one point, or pinnately netted, with veins branching along the entire length of the midvein.

2 Describe the major tissues of the leaf (epidermis, photosynthetic ground tissue, xylem, and phloem), and label them on a diagram of a leaf cross section (page 716).
  ❚ Upper and lower surfaces of the leaf blade are covered by an **epidermis.** A waxy **cuticle** coats the epidermis, enabling the plant to survive the dry conditions of a terrestrial existence.
  ❚ **Stomata** are small pores in the epidermis that permit gas exchange needed for photosynthesis. Each pore is surrounded by two **guard cells** that are often associated with special epidermal cells called **subsidiary cells.** Subsidiary

cells provide a reservoir of water and ions that move into and out of the guard cells as they change shape during stomatal opening and closing.

- **Mesophyll** consists of photosynthetic parenchyma cells. Mesophyll is divided into **palisade mesophyll,** which functions primarily for photosynthesis, and **spongy mesophyll,** which functions primarily for gas exchange.
- Leaf veins have **xylem** to conduct water and essential minerals to the leaf and **phloem** to conduct sugar produced by photosynthesis to the rest of the plant.

ThomsonNOW™ **Learn more about leaf tissues by clicking on the figure in ThomsonNOW.**

3 Compare leaf anatomy in eudicots and monocots (page 716).
- Monocot leaves have parallel venation, whereas eudicot leaves have netted venation.
- Some monocots (corn and other grasses) do not have mesophyll differentiated into distinct palisade and spongy layers.
- Some monocots (grasses, reeds, and sedges) have guard cells shaped like dumbbells, unlike the more common bean-shaped guard cells.

4 Relate leaf structure to its function of photosynthesis (page 716).
- Leaf structure is adapted for its primary function of **photosynthesis.** Most leaves have a broad, flattened blade that is quite efficient in collecting the sun's radiant energy.
- Stomata generally open during the day for gas exchange needed during photosynthesis and close at night to conserve water when photosynthesis is not occurring.
- The transparent epidermis allows light to penetrate into the middle of the leaf, where photosynthesis occurs.
- Air spaces in mesophyll tissue permit the rapid diffusion of CO_2 and water into, and oxygen out of, mesophyll cells.

5 Explain the role of blue light in the opening of stomata (page 722).
- **Blue light,** which is a component of sunlight, triggers the activation of proton pumps located in the guard cell plasma membrane. Blue light also triggers the synthesis of malic acid and the **hydrolysis** of starch.

6 Outline the physiological changes that accompany stomatal opening and closing (page 722).
- Protons (H^+) are pumped out of the guard cells. The protons are produced when malic acid ionizes. As protons leave the guard cells, an **electrochemical gradient** (a charge and concentration difference) forms on the two sides of the guard cell plasma membrane.
- The electrochemical gradient drives the uptake of potassium ions through **voltage-activated potassium channels** into the guard cells. Chloride ions are also taken into the guard cells through ion channels. These osmotically active ions increase the solute concentration in the guard cell

vacuoles. The resulting osmotic movement of water into the guard cells causes them to become turgid, forming a pore.
- As the day progresses, potassium ions slowly leave the guard cells and starch is hydrolyzed to sucrose, which increases in concentration in the guard cells. Stomata close when water leaves the guard cells as a result of a decline in the concentration of sucrose, an osmotically active solute. The sucrose is converted to starch, which is osmotically inactive.

ThomsonNOW™ **Watch stomata in action by clicking on the figure in ThomsonNOW.**

7 Discuss transpiration and its effects on plants (page 724).
- **Transpiration** is the loss of water vapor from aerial parts of plants. Transpiration occurs primarily through the stomata.
- The rate of transpiration is affected by environmental factors such as temperature, wind, and relative humidity.
- Transpiration appears to be both beneficial and harmful to the plant—that is, transpiration represents a trade-off between the CO_2 requirement for photosynthesis and the need for water conservation.

8 Distinguish between transpiration and guttation (page 724).
- **Guttation,** the release of liquid water from leaves of some plants, occurs through special structures when transpiration is negligible and available soil moisture is high. In contrast, transpiration is the loss of water vapor and occurs primarily through the stomata.

9 Define leaf *abscission*, explain why it occurs, and describe the physiological and anatomical changes that precede it (page 725).
- Leaf **abscission** is the loss of leaves that often occurs as winter approaches in temperate climates or at the beginning of the dry period in tropical climates with wet and dry seasons.
- Abscission is a complex process involving physiological and anatomical changes that occur prior to leaf fall. An **abscission zone** develops where the petiole detaches from the stem. Sugars, amino acids, and many essential minerals are transported from the leaves to other plant parts. Chlorophyll breaks down, and **carotenoids** and **anthocyanins** become evident.

10 List at least four examples of modified leaves, and give the function of each (page 726).
- **Spines** are leaves adapted to deter herbivores. Some **tendrils** are leaves modified for grasping and holding on to other structures (to support weak stems). **Bud scales** are leaves modified to protect delicate meristematic tissue or dormant buds. **Bulbs** are short, underground stems with fleshy leaves specialized for storage. Many plants adapted to arid conditions have succulent leaves for water storage. Carnivorous plants have leaves modified to trap insects.

TEST YOUR UNDERSTANDING

1. Plants with an alternate leaf arrangement have (a) blades divided into two or more leaflets (b) major veins that radiate out from one point (c) one leaf at each node (d) major veins branching off along the entire length of the midvein (e) two leaves at each node

2. The photosynthetic ground tissue in the middle of the leaf is called (a) cutin (b) mesophyll (c) the abscission zone (d) subsidiary cells (e) palisade and spongy stomata

3. The primary function of the spongy mesophyll is (a) reducing water loss from the leaf surface (b) changing the shape of the guard cells (c) supporting the leaf to prevent it from collapsing under its own weight (d) diffusing gases within the leaf (e) deterring herbivores

4. Gas exchange occurs through microscopic pores formed by two (a) subsidiary cells (b) abscission cells (c) mesophyll cells (d) guard cells (e) stipules

5. Most stomata are usually located in the _____ of the leaf. (a) upper epidermis (b) lower epidermis (c) cuticle (d) spongy mesophyll (e) palisade mesophyll

6. The thin, noncellular layer of wax secreted by the epidermis of leaves is the (a) stoma (b) subsidiary cell (c) trichome (d) bundle sheath (e) cuticle

7. The _____ encircles a vein. (a) palisade mesophyll (b) guard cell (c) bundle sheath (d) blade (e) cuticle

8. The _____ of a leaf vein transports water and dissolved minerals, whereas the _____ transports sugars produced by the leaf during photosynthesis. (a) xylem; phloem (b) xylem; bundle sheath (c) phloem; xylem (d) phloem; vein (e) vascular bundle; bundle sheath

9. Which of the following is *not* an adaptation of pine needles to conserve water? (a) less surface area exposed to the air than thin-bladed leaves (b) a relatively thick cuticle (c) sunken stomata (d) netted veins instead of parallel veins (e) both c and d are not adaptations of pine needles

10. Most of the water that a plant absorbs from the soil is lost by the process of (a) guttation (b) circadian rhythm (c) abscission (d) transpiration (e) photosynthesis

11. When transpiration is negligible, plants such as grasses exude excess water by (a) guttation (b) circadian rhythm (c) abscission (d) pumping H$^+$ out of and K$^+$ into guard cells (e) photosynthesis

12. At sunrise, the accumulation in the guard cells of the osmotically active substance _____ causes an inflow of water and the opening of the pore. (a) protons (b) starch (c) ATP synthase (d) sucrose (e) potassium ions

13. Stomatal opening is most pronounced in response to _____ light. (a) green (b) yellow (c) blue (d) ultraviolet (e) infrared

14. The seasonal detachment of leaves is known as (a) forest decline (b) transpiration (c) abscission (d) guttation (e) dormancy

15. Anatomically, the abscission zone where a petiole detaches from a stem consists of (a) thin-walled parenchyma cells with few fibers (b) thick-walled cork parenchyma cells (c) clusters of fibers and collenchyma strands (d) hard, pointed stipules (e) epidermal cells with sunken stomata

16. Modified leaves that enable a stem to climb are called _____, whereas modified leaves that cover the winter buds of a dormant woody plant are called _____. (a) spines; bud scales (b) bud scales; tendrils (c) tendrils; bud scales (d) tendrils; spines (e) carnivorous leaves; spines

17. There is a trade-off between photosynthesis and transpiration in leaves because (a) numerous stomatal pores provide both gas exchange for photosynthesis and openings through which water vapor escapes (b) a waxy layer, the cuticle, reduces water loss (c) blue light triggers an influx of potassium ions (K$^+$) into the guard cells (d) leaves of deciduous plants abscise as winter approaches in temperate climates (e) stomata are closed at night, although water continues to move into the roots by osmosis

CRITICAL THINKING

1. Suppose that you are asked to observe a micrograph of a leaf cross section and distinguish between the upper and lower epidermis. How would you make this decision?

2. Given that (a) xylem is located toward the upper epidermis in leaf veins and phloem is toward the lower epidermis and (b) the vascular tissue of a leaf is continuous with that of the stem, suggest one possible arrangement of vascular tissues in the stem that might account for the arrangement of vascular tissue in the leaf.

3. What might be some of the advantages of a plant having a few large leaves? What might be some disadvantages? What might be some advantages of having many small leaves? What disadvantages might this entail? How would your answers differ for plants growing in a humid environment compared to those in a desert?

4. Briefly explain why research on the molecular mechanism of stomatal closure might be of future use in agriculture.

5. **Evolution Link.** Why did natural selection favor the evolution of seasonal leaf abscission in woody flowering plants living in colder climates? What adaptations enable conifers to survive these climates without leaf abscission?

 Additional questions are available in ThomsonNOW at www.thomsonedu.com/login

Stems and Transport in Vascular Plants

Michael Fairchild/Peter Arnold, Inc.

Baobab tree. Baobabs (*Adansonia digitata*), which are native to Africa, Madagascar, and Australia, store large volumes of water and starch in their massive trunks. Baobab trees are relatively short (growing to 18 m, or 60 ft), but their trunks can be as much as 9 m (30 ft) in diameter. Photographed in Namibia, with children around the tree as a size reference.

KEY CONCEPTS

Primary tissues (epidermis, cortex, pith, xylem, and phloem) of stems develop from shoot apical meristems.

Secondary tissues (wood and bark) of stems develop from two lateral meristems, vascular cambium and cork cambium.

The concept of water potential explains the direction of water flow into, through, and out of a plant.

According to the tension–cohesion model, transpiration pulls water up through the stem as water evaporates from leaves by transpiration.

According to the pressure–flow hypothesis, sucrose is transported in phloem sap from the source, where the sugar is loaded into phloem, to the sink, where the sugar is removed from phloem.

A vegetative (not sexually reproductive) vascular plant has three parts: roots, leaves, and stems. As discussed in Chapter 32, roots serve to anchor the plant and absorb materials from the soil, whereas leaves are primarily for photosynthesis, converting radiant energy into the chemical energy of carbohydrate molecules. Stems, the focus of this chapter, link a plant's roots to its leaves and are usually located aboveground, although many plants have underground stems. Stems exhibit varied forms, ranging from rope-like vines to massive tree trunks. They can be either herbaceous, with soft, nonwoody tissues; or woody, with extensive hard tissues of wood and bark.

Stems perform three main functions in plants. First, stems of most species support leaves and reproductive structures. The upright position of most stems and the arrangement of the leaves on them allow each leaf to absorb light for use in photosynthesis. Reproductive structures (flowers and fruits) are located on stems in areas accessible to insects, birds, and air currents, which transfer pollen from flower to flower and help disperse seeds and fruits.

Second, stems provide internal transport. They conduct water and dissolved minerals (inorganic nutrients) from the roots, where these materials are absorbed from the soil, to leaves and other plant parts. Stems also conduct the sugar produced in leaves by photosynthesis to roots and other parts of the plant. Remember, however, that stems are not the only plant organs that conduct materials. The vascular system is continuous throughout all parts of a plant, and conduction occurs in roots, stems, leaves, and reproductive structures.

Third, stems produce new living tissue. They continue to grow throughout a plant's life, producing *buds* that develop into stems with new leaves and/or reproductive structures. In addition to the main functions of support, conduction, and production of new stem tissues, stems of some species are modified for asexual reproduction (see Chapter 36) or, if green, to manufacture sugar by photosynthesis. Also, some stems are specialized to store starch (see photograph on previous page). ∎

EXTERNAL STEM STRUCTURE IN WOODY TWIGS

Learning Objective

1   Describe the external features of a woody twig.

Although stems exhibit great variation in structure and growth, they all have **buds,** which are embryonic shoots. A **terminal bud** is the embryonic shoot located at the tip of a stem. The dormant (not actively growing) apical meristem of a terminal bud is covered and protected by an outer layer of **bud scales,** which are modified leaves (see Fig. 33-13c). **Axillary buds,** also called **lateral buds,** are located in the axils of a plant's leaves (see Fig. 33-1). An *axil* is the upper angle between a leaf and the stem to which it is attached. When terminal and axillary buds grow, they form branches that bear leaves and/or flowers. The area on a stem where each leaf is attached is called a **node,** and the region between two successive nodes is an **internode.**

To demonstrate certain structural features of the stem, we can use a woody twig of a deciduous tree that has shed its leaves, as shown in ∎ Figure 34-1. Bud scales cover the terminal bud and protect its delicate apical meristem during dormancy. When the bud resumes growth, the bud scales covering the terminal bud fall off, leaving **bud scale scars** on the stem where they were attached. Because temperate-zone woody plants form terminal buds at the end of each year's growing season, the number of sets of bud scale scars on a twig indicates its age. A **leaf scar** shows where each leaf was attached on the stem; the pattern of leaf scars can be used to determine leaf arrangement on a stem—alternate, opposite, or whorled (see Fig. 33-2b). The vascular (conducting) tissue that extends from the stem out into the leaf forms **bundle scars** within a leaf scar. Axillary buds may be found above the leaf scars. Also, the bark of a woody twig has **lenticels,** sites of loosely arranged cells that allow oxygen to diffuse into the interior of the woody stem. Lenticels look like tiny specks on the bark of a twig.

Review

∎   What is the difference between terminal and axillary buds?
∎   What is the function of bud scales? Of lenticels?
∎   How can you tell the age of a woody twig?

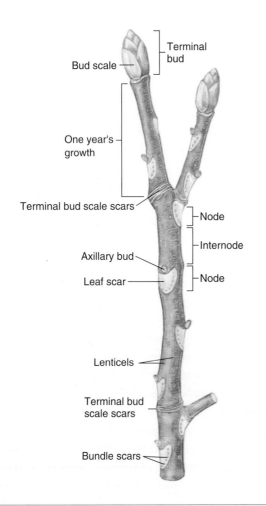

Bud scale
Terminal bud
One year's growth
Terminal bud scale scars
Node
Internode
Axillary bud
Leaf scar
Node
Lenticels
Terminal bud scale scars
Bundle scars

Figure 34-1 *Animated*   External structure of a woody twig in its winter condition

The age of a woody twig can be determined by the number of sets of bud scale scars (do not count side branches). How old is this twig?

STEM GROWTH AND STRUCTURE

Learning Objectives

2   Label cross sections of herbaceous eudicot and monocot stems, and describe the functions of each tissue.
3   Name the two lateral meristems, and describe the tissues that arise from each.
4   Outline the transition from primary growth to secondary growth in a woody stem.

You may recall from Chapter 32 that plants have two different types of growth. **Primary growth** is an increase in the length of a plant and occurs at **apical meristems** located at the tips of roots and shoots and also within the buds of stems. **Secondary growth** is an increase in the girth (thickness) of a plant as a result of the activity of **lateral meristems** located within stems and roots. The new tissues formed by the lateral meristems are called *secondary tissues* to distinguish them from *primary tissues* produced by apical meristems.

All plants have primary growth; some plants have both primary and secondary growth. Stems with only primary growth

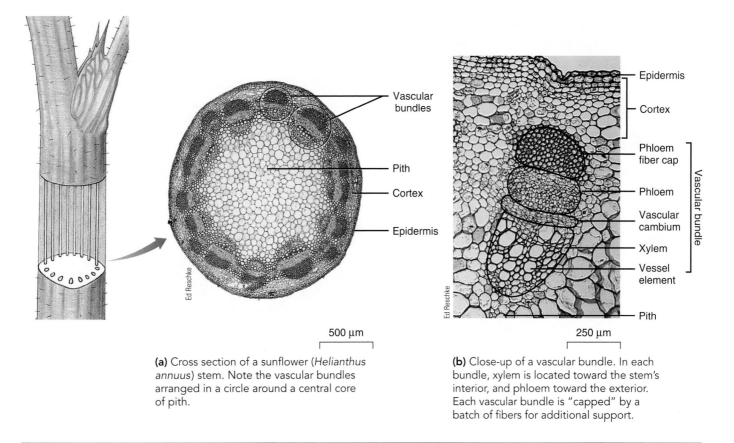

(a) Cross section of a sunflower (*Helianthus annuus*) stem. Note the vascular bundles arranged in a circle around a central core of pith.

500 µm

(b) Close-up of a vascular bundle. In each bundle, xylem is located toward the stem's interior, and phloem toward the exterior. Each vascular bundle is "capped" by a batch of fibers for additional support.

250 µm

Figure 34-2 *Animated* LMs of a herbaceous eudicot stem

are herbaceous, whereas those with both primary and secondary growth are woody.[1] A woody plant increases in length by primary growth at the tips of its stems and roots, whereas its older stems and roots farther back from the tips increase in girth by secondary growth. In other words, at the same time that secondary growth is adding wood and bark, thereby causing the stem to thicken, primary growth is increasing the length of the stem.

Herbaceous eudicot and monocot stems differ in internal structure

Although considerable structural variation exists in stems, they all possess an outer protective covering (epidermis or periderm), one or more types of ground tissue, and vascular tissues (xylem and phloem). Let us first consider the structure of herbaceous eudicot stems and then of monocot stems.

Vascular bundles of herbaceous eudicot stems are arranged in a circle in cross section

A young sunflower stem is a representative herbaceous eudicot stem that exhibits primary growth (❙ Fig. 34-2). Its outer covering, the **epidermis,** provides protection in herbaceous stems, as it does in leaves and herbaceous roots (see Table 32-4 and

Fig. 32-6). The **cuticle,** a waxy layer of *cutin,* covers the stem epidermis and reduces water loss from the stem surface. **Stomata** permit gas exchange. (Recall from Chapter 33 that a cuticle and stomata are also associated with the leaf epidermis.)

Inside the epidermis is the **cortex,** a cylinder of ground tissue that may contain parenchyma, collenchyma, and sclerenchyma cells (see Table 32-2 and Fig. 32-4). As might be expected from the various types of cells that it contains, the cortex in herbaceous eudicot stems can have several functions, such as photosynthesis, storage, and support. If a stem is green, photosynthesis occurs in chloroplasts of cortical parenchyma cells. Parenchyma in the cortex also stores starch (in amyloplasts) and crystals (in vacuoles). Collenchyma and sclerenchyma in the cortex confer strength and structural support for the stem.

The vascular tissues provide conduction and support. In herbaceous eudicot stems, the vascular tissues are located in bundles that, when viewed in cross section, are arranged in a circle. However, viewed lengthwise, these bundles extend as long strands throughout the length of a stem and are continuous with vascular tissues of both roots and leaves.

Each vascular bundle contains both **xylem,** which transports water and dissolved minerals from roots to leaves, and **phloem,** which transports dissolved sugar (see Table 32-3 and Fig. 32-5). Xylem is located on the inner side of the vascular bundle, and phloem is found toward the outside. Sandwiched between xylem and phloem in some herbaceous stems is a single layer of cells called the *vascular cambium,* a lateral meristem responsible for secondary growth (discussed later).

[1]Certain herbaceous stems, such as geranium and sunflower, also have a limited amount of secondary growth.

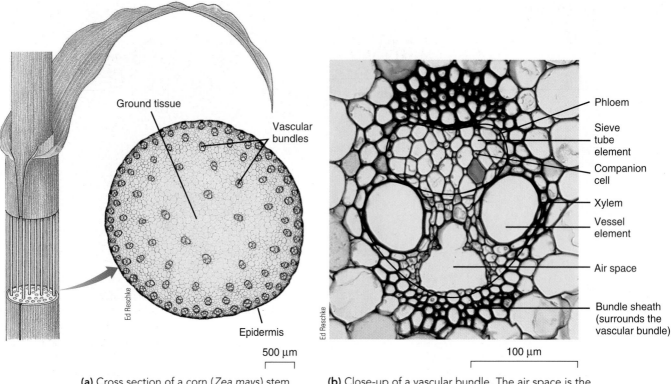

(a) Cross section of a corn (*Zea mays*) stem shows vascular bundles scattered throughout ground tissue.

(b) Close-up of a vascular bundle. The air space is the site where the first xylem elements were formed and later disintegrated. The entire bundle is enclosed in a bundle sheath of sclerenchyma for additional support.

Figure 34-3 *Animated* LMs of a monocot stem

Because most stems support the aerial plant body, they are much stronger than roots. The thick walls of tracheids and vessel elements in xylem help support the plant. Fibers also occur in both xylem and phloem, although they are usually more extensive in phloem. These fibers add considerable strength to the herbaceous stem. In sunflowers and certain other herbaceous eudicot stems, phloem contains a cluster of fibers toward the outside of the vascular bundle, called a **phloem fiber cap,** that helps strengthen the stem. The phloem fiber cap is not present in all herbaceous eudicot stems.

The **pith** is a ground tissue at the center of the herbaceous eudicot stem that consists of large, thin-walled parenchyma cells that function primarily in storage. Because of the arrangement of the vascular tissues in bundles, there is no distinct separation of cortex and pith between the vascular bundles. The areas of parenchyma between the vascular bundles are often referred to as **pith rays.**

Vascular bundles are scattered throughout monocot stems

An epidermis with its waxy cuticle covers monocot stems, such as the herbaceous stem of corn. As in herbaceous eudicot stems, the vascular tissues run in strands throughout the length of a stem. In cross section the vascular bundles contain xylem toward the inside and phloem toward the outside. In contrast with herbaceous eudicots, however, vascular bundles of monocots are not arranged in a circle but are scattered throughout the stem (❙ Fig. 34-3). Each vascular bundle is enclosed in a bundle sheath of supporting sclerenchyma cells. The monocot stem does not have distinct areas of cortex and pith. The ground tissue in which the vascular tissues are embedded performs the same functions as cortex and pith in herbaceous eudicot stems.

Monocot stems do not possess lateral meristems (vascular cambium and cork cambium) that give rise to secondary growth. Monocots have primary growth only and do not produce wood and bark. Although some treelike monocots (such as palms) attain considerable size, they do so by a modified form of primary growth in which parenchyma cells divide and enlarge. Stems of some monocots (such as bamboo and palm) contain a great deal of sclerenchyma tissue, which makes them hard and woodlike in appearance.

Woody plants have stems with secondary growth

Woody plants undergo secondary growth, an increase in the girth of stems and roots. Secondary growth occurs as a result of the activity of two lateral meristems: vascular cambium and cork cambium. Among flowering plants, only woody eudicots (such as apple, hickory, and maple) have secondary growth. Cone-bearing gymnosperms (such as pine, juniper, and spruce) also have secondary growth.

Cells in the **vascular cambium** divide and produce two conducting and supporting tissues: secondary xylem (wood) to replace primary xylem, and secondary phloem (inner bark) to replace primary phloem. Primary xylem and primary phloem are not able to transport materials indefinitely and so are replaced in plants that have extended life spans.

Cells of the outer lateral meristem, called **cork cambium,** divide and produce cork cells and cork parenchyma. Cork cambium and the tissues it produces are collectively referred to as **periderm** (outer bark), which functions as a replacement for the epidermis (see the last LM in Table 32-4).

Vascular cambium gives rise to secondary xylem and secondary phloem

Primary tissues in woody eudicot stems are organized like those in herbaceous eudicot stems, with the vascular cambium a thin layer of cells sandwiched between xylem and phloem in the vascular bundles. Once secondary growth begins, the internal structure of a stem changes considerably (Fig. 34-4). Although vascular cambium is not initially a continuous cylinder of cells (because the vascular bundles are separated by pith rays), it becomes continuous when production of secondary tissues begins. This continuity develops because certain parenchyma cells in each pith ray retain the ability to divide. These cells connect to vascular cambium cells in each vascular bundle, forming a complete ring of vascular cambium.

Cells in the vascular cambium divide and produce daughter cells in two directions. The cells formed from the dividing vascular cambium are located either *inside* the ring of vascular cambium (to become secondary xylem, or wood) or *outside* it (to become secondary phloem, or inner bark) (Fig. 34-5). When a cell in the vascular cambium divides tangentially (inward or outward), one daughter cell remains meristematic; that is, it remains part of the vascular cambium. The other cell may divide again several times, but eventually it stops dividing and develops into mature secondary tissue. Thus, vascular cambium is a thin layer

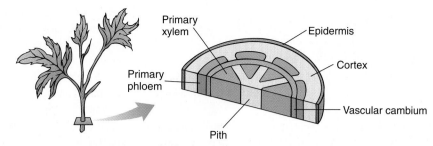

(a) At the onset of secondary growth, vascular cambium arises in the parenchyma between the vascular bundles (that is, in the pith rays), forming a cylinder of meristematic tissue (*blue circle in cross section*).

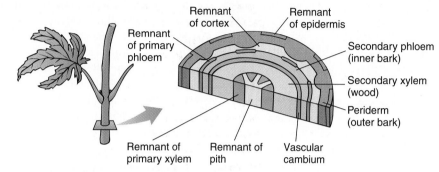

(b) Vascular cambium begins to divide, forming secondary xylem on the inside and secondary phloem on the outside.

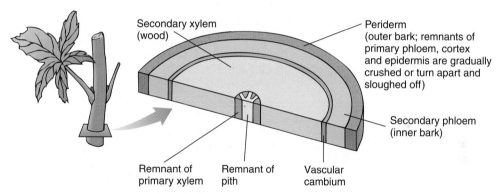

(c) A young woody stem. Vascular cambium produces significantly more secondary xylem than secondary phloem.

Figure 34-4 *Animated* Development of secondary growth

Vascular cambium and the tissues it produces are shown in cross section; cork cambium is not depicted. (The figures change in scale because of space limitations; pith and primary xylem are actually the same size in all three diagrams, but the change in scale makes those tissues appear to shrink in each succeeding diagram.)

of cells sandwiched between the wood and inner bark, the two tissues it produces (Fig. 34-6).

As the stem increases in circumference, the number of cells in the vascular cambium also increases. This increase in cells occurs by an occasional radial division of a vascular cambium cell, at right angles to its normal direction of division. In this case, both daughter cells remain meristematic.

What happens to the original primary tissues of a stem once secondary growth develops? As a stem increases in thickness, the orientation of the original primary tissues changes. For example,

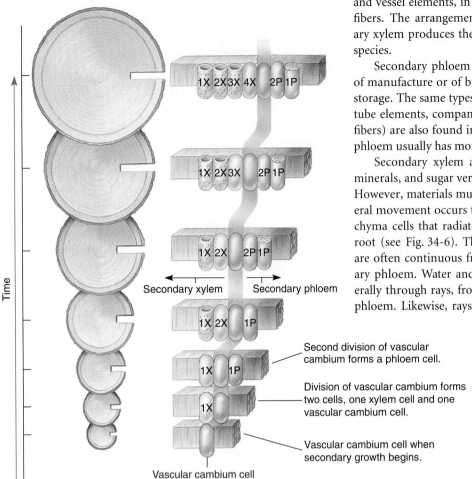

Figure 34-5 *Animated* Development of secondary xylem and secondary phloem

To study the figure, which shows a radial view of a dividing vascular cambium cell, start at the bottom and move up. Note that vascular cambium (*the blue cell*) divides in two directions, forming secondary xylem (X) to the inside and secondary phloem (P) to the outside. These cells, which are numbered in the order in which they are produced, differentiate to form the mature cell types associated with xylem and phloem. As secondary xylem accumulates, vascular cambium "moves" outward, and the woody stem increases in diameter.

Labels in figure:

1X 2X 3X 4X 2P 1P

1X 2X 3X 2P 1P

1X 2X 2P 1P

← Secondary xylem   Secondary phloem →

1X 2X 1P

1X 1P — Second division of vascular cambium forms a phloem cell.

1X — Division of vascular cambium forms two cells, one xylem cell and one vascular cambium cell.

Vascular cambium cell when secondary growth begins.

Vascular cambium cell

Time

---

secondary xylem and secondary phloem are laid down between the primary xylem and primary phloem within each vascular bundle. Therefore, as vascular cambium forms secondary tissues, the primary xylem and primary phloem in each vascular bundle become separated from one another. The primary tissues located outside the cylinder of secondary growth (that is, primary phloem, cortex, and epidermis) are subjected to the mechanical pressures produced by secondary growth and are gradually crushed or torn apart and sloughed off.

Secondary tissues replace the primary tissues in function. Secondary xylem conducts water and dissolved minerals from roots to leaves in the woody plant. It contains the same types of cells found in primary xylem: water-conducting tracheids and vessel elements, in addition to xylem parenchyma cells and fibers. The arrangement of the different cell types in secondary xylem produces the distinctive wood characteristics of each species.

Secondary phloem conducts dissolved sugar from its place of manufacture or of breakdown of starch to a place of use and storage. The same types of cells found in primary phloem (sieve tube elements, companion cells, phloem parenchyma cells, and fibers) are also found in secondary phloem, although secondary phloem usually has more fibers than primary phloem.

Secondary xylem and secondary phloem transport water, minerals, and sugar vertically throughout the woody plant body. However, materials must also move horizontally (laterally). Lateral movement occurs through **rays,** which are chains of parenchyma cells that radiate from the center of the woody stem or root (see Fig. 34-6). The vascular cambium forms rays, which are often continuous from the secondary xylem to the secondary phloem. Water and dissolved minerals are transported laterally through rays, from the secondary xylem to the secondary phloem. Likewise, rays form pathways for the lateral transport of dissolved sugar, from the secondary phloem to the secondary xylem, and of waste products to the center, or heart, of the tree (discussed later).

Cork cambium produces periderm

Cork cambium, which usually arises from parenchyma cells in the outer cortex, produces **periderm,** the functional replacement for the epidermis. Cork cambium is either a continuous cylinder of dividing cells (similar to vascular cambium) or a series of overlapping arcs of meristematic cells that form from parenchyma cells in successively deeper layers of the cortex and, eventually, secondary phloem. Variation in cork cambia and their rates of division explain why the outer bark of some tree species is fissured (as in bur oak), rough and shaggy (shagbark hickory), scaly (Norway pine), or smooth and peeling (paper birch).

As is true of vascular cambium, cork cambium divides to form new tissues in two directions—to its inside and its outside. Cork cells, formed to the outside of cork cambium, are dead at maturity and have walls that contain layers of *suberin* and waxes, making them waterproof. These cork cells protect the woody stem against mechanical injury, mild fires, attacks by insects and fungi, temperature extremes, and water loss. To its inside, cork cambium sometimes forms cork parenchyma cells that store water and starch granules. Cork parenchyma is only one to several cells thick, much thinner than the cork cell layer.

Cork cells are impermeable to water and gases, yet the living internal cells of the woody stem require oxygen and must be able to exchange gases with the surrounding atmosphere. As a stem thickens from secondary growth, the epidermis, including stomata that exchange gases for the herbaceous stem, dies. Stomata are replaced by lenticels, which permit gas exchange through the periderm (█ Fig. 34-7).

Common terms associated with wood are based on plant structure

If you have ever examined different types of lumber, you may have noticed that some trees have wood with two different colors (Fig. 34-8). The functional secondary xylem, that is, the part that conducts water and dissolved minerals, is the *sapwood,* a thin layer of younger, lighter-colored wood that is closest to the bark. *Heartwood,* the older wood in the center of the tree, is typically a brownish red. A microscopic examination of heartwood reveals that its vessels and tracheids are plugged with pigments, tannins, gums, resins, and other materials. Therefore, heartwood no longer functions in conduction but instead functions as a storage site for waste products. Because it is denser than sapwood, heartwood provides structural support for trees. Some evidence suggests that heartwood is also more resistant to decay.

Almost everyone has heard of hardwood and softwood. Botanically speaking, *hardwood* is the wood of flowering plants and *softwood* is the wood of conifers (conebearing gymnosperms). The wood of pine and other conifers typically lacks fibers (with their thick secondary cell walls) and vessel elements; the conducting cells in gymnosperms are tracheids. These cell differences generally make conifer wood softer than the wood of flowering plants, although there is a substantial variation from one species to another. The balsa tree, for example, is a flowering plant whose extremely light, soft wood is used to fashion airplane models.

Woody plants that grow in temperate climates where there is a growing period (during spring and summer) and a dormant period (during winter) exhibit *annual rings,* concentric circles found in cross sections of wood. To determine the age of a woody stem in the temperate zone, simply count the annual rings. In the tropics, environmental conditions, particularly seasonal or year-round precipitation patterns, determine the presence or absence of rings, so rings are not a reliable method of determining the ages of most tropical trees.

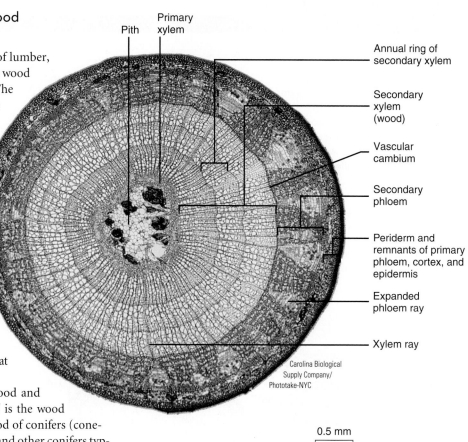

0.5 mm

Carolina Biological Supply Company/Phototake-NYC

Figure 34-6 Three-year-old stem in cross section

LM of entire cross section of basswood (*Tilia americana*) stem. Note the location of the vascular cambium between the secondary xylem (wood) and secondary phloem (inner bark).

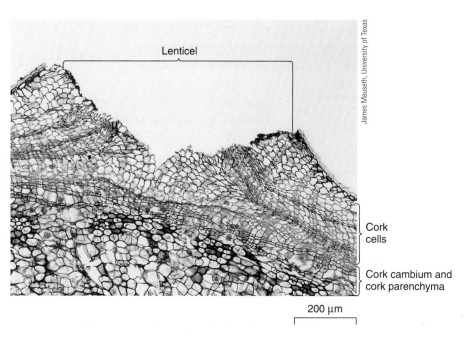

James Mauseth, University of Texas

200 μm

Figure 34-7 LM of stem periderm, showing a lenticel

The epidermis has ruptured because of the proliferation of loosely arranged cork cells in the lenticel. From the bark of a calico flower (*Aristolochia elegans*) stem.

TREE-RING ANALYSIS

In temperate climates, the number of annual rings indicates the age of a tree. The size of each ring varies depending on local weather conditions. Sometimes the variation can be attributed to a single environmental factor, and similar patterns appear in the rings of different species over a large geographic area. For example, years with adequate precipitation produce wider growth rings, and years of drought produce narrower rings.

It is possible to study ring sequences that are several thousand years old by constructing a *master chronology*, a complete sample of rings dating back as far as possible (see figure). First, a small core of wood is bored out of the trunk of a living tree. The oldest rings (toward the center of the tree) are matched with the youngest rings (toward the outside) of an older tree, perhaps a dead one in the forest. The master chronology is extended back in time by overlapping matching ring sequences of successively older and older sections of wood, even those found in prehistoric dwellings. Currently, the longest master chronology—of bristlecone pines in the western United States—goes back almost 9000 years.

Dendrochronology, the study of both visible and microscopic details of tree rings, has been used extensively in several fields. Tree-ring analysis has helped date Native American prehistoric sites in the Southwest. For example, using tree-ring analysis, scientists have determined that the Cliff Palace in Mesa Verde National Park dates back to the year 1073. Tree-ring analysis is also useful in ecology (to study changes in a forest community over time), environmental science (to study the effects of air pollution on tree growth), and geology (to date earthquakes and volcanic eruptions).

Climatologists are increasingly using tree-ring data to study past climate patterns. Annual rings of certain tree species that grow at high elevations are sensitive to yearly temperature variations; the rings of these trees are wider in warm years and narrower in cool years. Studying tree rings over long periods helps researchers determine the natural pattern of global temperature fluctuations. This information is particularly important because of concerns about the human influence on global climate.

Although scientists generally agree that Earth has warmed in recent decades, they are not sure how much of the recent warming is the result of human influence rather than natural climate variability. To help answer this vital question, European scientists are currently developing a 10,000-year master chronology that will help reconstruct annual temperatures across northern Europe and Asia since the end of the last Ice Age.

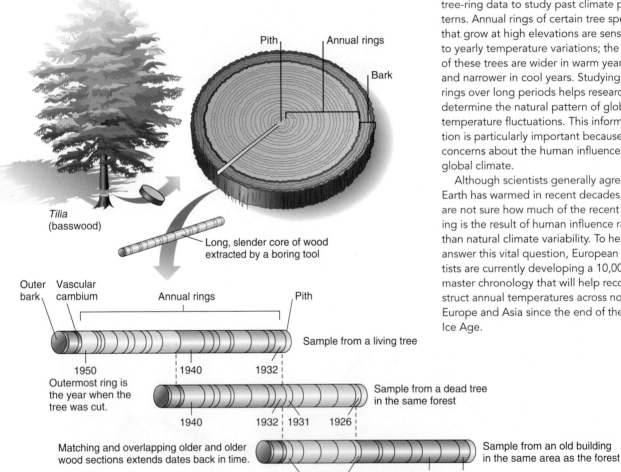

Animated Tree-ring dating

A master chronology is developed using progressively older pieces of wood from the same geographic area. By matching the rings of a wood sample of unknown age to the master chronology, investigators can accurately determine the age of the sample. Ring matching, once a tediously painstaking job, today is usually performed by computers.

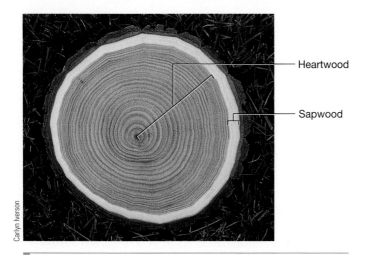

Carlyn Iverson

— Heartwood

— Sapwood

Figure 34-8 *Animated* Heartwood and sapwood

The wood of older trees consists of a dense, central heartwood and an outer layer of sapwood. The sapwood is the functioning xylem that conducts water and dissolved minerals. The annual rings in the heartwood are very conspicuous.

Examination of annual rings with a magnifying lens reveals no actual "ring," or line, separating one year's growth from the next. The appearance of a ring in cross section is due to differences in cell size and cell wall thickness between secondary xylem formed at the end of the preceding year's growth and that formed at the beginning of the following year's growth. In the spring, when water is plentiful, wood formed by vascular cambium has large-diameter conducting cells (tracheids and vessel elements) and few fibers and is appropriately called *springwood* or *early wood*. As summer progresses and water becomes less plentiful, the wood formed, known as *summerwood* or *late wood,* has narrower conducting cells and many fibers. It is this difference in cell size between the summerwood of one year and the springwood of the following year that gives the appearance of rings (❚ Fig. 34-9). A great deal of information about climate in past times can be learned from the study of annual rings of ancient trees (see *Focus On: Tree-Ring Analysis*).

As a woody stem increases in girth over the years, the branches that it bears grow along with it as long as they are alive. If a branch dies, it no longer continues to grow with the stem. In time, as the stem increases in girth, it surrounds the base of the dead branch. The basal portion of an embedded dead branch is called a *knot*. It is possible for a knot to contain bark as well as wood. The presence of knots in wood reduces its commercial value, except for ornamental purposes.

Review

❚ How does the arrangement of vascular tissue differ in primary eudicot stems and monocot stems in cross section?

❚ What is the difference between vascular cambium and cork cambium? What tissues arise from vascular cambium? From cork cambium?

❚ What happens to the primary tissues of a stem when secondary growth occurs?

❚ How do growth rings form in woody stems?

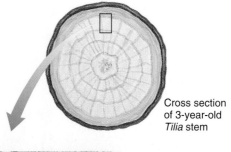

Cross section of 3-year-old *Tilia* stem

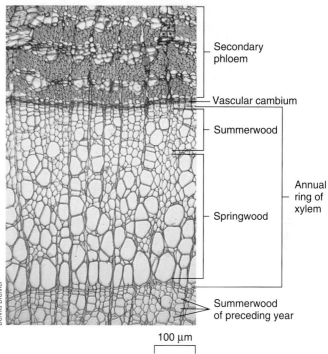

Dennis Drenner

Secondary phloem

Vascular cambium

Summerwood

Springwood

Annual ring of xylem

Summerwood of preceding year

100 µm

Figure 34-9 LM of a portion of a basswood (*Tilia americana*) stem cross section

One annual ring, or growth increment, is shown. Note the differences in cell size between the vessel elements of springwood and summerwood. The pink cells in the secondary phloem are fibers.

TRANSPORT IN THE PLANT BODY

Learning Objectives

5 Describe the pathway of water movement in plants.

6 Define *water potential*.

7 Explain the roles of tension–cohesion and root pressure as mechanisms responsible for the rise of water and dissolved minerals in xylem.

8 Describe the pathway of sugar translocation in plants.

9 Discuss the pressure–flow hypothesis of sugar translocation in phloem.

Now that we have discussed stem structure and primary and secondary growth, we examine internal transport in the vascular system of the plant (❚ Fig. 34-10). Roots obtain water and dissolved minerals from the soil. Once inside roots, these materi-

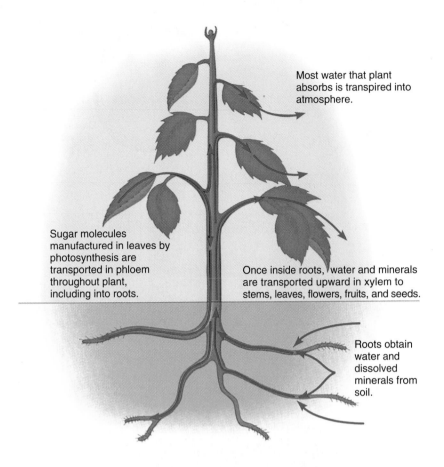

Most water that plant absorbs is transpired into atmosphere.

Sugar molecules manufactured in leaves by photosynthesis are transported in phloem throughout plant, including into roots.

Once inside roots, water and minerals are transported upward in xylem to stems, leaves, flowers, fruits, and seeds.

Roots obtain water and dissolved minerals from soil.

Figure 34-10 An overview of transport in vascular plants
Xylem transport is explained on the right side of the figure (start at the bottom and work upward), whereas phloem transport is on the left side.

als are transported upward to stems, leaves, flowers, fruits, and seeds. Furthermore, sugar molecules are transported from their place of manufacture or of breakdown of starch throughout the plant body, including into the subterranean roots. Water and dissolved minerals are transported from roots to other parts of the plant in xylem, whereas dissolved sugar is **translocated** in phloem.

Xylem transport and phloem translocation do not resemble the movement of materials in animals, because in plants nothing *circulates* in a system of vessels. Water and minerals, transported in xylem, travel in one direction only (upward), whereas translocation of dissolved sugar may occur upward or downward in separate phloem cells. In addition, xylem transport and phloem translocation differ from internal circulation in animals because movement in both xylem and phloem is driven largely by natural physical processes rather than by a pumping organ, or heart.

How do materials travel in the continuous system of the plant's vascular tissues? We first examine water and its movement through the plant, and later we discuss the translocation of dissolved sugar.

Water and minerals are transported in xylem

Water initially moves horizontally into roots from the soil, passing through several tissues until it reaches xylem. Once the water moves into the tracheids and vessel elements of root xylem, it travels upward through a continuous network of these hollow, dead cells from root to stem to leaf. Dissolved minerals are carried along passively in the water. The plant does not expend any energy of its own to transport water, which moves as a result of natural physical processes. The transport of xylem sap is the most rapid of any movement of materials in plants (❙ Table 34-1).

How does water move to the tops of plants? It is either pushed up from the bottom of the plant or pulled up at the top of the plant. Although plants use both mechanisms, current evidence indicates that most water is transported through xylem by being *pulled* to the top of the plant.

Water movement can be explained by a difference in water potential

To understand how water moves, it is helpful to introduce **water potential,** which is defined as the **free energy** of water (see Chapter 7). Water potential is important in plant physiology because it is a measure of a cell's ability to absorb water by **osmosis** (see Chapter 5). Water potential also provides a measure of water's tendency to evaporate from cells.

The water potential of pure water is conventionally set at 0 megapascals (MPa), because it cannot be measured directly. (A megapascal is a unit of pressure equal to about 10 atmospheres, or 145.1 lb/in^2.) However, botanists can measure differences in the free energy of water molecules in different situations. When solutes dissolve in water, the free energy of water decreases. Sol-

TABLE 34-1

Xylem and Phloem Transport Rates in Selected Plants

Plant	Maximum Rate in Xylem (cm/min)	Maximum Rate in Phloem (cm/min)
Conifer	2	0.8
Woody eudicot	73	2
Herbaceous eudicot or monocot	100	2.8–11
Herbaceous vine	250	1.2

Source: Adapted from J. D. Mauseth, *Botany: An Introduction to Plant Biology,* 2nd ed., Philadelphia, Saunders College Publishing, 1995.

Note: Xylem and phloem rates are from different plants within each general group and should be used for comparative purposes only.

Transpiration is the driving force of the tension–cohesion model.

① This model hypothesizes that water vapor transpires from surfaces of leaf mesophyll cells to drier atmosphere through stomata. This produces tension that pulls water out of leaf xylem toward mesophyll cells (MPa = megapascal).

② Cohesion of water molecules, caused by hydrogen bonding, allows unbroken columns of water to be pulled up narrow vessels and tracheids of stem xylem.

③ This in turn pulls water up root xylem, forming continuous column of water from root xylem to stem xylem to leaf xylem. Movement of water upward in root produces pull that causes soil water to diffuse into root.

Leaf vein
Leaf = −1.5 MPa
Stoma
Atmosphere = −80 MPa
Stem = −0.7 MPa
Root xylem
Soil water = −0.1 MPa
Root = −0.4 MPa
Coleus

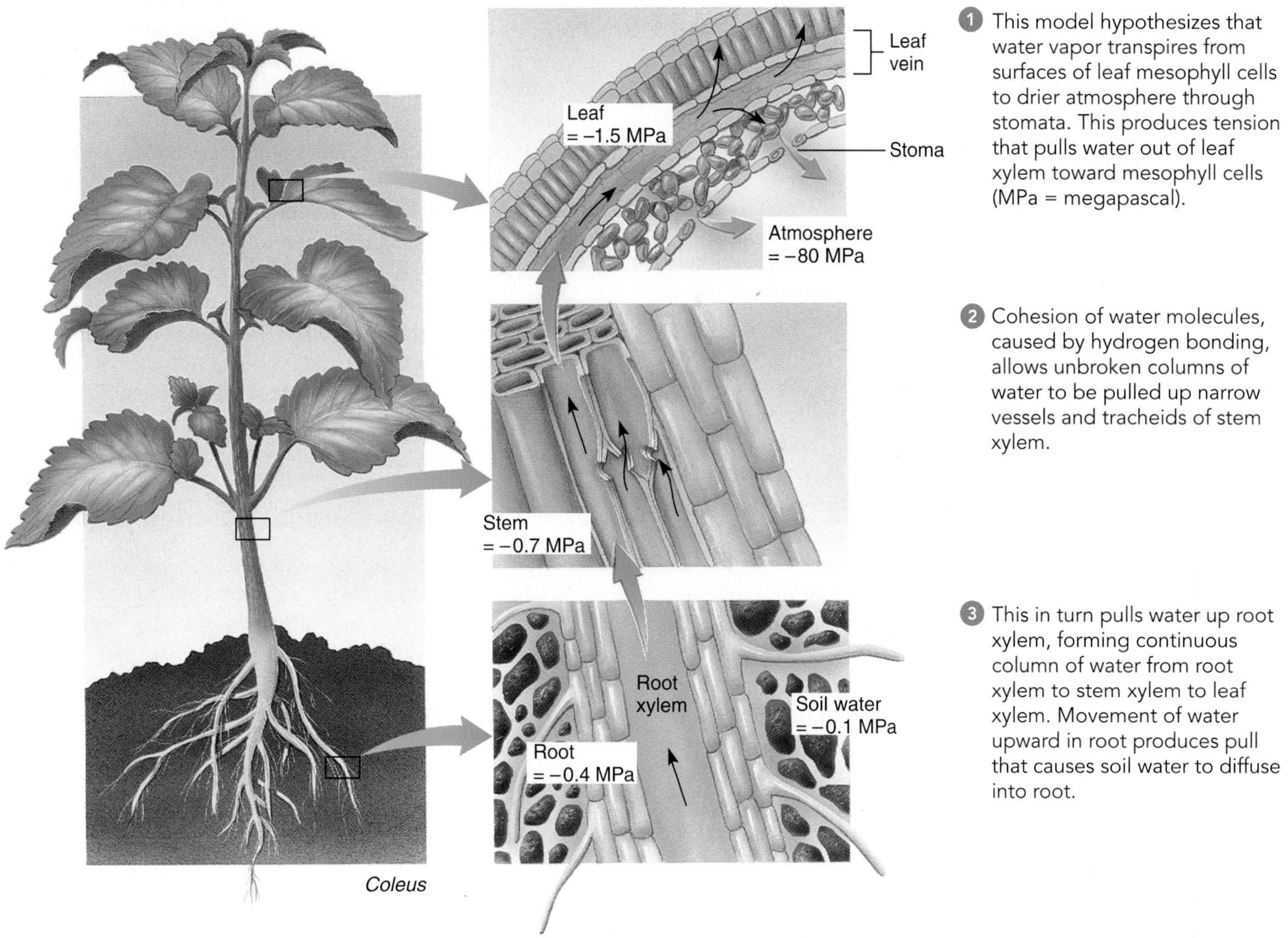

Figure 34-11 *Animated* The tension–cohesion model

utes induce **hydration,** in which water molecules surround ions and polar molecules, keeping them in solution by preventing them from coming together (see Fig. 2-10). The association of water molecules with hydrated molecules and ions reduces the motion of water molecules, decreasing their free energy. Thus, *dissolved solutes lower the water potential to a negative number. Water moves from a region of higher (less negative) water potential to a region of lower (more negative) water potential.*

The water potential of the soil varies, depending on how much water it contains. When a soil is extremely dry, its water potential is very low (very negative). When a soil is moister, its water potential is higher, although it still has a negative value because dissolved minerals are present in dilute concentrations.

The water potential in root cells is also negative because of the presence of dissolved solutes. Roots contain more dissolved materials than soil water does, unless the soil is extremely dry. This means that under normal conditions the water potential of the

root is more negative than the water potential of the soil. Thus, water moves by osmosis from the soil into the root.

According to the tension–cohesion model, water is pulled up a stem

In 1896, Irish botanist Henry Dixon proposed the tension–cohesion model to explain the ascent of water against the force of gravity. Although initially met with widespread skepticism in the scientific community, this model has withstood the test of time. According to the **tension–cohesion model,** also known as the **transpiration–cohesion model,** water is pulled up the plant as a result of a *tension* produced at the top of the plant (▌Fig. 34-11). This tension, which resembles that produced when drinking a liquid through a straw, is caused by the evaporative pull of transpiration. Recall from Chapter 33 that **transpiration** is the evaporation of water vapor from plants.

Most water loss from transpiration takes place through stomata, the numerous microscopic pores present on leaf and stem surfaces. The tension extends from leaves, where most transpiration occurs, down the stems and into the roots. It draws water up stem xylem to leaf cells that have lost water as a result of transpiration and pulls water from root xylem into stem xylem. As water is pulled upward, additional water from the soil is drawn into the roots. Thus, the pathway of water movement is as follows:

Soil ⟶ root tissues (epidermis, cortex, and so forth) ⟶ root xylem ⟶ stem xylem ⟶ leaf xylem ⟶ leaf mesophyll ⟶ atmosphere

This upward pulling of water is possible only as long as there is an unbroken column of water in xylem throughout the plant. Water forms an unbroken column in xylem because of the *cohesiveness* of water molecules. Recall from Chapter 2 that water molecules are **cohesive,** that is, strongly attracted to one another, because of **hydrogen bonding.** In addition, the **adhesion** of water to the walls of xylem cells, also the result of hydrogen bonding, is an important factor in maintaining an unbroken column of water. Thus, the cohesive and adhesive properties of water enable it to form a continuous column that can be pulled up through the xylem.

The movement of water in xylem by the tension–cohesion mechanism can be explained in terms of water potential. The atmosphere has an extremely negative water potential. For example, air with a relative humidity of 50% has a water potential of -100 MPa; even moist air at a relative humidity of 90% has a negative water potential of -13 MPa. Thus, *there is a water potential gradient from the least negative (the soil) up through the plant to the most negative (the atmosphere).* This gradient literally pulls the water from the soil up through the plant.

Although the tension–cohesion model was first proposed toward the end of the 19th century, conclusive experimental evidence to support this mechanism was not obtained until the late 1990s and early 2000s. At that time, several research groups made direct measurements of the large negative pressure that exists in xylem, indicating that the water potential gradients in root, stem, and leaf xylem are adequate to explain the observed movement of water.

Is the tension–cohesion model powerful enough to explain the rise of water in the tallest plants? Plant biologists have studied this question for years. For example, a 2004 study of five of the eight tallest trees in the world (all redwoods, or *Sequoia sempervirens*) concluded that the tension produced by transpiration is strong enough to pull water upward to a maximum height of 130 m (422 ft). Because the height of the tallest known living tree, located in the Humboldt Redwoods State Park in California, is about 113 m (367 ft), the tension–cohesion model easily accounts for the transport of water. Currently, most botanists consider the tension–cohesion model to be the dominant mechanism of xylem transport in most plants.

Root pressure pushes water from the root up a stem

In the less important mechanism for water transport, known as **root pressure,** water that moves into roots from the soil is *pushed* up through xylem toward the top of the plant. Root pressure occurs because mineral ions that are actively absorbed from the soil are pumped into the xylem, decreasing its water potential. This accumulation of ions has an osmotic effect, causing water to move into xylem cells from surrounding root cells. In turn, water moves into roots by osmosis because of the difference in water potential between the soil and root cells. The accumulation of water in root tissues produces a positive pressure (as high as $+0.2$ MPa) that forces the water up through the root xylem into the shoot.

Guttation, a phenomenon in which liquid water is forced out through special openings in the leaves (see Chapter 33), results from root pressure. However, root pressure is not strong enough to explain the rise of water to the tops of coastal redwoods and other tall trees. Root pressure exerts an influence in smaller plants, particularly in the spring when the soil is quite wet, but it clearly does not cause water to rise 100 m (330 ft) or more in the tallest plants. Furthermore, root pressure does not occur to any appreciable extent in summer (when water is often not plentiful in soil), yet water transport is greatest during hot summer days.

Sugar in solution is translocated in phloem

The sugar produced during photosynthesis is converted into sucrose (common table sugar), a disaccharide composed of one molecule of glucose and one of fructose (see Fig. 3-8b), before being loaded into phloem and translocated to the rest of the plant. Sucrose is the predominant photosynthetic product carried in phloem. Phloem sap also contains much smaller amounts of other materials, such as amino acids, organic acids, proteins, hormones, certain minerals, and sometimes disease-causing plant viruses. Translocation of phloem sap is not as rapid as xylem transport (see Table 34-1).

Fluid within phloem tissue moves both upward and downward. Sucrose is translocated in individual sieve tubes from a *source,* an area of excess sugar supply (usually a leaf), to a *sink,* an area of storage (as insoluble starch) or of sugar use, such as roots, apical meristems, fruits, and seeds.

The pressure–flow hypothesis explains translocation in phloem

Current experimental evidence supports the translocation of dissolved sugar in phloem by the **pressure–flow hypothesis,** which was first proposed in 1926 by the German scientist Ernst Münch. The pressure–flow hypothesis states that solutes (such as dissolved sugars) move in phloem by means of a pressure gradient—that is, a difference in pressure. The pressure gradient exists between the source, where the sugar is loaded into phloem, and the sink, where the sugar is removed from phloem.

At the source, the dissolved sucrose is moved from a leaf's mesophyll cells, where it was manufactured, into the companion cells, which load it into the sieve tube elements of phloem. This loading occurs by active transport, a process that requires adenosine triphosphate (ATP) (❙ Fig. 34-12). The ATP supplies energy to pump protons out of the sieve tube elements, producing a proton gradient that drives the uptake of sugar through specific channels by the **cotransport** of protons back into the sieve tube

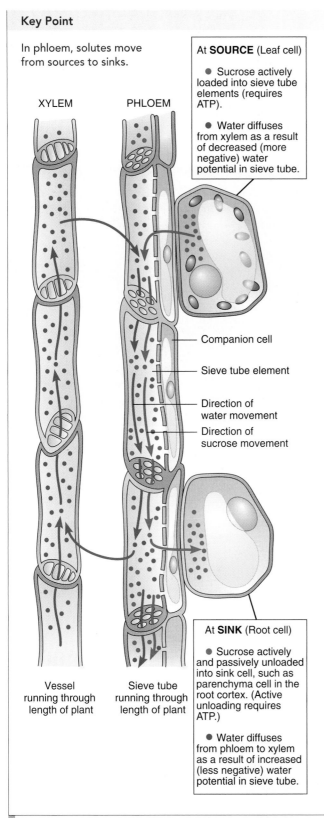

Key Point

In phloem, solutes move from sources to sinks.

XYLEM   PHLOEM

At SOURCE (Leaf cell)

● Sucrose actively loaded into sieve tube elements (requires ATP).

● Water diffuses from xylem as a result of decreased (more negative) water potential in sieve tube.

Companion cell

Sieve tube element

Direction of water movement

Direction of sucrose movement

Vessel running through length of plant

Sieve tube running through length of plant

At SINK (Root cell)

● Sucrose actively and passively unloaded into sink cell, such as parenchyma cell in the root cortex. (Active unloading requires ATP.)

● Water diffuses from phloem to xylem as a result of increased (less negative) water potential in sieve tube.

Figure 34-12 *Animated* The pressure–flow hypothesis

Sugar is actively loaded into the sieve tube element at the source. As a result, water diffuses from the xylem into the sieve tube element. At the sink, the sugar is actively or passively unloaded, and water diffuses from the sieve tube element into the xylem. The pressure gradient within the sieve tube, from source to sink, causes translocation from the area of higher turgor pressure (the source) to the area of lower turgor pressure (the sink).

elements (an example of a linked cotransport system, discussed in Chapter 5). The sugar therefore accumulates in the sieve tube element. The increase in dissolved sugars in the sieve tube element at the source—a concentration that is 2 to 3 times as great as in surrounding cells—decreases (makes more negative) the water potential of that cell. As a result, water moves by osmosis from the xylem cells into the sieve tubes, increasing the **turgor pressure** (hydrostatic pressure) inside them. Thus, phloem loading at the source occurs as follows:

Proton pump moves H^+ out of sieve tube element $\longrightarrow$ sugar is actively transported into sieve tube element $\longrightarrow$ water diffuses from xylem into sieve tube element $\longrightarrow$ turgor pressure increases within sieve tube

At its destination (the sink), sugar is unloaded by various mechanisms, both active and passive, from the sieve tube elements. With the loss of sugar, the water potential in the sieve tube elements at the sink increases (becomes less negative). Therefore, water moves out of the sieve tubes by osmosis and into surrounding cells where the water potential is more negative. Most of this water diffuses back to the xylem to be transported upward. This water movement decreases the turgor pressure inside the sieve tubes at the sink. Thus, phloem unloading at the sink proceeds as follows:

Sugar is transported out of sieve tube element $\longrightarrow$ water diffuses out of sieve tube element and into xylem $\longrightarrow$ turgor pressure decreases within sieve tube

The pressure–flow hypothesis explains the movement of dissolved sugar in phloem by means of a pressure gradient. The difference in sugar concentrations between the source and the sink causes translocation in phloem as water and dissolved sugar flow along the pressure gradient. This pressure gradient pushes the sugar solution through phloem much as water is forced through a hose.

The actual translocation of dissolved sugar in phloem does not require metabolic energy. However, the loading of sugar at the source and the active unloading of sugar at the sink require energy derived from ATP to move the sugar across cell membranes by active transport.

Although the pressure–flow hypothesis adequately explains current data on phloem translocation, much remains to be learned about this complex process. Phloem translocation is difficult to study in plants. Because phloem cells are under pressure, cutting into phloem to observe it releases the pressure and causes the contents of the sieve tube elements (the phloem sap) to exude and mix with the contents of other severed cells that are also unavoidably cut. In the 1950s, scientists developed a unique research tool to avoid contaminating the phloem sap: aphids, which are small insects that insert their mouthparts into phloem sieve tubes for feeding (❚ Fig. 34-13). The pressure in the punctured phloem drives the sugar solution through the aphid's mouthpart into its digestive system. When the aphid's mouthpart is severed from its body by a laser beam, the sugar solution continues to flow through the mouthpart at a rate proportional to the pressure in phloem. This rate can be measured, and the effects on phloem

QUESTION: How can phloem sap be studied without cutting nonphloem cells that would contaminate the sap?

HYPOTHESIS: An aphid mouthpart can be used to penetrate a single sieve tube.

EXPERIMENT: After allowing aphids to insert their mouthparts into phloem of a stem, researchers anesthetized the feeding aphids with CO_2 and used a laser to cut their bodies away from their mouthparts. The mouthparts remained in the phloem and functioned like miniature pipes.

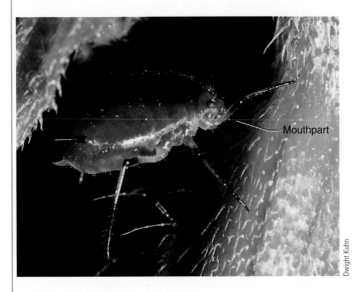

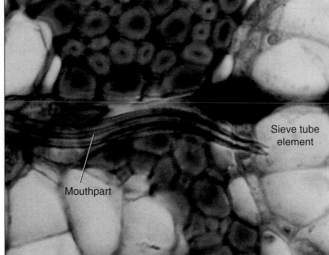

(a) Aphid feeding on a stem.

(b) LM of phloem cells, showing an aphid mouthpart penetrating a sieve tube element.

RESULTS AND CONCLUSION: About 1 mm³ of phloem sap exuded from each severed mouthpart per hour for several days, so researchers were able to collect and analyze the composition of the sap.

Figure 34-13 Collecting and analyzing phloem sap

This method, first used in the 1950s by insect physiologists studying aphids, was quickly adopted by plant physiologists studying phloem translocation.

transport of different environmental conditions—varying light intensities, darkness, and mineral deficiencies, for example—can be ascertained.

The identity and proportions of translocated substances can also be determined using severed aphid mouthparts. This technique has verified that in most plant species the sugar sucrose is the primary carbohydrate transported in phloem; however, some species transport other sugars, such as raffinose, or sugar alcohols, such as sorbitol.

Review

- How does the direction of transport differ in xylem and phloem?
- What is water potential? How is the movement of water related to water potential?
- How does the tension–cohesion model explain the rise of water in the tallest trees?
- How does the pressure–flow hypothesis explain sugar movement in phloem? Include in your answer the activities at source and sink.

SUMMARY WITH KEY TERMS

Learning Objectives

1 Describe the external features of a woody twig (page 732).
- Woody twigs demonstrate the external structure of stems. **Buds** are undeveloped embryonic shoots. A **terminal**

bud is located at the tip of a stem, whereas **axillary buds (lateral buds)** are located in leaf axils. A dormant bud is covered and protected by **bud scales**. When the bud resumes growth, bud scales covering the bud fall off, leaving **bud scale scars.**

- The area on a stem where each leaf is attached is called a **node,** and the region of a stem between two successive nodes is an **internode.** A **leaf scar** shows where each leaf was attached to the stem. **Bundle scars** are the areas within a leaf scar where the vascular tissue extended from the stem to the leaf. **Lenticels** are sites of loosely arranged cells that allow oxygen to diffuse into the interior of a woody stem.

2 Label cross sections of herbaceous eudicot and monocot stems, and describe the functions of each tissue (page 732).

- Herbaceous stems possess an epidermis, vascular tissue, and either ground tissue or cortex and pith. The **epidermis** is a protective layer covered by a water-conserving **cuticle. Stomata** permit gas exchange. **Xylem** conducts water and dissolved minerals, and **phloem** conducts dissolved sugar. The **cortex, pith,** and ground tissue function primarily for storage.

- All herbaceous stems have the same basic tissues, but their arrangement varies. Herbaceous eudicot stems have the vascular bundles arranged in a circle (in cross section) and have a distinct cortex and pith. Monocot stems have vascular bundles scattered in ground tissue.

ThomsonNOW™ **Learn more about eudicot and monocot stems by clicking on the figures in ThomsonNOW.**

3 Name the two lateral meristems, and describe the tissues that arise from each (page 732).

- **Vascular cambium** is the lateral meristem that produces secondary xylem (wood) and secondary phloem (inner bark).

- **Cork cambium** produces **periderm,** which consists of cork parenchyma and cork cells. Cork cells are the functional replacement for epidermis in a woody stem. Cork parenchyma functions primarily for storage in a woody stem.

4 Outline the transition from primary growth to secondary growth in a woody stem (page 732).

- **Secondary growth** (the production of the secondary tissues, wood and bark) occurs in some flowering plants (woody eudicots) and in all cone-bearing gymnosperms. During secondary growth, the vascular cambium divides in two directions to form secondary xylem (to the inside) and secondary phloem (to the outside). The primary xylem and primary phloem in the original vascular bundles become separated as secondary growth proceeds.

ThomsonNOW™ **Learn more about secondary growth by clicking on the figures in ThomsonNOW.**

5 Describe the pathway of water movement in plants (page 739).

- Water and dissolved minerals move from the soil into root tissues (epidermis, cortex, and so forth). Once in root xylem, water and minerals move upward, from root xylem to stem xylem to leaf xylem. Much of the water entering the leaf exits leaf veins and passes into the atmosphere.

6 Define *water potential* (page 739).

- **Water potential** is a measure of the **free energy** of water. Pure water has a water potential of 0 megapascals, whereas water with dissolved solutes has a negative water potential. Water moves from an area of higher (less negative) water potential to an area of lower (more negative) water potential.

7 Explain the roles of tension–cohesion and root pressure as mechanisms responsible for the rise of water and dissolved minerals in xylem (page 739).

- The **tension–cohesion model** explains the rise of water in even the tallest plants. The evaporative pull of **transpiration** causes tension at the top of the plant. This tension is the result of a water potential gradient that ranges from the slightly negative water potentials in the soil and roots to the very negative water potentials in the atmosphere. As a result of the **cohesive** and **adhesive** properties of water, the column of water pulled up through the plant remains unbroken.

ThomsonNOW™ **Learn more about the tension–cohesion model by clicking on the figure in ThomsonNOW.**

- **Root pressure,** caused by the movement of water into roots from the soil as a result of the active absorption of mineral ions from the soil, helps explain the rise of water in smaller plants, particularly when the soil is wet. Root pressure pushes water up through xylem.

8 Describe the pathway of sugar translocation in plants (page 739).

- Dissolved sugar is **translocated** upward or downward in phloem, from a source (an area of excess sugar, usually a leaf) to a sink (an area of storage or of sugar use, such as roots, apical meristems, fruits, and seeds). Sucrose is the predominant sugar translocated in phloem.

9 Discuss the pressure–flow hypothesis of sugar translocation in phloem (page 739).

- Movement of materials in phloem is explained by the **pressure–flow hypothesis.** Companion cells actively load sugar into the sieve tubes at the source; ATP is required for this process. The ATP supplies energy to pump protons out of the sieve tube elements. The proton gradient drives the uptake of sugar by the cotransport of protons back into the sieve tube elements. Sugar therefore accumulates in the sieve tube element, causing the movement of water into the sieve tubes by **osmosis.**

ThomsonNOW™ **Learn more about how sugar travels by clicking on the figure in ThomsonNOW.**

- Companion cells actively (requiring ATP) and passively (not requiring ATP) unload sugar from the sieve tubes at the sink. As a result, water leaves the sieve tubes by osmosis, decreasing the **turgor pressure** (hydrostatic pressure) inside the sieve tubes.

- The flow of materials between source and sink is driven by the turgor pressure gradient produced by water entering phloem at the source and water leaving phloem at the sink.

TEST YOUR UNDERSTANDING

1. The three main functions of stems are (a) support, conduction, and photosynthesis (b) support, anchorage in soil, and production of new living tissues (c) conduction, production of new living tissues, and sexual reproduction (d) conduction, asexual reproduction, and sexual reproduction (e) support, conduction, and production of new living tissues

2. Stems have undeveloped embryonic shoots called (a) lenticels (b) buds (c) vines (d) phloem fiber caps (e) periderm

3. Axillary buds are located (a) at the tips of stems (b) in unusual places, such as on roots (c) in the region between two successive nodes (d) in the upper angle between a leaf and the stem to which it is attached (e) within the loosely arranged cells of lenticels

4. The protective outer layer of cells covering herbaceous stems is the (a) periderm (b) cork cambium (c) lateral meristem (d) epidermis (e) bud scale

5. Ground tissue in monocot stems performs the same functions as _____ and _____ in herbaceous eudicot stems. (a) phloem; xylem (b) cork cambium; vascular cambium (c) epidermis; periderm (d) primary xylem; secondary xylem (e) cortex; pith

6. Which of the following statements is *false?* (a) primary growth is an increase in the length of a plant (b) primary growth occurs at both apical and lateral meristems (c) all plants have primary growth (d) herbaceous stems have primary growth, whereas woody stems have both primary and secondary growth (e) buds are embryonic shoots that contain apical meristems

7. The two lateral meristems responsible for secondary growth are (a) phloem and xylem (b) cork cambium and vascular cambium (c) epidermis and periderm (d) primary xylem and secondary xylem (e) cortex and pith

8. Cork cambium and the tissues it produces are collectively called (a) periderm (b) lenticels (c) cortex (d) epidermis (e) wood

9. Horizontal movement of materials in woody plants occurs in (a) bud scales (b) cortex (c) rays (d) lenticels (e) pith rays

10. The older, darker wood in the center of a tree trunk is commonly called (a) hardwood (b) softwood (c) sapwood (d) heartwood (e) cork

11. Each annual ring in a section of wood represents 1 year's growth of (a) primary xylem (b) secondary xylem (c) primary xylem or secondary xylem in alternate years (d) primary phloem (e) secondary phloem

12. Water potential is (a) the formation of a proton gradient across a cell membrane (b) the transport of a watery solution of sugar in phloem (c) the transport of water in both xylem and phloem (d) the removal of sucrose at the sink, causing water to move out of the sieve tubes (e) the free energy of water in a particular situation

13. Which of the following is a mechanism of water movement in xylem that is responsible for guttation: (a) pressure–flow hypothesis (b) tension–cohesion (c) root pressure (d) active transport of potassium ions into guard cells (e) transpiration

14. Which of the following is a mechanism of water movement in xylem that combines the evaporative pull of transpiration with the cohesive and adhesive properties of water? (a) pressure–flow (b) tension–cohesion (c) root pressure (d) active transport of potassium ions into guard cells (e) guttation

15. Which of the following is a mechanism of phloem transport in which dissolved sugar is moved by means of a pressure gradient that exists between the source and the sink? (a) pressure–flow (b) tension–cohesion (c) root pressure (d) active transport of potassium ions into guard cells (e) guttation

16. How does increasing solute concentration affect water potential? (a) water potential becomes more positive (b) water potential becomes more negative (c) water potential becomes more positive under certain conditions and more negative under other conditions (d) water potential is not affected by solute concentration (e) water potential is always zero when solutes are dissolved in water

17. Label the various tissues, give at least one function for each tissue, and identify the stem as a herbaceous eudicot, monocot, or woody plant. Use Figure 34-2 to check your answers.

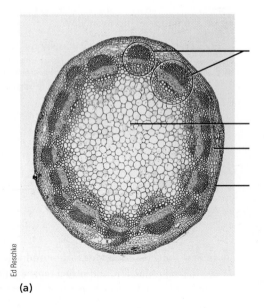

(a)

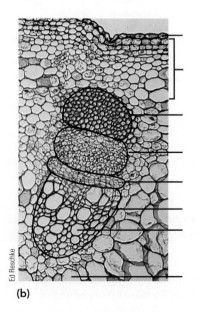

(b)

1. When secondary growth is initiated, certain cells become meristematic and begin to divide. Could a mature tracheid ever do this? A sieve tube element? Why or why not?

2. Why does the wood of many tropical trees lack annual rings? Why does the wood of other tropical trees possess annual rings?

3. Why is hardwood more desirable than softwood for making furniture? Explain your answer, based on the structural differences between hardwood and softwood.

4. Why should you cut off a few inches from the stem ends of cut flowers before placing them in water? Base your answer on what you have learned about the tension–cohesion model.

5. When a strip of bark is peeled off a tree branch, what tissues are usually removed?

6. Could a tree grow to a height of 150 m (488 ft)? Why or why not?

7. **Evolution Link.** Like stems in general, some vines are herbaceous and others are woody. Tropical rain forests have a greater diversity of vines than in any other environment on Earth, and most of these vines are woody. Develop a hypothesis to explain why natural selection has favored the evolution of more species of woody vines (as opposed to herbaceous vines) in tropical rain forests.

Additional questions are available in ThomsonNOW at www.thomsonedu.com/login

Roots and Mineral Nutrition

R. Calentine/Visuals Unlimited

Storage roots. Carrots (*Daucus carota*) are biennials (plants that live for 2 years). During the first year's growth, food is stored in the fleshy root system; during the second year, the shoot elongates and produces flowers. Carrots and other root crops are important sources of human food.

Branching underground root systems are often more extensive than a plant's aerial parts. The roots of a corn plant, for example, may grow to a depth of 2.5 m (about 8 ft) and spread outward 1.2 m (4 ft) from the stem. Desert-dwelling tamarisk (*Tamarix*) trees reportedly have roots that grow to a depth of 50 m (163 ft) to tap underground water. The extent of a plant's root depth and spread varies considerably among different species and even among different individuals in the same species.

Because roots are usually underground and out of sight, people do not always appreciate the important functions they perform. First, as anyone who has ever pulled weeds can attest, roots anchor a plant securely in the soil. A plant needs a solid foundation from which to grow. Firm anchorage is essential to a plant's survival so that the stem remains upright, enabling leaves to absorb sunlight effectively. Second, roots absorb water and dissolved minerals (inorganic nutrients) such as nitrates, phosphates, and sulfates, which are necessary for synthesizing important organic molecules. These dissolved minerals are then transported throughout the plant in the xylem.

Many roots perform the function of storage. Carrots (see photograph), sweet potatoes, cassava, and other root crops are important sources of human food. Surplus sugars produced in the leaves by photosynthesis are transported in the phloem to the roots for food storage (usually as starch or sucrose) until needed. Other plants, particularly those living in arid regions, possess storage roots adapted to store water.

In certain species, roots are modified for functions other than anchorage, absorption, conduction, and storage. Roots specialized to perform uncommon functions are discussed later in the chapter. ∎

KEY CONCEPTS

Roots anchor the plant in the soil, absorb water and minerals, conduct these materials to the rest of the plant body, and store carbon compounds and water.

Primary tissues (epidermis, cortex, endodermis, pericycle, xylem, and phloem) of roots develop from root apical meristems. Secondary tissues (wood and bark) of woody roots develop from lateral meristems.

The endodermis absorbs mineral ions from the soil solution by an active, energy-requiring process.

Many roots form associations with soil bacteria and fungi.

Soil, the layer of Earth's crust that has been modified by contact with weather, wind, water, and organisms, contains most of the essential elements required by plants.

ROOT STRUCTURE AND FUNCTION

Learning Objectives

1   Distinguish between taproot and fibrous root systems.
2   Label cross sections of a primary eudicot root and a monocot root, and describe the functions of each tissue.
3   Trace the pathway of water and mineral ions from the soil through the various root tissues, and distinguish between the symplast and apoplast.
4   Discuss the structure of roots with secondary growth.
5   Describe at least three roots that are modified to perform uncommon functions.

Two types of root systems, a taproot system and a fibrous root system, occur in plants (❙ Fig. 35-1). A **taproot** system consists of one main root that formed from the seedling's enlarging **radicle,** or embryonic root. Many lateral roots of various sizes branch out of a taproot. Taproots are characteristic of many eudicots and gymnosperms. A dandelion is a good example of a common herbaceous plant with a taproot system. A few trees, such as hickory, retain their taproots, which become quite massive as the plants age. Most trees, however, have taproots when young and later develop large, shallow, lateral roots from which other roots branch off and grow downward.

A **fibrous root** system has several to many roots of similar size developing from the end of the stem, with lateral roots branching off these roots. Fibrous root systems form in plants that have a short-lived embryonic root. The roots originate first from the base of the embryonic root and later from stem tissue. The main roots of a fibrous root system do not arise from pre-existing roots but from the stem; such roots are called **adventitious roots.** Adventitious organs occur in an unusual location, such as roots that develop on a stem, or buds that develop on roots. Onions, crabgrass, and other monocots have fibrous root systems.

Taproot and fibrous root systems are adapted to obtain water in different sections of the soil. Taproot systems often extend down into the soil to obtain water located deep underground, whereas fibrous root systems, which are located relatively close to the soil surface, are adapted to obtain rainwater from a larger area as it drains into the soil.

Roots have root caps and root hairs

Because of the need to adapt to the soil environment instead of the atmospheric environment, roots have several structures, such as root caps and root hairs, that shoots lack. Although stems and leaves have various types of hairs, they are distinct from root hairs in structure and function.

Each root tip is covered by a **root cap,** a protective, thimble-like layer many cells thick that covers the delicate root **apical meristem** (❙ Fig. 35-2a; also see Fig. 32-7). As the root grows, pushing its way through the soil, cells of the root cap are sloughed off by the frictional resistance of the soil particles and replaced by new cells formed by the root apical meristem. The root cap cells secrete lubricating polysaccharides that reduce friction as the root passes through the soil. The root cap also appears to be involved in orienting the root so that it grows downward (see discussion of gravitropism in Chapter 37). When a root cap is removed, the root apical meristem grows a new cap. However, until the root cap has regenerated, the root grows randomly rather than in the direction of gravity.

Root hairs are short-lived tubular extensions of epidermal cells located just behind the growing root tip. Root hairs continually form in the area of cell maturation closest to the root tip to replace those that are dying off at the more mature end of the root-hair zone (❙ Fig. 35-2b; also see Fig. 32-7). Each root hair is short (typically less than 1 cm, or 0.4 in, in length), but they are quite numerous. Root hairs greatly increase the absorptive capacity of roots by increasing their surface area in contact with moist soil. Soil particles are coated with a microscopically thin layer of water in which minerals are dissolved. The root hairs establish an intimate contact with soil particles, which allows efficient absorption of water and minerals.

Unlike stems, roots lack nodes and internodes and do not usually produce leaves or buds. Although herbaceous roots have certain primary tissues (such as epidermis, xylem, phloem, cortex, and pith) found in herbaceous stems, these tissues are arranged quite differently.

The arrangement of vascular tissues distinguishes the roots of herbaceous eudicots and monocots

Although considerable variation exists in herbaceous eudicot and monocot roots, they all have an outer protective covering (epidermis), a cortex for storage of starch and other organic mol-

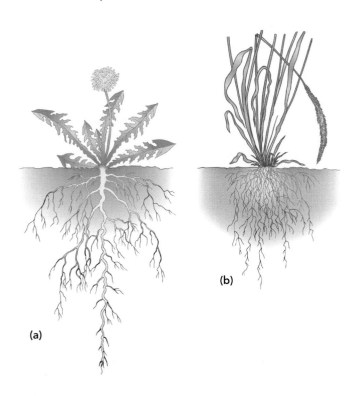

(b)

(a)

❙ **Figure 35-1** *Animated* Root systems

(a) A taproot system develops from the embryonic root in the seed. **(b)** The roots of a fibrous root system are adventitious and develop from stem tissue.

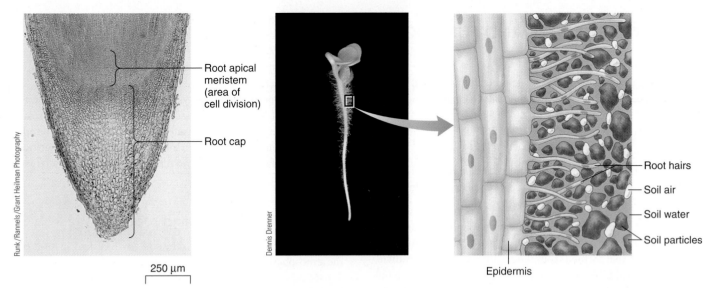

(a) Root cap. LM of an oak (*Quercus* sp.) root tip showing its root cap. The root apical meristem is protected by the root cap.

250 µm

Labels on image (a): Root apical meristem (area of cell division); Root cap

Photo credit: Runk/Rannels/Grant Heilman Photography

(b) Root hairs. Each delicate hair is an extension of a single cell of the root epidermis. Root hairs increase the surface area of the root in contact with the soil. This radish (*Raphanus sativus*) seedling is approximately 5 cm (2 in) long.

Labels on image (b): Root hairs; Soil air; Soil water; Soil particles; Epidermis

Photo credit: Dennis Drenner

Figure 35-2 *Animated* Structures unique to roots

---

ecules, and vascular tissues for conduction. Let us first consider the structure of herbaceous eudicot roots.

In most herbaceous eudicot roots, the central core of vascular tissue lacks pith

The buttercup root is a representative eudicot root with primary growth (❚ Fig. 35-3). Like other parts of this herbaceous eudicot, a single layer of protective tissue, the **epidermis,** covers its roots. The root hairs are a modification of the root epidermis that enables it to absorb more water from the soil. The root epidermis does not secrete a thick, waxy cuticle in the region of root hairs, because this layer would impede the absorption of water from the soil. Both the lack of a cuticle and the presence of root hairs increase absorption.

Beginning with the root hairs, most of the water that enters the root moves along the cell walls rather than enters the cells. One of the major components of cell walls is cellulose, which absorbs water as a sponge does. An example of the absorptive properties of cellulose is found in cotton balls, which are almost pure cellulose.

The **cortex,** which is composed primarily of loosely packed **parenchyma** cells, composes the bulk of a herbaceous eudicot root. Roots usually lack supporting **collenchyma** cells, probably because the soil supports the root, although roots as they age may develop some **sclerenchyma** (another supporting tissue; see Chapter 32). The primary function of the root cortex is storage. A microscopic examination of the parenchyma cells that form the cortex often reveals numerous amyloplasts (see Figs. 35-3b and 3-9a), which store starch. Starch, an insoluble carbohydrate composed of glucose subunits, is the most common form of stored

energy in plants. When used at a later time, these reserves provide energy for such activities as growth and cell replacement following an injury.

The large intercellular (between-cell) spaces, a common feature of the root cortex, provide a pathway for water uptake and allow for aeration of the root. The oxygen that root cells need for aerobic respiration diffuses from air spaces in the soil into the intercellular spaces of the cortex and from there into the cells of the root.

The water and dissolved minerals that enter the root cortex from the epidermis move in solution along two pathways: the symplast and apoplast (❚ Fig. 35-4). The **symplast** is the continuum of living cytoplasm, which is connected from one cell to the next by cytoplasmic bridges called **plasmodesmata** (see Fig. 5-27). Some dissolved mineral ions move from the epidermis through the cortex via the symplast. The **apoplast** consists of the interconnected porous cell walls of a plant, along which water and mineral ions move freely. The water and mineral ions can diffuse across the cortex without ever crossing a plasma membrane to enter a living cell.

The inner layer of the cortex, the **endodermis,** regulates the movement of water and minerals that enter the xylem in the root's center. Structurally, the endodermis differs from the rest of the cortex. Endodermal cells fit snugly against each other, and each has a special bandlike region, called a **Casparian strip** (❚ Fig. 35-5), on its radial (side) and transverse (upper and lower) walls. Casparian strips contain *suberin,* a fatty material that is waterproof. (Recall from Chapter 34 that suberin is also the waterproof material in cork cell walls.)

Until the endodermis is reached, most of the water and dissolved minerals have traveled along the apoplast and therefore

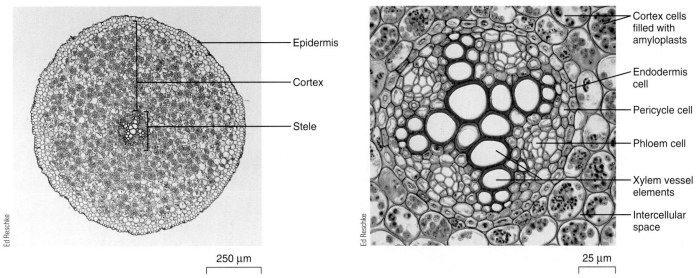

(a) Eudicot root. Cortex composes the bulk of herbaceous eudicot roots. Note the X-shaped xylem in the center of the root.

Epidermis

Cortex

Stele

250 µm

Ed Reschke

(b) Stele of eudicot root. Surrounding the solid core of vascular tissues is a single layer of pericycle, which is meristematic in growing roots.

Cortex cells filled with amyloplasts

Endodermis cell

Pericycle cell

Phloem cell

Xylem vessel elements

Intercellular space

25 µm

Ed Reschke

Figure 35-3 *Animated* LMs of cross sections of a herbaceous eudicot root

Shown is a buttercup (*Ranunculus*) root.

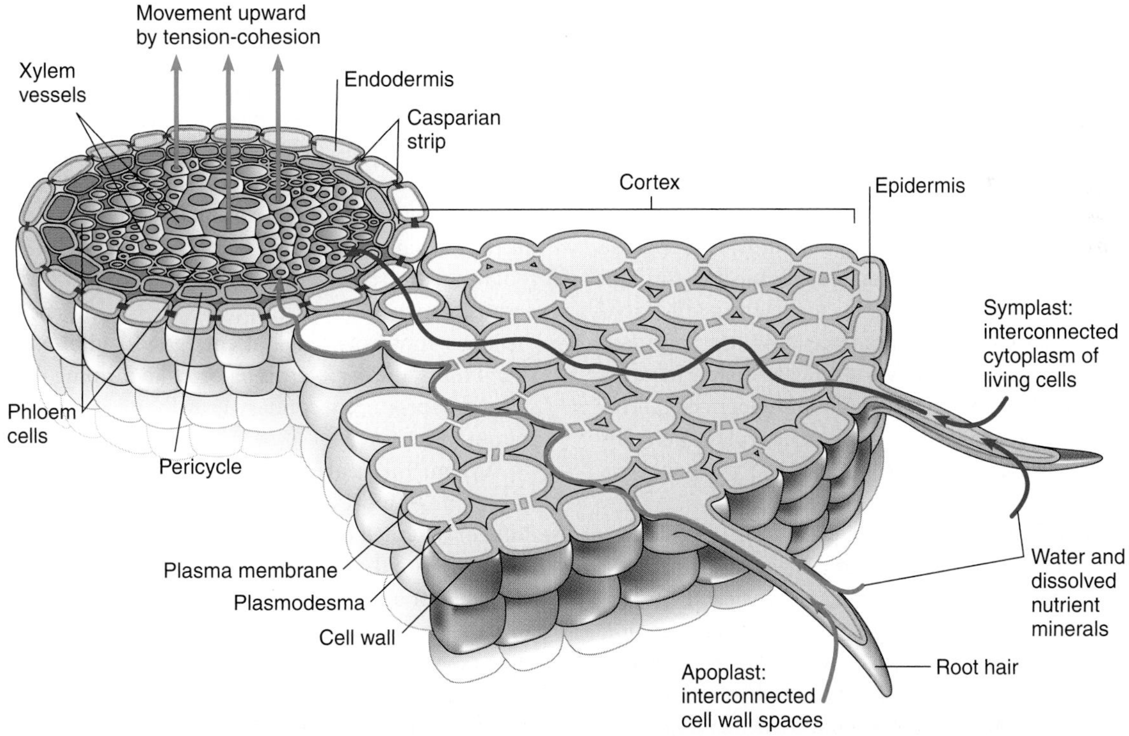

Movement upward by tension-cohesion

Xylem vessels

Endodermis

Casparian strip

Cortex

Epidermis

Symplast: interconnected cytoplasm of living cells

Phloem cells

Pericycle

Plasma membrane

Plasmodesma

Cell wall

Apoplast: interconnected cell wall spaces

Water and dissolved nutrient minerals

Root hair

Figure 35-4 *Animated* Pathways of water and dissolved minerals in the root

Water and dissolved minerals travel from cell to cell along the interconnected porous cell walls (the apoplast) or from one cell's cytoplasm to another through plasmodesmata (the symplast). On reaching the endodermis, water and minerals can continue to move into the root's center only if they pass through a plasma membrane and enter the cytoplasm of an endodermal cell. The Casparian strip blocks the passage of water and minerals along the cell walls between adjoining endodermal cells.

The endodermis regulates the kinds of minerals that are absorbed from the soil and conducted to the rest of the plant body.

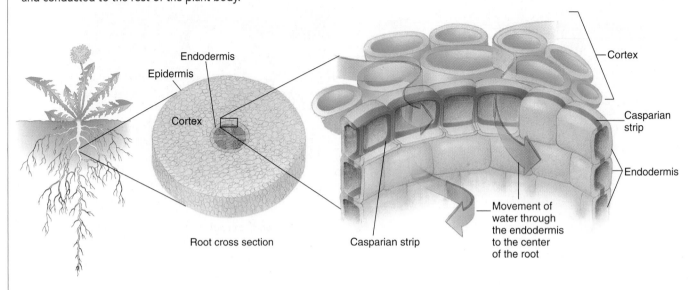

Root cross section

Casparian strip

Movement of water through the endodermis to the center of the root

Cortex

Casparian strip

Endodermis

Endodermis

Epidermis

Cortex

Figure 35-5 *Animated* Endodermis and mineral uptake

Note the Casparian strip around the radial and transverse walls that prevents water and dissolved minerals from passing into the stele along endodermal cell walls. To reach the vascular tissues, water and dissolved minerals must pass through the plasma membranes of endodermal cells.

have not passed through a plasma membrane or entered the cytoplasm of a root cell. However, the waterproof Casparian strip on the radial and transverse walls of the endodermal cells prevents water and minerals from continuing to move passively along the cell walls. For substances to pass farther into the root interior, they must move from the cell walls into the cytoplasm of the endodermal cells. Water enters by osmosis, both directly across the lipid bilayer of the endodermal plasma membrane and through **aquaporins,** integral membrane proteins that facilitate the rapid transport of water across the membrane. As in water movement across the lipid bilayer, water moves through aquaporins only along the osmotic gradient, from a region of higher concentration to a region of lower concentration. The key role of aquaporins in the uptake of water by root cells is currently under active study.

Minerals enter the endodermal cells by passing through carrier proteins in their plasma membranes. In **carrier-mediated active transport,** the mineral ions are pumped *against* their concentration gradient—that is, from an area of *low* concentration of that mineral in the soil solution to an area of *high* concentration in the plant's cells. One of many reasons why root cells require sugar and oxygen for aerobic respiration is that this active transport requires the expenditure of energy, usually in the form of ATP. From the endodermis, water and mineral ions enter the root xylem (botanists do not know precisely how this is done) and are conducted to the rest of the plant.

At the center of a eudicot primary root is the **stele,** or **vascular cylinder,** a central cylinder of vascular tissues (see Fig. 35-3a).

The outermost layer of the stele is the **pericycle,** which is just inside the endodermis. The pericycle consists of a single layer of parenchyma cells that give rise to multicellular *lateral roots,* also called branch roots (❙ Fig. 35-6). Lateral roots originate when cells in a portion of the pericycle start dividing. As it grows, the lateral root pushes through several layers of root tissue (endodermis, cortex, and epidermis) before entering the soil. Each lateral root has all the structures and features—root cap, root hairs, epidermis, cortex, endodermis, pericycle, xylem, and phloem—of the larger root from which it emerges. In addition to producing lateral roots, the pericycle is involved in forming the lateral meristems that produce secondary growth in woody roots (discussed later in the chapter).

Xylem, the centermost tissue of the stele, often has two, three, four, or more extensions, or "xylem arms" (see Fig. 35-3b). **Phloem** is located in patches between the xylem arms. The xylem and phloem of the root have the same functions and kinds of cells as in the rest of the plant: Water and dissolved minerals are conducted in **tracheids** and **vessel elements** of xylem, and dissolved sugar (sucrose) is conducted in **sieve tube elements** of phloem.

After passing through the endodermal cells, water enters the root xylem, often at one of the xylem arms. Up to this point the pathway of water has been horizontal from the soil into the center of the root:

Root hair/epidermis ⟶ cortex ⟶ endodermis ⟶
pericycle ⟶ root xylem

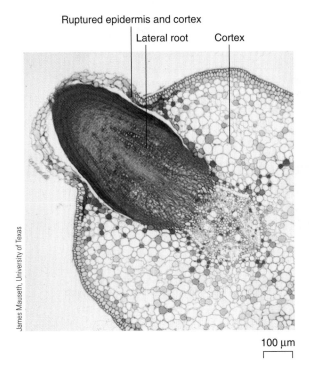

Ruptured epidermis and cortex
Lateral root   Cortex

100 μm

James Mauseth, University of Texas

Figure 35-6   LM of a lateral root

Lateral roots originate at the pericycle.

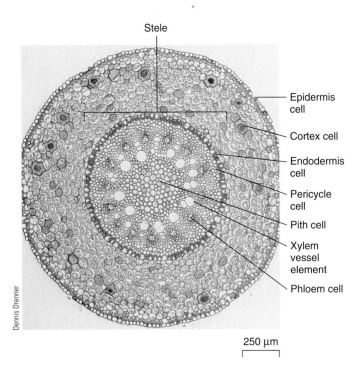

Stele

Epidermis cell

Cortex cell

Endodermis cell

Pericycle cell

Pith cell

Xylem vessel element

Phloem cell

250 μm

Dennis Drenner

Figure 35-7   LM of a cross section of a monocot root

Shown is a greenbrier (*Smilax*) root. As in herbaceous eudicot roots, the cortex of a monocot root is extensive.

Once water enters the xylem, it is transported upward through root xylem into stem xylem and from there to the rest of the plant.

One direction of phloem conduction is from the leaves, where sugar is made by photosynthesis, to the root, where sugar is used for the growth and maintenance of root tissues or stored, usually as starch. Another direction of phloem conduction is from the root, where sugar is stored as starch, to other parts of the plant, where sugar is used for growth and maintenance of tissues.

The *vascular cambium,* which gives rise to secondary tissues in woody plants, is sandwiched between the xylem and phloem. Because it has an inner core of vascular tissue, the primary eudicot root lacks **pith,** a ground tissue found in the centers of many stems and roots.

Xylem does not form the central tissue in some monocot roots

Monocot roots vary considerably in internal structure, compared with eudicot roots. The layers found in some monocot roots, starting from the outside, are epidermis, cortex, endodermis, and pericycle (▌Fig. 35-7). Unlike the xylem in herbaceous eudicot roots, the xylem in many monocot roots does not form a solid cylinder in the center. Instead, the phloem and xylem are in separate, alternating bundles arranged around the central pith, which consists of parenchyma cells.

Because virtually no monocots have secondary growth, no vascular cambium exists in monocot roots. Despite their lack of secondary growth, long-lived monocots, such as palms, may have

thickened roots produced by a modified form of primary growth in which parenchyma cells in the cortex divide and enlarge.

Woody plants have roots with secondary growth

Plants that produce stems with secondary growth also produce roots with secondary growth. These plants—gymnosperms and woody eudicots—have primary growth at apical meristems and secondary growth at lateral meristems. The production of secondary tissues occurs some distance back from the root tips and results from the activity of the same two lateral meristems found in woody stems: the vascular cambium and the cork cambium. Major roots of trees are often massive and have both wood and bark. In temperate climates, the wood of both roots and stems exhibits annual rings in cross section.

Before secondary growth starts in a root, the **vascular cambium** is sandwiched between the primary xylem and the primary phloem (▌Fig. 35-8 **1**). At the onset of secondary growth, the vascular cambium extends out to the pericycle, which develops into vascular cambium opposite the xylem arms. As a result, the pericycle links the separate sections of vascular cambium so that the vascular cambium becomes a continuous, noncircular loop of cells in cross section (▌Fig. 35-8 **2**). As the vascular cambium divides to produce secondary tissues, it eventually forms a cylinder of vascular cambium that continues to divide, producing secondary xylem (wood) to the inside and secondary phloem (inner bark) to the outside (▌Fig. 35-8 **3** and **4**). The root increases in

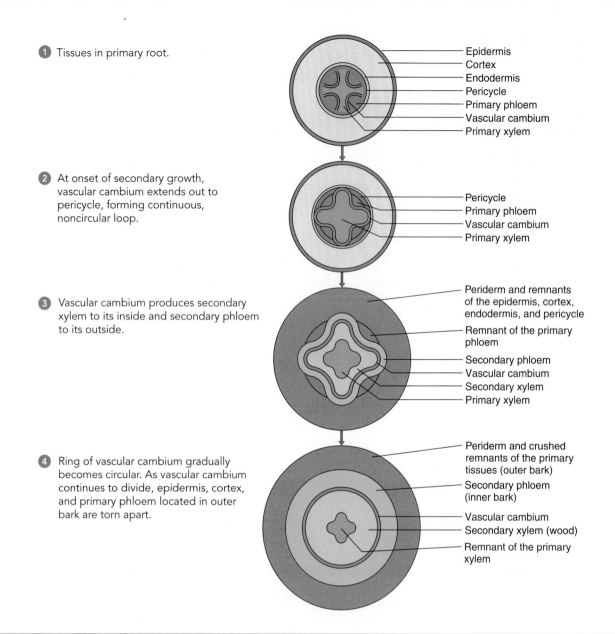

1 Tissues in primary root.

Epidermis
Cortex
Endodermis
Pericycle
Primary phloem
Vascular cambium
Primary xylem

2 At onset of secondary growth, vascular cambium extends out to pericycle, forming continuous, noncircular loop.

Pericycle
Primary phloem
Vascular cambium
Primary xylem

3 Vascular cambium produces secondary xylem to its inside and secondary phloem to its outside.

Periderm and remnants of the epidermis, cortex, endodermis, and pericycle
Remnant of the primary phloem
Secondary phloem
Vascular cambium
Secondary xylem
Primary xylem

4 Ring of vascular cambium gradually becomes circular. As vascular cambium continues to divide, epidermis, cortex, and primary phloem located in outer bark are torn apart.

Periderm and crushed remnants of the primary tissues (outer bark)
Secondary phloem (inner bark)
Vascular cambium
Secondary xylem (wood)
Remnant of the primary xylem

Figure 35-8 Development of secondary vascular tissues in a primary root

The figures are not drawn to scale because of space limitations; the primary xylem is actually the same size in all four diagrams, but differences in scale make the xylem appear different sizes.

girth (thickness), and the vascular cambium continues to move outward.

The epidermis, cortex, endodermis, and primary phloem are gradually torn apart as the root increases in girth. The root epidermis is replaced by **periderm,** composed of cork cells and cork parenchyma, both produced by the **cork cambium** (the last figure in Table 32-4 shows an LM of periderm). The cork cambium in the root initially arises from regions in the pericycle.

Some roots are specialized for unusual functions

Adventitious roots often arise from the nodes of stems. Many aerial adventitious roots are adapted for functions other than anchorage, absorption, conduction, or storage. **Prop roots** are

adventitious roots that develop from branches or a vertical stem and grow downward into the soil to help support the plant in an upright position (❙ Fig. 35-9a). Prop roots are more common in monocots than in eudicots. Corn and sorghum, both monocots, are herbaceous plants that produce prop roots. Many tropical and subtropical eudicot trees, such as red mangrove and banyan, also produce prop roots.

The roots of many tropical rainforest trees are shallow and concentrated near the surface in a mat only a few centimeters (an inch or so) thick. The root mat catches and absorbs almost all minerals released from leaves by decomposition. Swollen bases or braces called **buttress roots** hold the trees upright and aid in the extensive distribution of the shallow roots (❙ Fig. 35-9b).

In swampy or tidal environments where the soil is flooded or waterlogged, some roots grow upward until they are above

Linda R. Berg

(a) Prop roots are adventitious roots that arise near the base of the stem and provide additional support. Screw pine (*Pandanus*) has an elaborate set of aerial prop roots. Photographed in Kauai, Hawaii.

Prop roots

John Arnaldi

Buttress roots

(b) Tropical rainforest trees typically possess buttress roots that support them in the shallow, often wet soil. Shown are buttress roots on Australian banyan (*Ficus macrophylla*). Photographed at Selby Gardens in Sarasota, Florida.

Robert and Linda Mitchell

Pneumatophores

(c) White mangrove (*Laguncularia racemosa*) produces pneumatophores, shown protruding from the wet mud in the foreground. Pneumatophores may provide oxygen for roots buried in anaerobic (oxygen-deficient) soil. Photographed in Isla del Carmen, Mexico.

John Arnaldi

Aerial roots

(d) The moth orchid (*Phalaenopsis* hybrid) has photosynthetic aerial roots.

Figure 35-9 Specialized roots

the high-tide level. Even though roots live in the soil, they still require oxygen for aerobic respiration. Flooded soils are depleted of oxygen, so these aerial "breathing" roots, known as **pneumatophores,** may assist in getting oxygen to the submerged roots (❚ Fig. 35-9c). Pneumatophores, which also help anchor the plant, have a well-developed system of internal air spaces that is continuous with the submerged parts of the root, presumably allowing gas exchange. Black mangrove, white mangrove, and bald cypress are examples of plants with pneumatophores.

Climbing plants and **epiphytes,** which are plants that grow attached to other plants, have aerial roots that anchor the plant to the bark, branch, or other surface on which they grow. Some epiphytes have aerial roots specialized for functions other than anchorage. Certain epiphytic orchids, for example, have photosynthetic roots (❚ Fig. 35-9d). Epiphytic roots may absorb moisture as well.

Some parasitic epiphytes, such as mistletoe, have modified roots that penetrate the host-plant tissues and absorb water. Another plant that starts its life as an epiphyte is the strangler fig, which produces long roots that eventually reach the ground and anchor the plant (now a tree rather than an epiphyte) in the soil. The tree on which the strangler fig originally grew is often killed as the strangler fig grows around it, competing with it for light and other resources and crushing its secondary phloem.

Review

❚ What are the advantages of a taproot system? Of a fibrous root system?

❚ If you were examining a cross section of a primary root of a flowering plant, how would you determine whether it was a eudicot or a monocot?

❚ What is the symplast? The apoplast?

❚ How does a herbaceous eudicot root develop secondary tissues?

❚ What are the functions of each of the following: (1) prop roots, (2) buttress roots, and (3) pneumatophores?

ROOT ASSOCIATIONS WITH FUNGI AND BACTERIA

Learning Objective

6 List and describe two mutualistic relationships between roots and other organisms.

The roots of terrestrial plant species may form **mutualistic**—that is, mutually beneficial—relationships with certain soil microorganisms: *mycorrhizal fungi* and *rhizobial bacteria*. These mutualistic relationships provide plants with access to essential minerals that are in limited supply in the soil.

Mycorrhizae facilitate the uptake of essential minerals by roots

Subterranean associations between roots and certain fungi, known as **mycorrhizae,** permit the transfer of materials (such as carbon compounds produced by photosynthesis) from roots to the fungus. At the same time, essential minerals, such as phos-

phorus, move from the fungus to the roots of the host plant. The threadlike body of the fungal partner extends into the soil, extracting minerals well beyond the reach of the plant's roots. In some mycorrhizae, the fungal mycelium encircles the root like a sheath; in others, the fungus penetrates cell walls of the root cortex and forms branching **arbuscules** (❚ Fig. 35-10; also see Fig. 26-11).

The mycorrhizal relationship is an ancient one: some of the earliest known plant fossils (about 400 million years old) contain mycorrhizal fungi in the roots. The mycorrhizal relationship is mutually beneficial because when mycorrhizae are not present, neither the fungus nor the plant grows as well (see Fig. 26-19). The hyphal network of mycorrhizae appears to interconnect different plant species in the community so that carbon compounds may flow from one plant to another through their mutual fungal partner.

Rhizobial bacteria fix nitrogen in the roots of leguminous plants

Certain nitrogen-fixing bacteria, collectively called **rhizobia,** form associations with the roots of leguminous plants—clover, peas, and soybeans, for example. **Nodules** (swellings) that house millions of the rhizobia develop on the roots (see Fig. 54-9). Evidence indicates that the rhizobial relationship evolved much later than the mycorrhizal relationship; unlike mycorrhizae, which infect most plant species, rhizobia almost exclusively infect plants of the legume family, which evolved about 70 million years ago. (A few other plant–bacteria associations that fix nitrogen have also evolved.)

As with mycorrhizae, the association between nitrogen-fixing bacteria and the roots of legumes is mutually beneficial. Bacteria receive photosynthetic products from plants while helping them meet their nitrogen requirements by producing ammonia (NH_3) from atmospheric nitrogen.

For several decades, biologists have been studying the molecular basis of associations between plants and nitrogen-fixing bacteria. The two partners in this association use **cell signaling,** that is, a molecular dialogue back and forth between them, to establish contact and develop nodules. When soil contains a low level of nitrogen, legume roots secrete chemical attractants (flavonoids) toward which the rhizobial bacteria swim.

The initial signal molecules produced by rhizobial bacteria are called *nodulation factors,* or *nod factors.* These nod factors help root-hair cells in the roots of leguminous plants recognize the presence of the bacteria; recognition is the first step in the infection by rhizobia and the development of root nodules. The nod factor apparently attaches to a plant receptor on the root-hair plasma membrane. This attachment triggers an elaborate **signal transduction** cascade within root cells that alters gene expression and metabolism, resulting in nodule formation (❚ Fig. 35-11).

Some of the same plant genes are involved in the initiation of both mycorrhizal and rhizobial associations

Because both mycorrhizae and rhizobia infect plant roots, some biologists have hypothesized that the genes and signal transduc-

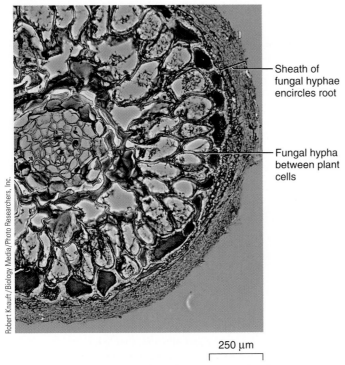

Sheath of
fungal hyphae
encircles root

Fungal hypha
between plant
cells

Robert Knauft/Biology Media/Photo Researchers, Inc.

250 μm

(a) LM of ectomycorrhizae, fungal associations that form a
sheath around the root. The fungal hyphae penetrate the
root between cortical cells but do not enter the cells.

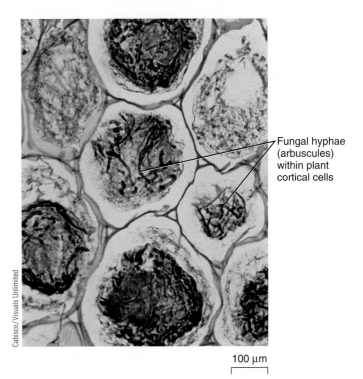

Fungal hyphae
(arbuscules)
within plant
cortical cells

Cabisco/Visuals Unlimited

100 μm

(b) LM of endomycorrhizae, fungal associations in which the
fungal hyphae penetrate root cells of the cortex and form
branched arbuscules (absorbing organs) within the cells to aid
in delivering and receiving nutrients. Roots of most vascular
plant species are colonized by endomycorrhizae.

Figure 35-10 Mycorrhizae

Mycorrhizae enhance plant growth by providing essential minerals to the roots.

tion components involved in mycorrhizal relationships may have
been recruited for the rhizobial relationships that evolved much
later. There is experimental evidence to support this hypothesis.
Some of the same genes that trigger infection and changes in root
development to form rhizobial nodules are also involved in establishing mycorrhizal infections. Evidence includes several legume mutants that biologists have identified; these mutants lack
the ability to develop nodules for rhizobial bacteria and cannot
form mycorrhizal associations. Molecular findings also support
the hypothesis.

Review

- What are mycorrhizal fungi?
- What are rhizobial bacteria?
- How are mycorrhizae and root nodules similar? How are they
  different?

THE SOIL ENVIRONMENT

Learning Objectives

7 Describe the roles of weathering, organisms, climate, and
topography in soil formation.

8 List the four components of soil, and give the ecological
significance of each.

9 Describe how roots absorb positively charged mineral ions by
the process of cation exchange.

10 Distinguish between macronutrients and micronutrients.

11 Explain the impacts of mineral depletion and soil erosion on
plant growth.

We now examine the soil environment in which most roots live.
Soil is a relatively thin layer of Earth's crust that has been modified
by the natural actions of weather, wind, water, and organisms. It
is easy to take soil for granted. We walk on and over it throughout
our lives but rarely stop to think about how important it is to our
survival. Vast numbers and kinds of organisms colonize soil and
depend on it for shelter and food. Most plants anchor themselves
in soil, and from it they receive water and minerals. Almost all
of the elements essential for plant growth are obtained directly
from the soil. With few exceptions, plants cannot survive on their
own without soil, and because we depend on plants for our food,
neither could humans exist without soil.

Soils are generally formed from rock (called *parent material*)
that is gradually broken down, or fragmented, into smaller and
smaller particles by biological, chemical, and physical **weathering processes.** Two important factors that work together in the
weathering of rock are climate and organisms. When plant roots
and other organisms living in the soil respire, they produce carbon dioxide (CO_2), which diffuses into the soil and reacts with
soil water to form carbonic acid (H_2CO_3). Soil organisms such

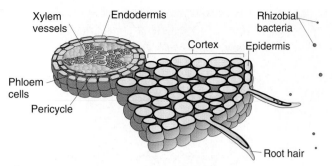

Xylem vessels — Endodermis — Rhizobial bacteria

Cortex — Epidermis

Phloem cells

Pericycle

Root hair

1 Root secretes chemical that attracts rhizobia.

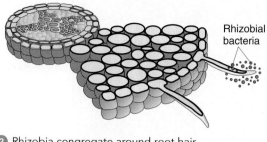

Rhizobial bacteria

2 Rhizobia congregate around root hair.

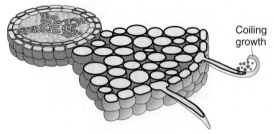

Coiling growth

3 Rhizobia secrete chemicals that cause root hair to coil around them.

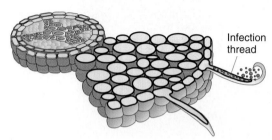

Infection thread

4 Rhizobia enter root hair and move through infection thread (tube surrounded by plasma membrane) that extends down root-hair cell.

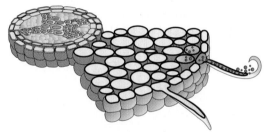

5 Rhizobia move from infection thread into cells in cortex; each bacterium is surrounded by a membrane.

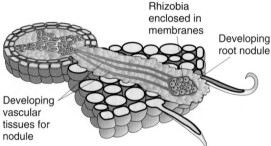

Rhizobia enclosed in membranes

Developing root nodule

Developing vascular tissues for nodule

6 Chemicals produced by rhizobia induce cell division and formation of root nodule.

Figure 35-11 Steps in the establishment of a relationship between rhizobial bacteria and leguminous plants

as lichens produce other kinds of acids.[1] These acids etch tiny cracks, or fissures, in the rock surface; water then seeps into these cracks. If the parent material is located in a temperate climate, the alternate freezing and thawing of the water during winter cause the cracks to enlarge, breaking off small pieces of rock. Small plants can then become established and send their roots into the larger cracks, fracturing the rock further.

Topography, a region's surface features, such as the presence or absence of mountains and valleys, is also involved in soil formation. Steep slopes often have little or no soil on them, because the soil and rock are continually transported down the slopes by gravity. Runoff from precipitation tends to amplify erosion on steep slopes. Moderate slopes, in contrast, may encourage the formation of deep soils.

---

[1] A lichen is a dual organism composed of a fungus and a phototroph (photosynthetic organism) (see Chapter 26).

The disintegration of solid rock into finer and finer mineral particles in the soil takes an extremely long time, sometimes thousands of years. Soil forms constantly as the weathering of parent material beneath soil that is already formed continues to add new soil.

Soil is composed of inorganic minerals, organic matter, air, and water

Soil comprises four distinct components—inorganic mineral particles (which make up about 45% of a typical soil), organic matter (about 5%), water (about 25%), and air (about 25%). The plants, animals, fungi, and microorganisms that inhabit soil interact with it, and minerals are continually cycled from the soil to organisms, which use them in their biological processes. When the organisms die, bacteria and other soil organisms decompose the remains, returning the minerals to the soil.

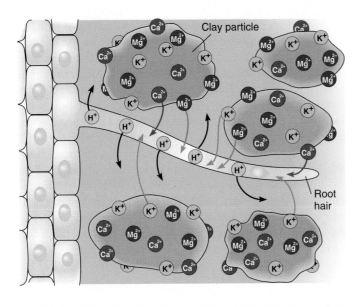

Figure 35-12 Cation exchange

Negatively charged clay particles bind to positively charged mineral cations, holding them in the soil. Roots pump out protons (H$^+$), which are exchanged for the cations, facilitating their absorption.

The inorganic mineral particles, which come from weathered rock, provide anchorage and essential minerals for plants, as well as pore space for water and air. Because different rocks consist of different minerals, soils vary in mineral composition and chemical properties. Rocks rich in aluminum form acidic soils, for example, whereas rocks that contain silicates of magnesium and iron form soils that may be deficient in calcium, nitrogen, and phosphorus. Also, soils formed from the same kind of parent material may not develop in the same way, because other factors, such as weather, topography, and organisms, differ.

The *texture,* or structural characteristic, of a soil is determined by the percentages (by weight) of the different-sized inorganic mineral particles—sand, silt, and clay—in it. The size assignments for sand, silt, and clay give soil scientists a way to classify soil texture. Particles larger than 2 mm in diameter, called gravel or stones, are not considered soil particles, because they do not have any direct value to plants. The largest soil particles are called *sand* (0.02 to 2 mm in diameter); the medium-sized particles, *silt* (0.002 to 0.02 mm in diameter); and the smallest particles, *clay* (less than 0.002 mm in diameter). Sand particles are large enough to be seen easily with the eye, and silt particles (about the size of flour particles) are barely visible. Most individual clay particles are too small to be seen with an ordinary light microscope; they can be seen only under an electron microscope.

A soil's texture affects many of that soil's properties, in turn influencing plant growth. The clay component of a soil is particularly important in determining many of its characteristics because clay particles have the greatest surface area of all soil particles. If the surface areas of about 450 g (1 lb) of clay particles were laid out side by side, they would occupy 1 hectare (2.5 acres).

Each clay particle has predominantly negative electric charges on its outer surface that attract and reversibly bind **cations,** which are positively charged mineral ions such as potassium (K$^+$) and

magnesium (Mg^{2+}). Because many cations are essential for plant growth, cation adhesion to soil particles is an important aspect of soil fertility. Roots secrete protons (hydrogen ions, H$^+$), which are exchanged for other positively charged mineral ions adhering to the surface of soil particles, in a process known as **cation exchange.** These "freed" ions and the water that forms a film around the soil particles are absorbed by the plant's roots (Fig. 35-12).

In contrast, **anions,** which are negatively charged mineral ions, are repelled by the negative surface charges of clay particles and tend to remain in solution. Anions such as nitrate (NO$_3^-$) are often washed out of the root zone by water moving through the soil.

Soil always contains a mixture of different-sized particles, but the proportions vary from one soil to another. A *loam,* which is an ideal agricultural soil, has an optimum combination of different soil particle sizes: it contains approximately 40% each of sand and silt and about 20% of clay. Generally, the larger particles provide structural support, aeration, and permeability to the soil, whereas the smaller particles bind into aggregates, or clumps, and hold minerals and water. Soils with larger proportions of sand are not as desirable for most plants, because they do not hold water and mineral ions well. Plants grown in such soils are more susceptible to drought and mineral deficiencies. Soils with larger proportions of clay are also not desirable for most plants because they provide poor drainage and often do not provide enough oxygen. Clay soils used in agriculture tend to compact, which reduces the number of soil spaces that can be filled by water and air.

A soil's organic matter consists of the wastes and remains of soil organisms

The organic matter in soil is composed of litter (dead leaves and branches on the soil's surface); droppings (animal dung); and the dead remains of plants, animals, and microorganisms in various stages of decomposition. Organic matter is decomposed by microorganisms, particularly bacteria and fungi, that inhabit the soil. During decomposition, essential mineral ions are released into the soil, where they may be bound by soil particles or absorbed by plant roots. Organic matter increases the soil's water-holding capacity by acting much like a sponge. For this reason gardeners often add organic matter to soils, especially sandy soils, which are naturally low in organic matter.

The partly decayed organic portion of the soil is referred to as **humus.** Humus, which is not a single chemical compound but a mix of many organic compounds, binds mineral ions and holds water. On average, humus persists in agricultural soil for about 20 years. Certain components of humus may persist in the soil for hundreds of years. Although humus is somewhat resistant to decay, a succession of microorganisms gradually reduces it to carbon dioxide, water, and minerals.

About 50% of soil volume is composed of pore spaces

Soil has numerous pore spaces of different sizes around and among the soil particles. Pore spaces occupy roughly 50% of a soil's volume and are filled with varying proportions of air and water. Both soil air and soil water are necessary to produce a moist but aerated soil that sustains plants and other soil-dwelling organisms. Water is usually held in the smaller pores, whereas air is found in the larger pores. After a prolonged rain, almost all the pore spaces may be filled with water, but water drains rapidly from the larger pore spaces, drawing air from the atmosphere into those spaces.

Soil air contains the same gases as atmospheric air, although they are usually present in different proportions. As a result of respiration by soil organisms, there is less oxygen and more carbon dioxide in soil air than in atmospheric air. (Aerobic respiration uses oxygen and produces carbon dioxide.) Among the important gases in soil are oxygen (O_2), required by soil organisms for aerobic respiration; nitrogen (N_2), used by nitrogen-fixing bacteria; and carbon dioxide (CO_2), involved in soil weathering.

Soil water originates as precipitation, which drains downward, or as groundwater (water stored in porous underground rock), which rises upward from the water table (the uppermost level of groundwater). Soil water contains low concentrations of dissolved minerals that enter the roots of plants when they absorb water. Water not bound to soil particles or absorbed by roots percolates (moves down) through soil, carrying dissolved minerals with it. The removal of dissolved materials from soil by percolating water is called **leaching.** The deposition of leached material in the lower layers of soil is known as **illuviation.** Iron and aluminum compounds, humus, and clay are some illuvial materials that can gather in the subsurface portion of the soil. Some substances completely leach out of the soil because they are so soluble that they migrate down into the groundwater. It is also possible for water that is moving *upward* through the soil to carry dissolved materials with it.

The organisms living in the soil form a complex ecosystem

A single teaspoon of fertile agricultural soil may contain millions of microorganisms, such as bacteria, fungi, algae, protozoa, and microscopic nematodes and other worms. Many other organisms also colonize the soil ecosystem, including plant roots, earthworms, insects, moles, snakes, and groundhogs (Fig. 35-13). Most numerous in soil are bacteria, which number in the hundreds of millions per gram of soil. Scientists have identified about 170,000 species of soil organisms, but thousands remain to be identified. Little is known about the roles of soil organisms, in part because it is hard to study their activities under natural conditions.

Worms are some of the most important organisms living in soil. Earthworms, probably one of the most familiar soil inhabitants, ingest soil and obtain energy and raw materials by digesting humus. *Castings,* bits of soil that have passed through the gut of an earthworm, are deposited on or near the soil surface. In this way, minerals from deeper layers are brought to upper layers. Earthworm tunnels serve to aerate the soil, and the worms' waste products and corpses add organic material to the soil.

Ants live in the soil in enormous numbers, constructing tunnels and chambers that help aerate it. Members of soil-dwelling ant colonies forage on the surface for bits of food, which they carry back to their nests. Not all this food is eaten, however, and its eventual decomposition helps increase the organic matter in the soil. Many ants are also indispensable in plant reproduction because they bury seeds in the soil (discussed in Chapter 36).

Soil pH affects soil characteristics and plant growth

Acidity is measured using the pH scale, which extends from 0 (extremely acidic) through 7.0 (neutral) to 14.0 (extremely alkaline). The pH of most soils ranges from 4.0 to 8.0, but some soils are outside this range. The soil of the Pygmy Forest in Men-

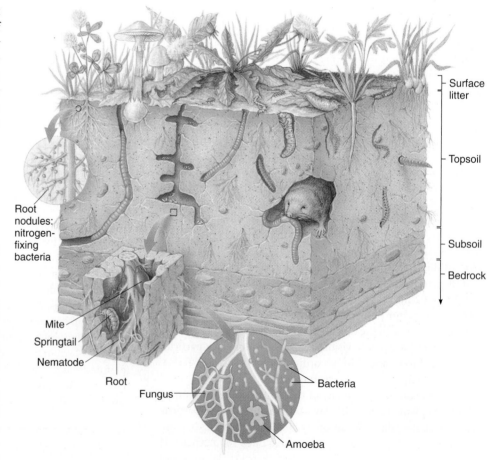

Figure 35-13 Diversity of life in fertile soil

Plants, algae, protozoa, fungi, bacteria, earthworms, flatworms, roundworms, insects, spiders and mites, and burrowing animals such as moles and groundhogs live in soil.

docino County, California, is extremely acidic (pH 2.8 to 3.9). At the other extreme, certain soils in Death Valley, California, have a pH of 10.5.

Plants are affected by soil pH, partly because the solubility of certain minerals varies with differences in pH. Soluble minerals can be absorbed by the plant, whereas insoluble forms cannot. At a low pH, for example, aluminum and manganese in soil water are more soluble and are sometimes absorbed by the roots in toxic concentrations. At a higher pH, certain mineral salts essential for plant growth, such as calcium phosphate, become less soluble and thus less available to plants.

Soil pH also affects the leaching of minerals. An acidic soil has less ability to bind positively charged ions to it because the soil particles also bind the abundant protons (❙ Fig. 35-14). As a result, certain mineral ions essential for plant growth, such as potassium (K^+), are leached more readily from acidic soil. The optimum soil pH for most plant growth is 6.0 to 7.0, because most essential minerals are available to plants in that pH range.

Acid precipitation, a type of air pollution in which sulfuric and nitric acids produced by human activities fall to the ground as acid rain, sleet, snow, or fog, can seriously decrease soil pH. Acid precipitation is one of several factors implicated in **forest decline,** the gradual deterioration, and often death, of trees that has been observed in many European and North American forests. Forest decline may be partly the result of soil changes, such as leaching of essential cations, caused by acid precipitation. In central European forests that have experienced forest decline, for example, a strong correlation exists between forest damage and soil chemistry altered by acid precipitation.

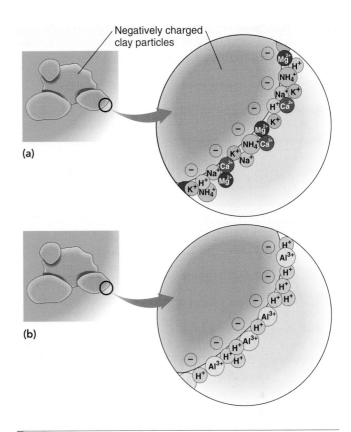

Figure 35-14   How acid alters soil chemistry

(a) In normal soil, positively charged mineral ions are attracted to the negatively charged soil particles. **(b)** In acidified soil, protons displace the cations. Aluminum ions released from inorganic mineral particles when the soil becomes acidified also adhere to soil particles.

Soil provides most of the minerals found in plants

More than 90 naturally occurring elements exist on Earth, and more than 60 of these, including elements as common as carbon and as rare as gold, have been found in plant tissues. Not all these elements are essential for plant growth, however.

Nineteen elements have been found essential for most, if not all, plants (❙ Table 35-1). Ten of these are required in fairly large quantities (greater than 0.05% dry weight) and are therefore known as **macronutrients.** Macronutrients include carbon, hydrogen, oxygen, nitrogen, potassium, calcium, magnesium, phosphorus, sulfur, and silicon. The remaining nine **micronutrients** are needed in trace amounts (less than 0.05% dry weight) for normal plant growth and development. Micronutrients include chlorine, iron, boron, manganese, sodium, zinc, copper, nickel, and molybdenum.

Four of the 19 elements—carbon, oxygen, hydrogen, and nitrogen—come directly or indirectly from soil water or from gases in the atmosphere. Carbon is obtained from carbon dioxide in the atmosphere during photosynthesis. Oxygen is obtained from atmospheric oxygen (O_2) and water (H_2O). Water also supplies hydrogen to the plant. Plants absorb their nitrogen from the soil as ions of nitrogen salts, but the nitrogen in nitrogen salts ultimately comes from atmospheric nitrogen (N_2). The remaining 15 essential elements are obtained from the soil as dissolved

mineral ions. Their ultimate source is the parent material from which the soil was formed.

Let us examine the main functions of essential elements. Carbon, hydrogen, and oxygen are found as part of the structure of all biologically important molecules, including lipids, carbohydrates, nucleic acids, and proteins. Nitrogen is part of proteins, nucleic acids, and chlorophyll.

Potassium, which plants use in fairly substantial amounts, is not found in a specific organic compound in plant cells. Instead, it remains as free K^+ and plays a key physiological role in maintaining the turgidity of cells because it is osmotically active. The presence of K^+ in cytoplasm causes water to pass through the plasma membrane into the cell by osmosis. Potassium is also involved in the opening of stomata.

Calcium plays a key structural role as a component of the middle lamella (the cementing layer between cell walls of adjacent plant cells). Calcium ions (Ca^{2+}) are also important in a number of physiological roles in plants, such as altering membrane permeability and as **second messengers** in various cell signaling responses (see Chapter 6).

Magnesium is critical for plants because it is part of the chlorophyll molecule. Phosphorus is a component of nucleic acids, phospholipids (part of cell membranes), and energy transfer molecules such as ATP. Sulfur is essential because it is found in certain amino acids and vitamins. Many plants accumulate sili-

TABLE 35-1

Functions of Elements Required by Most Plants

Element	Taken Up As	Major Functions
Macronutrients		
Carbon	CO_2	Component of carbohydrate, lipid, protein, and nucleic acid molecules
Hydrogen	H_2O	Component of carbohydrate, lipid, protein, and nucleic acid molecules
Oxygen	CO_2, H_2O	Component of carbohydrate, lipid, protein, and nucleic acid molecules
Nitrogen	NO_3^-, NH_4^+	Component of proteins, nucleic acids, chlorophyll, certain coenzymes
Potassium	K^+	Osmotic and ionic balance; opening of stomata; enzyme activator (for 40+ enzymes)
Calcium	Ca^{2+}	In cell walls; involved in membrane permeability; enzyme activator; second messenger
Magnesium	Mg^{2+}	In chlorophyll; enzyme activator in carbohydrate metabolism
Phosphorus	HPO_4^{2-}, $H_2PO_4^-$	In nucleic acids, phospholipids, ATP (energy transfer compound)
Sulfur	SO_4^{2-}	In certain amino acids and vitamins
Silicon	SiO_3^{2-}	In cell walls
Micronutrients		
Chlorine	Cl^-	Ionic balance; involved in photosynthesis
Iron	Fe^{2+}, Fe^{3+}	In enzymes and electron transport molecules involved in photosynthesis, respiration, and nitrogen fixation
Boron	$H_2BO_3^-$	In cell walls; involved in nucleic acid metabolism and in cell growth
Manganese	Mn^{2+}	In enzymes involved in respiration and nitrogen metabolism; required for photosynthesis
Sodium	Na^+	Involved in photosynthesis; substitutes for potassium in osmotic and ionic balance
Zinc	Zn^{2+}	In enzymes involved in respiration and nitrogen metabolism
Copper	Cu^+, Cu^{2+}	In enzymes involved in photosynthesis
Nickel	Ni^{2+}	In enzymes (urease) involved in nitrogen metabolism
Molybdenum	MoO_4^{2-}	In enzymes involved in nitrogen metabolism

con in their cell walls and intercellular spaces. Silicon enhances the growth and fertility of some species and may help reinforce cell walls.

Chlorine and sodium are micronutrients that help maintain cell turgor. In addition to this osmotic role, chloride (Cl^-) and sodium (Na^+) ions are essential for photosynthesis. Six of the micronutrients (iron, manganese, zinc, copper, nickel, and molybdenum) are associated with various plant enzymes, often as **cofactors,** and are involved in certain enzymatic reactions. Boron, present in cell walls, is also involved in nucleic acid metabolism and in cell growth.

How do biologists determine whether an element is essential?

It is impossible to conduct mineral nutrition experiments by growing plants in soil, because soil is too complex and contains too many elements. Therefore, nutritional studies require special methods. One of the most useful techniques to test whether an element is essential is **hydroponics,** which is the growing of plants in aerated water to which mineral salts have been added. Hydroponics has commercial applications in addition to its scientific use (Fig. 35-15).

If biologists suspect that a particular element is essential for plant growth, they grow plants in a nutrient solution that contains all known essential elements *except* the one in question. If plants grown in the absence of that element cannot develop normally or complete their life cycle, the element may be essential.

Additional criteria are used to confirm whether an element is essential. For example, it must be demonstrated that the element has a direct effect on the plant's metabolism and that the element is essential for a wide variety of plants.

Soil can be damaged by human mismanagement

Soil is a valuable natural resource on which humans depend for food. Many human activities generate or aggravate soil problems, including mineral depletion, soil erosion, and accumulation of salt.

Mineral depletion may occur in soils that are farmed

In a natural ecosystem, the essential minerals removed from the soil by plants are returned when the plants or the animals that eat them die and are decomposed. An agricultural system disrupts this pattern of nutrient cycling when the crops are harvested. Plant material containing minerals is removed from the nutrient cycle, and the harvested crops fail to decay and release their nutrients back to the soil. Over time, farmed soil inevitably loses its fertility, that is, its ability to produce abundant crops. Homeowners often mow their lawns and remove the clippings, similarly preventing decomposition and cycling of minerals that were in the grass blades.

Figure 35-15 Hydroponically grown lettuce

Lettuce is one of several hydroponic crops grown commercially. The white boards support the plants, which are not anchored in soil. A chemically defined liquid solution of minerals trickles over their roots, which also have access to atmospheric oxygen.

Figure 35-16 Soil erosion in an open field

Removal of plant cover exposes bare soil to erosion in heavy rains. Precipitation can cause gullies to enlarge rapidly, because they provide channels for the runoff of water.

The essential material (water, sunlight, or some essential element) that is in shortest supply usually limits plant growth. This phenomenon is sometimes called the concept of **limiting resources.** The three elements that are most often limiting resources for plants are nitrogen, phosphorus, and potassium. To sustain the productivity of agricultural soils, farmers periodically add fertilizers to depleted soils to replace the minerals that limit plant growth.

Soil erosion is the loss of soil from the land

Water, wind, ice, and other agents cause **soil erosion,** the wearing away or removal of soil from the land. Water and wind are particularly effective in moving soil from one place to another. Rainfall loosens soil particles that can then be transported by moving water (Fig. 35-16). Wind loosens soil and blows it away, particularly if the soil is barren and dry. Soil erosion is a natural process that can be greatly accelerated by human activities.

Soil erosion is a national and international problem that does not make the headlines very often. The U.S. Department of Agriculture estimates that approximately 4.2 billion metric tons (4.6 billion tons) of topsoil are eroded each year from U.S. croplands and rangelands. Erosion causes a loss in soil fertility because essential minerals and organic matter that are a part of the soil are removed. As a result of these losses, the productivity of eroded agricultural soils declines, and more fertilizer must be used to replace the nutrients lost to erosion.

Humans often accelerate soil erosion through poor soil management practices. Agriculture is not the only culprit, because the removal of natural plant communities during surface mining, unsound logging practices (such as clear-cutting large areas of forest), and the construction of roads and buildings also accelerate erosion.

Soil erosion has an impact on other natural resources. Sediment that enters streams, rivers, and lakes degrades water quality and fish habitats. Pesticides and fertilizer residues in sediment may add pollutants to the water. Also, when forests are removed within the watershed of a hydroelectric power facility, accelerated soil erosion fills the reservoir behind the dam with sediment much faster than usual. This process reduces electricity production at that facility.

Sufficient plant cover reduces the amount of soil erosion: leaves and stems cushion the impact of rainfall, and roots help hold the soil in place. Although soil erosion is a natural process, abundant plant cover makes it negligible in many natural ecosystems.

Salt accumulates in soil that is improperly irrigated

Although irrigation improves the agricultural productivity of arid and semiarid lands, it sometimes causes salt to accumulate in the soil, a process called **salinization.** In a natural scenario, as a result of precipitation runoff, rivers carry dissolved salts away. Irrigation water, however, normally soaks into the soil and does not run off the land into rivers, so when water evaporates, the salt remains behind and accumulates in the soil. Salty soil results in a decline in productivity and, in extreme cases, renders the soil completely unfit for crop production.

Most plants cannot obtain all the water they need from salty soil, because a water balance problem exists: water moves by osmosis *out* of plant roots and into the salty soil. Obviously, most plants cannot survive under these conditions (see Fig. 5-14). Plant species that thrive in saline soils have special adaptations that enable them to tolerate the high amount of salt. Black mangroves, for example, excrete excess salt through their leaves.

Most crops, unless they have been genetically selected to tolerate high salt, are not productive in saline soil. Research on the molecular aspects of salt tolerance suggests that mutations in membrane transport proteins confer salt tolerance. Biologists genetically engineered a salt-tolerant variety of *Arabidopsis* to overexpress a single gene that codes for a sodium transport protein in the vacuolar membrane. These plants can store large quantities of sodium in their vacuoles, thereby tolerating saline soils.

Review

▌ What are the four components of soil, and how is each important to plants?

▌ What is cation exchange?

▌ How do weathering processes convert rock into soil?

▌ What are macronutrients and micronutrients?

▌ How are minerals lost from the soil?

SUMMARY WITH KEY TERMS

Learning Objectives

1 Distinguish between taproot and fibrous root systems (page 749).

▌ A **taproot** system has one main root (formed from the radicle), from which many lateral roots extend.

▌ A **fibrous root** system has several to many **adventitious roots** of the same size developing from the end of the stem. Lateral roots branch from these adventitious roots.

ThomsonNOW™ **Learn more about root systems by clicking on the figure in ThomsonNOW.**

2 Label cross sections of a primary eudicot root and a monocot root, and describe the functions of each tissue (page 749).

▌ Primary roots have an epidermis, ground tissues (cortex and, in certain roots, pith), and vascular tissues (xylem and phloem). Each root tip is covered by a **root cap,** a protective layer that covers the delicate root **apical meristem** and may orient the root so that it grows downward.

▌ **Epidermis** protects the root. **Root hairs,** short-lived extensions of epidermal cells, aid in absorption of water and dissolved minerals.

▌ **Cortex** consists of parenchyma cells that often store starch. **Endodermis,** the innermost layer of the cortex, regulates the movement of water and minerals into the root xylem. Cells of the endodermis have a **Casparian strip** around their radial and transverse walls that is impermeable to water and dissolved minerals. Minerals are actively transported through carrier proteins in the plasma membranes of endodermal cells.

▌ Pericycle, xylem, and phloem collectively make up the root's **stele,** or **vascular cylinder. Pericycle** gives rise to lateral roots and lateral meristems. **Xylem** conducts water and dissolved minerals; **phloem** conducts dissolved sugar.

▌ Xylem of a herbaceous eudicot root forms a solid core in the center of the root. The center of a monocot root often consists of **pith** surrounded by a ring of alternating bundles of xylem and phloem. Monocot roots lack a **vascular cambium** and do not have secondary growth.

3 Trace the pathway of water and mineral ions from the soil through the various root tissues, and distinguish between the symplast and apoplast (page 749).

▌ As water and dissolved mineral ions move from the soil into the root, they pass through the following tissues: root hair/epidermis ⟶ cortex ⟶ endodermis ⟶ pericycle ⟶ root xylem.

▌ Water and dissolved minerals move through the epidermis and cortex along one of two pathways, the **apoplast** (along the interconnected porous cell walls) or the **symplast** (from one cell's cytoplasm to the next through **plasmodesmata**).

ThomsonNOW™ **Explore the pathways of water and dissolved minerals in the root by clicking on the figures in ThomsonNOW.**

4 Discuss the structure of roots with secondary growth (page 749).

▌ Roots of gymnosperms and woody eudicots develop secondary tissues (wood and bark). The production of secondary tissues is the result of the activity of two lateral meristems, the vascular cambium and cork cambium. The vascular cambium produces secondary xylem (wood) and secondary phloem (inner bark). The **cork cambium** produces **periderm** (outer bark).

5 Describe at least three roots that are modified to perform uncommon functions (page 749).

▌ **Prop roots** develop from branches or from a vertical stem and grow downward into the soil to help support certain plants in an upright position. **Buttress roots** are swollen bases or braces that support certain tropical rainforest trees that have shallow root systems.

▌ **Pneumatophores** are aerial "breathing" roots that may assist in getting oxygen to submerged roots.

▌ Certain epiphytes have roots modified to photosynthesize, absorb moisture, or, if parasitic, penetrate host tissues.

6 List and describe two mutualistic relationships between roots and other organisms (page 756).

▌ **Mycorrhizae** are mutually beneficial associations between roots and soil fungi.

▌ Root **nodules** are swellings that develop on roots of leguminous plants and house millions of **rhizobia** (nitrogen-fixing bacteria).

7 Describe the roles of weathering, organisms, climate, and topography in soil formation (page 757).

▌ Factors influencing soil formation include parent material (type of rock), climate, organisms, the passage of time, and topography. Most soils are formed from parent material that is broken into smaller and smaller particles by **weathering processes.** Climate and organisms work together in weathering rock.

▌ Soil organisms such as plants, algae, fungi, worms, insects, spiders, and bacteria are important not only in forming soil but also in cycling minerals.

▌ Topography, a region's surface features, affects soil formation. Steep slopes have little or no soil on them, whereas moderate slopes often have deep soils.

8 List the four components of soil, and give the ecological significance of each (page 757).

▌ Soil is composed of inorganic minerals, organic matter, air, and water. Inorganic minerals provide anchorage and essential minerals for plants. Organic matter increases the soil's water-holding capacity and, as it decomposes, releases essential minerals into the soil. Soil air provides oxygen for soil organisms to use during aerobic respiration. Soil water provides water and dissolved minerals to plants and other organisms.

9 Describe how roots absorb positively charged mineral ions by the process of cation exchange (page 757).

   ▮ **Cations,** positively charged mineral ions, are attracted and reversibly bound to clay particles, which have predominantly negative charges on their outer surfaces. In **cation exchange,** roots secrete protons (H^+), which are exchanged for other positively charged mineral ions, freeing them into the soil water to be absorbed by roots.

10 Distinguish between macronutrients and micronutrients (page 757).

   ▮ Plants require 19 essential elements for normal growth. Ten elements are **macronutrients:** carbon, hydrogen, oxy-gen, nitrogen, potassium, calcium, magnesium, phosphorus, sulfur, and silicon. Macronutrients are required in fairly large quantities.

   ▮ Nine elements are **micronutrients:** chlorine, iron, boron, manganese, sodium, zinc, copper, nickel, and molybdenum. Micronutrients are needed in trace amounts.

11 Explain the impacts of mineral depletion and soil erosion on plant growth (page 757).

   ▮ Mineral depletion may occur in soils that are farmed because the natural pattern of nutrient cycling is disrupted when crops are harvested (and not allowed to decompose into the soil).

   ▮ **Soil erosion** is the removal of soil from the land by the actions of agents such as water and wind. Erosion causes a loss in soil fertility, because essential minerals and organic matter that are a part of the soil are removed.

TEST YOUR UNDERSTANDING

1. One main root, formed from the enlarging embryonic root, with many smaller lateral roots branching off it is a(an) (a) fibrous root system (b) adventitious root system (c) taproot system (d) storage root system (e) prop root system

2. Roots produced at unusual places on the plant are (a) fibrous (b) adventitious (c) taproots (d) rhizobial (e) mycorrhizae

3. Root hairs (a) cover and protect the delicate root apical meristem (b) increase the absorptive capacity of roots (c) secrete a waxy cuticle (d) orient the root so it grows downward (e) store excess sugars produced in the leaves

4. Certain plants adapted to flooded soil produce aerial "breathing" roots known as (a) fibrous roots (b) pneumatophores (c) mycorrhizae (d) nodules (e) prop roots

5. Unlike stems, roots produce (a) nodes and internodes (b) root caps and internodes (c) axillary buds and root hairs (d) terminal buds and axillary buds (e) root caps and root hairs

6. The waterproof region around the radial and transverse walls of endodermal cells is the (a) Casparian strip (b) pericycle (c) apoplast (d) symplast (e) pneumatophore

7. The apoplast is (a) a layer of cells that surround the vascular region in roots (b) the layer of cells just inside the endodermis (c) a system of interconnected plant cell walls through which water moves (d) the central cylinder of the root that comprises the vascular tissues (e) a continuum of cytoplasm of many cells, all connected by plasmodesmata

8. Plants obtain positively charged mineral ions from clay particles in the soil by cation exchange, in which (a) roots passively absorb the positively charged mineral ions (b) mineral ions flow freely along porous cell walls (c) roots secrete protons, which free other positively charged mineral ions to be absorbed by roots (d) the Casparian strip effectively blocks the passage of water and mineral ions along the endodermal cell wall (e) a well-developed system of internal air spaces in the root allows both gas exchange and cation exchange

9. The cell layer from which lateral roots originate is the (a) epidermis (b) cortex (c) endodermis (d) pericycle (e) vascular cambium

10. The center of a herbaceous eudicot root is composed of _____, whereas the center of a monocot root is composed of _____. (a) pith; cortex (b) xylem; phloem (c) phloem; xylem (d) xylem; pith (e) pith; xylem

11. Mutually beneficial associations between certain soil fungi and the roots of most plant species are called (a) mycorrhizae (b) pneumatophores (c) nodules (d) rhizobia (e) humus

12. Which of the following statements about soil is *true?* (a) pore spaces are always filled with about 50% air and 50% water (b) a single teaspoon of fertile agricultural soil may contain up to several hundred living microorganisms (c) the texture of a soil is determined by the soil's pH (d) a soil's organic matter includes litter, droppings, and the dead remains of plants, animals, and microorganisms (e) soil formation is unaffected by a region's climate or topography

13. The technique of growing plants in aerated water containing dissolved mineral salts is known as (a) hydration (b) hydroponics (c) hydrophilic (d) hydrostatic (e) hydrolysis

14. Carbon, hydrogen, oxygen, nitrogen, potassium, calcium, magnesium, phosphorus, sulfur, and silicon are collectively known as (a) micronutrients (b) microvilli (c) micronuclei (d) macronuclei (e) macronutrients

15. In roots of woody plants, (a) xylem does not form the central tissue of the root (b) the cortex composes the bulk of the root (c) the vascular cambium gives rise to secondary xylem and secondary phloem (d) the pericycle gives rise to the apical meristem (e) secondary growth occurs despite the lack of a vascular cambium

16. Corn, sorghum, red mangrove, and banyan are plants that have (a) prop roots (b) buttress roots (c) pneumatophores (d) storage roots (e) root nodules

Roots and Mineral Nutrition   **765**

1. A mesquite root is found penetrating a mine shaft about 46 m (150 ft) below the surface of the soil. How could you determine *when* the root first grew into the shaft? (*Hint:* Mesquite is a woody plant.)

2. How would you distinguish between a root hair and a small lateral root?

3. You are given a plant part that was found growing in the soil and are asked to determine whether it is a root or an underground stem. How would you identify the structure without a microscope? With a microscope?

4. How would you design an experiment to determine whether gold is essential for plant growth? What would you use for an experimental control?

5. Why does overwatering a plant often kill it?

6. Explain why, once secondary growth has occurred, that portion of the root is no longer involved in absorption. Where does absorption of water and dissolved minerals occur in plants that have roots with secondary growth?

7. **Evolution Link.** A barrel cactus that is 60 cm tall and 30 cm in diameter has roots more than 3 m long. However, all the plant's roots are found in the soil at a depth of 5 to 15 cm. What possible adaptive value does such a shallow root system confer on a desert plant?

Additional questions are available in ThomsonNOW at www.thomsonedu.com/login

Reproduction in Flowering Plants

Flower of the common chickweed. This plant (*Stellaria media*) is a weedy annual native to Europe but widespread in North America. Each flower has five green sepals, five deeply notched yellow petals, three to five pollen-bearing stamens (this flower has three), and a single pistil.

KEY CONCEPTS

The flower is the site of sexual reproduction in angiosperms.

A typical flower consists of four whorls: sepals, petals, stamens, and carpels.

Pollen grains are transported to stigmas by a variety of agents, such as animals and wind.

Double fertilization results in a plant embryo and endosperm.

The seed is a mature ovule, and the fruit is a mature ovary.

Many vegetative organs (roots, stems, and leaves) have evolved modifications for reproducing asexually.

Flowering plants, or **angiosperms,** include about 300,000 species and are the largest, most successful group of plants. You may have admired flowers for their fragrances as well as their appealing colors and varied shapes (see photograph). The biological function of flowers is sexual reproduction. Their colors, shapes, and fragrances increase the likelihood that pollen grains, which produce sperm cells, will be carried from one plant to another. Sexual reproduction in plants includes *meiosis* and the fusion of reproductive cells—egg and sperm cells, collectively called **gametes.** The union of gametes, called *fertilization*, occurs within the flower's ovary.

Sexual reproduction offers the advantage of new gene combinations, not found in either parent, that may make an individual plant better suited to its environment. These new combinations result from the crossing-over and independent assortment of chromosomes that occur during meiosis, before the production of egg and sperm cells (see Chapter 11).

Many flowering plants also reproduce asexually. Asexual reproduction often does not involve the formation of flowers, seeds, and fruits. Instead, offspring generally form when a vegetative organ (such as a stem, root, or leaf) expands, grows, and then becomes separated from the rest of the plant, often by the death of tissues. Because asexual reproduction requires only one parent and no meiosis or fusion of gametes occurs, the offspring of asexual reproduction are virtually genetically identical to one another and to the parent plant from which they came.[1]

---

[1]Somatic mutations can result in some variability among asexually derived offspring.

This chapter examines both sexual and asexual reproduction in flowering plants, including floral adaptations that are important in pollination; seed and fruit structure and dispersal; germination and early growth; and several kinds of asexual reproduction. We conclude with a discussion of the evolutionary advantages and disadvantages of sexual and asexual reproduction. ■

THE FLOWERING PLANT LIFE CYCLE

Learning Objectives

1   Describe the functions of each part of a flower.
2   Identify where eggs and pollen grains are formed within the flower.

In Chapters 27 and 28 you learned that angiosperms and other plants undergo a cyclic **alternation of generations** in which they spend a portion of their life cycle in a multicellular haploid stage and a portion in a multicellular diploid stage. The haploid portion, called the **gametophyte generation,** gives rise to gametes by mitosis. When two gametes fuse during **fertilization,** the diploid portion of the life cycle, called the **sporophyte generation,** begins. The sporophyte generation produces haploid spores by meiosis. Each spore has the potential to give rise to a gametophyte plant, and the cycle continues.

In flowering plants the diploid sporophyte generation is larger and nutritionally independent. The haploid gametophyte generation, which is located in the flower, is microscopic and nutritionally dependent on the sporophyte. We say more about alternation of generations in flowering plants after our discussion of the role of flowers as reproductive structures. It may be helpful for you to review Figure 28-13, which shows the main stages in the flowering plant life cycle.

Darwin Dale/Photo Researchers, Inc.

(a)

Figure 36-1   *Animated*   Floral structure

(a) An *Arabidopsis thaliana* flower. **(b)** Cutaway view of an *Arabidopsis* flower. Each flower has four sepals (*two are shown*), four petals (*two are shown*), six stamens, and one long pistil. Four of the stamens are long, and two are short (*two long and two short are shown*). Pollen grains develop within sacs in the anthers. In *Arabidopsis,* the compound pistil consists of two carpels that each contain numerous ovules.

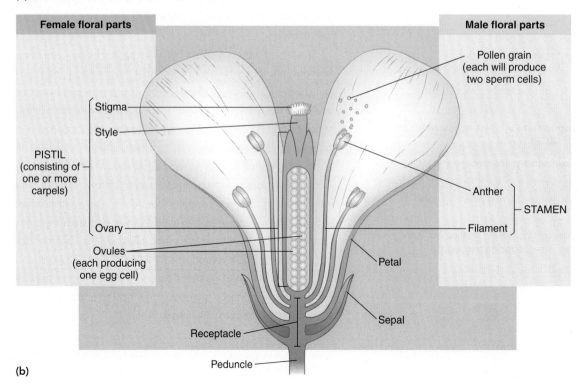

(b)

www.thomsonedu.com/biology/solomon

Flowers develop at apical meristems

How does a plant "know" it is time to start forming flowers? Correct timing in the switch from vegetative to reproductive development is crucial to ensure reproductive success. What happens, for example, if a plant flowers so late in the season that it does not have enough time to set seed before winter? A variety of environmental cues, such as temperature and day length, ensure proper timing, and different species are adapted to respond to distinct environmental cues. These environmental signals interact with a variety of plant hormones and developmental pathways.

In recent years some biologists have focused on the molecular aspects of developmental pathways that initiate flowering in the model organism *Arabidopsis*. When environmental conditions induce flowering, many genes are activated or inactivated. One gene, the *Flowering Locus C (FLC)* gene, codes for a **transcription factor** that represses flowering. Another gene, called *Flowering Locus D (FLD)* gene, codes for a protein that removes acetyl groups from histones in the chromatin where the *FLC* gene is located. When deacetylation occurs, the *FLC* gene is not transcribed (that is, the repressive transcription factor is not produced), and the shoot apical meristem undergoes a transition from vegetative growth to reproductive growth. It is intriguing that the plant FLD protein is homologous to a mammalian protein that also removes acetyl groups from chromatin.

Other genes are also involved in the initiation of flowering, and this area remains a focus of research interest. In Chapter 37 we discuss further the initiation of flowering.

Each part of a flower has a specific function

Flowers are reproductive shoots, usually consisting of four kinds of organs—sepals, petals, stamens, and carpels—arranged in whorls (circles) on the end of a flower stalk (❚ Fig. 36-1; also see Fig. 28-11). In flowers with all four organs, the normal order of whorls from the flower's periphery to the center (or from the flower's base upward) is as follows:

Sepals ⟶ petals ⟶ stamens ⟶ carpels

The tip of the stalk enlarges to form a **receptacle** on which some or all of the flower parts are borne. All four floral parts are important in the reproductive process, but only the stamens (the "male" organs) and carpels (the "female" organs) participate directly in sexual reproduction—sepals and petals are sterile.

Sepals, which constitute the outermost and lowest whorl on a floral shoot, cover and protect the flower parts when the flower is a bud. Sepals are leaflike in shape and form and are often green. Some sepals, such as those in lily flowers, are colored and resemble petals (❚ Fig. 36-2). The collective term for all the sepals of a flower is **calyx.**

The whorl just inside and above the sepals consists of **petals,** which are broad, flat, and thin (like sepals and leaves) but tremendously varied in shape and frequently brightly colored, which attracts pollinators. Petals play an important role in ensuring that sexual reproduction will occur. Sometimes petals fuse to form a tube or other floral shape. The collective term for all the petals of a flower is **corolla.**

Figure 36-2 Lily flower

In lily (*Lilium*) the three sepals and three petals are similar in size and color.

Just inside and above the petals are the **stamens,** the male reproductive organs. Each stamen has a thin stalk, called a **filament,** at the top of which is an **anther,** a saclike structure in which **pollen grains** form. For sexual reproduction to occur, pollen grains must be transferred from the anther to the female reproductive structure (the carpel), usually of another flower of the same species. At first, each pollen grain consists of two cells surrounded by a tough outer wall. One cell, the **generative cell,** divides mitotically to form two nonflagellate male gametes, known as *sperm cells.* The other cell, the **tube cell,** produces a **pollen tube,** through which the sperm cells travel to reach the ovule.

One or more **carpels,** the female reproductive organs, are located in the center or top of most flowers. Carpels bear **ovules,** which are structures with the potential to develop into seeds. The carpels of a flower may be separate or fused into a single structure. The female part of the flower, often called a **pistil,** may be a single carpel (a *simple pistil*) or a group of fused carpels (a *compound pistil*) (see Fig. 28-12). Each pistil has three sections: a **stigma,** on which the pollen grains land; a **style,** a necklike structure through which the pollen tube grows; and an **ovary,** a juglike structure that contains one or more ovules and can develop into a fruit.

Female gametophytes are produced in the ovary, male gametophytes in the anther

Before we proceed, it may be helpful to relate the stages in alternation of generations to floral structure. As discussed in Chapters 27 and 28, angiosperms and certain other plants are heterosporous and produce two kinds of spores: megaspores and microspores (❚ Fig. 36-3).

Each young ovule within an ovary contains a diploid cell, the *megasporocyte,* which undergoes meiosis to produce four haploid **megaspores.** Three of these usually disintegrate, and the fourth, the functional megaspore, divides mitotically to produce a multicellular **female gametophyte,** also called an *embryo sac.* The fe-

In alternation of generations in flowering plants, the female and male gametophytes are microscopic and nutritionally dependent on the sporophyte plant.

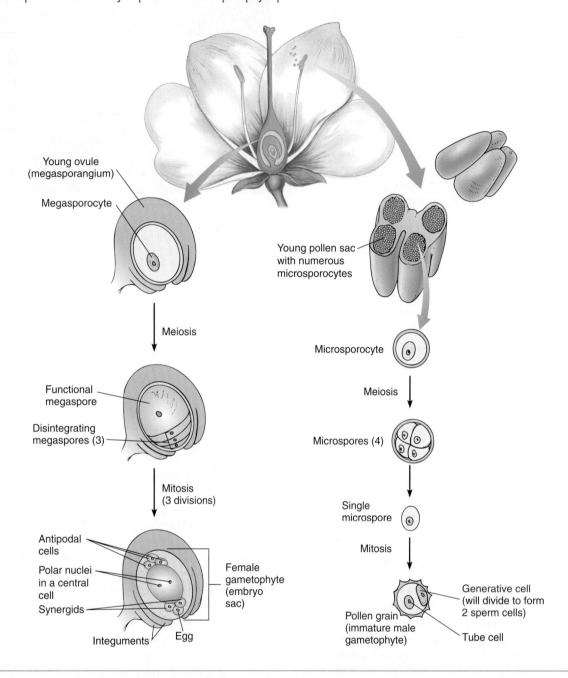

Figure 36-3 *Animated* Development of female and male gametophytes

(*Left side*) The female gametophyte, or embryo sac, develops within the ovule. (*Right side*) The immature male gametophytes, or pollen grains, develop within pollen sacs in the anthers. Each male gameto-phyte becomes mature when its generative cell divides mitotically to produce two sperm cells.

male gametophyte, which is embedded in the ovule, typically contains seven cells with eight haploid nuclei. Six of these cells, including the egg cell (the female gamete), contain a single nucleus each; a large central cell has two nuclei, called **polar nuclei.** The egg and both polar nuclei participate directly in fertilization.

Pollen sacs within the anther contain numerous diploid cells called *microsporocytes,* each of which undergoes meiosis to produce four haploid cells called **microspores.** Each microspore divides mitotically to produce an immature male gametophyte, also called a pollen grain, that consists of two cells, the tube cell and the generative cell. The pollen grain becomes mature when its generative cell divides to form two nonmotile sperm cells.

Review

- How do petals differ from sepals? How are they similar?
- How do stamens differ from carpels? How are they similar?
- What are female gametophytes, and where are they formed?
- What are male gametophytes, and where are they formed?

POLLINATION

Learning Objectives

3 Compare the evolutionary adaptations that characterize flowers pollinated in different ways (by insects, birds, bats, and wind).

4 Define *coevolution*, and give examples of ways that plants and their animal pollinators have affected one another's evolution.

Before fertilization can occur, pollen grains must travel from the anther (where they form) to the stigma. The transfer of pollen grains from anther to stigma is known as **pollination.** Plants are *self-pollinated* if pollination occurs within the same flower or a different flower on the same individual plant. When pollen grains are transferred to a flower on another individual of the same species, the plant is *cross-pollinated.* Flowering plants accomplish pollination in a variety of ways. Beetles, bees, flies, butterflies, moths, wasps, and other insects pollinate many flowers. Animals such as birds, bats, snails, and small nonflying mammals (rodents, primates, and marsupials) also pollinate plants. Wind is an agent of pollination for certain flowers; and water, for a few aquatic flowers.

Many plants have mechanisms to prevent self-pollination

In plant sexual reproduction, the two gametes that unite to form a zygote may be from the same parent or from two different parents. The combination of gametes from two different parents increases the variation in offspring, and this variation may confer a selective advantage. Some offspring, for example, may be able to survive environmental changes better than either parent can.

Plants have evolved a variety of mechanisms that prevent self-pollination and thus prevent **inbreeding,** which is the mating of genetically similar individuals. Inbreeding can increase the concentration of harmful genes in the offspring. Some plants, such as asparagus and willow, have separate male and female individuals; the male plants have staminate flowers that lack carpels, and the female plants have pistillate flowers that lack stamens. Other species have flowers with both stamens and pistils, but the pollen is shed from a given flower either before or after the time when the stigma of that flower is receptive to pollen. These characteristics promote **outcrossing** (also called outbreeding), which is the mating of dissimilar individuals.

Many species have genes for **self-incompatibility,** a genetic condition in which the pollen is ineffective in fertilizing the same flower or other flowers on the same plant. In other words, an individual plant can identify and reject its own pollen. Genes for self-incompatibility usually inhibit the growth of the pollen tube in the stigma and style, thereby preventing delivery of sperm cells to the ovules. Self-incompatibility, which is more common in wild species than in cultivated plants, ensures that reproduction occurs only if the pollen comes from a genetically different individual.

In plants such as oilseed rape, self-incompatibility is based on a high degree of variation at a particular locus called the S locus; the many alleles at this locus are designated S_1, S_2, S_3, S_4, and so on. Here is an example of how the S locus blocks self-fertilization: A plant with the genotype S_1S_2 produces pollen grains that land on a stigma of another plant with the genotype S_1S_3. In this case, the presence of the S_1 allele in both the pollen and stigma triggers a self-recognition signaling cascade in surface cells of the stigma. As a result, the stigma cells do not undergo changes that allow the pollen grain to grow a pollen tube. Therefore, fertilization does not occur.

The molecular basis of self-incompatibility in *Arabidopsis* and related plants is an area of active scientific interest. *Arabidopsis* can self-pollinate. Biologists have determined that the genes involved in self-incompatibility and outcrossing exist in *Arabidopsis* but have undergone mutations so that they are no longer functional. Thus, it appears that self-incompatibility in these plants is the ancestral (normal) condition.

Flowering plants and their animal pollinators have coevolved

Animal pollinators and the plants they pollinate have had such close, interdependent relationships over time that they have affected the evolution of certain physical and behavioral features in one another. The term **coevolution** describes such reciprocal adaptation, in which two species interact so closely that they become increasingly adapted to one another as they each undergo evolutionary change by *natural selection*. We now examine some of the features of flowers and their pollinators that may be the products of coevolution.

Flowers pollinated by animals have various features to attract their pollinators, including showy petals (a visual attractant) and scent (an olfactory attractant). One reward for the animal pollinator is food. Some flowers produce nectar, a sugary solution, in special floral glands called *nectaries*. Pollinators use nectar as an energy-rich food. Pollen grains are also a protein-rich food for many animals. As they move from flower to flower searching for food, pollinators inadvertently carry along pollen grains on their body parts, helping the plants reproduce sexually (Fig. 36-4).

Biologists estimate that insects pollinate about 70% of all flowering plant species. Bees are particularly important as pollinators of crop plants. Crops pollinated by bees provide about 30% of human food. Plants pollinated by insects often have blue or yellow petals (see Fig. 36-4a). The insect eye does not see color

(a) A honeybee (*Apis mellifera*) pollinates a Scotch broom flower (*Cytisus scoparius*). When the bee presses in to obtain nectar, the stamens spring out, dropping pollen grains on the bee's body.

(b) A ruby-throated hummingbird (*Archilochus colubris*) obtains nectar from a trumpet vine flower (*Campsis radicans*). The pollen grains on the bird's feathers are carried to the next plant.

Figure 36-4 Animal pollinators

(c) A lesser longnosed bat (*Leptonycteris curasoae*) obtains nectar from a cardon cactus flower (*Pachycereus*) and transfers pollen as it moves from flower to flower. Bats pollinate several hundred species of plants.

the same way the human eye does. Most insects see well in the violet, blue, and yellow range of visible light but do not perceive red as a distinct color. So flowers pollinated by insects are not usually red. Insects also see in the ultraviolet range, wavelengths that are invisible to the human eye. Insects see ultraviolet radiation as a color called *bee's purple*. Many flowers have dramatic UV markings that may or may not be visible to humans but that direct insects to the center of the flower where the pollen grains and nectar are (❚ Fig. 36-5).

Insects have a well-developed sense of smell, and many insect-pollinated flowers have a strong scent that may be pleasant or foul to humans. The carrion plant, for example, is pollinated by flies and smells like the rotting flesh in which flies lay their eggs. As flies move from one reeking flower to another looking for a place to lay their eggs, they transfer pollen grains.

Birds such as hummingbirds are important pollinators (see Fig. 36-4b). Flowers pollinated by birds are usually red, orange, or yellow, because birds see well in this range of visible light. Because most birds do not have a strong sense of smell, bird-pollinated flowers usually lack a scent.

Bats, which feed at night and do not see well, are important pollinators, particularly in the tropics (see Fig. 36-4c). Bat-pollinated flowers bloom at night and have dull white petals and a strong nighttime output of scent that usually smells like fermented fruit. Nectar-feeding bats are attracted to the flowers by the scent; they lap up the nectar with their long, extendible tongues. As they move from flower to flower, they transfer pollen grains. An unusual adaptation that encourages pollination by bats has evolved in the tropical vine *Mucana holtonii*. When the pollen in a given flower is mature, a concave petal lifts up. The petal bounces the echo from the bat's echo-locating calls back to the bat, helping it find the flower.

Specialized features such as petals, scent, and nectar to attract pollinators have coevolved with adaptations of animal pollinators. Specialized body parts and behaviors adapted animals to aid pollination as they obtain nectar and pollen grains as a reward. For example, coevolution has selected for bumblebees' hairy bodies, which catch and hold the sticky pollen grains for transport from one flower to another.

Coevolution may have led to the long, curved beaks of the 'i'iwi, one of the Hawaiian honeycreepers, that inserts its beak into tubular flowers to obtain nectar (see Figs. 1-11b and 20-16). The long, tubular corolla of the flowers that 'i'iwis visit probably also came about through coevolution. During the 20th century, some of the tubular flower species (such as lobelias) became rare, largely as a result of grazing by nonnative cows and feral goats, and about 25% of lobelia species have become extinct. The 'i'iwi now feeds largely on the flowers of the 'ohi'a tree, which lacks petals, and the 'i'iwi bill appears to be adapting to this change in feeding preference. Comparison of the bills of 'i'iwi museum

(a) The evening primrose (*Oenothera* sp.) as seen by the human eye is solid yellow.

(b) The same flower viewed under ultraviolet radiation provides clues about how the insect eye perceives it. The light blue portions of the petals appear purple to a bee, whereas the dark blue inner parts appear yellow.

Figure 36-5 *Animated* Ultraviolet markings on insect-pollinated flowers

These markings draw attention to the center of the flower, where the pollen grains and nectar are located.

Figure 36-6 An orchid flower that mimics bees, thereby accomplishing pollination

The woodcock orchid (*Ophrys scolopax*) attracts male bees with the similarity of its flowers to female bees and the production of a scent to which the males are irresistibly attracted. The woodcock orchid is native to southern Europe.

specimens collected in 1902 with the bills of live birds captured in the 1990s showed that 'i'iwi bills were about 3% shorter in the 1990s than in 1902.

Plants and the behavior of their animal pollinators have co-evolved, sometimes in bizarre and complex ways. The flowers of certain orchids (*Ophrys*) resemble female bees in coloring and shape (Fig. 36-6). The unpollinated flowers secrete a scent similar to that produced by female bees, and the males are irresistibly attracted to it. The resemblance between *Ophrys* flowers and female bees is so strong that male bees mount the flowers and try to copulate with them. During this *pseudocopulation,* a pollen sac usually attaches to the back of the bee. When the frustrated bee departs and tries to copulate with another orchid flower, pollen grains are transferred to that flower. Interestingly, once a flower has been pollinated, it emits a scent like that released by female bees that have already mated. Male bees have no interest in visit-

ing flowers that have been pollinated (just as they lose interest in female bees that have already been inseminated).

A single mutation in a gene for flower color can result in a shift in animal pollinators

Biologists used two species of monkey flowers, both native to western North America, to determine if a single gene mutation would have any effect on animal pollinators. The wild-type *Mimulus lewisii* has violet-pink flowers and is pollinated by bumblebees, whereas the closely related *M. cardinalis* has orange-red flowers and is pollinated by hummingbirds. Alleles at the same locus in both species affect flower color.

When the researchers transferred the allele for orange-red flowers from *M. cardinalis* to *M. lewisii,* the resulting *M. lewisii* flowers were yellow-orange instead of pink. These flowers were visited by 68 times as many hummingbirds as the wild-type *M. lewisii* flowers (Fig. 36-7). A shift in pollinators also occurred when *M. cardinalis* plants received the allele for pink flowers from *M. lewisii.* These *M. cardinalis* flowers were a dark pink in color and had 74 times as many bumblebee visits as the wild-type *M. cardinalis* flowers.

Biologists think that hummingbird-pollinated flowers arose from insect-pollinated flowers many times during the course of flowering plant evolution. These data suggest that a single gene

QUESTION: Can a single-gene mutation in flower color affect animal pollinators?

HYPOTHESIS: If flower color is changed by a mutation, pollinator preferences for the resulting flower may change.

EXPERIMENT: *Mimulus lewisii* has violet-pink flowers and is pollinated by bumblebees; the closely related *M. cardinalis* has orange-red flowers and is pollinated by hummingbirds. Alleles at the same locus in both species affect flower color.

Researchers transferred the allele for orange-red flowers from *M. cardinalis* to *M. lewisii*, which then produced yellow-orange petals. They also transferred the allele for pink flowers from *M. lewisii* to *M. cardinalis*, which then produced dark pink petals.

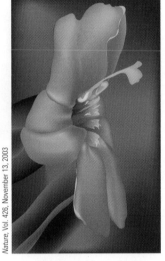

Wild-type *M. lewisii*   Wild-type *M. cardinalis*   Altered *M. lewisii*   Altered *M. cardinalis*

RESULTS AND CONCLUSION: The modified *M. lewisii* flowers were yellow-orange in color and had 68 times as many hummingbird visits as the unaltered *M. lewisii* flowers. The modified *M. cardinalis* flowers had dark pink petals and were visited by 74 times as many bumblebees as the unaltered *M. cardinalis* flowers. Thus, a single mutation in a gene for flower color resulted in a shift in animal pollinators.

Figure 36-7 Studying how pollinator preferences may have evolved

Biologists H. D. Bradshaw Jr. from the University of Washington and Douglas Schemske from Michigan State University used monkey flow-ers to determine if a single gene mutation would have any effect on animal pollinators.

mutation affecting petal color could have been partly responsible for the pollinator shift that resulted in the evolution of two species, *M. lewisii* and *M. cardinalis,* from their insect-pollinated common ancestor.

Some flowering plants depend on wind to disperse pollen

Some flowering plants, such as grasses, ragweed, maples, and oaks, are pollinated by wind. Wind-pollinated plants produce many small, inconspicuous flowers (▮ Fig. 36-8). They do not produce large, colorful petals, scent, or nectar. Some have large,

feathery stigmas, presumably to trap wind-borne pollen grains. Because wind pollination is a hit-or-miss affair, the likelihood of a particular pollen grain landing on a stigma of the same species of flower is slim. Wind-pollinated plants produce large quantities of pollen grains, which increase the likelihood that some pollen grains will land on the appropriate stigma.

Review

▮ A flower has yellow petals, a pleasant scent, and sugary nectar. How is it probably pollinated?

▮ A flower has small, inconspicuous flowers without petals, a scent, or nectar. How is it probably pollinated?

▮ What is coevolution?

FERTILIZATION AND SEED/FRUIT DEVELOPMENT

Learning Objectives

5 Distinguish between pollination and fertilization.

6 Trace the stages of embryo development in flowering plants, and list and define the main parts of seeds.

7 Explain the relationships among the following: ovules, ovaries, seeds, and fruits.

8 Distinguish among simple, aggregate, multiple, and accessory fruits.

Once pollen grains have been transferred from anther to stigma, the tube cell, one of the two cells in the pollen grain, grows a thin pollen tube down through the style and into an ovule in the ovary. How does the pollen tube "know" where to grow? Scientists have found that molecular signals from the female gametophyte guide the growing pollen tube toward the ovule. A small protein appears to be the chemical attractant in corn, but this compound or something like it has yet to be identified in other plants.

Dr. Jeremy Burgess /Science Photo Library/Photo Researchers, Inc.

Figure 36-8 Wind pollination

Each cluster of male oak (*Quercus*) flowers dangles from a tree branch and sheds a shower of pollen when the wind blows. These flowers lack petals.

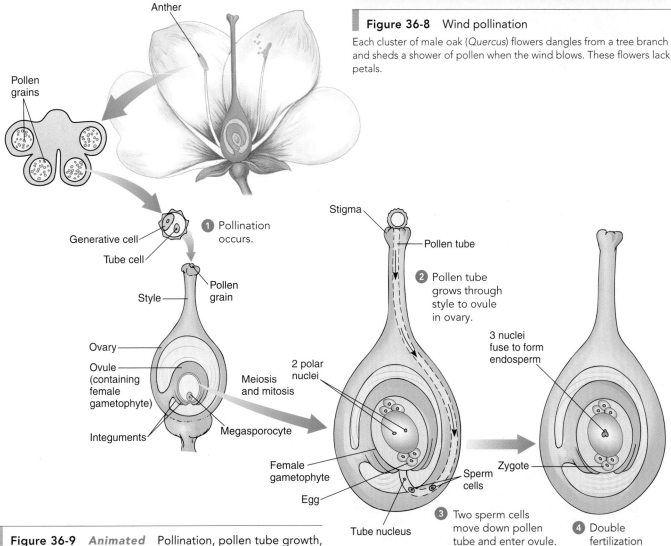

Figure 36-9 *Animated* Pollination, pollen tube growth, and double fertilization

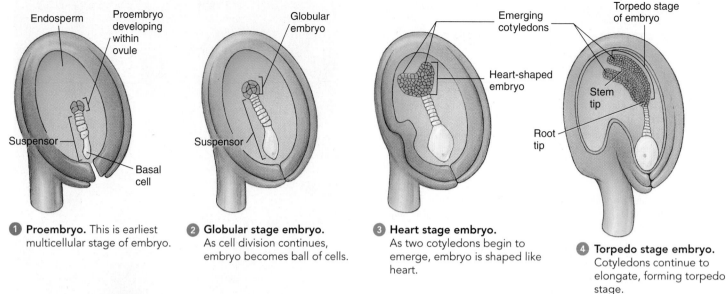

1 **Proembryo.** This is earliest multicellular stage of embryo.

2 **Globular stage embryo.** As cell division continues, embryo becomes ball of cells.

3 **Heart stage embryo.** As two cotyledons begin to emerge, embryo is shaped like heart.

4 **Torpedo stage embryo.** Cotyledons continue to elongate, forming torpedo stage.

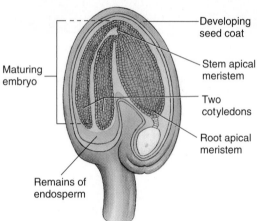

5 **Maturing embryo within seed.** Food originally stored in endosperm has been almost completely depleted during embryonic growth and development, and most of food for embryonic plant is now stored in cotyledons.

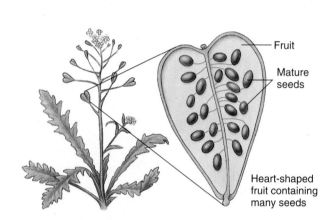

6 **Fruit.** Longitudinal section through fruit of shepherd's purse reveals numerous tiny seeds, each containing mature embryo. Each seed developed from ovule.

Figure 36-10 *Animated* Embryonic development in shepherd's purse (*Capsella bursa-pastoris*)

The ovule is shown apart from the ovary in parts **1** to **5** .

Once a pollen tube penetrates the ovule, the attracting signals are no longer produced. As a result, only one pollen tube enters each female gametophyte. The second cell within the pollen grain divides to form two male gametes (the sperm cells), which move down the pollen tube and enter the ovule (Fig. 36-9 on page 775).

A unique double fertilization process occurs in flowering plants

The egg within the ovule unites with one of the sperm cells, forming a zygote (fertilized egg) that develops into an embryonic plant contained in the future seed. The two polar nuclei in the central cell of the ovule fuse with the second sperm cell to form the first cell of the triploid ($3n$) **endosperm,** the tissue with nutritive and hormonal functions that surrounds the developing embryonic

plant in the seed. This process, in which two separate cell fusions occur, is called **double fertilization.** It is, with few exceptions, unique to flowering plants. (A type of double fertilization has been reported in two gymnosperm species, *Ephedra nevadensis* and *Gnetum gnemon.*)

After double fertilization, the ovule develops into a *seed* and the surrounding ovary develops into a *fruit* (see discussion in Chapter 28).

Embryonic development in seeds is orderly and predictable

Flowering plants package a young plant embryo, complete with stored nutrients, in a compact **seed,** which develops from the ovule after fertilization. The nutrients in seeds are not only used by *germinating* plant embryos but also eaten by animals, including humans. Development of the embryo and endosperm following fertilization is possible because of the constant flow of nutrients into the developing seed from the parent plant.

Cell divisions of the fertilized egg to form a multicellular embryo proceed in several ways in flowering plants. We describe eudicot embryonic development; monocot embryonic development is similar in the early stages.

The two cells (basal cell and apical cell) that are formed as a result of the first division of the fertilized egg establish *polarity,* or direction, in the embryo. The large **basal cell** (located toward the outside of the ovule) typically develops into a **suspensor,** an embryonic tissue that anchors the developing embryo and aids in nutrient uptake from the endosperm. The **apical cell** (toward the inside of the ovule) develops into the plant embryo. Scientists are currently studying molecular differences between the apical cell and basal cell to determine the initial molecular signals that are involved in establishing polarity.

Initially, the apical cell divides to form a small cluster of cells, called a *proembryo* (▌Fig. 36-10). As cell division continues, a sphere of cells, often called a *globular embryo,* develops. Cells begin to develop into specialized tissues during this stage. When the eudicot embryo starts to develop its two cotyledons (seed leaves), it has two lobes and resembles a heart; this is often called the *heart stage.* During the *torpedo stage,* the embryo continues to grow as the cotyledons elongate. As the embryo enlarges, it often curves back on itself and crushes the suspensor.

The mature seed contains an embryonic plant and storage materials

A mature seed contains an embryonic plant and food (stored in the cotyledons or endosperm), surrounded by a tough, protective **seed coat** derived from the **integuments,** which are the outermost layers of an ovule. The seed, in turn, is enclosed within a fruit.

The mature embryo within the seed consists of a short embryonic root, or **radicle;** an embryonic shoot; and one or two seed leaves, or **cotyledons** (▌Fig. 36-11). Monocots have a single cotyledon, whereas eudicots have two. The short portion of the embryonic shoot connecting the radicle to one or two cotyledons is the **hypocotyl.** The shoot apex, or terminal bud, located above the point of attachment of the cotyledon(s) is the **plumule.** After the radicle, hypocotyl, cotyledon(s), and plumule have formed, the young plant's development is arrested, usually by desiccation (drying out) or **dormancy** (a temporary state of arrested physiological activity). When conditions are right for continuing the developmental program, the seed **germinates,** or sprouts, and the embryo resumes growth.

Because the embryonic plant is nonphotosynthetic, it must be nourished during germination until it becomes photosynthetic and therefore self-sufficient. The cotyledons of many plants function as storage organs and become large, thick, and fleshy as they absorb the food reserves (starches, oils, and proteins) initially produced as endosperm. Seeds that store nutrients in cotyledons have little or no endosperm at maturity. Examples

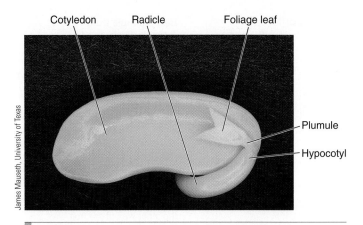

Cotyledon Radicle Foliage leaf

Plumule

Hypocotyl

James Mauseth, University of Texas

▌ **Figure 36-11** Seed structure

A bean seed has been dissected—its seed coat and one of its two cotyledons were removed—to show the radicle, hypocotyl, remaining cotyledon, and plumule with its foliage leaf at the shoot apex. This particular seed had begun germinating, so its radicle is larger than in an ungerminated seed.

of such seeds are peas, beans, squashes, sunflowers, and peanuts. Other plants—wheat and corn, for example—have thin cotyledons that function primarily to help the young plant digest and absorb food stored in the endosperm. (See Chapter 37 for a discussion of seed germination and early growth.)

Fruits are mature, ripened ovaries

After double fertilization takes place within the ovule, the ovule develops into a seed, and the ovary surrounding it develops into a **fruit.** For example, a pea pod is a fruit, and the peas within it are seeds. A fruit may contain one or more seeds; some orchid fruits contain several thousand to a few million seeds! Fruits provide protection for the enclosed seeds and sometimes aid in their dispersal.

There are several types of fruits; their differences result from variations in the structure or arrangement of the flowers from which they were formed. The four basic types of fruits—simple fruits, aggregate fruits, multiple fruits, and accessory fruits—are summarized in ▌Figure 36-12.

Most fruits are simple fruits. A **simple fruit** develops from a single ovary, which may consist of a single carpel or several fused carpels. At maturity, simple fruits may be fleshy or dry. Two examples of simple, fleshy fruits are berries and drupes. A **berry** is a fleshy fruit that has soft tissues throughout and contains few to many seeds; a blueberry is a berry, as are grapes, cranberries, bananas, and tomatoes. Many so-called berries do not fit the botanical definition. Strawberries, raspberries, and mulberries, for example, are not berries; these three fruits are discussed shortly.

A **drupe** is a simple, fleshy or fibrous fruit that contains a hard stone (pit) surrounding a single seed. Examples of drupes include peaches, plums, olives, avocados, and almonds. The almond shell is actually the stone, which remains after the rest of the fruit has been removed.

Many simple fruits are dry at maturity; some of these split open, usually along seams, called *sutures,* to release their seeds. A

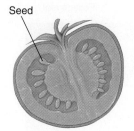

Berry (simple fruit)
A simple, fleshy fruit in which the fruit wall is soft throughout.

Tomato (*Lycopersicon lycopersicum*)

Seed

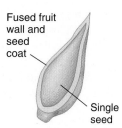

Caryopsis (simple fruit)
A simple, dry fruit in which the fruit wall is fused to the seed coat.

Wheat (*Triticum*)

Fused fruit wall and seed coat

Single seed

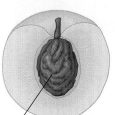

Drupe (simple fruit)
A simple, fleshy fruit in which the inner wall of the fruit is a hard stone.

Peach (*Prunus persica*)

Single seed inside stone

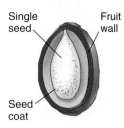

Achene (simple fruit)
A simple, dry fruit in which the fruit wall is separate from the seed coat.

Sunflower (*Helianthus annuus*)

Single seed

Fruit wall

Seed coat

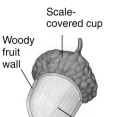

Nut (simple fruit)
A simple, dry fruit that has a stony wall, is usually large, and does not split open at maturity.

Oak (*Quercus*)

Scale-covered cup

Woody fruit wall

Single seed

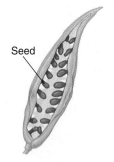

Follicle (simple fruit)
A simple, dry fruit that splits open along one suture to release its seeds; fruit is formed from ovary that consists of a single carpel.

Milkweed (*Asclepias syriaca*)

Seed

Aggregate fruit
A fruit that develops from a single flower with several to many pistils (i.e., carpels are not fused into a single pistil).

Blackberry (*Rubus*)

Seed

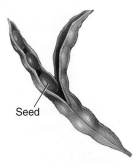

Legume (simple fruit)
A simple, dry fruit that splits open along two sutures to release its seeds; fruit is formed from ovary that consists of a single carpel.

Green bean (*Phaseolus vulgaris*)

Seed

Seed

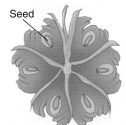

Multiple fruit
A fruit that develops from the ovaries of a group of flowers.

Mulberry (*Morus*)

Seed

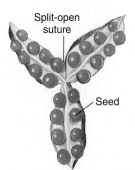

Capsule (simple fruit)
A simple, dry fruit that splits open along two or more sutures or pores to release its seeds; fruit is formed from ovary that consists of two or more carpels.

Iris (*Iris*)

Split-open suture

Seed

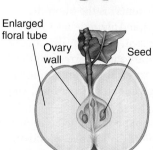

Accessory fruit
A fruit composed primarily of nonovarian tissue (such as the receptacle or floral tube).

Apple (*Malus sylvestris*)

Enlarged floral tube

Ovary wall

Seed

Figure 36-12 *Animated* Fruit types

Fruits are botanically classified into four groups—simple, aggregate, multiple, and accessory—based on their structure and mechanism of seed dispersal.

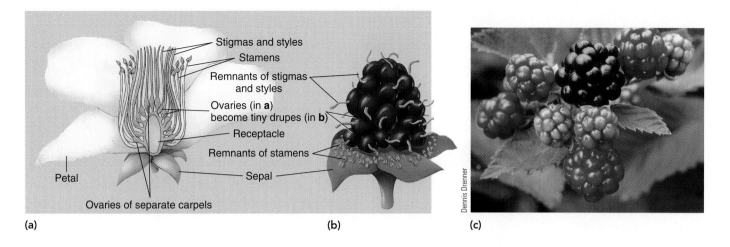

(a) **(b)** **(c)**

Dennis Drenner

Stigmas and styles
Stamens
Remnants of stigmas and styles
Ovaries (in **a**) become tiny drupes (in **b**)
Receptacle
Remnants of stamens
Sepal
Petal
Ovaries of separate carpels

Figure 36-13 An aggregate fruit

(a) Cutaway view of a blackberry (*Rubus*) flower, showing the many separate carpels in the center of the flower. **(b)** A developing blackberry fruit is an aggregate of tiny drupes. The little "hairs" on the blackberry are remnants of stigmas and styles. **(c)** Developing blackberry fruits at various stages of maturity.

milkweed pod is an example of a **follicle,** a simple, dry fruit that develops from a single carpel and splits open along one suture to release its seeds. A **legume** is a simple, dry fruit that develops from a single carpel and splits open along two sutures. Pea pods are legumes, as are green beans, although both are generally harvested before the fruit has dried out and split open. Pea seeds are usually removed from the fruit and consumed, whereas in green beans the entire fruit and seeds are eaten. A **capsule** is a simple, dry fruit that develops from two or more fused carpels and splits open along two or more sutures or pores. Iris, poppy, and cotton fruits are capsules.

Other simple, dry fruits, such as **caryopses** (sing., *caryopsis*), or **grains,** do not split open at maturity. Each caryopsis contains a single seed. Because the seed coat is fused to the fruit wall, a caryopsis looks like a seed rather than a fruit. Kernels of corn and wheat are fruits of this type.

An **achene** is similar to a caryopsis in that it is simple and dry, does not split open at maturity, and contains a single seed. However, the seed coat of an achene is not fused to the fruit wall. Instead, the single seed is attached to the fruit wall at one point only, permitting an achene to be separated from its seed. The sunflower fruit is an example of an achene. One can peel off the fruit wall (the shell) to reveal the sunflower seed within.

Nuts are simple, dry fruits that have a stony wall and do not split open at maturity. Unlike achenes, nuts are usually large, single seeded, and often derived from a compound ovary. Examples of nuts include chestnuts, acorns, and hazelnuts. Many so-called nuts do not fit the botanical definition. Peanuts and Brazil nuts, for example, are seeds, not nuts.

Aggregate fruits are a second main type of fruit. An **aggregate fruit** is formed from a single flower that contains several to many separate (free) carpels (❙ Fig. 36-13). After fertilization, each ovary from each individual carpel enlarges. As they enlarge, the ovaries may fuse to form a single fruit. Raspberries, blackberries, and magnolia fruits are examples of aggregate fruits.

A third type is the **multiple fruit,** formed from the ovaries of many flowers that grow in proximity on a common floral stalk.

The ovary from each flower fuses with nearby ovaries as it develops and enlarges after fertilization. Pineapples, figs, and mulberries are multiple fruits (❙ Fig. 36-14).

Accessory fruits are the fourth type. They differ from other fruits in that plant tissues in addition to ovary tissue make up the fruit. For example, the edible portion of a strawberry is the red,

Single female flower

Inflorescence (a cluster of flowers on a common floral stalk)

Multiple fruit

Figure 36-14 A multiple fruit

Mulberry (*Morus*) is formed from the ovaries and fleshy tepals (petals and sepals) of many flowers that fused to become a multiple fruit. Mulberry flowers are imperfect and contain either stamens or carpels. Also shown is an enlarged inflorescence of female flowers from which the mulberry fruit develops.

Reproduction in Flowering Plants   **779**

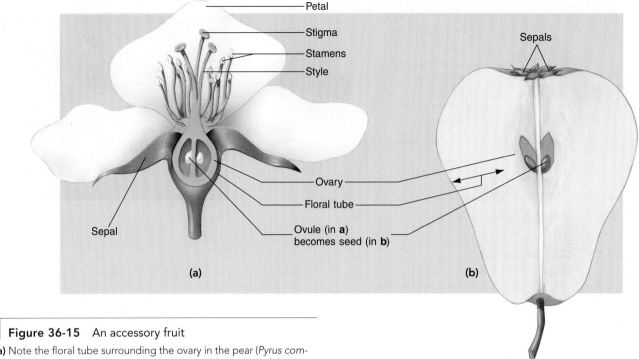

Petal
Stigma
Stamens
Style
Sepals
Ovary
Floral tube
Sepal
Ovule (in **a**)
becomes seed (in **b**)

(a)   **(b)**

Figure 36-15   An accessory fruit

(a) Note the floral tube surrounding the ovary in the pear (*Pyrus communis*) flower. This tube becomes the major edible portion of the pear. **(b)** Longitudinal section through a pear showing the fruit tissue, which derives from both the floral tube and the ovary.

fleshy receptacle. Apples and pears are also accessory fruits; the outer part of each of these fruits is an enlarged *floral tube,* consisting of receptacle tissue along with portions of the calyx, that surrounds the ovary (❚ Fig. 36-15).

Seed dispersal is highly varied

Wind, animals, water, and explosive dehiscence disperse the various seeds and fruits of flowering plants. Effective methods of seed dispersal have made it possible for certain plants to expand their geographic range. In some cases the seed is the actual agent of dispersal, whereas in others the fruit performs this role. In tumbleweeds, such as Russian thistle, the entire plant is the agent of dispersal because it detaches and blows across the ground, scattering seeds as it bumps along. Tumbleweeds are lightweight and are sometimes blown many kilometers by the wind.

Wind is responsible for seed dispersal in many plants. Plants such as maple trees have winged fruits adapted for wind dispersal. Light, feathery plumes enable other seeds or fruits to be transported by wind, often for considerable distances. Both dandelion fruits and milkweed seeds have this type of adaptation (❚ Fig. 36-16a).

Some plants have special structures that aid in dispersal of their seeds and fruits by animals. The spines and barbs of burdock burs and similar fruits catch in animal fur and fall off as the animal moves about (❚ Fig. 36-16b).

Fleshy, edible fruits are also adapted for animal dispersal. Birds, bats, primates, grazing ruminants, and ants are common dispersal agents. These animals are attracted to the fruit by its color, location, odor, and taste. As the animal eats these fruits, it either discards or swallows the seeds. Many seeds that are swallowed have thick seed coats and are not digested; instead, they pass through the digestive tract and are deposited in the animal's feces some distance from the parent plant. In fact, some seeds do not germinate unless they have passed through an animal's digestive tract; the animal's digestive juices probably aid germination by helping break down the seed coat. Some edible fruits apparently contain chemicals that function as laxatives to speed seeds through an animal's digestive tract; the less time these seeds spend in the gut, the more likely they are to germinate.

Animals such as squirrels and many bird species also help disperse acorns and other fruits and seeds by burying them for winter use. Many buried seeds are never used by the animal and germinate the following spring. Ants collect the seeds of many plants and take them underground to their nests. Ants disperse and bury seeds for hundreds of plant species in almost every terrestrial environment, from northern coniferous forests to tropical rain forests to deserts.

Both ants and flowering plants benefit from their association. The ants ensure the reproductive success of the plants whose seeds they bury, and the plants supply food to the ants. A seed that is collected and taken underground by ants often contains a special structure called an *elaiosome,* or *oil body,* that protrudes from the seed (❚ Fig. 36-17). Elaiosomes are a nutritious food for ants, which carry seeds underground before removing the elaiosome. Once an elaiosome is removed from a seed, the ants discard the undamaged seed in an underground refuse pile, which happens to be rich in organic material (such as ant droppings and dead ants) and minerals (inorganic nutrients) required by young seedlings. Thus, ants not only bury the seeds away from animals that might eat them but also place the seeds in rich soil that is ideal for germination and seedling growth.

The coconut is an example of a fruit adapted for dispersal by water. The coconut has air spaces that make it buoyant and capable of being carried by ocean currents for thousands of kilometers. When it washes ashore, the seed may germinate and grow into a coconut palm tree.

Some seeds are not dispersed by wind, animals, or water. Such seeds are found in fruits that use *explosive dehiscence,* in which the fruit bursts open suddenly and quite often violently, forcibly discharging its seeds (❙ Fig. 36-18). These fruits burst open due to pressure caused by differences in *turgor* (hydrostatic pressure) as the cells in the fruit dry out. The fruits of plants such as touch-me-not and bitter cress split open so explosively that seeds are scattered a meter or more.

(a) The feathery plumes of milkweed (*Asclepias syriaca*) seeds make them buoyant for dispersal by wind.

(b) Burdock (*Arctium minus*) burs (the hooked fruits) are carried away from the parent plant after sticking to bird feathers, mammal fur, or human clothing.

Figure 36-16 Seed (and fruit) dispersal

Review

- What is the difference between pollination and fertilization? Which process occurs first?
- What are the main stages of eudicot embryonic development?

- What is a fruit?
- How are the following fruits distinguished: simple, aggregate, multiple, and accessory?
- What are some characteristics of animal-dispersed seeds and fruits?

Figure 36-17 Dispersal of seeds by ants

The brown part of each bloodroot (*Sanguinaria canadensis*) seed is the seed proper, and the white part is the elaiosome, or oil body. The seeds have been placed along the midvein of an oak leaf to indicate scale.

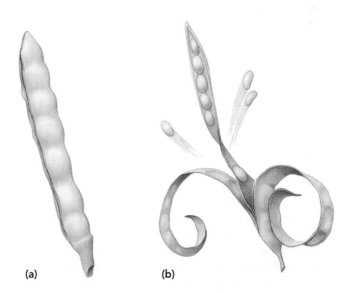

(a)   (b)

Figure 36-18 Explosive dehiscence

(a) An intact fruit of bitter cress (*Cardamine pratensis*) before it has opened. **(b)** The fruit splits open with explosive force, flinging the seeds some distance from the plant.

GERMINATION AND EARLY GROWTH

Learning Objective

9   Summarize the influence of internal and environmental factors on the germination of seeds.

Pollination and fertilization are followed by seed and fruit development in flowering plants. Each seed develops from an ovule and contains an embryonic plant and food to provide nourishment for the embryo during germination. **Germination**—the process of a seed sprouting—and the growth of young seedlings into mature plants are aspects of growth and development. Within a given species, the precise requirements for seed germination represent evolutionary adaptations that protect the young seedlings from adverse environmental conditions. Environmental cues, such as the presence of water and oxygen, proper temperature, and sometimes the presence of light penetrating the soil surface, influence whether or not a seed germinates.

No seed germinates unless it has absorbed water. The embryo in a mature seed is dehydrated, and a watery environment is necessary for active metabolism. When a seed germinates, its metabolic machinery is turned on, and numerous materials are synthesized and others degraded. Therefore, water is an absolute requirement for germination. The absorption of water by a dry seed is **imbibition.** As a seed imbibes water, it often swells to several times its original, dry size (Fig. 36-19). Cells imbibe water by adhesion of water onto and into materials such as cellulose, pectin, and starches within the seed.

Germination and subsequent growth require a great deal of energy. Because young plants obtain this energy by converting the energy of food molecules stored in the seed's endosperm or cotyledons to ATP during aerobic respiration, much oxygen is usually needed during germination. (Some plants, such as rice, grow in flooded soil where oxygen is absent and carry out *alcohol fermentation* during the early stages of germination and seedling growth.)

Temperature is another environmental factor that affects germination. Each species has an optimal, or ideal, temperature at which the germination percentage is highest. For most plants, the optimal germination temperature is between 25°C and 30°C (77°F and 86°F). Some seeds, such as those of apples, require prolonged exposure to low temperatures before their seeds break dormancy and germinate. Some of the environmental factors needed for germination help ensure the survival of the young plant. The requirement of a prolonged low-temperature period ensures that seeds adapted to temperate climates germinate in the spring rather than the fall.

Some plants, especially those with tiny seeds, such as lettuce, require light for germination. A light requirement ensures that a tiny seed germinates only if it is close to the soil surface. If such a seed were to germinate several centimeters below the soil surface, it might not have enough food reserves to grow to the surface. But if this light-dependent seed remains dormant until the soil is disturbed and it is brought to the surface, it has a much greater likelihood of survival.

Figure 36-19   Imbibition

Pinto bean (*Phaseolus vulgaris*) seeds before imbibition (*left*) and after (*right*). Dry seeds imbibe water before they germinate.

Some seeds do not germinate immediately

A mature seed is often dormant and may not germinate immediately even if growing conditions are ideal. In certain seeds, internal factors, which are under genetic control, prevent germination even when all external conditions are favorable. Many seeds are dormant because certain chemicals are present or absent or because the seed coat restricts germination. For example, the seeds of many desert plants contain high concentrations of *abscisic acid* (discussed in Chapter 37), which inhibits germination under unfavorable conditions. Abscisic acid is washed out only when rainfall is sufficient to support the plant's growth after the seed germinates.

Some seeds, such as certain legumes, have extremely hard, thick seed coats that prevent water and oxygen from entering, thereby inducing dormancy. *Scarification,* the process of scratching or scarring the seed coat (physically with a knife or chemically with an acid) before sowing it, induces germination in these plants. Scarification in nature occurs, for example, when these seeds pass through the digestive tracts of animals or when the seed coats are partially digested by soil bacteria.

Eudicots and monocots exhibit characteristic patterns of early growth

Once conditions are right for germination, the first part of the plant to emerge from the seed is the radicle, or embryonic root. As the root grows and forces its way through the soil, it encounters considerable friction from soil particles. A root cap protects the delicate apical meristem of the root tip.

The shoot (the stem with its leaves and reproductive structures) is next to emerge from the seed. Stem tips are not protected by a structure comparable to a root cap, but plants have ways to protect the delicate stem tip as it grows through the soil to the surface. The stem of a bean seedling (a eudicot), for instance, curves over to form a hook so that the stem tip and

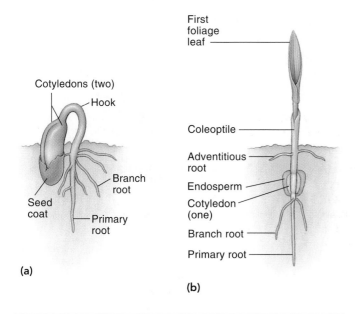

First
foliage
leaf

Cotyledons (two)

Hook

Coleoptile

Adventitious
root

Endosperm

Cotyledon
(one)

Branch
root

Seed
coat

Primary
root

Branch root

Primary root

(a)

(b)

Figure 36-20 *Animated* Germination and seedling growth

(a) Common bean, a eudicot. The hook in the stem protects the delicate stem tip as it grows through the soil. Once the shoot emerges from the soil, the hook straightens. **(b)** Corn, a monocot. The coleoptile, a sheath of cells, emerges first from the soil. The shoot and leaves grow through the middle of the coleoptile. (From Cecie Starr and Ralph Taggart, *Biology, Unity and Diversity of Life,* 9th ed., p. 547. Art by Raychel Ciemma.)

cotyledons are actually *pulled up* through the soil (∎ Fig. 36-20a). Corn and other grasses (monocots) have a special sheath of cells called a **coleoptile** that surrounds and protects the young shoot (∎ Fig. 36-20b). First, the coleoptile pushes up through the soil, and then the leaves and stem grow through the tip of the coleoptile.

Review

∎ What factors influence the germination of seeds?

∎ Why are plant growth and development so sensitive to environmental cues?

ASEXUAL REPRODUCTION IN FLOWERING PLANTS

Learning Objectives

10 Explain how the following structures may be used to propagate plants asexually: rhizomes, tubers, bulbs, corms, stolons, plantlets, and suckers.

11 Define *apomixis.*

Flowering plants have many kinds of asexual reproduction, several of which involve modified stems: rhizomes, tubers, bulbs, corms, and stolons. A **rhizome** is a horizontal underground stem that may or may not be fleshy. Fleshiness indicates that the rhizome is used for storing food materials, such as starch

(∎ Fig. 36-21a). Although rhizomes resemble roots, they are really stems, as indicated by the presence of scalelike leaves, buds, nodes, and internodes. (Roots have none of these features.) Rhizomes frequently branch in different directions. Over time, the old portion of the rhizome dies, and the two branches eventually separate to become distinct plants. Irises, bamboos, ginger, and many grasses are examples of plants that reproduce asexually by forming rhizomes.

Some rhizomes produce greatly thickened ends called **tubers,** which are fleshy underground stems enlarged for food storage. When the attachment between a tuber and its parent plant breaks, often as a result of the death of the parent plant, the tuber grows into a separate plant. Potatoes and elephant's ear (*Caladium*) are examples of plants that produce tubers (∎ Fig. 36-21b). The "eyes" of a potato are axillary buds, evidence that the tuber is an underground stem rather than a storage root such as a sweet potato or carrot.

A **bulb** is a modified underground bud in which fleshy storage leaves are attached to a short stem (∎ Fig. 36-21c). A bulb is globose (round) and covered by paperlike bulb scales, which are modified leaves. It frequently forms axillary buds that develop into small daughter bulbs (bulblets). These new bulbs are initially attached to the parent bulb, but when the parent bulb dies and rots away, each daughter bulb can become established as a separate plant. Lilies, tulips, onions, and daffodils are some plants that form bulbs.

A **corm** is a very short, erect underground stem that superficially resembles a bulb (∎ Fig. 36-21d). Unlike the bulb, whose food is stored in underground leaves, the corm's storage organ is a thickened underground stem covered by papery scales (modified leaves). Axillary buds frequently give rise to new corms; the death of the parent corm separates these daughter corms, which then become established as separate plants. Familiar garden plants that produce corms include crocus, gladiolus, and cyclamen.

Stolons, or **runners,** are horizontal, aboveground stems that grow along the surface and have long internodes (∎ Fig. 36-21e). Buds develop along the stolon, and each bud gives rise to a new shoot that roots in the ground. When the stolon dies, the daughter plants live separately. The strawberry plant produces stolons.

Some plants form detachable **plantlets** (small plants) in notches along their leaf margins. *Kalanchoe,* whose common name is "mother of thousands," has meristematic tissue that gives rise to an individual plantlet at each notch in the leaf (∎ Fig. 36-22). When these plantlets reach a certain size, they drop to the ground, root, and grow.

Some plants reproduce asexually by producing **suckers,** aboveground shoots that develop from adventitious buds on roots (∎ Fig. 36-23). Each sucker grows additional roots and becomes an independent plant when the parent plant dies. Examples of plants that form suckers include black locust, pear, apple, cherry, blackberry, and aspen. A quaking aspen (*Populus tremuloides*) colony in the Wasatch Mountains of Utah contains at least 47,000 tree trunks formed from suckers that can be traced back to a single individual; this massive "organism" occupies almost 43 hectares (106 acres). Some weeds, such as field bindweed, produce many suckers. These plants are difficult to control, because

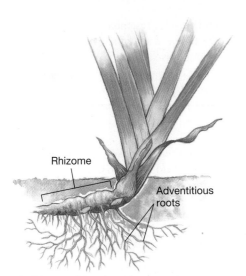

(a) Rhizome. Irises have horizontal underground stems called rhizomes. New aerial shoots arise from buds that develop on the rhizome.

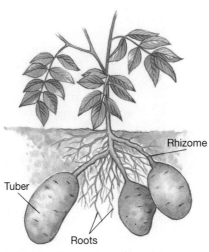

(b) Tuber. Potato plants form rhizomes, which enlarge into tubers (the potatoes) at the ends.

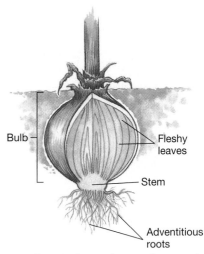

(c) Bulb. A bulb is a short underground stem to which overlapping, fleshy leaves are attached; most of the bulb consists of leaves.

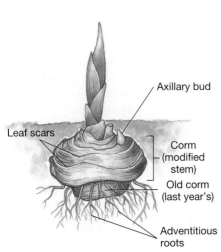

(d) Corm. A corm is an underground stem that is almost entirely stem tissue surrounded by a few papery scales.

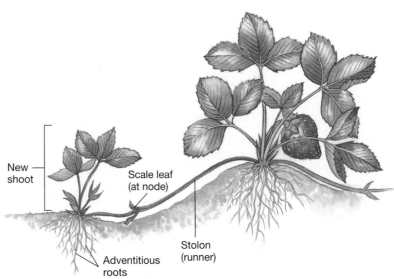

(e) Stolon. Strawberries reproduce asexually by forming stolons, or runners. New plants (shoots and roots) are produced at every other node.

Figure 36-21 Modified stems

pulling the plant out of the soil seldom removes all the roots, which can grow as deep as 3 m (10 ft). In fact, in response to wounding, the roots produce additional suckers, which can be a considerable nuisance to humans.

Apomixis is the production of seeds without the sexual process

Some flowering plants produce embryos in seeds without meiosis and the fusion of gametes. This asexual process is **apomixis.** For example, an embryo may develop from a diploid cell in the ovule rather than from a diploid zygote that forms from the union of two haploid gametes. Seed production by apomixis is a form of asexual reproduction. Because there is no fusion of gametes, the embryo is virtually genetically identical to the maternal genotype. However, the advantage of apomixis over other methods of asexual reproduction is that the seeds and fruits produced by apomixis can be dispersed by methods associated with sexual reproduction.

Apomixis, which occurs in various species of more than 40 angiosperm families, is often found in plants that are polyploid with varying degrees of sterility. Apomixis allows these sterile plants to reproduce and therefore survive. Examples of plants

Figure 36-22 Plantlets

The "mother of thousands" (*Kalanchoe*) produces detachable plantlets along the margins of its leaves. The young plantlets drop off and root in the ground.

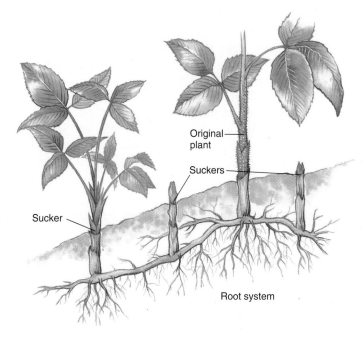

Original plant

Suckers

Sucker

Root system

Figure 36-23 Suckers in various stages of development

All the suckers are connected to one another through the root system.

that can reproduce by apomixis include dandelions, citrus trees, mango, blackberries, garlic, and certain grasses.

Plant biologists are actively engaged in identifying the genes involved in asexual reproduction by apomixis. If researchers can transfer these genes to crop plants that possess superior traits such as high yield, they may be able to develop apomictic crop plants whose traits do not get lost or diluted during the genetic shuffling of sexual reproduction.

Review

■ How are rhizomes and tubers involved in asexual reproduction?

■ How are corms and bulbs similar? How are they different?

■ What is apomixis? How is the ability to reproduce by apomixis advantageous?

A COMPARISON OF SEXUAL AND ASEXUAL REPRODUCTION

Learning Objective

12 State the differences between sexual reproduction and asexual reproduction, and discuss the evolutionary advantages and disadvantages of each.

Sexual reproduction and asexual reproduction are suited for different environmental circumstances. As you know, sexual reproduction results in offspring that are genetically different from the parents; that is, the parental genotypes are not preserved in the offspring. This genetic diversity of offspring may be selectively advantageous, particularly in an unstable, or changing, environment. (We define *environment* broadly to include all external conditions, both living and nonliving, that affect an organism.) If a plant species that reproduces sexually (and is therefore genetically diverse) is exposed to increasing annual temperatures as a result of global warming, for example, some of the individuals

may be more fit than either the parents or other offspring. The genetic diversity from sexual reproduction may also let the individuals of a species exploit new environments, thereby expanding their range.

You have seen that asexual reproduction results in offspring that are virtually genetically identical to the parent; that is, the parental genotype is preserved. Assuming the parent is well adapted to its environment (that is, the parent has a favorable combination of alleles), this genetic similarity may be selectively advantageous if the environment remains stable (unchanging) for several generations. None of the offspring of asexual reproduction is more fit than the parent, but neither is any of them less fit.

Despite the apparent advantages of asexual reproduction, most plant species whose reproduction is primarily asexual occasionally reproduce sexually. Even in what appears a stable environment, plants are exposed to changing selective pressures, such as changes in the number and kinds of predators and parasites, the availability of food, competition from other species, and climate. Sexual reproduction permits species whose reproduction is primarily asexual to increase their genetic variability so at least some individuals are adapted to the changing selective pressures in a stable environment.

Sexual reproduction has some disadvantages

Although the genetic diversity produced by sexual reproduction is advantageous to a species' long-term survival, sexual reproduction is a "costly" form of reproduction. Sexual reproduction requires both male and female gametes, which must meet for reproduction to occur. The many adaptations of flowers for different modes of pollination represent one cost of sexual reproduction.

Sexual reproduction produces some individuals with genotypes that are well adapted to the environment, but it also produces some individuals that are less well adapted. Therefore, sexual reproduction is usually accompanied by high death rates among offspring, particularly when selective pressures are strong. As discussed in Chapter 18, however, this aspect of sexual reproduction is an important part of evolution by natural selection.

Every biological process involves trade-offs, and sexual reproduction is no exception. Although sexual reproduction has its costs, the adaptive advantages of sexual reproduction clearly outweigh any disadvantages.

Review

- How does sexual reproduction in plants differ from asexual reproduction?
- What is an adaptive advantage of sexual reproduction? Why is sexual reproduction considered costly?

SUMMARY WITH KEY TERMS

Learning Objectives

1 Describe the functions of each part of a flower (page 768).
- **Sepals** cover and protect the flower parts when the flower is a bud. **Petals** play an important role in attracting animal pollinators to the flower.
- **Stamens** produce pollen grains. Each stamen consists of a thin stalk (the **filament**) attached to a saclike structure (the **anther**).
- The **carpel** is the female reproductive unit. A **pistil** may consist of a single carpel or a group of fused carpels. Each pistil has three sections: a **stigma**, on which the pollen grains land; a **style**, through which the pollen tube grows; and an **ovary** that contains one or more **ovules**.

ThomsonNOW™ **Learn more about flower structure by clicking on the figure in ThomsonNOW.**

2 Identify where eggs and pollen grains are formed within the flower (page 768).
- Pollen forms within pollen sacs in the anther. Each **pollen grain** contains two cells. One generates two sperm cells, and the other produces a **pollen tube** through which the sperm cells reach the ovule.
- An egg and two **polar nuclei**, along with several other nuclei, are formed in the ovule. Both egg and polar nuclei participate directly in fertilization.

ThomsonNOW™ **Watch plant reproduction unfold by clicking on the figure in ThomsonNOW.**

3 Compare the evolutionary adaptations that characterize flowers pollinated in different ways (by insects, birds, bats, and wind) (page 771).
- Flowers pollinated by insects are often yellow or blue and have a scent. Bird-pollinated flowers are often yellow, orange, or red and do not have a strong scent. Bat-pollinated flowers often have dusky white petals and are scented. Plants pollinated by wind often have smaller petals or lack petals altogether and do not produce a scent or nectar; wind-pollinated flowers make large amounts of pollen.

4 Define *coevolution,* and give examples of ways that plants and their animal pollinators have affected one another's evolution (page 771).
- **Coevolution** is reciprocal adaptation caused by two different species (such as flowering plants and their animal pollinators) forming an interdependent relationship and affecting the course of one another's evolution. For example, flowers with large, showy petals and scent have evolved in some plants, whereas hairy bodies that catch and hold sticky pollen grains have evolved in bees.

5 Distinguish between pollination and fertilization (page 775).
- **Pollination** is the transfer of pollen grains from anther to stigma. After pollination, **fertilization,** the fusion of gametes, occurs.
- Flowering plants undergo **double fertilization.** In the ovule, the egg fuses with one sperm cell, forming a zygote (fertilized egg) that eventually develops into a multicellular embryo in the **seed.** The two polar nuclei fuse with the second sperm cell, forming a triploid nutritive tissue called **endosperm.**

ThomsonNOW™ **Learn more about double fertilization by clicking on the figures in ThomsonNOW.**

6 Trace the stages of embryo development in flowering plants, and list and define the main parts of seeds (page 775).
- A eudicot embryo develops in the seed in an orderly fashion, from proembryo to globular embryo to the heart stage to the torpedo stage.
- A mature seed contains both a young plant embryo and nutritive tissue—stored in the endosperm or in the **cotyledons** (seed leaves)—for use during germination. A tough, protective **seed coat** surrounds the seed.

ThomsonNOW™ **Watch embryonic development in flowering plants by clicking on the figure in ThomsonNOW.**

7 Explain the relationships among the following: ovules, ovaries, seeds, and fruits (page 775).
- Ovules are structures with the potential to develop into seeds, whereas ovaries are structures with the potential to develop into fruits. Seeds are enclosed within **fruits,** which are mature, ripened ovaries.

8 Distinguish among simple, aggregate, multiple, and accessory fruits (page 775).
- **Simple fruits** develop from a single ovary that consists of one carpel or several fused carpels. **Aggregate fruits** develop from a single flower with many separate ovaries. **Multiple fruits** develop from the ovaries of many flowers growing in close proximity on a common axis. In **accessory fruits,** the major part of the fruit consists of tissue other than ovary tissue.

9 Summarize the influence of internal and environmental factors on the germination of seeds (page 782).
- **Germination** is the process of seed sprouting. Internal factors affecting whether a seed germinates include the maturity of the embryo; the presence or absence of chemical inhibitors; and the presence or absence of hard, thick seed coats.

■ External environmental factors that may affect germination include requirements for oxygen, water, temperature, and light. For example, before germinating, dry seeds absorb water by **imbibition.**

ThomsonNOW™ **Learn more about germination and seedling growth by clicking on the figure in ThomsonNOW.**

10 Explain how the following structures may be used to propagate plants asexually: rhizomes, tubers, bulbs, corms, stolons, plantlets, and suckers (page 783).

■ Rhizomes, tubers, bulbs, corms, and stolons are stems specialized for asexual reproduction. A **rhizome** is a horizontal underground stem. A **tuber** is a fleshy underground stem enlarged for food storage. A **bulb** is a modified underground bud with fleshy storage leaves attached to a short stem. A **corm** is a short, erect underground stem covered by papery scales. A **stolon** is a horizontal aboveground stem with long internodes.

■ Some leaves have meristematic tissue along their margins and give rise to detachable **plantlets.**

■ Roots may develop adventitious buds that develop into **suckers.** Suckers produce additional roots and may give rise to new plants.

11 Define *apomixis* (page 783).

■ **Apomixis** is the production of seeds and fruits without sexual reproduction.

12 State the differences between sexual reproduction and asexual reproduction, and discuss the evolutionary advantages and disadvantages of each (page 785).

■ Sexual reproduction involves the union of two gametes; the offspring produced by sexual reproduction are genetically variable. Asexual reproduction involves the formation of offspring without the fusion of gametes; the offspring are virtually genetically identical to the single parent.

■ The parental genotypes are not preserved in the offspring of sexual reproduction. Genetic diversity among offspring produced by sexual reproduction may let individuals survive in a changing environment or exploit new environments. Sexual reproduction is costly, because both male and female gametes must be produced and must meet.

■ The parental genotype is preserved in asexual reproduction. Genetic similarity may be advantageous if the environment is stable. Most plant species whose reproduction is primarily asexual occasionally reproduce sexually, thereby increasing their genetic variability.

TEST YOUR UNDERSTANDING

1. In flowering plants, the _____ is/are large (multicellular) and nutritionally independent. (a) gametes (b) microspores (c) megaspores (d) mature gametophyte (e) mature sporophyte

2. The normal order of whorls from the flower's periphery to the center is (a) sepals ⟶ petals ⟶ carpels ⟶ stamens (b) stamens ⟶ carpels ⟶ sepals ⟶ petals (c) sepals ⟶ petals ⟶ stamens ⟶ carpels (d) petals ⟶ carpels ⟶ stamens ⟶ sepals (e) carpels ⟶ stamens ⟶ petals ⟶ sepals

3. A pistil consists of (a) stigma, style, and stamen (b) anther and filament (c) sepal and petal (d) stigma, style, and ovary (e) radicle, hypocotyl, and plumule

4. The petals of a flower are collectively called a (a) calyx (b) capsule (c) carpel (d) cotyledon (e) corolla

5. The transfer of pollen grains from anther to stigma is (a) fertilization (b) double fertilization (c) pollination (d) germination (e) apomixis

6. The observation that insects with long mouthparts pollinate long, tubular flowers and insects with short mouthparts pollinate flowers with short corollas is explained by (a) coevolution (b) germination (c) double fertilization (d) apomixis (e) explosive dehiscence

7. The process of _____ in flowering plants involves one sperm cell fusing with an egg cell and one sperm cell fusing with two polar nuclei. (a) coevolution (b) germination (c) double fertilization (d) apomixis (e) pollination

8. The nutritive tissue in the seeds of flowering plants that is formed from the union of a sperm cell and two polar nuclei is called the (a) plumule (b) endosperm (c) cotyledon (d) hypocotyl (e) radicle

9. The _____ is a multicellular structure that anchors the embryo and aids in nutrient uptake from the endosperm. (a) proembryo (b) ovule (c) suspensor (d) cotyledon (e) pollen tube

10. After fertilization, the ovule develops into a _____ and the ovary develops into a _____. (a) fruit; seed (b) seed; fruit (c) calyx; corolla (d) corolla; calyx (e) follicle; legume

11. In plants that lack endosperm in their mature seeds, the cotyledons function to (a) enclose and protect the seed (b) aid in seed dispersal (c) serve as an absorptive embryonic root (d) store food reserves (e) attach the embryo within the ovule

12. _____ fruits develop from many ovaries of a single flower, whereas _____ fruits develop from the ovaries of many separate flowers. (a) multiple; accessory (b) simple; accessory (c) aggregate; multiple (d) accessory; aggregate (e) simple; multiple

13. Apples, strawberries, and pears are examples of what kind of fruit? (a) accessory (b) simple (c) multiple (d) aggregate (e) legume

14. A horizontal underground stem that may or may not be fleshy and that is often specialized for asexual reproduction is called a (a) stolon (b) bulb (c) corm (d) rhizome (e) tuber

15. Place the following events in the correct order.

1. pollen tube grows into ovule 2. insect lands on flower to drink nectar 3. embryo develops within the seed 4. fertilization occurs 5. pollen carried by insect contacts stigma

(a) 2, 5, 1, 4, 3 (b) 1, 4, 2, 5, 3 (c) 3, 2, 5, 1, 4 (d) 5, 1, 3, 4, 2 (e) 2, 5, 4, 3, 1

1. Sketch the kinds of flowers that form simple, aggregate, multiple, and accessory fruits.

2. Using what you have learned in this chapter, speculate whether it is more likely that offspring of asexual reproduction develop in close proximity to or widely dispersed from the parent plant. Explain your reasoning. How could you design an experiment to test your hypothesis?

3. Which type of reproduction, sexual or asexual, might be more beneficial in the following circumstances, and why: (a) a perennial (plant that lives more than two years) in a stable environment; (b) an annual (plant that lives one year) in a rapidly changing environment; and (c) a plant adapted to an extremely narrow climate range?

4. Using what you have learned in this chapter, offer an explanation of why telephone poles and wires strung across grassy fields or plains often have tree seedlings growing under them.

5. **Evolution Link.** Is seed dispersal by ants an example of co-evolution? Why or why not?

 Additional questions are available in ThomsonNOW at www.thomsonedu.com/login

Plant Growth and Development

Dwight R. Kuhn

Black-eyed Susan. This plant (*Rudbeckia hirta*), which grows to 0.9 m (3 ft), produces flowers in response to the shortening nights of spring and early summer.

KEY CONCEPTS

Growth and development are influenced by a plant's internal conditions and signals from its external environment.

A plant may respond to an external stimulus, such as light, gravity, or touch, by a directional growth response, or tropism. Study of tropisms has helped biologists elucidate links between the external environment and internal signals such as hormones.

Hormones are chemical signals responsible for coordinating and regulating many aspects of plant development.

Although there are many plant hormones, five have been well-characterized: auxins, gibberellins, cytokinins, ethylene, and abscisic acid.

Plants have different receptors that detect various colors of light. Phytochrome detects red and far-red light, which affect several aspects of development, including the timing of flowering.

The ultimate control of plant growth and development, which includes all the changes that take place during the entire life of an individual, is genetic. If the genes required for development of a particular trait, such as the shape of a leaf, the color of a flower, or the type of root system, are not present, that characteristic does not develop. When a particular gene is present, its expression—that is, how it exhibits itself as an observable feature of an organism—is determined by several factors, including signals from other genes and from the environment. The location of a cell in the young plant body also has a profound effect on gene expression during development. Chemical signals from adjacent cells help a cell "perceive" its location within the plant body. Each cell's spatial environment helps determine what that cell ultimately becomes.

Growth and development, including a plant's responses to various changes in its environment, are controlled by plant *hormones*, organic compounds that are present in very low concentrations in plant tissues and that act as chemical signals between cells. Environmental cues, such as changing day length and variations in precipitation and temperature, exert an important influence on gene expression and hormone production, as they do on all aspects of plant growth and development.

For example, the initiation of sexual reproduction is often under environmental control, particularly in temperate latitudes, and plants switch to reproductive growth after receiving appropriate signals from the environment. Many flowering plants are sensitive to changes in the relative amounts of daylight and darkness that accompany the changing seasons, and these plants flower in response to those changes (see photograph on page 789). Other plants have temperature requirements that induce sexual reproduction. Thus, plants continually perceive information from the environment and use this information to help regulate normal growth and development. We consider all of these aspects of growth and development in this chapter. ∎

TROPISMS

Learning Objective

1   Describe phototropism, gravitropism, and thigmotropism.

Plants exhibit movements in response to environmental stimuli such as light, gravity, and touch. A plant may respond to such an external stimulus by directional growth—that is, the direction of growth depends on the direction of the stimulus. Such a directional growth response, called a **tropism,** results in a change in the position of a plant part. Tropisms are irreversible and may be positive or negative, depending on whether the plant grows toward the stimulus (a positive tropism) or away from it (a negative tropism). Tropisms are under hormonal control, which is discussed later in the chapter.

Phototropism is the directional growth of a plant caused by light (∎ Fig. 37-1). Most growing shoot tips exhibit positive phototropism by bending (growing) toward light, something you may have observed if you place houseplants near a sunny window. This growth response increases the likelihood that stems and leaves receive adequate light for photosynthesis. The bending response of phototropism is triggered by blue light with wavelengths less than 500 nm. (You may recall from Chapter 33 that blue light also induces stomata to open.)

For a plant or any organism to have a biological response to light, it must contain a light-sensitive substance, called a *photoreceptor,* to absorb the light. The photoreceptor that absorbs blue light and triggers the phototropic response and other blue-light responses (such as stomatal opening) is a family of yellow pigments called **phototropins.** Phototropins are light-activated **kinases,** enzymes that transfer phosphate groups. There is evidence that phototropins become phosphorylated—that is, a phosphate

Figure 37-1 *Animated* Phototropism

Stems of corn (*Zea mays*) seedlings grow in the direction of light and therefore exhibit positive phototropism. The bending is caused by greater elongation on the shaded side of a stem than on the lighted side.

group is added—in response to blue light. Thus, phosphorylation is an early step in the blue light–signaling pathway.

Growth in response to the direction of gravity is called **gravitropism.** Most stem tips exhibit negative gravitropism by growing away from Earth's center, whereas most root tips exhibit positive gravitropism (∎ Fig. 37-2). The root cap is the site of gravity perception in roots; when the root cap is removed, the root con-

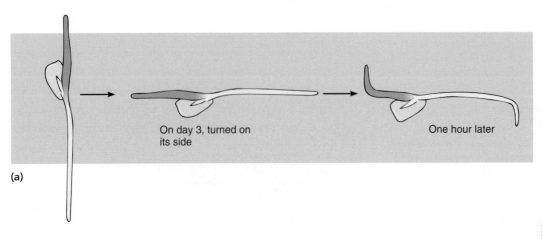

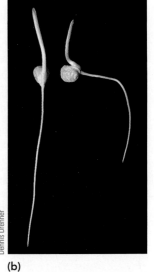

Figure 37-2 *Animated* Gravitropism

(a) A corn (*Zea mays*) seedling was turned on its side 3 days after germinating. In 1 hour the root and shoot tips had curved. **(b)** In 24 hours (*seedling on the right*), the new root growth was downward (positive gravitropism) and the new shoot growth was upward (negative gravitropism). A control seedling germinated at the same time is on the left.

tinues to grow, but it loses any ability to perceive gravity. Special cells in the root cap possess starch-containing **amyloplasts** that collect toward the bottom of the cells in response to gravity, and these amyloplasts may initiate at least some of the gravitropic response. If the root is put in a different position, as when a potted plant is laid on its side, the amyloplasts tumble to a new position, always settling in the direction of gravity. The gravitropic response (bending) occurs shortly thereafter and involves the hormone *auxin* (discussed later in the chapter). Despite the movement of amyloplasts in response to gravity, researchers question their role in gravitropism. A mutant *Arabidopsis* plant that lacks amyloplasts in its root cap cells still responds gravitropically when placed on its side, indicating that roots do not necessarily need amyloplasts to respond to gravity. Ongoing research may clarify how roots perceive gravity.

Thigmotropism is growth in response to a mechanical stimulus, such as contact with a solid object. The twining or curling growth of tendrils or stems, which helps attach a climbing plant such as a vine to some type of support, is an example of thigmotropism (see Fig. 33-13b).

Review

- What is phototropism? How does blue light trigger the phototropic response?
- How do phototropism and gravitropism differ?

PLANT HORMONES AND DEVELOPMENT

Learning Objectives

2   Describe a general mechanism of action for plant hormones, using auxin as your example.
3   Describe early auxin experiments involving phototropism.

4   List several ways that each of these hormones affects plant growth and development: auxins, gibberellins, cytokinins, ethylene, and abscisic acid.
5   Summarize the activities of these plant hormones and hormone-like signaling molecules: brassinosteroids, jasmonates, salicylic acid, systemin, and oligosaccharins.

A plant **hormone** is an organic compound that acts as a chemical signal eliciting a variety of responses that regulate growth and development. The study of plant hormones is challenging because hormones are effective in extremely small concentrations (less than 10^{-6} mol/L). In addition, the effects of different plant hormones overlap, and it is difficult to determine which hormone, if any, is the primary cause of a particular response. Plant hormones may also stimulate a response at one concentration and inhibit that same response at a different concentration.

Plant hormones are different from animal hormones in several ways. For example, most plant hormones are not produced in one part of the plant and transported over long distances to another part, where they cause a particular response; instead, the effects of plant hormones often occur close to where hormones are produced. Also, plant hormones are generally small molecules, not large, complex molecules like many animal hormones.

For many years, biologists studied five major classes of plant hormones: auxins, gibberellins, cytokinins, ethylene, and abscisic acid. More recently, researchers have uncovered compelling evidence for a variety of signaling molecules, such as brassinosteroids, jasmonates, salicylic acid, systemin, and oligosaccharins. (❙ Table 37-1 summarizes plant hormones and hormone-like signaling molecules discussed in this chapter.)

Plant hormones act by signal transduction

Researchers have used molecular genetic techniques to better understand the biology of plant hormones. *Arabidopsis* mutants are particularly useful. For example, different mutants have defects

TABLE 37-1

Plant Hormones and Signaling Molecules

Hormone	Site of Production	Principal Actions
Auxins (e.g., IAA)	Shoot apical meristem, young leaves, seeds	Stem elongation, apical dominance, root initiation, fruit development
Gibberellins (e.g., GA$_3$)	Young leaves and shoot apical meristems, embryo in seed	Seed germination, stem elongation, flowering, fruit development
Cytokinins (e.g., zeatin)	Roots	Cell division, delay of leaf senescence, inhibition of apical dominance, flower development, embryo development, seed germination
Ethylene	Stem nodes, ripening fruit, damaged or senescing tissue	Fruit ripening, responses to environmental stressors, seed germination, maintenance of apical hook on seedlings, root initiation, senescence and abscission in leaves and flowers
Abscisic acid	Almost all cells that contain plastids (leaves, stems, roots)	Seed dormancy, responses to water stress
Brassinosteroids (e.g., brassinolide)	Shoots (leaves and flower buds), seeds, fruits	Light-mediated gene expression, cell division, cell elongation, seed germination, vascular development
Jasmonates (e.g., jasmonic acid)	Leaves? Probably many tissues	Initiation of defenses against predators or disease organisms
Salicylic acid	Wound (site of infection)	Resistance to disease organisms
Systemin	Wound (site of herbivore or pathogen attack)	Initiation of defenses against predators (herbivores) or disease organisms
Oligosaccharins	Unknown	May function in normal cell growth and development; defense responses to disease organisms

Many plant hormones activate genes through a signal cascade involving a series of enzymes.

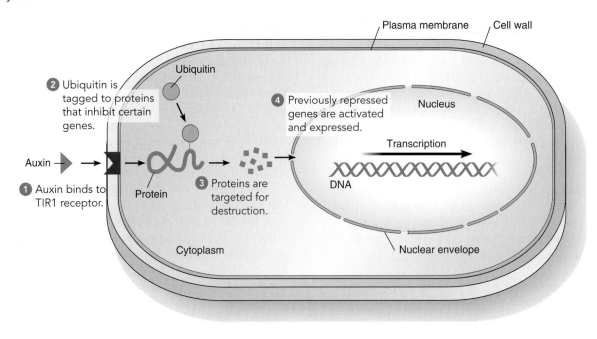

Figure 37-3 General mechanism of action for the hormone auxin

The numbered steps are explained in the text.

in hormone synthesis, hormone transport, signal reception, or signal transduction. In **signal transduction,** a receptor converts an extracellular signal into an intracellular signal that causes some change in the cell. Such mutants enable plant biologists to identify and clone genes involved in these aspects of hormone biology. Studying the mutant phenotypes helps plant biologists establish connections between the mutant genes and specific physiological activities involved in growth and development.

Using this research, biologists are elucidating the general mechanism of plant hormone action. It appears that many plant hormones bind to **enzyme-linked receptors** located in the plasma membrane; the hormone binds to the receptor, where it triggers an enzymatic reaction of some sort.

Let us consider a specific example that involves the plant hormone *auxin,* which is known to cause rapid changes in gene expression (Fig. 37-3). As shown in step ①, the receptor for auxin is located in the plasma membrane and has a three-dimensional shape that binds to the auxin molecule. The binding of auxin to its receptor catalyzes the attachment of the molecule *ubiquitin* to repressor proteins that inhibit certain genes (step ②). Whenever ubiquitin is attached to protein molecules, the cell targets those molecules for destruction (step ③). As a result, the genes that were repressed by those repressor proteins are now activated, resulting in changes in cell growth and development (step ④).

The identification of the receptor for auxin was reported in 2005 by two groups working independently. This receptor, called *transport inhibitor response 1 (TIR1)* protein, is an **F-box protein.** The *F-box* is a short sequence of amino acids found in molecules

that catalyze the addition of ubiquitin tags to proteins targeted for destruction. Interestingly, all animals, plants, and fungi examined to date use ubiquitin to target proteins for destruction, but bacteria do not. This suggests that the use of ubiquitin tags evolved early in eukaryote evolution and was conserved as the various groups of eukaryotes diversified. Plants have about 700 F-box proteins, but not much is known about most of them. Future research may identify other plant hormones and signaling molecules that use F-box proteins as receptors.

Auxins promote cell elongation

Charles Darwin, the British naturalist best known for developing the theory of natural selection to explain evolution, also provided the first evidence for the existence of auxins. The experiments that Darwin and his son Francis performed in the 1870s involved positive phototropism, the directional growth of plants toward light. The plants they used were newly germinated canary grass seedlings. As in all grasses, the first part of a canary grass seedling to emerge from the soil is the coleoptile, a protective sheath that encircles the stem. When coleoptiles are exposed to light from only one direction, they bend toward the light. The bending occurs below the tip of the coleoptile.

The Darwins tried to influence this bending in several ways (Fig. 37-4). For example, they covered the tip of the coleoptile as soon as it emerged from the soil. When they covered that part of the coleoptile above where the bend would be expected

QUESTION: What part of the grass coleoptile depends on light for phototropic growth?

HYPOTHESIS: The coleoptile tip is required for the growth (bending) of the coleoptile shaft that accompanies exposure to light coming from one direction.

EXPERIMENT: Some canary grass coleoptiles were uncovered, some were covered only at the tip, some had the tip removed, and some were covered everywhere but at the tip (left). The covers were impervious to light.

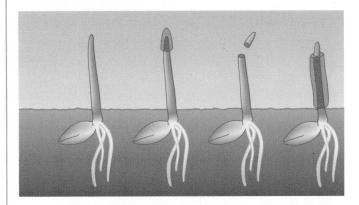 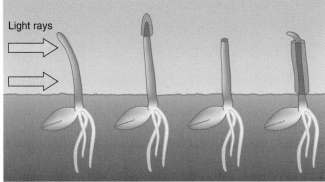

RESULTS AND CONCLUSION: After exposure to light coming from one direction, the uncovered plants and the plants with uncovered tips grew toward the light (right). The plants with tips covered or removed did not bend toward light. The Darwins concluded that some substance was produced in the tip and transmitted to the lower part, causing it to bend.

Figure 37-4 The Darwins' phototropism experiments

to occur, the plants did not bend. On other plants, they removed the coleoptile tip and found bending did not occur. When the bottom of the coleoptile where the curvature would occur was shielded from the light, the coleoptile bent toward light. From these experiments, the Darwins concluded that "some influence is transmitted from the upper to the lower part, causing it to bend."

In the 1920s, Frits Went, a young Dutch scientist, isolated the phototropic hormone from oat coleoptiles. He removed the coleoptile tips and placed them on tiny blocks of agar for a period of time. When he put one of these agar blocks squarely on a decapitated coleoptile, normal growth resumed. When he placed one of these agar blocks to one side of the tip of a decapitated coleoptile in the dark, bending occurred (❚ Fig. 37-5). The results indicated that the substance had diffused from the coleoptile tip into the agar and, later, from the agar into the decapitated coleoptile. Went named this substance **auxin** (from the Greek *aux,* "enlarge" or "increase"). The purification and elucidation of the primary auxin's chemical structure were accomplished in the mid-1930s by a research team led by U.S. biologist Kenneth Thimann at the California Institute of Technology.

Auxin is a group of several natural (and artificial) plant hormones; the most common and physiologically important auxin is **indoleacetic acid (IAA).** The movement of auxin in the plant is said to be *polar,* or unidirectional. Auxin moves downward along the shoot–root axis from its site of production, usually the shoot

apical meristem. Young leaves and seeds are also sites of auxin production.

Auxin's most characteristic action is promotion of cell elongation in stems and coleoptiles. This effect, apparently exerted by acidification of cell walls, increases their plasticity and thus enables them to expand under the force of the cell's internal turgor pressure. Auxin's effect on cell elongation also provides an explanation for phototropism. When a plant is exposed to a light from only one direction, some of the auxin migrates laterally to the shaded side of the stem before moving down the stem by polar transport. Because of the greater auxin concentration on the shaded side of the stem, the cells there elongate more than the cells on the light side, and the stem bends toward the light (❚ Fig. 37-6). Auxin is also involved in gravitropism and thigmotropism.

Auxin exerts other effects on plants. For example, some plants tend to branch out very little when they grow. Growth in these plants occurs almost exclusively from the apical meristem rather than from axillary buds, which do not develop as long as the terminal bud is present. Such plants are said to exhibit **apical dominance,** the inhibition of axillary bud growth by the apical meristem. In plants with strong apical dominance, auxin produced in the apical meristem inhibits axillary buds near the apical meristem from developing into actively growing shoots. When the apical meristem is pinched off, the auxin source is removed and axillary buds grow to form branches. Apical dominance is

Key Experiment

QUESTION: Is there a chemical substance responsible for elongation of coleoptiles, and can it be isolated?

HYPOTHESIS: The factor responsible for coleoptile growth is a diffusible chemical that can be isolated from coleoptile tips.

EXPERIMENT: Coleoptile tips were placed on agar blocks for a period of time **(a)**. The agar block was transferred to a decapitated coleoptile. It was placed off-center, and the coleoptile was left in darkness **(b)**.

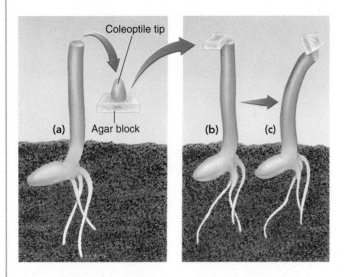

RESULTS AND CONCLUSION: The coleoptile bent **(c)**, indicating that a chemical moved from the original coleoptile tip to the agar block and from there to the same side of the decapitated coleoptile, causing that side to elongate.

Figure 37-5 *Animated* Isolating auxin from coleoptiles

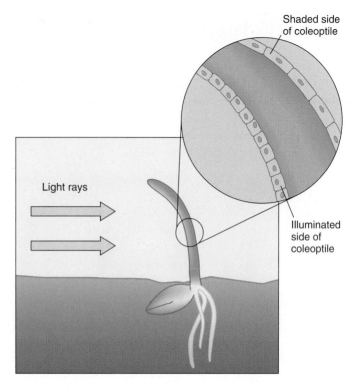

Figure 37-6 Phototropism and the unequal distribution of auxin

Auxin travels down the side of the stem or coleoptile *away* from the light, causing cells on the shaded side to elongate. Therefore, the stem or coleoptile bends toward light.

often quickly re-established, however, as one branch begins to inhibit the growth of others. Other hormones (ethylene and cytokinin, both discussed later) are also involved in apical dominance. As with many physiological activities, the changing ratios of these hormones may be the factor responsible for apical dominance.

Auxin produced by developing seeds stimulates the development of the fruit. When auxin is applied to certain flowers in which fertilization has not occurred (and, therefore, in which seeds are not developing), the ovary enlarges and develops into a seedless fruit. Seedless tomatoes are produced in this manner.[1] Auxin is not the only hormone involved in fruit development, however.

Some manufactured, or synthetic, auxins have structures similar to that of IAA. The synthetic auxin naphthalene acetic acid stimulates root development on stem cuttings and is used

for asexual propagation, particularly of woody plants with horticultural importance (■ Fig. 37-7). The synthetic auxins 2,4-D and 2,4,5-T have been used as selective herbicides (weed killers). These compounds kill plants with broad leaves but, for reasons not completely understood, do not kill grasses. Both herbicides are similar in structure to IAA and disrupt the plants' normal growth processes. Because many of the world's most important crops are grasses (such as wheat, corn, and rice), 2,4-D and 2,4,5-T can kill broadleaf weeds that compete with these crops. The use of 2,4,5-T is no longer allowed in the United States, however, because of its association with dioxins, a group of mildly to very toxic compounds formed as by-products during the manufacture of 2,4,5-T.

Gibberellins promote stem elongation

In the 1920s, a Japanese biologist was studying a disease of rice in which the young rice seedlings grew extremely tall and spindly, fell over, and died. The cause of the disease was a fungus (*Gibberella fujikuroi*) that produces a chemical substance named **gibberellin.** Not until after World War II did scientists in Europe and North America learn of the work done by the Japanese scientists. During the 1950s and 1960s, studies in the United States and Great Britain showed that gibberellins are produced by healthy plants as well as by the fungus.

---

[1]Not all seedless fruits are produced by treatment with auxin. In Thompson seedless grapes, fertilization occurs but the embryos abort, and therefore the seeds fail to develop. Thompson seedless grapes are sprayed with the hormone gibberellin to increase berry size.

Figure 37-7 Auxin and root development on stem cuttings

(*Left*) Many adventitious roots developed on a honeysuckle (*Lonicera fragrantissima*) cutting placed in a solution with a high concentration of synthetic auxin. (*Middle*) Fewer roots developed in a lower auxin concentration. (*Right*) The cutting placed in water (no auxin) served as a control and did not form roots in the same period.

Gibberellins are hormones involved in many normal plant functions. The symptoms of the seedling disease are caused by an abnormally high gibberellin concentration in the plant tissue (because both plant and fungus produce gibberellin). Currently, dozens of naturally occurring gibberellins are known, although many are probably inactive precursors; there are no synthetic gibberellins.

Gibberellins promote stem elongation in many plants. When gibberellin is applied to a plant, particularly certain dwarf varieties, this elongation may be spectacular. Some corn and pea plants that are dwarfs as a result of one or more mutations grow to a normal height when treated with gibberellin. Short-stemmed, high-yielding varieties of wheat have short stems because they have a reduced response to gibberellin. These varieties put less of their resources into stem height and more resources into grain production. Gibberellins are also involved in **bolting,** the rapid elongation of a floral stalk that occurs naturally in many plants when they initiate flowering (Fig. 37-8).

Gibberellins cause stem elongation by stimulating cells to divide as well as to elongate. The actual mechanism of cell elongation differs from that caused by auxin, however. Recall that IAA-induced cell elongation involves the acidification of the cell wall. In gibberellin-induced cell elongation, cell wall acidification does not occur; biologists do not yet know how gibberellins modify cell wall properties to cause cell elongation.

Gibberellins affect several reproductive processes in plants. They stimulate flowering, particularly in long-day plants (discussed later in the chapter). In addition, gibberellins substitute for the low temperature that biennials require before they begin flowering. If gibberellins are applied to biennials during their first year of growth, flowering occurs without exposure to a period of low temperature. Gibberellins, like auxin, affect fruit development. Agriculturalists apply gibberellins to several varieties of grapes to produce larger fruits.

Gibberellins are also involved in seed germination in certain plants. In a classic experiment involving barley seed germination,

Figure 37-8 Gibberellin and stem elongation

Like many biennials, Indian blanket (*Gaillardia pulchella*) grows as a rosette, which is a circular cluster of leaves close to the ground, during its first year (*left*). It then bolts when it initiates flowering in the second year (*right*). The plant in bloom grows to 0.6 m (2 ft).

researchers showed that the release of gibberellin from the embryo triggers the synthesis of α-amylase, an enzyme that digests starch in the endosperm. As a result, glucose becomes available for absorption by the embryo. Although enzymes mobilize starch reserves in many types of seeds, gibberellin control of seed enzymes appears restricted to cereals and other grasses. In addition to mobilizing food reserves in newly germinated grass seeds, application of gibberellins substitutes for low-temperature or light requirements for germination in seeds of plants such as lettuce, oats, and tobacco.

Cytokinins promote cell division

During the 1940s and 1950s, researchers were trying to find substances that might induce plant cells to divide in **tissue culture,** a technique in which cells are isolated from plants and grown in a nutrient medium (see *Focus On: Cell and Tissue Culture*). They discovered that cells would not divide without a substance found in coconut milk. Because coconut milk has a complex chemical composition, investigators did not chemically identify the division-inducing substance for some time. Finally, researchers isolated an active substance from a different source, aged DNA from herring sperm. They called it **cytokinin** because it induces cytokinesis, or cytoplasmic division. In 1963, researchers identified the first plant cytokinin, zeatin, in corn. Since then, similar molecules have been identified in other plants. Biologists have also synthesized several cytokinins. Cytokinins are structurally similar to adenine, a purine base that is part of DNA and RNA molecules.

CELL AND TISSUE CULTURE

Cells can be isolated from certain plants and grown in a chemically defined, sterile nutrient medium. In initial experiments with such cultures, plant cells could be kept alive, but they did not divide. Researchers later discovered that adding certain natural materials, such as the liquid endosperm of coconut, also known as coconut milk, induced cells to divide in culture. By the late 1950s, plant cells from a variety of sources could be cultured successfully, dividing to produce a mass of disorganized, relatively undifferentiated cells, or callus.

In 1958, F. C. Steward, a plant physiologist at Cornell University, succeeded in generating an entire carrot plant from a single callus cell derived from a carrot root (see Fig. 17-2). This demonstrated that each plant cell contains a genetic blueprint for all features of an entire organism. His work also showed that an entire plant can be grown from a single cell, provided the proper genes are expressed at the appropriate times.

Since Steward's pioneering work, biologists have successfully cultured many plants by using a variety of cell sources. Plants have been regenerated from different tissues, organ explants (excised organs or parts such as root apical meristems), and single cells. Cell and tissue culture techniques help answer many fundamental questions involving growth and development in plants. These techniques also have great practical potential. Using tissue culture, researchers can regenerate many genetically identical plants from the cells of a single, genetically superior plant. Many kinds of plants—from orchids and African violets to coastal redwoods—have been cultured in this way.

It is also possible to alter the genetic composition of a cell while it is in culture and then have these changes expressed in the whole plant following regeneration. Thus, tissue culture provides a valuable tool to genetic engineers who wish to introduce desirable new traits, such as better nutritional properties, into crop species such as rice (see Fig. 15-17b).

Cytokinins promote cell division and differentiation of young, relatively unspecialized cells into mature, more specialized cells in intact plants. They are a required ingredient in any plant tissue culture medium and must be present for cells to divide. In tissue culture, cytokinins interact with auxin during the formation of plant organs such as roots and stems (█ Fig. 37-9). For example, in tobacco tissue culture a high ratio of auxin to cytokinin induces root formation, whereas a low ratio of auxin to cytokinin induces shoot formation.

Cytokinins and auxin also interact in the control of apical dominance. Here their relationship is antagonistic: auxin inhibits the growth of axillary buds, and cytokinin promotes their growth.

One effect of cytokinins on plant cells is to delay the aging process. Plant cells, like all living cells, go through a natural aging process known as **senescence.** Senescence is accelerated in cells of plant parts that are cut, such as flower stems. Botanists think that plants must have a continual supply of cytokinins from the roots. Cut stems, of course, lose their source of cytokinins and therefore age rapidly. When cytokinins are sprayed on leaves of a cut stem of many species, they remain green, whereas unsprayed leaves turn yellow and die.

Despite their involvement in regulating many aspects of plant growth and development, cytokinins currently have few commercial applications other than plant tissue culture. However, in 1995, molecular biologists at the University of Wisconsin combined a promoter from a gene activated during normal senescence with a gene that encodes an enzyme involved in cytokinin synthesis. (Recall from Chapter 13 that the **promoter** is the nucleotide sequence in DNA to which RNA polymerase attaches to begin transcription.) The leaves of transgenic tobacco plants that contained this recombinant DNA produced more cytokinin and therefore lived longer and continued to photosynthesize

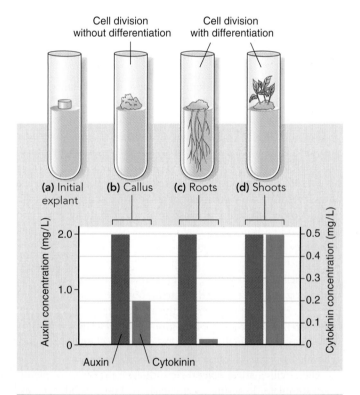

Figure 37-9 Auxin–cytokinin interactions in tissue culture

Varying amounts of auxin and cytokinin in the culture media produce different growth responses. **(a)** The initial explant is a small piece of sterile tissue from the pith of a tobacco stem, which is placed on nutrient agar. **(b)** Nutrient agar containing 2.0 mg/L of auxin and 0.2 mg/L of cytokinin causes cells to divide and form a clump of undifferentiated tissue called a callus. **(c)** Agar with 2.0 mg/L of auxin and 0.02 mg/L of cytokinin (high ratio of auxin to cytokinin) stimulates root growth. **(d)** Agar with 2.0 mg/L of auxin and 0.5 mg/L of cytokinin (low ratio of auxin to cytokinin) stimulates shoot growth.

Figure 37-10 Cytokinin synthesis and delay of senescence

The tobacco (*Nicotiana tabacum*) plant on the left was genetically engineered to produce additional cytokinin as it aged, whereas the tobacco plant of the same age on the right served as a control. Note the extensive senescence and death of older leaves on the control plant. Depending on the variety, tobacco grows 0.9 to 3 m (3 to 10 ft) tall.

(Fig. 37-10). This molecular technology has the potential to increase the longevity and productivity of certain crops.

Ethylene promotes abscission and fruit ripening

During the early 20th century, scientists observed that the gas **ethylene** (C_2H_4) has several effects on plant growth, but not until 1934 did they demonstrate that plants produce ethylene. This natural hormone influences many plant processes. Ethylene inhibits cell elongation, promotes seed germination, promotes apical dominance, and is involved in plant responses to wounding or invasion by disease-causing microorganisms.

Ethylene also has a major role in many aspects of senescence, including fruit ripening. As a fruit ripens, it produces ethylene, which triggers an acceleration of the ripening process. This induces the fruit to produce more ethylene, which further accelerates ripening. The expression "one rotten apple spoils the barrel" is true. A rotten apple is one that is overripe and produces large amounts of ethylene, which diffuses and triggers the ripening process in nearby apples. Humans use ethylene commercially to uniformly ripen bananas and tomatoes. These fruits are picked while green and shipped to their destination, where they are exposed to ethylene before they are delivered to grocery stores (Fig. 37-11).

Plants growing in a natural environment encounter rain, hail, wind, and contact with passing animals. All these mechanical stressors alter their growth and development, making them shorter and stockier than plants grown in a greenhouse. Ethylene regulates such developmental responses to mechanical stimuli, known as **thigmomorphogenesis.** Plants that are mechanically disturbed produce additional ethylene, which in turn inhibits stem elongation and enhances cell wall thickening in supporting cells (collenchyma and sclerenchyma). These changes are adaptive because shorter, thicker stems are less likely to be damaged by mechanical stressors.

Ethylene is involved in leaf abscission, which is actually influenced by two antagonistic hormones, ethylene and auxin. As a leaf ages (as autumn approaches, for deciduous trees in temperate climates), the level of auxin in the leaf decreases. Concurrently, cells in the abscission layer at the base of the petiole (where the leaf will break away from the stem) begin producing ethylene.

Abscisic acid promotes seed dormancy

In 1963, two different research teams discovered **abscisic acid.** Despite its name, abscisic acid does not induce abscission in most plants. Instead, abscisic acid is involved in a plant's response to stress and in seed **dormancy,** a temporary state of arrested physiological activity. As an environmental stress hormone, abscisic acid particularly promotes changes in plant tissues that are water stressed. (Recall that ethylene also affects plant responses to certain stressors, such as mechanical stressors and wounding.)

Figure 37-11 Ethylene and fruit ripening

Both boxes of tomatoes were picked at the same time, while green. The tomatoes in the box on the right were exposed to an atmosphere containing 100 ppm of ethylene for 3 days.

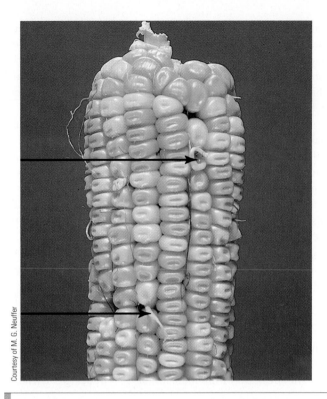

Courtesy of M. G. Neuffer

Figure 37-12 Abscisic acid and seed germination

In a corn (*Zea mays*) mutant that does not produce abscisic acid, some of the kernels have germinated while still on the ear, producing roots (*arrows*).

Botanists best understand the effect of abscisic acid on plants suffering from water stress. The level of abscisic acid increases dramatically in the leaves of plants exposed to severe drought conditions. The high level of abscisic acid in the leaves activates a signal transduction process that leads to the closing of stomata. The closing of stomata saves the water that the plant would normally lose by transpiration, thereby increasing the plant's likelihood of survival. As knowledge of abscisic acid signaling in guard cells increases, botanists hope to use this information to engineer crops and horticultural plants that are resistant to drought.

The onset of winter is also a type of stress on plants. A winter adaptation that involves abscisic acid is dormancy in seeds. Many seeds have high levels of abscisic acid in their tissues and do not germinate until the abscisic acid washes out. In a corn mutant unable to synthesize abscisic acid, the seeds germinate as soon as the embryos are mature, even while attached to the ear (Fig. 37-12).

Abscisic acid is not the only hormone involved in seed dormancy. For example, addition of gibberellin reverses the effects of dormancy. In seeds, the level of abscisic acid decreases during the winter, and the level of gibberellin increases. Cytokinins are also implicated in breaking dormancy. Once again, you see that a single physiological activity such as seed dormancy may be controlled in plants by the interaction of several hormones. The plant's actual response may result from changing ratios of hormones rather than the effect of each individual hormone.

Additional signaling molecules affect growth and development, including plant defenses

Biologists continue to discover new plant hormones and hormone-like signaling molecules. Many of these signaling molecules are involved in defensive responses of plants to disease organisms and insects. Here we briefly consider five groups—the brassinosteroids, jasmonates, salicylic acid, systemin, and oligosaccharins.

Brassinosteroids are steroid hormones found in plants

Although steroid hormones have crucial roles in animals, biologists are just beginning to understand the roles of these hormones in plants. The **brassinosteroids (BRs)** are a group of steroids that function as plant hormones. BRs are involved in several aspects of growth and development. *Arabidopsis* mutants that cannot synthesize BRs are dwarf plants with reduced fertility. Researchers reverse this defect by applying BR. Studies of these mutants suggest that the brassinosteroids are involved in multiple developmental processes, such as cell division, cell elongation, light-induced differentiation, seed germination, and vascular development.

Parts of the BR signaling pathway have been characterized. BRs bind to receptors in the plant plasma membrane (unlike steroid hormones in animals, which enter the cell and bind to receptors in the cytoplasm or nucleus). The binding of BR to its receptor triggers a signaling cascade that alters gene expression in the cell.

Jasmonates are derivatives of fatty acids

Jasmonates affect several plant processes, such as pollen development, root growth, fruit ripening, and senescence. These lipid-derived plant hormones, which are structurally similar to *prostaglandins* in animals, are also produced in response to the presence of insect pests and disease-causing organisms.

Jasmonates trigger the production of enzymes that confer an increased resistance against herbivorous (plant-eating) insects. For example, caterpillar-infested tomato plants release volatile jasmonates into the air that attract natural caterpillar enemies, such as parasitic wasps that lay their eggs on the caterpillar's bodies. When the eggs hatch, the wasp larvae consume, and ultimately kill, the host insects. In one study, treating plants with jasmonic acid increased parasitism on caterpillar pests twofold over control fields in which the plants did not have jasmonic acid applications. Jasmonates may be of practical value in controlling certain insect pests without the use of chemical pesticides.

About 20 different jasmonates have been identified and characterized. Biologists have made progress in identifying key aspects of the jasmonate signaling pathway, although important steps are not yet known. Interestingly, the jasmonate signaling pathway includes ubiquitin, which was discussed earlier in the chapter in relation to auxin (see Fig. 37-3). As in the auxin signaling pathway, the ubiquitin tags in the jasmonate signaling pathway may cause degradation of as-yet-unidentified transcription repressors, thereby altering gene expression.

Salicylic acid is a phenolic compound

For centuries people chewed willow (*Salix*) bark to treat headaches and other types of pain. **Salicylic acid** was first extracted from willow bark and is chemically related to aspirin (acetylsalicylic acid). More recently, biologists have shown that salicylic acid is a signaling molecule that helps plants defend against insect pests and pathogens such as viruses. When a plant is under attack, the concentration of salicylic acid increases, and it spreads systemically throughout the plant. Biologists think that salicylic acid binds to a cell receptor, triggering a signal transduction pathway that switches on genes. These genes code for proteins that establish **systemic acquired resistance** to fight infection and promote wound healing.

Plants also use salicylic acid to signal nearby plants. Tobacco plants infected with tobacco mosaic virus release into the air a volatile form of salicylic acid known as *methyl salicylate*, or *oil of wintergreen*. When nearby healthy plants receive the airborne chemical signal, they begin synthesizing antiviral proteins that enhance their resistance to the virus.

Systemin is a small polypeptide

Although many animal hormones are polypeptides, the first plant polypeptide with hormonal properties was not isolated until 1991. This polypeptide, **systemin,** consists of 18 amino acids and is transported systemically throughout the plant in response to wounding by insects. Systemin may stimulate natural defense mechanisms at extremely low concentrations, as low as one part per trillion. Systemin may trigger the plant to activate genes that produce *protease inhibitors,* molecules that disrupt insect digestion, curbing leaf damage done by caterpillars and other herbivorous insects. The discovery of systemin in tomato leaves prompted a flurry of research in search of polypeptide regulators and resulted in the discovery of additional polypeptides in other plants.

Oligosaccharins are composed of sugar residues

Oligosaccharins are carbohydrate fragments that consist of short, branched chains of sugar molecules. They are present in extremely small quantities in cells and active at much lower concentrations (100 to 1000 times as low) than hormones such as auxin. Different oligosaccharins may have distinct functions. Some trigger the production of **phytoalexins** (from the Greek *phyto,* "plant," and *alexi,* "to ward off"), antimicrobial compounds that limit the spread of plant pathogens such as fungi. Other oligosaccharins inhibit flowering and induce vegetative growth.

Progress is being made in identifying the elusive flower-promoting signal

Experiments in which different tobacco species are grafted together indicate that unidentified flower-promoting and flower-inhibiting substances may exist. *Nicotiana silvestris* is a *long-day* tobacco plant (it requires a short night to flower); a variety of *N. tabacum* is a *day-neutral* tobacco plant (seasonal changes do

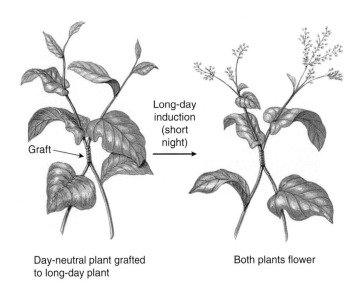

Figure 37-13 Evidence for the existence of a flower-promoting substance

When a long-day tobacco plant (*Nicotiana silvestris*) is grafted to a day-neutral tobacco plant (*N. tabacum*) and both plants are exposed to a long-day–short-night regimen, they both flower. The day-neutral plant flowers sooner than it normally would, presumably because a flower-promoting substance passes from the long-day plant to the day-neutral one through the graft.

not affect when it flowers). When a long-day tobacco is grafted to a day-neutral tobacco and exposed to short nights, both plants flower (Fig. 37-13). The day-neutral tobacco plant flowers sooner than it normally would.

Biologists hypothesize that a flower-promoting substance, **florigen,** may be induced in the long-day plant and transported to the day-neutral plant through the graft union, causing the day-neutral tobacco plant to flower sooner than expected. An intact plant may produce florigen in the leaves and transport it in the phloem to the shoot apical meristem. There, it induces a transition from vegetative to reproductive development—that is, to a meristem that produces flowers.

When a botanist grafts a long-day tobacco to a day-neutral tobacco and exposes them to long nights, neither plant flowers. As long as these conditions continue, the day-neutral plants do not flower even when they would normally do so. In this case, the long-day tobacco may produce a flower-inhibiting substance that is transported to the day-neutral tobacco through the graft union. This substance prevents the day-neutral tobacco from flowering.

Biologists have not yet isolated and chemically characterized all flower promoters (florigen) or flower inhibitors. In 2005, however, plant biologists working with the model plant *Arabidopsis* reported the identification of an important component of the florigen signal. A gene known as *Flowering Locus T (FT)* is activated in *Arabidopsis* leaves, and the mRNA transcript is expressed. This *FT* mRNA travels in the phloem from the leaf to the shoot apical meristem, where it is translated into FT protein. The FT protein interacts with **transcription factors** to alter gene expression, thereby initiating flower development at the shoot apex.

Review

■ How are signal reception, signal transduction, and altered cell activity part of the general mechanism of action for auxin?

■ How is auxin involved in phototropism?

■ Which hormones are involved in each of the following physiological processes: (1) seed germination, (2) stem elongation, (3) fruit ripening, (4) leaf abscission, and (5) seed dormancy?

■ How does salicylic acid help plants defend against insects and viruses?

LIGHT SIGNALS AND PLANT DEVELOPMENT

Learning Objectives

6 Explain how varying amounts of light and darkness induce flowering.

7 Describe the role of phytochrome in flowering, including a brief discussion of phytochrome signal transduction.

8 Define *circadian rhythm,* and give an example.

9 Distinguish between phytochrome and cryptochrome.

Photoperiodism is any response of a plant to the relative lengths of daylight and darkness. Initiation of flowering at the shoot apical meristem is one of several physiological activities that are photoperiodic in many plants. Plants are classified into four main groups—short-day, long-day, intermediate-day, and day-neutral—on the basis of how photoperiodism affects their transition from vegetative growth to flowering.

Short-day plants (also called **long-night plants**) flower when the night length is equal to or greater than some critical period (■ Fig. 37-14a). Thus, short-day plants detect the lengthening nights of late summer or fall. There are two kinds of short-day plants: qualitative and quantitative. In *qualitative short-day plants,* flowering occurs only in short days; in *quantitative short-day plants,* flowering is accelerated by short days. The initiation of flowering in short-day plants is not due to the shorter period of daylight but to the long, uninterrupted period of darkness. The minimum critical night length varies considerably from one plant species to another but falls between 12 and 14 hours for many. Examples of short-day plants are florist's chrysanthemum, cocklebur, and poinsettia, which typically flower in late summer or fall. Poinsettias, for example, typically initiate flower buds in early October in the Northern Hemisphere and flower about 8 to 10 weeks later; hence, their traditional association with Christmas.

Long-day plants (also called **short-night plants**) flower when the night length is equal to or less than some critical period (■ Fig. 37-14b). In *qualitative long-day plants,* flowering occurs only in long days; in *quantitative long-day plants,* long days accelerate flowering. Long-day plants, such as spinach, black-eyed Susan (see photograph in chapter introduction) and the model

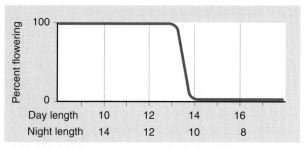

(a) Short-day plants flower when the night length is equal to or exceeds a certain critical length in a 24-hour period.

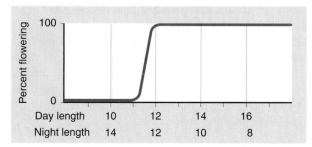

(b) Long-day plants flower when the night length is equal to or less than a certain critical length in a 24-hour period. The critical length varies among different species; short-day plants may require a shorter night length than long-day plants, as shown.

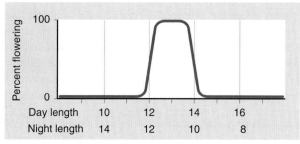

(c) Intermediate-day plants have a narrow night-length requirement.

Figure 37-14 Generalized photoperiodic responses in short-day, long-day, and intermediate-day plants

research plant *Arabidopsis thaliana,* flower when they detect the shortening nights of spring and early summer.

Intermediate-day plants do not flower when night length is either too long or too short (■ Fig. 37-14c). Sugarcane and coleus are intermediate-day plants. These plants flower when exposed to days and nights of intermediate length.

Some plants, called **day-neutral plants,** do not initiate flowering in response to seasonal changes in the period of daylight and darkness but instead respond to some other type of stimulus, external or internal. Cucumber, sunflower, corn, and onion are examples of day-neutral plants. Many of these plants originated in the tropics, where day length does not vary appreciably during the year. In contrast, short-day, long-day, and intermediate-day plants are temperate or subtropical (midlatitude species).

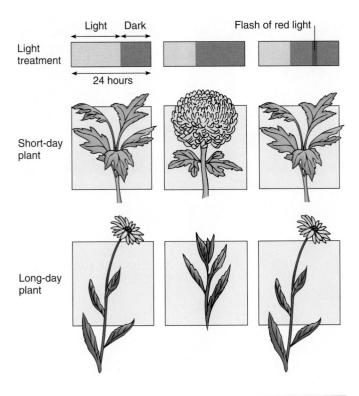

Light treatment

Light | Dark

Flash of red light

24 hours

Short-day plant

Long-day plant

Figure 37-15 *Animated* Photoperiodic responses of short-day and long-day plants

A short-day plant flowers when it is grown under long-night conditions (*top middle*), but it does not flower when exposed to a long night interrupted with a brief flash of red light (*top right*). A long-day plant does not flower when grown under long-night conditions (*bottom middle*) unless the long night is interrupted with a brief flash of red light (*bottom right*).

Plant biologists have experimented with the effects of various light regimens on flowering. ▌Figure 37-15 shows how flowering in long-day and short-day plants is affected by different light treatments, including a short-day–long-night regimen with a *night break*, a short burst of light in the middle of the night.

Phytochrome detects day length

The main photoreceptor for photoperiodism and many other light-initiated plant responses (such as germination and seedling establishment) is **phytochrome,** a family of about five blue-green pigment proteins, each of which is coded for by a different gene. A mixture of phytochrome proteins is present in cells of all vascular plants examined so far. For example, five members of the phytochrome family, designated phyA, phyB, phyC, phyD, and phyE, occur in *A. thaliana.*

Much of the current knowledge of phytochrome in *Arabidopsis* is based on various mutant plants that do not express a specific phytochrome gene, such as the gene that codes for phyA. By studying the physiological response of plants that do not produce an individual phytochrome, biologists have concluded that the individual forms of phytochrome have both unique and overlapping functions. PhyB appears to exert its influence at all stages of the plant life cycle, whereas the other forms of phytochrome have narrower functions at specific stages in the life cycle.

PhyA and phyB may have antagonistic (opposite) effects on flowering. In long-day plants, phyB may inhibit flowering and phyA may induce flowering. Flowering occurs more rapidly in long-day plants with mutations in the gene that codes for phyB; such mutations reduce or eliminate the production of phyB. In contrast, flowering is delayed or prevented in long-day plants with mutations in the gene that codes for phyA.

Each member of the phytochrome family exists in two forms and readily converts from one form to the other after absorption of light of specific wavelengths. One form, designated **Pr** (for *red*-absorbing *p*hytochrome), strongly absorbs light with a relatively short red wavelength (660 nm). In the process, the shape of the molecule changes to the second form of phytochrome, **Pfr,** so designated because it absorbs *far-r*ed light, which is light with a relatively long red wavelength (730 nm) (▌Fig. 37-16). When it absorbs far-red light, Pfr reverts to the original form, Pr. Pfr is the active form of phytochrome, triggering or inhibiting physiological responses such as flowering.

What does a pigment that absorbs red light and far-red light have to do with daylight and darkness? Sunlight consists of various amounts of the entire spectrum of visible light, in addition to ultraviolet and infrared radiation. Because sunlight contains more

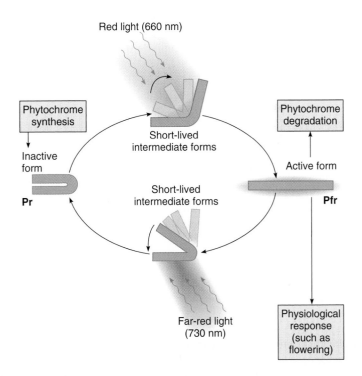

Figure 37-16 *Animated* Phytochrome

Each member of the phytochrome family occurs in two forms, designated Pr and Pfr, and readily converts from one form to the other. Red light (660 nm) converts Pr to Pfr, and far-red light (730 nm) converts Pfr to Pr.

red than far-red light, however, when a plant is exposed to sunlight, the level of Pfr increases. During the night, the level of Pfr slowly decreases as Pfr, which is less stable than Pr, is degraded.

The importance of phytochrome to plants cannot be overemphasized. Timing of day length and darkness is the most reliable way for plants to measure the change from one season to the next. This measurement, which synchronizes the stages of plant development, is crucial for survival, particularly in environments where the climate has an annual pattern of favorable and unfavorable seasons.

Competition for sunlight among shade-avoiding plants involves phytochrome

Plants sense the proximity of nearby plants, which are potential competitors, and react by changing the way they grow and develop. Many plants, from small herbs to large trees, compete for light, a response known as **shade avoidance,** in which plants tend to grow taller when closely surrounded by other plants. If successful, the shade-avoiding plant projects its new growth into direct sunlight and thereby increases its chances of survival.

Since the 1970s, botanists have recognized the environmental factor that triggers shade avoidance: plants perceive changes in the ratio of red to far-red light that result from the presence of nearby plants. The leaves of neighboring plants absorb much more red light than far-red light. (The green pigment chlorophyll strongly absorbs red light during photosynthesis.) In a densely plant-populated area, the ratio of red light to far-red light (r/fr) decreases, affecting the equilibrium between the Pr and Pfr forms, particularly of phyB. This signal triggers a series of responses that cause the shade-avoiding plant, which is adapted to full-light environments, to grow taller or flower earlier. ▋Figure 37-17 shows an interesting application of additional (reflected) far-red light without the reduction in incoming sunlight that occurs in shade avoidance.

When a plant is using many of its resources for stem elongation, it has fewer resources for new leaves and branches, storage tissues, and reproductive tissues. However, for a shade-avoiding plant that is shaded by its neighbors, a rapid increase in stem length is advantageous, because once this plant is taller than its neighbors, it obtains a larger share of unfiltered sunlight.

Phytochrome is involved in other responses to light, including germination

Phytochrome is involved in the light requirement that some seeds have for germination. Seeds with a light requirement must be exposed to light containing red wavelengths. Exposure to red light converts Pr to Pfr, and germination occurs. Many temperate forest species with small seeds require light for germination. (Larger seeds generally do not have a light requirement.) This adaptation enables the seeds to germinate at the optimal time. During early spring, sunlight, including red light, penetrates the

Cary Wolinsky

Figure 37-17 Colored mulches and plant growth processes

The red mulch reflects far-red light from the ground to the tomato plants. The plants, absorbing the extra far-red light, react as if nearby plants are present. Their aboveground growth is greater, and their fruits ripen earlier. Manipulating mulch colors also modifies yield, flavor, and nutrient content.

bare branches of overlying deciduous trees and reaches the soil between the trees. As spring temperatures warm, the seeds on the soil absorb red light and germinate. During their early growth, the newly germinated seedlings do not have to compete with the taller trees for sunlight.

Other physiological functions under the influence of phytochrome include sleep movements in leaves (discussed later); shoot dormancy; leaf abscission; and pigment formation in flowers, fruits, and leaves.

Phytochrome acts by signal transduction

Some phytochrome-induced responses are very rapid and short-term, whereas others are slower and long-term. The rapid responses probably involve reversible changes in the properties of membranes, altering cell ionic balances. Red light, for example, causes potassium ion (K^+) channels to open in cell membranes involved in plant movements caused by changes in turgor. Exposure to far-red light causes the K^+ channels to close. Slower phytochrome-induced responses involve the selective regulation of transcription of numerous genes. For example, phytochrome activates transcription of the gene for the small subunit of *rubisco*, an enzyme involved in photosynthesis.

The active form of phytochrome moves from the cytoplasm into the nucleus, where it affects gene expression by interacting with transcription factors such as PIF3.

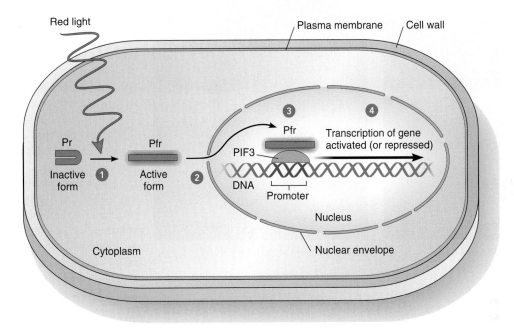

Figure 37-18 Phytochrome signal transduction

The numbered steps are explained in the text.

Each phytochrome molecule consists of a protein attached to a light-absorbing photoreceptor. Biologists hypothesize that the absorption of light by the photoreceptor portion of phytochrome elicits a change in the shape of the larger protein component. This change, in turn, triggers one or more signal transduction pathways.

One important research tool in studying phytochrome signal transduction is mutant plants that respond as if they were exposed to a particular light stimulus even if they were not. Research indicates that the active form of phytochrome moves from the cytoplasm into the nucleus, where it affects gene expression by activating a transcription factor. When activated, the transcription factor, which binds to the promoter of a gene, either turns on or represses transcription that leads to protein synthesis. Biologists have identified **phytochrome-interacting factor (PIF-3),** a transcription factor involved in phytochrome signaling.

The pathway in phytochrome signal transduction with PIF-3 is relatively simple (❙ Fig. 37-18). In step ❶, inactive phytochrome (Pr) in the cytoplasm absorbs red light and is converted into the active form, Pfr, which (step ❷) moves into the nucleus. There, phytochrome (step ❸) binds to the transcription factor PIF3 (which is already bound to the promoter) and (step ❹) activates (or represses) the transcription of light-responsive genes.

The signal transduction pathway is shut down by far-red light, which is absorbed by the Pfr in the Pfr–PIF3 complex in the nucleus. When Pfr is converted to Pr, it dissociates from PIF3.

The signal transduction pathway just described is not the end of the story. Biologists think that there are probably other pathways by which phytochrome regulates light-responsive gene expression.

Light influences circadian rhythms

Almost all organisms, including plants, animals, fungi, eukaryotic microorganisms, and many prokaryotes, appear to have an internal timer, or biological clock, that approximates a 24-hour cycle, the time it takes for Earth to rotate around its own axis. These internal cycles are known as **circadian rhythms** (from the Latin *circum,* "around," and *diurn,* "daily"). Circadian rhythms help an organism detect the time of day, whereas photoperiodism enables a plant to detect the time of year.

Why do plants and other organisms exhibit circadian rhythms? Predictable environmental changes, such as sunrise and sunset, occur during the course of each 24-hour period. These predictable changes may be important to an individual organism, causing it to change its physiological activities or its behavior (in the case of animals). Researchers think that circadian rhythms help an organism synchronize repeated daily activities so that they occur at the appropriate time each day. If, for example, an insect-pollinated flower does not open at the time of day that pollinating insects are foraging for food, reproduction will be unsuccessful.

(a) Leaf position at noon in a bean (*Phaseolus vulgaris*) seedling.

(b) Leaf position at midnight.

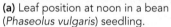

Figure 37-19 Sleep movements

Circadian rhythms in plants affect such biological events as gene expression, the timing of photoperiodism (that is, of seasonal reproduction), the rate of photosynthesis, and the opening and closing of stomata. **Sleep movements** observed in the common bean and other plants are another example of a circadian rhythm (❙ Fig. 37-19). During the day, bean leaves are horizontal, possibly for optimal light absorption, but at night the leaves fold down or up, a movement that orients them perpendicular to their daytime position. The biological significance of sleep movements is unknown at this time.

When constant environmental conditions are maintained, circadian rhythms such as sleep movements repeat every 20 to 30 hours, at least for several days. In nature, the rising and setting of the sun reset the biological clock so that the cycle repeats every 24 hours. What happens if a plant's circadian clock is mutated so that it is not resynchronized by the day–night cycle? In 2005, biologists at the University of Cambridge in England reported the identification of mutants of the model plant *Arabidopsis* in which the circadian clock is not synchronized to match the external day–night cycle. The mutant plants were found to contain less chlorophyll, fix less carbon by photosynthesis, grow more slowly, and have less of a competitive advantage than *Arabidopsis* plants with a normal circadian clock.

For many plants, two photoreceptors—the red light–absorbing phytochrome and the blue/ultraviolet-A light–absorbing **cryptochrome**—are implicated in resetting the biological clock. Certain amino acid sequences of the protein portion of phytochrome are homologous with amino acid sequences of clock proteins in fruit flies, fungi, mammals, and bacteria; this molecular evidence strongly supports the circadian-clock role of phytochrome. The evidence for cryptochrome as a clock protein is also convincing. First discovered in plants, cryptochrome counterparts are found in the fruit fly and mouse biological-clock proteins. Possibly both photoreceptors are involved in resetting the biological clock in plants; researchers have evidence that phytochrome and cryptochrome sometimes interact to regulate similar responses.

Review

❙ What is phytochrome? What are two roles of phytochrome?
❙ What are the steps in phytochrome signal transduction?
❙ In what process is cryptochrome involved?

SUMMARY WITH KEY TERMS

Learning Objectives

1 Describe phototropism, gravitropism, and thigmotropism (page 790).

❙ **Tropisms** are directional growth responses. **Phototropism** is growth in response to the direction of light. **Gravitropism** is growth in response to the influence of gravity. **Thigmotropism** is growth in response to contact with a solid object.

ThomsonNOW **Learn more about phototropism and gravitropism by clicking on the figures in ThomsonNOW.**

2 Describe a general mechanism of action for plant hormones, using auxin as your example (page 791).

❙ Plants produce and respond to **hormones,** organic compounds that act as highly specific chemical signals to elicit a variety of responses that regulate growth and development.

❙ Many plant hormones bind to **enzyme-linked receptors** located in the plasma membrane; the hormone binds to the receptor, triggering an enzymatic reaction of some sort. The receptor for auxin, which is located in the plasma membrane, binds to auxin. This catalyzes the attachment of ubiquitin to repressor proteins, targeting those molecules for destruction. When genes repressed by those proteins are then activated, changes in cell growth and development result.

3 Describe early auxin experiments involving phototropism (page 791).

❙ Darwin and his son performed phototropism experiments in the 1870s on grass seedlings. When they covered the coleoptile tip, the plant did not bend. When they removed the coleoptile tip, bending did not occur. When the bottom of the coleoptile was shielded from the light, the coleoptile bent. The Darwins concluded that some substance is transmitted from the upper to the lower part that causes the plant to bend.

- In the 1920s, Frits Went isolated the phototropic hormone from oat coleoptiles by removing the coleoptile tips and placing them on agar blocks. Normal growth resumed when he put one of these agar blocks squarely on a decapitated coleoptile; the substance had diffused from the coleoptile tip into the agar and, later, from the agar into the decapitated coleoptile. Went named this substance auxin.

ThomsonNOW™ **See the auxin experiments in action by clicking on the figure in ThomsonNOW.**

4 List several ways that each of these hormones affects plant growth and development: auxins, gibberellins, cytokinins, ethylene, and abscisic acid (page 791).

- **Auxin** is involved in cell elongation; tropisms; **apical dominance,** the inhibition of axillary buds by the apical meristem; and fruit development. Auxin also stimulates root development on stem cuttings.

- **Gibberellins** are involved in stem elongation, flowering, and germination.

- **Cytokinins** promote cell division and differentiation; delay **senescence,** the natural aging process; and interact with auxin and ethylene in apical dominance. Cytokinins induce cell division in **tissue culture,** a technique in which cells are isolated from plants and grown in a nutrient medium.

- **Ethylene** plays a role in ripening fruits; apical dominance; leaf abscission; wound response; **thigmomorphogenesis,** a developmental response to mechanical stressors such as wind; and senescence.

- **Abscisic acid** is an environmental stress hormone involved in stomatal closure caused by water stress and in seed **dormancy,** a temporary state of reduced physiological activity.

5 Summarize the activities of these plant hormones and hormone-like signaling molecules: brassinosteroids, jasmonates, salicylic acid, systemin, and oligosaccharins (page 791).

- Plant steroids known as **brassinosteroids** are involved in several aspects of plant growth and development, such as cell division, cell elongation, light-induced differentiation, seed germination, and vascular development.

- **Jasmonates** affect several plant processes, such as pollen development, root growth, fruit ripening, and senescence. They are also produced in response to the presence of insect pests and disease-causing organisms.

- **Salicylic acid** triggers **systemic acquired resistance** that helps defend plants against pathogens and insect pests.

- **Systemin,** a polypeptide with hormonal properties, stimulates a natural defense mechanism in which the plant produces molecules that disrupt insect digestion.

- **Oligosaccharins,** short branched chains of sugar molecules, inhibit flowering and stimulate vegetative growth.

6 Explain how varying amounts of light and darkness induce flowering (page 800).

- **Photoperiodism** is any response of plants to the duration and timing of light and dark. Flowering is a photoperiodic response in many plants. **Short-day plants** detect the lengthening nights of late summer or fall and flower at that time. **Long-day plants** detect the shortening nights of spring and early summer and flower at that time. **Intermediate-day plants** flower when exposed to days and nights of intermediate length.

ThomsonNOW™ **Learn more about photoperiodism by clicking on the figure in ThomsonNOW.**

7 Describe the role of phytochrome in flowering, including a brief discussion of phytochrome signal transduction (page 800).

- The photoreceptor in photoperiodism is **phytochrome,** a family of about five blue-green pigments. Each type of phytochrome has two forms, **Pr** and **Pfr,** named by the wavelength of light they absorb. Pfr is the active form, triggering or inhibiting physiological responses such as flowering, **shade avoidance,** and a light requirement for germination.

- The pathway in phytochrome **signal transduction** begins when inactive phytochrome in the cytoplasm absorbs red light and is converted into the active form, Pfr, which moves into the nucleus. There, phytochrome binds to the transcription factor PIF3 (phytochrome-interacting factor) and activates or represses the transcription of light-responsive genes.

ThomsonNOW™ **Learn more about phytochrome by clicking on the figure in ThomsonNOW.**

8 Define *circadian rhythm,* and give an example (page 800).

- A **circadian rhythm** is a regular period in an organism's growth or activities that approximates the 24-hour day and is reset by the rising and setting of the sun. Two examples of circadian rhythms are the opening and closing of stomata and **sleep movements.**

9 Distinguish between phytochrome and cryptochrome (page 800).

- Both phytochrome and **cryptochrome** are photoreceptors that sometimes interact to regulate similar responses, such as resetting the biological clock. Phytochrome strongly absorbs red light, whereas cryptochrome absorbs blue and ultraviolet-A light.

TEST YOUR UNDERSTANDING

1. In the signal transduction process for the hormone auxin, the molecule ubiquitin (a) absorbs blue light (b) becomes phosphorylated (c) tags certain proteins for destruction (d) interacts antagonistically with gibberellins (e) binds to a receptor in the plant cell's plasma membrane

2. When you prune shrubs to make them "bushier"—that is, to prevent apical dominance—you are affecting the distribution and action of which plant hormone? (a) auxin (b) jasmonic acid (c) systemin (d) ethylene (e) abscisic acid

3. A synthetic _____ known as 2,4-D is used as a selective herbicide. (a) auxin (b) gibberellin (c) cytokinin (d) ethylene (e) abscisic acid

4. Research on a fungal disease of rice provided the first clues about the plant hormone (a) auxin (b) gibberellin (c) cytokinin (d) ethylene (e) abscisic acid

5. This plant hormone interacts with auxin during the formation of plant organs in tissue culture. (a) florigen (b) gibberellin (c) cytokinin (d) ethylene (e) abscisic acid

6. The plant hormone _____ delays senescence, whereas the plant hormone _____ promotes senescence. (a) cytokinin; auxin (b) auxin; cytokinin (c) cytokinin; ethylene (d) abscisic acid; ethylene (e) gibberellin; auxin

7. The stress hormone that helps plants respond to drought is (a) auxin (b) gibberellin (c) cytokinin (d) ethylene (e) abscisic acid

8. This hormone promotes seed dormancy. (a) auxin (b) gibberellin (c) cytokinin (d) ethylene (e) abscisic acid

9. Which signaling molecule triggers the release of volatile substances that attract parasitic wasps to plant-eating caterpillars? (a) brassinosteroid (b) jasmonic acid (c) systemin (d) salicylic acid (e) oligosaccharin

10. The three classes of photoreceptors that enable plants to sense the presence and duration of light are (a) phytochrome, cryptochrome, and circadian rhythms (b) chlorophyll, photosynthesis, and photoperiodism (c) phytochrome, photoperiodism, and chlorophyll (d) phytochrome, cryptochrome, and phototropin (e) phototropin, photosynthesis, and photoperiodism

11. A plant's response to the relative amounts of daylight and darkness is (a) apical dominance (b) bolting (c) gravitropism (d) photoperiodism (e) phototropism

12. Pfr, the active state of _____, forms when red light is absorbed. (a) photosystem I (b) phytochrome (c) far-red light (d) phototropin (e) cryptochrome

13. Which of the following represents the correct order in the phytochrome signal transduction pathway?

1. red light 2. light-responsive gene is switched on (or off) 3. movement of Pfr to nucleus 4. conversion of Pr to Pfr 5. formation of PFr–PIF3 complex that is bound to promoter region

(a) 1, 3, 5, 4, 2 (b) 1, 5, 3, 2, 4 (c) 1, 2, 3, 4, 5 (d) 1, 4, 3, 2, 5 (e) 1, 4, 3, 5, 2

14. Which of the following statements about phytochrome is *incorrect?* (a) phytochrome is the main photoreceptor for photoperiodism (b) phytochrome is a family of about five blue-green pigment proteins, each coded by a different gene (c) most of our knowledge of phytochrome function is based on mutant corn (*Zea mays*) plants (d) each member of the phytochrome family exists in two forms: Pr and Pfr (e) phytochrome helps plants sense the presence of nearby plants with which they must compete for light

15. Which photoreceptor(s) is/are implicated in resetting the biological clock? (a) phytochrome only (b) cryptochrome only (c) phototropin only (d) cryptochrome and gibberellin (e) phytochrome and cryptochrome

16. The orientation of the growth of a plant according to the direction of light is called _____, whereas the twining of tendrils is an example of _____. (a) thigmomorphogenesis; gravitropism (b) photoperiodism; thigmomorphogenesis (c) phototropism; gravitropism (d) phototropism; thigmotropism (e) photoperiodism; thigmotropism

CRITICAL THINKING

1. Predict whether flowering in a short-day plant with a minimum critical night length of 14 hours would be expected to occur in the following situations. Explain each answer. (a) The plant is exposed to 15 hours of daylight and 9 hours of uninterrupted darkness. (b) The plant is exposed to 9 hours of daylight and 15 hours of darkness. (c) The plant is exposed to 9 hours of daylight and 15 hours of darkness, with a 10-minute exposure to red light in the middle of the night.

2. If you transplanted the short-day plant discussed in question 1 to the tropics, would it flower? Explain your answer.

3. **Evolution Link.** What adaptive advantages are conferred on a plant whose stems are positively phototropic and whose roots are positively gravitropic?

4. **Evolution Link.** Explain why a plant with a mutation in one of its phytochrome genes that affects flowering time would be at an evolutionary disadvantage.

5. **Analyzing Data.** Examine Figure 37-14a. When the day length is 10 hours and the night length is 14 hours, what percent of the plants flower? When the day length is 16 hours and the night length is 8 hours, what percent of the plants flower? What is the critical night length for this plant?

Additional questions are available in ThomsonNOW at www.thomsonedu.com/login

38

Animal Structure and Function: An Introduction

Fritz Polking/Dembinsky Photo Associates

Larger body size does not mean bigger cells. The cells of the Yacare caiman (*Caiman yacare*) and the flambeau butterfly (*Dryas julia*) on its head are all about the same size. The caiman is larger because its genes specify that its body should consist of a larger number of cells.

KEY CONCEPTS

Structure and function are closely linked at every level of organization.

The main types of tissues in a complex animal are epithelial, connective, muscle, and nervous.

Tissues and organs form the 11 main organ systems of a complex animal.

Homeostatic mechanisms are responsible for the body's automatic tendency to maintain a relatively stable internal environment.

Thermoregulation contributes to homeostasis.

Animal groups are dramatically diverse with radically different body structures. For example, consider how different the caiman and the butterfly are, not only in size but also in body form and lifestyle (see photograph). Despite their differences, animal groups share many characteristics. One of these characteristics is their relatively large size.

Why *are* most animals larger than bacteria, protists, and fungi? The answer may be related to *ecological niches*, which are the functional roles of a species within a community. By the time animals evolved, bacteria, protists, and fungi already occupied most available ecological niches. For new species to succeed, they had to displace others from a niche or adapt to a new one. Success in a new niche required a new body plan, and new body plans often involved larger size. Increased size provided more opportunity for capturing food. Predators are typically larger than their prey.

To grow larger than their bacterial and protist competitors, animals had to be multicellular. Recall that the size of a single cell is limited by the ratio of its surface area (plasma membrane) to its volume (see Chapter 4). The plasma membrane needs to be large enough relative to the cell's volume to permit passage of materials into and out of the cell so that the conditions necessary for life can be maintained. In a multicellular animal, each cell has a large enough surface area–to-volume ratio to effectively regulate its internal environment. Individual cells live and die and are replaced while the organism continues to maintain itself and thrive. The number of cells, not their individual sizes, is mainly responsible for the size of an animal.

In unicellular organisms, such as bacteria and many protists, the single cell carries on all the activities necessary for life. Recall that unicellular and small, flat organisms depend on diffusion for many life processes, including gas exchange and disposal of metabolic wastes. One reason they can be small is that they do not require complex organ systems.

In Chapter 1 we discussed the levels of organization of multicellular organisms. Recall that cells organize to form *tissues,* and tissues associate to form *organs* such as the heart or stomach. Groups of tissues and organs form the *organ systems* of a complex *organism.* Billions of cells may be organized to form the tissues, organs, and organ systems of a complex animal. **Anatomy** is the study of an organism's structure. **Physiology** is the study of how the body functions. Structure and function are closely linked at every level of organization.

Cells, tissues, organs, and organ systems work together to maintain appropriate conditions in the body. The tendency to maintain a relatively constant internal environment is called *homeostasis,* and the processes that accomplish the task are *homeostatic mechanisms.* For example, in mammals, nervous, endocrine, and cardiovascular systems work together to regulate body temperature.

In this chapter we focus on the basic form and function of animals. We describe the types and functions of tissues as well as organ systems. We discuss the important concept of homeostasis, using regulation of body temperature as an example. In the following chapters, we discuss how animals carry out life processes and how organ systems work together to maintain homeostasis of the animal as a whole. ■

▌ TISSUES

Learning Objectives

1 Compare the structure and function of the four main kinds of animal tissues: epithelial, connective, muscle, and nervous.
2 Compare the structure and function of the main types of epithelial tissue.
3 Compare the main types of connective tissue, and summarize their functions.
4 Contrast the three types of muscle tissue and their functions.
5 Relate the structure of the neuron to its function.

In a multicellular organism, cells specialize to perform specific tasks. A **tissue** consists of a group of closely associated, similar cells that carry out specific functions. Biologists classify animal tissues as epithelial, connective, muscle, or nervous. Classification of tissues depends on their structure and origin. Each kind of tissue is composed of cells with characteristic sizes, shapes, and arrangements; and each type of tissue is specialized to perform a specific function or group of functions. For example, some tissues are specialized to transport materials, whereas others contract, enabling the animal to move. Still others secrete hormones that regulate metabolic processes. As we discuss each tissue type, notice the relationship between its form and its function.

Epithelial tissues cover the body and line its cavities

Epithelial tissue (also called **epithelium**) consists of cells fitted tightly together to form a continuous layer, or sheet, of cells. One surface of the sheet is typically exposed because it covers the body (outer layer of the skin) or lines a cavity, such as the *lumen* (the cavity in a hollow organ) of the intestine. The other surface of an epithelial layer attaches to the underlying tissue by a noncellular **basement membrane** consisting of tiny fibers and nonliving polysaccharide material that the epithelial cells produce.

Epithelial tissue forms the outer layer of the skin and the linings of the digestive, respiratory, excretory, and reproductive tracts. As a result, everything that enters or leaves the body must cross at least one layer of epithelium. Food taken into the mouth and swallowed is not really "inside" the body until it is absorbed through the epithelium of the gut and enters the blood. To a large extent, the permeabilities of the various epithelial tissues regulate the exchange of substances between the different parts of the body, as well as between the animal and the external environment.

Epithelial tissues perform many functions, including protection, absorption, secretion, and sensation. The epithelial layer of the skin, the **epidermis,** covers the entire body and protects it from mechanical injury, chemicals, bacteria, and fluid loss. The epithelial tissue lining the digestive tract absorbs nutrients and water into the body. Some epithelial cells form **glands** that secrete cell products such as hormones, enzymes, or sweat. Other epithelial cells are sensory receptors that receive information from the environment. For example, epithelial cells in taste buds and in the nose specialize as chemical receptors.

▌ Table 38-1 illustrates the main types of epithelial tissue, indicates their locations in the body, and describes their functions. We can distinguish three types of epithelial cells on the basis of shape. **Squamous** epithelial cells are thin, flat cells shaped like flagstones. **Cuboidal** epithelial cells are short cylinders that from the side appear cube shaped, like dice. Actually, each cuboidal cell is typically hexagonal in cross section, making it an eight-sided polyhedron.

Columnar epithelial cells look like columns or cylinders when viewed from the side. The nucleus is usually located near the base of the cell. Viewed from above or in cross section, these cells often appear hexagonal. On its free surface, a columnar epithelial cell may have cilia that beat in a coordinated way, moving materials over the tissue surface. Most of the upper respiratory tract is lined with ciliated columnar epithelium that moves particles of dust and other foreign material away from the lungs.

Epithelial tissue is also classified by the number of layers. **Simple epithelium** is composed of one layer of cells. It is usually located where substances are secreted, excreted, or absorbed or where materials diffuse between compartments. For example, simple squamous epithelium lines the air sacs in the lungs. The structure of this thin tissue is wonderfully suited to permit diffusion of gases in and out of air sacs.

Stratified epithelium, which has two or more layers, is found where protection is required. Stratified squamous epithelium, which makes up the outer layer of your skin, continuously regenerates as it is sloughed off during normal wear and tear. The cells of **pseudostratified epithelium** falsely appear layered. Although all its cells rest on a basement membrane, not every cell extends to the exposed surface of the tissue. This arrangement gives the impression of two or more cell layers. Some of the respiratory passageways are lined with pseudostratified epithelium equipped with cilia.

The lining of blood and lymph vessels is called **endothelium.** Endothelial cells have a different embryonic origin from "true" epithelium. However, these cells are structurally similar to squamous epithelial cells and can be included in that category.

A gland consists of one or more epithelial cells specialized to produce and secrete a product such as sweat, milk, mucus, wax, saliva, hormones, or enzymes (▌ Fig. 38-1). Epithelial tissue lining the cavities and passageways of the body typically has some specialized mucus-secreting cells called **goblet cells.** The mucus lubricates these surfaces, offers protection, and facilitates the movement of materials.

Glands are classified as exocrine or endocrine. **Exocrine glands,** like goblet cells and sweat glands, secrete their products onto a free epithelial surface, typically through a duct (tube). **Endocrine glands** lack ducts. These glands release their products, called **hormones,** into the **interstitial fluid** (tissue fluid) or blood; hormones are typically transported by the cardiovascular system. (Endocrine glands are discussed in Chapter 48.)

An **epithelial membrane** consists of a sheet of epithelial tissue and a layer of underlying connective tissue. Types of epithelial membranes include mucous membranes and serous membranes. A **mucous membrane,** or mucosa, lines a body cavity that opens to the outside of the body, such as the digestive or respiratory tract. The epithelial layer secretes mucus that lubricates the tissue and protects it from drying.

A **serous membrane** lines a body cavity that does not open to the outside of the body. It consists of simple squamous epithelium over a thin layer of loose connective tissue. This type of membrane secretes fluid into the cavity it lines. Examples of serous membranes are the pleural membranes lining the pleural cavities around the lungs and the pericardial membranes lining the pericardial cavity around the heart.

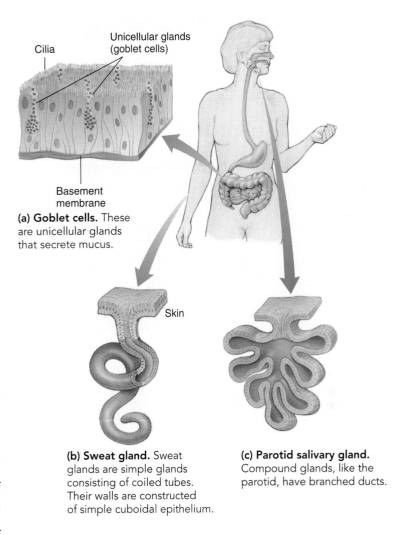

(a) Goblet cells. These are unicellular glands that secrete mucus.

(b) Sweat gland. Sweat glands are simple glands consisting of coiled tubes. Their walls are constructed of simple cuboidal epithelium.

(c) Parotid salivary gland. Compound glands, like the parotid, have branched ducts.

Figure 38-1 Glands

A gland consists of one or more epithelial cells.

Connective tissues support other body structures

Almost every organ in the body has a framework of **connective tissue** that supports and cushions it. Typically, connective tissues contain relatively few cells. Its cells are embedded in an extensive **intercellular substance** consisting of threadlike, microscopic **fibers** scattered throughout a **matrix,** a thin gel of polysaccharides that the cells secrete. The nature and function of each kind of connective tissue are determined in part by the structure and properties of the intercellular substance.

Connective tissue typically contains three types of fibers: collagen, elastic, and reticular. **Collagen fibers,** the most numerous type, are made of **collagens,** a group of fibrous proteins found in all animals (see Fig. 3-22b). Collagens are the most abundant proteins in mammals, accounting for about 25% of their total protein mass. Collagen is very tough (meat is tough because of its collagen content). The tensile strength (ability to stretch without tearing) of collagen fibers is comparable to that of steel. Collagen

TABLE 38-1

Epithelial Tissues

Nuclei of squamous epithelial cells

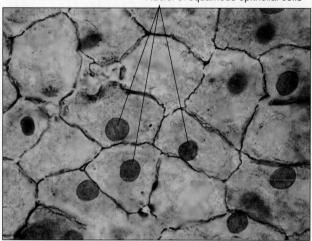

LM of simple squamous epithelium.

Simple Squamous Epithelium

Main Locations
Air sacs of lungs; lining of blood vessels

Functions
Passage of materials where little or no protection is needed and where diffusion is major form of transport

Description and Comments
Cells are flat and arranged as single layer

25 µm

Nuclei of cuboidal epithelial cells   Lumen of tubule

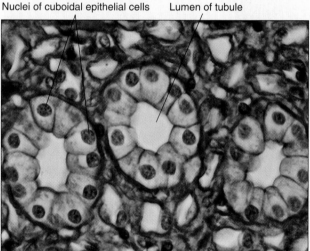

LM of simple cuboidal epithelium.

Simple Cuboidal Epithelium

Main Locations
Linings of kidney tubules; gland ducts

Functions
Secretion and absorption

Description and Comments
Single layer of cells; LM shows cross section through tubules; from the side each cell looks like a short cylinder; some have microvilli for absorption

25 µm

Goblet cell   Nuclei of culumnar cells

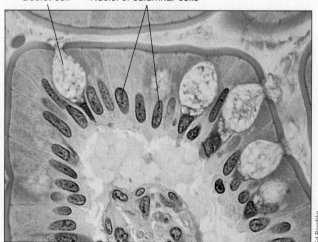

LM of simple columnar epithelium.

Simple Columnar Epithelium

Main Locations
Linings of much of digestive tract and upper part of respiratory tract

Functions
Secretion, especially of mucus; absorption; protection; movement of layer of mucus

Description and Comments
Single layer of columnar cells; sometimes with enclosed secretory vesicles (in goblet cells); highly developed Golgi complex; often ciliated

25 µm

TABLE 38-1

Continued

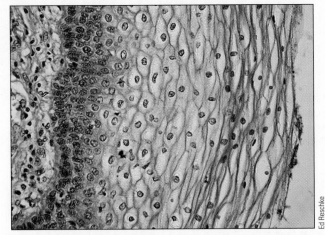

LM of stratified squamous epithelium.

Stratified Squamous Epithelium

Main Locations
Skin; mouth lining; vaginal lining

Functions
Protection only; little or no absorption or transit of materials; outer layer continuously sloughed off and replaced from below

Description and Comments
Several layers of cells, with only the lower ones columnar and metabolically active; division of lower cells causes older ones to be pushed upward toward surface, becoming flatter as they move

50 μm

Basement membrane   Cilia

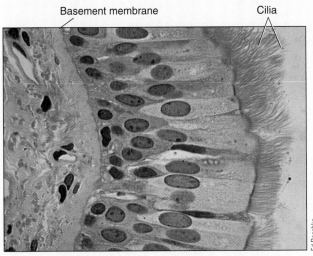

LM of pseudostratified columnar epithelium, ciliated.

Pseudostratified Epithelium

Main Locations
Some respiratory passages; ducts of many glands

Functions
Secretion; protection; movement of mucus

Description and Comments
Ciliated, mucus-secreting, or with microvilli; comparable in many ways to columnar epithelium except that not all cells are the same height; so, though all cells contact the same basement membrane, the tissue appears stratified

25 μm

fibers are wavy and flexible, allowing them to remain intact when tissue is stretched.

Elastic fibers branch and fuse to form networks. They can be stretched by a force and then (like a stretched rubber band) return to their original size and shape when the force is removed. Elastic fibers, composed of the protein elastin, are an important component of structures that must stretch.

Reticular fibers are very thin, branched fibers that form delicate networks joining connective tissues to neighboring tissues. Reticular fibers consist of collagen and some glycoprotein.

The cells of different kinds of connective tissues differ in their shapes and structures and in the kinds of fibers and matrices they secrete. **Fibroblasts** are connective tissue cells that produce the fibers, as well as the protein and carbohydrate complexes, of the matrix. Fibroblasts release protein components that become arranged to form the characteristic fibers. These cells are especially active in developing tissues and are important in healing wounds.

As tissues mature, the number of fibroblasts decreases and they become less active. **Macrophages,** the body's scavenger cells, commonly wander through connective tissues, cleaning up cell debris and phagocytosing foreign matter, including bacteria.

Some of the main types of connective tissue are (1) loose and dense connective tissues; (2) elastic connective tissue; (3) reticular connective tissue; (4) adipose tissue; (5) cartilage; (6) bone; and (7) blood, lymph, and tissues that produce blood cells. These tissues vary widely in their structural details and in the functions they perform (▌Table 38-2).

Loose connective tissue is the most widely distributed connective tissue in the vertebrate body. This thin filling between body parts serves as a reservoir for fluid and salts. Nerves, blood vessels, and muscles are wrapped in this tissue. Together with adipose tissue, loose connective tissue forms the subcutaneous (below the skin) layer that attaches skin to the muscles and other structures beneath. Loose connective tissue consists of fibers

TABLE 38-2

Connective Tissues

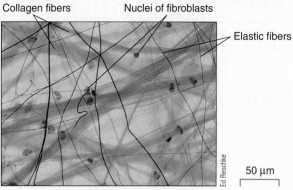

Collagen fibers   Nuclei of fibroblasts   Elastic fibers

50 µm

Ed Reschke

LM of loose connective tissue.

Loose Connective Tissue

Main Locations
Everywhere that support must be combined with elasticity, such as subcutaneous tissue (the layer of tissue beneath the dermis of the skin)

Functions
Support; reservoir for fluid and salts

Description and Comments
Fibers produced by fibroblast cells embedded in semifluid matrix and mixed with miscellaneous other cells

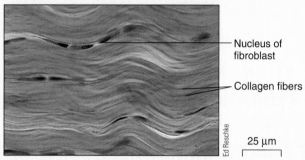

Nucleus of fibroblast

Collagen fibers

25 µm

Ed Reschke

LM of dense connective tissue.

Dense Connective Tissue

Main Locations
Tendons; many ligaments; dermis of skin

Functions
Support; transmission of mechanical forces

Description and Comments
Collagen fibers may be regularly or irregularly arranged

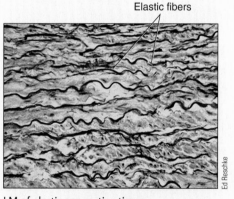

Elastic fibers

50 µm

Ed Reschke

LM of elastic connective tissue.

Elastic Connective Tissue

Main Locations
Structures that must both expand and return to their original size, such as lung tissue and large arteries

Function
Confers elasticity

Description and Comments
Branching elastic fibers interspersed with fibroblasts

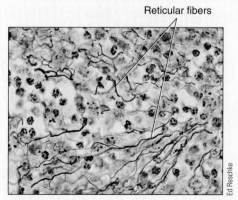

Reticular fibers

50 µm

Ed Reschke

LM of reticular connective tissue.

Reticular Connective Tissue

Main Locations
Framework of liver; lymph nodes; spleen

Function
Support

Description and Comments
Consists of interlacing reticular fibers

TABLE 38-2

Continued

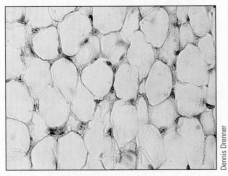

50 μm

Dennis Drenner

LM of adipose tissue.

Adipose Tissue

Main Locations
Subcutaneous layer; pads around certain internal organs

Functions
Food storage; insulation; support of such organs as mammary glands, kidneys

Description and Comments
Fat cells are star shaped at first; fat droplets accumulate until typical ring-shaped cells are produced

Chondrocytes   Lacuna   Intercellular substance

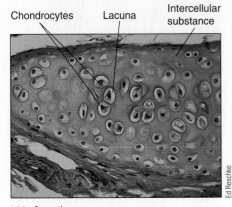

50 μm

Ed Reschke

LM of cartilage.

Cartilage

Main Locations
Supporting skeletons in sharks and rays; ends of bones in mammals and some other vertebrates; supporting rings in walls of some respiratory tubes; tip of nose; external ear

Function
Flexible support

Description and Comments
Cells (chondrocytes) separated from one another by intercellular substance; cells occupy lacunae

Lacunae   Haversian canal   Matrix

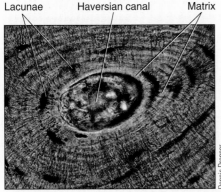

50 μm

Dennis Drenner

LM of bone.

Bone

Main Locations
Forms skeletal structure in most vertebrates

Functions
Support and protection of internal organs; calcium reservoir; skeletal muscles attach to bones

Description and Comments
Osteocytes in lacunae; in compact bone, lacunae embedded in lamellae, concentric circles of matrix surrounding Haversian canals

Red blood cells   White blood cells

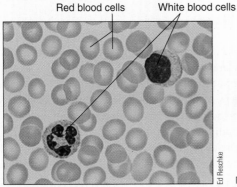

25 μm

Ed Reschke

LM of blood.

Blood

Main Locations
Within heart and blood vessels of circulatory system

Functions
Transports oxygen, nutrients, wastes, and other materials

Description and Comments
Consists of cells dispersed in fluid intercellular substance (plasma)

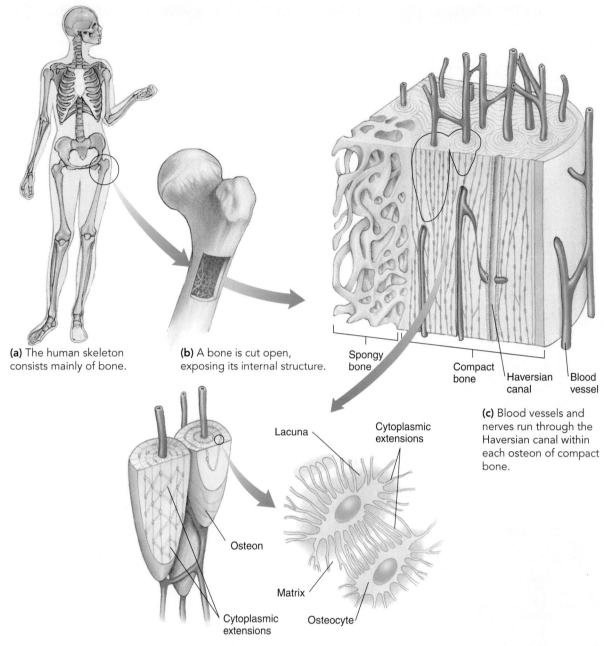

(a) The human skeleton consists mainly of bone.

(b) A bone is cut open, exposing its internal structure.

Spongy bone

Compact bone

Haversian canal

Blood vessel

(c) Blood vessels and nerves run through the Haversian canal within each osteon of compact bone.

Lacuna

Cytoplasmic extensions

Osteon

Matrix

Cytoplasmic extensions

Osteocyte

(d) The bone matrix is rigid and hard. Osteocytes become trapped within lacunae but communicate with one another by way of cytoplasmic extensions that extend through tiny canals.

Figure 38-2 Bone

running in all directions through a semifluid matrix. Its flexibility permits the parts it connects to move.

Dense connective tissue, found in the dermis (lower layer) of the skin, is very strong, though somewhat less flexible than loose connective tissue. Collagen fibers predominate. **Tendons,** the cords that connect muscles to bones, and **ligaments,** the cables that connect bones to one another, consist of dense connective tissue in which collagen bundles are arranged in a definite pattern.

Elastic connective tissue consists mainly of bundles of parallel elastic fibers. This tissue is found in structures that must ex-

pand and then return to their original size, such as lung tissue and the walls of large arteries.

Reticular connective tissue is composed mainly of interlacing reticular fibers. It forms a supporting internal framework in many organs, including the liver, spleen, and lymph nodes.

The cells of **adipose tissue** store fat and release it when fuel is needed for cellular respiration. Adipose tissue is found in the subcutaneous layer and in tissue that cushions internal organs.

The supporting skeleton of a vertebrate is made of cartilage, or of both cartilage and bone. **Cartilage** is the supporting skeleton in the embryonic stages of all vertebrates. In most vertebrates, bone replaces cartilage during development. However, cartilage remains in some supporting structures. For example, in humans cartilage is found in the external ear, the supporting rings in the walls of the respiratory passageways, the tip of the nose, the ends of some bones, and the discs that serve as cushions between the vertebrae.

Cartilage is firm yet elastic. Its cells, called **chondrocytes,** secrete a hard, rubbery matrix around themselves and also secrete collagen fibers, which become embedded in the matrix and strengthen it. Chondrocytes eventually come to lie, singly or in groups of two or four, in small cavities in the matrix called **lacunae.** These cells remain alive and are nourished by nutrients and oxygen that diffuse through the matrix. Cartilage tissue lacks nerves, lymph vessels, and blood vessels.

Bone, the main vertebrate skeletal tissue, is like cartilage in that it consists mostly of matrix material. The bone cells, called **osteocytes,** are contained within lacunae. Osteocytes secrete and maintain the matrix (▐ Fig. 38-2). Unlike cartilage, however, bone is a highly vascular tissue, with a substantial blood supply. Osteocytes communicate with one another and with capillaries by tiny channels (*canaliculi*) that contain long cytoplasmic extensions of the osteocytes.

A typical bone has an outer layer of *compact bone* surrounding a filling of *spongy bone.* Compact bone consists of spindle-shaped units called **osteons.** Within each osteon, osteocytes are arranged in concentric layers of matrix called *lamellae.* In turn, the lamellae surround central microscopic channels known as **Haversian canals,** through which capillaries and nerves pass.

Bones are amazingly light and strong. Calcium salts of bone render the matrix very hard, and collagen prevents the bony matrix from being overly brittle. Most bones have a large, central **marrow cavity** that contains a spongy tissue called *marrow.* Yellow marrow consists mainly of fat. Red marrow is the connective tissue in which blood cells are produced. We discuss bone in more detail in Chapter 39.

Blood and **lymph** are circulating tissues that help other parts of the body communicate and interact. Like other connective tissues, they consist of specialized cells dispersed in an intercellular substance. In mammals, blood consists of **red blood cells, white blood cells,** and **platelets,** all suspended within **plasma,** the liquid, noncellular part of the blood. In humans and other vertebrates, red blood cells contain the respiratory pigment that transports oxygen. White blood cells defend the body against disease-causing microorganisms (see Chapter 44). Platelets, small fragments broken off from large cells in the bone marrow, play a key role in blood clotting. Plasma consists of water, proteins, salts, and a variety of soluble chemical messengers such as hormones that it transports from one part of the body to another. We discuss blood in Chapter 43.

Muscle tissue is specialized to contract

Most animals move by contracting the long, cylindrical or spindle-shaped cells of **muscle tissue.** Each muscle cell is called a **muscle fiber** because of its length. A muscle fiber contains many thin, longitudinal, parallel contractile units called **myofibrils.** Two proteins, **myosin** and **actin,** are the chief components of myofibrils and play a key role in contraction of muscle fibers.

Some invertebrates have skeletal and smooth muscle. Vertebrates have three types of muscle tissue: skeletal, cardiac, and smooth (▐ Table 38-3). **Skeletal muscle** makes up the large muscle masses attached to the bones of the body. Skeletal muscle fibers are very long, and each fiber has many nuclei. The nuclei of skeletal muscle fibers are also unusual in their position. They lie just under the plasma membrane, which frees the entire central part of the skeletal muscle fiber for the myofibrils. This adaptation appears to increase the efficiency of contraction. When skeletal muscles contract, they move parts of the body. Whereas skeletal muscle fibers are generally under voluntary control, you

TABLE 38-3			
Muscle Tissues			
	Skeletal	**Cardiac**	**Smooth**
Location	Attached to skeleton	Walls of heart	Walls of stomach, intestines, etc.
Type of control	Voluntary	Involuntary	Involuntary
Shape of fibers	Elongated, cylindrical, blunt ends	Elongated, cylindrical, fibers that branch and fuse	Elongated, spindle shaped, pointed ends
Striations	Present	Present	Absent
Number of nuclei per fiber	Many	One or two	One
Position of nuclei	Peripheral	Central	Central
Speed of contraction	Most rapid	Intermediate (varies)	Slowest
Resistance to fatigue (with repetitive contraction)	Least	Intermediate	Greatest

Skeletal muscle fibers

Cardiac muscle fibers

Smooth muscle fibers

UNWELCOME TISSUES: CANCERS

A **neoplasm** ("new growth"), or **tumor,** is an abnormal mass of cells. A neoplasm may be benign or malignant (cancerous). A **benign** ("kind") tumor tends to grow slowly, and its cells stay together. Because benign tumors form masses with distinct borders, they can usually be removed surgically.

A **malignant** ("wicked") **neoplasm,** or **cancer,** usually grows much more rapidly and invasively than a benign tumor. In fact, two basic defects in behavior that characterize most cancer cells are rapid multiplication and abnormal relations with neighboring cells. Unlike normal cells, which respect one another's boundaries and form tissues in an orderly, organized manner, cancer cells grow helter-skelter on one another and infiltrate normal tissues. They apparently no longer receive or respond appropriately to signals from surrounding cells; communication is lacking (see figure).

In Chapter 17 you learned that cancer results from abnormal expression of specific genes critical for cell division. When a cancer cell multiplies, all the cells derived from it are also abnormal. Unlike the cells of benign tumors, cancer cells do not retain normal structural features. Cancers that develop from connective tissues or muscle are referred to as **sarcomas.** Those that originate in epithelial tissue are called **carcinomas.** Most human cancers are epithelial.

Death from cancer results from **metastasis,** a migration of cancer cells through blood or lymph channels to other parts of the body. Once there, cancer cells multiply, forming new malignant neoplasms that interfere with the normal functions of the tissues being invaded. Cancer often spreads so rapidly and extensively that surgeons cannot locate or remove all the malignant masses.

Some neoplasms grow to several millimeters in diameter and then enter a dormant stage, which may last for months or even years. Solid tumors, which account for more than 85% of cancer deaths, require blood vessels to ensure delivery of nourishment and oxygen. Eventually,

cancer cells release a chemical substance that stimulates nearby blood vessels to develop new capillaries. These blood vessels grow into the abnormal mass of cells. Nourished by its new blood supply, the neoplasm begins to grow rapidly. Newly formed blood vessels have leaky walls that provide a route for metastasis. Malignant cells enter the blood through these walls and are transported to new sites.

Cancer is the second leading cause of death in the United States. One in three people in the United States develops cancer at some time in his or her life. Currently, the key to survival is early diagnosis and treatment with some combination of surgery, hormonal treatment, radiation therapy, and **chemotherapy,** which may include drugs that suppress mitosis. Many new treatments are under investigation, including agents that inhibit the development of new blood vessels. A recently discovered molecular link between chronic inflammation and cancer is also inspiring new approaches to cancer treatment. Cancer is actually a large family of closely related diseases (there are hundreds of distinct varieties), so treatment must be tailored to the particular type of cancer.

Alleles of some genes appear to affect an individual's level of tolerance to **carcinogens,** cancer-producing agents. More than 80% of cancer cases are thought to be triggered by carcinogens in the environment. You can decrease risk of developing cancer by following these recommendations:

1. Do not smoke or use tobacco. Smoking is responsible for more than 80% of lung cancer cases, and it increases the risk for many other cancers.

2. Avoid prolonged exposure to the sun. When in the sun, use sunscreen or sunblock. Exposure to the sun is responsible for almost all the 400,000 cases of skin cancer reported each year in the United States.

3. Eat a healthy diet, including fruits and vegetables. Although research results are not clear, it appears prudent to

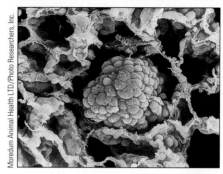

Moredum Animal Health LTD/Photo Researchers, Inc.

50 μm

Lung cancer
When the cancer cells shown in this color-enhanced SEM of human lung tissue multiply rapidly and invade normal tissues, they interfere with normal function. A mass of malignant cells (*pink*) occupies the air sac in the center, and cancer cells that have separated from the main tumor can be seen in other air sacs. Microvilli on the surface of the cancer cells give them a fuzzy appearance.

increase the fiber content of your diet and avoid high-fat, smoked, salt-cured, and nitrite-cured foods.

4. Avoid unnecessary exposure to X-rays.

5. Women should examine their breasts each month, have regular mammograms, and obtain annual Papanicolaou (Pap) and human papilloma virus (HPV) tests. Certain HPV strains can cause cervical cancer; vaccines are now available.

6. Men should regularly examine their testes and have prostate examinations yearly after age 50. They should also have the prostate-specific antigen (PSA) blood test. These routine tests detect cancer at an early, more treatable stage.

7. Beginning at age 50, both men and women should be screened for colorectal cancer. Detection and removal of polyps (benign growths that can become malignant) can prevent cancer.

do not normally contract your cardiac and smooth muscle fibers at will.

Light microscopy shows that both skeletal and cardiac fibers have alternating light and dark transverse stripes, or **striations,** that change their relative sizes during contraction. Striated muscle

fibers contract rapidly but cannot remain contracted for a long period. They must relax and rest momentarily before contracting again. (Muscle contraction is discussed in Chapter 39.)

Cardiac muscle is the main tissue of the heart. When this muscle contracts, the heart pumps the blood. The fibers of car-

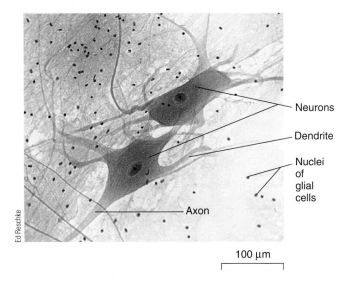

Ed Reschke

Neurons

Dendrite

Nuclei of glial cells

Axon

100 μm

Figure 38-3 LM of nervous tissue

Neurons transmit impulses. Glial cells support and nourish neurons.

diac muscle join end to end, and they branch and rejoin to form complex networks. One or two nuclei lie within each fiber. A characteristic feature of cardiac muscle tissue is the presence of *intercalated discs,* specialized junctions where the fibers join.

Smooth muscle occurs in the walls of the digestive tract, uterus, blood vessels, and many other internal organs. Contraction of smooth muscle allows an organ to perform some function. When smooth muscle in the wall of the digestive tract contracts, food is moved through the digestive tract. When smooth muscle in the walls of arterioles (small arteries) contract, the blood vessel constricts, raising blood pressure. Each spindle-shaped smooth muscle fiber contains a single, central nucleus.

Nervous tissue controls muscles and glands

Nervous tissue consists of neurons and glial cells. **Neurons** are specialized for receiving and transmitting signals. **Glial cells** support and nourish the neurons (❚ Fig. 38-3).

A typical neuron has a **cell body** containing the nucleus as well as two types of cytoplasmic extensions (discussed in Chapter 40). **Dendrites** are cytoplasmic extensions specialized for receiving signals and transmitting them to the cell body. The single **axon** transmits signals, called *nerve impulses,* away from the cell body. Axons range in length from 1 or 2 mm to more than a meter. Those extending from the spinal cord down the arm or leg in a human, for example, may be a meter or more in length.

Certain neurons receive signals from the external or internal environment and transmit them to the spinal cord and brain. Other neurons relay, process, or store information. Still others transmit signals from the brain and spinal cord to the muscles and glands. Neurons communicate at junctions called **synapses.** A **nerve** consists of a great many neurons bound together by connective tissue.

In this chapter, we have focused on normal tissues. For a discussion of some abnormal tissues, see *Focus On: Unwelcome Tissues: Cancers.*

Review

❚ What are the main differences in structure and function between epithelial tissue and connective tissue?

❚ What type of tissue lines the air sacs of the lungs? How is its structure adapted to its function?

❚ How is the structure of adipose tissue adapted to its function? Where is it found in the body?

❚ What are some differences between the three types of muscle tissue? How is the structure of each type adapted for its function?

❚ How is the neuron uniquely adapted for its function?

ORGANS AND ORGAN SYSTEMS

Learning Objectives

6 Briefly describe the organ systems of a mammal, and summarize the homeostatic actions of each organ system.

7 Define *homeostasis,* and contrast negative and positive feedback systems.

Tissues associate to form **organs.** Although an animal organ may be composed mainly of one type of tissue, other types are needed to support, protect, provide a blood supply, and transmit information. For example, the heart is mainly cardiac muscle tissue, but its chambers are lined with endothelium and its walls contain blood vessels made of endothelium, smooth muscle, and connective tissue. The heart also has nerves that transmit information and help regulate the rate and strength of its contractions.

An organized group of tissues and organs that together perform a specialized set of functions make up an **organ system.** Working together in a very coordinated way, organ systems perform the functions required by the **organism.** We can identify 11 major organ systems that work together to carry out the physiological processes of a mammal: **integumentary, skeletal, muscular, digestive, cardiovascular, immune (lymphatic), respiratory, urinary, nervous, endocrine,** and **reproductive systems.** ❚ Table 38-4 and ❚ Figure 38-4 summarize their principal organs and functions.

The body maintains homeostasis

To survive and function, animals must regulate the composition of the fluids that bathe their cells. They must maintain the appropriate concentration of nutrients, oxygen, and other gases, ions, and compounds needed for metabolism at all times. Animals must also maintain pH and internal temperature within relatively narrow limits.

Cells, tissues, organs, and organ systems work together to maintain appropriate conditions in the body. The balanced internal environment is referred to as **homeostasis,** and the control processes that maintain these conditions are **homeostatic mechanisms.** Homeostasis is a basic concept in physiology. First coined by U.S. physiologist Walter Cannon, the word *homeostasis* is derived from the Greek *homoios,* meaning "same," and *stasis,* "standing." Actually, the internal environment never really stays the same. It is a dynamic equilibrium in which conditions are maintained within narrow limits, which we call a **steady state.**

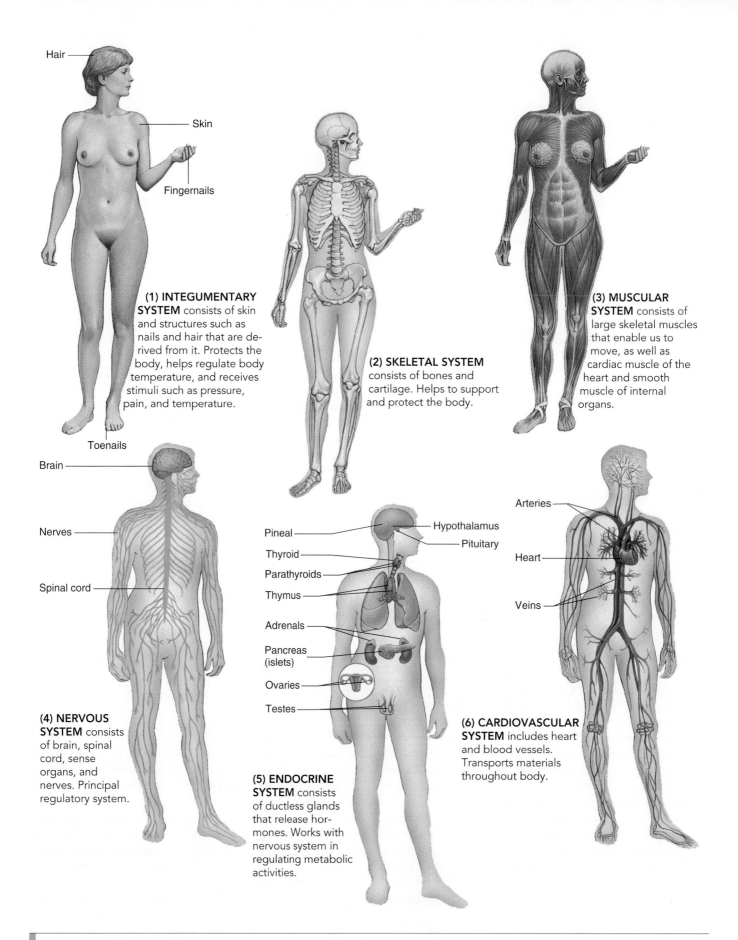

Hair

Skin

Fingernails

(1) INTEGUMENTARY SYSTEM consists of skin and structures such as nails and hair that are derived from it. Protects the body, helps regulate body temperature, and receives stimuli such as pressure, pain, and temperature.

Toenails

(2) SKELETAL SYSTEM consists of bones and cartilage. Helps to support and protect the body.

(3) MUSCULAR SYSTEM consists of large skeletal muscles that enable us to move, as well as cardiac muscle of the heart and smooth muscle of internal organs.

Brain

Nerves

Spinal cord

(4) NERVOUS SYSTEM consists of brain, spinal cord, sense organs, and nerves. Principal regulatory system.

Pineal

Hypothalamus

Pituitary

Thyroid

Parathyroids

Thymus

Adrenals

Pancreas (islets)

Ovaries

Testes

(5) ENDOCRINE SYSTEM consists of ductless glands that release hormones. Works with nervous system in regulating metabolic activities.

Arteries

Heart

Veins

(6) CARDIOVASCULAR SYSTEM includes heart and blood vessels. Transports materials throughout body.

Figure 38-4 *Animated* The 11 principal organ systems of the human body

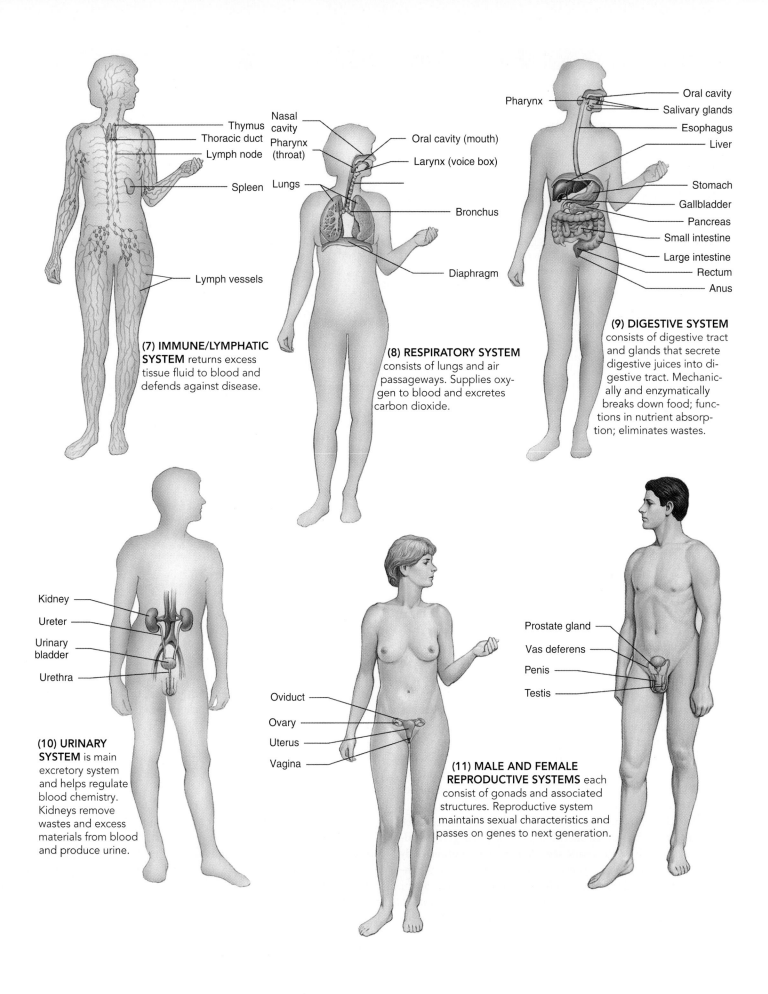

(7) IMMUNE/LYMPHATIC SYSTEM returns excess tissue fluid to blood and defends against disease.

Thymus
Thoracic duct
Lymph node
Spleen
Lymph vessels

(8) RESPIRATORY SYSTEM consists of lungs and air passageways. Supplies oxygen to blood and excretes carbon dioxide.

Nasal cavity
Pharynx (throat)
Lungs
Oral cavity (mouth)
Larynx (voice box)
Bronchus
Diaphragm

(9) DIGESTIVE SYSTEM consists of digestive tract and glands that secrete digestive juices into digestive tract. Mechanically and enzymatically breaks down food; functions in nutrient absorption; eliminates wastes.

Pharynx
Oral cavity
Salivary glands
Esophagus
Liver
Stomach
Gallbladder
Pancreas
Small intestine
Large intestine
Rectum
Anus

(10) URINARY SYSTEM is main excretory system and helps regulate blood chemistry. Kidneys remove wastes and excess materials from blood and produce urine.

Kidney
Ureter
Urinary bladder
Urethra

(11) MALE AND FEMALE REPRODUCTIVE SYSTEMS each consist of gonads and associated structures. Reproductive system maintains sexual characteristics and passes on genes to next generation.

Oviduct
Ovary
Uterus
Vagina

Prostate gland
Vas deferens
Penis
Testis

TABLE 38-4

The Mammalian Organ Systems and Their Functions

System	Components	Functions	Homeostatic Ability
Integumentary	Skin, hair, nails, sweat glands	Covers and protects body	Sweat glands help control body temperature; as a barrier, the skin helps maintain a steady state
Skeletal	Bones, cartilage, ligaments	Supports and protects body; provides for movement and locomotion; stores calcium	Helps maintain constant calcium level in blood
Muscular	Skeletal muscle, cardiac muscle, smooth muscle	Moves parts of skeleton; provides locomotion; moves internal materials	Ensures vital functions requiring movement, e.g., cardiac muscle circulates the blood
Digestive	Mouth, esophagus, stomach, intestine, liver, pancreas, salivary glands	Ingests and digests food; absorbs nutrients into blood	Maintains adequate supplies of fuel molecules and building materials
Cardiovascular	Heart, blood vessels, blood	Transports materials from one part of body to another	Distributes nutrients, oxygen, hormones to cells; transports wastes from cells to excretory organs; helps maintain water and salt balance of tissues
Immune/lymphatic	Lymphatic vessels, lymph, lymph nodes, spleen, thymus, tonsils	Collects and transports tissue fluid to blood; absorbs lipids from digestive tract and transports them to circulatory system; defends body against organisms that cause disease	Helps maintain water and salt balance of tissues; helps prevent disease
Respiratory	Lungs, trachea, and other air passageways	Exchanges gases between blood and external environment	Maintains adequate blood oxygen content and helps regulate blood pH; removes carbon dioxide from the blood
Urinary	Kidney, bladder, and associated ducts	Excretes metabolic wastes; removes excessive substances from blood	Helps regulate volume and composition of blood and body fluids
Nervous	Nerves and sense organs; brain and spinal cord	Receives stimuli from external and internal environment; conducts impulses; integrates activities of other systems	Principal regulatory system
Endocrine	Ductless glands (e.g., pituitary, adrenal, thyroid) and tissues that secrete hormones	Regulates blood chemistry and many body functions	Regulates metabolic activities and blood levels of various substances
Reproductive	Testes, ovaries, and associated structures	Sexual reproduction	Maintains sexual characteristics

Stressors, changes in the internal or external environment that affect normal conditions within the body, continuously challenge homeostasis. An internal condition that moves out of its homeostatic range (either too high or too low) causes **stress.** An organism functions effectively because homeostatic mechanisms continuously operate to manage stress.

Mammals are superb **regulators.** They have complex homeostatic mechanisms that maintain relatively constant internal conditions despite changes in the outside environment. Many animals are **conformers** for certain environmental conditions. Certain internal states vary with changes in the environment. For example, most marine invertebrates conform to the salinity of the surrounding sea water.

How do homeostatic mechanisms work? Many are **feedback systems,** sometimes called *biofeedback systems.* In **negative feedback,** a change in some steady state (for example, normal body temperature) triggers a response that opposes the change. A **sensor** detects a change, a deviation from the desired condition, or **set point.** The sensor signals an **integrator,** or control center. Based on the input of the sensor, the integrator activates homeo-static mechanisms that restore the steady state (❙ Fig. 38-5). The response counteracts the inappropriate change, thereby restoring the steady state

Note that in a negative feedback system, the response of the integrator is *opposite* (negative) to the output of the sensor. When some condition varies too far from the steady state (either too high or too low), a control system using negative feedback brings the condition back to the steady state. For example, when the glucose concentration in the blood *decreases* below its homeostatic level, negative feedback systems *increase* its concentration. Most homeostatic mechanisms in the body are negative feedback systems.

Let us discuss two specific examples: heat regulation in your home and in your body. In your home, you might set a particular room temperature on your thermostat. If the temperature in the room falls, the thermometer in the thermostat acts as a sensor that detects a change, or deviation, from the set point (❙ Fig. 38-6a). The thermometer sends a signal to the thermostat, which then acts as an integrator, or control center. The thermostat compares the sensor's input with the set point. The thermo-

Key Point

In a negative feedback system, the response of the integrator is opposite to the input of the sensor.

HOMEOSTASIS

Stressor

5 Normal condition (set point) restored.

1 Stressor causes deviation from set point.

4 Integrator activates effectors (homeostatic mechanisms).

2 Sensor detects change from set point.

3 Sensor signals integrator (control center).

Figure 38-5 *Animated* Negative feedback

stat then signals the furnace, which is the **effector** in this system. An effector is the device, organ, or process that helps restore the desired condition. The furnace increases its heat output, a corrective response that brings the room temperature back to the set

point. The thermometer no longer detects a change from the set point, so the thermostat and furnace are switched off.

The temperature-regulating system in the human body is somewhat similar. A very simplified illustration of this negative feedback system is shown in ❙ Figure 38-6b. When body temperature decreases below normal limits, specialized nerve cells (sensors) signal the temperature-regulating center in the hypothalamus of the brain (the integrator). The integrator activates effectors that bring the temperature back to the set point. The return to normal temperature signals the sensors, and the temperature-regulating center switches off the effectors.

A few **positive feedback systems** also operate in the body. In these systems a deviation from the steady state sets off a series of changes that intensify (rather than reverse) the changes. Although many positive feedback mechanisms are beneficial, they do not maintain homeostasis. For example, a positive feedback cycle operates during the birth of a baby. As the baby's head pushes against the cervix (lower part of uterus), a reflex action causes the uterus to contract. The contraction forces the head against the cervix again, resulting in another contraction, and the positive feedback cycle repeats again and again until the baby is born. Some positive feedback sequences, such as those that deepen circulatory shock following severe hemorrhage, can disrupt steady states and lead to death. A simplified example is shown in ❙ Figure 38-7.

Homeostatic mechanisms maintain the internal environment within the physiological limits that support life. As you continue your study of animal processes, you will learn many ways in which organ systems interact to maintain the steady state of the organism. All organ systems participate in these regulatory processes, but the nervous and endocrine systems play major roles.

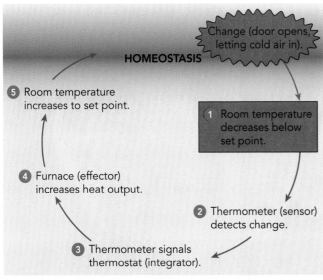

HOMEOSTASIS

Change (door opens, letting cold air in).

5 Room temperature increases to set point.

1 Room temperature decreases below set point.

4 Furnace (effector) increases heat output.

2 Thermometer (sensor) detects change.

3 Thermometer signals thermostat (integrator).

(a) Regulation of room temperature.

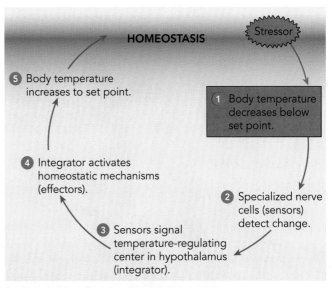

HOMEOSTASIS

Stressor

5 Body temperature increases to set point.

1 Body temperature decreases below set point.

4 Integrator activates homeostatic mechanisms (effectors).

2 Specialized nerve cells (sensors) detect change.

3 Sensors signal temperature-regulating center in hypothalamus (integrator).

(b) Regulation of body temperature.

❙ **Figure 38-6** Negative feedback in temperature regulation

Note that the diagram in (b) is highly simplified. Figure 38-9 shows greater detail.

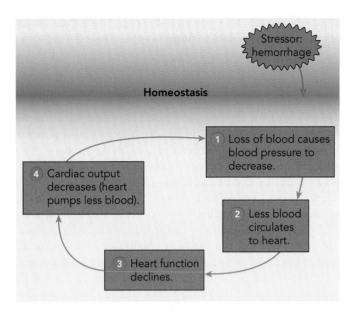

Homeostasis

Stressor: hemorrhage

1. Loss of blood causes blood pressure to decrease.
2. Less blood circulates to heart.
3. Heart function declines.
4. Cardiac output decreases (heart pumps less blood).

Figure 38-7   Positive feedback

In positive feedback, the changes that occur amplify the deviation from the set point. Loss of blood causes blood pressure to decrease. Less blood reaches the heart, so heart function decreases. The resulting decrease in cardiac output further decreases blood pressure, bringing about conditions farther from homeostasis.

In the next section, we discuss some homeostatic mechanisms that help regulate body temperature.

Review

- What are the main functions of each organ system?
- What is the basic difference between negative and positive feedback mechanisms?

- How do each of the following organ systems help maintain homeostasis: (1) respiratory, (2) urinary, and (3) endocrine? (Consult Table 38-4 for help.)

REGULATING BODY TEMPERATURE

Learning Objectives

8   Compare the costs and benefits of ectothermy.
9   Compare the costs and benefits of endothermy, and describe strategies animals use to adjust to challenging temperature changes.

Many animals have elaborate homeostatic mechanisms for regulating body temperature. Some are physiological, and others are structural or behavioral. **Thermoregulation** is the process of maintaining body temperature within certain limits despite changes in the surrounding temperature. Animals produce heat as a by-product of metabolic activities. Body temperature is determined by the rate at which heat is produced and the rate at which heat is lost to, or gained from, the outside environment.

The strategies for maintaining body temperature that are available to an animal may restrict the type of environment it can inhabit. Each species has an optimal environmental temperature range. Some animals, such as snowshoe hares, snowy owls, and weasels, can survive in cold arctic regions. Others, such as the

(a) Ectotherm. This marine iguana (*Amblyrhynchus cristatus*) increases its body temperature by sunning itself. Photographed in Galápagos Islands.

Breck P. Kent/Animals Animals/Earth Scenes

Frans Lanting/Minden Pictures

(b) Endotherm. The body temperature of this baby sooty tern (*Sterna fuscata*) is reduced as heat leaves the body through its open mouth. Photographed in Hawaii.

Figure 38-8   Behavioral adaptations for thermoregulation

Cape ground squirrel, which inhabits South Africa, are adapted to hot tropical climates. Although some animals can survive at temperature extremes, most survive only within moderate temperature ranges.

Ectotherms absorb heat from their surroundings

Ectotherms are animals that depend on the environment for their body heat. Their body temperature is determined mainly by the changing temperature of their surroundings. Most of the heat for their thermoregulation comes from the sun. You may be surprised to learn that most animals are ectotherms.

A benefit of ectothermy is that very little energy is used to maintain a high metabolic rate. In fact, an ectotherm's metabolic rate tends to change with the weather. As a result, ectotherms have a much lower daily energy expenditure than do endotherms. They survive on less food and convert more of the energy in their food to growth and reproduction. Ectothermy also has costs. One disadvantage is that daily and seasonal temperature conditions may limit activity.

Many ectotherms use behavioral strategies to adjust body temperature. For example, lizards may keep warm by burrowing in the soil at night. During the day, lizards take in heat by basking in the sun, orienting their bodies to expose the maximum surface area to the sun's rays (Fig. 38-8a). Many animals migrate to warmer climates during the winter. Another behavioral strategy for regulating temperature is *hibernation*.

Endotherms derive heat from metabolic processes

Birds and mammals, as well as some species of fish (such as tuna and some sharks) and some insects, are **endotherms,** which means they have homeostatic mechanisms that maintain body temperature despite changes in the external temperature. The most important benefit of endothermy is the high metabolic rate, which can be as much as 6 times higher than that of ectotherms. Their constant body temperature allows a higher rate of enzyme activity than is possible for ectotherms living in the same habitat. Endotherms respond more rapidly to internal and external stimuli and can be active even in low winter temperatures. However, these animals must pay the high energy cost of thermoregulation during times when they are inactive. You must maintain your body temperature even when you are asleep.

Endotherms have structural adaptations for maintaining body temperature. For example, the insulating feathers of birds, hair of mammals, and insulating layers of fat in birds and mammals reduce heat loss from the body. Birds and mammals also have behavioral adaptations for maintaining body temperature. The Cape ground squirrel positions its tail to shade its body from the direct rays of the sun. Elephants spray themselves with cool water.

Some insects use a combination of structural, behavioral, and physiological mechanisms to regulate body temperature. The

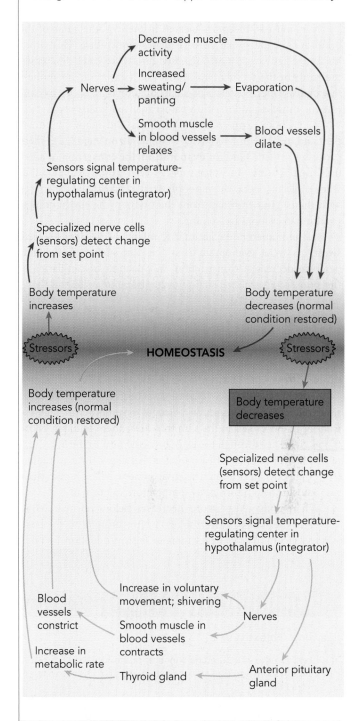

Figure 38-9 *Animated* Regulation of temperature in the human body

"furry" body of the moth helps conserve body heat. When a moth prepares for flight, it contracts its flight muscles with little movement of its wings. The metabolic heat generated enables the moth to sustain the intense metabolic activity needed for flight.

Endotherms have a variety of physiological mechanisms for maintaining temperature homeostasis. They regulate heat production and regulate heat exchange with the environment. Most of their body heat comes from their own metabolic processes (see *Focus On: Electron Transport and Heat*, in Chapter 8).

In mammals, receptors located in the hypothalamus of the brain and in the spinal cord regulate temperature. Heat from metabolic activities can be increased directly or can be increased indirectly by the action of hormones (such as thyroid hormones) that increase metabolic rate. Heat production is increased by contracting muscles, and in cold weather many animals shiver.

When body temperature rises, birds and many mammals pant and some mammals sweat (■ Fig. 38-8b on page 822). These processes provide fluid for **evaporation,** the conversion of a liquid, such as sweat, to water vapor. Recall from Chapter 2 that when molecules enter the vapor phase, they take their heat energy with them. Heat transfers from the body to the surroundings, resulting in evaporative cooling.

In the human body, about 2.5 million sweat glands secrete sweat. As sweat evaporates from the skin surface, body temperature decreases. Constriction and dilation of capillaries in the skin are also homeostatic mechanisms for regulating body temperature.

In humans, information about body temperature is sent to the temperature-regulating center in the hypothalamus (■ Fig. 38-9 on page 823). When your body temperature rises above normal, specialized nerve cells signal the temperature-regulating center in the hypothalamus. This center sends messages, by way of nerves to the sweat glands, that increase sweat secretion. At the same time, the hypothalamus sends messages to smooth muscle in the walls of blood vessels in the skin that cause them to dilate. More blood circulates through the skin, bringing heat to the body surface. The skin acts as a heat radiator that allows heat to radiate from the body surface into the environment. These homeostatic mechanisms help return body temperature to normal.

When body temperature decreases below normal, blood vessels in the skin constrict, so less heat is brought to the body surface. Hormonal messages increase heat production by body tissues. Nerves signal muscles to shiver or let you move muscles voluntarily to increase body temperature.

Many animals adjust to challenging temperature changes

Animals adjust to seasonal changes, a process called **acclimatization.** A familiar example is the thickening of a dog's coat in the winter. As water temperature decreases during fall and winter, a trout's enzyme systems decrease their level of activity, allowing the trout to remain active at the lowest metabolic cost.

When stressed by cold, many small endotherms sink into **torpor,** a state in which metabolic rate decreases, sometimes dramatically. Heart and respiratory rates decrease, and animals are less responsive to external stimulation. In endotherms, body temperature decreases below normal levels, an adaptive hypothermia. The decrease in body temperature is brought about by a decrease in the temperature set point in the hypothalamus. Daily torpor occurs in some small endotherms, for example, hummingbirds and shrews.

Hibernation is long-term torpor in response to winter cold and scarcity of food. Animals that hibernate store unsaturated fats as energy sources. **Estivation** is a state of torpor caused by lack of food or water during periods of high temperature. During estivation some mammals retreat to their burrows. The cactus mouse enters a state of hibernation during the winter in response to cold and scarcity of food. In summer, it estivates in response to lack of either food or water.

Review

- How do ectotherms adjust body temperature? What are some costs and benefits of ectothermy?
- What are some costs and benefits of endothermy?
- How is body temperature regulated in humans? Draw a diagram to illustrate your answer.
- What is acclimatization? Give an example.
- How do hibernation and estivation differ?

SUMMARY WITH KEY TERMS

Learning Objectives

1 Compare the structure and function of the four main kinds of animal tissues: epithelial, connective, muscle, and nervous (page 808).

- A **tissue** consists of a group of similarly specialized cells that associate to perform one or more functions.
- **Epithelial tissue (epithelium)** forms a continuous layer, or sheet, of cells covering a body surface or lining a body cavity. Epithelial tissue functions in protection, absorption, secretion, or sensation.
- In contrast, **connective tissue** consists of relatively few cells separated by **intercellular substance,** composed of **fibers** scattered throughout a **matrix.** The intercellular substance contains **collagen fibers, elastic fibers,** and/or **reticular fibers.** Connective tissue contains specialized

cells such as **fibroblasts** and **macrophages.** Connective tissue joins other tissues of the body, supports the body and its organs, and protects underlying organs.

- **Muscle tissue** consists of cells specialized to contract. Each cell is an elongated muscle fiber containing many contractile units called **myofibrils.**
- **Nervous tissue** consists of elongated cells called **neurons,** specialized for transmitting impulses, and **glial cells,** which support and nourish the neurons.

2 Compare the structure and function of the main types of epithelial tissue (page 808).

- Epithelial cells may be **squamous, cuboidal,** or **columnar** in shape. Epithelial tissue may be **simple, stratified,** or **pseudostratified** (summarized in Table 38-1).

■ Simple squamous epithelium lines blood vessels and the air sacs in the lungs; it permits exchange of materials by diffusion. Simple cuboidal and columnar epithelia line passageways and are specialized for secretion and absorption. Stratified squamous epithelium forms the outer layer of the skin and lines passageways into the body; it provides protection. Pseudostratified epithelium also lines passageways and protects underlying tissues.

■ Some epithelial tissue is specialized to form **glands.** **Goblet cells** are unicellular glands that secrete mucus. Goblet cells are **exocrine glands** that secrete their product through a duct onto an exposed epithelial surface. In contrast, **endocrine glands** release hormones into the **interstitial fluid** or blood.

■ An **epithelial membrane** consists of a sheet of epithelial tissue and a layer of underlying connective tissue. A **mucous membrane** lines a cavity that opens to the outside of the body. A **serous membrane** lines a body cavity that does not open to the outside.

3 Compare the main types of connective tissue, and summarize their functions (page 808).

■ The cells of **connective tissue** are embedded in an **intercellular substance** that consists of microscopic **collagen fibers, elastic fibers,** and **reticular fibers** (thin branched fibers) scattered through a **matrix,** a thin gel of polysaccharides. **Loose connective tissue** consists of fibers running in various directions through a semifluid matrix. This flexible tissue forms a covering for nerves, blood vessels, and muscles. **Dense connective tissue** is strong but less flexible than loose connective tissue. Its collagen fibers are arranged in a definite pattern. Dense connective tissue forms **tendons,** cords that connect muscles to bones, and **ligaments,** cables that connect bones to one another.

■ **Elastic connective tissue,** which consists of bundles of parallel elastic fibers, is found in lung tissue and in walls of large arteries. **Reticular connective tissue,** which consists of interlacing reticular fibers, forms a supporting framework for many organs.

■ **Adipose tissue,** which consists of fat cells, is found along with loose connective tissue in subcutaneous tissue.

■ **Cartilage** consists of **chondrocytes** that lie in **lacunae,** small cavities in a hard matrix. **Osteocytes** secrete and maintain the matrix of **bone.** Unlike cartilage, bone is quite vascular. Cartilage and bone form the skeleton of vertebrates.

■ **Blood** and **lymph** are circulating tissues that help various parts of an animal communicate with one another. The intercellular substances of blood and lymph are more fluid than those of other tissues.

4 Contrast the three types of muscle tissue and their functions (page 808).

■ **Skeletal muscle** is **striated** and under voluntary control. Each elongated, cylindrical **muscle fiber** has several nuclei. When skeletal muscles contract, they move parts of the body.

■ **Cardiac muscle** is also striated, but its contractions are involuntary. Its elongated, cylindrical fibers branch and fuse; each fiber has one or two central nuclei. When this muscle contracts, the heart pumps the blood.

■ **Smooth muscle** contracts involuntarily. Its elongated, spindle-shaped fibers lack striations. Each fiber has a single central nucleus. Smooth muscle is responsible for movement of body organs; for example, it pushes food through the digestive tract.

5 Relate the structure of the neuron to its function (page 808).

■ The elongated neuron is adapted for receiving and transmitting information. **Dendrites** receive signals and transmit them to the **cell body.** The **axon** transmits signals away from the cell body to other neurons or to a muscle or gland. A **synapse** is a junction between neurons.

6 Briefly describe the organ systems of a mammal, and summarize the homeostatic actions of each organ system (page 817).

■ Tissues and **organs** work together, forming **organ systems.** In mammals, 11 organ systems work together to carry out the functions required by the **organism.** These include the **integumentary, skeletal, muscular, digestive, cardiovascular, immune (lymphatic), respiratory, urinary, nervous, endocrine,** and **reproductive systems.** Each organ system functions to maintain homeostasis (summarized in Table 38-4).

ThomsonNOW Watch body systems work together by clicking on the figure in ThomsonNOW.

7 Define *homeostasis*, and contrast negative and positive feedback systems (page 817).

■ **Homeostasis** is the balanced internal environment, or steady state. The control processes that maintain these conditions are **homeostatic mechanisms.**

■ Homeostasis is actually a dynamic equilibrium maintained by **negative feedback systems.** When a stressor causes a change in some steady state, a response is triggered that opposes the change. A **sensor** detects a change, a deviation from the desired condition, or **set point.** The sensor signals an **integrator,** or control center. Based on the output of the sensor, the integrator activates one or more **effectors,** organs or processes that restore the steady state.

■ In a **positive feedback system,** a deviation from the steady state sets off a series of changes that intensify (rather than reverse) the changes.

8 Compare the costs and benefits of ectothermy (page 822).

■ Animals have structural, behavioral, and physiological strategies for **thermoregulation,** the process of maintaining body temperature within certain limits despite changes in the surrounding temperature. In **ectotherms,** body temperature depends to a large extent on the temperature of the environment. Many ectotherms use behavioral strategies to adjust body temperatures.

■ A benefit of ectothermy is that very little energy is used to maintain the metabolic rate. As a result, ectotherms can survive on less food. A disadvantage is that activity may be limited by daily and seasonal temperature conditions.

9 Compare the costs and benefits of endothermy, and describe strategies animals use to adjust to challenging temperature changes (page 822).

■ **Endotherms** have homeostatic mechanisms for regulating body temperature within a narrow range.

■ The most important benefit of endothermy is a high metabolic rate. Another advantage is that constant body temperature allows a higher rate of enzyme activity. Many endotherms are active even in low winter temperatures. A disadvantage of endothermy is its high energy cost.

■ **Acclimatization** is the process of adjustment to seasonal changes.

■ When challenged by a drop in surrounding temperature, many small endotherms enter a state of torpor. **Torpor** is an adaptive hypothermia; body temperature is maintained below normal levels. **Hibernation** is long-term torpor in response to winter cold. **Estivation** is torpor caused by lack of food or water during summer heat.

ThomsonNOW Explore negative feedback and temperature regulation in humans and other animals by clicking on the figures in ThomsonNOW.

1. In animals, (a) organ systems are always present (b) cells specialize to perform specific functions (c) small size provides more opportunity for capturing food (d) organelles form organ systems (e) epithelial tissue contains glial cells

2. A group of closely associated cells that carry out specific functions is a(an) (a) colony (b) tissue (c) organ system (d) organelle (e) homeostatic mechanism

3. Which tissue consists of cells that fit tightly together to form a sheet of cells? (a) muscle tissue (b) nervous tissue (c) connective tissue (d) epithelial tissue (e) reticular tissue

4. Epithelial cells that produce and secrete a product into a duct form a(an) (a) endocrine gland (b) exocrine gland (c) pseudostratified membrane (d) endocrine or exocrine gland (e) serous membrane

5. A serous membrane (a) lines a body cavity that does not open to the outside of the body (b) consists of loose connective tissue (c) has a framework of reticular connective tissue (d) covers the body (e) lines the digestive and respiratory systems

6. The most numerous fibers in connective tissue are (a) reticular (b) myofibrils (c) collagen (d) elastic (e) glial

7. Dense connective tissue is likely to be found in (a) the outer layer of skin (b) tendons (c) bone (d) the air sacs of lungs (e) the framework of the lymph nodes

8. Osteocytes are most likely to be found in (a) the outer layer of skin (b) bone (c) cartilage (d) lungs (e) blood

9. Tissue that contains fibroblasts and a great deal of intercellular substance is (a) connective tissue (b) muscle tissue (c) nervous tissue (d) pseudostratified epithelium (e) simple squamous epithelium

10. Tissue that contracts and is striated and involuntary is (a) connective tissue (b) smooth muscle (c) skeletal muscle (d) pseudostratified (e) cardiac muscle

11. Glial cells are characteristic of (a) muscle that is striated and voluntary (b) cardiac muscle tissue (c) nervous tissue (d) stratified epithelium (e) loose connective tissue

12. Tissue that has a hard, rubbery matrix and lacks blood vessels is (a) integumentary tissue (b) bone (c) nervous tissue (d) stratified epithelium (e) cartilage

13. The contractile elements in muscle tissue are (a) myofibrils (b) elastic (c) collagen (d) lacunae (e) dendrites

14. Which system consists of glands that secrete hormones? (a) integumentary (b) skeletal (c) nervous (d) endocrine (e) exocrine

15. Which system has the homeostatic function of helping maintain a constant calcium level in the blood? (a) integumentary (b) skeletal (c) nervous (d) reproductive (e) exocrine

16. Which system has the homeostatic function of helping regulate volume and composition of blood and body fluids? (a) integumentary (b) muscular (c) reproductive (d) urinary (e) exocrine

17. Many homeostatic functions are maintained by (a) negative feedback systems (b) positive feedback systems (c) set points (d) stressors (e) exocrine glands

18. An ectotherm (a) may use behavioral strategies to help adjust body temperature (b) has a variety of homeostatic mechanisms to precisely regulate body temperature (c) depends on sensors in the hypothalamus to regulate temperature (d) has a higher rate of enzyme activity than a typical endotherm (e) must expend more energy on thermoregulation than an endotherm

19. The state of torpor caused in some animals by lack of food or water during periods of high temperature is known as (a) positive feedback (b) hibernation (c) endothermy (d) ectothermy (e) estivation

1. Imagine that all the epithelium in an animal, such as a human, suddenly disappeared. What effects might this have on the body and its ability to function?

2. What would connective tissue be like if it had no intercellular substance? What effect would the absence of intercellular substance have on the body?

3. A high concentration of carbon dioxide in the blood and interstitial fluid results in more rapid breathing. Explain this observation in terms of homeostasis.

4. **Evolution Link.** What are some adaptive benefits of multicellularity?

5. **Evolution Link.** From an evolutionary perspective, why do you think most animals are ectotherms? (*Hint:* What are some benefits of ectothermy?)

Additional questions are available in ThomsonNOW at www.thomsonedu.com/login

Protection, Support, and Movement

Stephen Dalton

Grasshopper in midjump. The grasshopper (*Tropidacris dux*) uses powerful muscles in its hind legs to catapult itself into the air. Insect muscles attach to the exoskeleton.

KEY CONCEPTS

Many structures and processes have evolved in animals for protection, support, and movement.

Epithelial coverings protect underlying tissues and may be specialized for sensory, respiratory, or other functions.

Skeletal systems, whether they are hydrostatic skeletons, exoskeletons, or endoskeletons, support and protect the body and transmit mechanical forces important in movement.

Muscles contract; in most animals they move body parts by pulling on them.

During muscle contraction, energy from ATP is used to slide muscle filaments so that the muscle shortens.

In Chapter 38 we described animal tissues, organs, and systems, laying the foundation for the more detailed discussions in this and later chapters. Here we focus on epithelial coverings, skeletal systems, and muscle—structures that are closely interrelated in function. In our survey of the animal kingdom (Chapters 29 to 31), we described many adaptations involving these structures. For example, we discussed how the tough exoskeleton contributes to the great biological success of insects. In this chapter we compare these systems in several animal groups and then focus on their structures and functions in mammals.

In addition to protecting underlying tissues, the *epithelial coverings* of animals may be specialized to exchange gases, excrete wastes, regulate temperature, or secrete substances such as peptides that kill bacteria, mucus, or poison. Most animals are also protected by a *skeletal system*. Whether it is a fluid-filled compartment, a shell, or a system of bones, the skeletal system provides support for the body and typically protects internal organs. In vertebrates, bones store calcium and are important in maintaining homeostatic levels of calcium in the blood.

The skeletal system and *muscular system* work together to produce movement. Muscle tissue is specialized to contract. Although a few animals remain rooted to one spot, sweeping their surroundings with tentacles, **locomotion**—movement from place to place in the environment—is characteristic of most animals. In fact, for most animals dependable, rapid, and responsive movement is key to finding food and escaping from predators. Animals creep, walk, run, jump, swim, or fly (see photograph). Muscles responsible for locomotion anchor to the skeleton, which gives them something firm on which to act. In most vertebrates, bones serve as levers that transmit the force necessary to move various parts of the body.

Muscle powers both locomotion and the physiological actions necessary to maintain homeostasis. Many animals have internal circulating fluids, pumped by hearts and contained by hollow vessels that maintain their pressure with gentle squeezing. Most have digestive systems that push food along with peristaltic contractions. In complex animals, effective movement depends on the cooperation and interactions of several body systems, including the muscular, skeletal, nervous, circulatory, respiratory, and endocrine systems. ∎

EPITHELIAL COVERINGS

Learning Objectives

1 Compare the functions of the external epithelium of invertebrates and vertebrates.
2 Relate the structure of vertebrate skin to its functions.

Epithelial tissue covers all external and internal surfaces of the animal body. The structure and functions of the epithelial covering are adapted to the animal's environment and lifestyle. In both invertebrates and vertebrates, epithelial coverings protect the body. These coverings may also be specialized for secretion or gas exchange and typically contain receptors that receive sensory signals from the environment.

Invertebrate epithelium may function in secretion or gas exchange

In many invertebrates, the outer epithelium is specialized to secrete protective or supportive layers of nonliving material. In insects and many other animals, the secreted material forms an outer covering called a **cuticle.** Some animals, including corals and mollusks, secrete a shell made mainly of calcium carbonate. Epithelial cells may be modified as sensory cells that are selectively sensitive to light; chemical stimuli; or mechanical stimuli, such as contact with an object, or pressure. In many species, the epithelium contains cells that secrete lubricants or adhesives. Such cells may release odorous secretions used for communication or for marking trails. In other species, secretory cells produce toxins used for offense or defense. The earthworm epithelium secretes a lubricating mucus that promotes efficient diffusion of gases across the body wall and reduces friction during movement through the soil.

Vertebrate skin functions in protection and temperature regulation

The **integumentary system** of vertebrates includes the skin and structures that develop from it. In many fishes, in the African ant-eating pangolin (a mammal; see Fig. 18-12c), and in some reptiles, skin has developed into a set of scales formidable enough to be considered armor. Even human skin has considerable strength. In many amphibians, gas exchange takes place through the skin.

Derivatives of skin differ considerably among vertebrates. Fishes have bony or toothlike scales. Amphibians have naked skin covered with mucus, and some species are equipped with poison glands. Reptiles have epidermal scales, mammals have hair, and birds have feathers that provide even more effective insulation than fur. The feathers of birds and the fur of mammals are insulating structures that trap heat in the body, which helps maintain body temperature. Skin and its derivatives are often brilliantly colored in connection with courtship rituals, territorial displays, and other kinds of communication.

In mammals, structures derived from skin include claws (modified as fingernails and toenails in primates), hair, sweat glands, oil (sebaceous) glands, and several types of sensory receptors that give mammals the ability to feel pressure, temperature, and pain. The skin of mammals also contains mammary glands, specialized in females for secretion of milk.

In humans and other mammals, oil glands empty via short ducts into hair follicles. A **hair follicle** is the part of a hair below the skin surface, together with the surrounding epithelial tissue. Oil glands secrete **sebum,** a mixture of fats and waxes that inhibits the growth of harmful bacteria.

Although the skin of fishes, frogs, pelicans, and elephants appears very different, the basic structure of skin is the same in all vertebrates. The outer layer of skin, the **epidermis,** is a waterproof protective barrier. The epidermis consists of several strata, or sublayers. The deepest is **stratum basale,** and the most superficial is **stratum corneum** (▌ Fig. 39-1). Pigment cells in stratum basale and in the dermis below produce **melanin,** a pigment that contributes to the color of the skin. In stratum basale, cells divide and are pushed outward as other cells are produced below them. Epidermal cells mature as they move toward the skin surface.

As they move toward the body surface, epidermal cells manufacture **keratin,** an elaborately coiled protein that gives the skin considerable mechanical strength and flexibility. Keratin is insoluble and serves as a diffusion barrier for the body surface. As epidermal cells move through stratum corneum, they die. When they reach the outer surface of the skin, they wear off and must be continuously replaced.

Beneath the epidermis lies the **dermis** (see Fig. 39-1), a dense, fibrous connective tissue made up mainly of collagen fibers. Collagen imparts strength and flexibility to the skin. The dermis also contains blood vessels that nourish the skin and sensory receptors for touch, pain, and temperature. Hair follicles and, in most mammals, sweat glands are embedded in the dermis. In birds and

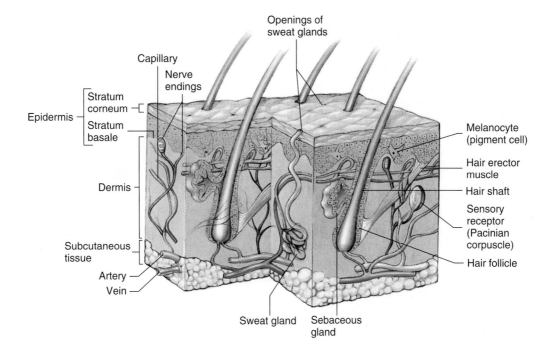

Figure 39-1 *Animated* Human skin

The integumentary system of humans and other mammals is a complex organ system consisting of skin and structures such as hair and nails that are derived from the skin. Stratum basale consists of a single layer of dividing epithelial cells.

mammals, the dermis rests on a layer of **subcutaneous tissue** composed mainly of adipose tissue that insulates the body from outside temperature extremes.

In humans, exposure to ultraviolet (UV) radiation, the short, invisible rays from the sun, causes the epidermis to thicken and stimulates pigment cells in the skin to produce melanin at an increased rate. Melanin is an important protective screen against the sun because it absorbs some of the harmful UV rays. An increase in melanin causes the skin to darken. The suntan so prized by sun worshipers is actually a sign that the skin has been exposed to too much UV radiation. When the melanin cannot absorb all the UV rays, the skin becomes inflamed, or sunburned. Dark-skinned people have more melanin and so suffer less sunburn, wrinkling, and skin cancer but still require protection from UV rays.

UV radiation damages DNA, causing mutations that lead to malignant transformation of cells (see Chapter 17). Most cases of skin cancer are caused by excessive, chronic exposure to UV radiation. The incidence of *malignant melanoma,* the most lethal type of skin cancer, is increasing faster than any other type of cancer.

Review

- How are the body coverings of invertebrates and vertebrates alike? How are they different?
- How is the structure of vertebrate skin related to its specific functions?
- What properties does keratin give human skin?
- What is the function of melanin?

SKELETAL SYSTEMS

Learning Objectives

3 Compare the structure and functions of different types of skeletal systems, including the hydrostatic skeleton, exoskeleton, and endoskeleton.

4 Describe the main divisions of the vertebrate skeleton and the bones that make up each division.

5 Describe the structure of a typical long bone, and differentiate between endochondral and intramembranous bone development.

6 Compare the main types of vertebrate joints.

In addition to having an epithelial covering, many animals are protected by a **skeletal system** that forms the framework of the body. The skeleton also functions in locomotion. Muscles typically act on hard structures such as chitin or bone. These skeletal structures transmit and transform mechanical forces generated by muscle contraction into the variety of motions that animals make. Three types of skeletons are hydrostatic skeletons, exoskeletons, and endoskeletons.

In hydrostatic skeletons, body fluids transmit force

Many soft-bodied invertebrates have **hydrostatic skeletons** made of fluid-filled body compartments. Imagine an elongated balloon full of water. If you pull on the balloon, it lengthens and becomes thinner. It does the same if you squeeze it. Conversely, if you push the ends toward the center, the balloon shortens and thickens. Many animals, including most cnidarians, flatworms, annelids, and roundworms, have hydrostatic skeletons that work something like a balloon filled with water. Fluid in a closed compartment of the body is held under pressure. When muscles in the compartment wall contract, they push against the tube of fluid. Because fluids cannot be compressed, the force is transmitted through the fluid, changing the shape and movement of the body.

In *Hydra* and other cnidarians, cells of the two body layers can contract. The contractile cells in the outer epidermal layer lie longitudinally, whereas the contractile cells of the inner layer (the gastrodermis) are arranged circularly around the central body axis (❙ Fig. 39-2). The two groups of cells work in **antagonistic** fashion: what one can do, the other can undo. When the epidermal (longitudinal) layer contracts, the hydra shortens. Due to the fluid in the gastrovascular cavity, force is transmitted so that the hydra thickens as well. In contrast, when the inner (circular) layer contracts, the hydra thins and its fluid contents force it to lengthen.

Mechanically, the hydra is a bag of fluid. The fluid acts as a hydrostatic skeleton because it transmits force when the contractile cells contract against it. (Although technically not a closed compartment, the gastrovascular cavity functions as a hydrostatic skeleton because its opening is small.) Hydrostatic skeletons permit only crude mass movements of the body or its appendages. Delicate movements are difficult, because force tends to be transmitted equally in all directions throughout the entire fluid-filled body of the animal. For example, it is not easy for the hydra to thicken one part of its body while thinning another.

The more sophisticated hydrostatic skeleton of the annelid worm lets it move with more flexibility. An earthworm's body consists of a series of segments divided by transverse partitions, or septa (see Fig. 30-12). The septa isolate portions of the body cavity and the coelomic fluid they contain. This structure allows the hydrostatic skeletons of each segment to be largely independent of one another. Thus, contraction of the circular muscle in the elongating anterior end does not interfere with the action of the longitudinal muscle in the segments at the posterior end.

You can find some examples of hydrostatic skeletons in invertebrates equipped with shells or endoskeletons, and even in vertebrates with endoskeletons of cartilage or bone. For example, sea stars and sea urchins move their tube feet by an ingenious adaptation of the hydrostatic skeleton (see Chapter 31). And the human penis becomes erect and stiff because of the turgidity of pressurized blood in its internal spaces.

Mollusks and arthropods have nonliving exoskeletons

In most animals, the skeleton is a lifeless shell, or **exoskeleton,** deposited atop the outer epithelial covering. In mollusks, the exoskeleton is a calcium carbonate shell secreted by the mantle, a thin sheet of epithelial tissue that extends from the body wall. The

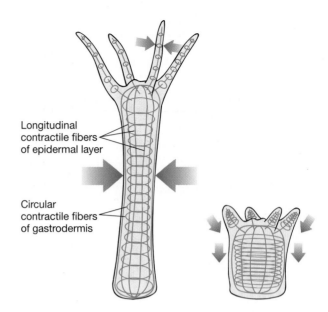

Longitudinal contractile fibers of epidermal layer

Circular contractile fibers of gastrodermis

(a) Contraction of circular contractile fibers elongates the body.

(b) Contraction of longitudinal fibers shortens the body.

Figure 39-2 Hydrostatic skeleton

In *Hydra,* fluid in the gastrovascular cavity transmits force when contractile cells in the body wall contract against it. The longitudinally arranged contractile cells of *Hydra* are antagonistic to the cells arranged in circles around the body axis.

exoskeleton provides protection, a retreat used in emergencies, with the bulk of the naked, tasty body exposed at other times.

Exoskeletons of arthropods serve not only to protect but also to transmit forces. In this respect they are comparable to the skeletons of vertebrates. The arthropod exoskeleton is a nonliving *cuticle* that contains the polysaccharide **chitin.** Although it is a continuous, one-piece sheath covering the entire body, the arthropod exoskeleton varies greatly in thickness and flexibility. Large, thick, inflexible plates are separated from one another by thin, flexible joints arranged segmentally. Enough joints are provided to make the arthropod's body as flexible as those of many vertebrates. The arthropod exoskeleton is adapted to a vast variety of lifestyles, and parts of it are modified to function as specialized tools or weapons.

A disadvantage of the rigid arthropod exoskeleton is that to accommodate growth, an arthropod must **molt,** that is, shed its exoskeleton and replace it with a new, larger one (❙ Fig. 39-3). (This process is also called *ecdysis.*) During molting the animal is weak and vulnerable to predators.

Internal skeletons are capable of growth

You are probably most familiar with the **endoskeleton** of echinoderms and chordates. This living internal skeleton consists of plates or shafts of calcium-impregnated tissue (such as cartilage or bone). Composed of living tissue, the endoskeleton grows along with the animal as a whole. The echinoderm endoskeleton consists of spines and plates of calcium salts embedded in the

THE NEUROBIOLOGY OF TRAUMATIC EXPERIENCE

In recent years, the 9/11 attack on the World Trade Center, civilian deaths by suicide bombers, the experiences of hundreds of thousands of troops at war in Iraq, and a host of natural disasters have heightened our awareness of the effects of traumatic experiences. A *traumatic experience* is an event that causes intense fear, helplessness, or horror and that overwhelms normal coping and defense mechanisms. Whether you are a combat veteran; a survivor of childhood abuse; or the victim of a robbery, rape, hurricane, or domestic violence, you may have experienced the effects of trauma.

Most of us have strategies for processing moderately disturbing experiences. For example, if you were one of the millions of people who were glued to their television sets on 9/11 or watched as people were wounded in Iraq, you may have been disturbed by the events you witnessed. You probably processed what you saw and heard by thinking and talking about it, perhaps even writing about it. You may have dreamed about your experience. As the brain actively reviews and sorts out a disturbing experience, we make sense of what happened, its emotional intensity decreases, and we store the memory along with other memories of more ordinary past events. The memory of the uncomfortable experience fades in importance.

Traumatic events are more difficult to process. If you are one of the survivors who actually ran for your life as the World Trade Center fell, or if you are a combat veteran, your body's response to the danger is more intense. Years later, your brain may still be secreting abnormally high amounts of stress hormones and neurotransmitters, and you may remain on red alert for possible danger.

How does the brain respond in a traumatic experience, and how does the experience affect the brain? When danger threatens, the amygdala sends messages to both the hypothalamus (which signals the autonomic division of the nervous system and the endocrine system) and the neocortex (which allows us to be conscious of our experience). The amygdala is programmed to remember the smells, sounds, and sensations that are part of the experience. Until the memories of the experience are fully processed, similar smells, sounds, and sensations remind us of the traumatic event and trigger the body to prepare for danger. The traumatized individual experiences fear and anxiety. Some trauma survivors develop posttraumatic stress disorder (PTSD), a condition in which they experience (1) intrusive thoughts, images, sensory experiences, memories, and dreams; (2) an urge to avoid reminders of the traumatic event; and (3) physiological hyperarousal, a condition in which the body remains on high-alert status, continuously scanning the environment for potential danger.

Because they are so overwhelming, traumatic memories are very difficult to process. They cause us so much discomfort and anxiety that we tend to avoid them rather than intentionally focus on them and sort them out. As a result, traumatic memories seem to stay "stuck" in the limbic system, and when triggered, the experience is replayed with its original emotional intensity (a flashback).

Several neuroimaging studies have shown an association between prolonged trauma (for example, severe child abuse) and long-term changes in the brain. These changes include EEG abnormalities, significantly smaller total brain and cerebral volumes, decreased right–left hemisphere integration, decreased size of the corpus callosum, and decreased volume of the hippocampus and amygdala.

Neuroimaging studies have also demonstrated changes in brain function, including increased amygdalar response and decreased cortical response. Investigators have used functional magnetic resonance imaging (fMRI) to measure amygdalar response to masked, fearful face targets. They found exaggerated amygdalar responses in subjects with PTSD. The investigators suggested that these intense responses occurred because the normal mediating influence of the medial prefrontal cortex was absent.

When researchers presented subjects with reminders of their traumatic experience, they observed decreased activity in the anterior cingulate cortex (ACC). This area, part of the medial prefrontal cortex, normally inhibits the amygdala so that it does not respond too strongly. When the ACC does not function normally, cerebral blood flow to the amygdala increases (and can be seen on neuroimages). The amygdala overresponds, and the individual experiences anxiety, distress, and physiological hyperarousal, which may lead to intense emotional responses that are rooted in the traumatic experience.

Broca's area, a region in the posterior part of the left frontal cortex, is also affected by trauma. Broca's area is critically important in generating words and thus in expressing language. Researchers have shown that when subjects are exposed to accounts of their traumatic experiences, activity in Broca's area decreases. This deactivation appears to be the physiological basis for the difficulty that trauma survivors have describing their experience in words. Trauma survivors have described flashbacks of their experience by saying that they are in a state of "speechless terror." Without words, it is difficult to process and resolve a traumatic experience.

Given these and other studies, we can conclude that traumatic experience can cause both structural and functional changes in the brain and that many trauma survivors respond strongly to triggers that remind them of their experience. The exaggerated responses of the amygdala to harmless stimuli perceived as threatening are dissociated from cortical (medial prefrontal) activation. Responses generated by the limbic system are rooted in emotion rather than in rationality and judgment.

---

ment and future behavior. Healthy attachment allows the baby to express emotions and provides the basis for learning how to regulate emotions. Attachment also determines how vulnerable (or resilient) the individual will be to traumatic experiences later in life.

Researchers first discovered in the 1950s that the limbic system is part of an important motivational system when they implanted electrodes in certain areas of the brains of laboratory animals. They found that a rat quickly learns to press a lever that

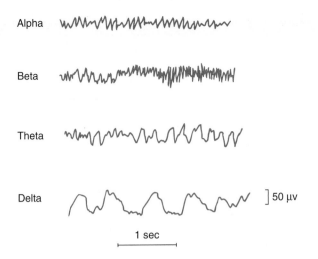

Alpha

Beta

Theta

Delta] 50 μv

1 sec

Figure 41-14 EEGs showing electrical activity in the brain

The regular waves characteristic of the relaxed state are alpha waves. During mental activity, rapid, low-amplitude beta waves are dominant. Recordings made in various stages of sleep show theta and delta waves, which have a lower frequency and greater amplitude.

stimulates a **reward center** in the brain. In fact, the rat will press the lever as many as 5000 times per hour, choosing this self-stimulation over food and water, until it drops from exhaustion.

We now know that the brain has reward centers that give us pleasure when we carry out vital activities such as eating, drinking, and sexual activity. These centers are important in experiencing emotion and in motivation. When stimulated, they activate reward circuits that allow us to feel pleasure in response to certain experiences.

The **mesolimbic dopamine pathway,** a major reward circuit, includes two important areas in the midbrain, the substantia nigra and the adjacent ventral tegmental area. These areas, which extend into behavioral control centers in the limbic system, contain the largest group of neurons that release dopamine in the brain. Dopamine neurons are activated by novel stimuli associated with reward or pleasure. These neurons signal us about surprising events or stimuli that predict rewards. Dopamine may motivate us to act when something important is happening.

As researchers continue to unravel the mechanisms of dopamine pathways, we may better understand such disorders as attention deficit disorder (ADD) and schizophrenia, conditions associated with excessive amounts of dopamine and other neurotransmitters in the brain. People with these disorders have difficulty filtering out sensory stimuli. The presence of excess dopamine may drive them to divert their attention to so many sensory stimuli that they have difficulty focusing on the most important stimuli.

Review

- How is the human CNS protected? What is the function of the cerebrospinal fluid?
- What are two main functions of the vertebrate spinal cord?
- What are the functions of each of the major lobes of the cerebrum?
- What is the function of the reticular activating system? of the limbic system?

INFORMATION PROCESSING

Learning Objectives

9 Relate neural plasticity to synaptic plasticity, and give three examples of neural plasticity.

10 Summarize how the brain processes information.

Neurobiologists are beginning to understand how the nervous system functions—how neurons signal one another; connect; and carry out basic functions such as regulating heart rate, respiration, and sleep–wake cycles; as well as how we learn and remember. An interesting question has been, To what extent are these functions **hard-wired,** that is, preset? Even animals with very simple nervous systems can learn. They learn to repeat behaviors associated with reward and to avoid behaviors that cause pain. Such changes in behavior are possible because of **neural plasticity,** the ability of the nervous system to change in response to experience. Such change modifies structure and function.

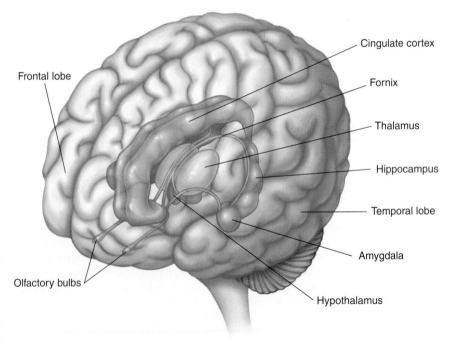

Figure 41-15 The limbic system

The limbic system consists of a ring of forebrain structures that surround the brain stem and are interconnected by complex neural pathways. This system is critical in regulation of emotional aspects of behavior and in motivation.

Many areas of the brain once thought to be hard-wired are now known to be flexible and capable of change. Familiar examples of neural plasticity in human motor skills include learning to walk, ride a bicycle, or catch a baseball. At first you were probably clumsy, but with practice your performance became smoother and more precise. For that to happen, changes must occur in the brain. Similarly, the ability to learn languages, solve problems, and perform scientific experiments depends on neural plasticity.

What are the mechanisms by which neurons signal one another to learn diverse activities, such as skilled movement and the study of biology? Canadian psychologist Donald O. Hebb proposed in 1949 that when two neurons connected by a *synapse* (that is, presynaptic and postsynaptic neurons) are active simultaneously, the synapse is strengthened. In 1973, British researchers T. V. P. Bliss and T. J. Lomo found experimental evidence for Hebb's hypothesis. Investigating rabbit brains, these researchers applied a series of high-frequency electrical stimuli to neurons in the hippocampus (a region that converts information into memories). The result was a long-term increase in synaptic strength. We can also credit Hebb for the axiom "neurons that fire together, wire together." Experience influences how neural circuits in our brains organize themselves.

Because neurons signal one another at synapses, these junctions between neurons are currently the focus of intense research. Neural plasticity may, in fact, be **synaptic plasticity,** because changes that occur during learning and remembering appear to take place at synapses. As the brain matures, it loses some of its plasticity. Neurobiologists are looking for molecules that inhibit the brain's ability to change.

Learning involves the storage of information and its retrieval

Learning is the process by which we acquire knowledge or skills as a result of experience. For learning to occur, we must be able to remember what we experience. **Memory** is the process of encoding, storing, and retrieving information or learned skills.

Implicit memory, also called *procedural memory,* is unconscious memory for perceptual and motor skills, such as riding a bicycle. Implicit memory is about "how" to do something and involves repeating the behavior until it becomes routine. **Explicit memory,** also called *declarative memory,* involves factual knowledge of people, places, or objects and requires conscious recall of the information. The hippocampus is critical in forming and retrieving explicit memories.

Information processing involves short- and long-term memory

At any moment you are bombarded with thousands of bits of sensory information. At this very moment, your eyes are receiving information about the words on this page, the objects around you, and the intensity of the light in the room. At the same time you may be hearing a variety of sounds—music, your friends talking in the next room, the hum of an air conditioner. Your olfactory epithelium may sense cologne or the smell of coffee. Maybe you are eating as you read. Sensory receptors in your hands may be receiving information regarding the weight and position of your book. Most sensory stimuli are not important to remember and so are filtered out.

When we select information to remember, we must process it. When we begin to process information, we *encode* it. We recognize patterns and meaningfully associate the stimuli to past experience or knowledge. Information processing involves both **short-term memory,** which lasts only minutes, and **long-term memory,** which may last for many years.

Typically, we can hold only about seven chunks of information (a chunk is some unit such as a word, syllable, or number) at a time in short-term memory. Short-term memory lets us recall information for a few minutes. When we look up a phone number, for example, we usually remember it only long enough to dial. If we need the same number an hour later, most of us have to look it up again. One hypothesis of short-term memory suggests that reverberating circuits are involved (see Fig. 40-13). A memory circuit may continue to reverberate for several minutes until it fatigues or until new signals interfere with the old.

The hippocampus temporarily holds new information and may integrate various aspects of an experience, including odors, sounds, and other information. **Memory consolidation,** which depends on the hippocampus, allows memories to be *stored* for long periods. Consolidation involves expression of genes and protein synthesis and depends on new neural connections at synapses. Time (several minutes, hours, or even longer) is required for the brain to *consolidate* a new memory.

When information is selected for long-term storage, the brain rehearses the material and then stores it in long-term memory in association with similar memories. We integrate new information with other knowledge already stored in the brain.

If a person suffers a brain concussion, memory of what happened immediately before the incident may be completely lost. This is known as *retrograde amnesia.* When the hippocampus is damaged, short-term memory may be unimpaired and the patient can still recall information stored in the past. However, new short-term memories can no longer be converted to long-term memories.

Where are memories stored? When large areas of the mammalian cerebral cortex are destroyed, information is lost somewhat in proportion to the extent of lost tissue. No specific area can be labeled the "memory bank." Rather, memories seem to be stored within many areas of the brain. For example, visual memories may be stored in the visual centers of the occipital lobes, and auditory memories may be stored in the temporal lobes. Researchers think that memories are integrated in many areas of the brain, including the amygdala and the hippocampus, association areas of the cerebral cortex, and the thalamus and hypothalamus. *Wernicke's area* in the temporal lobe has been identified as a very important association area for complex thought processes. Neurons within the association areas form interconnected pathways that permit complicated reverberation.

Retrieval of information stored in long-term memory is of considerable interest—especially to students! The challenge is to find information when you need it. When you seem to forget something, the problem may be that you have not searched effectively for the memory. Information retrieval can be improved by careful encoding, for example, by forming strong associations between items.

Neurobiological changes occur during learning

Short-term memory involves changes in the neurotransmitter receptors of postsynaptic neurons. These changes strengthen synaptic connections. The receptors are linked by second messengers (for example, cyclic AMP) to ion channels in the plasma membrane.

In some types of learning, changes take place in presynaptic terminals or postsynaptic neurons that permanently enhance or inhibit the transmission of impulses. In some cases, specific neurons may become more sensitive to neurotransmitter. Neurons typically transmit action potentials in bursts, and the amount of neurotransmitter released by each action potential may increase or decrease.

Repeated electrical stimulation of neurons causes a functional change at synapses. When a presynaptic neuron continues to transmit action potentials at a high rate for a minute or longer, there is a long-lasting strengthening of the connection between the presynaptic and postsynaptic neurons. This increased strength of the synaptic connection is known as **long-term potentiation (LTP)**. *Potentiation* means "to strengthen or make more potent." In contrast, low-frequency stimulation of neurons results in a long-lasting *decrease* in the strength of their synaptic connections, called **long-term depression (LTD)**. Information storage and forgetting depend on the strengthening and weakening of synaptic connections brought about by LTP and LTD.

Induction of LTP and LTD requires activation of two types of receptors: **NMDA receptors** and **AMPA receptors.** These receptors, present on the plasma membranes of postsynaptic neurons, control the passage of calcium ions into neurons. NMDA receptors respond to the neurotransmitter *glutamate* by opening Ca^{2+} channels. However, when the postsynaptic neuron is at its resting potential, NMDA ion channels are blocked by Mg^{2+}.

A model for the mechanism of LTP is illustrated in Figure 41-16. A presynaptic neuron releases glutamate, which binds with AMPA receptors. The postsynaptic neuron becomes depolarized. If depolarization of the postsynaptic neuron is sufficiently strong, Mg^{2+} move away from the NMDA receptors, unblocking them. Glutamate can then bind with these receptors, opening Ca^{2+} channels and letting Ca^{2+} move into the cell. Calcium ions appear to be an important trigger for LTP.

Calcium ions act as second messengers that initiate long-term changes ultimately responsible for LTP. For example, calcium ions activate a Ca^{2+}-dependent second-messenger pathway that results in the insertion of more AMPA receptors in the postsynaptic membrane. This is important because additional AMPA receptors increase sensitivity to glutamate. More EPSPs are produced, which strengthens the synapse (that is, helps maintain the LTP).

Calcium ions also activate a pathway that leads to release of a **retrograde signal,** a signal that moves backward, from the postsynaptic neuron to the presynaptic neuron. The soluble gas *nitric oxide (NO)* has been identified as a retrograde signal. This signal enhances neurotransmitter release by the presynaptic neuron. Note that this is a positive feedback loop that strengthens the connection between the two neurons.

Long-term memory involves gene expression

Gene expression and protein synthesis take place during the process of establishing long-term memory. This process involves slower, but longer-lasting, changes in synaptic connections. Long-term memory depends on activated receptors linked to G proteins. Cyclic AMP acts as a second messenger.

A high level of cyclic AMP activates a protein kinase that enters the nucleus, leading to gene activation and protein synthesis. In this process, protein kinase phosphorylates a transcription factor known as **CREB** (for *cyclic AMP response element binding protein*). CREB then turns on the transcription process of certain genes. CREB has been shown to be a signaling molecule in the memory pathway in many animals, including fruit flies and mice. The molecules and processes involved in learning and memory have been highly conserved during evolution.

Experience affects development and learning

Many studies have demonstrated neural plasticity in rats and other laboratory animals exposed to enriched environments. In contrast with rats housed in standard cages and provided with the basic necessities, those exposed to enriched environments are given toys and other stimulating objects, as well as the opportunity to socially interact with other rats. Animals reared in an enriched environment exhibit increased synaptic contacts and process and remember information more quickly than animals lacking such advantages. Mice exposed to enriched environments develop significantly greater numbers of neurons in the hippocampus and learn mazes faster than control animals.

Early environmental stimulation can also enhance the development of motor areas in the brain. For example, the brains of rats encouraged to exercise become slightly heavier than those of control animals. Characteristic changes occur within the cerebellum, including the development of larger dendrites.

Apparently, during early life certain critical or sensitive periods of nervous system development occur that are influenced by environmental stimuli. For example, when the eyes of young mice first open, neurons in the visual cortex develop large numbers of dendritic spines (structures on which synaptic contact takes place). If the animals are kept in the dark and deprived of visual stimuli, fewer dendritic spines form. If the mice are exposed to light later in life, some new dendritic spines form but never as many as develop in a mouse reared in a normal environment.

Studies linking the development of the brain with environmental experience have demonstrated that early stimulation is important for the sensory, motor, and intellectual development of children. Such studies led to a greatly expanded educational

When glutamate binds to NMDA receptors, Ca^{2+} enter the postsynaptic neuron and activate intracellular signaling systems that lead to changes responsible for LTP.

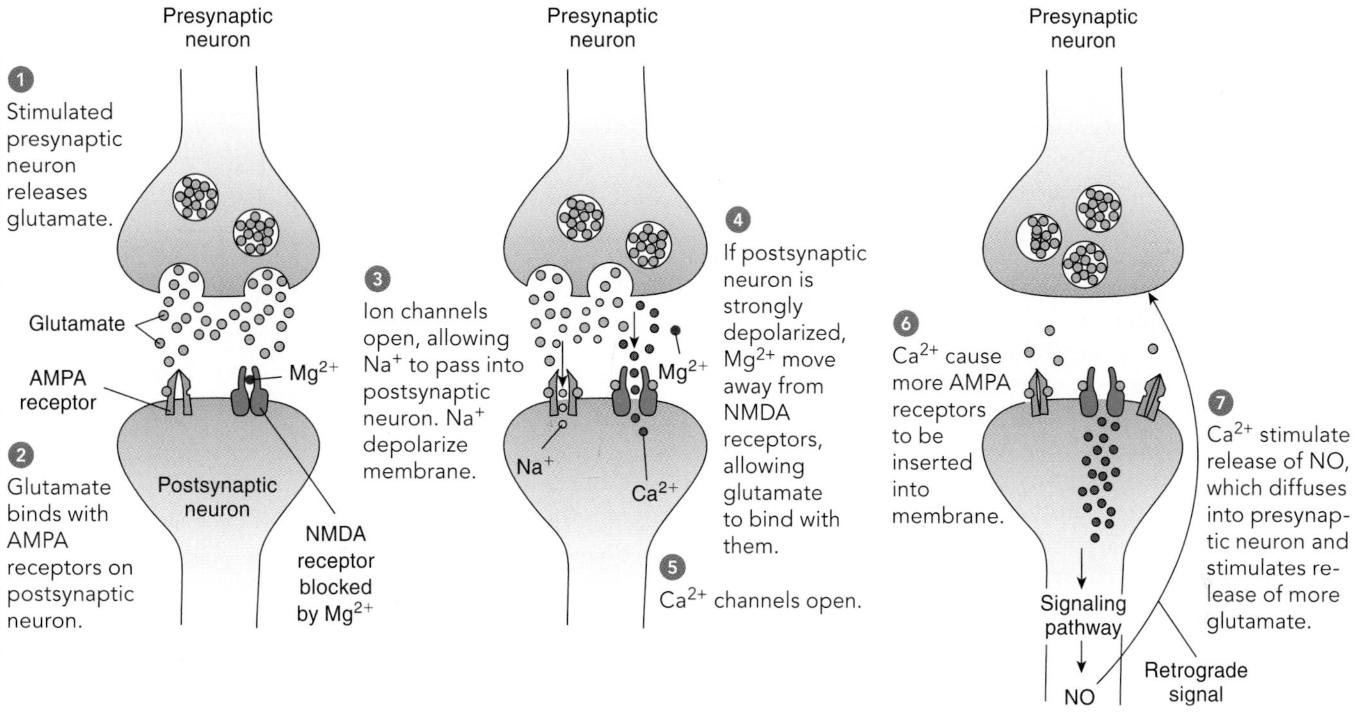

Presynaptic neuron

1 Stimulated presynaptic neuron releases glutamate.

Glutamate

AMPA receptor

Mg^{2+}

2 Glutamate binds with AMPA receptors on postsynaptic neuron.

Postsynaptic neuron

NMDA receptor blocked by Mg^{2+}

Presynaptic neuron

3 Ion channels open, allowing Na^+ to pass into postsynaptic neuron. Na^+ depolarize membrane.

Mg^{2+}

Na^+

Ca^{2+}

4 If postsynaptic neuron is strongly depolarized, Mg^{2+} move away from NMDA receptors, allowing glutamate to bind with them.

5 Ca^{2+} channels open.

Presynaptic neuron

6 Ca^{2+} cause more AMPA receptors to be inserted into membrane.

Signaling pathway

NO

Retrograde signal

7 Ca^{2+} stimulate release of NO, which diffuses into presynaptic neuron and stimulates release of more glutamate.

(a) Activation of AMPA receptors. Glutamate released by a presynaptic neuron binds with AMPA receptors on the postsynaptic neuron. Note that NMDA receptors are blocked by Mg^{2+}.

(b) Activation of NMDA receptors. The combination of strong stimulation and glutamate binding activates NMDA receptors.

(c) Long-term potentiation. Calcium ions activate pathways, leading to changes that are ultimately responsible for LTP.

Figure 41-16 A proposed model for the mechanism of long-term potentiation (LTP)

toy market and to widespread acceptance of early-education programs. Researchers have also suggested that continuing environmental stimulation is needed to maintain the status of the cerebral cortex in later life.

Language involves comprehension and expression

Language is a form of communication in which we use words to symbolize objects and convey ideas. In most people, the areas of the brain responsible for language are located only in the left hemisphere. **Wernicke's area,** located in the temporal lobe, is an important center for language comprehension (see Fig. 41-12). This area helps us recognize and interpret both spoken and written words. Wernicke's area also helps us formulate the choice and sequence of words and transfers information to **Broca's area.** Located near motor areas in the left frontal lobe, Broca's area controls our ability to speak. Broca's area is also important in language processing and speech comprehension.

Review

- What is synaptic plasticity?
- How is information processed? How might short-term memory be adaptive?

THE PERIPHERAL NERVOUS SYSTEM

Learning Objectives

11 Describe the organization of the peripheral nervous system, and compare its somatic and autonomic divisions.

12 Contrast the sympathetic and parasympathetic divisions of the autonomic system, and give examples of the effects of these systems on specific organs such as the heart and intestine.

The peripheral nervous system consists of the sensory receptors, the nerves that link these receptors with the central nervous system, and the nerves that link the CNS with effectors (muscles and glands). The somatic division of the PNS helps the body respond to changes in the external environment. The nerves and receptors that maintain homeostasis despite internal changes make up the autonomic division.

The somatic division helps the body adjust to the external environment

The **somatic division** of the PNS includes the receptors that react to changes in the external environment, the sensory neurons that inform the CNS of those changes, and the motor neurons that adjust the positions of the skeletal muscles that help maintain the body's posture and balance. In mammals, 12 pairs of **cranial nerves** emerge from the brain. They transmit information regarding the senses of smell, sight, hearing, and taste from the special sensory receptors, as well as information from the general sensory receptors, especially in the head region. For example, cranial nerve II, the optic nerve, transmits signals from the retina of the eye to the brain. The cranial nerves also bring orders from the CNS to the voluntary muscles that control movements of the eyes, face, mouth, tongue, pharynx, and larynx. Cranial nerve VII, the facial nerve, transmits signals to the muscles used in facial expression and to the salivary glands.

In humans, 31 pairs of **spinal nerves** emerge from the spinal cord. Named for the general region of the vertebral column from which they originate, they comprise 8 pairs of cervical, 12 pairs of thoracic, 5 pairs of lumbar, 5 pairs of sacral, and 1 pair of coccygeal spinal nerves. The ventral branches of several spinal nerves form tangled networks called *plexi* (sing., *plexus*). Within a plexus, the fibers of a spinal nerve may separate and then regroup with fibers that originated in other spinal nerves. Thus, nerves emerging from a plexus consist of neurons from several different spinal nerves.

The autonomic division regulates the internal environment

The **autonomic division** helps maintain homeostasis in the internal environment. For example, it regulates the heart rate and helps maintain a constant body temperature. The autonomic system works automatically and without voluntary input. Its effectors are smooth muscle, cardiac muscle, and glands. Like the somatic system, it is functionally organized into reflex pathways. Receptors within the viscera relay information via afferent nerves to the CNS. The information is integrated at various levels. Then the decision is transmitted along efferent nerves to the appropriate muscles or glands.

Afferent and efferent neurons of the autonomic division lie within cranial or spinal nerves. For example, afferent fibers of cranial nerve X, the vagus nerve, transmit signals from many organs of the chest and upper abdomen to the CNS. Its efferent fibers transmit signals from the brain to the heart, stomach, small intestine, and several other organs.

The efferent portion of the autonomic division is subdivided into **sympathetic** and **parasympathetic systems.** Many organs are innervated by both types of nerves. In general, sympathetic

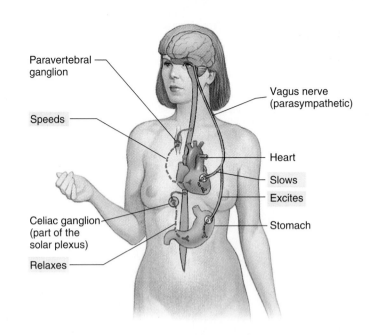

Figure 41-17 Dual innervation of the heart and stomach

A sympathetic nerve increases the heart rate, whereas the vagus nerve, a parasympathetic nerve, slows the heartbeat. In contrast, sympathetic nerves slow contractions of the stomach and intestine, whereas the vagus nerve stimulates contractions of the digestive organs. Sympathetic nerves are shown in red; postganglionic fibers are shown as dotted lines.

and parasympathetic systems have *opposite* effects (❚ Figs. 41-17 and 41-18). For example, the heart rate is speeded up by messages from sympathetic nerve fibers and slowed by impulses from its parasympathetic nerve fibers. In many cases sympathetic nerves operate to stimulate organs and to mobilize energy, especially in response to stress; whereas the parasympathetic nerves influence organs to conserve and restore energy, particularly during quiet, calm activities.

Instead of using a single efferent neuron, as in the somatic system, the autonomic system uses a relay of two neurons between the CNS and the effector. The first neuron, called the **preganglionic neuron,** has a cell body and dendrites within the CNS. Its axon, part of a peripheral nerve, ends by synapsing with a **postganglionic neuron.** The dendrites and cell body of the postganglionic neuron are in a ganglion outside the CNS. Its axon terminates near or on the effector. The sympathetic ganglia are paired, and a chain of them, the **paravertebral sympathetic ganglion chain,** runs on each side of the spinal cord from the neck to the abdomen. Some sympathetic preganglionic neurons do not end in these ganglia but pass on to **collateral ganglia** in the abdomen, close to the aorta and its major branches. Parasympathetic preganglionic neurons synapse with postganglionic neurons in **terminal ganglia** near or within the walls of the organs they innervate.

The sympathetic and parasympathetic systems also differ in the neurotransmitters they release at the synapse with the effector. Both preganglionic and postganglionic parasympathetic neurons secrete acetylcholine. Sympathetic postganglionic neurons

Most organs are innervated by both sympathetic and parasympathetic nerves.

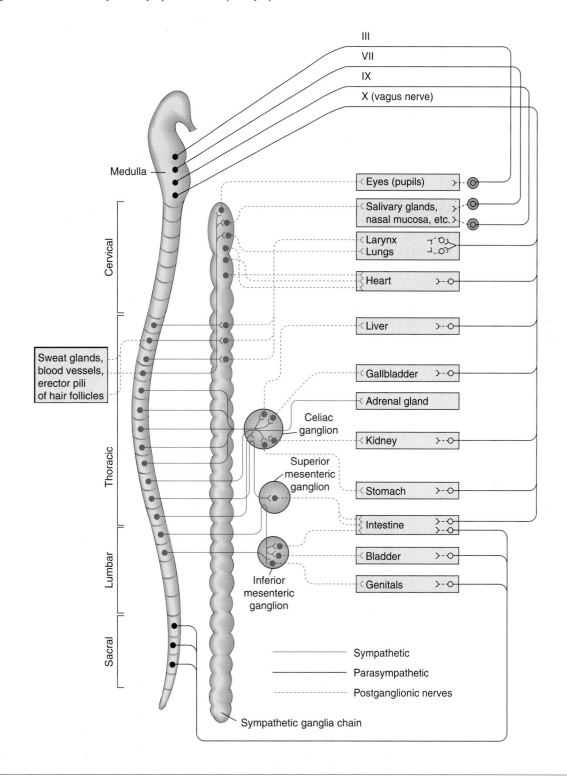

Figure 41-18 The sympathetic and parasympathetic systems

Complex as it appears, this diagram has been greatly simplified. Red lines represent sympathetic nerves, black lines represent parasympathetic nerves, and dotted lines represent postganglionic nerves.

release norepinephrine (although preganglionic sympathetic neurons secrete acetylcholine).

The autonomic division got its name from the original idea that it was independent of the CNS, that is, autonomous. Neurobiologists have disproved this and shown that the hypothalamus and many other parts of the CNS help regulate the autonomic system. Although the autonomic system usually functions automatically, its activities can be consciously influenced. Biofeedback provides a person with visual or auditory evidence concerning the status of an autonomic body function. For example, a tone may be sounded when blood pressure reaches a desirable level. Using such techniques, individuals have learned to control autonomic activities such as blood pressure, brain wave pattern, heart rate, and blood sugar level. Even certain abnormal heart rhythms can be consciously modified.

Review

- What are some functions of the PNS?
- What are some differences between the somatic division and the autonomic division?
- What are some differences in structure and function of the sympathetic system and the parasympathetic system?

States were current users of illicit drugs in 2004, meaning they used an illicit drug at least once during the 30 days prior to the interview. More than 9 million children live with a parent who is dependent on alcohol and/or illicit drugs. ▌Table 41-3 lists several commonly used and abused drugs and gives their effects. Also see *Focus On: Alcohol: The Most Abused Drug.*

Habitual use of almost any mood-altering drug can result in **psychological dependence,** in which the user becomes emotionally dependent on the drug. When deprived of it, the user craves the feeling of euphoria (well-being) the drug induces. Some drugs induce **tolerance** after several days or weeks. Response to the drug decreases, and greater amounts are required to obtain the desired effect. Tolerance often occurs because the liver cells are stimulated to produce more of the enzymes that metabolize and inactivate the drug. It can also occur when the number of postsynaptic receptors that bind with the drug decrease (*down-regulation*).

Drug addiction is a serious societal problem that involves compulsive use of a drug despite negative health consequences and negative effects on the ability to function socially and occupationally. Use of some drugs, such as heroin, tobacco, alcohol, and barbiturates, may also result in physical dependence, in which physiological changes occur that make the user dependent

EFFECTS OF DRUGS ON THE NERVOUS SYSTEM

Learning Objective

13 Discuss the biological actions and effects on mood of the following types of drugs: alcohol, antidepressants, barbiturates, anti-anxiety drugs, antipsychotic drugs, opiates, stimulants, hallucinogens, and marijuana.

About 25% of all prescribed drugs are taken to alter psychological conditions, and almost all the commonly abused drugs affect mood. Many act by changing the levels of neurotransmitters within the brain. In particular, levels of norepinephrine, serotonin, and dopamine influence mood.

According to the Substance Abuse and Mental Health Services Administration, an estimated 22.5 million individuals in the United States suffer from substance abuse or substance dependence. The report states that 19.1 million people in the United

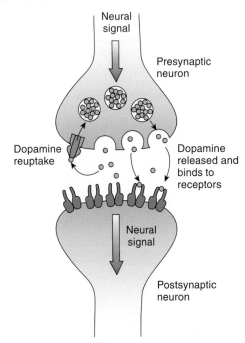

(a) Reuptake of the neurotransmitter dopamine. Dopamine binds with a transporter (*green*) that shuttles it back into the presynaptic terminal.

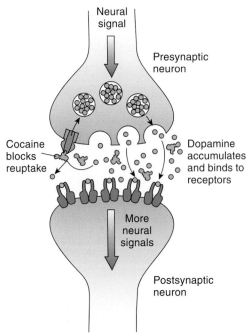

(b) Cocaine blocks the reuptake of dopamine by binding competitively with the dopamine reuptake transporter. As a result, more dopamine is available to bind with postsynaptic receptors.

Figure 41-19 Mechanism of cocaine action

ALCOHOL: THE MOST ABUSED DRUG

After tobacco, alcohol is the leading cause of premature death in the United States. It is linked to approximately 100,000 deaths every year and costs our society more than $185 billion annually. According to the 2004 National Survey on Drug Use and Health, an estimated 16.7 million Americans (6.9%) engaged in binge drinking (five or more drinks) on five or more days during the month preceding the survey. The highest prevalence of binge and heavy drinking was for young adults aged 18 to 25 years.

More than half of men and women in the United States report that one or more of their close relatives have a drinking problem. Alcohol abuse is not limited to adults. About 4.6 million adolescents, or nearly one of every three high school students, experience negative consequences from alcohol use, including difficulty with parents, poor performance at school, and trouble with the law.

Alcohol abuse results in physiological, psychological, and social impairment and has serious negative consequences for family, friends, and society. Alcohol abuse has been linked to the following statistics:

- More than 40% of all traffic fatalities
- More than 50% of violent crimes and more than 60% of child-abuse and spouse-abuse cases
- More than 50% of suicides
- More than 15,000 babies born each year with serious birth defects because their mothers drank alcohol during

pregnancy; fetal alcohol syndrome is the leading cause of preventable mental retardation in the United States

- Increased risk for certain cancers, especially those of the liver, esophagus, throat, and larynx

A single drink, that is, 12 oz of beer, 5 oz of wine, or 1.5 oz of 80-proof liquor, results in a blood-alcohol concentration of approximately 20 mg/dL (milligrams per deciliter). This represents about 0.5 oz of pure alcohol in the blood. Alcohol accumulates in the blood because absorption occurs more rapidly than do oxidation and excretion. Every cell in the body can take in alcohol from the blood. At first the drinker may feel stimulated, but alcohol actually depresses the CNS.

As blood-alcohol concentration rises, information processing, judgment, memory, sensory perception, and motor coordination all become progressively impaired. Depression and drowsiness generally occur. Contrary to popular belief, alcohol decreases sexual performance. Some individuals become loud, angry, or violent. A blood-alcohol concentration of 80 mg/dL (or 0.08 gm/100 mL) legally defines driving while intoxicated (DWI). A 170-pound man typically reaches this level by drinking four drinks in an hour on an empty stomach.

Alcohol metabolism occurs at a fixed rate in the liver, so only time, not coffee, decreases its effects. Because alcohol inhibits water reabsorption in the kidneys, more fluid is excreted as urine than is consumed. This results in dehydration that,

together with low blood sugar level, may cause the stupor produced by excessive drinking.

In chronic drinkers, cells of the CNS adapt to the presence of the drug. This causes *tolerance* (more and more alcohol is needed to experience the same effect) and physical dependence. Abrupt withdrawal can result in sleep disturbances, severe anxiety, tremors, seizures, hallucinations, and psychoses.

The *A1* allele of the D2 dopamine receptor gene increases the risk for alcoholism. Individuals with this allele have fewer D2 dopamine receptors in their brain. Alcohol, like cocaine, heroin, amphetamines, and nicotine, enhances dopamine activity in the *mesolimbic dopamine pathway* of the brain. Individuals with the *A1* allele may compensate for their low dopamine levels by using alcohol and other drugs. However, the mechanisms underlying alcohol addiction are complex. Evidence suggests that many neurotransmitters are involved, including glutamate, serotonin, GABA, and opioid peptides. Recently, researchers used microarray techniques to identify a number of genes that may be associated with alcohol dependence.

Treatment for alcohol problems includes various forms of psychotherapy and use of drugs such as Disulfiram (Antabuse), which causes immediate symptoms of a "hangover" after alcohol intake. The group support offered by Alcoholics Anonymous (AA) has proved effective for many struggling with alcohol dependence.

on the drug. For example, when heroin or alcohol is withheld, the addict suffers characteristic withdrawal symptoms. Physical addiction can occur when a drug, for example, morphine, has components similar to substances that body cells normally manufacture on their own. Some highly addictive drugs, such as crack cocaine and methamphetamine, do not cause serious withdrawal symptoms.

The neurophysiological mechanisms for drug addiction involve the *mesolimbic dopamine pathway*. Neurons in this pathway signal neurons in an area of the limbic system (the nucleus accumbens). Effects on neurons in the mesolimbic dopamine pathway have been demonstrated in nicotine, heroin, amphetamine,

and cocaine addiction. For example, amphetamines stimulate dopamine release. Cocaine blocks the reuptake of dopamine into presynaptic neurons, thus prolonging the stimulation (Fig. 41-19). Amphetamines or cocaine can increase dopamine concentration in the limbic system by up to 1000 times normal amounts! Prolonged drug use alters gene expression and changes neuron structure.

Review

- How does each of the following drugs affect the CNS: (1) alcohol, (2) antipsychotic drugs, (3) antidepressants, (4) amphetamines, and (5) MDMA ("Ecstasy")?

TABLE 41-3

Effects of Some Commonly Used Drugs

Name of Drug	Effect on Mood	Actions on Body	Side Effects/Dangers Associated with Abuse
Antidepressants			
Tricyclic antidepressants (e.g., Elavil, Anafranil)	Elevate mood; relieve depression; used to treat obsessive-compulsive behavior	Most block reuptake of biogenic amines (especially norepinephrine), increasing their concentration in synapses	Sedation, weight gain, sexual dysfunction
Selective serotonin reuptake inhibitors (SSRIs) (e.g., Prozac, Zoloft, Lexapro)	Relieve depression; used to treat obsessive-compulsive behavior	Block serotonin reuptake	Nausea, headache, insomnia, anxiety
MAO inhibitors (e.g., Parnate, Nardil)	Relieve depression	Block enzymatic breakdown of biogenic amines, increasing their concentration in synapses	Liver toxicity, excessive CNS stimulation; overdose may affect blood pressure, cause hallucinations
Anti-anxiety Medications			
Benzodiazepines (e.g., Xanax, Valium, Librium)	Sedation; induce sleep	Bind to GABA receptors on postsynaptic neuron; increase effectiveness of GABA in opening chloride channels, causing hyperpolarization; cause relaxation of skeletal muscles	Drowsiness, confusion; psychological dependence, addiction; effects additive with alcohol
Barbiturates ("downers") (e.g., phenobarbital)	Sedation; induce sleep	Bind to receptor adjacent to GABA receptor on chloride channels; chloride channels open so more chloride ions move in, causing hyperpolarization	Drowsiness, confusion; psychological dependence, tolerance, addiction; severe CNS depression, resulting in coma and death (especially lethal in combination with alcohol)
Antipsychotic Medications			
Phenothiazines (e.g., Thorazine, Mellaril, Stelazine)	Relieve symptoms of schizophrenia; reduce impulsive and aggressive behavior	Block dopamine receptors	Muscle spasms, shuffling gait, and other symptoms of Parkinson's disease
Newer antipsychotic medications (e.g., Abilify, Risperdal)	Similar to phenothiazines, but also increase motivation	Block dopamine receptors and other neurotransmitter receptors	Fewer side effects than with the phenothiazines
Narcotic Analgesics			
Opiates (e.g., morphine, codeine, Demerol, heroin)	Euphoria; sedation; relieve pain	Mimic actions of endorphins; bind to opiate receptors	Depress respiration; constrict pupils; impair coordination; tolerance, psychological dependence, addiction; convulsions, death from overdose
Cocaine	Brief, intense high with euphoria, followed by fatigue and depression	Stimulates release and inhibits reuptake of norepinephrine and dopamine, leading to CNS stimulation followed by depression; autonomic stimulation; dilation of pupils; local anesthesia	Mental impairment, convulsions, hallucinations, unconsciousness; death from overdose
Marijuana	Euphoria	Impairs coordination; impairs depth perception and alters sense of timing; impairs short-term memory (probably by decreasing acetylcholine levels in the hippocampus); inflames eyes; causes peripheral vasodilation	In large doses, sensory distortions, hallucinations, evidence of lowered sperm counts and testosterone levels
Amphetamines			
"Uppers," "pep pills," "crystal meth" (e.g., Dexedrine); methamphetamine is a popular club drug	Euphoria, stimulation, hyperactivity	Stimulate release and block reuptake of dopamine and norepinephrine; enhance flow of impulses in RAS, leading to increased heart rate, blood pressure; pupil dilation	Tolerance, physical dependence; hallucinations; increased blood pressure; psychotic episodes; death from overdose

TABLE 41-3

(Continued)

Name of Drug	Effect on Mood	Actions on Body	Side Effects/Dangers Associated with Abuse
Amphetamines (*continued*)			
Methylphenidate Amphetamine-like prescription stimulant commonly used to treat attention deficit hyperactivity disorder (ADHD) (e.g., Ritalin, Concerta)	CNS stimulation, but has calming effect on many children and adults who have ADHD; increases focus	Increases heart rate, blood pressure, and body temperature; like amphetamines and cocaine, blocks dopamine transporter so increases dopamine concentration in synapses	Not addictive at prescribed doses; used illegally and abused by some high school and college students who use them in higher doses than would be prescribed; can cause irregular heartbeat and respiration, hallucinations, delusions, and convulsions; abuse can lead to tolerance and physical addiction
MDMA (3,4-methylenedioxymethamphetamine) "Ecstasy" (most popular of club drugs)	Stimulation, sense of pleasure, self-confidence, feelings of closeness to others; mild hallucinogenic	CNS stimulant; damages serotonin pathways in rat and monkey brains	Hyperthermia (excessive body heat); overdose causes rapid heartbeat, high blood pressure, panic attacks, and seizures; long-term neurological changes; cognitive impairment
Rohypnol (flunitrazepam) "Date rape" drug, "roofies," "roach"	Sedative-hypnotic	Muscle relaxation; amnesia	Physical and psychological dependence; withdrawal may cause seizures; may be lethal when mixed with alcohol and/or other depressants
PCP (phencyclidine) "Angel dust," "ozone"	Feelings of strength, power, invulnerability, and a numbing effect on the mind	Increase in blood pressure and heart rate; shallow respiration; flushing, profuse sweating; numbness of extremities; muscular incoordination; changes in body awareness, similar to those associated with alcohol intoxication	Psychological dependence, craving; in adolescents, interferes with hormones related to growth and development; may interfere with learning; causes hallucinations, delusions, disordered thinking; high doses cause drop in blood pressure and heart rate, seizures, coma, and death; chronic use causes memory loss, difficulties with speech and thinking, and depression
LSD (lysergic acid diethylamide)	Overexcitation; sensory distortions, hallucinations	Alters levels of neurotransmitters in brain; potent CNS stimulator; dilates pupils, sometimes unequally; increases heart rate; raises blood pressure	Irrational behavior
Methaqualone (e.g., Quaalude, Sopor)	Hypnotic	Depresses CNS; depresses certain spinal reflexes	Tolerance, physical dependence; convulsions, death
Caffeine	Increases mental alertness; decreases fatigue and drowsiness	Acts on cerebral cortex; relaxes smooth muscle; stimulates cardiac and skeletal muscle; increases urine volume	Very large doses stimulate centers in the medulla (may slow the heart); toxic doses may cause convulsions
Nicotine	Lessens psychological tension	Stimulates sympathetic nervous system; increases dopamine in mesolimbic pathway	Tolerance, physical dependence; stimulates development of atherosclerosis by stimulating synthesis of lipid in arterial wall
Alcohol	Euphoria, relaxation, release of inhibitions	Depresses CNS; impairs vision, coordination, judgment; lengthens reaction time	Physical dependence; damage to pancreas and liver (e.g., cirrhosis); brain damage

Learning Objectives

1 Contrast nerve nets and radial nervous systems with bilateral nervous systems (page 866).

■ Nerve nets and radial nervous systems are typical of radially symmetrical invertebrates. Bilateral nervous systems are characteristic of bilaterally symmetrical animals.

■ Cnidarians have a **nerve net** of nerve cells scattered throughout the body; they have no central control organ. Echinoderms typically have a **radial nervous system** consisting of a nerve ring and nerves that extend to various parts of the body.

■ In a **bilateral nervous system,** nerve cells concentrate to form **nerves, nerve cords, ganglia,** and **brain;** typically, sense organs are concentrated in the head region. An increased number of interneurons and more complex synaptic contacts permit a wide range of responses.

■ In planarian flatworms the bilateral nervous system includes **cerebral ganglia** and, usually, two solid ventral nerve cords connected by transverse nerves. Annelids and arthropods typically have a ventral nerve cord and numerous ganglia. The cerebral ganglia of arthropods have specialized regions. Octopods and other cephalopod mollusks have highly developed nervous systems with neurons concentrated in a central region.

ThomsonNOW **Learn more about various nervous systems by clicking on the figures in ThomsonNOW.**

2 Identify trends in the evolution of the invertebrate nervous system (page 866).

■ Trends in nervous system evolution include increased numbers and concentration of nerve cells; specialization of function; increased number of association neurons; more complex synaptic contacts; and **cephalization,** formation of a head.

3 Describe the two main divisions of the vertebrate nervous system (page 867).

■ The vertebrate nervous system consists of the **central nervous system (CNS)** and **peripheral nervous system (PNS).** The CNS consists of the brain and dorsal, tubular **spinal cord.** The PNS consists of sensory receptors and nerves.

ThomsonNOW **Learn more about the vertebrate nervous system by clicking on the figure in ThomsonNOW.**

4 Trace the development of the principal vertebrate brain regions from the hindbrain, midbrain, and forebrain; and compare the brains of fishes, amphibians, reptiles, birds, and mammals (page 868).

■ In the vertebrate embryo, the brain and spinal cord arise from the **neural tube.** The anterior end of the tube differentiates into **hindbrain, midbrain,** and **forebrain.**

■ The hindbrain subdivides into the **metencephalon** and **myelencephalon.** The myelencephalon develops into the **medulla,** which contains vital centers and other reflex centers. The **fourth ventricle,** the cavity of the medulla, communicates with the central canal of the spinal cord. The metencephalon gives rise to the **cerebellum,** which is responsible for muscle tone, posture, and equilibrium, and to the **pons,** which connects various parts of the brain.

■ The midbrain is the largest part of the brain in fishes and amphibians. It is their main association area, linking sensory input and motor output. In reptiles, birds, and

mammals, the midbrain serves as a center for visual and auditory reflexes. The medulla, pons, and midbrain make up the **brain stem.**

■ The forebrain differentiates to form the **diencephalon** and **telencephalon.** The diencephalon develops into the thalamus and hypothalamus. The **thalamus** is a relay center for motor and sensory information. The **hypothalamus** controls autonomic functions; links nervous and endocrine systems; controls temperature, appetite, and fluid balance; and is involved in some emotional and sexual responses.

■ The telencephalon develops into the cerebrum and **olfactory bulbs.** In most vertebrates the **cerebrum** is divided into right and left **hemispheres.** In fishes and amphibians, the cerebrum mainly integrates incoming sensory information. In birds, the corpus striatum controls complex behavior patterns, such as flying and singing.

■ In mammals, the **neocortex** accounts for a large part of the **cerebral cortex,** the **gray matter** of the brain. The cerebrum has complex association functions.

ThomsonNOW **Explore the early development of the vertebrate nervous system and the evolution of the vertebrate brain by clicking on the figures in ThomsonNOW.**

5 Describe the structure and functions of the human spinal cord (page 871).

■ The human brain and spinal cord are protected by bone and three **meninges**—the **dura mater, arachnoid,** and **pia mater;** the brain and spinal cord are cushioned by **cerebrospinal fluid (CSF).**

■ The spinal cord transmits impulses to and from the brain and controls many **reflex actions.** The spinal cord consists of **ascending tracts,** which transmit information to the brain, and **descending tracts,** which transmit information from the brain. Its gray matter contains nuclei that serve as reflex centers.

ThomsonNOW **Learn more about the spinal cord by clicking on the figure in ThomsonNOW.**

■ A **withdrawal reflex** involves sensory receptors; sensory neurons, interneurons, and motor neurons; and effectors, such as muscles.

6 Describe the structure and functions of the human cerebrum (page 871).

■ The human cerebral cortex consists of gray matter, which forms folds or **convolutions.** Deep furrows between the folds are called **fissures.**

■ The cerebrum is functionally divided into **sensory areas** that receive incoming sensory information; **motor areas** that control voluntary movement; and **association areas** that link sensory and motor areas and are responsible for learning, language, thought, and judgment. The cerebrum consists of lobes, including the **frontal lobes, parietal lobes, temporal lobes,** and **occipital lobes.**

■ The **white matter** of the cerebrum lies beneath the cerebral cortex. The **corpus callosum,** a large band of white matter, connects right and left hemispheres. The **basal ganglia,** a cluster of nuclei within the white matter, are important centers for motor function.

ThomsonNOW **Learn more about the human brain by clicking on the figures in ThomsonNOW.**

7 Describe the sleep–wake cycle, and contrast rapid-eye-movement (REM) sleep and non-REM sleep (page 871).

- Brain activity cycles in a sleep–wake pattern that is regulated by the hypothalamus and brain stem. **Alpha wave** patterns are characteristic of relaxed states. **Beta wave** patterns accompany heightened mental activity; and slower-frequency, higher-amplitude **theta** and **delta waves** are characteristic of non-REM sleep.
- The **reticular activating system (RAS)** is an arousal system; its neurons filter sensory input, selecting which information is transmitted to the cerebrum.
- Electrical activity of the cerebral cortex and metabolic rate slow during **non-REM sleep**, whereas dreaming characterizes **REM sleep**.
- The body's main biological clock—the **suprachiasmatic nucleus**—receives information about light and dark and apparently transmits it to other nuclei that regulate sleep. When sleep centers are activated, they release **serotonin**.

8 Describe the actions of the limbic system (page 871).

- The **limbic system** of the brain affects the emotional aspects of behavior, motivation, sexual activity, autonomic responses, and biological rhythms. The **hippocampus** is important in categorizing information and consolidating memories. The **amygdala** evaluates incoming information and signals danger.

9 Relate neural plasticity to synaptic plasticity, and give three examples of neural plasticity (page 880).

- Changes in behavior are possible because of **neural plasticity,** the ability of the nervous system to change in response to experience. Examples include learning to walk, ride a bicycle, or speak a new language. **Synaptic plasticity** is the ability of the nervous system to modify synapses, which allows learning and remembering. Synaptic plasticity appears to be the major form of neural plasticity.

10 Summarize how the brain processes information (page 880).

- **Learning** is the process by which we acquire knowledge or skills as a result of experience. **Memory** is the process by which information is encoded, stored, and retrieved. **Implicit memory** is unconscious ("how-to") memory for perceptual and motor skills. **Explicit memory** is factual memory of people, places, or objects.
- **Short-term memory** lets us recall information, such as a telephone number, for a few minutes. Information can

be transferred from short-term memory to **long-term memory.**

- Learning and memory involve long-term functional changes at synapses. Known as **long-term potentiation (LTP),** such changes involve increased sensitivity to an action potential by a postsynaptic neuron. **Long-term depression (LTD)** is a long-lasting decrease in the strength of synaptic connections. Long-term memory storage involves gene activation. The transcription factor **CREB** is an important signaling molecule in the memory pathway. The physical structure and chemistry of the brain can be altered by environmental experience.

11 Describe the organization of the peripheral nervous system, and compare its somatic and autonomic divisions (page 883).

- The PNS consists of sensory receptors and nerves, including the **cranial nerves** and **spinal nerves** and their branches. The **somatic division** of the PNS responds to changes in the external environment. The **autonomic division** regulates the internal activities of the body.

12 Contrast the sympathetic and parasympathetic divisions of the autonomic system, and give examples of the effects of these systems on specific organs such as the heart and intestine (page 883).

- The **sympathetic system** permits the body to respond to stressful situations. The **parasympathetic system** influences organs to conserve and restore energy. Many organs are innervated by sympathetic and parasympathetic nerves, which function in opposite ways. For example, the sympathetic system increases heart rate, whereas the parasympathetic system decreases heart rate.

13 Discuss the biological actions and effects on mood of the following types of drugs: alcohol, antidepressants, barbiturates, anti-anxiety drugs, antipsychotic drugs, opiates, stimulants, hallucinogens, and marijuana (page 886).

- Many drugs alter mood by increasing or decreasing the concentrations of specific neurotransmitters within the brain. See Table 41-3 for a summary of these drugs.
- Habitual use of mood-altering drugs can result in **psychological dependence** or **drug addiction.** Many drugs induce **tolerance,** in which the body's response to the drug decreases so that greater amounts are needed to obtain the desired effect.

TEST YOUR UNDERSTANDING

1. A radially symmetrical animal such as *Hydra* is likely to have (a) a forebrain (b) a nerve net (c) cerebral ganglia (d) a ventral nerve cord (e) cerebral ganglia and a nerve net

2. In vertebrate embryos the brain develops from the (a) spinal cord (b) sympathetic nervous system (c) parasympathetic nervous system (d) neural tube (e) forebrain

3. Which part of the brain maintains posture, muscle tone, and equilibrium? (a) cerebrum (b) medulla (c) cerebellum (d) neocortex (e) thalamus

4. Which part of the brain controls autonomic functions and regulates body temperature? (a) cerebrum (b) hypothalamus (c) cerebellum (d) pons (e) thalamus

5. In a withdrawal reflex, following reception a signal is transmitted by (a) a motor neuron to an association neuron in the CNS (b) an association neuron in the CNS to an afferent

neuron (c) an afferent neuron in the CNS to a motor neuron (d) a sensory neuron to an interneuron in the CNS (e) a sensory neuron to an interneuron in the PNS

6. Association areas in the human brain are concentrated in the (a) cerebral cortex (b) medulla (c) ventricle (d) hippocampus (e) meninges

7. The human brain is protected by (a) meninges, cerebrospinal fluid, and skull bones (b) meninges and skull bones only (c) dura mater and fourth ventricle (d) pia mater and skull bones (e) arachnoid, pia mater, cerebrospinal fluid, and ganglia

8. Which of the following is *not* a function of the spinal cord? (a) controls many reflex actions (b) transmits information to the brain (c) transmits information from the brain (d) regulates sleep–wake cycles (e) controls the withdrawal reflex

9. The most prominent part of the amphibian brain is the (a) midbrain (b) medulla (c) cerebellum (d) neocortex (e) cerebrum

10. The visual centers are located in the (a) parietal lobes (b) thalamus (c) occipital lobes (d) limbic lobes (e) frontal lobes

11. As you are answering these questions, what is the predominant type of wave your brain is emitting? (a) REM (b) alpha (c) theta (d) delta (e) beta

12. Implicit memory is (a) short-term memory (b) long-term memory (c) factual knowledge of people, places, or objects (d) unconscious memory for perceptual or motor skills (e) learning that depends on long-term depression (LTD)

13. Long-term potentiation (LTP) (a) is mainly the responsibility of cranial nerve X (b) is a long-lasting increase in synaptic strength (c) is associated with short-term memory (d) is a long-lasting decrease in synaptic strength (e) occurs mainly during REM sleep

14. The heart rate is slowed by (a) sympathetic nerves (b) parasympathetic nerves (c) corpus callosum (d) both sympathetic and parasympathetic nerves (e) the hippocampus working together with sympathetic nerves

15. After taking a mood-altering drug for several weeks, a patient notices it no longer works as effectively. This is an example of (a) psychological dependence (b) withdrawal (c) addiction (d) tolerance (e) neurotransmitter increases

16. Amphetamines (a) affect neurons in the mesolimbic dopamine pathway (b) decrease the amount of dopamine secreted (c) do not result in tolerance (d) are sometimes referred to as *downers* (e) are CNS depressants

17. Label the diagram. Use Figure 41-11a to check your answers.

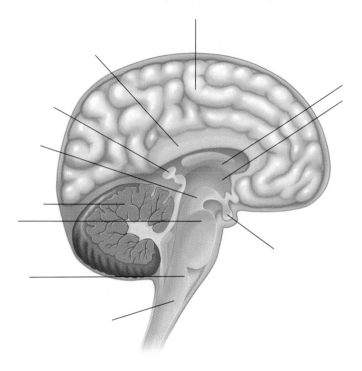

CRITICAL THINKING

1. Imagine that you have just become a parent. What kinds of things can you do to enhance the development of your child's intellectual abilities?

2. What is the adaptive function of the limbic system when you are in danger?

3. Hypothesize a possible relationship between inheritance and intelligence based on gene activation during information processing.

4. **Evolution Link.** What general trends can you identify in the evolution of the vertebrate brain?

5. **Evolution Link.** CREB has been shown to be a signaling molecule in the memory pathway in many animals, including fruit flies and mice. What does this suggest about the evolution of learning and memory?

Additional questions are available in ThomsonNOW at www.thomsonedu.com/login

Sensory Systems

Merlin D. Tuttle/Bat Conservation International/Photo Researchers, Inc.

Bats are guided by echoes.
This California leaf-nosed bat
(*Macrotus californicus*) has
located an insect.

KEY CONCEPTS

Sensory receptors detect information about changes in the internal or external environment.

Steps that take place in sensory processing include sensory reception, energy transduction, transmission of signals, and interpretation in the brain.

Mechanoreceptors respond to touch, pressure, gravity, stretch, or movement.

Most animals have mechanoreceptors that detect motion and position; in vertebrates, the vestibular apparatus in the inner ear helps maintain equilibrium.

In vertebrates, the organ of Corti within the cochlea of the inner ear contains auditory receptors.

Most organisms have chemoreceptors for taste and smell that detect chemical substances in food, water, and air.

Almost all organisms respond to light; in vertebrates, the retina contains the photoreceptor cells, rods and cones.

Have you heard dolphins make clicking sounds? Humans can hear only a limited number of the wide range of sound frequencies that dolphins emit. Dolphins, bats, and a few other vertebrates detect distant objects by *echolocation*, sometimes called biosonar. Echolocation works something like radar but uses sound rather than radio waves. For example, a bat emits high-pitched sounds that bounce off objects in its path and echo back. By rapidly responding to the echo, the bat skillfully avoids obstacles and captures prey (see photograph). Using echolocation, bats and dolphins determine the size, shape, texture, and density of an object, as well as its location. Certain moths and other insect prey have evolved the ability to detect sounds emitted by some foraging bats. Through the course of evolution, some species of bats have adapted to this ability of their prey and now emit even higher-frequency sounds that their prey cannot detect.

By informing an animal about its environment, *sensory systems* are the link between the animal and the outside world. The kinds of sensory receptors an animal has and the way in which the brain *perceives* incoming sensory information determine what the animal's world is like. We humans live in a world of rich colors, numerous shapes, and varied sounds. But we cannot hear the high-pitched whistles that are audible to dogs and cats or the ultrasonic echoes by which bats navigate. Nor do we ordinarily recognize our friends by their distinctive odors! And although vision is our dominant and most refined sense, we are blind to the ultraviolet (UV) hues that light up the world for insects.

We are most familiar with the senses of sight, hearing, smell, taste, and touch, but there are other senses. In fact, touch is a compound sense that allows us to detect pressure, pain, and temperature. Some sensory receptors allow us to sense balance, muscle tension, and joint position, so we have an awareness of our bodies. Birds and some other animals can detect changes in magnetic fields. Sharks, rays, and several other species of fishes detect electric fields.

In this chapter we will first examine how sensory systems function. Next we will introduce the major types of sensory receptors. We then discuss several types of sensory receptors with emphasis on auditory and visual receptors. ∎

HOW SENSORY SYSTEMS WORK

Learning Objective

1   Describe sensory system function, including sensory reception, energy transduction, receptor potential, sensory adaptation, and perception.

Sensory receptors detect information about changes in the internal or external environment. These receptors consist of specialized neuron endings or specialized cells in close contact with neurons. Sensory receptors, along with other types of cells, make up complex **sense organs,** such as eyes, ears, nose, and taste buds. A human taste bud, for example, consists of modified epithelial cells that detect chemicals dissolved in saliva.

Several steps take place in sensory processing, including sensory reception, energy transduction, transmission of the signal, and interpretation in the brain. With minor variations, this is how all sensory systems operate.

Sensory receptors receive information

Sensory receptors receive stimuli from the external or internal environment. In the process of reception, they absorb a small amount of energy from some stimulus. Each kind of sensory receptor is especially sensitive to one particular form of energy. The photoreceptors of the human eye are stimulated by an extremely faint beam of light, and taste receptors are stimulated by a minute amount of a chemical compound.

Sensory receptors transduce energy

Sensory receptors *transduce,* or convert, the energy of the stimulus into electrical signals, the information currency of the nervous system. This process is known as **energy transduction.** When unstimulated, a sensory receptor maintains a *resting potential,* that is, a difference in charge between the inside and outside of the cell (see Chapter 40). Transduction couples a stimulus with the opening or closing of ion channels in the plasma membrane of sensory receptors. The permeability of the plasma membrane to various ions is altered.

A change in ion distribution causes a change in voltage across the membrane. If the difference in charge increases, the receptor becomes hyperpolarized. If the potential decreases, the receptor becomes depolarized. A change in membrane potential is a **receptor potential.** A receptor potential does not directly trigger an action potential. Like an excitatory postsynaptic potential, or EPSP (see Chapter 40), the receptor potential is a **graded response** in which the magnitude of change depends on the energy of the stimulus.

The function of a receptor potential is to generate action potentials that transmit information to the central nervous system (CNS). If the receptor is a separate cell, receptor potentials stimulate the release of a neurotransmitter, which flows across the synapse and binds to sites on sensory neurons (∎ Fig. 42-1). When a sensory neuron becomes sufficiently depolarized to reach its threshold level, an action potential is generated.

Many receptors are specialized neurons rather than separate cells. In these neurons, the specialized region of the plasma membrane that transduces energy does not generate action potentials. Current generated by receptor potentials flows to a region along the axon where an action potential can be generated.

As we discussed in Chapter 40, **sensory neurons,** also known as *afferent neurons,* transmit information from receptors to the CNS. The action potential travels along the axon of the sensory neuron to the CNS. In summary,

Stimulus (such as light energy) ⟶ receptor transduces energy of stimulus into electrical energy ⟶ receptor potential ⟶ action potential in sensory neuron ⟶ signal transmitted to CNS

Sensory input is integrated at many levels

Sensory integration begins in the receptor itself. For example, receptor potentials produced by various stimuli are integrated by summation. Integration also occurs in the spinal cord and at various locations in the brain.

Sensory receptors adapt to stimuli

Many sensory receptors do not continue to respond at the initial rate, even if the stimulus continues at the same intensity. This process, called **sensory adaptation,** occurs for two reasons. First, during a sustained stimulus, the receptor sensitivity decreases and produces a smaller receptor potential (resulting in a lower frequency of action potentials in the sensory neurons). Second, changes take place at synapses in the neural pathway activated by the receptor. For example, the release of neurotransmitter from a presynaptic terminal may decrease in response to a series of action potentials.

Some receptors, such as those for pain or cold, adapt so slowly that they continue to trigger action potentials as long as

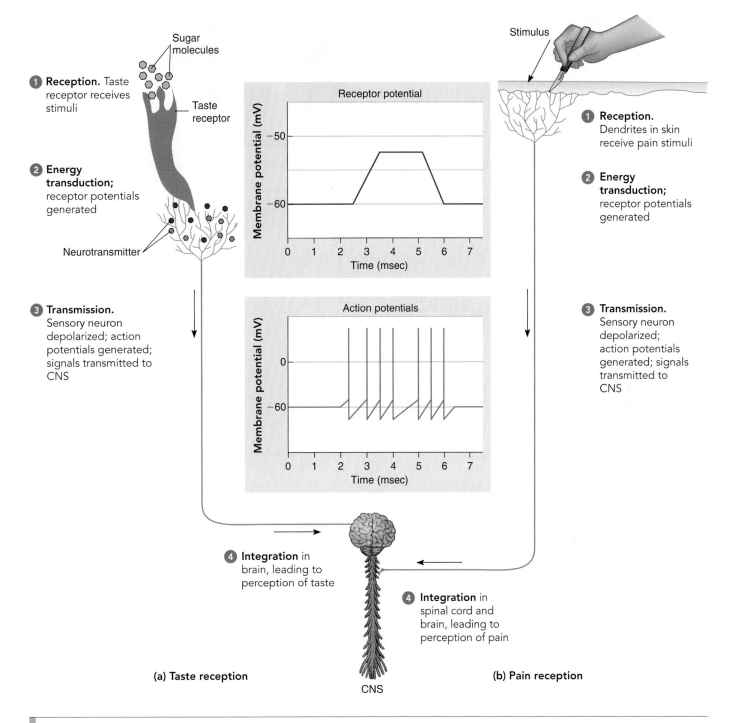

Figure 42-1 How sensory systems work

Sensory receptors receive stimuli, transduce stimulus energy to electrical energy, and thus generate receptor potentials. Action potentials are generated in sensory neurons that transmit signals to the CNS, where they are integrated.

the stimulus persists. Other receptors adapt rapidly, permitting us to ignore persistent unpleasant or unimportant stimuli. For example, when you first pull on a pair of tight jeans, your pressure receptors let you know that you are being squished, and you may feel uncomfortable. Soon, though, these receptors adapt, and you hardly notice the sensation of the tight fit. In the same way, people quickly adapt to odors that at first seem to assault their senses. Sensory adaptation enables an animal to discriminate between unimportant background stimuli that can be ignored and new or important stimuli that require attention.

Sensation depends on transmission of a coded message

How do you know whether you are seeing a blue sky, tasting a doughnut, or hearing a note played on a piano? All action potentials are qualitatively the same. Light of the wavelength 450 nm (blue), sugar molecules (sweet), and sound waves of 440 hertz (Hz) (A above middle C) all cause transmission of similar action potentials. Our ability to differentiate stimuli depends on both the sensory receptor itself and on the brain. We can distinguish the color of a blue sky from the scent of cologne; a sweet taste from a light breeze; or the sound of a piano from the heat of the sun because cells of each sensory receptor are connected to specific neurons in particular parts of the brain. Because a receptor normally responds to only one type of stimulus (for example, light), the brain interprets a message arriving from a particular receptor as meaning that a certain type of stimulus occurred (such as a flash of color).

When stimulated, a sensory receptor initiates what might be considered a coded message, composed of action potentials transmitted by sensory neurons. This coded message is later decoded in the brain. Impulses from the sensory receptor may differ in the (1) total number of sensory neurons transmitting the signal, (2) specific neurons transmitting action potentials and their targets, (3) total number of action potentials transmitted by a given neuron, and (4) frequency of the action potentials transmitted by a given fiber. The intensity of a stimulus is coded by the frequency of action potentials fired by sensory neurons during the stimulus. A strong stimulus results in a greater depolarization of the receptor membrane, so the sensory neuron fires action potentials with greater frequency than would a weak stimulus. This variation is possible because as a graded response, the receptor potential can vary in magnitude.

The difference in sound intensity between the gentle rustling of leaves and a clap of thunder depends on the number of neurons transmitting action potentials as well as the frequency of the action potentials transmitted by each neuron. Just how the sensory receptor initiates different codes and how the brain analyzes and interprets them to produce various sensations are not yet completely understood.

Sensation takes place in the brain. Interpretation of the message and the type of sensation depends on which interneurons receive the message. The rods and cones of the eye do not see. When stimulated, they send a message to the brain that interprets the signals and translates them into a rainbow, an elephant, or a child. Artificial (for example, electrical) stimulation of brain centers can also result in sensation. In contrast, many sensory messages never give rise to sensations at all. For example, certain chemoreceptors sense internal changes in the body but never stir our consciousness.

Sensory perceptions are constructed by the mind

The brain interprets sensations by converting them to perceptions of stimuli. Sensory **perception** is the process of selecting, interpreting, and organizing sensory information. Sensory perceptions are constructed by the mind by comparing present sensory experience with our memories of past experiences. You may perceive the incoming visual and auditory input of a hissing snake very differently than does your friend or a herpetologist who studies snakes. Perceptions are also influenced by our state of mind at the time we receive sensory information. Optical il-

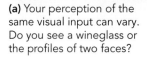

(a) Your perception of the same visual input can vary. Do you see a wineglass or the profiles of two faces?

(b) Do you "see" a white square? If so, your brain is completing a familiar pattern suggested by the right-angle wedges taken out of each of the four red circles.

Figure 42-2 Visual perception

lusions demonstrate that the brain can interpret sensory stimuli in various ways (▮ Fig. 42-2a). The brain also modifies stimuli to make them more complete, familiar, or logical (▮ Fig. 42-2b).

Review

▮ What is a receptor potential?

▮ What is the function of sensory adaptation?

▮ How is sensory information integrated?

TYPES OF SENSORY RECEPTORS

Learning Objective

2 Classify sensory receptors according to the location of stimuli to which they respond and according to the types of energy they transduce.

Sensory receptors can be classified according to the location of the stimuli to which they respond. **Exteroceptors** receive stimuli from the outside environment, enabling an animal to know and explore the world, search for food, find and attract a mate, recognize friends, and detect enemies. **Interoceptors** are sensory receptors within body organs that detect changes in pH, osmotic pressure, body temperature, and the chemical composition of the blood. You are not usually aware of messages sent to the CNS by these receptors as they work continuously to maintain homeostasis. You become aware of their activity when they signal certain internal conditions such as thirst, hunger, nausea, pain, and orgasm.

We can also classify sensory receptors based on the type of energy they transduce (▮ Table 42-1). *Thermoreceptors* respond to heat and cold. *Electroreceptors* sense differences in electrical potential. Some electroreceptors are sensitive enough to detect Earth's magnetic field. *Nociceptors* (pain receptors) respond to mechanical, temperature, and other stimuli that could be damaging. *Mechanoreceptors* transduce mechanical energy—touch, pressure, gravity, stretching, and movement. These receptors

TABLE 42-1

Classification of Receptors by Type of Energy They Transduce

Type of Receptor	Type of Energy Transduced	Examples
Thermoreceptors	Heat	Temperature receptors in blood-sucking insects and ticks; pit organs in pit vipers; nerve endings and receptors in skin and tongues of many animals
Electroreceptors and electromagnetic receptors	Electrical; receptors sense differences in electrical potential; electromagnetic receptors detect magnetic fields	Electrical currents in water used to navigate by many fish and some amphibian species; magnetic fields used for orientation and migration
Nociceptors (pain receptors)	Mechanical; physical force such as strong touch, pressure; heat, temperature extremes; damaging chemicals	Neuron endings in skin and other tissues

Mechanoreceptors	Mechanical; change shape as a result of being pushed or pulled	Tactile receptors (free nerve endings, Merkel discs, Meissner corpuscles, Ruffini corpuscles, Pacinian corpuscles); respond to touch and pressure
		
		Proprioceptors; respond to movement and body position
		    Muscle spindles; respond to muscle contraction
		    Golgi tendon organs; respond to stretch of a tendon
		    Joint receptors; respond to movement in ligaments
		
		Statocysts in invertebrates; have hair cells that respond to gravity
		
		Lateral line organs in fish; detect vibrations in the water; respond to waves and currents
		
		Vestibular apparatus
		    Hair cells in saccule and utricle; respond to gravity, linear acceleration
		    Hair cells in semicircular canals; respond to angular acceleration
		Hair cells in organ of Corti in cochlea; respond to pressure waves (sound)

Chemoreceptors	Specific chemical compounds	Taste buds; olfactory epithelium
Photoreceptors	Light	Eyespots; ommatidia in compound eye of arthropods; rods and cones in retina of vertebrates

convert mechanical forces directly into electrical signals. *Chemoreceptors* transduce the energy of certain chemical compounds, and *photoreceptors* transduce light energy.

Review

▌ Name five kinds of sensory receptors based on the type of energy they transduce. What is the specific function of each?

▌ THERMORECEPTORS

Learning Objective

3   Identify three functions of thermoreceptors.

Thermoreceptors respond to heat and cold. Mosquitoes, ticks, and other blood-sucking arthropods use thermoreception in their search for an endothermic host. Some have temperature receptors on their antennae that are sensitive to changes of less than 0.5°C. At least two types of snakes—pit vipers and boas—use thermoreceptors to locate their prey (▌ Fig. 42-3).

In mammals, which are endothermic, free nerve endings (and perhaps, specialized receptors) in the skin and tongue detect temperature changes in the outside environment. However,

Carmela Lesczynski/Animals Animals

Figure 42-3   Thermoreception

The pit organ of this bamboo viper (*Trimeresurus stejnegeri*) is a sense organ located between each eye and nostril. The pit organ of many snakes can detect the heat from an endothermic animal up to a distance of 1 to 2 m.

pain receptors sense extreme temperatures that pose a threat to the body. Thermoreceptors in the hypothalamus detect internal changes in temperature and receive and integrate information from receptors on the body surface. The hypothalamus then initiates homeostatic mechanisms that ensure a constant body temperature.

Review

▮ What are the functions of thermoreceptors in mosquitoes, snakes, and mammals?

ELECTRORECEPTORS AND ELECTROMAGNETIC RECEPTORS

Learning Objective

4 Describe the functions of electroreceptors and electromagnetic receptors.

Electroreceptors sense differences in electrical potential. Some predatory species of sharks, rays, and bony fishes detect electric fields generated in the water by the muscle activity of their prey. Several groups of fishes have electric organs, specialized muscle or nerve cells that emit electrical signals and receive a feedback signal. In species that produce a weak electric current, electroreceptors may help in navigation. This mechanism is particularly useful in murky water, where visibility and olfaction are poor.

Electroreception also appears important in communication, for example, in recognizing a potential mate. Males and females have different frequencies of electric discharge. A few fishes, such as electric eels or electric rays, have electric organs in their heads capable of delivering powerful shocks (up to several hundred volts) that stun prey or predators. Weaker electrical signals are emitted continuously to help in assessing the environment, orienting themselves, and communicating.

Some electroreceptors, called **electromagnetic receptors,** are sensitive enough to detect Earth's magnetic field. When a shark, ray, or skate swims across one of Earth's magnetic field lines, its electromagnetic receptors detect changes that occur in electric currents. This information, which is integrated in the brain, helps the animal orient itself. Many other organisms, including some bacteria, insects, amphibians, fishes, reptiles, birds, and a few mammals, can also detect magnetic fields and use this information to orient themselves. Migratory birds and sea turtles are well-known examples of animals that use magnetic fields to navigate (discussed further in Chapter 51).

Review

▮ What are two functions of electroreceptors?
▮ What is the function of electromagnetic receptors?

NOCICEPTORS

Learning Objective

5 Describe the functions of nociceptors, and identify the roles of glutamate and substance P.

Pain receptors, called **nociceptors** (from the Latin *nocere,* meaning "to injure"), are free nerve endings (dendrites) of certain sensory neurons found in almost every tissue. Three main types of nociceptors have been identified. Mechanical nociceptors respond to strong tactile stimuli such as cutting, crushing, or pinching. Thermal nociceptors respond to temperature extremes (temperatures above 45°C or below 5°C). Other nociceptors respond to a variety of damaging stimuli, including certain chemicals.

When stimulated, nociceptors transmit signals through sensory neurons to interneurons in the spinal cord. The sensory neurons release the neurotransmitter **glutamate,** as well as several neuropeptides, including **substance P,** that enhance and prolong the actions of glutamate. The interneuron transmits the message to the opposite side of the spinal cord and then upward to the thalamus, where pain perception begins. From the thalamus, impulses are sent into the parietal lobes and several other cortical regions, including areas of the limbic system where the emotional aspects of the pain are processed. When pain signals reach the cerebrum, we become fully aware of the pain and can evaluate the situation.

Review

▮ What is the function of each main type of nociceptor?
▮ What is the function of substance P?

MECHANORECEPTORS

Learning Objectives

6 Compare the functions and mechanisms of action of the following mechanoreceptors: tactile receptors, proprioceptors, statocysts, hair cells, and lateral line organs.

7 Compare the structure and function of the saccule and utricle with that of the semicircular canals in maintaining equilibrium.

8 Trace the path taken by sound waves through the structures of the ear, and explain how the organ of Corti functions as an auditory receptor.

Mechanoreceptors are activated when they change shape as a result of being mechanically pushed or pulled. They transduce mechanical energy, permitting animals to feel, hear, and maintain balance. These receptors provide information about the shape, texture, weight, and topographic relations of objects in the external environment. Some mechanoreceptors are extremely sensitive. For example, alligators have pressure receptors on their faces that can detect the movement in the surrounding water caused by a single drop of falling water.

Certain mechanoreceptors enable an organism to maintain its body position with respect to gravity (for us, head up and feet down; for a dog, dorsal side up and ventral side down; for a tree sloth, ventral side up and dorsal side down). When displaced from its normal position, the animal quickly adjusts its body to reassume normal orientation. Mechanoreceptors continuously send information to the CNS regarding the position and movements of the body. Mechanoreceptors also provide information about the operation of internal organs. For example, they inform us about the presence of food in the stomach, feces in the rectum, urine in the bladder, and a fetus in the uterus.

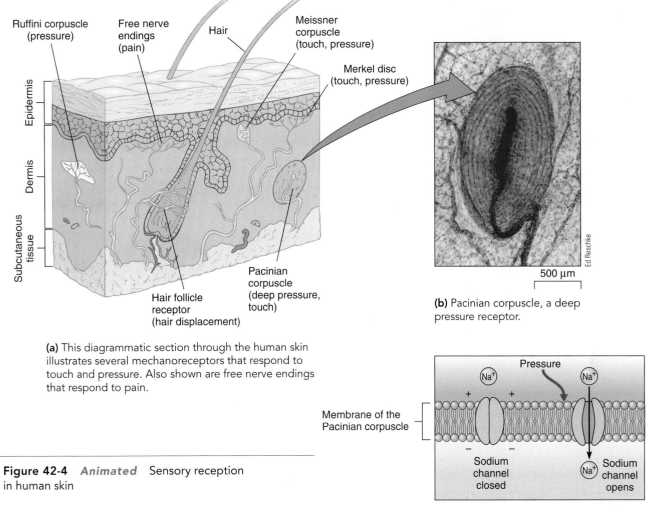

Ruffini corpuscle
(pressure)

Free nerve
endings
(pain)

Hair

Meissner
corpuscle
(touch, pressure)

Merkel disc
(touch, pressure)

Epidermis

Dermis

Subcutaneous
tissue

Hair follicle
receptor
(hair displacement)

Pacinian
corpuscle
(deep pressure,
touch)

(a) This diagrammatic section through the human skin illustrates several mechanoreceptors that respond to touch and pressure. Also shown are free nerve endings that respond to pain.

500 μm

Ed Reschke

(b) Pacinian corpuscle, a deep pressure receptor.

Pressure

Na^+

Na^+

Membrane of the
Pacinian corpuscle

+   +

−   −

Na^+

Sodium
channel
closed

Sodium
channel
opens

(c) The mechanical forces these receptors sense cause Na^+ channels to open, depolarizing the axon.

▌ **Figure 42-4** *Animated* Sensory reception in human skin

Touch receptors are located in the skin

The simplest mechanoreceptors are free nerve endings in the skin. They detect touch, pressure, and pain when stimulated by objects that contact the body surface. In many invertebrates and vertebrates, **tactile** (touch) **receptors** lie at the base of a hair or bristle. These receptors may sense the body's orientation in space with respect to gravity. They may also detect air and water vibrations and contact with other objects. The tactile receptor is stimulated indirectly when the hair is bent or displaced. A receptor potential develops, and a few action potentials may be generated. This type of receptor responds only when the hair is moving. Even though the hair may be maintained in a displaced position, the receptor is not stimulated unless there is motion.

Thousands of specialized tactile receptors are located in mammalian skin (▌ Fig. 42-4). **Merkel discs** sense touch and pressure. They adapt slowly, permitting us to know that an object continues to touch the skin. Three types of mechanoreceptors in the skin have encapsulated endings: Meissner corpuscles, Ruffini corpuscles, and Pacinian corpuscles. **Meissner corpuscles** are sensitive to light touch and pressure and adapt quickly to a sustained stimulus. **Ruffini corpuscles,** which adapt very slowly, inform us of heavy, continuous touch and pressure.

Pacinian corpuscles are sensitive to deep pressure that causes rapid movement of the tissues. They are especially sensitive to stimuli that vibrate. The Pacinian corpuscle consists of a neuron ending surrounded by concentric connective tissue layers interspersed with fluid. Compression causes displacement of the layers, which stimulates the axon. Even though the displacement is maintained under steady compression, the receptor potential rapidly falls to zero, and action potentials cease—an excellent example of sensory adaptation.

Proprioceptors help coordinate muscle movement

Proprioceptors help animals maintain postural relations—the position of one part of the body with respect to another. Located within muscles, tendons, and joints, proprioceptors continuously respond to tension and movement. With their continuous input, an animal can perceive the positions of its arms, legs, head,

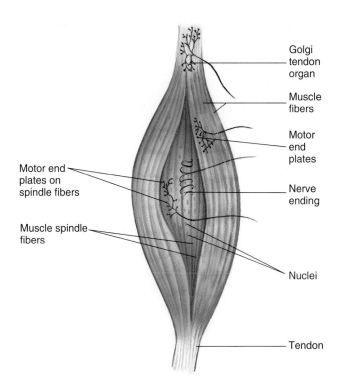

Figure 42-5 Proprioceptors

Muscle spindles detect muscle movement. A muscle spindle consists of an elongated bundle of specialized muscle fibers. Golgi tendon organs respond to tension in muscles and their associated tendons.

Labels for Figure 42-5:
- Golgi tendon organ
- Muscle fibers
- Motor end plates
- Nerve ending
- Motor end plates on spindle fibers
- Muscle spindle fibers
- Nuclei
- Tendon

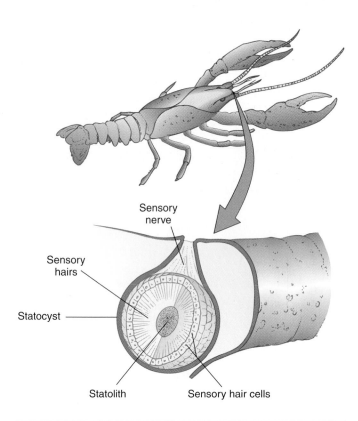

Figure 42-6 A statocyst

Many invertebrates have statocysts, receptors that sense gravitational force and provide information about orientation of the body with respect to gravity.

Labels for Figure 42-6:
- Sensory nerve
- Sensory hairs
- Statocyst
- Statolith
- Sensory hair cells

and other body parts, along with the orientation of its body as a whole. This information is essential for all forms of locomotion and for all coordinated and skilled movements, from spinning a cocoon to completing a reverse one-and-a-half dive with twist. With the help of proprioceptors, we carry out activities such as dressing or playing the piano even in the dark.

Vertebrates have three main types of proprioceptors: **muscle spindles,** which detect muscle movement (Fig. 42-5); **Golgi tendon organs,** which respond to tension in contracting muscles and in the tendons that attach muscle to bone; and **joint receptors,** which detect movement in ligaments. Impulses from proprioceptors help coordinate the contractions of the several distinct muscles involved in a single movement. Without such receptors, complicated, skilled acts would be impossible. Proprioceptors are also important in maintaining balance.

Proprioceptors are probably more numerous and more continually active than any of the other sensory receptors, although we are less aware of them than of most of the others. As long as the stimulus is present, receptor potentials are maintained (although not at constant magnitude) and action potentials continue to be generated. Proprioceptors continuously send signals to inform the brain about position.

The mammalian muscle spindle, one of the more versatile stretch receptors, helps maintain muscle tone. It is a bundle of specialized muscle fibers, with a central region encircled by sen-

sory neuron endings. These neurons continue to transmit signals for a prolonged period, in proportion to the degree of stretch.

Many invertebrates have gravity receptors called statocysts

Many invertebrates—from jellyfish to crayfish—have gravity receptors called **statocysts** (Fig. 42-6). Statocysts are the simplest organs of equilibrium. A statocyst is basically an infolding of the epidermis lined with receptor cells, called sensory hair cells, equipped with sensory hairs (not true hairs). The cavity contains one or more **statoliths,** tiny granules of loose sand grains or calcium carbonate held together by an adhesive material secreted by cells of the statocyst. Normally the particles are pulled downward by gravity and stimulate the sensory cells.

When the animal changes position, the statocyst tilts and the statolith moves in the same direction. This movement bends the sensory hairs. Each sensory hair cell responds maximally when the animal is at a particular position with respect to gravity. The mechanical displacement results in receptor potentials and action potentials that inform the CNS of the change in position. By "knowing" which sensory cells are firing, the animal knows where "down" is and can correct any abnormal orientation.

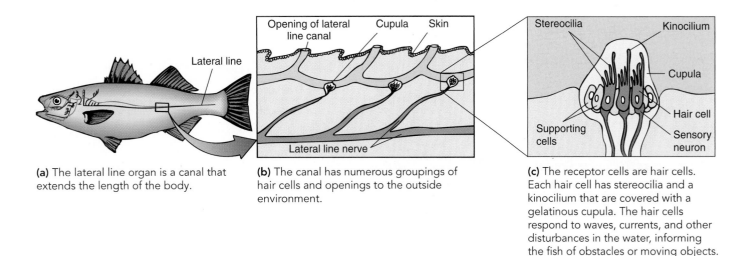

(a) The lateral line organ is a canal that extends the length of the body.

(b) The canal has numerous groupings of hair cells and openings to the outside environment.

(c) The receptor cells are hair cells. Each hair cell has stereocilia and a kinocilium that are covered with a gelatinous cupula. The hair cells respond to waves, currents, and other disturbances in the water, informing the fish of obstacles or moving objects.

Figure 42-7 Lateral line organ

In a classic experiment, the function of the statocyst was demonstrated by substituting iron filings for sand grains in the statocysts of crayfish. The force of gravity was overcome by holding magnets above the animals. The iron filings were attracted upward toward the magnets, and the crayfish began to swim upside down in response to the new information provided by their gravity receptors.

Hair cells are characterized by stereocilia

Vertebrate **hair cells** are mechanoreceptors that detect movement. They are the sensory receptors, located in the lateral line of fishes, that detect water movements (Fig. 42-7). Hair cells help maintain position and equilibrium and are important in hearing.

The surface of a vertebrate hair cell typically has a single long **kinocilium** and many shorter **stereocilia,** hairlike projections that increase in length from one side of the hair cell to the other (see Fig. 42-7c). A kinocilium is a true cilium with a 9 × 2 arrangement of microtubules (see Chapter 4). In contrast, stereocilia are not really cilia. They are microvilli that contain actin filaments.

Mechanical stimulation of the stereocilia causes voltage changes. Mechanical stimulation in one direction causes depolarization and release of neurotransmitter, whereas mechanical stimulation in the opposite direction results in hyperpolarization. However, hair cells have no axons and do not produce their own action potentials. Rather, they release neurotransmitters that depolarize associated neurons.

Lateral line organs supplement vision in fishes

Lateral line organs of fishes and aquatic amphibians detect vibrations in the water. They inform the animal of obstacles in its way and of moving objects such as prey, enemies, and others in its school. Typically, the lateral line organ is a long canal running the length of the body and continuing into the head. The canal is lined with sensory hair cells. The tips of the stereocilia are enclosed by a **cupula,** a mass of gelatinous material secreted by the hair cells.

Pressure waves, currents, and even slight movement in the water cause vibrations in the lateral line organ. As the water moves the cupula, the stereocilia bend. This changes the membrane potential of the hair cell, which may then release neurotransmitter. If an associated sensory neuron is sufficiently stimulated, an action potential may be dispatched to the CNS.

The vestibular apparatus maintains equilibrium

When you think of the ear, you probably think of hearing. However, in vertebrates its main function is to help maintain equilibrium. Many vertebrates do not have outer or middle ears, but all have inner ears (Fig. 42-8). The mammalian **inner ear** includes organs of equilibrium equipped with hair cells that sense the position of the body with respect to gravity.

The inner ear is a group of interconnected canals and sacs, called the **labyrinth,** which includes a membranous labyrinth that fits inside a bony labyrinth. In mammals the membranous labyrinth has two saclike chambers—the **saccule** and the **utricle**—and three **semicircular canals.** Collectively, the saccule, utricle, and semicircular canals are called the **vestibular apparatus** (see Fig. 42-8b). Destruction of these organs leads to a considerable loss of the sense of equilibrium. A pigeon whose vestibular apparatus has been destroyed cannot fly but in time can relearn how to maintain equilibrium by using visual stimuli.

The sensory cells of the saccule and utricle are hair cells similar to those of the lateral line organ (Fig. 42-9). The stereocilia projecting from the hair cells are covered by a gelatinous cupula

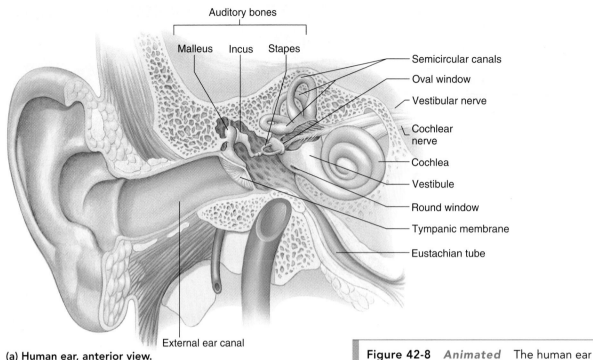

(a) **Human ear, anterior view.**

Auditory bones

Malleus   Incus   Stapes

Semicircular canals

Oval window

Vestibular nerve

Cochlear nerve

Cochlea

Vestibule

Round window

Tympanic membrane

Eustachian tube

External ear canal

Figure 42-8 *Animated* The human ear

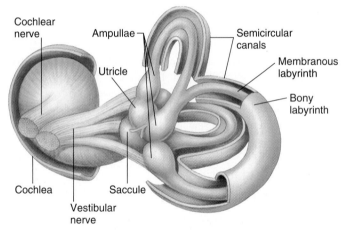

Cochlear nerve

Ampullae

Utricle

Cochlea

Saccule

Vestibular nerve

Semicircular canals

Membranous labyrinth

Bony labyrinth

(b) **Vestibular apparatus** consists of saccule, utricle, and semi-circular canals. The utricle and saccule are best seen in the posterior view shown here.

in which calcium carbonate ear stones, called **otoliths,** are embedded. The hair cells in the saccule and utricle lie in different planes.

Normally, the pull of gravity causes the otoliths to press against the stereocilia, stimulating them to initiate impulses. Sensory neurons at the bases of the hair cells transmit these signals to the brain. When the head is tilted or in linear acceleration (increasing speed when the body is moving in a straight line), otoliths press on the stereocilia of different cells, deflecting them. Deflection of the stereocilia toward the kinocilium depolarizes the hair cell. Deflection in the opposite direction hyperpolarizes the hair cell. Thus, depending on the direction of movement, the hair cells release more or less neurotransmitter. The brain in-

terprets the neural messages so the animal is aware of its full body position relative to the ground regardless of how its head is positioned.

Information about turning movements, referred to as angular acceleration, is provided by the three semicircular canals. Each canal, a hollow ring connected with the utricle, lies at right angles to the other two and is filled with fluid called **endolymph.** At one of the openings of each canal into the utricle, there is a small, bulblike enlargement, the **ampulla.** Within each ampulla lies a clump of hair cells called a **crista** (pl., *cristae*), similar to the groups of hair cells in the utricle and saccule. No otoliths are present. The stereocilia of the hair cells of the cristae are stimulated by movements of the endolymph in the canals (❚ Fig. 42-10).

When the head is turned, there is a lag in the movement of the fluid within the canals. The stereocilia move in relation to the fluid and are stimulated by its flow. This stimulation produces not only the consciousness of rotation but also certain reflex movements in response to it. These reflexes cause the eyes and head to move in a direction opposite that of the original rotation. Because the three canals are in three different planes, head movement in any direction stimulates fluid movement in at least one of the canals.

We humans are used to movements in the horizontal plane but not to vertical movements (parallel to the long axis of the upright body). The motion of an elevator or of a ship pitching in a rough sea stimulates the semicircular canals in an unusual way and may cause seasickness or motion sickness, with resultant nausea or vomiting. When a seasick person lies down, the movement stimulates the semicircular canals in a more familiar way and nausea is less likely to occur.

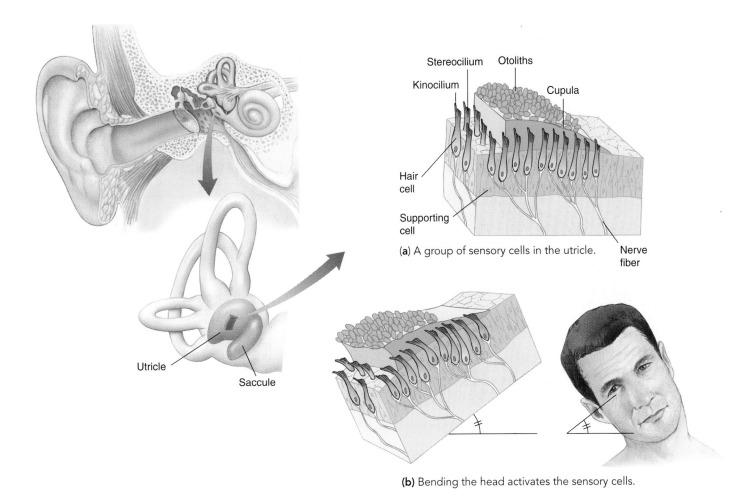

(a) A group of sensory cells in the utricle.

(b) Bending the head activates the sensory cells.

Figure 42-9 *Animated* Function of the saccule and utricle in maintaining equilibrium

The saccule and utricle sense linear acceleration, allowing us to know our position relative to the ground. Compare the positions of the otoliths and hair cells in **(a)** with those in **(b)**. Changes in head position cause the force of gravity to distort the cupula, which in turn distorts the stereocilia of the hair cells. The hair cells respond by releasing neurotransmitters. Impulses are transmitted along the vestibulocochlear nerve to the brain.

Auditory receptors are located in the cochlea

Many arthropods and most vertebrates have sound receptors, but for many of them hearing does not seem to be a sensory priority. It is important in tetrapods, however, and both birds and mammals have a highly developed sense of hearing. Their auditory receptors, located in the **cochlea** of the inner ear, contain mechanoreceptor hair cells that detect pressure waves.

The cochlea is a spiral tube that resembles a snail's shell (▌Fig. 42-11). If the cochlea were uncoiled, you would see that it consists of three canals separated from one another by thin membranes. The canals come almost to a point at the apex. Two of these canals, the **vestibular canal** and the **tympanic canal,** connect at the apex of the cochlea and are filled with a fluid known as **perilymph.** The middle canal, the **cochlear duct,** is filled with endolymph and contains the auditory organ, the **organ of Corti.**

Each organ of Corti contains about 18,000 hair cells, in rows that extend the length of the coiled cochlea. The stereocilia of each hair cell extend into the cochlear duct. The hair cells rest on the **basilar membrane,** which separates the cochlear duct from the tympanic canal. Another membrane, the **tectorial membrane,** overhangs and is in contact with the hair cells.

In terrestrial vertebrates, sound waves in the air are transformed into pressure waves in the cochlear fluid. In the human ear, for example, sound waves pass through the **external auditory canal** and vibrate the **tympanic membrane,** or eardrum, which separates the outer ear and the middle ear. The vibrations are transmitted across the middle ear by three tiny bones—the **malleus, incus,** and **stapes** (or hammer, anvil, and stirrup, so called because of their shapes). The malleus is in contact with the eardrum, and the stapes is in contact with a thin, membranous region of the cochlea called the **oval window.** The **eustachian tube,** which connects the middle ear with the pharynx (throat

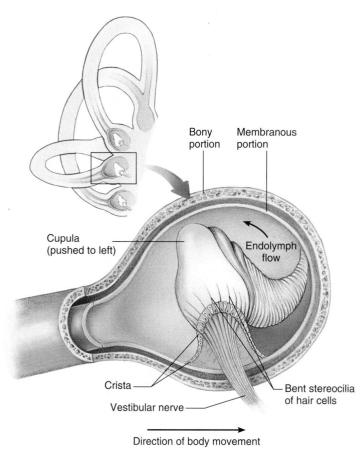

Bony portion

Membranous portion

Cupula (pushed to left)

Endolymph flow

Crista

Bent stereocilia of hair cells

Vestibular nerve

Direction of body movement

Figure 42-10 Semicircular canals and equilibrium

When the head changes its rate of rotation, endolymph within the ampulla of the semicircular canal distorts the cupula. The stereocilia of the hair cells bend, increasing the frequency of action potentials in sensory neurons. Information is transmitted to the brain via the vestibular nerve.

region), helps equalize pressure between the middle ear and the atmosphere.

The malleus, incus, and stapes act as three interconnected levers that amplify vibrations. A small movement in the malleus causes a larger movement in the incus and a very large movement in the stapes. The vibrations pass through the oval window to the perilymph in the vestibular canal. If sound waves were conducted directly from air to the oval window, much energy would be lost. The middle ear functions to couple sound waves in the air with the pressure waves conducted through the fluid in the cochlea.

Because liquids cannot be compressed, the oval window could not cause movement of the fluid in the vestibular canal if there were no escape valve for the pressure. This valve is provided by the membranous **round window** at the end of the tympanic canal. The pressure wave presses on the membranes separating the three ducts, is transmitted to the tympanic canal, and makes the round window bulge. The movements of the basilar membrane

produced by these pulsations cause the stereocilia of the organ of Corti to rub against the overlying tectorial membrane.

Stereocilia are bent by their contact with the tectorial membrane. As a result, ion channels in the plasma membrane of the hair cells open. Calcium ions move into the hair cell, causing the release of *glutamate*. This neurotransmitter binds to receptors on sensory neurons that synapse with each hair cell, leading to depolarization of these sensory neurons. Axons of the sensory neurons join to form the **cochlear nerve,** a component of the vestibulocochlear nerve (cranial nerve VIII; also referred to as the *auditory nerve*). We can summarize the sequence of events involved in hearing as follows:

Sound waves enter external auditory canal ⟶ tympanic membrane vibrates ⟶ malleus, incus, and stapes amplify vibrations ⟶ oval window vibrates ⟶ vibrations are conducted through fluid ⟶ basilar membrane vibrates ⟶ hair cells in organ of Corti are stimulated ⟶ cochlear nerve transmits impulses to brain

Sounds differ in pitch, loudness, and tone quality. **Pitch** depends on frequency of sound waves, or number of vibrations per second, and is expressed as hertz (Hz). Low-frequency vibrations result in the sensation of low pitch, whereas high-frequency vibrations result in the sensation of high pitch. Sounds of a given frequency set up resonance waves in the cochlear fluid that cause a particular section of the basilar membrane to vibrate. High frequencies are detected by hair cells located near the base of the cochlea, whereas low frequencies are sensed by hair cells near the apex of the cochlea. The brain infers the pitch of a sound from the particular hair cells that are stimulated.

The human ear typically registers sound frequencies between about 20 and 20,000 Hz (also expressed as 20 kilohertz), and is most sensitive to sounds between 1000 and 4000 Hz (❚ Fig. 42-12). Comparing the energy of audible sound waves with the energy of visible light waves, your ear is 10 times as sensitive as your eye. Dogs can hear sounds up to 40,000 Hz, and bats can detect frequencies as high as 100,000 Hz.

Loud sounds cause resonance waves of greater amplitude (height). The hair cells are more intensely stimulated, and the cochlear nerve transmits a greater number of impulses per second. Variations in tone quality, or timbre, such as when an oboe, a cornet, and a violin play the same note, depend on the number and kinds of overtones, or harmonics, produced. Many hair cells along the basilar membrane are stimulated and vibrate simultaneously, in addition to the main vibration common to all three instruments. Thus, differences in tone quality are recognized in the pattern of the hair cells stimulated.

Deafness may be caused by injury to, or malformation of, either the sound-transmitting mechanism of the outer, middle, or inner ear or the sound-perceiving mechanism of the inner ear. Efferent neurons from the brain stem protect the hair cells of the organ of Corti by dampening their response. Even so, exposure to high-intensity sound, such as heavily amplified music, damages hair cells in the organ of Corti.

Key Point

The hair cells in the organs of Corti in the cochlea are the receptors for hearing.

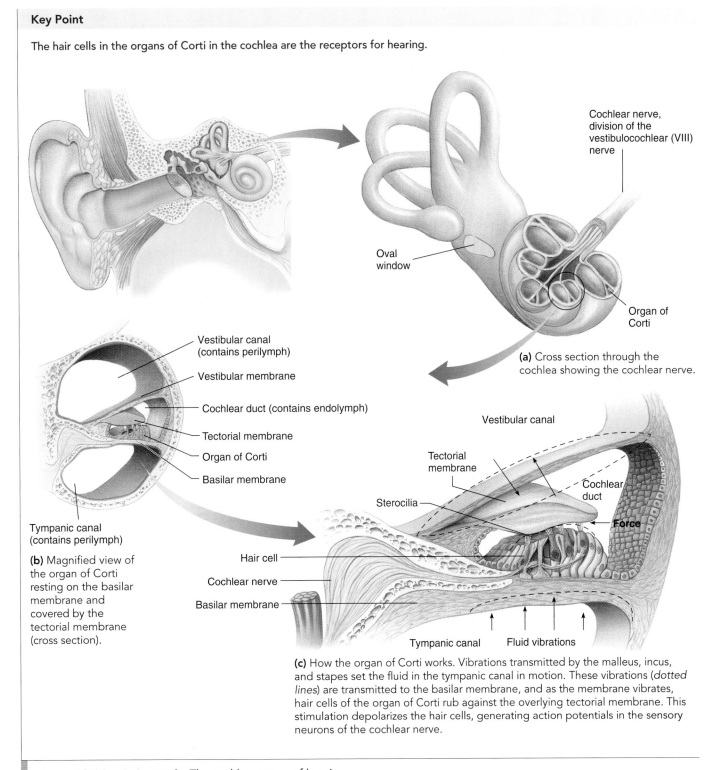

Cochlear nerve, division of the vestibulocochlear (VIII) nerve

Oval window

Organ of Corti

(a) Cross section through the cochlea showing the cochlear nerve.

Vestibular canal (contains perilymph)

Vestibular membrane

Cochlear duct (contains endolymph)

Tectorial membrane

Organ of Corti

Basilar membrane

Tympanic canal (contains perilymph)

(b) Magnified view of the organ of Corti resting on the basilar membrane and covered by the tectorial membrane (cross section).

Vestibular canal

Tectorial membrane

Sterocilia

Cochlear duct

Force

Hair cell

Cochlear nerve

Basilar membrane

Tympanic canal

Fluid vibrations

(c) How the organ of Corti works. Vibrations transmitted by the malleus, incus, and stapes set the fluid in the tympanic canal in motion. These vibrations (*dotted lines*) are transmitted to the basilar membrane, and as the membrane vibrates, hair cells of the organ of Corti rub against the overlying tectorial membrane. This stimulation depolarizes the hair cells, generating action potentials in the sensory neurons of the cochlear nerve.

Figure 42-11 *Animated* The cochlea, organ of hearing

Review

- Which sensory receptors permit you to perform actions such as getting dressed or finding your way into bed in the dark?
- How do hair cells work?
- What are otoliths, and what is their role in maintaining equilibrium?
- What is the sequence of events involved in hearing?

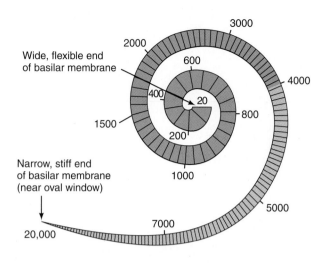

Figure 42-12 Distinguishing pitch

The basilar membrane is shown here coiled as it would be within the cochlea. Various parts of the basilar membrane vibrate in response to specific frequencies. The narrow, stiff end of the basilar membrane responds to higher frequencies, whereas the wide, flexible end responds to lower frequencies. Pitch depends on frequency of sound waves expressed as hertz (Hz). The human ear is most sensitive to sounds between 1000 and 4000 Hz.

CHEMORECEPTORS

Learning Objective

9 Compare the structure and function of the receptors of taste and smell, and describe the function of the vomeronasal organ.

Two highly sensitive chemoreceptive systems have evolved for **gustation,** or taste, and **olfaction,** or smell. These chemical senses, found throughout the animal kingdom, allow animals to detect chemical substances in food, water, and air. Evaluating such chemical cues allows animals to find food and mates and to avoid predators. Members of the same species use chemoreception to communicate about these critical issues.

For terrestrial vertebrates, the sense of smell involves gaseous substances that reach olfactory receptors through the air. The sense of taste involves materials dissolved in water (or saliva) in the mouth. For aquatic animals, these distinctions blur. Does a catfish smell or taste the water?

The organs of taste and smell in insects are sensory hairs called **sensilla** (sing., *sensillum*), which project from the end of the leg and from the mouthparts. In ants, bees, and some other insect species, receptors located on the antennae sense odors. In the blowfly, each sensillum contains four chemoreceptors (❚ Fig. 42-13). Dendrites from each cell extend to an opening at the tip, and axons transmit sensory signals to the brain. Recent findings suggest that insects can distinguish among various bitter tastes more precisely than mammals can. This ability is important in discriminating tastes associated with spoiled, bacteria-laden foods or toxins.

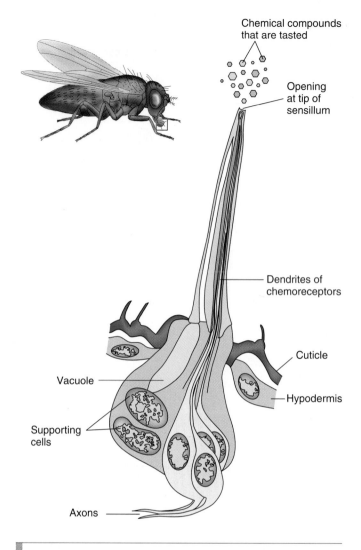

Figure 42-13 An insect taste receptor

Insects respond to a wide variety of chemical substances found in foods. Each sensillum in the blowfly contains four chemoreceptors. Dendrites that extend to an opening at the tip sense taste. The other end of each chemoreceptor is an axon that transmits sensory signals to the brain.

Taste receptors detect dissolved food molecules

The ability to discriminate among tastes has survival value. For example, foods with a high caloric value often taste sweet, whereas poisons are generally bitter. Mammals sense taste with **taste buds** located in the mouth. In humans, taste buds lie mainly in tiny elevations, or papillae, on the tongue. Each of the thousands of taste buds is an oval epithelial capsule containing about 100 taste receptor cells interspersed with supporting cells (❚ Fig. 42-14).

The plasma membrane at the tip of each taste receptor cell has microvilli that extend into a taste pore on the surface of the tongue, where they are bathed in saliva. The taste receptors detect chemical substances dissolved in saliva. Certain molecules, such as those perceived as sweet, activate a signal transduction pro-

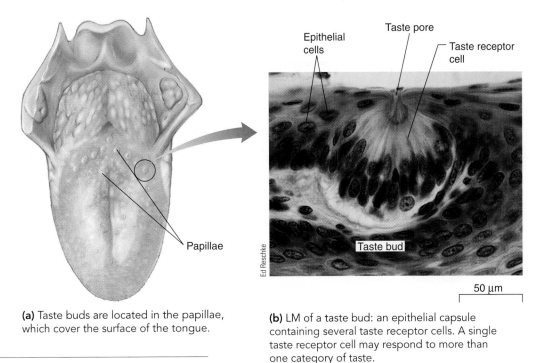

(a) Taste buds are located in the papillae, which cover the surface of the tongue.

(b) LM of a taste bud: an epithelial capsule containing several taste receptor cells. A single taste receptor cell may respond to more than one category of taste.

50 µm

Ed Reschke

Figure 42-14 *Animated* Taste buds

cess involving a G protein (see Fig. 42-14c). Adenylyl cyclase activity increases, elevating cyclic AMP levels. A protein kinase is activated that phosphorylates and closes K⁺ channels. This decrease in K⁺ permeability sets up a depolarizing receptor potential. Action potentials are then generated in sensory neurons that synapse with the taste receptor cell. One sensory neuron can innervate several taste buds.

Traditionally, four basic tastes have been recognized: sweet, sour, salty, and bitter. In 2000, physiologists reported a fifth taste, glutamate. Although the idea of a fifth taste is still somewhat controversial, it was first suggested almost a century ago by the Japanese physiologist Kikunae Ikeda. In 1908, Ikeda identified glutamate as the compound that triggers the taste he called umami, responsible for the savoriness of certain aged cheeses, soy sauce, anchovy, and some types of seafood.

Flavor depends on the four or five basic tastes in combination with smell, texture, and temperature. Smell affects flavor because odors pass from the mouth to the nasal chamber. No doubt you have observed that when your nose is congested, food seems to have little "taste." The taste buds are not affected, but the blockage of nasal passages severely reduces the participation of olfactory reception in the composite sensation of flavor.

Awareness of the genetic component of taste can be traced to 1931 when Arthur L. Fox, a chemist at the DuPont Company,

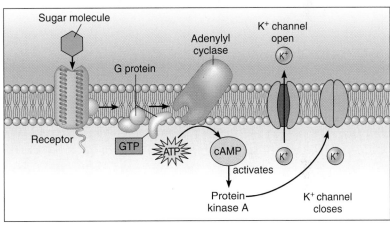

1 Sugar molecule binds with receptor in plasma membrane of taste receptor cell.

2 G protein is activated, and activates adenylyl cyclase.

3 ATP is converted to cyclic AMP (cAMP).

4 Cyclic AMP activates a protein kinase that closes K⁺ channels.

(c) A sugar molecule activates a signal transduction process that leads to a decrease in K⁺ permeability. The membrane becomes depolarized, resulting in action potentials in sensory neurons.

synthesized a compound called *phenylthiocarbamide (PTC)*. Some PTC blew into the air and was inhaled by a colleague, who experienced it as very bitter. Because Fox himself could not taste it, he became intrigued and asked other people to taste it. Fox found that about 25% of people were nontasters. Everyone else experienced PTC as bitter. Further research showed that the ability to taste PTC is inherited as a dominant trait.

Since that time, taste researchers have found that some tasters are especially sensitive to bitter tastes. In fact, about 25% of

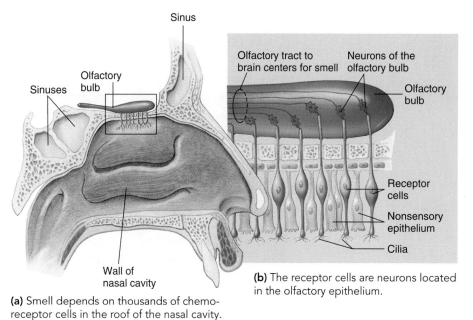

(a) Smell depends on thousands of chemo-receptor cells in the roof of the nasal cavity.

(b) The receptor cells are neurons located in the olfactory epithelium.

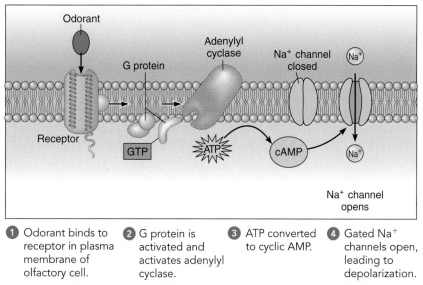

① Odorant binds to receptor in plasma membrane of olfactory cell.

② G protein is activated and activates adenylyl cyclase.

③ ATP converted to cyclic AMP.

④ Gated Na⁺ channels open, leading to depolarization.

(c) An odorant activates a signal transduction process that leads to an increase in Na⁺ permeability. The membrane becomes depolarized, resulting in transmission of action potentials to the brain.

Figure 42-15 *Animated* Olfactory epithelium

the U.S. population are supertasters, 50% are regular tasters, and 25% are nontasters. Supertasters avoid broccoli, brussels sprouts, cabbage, and many other vegetables and fruits that taste bitter. These foods contain flavonoids and other compounds that are thought to protect against cancer. Continuing investigation of taste preferences and their nutritional consequences may lead to more effective approaches to improving nutrition and to a better understanding of the relative importance of genetics and learning in food selection.

The olfactory epithelium is responsible for the sense of smell

Most invertebrates depend on **olfaction,** the detection of odors, as their main sensory modality. Recent findings indicate that the mechanisms for chemoreception have been highly conserved throughout the animal kingdom.

In terrestrial vertebrates, olfaction occurs in the nasal epithelium. In humans, the **olfactory epithelium** is found in the roof of the nasal cavity (❙ Fig. 42-15). It contains about 100 million olfactory receptor cells with ciliated tips. The long, nonmotile cilia extend into a layer of mucus on the epithelial surface of the nasal passageway. Receptor molecules on the cilia bind with compounds dissolved in the mucus. The other end of each olfactory receptor cell is an axon that projects directly to the brain. These axons make up the **olfactory nerve** (cranial nerve I), which extends to the olfactory bulb in the brain. From there, information is transmitted to the **olfactory cortex,** which is in the limbic system, a part of the brain also associated with emotional behavior. (Odors are often associated with feelings and memories.)

When an *odorant,* a molecule that can be smelled, binds with a receptor on a cilium of an olfactory receptor cell, a signal transduction process is initiated (see Fig. 42-15c). A G protein is activated, leading to the synthesis of cyclic AMP, which opens gated channels in the plasma membrane. These channels permit Na⁺ and other cations to enter the cell, causing depolarization, which is the receptor potential. The number of odorous molecules determines the intensity, which in turn determines the magnitude of the receptor potential.

Humans can detect at least seven groups of odors: camphor, musk, floral, peppermint, ethereal, pungent, and putrid. About 1000 genes code for 1000 types of olfactory receptors. Each odor consists of several component chemical groups, and each type of receptor may bind with a particular component. The combination of receptors activated determines what odor you perceive. You are capable of perceiving about 10,000 scents.

The olfactory receptors respond to remarkably small amounts of a substance. For example, most people can detect ionone, the synthetic substitute for the odor of violets, when it is present in a concentration of only 1 part to more than 30 billion parts of air. Smell is perhaps the sense that adapts most quickly. The olfactory receptors adapt about 50% in the first second or so after stimulation, so even offensively odorous air may seem odorless after only a few minutes.

Many animals communicate with pheromones

Animals within many species communicate with one another by releasing **pheromones,** small, volatile molecules that are secreted into the environment. For example, female moths release pheromones that attract males. Dogs and wolves use pheromones to mark territory.

Mammals have specialized chemoreceptor cells that detect pheromones. These cells make up the **vomeronasal organ** located in the epithelium of the nose (separate from the main olfactory epithelium). The vomeronasal sensory neurons send signals to areas of the hypothalamus, which serves as a link with the endocrine system. (We discuss pheromones further in Chapter 51.)

Review

▮ How is signal transduction similar in taste buds and olfactory receptor cells? How is it different?

▮ PHOTORECEPTORS

Learning Objectives

10 Contrast simple eyes, compound eyes, and vertebrate eyes.
11 Describe the functions of each structure of the verterate eye.
12 Describe the events that take place in human vision; compare the two types of photoreceptors, and describe the signal transduction pathway as part of your explanation.

Most animals have photoreceptors that use pigments to absorb light energy. **Rhodopsins** are the photopigments in the eyes of cephalopod mollusks (squids and octopods), arthropods, and vertebrates. Light energy striking a light-sensitive receptor cell containing these pigments triggers chemical changes in the pigment molecules that result in receptor potentials.

Invertebrate photoreceptors include eyespots, simple eyes, and compound eyes

The simplest light-sensitive structures in animals are found in certain cnidarians and flatworms (▮ Fig. 42-16). They are **eyespots,** called **ocelli,** that detect light but do not form images. Eyespots

are often bowl-shaped clusters of light-sensitive cells within the epidermis. They may detect the direction of the light source and distinguish light intensity.

Effective image formation requires a more complex **eye,** usually with a lens. A **lens** is a structure that concentrates light on a group of photoreceptors. Vision also requires a brain that can interpret the action potentials generated by the photoreceptors. The brain integrates information about movement, brightness, location, position, and shape of the visual stimulus. Two fundamentally different types of eyes evolved: the compound eye of arthropods and the camera eye of vertebrates and cephalopod mollusks.

The **compound eyes** of crustaceans and insects differ structurally and functionally from vertebrate eyes. The surface of a compound eye appears faceted, which means "having many faces," like a diamond (▮ Fig. 42-17). Each **facet** is the convex *cornea* of one of the eye's visual units, called **ommatidia** (sing., *ommatidium*). The number of ommatidia varies with the species. For example, each eye of certain crustaceans has only 20 ommatidia, whereas the dragonfly eye has as many as 28,000.

The optical part of each ommatidium includes a biconvex lens and a **crystalline cone.** The lens and crystalline cone focus light onto photoreceptor cells called **retinular cells.** These cells have a light-sensitive membrane made up of microvilli that contain rhodopsin. The membranes of several retinular cells may fuse to form a central rod-shaped **rhabdome** that is sensitive to light.

Compound eyes do not perceive form well. Although the lens system of each ommatidium is adequate to focus a small inverted image, there is little evidence that the animal actually perceives them as images. However, all the ommatidia together produce a composite image, or **mosaic** picture. Each ommatidium, in gath-

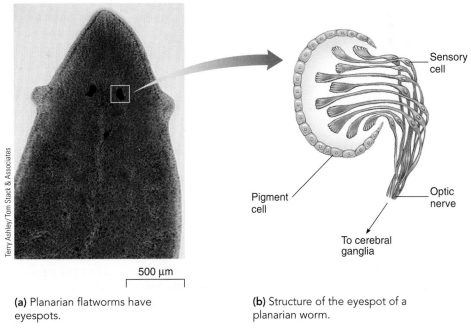

(a) Planarian flatworms have eyespots.

500 μm

Terry Ashley/Tom Stack & Associates

(b) Structure of the eyespot of a planarian worm.

Sensory cell

Pigment cell

Optic nerve

To cerebral ganglia

Figure 42-16 Eyespots

The simplest light-sensitive structures are the ocelli, or eyespots, found in certain invertebrates.

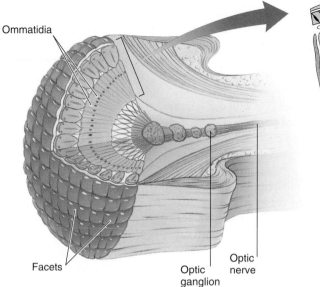

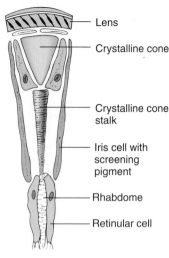

(a) SEM of Mediterranean fruit fly (*Ceratitis capitata*) showing its prominent compound eyes.

(b) Structure of compound eye showing several ommatidia. The eye registers changes in light and shade, permitting the animal to detect movement.

(c) Structure of an ommatidium. The rhabdome is the light-sensitive core of the ommatidium.

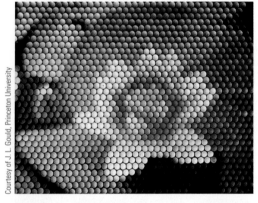

(d) A bee's-eye view of a flower. This photograph was taken using an optical device to achieve an approximate simulation of how the bee might see this flower. However, the bee would see UV light rather than red, and circles would be vertically elongated ellipses.

Figure 42-17 Compound eyes

ering a point of light from a narrow sector of the visual field, is in fact sampling a mean intensity from that sector. All these points of light taken together form a mosaic picture (see Fig. 42-17d).

Arthropod eyes usually adapt to different intensities of light. A sheath of pigmented cells envelops each ommatidium, and screening pigments are present in *iris cells* and in retinular cells. In nocturnal and crepuscular (active at dusk) insects and many crustaceans, pigment migrates proximally and distally within each pigmented cell. When the pigment is in the proximal position, each ommatidium is shielded from its neighbor, and only light entering directly along its axis can stimulate the receptors.

When the pigment is in the distal position, light striking at any angle can pass through several ommatidia and stimulate many retinal units. As a result, sensitivity is increased in dim light and the eye is protected from excessive stimulation in bright light. Pigment migration is under neural control in insects and under hormonal control in crustaceans. In some species it follows a daily rhythm.

Although the compound eye can form only coarse images, it compensates by following flickers to higher frequencies. Flies can detect up to about 265 flickers per second. In contrast, the human eye can detect only 45 to 53 flickers per second; for us, flickering lights fuse above these values, so we see light provided by an ordinary bulb as steady and the movement in motion pictures as smooth. To an insect, both room lighting and motion pictures must flicker horribly. The insect's high critical flicker fusion threshold permits immediate detection of even slight movement by prey or enemy. The compound eye is an important adaptation to the arthropod's way of life.

Compound eyes differ from our eyes in another respect. They are sensitive to wavelengths of light in the range from red to ultraviolet (UV). Accordingly, an insect can see UV light well, and its world of color is very different from ours. Because they reflect UV light to various degrees, flowers that appear identically colored to humans may appear strikingly dissimilar to insects (see Fig. 36-5).

Vertebrate eyes form sharp images

The position of the eyes in the front of the head of primates and certain birds allows both eyes to focus on the same object. The

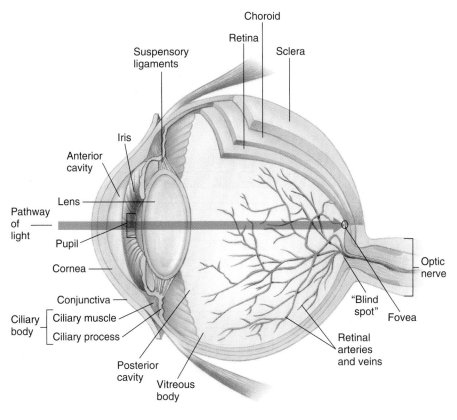

Figure 42-18 *Animated* Structure of the human eye

Light passes through the human eye to photoreceptor cells in the retina. In this lateral view, the eye is shown partly sectioned along the sagittal plane to expose its internal structures.

light-sensitive film used in a camera. Outside the retina is the **choroid layer,** a sheet of cells filled with black pigment that absorbs extra light. This mechanism prevents light from being reflected back into the photoreceptors, which would cause blurring of images. (Cameras are also black on the inside.) The choroid is rich in blood vessels that supply the retina.

The outer coat of the eyeball, called the **sclera,** is a tough, opaque, curved sheet of connective tissue that protects the inner structures and helps maintain the rigidity of the eyeball. On the front surface of the eye, this sheet becomes the thinner, transparent **cornea,** through which light enters. The cornea serves as a fixed lens that focuses light.

The lens of the eye is a transparent, elastic ball just behind the iris. It bends the light rays coming in and brings them to a focus on the retina. The lens is aided by the curved surface of the cornea and by the refractive properties (ability to bend light rays) of the liquids inside the eyeball. The **anterior cavity** between the cornea and the lens is filled with a watery substance, the **aqueous fluid.** The larger **posterior cavity** between the lens and the retina is filled with a more viscous fluid, the **vitreous body.** Both fluids are important in maintaining the shape of the eyeball by providing an internal fluid pressure.

At its anterior margin, the choroid is thick and projects medially into the eyeball, forming the **ciliary body,** which consists of ciliary processes and the ciliary muscle. The ciliary processes are glandlike folds that project toward the lens and secrete the aqueous fluid.

We focus a camera by changing the distance between the lens and the film. The eye has the power of **accommodation,** the ability to change focus for near or far vision by changing the shape of the lens (❚ Fig. 42-19). This accommodation is accom-

overlap in information they receive results in the same visual information striking the two retinas (light-sensitive areas) at the same time. This **binocular vision** is important in judging distance and in depth perception.

The vertebrate eye can be compared to an old-fashioned camera. An adjustable lens can be focused for different distances, and a diaphragm, called the **iris,** regulates the size of the light opening, called the **pupil** (❚ Fig. 42-18). The **retina** corresponds to the

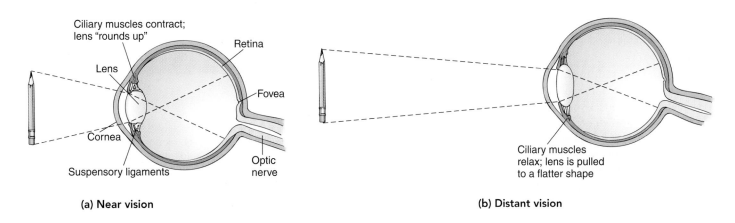

(a) Near vision

(b) Distant vision

Figure 42-19 *Animated* Accommodation

How the eye changes focus for **(a)** near and **(b)** distant vision.

plished by the **ciliary muscle,** a part of the ciliary body. To focus on close objects, the ciliary muscle contracts, making the elastic lens rounder. To focus on more distant objects, the ciliary muscle relaxes and the lens assumes a flattened (ovoid) shape.

The amount of light entering the eye is regulated by the iris, a ring of smooth muscle that appears as blue, green, gray, or brown depending on the amount and nature of pigment present. The iris consists of two mutually antagonistic sets of muscle fibers. One set is arranged circularly and contracts to *decrease* the size of the pupil. The other is arranged radially and contracts to *increase* the size of the pupil.

Each eye also has six muscles that extend from the surface of the eyeball to various points in the bony socket. These muscles enable the eye as a whole to move and be oriented in a given direction. Cranial nerves innervate the muscles in such a way that the eyes normally move together and focus on the same area.

The retina contains light-sensitive rods and cones

The light-sensitive structure in the vertebrate eye is the retina, which lines the posterior two thirds of the eyeball and covers the choroid. The retina, which is composed of 10 layers, contains the photoreceptor cells called, according to shape, **rods** and **cones.** In both rod and cone cells, infoldings of the plasma membrane form stacks of membranous discs that contain the photopigments. The discs greatly increase the light-absorbing surface.

The human eye has about 125 million rods and 6.5 million cones. Rods function in dim light so that we can detect shape and movement. They are not sensitive to colors. Because the rods are more numerous in the periphery of the retina, you can see an object better in dim light if you look slightly to one side of it (allowing the image to fall on the rods).

Cones respond to light at higher levels of intensity, such as daylight, and they allow us to perceive fine detail. Cones are responsible for color vision; they are differentially sensitive to different wavelengths (colors) of light. In primates and certain hunting birds (such as eagles), the cones are most concentrated in the **fovea,** a small, depressed area in the center of the retina. The fovea is the region of sharpest vision because it has the greatest density of receptor cells and because the retina is thinner there.

Light must pass through several layers of connecting neurons in the retina to reach the rods and cones (❙ Fig. 42-20). This arrangement lets the rods and cones contact a layer of pigmented epithelium that provides retinal, a component of rhodopsin. The retina has five main types of neurons: (1) Photoreceptors (rods and cones) synapse on (2) **bipolar cells,** which make synaptic contact with (3) **ganglion cells.** Two types of lateral interneurons are the (4) **horizontal cells,** which receive information from the photoreceptor cells and send it to bipolar cells, and (5) **amacrine cells,** which receive messages from the bipolar cells and send signals back to the bipolar cells or to ganglion cells (❙ Fig. 42-21). Interestingly, one group of ganglion cells project to the suprachiasmatic nucleus, the body's main biological clock. These gan-

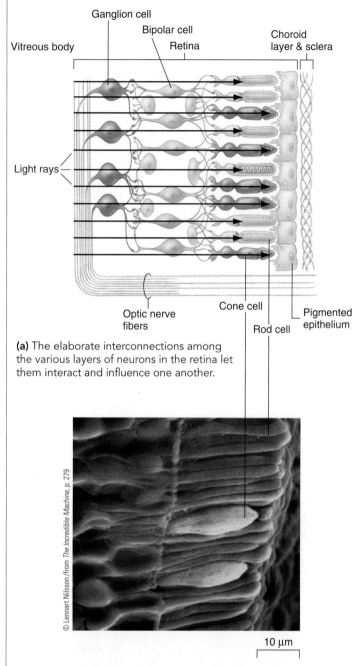

(a) The elaborate interconnections among the various layers of neurons in the retina let them interact and influence one another.

(b) Rods (*red elongated structures*) and two cones (*shorter, thicker, yellow structures*) are seen in this SEM. The elongated rods permit you to see shape and movement, whereas the shorter cones allow you to view the world in color.

Figure 42-20 *Animated* Organization of the retina

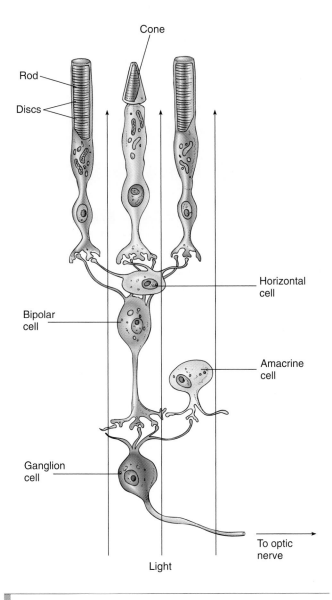

Figure 42-21 Neural pathway in the retina

The photoreceptor cells (rods and cones) are located in the back of the retina. They synapse with bipolar and horizontal cells. The bipolar cells synapse with ganglion cells and amacrine cells. Axons of ganglion cells make up the optic nerve.

glion cells contain a light-sensitive pigment (melanopsin) that is important in nonvisual responses to light.

The axons of the ganglion cells extend across the surface of the retina and unite to form the **optic nerve.** The area where the optic nerve passes out of the eyeball, the **optic disk,** is known as the *blind spot;* because it lacks rods and cones, images falling on it cannot be perceived.

Here is a simplified summary of the visual pathway:

Light passes through cornea ⟶ through aqueous fluid ⟶ through lens ⟶ through vitreous body ⟶ image forms on photoreceptor cells in retina ⟶ signal bipolar cells ⟶ signal ganglion cells ⟶ optic nerve transmits signals to thalamus ⟶ integration by visual areas of cerebral cortex

A chemical change in rhodopsin leads to the response of a rod to light

How is light converted into the neural signals that transmit information about environmental stimuli into pictures in the brain? Rod cells are so sensitive to light that they can respond to a single photon. Rhodopsin in the rod cells and some very closely related photopigments in the cone cells are responsible for the ability to see. Rhodopsin consists of *opsin,* a large protein that is chemically joined with *retinal,* an aldehyde of vitamin A (see Fig. 3-14). Two isomers of retinal exist: the *cis* form, which is folded, and the *trans* form, which is straight.

In the dark, opsin binds to retinal in the *cis* form. Cyclic GMP (cyclic guanosine monophosphate, a molecule similar to cyclic AMP) opens nonspecific channels that permit passage of Na^+ and other cations into the rod cell (Fig. 42-22). This process depolarizes the rod cell, which releases the neurotransmitter glutamate. The glutamate hyperpolarizes the membrane of the bipolar cell so it does not transmit messages.

Note that the photoreceptor differs from other neurons in that the ion channels in its membrane are normally open; it is depolarized and continuously releases neurotransmitter. The steady flow of ions into the cell when the environment is dark is called the *dark current.* Another unusual characteristic of photoreceptors, and of bipolar cells, is that they do not produce action potentials. Their release of neurotransmitter is graded, regulated by the extent of depolarization.

When light strikes rhodopsin, light energy is transduced. This process can be considered in three stages:

1. Light activates rhodopsin. Light transforms *cis*-retinal to *trans*-retinal. Rhodopsin changes shape and breaks down into its components, opsin and retinal. This is the light-dependent process in vision.

2. Rhodopsin is part of a signal transduction pathway. When it changes shape, it binds with **transducin,** a G protein. Transducin activates an esterase that hydrolyzes cyclic GMP (cGMP) to GMP, reducing the concentration of cGMP.

3. When the concentration of cGMP in the rod decreases, its Na^+ channels begin to close. Fewer cations pass into the rod cell, and it becomes more negative (hyperpolarized). The rod cell then releases less glutamate. Thus, light decreases the number of neural signals from the rod cells.

The release of glutamate normally hyperpolarizes the membrane of the bipolar cell, so a decrease in glutamate release results in its depolarization. The depolarized bipolar cell increases its release of a neurotransmitter, which typically stimulates the ganglion cell.

When you move from bright daylight to a dark room, you experience "blindness" for a few moments while your eyes adapt. During exposure to bright light, rhodopsin breaks down, decreasing the sensitivity of your eyes. **Dark adaptation** occurs as enzymes restore rhodopsin to a form in which it can respond to light. In dim light, vision depends on the rods. Conversely, when you move from the dark to bright light, your eyes are very sensitive and you may feel "blinded" by the light. **Light adapta-**

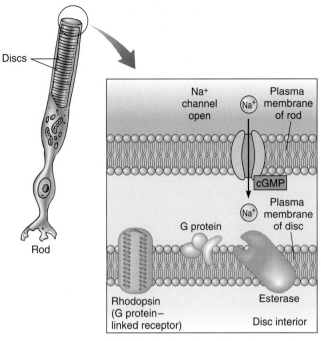

Discs

Rod

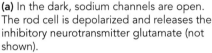

Na⁺ channel open

Na⁺

Plasma membrane of rod

cGMP

Plasma membrane of disc

Na⁺

G protein

Rhodopsin (G protein–linked receptor)

Esterase

Disc interior

(a) In the dark, sodium channels are open. The rod cell is depolarized and releases the inhibitory neurotransmitter glutamate (not shown).

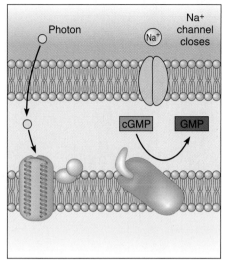

Photon

Na⁺ channel closes

Na⁺

cGMP

GMP

(b) Light makes rhodopsin change shape, activating a signal transduction pathway. The activation of a G protein leads to hyperpolarization and a decrease in inhibitory signals from the rod cell.

Figure 42-22 The biochemistry of vision

The discs of the rod are infoldings of the plasma membrane. In the dark the rod cell is depolarized; in the light it becomes hyperpolarized.

tion occurs as some of the photopigments break down and the sensitivity of your eyes decreases.

Color vision depends on three types of cones

Many invertebrates and at least some animals in each vertebrate class have color vision. Color perception depends on cones. Most mammals have two types of cones, but humans and other primates have three types: blue, green, and red. Each contains a slightly different photopigment. Although the retinal portion of the pigment molecule is the same as in rhodopsin, the opsin protein differs slightly in each type of photoreceptor.

Each type of cone responds to light within a considerable range of wavelengths but is named for the ability of its pigment to absorb a particular wavelength more strongly than other cones do. For example, red light can be absorbed by all three types of cones, but those cones most sensitive to red act as red receptors. By comparing the relative responses of the three types of cones, the brain can detect colors of intermediate wavelengths. Color blindness occurs when there is a deficiency of one or more of the three types of cones. This is usually an inherited X-linked condition (see Fig. 11-16).

Integration of visual information begins in the retina

The size, intensity, and location of light stimuli determine initial processing in the retina. Rods and cones send signals directly to bipolar cells, and bipolar cells send signals directly to ganglion cells. However, as illustrated in Figure 42-21, photoreceptor cells can also transmit signals to horizontal cells, and bipolar cells can transmit signals to amacrine cells. Ganglion cells are inhibited by amacrine cells and can be inhibited indirectly by horizontal cells as a result of their action on bipolar cells. Thus, the horizontal and amacrine cells integrate signals laterally.

Bipolar, horizontal, and amacrine cells combine signals from several photoreceptors. Each ganglion cell has a **receptive field,** a specific group of photoreceptors that light must strike for the ganglion cell to be stimulated. Ganglion cells are activated or inhibited depending on which photoreceptor cells have been stimulated.

The pattern of neuron firing in the retina is very important. The signals sent to ganglion cells depend on spatial patterns and timing of the light striking the retina. Ganglion cells receive signals about specific types of visual stimuli, such as color, brightness, and motion, and transmit this information about the character-

914 Chapter 42

www.thomsonedu.com/biology/solomon

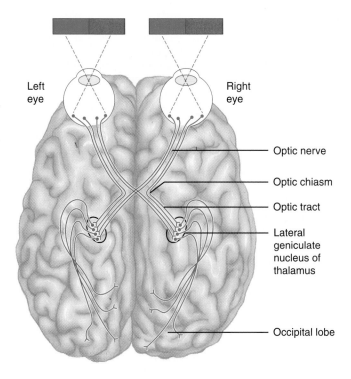

Left eye

Right eye

Optic nerve

Optic chiasm

Optic tract

Lateral geniculate nucleus of thalamus

Occipital lobe

Figure 42-23 *Animated* Neural pathway for transmission of visual information

Axons of ganglion cells form the optic nerves. The optic nerves cross, forming the optic chiasm. Many optic nerve fibers end in the lateral geniculate nuclei of the thalamus. From there, signals are sent to the visual cortex in the occipital lobe. Brain is viewed from above with overlying structures removed.

istics of a visual image. Ganglion cells produce action potentials, in contrast with rods and cones and most other neurons of the retina, which produce only graded potentials.

The retina begins the construction of visual images, but those images are interpreted in the brain. Axons of more than a million

ganglion cells form the optic nerves that transmit information to the brain by complex, encoded signals. The optic nerves cross in the floor of the hypothalamus, forming an X-shaped structure, the **optic chiasm** (Fig. 42-23). Some of the axons of the optic nerves cross over and then extend to the opposite side of the brain.

Axons of the optic nerves end in the **lateral geniculate nuclei** of the thalamus. From there, neurons send signals to the **primary visual cortex** in the occipital lobe of the cerebrum. The lateral geniculate nucleus controls which information is sent to the visual cortex. Neurons in the reticular activating system are involved in this integration. Information is transmitted from the primary visual cortex to other cortical areas for further processing.

Different types of ganglion cells receive information about different aspects of a visual stimulus. As a result, information from each point in the retina is transmitted in parallel by several types of ganglion cells. These cells project to different kinds of neurons in the lateral geniculate nucleus and primary visual cortex.

Neurobiologists have not yet discovered all the mechanisms by which the brain makes sense out of the visual information it receives. We know that a large part of the association areas of the cerebrum are involved in integrating visual input. The neurons of the visual cortex are organized as a map of the external visual field. They are also organized into columns. Each column of neurons receives signals originating from light entering either the right or left eye. Columns are also organized by orientation of light. Neurons in a column receive signals originally triggered by light with the same orientation. However, these signals originate from different locations in the retina.

Review

■ How does the insect's compound eye compare with the vertebrate eye?

■ What happens when light strikes rhodopsin? (Include a description of the signal transduction pathway in your answer.)

■ What is the sequence of neural signaling in the retina?

■ What is meant by the statement "Vision happens mainly in the brain?"

SUMMARY WITH KEY TERMS

Learning Objectives

1 Describe sensory system function, including sensory reception, energy transduction, receptor potential, sensory adaptation, and perception (page 894).

■ **Sensory receptors** are neuron endings or specialized receptor cells in close contact with neurons. **Sense organs** consist of sensory receptors and accessory cells.

■ Sensory receptors absorb energy, transduce that energy—**energy transduction**—into electrical energy, and produce **receptor potentials,** which are depolarizations or hyperpolarizations of the membrane. Receptor potentials are **graded responses.**

■ Sensory receptors transmit coded signals to the CNS through **sensory neurons. Sensory adaptation,** the decrease in frequency of action potentials in a sensory

neuron even when the stimulus is maintained, decreases response to that stimulus.

■ The brain interprets sensations by converting them to perceptions of stimuli. Sensory **perception** is the process of selecting, interpreting, and organizing sensory information.

ThomsonNOW™ **Learn more about sensory reception in human skin by clicking on the figure in ThomsonNOW.**

2 Classify sensory receptors according to the location of stimuli to which they respond and according to the types of energy they transduce (page 896).

■ Sensory receptors can be classified according to the location of the stimuli to which they respond. **Exteroceptors** receive stimuli from the outside environment. **Interoceptors** detect changes within the body, such as changes in pH, osmotic pressure, body temperature, and the chemical composition of the blood.

- Sensory receptors can be classified according to the types of energy they transduce. Thermoreceptors respond to heat and cold. Electroreceptors sense differences in electrical potential. Some electromagnetic receptors can detect Earth's magnetic field. Nociceptors (pain receptors) respond to mechanical, temperature, and other stimuli that could be damaging. Mechanoreceptors transduce mechanical energy—touch, pressure, gravity, stretching, and movement. Chemoreceptors transduce certain chemical compounds, and photoreceptors transduce light energy.

3 Identify three functions of thermoreceptors (page 897).

- **Thermoreceptors** transduce thermal energy; they respond to heat and cold. Some parasites use thermoreceptors to find an endothermic host. Some animals use thermoreceptors to locate endothermic prey. In endothermic animals, thermoreceptors provide cues about body temperature.

4 Describe the functions of electroreceptors and electromagnetic receptors (page 898).

- **Electroreceptors,** used by predatory fishes to detect prey, respond to electrical stimuli.

- Some electroreceptors, called **electromagnetic receptors,** are sensitive enough to detect Earth's magnetic field. Some animals use information about magnetic fields to orient themselves. Many migratory animals use magnetic fields to navigate.

5 Describe the functions of nociceptors, and identify the roles of glutamate and substance P (page 898).

- **Nociceptors** are pain receptors; they are free nerve endings of certain sensory neurons that respond to strong tactile stimuli, temperature extremes, and certain chemicals.

- Sensory neurons that transmit pain signals release the neurotransmitters **glutamate** and **substance P.** Substance P enhances and prolongs the actions of glutamate.

6 Compare the functions and mechanisms of action of the following mechanoreceptors: tactile receptors, proprioceptors, statocysts, hair cells, and lateral line organs (page 898).

- **Mechanoreceptors** transduce mechanical energy, permitting animals to feel, hear, and maintain balance. Mechanoreceptors respond to touch, pressure, gravity, stretch, or movement. They are activated when they change shape as a result of being mechanically pushed or pulled.

- **Tactile receptors** in the skin respond to mechanical displacement of hairs or to displacement of the receptor cells themselves. The **Pacinian corpuscle** is a tactile receptor that responds to touch and pressure.

- **Proprioceptors** allow an animal to perceive orientation of its body and the positions of its parts. **Muscle spindles, Golgi tendon organs,** and **joint receptors** are proprioceptors that continuously respond to tension and movement.

- **Statocysts,** gravity receptors found in many invertebrates, are simple organs of equilibrium. A statocyst consists of a chamber lined with ciliated mechanoreceptors; the chamber contains **statoliths,** granules of sand grains or calcium carbonate that change position as the animal moves. As statoliths move, sensory hairs bend and different sensory cells are stimulated. This mechanical displacement results in receptor potentials and action potentials that inform the CNS of the change in position.

- Vertebrate **hair cells** detect movement; they are found in the lateral line of fishes, the vestibular apparatus, semicircular canals, and cochlea. Each hair cell has a single **kinocilium,** which is a true cilium, and **stereocilia,** which are microvilli that contain actin filaments.

- **Lateral line organs** supplement vision in fishes and some amphibians by informing the animal of moving objects or objects in its path.

7 Compare the structure and function of the saccule and utricle with that of the semicircular canals in maintaining equilibrium (page 898).

- The vertebrate **inner ear** consists of a **labyrinth** of fluid-filled chambers and canals that help maintain equilibrium. The **vestibular apparatus** in the upper part of the labyrinth consists of the **saccule, utricle,** and **semicircular canals.**

- The saccule and utricle contain **otoliths** that change position when the head is tilted or when the body is moving in a straight line. The otoliths stimulate hair cells that send signals to the brain, enabling the animal to perceive the direction of gravity.

- The semicircular canals inform the brain about turning movements. Clumps of hair cells, called **cristae,** are located within each bulblike enlargement, called an **ampulla.** Cristae are stimulated by movements of the **endolymph,** a fluid that fills each canal.

8 Trace the path taken by sound waves through the structures of the ear, and explain how the organ of Corti functions as an auditory receptor (page 898).

- In terrestrial vertebrates, the **organ of Corti** within the **cochlea** contains auditory receptors.

- Sound waves pass through the **external auditory canal** and cause the **tympanic membrane** (eardrum) to vibrate. The ear bones—**malleus, incus,** and **stapes**—transmit and amplify the vibrations through the middle ear.

- Vibrations pass through the **oval window** to fluid within the vestibular canal. Pressure waves press on the membranes that separate the three ducts of the cochlea.

- The bulging of the **round window** serves as an escape valve for the pressure. The pressure waves cause movements of the **basilar membrane.** These movements stimulate the hair cells of the organ of Corti by rubbing them against the overlying **tectorial membrane.**

- Neural impulses are initiated in the dendrites of neurons that lie at the base of each hair cell and are transmitted by the **cochlear nerve** to the brain.

ThomsonNOW™ **Watch the human ear in action by clicking on the figures in ThomsonNOW.**

9 Compare the structure and function of the receptors of taste and smell, and describe the function of the vomeronasal organ (page 906).

- **Chemoreceptors** transduce certain kinds of chemical compounds, allowing **gustation,** or taste, and **olfaction,** or smell.

- In vertebrates, taste receptor cells are specialized epithelial cells in **taste buds.** The **olfactory epithelium** contains specialized olfactory cells with axons that extend to the brain as fibers of the **olfactory nerves.**

- When a molecule binds with a receptor on a taste receptor cell or an olfactory receptor cell, a signal transduction process involving a G protein is initiated.

- Receptors that make up the **vomeronasal organ,** located in the nasal epithelium of mammals, detect **pheromones,** small, volatile signaling molecules released by animals.

ThomsonNOW™ **Learn more about taste buds and olfactory epithelium by clicking on the figures in ThomsonNOW.**

10 Contrast simple eyes, compound eyes, and vertebrate eyes (page 909).
- **Photoreceptors** transduce light energy and serve as the sensory receptors in eyespots and eyes.
- **Eyespots,** or **ocelli,** found in cnidarians and flatworms, detect light but do not form images.
- The **compound eye** of insects and crustaceans consists of visual units called **ommatidia,** which collectively produce a **mosaic** image. Each ommatidium has a transparent **lens** and a **crystalline cone** that focus light onto receptor cells called **retinular cells.**
- We can compare the vertebrate eye to an old-fashioned camera.

11 Describe the functions of each structure of the vertebrate eye (page 909).
- In the human eye, light enters through the **cornea,** is focused by the lens, and produces an image on the **retina.** The **iris** regulates the amount of light that can enter.

ThomsonNOW™ **Learn more about the human eye and the organization of the retina by clicking on the figures in ThomsonNOW.**

12 Describe the events that take place in human vision; compare the two types of photoreceptors, and describe the signal transduction pathway as part of your explanation (page 909).

- The retina contains the photoreceptor cells: **rods,** which function in dim light and form images in black and white; and **cones,** which function in bright light and permit color vision.
- The retina also contains **bipolar cells** that send signals to **ganglion cells.** Two types of lateral interneurons integrate information: **Horizontal cells** receive signals from the rods and cones and send signals to bipolar cells, and **amacrine cells** receive signals from bipolar cells and send signals back to bipolar cells and to ganglion cells.
- When the environment is dark, ion channels in the plasma membranes of rod cells are open and the cells are depolarized; they release glutamate, which hyperpolarizes the membranes of bipolar cells so they do not send signals.
- When light strikes the photopigment **rhodopsin** in the rod cells, its retinal portion changes shape and initiates a signal transduction process that involves **transducin.** This G protein activates an esterase that hydrolyzes cGMP, reducing its concentration. As a result, ion channels close and the membrane becomes hyperpolarized. The rod cells release less glutamate, and fewer signals are transmitted. As a result, bipolar cells become depolarized and release neurotransmitter that stimulates ganglion cells.
- Axons of the ganglion cells make up the **optic nerves.** The optic nerves transmit information to the **lateral geniculate nuclei** in the thalamus. From there, neurons project to the **primary visual cortex** and then to integration centers in the cerebral cortex.

ThomsonNOW™ **See how vision disorders change the eye by clicking on the figure in ThomsonNOW.**

TEST YOUR UNDERSTANDING

1. Which of the following is *not* true of a sensory receptor? (a) detects a stimulus in the environment (b) converts energy of the stimulus into electrical energy (c) produces a receptor potential (d) produces a graded response (e) interprets sensory stimuli

2. A sensory receptor absorbs energy from some stimulus. The next step is (a) release of neurotransmitter (b) transmission of an action potential (c) energy transduction (d) transmission of a receptor potential (e) sensory adaptation

3. Interoceptors (a) help maintain homeostasis (b) are located in the skin (c) are photoreceptors (d) adapt rapidly (e) convert absorbed energy directly into action potentials

4. Which of the following are *not* correctly matched? (a) mechanoreceptors—touch, pressure (b) electroreceptors—voltage (c) photoreceptors—light (d) chemoreceptors—gravity (e) nociceptors—extreme temperature

5. A Pacinian corpuscle (a) is a mechanoreceptor (b) responds to heat (c) responds to pressure (d) a, b, and c (e) a and c

6. Which of the following is *not* located in the vertebrate inner ear? (a) vestibular apparatus (b) cochlea (c) malleus (d) organ of Corti (e) utricle

7. The auditory receptors are located in the (a) utricle (b) organ of Corti (c) vestibular apparatus (d) saccule (e) semicircular canals

8. Which of the following is/are associated with informing the brain about turning movements? (a) semicircular canals (b) saccule (c) lymph (d) otoliths (e) utricle

9. Which of the following receptors do *not* have hair cells? (a) lateral line organ (b) saccule (c) utricle (d) semicircular canals (e) rods

10. In the process of hearing, the basilar membrane vibrates. Which event occurs next? (a) tympanic membrane vibrates (b) bones in middle ear amplify and conduct vibrations (c) cochlear nerve transmits impulses to organ of Corti (d) hair cells in organ of Corti are stimulated (e) vibrations conducted to chemoreceptors

11. In olfaction, (a) pheromones stimulate the taste buds (b) the number of odorous molecules determines the magnitude of the receptor potential (c) chemical signals are converted directly into electrical signals (d) a G protein is activated and leads to the closing of gated channels (e) the concentration

of cyclic AMP increases, leading to hyperpolarization of the receptor

12. The vomeronasal organ (a) detects pheromones (b) is located at the back of the tongue (c) sends signals directly to the thalamus (d) releases pheromones (e) detects signals from predators

13. Mechanical stimulation of stereocilia (a) decreases the number of action potentials (b) inhibits glutamate secretion (c) causes voltage changes (d) degrades opiates (e) causes rhodopsin release

14. In the human visual pathway, after light passes through the cornea, it (a) stimulates ganglion cells (b) passes through the lens (c) sends signals through the optic nerve (d) depolarizes horizontal cells (e) hyperpolarizes rod cells

15. Cones (a) are most concentrated in the ganglion area (b) are more numerous than rods (c) are responsible for vision in bright light (d) are found in all animals with photoreceptors (e) are a type of bipolar cell

16. Rhodopsin (a) is concentrated in bipolar cells (b) binds with a G protein (c) changes shape when the membrane of the cone is depolarized (d) sends signals to the lateral geniculate nuclei (e) activates substance P

CRITICAL THINKING

1. If all neurons transmit the same type of message, how do you know the difference between sound and light? How are you able to distinguish between an intense pain and a mild one? How are these discriminations adaptive?

2. Connoisseurs can recognize many varieties of cheese or wine by "tasting." How can this be, when there are only a few types of taste receptors? How might it be adaptive for organisms to have only four or five basic tastes?

3. Design a synaptic mechanism by which the same neurotransmitter has an excitatory effect on one bipolar cell and an inhibitory effect on another.

4. **Evolution Link.** Proteins known as opsins are present in the photoreceptors of all light-sensitive receptors that have been studied. What does this suggest about the relationships of organisms as diverse as amoebas, flatworms, crayfish, and elephants?

5. **Evolution Link.** Once the ability to produce images evolved, eyes evolved relatively rapidly. Hypothesize why eyes evolved quickly and are now present in the majority of animal groups.

 **Additional questions are available in ThomsonNOW at www.thomsonedu.com/login**

Internal Transport

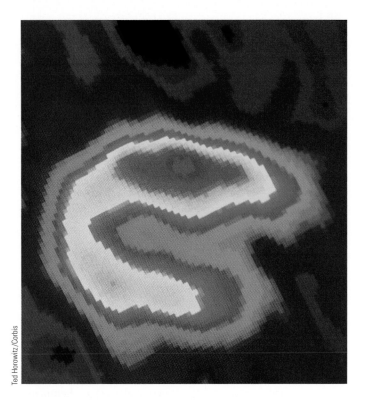

Ted Horowitz/Corbis

PET scan visualizing the left ventricle of the human heart. The scan measures metabolic activity based on glucose levels in the cardiac muscle tissue, providing information about the condition of the heart. (The color progression from least to most active is blue, green, yellow, red.)

KEY CONCEPTS

A circulatory system typically consists of blood, a heart, and a system of blood vessels or spaces through which blood circulates.

Arthropods and most mollusks have an open circulatory system in which blood bathes the tissues directly; some invertebrates and all vertebrates have a closed circulatory system in which blood flows through a continuous circuit of blood vessels.

The vertebrate circulatory system transports nutrients, oxygen, wastes, and hormones; helps maintain fluid balance, appropriate pH, and body temperature; and defends the body against disease.

During the evolution of terrestrial vertebrates, adaptations in circulatory system structures separated oxygen-rich from oxygen-poor blood. The four-chambered hearts and double circuits of endothermic birds and mammals completely separate oxygen-rich blood from oxygen-poor blood.

The lymphatic system helps maintain fluid homeostasis by returning interstitial fluid to the blood.

Most cells need a continuous supply of nutrients and oxygen and the removal of waste products. In very small animals, these metabolic needs are met by simple **diffusion,** the net movement of particles from a region of higher concentration to a region of lower concentration as a result of random motion. A molecule can diffuse 1 micrometer (μm) in less than 1 millisecond (msec), so diffusion is adequate over microscopic distances. In invertebrates that are only a few cells thick, diffusion is an effective mechanism for distributing materials to and from their cells. Consequently, many small, aquatic invertebrates have no circulatory system. Fluid between the cells, called **interstitial fluid,** or tissue fluid, bathes the cells and provides a medium for diffusion of oxygen, nutrients, and wastes.

How much time diffusion requires increases with the square of the distance over which diffusion occurs. A cell that is 10 μm away from its oxygen (or nutrient) supply can receive oxygen by diffusion in about 50 msec, but a cell that is 1000 μm (1 mm) away from its oxygen supply would have to wait several minutes and could not survive if it had to depend on diffusion alone. In animals that are many cells thick, specialized **circulatory systems** transport oxygen, nutrients, hormones, and other materials to the interstitial fluid surrounding all the cells and remove metabolic wastes. A circulatory system reduces the diffusion distance that needed materials must travel. In most animals, a circulatory system interacts with every other organ system in the body.

The positron-emission tomography (PET) scan shown here visualizes the left ventricle of the human heart. The human circulatory system, known as the **cardiovascular system,** is the focus of extensive research because cardiovascular disease is the leading cause of death in the United States and most other industrial societies. A major risk factor for cardiovascular disease is elevated levels of cholesterol and low-density lipoprotein (LDL) in the blood. In contrast, high-density lipoprotein (HDL) seems to play a protective role, removing excess cholesterol from the blood and tissues. Cells in the liver and certain other organs bind with the HDL, remove the cholesterol, and use it in synthesizing needed compounds.

Researchers have identified a gene in mice that codes for an HDL receptor. When investigators knock out this gene, cholesterol levels more than double. As new receptors and mechanisms involved in lipid transport and metabolism are discovered, they serve as targets for new drugs and other treatments for cardiovascular disease. ∎

TYPES OF CIRCULATORY SYSTEMS

Learning Objective

1   Compare and contrast internal transport in animals with no circulatory system, those with an open circulatory system, and those with a closed circulatory system.

No specialized circulatory structures are present in sponges, cnidarians (such as jellyfish), ctenophores (comb jellies), flatworms, or nematodes (roundworms). In cnidarians, the central gastrovascular cavity serves as a circulatory organ as well as a digestive organ (∎ Fig. 43-1a). As the animal stretches and contracts, body movements stir up the contents of the gastrovascular cavity and help distribute nutrients.

The flattened body of the flatworm permits effective gas exchange by diffusion (∎ Fig. 43-1b). Its branched gastrovascular cavity brings nutrients close to all the cells. As in cnidarians, circulation is aided by contractions of the body wall muscles, which agitate the fluid in the gastrovascular cavity. The branching excretory system of planarians provides for internal transport of metabolic wastes that are then expelled from the body.

Fluid in the body cavity of nematodes and other pseudocoelomate animals helps circulate materials. Nutrients, oxygen, and wastes dissolve in this fluid and diffuse through it to and from the cells. Body movements of the animal move the fluid and thus distribute these materials.

Larger animals require a circulatory system to efficiently distribute materials. A circulatory system typically has the following components: (1) **blood,** a connective tissue consisting of cells and cell fragments dispersed in fluid known as *plasma;* (2) a pumping organ, generally a **heart;** and (3) a system of **blood vessels** or spaces through which blood circulates. Two main types of circulatory systems are open and closed systems.

Many invertebrates have an open circulatory system

Arthropods and most mollusks have an **open circulatory system,** in which the heart pumps blood into vessels that have open ends. Their blood and interstitial fluid are indistinguishable and are collectively referred to as **hemolymph.** This fluid spills out of the open ends of the blood vessels, filling large spaces, called *sinuses.* The sinuses make up the **hemocoel** (blood cavity), which is not part of the coelom. (In arthropods and mollusks, the coelom is reduced.) The hemolymph bathes the cells of the body directly. Hemolymph re-enters the circulatory system through openings in the heart (in arthropods) or through open-ended vessels that lead to the gills (in mollusks).

In the open circulatory system of most mollusks, the heart typically has three chambers (∎ Fig. 43-2a). The two *atria* receive hemolymph from the gills. Then the single *ventricle* pumps oxygen-rich hemolymph into blood vessels that conduct it into the large sinuses of the hemocoel. After bathing the body cells, the hemolymph passes into vessels that lead to the gills, where it is recharged with oxygen. The hemolymph then returns to the heart.

Some mollusks, as well as arthropods, have a hemolymph pigment, **hemocyanin,** containing copper that transports oxygen. When oxygenated, hemocyanin is blue and imparts a bluish color to the hemolymph of these animals (the original blue bloods!).

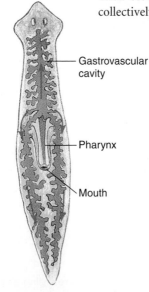

(a) In *Hydra* and other cnidarians, nutrients circulate through the gastrovascular cavity and come in contact with the inner layer of body cells. Nutrients diffuse the short distance to the outer layer of cells.

Gastrovascular cavity

Pharynx

Mouth

(b) In planarian flatworms, the branched gastrovascular cavity allows nutrients to come within close proximity of most body cells.

Figure 43-1   Invertebrates with no circulatory system

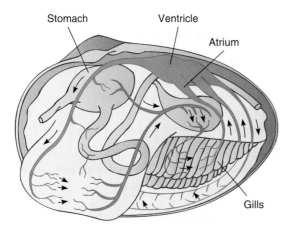

Stomach   Ventricle   Atrium   Gills

(a) In most mollusks, hemolymph enters vessels that conduct it to the gills, where it is recharged with oxygen. Hemolymph then returns to the heart.

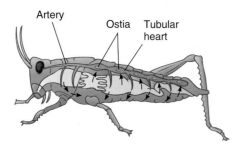

Artery   Ostia   Tubular heart

(b) In arthropods, a tubular heart pumps hemolymph into arteries that deliver it to the sinuses of the hemocoel. After circulating, hemolymph re-enters the heart through ostia in the heart wall.

Figure 43-2 *Animated* Open circulatory systems

In mollusks and arthropods, a heart pumps the blood into arteries that end in sinuses of the hemocoel. Hemolymph circulates through the hemocoel.

In arthropods, a tubular heart pumps hemolymph into arteries, blood vessels that deliver it to the sinuses of the hemocoel (∥ Fig. 43-2b). Hemolymph then circulates through the hemocoel, eventually returning to the pericardial cavity surrounding the heart. Hemolymph enters the heart through ostia, tiny openings equipped with valves that prevent backflow. The rate of hemolymph circulation increases when the animal moves. Thus, when an animal is active and most needs nutrients for fuel, its own movement ensures effective circulation.

In crayfish and other crustaceans, gas exchange takes place as hemolymph circulates through the gills. However, an open circulatory system cannot provide enough oxygen to maintain the active lifestyle of insects. Hemolymph mainly distributes nutrients and hormones in insects. Oxygen diffuses directly to the cells through a system of air tubes (tracheae) that make up the respiratory system (see Figure 45-2).

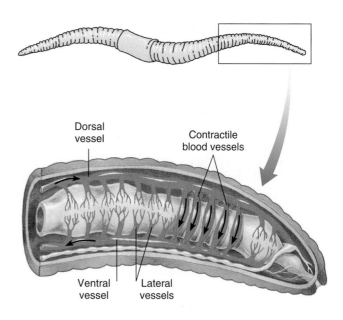

Dorsal vessel   Contractile blood vessels   Ventral vessel   Lateral vessels

Figure 43-3 *Animated* Closed circulatory system of the earthworm

Blood circulates through a continuous system of blood vessels. Five pairs of contractile blood vessels deliver blood from the dorsal vessel to the ventral vessel.

Some invertebrates have a closed circulatory system

Annelids, some mollusks (cephalopods), and echinoderms have a **closed circulatory system.** In them, blood flows through a continuous circuit of blood vessels. The walls of the smallest blood vessels, the *capillaries,* are thin enough to permit diffusion of gases, nutrients, and wastes between blood in the vessels and the interstitial fluid that bathes the cells.

A rudimentary closed circulatory system characterizes proboscis worms (phylum Nemertea). This system consists of a complete network of blood vessels but no heart. Blood flow depends on movements of the animal and on contractions in the walls of large blood vessels.

Earthworms and other annelids have a closed circulatory system (∥ Fig. 43-3). Two main blood vessels extend lengthwise in the body. The ventral vessel conducts blood posteriorly, and the dorsal vessel conducts blood anteriorly. Dorsal and ventral vessels are connected by lateral vessels in every segment. Branches of the lateral vessels deliver blood to the surface, where it is oxygenated. In the anterior part of the worm, five pairs of contractile blood vessels (sometimes called "hearts") connect dorsal and ventral vessels. Contractions of these paired vessels and of the dorsal vessel, as well as contraction of the body wall muscles, circulate the blood. Earthworms have **hemoglobin,** the same red pigment that transports oxygen in vertebrate blood. However, their hemoglobin is not contained within red blood cells but is dissolved in the blood plasma.

Although other mollusks have an open circulatory system, the fast-moving cephalopods, such as the squid and octopus, require a more efficient means of internal transport. They have a closed system made even more effective by accessory "hearts" at the base of the gills, which speed passage of blood through the gills.

Vertebrates have a closed circulatory system

All vertebrates have a ventral, muscular heart that pumps blood into a closed system of blood vessels. **Capillaries,** the tiniest blood vessels, have very thin walls that permit exchange of materials between blood and interstitial fluid. The vertebrate circulatory system consists of heart, blood vessels, blood, lymph, lymph vessels, and associated organs such as the thymus, spleen, and liver. The system performs several functions:

1. Transports nutrients from the digestive system and from storage depots to each cell

2. Transports oxygen from respiratory structures (gills or lungs) to the cells

3. Transports metabolic wastes from each cell to organs that excrete them

4. Transports hormones from endocrine glands to target tissues

5. Helps maintain fluid balance

6. Helps distribute metabolic heat within the body, which helps maintain a constant body temperature in endothermic animals

7. Helps maintain appropriate pH

8. Defends the body against invading microorganisms

Review

▮ How are nutrients and oxygen transported to the body cells in a hydra, flatworm, earthworm, insect, and frog?

▮ How does an open circulatory system differ from a closed circulatory system?

▮ What are five functions of the vertebrate circulatory system?

VERTEBRATE BLOOD

Learning Objectives

2 Compare the structure and function of plasma, red blood cells, white blood cells, and platelets.

3 Summarize the sequence of events involved in blood clotting.

In vertebrates, blood consists of a pale yellowish fluid called **plasma,** in which red blood cells, white blood cells, and platelets are suspended (▮ Fig. 43-4 and ▮ Table 43-1). In humans the total circulating blood volume is 5 L (5.3 qt) in an adult female and about 5.5 L (5.8 qt) in an adult male. About 55% of the blood volume is plasma. The remaining 45% is made up of blood cells and platelets. Because cells and platelets are heavier than plasma, they can be separated from it by centrifugation. Plasma does not separate from blood cells in the body, because the blood is constantly mixed as it circulates in the blood vessels.

Plasma is the fluid component of blood

Plasma consists of water (about 92%), proteins (about 7%), salts, and a variety of materials being transported, such as dissolved gases, nutrients, wastes, and hormones. Plasma is in dynamic equilibrium with the interstitial fluid bathing the cells and with the intracellular fluid. As blood passes through the capillaries, substances continuously move into and out of the plasma. Any deviation from its normal composition signals one or more organs to restore homeostasis.

Plasma contains several kinds of **plasma proteins,** each with specific properties and functions: **fibrinogen;** alpha, beta, and gamma **globulins;** and **albumin.** Fibrinogen is one of the proteins involved in the clotting process. When the proteins involved in blood clotting have been removed from the plasma, the remaining liquid is called **serum.** Alpha globulins include certain

TABLE 43-1

Cell Components of Blood

	Normal Range	Function	Pathology
Red Blood Cells (RBCs)	Male: 4.2–5.4 million/μL Female: 3.6–5.0 million/μL	Oxygen transport; carbon dioxide transport	Too few: anemia Too many: polycythemia
Platelets	150,000–400,000/μL	Essential for clotting	Clotting malfunctions; bleeding; easy bruising
White Blood Cells (WBCs)	5000–10,000/μL		
Neutrophils	About 60% of WBCs	Phagocytosis	Too many: may be due to bacterial infection, inflammation, myelogenous leukemia
Eosinophils	1%–3% of WBCs	Play some role in allergic response	Too many: may result from allergic reaction, parasitic infestation
Basophils	1% of WBCs	May play role in prevention of inappropriate clotting	
Lymphocytes	25%–35% of WBCs	Produce antibodies; destroy foreign cells	Atypical lymphocytes present in infectious mononucleosis; too many may be due to lymphocytic leukemia, certain viral infections
Monocytes	6% of WBCs	Differentiate to form macrophages	May increase in monocytic leukemia and fungal infections

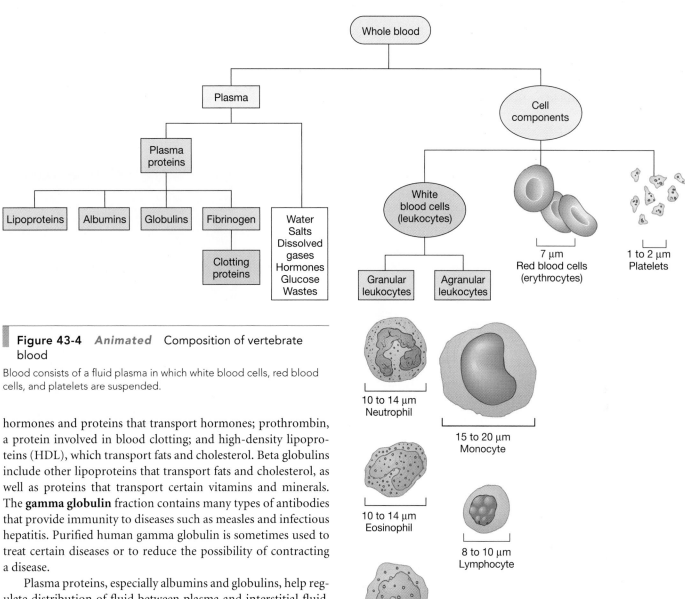

Figure 43-4 *Animated* Composition of vertebrate blood

Blood consists of a fluid plasma in which white blood cells, red blood cells, and platelets are suspended.

hormones and proteins that transport hormones; prothrombin, a protein involved in blood clotting; and high-density lipoproteins (HDL), which transport fats and cholesterol. Beta globulins include other lipoproteins that transport fats and cholesterol, as well as proteins that transport certain vitamins and minerals. The **gamma globulin** fraction contains many types of antibodies that provide immunity to diseases such as measles and infectious hepatitis. Purified human gamma globulin is sometimes used to treat certain diseases or to reduce the possibility of contracting a disease.

Plasma proteins, especially albumins and globulins, help regulate distribution of fluid between plasma and interstitial fluid. Too large to pass readily through the walls of blood vessels, these proteins contribute to the blood's osmotic pressure, which helps maintain an appropriate blood volume. Plasma proteins (along with the hemoglobin in the red blood cells) are also important acid–base buffers. They help maintain the pH of the blood within a narrow range—near its normal, slightly alkaline pH of 7.4.

Red blood cells transport oxygen

Erythrocytes, informally called **red blood cells (RBCs),** are highly specialized for transporting oxygen. In most vertebrates except mammals, circulating RBCs have nuclei. For example, birds have large, oval, nucleated RBCs. In mammals, the nucleus is ejected from the RBC as the cell develops. Each mammalian RBC is a flexible, biconcave disc 7 to 8 μm in diameter and 1 to 2 μm thick. An internal elastic framework maintains the disc shape and permits the cell to bend and twist as it passes through blood vessels even smaller than its own diameter. Its biconcave shape provides a high ratio of surface area to volume, allowing efficient diffusion of oxygen and carbon dioxide into and out of

the cell. In an adult human, about 30 trillion RBCs circulate in the blood, approximately 5 million per μL.

Erythrocytes are produced within the red bone marrow of certain bones: vertebrae, ribs, breastbone, skull bones, and long bones. As an RBC develops, it produces great quantities of hemoglobin, the oxygen-transporting pigment that gives vertebrate blood its red color. (Oxygen transport is discussed in Chapter 45.) The life span of a human RBC is about 120 days. As blood circulates through the liver and spleen, phagocytic cells remove worn-out RBCs from the circulation. These RBCs are then disassembled, and some of their components are recycled. In the human body, more than 2.4 million RBCs are destroyed every second, so an equal number must be produced in the bone marrow to replace them. Red blood cell production is regulated by the

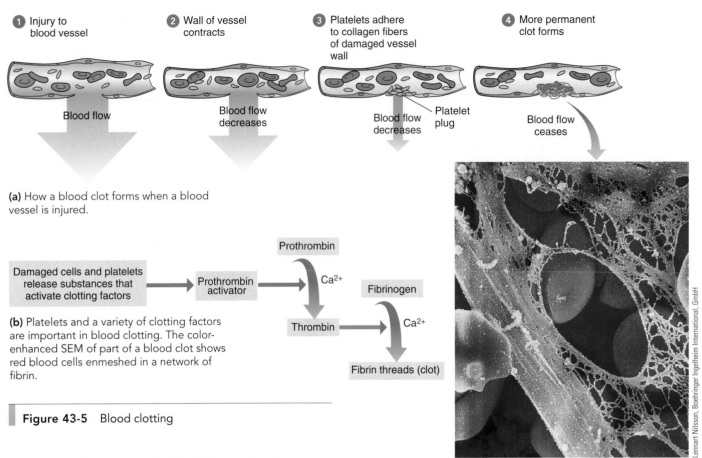

1 Injury to blood vessel

2 Wall of vessel contracts

3 Platelets adhere to collagen fibers of damaged vessel wall

4 More permanent clot forms

Blood flow

Blood flow decreases

Blood flow decreases — Platelet plug

Blood flow ceases

(a) How a blood clot forms when a blood vessel is injured.

Damaged cells and platelets release substances that activate clotting factors → Prothrombin activator

Prothrombin → Ca^{2+} → Thrombin

Fibrinogen → Ca^{2+} → Fibrin threads (clot)

(b) Platelets and a variety of clotting factors are important in blood clotting. The color-enhanced SEM of part of a blood clot shows red blood cells enmeshed in a network of fibrin.

Figure 43-5 Blood clotting

5 µm

hormone **erythropoietin,** which the kidneys release in response to a decrease in oxygen.

Anemia is a deficiency in hemoglobin (often accompanied by a decrease in the number of RBCs). When hemoglobin is insufficient, the amount of oxygen transported is inadequate to supply the body's needs. An anemic person may complain of feeling weak and may become easily fatigued. Three general causes of anemia are (1) loss of blood from hemorrhage or internal bleeding; (2) decreased production of hemoglobin or red blood cells, as in iron-deficiency anemia or pernicious anemia (which can be caused by vitamin B_{12} deficiency); and (3) increased rate of RBC destruction—the **hemolytic anemias,** such as sickle cell anemia (see Chapter 16).

White blood cells defend the body against disease organisms

The **leukocytes,** or **white blood cells (WBCs),** are specialized to defend the body against harmful bacteria and other microorganisms. Leukocytes are amoeba-like cells capable of independent movement. Some types routinely slip through the walls of blood vessels and enter the tissues. Human blood contains five kinds of leukocytes, classified as either granular or agranular (see Fig. 43-4). Both types are manufactured in the red bone marrow.

The **granular leukocytes** are characterized by large, lobed nuclei and distinctive granules in their cytoplasm. The three varieties of granular leukocytes are the neutrophils, eosinophils, and basophils. **Neutrophils,** the principal phagocytic cells in the

blood, are especially adept at seeking out and ingesting bacteria. They also phagocytose dead cells, a cleanup task especially demanding after injury or infection. Most granules in neutrophils contain enzymes that digest ingested material.

Eosinophils have large granules that stain bright red with eosin, an acidic dye. The lysosomes of these WBCs contain enzymes such as oxidases and peroxidases, which suggests that they function in detoxifying foreign proteins and other substances. Eosinophils increase in number during allergic reactions and during parasitic (for example, tapeworm) infestations.

Basophils exhibit deep blue granules when stained with basic dyes. Like eosinophils, these cells play a role in allergic reactions. Basophils do not contain lysosomes. Granules in their cytoplasm contain **histamine,** a substance that dilates blood vessels and makes capillaries more permeable. Basophils release histamine in injured tissues and in allergic responses. Other basophil granules contain **heparin,** an anticoagulant that helps prevent blood from clotting inappropriately within the blood vessels.

Agranular leukocytes lack large, distinctive granules, and their nuclei are rounded or kidney shaped. Two types of agranular leukocytes are lymphocytes and monocytes. Some **lymphocytes** are specialized to produce antibodies, whereas others directly attack foreign invaders such as bacteria or viruses (discussed in Chapter 44).

Monocytes are the largest WBCs, reaching 20 μm in diameter. After circulating in the blood for about 24 hours, a monocyte leaves the circulation and completes its development in the tissues. The monocyte greatly enlarges and becomes a **macrophage,** a giant scavenger cell. All macrophages found in the tissues develop this way. Macrophages voraciously engulf bacteria, dead cells, and debris.

Human blood normally has about 7000 WBCs per μL of blood (only 1 for every 700 RBCs). During bacterial infections the number may rise sharply, so a WBC count is useful in diagnosis. The proportion of each kind of WBC is determined by a differential WBC count. The normal distribution of WBCs is indicated in Table 43-1.

Leukemia is a form of cancer in which any one of the various kinds of WBCs multiplies rapidly within the bone marrow. Many of these cells do not mature, and their large numbers crowd out developing RBCs and platelets, leading to anemia and impaired clotting. A common cause of death from leukemia is internal hemorrhaging, especially in the brain. Another frequent cause of death is infection; although the WBC count may rise dramatically, the cells are immature and abnormal and cannot defend the body against disease organisms.

Platelets function in blood clotting

In most vertebrates other than mammals, the blood contains small, oval, nucleated cells called **thrombocytes** that function in blood clotting. Mammals have **platelets,** tiny spherical or disc-shaped bits of cytoplasm that lack nuclei. About 300,000 platelets per μL are present in human blood. Platelets are pinched off from very large cells in the bone marrow. Thus, a platelet is not a whole cell but a fragment of cytoplasm enclosed by a membrane.

Platelets function in blood clotting and may also stimulate the immune system. When a blood vessel is cut, it constricts to reduce blood loss. Platelets physically patch the break by sticking to the rough, cut edges of the vessel. As platelets begin to gather, they release substances that attract other platelets. The platelets become sticky and adhere to collagen fibers in the blood vessel wall. Within about 5 minutes after injury, they form a platelet plug, or temporary clot.

At the same time that the temporary clot forms, a stronger, more permanent clot begins to develop. More than 30 chemical substances interact in this complex process. The series of reactions that leads to clotting is triggered when one of the clotting factors in the blood is activated by contact with the injured tissue. In **hemophilia,** one clotting factor is absent, as a result of an inherited genetic mutation (see Chapter 16).

Prothrombin, a plasma protein manufactured in the liver, requires vitamin K for its production. In the presence of clotting factors, calcium ions, and compounds released from platelets, prothrombin is converted to **thrombin.** Then thrombin catalyzes the conversion of the soluble plasma protein *fibrinogen* to an insoluble protein, **fibrin.** Once formed, fibrin polymerizes, producing long threads that stick to the damaged surface of the blood vessel and form the webbing of the clot. These threads trap blood cells and platelets, which help strengthen the clot. The clotting process is summarized in ▌ Figure 43-5.

Review

▌ What are the functions of the main groups of plasma proteins?

▌ What is the function of red blood cells? Of neutrophils?

▌ What are the major steps in blood clotting?

VERTEBRATE BLOOD VESSELS

Learning Objective

4   Compare the structure and function of different types of blood vessels, including arteries, arterioles, capillaries, and veins.

The vertebrate circulatory system includes three main types of blood vessels: arteries, capillaries, and veins (▌ Fig. 43-6). An **artery** carries blood away from a heart chamber, toward other tissues. When an artery enters an organ, it divides into many smaller branches called **arterioles.** The arterioles deliver blood into the microscopic capillaries. After blood circulates through an organ, capillaries merge to form **veins** that channel the blood back toward the heart.

The wall of an artery or vein has three layers (see Fig. 43-6b). The innermost layer, which lines the blood vessel, consists mainly of **endothelium,** a tissue that resembles squamous epithelium (see Chapter 38). The middle layer is connective tissue and smooth muscle cells, and the outer coat is connective tissue rich in elastic and collagen fibers.

Gases and nutrients cannot pass through the thick walls of arteries and veins. Materials are exchanged between the blood and interstitial fluid bathing the cells through the capillary walls, which are only one cell thick. Capillary networks in the body are so extensive that at least one of these tiny vessels is located close to every cell in the body. The total length of all capillaries in the body has been estimated to be more than 96,000 km!

Smooth muscle in the arteriole wall can constrict (**vasoconstriction**) or relax (**vasodilation**), changing the radius of the arteriole. Such changes help maintain appropriate blood pressure and help control the volume of blood passing to a particular tissue. Changes in blood flow are regulated by the nervous system in response to the metabolic needs of the tissue, as well as by the demands of the body as a whole. For example, when a tissue is metabolizing rapidly, it needs more nutrients and oxygen. During exercise, arterioles within skeletal muscles dilate, increasing by more than 10-fold the amount of blood flowing to these muscle tissues.

If all your blood vessels dilated at the same time, you would not have enough blood to fill them completely. Normally your liver, kidneys, and brain receive the lion's share of blood. However, if an emergency suddenly occurred requiring rapid action, your blood would be rerouted quickly in favor of heart and muscles. At such a time the digestive system and kidneys can do with less blood, because they are not crucial in responding to the crisis.

The small vessels that directly link arterioles with venules (small veins) are **metarterioles.** The true capillaries branch off from the metarterioles and then rejoin them (▌ Fig. 43-7). True

Figure 43-6 *Animated* Blood flow through blood and lymphatic vessels

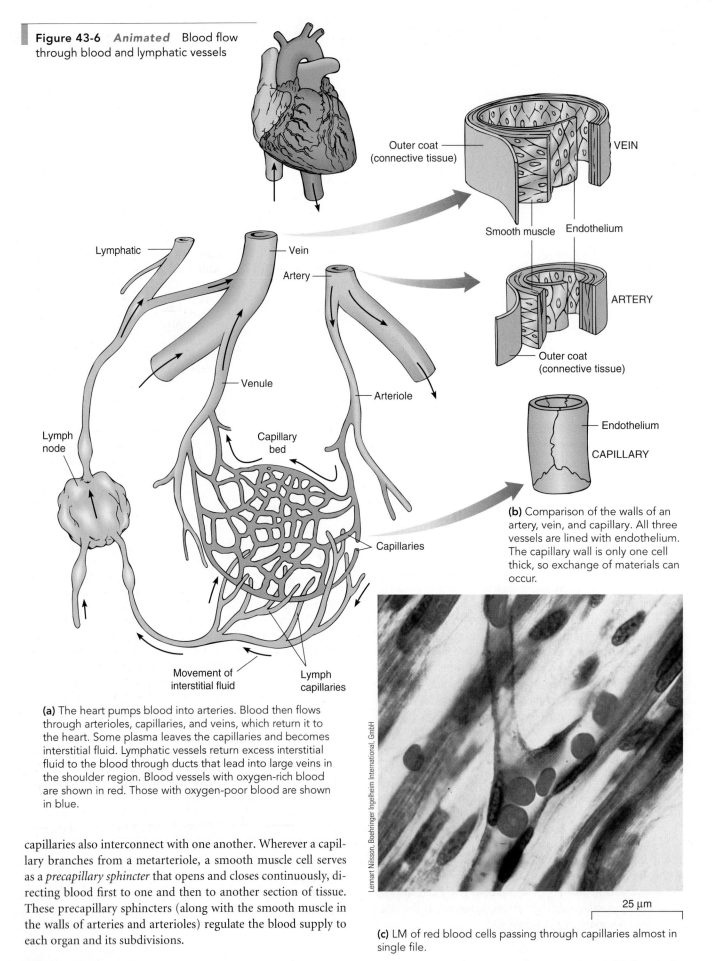

Lymphatic

Vein

Artery

Venule

Arteriole

Lymph node

Capillary bed

Capillaries

Movement of interstitial fluid

Lymph capillaries

Outer coat (connective tissue)

VEIN

Smooth muscle

Endothelium

ARTERY

Outer coat (connective tissue)

Endothelium

CAPILLARY

(b) Comparison of the walls of an artery, vein, and capillary. All three vessels are lined with endothelium. The capillary wall is only one cell thick, so exchange of materials can occur.

(a) The heart pumps blood into arteries. Blood then flows through arterioles, capillaries, and veins, which return it to the heart. Some plasma leaves the capillaries and becomes interstitial fluid. Lymphatic vessels return excess interstitial fluid to the blood through ducts that lead into large veins in the shoulder region. Blood vessels with oxygen-rich blood are shown in red. Those with oxygen-poor blood are shown in blue.

25 μm

Lennart Nilsson, Boehringer Ingelheim International, GmbH

(c) LM of red blood cells passing through capillaries almost in single file.

capillaries also interconnect with one another. Wherever a capillary branches from a metarteriole, a smooth muscle cell serves as a *precapillary sphincter* that opens and closes continuously, directing blood first to one and then to another section of tissue. These precapillary sphincters (along with the smooth muscle in the walls of arteries and arterioles) regulate the blood supply to each organ and its subdivisions.

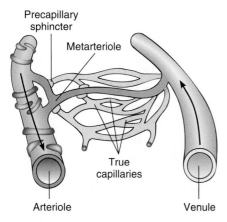

Precapillary sphincter

Metarteriole

True capillaries

Arteriole   **Venule**

(a) Sphincters closed. When a tissue is inactive, only its metarterioles are open.

Precapillary sphincter

Metarteriole

True capillaries

Arteriole   **Venule**

(b) Sphincters open. When the tissue becomes active, decreased oxygen tension in the tissue relaxes the precapillary sphincters, and the capillaries open. This process increases the blood supply and thus the delivery of nutrients and oxygen to the active tissue.

Figure 43-7   Blood flow through a capillary network

As a tissue becomes active, the pattern of blood flow through its capillary networks changes.

Review

- Compare the functions of arteries, capillaries, and veins.
- How do arterioles function in maintaining homeostasis?

EVOLUTION OF THE VERTEBRATE CARDIOVASCULAR SYSTEM

Learning Objective

5   Trace the evolution of the vertebrate cardiovascular system from fish to mammal.

The vertebrate cardiovascular system became modified in the course of evolution, as the site of gas exchange shifted from gills to lungs and as certain vertebrates became active, endothermic animals with higher metabolic rates. The vertebrate heart has one or two **atria** (sing., *atrium*), chambers that receive blood returning from the tissues, and one or two **ventricles** that pump blood into the arteries (❚ Fig. 43-8). Vertebrates in some groups have additional chambers.

The fish heart has one atrium and one ventricle. The atrium pumps blood into the ventricle, which pumps blood into a single circuit of blood vessels. Blood is oxygenated as it passes through capillaries in the gills. After blood circulates through the gill capillaries, its pressure is low, so blood passes very slowly to the other organs. The fish's swimming movements facilitate circulation. Blood returning to the heart has a low oxygen content. A thin-walled sinus venosus receives blood returning from the tissues and pumps it into the atrium.

In amphibians, blood flows through a double circuit: the **pulmonary circulation** and the **systemic circulation.** Oxygen-rich

and oxygen-poor blood are kept somewhat separate. The amphibian heart has two atria and one ventricle. A *sinus venosus* collects oxygen-poor blood returning from the veins and pumps it into the right atrium. Blood returning from the lungs passes directly into the left atrium.

Both atria pump into the single ventricle, but oxygen-poor blood is pumped out of the ventricle before oxygen-rich blood enters it. Blood passes into an artery, the *conus arteriosus,* equipped with a fold that helps keep the blood separate. Much of the oxygen-poor blood is directed into the pulmonary circulation, which delivers it to the lungs and skin, where it is recharged with oxygen. The systemic circulation delivers oxygen-rich blood into arteries that conduct it to the various tissues of the body.

Most reptiles also have a double circuit of blood flow, made more efficient by a wall that partly divides the ventricle. Mixing of oxygen-rich and oxygen-poor blood is minimized by the timing of contractions of the left and right sides of the heart and by pressure differences. In crocodilians (crocodiles and alligators), the wall between the ventricles is complete, so the heart consists of two atria and two ventricles. Thus, a four-chambered heart first evolved among the reptiles.

Unlike birds and mammals, amphibians and reptiles do not ventilate their lungs continuously. Therefore, it would be inefficient to pump blood through the pulmonary circulation continuously. The shunts between the two sides of the heart allow blood to be distributed to the lungs as needed.

In all birds and mammals (and in crocodilians), the septum (wall) between the ventricles is complete. Biologists hypothesize that the completely divided heart evolved twice during the course of vertebrate evolution; first in the crocodilian–bird clade and then independently in mammals. The septum (wall) between the ventricles prevents oxygen-rich blood in the left chamber from mixing with oxygen-poor blood in the right chamber. The conus arteriosus has split and become the base of the **aorta** (the largest artery) and the pulmonary artery. No sinus venosus is present as a separate chamber, but a vestige remains as the *sinoatrial node* (the pacemaker).

Complete separation of the right and left sides of the heart requires blood to pass through the heart twice each time it tours the body. The complete double circuit allows birds and mammals to maintain relatively high blood pressures in the systemic circulation and modest pressures in the pulmonary circulation. The higher pressure is necessary for efficient circulation of blood through the body. However, the delicate air sacs and capillaries of the lungs would be damaged by this pressure.

The double circulatory circuit delivers materials to the tissues rapidly and efficiently. Because their blood contains more oxygen

Key Point

The four-chambered heart and double circuit that separates oxygen-rich from oxygen-poor blood are important adaptations that evolved as vertebrates diversified and some became active, terrestrial, endothermic animals.

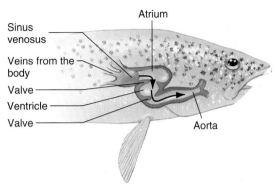

(a) Fishes. The single atrium and ventricle of the fish heart are part of a single circuit of blood flow.

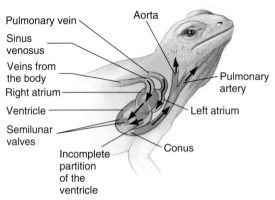

(c) Reptiles. The reptilian heart has two atria and two ventricles. In most reptiles, the wall separating the ventricles is incomplete, so blood from the right and left chambers mixes to some extent. In crocodiles and alligators, the septum is complete and the heart consists of four separate chambers.

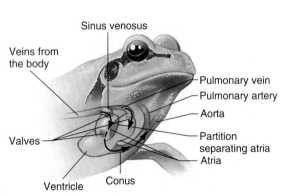

(b) Amphibians. In amphibians, the heart consists of two atria and one ventricle, and blood flows through a double circuit.

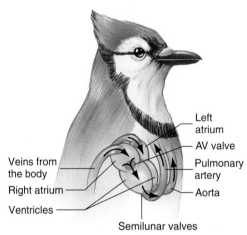

(d) Birds and mammals. In birds and mammals, the two atria and two ventricles separate oxygen-rich blood from oxygen-poor blood.

Figure 43-8 Evolution of the vertebrate cardiovascular system

per unit volume and circulates more rapidly than in other vertebrates, the tissues of birds and mammals receive more oxygen. As a result, these animals can maintain a higher metabolic rate and a constant, high body temperature even in cold surroundings.

The pattern of blood circulation in birds and mammals can be summarized as follows:

Veins (conduct blood from organs) ⟶ right atrium ⟶ right ventricle ⟶ pulmonary arteries ⟶ capillaries in the lungs ⟶ pulmonary veins ⟶ left atrium ⟶ left ventricle ⟶ aorta ⟶ arteries (conduct blood to organs) ⟶ arterioles ⟶ capillaries

Review

■ What are some of the major adaptations that occurred in the evolution of the vertebrate cardiovascular system?

THE HUMAN HEART

Learning Objectives

6 Describe the structure and function of the human heart. (Include the heart's conduction system in your answer.)

7 Trace the events of the cardiac cycle, and relate normal heart sounds to these events.

8 Define *cardiac output,* describe how it is regulated, and identify factors that affect it.

Not much bigger than a fist and weighing less than a pound, the human heart is a remarkable organ that beats about 2.5 billion times in an average lifetime, pumping about 300 million L (80 million gal) of blood. To meet the body's changing needs, the heart can vary its output from 5 L to more than 20 L of blood per minute.

Your heart is a hollow, muscular organ located in the chest cavity directly under the breastbone. Its wall consists mainly of cardiac muscle attached to a framework of collagen fibers. The **pericardium,** a tough connective tissue sac, encloses the heart. A smooth layer of endothelium covers the inner surface of the pericardium and the outer surface of the heart. Between these two surfaces is a small **pericardial cavity** filled with fluid, which reduces friction to a minimum as the heart beats.

A wall, or **septum,** separates the right atrium and ventricle from the left atrium and ventricle. Between the atria, the wall is known as the **interatrial septum;** between the ventricles, the **interventricular septum.** A shallow depression, the **fossa ovalis,** on the interatrial septum marks the place where an opening, the **foramen ovale,** was located in the fetal heart. In the fetus, the foramen ovale lets the blood flow directly from right to left atrium, so very little passes to the nonfunctional lungs. A small muscular pouch, called an **auricle,** lies at the upper surface of each atrium.

To keep blood from flowing backward, the heart has valves that close automatically (▌Fig. 43-9). The valve between the right atrium and right ventricle is called the **right atrioventricular (AV) valve,** or **tricuspid valve.** The **left AV valve** (between the left atrium and left ventricle) is the **mitral valve,** or **bicuspid valve.** The AV valves are held in place by stout cords, or "heartstrings," the **chordae tendineae.** These cords attach the valves to the papillary muscles that project from the walls of the ventricles.

When blood returning from the tissues fills the atria, blood pressure on the AV valves forces them to open into the ventricles, filling the ventricles with blood. As the ventricles contract, blood is forced back against the AV valves, pushing them closed. Contraction of the papillary muscles and tensing of the chordae tendineae prevent the valves from opening backward into the atria. These valves are like swinging doors that open in only one direction.

Semilunar valves (named for their flaps, which are shaped like half-moons) guard the exits from the heart. The semilunar valve between the left ventricle and the aorta is the **aortic valve,** and the one between the right ventricle and the pulmonary artery is the **pulmonary valve.** When blood passes out of the ventricles, the flaps of the semilunar valves are pushed aside and offer no resistance to blood flow. But when the ventricles are relaxing and filling with blood from the atria, blood pressure in the arteries is higher than in the ventricles. Blood then fills the pouches of the valves, stretching them across the artery so blood cannot flow back into the ventricle.

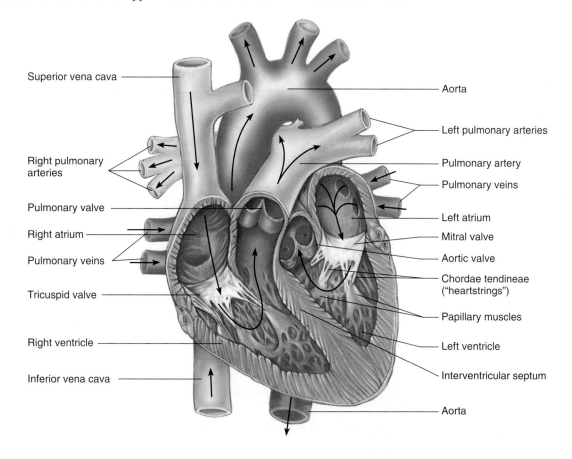

Superior vena cava

Right pulmonary arteries

Pulmonary valve

Right atrium

Pulmonary veins

Tricuspid valve

Right ventricle

Inferior vena cava

Aorta

Left pulmonary arteries

Pulmonary artery

Pulmonary veins

Left atrium

Mitral valve

Aortic valve

Chordae tendineae ("heartstrings")

Papillary muscles

Left ventricle

Interventricular septum

Aorta

▌ **Figure 43-9** *Animated* Section through the human heart showing the valves

Note the right and left atria, which receive blood, and the right and left ventricles, which pump blood into the arteries. Arrows indicate the pattern of blood flow.

Each heartbeat is initiated by a pacemaker

Horror films sometimes feature a scene in which a heart cut out of a human body continues to beat. Scriptwriters of these tales actually have some factual basis for their gruesome fantasies, because a heart carefully removed from the body does continue to beat for many hours if kept in a nutritive, oxygenated fluid. This is possible because the contractions of cardiac muscle begin within the muscle itself and can occur independently of any nerve supply (▌Fig. 43-10a).

At their ends cardiac muscle cells are joined by dense bands called **intercalated discs** (▌Fig. 43-10b and c). Each disc is a type of *gap junction* (see Chapter 5) in which two cells connect through pores. This type of junction is of great physiological importance, because it offers very little resistance to the passage of an action potential. Ions move easily through the gap junctions, allowing the entire atrial (or ventricular) muscle mass to contract as one giant cell.

Compared with skeletal muscle that has action potentials typically lasting 1 to 2 msec, the action potentials of cardiac muscle are much longer, several hundred milliseconds. Voltage-activated Ca^{2+} channels open during depolarization of cardiac muscle fibers. Entry of Ca^{2+} contributes to the longer depolarization time. Another factor is a type of K^+ channel that stays open when the cell is at its resting potential but closes during depolarization, lengthening depolarization time by decreasing the permeability of the membrane to K^+. From experiments on isolated cardiac muscle fibers, investigators determined that spontaneous contraction results from the combination of a slow decrease in K^+ permeability and a slow increase in Na^+ and Ca^{2+} permeability.

A specialized conduction system ensures that the heart beats in a regular and effective rhythm. Each beat is initiated by the **pacemaker,** also called the **sinoatrial (SA) node.** The SA node is a small mass of specialized cardiac muscle in the posterior wall of the right atrium near the opening of a large vein, the superior vena cava. The action potential in the SA node is triggered mainly by the opening of Ca^{2+} channels. Ends of the SA node fibers fuse with ordinary atrial muscle fibers, so each action potential spreads through both atria, producing atrial contraction.

One group of atrial muscle fibers conducts the action potential directly to the **atrioventricular (AV) node,** located in the right atrium along the lower part of the septum. Here transmission is delayed briefly so that the atria finish contracting before the ventricles begin to contract. From the AV node the action potential spreads into specialized muscle fibers that make up the **AV bundle.** The AV bundle divides, sending branches into each ventricle. Fibers of the bundle branches divide further, eventually forming small *Purkinje fibers.* When an impulse reaches the ends of the Purkinje fibers, it spreads through the ordinary cardiac muscle fibers of the ventricles.

SA node ⟶ atrial muscle fibers (atria contract) ⟶ AV node ⟶ AV bundle ⟶ right and left bundle branches ⟶ Purkinje fibers ⟶ ventricular muscle fibers (ventricles contract)

As each wave of contraction spreads through the heart, electric currents flow into the tissues surrounding the heart and onto

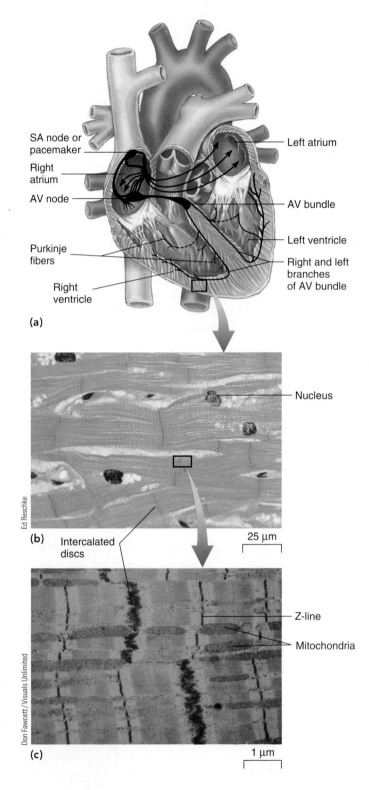

(a)

(b)

Ed Reschke

25 μm

Don Fawcett / Visuals Unlimited

(c)

1 μm

▌ **Figure 43-10** *Animated* Conduction system of the heart

(a) The sinoatrial (SA) node initiates each heartbeat. The action potential spreads through the muscle fibers of the atria, producing atrial contraction. Transmission is briefly delayed at the atrioventricular (AV) node before the action potential spreads through specialized muscle fibers into the ventricles. **(b)** LM of cardiac muscle. **(c)** TEM of cardiac muscle.

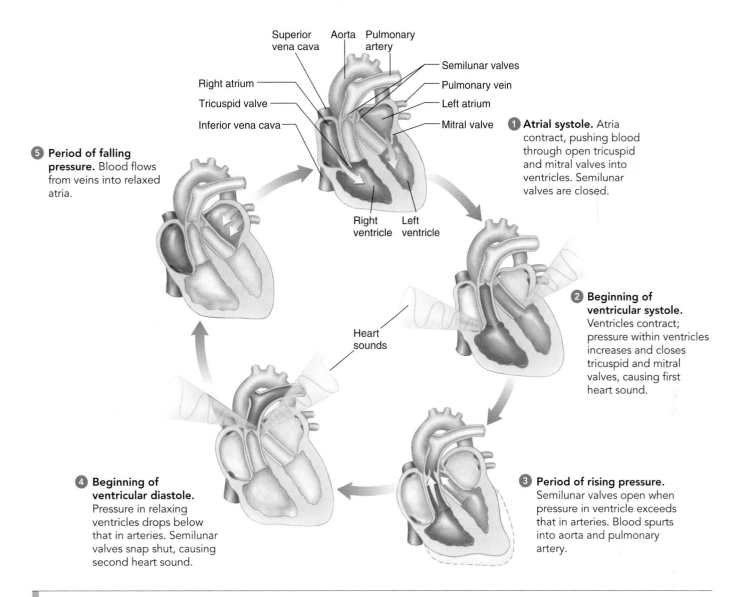

Superior vena cava
Aorta
Pulmonary artery
Semilunar valves
Right atrium
Pulmonary vein
Tricuspid valve
Left atrium
Inferior vena cava
Mitral valve
Right ventricle
Left ventricle

1 Atrial systole. Atria contract, pushing blood through open tricuspid and mitral valves into ventricles. Semilunar valves are closed.

2 Beginning of ventricular systole. Ventricles contract; pressure within ventricles increases and closes tricuspid and mitral valves, causing first heart sound.

3 Period of rising pressure. Semilunar valves open when pressure in ventricle exceeds that in arteries. Blood spurts into aorta and pulmonary artery.

4 Beginning of ventricular diastole. Pressure in relaxing ventricles drops below that in arteries. Semilunar valves snap shut, causing second heart sound.

5 Period of falling pressure. Blood flows from veins into relaxed atria.

Heart sounds

Figure 43-11 *Animated*   The cardiac cycle

The cycle comprises contraction of both atria followed by both ventricles. White arrows indicate the direction of blood flow; dotted lines indicate the change in size as contraction occurs.

the body surface. By placing electrodes on the body surface on opposite sides of the heart, a physician can amplify and record the electrical activity. The graph produced is called an **electrocardiogram** (**EKG** or **ECG**). Abnormalities in the EKG indicate disorders in the heart or its rhythm. For example, in **heart block,** impulse transmission is delayed or blocked at some point in the conduction system. **Artificial pacemakers** can help patients with severe heart block. The pacemaker is implanted beneath the skin and its electrodes connected to the heart. This device provides continuous rhythmic impulses that avoid the block and drive the heartbeat.

Each minute the heart beats about 70 times. One complete heartbeat takes about 0.8 second and is referred to as a **cardiac cycle.** That portion of the cycle in which contraction occurs is known as **systole;** the period of relaxation is **diastole.** Figure 43-11 shows the sequence of events that occur during one cardiac cycle.

You can measure your heart rate by placing a finger over the radial artery in your wrist or the carotid artery in your neck and counting the pulsations for 1 minute. **Arterial pulse** is the alternate expansion and recoil of an artery. Each time the left ventricle pumps blood into the aorta, the elastic wall of the aorta expands to accommodate the blood. This expansion moves in a wave down the aorta and the arteries that branch from it. When this pressure wave passes, the elastic arterial wall snaps back to its normal size.

When you listen to the heartbeat with a stethoscope, you can hear two main heart sounds, "lub-dup," which repeat rhythmically. These sounds result from the heart valves closing. When the valves close, they cause turbulence in blood flow that sets up vibrations in the walls of the heart chambers. The first heart sound, "lub," is low-pitched, not very loud, and fairly long-lasting. It is caused mainly by the closing of the AV (mitral and tricuspid) valves and marks the beginning of ventricular systole. The "lub"

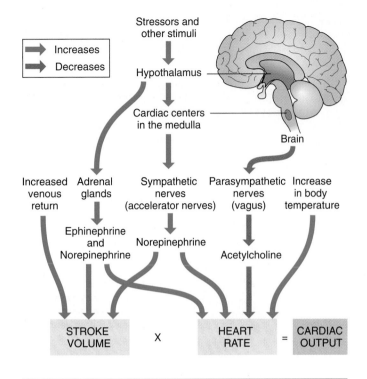

Stressors and
other stimuli

Hypothalamus

Cardiac centers
in the medulla

Brain

Increased venous return | Adrenal glands | Sympathetic nerves (accelerator nerves) | Parasympathetic nerves (vagus) | Increase in body temperature

Ephinephrine and Norepinephrine | Norepinephrine | Acetylcholine

STROKE VOLUME X HEART RATE = CARDIAC OUTPUT

Figure 43-12 Some factors that influence cardiac output

sound is quickly followed by the higher-pitched, louder, sharper, and shorter "dup" sound. Heard almost as a quick snap, the "dup" marks the closing of the semilunar valves and the beginning of ventricular diastole.

The quality of these sounds tells a discerning physician much about the state of the valves. For example, when a valve does not close tightly, blood may flow backward. When the semilunar valves are injured, a soft, hissing noise ("lub-shhh") known as a **heart murmur** is heard in place of the normal sound. Valve deformities are sometimes present at birth, or they may result from certain diseases, such as rheumatic fever or syphilis. Diseased or deformed valves can be surgically repaired or replaced with artificial valves.

The nervous system regulates heart rate

Although the heart can beat independently, its rate is, in fact, carefully regulated by the nervous and endocrine systems (Fig. 43-12). Sensory receptors in the walls of certain blood vessels and heart chambers are sensitive to changes in blood pressure. When stimulated, the receptors send messages to **cardiac centers** in the medulla of the brain. These cardiac centers govern two sets of autonomic nerves that pass to the SA node: parasympathetic and sympathetic nerves.

Parasympathetic and sympathetic nerves have opposite effects on heart rate (see Fig. 41-17). Parasympathetic nerves release the neurotransmitter *acetylcholine,* which slows the heart.

Acetylcholine slows the rate of depolarization by increasing the membrane's permeability to K^+ (Fig. 43-13a). Sympathetic nerves release norepinephrine, which speeds the heart rate and increases the strength of contraction. Norepinephrine stimulates Ca^{2+} channel opening during depolarization (Fig. 43-13b).

Both norepinephrine and acetylcholine act indirectly on ion channels. They activate a *signal transduction* process involving a G protein. Norepinephrine binds to *beta-adrenergic receptors,* one of the two main types of adrenergic receptors. These receptors are targeted by *beta blockers,* drugs that block the actions of norepinephrine on the heart and are used clinically in treating hypertension and other types of heart disease.

In response to physical and emotional stressors, the adrenal glands release epinephrine and norepinephrine, which speed the heart. An elevated body temperature also increases heart rate. During fever, the heart may beat more than 100 times per minute. As you might expect, heart rate decreases when body temperature is lowered. This is the reason that physicians may deliberately lower a patient's temperature during heart surgery.

Stroke volume depends on venous return

The volume of blood one ventricle pumps during one beat is the **stroke volume.** Stroke volume depends mainly on **venous return,** the amount of blood the veins deliver to the heart. According to **Starling's law of the heart,** if the veins deliver more blood to the heart, the heart pumps more blood (within physiological limits). When extra amounts of blood fill the heart chambers, the cardiac muscle fibers stretch more and contract with greater force, pumping a larger volume of blood into the arteries. Norepinephrine released by sympathetic nerves and epinephrine released by the adrenal glands during stress also increase the force of contraction of cardiac muscle fibers.

Cardiac output varies with the body's need

By multiplying the stroke volume by the number of times the left ventricle beats per minute, we can compute the **cardiac output (CO).** The CO is the volume of blood pumped by the left ventricle into the aorta in 1 minute. For example, in a resting adult the heart may beat about 72 times per minute and pump about 70 mL of blood with each contraction.

CO = stroke volume × heart rate (number of ventricular contractions per minute)

= 70 mL/stroke × 72 strokes/min

= 5040 mL/min (about 5 L/min)

Cardiac output varies with changes in either stroke volume or heart rate (see Fig. 43-12).When stroke volume increases, CO increases. The CO varies dramatically with the changing needs of the body. During stress or heavy exercise, the normal heart can increase its CO fivefold so that it pumps 20 to 30 L of blood per minute.

① Parasympathetic neuron releases acetylcholine.

② Acetylcholine binds with receptors on plasma membrane of cardiac muscle.

③ Receptor activates G protein.

④ G protein binds with K⁺ channel, opening it.

⑤ K⁺ leaves the cell, hyperpolarizing the membrane. Action potentials occur more slowly.

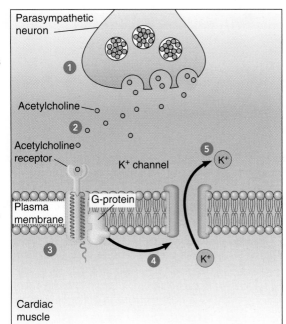

(a) Parasympathetic action on cardiac muscle

① Sympathetic neuron releases norepinephrine.

② Norepinephrine binds with receptors on the plasma membrane of cardiac muscle.

③ Receptor activates G protein.

④ G protein activates adenylyl cyclase, which converts ATP to cyclic AMP.

⑤ Cyclic AMP activates protein kinase.

⑥ Protein kinase phosphorylates Ca²⁺ channels so that they open more easily when neuron is depolarized. Action potentials occur more rapidly.

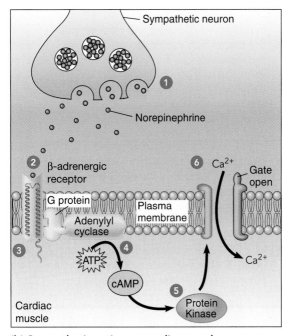

(b) Sympathetic action on cardiac muscle

Figure 43-13 Actions of sympathetic and parasympathetic neurons on cardiac muscle cells
Neurotransmitters released by sympathetic and parasympathetic neurons initiate a signal transduction process.

Review

■ What factors influence cardiac output?
■ What sequence of events occurs during cardiac conduction?
■ How is the heart regulated? (Include a description of the actions of acetylcholine and norepinephrine.)

BLOOD PRESSURE

Learning Objective

9 Identify factors that determine and regulate blood pressure, and compare blood pressure in different types of blood vessels.

Blood pressure is the force exerted by the blood against the inner walls of the blood vessels. It is determined by CO, blood volume, and the resistance to blood flow (▌Fig. 43-14a). When CO increases, the increased blood flow causes a rise in blood pressure. When CO decreases, the decreased blood flow causes a fall in blood pressure. If the volume of blood is reduced by hemorrhage or by chronic bleeding, the blood pressure drops. In contrast, an increase in blood volume results in an increase in blood pressure. For example, a high dietary intake of salt causes water retention. This increases blood volume and raises blood pressure.

Blood flow is impeded by resistance; when the resistance to flow increases, blood pressure rises. **Peripheral resistance** is the resistance to blood flow caused by blood viscosity and by friction between blood and the blood vessel wall. In the blood of a healthy person, viscosity remains fairly constant and is only a minor influence on changes in blood pressure. More important is the friction between blood and the blood vessel wall. The length and diameter of a blood vessel determine the amount of surface area in contact with the blood. Blood vessel length does not change, but the diameter, especially of an arteriole, does. A small change in blood vessel diameter causes a big change in blood pressure. Constriction of blood vessels raises blood pressure; dilation lowers blood pressure.

Blood pressure in arteries rises during systole and falls during diastole. The National Institutes of Health defines normal blood pressure as a systolic pressure of less than 120 and a diastolic pressure of less than 80. An example of a normal blood pressure, measured in the upper arm with a sphygmomanometer, is 110/73 mm of mercury, abbreviated mm Hg. Systolic pressure is indicated by the first number, diastolic by the second.

If you have a blood pressure of 120 to 139 systolic over 80 to 89 diastolic, you are considered *prehypertensive,* and you need to modify your lifestyle to prevent cardiovascular disease and

CARDIOVASCULAR DISEASE

Cardiovascular disease is the leading cause of death in the United States and in most other industrialized countries. Death often results from some complication of **atherosclerosis,**[1] a progressive disease in which the walls of certain arteries are damaged, inflamed, and narrowed as a result of lipid deposits in their walls. Although it can affect almost any artery, the disease most often develops in the aorta and in the coronary and cerebral arteries. When it occurs in the cerebral arteries, less blood is delivered to the brain. This condition can lead to a **cerebrovascular accident (CVA),** commonly referred to as a stroke.

The mechanisms leading to atherosclerosis are not completely understood. A widely accepted hypothesis is that the disease process is a "response to injury." Studies indicate that the injury commonly occurs when excess low-density lipoprotein (LDL), known as "bad" cholesterol, in the blood triggers inflammation in the inner layer of the arterial wall. (Inflammation, discussed in Chapter 44, is one of the body's most important responses to infection or injury.) Other suspected causes are infectious agents and toxins, including those from cigarette smoke.

When excess LDLs are present, they become oxidized and adhere to the wall of the blood vessel. Macrophages take up LDL cholesterol and stimulate a series of events that promotes inflammation. Underlying smooth muscle fibers then migrate to the inner lining of the artery and

---

[1]Atherosclerosis is a type of arteriosclerosis that affects large and medium-sized arteries. *Arteriosclerosis* is a general term for thickening of the walls of arteries.

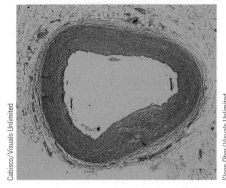

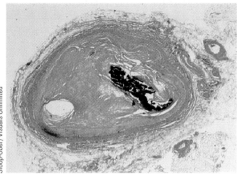

500 μm

500 μm

(a) Normal coronary artery.

(b) This artery is almost completely blocked with atherosclerotic plaque.

Progression of atherosclerosis. LMs of cross sections through two arteries showing changes that take place in atherosclerosis.

multiply there. Collagen and other connective tissue fibers also proliferate, reducing the elasticity of the artery wall. The muscle fibers and collagen form a fibrous cap of tissue over the injured area of the artery. The buildup of fatty substances, smooth muscle fibers, collagen, and sometimes, calcium is called **atherosclerotic plaque.**

As atherosclerotic plaque develops, arteries lose their ability to stretch when filled with blood. As the walls thicken, the diameter of the artery decreases, and it gradually becomes occluded (blocked), as shown in the figure. The occluded vessels deliver less blood to the tissues, which then become **ischemic,** or lacking in blood. Now the tissue is deprived of an adequate oxygen and nutrient supply.

In **ischemic heart disease,** the increased need for oxygen during exercise or emotional stress results in the pain characteristic of **angina pectoris.** Nitroglycerin pills or a nitroglycerin patch dilates veins, decreasing venous return. Cardiac output is lowered so the heart is not working as hard and requires less oxygen. Nitroglycerin also dilates the coronary arteries slightly, allowing more blood to reach the heart muscle. *Calcium ion channel blockers* are drugs that slow the heart by inhibiting the passage of Ca^{2+} into cardiac muscle fibers. They also dilate the coronary arteries.

Myocardial infarction (MI), or heart attack, is a very serious, often fatal, consequence of cardiovascular disease. MI is the leading cause of death and disability

---

stroke. Lifestyle changes that will reduce your risk include exercising, losing excess weight, following a heart-healthy diet, reducing salt intake, limiting alcohol, and not smoking.

If your systolic pressure consistently measures 140 mm Hg or higher or your diastolic pressure measures 90 mm Hg or higher, you have **hypertension,** or high blood pressure. Hypertension is a risk factor for atherosclerosis and other cardiovascular disease (see *Focus On: Cardiovascular Disease*). In hypertension, there is usually increased vascular resistance, especially in the arterioles and small arteries. The heart's workload increases because it must pump against this greater resistance. If this condition persists, the left ventricle enlarges and may deteriorate in function. Heredity, aging, and ethnicity contribute to the development of hypertension.

Blood pressure varies in different blood vessels

As you might guess, blood pressure is greatest in the large arteries, decreasing as blood flows away from the heart and through the smaller arteries and capillaries (▮ Fig. 43-14b). By the time blood enters the veins, its pressure is very low, even approaching zero. Flow rate can be maintained in veins at low pressure, because they are low-resistance vessels. Their diameter is larger than that of corresponding arteries, and their walls have little smooth muscle. Flow of blood through veins depends on several factors, including skeletal muscle movement, which compresses veins. Most veins larger than 2 mm (0.08 in) in diameter that conduct blood

in the United States. In fact, for about 65% of men and almost 50% of women, the first symptom of atherosclerotic cardiovascular disease is MI or sudden cardiac death, death within one hour of onset of the symptoms. MI often results from a sudden decrease in coronary blood supply. The part of cardiac muscle deprived of oxygen dies within a few minutes and is then referred to as an **infarct.** The sudden decrease in blood supply often occurs when inflammation causes the fibrous cap over the atherosclerotic plaque to break open. Platelets adhere to the roughened arterial wall and initiate clotting.

A **thrombus,** a clot that forms within a blood vessel or within the heart, can block a sizable branch of a coronary artery. Blood flow to a portion of heart muscle is impeded or completely halted. When such blockage, referred to as a *coronary occlusion,* prevents blood flow to a large region of cardiac muscle, the heart may stop beating; that is, **cardiac arrest** occurs, and death can follow within moments. If only a small part of the heart is affected, however, the heart may continue to function. Cells in the region deprived of oxygen die and are replaced by scar tissue.

An episode of ischemia may trigger a fatal arrhythmia such as **ventricular fibrillation,** a condition in which the ventricles contract very rapidly without actually pumping blood. The pulse may stop, and blood pressure may fall precipitously. Ventricular fibrillation has been linked with about 65% of cardiac arrests. The only effective treatment for fibrillation is **defibrillation** with electric shock. The shock appears to depolarize every muscle fiber in the heart so that its timing mechanism can reset.

Patients with progressive cardiovascular disease can be treated with *coronary bypass surgery* in which veins from another location in the patient's body are grafted around occluded coronary arteries. The newly positioned blood vessels restore adequate blood flow to the affected area. Another procedure, *coronary angioplasty,* involves inserting a small balloon into an occluded coronary artery. Inflating the balloon breaks up the plaque in the arterial wall and thus widens the vessel.

Approaches to preventing and treating cardiovascular disease include use of anti-inflammatory agents and statins, medications that lower cholesterol level. A molecular marker of inflammation called C-reactive protein (CRP) is used to assess the risk of heart disease. Several major *modifiable* risk factors for cardiovascular disease have been identified:

1. *Elevated LDL–cholesterol levels in the blood:* Elevation of LDL–cholesterol levels is associated with diets rich in total calories, total fats, saturated fats, and cholesterol.

2. *Hypertension:* The higher the blood pressure is, the greater the risk. The hormone angiotensin II, which at abnormally high levels is thought to contribute to hypertension, also stimulates inflammation.

3. *Cigarette smoking:* The risk of developing atherosclerosis is 2 to 6 times as great in smokers as in nonsmokers and is directly proportional to the number of cigarettes smoked daily.

Components of cigarette smoke cause oxidants to form, which may contribute to the inflammation by increasing oxidation of the altered LDL in the arterial wall.

4. *Diabetes mellitus:* In this endocrine disorder, glucose is not metabolized normally. The body shifts to fat metabolism, and there is a marked increase in circulating lipids, leading to atherosclerosis. Impaired glucose tolerance is also a risk factor.

5. *Physical inactivity:* One in four adults in the United States has a sedentary lifestyle and does not engage in regular exercise. Physical inactivity contributes to more than one third of the approximately 500,000 deaths related to heart disease in the United States each year. Exercise reduces the concentrations of triacylglycerols (triglycerides) and cholesterol in the blood (these lipids have been associated with heart disease). At the same time, exercise increases the concentration of high-density lipoproteins (HDLs), which protect against heart disease.

6. *Obesity:* Obesity can affect cholesterol levels and increase the risk of hypertension and diabetes.

The risk of developing cardiovascular disease also increases with age. Other probable risk factors currently being studied are hereditary predisposition, hormone levels, stress and behavior patterns, and dietary factors.

against the force of gravity are equipped with *valves* to prevent backflow (❙ Fig. 43-15). Such valves usually consist of two cusps formed by inward extensions of the vein wall.

When a person stands perfectly still for a long time, as when a soldier stands at attention or a store clerk stands at a cash register, blood tends to pool in the veins. When fully distended, veins can accept no more blood from the capillaries. Pressure in the capillaries increases, and large amounts of plasma are forced out of the circulation through the thin capillary walls. Within just a few minutes, as much as 20% of the blood volume is lost from the circulation—with drastic effect. Arterial blood pressure falls dramatically, reducing blood flow to the brain. Sometimes the resulting lack of oxygen in the brain causes fainting, a protective response. Lying in a prone position increases blood supply to the brain. In fact, lifting a person who has fainted to an upright position can result in circulatory shock and even death.

Blood pressure is carefully regulated

Each time you get up from a horizontal position, your blood pressure changes. Several mechanisms interact to maintain normal blood pressure so that you do not faint when you get out of bed each morning or change position during the day. When blood pressure decreases, sympathetic nerves to the blood vessels stimulate vasoconstriction, causing the pressure to rise again.

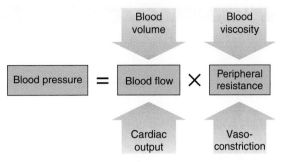

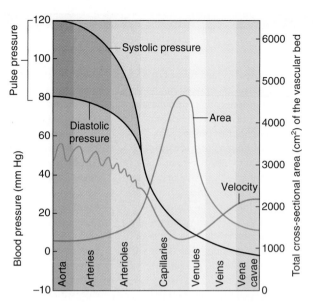

(a) Blood pressure depends on blood flow and resistance to that flow. A variety of factors affect blood flow and resistance.

Figure 43-14 *Animated* Blood pressure

Area in graph refers to the surface area of blood vessels in contact with blood.

(b) Blood pressure varies in different types of blood vessels. Systolic and diastolic variations in arterial blood pressures are shown. Note that the venous pressure drops below zero (below atmospheric pressure) near the heart.

Baroreceptors, specialized receptors in the walls of certain arteries and in the heart wall, are sensitive to changes in blood pressure. When an increase in blood pressure stretches the baroreceptors, messages are sent to the cardiac and vasomotor centers in the medulla of the brain. The cardiac center stimulates parasympathetic nerves that slow the heart, lowering blood pressure. The vasomotor center inhibits sympathetic nerves that constrict arterioles; this action causes vasodilation, which also lowers blood pressure. These neural reflexes continuously work in a complementary way to maintain blood pressure within normal limits.

Several hormones are also involved in regulating blood pressure (discussed in more detail in Chapters 47 and 48). In response to low blood pressure, the kidneys release **renin,** which activates the **renin–angiotensin–aldosterone pathway.** Renin acts on a plasma protein (angiotensinogen), triggering a cascade of reactions that produces the hormone **angiotensin II,** a powerful vasoconstrictor. Vasoconstriction *increases* blood pressure, restoring homeostasis. Angiotensin II also acts indirectly to maintain blood pressure by increasing the synthesis and release of the hormone **aldosterone** by the adrenal glands. Aldosterone increases retention of Na^+ by the kidneys, resulting in greater fluid retention and increased blood volume.

When the body becomes dehydrated, the osmotic concentration of the blood increases. In response, the posterior lobe of the pituitary gland releases **antidiuretic hormone (ADH).** ADH increases reabsorption of water in the kidneys (and only a small volume of concentrated urine is produced). Blood volume increases, increasing blood pressure and restoring homeostasis.

When blood volume increases, the atria of the heart release a hormone called **atrial natriuretic peptide (ANP),** which increases sodium excretion. As a result, blood pressure *decreases.* Nitric oxide also helps regulate blood pressure by causing vasodilation and thus decreased blood pressure.

Review

■ What is peripheral resistance? How does it affect blood pressure?

■ Blood pressure is low in capillaries. How does this help retain fluid in the circulation?

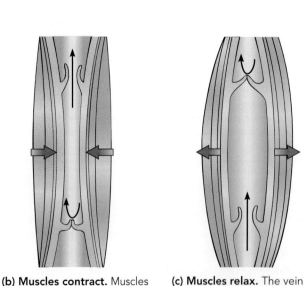

(a) Resting condition.

(b) Muscles contract. Muscles bulge, compressing veins and forcing blood toward the heart. The lower valve prevents backflow.

(c) Muscles relax. The vein expands and fills with blood from below. The upper valve prevents backflow.

Figure 43-15 *Animated* Venous blood flow

Contraction of skeletal muscles helps move blood through the veins.

THE PATTERN OF CIRCULATION

Learning Objective

10 Trace a drop of blood through the pulmonary and systemic circulations, naming in sequence each structure through which it passes.

Most vertebrates other than fishes have a double circuit of blood vessels: (1) the pulmonary circulation connects the heart and lungs; and (2) the systemic circulation connects the heart with all the body tissues. You can trace this general pattern of circulation in ▌ Figure 43-16.

The pulmonary circulation oxygenates the blood

Blood from the tissues returns to the right atrium of the heart. This oxygen-poor blood, loaded with carbon dioxide, is pumped by the right ventricle into the pulmonary circulation. As it emerges from the heart, the large pulmonary trunk branches to form the **pulmonary arteries** that deliver blood to the lungs. These are the only arteries that carry oxygen-poor blood.

In the lungs the pulmonary arteries branch into smaller and smaller vessels, which finally give rise to extensive networks of pulmonary capillaries that surround the air sacs of the lungs. As blood circulates through the pulmonary capillaries, carbon dioxide diffuses out of the blood and into the air sacs. Oxygen from the air sacs diffuses into the blood so that by the time it enters the **pulmonary veins** leading back to the left atrium of the heart, the blood is charged with oxygen. Pulmonary veins are the only veins in the body that carry blood rich in oxygen.

In summary, blood flows through the pulmonary circulation in the following sequence:

> Right atrium ⟶ right ventricle ⟶ pulmonary arteries ⟶ pulmonary capillaries (in lungs) ⟶ pulmonary veins ⟶ left atrium

The systemic circulation delivers blood to the tissues

Blood entering the systemic circulation is pumped by the left ventricle into the *aorta,* the largest artery. Arteries that branch off from the aorta conduct blood to all regions of the body. Some of the principal branches include the **coronary arteries** to the heart wall itself, the **carotid arteries** to the brain, the **subclavian arteries** to the shoulder region, the **mesenteric artery** to the intestine, the **renal arteries** to the kidneys, and the **iliac arteries** to the legs (▌ Fig. 43-17). Each of these arteries gives rise to smaller and smaller vessels, somewhat like branches of a tree that divide until they form tiny twigs. Eventually blood flows into capillary networks within each tissue or organ.

Blood returning from the capillary networks within the brain passes through the **jugular veins.** Blood from the shoulders and

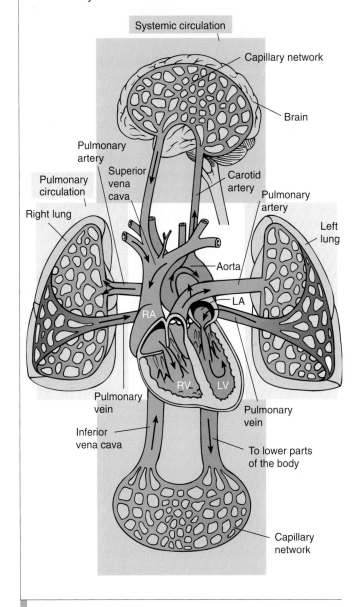

Figure 43-16 *Animated* Systemic and pulmonary circulation

In this highly simplified diagram, red represents oxygen-rich blood and blue represents oxygen-poor blood. Green screen highlights systemic circulation. Yellow screen highlights pulmonary circulation.

arms drains into the **subclavian veins.** These veins and others returning blood from the upper portion of the body merge to form a very large vein that empties blood into the right atrium. In humans this vein is called the **superior vena cava.** The **renal veins** from the kidneys, **iliac veins** from the lower limbs, **hepatic veins** from the liver, and other veins from the lower portion of the body return blood to the **inferior vena cava,** which delivers blood to the right atrium.

As an example of blood circulation through the systemic circuit, let us trace a drop of blood from the heart to the right leg and back to the heart:

Left atrium ⟶ left ventricle ⟶ aorta ⟶ right common iliac artery ⟶ smaller arteries in leg ⟶ capillaries in leg ⟶ small veins in leg ⟶ common iliac vein ⟶ inferior vena cava ⟶ right atrium

In most fishes, amphibians, and reptiles, the heart muscle is spongy and receives oxygen directly from the blood as it passes through the heart chambers. However, in birds and mammals the walls of the heart are too thick for nutrients and oxygen to reach all the muscle fibers by diffusion. Instead, the cardiac muscle has its own system of blood vessels. In humans, **coronary arteries** give rise to a network of capillaries within the heart wall. The **coronary veins** join to form a large vein, the **coronary sinus,** which empties directly into the right atrium.

Blood almost always travels from artery to capillary to vein to the heart. An exception occurs in the **hepatic portal system.** Instead of leading directly back to the heart (as most veins do), the hepatic portal vein delivers nutrients from the intestine to the liver. Within the liver, the hepatic portal vein gives rise to an extensive network of tiny blood sinuses. As blood courses through the hepatic sinuses, liver cells remove

nutrients and store them. Eventually liver sinuses merge to form hepatic veins, which deliver blood to the inferior vena cava.

Review

▌ What sequence of blood vessels and heart chambers would a red blood cell pass through on its way (1) from the inferior vena cava to the aorta and (2) from the renal vein to the renal artery?

▌ What is the function of the hepatic portal system? How does its sequence of blood vessels differ from that in most other circulatory routes?

THE LYMPHATIC SYSTEM

Learning Objective

11 Describe the structure and functions of the lymphatic system.

Vertebrates have in addition to the blood circulatory system an accessory circulatory system, the **lymphatic system** (▌ Fig. 43-18). The lymphatic system (1) collects and returns interstitial fluid to the blood, (2) launches immune responses that defend the body against disease organisms, and (3) absorbs lipids from the

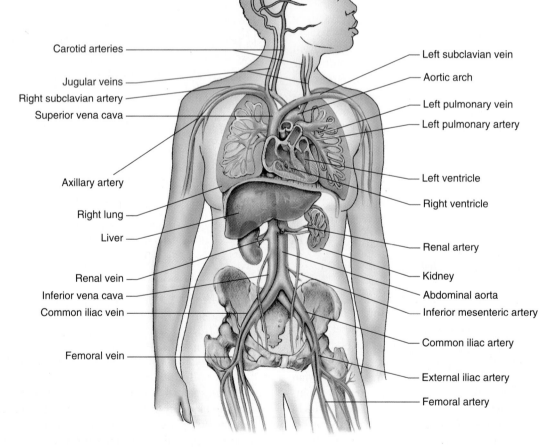

Carotid arteries
Jugular veins
Right subclavian artery
Superior vena cava

Axillary artery

Right lung
Liver

Renal vein
Inferior vena cava
Common iliac vein

Femoral vein

Left subclavian vein
Aortic arch
Left pulmonary vein
Left pulmonary artery

Left ventricle
Right ventricle

Renal artery
Kidney
Abdominal aorta
Inferior mesenteric artery

Common iliac artery

External iliac artery
Femoral artery

Figure 43-17 *Animated* Blood circulation through some of the principal arteries and veins of the human body

Blood vessels carrying oxygen-rich blood are red; those carrying oxygen-poor blood are blue.

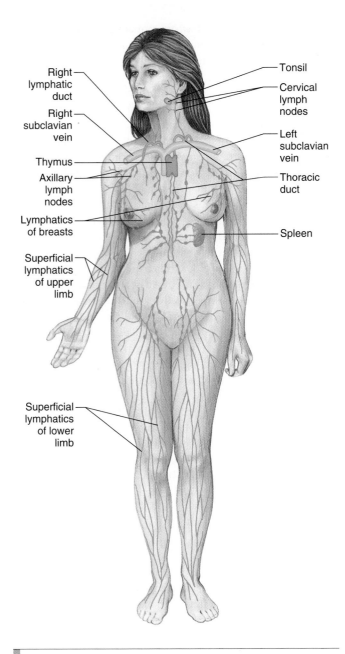

Figure 43-18 *Animated* Human lymphatic system

Note that the lymphatic vessels extend into most tissues of the body but the lymph nodes are clustered in certain regions. The right lymphatic duct drains lymph from the upper right quadrant of the body. The thoracic duct drains lymph from other regions of the body.

Labels (left figure): Right lymphatic duct; Right subclavian vein; Thymus; Axillary lymph nodes; Lymphatics of breasts; Superficial lymphatics of upper limb; Superficial lymphatics of lower limb; Tonsil; Cervical lymph nodes; Left subclavian vein; Thoracic duct; Spleen

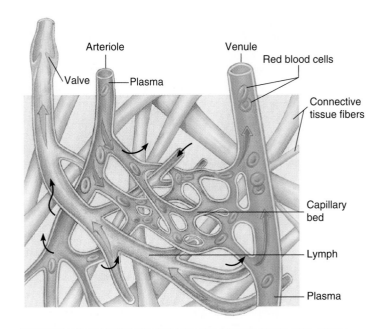

Figure 43-19 Lymph capillaries

Lymph capillaries drain excess interstitial fluid from the tissues. Note that blood capillaries are connected to vessels at both ends, whereas lymph capillaries (*shown in green*) are "dead-end streets." The arrows indicate direction of flow.

Labels (right figure): Arteriole; Valve; Plasma; Venule; Red blood cells; Connective tissue fibers; Capillary bed; Lymph; Plasma

digestive tract. In this section we focus on the first function. We discuss immunity in Chapter 44 and lipid absorption in Chapter 46.

The lymphatic system consists of lymphatic vessels and lymph tissue

The lymphatic system consists of (1) an extensive network of **lymphatic vessels,** or simply **lymphatics,** that conduct **lymph,** the clear, watery fluid formed from interstitial fluid; and (2) **lymph** **tissue,** a type of connective tissue with large numbers of lymphocytes. Lymph tissue is organized into small masses of tissue called **lymph nodes** and **lymph nodules.** The tonsils, thymus gland, and spleen, which consist mainly of lymph tissue, are also part of the lymphatic system.

Tiny "dead-end" capillaries of the lymphatic system extend into almost all body tissues (Fig. 43-19). Lymph capillaries join to form larger lymphatics (which you can think of as lymph veins). There are no lymph arteries.

Interstitial fluid enters lymph capillaries, where it becomes lymph. The lymph is conveyed into lymphatics, which at certain locations empty into lymph nodes. As lymph circulates through the lymph nodes, phagocytes filter out bacteria and other harmful materials. The lymph then flows into lymphatics that conduct it away from the lymph node. Lymphatics from all over the body conduct lymph toward the shoulder region. These vessels join the circulatory system at the base of the subclavian veins by way of ducts: the **thoracic duct** on the left side and the **right lymphatic duct** on the right.

Tonsils are masses of lymph tissue under the lining of the oral cavity and throat. (When enlarged, the pharyngeal tonsils in back of the nose are called **adenoids.**) Tonsils help protect the respiratory system from infection by destroying bacteria and other foreign matter that enter the body through the mouth or nose. Unfortunately, tonsils are sometimes overcome by invading bacteria and become the site of frequent infection themselves.

Some vertebrates, such as frogs, have lymph "hearts," which pulsate and squeeze lymph along. However, in mammals the walls of lymphatic vessels themselves pulsate. Valves within the lymphatics prevent the lymph from flowing backward. When

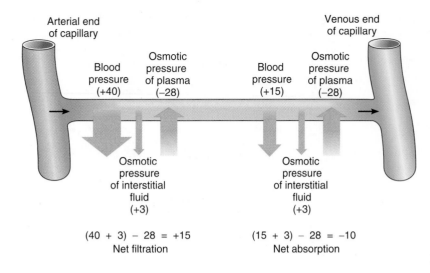

$$(40 + 3) - 28 = +15$$
Net filtration

$$(15 + 3) - 28 = -10$$
Net absorption

Figure 43-20 Fluid movement between blood and interstitial fluid

Blood hydrostatic pressure and osmotic pressures are responsible for fluid movement, and thus exchange of dissolved materials, between blood and interstitial fluid. At the arterial end of a capillary, blood pressure forces plasma out of the capillary. The osmotic pressure of the blood is an opposing force acting to draw fluid into the blood. Osmotic pressure of the interstitial fluid contributes to the net filtration pressure but does not change much between arterial and venous ends of the capillary. At the venous end of the capillary, fluid enters the blood because blood pressure is much lower. However, more fluid leaves the blood than returns, and the lymphatic system collects the excess interstitial fluid. The numbers given are hypothetical and represent millimeters of mercury. Net filtration is the total pressure moving fluid out of the capillary. Net absorption is the total pressure drawing fluid into the capillary.

muscles contract or when arteries pulsate, pressure on the lymphatic vessels increases lymph flow. The rate at which lymph flows is slow and variable, and the total lymph flow is about 100 mL per hour—far slower than the 5 L per min of blood flowing in the vascular system.

The lymphatic system plays an important role in fluid homeostasis

When blood enters a capillary network, it is under rather high pressure, so some plasma is forced out of the capillaries into the tissues. Once it leaves the blood vessels, this fluid is called *interstitial fluid*, or tissue fluid. It contains no red blood cells or platelets and only a few white blood cells. Its protein content is about one fourth that found in plasma, because proteins are too large to pass easily through capillary walls. However, smaller molecules dissolved in the plasma do pass out with the fluid leaving the

blood vessels. Thus, interstitial fluid contains glucose, amino acids, other nutrients, and oxygen, as well as a variety of salts. This nourishing fluid bathes all the cells.

The main force (*filtration pressure*) pushing plasma out of the blood is hydrostatic pressure, that is, the blood pressure against the capillary wall (Fig. 43-20). The osmotic pressure of the interstitial fluid adds to the filtration pressure. The principal opposing force is the osmotic pressure of the blood, which is greater than the osmotic pressure of the interstitial fluid and restrains fluid loss from the capillary.

At the venous ends of the capillaries the blood hydrostatic pressure is much lower, and the osmotic pressure of the blood draws fluid back into the capillary. However, not as much fluid is absorbed back into the circulation as is filtered out. Furthermore, protein does not return effectively into the venous capillaries and instead tends to accumulate in the interstitial fluid. These potential problems are so serious that without the lymphatic system, fluid balance in the body would be significantly disturbed within a few hours, and death would occur within about 24 hours. The lymphatic system preserves fluid balance by collecting some of the interstitial fluid (about 10%), including the protein that accumulates in it, and returning it to the circulation.

The walls of the lymph capillaries consist of endothelial cells that overlap slightly. When interstitial fluid accumulates, it presses against these cells, pushing them inward like tiny swinging doors that swing in only one direction. As fluid accumulates within the lymph capillary, these cell doors are pushed closed.

Obstruction of the lymphatic vessels causes **edema,** swelling from excessive accumulation of interstitial fluid. Lymphatic vessels can be blocked as a result of injury, inflammation, surgery, or parasitic infection. For example, when a breast is removed (mastectomy) because of cancer, lymph nodes in the underarm region may also be removed to prevent the spread of cancer cells. The disrupted lymph circulation makes the patient's arm swell. New lymphatic vessels develop within a few months, and the swelling slowly subsides.

Review

- What are the relationships among plasma, interstitial fluid, and lymph?
- How does the lymphatic system help maintain fluid balance?

SUMMARY WITH KEY TERMS

Learning Objectives

1   Compare and contrast internal transport in animals with no circulatory system, those with an open circulatory system, and those with a closed circulatory system (page 920).

- Small, simple invertebrates, such as sponges, cnidarians, and flatworms, depend on **diffusion** for internal transport.

Larger animals require a specialized **circulatory system,** which typically consists of **blood,** a **heart,** and a system of **blood vessels** or spaces through which blood circulates. In all animals, **interstitial fluid,** the tissue fluid between cells, brings oxygen and nutrients into contact with cells.

- Arthropods and most mollusks have an **open circulatory system** in which blood flows into a **hemocoel,** bathing the tissues directly.

- Some invertebrates and all vertebrates have a **closed circulatory system** in which blood flows through a continuous circuit of blood vessels.

- The vertebrate circulatory system consists of a muscular heart that pumps blood into a system of **arteries, capillaries,** and **veins.** This system transports nutrients, oxygen, wastes, and hormones; helps maintain fluid balance, appropriate pH, and body temperature; and defends the body against disease.

ThomsonNOW™ **Learn more about open and closed circulatory systems by clicking on the figures in ThomsonNOW.**

2 Compare the structure and function of plasma, red blood cells, white blood cells, and platelets (page 922).
- **Plasma** consists of water, salts, substances in transport, and **plasma proteins,** including albumins, globulins, and fibrinogen.
- **Red blood cells,** also called **erythrocytes,** transport oxygen and carbon dioxide. Red blood cells produce large quantities of **hemoglobin,** a red pigment that binds with oxygen.
- **White blood cells,** also called **leukocytes,** defend the body against disease organisms. **Lymphocytes** and **monocytes** are agranular white blood cells; **neutrophils, eosinophils,** and **basophils** are granular white blood cells.
- **Platelets** patch damaged blood vessels and release substances essential for blood clotting.

ThomsonNOW™ **Learn more about the composition of vertebrate blood by clicking on the figures in ThomsonNOW.**

3 Summarize the sequence of events involved in blood clotting (page 922).
- Damaged cells and platelets release substances that activate clotting factors. Prothrombin is converted to **thrombin,** which catalyzes the conversion of **fibrinogen** to an insoluble protein, **fibrin.** Fibrin forms long threads that make up the webbing of the clot.

4 Compare the structure and function of different types of blood vessels, including arteries, arterioles, capillaries, and veins (page 925).
- Arteries carry blood away from the heart; veins return blood to the heart.
- **Arterioles** constrict (**vasoconstriction**) and dilate (**vasodilation**) to regulate blood pressure and distribution of blood to the tissues.
- Capillaries are the thin-walled exchange vessels through which blood and tissues transfer materials.

5 Trace the evolution of the vertebrate cardiovascular system from fish to mammal (page 927).
- The vertebrate heart has one or two **atria,** which receive blood, and one or two **ventricles,** which pump blood into the arteries.
- The fish heart consists of a single atrium and ventricle that are part of a single circuit of blood flow.
- In terrestrial vertebrates, circulatory systems separate oxygen-rich from oxygen-poor blood; this allows the higher metabolic rate needed to support an active terrestrial lifestyle.
- Amphibians have two atria and a ventricle, and blood flows through a double circuit so that oxygen-rich blood is partly separated from oxygen-poor blood. Most reptiles have a wall that partly divides the ventricles, minimizing the mixing of oxygen-rich and oxygen-poor blood.

- The four-chambered hearts of birds and mammals separate oxygen-rich blood from oxygen-poor blood.

6 Describe the structure and function of the human heart. (Include the heart's conduction system in your answer.) (page 928).
- The heart is enclosed by a **pericardium** and has valves that prevent backflow of blood. The valve between the right atrium and ventricle is the **right atrioventricular (AV) valve,** or **tricuspid valve.** The valve between the left atrium and ventricle is the **mitral valve. Semilunar valves** guard the exits from the heart. Cardiac muscle fibers are joined by **intercalated discs.**
- The **sinoatrial (SA) node,** or **pacemaker,** initiates each heartbeat. A specialized electrical conduction system coordinates heartbeats.

ThomsonNOW™ **Learn more about heart anatomy by clicking on the figure in ThomsonNOW.**

ThomsonNOW™ **See cardiac conduction by clicking on the figure in ThomsonNOW.**

7 Trace the events of the cardiac cycle, and relate normal heart sounds to these events (page 928).
- One complete heartbeat makes up a **cardiac cycle.** Contraction occurs during **systole.** The period of relaxation is **diastole.** At the beginning of ventricular systole, the closing of the AV valves makes a low-pitched "lub" sound. The closing of the semilunar valves, the beginning of ventricular diastole, makes a short, loud "dup" sound.

ThomsonNOW™ **Watch the cardiac cycle by clicking on the figure in ThomsonNOW.**

8 Define *cardiac output,* describe how it is regulated, and identify factors that affect it (page 929).
- **Cardiac output (CO)** equals **stroke volume** times heart rate. Stroke volume depends on **venous return** and on neural messages and hormones, especially epinephrine and norepinephrine. According to **Starling's law of the heart,** the more blood delivered to the heart by the veins, the more blood the heart pumps.
- Heart rate is regulated mainly by the nervous system and is influenced by hormones and body temperature.

9 Identify factors that determine and regulate blood pressure, and compare blood pressure in different types of blood vessels (page 933).
- **Blood pressure** is the force blood exerts against the inner walls of the blood vessel. Blood pressure is greatest in the arteries and decreases as blood flows through the capillaries.
- Blood pressure depends on cardiac output, blood volume, and resistance to blood flow. **Peripheral resistance** is the resistance to blood flow caused by blood viscosity and by friction between blood and the blood vessel wall.
- **Baroreceptors** sensitive to blood pressure changes send messages to the cardiac and vasomotor centers in the medulla of the brain. When informed of an increase in blood pressure, the cardiac center stimulates parasympathetic nerves that slow heart rate, and the vasomotor center inhibits sympathetic nerves that constrict blood vessels. These actions reduce blood pressure.
- **Angiotensin II** is a hormone that raises blood pressure. **Aldosterone** helps regulate salt excretion, which affects blood volume and blood pressure.

ThomsonNOW™ **Learn more about determining blood pressure by clicking on the figure in ThomsonNOW.**

10 Trace a drop of blood through the pulmonary and systemic circulations, naming in sequence each structure through which it passes (page 937).

- The **pulmonary circulation** connects the heart and the lungs; the **systemic circulation** connects the heart and the tissues.
- In the pulmonary circulation, the right ventricle pumps blood into the **pulmonary arteries,** which go to the lungs. Blood circulates through pulmonary capillaries in the lung and then is conducted to the left atrium by a **pulmonary vein.**
- In the systemic circulation, the left ventricle pumps blood into the **aorta,** which branches into arteries leading to the body organs. After flowing through capillary networks within various organs, blood flows into veins that conduct it to the **superior vena cava** or **inferior vena cava,** which returns blood to the right atrium.
- The **coronary arteries** supply the heart muscle with blood. The **hepatic portal system** circulates nutrient-rich blood through the liver.

ThomsonNOW **Learn more about systemic and pulmonary circulation by clicking on the figures in ThomsonNOW.**

11 Describe the structure and functions of the lymphatic system (page 938).

- The **lymphatic system** collects interstitial fluid and returns it to the blood, so it plays an important role in homeostasis of fluids. The lymph system also defends the body against disease and absorbs lipids from the digestive tract.
- **Lymphatic vessels** conduct **lymph,** a clear fluid formed from interstitial fluid, to the **thoracic duct** and **right lymphatic duct** in the shoulder region; these ducts return lymph to the blood circulatory system. **Lymph nodes** are small masses of tissue that filter bacteria and harmful materials out of lymph.

ThomsonNOW **Learn more about the human lymphatic system by clicking on the figure in ThomsonNOW.**

TEST YOUR UNDERSTANDING

1. An open circulatory system (a) is found in flatworms (b) typically includes a hemocoel (c) has a continuous circuit of vessels with openings in the capillaries (d) is characteristic of vertebrates (e) is typically found in animals with a two-chambered heart

2. Which of the following is *not* a function of the vertebrate circulatory system? (a) helps maintain appropriate pH (b) transports nutrients, oxygen, and metabolic wastes (c) helps maintain fluid balance (d) produces hemocyanin (e) provides internal defense

3. Lipoproteins (a) are mainly transported in granular leukocytes (b) transport cholesterol (c) have been linked to clotting disorders (d) are associated with platelets (e) are stored in red blood cells

4. Which of the following are most closely associated with oxygen transport? (a) red blood cells (b) platelets (c) neutrophils (d) basophils (e) lymphocytes

5. Which of the following are most closely associated with blood clotting? (a) red blood cells (b) platelets (c) neutrophils (d) basophils (e) lymphocytes

6. In blood clotting, (a) thrombin $\longrightarrow$ prothrombin; fibrinogen $\longrightarrow$ fibrin (b) prothrombin $\longrightarrow$ thrombin; fibrin $\longrightarrow$ fibrinogen (c) prothrombin $\longrightarrow$ thrombin; fibrinogen $\longrightarrow$ fibrin (d) clotting factors $\longrightarrow$ platelets; thrombin $\longrightarrow$ fibrinogen (e) prothrombin $\longrightarrow$ thrombin; fibrinogen $\longrightarrow$ platelets

7. Blood vessels that carry blood away from the heart are (a) arteries (b) sinuses (c) veins (d) capillaries (e) arterioles and venules

8. Arterioles (a) help regulate blood pressure (b) help regulate distribution of blood to the tissues (c) deliver blood to arteries (d) a, b, and c (e) a and b

9. Which choice most accurately describes one sequence of blood flow? (a) right atrium $\longrightarrow$ right ventricle $\longrightarrow$ pulmonary artery (b) right atrium $\longrightarrow$ left atrium $\longrightarrow$ left ventricle $\longrightarrow$ aorta (c) left atrium $\longrightarrow$ left ventricle $\longrightarrow$ pulmonary artery (d) left ventricle $\longrightarrow$ left atrium $\longrightarrow$ aorta (e) right atrium $\longrightarrow$ right ventricle $\longrightarrow$ aorta

10. Which choice most accurately describes one sequence of blood flow? (a) pulmonary vein $\longrightarrow$ pulmonary artery $\longrightarrow$ right atrium (b) pulmonary artery $\longrightarrow$ left atrium $\longrightarrow$ left ventricle (c) pulmonary artery $\longrightarrow$ pulmonary capillaries $\longrightarrow$ pulmonary vein $\longrightarrow$ left atrium (d) left ventricle $\longrightarrow$ aorta $\longrightarrow$ pulmonary artery (e) pulmonary artery $\longrightarrow$ pulmonary capillaries $\longrightarrow$ pulmonary vein $\longrightarrow$ right atrium

11. A cardiac cycle (a) consists of one ventricular heartbeat (b) includes a systole (c) equals stroke volume times heart rate (d) includes a diastole (e) includes a systole and a diastole

12. Blood pressure is determined by (a) cardiac output (b) peripheral resistance (c) blood volume (d) a, b, and c (e) b and c

13. Lymph forms from (a) interstitial fluid (b) blood serum (c) plasma combined with protein (d) fluid released by lymph nodes (e) angiotensins

14. The valve between the right atrium and right ventricle is the (a) mitral valve (b) semilunar valve (c) tricuspid valve (d) pulmonary valve (e) aortic valve

15. Norepinephrine (a) slows heart rate (b) is released in cardiac muscle by parasympathetic nerves (c) causes K^+ channels in cardiac muscle to open (d) decreases stroke volume (e) causes Ca^{2+} channels in cardiac muscle to open

16. Baroreceptors (a) stimulate renin release (b) activate the renin–angiotensin–aldosterone pathway (c) stimulate sympathetic nerves (d) are stimulated by increased blood pressure (e) send messages to cardiac centers that increase blood pressure

17. Atherosclerosis (a) is associated with thickening of the walls of arteries (b) is associated with high concentrations of low-density lipoprotein (c) can lead to ischemic heart disease (d) a, b, and c (e) b and c

18. Label the diagram. See Figure 43-9 to check your answers.

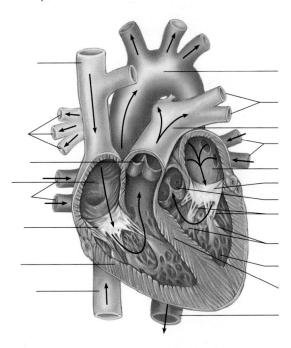

CRITICAL THINKING

1. How is the heart of the fish specifically adapted to its lifestyle?

2. When the nerves to the heart are cut, the heart rate increases to about 100 contractions per minute. What does this indicate about the regulation of the heart rate?

3. List five modifiable risk factors associated with the development of cardiovascular disease. Describe the disease process in atherosclerosis, and explain the association between atherosclerosis and ischemic heart disease. What happens in myocardial infarction?

4. **Evolution Link.** A heart divided into four chambers independently evolved in birds and mammals. What are the advantages of this adaptation?

Additional questions are available in ThomsonNOW at www.thomsonedu.com/login

The Immune System: Internal Defense

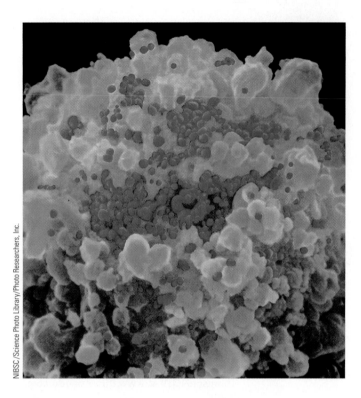

NIBSC/Science Photo Library/Photo Researchers, Inc.

A T cell infected with HIV. In this colorized SEM, an immune system T cell (*green*) is infected by human immunodeficiency virus (HIV; *red*).

KEY CONCEPTS

Nonspecific immune responses provide general and immediate protection against pathogens, some toxins, and cancer cells.

Specific immune responses include cell-mediated immunity and antibody-mediated immunity.

In cell-mediated immunity, specific lymphocytes known as T cells are activated and then release proteins that destroy cells infected with viruses or other intracellular pathogens.

In antibody-mediated immunity, specific lymphocytes known as B cells are activated; they multiply and differentiate into plasma cells, which produce antibodies.

Sometimes the immune system is hypersensitive, such as in allergic responses; immune responses directed at self tissue can cause autoimmune disease.

The **immune system,** our internal defense system, protects the body against disease-causing organisms and certain toxins. Derived from the Latin for "safe," the word *immune* refers to the early observation that when people recovered from smallpox and other serious infections, they were safe from contracting the same illnesses again.

Disease-causing organisms, or **pathogens,** include certain viruses, bacteria, fungi, and protozoa. Pathogens enter the body with air, food, and water; during copulation; and through wounds in the skin. The immune system recognizes pathogens and toxins and responds to eliminate them.

Sometimes pathogens overcome the body's internal defenses, and disease results. Some diseases, as well as certain genetic mutations, prevent or compromise immune function. HIV, the retrovirus that causes AIDS, infects T cells, an important component of the immune system (see photograph).

The immune system responds in ways that are clinically important. This system may overfunction, as in allergic reactions, may endanger the life of a fetus that is Rh incompatible, or may destroy the cells of organ transplants. In about 5% of adults in highly developed countries, certain immune responses are directed against self tissues, resulting in *autoimmune disease.*

Immunology, the study of internal defense systems of humans and other animals, is one of the most rapidly changing, challenging, and exciting fields of biomedical research. The immune system is a collection of many types of cells and tissues scattered throughout the body. Immune responses require communication among cells, or **cell signaling.** Cells of the immune system communicate directly by means of their surface molecules and indirectly by release of messenger molecules. Understanding the complex signaling systems of the immune system is a major focus of research.

One of the greatest accomplishments of immunologists has been the development of vaccines that prevent disease. Another has been developing techniques for successful tissue and organ transplantation. Sophisticated research tools, such as gene transfer, have enabled immunologists to expand their knowledge of the cells and molecules that interact to generate immune responses. These techniques have led to development of new approaches to the prevention and treatment of disease. Much has been learned, and many challenges lie ahead. ■

NONSPECIFIC AND SPECIFIC IMMUNITY: AN OVERVIEW

Learning Objectives

1 Distinguish between nonspecific and specific immune responses.
2 Compare, in general terms, the immune responses of invertebrates and vertebrates.

An **immune response** is the process of recognizing foreign or dangerous macromolecules and responding to eliminate them. Two main types of immune responses protect the body: nonspecific and specific (■ Fig. 44-1). **Nonspecific immune responses,** or *innate immunity,* provide immediate, general protection against pathogens, parasites, some toxins and drugs, and cancer cells. Nonspecific immune responses prevent most pathogens from entering the body and rapidly destroy those that do penetrate the outer defenses. For example, the cuticle or skin provides a physical barrier to pathogens that come in contact with an animal's body. Phagocytosis, another nonspecific defense, destroys bacteria that invade the body.

Specific immune responses, also referred to as *adaptive* or *acquired immunity,* are highly specific for distinct macromolecules. Any molecule that cells of the immune system specifically recognize as foreign is called an **antigen.** Proteins are the most powerful antigens, but some polysaccharides and lipids are also antigenic. Specific immune responses are directed toward particular antigens and typically include the production of **antibodies,** highly specific proteins that recognize and bind to specific antigens. An important characteristic of specific immune responses is **immunological memory,** which means that the system "remembers" foreign or dangerous molecules and responds more effectively to repeated encounters with the same molecules.

The immune system responds to danger signals

For more than 50 years, immunologists based their work on the hypothesis that internal defense depends on the animal's ability to distinguish between *self* and *nonself.* Such recognition is possible because each individual is biochemically unique. Cells have surface proteins different from those on the cells of other species or even other individuals of the same species. An animal's immune system recognizes its own cells and can identify those of

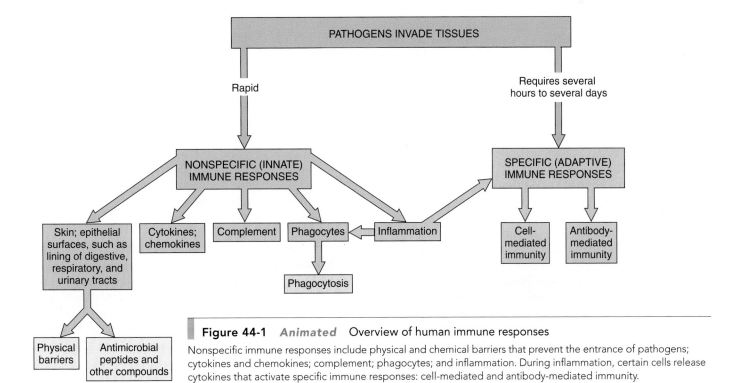

Figure 44-1 *Animated* Overview of human immune responses

Nonspecific immune responses include physical and chemical barriers that prevent the entrance of pathogens; cytokines and chemokines; complement; phagocytes; and inflammation. During inflammation, certain cells release cytokines that activate specific immune responses: cell-mediated and antibody-mediated immunity.

other organisms as foreign. Thus, the distinctive macromolecules of a pathogen that invades an animal stimulate the animal's defensive responses. A single bacterium may have from 10 to more than 1000 distinctive macromolecules on its surface.

Immunologists are aware of several limitations of the self–nonself hypothesis. For example, the immune system does not typically respond to foreign molecules that are harmless. The **danger model,** proposed by Polly Matzinger of the U.S. National Institutes of Health, hypothesizes that the immune system does more than distinguish between self and nonself. It responds to danger signals from injured tissues, such as proteins released when cell membranes are damaged. Most immunologists now agree that internal defense relies on a combination of factors, including the ability to identify foreign molecules *and* to respond to chemical clues from injured tissues.

Invertebrates launch nonspecific immune responses

Many invertebrates (cnidarians, annelids, and mollusks) are covered by mucus that traps and kills pathogens. Tough external skeletons, such as cuticles, shells, or chitinous exoskeletons, shield the bodies of nematodes, most mollusks, arthropods, and many other invertebrates.

All animal species studied have **phagocytes** that move through the body and engulf and destroy bacteria and other foreign matter. In mollusks, substances in the hemolymph (blood) enhance phagocytosis. Annelids are the only invertebrates known to have *natural killer cells,* a type of lymphocyte that kills cells infected by pathogens.

Researchers have described more than 800 **antimicrobial peptides,** chemicals that inactivate or kill pathogens. Investigators have identified antimicrobial peptides in all eukaryotes (including plants) that have been studied, suggesting an early common origin of these molecules. When researchers inject an antigen into an insect, as many as 15 different antimicrobial peptides are produced within a few hours. These chemicals are very effective in part because of their small size (a dozen or fewer amino acids), which facilitates their rapid production and diffusion.

What stimulates an animal to produce antimicrobial peptides and other immune defenses? Animal cells have receptors that recognize certain types of pathogen molecules and then signal the cell to produce antimicrobial peptides.

Like vertebrates, invertebrates distinguish between their own cells and those of other species. For example, sponge cells have specific glycoproteins on their surfaces that enable them to recognize their own species. When cells of two different species are mixed, they reassort according to species. When two different species of sponges are forced to grow in contact with each other, tissue is destroyed along the region of contact. Cnidarians (such as corals), annelids (such as earthworms), arthropods (such as insects), and echinoderms (such as sea stars) reject tissue grafted from other animals, even from the same species.

Certain cnidarians, arthropods, some echinoderms, and simple chordates (such as tunicates) appear to remember antigens for a short period. As mentioned earlier, immunological memory enables the body to respond more effectively when it encounters the same pathogens again. Although certain invertebrates demonstrate some specificity and memory, their immune responses are primarily nonspecific. Echinoderms and tunicates are the simplest animals known to have differentiated white blood cells that perform limited immune functions.

Vertebrates launch nonspecific and specific immune responses

Vertebrates carry out a variety of nonspecific immune responses (discussed in the next section). However, the sophisticated specific immune responses are a distinctive feature of these animals. The specialized lymphatic system that has evolved in vertebrates includes cells such as **lymphocytes,** white blood cells specialized to carry out immune responses, and organs such as lymph nodes (see Chapter 43). Specific immune responses, including lymphocytes and antibodies, first evolved in jawed vertebrates. In the discussion that follows, we focus on the human immune system, with references to immune mechanisms of other animals.

Review

■ What are two key ways in which specific immune responses differ from nonspecific immune responses?

■ In general, how do vertebrate immune responses differ from invertebrate responses?

NONSPECIFIC IMMUNE RESPONSES

Learning Objective

3 Describe nonspecific immune responses, including physical and chemical barriers; recognition by Toll-like receptors; the actions of cells, such as phagocytes and natural killer cells; cytokines and the proteins that make up the complement system; and the inflammatory response.

Physical and chemical barriers prevent most pathogens from entering the body. An animal's first line of defense is its outer covering—cuticle, shell, chitin, or skin—which blocks the entry of pathogens. Microorganisms that enter with food are usually destroyed by the acid secretions and enzymes of the stomach, which are effective chemical barriers. In addition, peptides, known as **defensins,** are produced by epithelium in the skin of many animals, including mammals. Also found in saliva, defensins kill a variety of bacteria and fungi.

Mucous membranes of the respiratory and reproductive tracts protect these passageways that open to the outside of the body. For example, pathogens that enter the body with inhaled air may be filtered out by hairs in the nose or trapped in the sticky mucous lining of the respiratory tract. Once trapped, they are destroyed by phagocytes. Mucus contains a substance called *mucin,* which chemically destroys invaders. **Lysozyme,** an enzyme found in many tissues, as well as in tears and other body fluids, attacks the cell walls of many gram-positive bacteria.

When pathogens breach the body's outer barriers, their chemical properties activate other nonspecific defense responses,

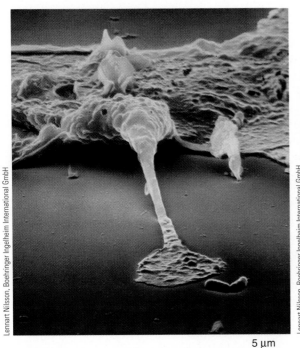

Lennart Nilsson, Boehringer Ingelheim International GmbH

(a) A macrophage extends a pseudopod toward an invading *Escherichia coli* bacterium that is already multiplying.

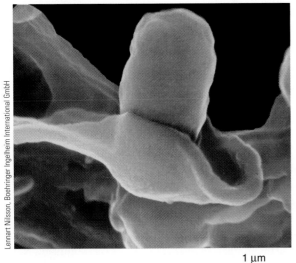

Lennart Nilsson, Boehringer Ingelheim International GmbH

(b) A bacterium is trapped within the engulfing pseudopod.

Figure 44-2 Phagocytosis

These colorized SEMs show macrophages, which are efficient warriors.

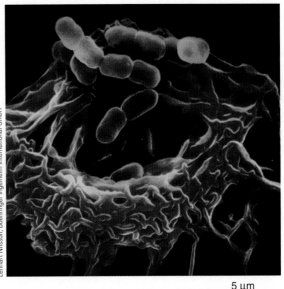

Lennart Nilsson, Boehringer Ingelheim International GmbH

(c) A macrophage takes in trapped bacteria along with its own plasma membrane. The macrophage plasma membrane will seal over the bacteria, and powerful lysosomal enzymes will destroy them.

Nonspecific immune responses depend on the action of several types of cells, including phagocytes and natural killer cells. These responses also involve numerous molecules, including *cytokines,* which are important signaling molecules, and proteins of the *complement system* that destroy pathogens. The activation and concentration of these cells and molecules produce inflammation, a major nonspecific immune response.

Phagocytes and natural killer cells destroy pathogens

Neutrophils, the most common type of white blood cell (see Chapter 43), and macrophages are the main phagocytes in the body. Recall that **phagocytosis** is a type of endocytosis in which cells engulf microorganisms, foreign matter, or other cells (see Fig. 5-21). A neutrophil can phagocytose about 20 bacteria before it becomes inactivated (perhaps by leaking lysosomal enzymes) and dies.

Macrophages are large phagocytes that develop from nongranular white blood cells called *monocytes.* A macrophage can phagocytose about 100 bacteria during its life span (Fig. 44-2). Some macrophages patrol the body's tissues, phagocytosing damaged cells or foreign matter (including bacteria); they also release antiviral agents. Other macrophages stay in one place and destroy bacteria that pass by. For example, air sacs in the lungs contain large numbers of macrophages that destroy foreign matter entering with inhaled air.

Vertebrate macrophages have Toll-like receptors that recognize certain PAMPs and respond by producing cytokines. For ex-

and most invaders are quickly eliminated. Nonspecific immune responses are activated by **Toll-like receptors,** cell-surface receptors on phagocytes and certain other types of cells. Toll-like receptors (homologous to a protein known as the Toll protein in *Drosophila*) recognize certain common molecular features of classes of pathogens called **pathogen-associated molecular patterns,** or **PAMPs.** PAMPs are shared by whole groups of viruses, bacteria, or fungi. Examples of PAMPs include the double-stranded RNA of certain viruses and peptidoglycan in gram-positive bacteria. When activated by PAMPs, Toll-like receptors activate phagocytes.

ample, when stimulated by a lipopolysaccharide found on gram-negative bacteria, macrophages release cytokines that enhance the inflammatory response.

Can bacteria counteract the phagocyte's attack? Many strategies have evolved in bacteria that protect them from their hosts. *Streptococcus pneumoniae,* for example, has cell walls or capsules that resist the action of the phagocyte's lysosomal enzymes. Other bacteria release enzymes that destroy lysosomal membranes. The powerful lysosomal enzymes then spill out into the cytoplasm and may destroy the phagocyte.

Natural killer (NK) cells are large, granular lymphocytes that originate in the bone marrow. They account for about 10% of circulating lymphocytes. NK cells are active against tumor cells and cells infected with some types of viruses. They destroy target cells by both nonspecific and specific (antibody-mediated) processes.

NK cells release cytokines, as well as perforins and granzymes, enzymes that destroy target cells. **Perforins** cause pores to form in the plasma membrane of the target cell, allowing granzymes to enter the cell. **Granzymes** then activate a cascade of reactions that cause the cell to self-destruct by **apoptosis,** programmed cell death. Several cytokines stimulate NK cell activity. When NK cell levels are high, resistance to certain cancers may be strengthened. Psychological stressors are thought to decrease NK cell activity and thus enhance tumor growth.

Cytokines and complement mediate immune responses

Cells of the immune system secrete a remarkable number of regulatory and antimicrobial peptides and proteins. We discuss two major groups of molecules that are critical in nonspecific defense responses: cytokines and complement.

Cytokines are important signaling molecules

Cytokines are a large, diverse group of peptides and proteins (most are glycoproteins) that serve as signals and perform regulatory functions. Just as the functions of nonspecific and specific immune responses overlap, the actions of cytokines of these subsystems also overlap. For example, cytokines produced by nonspecific cells such as macrophages can activate lymphocytes involved in specific immune responses.

Cytokines help regulate the intensity and duration of immune responses. They are also important in regulating many other biological processes, such as cell growth, repair, and cell activation. Cytokines bind to specific membrane receptors on target cells. These signaling molecules can act as *autocrine agents,* which affect the very cells that produce them, or as *paracrine agents,* which regulate the activity of nearby cells (see Chapter 48). When produced in large amounts, some cytokines act as endocrine agents (hormones); they circulate in the blood and affect distant tissues. Biologists name the types of cytokines for their function or origin. We discuss several examples, including *interferons, tumor necrosis factors, interleukins,* and *chemokines.*

When infected by viruses or other intracellular parasites (some types of bacteria, fungi, and protozoa), certain cells respond by secreting cytokines called **interferons.** Type I interferons are produced by either macrophages or *fibroblasts,* cells

that produce the fibers of connective tissues. Type I interferons inhibit viral replication and activate NK cells that have antiviral actions. Viruses produced in cells exposed to Type I interferons are not very effective at infecting other cells. Interestingly, even though Type I interferons are produced as part of the nonspecific immune response, once released, they can regulate the action of certain cells of the specific immune response. This interaction of nonspecific and specific immune mechanisms is characteristic of the power and complexity of the immune system.

Type II interferon (also called IFN-gamma) is critical in both nonspecific and specific immunity. In nonspecific immunity, Type II interferon stimulates macrophages to destroy tumor cells and host cells that have been infected by viruses. (The interferon activates gene transcription that encodes enzymes needed to produce antimicrobial compounds in lysosomes.)

Since their discovery in 1957, interferons have been the focus of a great deal of research. Pharmaceutical companies now use recombinant DNA techniques to produce large quantities of some interferons. The U.S. Food and Drug Administration (FDA) has approved interferons for treating several diseases, including hepatitis B and hepatitis C, genital warts, a type of leukemia, a type of multiple sclerosis, and AIDS-related Kaposi's sarcoma. Researchers are testing interferons in clinical trials for the treatment of HIV infection and several types of cancer.

Tumor necrosis factor (TNF) stimulates immune cells to initiate an inflammatory response in response to gram-negative bacteria and other pathogens. TNF is secreted mainly by macrophages and by certain lymphocytes called *T cells.* In severe infections, large amounts of TNF are produced. TNF is then circulated in the blood and can have widespread effects. For example, it acts on the hypothalamus, inducing *fever.* Sometimes infection by gram-negative bacteria, such as *Salmonella typhi,* results in the release of large amounts of TNF and other cytokines. This results in a cascade of reactions leading to *septic shock,* a potentially fatal condition that may involve high fever and malfunction of the circulatory system. Thus, cytokines can sometimes have harmful effects.

Interleukins are a diverse group of cytokines secreted mainly by macrophages and lymphocytes. Interleukins are numbered according to their order of discovery. They regulate interactions between white blood cells, such as lymphocytes and macrophages, and other cells. *Interleukin-1 (IL-1)* functions with TNF in mediating inflammation. *Interleukin-12 (IL-12)* stimulates NK cells, as well as T cells, to produce IFN-gamma. Some interleukins have widespread effects. During infection IL-1 can reset the body's thermostat in the hypothalamus, resulting in fever and its symptoms. The overproduction of certain interleukins has been associated with clinical disease, including atherosclerosis.

Chemokines, a large group of cytokines, are signaling molecules that attract, activate, and direct the movement of various cells of the immune system. For example, they regulate the migration of white blood cells from the blood to the tissues. Some chemokines are produced in response to infection and are mediators of the inflammatory response.

Complement leads to the destruction of pathogens

Cytokines produced by phagocytes can activate the *complement system.* **Complement,** so named because it "complements" the

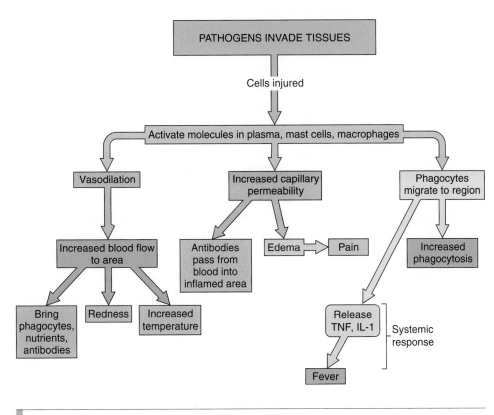

Figure 44-3 *Animated* The inflammatory response

action of other defensive responses, consists of more than 20 proteins present in plasma and other body fluids. Similarities in complement proteins in many species, including horseshoe crabs, sea urchins, tunicates, and mammals, suggest that these molecules evolved millions of years ago and have been conserved.

Normally, complement proteins are inactive until the body is exposed to an antigen. Certain pathogens activate the complement system directly. In other cases, the binding of an antigen and antibody stimulates activation. Complement activation involves a cascade of reactions; each component acts on the next in the series. Proteins of the complement system then work to destroy pathogens.

Complement proteins are activated against many antigens, and their actions are nonspecific: They (1) lyse viruses, bacteria, and other cells; (2) coat pathogens, making them less "slippery" so that macrophages and neutrophils phagocytose them more easily; (3) attract white blood cells to the site of infection, a process called *chemotaxis;* and (4) bind to specific receptors on cells of the immune system, stimulating specific actions, such as secreting regulatory molecules and enhancing the inflammatory response.

Inflammation is a protective response

The **inflammatory response,** or *inflammation,* begins immediately after pathogen invasion or physical injury (▌Fig. 44-3). Tissue injury activates a clotting factor in the blood plasma that turns on three different interconnecting molecular cascades in the plasma, including the clotting cascade. The reactions generate

molecules—for example, the peptide *bradykinin*—that mediate the inflammatory process. These plasma mediators dilate blood vessels and increase capillary permeability. Inflammation also activates the complement system.

Many cytokines signal white blood cells to launch the inflammatory response, and some cytokines help regulate the process. For example, large numbers of neutrophils migrate into the inflamed tissue within a few hours of tissue injury or infection. Neutrophils secrete chemokines and other cytokines. Macrophages and **mast cells** stationed in the tissues also respond rapidly to damaged tissue or infection. Mast cells release **histamine,** cytokines, and other compounds that dilate blood vessels in the affected area and increase capillary permeability. Mast cells and macrophages release signaling molecules that attract and activate additional neutrophils. In turn, neutrophils send signals that attract lymphocytes.

The clinical characteristics of inflammation are heat, redness, edema (swelling), and pain. The inflammatory response includes three main processes.

1. *Vasodilation.* Expanded blood vessel diameter brings more blood to the area. The increased blood flow warms the skin and reddens skin that contains little pigment.

2. *Increased capillary permeability.* Fluid and antibodies leave the circulation and enter the tissues. As the volume of interstitial fluid increases, *edema* occurs. The edema, along with the action of certain enzymes in the plasma, causes the pain that characterizes inflammation.

3. *Increased phagocytosis.* Increased blood flow brings large numbers of neutrophils and other phagocytic cells to the infected region. The phagocytes migrate out of the capillaries and into the infected tissues. One of the main functions of inflammation is increased phagocytosis.

Although the inflammatory response begins as a local response, sometimes the entire body becomes involved. **Fever,** a common clinical symptom of widespread inflammation, helps the body fight infection. Elevated body temperature increases phagocytosis and interferes with the growth and replication of microorganisms. Fever breaks down lysosomes, destroying cells infected by viruses. Fever also promotes the activity of certain lymphocytes. A short-term, low fever speeds recovery.

Review

▌ What are the main groups of cytokines? What is the function of each group?

▌ What processes are involved in the inflammatory response? What types of cells help mediate the inflammatory response?

SPECIFIC IMMUNE RESPONSES

Learning Objectives

4   Distinguish between cell-mediated and antibody-mediated immunity.

5   Describe the principal cells of the immune system, and summarize the function of the major histocompatibility complex.

While nonspecific immune responses destroy pathogens and prevent the spread of infection, the body mobilizes its specific immune responses. It takes several days to activate specific immune responses, but once in gear, these mechanisms are extremely effective. As discussed earlier, specific immune responses are precisely targeted to destroy specific antigens, and they have immunological memory. Two types of specific immune responses are *cell-mediated immunity* and *antibody-mediated immunity*.

Many types of cells are involved in specific immune responses

The primary types of cells of the immune system are shown in ▮ Figure 44-4. Two key cell types that participate in specific immune responses are lymphocytes and antigen-presenting cells. Phagocytes and many other cells important in nonspecific immune responses also participate in specific responses. **Eosinophils** are white blood cells that release proteins, including cytokines, from their granules. They also release proteins that are toxic to parasitic worms.

Lymphocytes are the principal warriors in specific immune responses

Three main types of lymphocytes are T lymphocytes, B lymphocytes, and NK cells. As already discussed, NK cells kill virally infected cells and tumor cells. **T lymphocytes,** or **T cells,** are responsible for **cell-mediated immunity.** T cells are the body's cellular soldiers. They travel to the site of infection and attack body cells infected by invading pathogens, as well as foreign cells (such as those introduced in tissue grafts or organ transplants). T cells also destroy cells altered by mutation (cancer cells).

B lymphocytes, or **B cells,** are responsible for **antibody-mediated immunity.** B cells mature into **plasma cells,** which produce specific antibodies. Antibodies bind to specific antigens, neutralizing them or marking them for destruction. Antibody-mediated immunity is one of the body's chief chemical defense strategies.

All lymphocytes develop from stem cells in the bone marrow. B cells complete their development in the adult bone marrow. T cells mature in the **thymus gland** (from which the *T* in their name derives). Large numbers of mature lymphocytes reside in lymph organs, including the spleen, lymph nodes, tonsils, and other lymph tissues strategically positioned throughout the body.

Although T cells and B cells are similar in appearance when viewed with a light microscope, sophisticated techniques such as fluorescence microscopy demonstrate that these cells can be differentiated by their unique cell-surface macromolecules. T cells and B cells have different functions and life histories and tend to locate in (or "home" to) separate regions of the spleen, lymph nodes, and other lymph tissues.

Millions of B cells are produced in the bone marrow daily. Each B cell is genetically programmed to encode a glycoprotein receptor that can bind with a specific type of antigen. When a B cell comes into contact with an antigen that binds to its receptors, the B cell becomes activated. B-cell activation is a complex process that typically requires the participation of a particular type of T cell.

Once activated, a B cell divides rapidly to form a clone of identical cells. The cloned B cells differentiate into plasma cells, which produce antibody. A plasma cell can produce more than 10 million molecules of antibody per hour! The antibody binds to the antigen that originally activated the B cells. Some activated B cells do not become plasma cells but instead become long-living **memory B cells** that continue to produce small amounts of antibody after the body overcomes an infection.

On their way from the bone marrow to the lymph tissues, T cells are processed in the thymus gland. The thymus makes T cells *immunocompetent*, capable of immunological response. As T cells move through the thymus, they divide many times and develop specific surface proteins with distinctive receptor sites. The T-cell population undergoes positive and negative selection. In *positive selection*, T cells are allowed to mature if they recognize self-antigens and bind with foreign antigens.

In *negative selection*, T cells in the thymus that react to self-antigens undergo apoptosis. Immunologists estimate that more than 90% of developing T cells are negatively selected. The remaining T cells differentiate and leave the thymus to take up residence in other lymph tissues or to launch immune responses in infected tissues. By selecting only appropriate T cells, the thymus gland ensures that T cells can distinguish between the body's own molecules and foreign antigens.

Most T cells in the thymus differentiate just before birth and during the first few months of postnatal life. If the thymus is removed before this processing takes place, an animal does not develop cell-mediated immunity. If the thymus is removed after that time, cell-mediated immunity is less seriously impaired.

T cells are distinguished by the **T-cell receptor (TCR),** which recognizes specific antigens. Two main populations of T cells develop in the thymus: T cytotoxic cells and T helper cells. **T cytotoxic (T_C) cells** are also known as *CD8 cells* because they have a glycoprotein designated CD8 on their plasma membrane surface. Known less formally as *killer T cells*, T_C cells recognize and destroy cells with foreign antigens on their surfaces. Among their target cells are virus-infected cells, cancer cells, and foreign tissue grafts.

T helper (T_H) cells, also known as *CD4 cells*, have a surface glycoprotein designated CD4. T_H cells are regulatory cells. They secrete cytokines that activate B cells and macrophages. After an infection, **memory T cells** (both T_C and helper) remain in the body.

Antigen-presenting cells activate T helper cells

Macrophages, dendritic cells, and B cells function as "professional" **antigen-presenting cells (APCs)** that display foreign antigens as well as their own surface proteins. APCs are inac-

Two main categories of immune system cells are lymphocytes (T cells and B cells) and antigen-presenting cells (macrophages and dendritic cells).

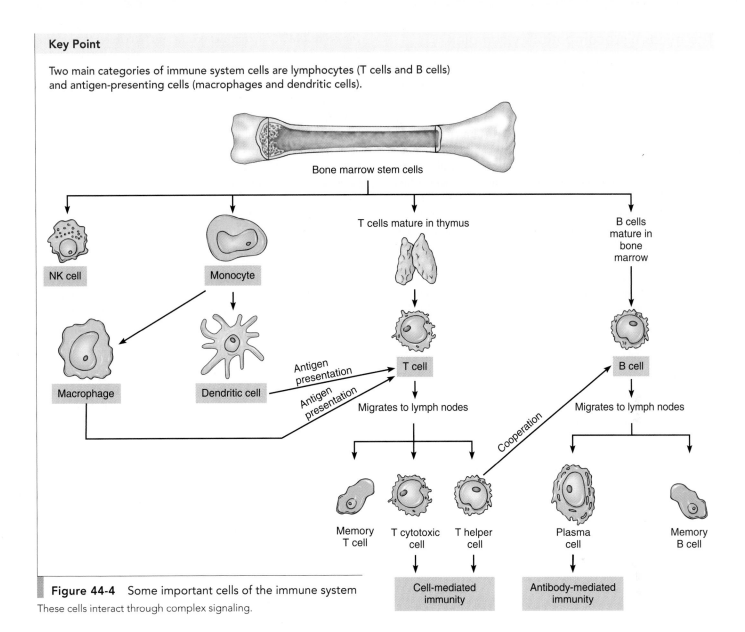

Figure 44-4 Some important cells of the immune system

These cells interact through complex signaling.

tive until their pattern-recognition receptors recognize PAMPs on pathogens. Once activated, an APC ingests the pathogen, and lysosomal enzymes degrade most, but not all, of the bacterial antigens. The APC displays fragments of the foreign antigens on its cell surface in association with a type of self-molecule (class II MHC, discussed later). The activated APC expresses additional signaling molecules, called **co-stimulatory molecules,** along with the displayed antigen. APCs present displayed antigens to T cells.

In addition to their function as APCs, macrophages secrete more than 100 different compounds, including cytokines and enzymes that destroy bacteria. When stimulated by bacteria, macrophages secrete interleukins that activate B cells and certain T cells. Interleukins also promote nonspecific immune responses, such as fever.

Biologists first identified **dendritic cells** in 1868 and then rediscovered them in 1973, but the small numbers of these cells made them hard to study until sophisticated techniques became

available in the 1990s. Their name derives from their long cytoplasmic extensions known as *dendrites* (not to be confused with the dendrites of neurons). Like macrophages, they develop from monocytes. Dendritic cells are strategically stationed in all tissues of the body that come into contact with the environment—the skin and the linings of the digestive, respiratory, urinary, and vaginal passageways.

When pathogens infect a tissue, PAMPs activate dendritic cells. Activated dendritic cells capture the pathogens (or their products) by phagocytosis or by receptor-mediated endocytosis (see Chapter 5). Guided by chemokines produced by cells in the lymph nodes and lymph vessels, dendritic cells make their way to the lymph nodes. They mature as they migrate. Mature dendritic cells break down antigens and display the fragments on their cell surface. Co-stimulatory molecules, along with the displayed antigen, attract and activate specific T cells capable of responding to the antigen. Thus, dendritic cells are specialized to process, transport, and present antigens to T cells.

The major histocompatibility complex is responsible for recognition of self

The ability of the vertebrate immune system to distinguish self from nonself depends largely on a cluster of closely linked genes known as the **major histocompatibility complex (MHC).** In humans, the MHC is also called the **HLA (human leukocyte antigen) complex.** The MHC genes are polymorphic (variable). Within the population there are multiple alleles, more than 40 for each locus. As a result, the cell-surface proteins for which they code are generally different in each individual. With so many possible combinations, no two people, except identical twins, are likely to have all the same MHC proteins on their cells. The more closely related two individuals are, the more MHC genes they have in common. Thus, a person's MHC proteins serve as a biochemical "fingerprint."

The MHC genes encode **MHC antigens,** or self-antigens, that differ in chemical structure, function, and tissue distribution. Class I MHC genes encode glycoproteins expressed on the surface of most nucleated cells. These antigens bind with foreign antigens from viruses or other pathogens within the cell to form a foreign antigen–class I MHC glycoprotein complex. The cell displays this complex on its surface and presents it to T_C cells. Thus, any infected cell can function as an antigen-presenting cell and can activate certain T_C cells.

Class II MHC genes encode glycoproteins expressed primarily on "professional" APCs—dendritic cells, macrophages, and B cells. These class II MHC antigens combine with foreign antigen from bacteria, and the cell presents the complex to T_H cells. Class III MHC genes encode many secreted proteins involved in the immune response, including components of the complement system and TNFs.

Review

- How are cell-mediated immunity and antibody-mediated immunity different?
- What is the function of antigen-presenting cells? Describe two types.
- What are two main types of T cells? What is the function of B cells?

CELL-MEDIATED IMMUNITY

Learning Objective

6 Describe the sequence of events in cell-mediated immunity.

T cells and APCs (mainly dendritic cells and macrophages) are responsible for cell-mediated immunity (❚ Fig. 44-5). T cells destroy cells infected with viruses and cells that have been altered in some way, such as cancer cells. They also destroy the cells of foreign grafts such as a transplanted kidney. There are two general populations of T cells: T_C cells and T_H cells. Each T_C cell expresses the CD8 molecule on its surface, as well as more than 50,000 identical T-cell receptors (TCRs) that bind to one specific type of antigen. Each T_H cell expresses the glycoprotein called CD4. Within each general category of T cell, there are thousands of different antigen specificities that can be recognized. T_C cells recognize antigens presented to them as part of a foreign anti-

gen–class I MHC glycoprotein complex. T_H cells recognize antigens presented to them as part of a foreign antigen–class II MHC glycoprotein complex.

How do T cells know which cells to target? T cells do not recognize antigens unless they are presented properly. When a virus (or other pathogen) infects a cell, some of the viral protein is broken down to peptides and displayed with class I MHC molecules on the cell surface. Only T_C cells with receptors that bind to the specific antigen–MHC class I complex become activated. Generally, fewer than 1 in 10,000 T cells have the same antigen specificity and can respond. T_C cell activation requires at least two signals in addition to the presented antigen: a co-stimulatory signal and an interleukin signal.

Once activated, a T_C cell increases in size and gives rise to a clone of many T_C cells. Activated T_C cells are the effector cells that make up the cell infantry. They leave the lymph nodes and make their way to the infected area, where they destroy target cells within seconds after contact. When a T_C cell combines with antigen on the surface of the target cell, it destroys the cell in much the same way that NK cells destroy their target cells. The T_C secretes perforins and granzymes that perforate the plasma membrane of the target cell and induce it to kill itself by apoptosis. After releasing cytotoxic substances, the T cell disengages itself from its victim cell and seeks out a new target.

The sequence of events in cell-mediated immunity can be summarized as follows:

> Virus invades body cell ⟶ infected cell displays foreign antigen–MHC class I antigen complex ⟶ specific T_C cell activated by this complex ⟶ clone of T_C cells produced ⟶ T_C cells migrate to area of infection ⟶ T_C cells release proteins that stimulate destruction of target cells

T_H cells are activated by the foreign antigen–MHC class II antigen complex displayed on the surface of APCs. Once activated, a T_H cell gives rise to a clone of T_H cells. Some T_H cells attract macrophages to the site of infection and promote the destruction of intracellular pathogens. Others function mainly in promoting antibody-mediated immunity.

Review

- How do T cytotoxic cells function in cell-mediated immunity?
- How do T helper cells function in cell-mediated immunity?

ANTIBODY-MEDIATED IMMUNITY

Learning Objectives

7 Summarize the sequence of events in antibody-mediated immunity, including the effects of antigen–antibody complexes on pathogens.
8 Describe the basic structure and function of an antibody, and explain the basis of antibody diversity.

B cells are responsible for antibody-mediated immunity. A given B cell can produce many copies of one specific antibody. Antibody molecules serve as cell-surface receptors that combine with antigens. Only a B cell displaying a matching receptor on its sur-

T cells and APCs are responsible for cell-mediated immunity.

1 T$_H$ cell is activated by specific foreign antigen–MHC complex presented by APC.

2 Activated T$_H$ cell increases in size and divides by mitosis.

3 Clone of competent T$_H$ cells and memory T cells produced.

Memory T cells

4 T$_H$ cells differentiate and release cytokines that act on other T cells and on macrophages.

5 T$_C$ cells are activated by foreign antigen-MHC complex displayed by infected cells. T$_C$ cells form clones (*not shown*) and migrate to tissues. They release proteins that destroy infected cells.

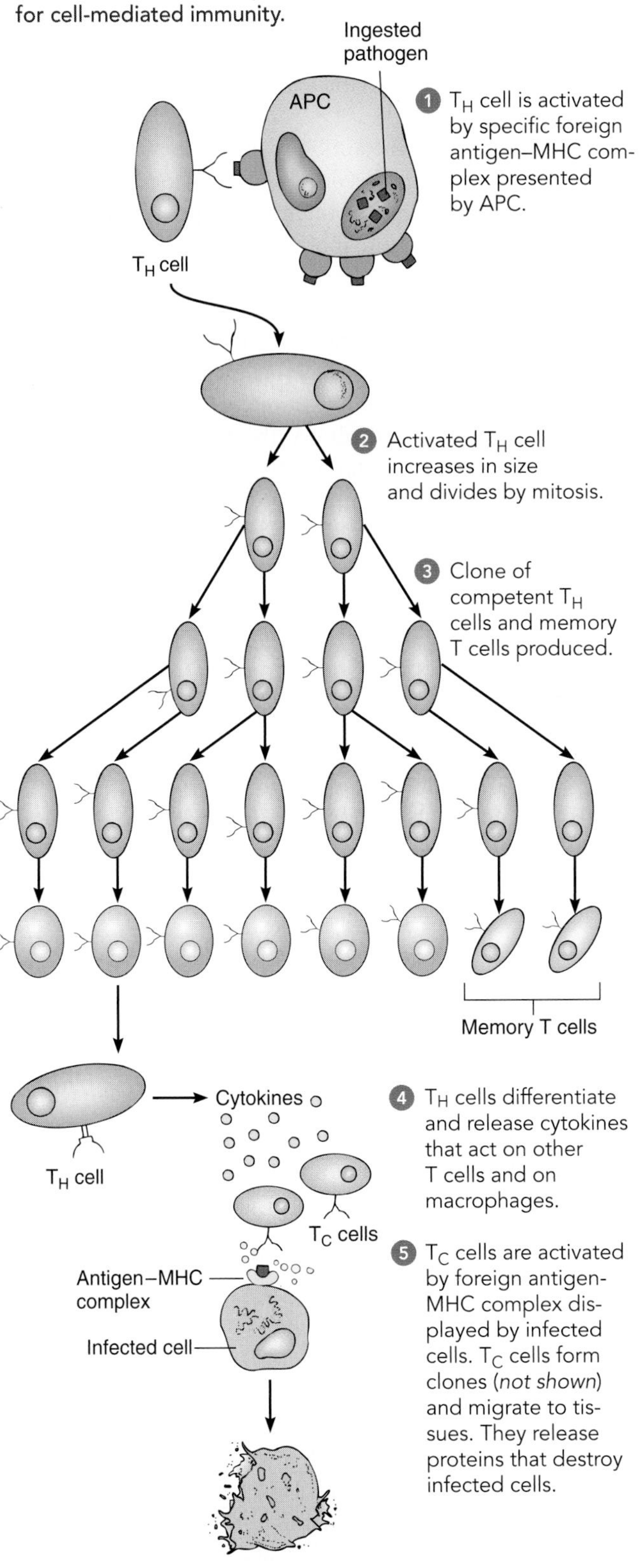

Figure 44-5 *Animated* Cell-mediated immunity

When activated by a foreign antigen–MHC complex presented by an APC and by cytokines, a T$_H$ cell divides and gives rise to a clone of T$_H$ cells. T$_C$ cells are activated by antigens presented on the surfaces of infected cells. T$_C$ cells also form a clone. Both types of T cells migrate to the site of infection. T$_H$ cells release cytokines that stimulate T$_C$ cells and activate macrophages so that they phagocytose pathogens. T$_C$ cells release proteins that destroy infected cells.

face can bind a particular antigen. Inside the B cell, the antigen is degraded into peptide fragments. The B cell then displays the peptide fragments together with MHC protein class II on its surface.

In most cases, the activation of B cells involves APCs and T$_H$ cells (❚ Fig. 44-6). T$_H$ cells stimulate B cells to divide and produce antibodies. However, the T$_H$ cell itself must first be activated. T$_H$ cells do not recognize an antigen that is presented alone. The antigen must be presented as part of a foreign antigen–class II MHC complex on the surface of an APC. When an APC displaying a foreign antigen–MHC complex contacts a T$_H$ cell with complementary T-cell receptors, a complicated interaction occurs. Multiple chemical signals are sent back and forth between cells. For example, the APC secretes interleukins, such as IL-1, that activate T$_H$ cells.

B cells serve as APCs to T cells. An activated T$_H$ cell binds with the foreign antigen–MHC complex on the B cell. The T$_H$ cell then releases interleukins, which together with antigen activate the B cell.

Once activated, a B cell divides by mitosis, giving rise to a clone of identical cells (❚ Fig. 44-7). This cell division in response to a specific antigen is known as *clonal selection* (discussed later). The cloned B cells mature into plasma cells that secrete antibodies specific to the antigen that activated the original B cell. It is important to remember that the specificity of the clone is determined *before* the B cell encounters the antigen.

Unlike T cells, most plasma cells do not leave the lymph nodes. Only the antibodies they secrete pass out of the lymph tissues and make their way via the lymph and blood to the infected area. The secreted antibody is a soluble form of the glycoprotein receptor of the activated B cell.

The sequence of events in antibody-mediated immunity can be summarized as follows:

Pathogen invades body ⟶ APC phagocytoses pathogen ⟶ foreign antigen–MHC complex displayed on APC surface ⟶ T$_H$ cell binds with foreign antigen–MHC complex ⟶ activated T$_H$ cell interacts with a B cell that displays the same antigen ⟶ B cell activated ⟶ clone of B cells ⟶ B cells differentiate, becoming plasma cells ⟶ plasma cells secrete antibodies ⟶ antibodies form complexes with pathogen ⟶ pathogen is destroyed

Some activated B cells do not differentiate into plasma cells but instead become memory B cells (discussed in a later section).

A typical antibody consists of four polypeptide chains

An antibody molecule, also called **immunoglobulin (Ig),** has two main functions: it combines with antigen, and it activates processes that destroy the antigen that binds to it. For example,

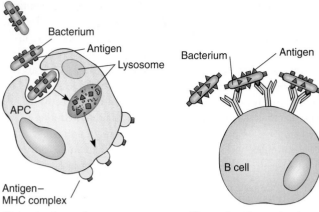

1a APC takes in bacterium, breaks it down, and presents foreign antigens on its surface in combination with MHC class II antigens.

1b B cell independently combines with foreign antigens on surface of bacterium, but B cell is typically not active until stimulated by T_H cells.

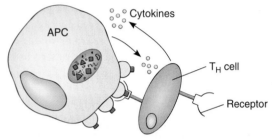

2 T_H cells are activated when their receptors combine with foreign antigen–MHC antigen complexes on APC. Signaling takes place via cytokines secreted by both cells.

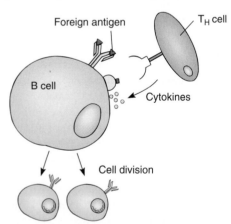

3 Activated T_H cell recognizes antigen–MHC complex on B cell. T_H cell secretes cytokines that activate B cell that has interacted with foreign antigen. B cell then divides, and subsequent divisions produce clone of identical B cells.

Figure 44-6 *Animated* B-cell activation

an antibody may stimulate phagocytosis. Note that an antibody does not destroy an antigen directly; rather, it *labels* the antigen for destruction.

Rodney Porter, of the University of Oxford in England, and Gerald Edelman, of Rockefeller University in New York, clarified the basic structure of the immunoglobulin molecule during the

1960s. Porter used the plant enzyme papain, a protease, to split Ig molecules into fragments. Based on his findings, Porter developed a working model of the structure of the Ig molecule and was the first to suggest that it is Y shaped. These researchers then constructed an accurate model of the antibody molecule. Porter and Edelman were awarded the 1972 Nobel Prize in Medicine for their contributions.

Two fragments of the antibody molecule bind antigen and are referred to as *Fab fragments* (*Fab* stands for *antigen-binding fragment*). The fragment that interacts with cells of the immune system is the *Fc fragment* (*c* indicates that this fragment crystallizes during cold storage). Many cells of the immune system have Fc receptors.

In a Y-shaped antibody molecule, the two arms of the Y, the *Fab*, or *antigen-binding fragments*, bind with antigen (❚ Fig. 44-8). This shape permits the antibody molecule to combine with two antigen molecules, which allows the formation of **antigen–antibody complexes.** The tail of the Y, called the *Fc fragment*, interacts with cells of the immune system, such as phagocytes, or binds with molecules of the complement system.

An antibody molecule consists of four polypeptide chains: two identical long chains, called *heavy chains,* and two identical short chains, called *light chains.* Each chain has a constant region and a variable region. In the **constant (C) region** of the heavy chains, the amino acid sequence is constant within a particular immunoglobulin class. You can think of the C region of the heavy chains as the handle portion of a door key. The C region is the tail of the Y part of the antibody molecule.

The **variable (V) region** has a unique amino acid sequence, like the pattern of bumps and notches on the part of a key that slides into a lock. At its variable regions, the antibody folds three-dimensionally, assuming a shape that enables it to combine with a specific antigen. When they meet, antigen and antibody fit together somewhat like a lock and key. They must fit in just the right way for the antibody to be effective (❚ Fig. 44-9).

The specific part of the antigen molecule that is recognized by an antibody or T-cell receptor is called an **antigenic determinant,** or *epitope.* An antigen may have many antigenic determinants on its surface; some have hundreds. Antigenic determinants often differ from one another, so several different kinds of antibodies can combine with a single antigen. A given antibody can bind with different strengths, or *affinities,* to different antigens. In the course of an immune response, higher-affinity antibodies are generated.

Antibodies are grouped in five classes

Antibodies are grouped in five classes, defined by unique amino acid sequences in the C regions of the heavy chains. The classes are named by using the abbreviation *Ig* for *immunoglobulin* and are designated IgG, IgM, IgA, IgD, and IgE. In humans, about 75% of the circulating antibodies belong to the **IgG** class; these are part of the gamma globulin fraction of the plasma. IgG and **IgM** share some functional properties. Both interact with macrophages and activate the complement system. They defend against many pathogens carried in the blood, including bacteria, viruses, and some fungi. IgM is the first antibody humans make in an immune response. IgG is the antibody that crosses the placenta and protects the fetus and, later, the newborn baby.

IgA, present in mucus, tears, saliva, and breast milk, prevents viruses and bacteria from attaching to epithelial surfaces. This immunoglobulin, which defends against inhaled or ingested pathogens, is strategically secreted into the respiratory, digestive, urinary, and reproductive tracts. **IgD** has a low concentration (less than 1%) in plasma. Along with IgM, it is an important immunoglobulin on the B-cell surface. IgD has a critical role in maturation of B cells and helps activate B cells following antigen binding.

IgE, which has an even lower plasma concentration, binds to mast cells, which contain potent signaling molecules, such as histamine. These chemicals are released when an antigen binds to IgE on a mast cell. Histamine triggers many allergy symptoms, including inflammation. IgE is also responsible for an immune response to invading parasitic worms.

Antigen–antibody binding activates other defenses

Antibodies mark a pathogen as foreign by combining with an antigen on its surface. Generally, several antibodies bind with several antigens, creating a mass of clumped antigen–antibody complexes. The combination of antigen and antibody activates several defensive responses.

1. The antigen–antibody complex may inactivate the pathogen or its toxin. For example, when an antibody attaches to its surface, a virus may lose the ability to attach to a host cell. Snake venom antitoxin contains antibodies that neutralize the toxins that enter the body with poisonous snakebites.

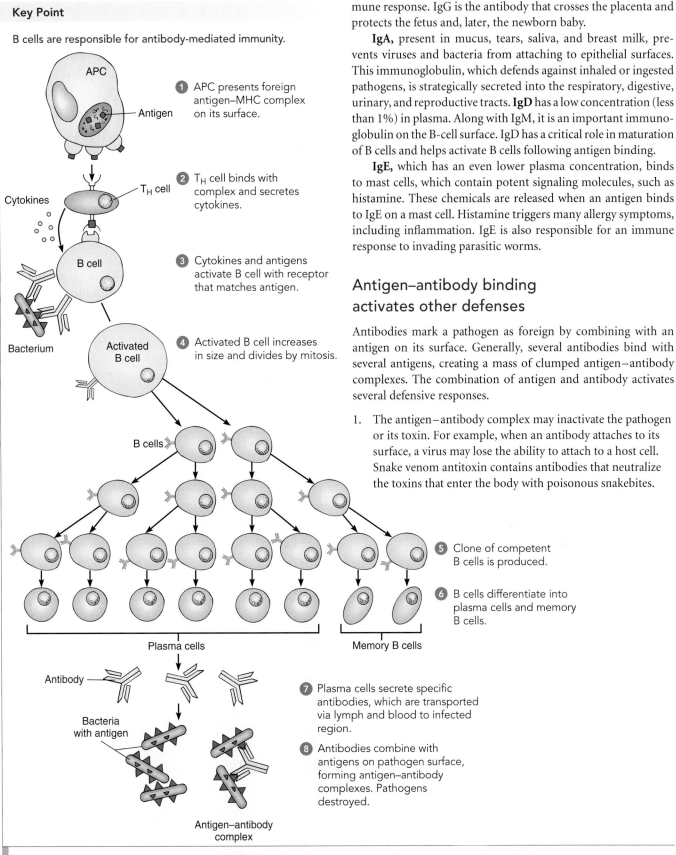

Key Point

B cells are responsible for antibody-mediated immunity.

1. APC presents foreign antigen–MHC complex on its surface.

2. T$_H$ cell binds with complex and secretes cytokines.

3. Cytokines and antigens activate B cell with receptor that matches antigen.

4. Activated B cell increases in size and divides by mitosis.

5. Clone of competent B cells is produced.

6. B cells differentiate into plasma cells and memory B cells.

7. Plasma cells secrete specific antibodies, which are transported via lymph and blood to infected region.

8. Antibodies combine with antigens on pathogen surface, forming antigen–antibody complexes. Pathogens destroyed.

Figure 44-7 *Animated* Antibody-mediated immunity

A specific B cell may be activated when a specific antigen binds to immunoglobulin receptors on the B-cell surface. Typically, the B cell also requires cytokine stimulation from nearby T$_H$ cells. Once activated, the B cell divides and produces a clone of B cells. These cells differentiate and become plasma cells that secrete antibodies. The plasma cells remain in the lymph tissues, but the antibodies are transported to the site of infection by the blood or lymph. After the infection has been overcome, memory B cells remain in the tissues.

The Immune System: Internal Defense   **955**

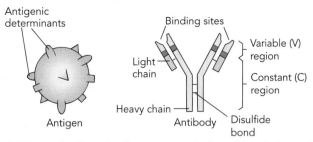

(a) The antibody molecule is composed of two light chains and two heavy chains joined by disulfide bonds. The constant (C) and variable (V) regions of the chains are labeled.

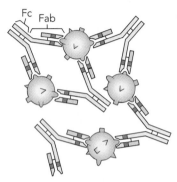

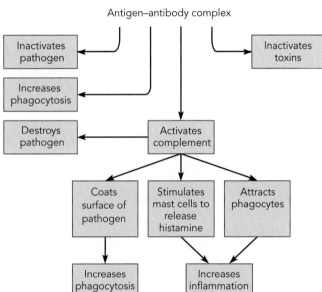

(b) The Fab portion of the antibody binds to antigen. The Fc portion of the molecule binds to a cell of the immune system (*not shown*). Antigen–antibody complexes directly inactivate pathogens and increase phagocytosis. They also activate the complement system.

Figure 44-8 *Animated* Antibody structure and function

Antibodies combine with antigens to form antigen–antibody complexes.

2. The antigen–antibody complex stimulates phagocytic cells to ingest the pathogen.

3. Antibodies of the IgG and IgM groups work mainly through the complement system. When antibodies combine with a

specific antigen on a pathogen, complement proteins destroy the pathogen. IgG molecules have an Fc fragment that binds Fc receptors; these receptors are expressed on many cells, including macrophages and neutrophils. When an antibody molecule is bound to a pathogen and the Fc portion of the antibody binds with an Fc receptor on a phagocyte, the pathogen is more easily destroyed.

The immune system responds to millions of different antigens

How can the immune system recognize every possible antigen, even those produced by newly mutated viruses never before encountered during the evolution of our species? In the 1950s, the Danish researcher Niels Jerne, of the Basel Institute for Immunology in Basel, Switzerland; David Talmage, of the University of Chicago; and Frank Macfarlane Burnet, of the Walter and Eliza Hall Institute for Medical Research in Melbourne, Australia, developed the **clonal selection** hypothesis. This hypothesis proposes that before a lymphocyte ever encounters an antigen, the

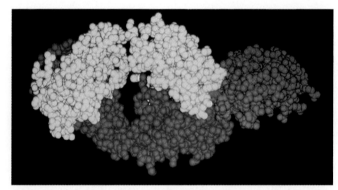

(a) A portion of the antigenic determinant (*shown in red*) fits into a groove in the antibody molecule.

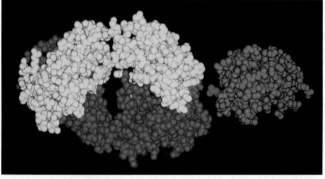

(b) The antigen–antibody complex has been pulled apart to show its structure.

Figure 44-9 An antigen–antibody complex

The components of an antigen–antibody complex fit together as shown in this computer simulation of their molecular structure. The antigen lysozyme is shown in green, the heavy chain of the antibody is shown in blue, and the light chain is shown in yellow.

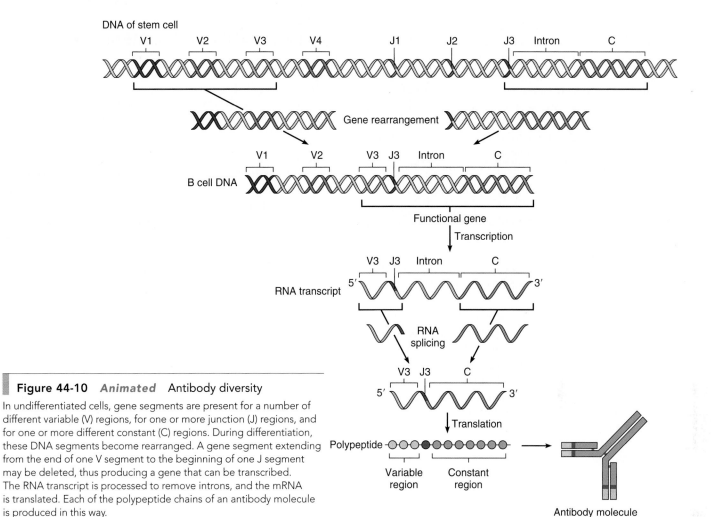

Figure 44-10 *Animated* Antibody diversity

In undifferentiated cells, gene segments are present for a number of different variable (V) regions, for one or more junction (J) regions, and for one or more different constant (C) regions. During differentiation, these DNA segments become rearranged. A gene segment extending from the end of one V segment to the beginning of one J segment may be deleted, thus producing a gene that can be transcribed. The RNA transcript is processed to remove introns, and the mRNA is translated. Each of the polypeptide chains of an antibody molecule is produced in this way.

lymphocyte has specific receptors for that antigen on its surface. (Jerne and Burnet later were awarded Nobel prizes in Medicine for their work in immunology.)

When an antigen binds to a matching receptor on a lymphocyte, it activates the lymphocyte, which then gives rise to a clone of cells with identical receptors. A problem with this hypothesis when it was initially proposed was its suggestion that our cells must contain millions of separate antibody genes, each coding for an antibody with a different specificity. Although each human cell has a large amount of DNA, it is not enough to provide a different gene to code for each of the millions of possible specific antibody molecules.

An explanation for how the immune system could have so many different specificities came later. In 1965, W. J. Dreyer, of the Institute of Biochemistry at the University of Zurich, Switzerland; and Joe Claude Bennett, of the University of Alabama at Birmingham, suggested that the C region and the V region of an immunoglobulin are encoded by two separate genes. Their hypothesis was at first rejected by many biologists because it contradicted the prevailing hypothesis that one gene codes for one polypeptide.

The technology needed to test Dreyer and Bennett's hypothesis was not immediately available, and it was not until 1976 that Susumu Tonegawa, of the Massachusetts Institute of Technology,

and his colleagues demonstrated that separate genes encode the V and C regions of immunoglobulins. Tonegawa further showed that three separate families of genes code for immunoglobulins and that each gene family contains a large number of DNA segments that code for V regions. Recombination of these DNA segments during the differentiation of B cells is responsible for antibody diversity. Tonegawa was awarded the 1987 Nobel Prize in Medicine for his work, which transformed the emerging science of immunology.

We now understand that in undifferentiated B cells, gene segments are present for a number of different V regions, for one or more junction (J) regions, and for one or more different C regions (▌Fig. 44-10). Rearrangement of these DNA segments produces an enormous number of potential combinations. Millions of different types of B (and T) cells are produced. By chance, one of those B cells may produce just the right antibody to destroy a pathogen that invades the body. To appreciate the capacity for antibody diversity, consider the diverse combinations of things you create in your everyday life. A familiar example is making an ice-cream sundae. Imagine the varied combinations that are possible using 10 flavors of ice cream, 6 types of sauce, and 15 kinds of toppings.

Researchers have identified additional sources of antibody diversity. For example, the DNA of the mature B cells that codes for the regions of immunoglobulins mutates very readily. These somatic mutations produce genes that code for slightly different antibodies.

We have used antibody diversity here as an example, but similar genetic mechanisms account for the diversity of T-cell receptors. Although we may actually use only a relatively few types of antibodies or T cells in a lifetime, the remarkable diversity of the immune system prepares it to attack most potentially harmful antigens that may invade the body.

Monoclonal antibodies are highly specific

Before 1975, the only method for obtaining antibodies for medicine or research was immunizing animals and collecting their blood. Then Cesar Milstein and Georges Kohler at the Laboratory of Molecular Biology in Cambridge, England, developed **monoclonal antibodies,** identical antibodies produced by cells cloned from a single cell.

One way to produce monoclonal antibodies in the laboratory is to inject mice with the antigen of interest, for example, an antigen from a particular bacterium. After the mice have produced antibodies to the antigen, their plasma cells are collected. However, these normal plasma cells survive in culture for only a few generations, limiting the amount of antibody that can be produced. In contrast, cancer cells live and divide in tissue culture indefinitely. B cells can be suspended in a culture medium together with lymphoma cells from other mice. (Lymphoma is cancer of lymphocytes.) Then the B cells and lymphoma cells are induced to fuse. They form hybrid cells, known as **hybridomas,** which have properties of the two "parent" cells. Hybridomas can be cultured indefinitely (a cancer cell property) and continue to secrete antibodies (a B-cell property).

Researchers select hybrid cells that are manufacturing the specific antibody needed and then clone them in a separate cell culture. Cells of this clone secrete large amounts of the particular antibody—thus the name *monoclonal antibodies.* Each type of monoclonal antibody is specific for a single antigenic determinant.

Immunologists also use recombinant DNA technology to produce monoclonal antibodies. Because of their purity and specificity, monoclonal antibodies have proved to be invaluable tools in modern biology. For example, a researcher may want to detect a specific molecule present in very small amounts in a mixture. The reaction of that molecule (the antigen) with a specific monoclonal antibody makes its presence known. Monoclonal antibodies are used in similar ways in various diagnostic tests. Home pregnancy tests make use of a monoclonal antibody that is specific for human chorionic gonadotropin (hCG), a hormone produced by a developing human embryo (see Chapters 49 and 50). Several monoclonal antibodies are being used clinically to treat cancer. The drug Herceptin is a monoclonal antibody used in treating a form of breast cancer.

Review

- How do antibodies recognize antigens?
- What are three ways that antigen–antibody complexes affect pathogens?

IMMUNOLOGICAL MEMORY

Learning Objectives

9 Describe the basis of immunological memory, and contrast secondary and primary immune responses.
10 Compare active and passive immunity, and give examples of each.

Memory B cells and memory T cells are responsible for long-term immunity. Following an immune response, memory T cells group strategically in many nonlymphatic tissues, including the lung, liver, kidney, and intestine. Many infections occur at such sites, and T cells stationed in those locations can respond quickly. In response to antigen, memory T cells rapidly become T_C cells that produce substances that kill invading cells.

Memory B cells have a "survival gene" that prevents apoptosis, the programmed cell death that is the eventual fate of plasma cells. Memory B cells continue to live and produce small amounts of antibody long after the body has overcome an infection. This circulating antibody is part of the body's arsenal of chemical weapons. If the same pathogen enters the body again, the antibody immediately targets it for destruction. At the same time, specific memory cells are stimulated to divide, producing new clones of plasma cells that produce the same antibody. The presence of circulating antibodies is used clinically to detect previous exposure to specific pathogens. For example, some HIV screening tests measure antibody to the virus that causes AIDS.

A secondary immune response is more effective than a primary response

The first exposure to an antigen stimulates a **primary immune response.** An infection, or an antigen injected into an animal, causes specific antibodies to appear in the blood plasma in 3 to 14 days. After antigen injection, there is a brief *latent period* during which the antigen is recognized and appropriate lymphocytes begin to form clones. A *logarithmic phase* follows, during which the antibody concentration rises rapidly for several days until it reaches a peak (▌ Fig. 44-11). IgM is the principal antibody synthesized during the primary immune response. Finally, there is a *decline phase,* during which the antibody concentration decreases to a very low level.

A second exposure to the same antigen, even years later, results in a **secondary immune response.** Because memory B cells bearing antibodies to that antigen (and also memory T cells with their specific T-cell receptors) persist for many years, the secondary immune response is much more rapid than the primary immune response and has a shorter latent period. Much less antigen is necessary to stimulate a secondary immune response than a primary response, and more antibodies are produced. In addition, the affinity of antibodies is generally much higher. The predominant antibody in a secondary immune response is IgG.

The body's ability to launch a rapid, effective response during a second encounter with an antigen explains why we do not usually suffer from the same infectious disease several times. For example, a person who contracts measles or chickenpox is unlikely

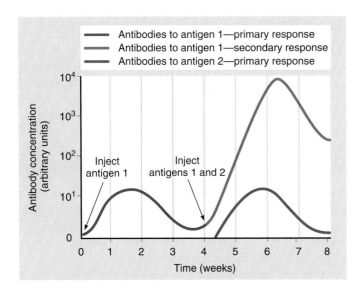

Figure 44-11 *Animated* Immunological memory

Antigen 1 was injected at day 0, and the immune response was assessed by measuring antibody levels to the antigen. At week 4, the primary immune response had subsided. Antigen 1 was injected again, together with a new protein, antigen 2. The secondary immune response was greater and more rapid than the primary response. It was also specific to antigen 1. A primary immune response was made to the newly encountered antigen 2.

to become reinfected in the future. When someone is exposed a second time, the immune system responds quickly, destroying the pathogens before they have time to multiply and cause symptoms of the disease. Booster shots of vaccine are given to elicit a secondary immune response, thus reinforcing immunological memory.

You may wonder how a person can get influenza (the flu) or a cold more than once. The reason is that there are many varieties of these viral diseases, each caused by a virus with slightly different antigens. For example, more than 100 different viruses cause the common cold, and new varieties of cold and flu viruses evolve continuously by mutation (a survival mechanism for them), which may result in changes in their surface antigens.

Even a slight change may prevent recognition by memory cells. Because the immune system is so specific, each different antigen is treated by the body as a new immunological challenge.

Immunization induces active immunity

We have been considering **active immunity,** immunity that develops following exposure to antigens. Active immunity can be naturally or artificially induced (▋ Table 44-1). If someone with chickenpox sneezes near you and you contract the disease, you develop active immunity naturally. Active immunity can also be artificially induced by **immunization,** that is, by exposure to a *vaccine.* When an effective vaccine is introduced into the body, the immune system actively develops clones of cells, produces antibodies, and develops memory cells.

The first vaccine was prepared in 1796 by British physician Edward Jenner against vaccinia, the cowpox virus. (The term *vaccination* was derived from the name of the cowpox virus.) Jenner's vaccine provided humans with immunity against the deadly disease smallpox, which closely resembles cowpox. Jenner had no knowledge of microorganisms or of immunology, and 100 years passed before French chemist Louis Pasteur began to develop scientific methods for preparing vaccines.

Pasteur showed that inoculations with preparations of attenuated (weakened) pathogens could be used to develop immunity against the virulent (infectious) form of the pathogen. However, not until 20th-century advances in immunology—for example, Burnet's clonal selection theory in 1957 and the discovery of T cells and B cells in 1965—did scientists gain a modern understanding of vaccines. Effective vaccination stimulates the body to launch an immune response against the antigens contained in the vaccine. Memory cells develop, and future encounters with the same pathogen are dealt with rapidly.

Microbiologists prepare effective vaccines in various ways. They can attenuate a pathogen so it loses its ability to cause disease. When pathogens are cultured for long periods in nonhuman cells, mutations adapt the pathogens to the nonhuman host so that they no longer cause disease in humans. This method is used to produce the Sabin polio, measles, mumps, and rubella vaccines. Tetanus and diphtheria vaccines are made from toxins

TABLE 44-1

Active and Passive Immunity

Type of Immunity	When Developed	Development of Memory Cells	Duration of Immunity
Active			
Naturally induced	After pathogens enter the body through natural encounters (e.g., a person with measles sneezes on you)	Yes	Many years
Artificially induced	After immunization with a vaccine	Yes	Many years
Passive			
Naturally induced	After transfer of antibodies from mother to developing baby	No	Few months
Artificially induced	After injection with gamma globulin	No	Few months

secreted by the respective pathogens. The toxin is altered so that it can no longer destroy tissues, but its antigenic determinants are still intact.

Researchers are developing **DNA vaccines** (or RNA vaccines) made from a part of the pathogen's genetic material. The DNA of the pathogen is altered so that it transfers genes that specify antigens. When injected into a patient, the altered DNA is taken up by cells and makes its way to the nucleus. The encoded antigens are manufactured and stimulate both cell-mediated and antibody-mediated immunity. Several DNA vaccines, including vaccines to prevent and treat HIV infection, are in clinical trials.

Passive immunity is borrowed immunity

To provide **passive immunity,** physicians inject people with antibodies actively produced by another organism. The serum or gamma globulin that contains these antibodies is obtained from humans or other animals. Nonhuman serums are less desirable, because nonhuman proteins can act as antigens, stimulating an immune response that may result in serum sickness.

Passive immunity is borrowed immunity. Its effects are temporary whether the immunity is artificially or naturally induced (see Table 44-1). Physicians use artificially induced passive immunity to boost the body's defenses temporarily against a particular disease. For example, when exposed to hepatitis A, a form of viral hepatitis spread through contaminated food or water, people at risk of infection are injected with gamma globulin containing antibodies to the hepatitis pathogen. However, gamma globulin injections offer protection for only a few weeks. Because the body has not actively launched an immune response, it has no memory cells and cannot produce antibodies to the pathogen. Once the injected antibodies are broken down, the immunity disappears.

Pregnant women confer naturally induced passive immunity on their developing babies by manufacturing antibodies for them. These maternal antibodies, of the IgG class, pass through the placenta (the organ of exchange between mother and developing fetus) and provide the fetus and newborn infant with a defense system until its own immune system matures. Babies who are breast-fed continue to receive immunoglobulins, particularly IgA, in their milk.

Review

■ What is the basis of immunological memory?

■ How is passive immunity different from active immunity?

▌ THE IMMUNE SYSTEM AND DISEASE

Learning Objective

11 Describe the body's response to cancer cells and HIV.

The immune system is generally very effective in defending the body against invading pathogens. However, many factors can impair immune function, including genetic mutations, malnutrition, sleep deprivation, pre-existing disease, stress, and virulent pathogens.

Cancer cells evade the immune system

Cancer is now the second-leading cause of death in Western countries. An estimated one person in three in the United States will develop cancer, and one in five will die of cancer. We have discussed cancer in previous chapters, including Chapters 17 and 38. Here we focus on the interaction between cancer cells and cells of the immune system and on new treatment strategies based on immune mechanisms.

Cancer cells are body cells that have been transformed in a way that alters their normal growth-regulating mechanisms. Every day, a few normal cells may be transformed into precancer cells in each of us in response to the sun's UV rays, X-rays, certain viruses, chemical carcinogens in the environment, and other unknown factors. How does the immune system respond? And why is immune response sometimes ineffective?

Human T cells recognize two main groups of cancer cell antigens: tumor-specific antigens and tumor-associated antigens. *Tumor-specific antigens* are unique to cancer cells. They are induced by certain viruses and by chemical and physical carcinogens, such as ultraviolet light. Most cancer antigens are *tumor-associated antigens,* which are not unique to cancer cells; normal cells express them during fetal development. Some tumor-associated antigens are not expressed after birth, whereas others are expressed at very low levels. High levels stimulate the immune system. Researchers have shown that oncogenes encode tumor-associated antigens (see Chapter 17).

Antigens on cancer cells induce both cell-mediated and antibody-mediated immune responses. NK cells and macrophages destroy cancer cells. Macrophages produce cytokines, including TNFs that inhibit tumor growth. Dendritic cells present antigens to T cells, stimulating them to produce interferons, which have an antitumor effect. T_C cells attack cancer cells directly and also produce interleukins, which attract and activate macrophages and NK cells (▌ Fig. 44-12).

Sometimes cancer cells evade the immune system and multiply unchecked. Some types of cancer cells can block T_C cells. Others dramatically decrease their expression of class I MHC molecules. Recall that T_C cells recognize only antigen associated with class I MHC, so a low level of these molecules prevents tumor destruction. Some cancer cells do not produce co-stimulatory molecules needed to activate T_C cells.

Conventional cancer treatments, such as chemotherapy and radiation, destroy normal cells as well as cancer cells. To treat cancer effectively, treatment strategies must be more specific, or even customized to the particular cancer. For example, cancer researchers are genetically engineering cancer cells to secrete cytokines that would stimulate the patient's immune response.

A few cancer drugs, such as Herceptin, are monoclonal antibodies. Herceptin binds with a growth-factor receptor that is present in excessive numbers on the cells of about 30% of metastatic breast cancers. Herceptin blocks the growth factors that would stimulate proliferation of the cells. More than 50 *angiogenesis inhibitors* are currently being tested. These drugs inhibit the development of blood vessels that tumors need, and although these inhibitors do not cure cancer, they slow its growth.

Researchers are developing cancer vaccines that contain cancer cells or antigens; they stimulate the patient's immune system

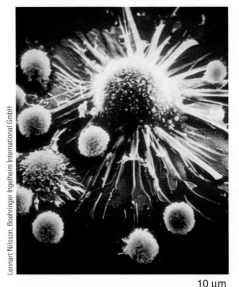

10 µm

(a) An army of T$_C$ cells surrounds a large cancer cell. The T$_C$ cells recognize the cancer cell as nonself because it displays altered or unique antigens on its surface.

10 µm

(b) Some T$_C$ cells elongate as they chemically attack the cancer cell and break down its plasma membrane.

10 µm

(c) The cancer cell has been destroyed. Only a collapsed fibrous cytoskeleton remains.

Figure 44-12　Colorized SEMs showing cancer cell destruction

to launch cell-mediated responses to tumor-associated antigens. These vaccines differ from traditional vaccines that are used to destroy pathogens *before* they cause disease. Several kinds of vaccines are being developed, including antigen vaccines, dendritic cell vaccines, DNA vaccines, and tumor cell vaccines.

Scientists make dendritic cell vaccines by taking dendritic cells from a cancer patient and exposing them to antigens from the patient's cancer cells. The combination of dendritic cells and antigens is then injected into the patient. The dendritic cells display the antigens to killer T cells, which then attack the cancer cells. DNA vaccines are made from bits of DNA from a patient's cancer cells. When injected into the patient, the vaccine stimulates increased antigen production. In response, the patient's immune system produces more T cells.

DNA microarrays detect patterns of gene expression in cancer cells. The microarray is a slide or chip that the technician can dot with DNA from thousands of genes (see Fig. 15-13). These genes are used as probes to determine which genes are active in cancer cells. Researchers use microarrays to identify cancer subtypes, information that can help identify and develop the most effective treatments.

Immunodeficiency disease can be inherited or acquired

The absence or failure of some component of the immune system can result in **immunodeficiency disease,** a condition that increases susceptibility to infection. *Inherited* immunodeficiencies have an estimated incidence of 1 per 10,000 births. *Severe combined immunodeficiency syndromes (SCIDs)* are X-linked and

autosomal recessive disorders that profoundly affect both cell-mediated immunity and antibody-mediated immunity, resulting in multiple infections. Babies born with SCID typically die by 2 years of age unless they are maintained in a protective bubble until they can be effectively treated, typically with bone marrow transplants.

In *DiGeorge syndrome,* the thymus is reduced or absent and the patient is deficient in T cells. Children born with this disorder are prone to serious viral infections. Treatment involves transplanting bone marrow or fetal thymus tissue.

Researchers use several animal models to study immunodeficiency. For example, the nude mouse (a hairless mutant mouse) does not develop a functional thymus. Because nude mice are deficient in mature T cells, they are used to study the effects of compromised cell-mediated immunity, as well as possible treatments. *Stem cell research* and advances in genetic engineering suggest new approaches to treating immunodeficiency (see Chapters 16 and 17).

Worldwide, the leading cause of *acquired* immunodeficiency in children is protein malnutrition. Lack of protein decreases T-cell numbers and decreases the ability to manufacture antibodies, which in turn increases the risk of contracting opportunistic infections. Another important cause of acquired immunodeficiency is chemotherapy administered to cancer patients.

HIV is the major cause of acquired immunodeficiency in adults

The **human immunodeficiency virus (HIV)** was first isolated in 1983 and was shown to be the cause of **acquired immunodefi-**

ciency syndrome (AIDS) in 1984. HIV, a retrovirus, has been studied more than any other virus. (Recall from Chapter 24 that a retrovirus is an RNA virus that uses its RNA as a template to make DNA with the help of reverse transcriptase.) Several different strains of the virus are known; HIV-1 is the most virulent form in humans.

The AIDS pandemic has claimed more than 27 million lives. HIV kills more than 3 million people every year, making it the fourth-highest cause of death globally. Epidemiologists estimate that more than 38 million adults and more than 2 million children worldwide are now infected with HIV. The General Assembly of the United Nations has called AIDS a "global emergency" and has outlined measures for mobilizing world resources to combat this disease. HIV/AIDS is more than a medical issue; it is a political, economic, and human rights threat.

HIV is transmitted mainly during sexual intercourse with an infected person or by direct exposure to infected blood or blood products. Sharing needles and/or syringes (usually for drug injection) with an infected person is a major risk factor. The virus is not spread by casual contact. People do not contract HIV by hugging, casual kissing, or using the same bathroom facilities. Friends and family members who have contact with AIDS patients are not more likely to become infected.

In the United States, 40,000 new HIV infections are occurring each year. Of these new infections, 70% occur in men and 30% in women. Half of all new infections in the United States occur in people 25 years old or younger. Heterosexual contact with infected individuals accounts for more than 75% of HIV infections in women and for an increasing number of cases in both men and women. Use of a latex condom during sexual intercourse provides some protection against HIV transmission, and use of a spermicide containing nonoxynol-9 may provide additional protection.

Up to one third of untreated pregnant women infected with HIV will transmit the virus to their babies. HIV can also be transmitted to babies through breastfeeding. Recent drug treatment protocols have reduced the rate of HIV transmission between mother and fetus. Effective blood-screening procedures now

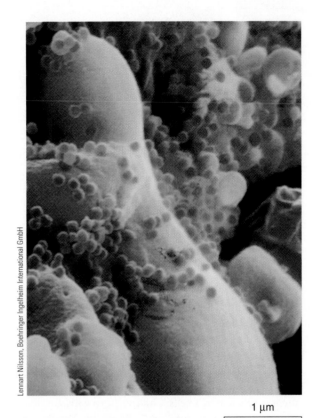

(a) HIV particles attack a T_H cell.

Figure 44-13 HIV infecting a T helper cell

In these colorized SEMs, HIV appears blue. The T helper cell (T_H) appears light green.

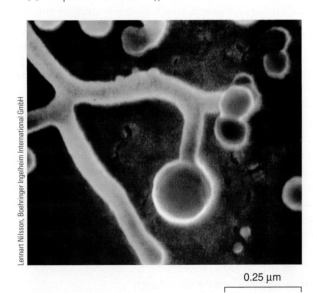

(b) HIV particles budding from the ends of the branched microvilli of a T_H cell.

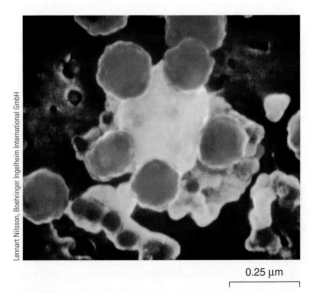

(c) An even higher magnification of HIV particles budding from a bleb (a cytoplasmic extension broader than a microvillus).

safeguard blood bank supplies, markedly reducing the risk of infection by blood transfusion in highly developed countries.

HIV destroys the host immune system

When HIV infects the body, an immune response is launched. Dendritic cells in the mucous membranes at sites of entry appear to be the first cells that HIV targets. When inflammation from another infection is present in the mucosa of the cervix or rectum, large numbers of dendritic cells are present, increasing the risk of HIV infection. Dendritic cells transport HIV from their point of entry to the lymph nodes.

HIV has a protein (gp120) on its outer envelope that attaches to CD4 on the surface of T_H cells, its main target (▌Fig. 44-13; also see chapter opening photograph). The virus also binds to a second receptor that normally binds chemokines. HIV enters the T_H cell and is reverse-transcribed to form DNA. Its DNA is then incorporated into the host genome. Over time HIV destroys T_H cells, resulting in a dramatic decrease in the T_H cell population. This decrease severely impairs the body's ability to resist infection. T_C cells appear to be the main cells that attack HIV, limiting viral replication and delaying the progress of the disease. Too often, the virus eventually wins the battle.

Although most individuals have no symptoms when first infected, about 15% of infected individuals experience mild flulike symptoms (fever and aching muscles) for a week or so. Some cells infected by HIV are not destroyed, and the virus may continue to replicate slowly for many years. After a time, a progression of symptoms occurs, including swollen lymph glands, night sweats, fever, and weight loss (▌Fig. 44-14).

In most HIV-infected individuals, the disease eventually progresses to AIDS, the final, lethal stage. By this time, lymph tissue has been destroyed and the blood T_H count has dropped below 200 cells/mm^3. In about one third of AIDS patients, the virus infects the nervous system and causes *AIDS dementia complex*. These patients exhibit progressive cognitive, motor, and behavioral dysfunction that typically ends in coma and death. As a consequence of immunosuppression, many AIDS patients develop and die from serious opportunistic infections or rare forms of cancer, such as Kaposi's sarcoma, a tumor of the endothelial cells lining blood vessels that causes purplish spots on the skin.

Researchers are actively working to develop effective treatments

Researchers throughout the world are searching for drugs that will successfully combat the AIDS virus. Because HIV often infects the central nervous system, an effective drug must cross the blood–brain barrier. **AZT** (azidothymidine), the first drug developed to treat HIV infection, can prolong the period prior to the onset of AIDS symptoms. AZT inhibits the action of reverse transcriptase, the enzyme the retrovirus uses to synthesize DNA. As a result, AZT blocks HIV replication. Without producing DNA, the virus cannot incorporate itself into the host cell's DNA. Unfortunately, the reverse transcriptase HIV uses to synthesize DNA makes many mistakes: about 1 in every 2000 nucleotides it incorporates is incorrect. By mutating in this way, viral strains resistant to AZT have evolved.

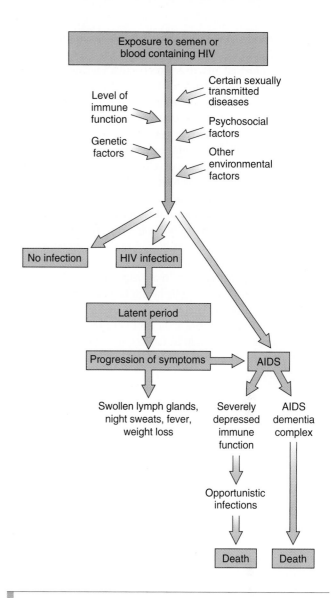

Figure 44-14 *Animated* The course of HIV infection

Exposure to semen or blood that contains HIV can lead to HIV infection. The infection is established in lymph nodes and other lymph tissues. During the latent period, chronic infection develops and the virus replicates at a low level. Eventually, viral replication may increase and lead to a progression of symptoms and to AIDS, the final, lethal stage. Although some exposed individuals apparently do not become infected, the risk increases with multiple exposures. Many factors determine whether a person exposed to HIV will develop AIDS.

Protease inhibitors block the viral enzyme protease, resulting in viral copies that cannot infect new cells. Protease inhibitors are used in combination with AZT and other reverse transcriptase inhibitors. Triple-combination treatment has been very effective for many AIDS patients, preventing opportunistic infections and prolonging life. In fact, combination treatment has led to a decline in AIDS incidence and mortality in the United States. Unfortunately, people in many parts of the world cannot afford these expensive drugs, and in impoverished areas, the AIDS epidemic continues to grow. In addition, the emergence of drug-resistant HIV viruses has been a serious problem.

Worldwide containment of the AIDS epidemic will require an effective vaccine that prevents the spread of HIV. Developing such a vaccine remains a daunting challenge for immunologists. More than 50 vaccines have been developed and tested clinically, but none has proved effective. An effective vaccine would have to counteract HIV's destruction of key cells needed to mount an immune response. Also, because mutations arise at a very high rate, new viral strains with new antigens evolve quickly. A vaccine would not be effective against new antigens and would quickly become obsolete. Some optimism has been generated by an apparently effective vaccine developed against SIV, the simian (monkey) virus that is closely related to HIV.

While immunologists work to develop a successful vaccine and more effective drugs to treat infected patients, massive educational programs have been launched to slow the spread of HIV. Educating the public that having multiple sexual partners increases the risk of AIDS, and teaching sexually active individuals the importance of "safe sex," may help contain the pandemic. Some have suggested that public health facilities should provide free condoms to those who are sexually active and free sterile hypodermic needles to those addicted to intravenous drugs. The cost of these measures would be far less than the cost of medical care for increasing numbers of AIDS patients and the toll in human suffering.

Review

■ How does the body defend itself against cancer?
■ How does HIV infect the body? How does it affect the body?
■ Why is it difficult to develop a vaccine against HIV?

HARMFUL IMMUNE RESPONSES

Learning Objective

12 Summarize the immunological basis of graft rejection, and describe the events that occur during hypersensitivity reactions, including Rh incompatibility, allergic reactions, and autoimmune diseases.

The immune system's ability to distinguish self from nonself sometimes interferes with clinical interventions. For example, efforts to save a patient's life with a blood transfusion or organ transplant can be disastrous if physicians ignore the effects of immune function. Sometimes the immune system malfunctions. **Hypersensitivity** is an exaggerated, damaging immune response to an antigen that is normally harmless, as occurs in an allergic reaction. Inappropriately directed or ineffective immune responses also result in disease states.

Graft rejection is an immune response against transplanted tissue

Skin can be successfully transplanted from one part of the same body to another or from one identical twin to another. However, when skin is taken from one person and grafted onto the body of a non-twin, it is rejected and sloughs off. Why? Recall that tissues from the same individual or from identical twins have identical MHC alleles and thus the same MHC antigens. Such tissues are compatible.

Because there are many alleles for each of the MHC genes, it is difficult to find identical matches. When a tissue or organ is taken from a donor and transplanted to the body of a non-twin host, several of the MHC antigens are likely to be different. The host's immune system regards the graft as foreign and launches an immune response called **graft rejection.** In the first stage of rejection, the *sensitization stage,* T_H cells and T_C cells recognize the antigens. In the second stage, the *effector stage,* T_C cells, the complement system, and cytokines secreted by T_H cells attack the transplanted tissue and can destroy it within a week.

Tissue from donor transplanted into body of recipient $\longrightarrow$ T_H cells and T_C cells recognize MHC antigens on transplant cells as foreign $\longrightarrow$ T_C cells launch immune response (graft rejection) $\longrightarrow$ T_H cells secrete cytokines, and T_C cells destroy transplant cells

Before transplants are performed, tissues from the patient and from potential donors must be typed and matched as closely as possible. Cell typing is somewhat similar to blood typing but is more complex. If all the MHC antigens are matched, the graft has about a 95% chance of surviving the first year. Unfortunately, not many people are lucky enough to have an identical twin to supply spare parts, so perfect matches are difficult to find. Furthermore, some organs, such as the heart, cannot be spared. Most organs to be transplanted, therefore, are removed from unrelated donors, often from patients who have just died.

To prevent graft rejection in less compatible matches, physicians use anti-rejection drugs and/or radiation to suppress the immune system. Unfortunately, these treatments make the transplant patient more vulnerable to pneumonia or other infections and increase the risk of certain types of cancer. If the patient survives the first few months, immunosuppressant drug dosages are reduced. Researchers are developing specific immunosuppression techniques, monoclonal antibodies that will target the specific lymphocytes that cause graft rejection.

Within the United States, thousands of patients are in need of organ transplants. Because the number of human donors does not meet this need, investigators are developing techniques for transplanting animal tissues and organs to humans. Challenges include risks of transmitting animal diseases to humans and graft rejection. Researchers are developing methods for genetically engineering pigs and other animals so that they do not produce antigens that stimulate immune responses in human recipients. These animals could then be cloned and used as donors of hearts, kidneys, and other organs. Effective artificial organs are also being developed.

The body has a few immunologically privileged locations where foreign tissue is accepted. For example, corneal transplants are highly successful because the cornea has almost no associated blood or lymphatic vessels and is thus out of reach of most lymphocytes. Furthermore, antigens in the corneal graft probably would not find their way into the circulatory system and therefore could not stimulate an immune response.

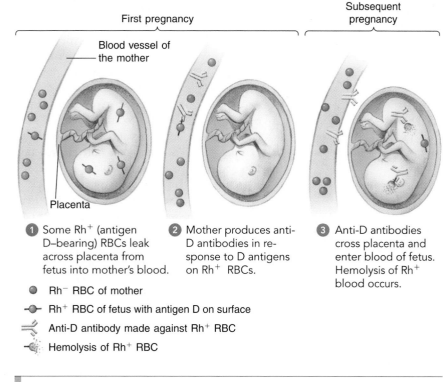

First pregnancy

Subsequent pregnancy

Blood vessel of the mother

Placenta

1 Some Rh⁺ (antigen D–bearing) RBCs leak across placenta from fetus into mother's blood.

2 Mother produces anti-D antibodies in response to D antigens on Rh⁺ RBCs.

3 Anti-D antibodies cross placenta and enter blood of fetus. Hemolysis of Rh⁺ blood occurs.

● Rh⁻ RBC of mother

◕ Rh⁺ RBC of fetus with antigen D on surface

⚹ Anti-D antibody made against Rh⁺ RBC

◔ Hemolysis of Rh⁺ RBC

Figure 44-15 *Animated* Rh incompatibility

When an Rh-negative woman produces Rh-positive offspring, her immune system can be sensitized. In subsequent pregnancies she may produce anti-D antibodies that cross the placenta and destroy fetal red blood cells.

Rh incompatibility can result in hypersensitivity

Named for the rhesus monkeys in whose blood it was first found, the Rh system consists of more than 40 kinds of Rh antigens, each referred to as an **Rh factor.** By far the most important of these factors is *antigen D.* About 85% of U.S. residents who are of Western European descent are Rh positive—they have antigen D on the surfaces of their red blood cells (in addition to the antigens of the ABO system and other blood group antigens). The 15% or so of this population who are Rh negative have no antigen D. Unlike the situation discussed in Chapter 11 for the ABO blood group, Rh-negative individuals do not naturally produce antibodies against antigen D (anti-D). However, they produce anti-D antibodies if they are exposed to Rh-positive blood. The allele coding for antigen D is dominant to the allele for the absence of antigen D. Hence, Rh-negative people are homozygous recessive, and Rh-positive people are heterozygous or homozygous dominant.

Although several kinds of maternal-fetal blood type incompatibilities are known, **Rh incompatibility** is the most serious (❙ Fig. 44-15). If a woman is Rh negative and the father of the fetus she is carrying is Rh positive, the fetus may also be Rh positive, having inherited the D allele from the father. Ordinarily no mixing of maternal and fetal blood occurs; nutrients, oxygen, and other substances are exchanged between these two circulatory systems across the placenta. However, late in pregnancy or dur-

ing the birth process, a small quantity of blood from the fetus may pass through a defect in the placenta.

The fetus's red blood cells, which bear antigen D, activate the mother's immune system, stimulating B cells to produce antibodies to antigen D. If the woman becomes pregnant again, she produces anti-D antibodies that cross the placenta and enter the fetal blood. They combine with antigen D molecules on the surface of the fetal red blood cells, causing hemolysis (cells rupture and release hemoglobin into the circulation). The ability to transport oxygen is reduced, and breakdown products of the released hemoglobin damage organs, including the brain. In extreme cases of this disease, known as **erythroblastosis fetalis,** so many fetal red blood cells are destroyed that the fetus may die.

When Rh-incompatibility problems are suspected, fetal blood can be exchanged by transfusion before birth, but this is a risky procedure. More commonly, Rh-negative women are treated just after childbirth (or at termination of pregnancy by miscarriage or abortion) with a preparation of anti-D antibodies known as RhoGAM. These antibodies clear the Rh-positive fetal red blood cells from the mother's blood very quickly, minimizing the chance for her own white blood cells to become sensitized and develop memory cells. The antibodies are also soon eliminated from her body. As a result, if she becomes pregnant again, her blood does not contain the anti-D that could harm the developing fetus.

Allergic reactions are directed against ordinary environmental antigens

About 20% of the U.S. population is plagued by an allergic disorder, such as allergic asthma or hay fever. A predisposition toward these disorders appears to be inherited. In **allergic reactions,** hypersensitivity results in the manufacture of antibodies against mild antigens, called **allergens,** that normally do not stimulate an immune response. Common environmental agents, such as house-dust mites, roach feces, pet dander, and pollen, can trigger allergic reactions in some individuals. In many kinds of allergic reactions, distinctive IgE immunoglobulins are produced.

Let us examine a common allergic reaction, a hay fever response to ragweed pollen (❙ Fig. 44-16). **1** Exposure to pollen causes *sensitization.* Macrophages degrade the allergen and present fragments of it to T cells. The activated T cells then stimulate B cells to become plasma cells and produce pollen-specific IgE. **2** These antibodies attach to receptors on mast cells. The Fc end of each IgE molecule attaches to a mast cell receptor; its V region is free to combine with the ragweed pollen allergen. **3** When a sensitized, allergic person inhales the microscopic pollen, **4** allergen molecules rapidly attach to the IgE on sensitized mast cells.

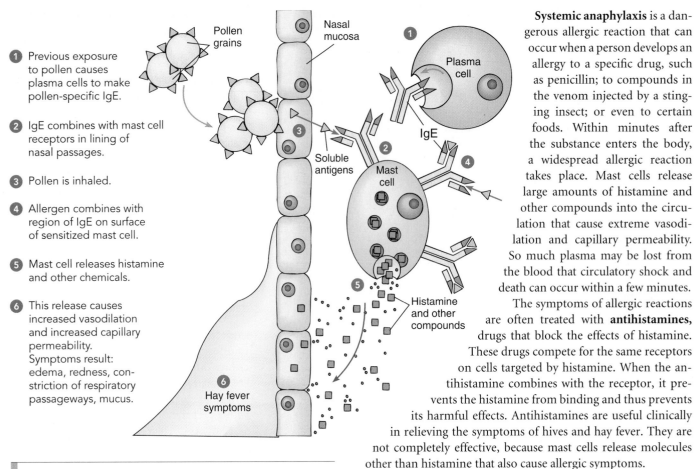

1. Previous exposure to pollen causes plasma cells to make pollen-specific IgE.

2. IgE combines with mast cell receptors in lining of nasal passages.

3. Pollen is inhaled.

4. Allergen combines with region of IgE on surface of sensitized mast cell.

5. Mast cell releases histamine and other chemicals.

6. This release causes increased vasodilation and increased capillary permeability. Symptoms result: edema, redness, constriction of respiratory passageways, mucus.

Figure 44-16 An allergic reaction

Pollen causes a common type of allergic reaction in many people.

5 This binding of allergen with IgE antibody stimulates the mast cell to release granules filled with histamine and other molecules that cause inflammation. 6 Blood vessels dilate, and capillaries become more permeable, leading to edema and redness. Such responses cause the nasal passages to become swollen and irritated. A runny nose, sneezing, watery eyes, and a general feeling of discomfort are common.

A prolonged allergic response occurs when neutrophils are attracted by chemical compounds released by the mast cells. Neutrophils migrate to the inflamed area and release compounds that prolong the allergic reaction.

In **allergic asthma,** an allergen–IgE response occurs in the bronchioles of the lungs. Mast cells release substances that cause smooth muscle to contract, and the airways in the lungs sometimes constrict for several hours, making breathing difficult. An antibody to IgE is being clinically tested as a treatment for allergic asthma.

Certain foods or drugs act as allergens in some people, causing a reaction in the walls of the digestive tract that leads to discomfort and diarrhea. The allergen may be absorbed and cause mast cells to release granules elsewhere in the body. When the allergen–IgE reaction takes place in the skin, the histamine released by mast cells causes the swollen red welts called **hives.**

Systemic anaphylaxis is a dangerous allergic reaction that can occur when a person develops an allergy to a specific drug, such as penicillin; to compounds in the venom injected by a stinging insect; or even to certain foods. Within minutes after the substance enters the body, a widespread allergic reaction takes place. Mast cells release large amounts of histamine and other compounds into the circulation that cause extreme vasodilation and capillary permeability. So much plasma may be lost from the blood that circulatory shock and death can occur within a few minutes.

The symptoms of allergic reactions are often treated with **antihistamines,** drugs that block the effects of histamine. These drugs compete for the same receptors on cells targeted by histamine. When the antihistamine combines with the receptor, it prevents the histamine from binding and thus prevents its harmful effects. Antihistamines are useful clinically in relieving the symptoms of hives and hay fever. They are not completely effective, because mast cells release molecules other than histamine that also cause allergic symptoms.

In an autoimmune disease, the body attacks its own tissues

During lymphocyte development, self-tolerance (self-recognition) is established so that lymphocytes do not attack tissues of their own body. However, some lymphocytes remain that have the potential to be *autoreactive,* that is, to launch an immune response against self tissues. Such autoreactivity can lead to a form of hypersensitivity known as *autoimmunity,* or **autoimmune disease,** in which immune cells react immunologically against the body's own cells. Some of the diseases that result from such failure in self-tolerance are rheumatoid arthritis, systemic lupus erythematosus (SLE), insulin-dependent diabetes, scleroderma, and multiple sclerosis.

In autoimmune diseases, antibodies and T cells attack the body's own tissues. In **rheumatoid arthritis,** T cells in the area of inflammation produce a cytokine that promotes inflammation. In addition, antibodies to a joint protein may generate antigen–antibody complexes that are deposited in the small joints. In **multiple sclerosis,** an autoimmune disease of the central nervous system, T_H cells attack self myelin antigens. MRI scans show the loss of myelin sheaths in multiple areas around axons in the brain and spinal cord that are normally myelinated. Scar tissue forms and compromises the ability of these damaged neurons to transmit signals.

Genetic risk factors are important in autoimmune diseases. Mutations that cause abnormal apoptosis or inadequate clearing of dead cells may result in danger signals that activate the immune system. Another important factor is that viral or bacterial infection often precedes the onset of an autoimmune disease. Some pathogens have evolved a tactic known as *molecular mimicry.* They trick the body by producing molecules that look like self molecules. For example, an adenovirus that causes respiratory and intestinal illness produces a peptide that mimics myelin protein. When the body launches responses to the adenovirus peptide, it may also begin to attack the similar self molecule, myelin.

To maintain homeostasis and health, the numerous types of cells, regulatory and signaling molecules, and the complex actions of the immune system must be carefully regulated. Specific T cells and B cells proliferate and respond to challenges by pathogens, but after an effective immune response the expanded populations of T cells and B cells must be decreased to normal numbers. Cells that are no longer needed die by apoptosis. Overactivation of critical components can result in inflammation and autoimmune disease. Underactivation allows pathogens to cause serious disease. Either situation can lead to death.

Review

- What is the immunological basis for graft rejection?
- What immunological events take place in a common type of allergic reaction, such as hay fever?
- What is an autoimmune disease? Give two examples.

SUMMARY WITH KEY TERMS

Learning Objectives

1 Distinguish between nonspecific and specific immune responses (page 945).
  - **Immune responses** are the body's defensive responses against **pathogens,** toxins, and other harmful agents. In immune responses, the body recognizes foreign or dangerous macromolecules and responds to eliminate them. **Nonspecific immune responses** provide general and immediate protection against pathogens, some toxins and drugs, and cancer cells.
  - **Specific immune responses** target distinct antigens, molecules recognized as foreign or dangerous by cells of the immune system. When it encounters antigens, the immune system produces **antibodies,** highly specific proteins that recognize and bind to specific antigens. Another defining feature of specific immunity is **immunological memory,** that is, the system "remembers" foreign or dangerous molecules and responds more strongly to repeated encounters with the same molecules.

2 Compare, in general terms, the immune responses of invertebrates and vertebrates (page 945).
  - Invertebrates depend on nonspecific immune responses, such as physical barriers (cuticle, skin); **phagocytosis;** and **antimicrobial peptides,** soluble molecules that destroy pathogens.
  - Vertebrates use both nonspecific and specific immune responses.

3 Describe nonspecific immune responses, including physical and chemical barriers; recognition by Toll-like receptors; the actions of cells, such as phagocytes and natural killer cells; cytokines and the proteins that make up the complement system; and the inflammatory response (page 946).
  - Nonspecific immune responses include physical barriers, such as the skin and the mucous linings of the respiratory and digestive tracts. Antimicrobial peptides are produced by epithelial membranes.
  - When pathogens break through these first-line defenses, other nonspecific defenses are activated by **Toll-like receptors,** cell-surface receptors on phagocytes and certain other types of cells. Toll-like receptors recognize certain common molecular features of classes of pathogens called **pathogen-associated molecular patterns,** or **PAMPs.**

  - **Phagocytes,** including **neutrophils** and **macrophages,** destroy bacteria. **Natural killer (NK) cells** destroy cells infected with viruses and foreign or altered cells such as tumor cells.
  - **Cytokines** are signaling proteins that regulate interactions between cells. Important groups are interferons, tumor necrosis factors, interleukins, and chemokines. **Interferons** inhibit viral replication and activate natural killer cells. **Tumor necrosis factor (TNF)** stimulates immune cells to initiate an inflammatory response. **Interleukins** help regulate interactions between lymphocytes and other cells of the body, help mediate inflammation, and can mediate fever. **Chemokines** attract, activate, and direct the movement of certain cells of the immune system.
  - **Complement** proteins lyse the cell wall of pathogens; coat pathogens, thereby enhancing phagocytosis; and attract white blood cells to the site of infection. These actions enhance the inflammatory response.
  - When pathogens invade tissues, they trigger an **inflammatory response,** which includes three main processes: vasodilation, which brings more blood to the area of infection; increased capillary permeability, which allows fluid and antibodies to leave the circulation and enter the tissues; and increased phagocytosis. In response to infection, **mast cells** release **histamine** and other compounds that cause vasodilation and increased capillary permeability.

ThomsonNOW™ Watch immune responses, including the inflammatory response, by clicking on the figures in ThomsonNOW.

4 Distinguish between cell-mediated and antibody-mediated immunity (page 950).
  - In **cell-mediated immunity,** specific T cells are activated; T cells release proteins that destroy cells infected with viruses or other intracellular pathogens.
  - In **antibody-mediated immunity,** specific B cells are activated; B cells multiply and differentiate into plasma cells, which produce antibodies.

5 Describe the principal cells of the immune system, and summarize the function of the major histocompatibility complex (page 950).

- Two main types of cells important in specific immune responses are lymphocytes and antigen-presenting cells. Lymphocytes develop from stem cells in the bone marrow. **T cells** are lymphocytes responsible for cell-mediated immunity. The **thymus gland** confers immunocompetence on T cells by making them capable of distinguishing between self and nonself. Three types of T cells are **T cytotoxic cells (T_C cells), T helper cells (T_H),** and **memory T cells.** T cells are distinguished by their **T-cell receptors (TCRs).**

- **B cells** are lymphocytes responsible for antibody-mediated immunity. B cells differentiate into **plasma cells,** which produce antibodies. Some activated B cells become **memory B cells,** which continue to produce antibodies after an infection has been overcome.

- **Antigen-presenting cells (APCs),** including dendritic cells, macrophages, and B cells, display foreign antigens as well as their own surface proteins. **Dendritic cells** are located in the skin and other tissues of the body that interact with the environment. They are specialized to process, transport, and present antigens.

- Immune responses depend on the **major histocompatibility complex (MHC),** a group of genes that encode MHC proteins. Class I MHC genes encode self antigens, glycoproteins expressed on the surface of most nucleated cells. Class II MHC genes encode glycoproteins expressed on APCs of the immune system. Class III MHC genes encode components of the complement system and TNFs.

6 Describe the sequence of events in cell-mediated immunity (page 952).

- In cell-mediated immunity, specific T cells are activated by a foreign antigen–MHC complex on the surface of an infected cell. A co-stimulatory signal and interleukins are also required. Activated T_C cells multiply and give rise to a clone. These cells migrate to the site of infection and destroy pathogen-infected cells. Activated T_H cells give rise to a clone of T_H cells, which secrete cytokines that activate B cells and macrophages.

7 Summarize the sequence of events in antibody-mediated immunity, including the effects of antigen–antibody complexes on pathogens (page 952).

- In antibody-mediated immunity, B cells are activated when they combine with antigen. Activation requires both an APC (such as a dendritic cell or macrophage) that has a foreign antigen–MHC complex displayed on its surface and a T_H cell that secretes interleukins.

- When activated B cells multiply, they give rise to clones of cells. The cloned cells differentiate and form plasma cells. Plasma cells produce specific antibodies, called **immunoglobulins (Ig),** in response to the specific antigens that activated them. An antibody combines with a specific antigen to form an **antigen–antibody complex,** which may inactivate the pathogen, stimulate phagocytosis, or activate the complement system.

ThomsonNOW™ **See cell-mediated and antibody-mediated immunity by clicking on the figures in ThomsonNOW.**

8 Describe the basic structure and function of an antibody, and explain the basis of antibody diversity (page 952).

- In a Y-shaped antibody, the two arms combine with antigen. An antibody molecule consists of four polypeptide chains: two identical heavy chains and two shorter light chains. Each chain has a **constant (C) region** and a **variable (V) region.**

- Rearrangement of DNA segments during the differentiation of B cells is the main factor responsible for antibody diversity; millions of different types of B (and T) cells are produced.

ThomsonNOW™ **Learn more about antibody structure, function, and diversity by clicking on the figures in ThomsonNOW.**

9 Describe the basis of immunological memory, and contrast secondary and primary immune responses (page 958).

- After an infection, memory B cells and memory T cells remain in the body. These cells are responsible for long-term immunity.

- The first exposure to an antigen stimulates a **primary immune response.** A second exposure to the same antigen evokes a **secondary immune response,** which is more rapid and more intense than the primary response.

10 Compare active and passive immunity, and give examples of each (page 958).

- **Active immunity** develops as a result of exposure to antigens; it may occur naturally after recovery from a disease or can be artificially induced by immunization with a vaccine.

- **Passive immunity** is a temporary condition that develops when an individual receives antibodies produced by another person or animal.

11 Describe the body's response to cancer cells and HIV (page 960).

- NK cells, macrophages, T cells, and other cells of the immune system recognize antigens on cancer cells and launch an immune response against them. Cancer cells evade the immune system by blocking T_C directly or by decreasing the class I MHC molecules on T_C cells.

- **Acquired immunodeficiency syndrome (AIDS)** is caused by the **human immunodeficiency virus (HIV),** a retrovirus. HIV destroys T helper cells, severely impairing immunity and placing the patient at risk for opportunistic infections.

12 Summarize the immunological basis of graft rejection, and describe the events that occur during hypersensitivity reactions, including Rh incompatibility, allergic reactions, and autoimmune diseases (page 964).

- Transplanted tissues have MHC antigens that stimulate **graft rejection,** an immune response in which T cells destroy the transplant.

- When an Rh-negative woman gives birth to an Rh-positive baby, she may develop anti-D antibodies. **Rh incompatibility** can then occur in future pregnancies.

- In an **allergic reaction,** an **allergen** stimulates the production of IgE, which combines with receptors on mast cells. The mast cells release histamine and other molecules that cause inflammation and other symptoms of allergy. **Systemic anaphylaxis** is a rapid, widespread allergic reaction that can lead to death.

- In **autoimmune diseases,** the body reacts immunologically against its own tissues.

TEST YOUR UNDERSTANDING

1. A molecule recognized as foreign by cells of the immune system is a(an) (a) antibody (b) antigen (c) immunoglobulin (d) interferon (e) cytokine

2. Nonspecific (innate) immune responses include (a) inflammation (b) antigen–antibody complexes (c) immunoglobulin action (d) complement and memory T cells (e) memory B cells

3. Invertebrate defense responses include (a) phagocytosis (b) antimicrobial peptides (c) ability to distinguish between self and nonself (d) a, b, and c (e) a and c

4. Cytokines (a) are regulatory Toll-like receptors (b) prevent the inflammatory response (c) include interferons and interleukins (d) are immunoglobulins (e) include interleukins and complement proteins

5. Which of the following is *not* an action of complement? (a) enhances phagocytosis (b) enhances inflammatory response (c) coats pathogens (d) lyses viruses (e) stimulates allergen release

6. Which of the following cells are antigen-presenting cells? (a) NK cells and monocytes (b) macrophages and plasma cells (c) dendritic cells and macrophages (d) mast cells and B cells (e) memory T cells and memory B cells

7. Which of the following cells are especially adept at destroying tumor cells? (a) NK cells (b) plasma cells (c) neutrophils (d) B cytotoxic cells (e) mast cells

8. Which of the following cells become immunologically competent after processing in the thymus gland? (a) NK cells (b) T cells (c) macrophages (d) B cells (e) plasma cells

9. Cells that have a surface marker called CD4 are (a) NK cells (b) T cytotoxic cells (c) T helper cells (d) B cells (e) plasma cells

10. The major histocompatibility complex (MHC) (a) encodes a group of cell-surface proteins (b) encodes certain antibodies (c) encodes Toll-like receptors (d) inhibits complement release from macrophages (e) consists of Y-shaped molecules

11. Which sequence most accurately describes antibody-mediated immunity?

    1. B cell divides and gives rise to clone 2. antibodies produced 3. cells differentiate and form plasma cells 4. activated T helper cell interacts with B cell displaying same antigen complex 5. B cell activated

    (a) 1, 2, 3, 4, 5 (b) 3, 2, 1, 4, 5 (c) 4, 5, 3, 2, 1 (d) 4, 5, 1, 3, 2 (e) 4, 3, 1, 2, 5

12. A typical antibody (a) is activated by APCs (b) has four identical heavy chains and four identical light chains (c) has IgG and IgD components (d) suppresses phagocyte actions (e) has a Y shape

13. Immunoglobulin A (a) recognizes pathogen-associated molecular patterns (b) combines with NK cells (c) prevents pathogens from attaching to epithelial surfaces (d) is found mainly on B-cell surfaces (e) is found mainly on T cells

14. When a person is exposed to the same antigen a second time, the response is (a) a secondary immune response (b) a more rapid response (c) a response mediated by dendritic cells (d) a, b, and c (e) a and b

15. Graft rejection (a) is an example of passive immunity (b) occurs in mild form after immunization (c) generates monoclonal antibodies (d) does not occur when tissue is transplanted from one identical twin to the other (e) is initiated by the thymus gland

16. In an allergic reaction, (a) the body is immunodeficient (b) an allergen binds with IgE (c) T helper cells release histamine (d) allergen stimulates graft rejection (e) mast cells are deactivated

17. HIV (a) is a retrovirus (b) destroys T cytotoxic cells (c) is attacked mainly by B cells (d) a, b, and c (e) none of the preceding

CRITICAL THINKING

1. Specificity, diversity, and memory are key features of the immune system. Giving specific examples, explain how each of these features is important.

2. Macrophages can be selectively destroyed in the body by the administration of a certain chemical. What would be the effects of such a loss of macrophages? Which do you think would have a greater effect on the immune system, loss of macrophages or loss of B cells?

3. What are the advantages of having MHC antigens? Disadvantages? What do you think would be the consequences of not having them?

4. A playmate in kindergarten exposes John and Jack to measles. John has been immunized against measles, but Jack has not received the measles vaccine. How are their immune responses different? Five years later, John and Jack are playing together when Judy, who has just been diagnosed with measles, sneezes on both of them. Compare their immune responses.

5. Imagine that you are a researcher developing new HIV treatments. What approaches might you take? What public policy decisions would you recommend that might help slow the spread of AIDS while new treatments or vaccines are being developed?

6. **Evolution Link.** Fishes have only one antibody class, IgM. Amphibians have two classes, and the number increases to five main classes in mammals. Hypothesize an adaptive advantage of having a greater variety of antibody groups.

Additional questions are available in ThomsonNOW at www.thomsonedu.com/login

45

Gas Exchange

Ventilation. The white-tailed deer (*Odocoileus virginanus*) and other terrestrial vertebrates ventilate their lungs by breathing air.

KEY CONCEPTS

Air has a higher concentration of molecular oxygen than water does, and animals require less energy to move air than to move water over a gas exchange surface; adaptations in terrestrial animals protect their respiratory surfaces from drying.

Adaptations for gas exchange include a thin, moist body surface; gills in aquatic animals; and tracheal tubes and lungs in terrestrial animals.

In mammals, oxygen and carbon dioxide are exchanged between alveoli and blood by diffusion; the pressure of a particular gas determines its direction and rate of diffusion.

Respiratory pigments combine with oxygen and transport it. Almost all of the oxygen in vertebrate blood is transported as oxyhemoglobin; carbon dioxide is transported mainly as bicarbonate ions.

Most animal cells require a continuous supply of oxygen for cellular respiration. Some cells, such as mammalian brain cells, may be damaged beyond repair if their oxygen supply is cut off for only a few minutes. To stay alive, animals must take in oxygen and release excess carbon dioxide. The exchange of gases between an organism and its environment is known as **respiration.**

Two phases of respiration are organismic and cellular respiration. During **organismic respiration,** oxygen from the environment is taken up by the animal and delivered to its individual cells. At the same time, carbon dioxide generated during cellular respiration is excreted into the environment. In **aerobic cellular respiration,** which takes place in mitochondria, oxygen is essential because it serves as the final electron acceptor in the mitochondrial electron transport chain (see Chapter 8). Carbon dioxide is a metabolic waste product of cellular respiration.

In small, aquatic organisms, such as sponges, hydras, and flatworms, gas exchange occurs entirely by simple **diffusion,** the passive movement of particles (atoms, ions, or molecules) from a region of higher concentration to a region of lower concentration, that is, down a concentration gradient. Most cells are in direct contact with the environment. Dissolved oxygen from the surrounding water diffuses into the cells, whereas carbon dioxide diffuses out of the cells and into the water. No specialized respiratory structures are needed.

Oxygen diffuses through tissues slowly, however. In an animal more than about 1 mm thick, oxygen cannot diffuse quickly enough through layers of cells to support life. Specialized respiratory structures, such as gills or lungs, are required to deliver oxygen to the cells or to a transport system and to facilitate excretion of carbon dioxide. Such respiratory systems, working with circulatory systems, provide the efficient intake and transport of oxygen necessary to support high metabolic rates. Animals with lungs carry on *ventilation* by breathing air (see photograph). ∎

ADAPTATIONS FOR GAS EXCHANGE IN AIR OR WATER

Learning Objective

1 Compare the advantages and disadvantages of air and water as mediums for gas exchange, and describe adaptations for gas exchange in air.

Gas exchange in air has certain advantages over gas exchange in water. Air contains a much higher concentration of molecular oxygen than water does. In addition, oxygen diffuses much faster through air than through water. Another advantage is that less energy is needed to move air than to move water over a gas exchange surface, because air is less dense and less viscous. A problem with gas exchange in air is the threat of desiccation. Animals that respire in air struggle continuously with water loss.

Gills are adapted for gas exchange in water, and the tracheal tubes of insects and lungs of vertebrates are respiratory structures adapted for gas exchange in air. Respiratory surfaces must be kept moist because oxygen and carbon dioxide are dissolved in the fluid that bathes the cells of the respiratory surfaces. Whether an animal makes its home on land or water, gas exchange takes place across a moist surface.

Adaptations have evolved that keep respiratory surfaces moist and minimize desiccation. For example, the lungs of air-breathing vertebrates are located deep within the body, not exposed as gills are. Air is humidified and brought to body temperature as it passes through the upper respiratory passageways, and expired air must again pass through these airways (providing opportunity for retaining water) before leaving the body. These adaptations protect the lungs from the drying and cooling effects of air.

Review

∎ What are some advantages of gas exchange in air over gas exchange in water?

TYPES OF RESPIRATORY SURFACES

Learning Objective

2 Describe the following adaptations for gas exchange: body surface, tracheal tubes, gills, and lungs.

Not only must respiratory structures be moist but they must also have thin walls through which diffusion can easily occur. Respiratory structures are generally richly supplied with blood vessels to facilitate transport and exchange of respiratory gases. Four main types of respiratory surfaces have evolved in animals: the animal's own body surface, tracheal tubes, gills, and lungs (∎ Fig. 45-1). Some animals use a combination of these adaptations.

Most respiratory systems carry on **ventilation,** that is, they actively move air or water over their respiratory surfaces. If the air or water supplying oxygen to the cells can be continuously renewed, more oxygen will be available. Sponges do this with flagella, setting up a current of water through the channels of their bodies. Most fishes gulp water, which then passes over their gills. Terrestrial vertebrates have lungs and breathe air. When mammals breathe, the diaphragm and other muscles move air in and out of the lungs.

The body surface may be adapted for gas exchange

Gas exchange occurs through the entire body surface in many animals, including nudibranch mollusks, most annelids, and some amphibians. These animals are small, with a high surface area–to-volume ratio. They also have a low metabolic rate that requires smaller quantities of oxygen per cell. In aquatic animals, the body surface is kept moist by the surrounding water. In terrestrial animals, the body secretes fluids that keep its surface moist. Many animals that exchange gases across the body surface also have gills or lungs.

Tracheal tube systems deliver air directly to the cells

The relatively inefficient open circulatory system of arthropods cannot supply these active animals with enough oxygen. They need a specialized respiratory system. In insects and some other arthropods (such as chilopods, diplopods, some mites, and some spiders), the respiratory system is a network of **tracheal tubes,** also called **tracheae** (∎ Fig. 45-2). Air enters the tracheal tubes through a series of up to 20 tiny openings called **spiracles** along the body surface. In some insects, especially large, active ones, muscles help ventilate the tracheae by pumping air in and out of the spiracles. For example, the grasshopper draws air in through the first four pairs of spiracles when the abdomen expands. Then the abdomen contracts, forcing air out through the last six pairs of spiracles.

Once inside the body, the air passes through a system of branching tracheal tubes, which extend to all parts of the animal. The tracheal tubes terminate in microscopic, fluid-filled tracheoles. Gases are exchanged between this fluid and the body cells. The tracheal system supplies enough oxygen to support the high metabolic rates required by many insects.

Earthworm

(a) Body surface. Small, multicellular animals exchange gases through the body surface.

Grasshopper

(b) Tracheal tubes. Insects and some other arthropods exchange gases through a system of tracheal tubes, or tracheae.

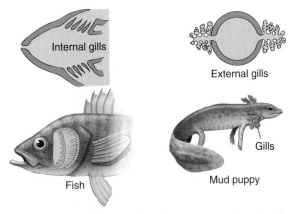

Internal gills

External gills

Fish

Gills

Mud puppy

(c) Gills. Most aquatic animals exchange gases through gills, thin structures that extend from the body. Gills can be internal or external.

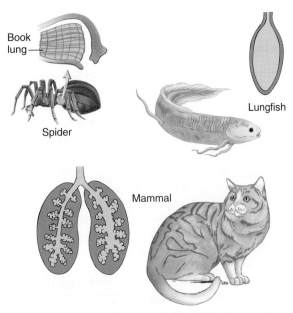

Book lung

Spider

Lungfish

Mammal

(d) Lungs. Lungs are adaptations for terrestrial gas exchange.

Figure 45-1 *Animated* Adaptations for gas exchange

Gills are the respiratory surfaces in many aquatic animals

Found mainly in aquatic animals, **gills** are moist, thin structures that extend from the body surface. They are supported by the buoyancy of water but tend to collapse in air. In many animals, the outer surface of the gills is exposed to water, whereas the inner side is in close contact with networks of blood vessels.

Sea stars and sea urchins have **dermal gills** that project from the body wall. Their ciliated epidermal cells ventilate the gills by beating a stream of water over them. Gases are exchanged between the water and the coelomic fluid inside the body by diffusion through the gills.

Various types of gills are found in some annelids, aquatic mollusks, crustaceans, fishes, and amphibians. Mollusk gills are folded, providing a large surface for respiration. In clams and other bivalve mollusks and in simple chordates, gills may also be adapted for trapping and sorting food. The rhythmic beating of cilia draws water over the gill area, and food is filtered out of the water while gases are exchanged. In mollusks, gas exchange also takes place through the mantle.

In chordates, gills are usually internal. A series of slits perforates the pharynx, and the gills lie along the edges of these gill slits (see Fig. 31-4). In bony fishes, the fragile gills are protected by an external bony plate, the **operculum.** In some fishes, movements of the jaw and operculum help pump water rich in oxygen through the mouth and across the gills. The water leaves through the gill slits.

Each gill in the bony fish consists of many **gill filaments,** which provide an extensive surface for gas exchange (▮ Fig. 45-3a and b). The filaments extend out into the water, which continuously flows over them. A capillary network delivers blood to the gill filaments, facilitating diffusion of oxygen and carbon dioxide between blood and water. This system is extremely efficient because blood flows in a direction opposite to the movement of the water. This arrangement, called a **countercurrent exchange system,** maximizes the difference in oxygen concentration between blood and water throughout the area where the two remain in contact (▮ Fig. 45-3c).

If blood and water flowed in the *same* direction, that is, *concurrent exchange,* the difference between the oxygen concentrations in blood (low) and water (high) would be very large initially and very small at the end (▮ Fig. 45-3d). The oxygen concentration in the water would decrease as the concentration in the blood increased. When the concentrations in the two fluids became equal, an equilibrium would be reached, and the net diffusion of oxygen would stop. Only about 50% of the oxygen dissolved in the water could diffuse into the blood.

In the countercurrent exchange system, however, blood low in oxygen comes in contact with water that is partly depleted of oxygen. Then, as the blood flowing through the capillaries becomes progressively richer in oxygen, the blood comes in contact with water with an increasingly higher concentration of oxygen. Thus, all along the capillaries, the diffusion gradient favors passage of oxygen from the water into the gill. A high rate of diffusion is maintained, ensuring that a very high percentage (more

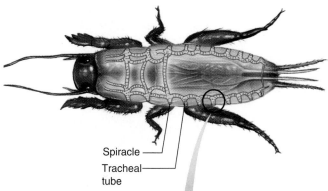

Spiracle
Tracheal
tube

(a) Location of spiracles and tracheal tubes. Air enters the system of tracheal tubes through openings called spiracles.

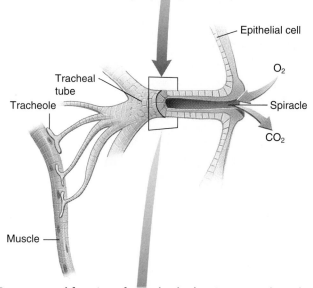

Epithelial cell

Tracheal
tube

Tracheole

O_2

Spiracle

CO_2

Muscle

(b) Structure and function of a tracheal tube. Air passes through a system of branching tracheal tubes that conduct oxygen to all cells of the insect.

Courtesy of Dr. James L. Nation and *Stain Technology*, Vol. 58, 1983

100 μm

(c) SEM of a mole cricket trachea. The wrinkles are a part of a long spiral that wraps around the tube, strengthening the tracheal wall somewhat as a spring strengthens the plastic hoses of many vacuum cleaners. The tracheal wall is composed of chitin.

Figure 45-2 Tracheal tubes

than 80%) of the available oxygen in the water diffuses into the blood.

Oxygen and carbon dioxide do not interfere with each other's diffusion, and they simultaneously diffuse in opposite directions. The reason is that oxygen is more concentrated outside the gills than inside but carbon dioxide is more concentrated inside the gills than outside. Thus, the same countercurrent exchange mechanism that ensures efficient inflow of oxygen also results in equally efficient outflow of carbon dioxide.

Terrestrial vertebrates exchange gases through lungs

Lungs are respiratory structures that develop as ingrowths of the body surface or from the wall of a body cavity such as the pharynx (throat region). For example, the **book lungs** of spiders are enclosed in an inpocketing of the abdominal wall. These lungs consist of a series of thin, parallel plates of tissue (like the pages of a book) filled with hemolymph (see Fig. 45-1d). The plates of tissue are separated by air spaces that receive oxygen from the outside environment through a spiracle. A different type of lung evolved in land snails and slugs. (These terrestrial mollusks lack gills.) Gas exchange takes place through a lung, which is a vascularized region of the mantle.

Fossil evidence suggests that early lobe-finned fishes had lungs somewhat similar to those of modern lungfishes. The Australian lungfish can use either its gills or its lungs, depending on the conditions in its environment. The gills of African and South American lungfishes degenerate with age, so adults must rise to the surface and exchange gases entirely through their lungs. Some paleontologists hypothesize that all early bony fishes may have had lungs or lunglike structures. Most modern bony fishes have no lungs, but nearly all have homologous **swim bladders** (see Chapter 31). By adjusting the amount of gas in its swim bladder, the fish can control its buoyancy.

Some amphibians do not have lungs. Among plethodontid (lungless) salamanders, for example, all gas exchange takes place in the pharynx or across the thin, wet skin. However, even though they depend mainly on their body surface for gas exchange, most amphibians have lungs (❚ Fig. 45-4). The lungs of salamanders are two long, simple sacs richly supplied with capillaries. Frogs and toads have ridges containing connective tissue on the inside of the lungs, somewhat increasing the respiratory surface.

The lungs of most reptiles are rather simple sacs; some folding of the wall increases the surface for gas exchange. Gas exchange is not very efficient and does not supply enough oxygen to sustain long periods of activity. In some turtles, lizards, and crocodiles, the lungs have subdivisions that give them a spongy texture.

Birds have the most efficient respiratory system of any living vertebrate. Very active, endothermic animals with high metabolic rates, birds require large amounts of oxygen to sustain flight and other activities. Their small, bright red lungs have extensions (usually nine) called **air sacs,** which reach into all parts of the body and even connect with air spaces in some of the bones. The air sacs act as bellows, drawing air into the system. Collapse of

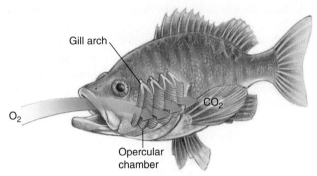

Gill arch

O_2

CO_2

Opercular chamber

(a) Location of gills. The gills lie under a bony plate, the operculum, which has been removed in this side view. The gills form the lateral wall of the pharyngeal cavity.

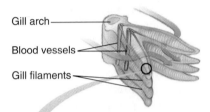

Gill arch

Blood vessels

Gill filaments

(b) Structure of a gill. Each gill consists of a cartilaginous gill arch to which two rows of leaflike gill filaments attach. As water flows past the gill filaments, blood circulates within them.

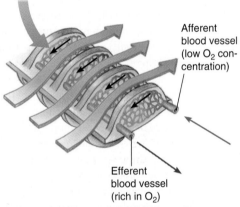

Afferent blood vessel (low O_2 concentration)

Efferent blood vessel (rich in O_2)

(c) Countercurrent flow. Each gill filament has many smaller extensions rich in capillaries. Blood entering the capillaries is deficient in oxygen. The blood flows through the capillaries in a direction opposite to that taken by the water. This countercurrent exchange system efficiently charges the blood with oxygen.

Figure 45-3 *Animated* Gills in bony fishes

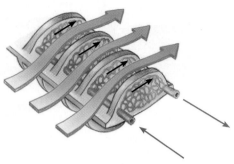

(d) Concurrent flow (hypothetical). If the system were concurrent, that is, if blood flowed through the capillaries in the same direction as the flow of the water, much less of the oxygen dissolved in the water could diffuse into the blood.

(e) Gills of the salmon.

the air sacs during exhalation forces air out. Gas exchange does not take place across the walls of the air sacs.

The high-performance bird respiratory system is arranged so that air flows in one direction through the lungs and is renewed during a two-cycle process (Fig. 45-5). Air entering the body passes into the posterior air sacs and the part of the lungs closest to these air sacs. When the bird exhales, that air flows into the lungs. At the second breath, the air flows from the lungs to the anterior air sacs. Finally, at the second exhalation, the air

leaves the body as another breath of air enters the lungs. Thus, a bird gets fresh air across its lungs through both inhalation and exhalation.

Bird lungs have tiny, thin-walled tubes called **parabronchi,** which are open at both ends. Gas exchange takes place across the walls of these tubes. Chickens and other weak-flying birds have about 400 parabronchi per lung; pigeons and other strong-flying birds have about 1800 parabronchi per lung. The direction of blood flow in the lungs is in a different direction from that of airflow through the parabronchi. This arrangement, similar in principle to the countercurrent exchange in the gills of fishes, increases the amount of oxygen that enters the blood. However, in birds the capillaries are oriented at right angles to the parabronchi rather than along their length. For this reason, this arrangement is referred to as *crosscurrent* rather than countercurrent.

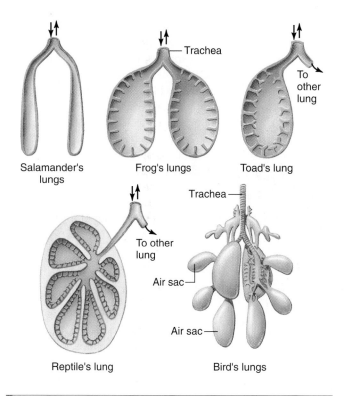

Review

■ Why are specialized respiratory structures necessary in a tadpole but not in a flatworm?

■ How does gas exchange differ among the following animals: (1) earthworm, (2) grasshopper, (3) fish, and (4) bird?

■ How does the countercurrent exchange system increase the efficiency of gas exchange between a fish's gills and blood?

Figure 45-4 *Animated* Evolution of vertebrate lungs

The surface area of the lung has increased during vertebrate evolution. Salamander lungs are simple sacs. Other amphibians and reptiles have lungs with small ridges or folds that help increase surface area. Birds have an elaborate system of lungs and air sacs. Mammalian lungs have millions of alveoli that increase the surface available for gas exchange (see Fig. 45-7).

THE MAMMALIAN RESPIRATORY SYSTEM

Learning Objectives

3　Trace the passage of oxygen through the human respiratory system from nostrils to alveoli.

4　Summarize the mechanics and the regulation of breathing in humans, and describe gas exchange in the lungs and tissues.

5　Explain the role of hemoglobin in oxygen transport, and identify factors that determine and influence the oxygen–hemoglobin dissociation curve.

6　Summarize the mechanisms by which carbon dioxide is transported in the blood.

7　Describe the physiological effects of hyperventilation and of sudden decompression when a diver surfaces too quickly from deep water.

The respiratory system of mammals consists of the lungs and a series of tubes through which air passes on its journey from the nostrils to the lungs and back (■ Fig. 45-6). The complex lungs have an enormous surface area. In the following sections we focus on the human respiratory system.

The airway conducts air into the lungs

A breath of air enters the body through the **nostrils** and flows through the **nasal cavities.** Air passing through the nose is filtered, moistened, and brought to body temperature. The nasal cavities

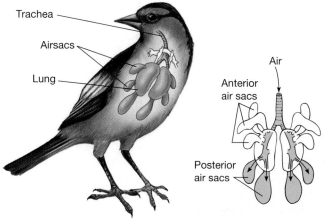

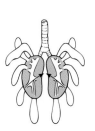

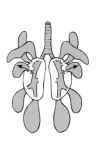

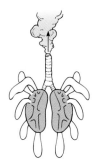

(a) Structure of the bird respiratory system.

(b) First inhalation. As the bird inhales, fresh air flows into the posterior air sacs (*blue*) and partly into the lungs (*not shown*).

(c) First exhalation. As the bird exhales, air from the posterior air sacs is forced into the lungs.

(d) Second inhalation. Air from the first breath moves into the anterior air sacs and partly into the lungs (*not shown*). Air from the second inhalation flows into the posterior air sacs (*pink*).

(e) Second exhalation. Most of the air from the first inhalation leaves the body, and air from the second inhalation flows into the lungs.

Figure 45-5 *Animated* Gas exchange in birds

The bird respiratory system includes lungs and air sacs. The bird's breathing process requires two cycles of inhalation and exhalation to support a one-way flow of air through the lungs.

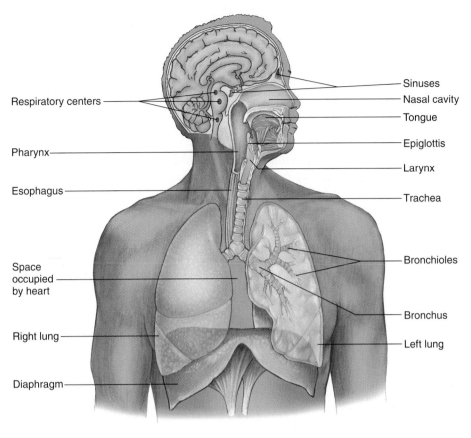

The trachea divides into two branches, the **bronchi** (sing., *bronchus*); one bronchus connects to each lung. Both trachea and bronchi are lined by a mucous membrane containing ciliated cells. These cells trap medium-sized particles that escape the cleansing mechanisms of nose and larynx. Mucus containing these particles is constantly beaten upward by the cilia to the pharynx, where it is periodically swallowed. This mechanism, functioning as a cilia-propelled elevator of mucus, helps keep foreign material out of the lungs.

Gas exchange occurs in the alveoli of the lungs

The lungs are large, paired, spongy organs occupying the *thoracic* (chest) *cavity*. The right lung is divided into three lobes, the left lung into two lobes. Each lung is covered with a **pleural membrane,** which forms a continuous sac that encloses the lung and becomes the lining of the thoracic cavity. The *pleural cavity* is the space between the pleural membranes. A film of fluid in the pleural cavity provides lubrication between the lungs and the chest wall.

Because the lung consists largely of air tubes and elastic tissue, it is a spongy, elastic organ with a very large internal surface area for gas exchange. Inside the lungs the bronchi branch, becoming smaller and more numerous. These branches give rise to more than 1 million tiny **bronchioles** in each lung. Each bronchiole ends in a cluster of tiny air sacs, the **alveoli** (sing., *alveolus*) (❚ Fig. 45-7). Each human lung contains more than 300 million alveoli, which provide an internal surface area the approximate size of a tennis court. Each alveolus is lined by an extremely thin, single layer of epithelial cells. Gases diffuse freely through the wall of the alveolus and into the capillaries that surround it. Only two thin cell layers, the epithelia of the alveolar wall and the capillary wall, separate the air in the alveolus from the blood. In summary, air passes through the following sequence of structures after it enters the body:

Nostrils ⟶ nasal cavities ⟶ pharynx ⟶ larynx ⟶ trachea ⟶ bronchi ⟶ bronchioles ⟶ alveoli

Figure 45-6 *Animated* The human respiratory system

The internal view of one lung illustrates a portion of its extensive system of air passageways. The muscular diaphragm forms the floor of the thoracic cavity. The respiratory centers in the brain regulate the rate of respiration.

are lined with a moist, ciliated epithelium rich in blood vessels. Inhaled dirt, bacteria, and other foreign particles are trapped in the stream of mucus produced by cells within the epithelium and pushed along toward the throat by the cilia. In this way, foreign particles are delivered to the digestive system, which can more effectively dispose of such materials than can the delicate lungs. A person normally swallows more than a pint of nasal mucus each day, and even more during an infection or allergic reaction.

The back of the nasal cavities is continuous with the throat region, or **pharynx.** Air finds its way into the pharynx whether one breathes through the nose or mouth. An opening in the floor of the pharynx leads into the **larynx.** Because the larynx contains the vocal cords, it is also referred to as the voice box. Cartilage embedded in its wall prevents the larynx from collapsing and makes it hard to the touch when felt through the neck.

During swallowing, a flap of tissue called the **epiglottis** automatically closes off the larynx so that food and liquid enter the esophagus rather than the lower airway. If this mechanism fails and foreign matter enters the sensitive larynx, a cough reflex expels the material. Despite these mechanisms, choking sometimes occurs.

From the larynx, air passes into the **trachea,** or windpipe, which is kept from collapsing by rings of cartilage in its wall.

Ventilation is accomplished by breathing

Breathing is the mechanical process of moving air from the environment into the lungs and of expelling air from the lungs. Inhaling air is called **inhalation** or *inspiration;* exhaling air is **exhalation** or *expiration.* The thoracic cavity is closed so no air can enter except through the trachea. (When the chest wall is punctured, for example, by a fractured rib or gunshot wound, air enters the pleural space and the lung collapses.)

During inhalation, the volume of the thoracic cavity is increased by the contraction of the **diaphragm,** the dome-shaped muscle that forms its floor. When the diaphragm contracts, it moves downward, thus increasing the volume of the thoracic cavity (Fig. 45-8). During forced inhalation, when a large volume of air is inhaled, the *external intercostal muscles* contract as well. This action moves the ribs upward, also increasing the volume of the thoracic cavity.

Because the lungs adhere to the walls of the thoracic cavity, when the volume of the thoracic cavity increases, the space within each lung also increases. The air in the lungs now has more space in which to move about. Air pressure in the lungs falls by 2 or 3 millimeters of mercury (mm Hg) below the air pressure outside the body. As a result of this pressure difference, air from the outside rushes in through the respiratory passageways and fills the lungs until the two pressures are equal once again.

Exhalation occurs when the diaphragm relaxes. The volume of the thoracic cavity decreases, raising the pressure in the lungs to 2 to 3 mm Hg above atmospheric pressure. The millions of distended air sacs partially deflate and expel the inhaled air. The pressure returns to normal, and the lung is ready for another inhalation. Thus, in inhalation the millions of alveoli fill with

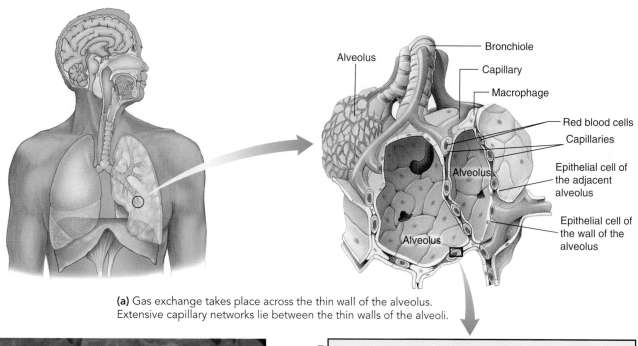

(a) Gas exchange takes place across the thin wall of the alveolus. Extensive capillary networks lie between the thin walls of the alveoli.

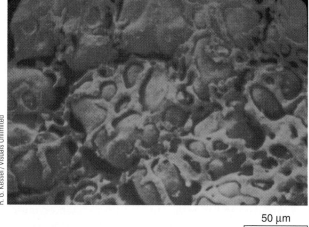

50 μm

(b) Color-enhanced SEM showing the capillary network surrounding a portion of several alveoli.

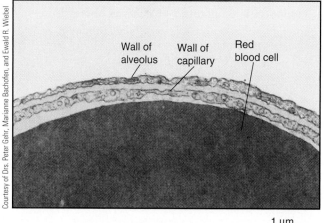

1 μm

(c) Color-enhanced TEM of a portion of a capillary and the wall of an alveolus. The dark structure extending through the capillary is part of a red blood cell. Notice the very short distance oxygen must diffuse to get from the air within the alveolus to the red blood cells that transport it to the body tissues.

Figure 45-7 *Animated* Structure of alveoli

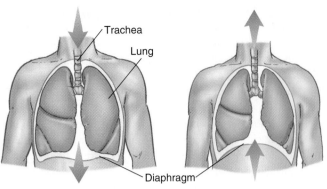

(a) Inhalation. The diaphragm contracts, increasing the volume of the thoracic cavity. Air moves into the lungs.

(b) Exhalation. The diaphragm relaxes, decreasing the volume of the thoracic cavity.

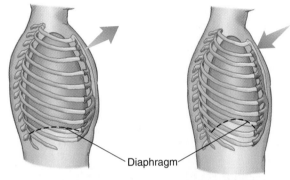

(c) Forced inhalation. The external intercostal muscles contract, pulling the rib cage upward and outward. This increases the front-to-back dimension of the chest and correspondingly increases the volume of the thoracic cavity.

(d) Forced exhalation. The internal intercostal muscles contract, pulling the rib cage downward and inward. The volume of the thoracic cavity decreases.

Figure 45-8 *Animated* Mechanics of breathing

Changes in position of the diaphragm in exhalation and inhalation change the volume of the thoracic cavity.

air like so many tiny balloons. Then, during exhalation, the air rushes out of the alveoli, partially deflating them.

During deep or forced inhalation, the *external intercostal muscles* also contract, pulling the rib cage upward and outward. This action further increases the volume of the thoracic cavity. During forced exhalation, muscles of the abdominal wall and the *internal intercostal muscles* contract, pushing the diaphragm up and the ribs down. This action decreases the volume of the thoracic cavity.

Some of the work in stretching the thorax and the lungs is necessary to stretch elastic connective tissue. Work is also required to overcome the cohesive force of water molecules associated with the pleural membranes. The forces between the water molecules produce a surface tension that resists stretching.

The work of breathing is reduced by **pulmonary surfactant,** a detergent-like phospholipid mixture secreted by specialized

epithelial cells in the lining of the alveoli. Pulmonary surfactant intersperses between the water molecules, reducing their cohesive force. This action markedly reduces the surface tension of the water, prevents the alveoli from collapsing, and reduces the energy required to stretch the lungs. Premature infants often cannot produce enough surfactant and thus suffer from respiratory distress syndrome. In these infants, high surface tension makes it difficult to inflate the lungs, and the alveoli collapse during expiration. Breathing is labored. These infants may be placed on respirators that help them breathe.

The quantity of respired air can be measured

The amount of air moved into and out of the lungs with each normal resting breath is called the **tidal volume.** The normal tidal volume is about 500 mL. The **vital capacity** is the maximum amount of air a person can exhale after filling the lungs to the maximum extent. Vital capacity is greater than tidal volume because the lungs are not completely emptied of stale air and filled with fresh air with each normal resting breath. The volume of air that remains in the lungs at the end of a normal expiration is the **residual capacity.**

Gas exchange takes place in the alveoli

The respiratory system delivers oxygen to the alveoli, but if oxygen remained in the lungs, all the other body cells would soon die. The vital link between alveolus and body cell is the circulatory system. Each alveolus serves as a tiny depot from which oxygen diffuses into blood brought close to the alveolar air by capillaries (Fig. 45-9).

Oxygen molecules efficiently pass by simple diffusion from the alveoli, where they are more concentrated, into the blood in the pulmonary capillaries, where they are less concentrated. At the same time, carbon dioxide moves from the blood, where it is more concentrated, to the alveoli, where it is less concentrated. Each gas diffuses through the single layer of cells lining the alveoli and the single layer of cells lining the capillaries.

Cellular respiration results in the continuous use of oxygen and production of carbon dioxide. Inhaled (atmospheric) air contains about 20.9% oxygen, but exhaled (alveolar) air contains only 14% oxygen. Because carbon dioxide is produced during cellular respiration, exhaled air contains 100 times as much (5.6%) carbon dioxide as inhaled air (.04% carbon dioxide).

The concentration of oxygen in the cells is lower than in the capillaries entering the tissues, and the concentration of carbon dioxide is higher in the cells than in the capillaries. As blood circulates through capillaries of a tissue such as brain or muscle, oxygen moves by simple diffusion from the blood to the cells, and carbon dioxide moves from the cells into the blood.

The factor that determines the direction and rate of diffusion is the pressure or tension of the particular gas. According to **Dalton's law of partial pressures,** in a mixture of gases the total pressure of the mixture is the sum of the pressures of the individual gases. Each gas exerts, independently of the others,

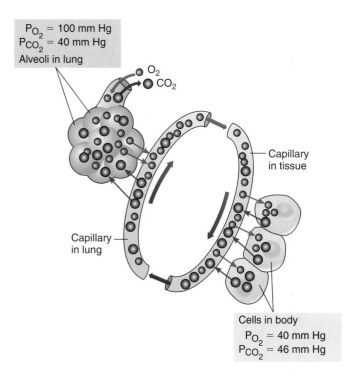

P_{O_2} = 100 mm Hg
P_{CO_2} = 40 mm Hg
Alveoli in lung

O_2
CO_2

Capillary in tissue

Capillary in lung

Cells in body
P_{O_2} = 40 mm Hg
P_{CO_2} = 46 mm Hg

Figure 45-9 *Animated* Gas exchange in the lungs and tissues

The concentration of oxygen is greater in the alveoli than in the pulmonary capillaries, so oxygen diffuses from the alveoli into the blood. Carbon dioxide is more concentrated in the blood than in the alveoli, so it diffuses out of the capillaries and into the alveoli. In the tissues, oxygen is more concentrated in the blood than in the body cells; it diffuses out of the capillaries into the cells. Carbon dioxide is more concentrated in the cells, so it diffuses out of the cells and moves into the blood. Note the differences in partial pressures of oxygen and carbon dioxide before and after gases are exchanged in the tissues.

a **partial pressure**—the same pressure it would exert if it were present alone. At sea level, the barometric pressure (the pressure of Earth's atmosphere) typically supports a column of mercury 760 mm high. Because oxygen makes up about 21% of the atmosphere, oxygen's share of that pressure is $0.21 \times 760 = 160$ mm Hg. Thus, 160 mm Hg is the partial pressure of atmospheric O_2, abbreviated P_{O_2}. In contrast, the partial pressure of atmospheric CO_2 is 0.3 mm Hg, abbreviated P_{CO_2}.

Fick's law of diffusion explains that the amount of oxygen or carbon dioxide that diffuses across the membrane of an alveolus depends on the differences in partial pressure on the two sides of the membrane and on the surface area of the membrane. The gas diffuses faster if the difference in pressure or the surface area increases.

Gas exchange takes place in the tissues

The partial pressure of oxygen in arterial blood is about 100 mm Hg. The P_{O_2} in the tissues is still lower, ranging from 0 to 40 mm Hg. Consequently, oxygen diffuses out of the capillaries and into the tissues. Not all the oxygen leaves the blood, however. The blood passes through the tissue capillaries too rapidly for equi-

librium to be reached. As a result, the partial pressure of oxygen in venous blood returning to the lungs is about 40 mm Hg. Thus, exhaled air has had only part of its oxygen removed—a good thing for those in need of mouth-to-mouth resuscitation!

Respiratory pigments increase capacity for oxygen transport

Respiratory pigments combine reversibly with oxygen and greatly increase the capacity of blood to transport it. **Hemocyanins** are copper-containing proteins dispersed in the hemolymph of many species of mollusks and arthropods. Without oxygen, these pigments are colorless. When oxygen combines with the copper, hemocyanins are blue.

Hemoglobin and **myoglobin** are the most common respiratory pigments in animals. Hemoglobin is the only type of pigment found in the blood of vertebrates. It is also present in many invertebrate species, including annelids, nematodes, mollusks, and arthropods. In some of these animals, the hemoglobin is dispersed in the plasma rather than confined to blood cells. Myoglobin is a form of hemoglobin found in muscle fibers.

In humans and other mammals, inhaled oxygen diffuses out of the alveoli and enters the pulmonary capillaries. Plasma in equilibrium with alveolar air can take up only 0.25 mL of oxygen per 100 mL. However, oxygen diffuses into red blood cells (RBCs) and combines with hemoglobin. The properties of hemoglobin permit whole blood to carry some 20 mL of oxygen per 100 mL. Hemoglobin transports almost 99% of the oxygen. The rest is dissolved in the plasma.

The term *hemoglobin* is actually a general name for a group of related compounds, all of which consist of an iron–porphyrin, or heme, group bound to a protein known as a *globin*. The protein portion of the molecule varies in size, amino acid composition, and physical properties among various species.

The protein portion of hemoglobin is composed of four peptide chains, typically two α and two β chains, each attached to a heme (iron–porphyrin) ring (see Fig. 3-22a). An iron atom is bound in the center of each heme ring. Hemoglobin has the remarkable property of forming a weak chemical bond with oxygen. An oxygen molecule can attach to the iron atom in each heme. In the lung (or gill), oxygen diffuses into the RBCs and combines with hemoglobin (Hb) to form **oxyhemoglobin (HbO$_2$)**. When combined with oxygen, hemoglobin is bright red; without oxygen, it appears dark red, imparting a purplish color to venous blood.

Because the chemical bond formed between the oxygen and the hemoglobin is weak, the reaction is readily reversible. As blood circulates through tissues where the oxygen concentration is low, the reaction proceeds to the left. Hemoglobin releases oxygen, which diffuses out of the blood and into the tissue cells.

$$Hb + O_2 \rightleftharpoons HbO_2$$

The maximum amount of oxygen that hemoglobin can transport is its **oxygen-carrying capacity.** The actual amount of oxygen bound to hemoglobin is the **oxygen content.** The ratio of O_2 content to oxygen-carrying capacity is the **percent O_2 satura-**

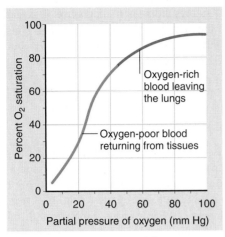

(a) Normal oxygen–hemoglobin dissociation curve. Look at the relationship between the partial pressure of oxygen and the percent O_2 saturation of hemoglobin. Oxygen binds to hemoglobin in the lungs and unloads from hemoglobin in the tissues.

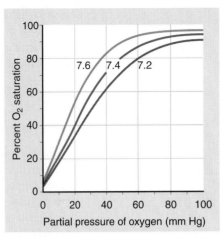

(b) Effect of pH on the oxygen–hemoglobin curve (Bohr effect). The normal pH of human blood is 7.4. Find the location on the horizontal axis where the partial pressure of oxygen is 40, and follow the line up through the curves. Notice that the saturation of hemoglobin with oxygen differs among the three curves, even though the partial pressure of oxygen is the same. Oxygen loading increases at higher pH (7.6). At lower pH (7.2), hemoglobin unloads more oxygen.

Figure 45-10 Oxygen–hemoglobin dissociation curves

tion of the hemoglobin. The percent O_2 saturation is highest in the pulmonary capillaries, where the concentration of oxygen is greatest. In the capillaries of the tissues, where there is less oxygen, oxyhemoglobin dissociates, releasing oxygen. There, the percent saturation of hemoglobin is correspondingly lower. The **oxygen–hemoglobin dissociation curve** shown in ▌ Figure 45-10a illustrates this relationship. As oxygen concentration increases, there is a progressive increase in the percentage of hemoglobin that is combined with oxygen. The ability of oxygen to combine with hemoglobin and be released from oxyhemoglobin is influenced by several factors in addition to percent O_2 saturation; these include pH, carbon dioxide concentration, and temperature.

Carbon dioxide formed in respiring tissue reacts with water in the plasma to form carbonic acid, H_2CO_3. For this reason, any increase in carbon dioxide concentration also increases the acidity (lowers the pH) of the blood. Oxyhemoglobin unloads oxygen more readily in an acidic environment than in an environment with normal pH. Displacement of the oxygen–hemoglobin dissociation curve by a change in pH is known as the **Bohr effect** (▌ Fig. 45-10b). Lactic acid released from active muscles also lowers blood pH and has a similar effect on the oxygen–hemoglobin dissociation curve—more oxygen is unloaded and available for the muscle cells.

Some carbon dioxide is transported by the hemoglobin molecule. Although carbon dioxide attaches to the hemoglobin molecule in a different way and at a different site than oxygen does, the attachment of a carbon dioxide molecule releases an oxygen molecule from the hemoglobin. The effect of carbon dioxide con-

centration on the oxygen–hemoglobin dissociation curve is important. In the capillaries of the lungs (or gills in fishes), carbon dioxide concentration is relatively low and oxygen concentration is high, so oxygen combines with a very high percentage of hemoglobin. In the capillaries of the tissues, carbon dioxide concentration is high and oxygen concentration is low, so hemoglobin readily unloads oxygen. The oxygen and carbon dioxide bound to hemoglobin in RBCs do not contribute to the partial pressures that govern diffusion. Loading and unloading of gases onto hemoglobin depend on partial pressures of the plasma and interstitial fluid (which is determined by the gases dissolved in these fluids).

Carbon dioxide is transported mainly as bicarbonate ions

Blood transports carbon dioxide in three forms. About 10% of carbon dioxide dissolves in plasma. Another 30% enters the RBCs and combines with hemoglobin. Because the bond between the hemoglobin and carbon dioxide is very weak, the reaction is readily reversible. Most carbon dioxide (about 60%) moves through plasma as **bicarbonate ions** (HCO_3^-).

In plasma, carbon dioxide slowly combines with water to form carbonic acid. This reaction proceeds much more rapidly inside RBCs as a result of the action of the enzyme **carbonic anhydrase** (▌ Fig. 45-11). Carbonic acid dissociates, forming hydrogen ions and bicarbonate ions.

$$\underset{\substack{\text{Carbon}\\\text{dioxide}}}{CO_2} + \underset{\text{Water}}{H_2O} \xrightarrow{\overset{\substack{\text{Carbonic}\\\text{anhydrase}}}{}} \underset{\substack{\text{Carbonic}\\\text{acid}}}{H_2CO_3} \longrightarrow \underset{\substack{\text{Hydrogen}\\\text{ion}}}{H^+} + \underset{\substack{\text{Bicarbonate}\\\text{ion}}}{HCO_3^-}$$

Most hydrogen ions released from carbonic acid combine with hemoglobin, which is a very effective buffer. Many of the bicarbonate ions diffuse into the plasma. The action of carbonic anhydrase in the RBCs maintains a diffusion gradient for carbon dioxide to move into and then out of RBCs. As the negatively charged bicarbonate ions move out of RBCs, chloride ions (Cl^-) in the plasma diffuse into the RBCs to replace them, a process known as the **chloride shift.** In the alveolar capillaries, CO_2 diffuses out of the plasma and into the alveoli. As the CO_2 concentration decreases, the reaction sequence just described reverses.

Any condition (such as emphysema) that interferes with the removal of carbon dioxide by the lungs can lead to *respiratory acidosis.* In this situation, carbon dioxide is produced more rapidly than it is excreted by the lungs. As a result, the concentration of

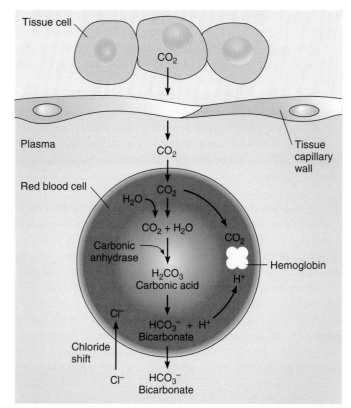

(a) In the tissues, carbon dioxide diffuses from cells into the plasma. Most of the carbon dioxide enters red blood cells, where the enzyme carbonic anhydrase rapidly converts it to carbonic acid. Carbonic acid dissociates and forms H^+ and bicarbonate. As bicarbonate ions move out into the plasma, chloride ions replace them. Hemoglobin combines with H^+, preventing a decrease in pH.

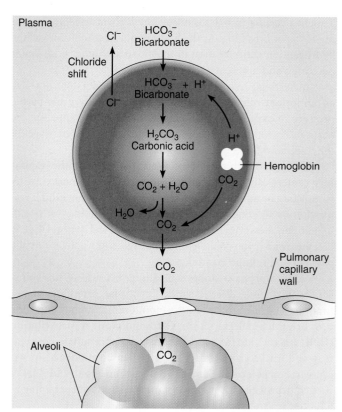

(b) In the lungs, carbon dioxide diffuses out of the plasma and into the alveoli. The processes described in **(a)** are reversed. Bicarbonate ions diffuse from the plasma into the red blood cells. H^+ released from hemoglobin combines with bicarbonate ions, forming carbonic acid. Carbon dioxide produced from the carbonic acid diffuses out of the blood and into the alveoli.

Figure 45-11 Carbon dioxide transport

carbonic acid in the blood increases. When blood pH falls below 7, the central nervous system becomes depressed and the individual becomes disoriented. Untreated respiratory acidosis can cause coma and death.

Breathing is regulated by respiratory centers in the brain

Breathing is a rhythmic, involuntary process regulated by **respiratory centers** in the brain stem (see Fig. 45-6). Groups of neurons in the medulla regulate the basic rhythm of breathing. These neurons send a burst of impulses to the diaphragm and external intercostal muscles, which causes them to contract. After several seconds, these neurons become inactive, the muscles relax, and exhalation occurs. Respiratory centers in the pons help control the transition from inspiration to exhalation. These centers can stimulate or inhibit the respiratory centers in the medulla. The cycle of activity and inactivity repeats itself so that at rest you breathe about 14 times per minute. Overdose of certain medica-

tions, such as barbiturates, depresses the respiratory centers and may lead to respiratory failure.

The basic rhythm of respiration changes in response to needs of the body. When you are playing a fast game of tennis, you need more oxygen than when you are studying biology. During exercise the rate of aerobic cellular respiration increases, producing more carbon dioxide. Your body must dispose of this carbon dioxide through increased ventilation.

Carbon dioxide concentration is the most important chemical stimulus for regulating the rate of respiration. Specialized **chemoreceptors** in the medulla and in the walls of the aorta and carotid arteries are sensitive to changes in arterial carbon dioxide concentration. When stimulated, they send impulses to the respiratory centers, which increase breathing rate.

The chemoreceptors in the walls of the aorta, called *aortic bodies,* and those in the walls of the carotid arteries, the *carotid bodies,* are sensitive to changes in hydrogen ion concentration and oxygen concentration, as well as to carbon dioxide levels. Recall that an increase in carbon dioxide concentration increases hydrogen ions from carbonic acid and lowers blood pH. Even a slight decrease in pH stimulates these chemoreceptors, leading to faster breathing. As the lungs remove carbon dioxide, the hydrogen ion concentration in the blood and other body fluids decreases, and homeostasis is re-established.

Interestingly, oxygen concentration generally does not play an important role in regulating respiration. Only if the partial pressure of oxygen falls markedly do the chemoreceptors in the aorta and carotid arteries become stimulated to send messages to the respiratory centers.

Although breathing is involuntary, the action of the respiratory centers can be consciously influenced for a short time by stimulating or inhibiting them. For example, you can inhibit respiration by holding your breath. You cannot hold your breath indefinitely, however, because eventually you feel a strong urge to breathe. Even if you could ignore this urge, you would eventually pass out and resume breathing.

People who have stopped breathing because of drowning, smoke inhalation, electric shock, or cardiac arrest can sometimes be sustained by mouth-to-mouth resuscitation until their own breathing reflexes return. **Cardiopulmonary resuscitation (CPR)** is a method for aiding victims who have suffered respiratory and cardiac arrest. CPR must be started immediately, because irreversible brain damage occurs within about 4 minutes of oxygen deprivation. A number of organizations offer training in CPR to the general public.

Hyperventilation reduces carbon dioxide concentration

Underwater swimmers and some Asian pearl divers undergo voluntary **hyperventilation** before going underwater. By making a series of deep inhalations and exhalations, they "blow off" CO_2, markedly reducing the CO_2 content of the alveolar air and of the blood. This action allows them to remain underwater longer before the urge to breathe becomes irresistible.

When hyperventilation continues for a long period, dizziness and sometimes unconsciousness may occur, because a certain concentration of carbon dioxide is needed in the blood to maintain normal blood pressure. (This mechanism operates by way of the vasoconstrictor center in the brain, which maintains the muscle tone of blood vessel walls.) Furthermore, if divers hold their breath too long, the low concentration of oxygen may result in unconsciousness and drowning.

High flying or deep diving can disrupt homeostasis

The barometric pressure decreases at progressively higher altitudes. Because the concentration of oxygen in the air remains at 21%, the partial pressure of oxygen decreases along with the barometric pressure. When a person moves to a high altitude, his or her body adjusts over time by producing a greater number of RBCs. Getting sufficient oxygen from the air becomes an ever-increasing problem at higher altitudes. All high-flying jets have airtight cabins pressurized to the equivalent of the barometric pressure at an altitude of about 2000 m (6600 ft). If a jet flying at 11,700 m (more than 38,000 ft) suddenly decompressed, the pilot would lose consciousness in about 30 seconds and become comatose in about 1 minute.

Shallow breathing, which occurs in anxiety and in many respiratory diseases, causes **hypoxia,** a deficiency of oxygen. Even *rapid,* shallow breathing results in hypoxia, because we do not clear out the stale air in the airway and ventilate the lung. Hypoxia causes drowsiness, mental fatigue, headache, and sometimes euphoria. The ability to think and make judgments is impaired, as is the ability to perform tasks requiring coordination.

In addition to the problems of hypoxia, a rapid decrease in barometric pressure causes **decompression sickness** (commonly known as the "bends" because those suffering from it bend over in pain). When the barometric pressure drops below the total pressure of all gases dissolved in the blood and other body fluids, the dissolved gases tend to come out of solution and form gas bubbles. A familiar example occurs each time you uncap a bottle of soda, reducing pressure in the bottle. Carbon dioxide is released from solution and bubbles out into the air. In the body, nitrogen has a low solubility in blood and tissues. When it comes out of solution, the bubbles formed may damage tissues and block capillaries, interfering with blood flow. The clinical effects of decompression sickness are pain, dizziness, paralysis, unconsciousness, and even death.

Decompression sickness is more common in scuba diving than in high-altitude flying. As a diver descends, the surrounding pressure increases tremendously—1 atmosphere (the atmospheric pressure at sea level, which equals 760 mm Hg) for each 10 m (about 32 ft). To prevent lung collapse, a diver must be supplied with air under pressure, exposing the lungs to very high alveolar gas pressures.

At sea level an adult human has about 1 L of nitrogen dissolved in the body, with about half in the fat and half in the body fluids. After a diver's body has become saturated with nitrogen at a depth of 100 m (325 ft), the body fluids contain about 10 L of nitrogen. To prevent this nitrogen from rapidly bubbling out of solution and causing decompression sickness, the diver must rise to the surface gradually, with stops at certain levels on the way up. These pauses allow nitrogen to be expelled slowly through the lungs.

Some mammals are adapted for diving

Some air-breathing mammals can spend long periods in the ocean depths without coming up for air. Dolphins, whales, seals, and beavers have structural and physiological adaptations that allow them to dive for food or to elude their enemies. With their streamlined bodies and forelimbs modified as fins or flippers, diving mammals perform impressive aquatic feats. The Weddell seal can swim under the ice at a depth of 596 m (1955 ft) for more than an hour without coming up for air. The enormous elephant seal, which measures about 5 m (16 to 18 ft) in length and weighs 2 to 4 tons, can plunge even deeper (Fig. 45-12). A female elephant seal can dive to depths of more than 1500 m (almost 5000 ft)

Figure 45-12 Deep diver

Elephant seals, like this northern elephant seal (*Mirounga angustiros-tris*), may be the deepest divers on Earth. Researchers have recorded their dives at depths of more than 1500 m (5000 ft).

in less than 20 minutes and stay beneath the surface for more than an hour.

Physiological adaptations, including ways to distribute and store oxygen, permit some mammals to dive deeply and remain underwater for long periods. Turtles and birds that dive depend on oxygen stored in their lungs; however, diving mammals do not take in and store extra air before a dive. In fact, seals exhale before they dive. With less air in their lungs, they are less buoyant. Their lungs collapse at about 50 to 70 m into their dive and then reinflate as they ascend, so their lungs do not function for most of the dive. These adaptations are thought to reduce the chance of decompression sickness, because with less air in the lungs there is less nitrogen in the blood to dissolve during the dive.

Seals have about twice the volume of blood, relative to their body weight, as nondiving mammals. Diving mammals also have high concentrations of myoglobin, which stores oxygen in muscles. These animals have up to 10 times as much myoglobin as do terrestrial mammals. The very large spleen typical of many diving mammals stores oxygen-rich red blood cells. Under pres-

sure (during a dive), the spleen is squeezed and releases these red blood cells into the circulation.

Diving mammals markedly reduce the energy expended in deep diving (more than 200 m) by gliding. Filmed video sequences of diving seals and whales show that they glide most of the way down. Gliding is possible because the animal's buoyancy decreases as the lungs gradually collapse and the amount of air in the lungs is reduced. Gravity then pulls them downward.

When a mammal dives to its limit, physiological mechanisms known collectively as the **diving reflex** are activated. Metabolic rate decreases by about 20%, which conserves oxygen. Breathing stops, and bradycardia (slowing of the heart rate) occurs. The heart rate may decrease to one tenth the normal rate, reducing the body's consumption of oxygen and energy. Blood is redistributed so that the heart and brain receive the greatest share; skin, muscles, digestive organs, and other internal organs can survive with less oxygen and receive less blood while an animal is submerged.

The diving reflex is present to some extent in humans, where it may act as a protective mechanism during birth, when an infant may be deprived of oxygen for several minutes. Cases of near drownings, especially of young children, have been documented in which the victim was submerged for as long as 45 minutes in very cold water before being rescued and resuscitated. Many of these survivors showed no brain damage. The shock of icy water slows heart rate, increases blood pressure, and shunts blood to internal organs of the body that most need oxygen (blood flow in arms and legs decreases). Metabolic rate decreases, so less oxygen is required.

Review

- What is the sequence of inhaled airflow through the respiratory structures in a mammal?
- What is the function of respiratory pigments? How do they work?
- Why does alveolar air differ in composition from atmospheric air? Explain.
- What physiological mechanisms bring about an increase in rate and depth of breathing during exercise? Why is such an increase necessary?

BREATHING POLLUTED AIR

Learning Objective

8   Describe the defense mechanisms that protect the lungs, and describe the effects of polluted air on the respiratory system.

Several defense mechanisms protect the delicate lungs from the harmful substances we breathe (Fig. 45-13). Hairs in the nostrils, the ciliated mucous lining in the nose and pharynx, and the cilia–mucus elevator of the trachea and bronchi trap foreign particles in inspired air. One of the body's most rapid defense responses to breathing dirty air is **bronchial constriction.** In this process, the bronchial tubes narrow, increasing the chance that inhaled particles will land on the sticky mucous lining. Unfortunately, bronchial constriction narrows the airway so less air

THE EFFECTS OF SMOKING

Tobacco smoke is by far the most important risk factor for lung cancer, the most common lethal cancer worldwide. Tobacco smoke is also an important risk factor for cardiovascular disease and many other diseases. In June 2006, U.S. Surgeon General Richard H. Carmona issued the annual *Report of the Surgeon General.** The report states once again that smoking is the single most preventable cause of death in the United States. More than 400,000 adults die each year of tobacco-related causes. Yet in the United States, an estimated 26% of men and 21% of women ages 18 and over continue to smoke. The direct medical costs of tobacco use amount to more than $75 billion annually.

The International Agency for Research on Cancer and several U.S. agencies have concluded that secondhand tobacco smoke is a known human carcinogen. The 2006 *Report of the Surgeon General* focuses on the effects of secondhand smoke. The report states that millions of children and adults in the United States are still exposed to secondhand smoke in their homes and workplaces. This exposure causes disease and premature death in children and adults who do not smoke. Babies and children exposed to secondhand smoke are at increased risk for sudden infant death syndrome (SIDS), acute respiratory infections, pneumonia, bronchitis, ear infections, and asthma. In adults, secondhand smoke causes cardio-

vascular disease and lung cancer. About 60% of nonsmokers in the United States exhibit biological evidence of exposure to secondhand smoke.

Nicotine is highly addictive. Like morphine, amphetamines, and cocaine, nicotine increases dopamine concentration and activates cells in the *nucleus accumbens*, an area at the base of the forebrain. This area helps integrate emotion, and dopamine may facilitate learning an association between the pleasurable effects of the drug and other stimuli, such as the smell of smoke. Nicotine, like cocaine and other abused drugs, causes long-lasting changes in the brain. Cell signaling and other mechanisms underlying drug addiction are very similar to those of learning and memory. Investigators are searching for drugs that block the dopamine release caused by nicotine and other addictive drugs.

Here are some facts about smoking:

- The life of a 30-year-old who smokes 15 cigarettes a day is shortened by an average of more than 5 years.

- If you smoke more than one pack per day, you are about 20 times as likely as a nonsmoker to develop lung cancer. According to the American Cancer Society, cigarette smoking causes more than 75% of all lung cancer deaths.

- If you smoke, you double your chances of dying from cardiovascular disease.

- If you smoke, you are 20 times as likely as a nonsmoker to develop chronic bronchitis and emphysema. About 90% of all deaths from COPD are caused by cigarette smoking.

- If you smoke, you have about 5% less oxygen circulating in your blood (because carbon monoxide binds to hemoglobin) than does a nonsmoker.

- If you smoke when you are pregnant, your baby will weigh about 6 ounces less at birth, and there is double the risk of miscarriage, stillbirth, and infant death.

- Infants whose parents smoke have double the risk of contracting pneumonia or bronchitis in their first year of life.

- There is no risk-free level of exposure to secondhand smoke. Nonsmokers who are exposed to secondhand smoke increase their risk of developing heart disease by 25% to 30% and lung cancer by 20% to 30%.

- When smokers quit smoking, their risk of dying from chronic obstructive pulmonary disease, cardiovascular disease, or cancer gradually decreases. (Precise changes in risk depend on the number of years the person smoked, the number of cigarettes smoked per day, the age of starting to smoke, and the number of years since quitting.)

- An estimated 70% of smokers (more than 33 million) want to quit, but only 2.5% per year succeed in quitting permanently. Nicotine replacement with gum, patches, or nasal spray has been shown to be effective as an aid to smoking cessation, especially when used with behavioral therapy. Certain antidepressants are used to reduce the craving for nicotine.

- Almost every American who takes up smoking is a teenager—6000 every day, more than 2 million every year. Of children who begin smoking, 10% start by the fourth grade, and nearly two thirds by the tenth grade. Teen smoking rates have increased every year since 1992.

---

*U.S. Department of Health and Human Services, The Health Consequences of Involuntary Exposure to Tobacco Smoke: A Report of the Surgeon General, U.S. Department of Health and Human Services, Centers for Disease Control and Prevention, National Center for Chronic Disease Prevention and Health Promotion, Office on Smoking and Health, 2006.

---

can reach the lungs, decreasing the amount of oxygen available to body cells. Fifteen puffs on a cigarette during a 5-minute period increases airway resistance as much as threefold, and this added resistance to breathing lasts more than 30 minutes. Chain-smokers and those who breathe heavily polluted air are in a state of chronic bronchial constriction.

Neither the smallest bronchioles nor the alveoli are equipped with mucus or ciliated cells. Foreign particles that get through other respiratory defenses and find their way into the alveoli may

be engulfed by macrophages. The macrophages may then accumulate in the lymph tissue of the lungs. Lung tissues of chronic smokers and those who work in dirty industries burning fossil fuel contain large blackened areas where carbon particles have been deposited (❚ Fig. 45-14).

Continued insult to the respiratory system results in disease. Chronic bronchitis, pulmonary emphysema, and lung cancer have been linked to cigarette smoking and breathing polluted air. More than 75% of patients with **chronic bronchitis** have

Figure 45-13 Urban air pollution

Industry spews tons of pollutants into the atmosphere. Unless expensive air pollution control equipment is installed, harmful gases, such as sulfur dioxide and nitrogen oxides, are released and contribute to acid precipitation.

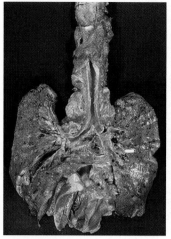

(a) Lungs and major bronchi of a nonsmoker.

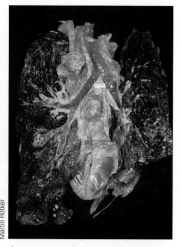

(b) Lungs and heart of a cigarette smoker. The dark spots in the lung tissue are particles of carbon, tar, and other substances that passed through the respiratory defenses and lodged in the lungs.

a history of heavy cigarette smoking (see *Focus On: The Effects of Smoking*). People with chronic bronchitis often develop **pulmonary emphysema,** a disease also most common in cigarette smokers. In this disorder, alveoli lose their elasticity, and walls between adjacent alveoli are destroyed. The surface area of the lung is so reduced that gas exchange is seriously impaired. Air is not expelled effectively, and stale air accumulates in the lungs.

The emphysema victim struggles for every breath, and still the body does not get enough oxygen. To compensate, the right ventricle of the heart pumps harder and becomes enlarged. Emphysema patients frequently die of heart failure.

Most patients with **chronic obstructive pulmonary disease (COPD),** a condition characterized by obstructed airflow, have both chronic bronchitis and emphysema. Asthma also contributes to COPD. People with asthma respond to inhaled stimuli with exaggerated bronchial constriction, and the airway is typically inflamed.

Figure 45-14 Effects of cigarette smoking

Cigarette smoking is also the main cause of **lung cancer.** More than 69 of the 4800 chemical compounds in tobacco smoke cause cancer in animals, including humans. These carcinogenic substances irritate the cells lining the respiratory passages and alter their metabolic balance. Normal cells are transformed into cancer cells, which may multiply rapidly and invade surrounding tissues.

Review

▌ What mechanisms does the human respiratory system have for getting rid of inhaled dirt?

▌ What happens when so much dirty air is inhaled that these mechanisms cannot function effectively?

SUMMARY WITH KEY TERMS

Learning Objectives

1 Compare the advantages and disadvantages of air and water as mediums for gas exchange, and describe adaptations for gas exchange in air (page 971).

   ▌ Air contains a higher concentration of molecular oxygen than does water, and oxygen diffuses more rapidly through air than through water. Air is less dense and less viscous than water, so less energy is needed to move air over a gas exchange surface.

   ▌ Terrestrial animals have adaptations that protect their respiratory surfaces from drying.

2 Describe the following adaptations for gas exchange: body surface, tracheal tubes, gills, and lungs (page 971).

   ▌ Small aquatic animals exchange gases by diffusion, requiring no specialized respiratory structures. Some

invertebrates, including most annelids, and a few vertebrates (many amphibians) exchange gases across the body surface.

   ▌ In insects and some other arthropods, air enters a network of **tracheal tubes,** or **tracheae,** through openings, called **spiracles,** along the body surface. Tracheal tubes branch and extend to all regions of the body.

   ▌ **Gills** are moist, thin projections of the body surface found mainly in aquatic animals. In chordates, gills are usually internal, located along the edges of the gill slits. In bony fishes an **operculum** protects the gills. A **countercurrent exchange system** maximizes diffusion of oxygen into the blood and diffusion of carbon dioxide out of blood.

   ▌ Animals carry on **ventilation,** the process of actively moving air or water over respiratory surfaces. Terrestrial vertebrates have **lungs** and some means of ventilating them. Amphibians and reptiles have lungs with only some ridges or folds that increase surface area.

■ In birds, the lungs have extensions, called **air sacs,** that act as bellows, drawing air into the system. Two cycles of inhalation and exhalation support a one-way flow of air through the lungs. Air flows from the outside into the posterior air sacs, to the lung, through the anterior air sacs, and then out of the body. Gas exchange takes place through the walls of the **parabronchi** in the lungs. A cross-current arrangement, in which blood flow is at right angles to the parabronchi, increases the amount of oxygen that enters the blood.

ThomsonNOW **Learn more about adaptations for gas exchange, including gills in bony fishes, vertebrate lungs, and the bird respiratory system, by clicking on the figures in ThomsonNOW.**

3 Trace the passage of oxygen through the human respiratory system from nostrils to alveoli (page 975).

■ The human respiratory system includes the lungs and a system of airways. A breath of air passes in sequence through the **nostrils, nasal cavities, pharynx, larynx, trachea, bronchi, bronchioles,** and **alveoli.** Each lung occupies a pleural cavity and is covered with a **pleural membrane.**

ThomsonNOW **Learn more about the human respiratory system by clicking on the figures in ThomsonNOW.**

4 Summarize the mechanics and the regulation of breathing in humans, and describe gas exchange in the lungs and tissues (page 975).

■ During breathing, the **diaphragm** contracts, expanding the chest cavity. The membranous walls of the lungs move outward along with the chest walls, lowering pressure within the lungs. Air from outside the body rushes in through the air passageways and fills the lungs until the pressure equals atmospheric pressure.

■ **Tidal volume** is the amount of air moved into and out of the lungs with each normal breath. **Vital capacity** is the maximum volume that can be exhaled after the lungs fill to the maximum extent. The volume of air that remains in the lungs at the end of a normal expiration is the **residual capacity.**

■ **Respiratory centers** in the medulla and pons regulate respiration. These centers are stimulated by **chemoreceptors** sensitive to an increase in carbon dioxide concentration. They also respond to an increase in hydrogen ions and to very low oxygen concentration.

■ Oxygen and carbon dioxide are exchanged between alveoli and blood by diffusion. The pressure of a particular gas determines its direction and rate of diffusion.

■ **Dalton's law of partial pressures** explains that in a mixture of gases, the total pressure of the mixture is the sum of the pressures of the individual gases. Thus, each gas in a mixture of gases exerts a **partial pressure,** the same pressure it would exert if it were present alone. The partial pressure of atmospheric oxygen, P_{O_2}, is 160 mm Hg at sea level.

■ According to **Fick's law of diffusion,** the greater the difference in pressure on the two sides of a membrane and the larger the surface area, the faster the gas diffuses across the membrane.

ThomsonNOW **See the breathing mechanisms in action by clicking on the figures in ThomsonNOW.**

5 Explain the role of hemoglobin in oxygen transport, and identify factors that determine and influence the oxygen–hemoglobin dissociation curve (page 975).

■ **Hemoglobin** is the respiratory pigment in the blood of vertebrates. Almost 99% of the oxygen in human blood is transported as **oxyhemoglobin (HbO_2).**

■ The maximum amount of oxygen that can be transported by hemoglobin is the **oxygen-carrying capacity.** The actual amount of oxygen bound to hemoglobin is the **oxygen content.** The **percent O_2 saturation,** the ratio of oxygen content to oxygen-carrying capacity, is highest in pulmonary capillaries, where oxygen concentration is greatest.

■ The **oxygen–hemoglobin dissociation curve** shows that as oxygen concentration increases, there is a progressive increase in the amount of hemoglobin that combines with oxygen. The curve is affected by pH, temperature, and CO_2 concentration.

■ Oxyhemoglobin dissociates more readily as carbon dioxide concentration increases, because carbon dioxide combines with water and produces carbonic acid, which lowers the pH. Displacement of the oxygen–hemoglobin dissociation curve by a change in pH is called the **Bohr effect.**

6 Summarize the mechanisms by which carbon dioxide is transported in the blood (page 975).

■ About 60% of the carbon dioxide in the blood is transported as bicarbonate ions. About 30% combines with hemoglobin, and another 10% is dissolved in plasma.

■ Carbon dioxide combines with water to form carbonic acid; the reaction is catalyzed by **carbonic anhydrase.** Carbonic acid dissociates, forming bicarbonate ions (HCO_3^-) and hydrogen ions (H^+).

■ Hemoglobin combines with H^+, buffering the blood. Many bicarbonate ions diffuse into the plasma and are replaced by Cl^-; this exchange is known as the **chloride shift.**

7 Describe the physiological effects of hyperventilation and of sudden decompression when a diver surfaces too quickly from deep water (page 975).

■ **Hyperventilation** reduces the concentration of carbon dioxide in the alveolar air and in the blood. A certain carbon dioxide concentration in the blood is needed to maintain normal blood pressure.

■ As altitude increases, barometric pressure (the pressure of Earth's atmosphere) falls, and less oxygen enters the blood. This situation can lead to **hypoxia,** or oxygen deficiency, which in turn can lead to loss of consciousness and death. Rapid decrease in barometric pressure can also cause **decompression sickness,** especially common among divers who ascend too rapidly from a dive.

■ Diving mammals have high concentrations of **myoglobin,** a pigment that stores oxygen, in their muscles. The **diving reflex,** a group of physiological mechanisms (including a decrease in metabolic rate), is activated when a mammal dives to its limit.

8 Describe the defense mechanisms that protect the lungs, and describe the effects of polluted air on the respiratory system (page 983).

■ The ciliated mucous lining of the nose, pharynx, trachea, and bronchi traps inhaled particles. Inhaling polluted air results in **bronchial constriction,** increased mucus secretion, damage to ciliated cells, and coughing. Breathing polluted air or inhaling cigarette smoke can cause **chronic bronchitis, pulmonary emphysema,** and **lung cancer.**

1. Breathing is an example of (a) countercurrent exchange (b) cellular respiration (c) ventilation (d) diffusion (e) gas exchange through the body surface

2. Which of the following is a benefit of gas exchange in air compared with gas exchange in water? (a) air has a higher concentration of molecular oxygen (b) oxygen diffuses more slowly in air (c) no energy is required for ventilation (d) moist respiratory surface is not needed (e) air is denser than water

3. Which of the following adaptations for gas exchange is most characteristic of insects? (a) lungs (b) tracheal tubes (tracheae) (c) parabronchi (d) air sacs (e) dermal gills

4. Which of the following are accurately matched? (a) bony fish—operculum (b) insect—alveoli (c) bird—spiracles (d) aquatic mammal—gill filaments (e) shark—dermal gills

5. Tracheal tubes (tracheae) (a) have moist surfaces throughout their length (b) are highly vascular (c) branch and extend to all the cells (d) are characteristic of many vertebrates (e) a, b, and c

6. The most efficient vertebrate respiratory system is that of (a) amphibians (b) birds (c) reptiles (d) mammals (e) humans

7. In a bird, the correct sequence for a breath of air is (a) anterior air sacs ⟶ posterior air sacs ⟶ lung (b) posterior air sacs ⟶ lung ⟶ anterior air sacs (c) parabronchi ⟶ posterior air sacs ⟶ anterior air sacs (d) posterior air sacs ⟶ alveoli ⟶ anterior air sacs (e) posterior air sacs ⟶ capillaries ⟶ cells

8. Respiratory pigments (a) combine reversibly with oxygen (b) are found mainly in vertebrates (c) all have a heme (porphyrin) group that combines with oxygen (d) diffuse into the air sacs (e) attach to the alveolar wall

9. Which sequence most accurately describes the sequence of airflow in the human respiratory system?

   1. pharynx 2. bronchus 3. trachea 4. larynx 5. alveolus 6. bronchiole

   (a) 4, 1, 3, 2, 5, 6 (b) 1, 4, 3, 2, 5, 6 (c) 4, 1, 3, 2, 6, 5 (d) 1, 4, 3, 2, 6, 5 (e) 1, 3, 4, 2, 6, 5

10. The amount of air moved in and out of the lungs with each normal resting breath is the (a) vital capacity (b) residual capacity (c) vital volume (d) partial pressure (e) tidal volume

11. According to Fick's law of diffusion, a gas will diffuse faster if the (a) surface is reduced (b) difference in pressure is increased (c) difference in pressure is decreased (d) pH is lowered (e) a and c

12. Oxygen in the blood is transported mainly (a) in combination with hemoglobin (b) as bicarbonate ions (c) as carbonic acid (d) dissolved in plasma (e) combined with carbon dioxide

13. The concentration of which substance is most important in regulating the rate of respiration? (a) chloride ions (b) oxygen (c) bicarbonate ions (d) nitrogen (e) carbon dioxide

14. When a diver ascends too rapidly, (a) bronchial constriction occurs (b) a diving reflex is activated (c) nitrogen rapidly bubbles out of solution in the body fluids (d) nitrogen hypoxia occurs (e) carbon dioxide bubbles damage the alveoli

15. Which of the following is *not* true of the diving reflex? (a) breathing stops (b) the heart slows (c) less blood is distributed to the muscles (d) metabolic rate increases by about 20% (e) energy consumption decreases

16. Pulmonary emphysema (a) results from chronic obstructive pulmonary disease (b) is uncommon in cigarette smokers (c) results from bronchial constriction (d) is sometimes referred to as the Bohr effect (e) is characterized by loss of elasticity of the alveolar walls

17. Label the figure. Use Figure 45-6 to check your answers.

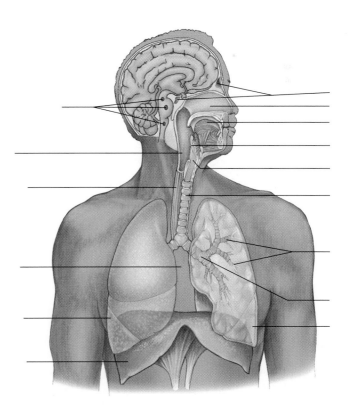

1. What problems would be faced by a terrestrial animal having gills instead of lungs?

2. **Evolution Link.** Under what conditions might it be advantageous for a fish to have lungs as well as gills? What function do the "lungs" of modern fishes serve?

3. **Evolution Link.** Aquatic mammals such as whales and dolphins use lungs rather than gills for gas exchange. Propose a hypothesis to explain this.

4. **Evolution Link.** What are the advantages of having millions of alveoli rather than a pair of simple, balloonlike lungs?

5. **Analyzing Data.** Look at Figure 45-10b. How would strenuous muscle activity affect the oxygen–hemoglobin dissociation curve? (*Hint:* How does muscular activity affect pH? See Chapter 39 if you need to review this concept.) What would be the advantage of the shift in the curve?

Additional questions are available in ThomsonNOW at www.thomsonedu.com/login

46

Processing Food and Nutrition

McMurray Photography

Obtaining nutrients. Spotted hyenas (*Crocuta crocuta*) and several species of birds of prey gathered around a dead elephant. Photographed in East Africa.

KEY CONCEPTS

Many animal adaptations are associated with mode of nutrition.

Food processing includes ingestion, digestion, absorption, and elimination.

Various parts of the vertebrate digestive system are specialized to perform specific functions; accessory glands (liver, pancreas, and salivary glands) secrete fluids and enzymes that are important in digestion.

Most animals require the same basic nutrients: carbohydrates, lipids, proteins, vitamins, and minerals.

Basal metabolic rate is the body's cost of metabolic living. When energy (kilocalories) input equals energy output, body weight remains constant.

Sponges and baleen whales feed on food particles suspended in the water. Giraffes eat leaves, and antelope graze on grasses. Frogs, lions, and some insects capture other animals. Although their choices of food and feeding mechanisms are diverse, all animals are **heterotrophs,** organisms that must obtain their energy and nourishment from the organic molecules manufactured by other organisms. What they eat, how they obtain food, and how they process and use food are the focus of this chapter.

Nutrients are substances in food that are used as energy sources to power the systems of the body, as ingredients to make compounds for metabolic processes, and as building blocks in the growth and repair of tissues. Obtaining nutrients is of such vital importance that both individual organisms and ecosystems are structured around the central theme of **nutrition,** the process of taking in and using food. An animal's body plan and its lifestyle are adapted to its particular mode of obtaining food. For example, spotted hyenas, which are formidable predators as well as scavengers, have massive jaws that let them consume an entire elephant carcass, including bones and hide (see photograph). Few nutrients are wasted.

Nutrition has been an important force in human evolution. Through natural selection, the human diet became more varied than the diets of other primates. Humans also became very efficient at obtaining food. Their higher-quality diet supported the evolution of a larger, more complex brain.

With only slight variations, all animals require the same basic nutrients: carbohydrates, lipids, proteins, vitamins, and minerals. Carbohydrates, lipids, and proteins are used as energy sources.

Eating too much of any of these nutrients can result in weight gain, whereas eating too few nutrients or an unbalanced diet can result in malnutrition and death. **Malnutrition,** or poor nutritional status, results from dietary intake that is either below or above required needs. In human populations, both undernutrition (particularly protein deficiency) and obesity (which results from overnutrition) are serious health problems. Food processing and nutrition are active areas of research. ■

NUTRITIONAL STYLES AND ADAPTATIONS

Learning Objective

1   Describe food processing, including ingestion, digestion, absorption, and egestion or elimination; and compare the digestive system of a cnidarian (such as *Hydra*) with that of an earthworm or vertebrate.

Feeding is the selection, acquisition, and ingestion of food. **Ingestion** is the process of taking food into the digestive cavity. In many animals, including vertebrates, ingestion includes taking food into the mouth and swallowing it. Most animals have a specialized digestive system that processes the food they eat. The process of breaking down food is called **digestion.** Because animals eat the macromolecules tailor-made by and for other organisms, they must break down these molecules and refashion them for their own needs. A hyena cannot incorporate the proteins and other organic compounds from an elephant carcass directly into its own cells. It must *mechanically digest* its food and then *chemically digest* it by enzymatic hydrolysis (see Chapter 3). During digestion, complex organic compounds are degraded into smaller molecular components. For example, proteins are broken down into their component amino acids.

Amino acids and other nutrients pass through the lining of the digestive tract and into the blood by **absorption.** Then the circulatory system transports the nutrients to the body cells, which use them to synthesize proteins and other organic compounds. Food that is not digested and absorbed is discharged from the body, a process called **egestion** or **elimination.**

Animals are adapted to their mode of nutrition

We classify animals as herbivores, carnivores, or omnivores on the basis of the type of food they typically eat (❙ Fig. 46-1). Many adaptations have evolved for each type of feeding. For example, the beaks of birds and the teeth of many vertebrates are specialized for cutting, tearing, or chewing food.

Animals that feed directly on producers are **herbivores,** or primary consumers. Most of what a herbivore eats is not efficiently digested and is eliminated from the body, almost unchanged, as waste. For this reason herbivores must eat large quantities of food to obtain the nourishment they need. Many herbivores—grasshoppers, locusts, elephants, and cattle, for example—spend a major part of their lives eating.

Animals cannot digest the cellulose of plant cell walls, and many adaptations have evolved for extracting nutrients from the plant material they eat. Many herbivores, including termites, cows, and horses, have a symbiotic relationship with bacteria that inhabit their digestive tracts. **Ruminants** (cattle, sheep, deer, giraffes) are hooved animals with a stomach divided into four chambers. Symbiotic bacteria and protists living in the first two chambers digest cellulose, splitting some of it into sugars, which are then used by the host and the bacteria themselves. The bacteria produce fatty acids during their metabolism, some of which are absorbed by the animal and serve as an important energy source. Food that is not sufficiently chewed clumps together, forming a cud. The cud is regurgitated into the animal's mouth, where it is mixed with saliva and chewed again. When the cud is reswallowed, the partly digested food is further broken down by the cow's own enzymes.

Herbivores are sometimes eaten by flesh-eating **carnivores,** which may also eat one another. Many carnivores (secondary and higher-level consumers in ecosystems) are predators, adapted for capturing and killing prey. Some carnivores seize their victims and swallow them alive and whole (see Fig. 46-1d). Others paralyze, crush, or shred their prey before ingesting it. Adaptations for this type of feeding include tentacles, claws, fangs, poison glands, and teeth.

Carnivorous mammals have well-developed canine teeth for stabbing their prey during combat. The digestive juice of the stomach breaks down proteins. Meat is more easily digested than plant food, and carnivore digestive tracts are shorter than those of herbivores.

Omnivores, such as pigs, humans, and some fishes, consume both plants and animals. Earthworms ingest large amounts of soil containing both animal and plant material. The blue whale, the largest animal, is a filter-feeder that strains out tiny algae and animals as it swims. Omnivores often have adaptations that help them distinguish among a wide range of smells and tastes and thereby select a variety of foods.

Animals can also be classified according to the mechanisms they use to feed. Many omnivores are **suspension feeders** that remove suspended food particles from the water they inhabit. Some animals expose a sticky, mucus-coated surface to flowing water; suspended particles adhere to the surface. For example, some echinoderms have tentacles coated with mucus. Others, like bivalve mollusks, filter water. Baleen whales use rows of hard plates (baleen) suspended from the roof of the mouth to filter small crustaceans.

Some animals feed on fluids by piercing and sucking. Mosquitoes have highly adapted structures for piercing skin and sucking blood. Birds that feed on pollen and nectar have long bills and tongues. The shape, size, and curve of the beak may be specialized for feeding on a particular type of flower (see Fig. 1-11). Bats that feed on nectar have a long tongue and reduced dentition (number of teeth).

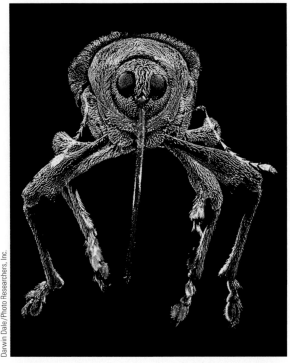

(a) The impressively long "snout" of the herbivorous acorn weevil (*Curculio*) is adapted both for feeding and for making a hole in the acorn through which it deposits an egg. When it has hatched, the larva feeds on the contents of the acorn seed.

(b) The herbivorous giant panda's (*Ailuropoda melanoleuca*) large, flat teeth and well-developed jaws and jaw muscles are adaptations for grinding high-fiber plant food.

(c) The mouth (*on left*) of the carnivorous long-nose butterfly fish (*Forcipiger longirostris*) is adapted for extracting small worms and crustaceans from tight spots in coral reefs.

(d) This carnivorous snake (*Dromicus*) is strangling a lava lizard (*Tropidurus*). A snake can swallow very large prey whole because of the structure of the jaws. The jaw dislocates and expands during swallowing.

Figure 46-1 Adaptations for obtaining and processing food

Some invertebrates have a digestive cavity with a single opening

Sponges, the simplest invertebrates, obtain food by filtering microscopic organisms from the surrounding water. Individual cells phagocytose the food particles, and digestion is *intracellular* within food vacuoles. Wastes are egested into the water that continuously circulates through the sponge body.

Most animals have a digestive cavity. Digestion within a cavity is more efficient than intracellular digestion because digestive enzymes can be released into one confined space, so less surface area is required. Cnidarians (such as hydras and jellyfish) and flatworms have a **gastrovascular cavity,** a central digestive cavity with a single opening. Cnidarians capture small aquatic animals with the help of their stinging cells and tentacles (❚ Fig. 46-2a). The mouth opens into the gastrovascular cavity. Cells lining this

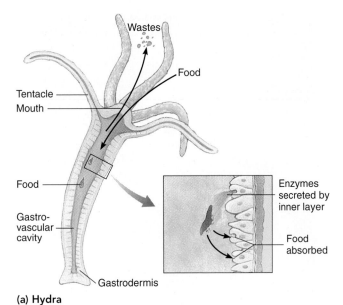

(a) Hydra

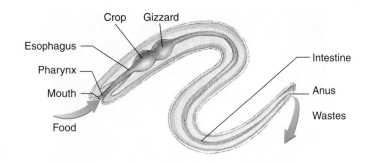

■ **Figure 46-3** Digestive tract with two openings

The earthworm, like most animals, has a complete digestive tract extending from mouth to anus. Various regions of the digestive tract are specialized to perform different food-processing functions.

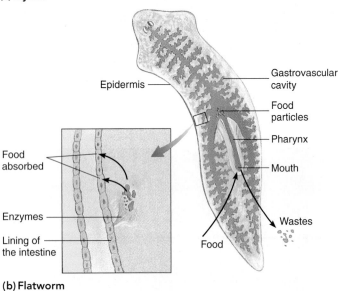

(b) Flatworm

■ **Figure 46-2** *Animated* Simple invertebrate digestive systems

(a) Hydras and **(b)** flatworms (planarians) have a gastrovascular cavity, a digestive tract with a single opening that serves as both mouth and anus.

digestive cavity secrete enzymes that break down proteins. Digestion continues intracellularly within food vacuoles, and digested nutrients diffuse into other cells. Body contractions promote egestion of undigested food particles through the mouth.

Free-living flatworms begin to digest their prey even before ingesting it. They extend their pharynx out through their mouth and secrete digestive enzymes onto the prey (■ Fig. 46-2b). After it is ingested, the food enters the branched gastrovascular cavity, where enzymes continue to digest it. Cells lining the gastrovascular cavity phagocytose partly digested food fragments, and digestion is completed within food vacuoles. As in cnidarians,

the flatworm digestive cavity has only one opening, so undigested wastes are egested through the mouth.

Most animal digestive systems have two openings

Most invertebrates, and all vertebrates, have a tube-within-a-tube body plan. The body wall forms the outer tube. The inner tube is a digestive tract with two openings, sometimes referred to as a complete digestive system (■ Fig. 46-3). Food enters through the mouth, and undigested food is eliminated through the anus. The mixing and propulsive movements of the digestive tract are referred to as **motility.** The propulsive activity characteristic of most regions of the digestive tract is **peristalsis,** waves of muscular contraction that push the food in one direction. More food can be taken in while previously eaten food is being digested and absorbed farther down the digestive tract. In a digestive tract with two openings, various regions of the tube are adapted to perform specific functions.

Review

■ How have carnivores adapted to their mode of nutrition?
■ How does food processing differ in earthworms and flatworms?

THE VERTEBRATE DIGESTIVE SYSTEM

Learning Objectives

2   Trace the pathway traveled by an ingested meal in the human digestive system, and describe the structure and function of each organ involved.
3   Describe the step-by-step digestion of carbohydrate, protein, and lipid.
4   Describe the structural adaptations that increase the surface area of the digestive tract.
5   Compare lipid absorption with absorption of other nutrients.

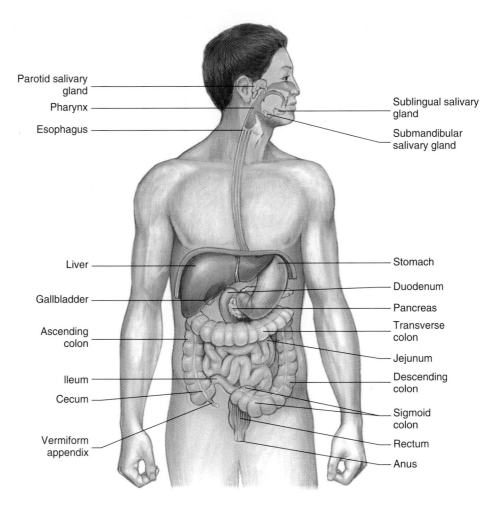

Figure 46-4 *Animated* Human digestive system

The human digestive tract is a long, coiled tube extending from mouth to anus. The small intestine consists of the duodenum, jejunum, and ileum. The large intestine includes the cecum, colon, rectum, and anus. Locate the three types of accessory glands: the liver, pancreas, and salivary glands.

Various regions of the vertebrate digestive tract are specialized to perform specific functions (❙ Fig. 46-4). Food passes in sequence through the following specialized regions:

Mouth ⟶ pharynx (throat) ⟶ esophagus ⟶ stomach ⟶ small intestine ⟶ large intestine ⟶ anus

The liver, pancreas, and, in terrestrial vertebrates, the salivary glands are accessory glands that secrete digestive juices into the digestive tract.

The wall of the digestive tract has four layers. Although various regions differ somewhat in structure, the layers are basically similar throughout the digestive tract (❙ Fig. 46-5). The **mucosa,** a layer of epithelial tissue and underlying connective tissue, lines the *lumen* (inner space) of the digestive tract. In the stomach and intestine, the mucosa is greatly folded to increase the secreting and absorbing surface. Surrounding the mucosa is the **submucosa,** a connective tissue layer rich in blood vessels, lymphatic vessels, and nerves.

A **muscle layer,** consisting of two sublayers of smooth muscle, surrounds the submucosa. In the inner sublayer, the muscle fibers are arranged circularly around the digestive tube. In the outer sublayer, the muscle fibers are arranged longitudinally. Below the level of the diaphragm, the outer connective tissue coat of the digestive tract is called the **visceral peritoneum.** By various folds it is connected to the **parietal peritoneum,** a sheet of connective tissue that lines the walls of the abdominal and pelvic cavities. The visceral and parietal peritonea enclose part of the coelom called the **peritoneal cavity.** Inflammation of the peritoneum, called **peritonitis,** can be very serious, because infection can spread along the peritoneum to most of the abdominal organs.

Food processing begins in the mouth

Imagine that you have just taken a big bite of a hamburger. The mouth is specialized for ingestion and for beginning the digestive process. Mechanical digestion begins as you bite, grind, and chew the meat and bun with your teeth. Unlike the simple, pointed teeth of fish, amphibians, and reptiles, the teeth of mammals vary in size and shape and are specialized to perform specific functions. The chisel-

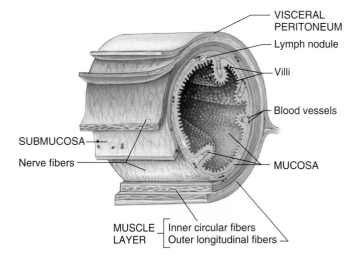

Figure 46-5 Wall of the digestive tract

From inside out, the layers of the wall are the mucosa, submucosa, muscle layer, and visceral peritoneum.

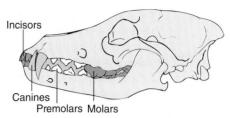

(a) Carnivore. Skull of a coyote showing the pointed incisors and canines, adaptations for ripping flesh.

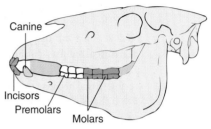

(b) Herbivore. Horses and other herbivores have incisors (and sometimes canines) adapted for cutting off bits of vegetation. Canines are absent in some herbivores. The broad, ridged surfaces of the molars are adapted for grinding plant material.

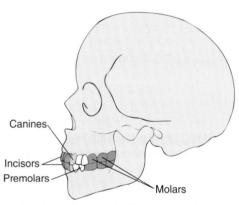

(c) Omnivore. The teeth of humans and other omnivores are adapted for chewing a variety of foods.

Figure 46-6 Teeth and diet

shaped **incisors** are used for biting, whereas the long, pointed **canines** are adapted for tearing food (Fig. 46-6). The flattened surfaces of the **premolars** and **molars** are specialized for crushing and grinding.

Each tooth is covered by **enamel,** the hardest substance in the body (Fig. 46-7). Most of the tooth consists of **dentin,** which resembles bone in composition and hardness. Beneath the dentin is the **pulp cavity,** a soft connective tissue containing blood vessels, lymph vessels, and nerves.

While food is being mechanically disassembled by the teeth, it is moistened by saliva. As some of the food molecules dissolve,

you can taste the food. Recall from Chapter 42 that taste buds are located on the tongue and other surfaces of the mouth. Three pairs of **salivary glands** secrete about a liter of saliva into the mouth cavity each day. Saliva contains an enzyme, **salivary amylase,** which begins the chemical digestion of starch into sugar.

The pharynx and esophagus conduct food to the stomach

After being chewed and fashioned into a lump called a **bolus,** the bite of food is swallowed—moved through the **pharynx** into the **esophagus.** The pharynx, or throat, is a muscular tube that serves as the hallway of both the respiratory system and the digestive system. During swallowing, a small flap of tissue, the **epiglottis,** closes the opening to the airway.

Waves of peristalsis sweep the bolus through the pharynx and esophagus toward the stomach (Fig. 46-8). Circular muscle fibers in the wall of the esophagus contract around the top of the bolus and push it downward. Almost at the same time, longitudinal muscles around the bottom of the bolus and below it contract, which shortens the tube.

When the body is in an upright position, gravity helps move the food through the esophagus, but gravity is not essential. Astronauts eat in its absence, and even if you are standing on your head, food will reach your stomach.

Food is mechanically and enzymatically digested in the stomach

The entrance to the large, muscular **stomach** is normally closed by a ring of muscle at the lower end of the esophagus. When a peristaltic wave passes down the esophagus, the muscle relaxes and the bolus enters the stomach (Fig. 46-9).When empty, the stomach is collapsed and shaped almost like a hot dog. Folds of the stomach wall, called **rugae,** give the inner lining a wrinkled appearance. As food enters, the rugae gradually smooth out, which expands the capacity of the stomach to more than a liter.

The stomach is lined with a simple, columnar epithelium that secretes large amounts of mucus. Tiny pits mark the entrances to the millions of **gastric glands,** which extend deep into the stomach wall. **Parietal cells** in the gastric glands secrete hydrochloric acid and *intrinsic factor,* a substance needed for adequate absorption of vitamin B_{12}. **Chief cells** in the gastric glands secrete **pepsinogen,** an inactive enzyme precursor. When it comes in contact with the acidic gastric juice in the stomach, pepsinogen is converted to **pepsin,** the main digestive enzyme of the stomach. Pepsin hydrolyzes proteins, converting them to short polypeptides.

Several protective mechanisms prevent the gastric juice from digesting the wall of the stomach. Cells of the gastric mucosa secrete an alkaline mucus that coats the stomach wall and neutralizes the acidity of the gastric juice along the lining. In addition, the epithelial cells of the lining fit tightly together, preventing gastric juice from leaking between them and into the tissue be-

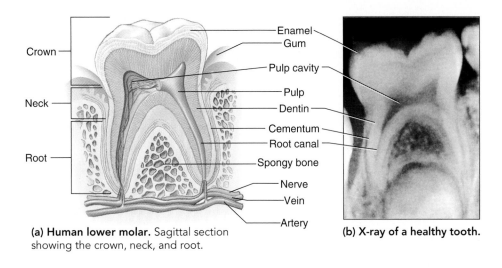

(a) Human lower molar. Sagittal section showing the crown, neck, and root.

Crown
Neck
Root

Enamel
Gum
Pulp cavity
Pulp
Dentin
Cementum
Root canal
Spongy bone
Nerve
Vein
Artery

(b) X-ray of a healthy tooth.

Figure 46-7 *Animated* Tooth structure

neath. If some of the epithelial cells become damaged, they are quickly replaced. In fact, about a half million of these cells are shed and replaced every minute!

Sometimes, these protective mechanisms malfunction and part of the stomach lining is digested, leaving an open sore, or **peptic ulcer.** Such ulcers often occur in the duodenum and sometimes in the lower part of the esophagus. The bacterium *Helicobacter pylori* has been implicated as a causative factor in ulcers (see Fig. 24-19). *Helicobacter pylori* infects the mucus-secreting cells of the stomach lining, decreasing the protective mucus, which can lead to peptic ulcers or cancer. This type of infection responds to antibiotic therapy.

What changes occur in a bite of hamburger during its 3- to 4-hour stay in the stomach? The stomach churns and chemically degrades the food so that it assumes the consistency of a thick soup; this partially digested food is called **chyme.** Protein digestion then degrades much of the hamburger protein to polypeptides. Digestion of the starch in the bun to small polysaccharides and maltose continues until salivary amylase is inactivated by the acidic pH of the stomach. Over a period of several hours, peristaltic waves release the chyme in spurts through the stomach exit, the **pylorus,** and into the small intestine.

Most enzymatic digestion takes place in the small intestine

Digestion of food is completed in the **small intestine,** and nutrients are absorbed through its wall. The small intestine, which is 5 to 6 m (about 17 ft) in length, has three regions: the **duodenum, jejunum,** and **ileum.** Most chemical digestion takes place in the duodenum, the first portion of the small intestine, not in the stomach. Bile from the liver and enzymes from the pancreas are released into the duodenum and act on the chyme. Then enzymes produced by the epithelial cells lining the duodenum catalyze the

final steps in the digestion of the major types of nutrients.

The lining of the small intestine appears velvety because of its millions of tiny fingerlike projections, the intestinal **villi** (sing., *villus*) (Fig. 46-10). The villi increase the surface area of the small intestine for digestion and absorption of nutrients. The intestinal surface is further expanded by **microvilli,** projections of the plasma membrane of the simple columnar epithelial cells of the villi. About 600 microvilli protrude from the exposed surface of each cell, giving the epithelial lining a fuzzy appearance when viewed with the electron microscope.

If the intestinal lining were smooth, like the inside of a water pipe, food would zip right through the intestine, and many valuable nutrients would not be absorbed. Folds in the wall of the intestine, the villi, and microvilli together increase the surface area of the small intestine by about 600 times. If we could unfold and spread out the lining of the small intestine of an adult human, its surface would approximate the size of a tennis court.

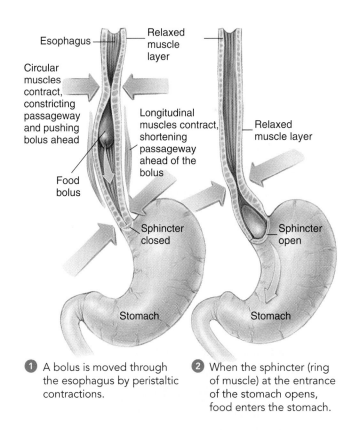

Esophagus
Relaxed muscle layer
Circular muscles contract, constricting passageway and pushing bolus ahead
Longitudinal muscles contract, shortening passageway ahead of the bolus
Relaxed muscle layer
Food bolus
Sphincter closed
Sphincter open
Stomach
Stomach

① A bolus is moved through the esophagus by peristaltic contractions.

② When the sphincter (ring of muscle) at the entrance of the stomach opens, food enters the stomach.

Figure 46-8 *Animated* Peristalsis

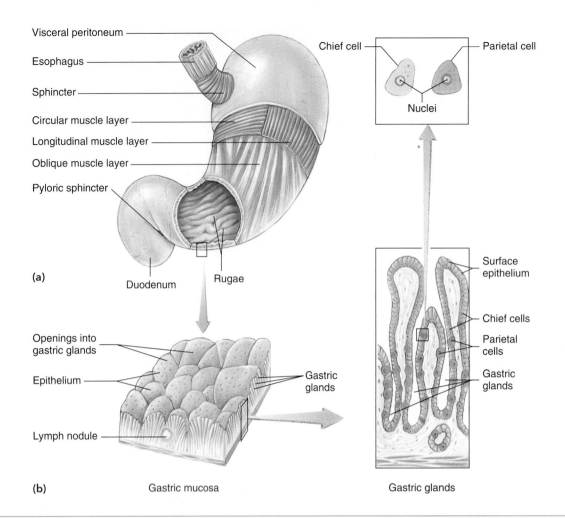

(a)

- Visceral peritoneum
- Esophagus
- Sphincter
- Circular muscle layer
- Longitudinal muscle layer
- Oblique muscle layer
- Pyloric sphincter
- Duodenum
- Rugae

Chief cell — Parietal cell — Nuclei

(b) Gastric mucosa

- Openings into gastric glands
- Epithelium
- Gastric glands
- Lymph nodule

Gastric glands

- Surface epithelium
- Chief cells
- Parietal cells
- Gastric glands

Figure 46-9   Structure of the stomach

From the esophagus, food enters the stomach, where it is mechanically and enzymatically digested.
(a) The wall of the stomach has been progressively removed to show muscle layers and rugae.
(b) Stomach lining and gastric glands.

The liver secretes bile

The **liver,** the largest internal organ and also one of the most complex organs in the body, lies in the upper-right part of the abdomen just under the diaphragm (❚ Fig. 46-11). The liver secretes **bile,** which mechanically digests fats by a detergent-like action (discussed in a later section). Bile consists of water, bile salts, bile pigments, cholesterol, salts, and lecithin (a phospholipid). Because it contains no digestive enzymes, bile does not enzymatically digest food. The pear-shaped **gallbladder** stores and concentrates bile and releases it into the duodenum as needed.

A single liver cell can carry on more than 500 separate, specialized metabolic activities! The liver performs these vital functions:

1. Secretes bile that mechanically digests fats
2. Helps maintain homeostasis by removing or adding nutrients to the blood
3. Converts excess glucose to glycogen and stores it
4. Converts excess amino acids to fatty acids and urea
5. Stores iron and certain vitamins
6. Detoxifies alcohol and other drugs and poisons

The pancreas secretes digestive enzymes

The **pancreas** is an elongated gland that secretes both digestive enzymes and hormones that help regulate the level of glucose in the blood. The cells that line the pancreatic ducts secrete an alkaline solution rich in bicarbonate ions. This pancreatic juice neutralizes the stomach acid in the duodenum and provides the optimal pH for action of the pancreatic enzymes. Pancreatic enzymes include **trypsin** and **chymotrypsin,** which digest polypeptides to dipeptides; **pancreatic lipase,** which degrades fats; **pancreatic amylase,** which breaks down almost all types of complex carbohydrates, except cellulose, to disaccharides; and **ribonuclease** and **deoxyribonuclease,** which split the nucleic acids ribonucleic acid (RNA) and deoxyribonucleic acid (DNA) to free nucleotides.

Nutrients are digested as they move through the digestive tract

Chyme moves through the digestive tract by peristalsis, mixing contractions, and motions of the villi. As nutrients in the chyme move through the small intestine, they come into contact with enzymes that digest them (❚ Table 46-1).

Carbohydrates are digested to monosaccharides

Polysaccharides, such as starch and glycogen, are important components of the food ingested by most animals. The glucose units of these large molecules are connected by glycosidic bonds linking carbon 4 (or 6) of one glucose molecule with carbon 1 of the adjacent glucose molecule. These bonds are hydrolyzed by **amylases** that digest polysaccharides to the disaccharide maltose. Although amylase can split the α-glycosidic linkages present in starch and glycogen, it cannot split the β-glycosidic linkages present in cellulose (see Figs. 3-9 and 3-10).

Amylase cannot split the bond between the two glucose units of maltose. Enzymes produced by the cells lining the small intestine break down disaccharides such as maltose to monosaccharides. **Maltase,**

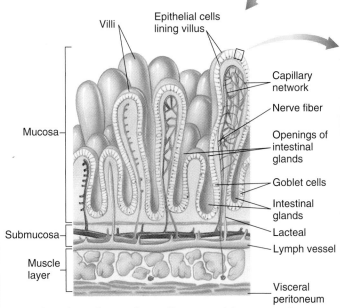

(a) SEM of a cross section of the small intestine.

500 μm

(b) Enlarged view of a small portion of the intestinal wall. Some of the villi have been opened to show the blood and lymph vessels within.

Villi
Epithelial cells lining villus
Capillary network
Nerve fiber
Openings of intestinal glands
Goblet cells
Intestinal glands
Lacteal
Lymph vessel
Visceral peritoneum
Mucosa
Submucosa
Muscle layer

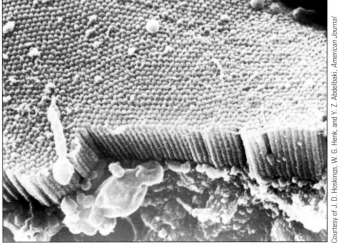

(c) SEM of the surface of an epithelial cell from the lining of the small intestine, showing microvilli. The epithelium has been cut vertically, allowing the microvilli to be viewed from the side as well as from above.

1 μm

Courtesy of J. D. Hoskings, W. G. Henk, and Y. Z. Abdelbaki, *American Journal of Veterinary Research*, 1982, Vol. 43, No. 10, pp. 1715–1720

G. Shih–R. Kessel / Visuals Unlimited

Figure 46-10 *Animated* Villi and microvilli

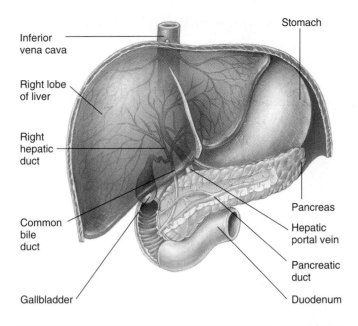

Inferior vena cava

Right lobe of liver

Right hepatic duct

Common bile duct

Gallbladder

Stomach

Pancreas

Hepatic portal vein

Pancreatic duct

Duodenum

Figure 46-11 The liver and pancreas

The gallbladder stores bile from the liver. Note the ducts that conduct bile to the gallbladder and the duodenum. The stomach has been displaced to expose the pancreas.

for example, splits maltose into two glucose molecules (see Fig. 3-8a). Hydrolysis occurs while the disaccharides are being absorbed through the epithelium of the small intestine.

Proteins are digested to amino acids

Several proteolytic enzymes are secreted into the digestive tract. Each breaks peptide bonds at one or more specific locations in a polypeptide chain. Trypsin, secreted in an inactive form by the pancreas, is activated by an enzyme called *enterokinase*. The trypsin then activates chymotrypsin and carboxypeptidase, as well as additional trypsin.

Pepsin, trypsin, and chymotrypsin break certain internal peptide bonds of proteins and polypeptides. Carboxypeptidase removes amino acids with free terminal carboxyl groups from the end of polypeptide chains. **Dipeptidases** released by the duodenum then split the small peptides to amino acids.

Fats are digested to fatty acids and monoacylglycerols

Lipids are usually ingested as large masses of triacylglycerols (also called triglycerides). They are digested mainly within the duodenum by pancreatic lipase. Like many other proteins, lipase is water soluble, but its substrates are not. Thus, the enzyme can attack only the fat molecules at the surface of a mass of fat. Bile salts act like detergents, reducing the surface tension of fats. Their action, called **emulsification,** breaks large masses of fat into smaller droplets. Emulsification greatly increases the surface area of fat exposed to the action of pancreatic lipase and so increases the rate of lipid digestion.

Conditions in the intestine are usually not optimal for the complete hydrolysis of lipids to glycerol and fatty acids. Consequently, the products of lipid digestion include monoacylglycerols (monoglycerides) and diacylglycerols (diglycerides) as well as glycerol and fatty acids. Undigested triacylglycerols also remain, some of which are absorbed without digestion.

Nerves and hormones regulate digestion

Most digestive enzymes are produced only when food is present in the digestive tract. Salivary gland secretion is controlled entirely by the nervous system, but secretion of other digestive

---

TABLE 46-1

Summary of Digestion

	Carbohydrates	Proteins	Lipids
Mouth	Polysaccharides ↓ Salivary amylase Maltose and small polysaccharides		
Stomach	Action continues until acidic pH inactivates salivary amylase	Protein ↓ Pepsin Short polypeptides	
Small intestine	Undigested polysaccharides ↓ Pancreatic amylase Maltose and other disaccharides ↓ Maltase, sucrase, lactase Monosaccharides	Polypeptides ↓ Trypsin, chymotrypsin Small polypeptides and peptides ↓ Carboxypeptidase, peptidases, dipeptidases Amino acids	Glob of fat ↓ Bile salts Emulsified fat droplets ↓ Pancreatic lipase Fatty acids and glycerol

TABLE 46-2

Some Hormones That Regulate Digestion

Hormone	Source	Target Tissue	Actions	Factors That Stimulate Release
Gastrin	Stomach (mucosa)	Stomach (gastric glands)	Stimulates gastric glands to secrete pepsinogen	Distention of the stomach by food; certain substances such as partially digested proteins and caffeine
Secretin	Duodenum (mucosa)	Pancreas	Signals secretion of sodium bicarbonate	Acidic chyme acting on mucosa of duodenum
		Liver	Stimulates bile secretion	
Cholecystokinin (CCK)	Duodenum (mucosa)	Pancreas	Stimulates release of digestive enzymes	Presence of fatty acids and partially digested proteins in duodenum
		Gallbladder	Stimulates emptying of bile	
Glucose-dependent insulinotropic peptide (GIP)	Duodenum (mucosa)	Pancreas	Stimulates insulin secretion	Presence of glucose in duodenum

juices is regulated by both nerves and hormones. The wall of the digestive tract contains dense networks of neurons. This so-called *enteric nervous system* continues to regulate many motor and secretory activities of the digestive system even if sympathetic and parasympathetic nerves to these organs are cut. Many neuropeptides present in the brain are also released by neurons in the digestive tract and help regulate digestion. For example, *substance P* stimulates smooth muscle contraction of the digestive tract, and *enkephalin* inhibits it (see Table 40-2).

Several hormones, including **gastrin, secretin, cholecystokinin (CCK),** and **glucose-dependent insulinotropic peptide (GIP),** help regulate the digestive system (▌ Table 46-2). These hormones are polypeptides secreted by endocrine cells in the mucosa of certain regions of the digestive tract. Investigators are studying several other messenger peptides that are important in regulating digestive activity.

As an example of the regulation of the digestive system, consider the secretion of gastric juice. Seeing, smelling, tasting, or even thinking about food causes the brain to send neural signals to the gastric glands in the stomach, stimulating them to secrete. In addition, when food distends the stomach, stretch receptors send neural messages to the medulla. The medulla then sends messages to endocrine cells in the stomach wall that secrete gastrin. This hormone is absorbed into the blood; it stimulates the stomach to release gastric juice and also stimulates gastric emptying and intestinal motility.

Absorption takes place mainly through the villi of the small intestine

Only a few substances—water, simple sugars, salts, alcohol, and certain drugs—are small enough to be absorbed through the stomach wall. Absorption of nutrients is primarily the job of the intestinal villi. As illustrated in Figure 46-10, the wall of a villus is a single layer of epithelial cells. Inside each villus is a network of capillaries and a central lymph vessel, called a **lacteal.**

To reach the blood (or lymph), a nutrient molecule must pass through an epithelial cell lining the intestine and through a cell lining a blood or lymph vessel. Absorption occurs by a combination of simple diffusion, facilitated diffusion, and active transport. Because glucose and amino acids cannot diffuse through the intestinal lining, they must be absorbed by active transport. Absorption of these nutrients is coupled with the active transport of sodium (see Chapter 5). Fructose is absorbed by facilitated diffusion.

Amino acids and glucose are transported directly to the liver by the **hepatic portal vein.** In the liver this vein divides into a vast network of tiny blood sinuses, vessels similar to capillaries. As the nutrient-rich blood moves slowly through the liver, nutrients and certain toxic substances are removed from the circulation.

The products of lipid digestion are absorbed by a different process and different route. After free fatty acids and monoacylglycerols enter an epithelial cell in the intestinal lining, they are reassembled as triacylglycerols in the smooth endoplasmic reticulum. The triacylglycerols, along with absorbed cholesterol and phospholipids, are packaged into protein-covered fat droplets, called **chylomicrons.**

After they are released into the interstitial fluid, chylomicrons enter the lacteal (lymph vessel) of the villus. Chylomicrons are transported in the lymph to the subclavian veins, where the lymph and its contents enter the blood. About 90% of absorbed fat enters the blood circulation in this indirect way. The rest, mainly short-chain fatty acids such as those in butter, are absorbed directly into the blood. After a meal rich in fats, the great number of chylomicrons in the blood may give the plasma a turbid, milky appearance for a few hours. Chylomicrons transport fats to the liver and other tissues.

Most of the nutrients in the chyme are absorbed by the time it reaches the end of the small intestine. What is left (mainly

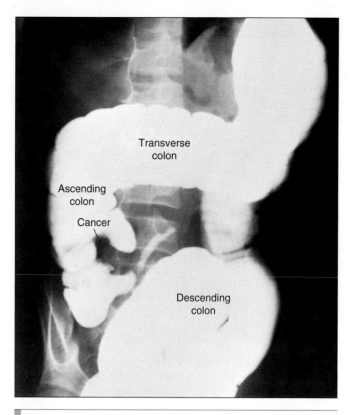

Figure 46-12   Colon cancer

In this radiographic view of the large intestine, cancer is evident as a mass that projects into the lumen of the colon. The large intestine has been filled with a suspension of barium sulfate, which makes irregularities in the wall visible.

waste) passes through a sphincter, the **ileocecal valve,** into the large intestine.

The large intestine eliminates waste

Undigested material, such as the cellulose of plant foods, along with unabsorbed chyme, passes into the **large intestine** (see Fig. 46-4). Although only about 1.3 m (about 4 ft) long, this part of the digestive tract is referred to as "large" because its diameter is greater than that of the small intestine. The small intestine joins the large intestine about 7 cm (2.8 in) from the end of the large intestine, forming a blind pouch, the **cecum.** The **vermiform appendix** projects from the end of the cecum. (Appendicitis is an inflammation of the appendix.) Herbivores such as rabbits have a large, functional cecum that holds food while bacteria digest its cellulose. In humans, the functions of the cecum and appendix are not known, and these structures are generally considered vestigial organs.

From the cecum to the **rectum** (the last portion of the large intestine), the large intestine is known as the **colon.** The regions of the large intestine are the cecum; ascending colon; transverse colon; descending colon; sigmoid colon; rectum; and **anus,** the opening for the elimination of wastes.

As chyme passes slowly through the large intestine, water and sodium are absorbed from it, and it gradually assumes the con-

sistency of normal feces. Bacteria inhabiting the large intestine are nourished by the last remnants of the meal and benefit their host by producing vitamin K and certain B vitamins that can be absorbed and used.

A distinction should be made between elimination and excretion. *Elimination* is the process of getting rid of digestive wastes—materials that have not been absorbed from the digestive tract and did not participate in metabolic activities. In contrast, *excretion* is the process of getting rid of metabolic wastes, which in mammals is mainly the function of the kidneys and lungs. The large intestine, however, does excrete bile pigments.

When chyme passes through the intestine too rapidly, **defecation** (expulsion of feces) becomes more frequent, and the feces are watery. This condition, called *diarrhea,* may be caused by anxiety, certain foods, or by pathogens that irritate the intestinal lining. Prolonged diarrhea results in loss of water and salts and leads to dehydration—a serious condition, especially in infants. *Constipation* results when chyme passes through the intestine too slowly. Because more water than usual is removed from the chyme, the feces may be hard and dry. Constipation is often caused by a diet deficient in fiber.

In Western countries, colorectal cancer (cancer of the colon and rectum) is the third most common cancer (∎ Fig. 46-12). Several factors contribute to the risk of developing colon cancer, including diet, smoking, family history, and physical inactivity. A diet high in red meat and low in fresh fruit, vegetables, poultry, and fish increases the risk.

Review

∎ Imagine that you have just taken a bite of a peanut butter sandwich. List in sequence the structures through which it passes in its journey through the digestive system. What happens in each structure?

∎ What is the function of villi?

∎ What are chylomicrons?

REQUIRED NUTRIENTS

Learning Objectives

6 Summarize the nutritional requirements for dietary carbohydrates, lipids, and proteins; and trace the fate of glucose, lipids, and amino acids after their absorption.

7 Describe the nutritional functions of vitamins, minerals, and phytochemicals.

Adequate amounts of essential nutrients are necessary for metabolic processes. Animals require carbohydrates, lipids, proteins, vitamins, and minerals. Nutritionists are also investigating plant compounds, called *phytochemicals,* that are important in nutrition. Although not considered a nutrient in a strict sense, water is a necessary dietary component. Enough fluid must be ingested to replace fluid lost in urine, sweat, feces, and breath.

Nutritionists consider new data and continually re-examine the *relative* amounts of various types of nutrients recommended for a healthy diet. According to the U.S. Department of Agriculture's *Dietary Guidelines for Americans 2005,* 45% to 65% of one's calories should come from carbohydrates, 20% to 35% from fat, and 10% to 35% from protein.

Recall from Chapter 1 that **metabolism** includes all the chemical processes that take place in the body. Metabolic processes include anabolism and catabolism. **Anabolism** includes the synthetic aspects of metabolism, such as the production of proteins and nucleic acids. **Catabolism** includes breakdown processes, such as hydrolysis. Nutritionists measure the energy value of food in kilocalories, or simply Calories. A Calorie, spelled with a capital *C,* is a **kilocalorie (kcal),** defined as the amount of heat required to raise the temperature of a kilogram of water one degree Celsius.

Carbohydrates provide energy

Sugars and starches are important sources of energy in the human diet. Most carbohydrates are ingested in the form of starch and cellulose, both polysaccharides composed of long chains of glucose subunits. (You may want to review the discussion of carbohydrates in Chapter 3.) Nutritionists refer to polysaccharides as **complex carbohydrates.** Foods rich in complex carbohydrates include rice, potatoes, corn, and other cereal grains.

When we eat an excess of carbohydrate-rich food, the liver cells become fully packed with glycogen and convert excess glucose to fatty acids and glycerol. Liver cells convert these compounds to triacylglycerols and send them to the fat depots of the body for storage.

Refined carbohydrates, such as white bread and white rice, are unhealthy because the refining process removes fiber and many vitamins and minerals. The refining process also produces a form of starch that the digestive system rapidly breaks down to glucose. The resulting rapid increase in glucose concentration in the blood stimulates the pancreas to release a large amount of insulin. This hormone stimulates the liver and muscles to remove glucose from the blood. Insulin lowers blood glucose level. When glucose and insulin levels are high, triglyceride levels increase and high-density lipoprotein (HDL; the good cholesterol) concentration decreases. These metabolic events can lead to cardiovascular disease and increase the risk of type 2 diabetes (discussed in Chapter 48).

Dietary intake of fiber decreases cholesterol concentration in the blood and is associated with a lower risk for cardiovascular disease and diabetes. **Fiber** is mainly a mixture of cellulose and other indigestible carbohydrates. We obtain dietary fiber by eating fruits, vegetables, and whole grains. The U.S. diet is low in fiber because of low intake of fruits and vegetables and use of refined flour. Increasing fiber in the diet has a variety of health benefits. Fiber stimulates the feeling of being satisfied (satiety) after food intake and thus is useful in treating obesity.

Lipids provide energy and are used to make biological molecules

Cells use ingested lipids to provide energy and to make a variety of lipid compounds, such as components of cell membranes, steroid hormones, and bile salts. We ingest about 98% of our dietary lipids in the form of triacylglycerols (triglycerides). (Recall from Chapter 3 that a triacylglycerol is a glycerol molecule chemically combined with three fatty acids; see Fig. 3-12b.) Triacylglycerols may be saturated, that is, fully loaded with hydrogen atoms, or their fatty acids may be monounsaturated (containing one double bond) or polyunsaturated (containing two or more double bonds).

Three polyunsaturated fatty acids (linoleic, linolenic, and arachidonic acids) are essential fatty acids that humans must obtain from food. Given these and sufficient nonlipid nutrients, the body can make all the lipid compounds (including fats, cholesterol, phospholipids, and prostaglandins) that it needs. The average U.S. diet provides far more cholesterol than the recommended daily maximum of 300 mg. High-cholesterol sources include egg yolks, butter, and meat.

How are lipids transported? Recall that chylomicrons transport lipids from the intestine to the liver and to other tissues. As chylomicrons circulate in the blood, the enzyme **lipoprotein lipase** breaks down triacylglycerols. The fatty acids and glycerol can then be taken up by the cells. What is left of the chylomicron, a remnant made up mainly of cholesterol and protein, is taken up by the liver.

Liver cells repackage cholesterol and triacylglycerols. These lipids are bound to proteins and transported as large molecular complexes, called **lipoproteins.** Some plasma cholesterol is transported by **high-density lipoproteins,** or **HDLs** ("good" cholesterol), but most is transported by **low-density lipoproteins,** or **LDLs** ("bad" cholesterol). LDLs deliver cholesterol to the cells. For cells to take in LDLs, a protein (apolipoprotein B) on the LDL surface must bind with a protein **LDL receptor** on the plasma membrane (see Fig. 5-23). After binding, the LDL enters the cell, and its cholesterol and other components are used. When cholesterol levels are high, HDLs collect the excess cholesterol and transport it to the liver. HDLs decrease risk for cardiovascular disease.

When needed, stored fats are hydrolyzed to fatty acids and released into the blood. Before cells can use these fatty acids as fuel, they are broken down into smaller compounds and combined with coenzyme A to form molecules of acetyl coenzyme A (acetyl CoA; ▌Fig. 46-13). Acetyl CoA enters the citric acid cycle (see Chapter 8). The conversion of fatty acids to acetyl CoA is accomplished in the liver by a process known as *β-oxidation.*

For transport to the cells, acetyl coenzyme A is converted into one of three types of *ketone bodies* (four-carbon ketones). Normally, the level of ketone bodies in the blood is low, but in certain abnormal conditions, such as starvation and diabetes mellitus, fat metabolism is tremendously increased. Ketone bodies are then produced so rapidly that their level in the blood becomes excessive, which may make the blood too acidic. Such disrupted pH balance can lead to death.

What is the relationship between fat and cholesterol intake and cardiovascular disease? Lipids play a key role in the development of atherosclerosis, a progressive disease in which the arteries become clogged with fatty material (see Chapter 43). LDLs are the main source of the cholesterol that builds up in the walls of arteries. The type of fat consumed, as well as other dietary and

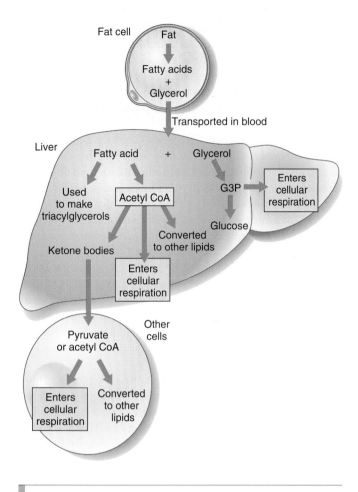

Figure 46-13 How the body uses fat

The liver converts glycerol and fatty acids to compounds that are used as fuel in cellular respiration. Recall from Chapter 8 that G3P is glyceraldehyde-3-phosphate.

lifestyle factors, is important. You can promote a healthy proportion of HDL to LDL by exercising regularly, following a healthy diet, maintaining appropriate body weight (obesity raises LDL and triacylglycerol levels), and not smoking cigarettes.

In general, animal foods are rich in both saturated fats and cholesterol, whereas most plant foods contain unsaturated fats and no cholesterol. Butter contains mainly saturated fats. Commonly used polyunsaturated vegetable oils are corn, soy, cottonseed, and safflower oils. Monounsaturated fats raise HDL level, as does stearic acid, the saturated fat in chocolate. Olive, canola, and peanut oils contain large amounts of monounsaturated fats. A few plant oils, including palm oil and coconut oil, are high in saturated fats. **Omega-3 fatty acids** (found in fish and some plant oils) decrease LDL levels and play other protective roles in decreasing the risk for coronary heart disease.

A diet high in saturated fats and cholesterol raises the blood cholesterol level by as much as 25%. Because polyunsaturated fats decrease the blood cholesterol level, many people now cook with vegetable oils rather than with butter and lard, drink skim milk rather than whole milk, and eat low-fat ice cream. Some use margarine instead of butter. However, during production of many margarines, vegetable oils are partially hydrogenated (some of the carbons accept hydrogen to become fully saturated). During hydrogenation, some double bonds are transformed from a *cis* arrangement to a *trans* arrangement (see Fig. 3-3b), forming ***trans* fatty acids;** the harder the margarine, the higher the *trans* fatty acid content. Health concerns have been raised because *trans* fatty acids increase LDL cholesterol in the blood and lower HDL. Many processed foods, including cakes, doughnuts, cookies, crackers, and potato chips, contain *trans* fatty acids. Although some fats, such as *trans* fatty acids and saturated fats, are unhealthy, others, such as omega-3 fatty acids and monounsaturated fat, are healthy. The Inuits in Greenland eat diets extremely high in fat, but because their dietary fat is rich in omega-3 fatty acids, they have a low incidence of heart disease.

Proteins serve as enzymes and as structural components of cells

Proteins are essential building blocks of cells, serve as enzymes, and are used to make compounds such as hemoglobin and myosin. Protein consumption is an index of a country's (or an individual's) economic status, because high-quality protein tends to be the most expensive and least available of the nutrients. In many parts of the world, protein poverty is one of the most pressing health problems.

Ingested proteins are degraded in the digestive tract to small peptides and amino acids. Of the 20 or so amino acids important in nutrition, approximately 9 (10 in children) cannot be synthesized by humans at all, or at least not in sufficient quantities to meet the body's needs. The diet must provide these **essential amino acids** (see Chapter 3).

Complete proteins, those that contain the most appropriate distribution of amino acids for human nutrition, are found in fish, meat, nuts, eggs, and milk. Some foods, such as gelatin and legumes (soybeans, beans, peas, peanuts), contain a high proportion of protein. However, they either do not contain all the essential amino acids, or they do not contain them in proper nutritional proportions. Most plant proteins are deficient in one or more essential amino acids. The healthiest sources of protein are fish, chicken, nuts, and legumes.

Amino acids circulating in the blood are taken up by cells and used for the synthesis of proteins. The liver removes excess amino acids from the circulation. Liver cells deaminate amino acids, that is, remove the amino group (▌ Fig. 46-14). During deamination, ammonia forms from the amino group. Ammonia, which is toxic at high concentrations, is converted to urea and excreted from the body. The remaining carbon chain of the amino acid (called a *keto acid*) may be converted into carbohydrate or lipid and used as fuel or stored. Thus, people who eat high-protein diets can gain weight if they eat too much.

Vitamins are organic compounds essential for normal metabolism

Vitamins are organic compounds required in the diet in relatively small amounts for normal biochemical functioning. Many are components of coenzymes (see Chapter 7). Nutritionists divide vitamins into two main groups. **Fat-soluble vitamins** include vitamins A, D, E, and K. **Water-soluble vitamins** are the B and C vitamins. Fruit and vegetables are rich sources of vitamins. ❚ Table 46-3 provides the sources, functions, and consequences of deficiency for most of the vitamins.

Health professionals debate the advisability of taking large amounts of certain specific vitamins, such as vitamin C to prevent colds or vitamin E to protect against vascular disease. Some studies suggest that vitamin A (found in yellow and green vegetables) and vitamin C (found in citrus fruit and tomatoes) help protect against certain forms of cancer. We do not yet understand all the biochemical roles played by vitamins or the interactions among various vitamins and other nutrients. We do know that, like vitamin deficiency, large overdoses of vitamins can be harmful. Moderate overdoses of the B and C vitamins are excreted in the urine, but surpluses of the fat-soluble vitamins are not easily excreted and can accumulate to harmful levels.

Minerals are inorganic nutrients

Minerals are inorganic nutrients ingested in the form of salts dissolved in food and water (❚ Table 46-4). We need certain minerals, including sodium, chloride, potassium, calcium, phosphorus, magnesium, and sulfur, in amounts of 100 mg or more daily. Several others, such as iron, copper, iodide, fluoride, and selenium, are **trace elements,** minerals that we require in amounts less than 100 mg per day.

Minerals are necessary components of body tissues and fluids. Salt content (about 0.9% in plasma) is vital in maintaining the fluid balance of the body, and because salts are lost from the body daily in sweat, urine, and feces, they must be replaced by dietary intake. Sodium chloride (common table salt) is the salt needed in largest quantity in blood and other body fluids. A deficiency results in dehydration. Iron is the mineral most often deficient in the diet. In fact, iron deficiency is one of the most widespread nutritional problems in the world.

Antioxidants protect against oxidants

Normal cell processes that require oxygen produce **oxidants,** highly reactive molecules such as free radicals, peroxides, and superoxides. **Free radicals,** molecules or ions with one or more unpaired electrons, are also generated by ionizing radiation, tobacco smoke, and other forms of air pollution. Oxidants damage DNA, proteins, and unsaturated fatty acids by snatching electrons. Damage to DNA causes mutations that lead to cancer, and injury to unsaturated fatty acids can damage cell membranes. Free radicals are thought to contribute to atherosclerosis by causing oxidation of LDL cholesterol. Oxidative damage to the body over many years contributes to the aging process.

Cells have **antioxidants** that destroy free radicals and other reactive molecules. Antioxidants in tissues include certain enzymes, for example, catalase and peroxidase. Their action requires minerals such as selenium, zinc, manganese, and copper. Certain vitamins—vitamin C, vitamin E, and vitamin A—have strong

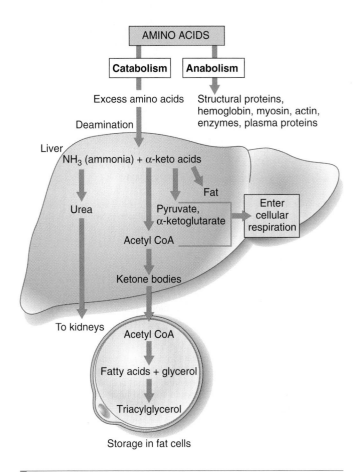

Figure 46-14 How the body uses protein

The liver plays a central role in protein metabolism. Deamination of amino acids and conversion of the amino groups to urea take place there. In addition, many proteins are synthesized in the liver. Note that you can gain weight on a high-protein diet, because excess amino acids can be converted to fat.

antioxidant activity. Vitamins A and E protect cell membranes from free radicals. In addition, a variety of phytochemicals are potent antioxidants (discussed in next section).

Nutritionists recommend that we increase the antioxidants in our diet by eating fruits, vegetables, and other foods high in antioxidants. We do not yet know whether antioxidant supplements are useful.

Phytochemicals play important roles in maintaining health

Diets rich in fruits and vegetables lower the incidence of heart disease and of certain types of cancer. Studies comparing diets in various countries suggest that intake of fruits and vegetables may be more important than differences in dietary fat. Yet nutritionists estimate that in the United States only about 1 in 11 people eats the daily recommended three to five servings of vegetables and two to three of fruit. A diet that includes all the essential nutrients does not provide the same health benefits as one rich in fruit and vegetables. The missing ingredients appear to be **phy-**

TABLE 46-3

The Vitamins

Vitamins and U.S. RDA*	Actions	Effect of Deficiency	Sources
Fat soluble			
Vitamin A, retinol 5000 IU[†]	Converted to retinal; essential for normal vision; essential for normal growth and differentiation of cells; reproduction; immunity	Growth retardation; night blindness; worldwide 250 million children are at risk of blindness from vitamin A deficiency	Liver, fortified milk, yellow and green vegetables such as carrots and broccoli
Vitamin D, calciferol 400 IU	Promotes calcium and phosphorus absorption from digestive tract; essential to normal growth and maintenance of bone	Weak bones; bone deformities; rickets in children, osteomalacia in adults	Fish oils, egg yolk, fortified milk, butter, margarine
Vitamin E, tocopherols 30 IU	Antioxidant; protects unsaturated fatty acids and cell membranes	Increased catabolism of unsaturated fatty acids so that not enough are available for maintenance of cell membranes; prevention of normal growth; nerve damage	Vegetable oils, nuts, leafy greens
Vitamin K, about 80 mcg[‡]	Synthesis of blood-clotting proteins	Prolonged blood-clotting time	Normally supplied by intestinal bacteria; leafy greens, legumes
Water soluble			
Vitamin C, ascorbic acid 60 mg	Collagen synthesis; antioxidant; needed for synthesis of some hormones and neurotransmitters; important in immune function	Scurvy (wounds heal very slowly, and scars become weak and split open; capillaries become fragile; bone does not grow or heal properly)	Citrus fruit, strawberries, tomatoes, leafy vegetables, cabbage
B-complex vitamins Vitamin B_1, thiamine 1.5 mg	Active form is a coenzyme in many enzyme systems; important in carbohydrate and amino acid metabolism	Beriberi (weakened heart muscle, enlarged right side of heart, nervous sytem and digestive tract disorders); common in alcoholics	Liver, yeast, whole and enriched grains, meat, green leafy vegetables
Vitamin B_2, riboflavin 1.7 mg	Used to make coenzymes (e.g., FAD) essential in cellular respiration	Dermatitis, inflammation and cracking at corners of mouth; confusion	Liver, milk, eggs, green leafy vegetables, enriched grains
Niacin, 20 mg	Component of important coenzymes (NAD^+ and $NADP^+$); essential to cellular respiration	Pellagra (dermatitis, diarrhea, mental symptoms, muscular weakness, fatigue)	Liver, chicken, tuna, milk, green leafy vegetables, enriched grains
Vitamin B_6, pyridoxine 2 mg	Derivative is coenzyme in many reactions in amino acid metabolism	Dermatitis, digestive tract disturbances; convulsions	Meat, whole grains, legumes, green leafy vegetables
Pantothenic acid 10 mg	Constituent of coenzyme A (important in cellular metabolism)	Deficiency extremely rare	Meat, whole grains, legumes
Folate 400 mcg	Coenzyme needed for nucleic acid synthesis and for maturation of red blood cells	A type of anemia; certain birth defects; increased risk of cardiovascular disease; deficiency in alcoholics, smokers, and pregnant women	Liver, legumes, dark green leafy vegetables, orange juice
Biotin 30 mcg	Coenzyme important in metabolism	—	Produced by intestinal bacteria; liver, chocolate, egg yolk
Vitamin B_{12}, 2.4 mcg	Coenzyme important in metabolism; contains cobalt	A type of anemia	Liver, meat, fish

*RDA is the recommended dietary allowance, established by the Food and Nutrition Board of the National Research Council, to maintain good nutrition for healthy adults.

[†]International Unit: the amount that produces a specific biological effect and is internationally accepted as a measure of the activity of the substance.

[‡]mcg = micrograms.

tochemicals, plant compounds that promote health. Many phytochemicals are antioxidants. For example, *lycopenes,* responsible for the red color of tomatoes, are powerful antioxidants.

Nutritionists have begun to intensively investigate phytochemicals. Some important classes of phytochemicals are carotenoids (which include vitamin A and lycopene) in deeply pigmented fruits and vegetables such as carrots and tomatoes; flavonoids in berries, herbs, vegetables, barley, soy, black and green tea, and cocoa, especially dark chocolate; isothiocyanates and indoles in broccoli and other cruciferous vegetables; and or-

ganosulfur compounds in garlic and onions. In Asian countries where diets are low in fat and high in soy and green tea, the incidence of breast, prostate, and colorectal cancer is low. In the United States, a national program known as 5 A Day for Better Health is committed to increasing the number of daily servings of fruits and vegetables to five or more.[1]

---

[1] The Centers for Disease Control and Prevention website (www.cdc.gov/) has information about this program and about other nutritional issues.

TABLE 46-4

Some Important Minerals and Their Functions

Mineral	Functions	Sources; Comments
Calcium	Main mineral of bones and teeth; essential for normal blood clotting, muscle function, nerve function, and regulation of cell activities	Milk and other dairy products, fish, green leafy vegetables; bones serve as calcium reservoir
Phosphorus	Performs more functions than any other mineral; structural component of bone; component of ATP, DNA, RNA, and phospholipids	Meat, dairy products, cereals
Sulfur	Component of many proteins and vitamins	High-protein foods such as meat, fish, legumes, nuts
Potassium	Principal positive ion within cells; important in muscle contraction and nerve function	Meat, milk, fruit, vegetables, grains
Sodium	Principal positive ion in interstitial fluid; important in fluid balance; neural transmission	Many foods, table salt; too much ingested in average American diet; excessive amounts may contribute to high blood pressure
Chloride	Principal negative ion of interstitial fluid; important in fluid balance and in acid–base balance	Many foods, table salt
Magnesium	Needed for normal muscle and nerve function	Nuts, whole grains, green leafy vegetables, seafood, chocolate
Copper	Component of several enzymes; essential for hemoglobin synthesis	Liver, eggs, fish, whole wheat flour, beans
Iodide	Component of thyroid hormones (hormones that increase metabolic rate); deficiency results in goiter (abnormal enlargement of thyroid gland)	Seafood, iodized salt, vegetables grown in iodine-rich soils
Manganese	Activates many enzymes	Whole-grain cereals, nuts, leafy green vegetables; poorly absorbed from intestine
Iron	Component of hemoglobin, myoglobin, important respiratory enzymes (cytochromes), and other enzymes essential to oxygen transport and cellular respiration; deficiency results in anemia and may impair cognitive function	Mineral most likely to be deficient in diet. Good sources: meat (especially liver), fish, nuts, egg yolk, legumes, dried fruit
Fluoride	Component of bones and teeth; makes teeth resistant to decay; excess causes tooth mottling	Fish; in areas where it does not occur naturally, fluoride may be added to municipal water supplies (fluoridation)
Zinc	Cofactor for at least 70 enzymes; helps regulate synthesis of certain proteins; needed for growth and repair of tissues; deficiency may impair cognitive function	Meat, fish, milk, yogurt, grains, vegetables
Selenium	Antioxidant (cofactor for a peroxidase that breaks down peroxides); appears to protect against prostate cancer in men	Seafood, eggs, meat, whole grains

Review

∎ How do complex carbohydrates affect the body?

∎ What is the function of glucose? Of essential amino acids?

∎ What is the function of high-density lipoproteins (HDLs)?

∎ What measures can you take to lower your risk for heart disease?

ENERGY METABOLISM

Learning Objectives

8 Contrast basal metabolic rate with total metabolic rate; write the basic energy equation for maintaining body weight, and describe the consequences of altering it in either direction.

9 In general terms, describe the effects of malnutrition, including both undernutrition and overnutrition.

10 Summarize current hypotheses about the regulation of food intake and energy homeostasis, including the roles of leptin and neuropeptide Y.

The amount of energy liberated by the body per unit time is a measure of its **metabolic rate.** Much of the energy expended by the body is ultimately converted to heat. Metabolic rate may be expressed either in kilocalories of heat energy expended per day or as a percentage above or below a standard normal level.

The **basal metabolic rate (BMR)** is the rate at which the body releases heat as a result of breaking down fuel molecules. BMR is the body's basic cost of living, that is, the rate of energy used during resting conditions. This energy is required to maintain body functions such as heart contraction, breathing, and kidney function. A person's **total metabolic rate** is the sum of his or her BMR and the energy used to carry on all daily activities. A laborer has a greater total metabolic rate than an account executive who is inactive most of the day and who does not exercise regularly.

An average-sized man who does not exercise and who sits at a desk all day expends about 2100 kcal daily (women expend less energy). If the food the individual eats each day also contains about the same number of kilocalories, the body will be in a state of energy balance; that is, energy input will equal energy output. This is an extremely important concept, because body weight remains constant when

$$\text{Energy input} = \text{energy output}$$

When energy output is greater than energy input, stored fat is burned and body weight decreases. People gain weight when they take in more energy in food than they expend in daily activity, in other words, when

$$\text{Energy input} > \text{energy output}$$

Undernutrition can cause serious health problems

Millions of people do not have enough to eat or do not eat a balanced diet. Individuals suffering from *undernutrition* are malnourished. They may feel weak, are easily fatigued, and are highly susceptible to infection as a result of depressed immune function. Iron, calcium, and vitamin A are commonly deficient, but essential amino acids are the nutrients most often deficient in the diet. Millions of people suffer from poor health and lowered resistance to disease because of protein deficiency. Children's physical and mental development are retarded when these essential building blocks of cells are not provided in the diet. Because these children's bodies cannot manufacture antibodies (which are proteins) and cells needed to fight infection, common childhood diseases, such as measles, whooping cough, and chickenpox, are often fatal in children suffering from protein malnutrition.

In young children, severe protein malnutrition results in the condition known as **kwashiorkor.** This Ashanti (West African) word means "first-second." It refers to the situation in which a first child is displaced from its mother's breast when a younger sibling is born. The older child is then given a diet of starchy cereal or cassava that is deficient in protein. Growth becomes stunted, muscles are wasted, and edema develops (as displayed by a swollen belly); the child becomes apathetic and anemic, with an impaired metabolism (Fig. 46-15). Without essential amino acids, digestive enzymes cannot be manufactured, so what little protein is ingested cannot be digested.

Obesity is a serious nutritional problem

Malnutrition can also result from overnutrition. **Obesity,** the excess accumulation of body fat, is a serious form of malnutrition that has become a problem of epidemic proportions in affluent societies. The World Health Organization considers obesity among the most important global health problems. Approximately 30% of U.S. adults are obese, and another 35% are

P. Pittet/United Nations Food and Agricultural Organization

Figure 46-15 Protein deficiency

Millions of children suffer from kwashiorkor, a disease caused by severe protein deficiency. Note the characteristic swollen belly, which results from fluid imbalance.

overweight. An estimated 20% of U.S. children and adolescents are overweight. Obesity contributes to about 300,000 deaths annually in the United States and is the second leading preventable cause of death (second only to smoking). It is a major risk factor for heart disease, diabetes mellitus, osteoarthritis, and certain types of cancer, including breast and colon cancers. In the well-known Framingham study of more than 2000 men, those who were 20% overweight had a significantly higher mortality rate from all causes.

Body mass index (BMI), now used worldwide as a measure of body size, is an index of weight in relation to height. It is calculated by dividing the square of the weight (kg^2) by height (m). The equivalent in the United States is 4.89 times the weight (lb) divided by the square of the height (ft^2). A person is considered obese if the BMI is 30 or more. Each of us appears to have a **set point,** or steady state, around which body weight is regulated. When BMI decreases below an individual's set point, energy-conserving mechanisms are activated and energy expenditure decreases.

Obesity can result from an increase in the size of fat cells or from an increase in the number of fat cells, or both. The number of fat cells in the adult is apparently determined mainly by the amount of fat stored during infancy and childhood. When we are overfed early in life, abnormally large numbers of fat cells are formed. Later in life, these fat cells may be fully stocked with excess lipids or may shrink in size, but they are always there. People with such increased numbers of fat cells are thought to be at greater risk for obesity than are those with normal numbers.

Although an estimated 40% to 70% of the factors involved in obesity are inherited, there is plasticity in the system that regulates body weight. High-calorie diets, overeating, and underexercising lead to obesity, and psychosocial factors influence these behaviors. A sedentary lifestyle combined with the ready availability of high-energy foods contributes to the problem. One researcher cleverly described the roots of the obesity problem as the combination of "computer chips and potato chips." For every 9.3 kcal of excess food taken into the body, about 1 g of fat is stored. (An excess of about 140 kcal, less than a typical candy bar, per day for a month results in a 1-lb weight gain.)

In their search for the causes of obesity, biologists are focusing on signaling pathways. Investigators have isolated a mouse gene (*ob* locus) responsible for an obese phenotype. Mice with a mutated allele of this gene apparently lack some weight-regulating substance and become grossly obese. The increased adipose tissue in these mice is part of a syndrome that parallels morbid obesity in humans, a condition in which an individual's body weight is 100 lb (46.5 kg) or more above normal.

Injections of the hormone **leptin,** the normal *ob* gene product, results in weight loss by obese and nonobese mice. When researchers inject leptin into grossly obese mice, the appetites of the mice decrease and their energy use increases. They lose weight, mainly from loss of body fat (■ Fig. 46-16). Leptin, produced in adipose tissue and by the stomach, signals centers in the brain about the status of energy stores in adipose tissue. The brain then adjusts feeding behavior and energy metabolism. An *ob* gene has been identified in humans, indicating that this gene has been conserved during evolution. Unfortunately, most obese humans are resistant to the actions of leptin; cell signaling is impaired, or leptin is not transported across the blood–brain barrier.

During the past few years, researchers have identified several gene mutations and signaling molecules that help regulate food intake and energy homeostasis. These signaling molecules form complex regulatory pathways involving genetic, neural, and endocrine mechanisms. Leptin and insulin are long-term regulators that stabilize the body's fat stores. During negative energy balance, the body's adipose tissue decreases, depressing both leptin and insulin secretion.

When leptin levels and food intake are low, as during dieting or starvation, the hypothalamus increases secretion of the neurotransmitter **neuropeptide Y (NPY).** This neuropeptide increases appetite and slows metabolism, actions that help restore energy homeostasis. Several other appetite regulators signal us to eat or to stop eating. When the stomach is empty, it secretes the peptide hormone *ghrelin,* which stimulates appetite. The gastrointestinal tract also releases several peptides, including

Figure 46-16 *Animated* Effect of leptin

Mutant *Ob* mice before (*left*) and after (*right*) treatment with leptin.

the hormones obestatin and cholecystokinin, that signal satiety and suppress food intake. Investigators are testing these signaling molecules for use as appetite suppressants.

The brain, especially the hypothalamus, processes information from leptin, insulin, and other signaling molecules and integrates this input with other information, such as the amount of nutrients circulating in the blood. The brain sends signals that adjust feeding behavior and metabolism so that homeostasis of stored fat and fuel metabolism is maintained. When food is available and energy stores are adequate, energy use increases and fat stores are mobilized. Energy intake and endogenous glucose production are inhibited.

As researchers discover the genetic and biochemical mechanisms that regulate energy metabolism, they provide potential targets for the development of pharmacological treatments for obesity. The following are among the many targets for drug action that are the focus of research: (1) reduce appetite so that food intake is decreased; (2) block absorption of fat; (3) uncouple metabolism from energy production so that food energy is dissipated as heat (see *Focus On: Electron Transport and Heat,* in Chapter 8); and (4) block receptors for molecules that send signals leading to increased energy storage.

Review

- Write an equation to describe energy balance, and explain what happens when the equation is altered in either direction.
- What is body mass index (BMI)?
- What is the function of leptin?

SUMMARY WITH KEY TERMS

Learning Objectives

1 Describe food processing, including ingestion, digestion, absorption, and egestion or elimination; and compare the digestive system of a cnidarian (such as *Hydra*) with that of an earthworm or vertebrate (page 990).
    - **Nutrition** is the process of taking in and using food.

- **Feeding** is the selection, acquisition, and **ingestion** of food, that is, taking food into the body. **Digestion** is the process of breaking down food mechanically and chemically. Nutrients pass through the lining of the digestive tract and into the blood by **absorption.** Food that is not digested and absorbed is discharged from the body by **egestion** or **elimination.**

- In cnidarians and flatworms, food is digested in the **gastrovascular cavity.** This cavity has only one opening that serves as both mouth and anus.
- In more complex invertebrates and in all vertebrates, the digestive tract is a complete tube with an opening at each end. Digestion takes place as food passes through the tube. Various parts of the digestive tract are specialized to perform specific functions.

ThomsonNOW™ **Explore various digestive systems by clicking on the figure in ThomsonNOW.**

2 Trace the pathway traveled by an ingested meal in the human digestive system, and describe the structure and function of each organ involved (page 992).

- Mechanical and enzymatic digestion of carbohydrates begins in the mouth. Mammalian teeth include **incisors** for biting, **canines** for tearing food, and **premolars** and **molars** for crushing and grinding. Three pairs of **salivary glands** secrete saliva, a fluid containing the enzyme **salivary amylase** that digests starch.
- As food is swallowed, it is propelled through the **pharynx** and **esophagus.** A **bolus** of food is moved along through the digestive tract by **peristalsis,** waves of muscular contraction that push food along.
- In the **stomach,** food is mechanically digested by vigorous churning, and proteins are enzymatically digested by the action of **pepsin** in the gastric juice. **Rugae** are folds in the stomach wall that expand as the stomach fills with food. **Gastric glands** secrete hydrochloric acid and **pepsinogen,** the precursor of pepsin.
- After several hours, a soup of partly digested food, called **chyme,** leaves the stomach through the **pylorus** and enters the **small intestine** in spurts. Most enzymatic digestion takes place in the **duodenum,** which produces several digestive enzymes and receives secretions from the liver and pancreas.
- The **liver** produces **bile,** which emulsifies fats. The **pancreas** releases enzymes that digest protein, lipid, and carbohydrate, as well as RNA and DNA. **Trypsin** and **chymotrypsin** digest polypeptides to dipeptides. **Pancreatic lipase** degrades fats, and **pancreatic amylase** digests complex carbohydrates.
- The **large intestine,** which consists of the **cecum, colon, rectum,** and **anus,** is responsible for eliminating undigested wastes. It also incubates bacteria that produce vitamin K and certain B vitamins.

ThomsonNOW™ **Learn more about human digestion by clicking on the figures in ThomsonNOW.**

3 Describe the step-by-step digestion of carbohydrate, protein, and lipid (page 992).

- Nutrients in chyme are enzymatically digested as they move through the digestive tract. Polysaccharides are digested to the disaccharide maltose by salivary and pancreatic amylases. **Maltase** in the small intestine splits maltose into glucose, the main product of carbohydrate digestion.
- Proteins are split by pepsin in the stomach and by proteolytic enzymes in the pancreatic juice. **Dipeptidases** split small peptides to amino acids.
- Lipids are emulsified by bile salts and then hydrolyzed by pancreatic lipase.

4 Describe the structural adaptations that increase the surface area of the digestive tract (page 992).

- The surface area of the small intestine is greatly expanded by folds in its wall; by the intestinal **villi,** elongated projections of the mucosa; and by **microvilli,** projections of the plasma membrane of the epithelial cells of the villi.

5 Compare lipid absorption with absorption of other nutrients (page 992).

- Nutrients are absorbed through the thin walls of the intestinal villi. The **hepatic portal vein** transports amino acids and glucose to the liver.
- Fatty acids and monoacylglycerols enter epithelial cells in the intestinal lining, where they are reassembled into triacylglycerols. They are packaged into **chylomicrons,** droplets that also contain cholesterol and phospholipids and are covered by a protein coat. The lymphatic system transports chylomicrons to the blood circulation.

6 Summarize the nutritional requirements for dietary carbohydrates, lipids, and proteins; and trace the fate of glucose, lipids, and amino acids after their absorption (page 1000).

- For a balanced diet, humans and other animals require carbohydrates, lipids, proteins, vitamins, and minerals. Most carbohydrates are ingested in the form of polysaccharides—starch and cellulose. Polysaccharides are referred to as **complex carbohydrates. Fiber** is mainly a mixture of cellulose and other indigestible carbohydrates. Carbohydrates are used mainly as an energy source. Glucose concentration in the blood is carefully regulated. Excess glucose is stored as glycogen and can also be converted to fat.
- Lipids are used to provide energy, to form components of cell membranes, and to synthesize steroid hormones and other lipid substances. Most lipids are ingested in the form of triacylglycerols. Fatty acids are converted to molecules of acetyl coenzyme A, which enter the citric acid cycle. Excess fatty acids are converted to triacylglycerol and stored as fat.
- Lipids are transported as large molecular complexes called **lipoproteins. Low-density lipoproteins (LDLs)** deliver cholesterol to the cells. **High-density lipoproteins (HDLs)** collect excess cholesterol and transport it to the liver.
- Proteins serve as enzymes and are essential structural components of cells. The best distribution of **essential amino acids** is found in the complete proteins of animal foods. Excess amino acids are deaminated by liver cells. Amino groups are converted to urea and excreted in urine; the remaining keto acids are converted to carbohydrate and used as fuel or converted to lipid and stored in fat cells.

7 Describe the nutritional functions of vitamins, minerals, and phytochemicals (page 1000).

- **Vitamins** are organic compounds required in small amounts for many biochemical processes. Many serve as components of coenzymes. **Fat-soluble vitamins** include vitamins A, D, E, and K. **Water-soluble vitamins** are the B and C vitamins.
- **Minerals** are inorganic nutrients ingested as salts dissolved in food and water. **Trace elements** are minerals required in amounts less than 100 mg per day.
- **Phytochemicals** are plant compounds that promote health. Many phytochemicals are **antioxidants** that destroy **oxidants,** which include **free radicals** and other reactive molecules that damage DNA, proteins, and unsaturated fatty acids by snatching electrons.

8 Contrast basal metabolic rate with total metabolic rate; write the basic energy equation for maintaining body weight, and describe the consequences of altering it in either direction (page 1005).

- **Basal metabolic rate (BMR)** is the body's cost of metabolic living. **Total metabolic rate** is the BMR plus the energy used to carry on daily activities.
- When energy (kilocalories) input equals energy output, body weight remains constant. When energy output is greater than energy input, body weight decreases.

When energy input exceeds energy output, body weight increases.

9 In general terms, describe the effects of malnutrition, including both undernutrition and overnutrition (page 1005).

▌ Millions of people suffer from undernutrition, a form of malnutrition that causes fatigue and depressed immune function. Essential amino acids are the nutrients most often deficient in the diet.

▌ In **obesity,** a serious form of malnutrition caused by overnutrition, an excess amount of fat accumulates in adipose tissues. Obesity is a major risk factor for heart disease, diabetes mellitus, and several other disorders. A person gains weight by taking in more energy, in the form of kilocalories, than is expended in activity.

10 Summarize current hypotheses about the regulation of food intake and energy homeostasis, including the roles of leptin and neuropeptide Y (page 1005).

▌ Researchers are identifying signaling molecules that make up the complex pathways that regulate food intake and energy metabolism. The hormone **leptin** is produced by fat cells in proportion to body fat mass; it signals the brain about the status of energy stores. **Neuropeptide Y (NPY),** a neurotransmitter produced in the hypothalamus, increases appetite and slows metabolism when leptin levels and food intake are low.

Thomson**NOW**   **Learn more about leptins and body weight by clicking on the figure in ThomsonNOW.**

TEST YOUR UNDERSTANDING

1. The process of taking in and using food is (a) nutrition (b) chemical digestion (c) egestion (d) ingestion (e) absorption

2. Animals that feed mainly on producers are (a) herbivores (b) secondary consumers (c) animals with gastrovascular cavities (d) carnivores (e) secondary consumers and carnivores

3. Teeth adapted for crushing and grinding are (a) incisors (b) canines (c) premolars and molars (d) incisors and premolars (e) canines and molars

4. The layer of tissue that lines the lumen of the digestive tract is the (a) muscle layer (b) visceral peritoneum (c) parietal peritoneum (d) mucosa (e) submucosa

5. Which of the following are accessory digestive glands? (a) salivary glands (b) pancreas (c) liver (d) a, b, and c (e) a and b

6. Arrange the following into the correct sequence.

   1. stomach 2. esophagus 3. pharynx 4. small intestine 5. colon

   (a) 2, 3, 1, 4, 5 (b) 3, 2, 1, 5, 4 (c) 3, 2, 5, 1, 4 (d) 2, 3, 1, 5, 4 (e) 3, 2, 1, 4, 5

7. Amylase is produced by the (a) liver and pancreas (b) stomach and pancreas (c) colon and salivary glands (d) liver and pancreas (e) pancreas and salivary glands

8. Pepsin is produced by the (a) liver (b) stomach (c) pancreas (d) duodenum (e) salivary glands

9. Which sequence most accurately describes the digestion of protein?

   1. dipeptide 2. amino acid 3. protein 4. polypeptide 5. glycerol

   (a) 4, 3, 1, 2 (b) 3, 4, 1, 2 (c) 3, 4, 1, 5 (d) 4, 3, 5, 1, 2 (e) 2, 1, 4, 5

10. The surface area of the small intestine is increased by (a) folds in its wall (b) villi (c) microvilli (d) a, b, and c (e) a and b

11. Lipids are transported from the intestine to the liver by (a) chylomicrons (b) HDLs (c) LDLs (d) glycerol transporters (e) leptin

12. Most vitamins are (a) inorganic compounds (b) components of coenzymes (c) used as fuel (d) electrolytes (e) required by herbivores and carnivores but not omnivores

13. When energy input is greater than energy output, (a) weight loss occurs (b) weight remains stable (c) weight gain occurs (d) leptin secretion is inhibited (e) the *ob* gene is deactivated

14. A hormone that stimulates gastric glands to secrete pepsinogen is (a) secretin (b) GIP (c) gastrin (d) cholecystokinin (CCK) (e) substance P

15. Neuropeptide Y (NPY) (a) increases appetite (b) increases metabolism (c) is encoded by the mutant *ob* gene (d) is a hormone produced by the pancreas (e) decreases bile production

16. Many phytochemicals (a) are harmful to the body (b) cause high cholesterol levels (c) inhibit gastrin (d) are used as energy sources (e) function as antioxidants

17. Label the diagram. Use Figure 46-4 to check your answers.

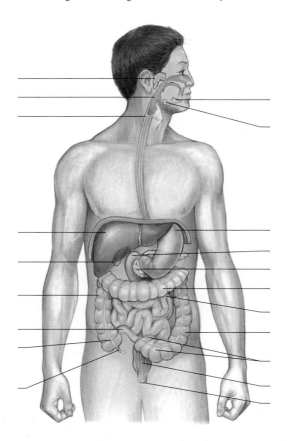

1. Design an experiment to test the hypothesis that the B vitamin pyridoxine is an essential nutrient in mice.

2. Why are proteolytic enzymes produced in an inactive form?

3. Investigators are unraveling the biochemical pathways that help regulate body weight. How might their work lead to a cure for obesity?

4. **Evolution Link.** If you were presented with an unfamiliar animal and asked to determine its nutritional lifestyle, how would you do so? (*Hint:* Consider adaptations.)

Additional questions are available in ThomsonNOW at www.thomsonedu.com/ login

Osmoregulation and Disposal of Metabolic Wastes

McMurray Photography

Most terrestrial animals inhabit areas near water sources. Zebra, wildebeest, and a variety of birds gather to drink at a lake in Ngorongoro Crater in Tanzania.

KEY CONCEPTS

Osmoregulation is the process by which organisms control the concentration of water and salt in the body so that their body fluids do not become too dilute or too concentrated.

Excretory systems have evolved that function in both osmoregulation and in disposal of metabolic wastes.

Freshwater, marine, and terrestrial animals have different adaptations to meet the challenges of these diverse environments.

The vertebrate kidney maintains water and electrolyte balance and excretes metabolic wastes.

The nephron is the functional unit of the vertebrate kidney.

Water, the most abundant molecule both in the cell and on Earth's surface, shapes life and the distribution of organisms on our planet. Ions are dissolved in water, and water is the medium in which most metabolic reactions take place. The volume and ionic composition of an animal's body fluids must be maintained within homeostatic limits. Natural selection has resulted in the evolution of a variety of homeostatic mechanisms that regulate the volume and composition of fluids in the internal environment. In this chapter we discuss some of these mechanisms.

Many small animals live in the ocean and get their food and oxygen directly from the sea water that surrounds them; they release waste products into the sea water. In larger animals and in most terrestrial animals, the fluid that bathes the cells serves as an internal sea.

Terrestrial animals have a continuous need to conserve water. Water loss from the body must be carefully regulated, and water lost must be replaced. Water is taken in with food and drink and is also produced in metabolic reactions. Like the animals in the photograph, most animals need a dependable source of water with which to replenish their body fluids, so they often inhabit areas near water sources. Two processes that help maintain fluid and electrolyte (salt) homeostasis in animals are *osmoregulation* and *excretion*, disposal of metabolic wastes. ∎

MAINTAINING FLUID AND ELECTROLYTE BALANCE

Learning Objective

1   Describe how the processes of *osmoregulation* and *excretion* contribute to fluid and electrolyte homeostasis.

Intracellular fluid, the fluid within cells, accounts for most of the body fluid. **Extracellular fluid,** the fluid outside the cells, includes **interstitial fluid** (the fluid between cells), lymph, and blood plasma (or hemolymph). In vertebrates, blood plasma, which is mainly water, transports nutrients, gases, waste products, and other materials throughout the body. Interstitial fluid forms from the blood plasma and bathes all the cells. Excess water evaporates from the body surface or is excreted by specialized structures.

Electrolytes are compounds such as inorganic salts, acids, and bases that form ions in solution. Electrolytes are very important solutes in body fluids. Many homeostatic mechanisms act to maintain fluid and **electrolyte balance.**

Recall from Chapter 5 that *osmosis* is the diffusion of water through a selectively permeable membrane. The net flow of water is from a more dilute solution (one with a lower solute concentration) to a less dilute solution (one with a higher solute concentration).

If the solute concentrations of two solutions are equal, they are *isotonic* to each other. If solution A has a greater solute concentration than solution B, solution A is *hypertonic* to solution B. Solution B, with a lower solute concentration relative to solution A, is *hypotonic* to solution A. Remember that the total solute concentration of a solution is responsible for the direction of water movement. The *osmotic pressure* of a solution is the pressure that must be exerted on the side of a selectively permeable membrane that contains the higher solute concentration to prevent net movement of water from the side containing the lower solute concentration.

An **osmol** is a unit of osmotic pressure equal to the molarity of the solution divided by the number of particles produced when the solute dissolves. For example, glucose dissolves to give only one kind of particle, so a mole of glucose in solution is one osmole. A mole of NaCl in solution produces two kinds of particles (Na^+ and Cl^-), so a mole of NaCl is two osmoles. Both types of ions affect the osmotic pressure of the solution. A *milliosmol* is 1/1000 of an osmol.

Osmolarity is a measure of the number of osmoles of solute per liter of solution. Solutions with the same osmolarity are described as *isosmotic*. If solution X has a higher osmolarity than solution Y, it is described as *hyperosmotic*. Solution Y would be described as *hypo-osmotic* relative to solution X.

Osmoregulation is the process by which organisms control the concentration of water and salt so that their body fluids do not become too dilute or too concentrated. **Excretion** is the process of ridding the body of metabolic wastes. **Excretory systems** have evolved that function in both osmoregulation and in disposal of metabolic wastes. Excretory systems rid the body of excess water and ions, metabolic wastes, and harmful substances.

As we will discuss, hormones are important signaling molecules in these regulatory processes.

Review

■   What is osmoregulation?
■   What is excretion?

METABOLIC WASTE PRODUCTS

Learning Objective

2   Contrast the benefits and costs of excreting ammonia, uric acid, or urea.

Metabolic wastes must be excreted so that they do not accumulate and reach concentrations that would disrupt homeostasis. The principal metabolic waste products produced by most animals are water, carbon dioxide, and **nitrogenous wastes,** those that contain nitrogen. Carbon dioxide is excreted mainly by respiratory structures (see Chapter 45). In terrestrial animals, some water is also lost from respiratory surfaces. Excretory organs, such as kidneys, remove and excrete most of the water and nitrogenous wastes.

Nitrogenous wastes include ammonia, uric acid, and urea. Recall that amino acids and nucleic acids contain nitrogen. During the metabolism of amino acids, the nitrogen-containing amino group is removed (in a process known as deamination) and converted to **ammonia** (▌Fig. 47-1). However, ammonia is highly toxic. Some aquatic animals excrete it into the surrounding water before it can build up to toxic concentrations in their tissues. A few terrestrial animals, including some snails and wood lice, vent it directly into the air. But many animals, humans included, convert ammonia to some less toxic nitrogenous waste such as uric acid or urea.

Uric acid is produced both from ammonia and by the breakdown of nucleotides from nucleic acids. Uric acid is insoluble in water and forms crystals that are excreted as a crystalline paste, so little fluid loss results. This is an important water-conserving adaptation in many terrestrial animals, including insects, certain reptiles, and birds. Also, because uric acid is not toxic and can be safely stored, its excretion is an adaptive advantage for species whose young begin their development enclosed in eggs.

Urea, the principal nitrogenous waste product of amphibians and mammals, is synthesized in the liver from ammonia and carbon dioxide by a sequence of reactions known as the **urea cycle.** Like the formation of uric acid, these reactions require specific enzymes and the input of energy by the cells. Compared to the energy cost of producing ammonia, that of producing urea and uric acid is high.

Urea has the advantage of being far less toxic than ammonia and can accumulate in higher concentrations without causing tissue damage; thus, it can be excreted in more concentrated form. Because urea is highly soluble, however, it is dissolved in water, and more water is needed to excrete urea than to excrete uric acid.

Review

■   What are the principal types of nitrogenous wastes? Give examples of animals that excrete each type.

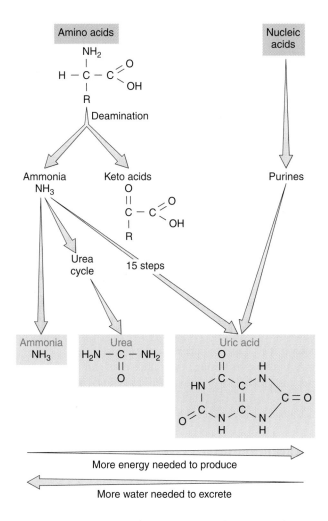

Figure 47-1 Formation of nitrogenous wastes

Deamination of amino acids and metabolism of nucleic acids produce nitrogenous wastes. Ammonia is the first metabolic product of deamination. Many aquatic animals excrete ammonia, but terrestrial animals convert it to urea and uric acid, which are less toxic. Most amphibians and mammals convert ammonia to urea via the urea cycle. Insects, many reptiles, and birds convert ammonia to uric acid. Energy is required to convert ammonia to urea and uric acid, but less water is required to excrete these wastes.

▌ What are the benefits of excreting nitrogenous wastes in the form of uric acid? In the form of urea?

OSMOREGULATION AND EXCRETION IN INVERTEBRATES

Learning Objectives

3 Compare osmoconformers and osmoregulators.
4 Describe protonephridia, metanephridia, and Malpighian tubules, and compare their functions.

The ocean is a stable environment, and its salt concentration does not vary much. The electrolyte concentration in the cells and body fluids of marine sponges and cnidarians is very similar to sea water, so these animals do not require specialized excretory structures. They expend little or no energy in excreting wastes because wastes simply diffuse from their cells to the external environment and are washed away by water currents. When water stagnates and currents do not wash away wastes, aquatic environments, such as coral reefs, are damaged by the waste accumulation.

The body fluids of most marine invertebrates are in osmotic equilibrium with the surrounding sea water. These animals are known as **osmoconformers,** because the concentration of their body fluids varies along with changes in the sea water. However, many osmoconforming marine animals still regulate some ions in their body fluids.

Coastal habitats, such as estuaries that contain brackish water, are much less stable environments than is the open ocean. Salt concentrations change frequently with shifting tides. Many invertebrates and vertebrates that inhabit these environments are **osmoregulators.** These animals have homeostatic mechanisms that maintain an optimal salt concentration in their tissues regardless of changes in the salt concentration of their surroundings.

In a coastal environment where fresh water enters the ocean, the water may have a lower salt concentration than do the body fluids of the animal. Water osmotically moves into the body, and salt diffuses out. An animal adapted to this environment has excretory structures that remove the excess water. Certain polychaete worms and the blue crab are among the animals that can be osmoconformers or osmoregulators depending on environmental conditions.

Dehydration is a constant threat to terrestrial animals. Because their fluid concentration is higher than that of the air around them, they tend to lose water by evaporation from both the body surface and respiratory surfaces and may also lose water as wastes are excreted. As animals moved onto the land, natural selection favored the evolution of structures and processes that conserve water.

Excretory systems help maintain fluid and electrolyte homeostasis by selectively adjusting the concentrations of salts and other substances in blood and other body fluids. Typically, an excretory system collects fluid, generally from the blood or interstitial fluid. It then adjusts the composition of this fluid by selectively returning needed substances to the body fluid. Finally, the body releases the adjusted excretory product containing excess or potentially toxic substances.

Nephridial organs are specialized for osmoregulation and/or excretion

Nephridial organs, or nephridia, are excretory structures that evolved in many invertebrates, including flatworms, nemerteans, rotifers, annelids, mollusks, and lancelets. Each nephridial organ consists of simple or branching tubes that typically open to the outside of the body through excretory pores, called *nephridiopores.* Two types of nephridial organs are protonephridia and metanephridia.

In flatworms and nemerteans, metabolic wastes pass through the body surface by diffusion, but these animals also have **proto-**

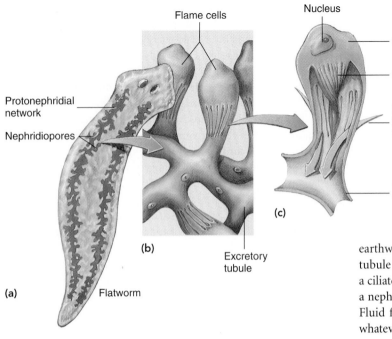

Figure 47-2 Protonephridia of a flatworm

(a) The protonephridia of a planarian, which function mainly in osmoregulation, form a system of branching tubules. (b) Interstitial fluid enters flame cells and passes through a series of tubules. Excess fluid leaves the body through nephridiopores. (c) A single flame cell.

nephridia (Fig. 47-2). These nephridial organs are composed of tubules with no internal openings. Their enlarged blind ends consist of **flame cells** with brushes of cilia, so named because their constant motion reminded early biologists of flickering flames. The flame cells lie in the interstitial fluid that bathes the body cells. Fluid enters the flame cells, and the beating of the cilia propels the fluid through the tubules. Excess fluid leaves the body through nephridiopores.

Most annelids and mollusks have more complex nephridial organs called **metanephridia** (Fig. 47-3). Each segment of an earthworm has a pair of metanephridia. A metanephridium is a tubule open at both ends. The inner end opens into the coelom as a ciliated funnel, and the outer end opens to the outside through a nephridiopore. Around each tubule is a network of capillaries. Fluid from the coelom passes into the tubule, bringing with it whatever it contains—glucose, salts, or wastes.

As fluid moves through the tubule, needed materials (such as water and glucose) are removed from the fluid by the tubule and are reabsorbed by the capillaries, leaving the wastes behind. In this way, urine is produced that contains concentrated wastes.

Malpighian tubules conserve water

The excretory system of insects and spiders consists of several hundred **Malpighian tubules** (Fig. 47-4). Malpighian tubules are slender extensions of the gut wall. Their blind ends lie in the

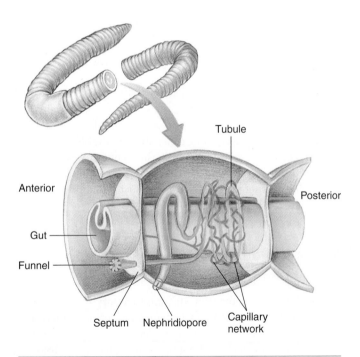

Figure 47-3 Metanephridium of an earthworm

Each metanephridium consists of a ciliated funnel opening into the coelom, a coiled tubule, and a nephridiopore opening to the outside. This 3-D internal view shows parts of three segments.

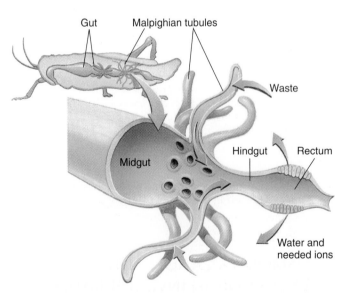

Figure 47-4 Malpighian tubules of an insect

The slender Malpighian tubules have blind ends that extend into the hemocoel. Their cells transfer uric acid and some ions from the hemolymph to the cavity of the tubule. Water follows by diffusion. The wastes are discharged into the gut. The epithelium lining the rectum (part of the hindgut) actively reabsorbs most of the water and needed ions.

hemocoel (blood cavity) and are bathed in hemolymph. Cells of the tubule wall actively transport uric acid, potassium ions, and some other substances from the hemolymph into the tubule lumen. Other solutes and water follow by diffusion. The Malpighian tubules empty into the gut. Water, some salts, and other solutes are reabsorbed into the hemolymph by a specialized epithelium in the rectum.

Uric acid, the major waste product, is excreted as a semidry paste with a minimum of water. Because Malpighian tubules effectively conserve body fluids, they have contributed to the success of insects in terrestrial environments.

Review

■ How are nephridial organs and Malpighian tubules alike? How are they different?

OSMOREGULATION AND EXCRETION IN VERTEBRATES

Learning Objectives

5   Relate the function of the vertebrate kidney to the success of vertebrates in a wide variety of habitats.
6   Compare adaptations for osmoregulation in freshwater fishes, marine bony fishes, sharks, marine mammals, and terrestrial vertebrates.

Vertebrates live successfully in a wide range of habitats—in fresh water, the ocean, tidal regions, and on land, even in extreme environments such as deserts. In response to the requirements of these diverse environments, adaptations have evolved for regulating salt and water content and for excreting wastes. The main osmoregulatory and excretory organ in most vertebrates is the **kidney.** The kidney excretes most nitrogenous wastes and helps maintain fluid balance by adjusting the salt and water content of the urine. The skin, lungs or gills, and digestive system also help maintain fluid balance and dispose of metabolic wastes.

Freshwater vertebrates must rid themselves of excess water

As fishes began to move into freshwater habitats about 470 million years ago (mya), there must have been strong selection for adaptations that promoted effective osmoregulation. Having body fluids more dilute than sea water is a major adaptation. However, the salt concentration of the body fluids of these fishes is still higher than that of the fresh water that surrounds them. They are hypertonic to their watery environment. As a result, water moves into the body, and they are in constant danger of becoming waterlogged.

Freshwater fishes are covered by scales and a mucous secretion that retard the passage of water into the body. However, water constantly enters through the gills and through the mouth (with food). How do these animals meet this challenge? Some water leaves the body through the gills. Furthermore, the kidneys of freshwater fishes are adapted to excrete large amounts of dilute

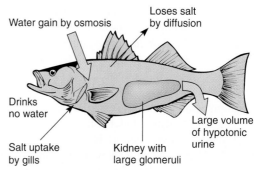

(a) Freshwater fishes live in a hypotonic medium. Water continuously enters the body, and salts diffuse out. These fishes excrete large quantities of dilute urine and actively transport salts in through the gills.

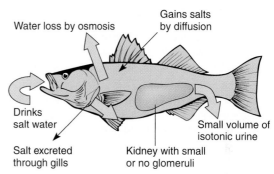

(b) Marine fishes live in a hypertonic medium. They lose water by osmosis. They gain salts from the sea water they drink and by diffusion. To compensate, the fish drinks water, excretes the salt, and produces a small volume of urine.

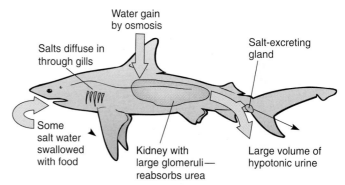

(c) Cartilaginous fishes (sharks and rays). The shark kidney reabsorbs urea in high enough concentration that its tissues become hypertonic to the surrounding medium. As a result, water enters the shark by osmosis, and the shark excretes a large quantity of dilute urine.

Figure 47-5   Osmoregulation in fishes

urine (■ Fig. 47-5a). The kidneys have large *glomeruli,* capillary clusters that filter the blood and produce urine.

Water entry is only part of the challenge of osmoregulation in freshwater fishes. These animals also tend to lose salts by diffusion through the gills into the surrounding water. To compensate, special gill cells have evolved that actively transport salts (mainly sodium chloride) from the water into the body.

The gills excrete most of the nitrogenous wastes of freshwater fishes, but the kidneys are also important excretory organs. Ammonia is the main nitrogenous waste; about 10% of nitrogenous wastes are excreted as urea.

Most amphibians are at least semiaquatic, and their mechanisms of osmoregulation are similar to those of freshwater fishes. They, too, produce large amounts of dilute urine. In its urine and through its skin, a frog can lose an amount of water equivalent to one third of its body weight in one day. Active transport of salt inward by special cells in the skin compensates for loss of salt through skin and in urine.

Marine vertebrates must replace lost fluid

Freshwater fishes adapted very successfully to their aquatic habitats. Thus, when some freshwater fishes returned to the sea about 200 mya, their blood and body fluids were less salty than (hypo-osmotic to) their surroundings. They lose water osmotically and take in salt. To compensate for fluid loss, many marine bony fishes drink sea water (▌ Fig. 47-5b). They retain the water and excrete salt by the action of specialized cells in their gills. The gills are also responsible for ammonia excretion. Very little urine is excreted by the kidneys, which have only small (or no) glomeruli.

Marine cartilaginous fishes (sharks and rays) have different osmoregulatory adaptations that allow them to tolerate the salt concentrations of their environment. These animals accumulate and tolerate urea (▌ Fig. 47-5c). Their tissues are adapted to function at concentrations of urea that would be toxic to most other animals. The high urea concentration makes the body fluids slightly hypertonic to sea water, resulting in a net inflow of water. Their well-developed kidneys excrete a large volume of urine. Excess salt is excreted by the kidneys and, in many species, by a rectal gland.

The heads of certain reptiles and marine birds have salt glands that excrete salt that enters the body with ingested sea water or in salty food. Salt glands are usually inactive; they function only in response to osmotic stress.

Whales, dolphins, and other marine mammals ingest sea water along with their food. Their kidneys produce concentrated urine, much saltier than sea water. This is an important physiological adaptation, especially for marine carnivores. The high-protein diet of these animals results in the production of large amounts of urea, which must be excreted in urine without losing much water.

Terrestrial vertebrates must conserve water

Adult amphibians generally inhabit moist environments. They excrete urea and reabsorb some water from the urinary bladder. Amniotes (reptiles, birds, and mammals) have more effective adaptations for life on land. Their skin minimizes water loss by evaporation, and many amniotes excrete uric acid, which requires very little water.

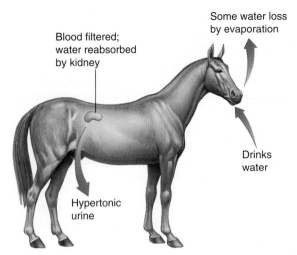

(a) The kidney of terrestrial vertebrates conserves water by reabsorbing it. Birds and mammals can produce concentrated urine.

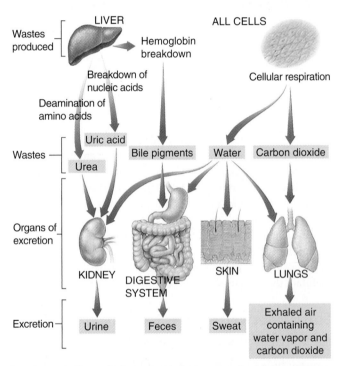

(b) Disposal of metabolic wastes in humans and other terrestrial mammals. To conserve water, mammals produce a small amount of hypertonic urine. Nitrogenous wastes are produced by the liver and transported to the kidneys. All cells produce carbon dioxide and some water during cellular respiration.

Figure 47-6 Excretory organs in terrestrial vertebrates

Because birds and mammals are endothermic (maintain a constant body temperature) and have a high rate of metabolism, they produce a relatively large volume of nitrogenous wastes. They have numerous adaptations, including very efficient kidneys, for conserving water. An extreme example is the desert-dwelling kangaroo rat, which obtains most of its water from its own metabolism (recall from Chapter 8 that aerobic respiration produces water). Its kidneys are so efficient that it loses little fluid as urine.

Birds conserve water by excreting nitrogen as uric acid and by efficiently reabsorbing water from the cloaca and intestine. Mammals excrete urea. Their kidneys produce very concentrated (hypertonic) urine.

In terrestrial vertebrates, the lungs, skin, and digestive system are important in osmoregulation and waste disposal (❙ Fig. 47-6). Most carbon dioxide is excreted by the lungs. In birds and mammals, some water is lost from the body as water vapor in exhaled air. Although primarily concerned with the regulation of body temperature, the sweat glands of humans and some other mammals excrete 5% to 10% of all metabolic wastes.

The liver produces both urea and uric acid, which are transported by the blood to the kidneys. Most of the bile pigments produced by the breakdown of red blood cells are normally excreted by the liver into the intestine. From the intestine they pass out of the body with the feces.

Review

- What type of osmoregulatory challenge is faced by marine fishes? By freshwater fishes? What mechanisms have evolved to meet these challenges?
- What adaptations in birds and mammals help meet the challenges of osmoregulation and metabolic waste disposal?

THE URINARY SYSTEM

Learning Objectives

7 Describe (or label on a diagram) the organs of the mammalian urinary system, and give the functions of each.

8 Describe (or label on a diagram) the structures of a nephron (including associated blood vessels), and give the functions of each structure.

9 Trace a drop of filtrate from Bowman's capsule to its release from the body as urine.

10 Describe the hormonal regulation of fluid and electrolyte balance by antidiuretic hormone (ADH), the renin–angiotensin–aldosterone system, and atrial natriuretic peptide (ANP).

The mammalian **urinary system** consists of the kidneys, the urinary bladder, and associated ducts. The overall structure of the human urinary system is shown in ❙ Figure 47-7. Located just below the diaphragm in the "small of the back," the kidneys look like a pair of giant, dark red lima beans, each about the size of a fist. Each kidney is covered by a connective tissue capsule (❙ Fig. 47-8). The outer portion of the kidney is the **renal cortex;** the inner portion is the **renal medulla.** The renal medulla contains 8 to 10 cone-shaped structures called **renal pyramids.** The tip of each pyramid is a **renal papilla.** Each papilla has several pores, the openings of **collecting ducts.**

As urine is produced, it flows from collecting ducts through a renal papilla and into the **renal pelvis,** a funnel-shaped chamber. Urine then flows into one of the paired **ureters,** ducts that connect each kidney with the **urinary bladder.** The urinary bladder is a remarkable organ capable of holding (with practice) up to 800 mL (about a pint and a half) of urine. Emptying the bladder changes it from the size of a small melon to that of a pecan. This feat is made possible by the smooth muscle and specialized epithelium of the bladder wall, which is capable of great shrinkage and stretching.

During **urination,** urine is released from the bladder and flows through the **urethra,** a duct leading to the outside of the body. In the male, the urethra is lengthy and passes through the penis. Semen, as well as urine, passes through the male urethra. In the female, the urethra is short and transports only urine. Its opening to the outside is just above the opening of the vagina. The length of the male urethra discourages bacterial invasions of the bladder. This length difference helps explain why bladder infections are more

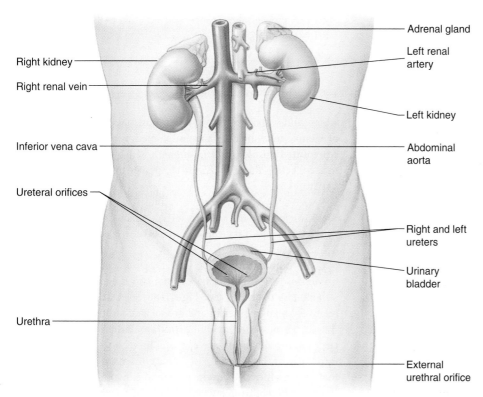

Right kidney

Right renal vein

Inferior vena cava

Ureteral orifices

Urethra

Adrenal gland

Left renal artery

Left kidney

Abdominal aorta

Right and left ureters

Urinary bladder

External urethral orifice

Figure 47-7 *Animated* The human urinary system

The kidneys produce urine, which passes through the ureters to the urinary bladder for temporary storage. The urethra then conducts urine from the bladder to the outside of the body through an opening, the external urethral orifice.

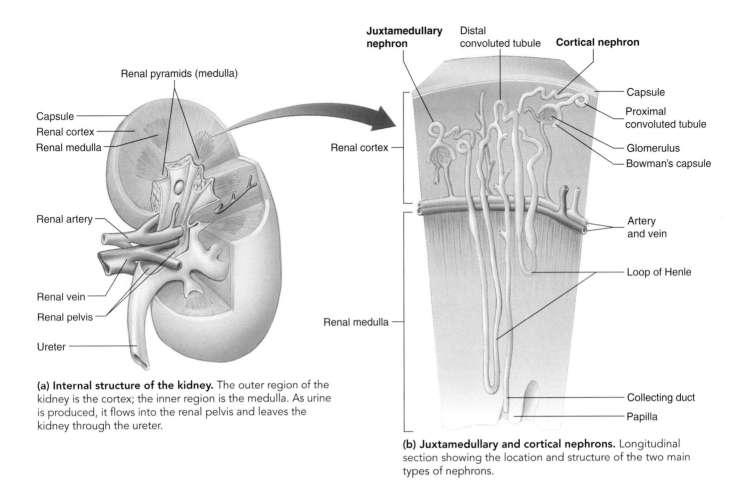

(a) Internal structure of the kidney. The outer region of the kidney is the cortex; the inner region is the medulla. As urine is produced, it flows into the renal pelvis and leaves the kidney through the ureter.

(b) Juxtamedullary and cortical nephrons. Longitudinal section showing the location and structure of the two main types of nephrons.

Figure 47-8 *Animated* Structure of the kidney

common in females than in males. In summary, urine flows through the following structures:

> Kidney (through renal pelvis) ⟶ ureter ⟶ urinary bladder
> ⟶ urethra

The nephron is the functional unit of the kidney

The principal function of the kidneys is to help maintain homeostasis by regulating fluid balance and excreting metabolic wastes. However, the kidneys have other functions as well. They produce the enzyme *renin*, which helps regulate fluid balance and blood pressure (discussed later in the chapter). They also produce at least two hormones: *erythropoietin*, which stimulates red blood cell production, and *1,25-dihydroxyvitamin D3*, which stimulates calcium absorption by the intestine.

Each kidney has more than 1 million functional units called **nephrons.** A nephron consists of a cuplike **Bowman's capsule** connected to a long, partially coiled **renal tubule** (❙ Fig. 47-9). Positioned within Bowman's capsule is a cluster of capillaries called a **glomerulus.** Three main regions of the renal tubule are the **proximal convoluted tubule** (or simply, *proximal tubule*),

which conducts the filtrate from Bowman's capsule; the **loop of Henle,** an elongated, hairpin-shaped portion; and the **distal convoluted tubule** (or *distal tubule*), which conducts the filtrate to a collecting duct. Thus, filtrate passes through the following structures:

> Bowman's capsule ⟶ proximal convoluted tubule ⟶ loop
> of Henle ⟶ distal convoluted tubule ⟶ collecting duct

The human kidney has two types of nephrons: the more numerous (85%) cortical nephrons and the more internal juxtamedullary nephrons (see Fig. 47-8b). **Cortical nephrons** have relatively small glomeruli and are located almost entirely within the cortex or outer medulla. **Juxtamedullary nephrons** have large glomeruli, and their very long loops of Henle extend deep into the medulla. The loop of Henle consists of a *descending limb* that receives filtrate from the proximal convoluted tubule and an *ascending limb,* through which the filtrate passes on its way to the distal convoluted tubule. The juxtamedullary nephrons contribute to the ability of the mammalian kidney to concentrate urine. Excretion of urine that is hypertonic to body fluids is an important mechanism for conserving water.

Blood is delivered to the kidney by the **renal artery.** Small branches of the renal artery give rise to **afferent arterioles** (*af-*

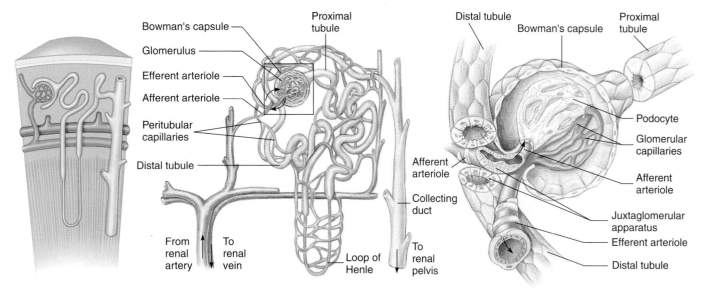

Bowman's capsule

Glomerulus

Efferent arteriole

Afferent arteriole

Peritubular capillaries

Distal tubule

Proximal tubule

From renal artery

To renal vein

Loop of Henle

Collecting duct

Afferent arteriole

To renal pelvis

(a) Location and the basic structure of a nephron. Urine forms by filtration of the blood in the glomerulus and by adjustment of the filtrate as it passes through the series of tubules that drain Bowman's capsule.

Distal tubule

Bowman's capsule

Proximal tubule

Podocyte

Glomerular capillaries

Afferent arteriole

Afferent arteriole

Juxtaglomerular apparatus

Efferent arteriole

Distal tubule

(b) Cutaway view of Bowman's capsule. Note that the distal tubule is adjacent to the afferent and efferent arterioles. The juxtaglomerular apparatus (discussed later in the chapter) is a small group of cells located in the walls of the tubule and arterioles.

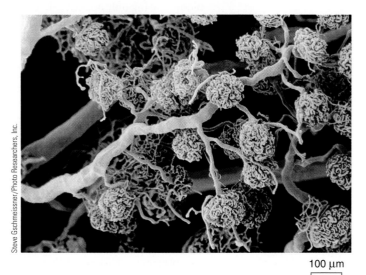

100 μm

(c) Low-power SEM of a portion of the kidney cortex. The tissue was treated to remove some structures and to show glomeruli and associated blood vessels.

Figure 47-9 Nephron structure

ferent means "to carry toward"). An afferent arteriole conducts blood into the capillaries that make up each glomerulus. As blood flows through the glomerulus, some of the plasma is forced into Bowman's capsule.

You may recall that in a typical circulatory route, capillaries deliver blood into veins. Circulation in the kidneys is an exception, because blood flowing from the glomerular capillaries next passes into an **efferent arteriole.** Each efferent arteriole conducts blood *away* from a glomerulus. The efferent arteriole delivers blood to a second capillary network, the **peritubular capillaries** surrounding the renal tubule.

Blood is filtered as it flows through the first set of capillaries, those of the glomerulus. The peritubular capillaries receive materials returned to the blood by the renal tubule. Blood from the peritubular capillaries enters small veins that eventually lead to the **renal vein.** In summary, blood circulates through the kidney in the following sequence:

Renal artery ⟶ afferent arterioles ⟶ capillaries of glomerulus ⟶ efferent arterioles ⟶ peritubular capillaries ⟶ small veins ⟶ renal vein

Urine is produced by filtration, reabsorption, and secretion

Urine, the watery discharge of the urinary system, is produced by a combination of three processes: filtration, reabsorption, and tubular secretion (Fig. 47-10).

Filtration is not selective with regard to ions and small molecules

Blood flows through the glomerular capillaries under high pressure, forcing more than 10% of the plasma out of the capillaries and into Bowman's capsule. **Filtration** is similar to the mechanism whereby interstitial fluid is formed as blood flows through other capillary networks in the body. However, blood flow through glomerular capillaries is at much higher pressure, so more plasma is filtered in the kidney.

Several factors contribute to filtration. First, the hydrostatic blood pressure in the glomerular capillaries is higher than in other

Key Point

The glomerulus filters blood. As the filtrate moves through the renal tubule, the filtrate's composition is adjusted by selective reabsorption and secretion.

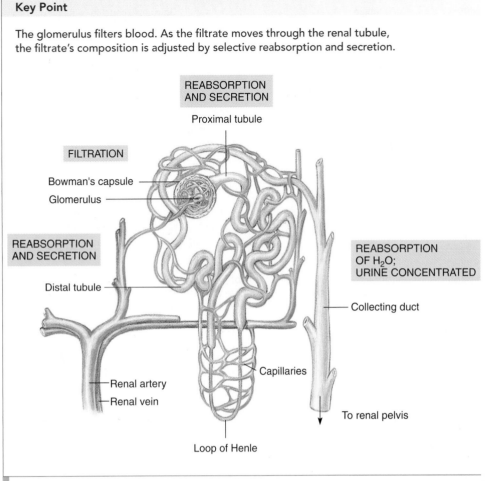

Figure 47-10 *Animated* General regions of filtration, reabsorption, and secretion
The adjusted filtrate is urine.

output. As plasma passes through the glomerulus, it loses more than 10% of its volume to the glomerular filtrate. The normal glomerular filtration rate amounts to about 180 L (about 45 gal) each 24 hours. This is 4.5 times the amount of fluid in the entire body! Common sense tells us that urine could not be excreted at that rate. Within a few moments, dehydration would become life threatening.

Reabsorption is highly selective

The threat to homeostasis posed by the vast amounts of fluid filtered by the kidneys is avoided by **reabsorption.** The renal tubules reabsorb about 99% of the filtrate into the blood, leaving only about 1.5 L to be excreted as urine during a 24-hour period. Reabsorption permits precise regulation of blood chemistry by the kidneys. Wastes, excess salts, and other materials remain in the filtrate and are excreted in the urine, whereas needed substances such as glucose and amino acids are returned to the blood. Each day the tubules reabsorb more than 178 L of water, 1200 g (2.6 lb) of salt, and about 250 g (0.5 lb) of glucose. Most of this, of course, is reabsorbed many times over.

The simple epithelial cells lining the renal tubule are well adapted for reabsorbing materials. Their abundant microvilli increase the surface area for reabsorption. These epithelial cells contain numerous mitochondria that provide the energy for running the cell pumps that actively transport materials.

Most (about 65%) of the filtrate is reabsorbed as it passes through the proximal convoluted tubule. Glucose, amino acids, vitamins, and other substances of nutritional value are entirely reabsorbed there. Many ions, including sodium, chloride, bicarbonate, and potassium, are partially reabsorbed. Some of these ions are actively transported; others are reabsorbed by diffusion. Reabsorption continues as the filtrate passes through the loop of Henle and the distal convoluted tubule. Then the filtrate is further concentrated as it passes through the collecting duct that leads to the renal pelvis.

Normally, substances that are useful to the body, such as glucose or amino acids, are reabsorbed from the renal tubules. If the concentration of a particular substance in the blood is high, however, the tubules may not be able to reabsorb it all. The maximum rate at which a substance can be reabsorbed is its **tubular transport maximum (Tm).** When that rate is reached, the binding sites are saturated on the membrane proteins that transport the substance. For example, the tubular load of glucose is about 125 mg per minute, and almost all of it is normally reabsorbed.

capillaries. This high pressure is mainly due to the high resistance to outflow presented by the efferent arteriole, which is smaller in diameter than the afferent arteriole (see Fig. 47-9b). A second factor contributing to the large amount of **glomerular filtrate** is the large surface area for filtration provided by the highly coiled glomerular capillaries. A third factor is the great permeability of the glomerular capillaries. Numerous small pores between the endothelial cells that form their walls make the glomerular capillaries more porous than typical capillaries.

The wall of Bowman's capsule in contact with the capillaries consists of specialized epithelial cells called **podocytes.** These cells have numerous cytoplasmic extensions called *foot processes* that cover most of the surfaces of the glomerular capillaries (❚ Fig. 47-11). Foot processes of adjacent podocytes are separated by narrow gaps called **filtration slits.** The porous walls of the glomerular capillaries and the filtration slits of the podocytes form a **filtration membrane** that permits fluid and small solutes dissolved in the plasma, such as glucose, amino acids, sodium, potassium, chloride, bicarbonate, other salts, and urea, to pass through and become part of the filtrate. This filtration membrane holds back blood cells, platelets, and most of the plasma proteins.

The total volume of blood passing through the kidneys is about 1200 mL per minute, or about one fourth of the entire cardiac

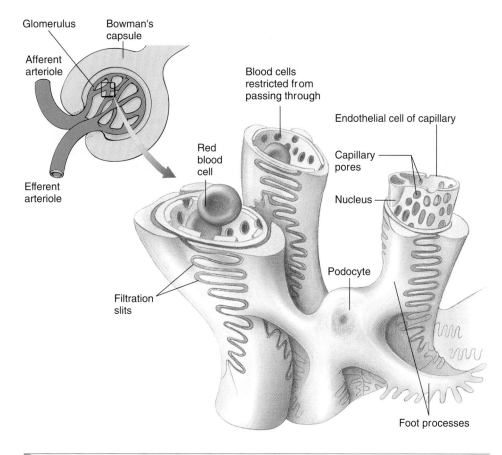

Figure 47-11 Filtration membrane of the kidney

The porous walls of the glomerular capillaries and the filtration slits between podocytes form a filtration membrane that is highly permeable to water and small molecules but restricts the passage of blood cells and large molecules.

However, in a person with uncontrolled diabetes mellitus, the concentration of glucose in the blood exceeds its Tm (glucose concentration in excess of 320 mg per minute). The excess glucose cannot be reabsorbed and is excreted in the urine (glucosuria), a symptom of this disorder (see Chapter 48).

Some substances are actively secreted from the blood into the filtrate

Tubular **secretion** is the passage of substances across the tubule epithelium in a direction opposite that of reabsorption. Secretion occurs mainly in the region of the distal convoluted tubule. Potassium, hydrogen ions, and ammonium ions, as well as some organic ions such as creatinine (a metabolic waste), are secreted into the filtrate. Certain drugs, such as penicillin, are also removed from the blood by secretion.

Secretion of hydrogen ions by the collecting ducts is an important homeostatic mechanism for regulating the pH of the blood. Carbon dioxide, which diffuses from the blood into the cells of the distal tubules and collecting ducts, combines with water, which produces carbonic acid. This acid then dissociates to form hydrogen ions and bicarbonate ions. When the blood becomes too acidic, more hydrogen ions are secreted into the urine:

$$CO_2 + H_2O \rightleftharpoons H_2CO_3 \rightleftharpoons H^+ + HCO_3^-$$

Potassium ion secretion is another important homeostatic mechanism. When K^+ concentration is too high, nerve impulses are not effectively transmitted, and the strength of muscle contraction decreases. The heart rhythm becomes irregular, and cardiac arrest can occur. When K^+ concentration exceeds its homeostatic level, K^+ are secreted from the blood into the renal tubules and then excreted in the urine. Secretion results partly from a direct effect of the K^+ on the tubules. In addition, the adrenal cortex increases its output of the hormone aldosterone, which further stimulates secretion of K^+.

Urine becomes concentrated as it passes through the renal tubule

We can survive with limited fluid intake because the kidneys can produce highly concentrated urine—more than 4 times as concentrated as blood. The osmolarity of human blood is about 300 milliosmols per liter (mOsm/L). The kidneys can produce urine with an osmolarity of about 1200 mOsm/L.

As the filtrate passes through various regions of the renal tubule, salt (NaCl) is reabsorbed into the interstitial fluid, and a salt concentration gradient is established (Figs. 47-12 and 47-13). The gradient is used to produce a concentrated urine.

When filtrate flows from Bowman's capsule to the proximal tubule, its osmolarity is the same as that of blood (about 300 mOsm/L). Water and salt are reabsorbed from the proximal tubule. Sodium ions are actively transported out of the proximal tubule, and chloride follows passively. As salt passes into the interstitial fluid, water follows osmotically.

The walls of the descending limb of the loop of Henle are relatively permeable to water but relatively impermeable to sodium and urea. The interstitial fluid has a high concentration of Na^+, so as the filtrate passes down the loop of Henle, water moves out by osmosis. This process concentrates the filtrate inside the loop of Henle.

The loop of Henle is specialized to highly concentrate sodium chloride in the interstitial fluid of the medulla. It maintains a highly hypertonic interstitial fluid in the medulla near the bottom of the loop, which in turn lets the kidneys produce concentrated urine. At the turn of the loop of Henle, the walls become more permeable to salt and less permeable to water. As the concentrated filtrate begins to move up the ascending limb (the thin region), salt diffuses out into the interstitial fluid. This contributes to the high salt concentration of the interstitial fluid in the medulla surrounding the loop of Henle. Farther along the

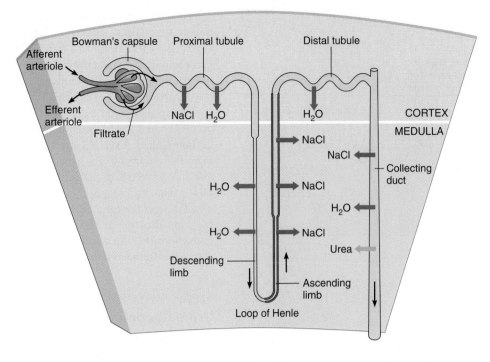

Figure 47-12 *Animated* Movement of water, ions, and urea through the renal tubule and collecting duct

Water passes out of the descending limb of the loop of Henle, leaving a more concentrated filtrate inside. The heavy outline along the ascending limb indicates that this region is relatively impermeable to water. NaCl diffuses out from the lower (*thin*) part of the ascending limb. In the upper (*thick*) part of the ascending limb, NaCl is actively transported into the interstitial fluid. The saltier the interstitial fluid becomes, the more water moves out of the descending limb. This process leaves a concentrated filtrate inside, so more salt passes out. Note that this is a positive feedback system. Some urea also moves out into the interstitial fluid through the collecting ducts. Water from the collecting ducts moves out osmotically into this hypertonic interstitial fluid.

ascending limb (the thick region), sodium is actively transported out of the tubule.

Because water passes out of the descending limb of the loop of Henle, the filtrate at the bottom of the loop has a high salt concentration. However, because salt (but not water) is removed in the ascending limb, by the time the filtrate moves into the distal tubule, its osmolarity may be the same as or even lower than that of blood.

As it passes through the distal tubule, the filtrate may become even more dilute. The distal tubule is relatively impermeable to water but actively transports salt out into the interstitial fluid. The filtrate passes from the renal tubule into a larger collecting duct that eventually empties into the renal pelvis.

Note the *counterflow* of fluid through the two limbs of the loop of Henle. Filtrate passing down through the descending limb is flowing in a direction opposite that of the filtrate moving upward through the ascending limb. The filtrate becomes concentrated as it moves down the descending limb and diluted as it moves up the ascending limb. This countercurrent mechanism helps maintain a high salt concentration in the interstitial fluid of the medulla. The hypertonic interstitial fluid draws water osmotically from the filtrate in the collecting ducts.

The inner medullary collecting ducts are permeable to urea, so some of the concentrated urea in the filtrate can diffuse out

into the interstitial fluid. The urea contributes to the high solute concentration of the inner medulla. This process helps concentrate urine.

The collecting ducts pass through the zone of very salty interstitial fluid. As the filtrate moves through the collecting duct, water passes osmotically into the interstitial fluid, where it is collected by capillaries. Sufficient water can leave the collecting ducts to produce highly concentrated urine. A hypertonic urine conserves water.

Some of the water that diffuses from the filtrate into the interstitial fluid is removed by the **vasa recta,** long, straight capillaries that extend from the efferent arterioles of the juxtamedullary nephrons. The vasa recta extend deep into the medulla, only to negotiate a hairpin curve and return to the cortical venous drainage of the kidney. These capillaries parallel the renal tubules.

Blood flows in opposite directions in the ascending and descending regions of the vasa recta, just as filtrate flows in opposite directions in the ascending and descending limbs of the loop of Henle. As a consequence of this countercurrent flow, much of the salt and urea that enter the blood through the descending region of the vasa recta leave again from the ascending region. As a result, the solute concentration of the blood leaving the vasa recta is only slightly higher than that of the blood entering. This mechanism helps maintain the high solute concentration of the interstitial fluid in the renal medulla.

Urine consists of water, nitrogenous wastes, and salts

By the time the filtrate reaches the renal pelvis, its composition has been precisely adjusted. The adjusted filtrate, called *urine*, consists of approximately 96% water, 2.5% nitrogenous wastes (mainly urea), 1.5% salts, and traces of other substances, such as bile pigments, that may contribute to the characteristic color and odor. Healthy urine is sterile and has been used to wash battlefield wounds when clean water was not available. However, when exposed to bacterial action, urine swiftly decomposes and forms ammonia and other products. Ammonia produces the diaper rash of infants.

The composition of urine yields many clues to body function and malfunction. **Urinalysis,** the physical, chemical, and microscopic examination of urine, is a very important diagnostic tool that has been used to monitor diabetes mellitus and many other disorders. Urinalysis is also extensively used in drug testing, be-

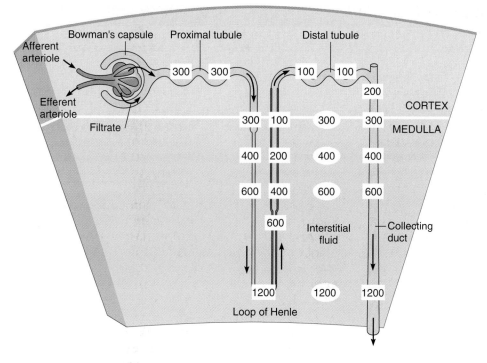

Figure 47-13 Concentration of the filtrate as it moves through the nephron

This figure shows the relative concentration of ions, mainly Na^+ and Cl^-, during formation of a very concentrated urine. Numbers indicate the concentration of salt in the filtrate and interstitial fluid expressed in milliosmols per liter. The more hypertonic a solution is, the higher its osmotic pressure. The very hypertonic interstitial fluid near the renal pelvis draws water osmotically from the filtrate in the collecting ducts. The heavy outline along the ascending loop indicates that this region is relatively impermeable to water.

cause breakdown products of some drugs can be identified in the urine for several days or weeks after the drugs are taken.

Hormones regulate kidney function

The kidneys maintain fluid volume and electrolyte concentration. They help maintain the concentration of Na^+ and K^+ in the blood within narrow limits and also regulate blood pH. Several hormones regulate these homeostatic functions of the kidneys (▮ Table 47-1).

ADH increases water reabsorption

The amount of urine produced depends on the body's need to retain or rid itself of water. When fluid intake is low, the body begins to dehydrate, and the blood volume decreases. As blood volume decreases, the concentration of salts dissolved in the blood becomes greater, causing an increase in osmotic pressure. Certain receptors in the hypothalamus are sensitive to this osmotic change. They signal the posterior lobe of the pituitary gland to release **antidiuretic hormone (ADH)**. This hormone is actually produced in the hypothalamus but is stored in the posterior pituitary and released as needed. A thirst center in the hypothalamus also responds to dehydration, stimulating an increase in fluid intake.

ADH makes the collecting ducts more permeable to water so that more water is reabsorbed. As a result, a small volume of concentrated urine is produced (▮ Fig. 47-14). ADH acts on aquaporin-2, a membrane protein that forms gated water channels in the wall of the collecting ducts (see Chapter 5). These channels allow water to pass rapidly through the plasma membrane.

When you drink a large volume of water, your blood becomes diluted and its osmotic pressure falls. Release of ADH by the pituitary gland decreases, lessening the amount of water reabsorbed from the collecting ducts. The kidneys produce a large volume of dilute urine.

In the disorder *diabetes insipidus* (not to be confused with the more common disorder, diabetes mellitus) the pituitary gland malfunctions and does not release enough ADH. Diabetes insipidus can also develop from an acquired insensitivity of the kidney to ADH. In diabetes insipidus, water is not efficiently reabsorbed from the ducts, so the body produces a large volume of urine. A person with severe diabetes insipidus may excrete up to 25 quarts of urine each day, a serious water loss. The affected individual becomes dehydrated and must drink almost continually to offset fluid loss. Injections of ADH or use of an ADH nasal spray can often control diabetes insipidus.

The renin–angiotensin–aldosterone pathway increases sodium reabsorption

Sodium is the most abundant extracellular ion, accounting for about 90% of all positive ions in the extracellular fluid. Several hormones work together to regulate sodium concentration. **Aldosterone,** which is secreted by the cortex of the adrenal glands, stimulates the distal tubules and collecting ducts to increase sodium reabsorption. When researchers remove the adrenal glands of experimental animals, too much sodium is excreted, leading to serious depletion of the extracellular fluid.

Aldosterone secretion is stimulated both by hormones and by a decrease in blood pressure (caused by a decrease in volume of blood and interstitial fluid). When blood pressure falls, cells of the **juxtaglomerular apparatus** secrete the enzyme **renin,** which activates the **renin–angiotensin–aldosterone pathway.** The juxtaglomerular apparatus is a small group of cells in the region where the renal tubule contacts the afferent and efferent arterioles (see Fig. 47-9b). Renin converts the plasma protein angiotensinogen to angiotensin I. *Angiotensin-converting enzyme (ACE)* converts angiotensin I into its active form, **angiotensin II,** a peptide hormone. Angiotensin II stimulates aldosterone secre-

TABLE 47-1

Hormonal Control of Kidney Function

Hormone	Source	Target Tissue	Actions	Factors That Stimulate Release
Antidiuretic hormone (ADH)	Produced in hypothalamus; released by posterior pituitary gland	Collecting ducts	Increases permeability of collecting ducts to water, increasing reabsorption and decreasing water excretion	Low fluid intake decreases blood volume and increases osmotic pressure of blood; receptors in hypothalamus stimulate posterior pituitary
Aldosterone	Adrenal glands (cortex)	Distal tubules and collecting ducts	Increases sodium reabsorption	Angiotensin II (when blood pressure decreases)
Angiotensin II	Produced from angiotensin I	Blood vessels and adrenal glands	Constricts blood vessels, which raises blood pressure; stimulates aldosterone secretion	Decrease in blood pressure causes renin secretion; renin catalyzes conversion of angiotensinogen to angiotensin I, which is then converted to angiotensin II by ACE
Atrial natriuretic peptide (ANP)	Atrium of heart	Afferent arterioles; collecting ducts	Dilates afferent arterioles; inhibits sodium reabsorption by collecting ducts; inhibits aldosterone secretion; decreases blood pressure	Stretching of atria due to increased blood volume

tion. ACE is produced by the endothelial cells in the walls of pulmonary capillaries.

Angiotensin II increases the synthesis and release of aldosterone. In addition, angiotensin II raises blood pressure directly by constricting blood vessels, stimulates the posterior pituitary to release ADH, and stimulates thirst. All of these actions help increase extracellular fluid volume and raise blood pressure. In individuals with hypertension, *ACE inhibitors* are sometimes used to block the production of angiotensin II. We can summarize the renin−angiotensin−aldosterone pathway as follows:

> Blood volume decreases ⟶ blood pressure decreases ⟶ cells of juxtaglomerular apparatus secrete renin ⟶ renin catalyzes conversion of angiotensinogen to angiotensin I ⟶ ACE catalyzes conversion of angiotensin I to angiotensin II ⟶ angiotensin II constricts blood vessels and stimulates aldosterone secretion ⟶ aldosterone increases sodium reabsorption ⟶ blood pressure increases

Atrial natriuretic peptide inhibits sodium reabsorption

Atrial natriuretic peptide (ANP), a hormone produced by the heart, increases sodium excretion and decreases blood pressure. ANP is stored in granules in atrial muscle cells. When Na^+ concentration increases, fluid is retained, and blood volume increases. Atrial muscle cells are stretched and respond by releasing ANP into the circulation. ANP dilates afferent arterioles, thereby increasing the glomerular filtration rate. ANP inhibits sodium reabsorption by the collecting ducts directly and also indirectly by inhibiting aldosterone secretion. ANP also reduces plasma aldosterone concentration by inhibiting renin release. These actions of ANP increase sodium excretion and urine output, which lowers blood volume and blood pressure. Note that the renin−angiotensin−aldosterone pathway and ANP work antagonisti-

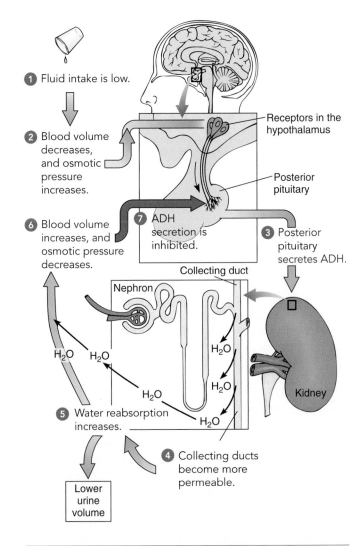

Figure 47-14 Regulation of urine volume by antidiuretic hormone (ADH)

When the body is dehydrated, the hormone ADH increases the permeability of the collecting ducts to water. More water is reabsorbed, and only a small volume of concentrated urine is produced.

cally in regulating fluid balance, electrolyte balance, and blood pressure.

> Blood volume increases ⟶ blood pressure increases ⟶ atria of heart stretched ⟶ atria release ANP ⟶ directly inhibits sodium reabsorption and inhibits aldosterone secretion (which also inhibits sodium reabsorption) ⟶ larger urine volume ⟶ blood volume decreases ⟶ blood pressure decreases

Review

■ Which structure(s) in the mammalian body is/are associated with each of the following processes: (1) urea formation; (2) urine formation; (3) temporary storage of urine; (4) conduction of urine out of the body?

■ Which part(s) of the nephron is/are associated with the following processes: (1) filtration; (2) reabsorption; (3) secretion?

■ Through what sequence of structures does a drop of filtrate pass as it moves from Bowman's capsule to the urinary bladder?

■ Through what sequence of blood vessels does a drop of blood pass as it is conducted to and from a nephron?

■ How does the kidney regulate the composition of the blood?

■ What are the actions of the renin–angiotensin–aldosterone pathway?

SUMMARY WITH KEY TERMS

Learning Objectives

1 Describe how the processes of *osmoregulation* and *excretion* contribute to fluid and electrolyte homeostasis (page 1012).

■ **Osmoregulation** is the active regulation of osmotic pressure of body fluids so that fluid and electrolyte homeostasis is maintained. **Excretion** is the process of ridding the body of metabolic wastes.

2 Contrast the benefits and costs of excreting ammonia, uric acid, or urea (page 1012).

■ The principal waste products of animal metabolism are water; carbon dioxide; and **nitrogenous wastes,** including ammonia, urea, and uric acid. **Ammonia** is toxic and is excreted mainly by aquatic animals.

■ **Urea** and **uric acid** are far less toxic than ammonia, but their synthesis requires energy. Urea excretion requires water. Uric acid can be excreted as a semisolid paste, a water-conserving adaptation.

3 Compare osmoconformers and osmoregulators (page 1013).

■ Most marine invertebrates are **osmoconformers**—the salt concentration of their body fluids varies with changes in the sea water. Some marine invertebrates, especially those inhabiting coastal habitats, are **osmoregulators** that maintain an optimal salt concentration despite changes in salinity of their surroundings.

4 Describe protonephridia, metanephridia, and Malpighian tubules, and compare their functions (page 1013).

■ **Nephridial organs** help maintain homeostasis by regulating the concentration of body fluids through osmoregulation and excretion of metabolic wastes. **Protonephridia,** nephridial organs found in flatworms and nemerteans, are tubules with no internal openings. Interstitial fluid enters their blind ends, which consist of **flame cells,** cells with brushes of cilia. Cilia propel fluid through the tubules; excess fluid exits through nephridiopores.

■ Most annelids and mollusks have nephridial organs called **metanephridia,** which are tubules open at both ends. As fluid from the coelom moves through the tubule, needed materials are reabsorbed by capillaries. Urine, containing wastes, exits the body through nephridiopores.

■ **Malpighian tubules,** extensions of the insect gut wall, have blind ends that lie in the hemocoel. Cells of the tubule actively transport uric acid and some other substances from the hemolymph into the tubule, and water follows by diffusion. The contents of the tubule pass into the gut. Water and some solutes are reabsorbed in the rectum. Malpighian tubules effectively conserve water and

have contributed to the success of insects as terrestrial animals.

5 Relate the function of the vertebrate kidney to the success of vertebrates in a wide variety of habitats (page 1015).

■ The vertebrate **kidney** excretes nitrogenous wastes and helps maintain fluid balance by adjusting the salt and water content of the urine. Freshwater, marine, and terrestrial habitats present different problems for maintaining internal fluid balance and for the excretion of nitrogenous wastes. The structure and function of the vertebrate kidney have adapted to meet the various osmotic challenges presented by these different habitats.

6 Compare adaptations for osmoregulation in freshwater fishes, marine bony fishes, sharks, marine mammals, and terrestrial vertebrates (page 1015).

■ Freshwater fishes take in water osmotically; they excrete a large volume of hypotonic urine.

■ Marine bony fishes lose water osmotically. They compensate by drinking sea water and excreting salt through their gills; they produce only a small volume of isotonic urine. Sharks and other marine cartilaginous fishes retain large amounts of urea, allowing them to take in water osmotically through the gills. They excrete a large volume of hypotonic urine. Marine mammals ingest sea water with their food. They produce a concentrated urine.

■ Terrestrial vertebrates must conserve water. Endotherms have a high metabolic rate and produce a large volume of nitrogenous wastes. They have numerous adaptations for conserving water, including efficient kidneys.

7 Describe (or label on a diagram) the organs of the mammalian urinary system, and give the functions of each (page 1017).

■ The **urinary system** is the principal excretory system in humans and other mammals. In mammals, the kidneys produce urine, which passes through the **ureters** to the **urinary bladder** for storage. During urination, the urine is released from the body through the **urethra.**

■ The outer portion of each kidney is the **renal cortex;** the inner portion is the **renal medulla.** The renal medulla contains 8 to 10 **renal pyramids.** The tip of each pyramid is a **renal papilla.** As urine is produced, it flows into **collecting ducts,** which empty through a renal papilla into a funnel-shaped chamber, the **renal pelvis.** Each kidney has more than 1 million functional units called **nephrons.**

8 Describe (or label on a diagram) the structures of a nephron (including associated blood vessels), and give the functions of each structure (page 1017).

■ Each nephron consists of a cluster of capillaries, called a **glomerulus,** surrounded by a **Bowman's capsule** that opens into a long, coiled **renal tubule.** The renal tubule consists of a **proximal convoluted tubule, loop of Henle,** and **distal convoluted tubule.**

■ **Cortical nephrons,** located almost entirely within the cortex or outer medulla, have small glomeruli. **Juxtamedullary nephrons** have large glomeruli and long loops of Henle that extend deep into the medulla. These nephrons are important in concentrating urine.

■ Blood flows from small branches of the **renal artery** to **afferent arterioles** and then to glomerular capillaries. Blood then flows into an **efferent arteriole** that delivers blood into a second set of capillaries, the **peritubular capillaries** that surround the renal tubule. Blood leaves the kidney through the **renal vein.**

9 Trace a drop of filtrate from Bowman's capsule to its release from the body as urine (page 1017).

■ Urine is produced by **filtration** of plasma, **reabsorption** of needed materials, and **secretion** of a few substances, such as potassium and hydrogen ions, into the renal tubule.

■ Plasma filters through the glomerular capillaries and into Bowman's capsule. The permeable walls of the capillaries and **filtration slits** between **podocytes,** specialized epithelial cells that make up the inner wall of Bowman's capsule, serve as a **filtration membrane.** Filtration is nonselective with regard to small molecules; glucose and other needed materials, as well as metabolic wastes, become part of the filtrate.

■ About 99% of the filtrate is reabsorbed from the renal tubules into the blood. Reabsorption is a highly selective process that returns usable materials to the blood but leaves wastes and excesses of other substances to be excreted in the urine. The maximum rate at which a substance can be reabsorbed is its **tubular transport maximum (Tm).**

■ In secretion, hydrogen ions, certain other ions, and some drugs are actively transported into the renal tubule to become part of the urine.

■ Production of concentrated urine depends on a high salt and urea concentration in the interstitial fluid of the kidney medulla. The interstitial fluid in the medulla has a concentration gradient in which the salt is most concentrated around the bottom of the loop of Henle. This gradient is maintained, in part, by salt reabsorption from various parts of the renal tubule. A counterflow of fluid through the two limbs of the loop of Henle concentrates filtrate as it moves down the descending loop and dilutes it as it moves up the ascending loop.

■ Water is drawn by osmosis from the filtrate as it passes through the collecting ducts. This process concentrates urine in the collecting ducts.

■ Some of the water that diffuses from the filtrate into the interstitial fluid is removed by the **vasa recta,** a system of capillaries that extend from the efferent arterioles.

■ **Urine** is a watery solution of nitrogenous wastes, excess salts, and other substances not needed by the body.

10 Describe the hormonal regulation of fluid and electrolyte balance by antidiuretic hormone (ADH), the renin–angiotensin–aldosterone system, and atrial natriuretic peptide (ANP) (page 1017).

■ When the body needs to conserve water, the posterior pituitary gland increases its release of **antidiuretic hormone (ADH).** The pituitary gland responds to an increase in osmotic concentration of the blood (caused by dehydration). ADH increases the permeability of the collecting ducts to water. As a result, more water is reabsorbed and only a small volume of urine is produced.

■ The renin–angiotensin–aldosterone pathway and atrial natriuretic peptide work antagonistically. When blood pressure decreases, cells of the **juxtaglomerular apparatus** secrete the enzyme **renin,** which activates a pathway leading to production of **angiotensin II.** This hormone stimulates aldosterone release. **Aldosterone** increases sodium reabsorption, and angiotensin II constricts arterioles; both actions raise blood pressure.

■ When blood pressure increases, **atrial natriuretic peptide (ANP)** increases sodium excretion and inhibits aldosterone secretion. These actions increase urine output and lower blood pressure.

TEST YOUR UNDERSTANDING

1. The process that maintains homeostasis of body fluids by keeping them from becoming too dilute or too concentrated is (a) excretion (b) elimination (c) osmoregulation (d) glomerular filtration (e) tubular secretion

2. The main nitrogenous waste product of insects and birds is (a) urea (b) uric acid (c) ammonia (d) carbon dioxide (e) nitrate

3. The main nitrogenous waste product of amphibians and mammals is (a) urea (b) uric acid (c) ammonia (d) carbon dioxide (e) nitrate

4. Osmoconformers (a) maintain an optimal salt concentration despite fluctuations in salt concentration of their surroundings (b) include many animals that inhabit coastal habitats (c) experience variation in concentration of their body fluids as changes occur in salinity of the sea water (d) typically have cells in their gills that remove salts from the surrounding water (e) a and b

5. Which of the following is *not* a correct pair? (a) protonephridia—flatworm (b) metanephridia—annelid (c) flame cell—flatworm (d) Malpighian tubule—mollusk (e) kidney—vertebrate

6. To compensate for fluid loss, many marine bony fishes (a) accumulate urea (b) have glands that excrete glucose (c) eat a low-protein diet (d) excrete a large volume of hypertonic urine (e) drink sea water

7. Arrange the following structures into an accurate sequence through which urine passes.

   1. urethra 2. urinary bladder 3. kidney 4. ureter

   (a) 4, 3, 2, 1 (b) 3, 4, 2, 1 (c) 1, 2, 3, 4 (d) 4, 2, 1, 3 (e) 3, 1, 2, 4

8. Arrange the following structures into an accurate sequence through which filtrate passes.

   1. proximal convoluted tubule 2. loop of Henle 3. collecting duct 4. distal convoluted tubule 5. Bowman's capsule

   (a) 5, 4, 3, 2, 1 (b) 3, 4, 2, 5, 1 (c) 1, 5, 2, 3, 4 (d) 5, 4, 2, 3, 1 (e) 5, 1, 2, 4, 3

9. The afferent arteriole delivers blood to the (a) renal artery (b) efferent arteriole (c) renal vein (d) capillaries of the glomerulus (e) peritubular capillaries

10. Which of the following does *not* contribute to the process of filtration? (a) high hydrostatic blood pressure in glomerular capillaries (b) large surface area for filtration (c) permeability of glomerular capillaries (d) active transport by epithelial cells lining renal tubules (e) podocytes

11. Tubular transport maximum is (a) the maximum concentration of a substance in the plasma that can be reabsorbed by the kidney (b) the most rapid rate at which urine can be transported through the ureter (c) the maximum rate at which a substance can be reabsorbed from the filtrate in the renal tubules (d) the maximum rate at which a substance can pass through the loop of Henle (e) the maximum rate at which a substance can be secreted into the filtrate

12. Which of the following does *not* contribute to the high salt concentration in the interstitial fluid in the medulla of the kidney? (a) reabsorption of salt from various regions of Bowman's capsule (b) diffusion of salt from the ascending limb of the loop of Henle (c) active transport of sodium from the upper part of the ascending limb (d) counterflow of fluid through the two limbs of the loop of Henle (e) diffusion of urea out of the collecting duct

13. Which of the following is *not* normally present in urine? (a) urea (b) glucose (c) salts (d) water (e) traces of bile pigments

14. Which is *not* true of ADH? (a) released by posterior lobe of the pituitary gland (b) increases water reabsorption (c) secretion increases when osmotic pressure in body increases (d) increases urine volume (e) secretion decreases when you drink a lot of water

15. Aldosterone (a) is released by the posterior pituitary gland (b) decreases sodium reabsorption (c) secretion is stimulated by an increase in blood pressure (d) is an enzyme that converts angiotensin into angiotensin II (e) secretion increases in response to angiotensin II

16. Label the diagram. Use Figure 47-9a to check your answers.

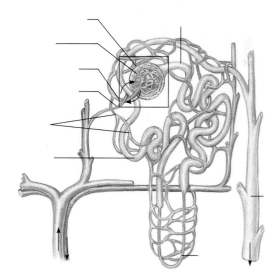

CRITICAL THINKING

1. The number of protonephridia in a planarian is related to the salinity of its environment. Planaria inhabiting slightly salty water develop fewer protonephridia, but the number quickly increases when the concentration of salt in the environment is lowered. Explain why.

2. What types of osmoregulatory challenges do humans experience? Explain. What mechanisms do we have to meet these challenges?

3. Why is glucose normally not present in urine? Why is it present in the urine of individuals with diabetes mellitus? Why do you suppose that people with diabetes mellitus experience an increased output of urine?

4. **Evolution Link.** The kangaroo rat's diet consists of dry seeds, and it drinks no water. What types of osmoregulatory adaptations might help this animal survive?

5. **Evolution Link.** Although the kidney is basically similar in all vertebrates, its structure and function varies somewhat among vertebrate taxa. What are some of the adaptations that have occurred during the evolution of the vertebrate kidney?

Additional questions are available in ThomsonNOW at www.thomsonedu.com/login

48

Endocrine Regulation

Shark Song/M. Kazmers/Dembinsky Photo Associates

Hormones regulate color changes in crustaceans. This juvenile Puget Sound king crab (*Lopholithodes mandtii*) changes color to blend with its background. Changes in the distribution of pigment in its cells result in color changes.

KEY CONCEPTS

Endocrine glands and tissues secrete hormones, chemical messengers that bind to specific receptors on (or in) target cells and regulate physiological processes.

The endocrine and nervous systems work closely together to regulate life processes and maintain homeostasis.

Most endocrine processes are regulated by negative feedback systems, often involving concentrations of specific ions or chemical compounds.

Small, lipid-soluble hormones enter target cells and activate genes; hydrophilic hormones bind to cell-surface receptors and initiate signal transduction, leading to metabolic changes in the cell.

Many invertebrate hormones are neurohormones secreted by neurons.

In vertebrates, the hypothalamus and pituitary gland work together to regulate growth, metabolism, reproduction, response to stress, and many other processes.

A caterpillar becomes a butterfly. A crustacean changes color to blend with its background (see photograph). A young girl develops into a woman. An adult copes with chronic stress. The **endocrine system** regulates growth and development, response to stress, reproduction, fluid balance, and many other physiological processes. The tissues and organs of the endocrine system secrete *hormones*, chemical messengers that signal other cells.

Endocrinology, the study of endocrine activity, is an active, exciting field of biomedical research. This branch of biology has its roots in experiments performed by German physiologist A. A. Berthold in the 1840s. Berthold removed the testes from young roosters and observed that the roosters' combs (a male secondary sex characteristic) did not grow as large as those in normal roosters. He then transplanted testes into some of the birds and observed that the combs grew to normal size. Berthold's methods became a model for subsequent studies in endocrinology and are still used by researchers today.

Investigators are actively studying mechanisms of hormone action. These studies include characterizing receptors and identifying the molecules involved in *signal transduction*. Some endocrinologists now use a reverse strategy for discovering new hormones and signaling pathways within the cell. They focus on "orphan" nuclear receptors, those for which ligands (the molecules that bind with them) are not yet known. Some of these "orphan" receptors are receptors for hormones that have not yet been identified. Using this strategy, researchers have identified intracellular signaling pathways for steroids, fatty acids, and several other compounds.

Interestingly, many signaling pathways share the same molecules even though they respond to different stimuli and activate different responses. Much remains to be learned about how similar pathways lead to many different, and very specific, cell actions.

In this chapter we discuss how the endocrine system regulates life processes. We focus on how hormones maintain homeostasis, and we examine how overproduction or deficiency of various hormones interferes with normal functioning. ■

AN OVERVIEW OF ENDOCRINE REGULATION

Learning Objectives

1 Compare endocrine with nervous system function, and describe how these systems work together to regulate body processes.
2 Summarize the regulation of endocrine action by negative feedback systems.
3 Identify four main chemical groups to which hormones are assigned, and give two examples for each group.

The endocrine system is a diverse collection of cells, tissues, and organs, including specialized **endocrine glands,** that produce and secrete **hormones,** chemical messengers that regulate many physiological processes. The term *hormone* is derived from a Greek word meaning "to excite." Hormones excite, or stimulate, changes in specific tissues. Endocrine glands differ from **exocrine glands** (such as sweat glands and gastric glands) that release their secretions into ducts. Endocrine glands have no ducts and secrete their hormones into the surrounding interstitial fluid or blood.

Endocrinologists have identified about 10 discrete endocrine glands. They have also discovered specialized cells in the digestive tract, heart, kidneys, and many other parts of the body that release hormones or hormone-like substances. As a result of these discoveries, the scope of endocrinology now includes the production and actions of chemical messengers produced by a wide variety of organs, tissues, and cells.[1]

Hormones are typically transported by the blood and produce a characteristic response only after they reach **target cells** and bind with specific receptors. Target cells, the cells influenced by a particular hormone, may be in another endocrine gland or in an entirely different type of organ, such as a bone or kidney. Target cells may be located far from the endocrine gland. For example, the vertebrate thyroid gland secretes hormones that stimulate metabolism in tissues throughout the body.

Several types of hormones may be involved in regulating the metabolic activities of a particular type of cell. In fact, many hormones produce a synergistic effect in which the presence of one hormone enhances the effects of another.

The endocrine system and nervous system interact to regulate the body

The endocrine system works closely with the nervous system to maintain homeostasis, the steady state of the body. Recall from

[1]Pheromones are chemical messengers that animals release to communicate with other animals of the same species. Because pheromones are generally produced by exocrine glands and do not regulate metabolic activities within the animal that produces them, most biologists do not classify them as hormones. Their role in regulating behavior is discussed in Chapter 51.

Chapters 40 and 41 that the nervous system responds rapidly to stimuli by transmitting electrical and chemical signals. Neurons signal muscle and gland cells, including endocrine cells. The endocrine system signals a much wider range of target cell types than the nervous system does.

In general, the endocrine system responds more slowly, but its responses are longer lasting. As we will learn, the nervous system helps regulate many endocrine responses. For example, when the body is threatened, the hypothalamus signals the adrenal glands to secrete the hormone epinephrine. The nervous system even produces some hormones (neurohormones). Nervous function and endocrine function further blur when we consider that the same signal molecule can function as either a neurotransmitter or a hormone, depending on its source. You may recall that norepinephrine is secreted at synapses by neurons in the brain and sympathetic system. Norepinephrine is also secreted into the blood as a hormone by the adrenal medulla, an endocrine organ.

Negative feedback systems regulate endocrine activity

Although present in minute amounts, more than 50 different hormones may be circulating in the blood of a vertebrate at any given time. Hormone molecules continuously move out of the circulation and bind with target cells. They are removed from the blood by the liver, which inactivates some hormones, and by the kidneys, which excrete them.

Hormone production is regulated by the nervous system and by other endocrine glands. Most endocrine action is regulated by **negative feedback systems** (discussed in Chapter 38). The parathyroid glands, located in the neck of tetrapod vertebrates, provide a good example of how negative feedback works in endocrine regulation (▌ Fig. 48-1). The parathyroid glands regulate the calcium concentration of the blood. When the calcium concentration is not within homeostatic limits, nerves and muscles cannot function properly. For instance, when too few calcium ions are present, neurons fire spontaneously, which causes muscle spasms. When calcium concentration varies too far from the steady state (either too high or too low), negative feedback systems restore homeostasis.

A decrease in the calcium concentration in the plasma signals the parathyroid glands to release more parathyroid hormone. This hormone increases the concentration of calcium in the blood. (The details of parathyroid action are discussed later in the chapter.) When the calcium concentration rises above normal limits, the parathyroid glands slow their output of hormone. Both responses are negative feedback systems. An increase in calcium concentration results in decreased release of parathyroid

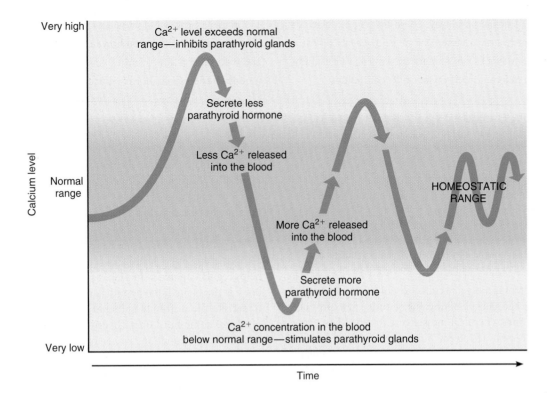

Very high

Ca²⁺ level exceeds normal
range—inhibits parathyroid glands

Secrete less
parathyroid hormone

Less Ca²⁺ released
into the blood

Calcium level

Normal
range

HOMEOSTATIC
RANGE

More Ca²⁺ released
into the blood

Secrete more
parathyroid hormone

Ca²⁺ concentration in the blood
below normal range—stimulates parathyroid glands

Very low

Time

Figure 48-1 Regulation by negative feedback

Whether the calcium concentration in the blood is too high or too low, negative feedback restores homeostasis. Parathyroid hormone increases calcium concentration by stimulating calcium release from bone and calcium reabsorption from the kidney tubules.

hormone, whereas a decrease in calcium leads to increased hormone secretion. In each case, the response counteracts the inappropriate change, restoring the steady state.

Hormones are assigned to four chemical groups

Although hormones are chemically diverse, they generally belong to one of four chemical groups: (1) fatty acid derivatives, (2) steroids, (3) amino acid derivatives, or (4) peptides or proteins. *Prostaglandins* and the juvenile hormones of insects are **fatty acid derivatives** (Fig. 48-2a). Prostaglandins are synthesized from arachidonic acid, a 20-carbon fatty acid. A prostaglandin has a five-carbon ring in its structure.

The molting hormone of insects is a **steroid hormone** (Fig. 48-2b). In vertebrates, the adrenal cortex, testis, ovary, and placenta secrete steroid hormones synthesized from cholesterol. Examples of steroid hormones are cortisol, secreted by the adrenal cortex; the male hormone testosterone secreted by the testis; and the female hormones progesterone and estrogens produced by the ovary. Athletes (and others) sometimes abuse synthetic hormones known as anabolic steroids (see *Focus On: Anabolic Steroids and Other Abused Hormones*).

Chemically, the simplest hormones are **amino acid derivatives** (Fig. 48-2c). The thyroid hormones (T_3 and T_4) are synthesized from the amino acid tyrosine and iodide. Epinephrine (also known as adrenaline) and norepinephrine (also known as noradrenaline), produced by the medulla of the adrenal gland, are also derived from tyrosine. Melatonin is synthesized from the amino acid tryptophan.

The water-soluble **peptide hormones** are the largest hormone group. (Many endocrinologists include protein hormones in this group.) **Neuropeptides** are a large group of signaling molecules produced by neurons. Oxytocin and antidiuretic hormone (ADH), produced in the hypothalamus, are short neuropeptides, each composed of nine amino acids (Fig. 48-2d). Seven of the amino acids are identical in the two hormones, but the actions of these hormones are quite different. The hormones glucagon (see Fig. 3-19), secretin, adrenocorticotropic hormone (ACTH), and calcitonin are somewhat longer peptides that consist of about 30 amino acids.

Insulin is a small protein that consists of two peptide chains joined by disulfide bonds. Growth hormone, thyroid-stimulating hormone, and the gonadotropic hormones, all secreted by the anterior lobe of the pituitary gland, are large proteins with molecular weights of 25,000 or more.

Review

■ What are target cells?

■ How do the nervous and endocrine systems work together?

■ How does negative feedback work?

■ How does the body respond when the concentration of calcium in the blood is below the homeostatic level?

■ What are four major chemical groups of hormones? Give an example of each.

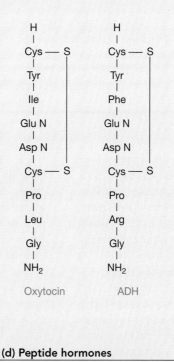

(a) Hormones derived from fatty acids

(b) Steroid hormones

Molting hormone (ecdysone)

Cortisol

Estradiol (principal estrogen)

(c) Amino acid derivatives

Norepinephrine

Epinephrine

Thyroid hormones

Thyroxine (T_4)

Triiodothyronine (T_3)

(d) Peptide hormones

Oxytocin

ADH

Figure 48-2 Major chemical groups of hormones

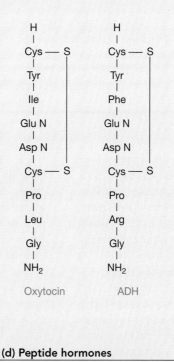

TYPES OF ENDOCRINE SIGNALING

Learning Objective

4 Compare four types of endocrine signaling.

Hormones signal their target cells by classical endocrine signaling, neuroendocrine signaling, autocrine signaling, and paracrine signaling. In **classical endocrine signaling,** hormones are secreted by endocrine glands and are transported by the blood to target cells (▌Fig. 48-3a). Steroid hormones and thyroid hormones are transported bound to plasma proteins. The water-soluble peptide hormones are dissolved in the plasma. Hormones move out of the blood with plasma and are present in interstitial fluid.

Neurohormones are transported in the blood

Certain neurons, known as **neuroendocrine cells,** are an important link between the nervous and endocrine systems. In **neuroendocrine signaling,** neuroendocrine cells produce **neurohormones** that are transported down axons and released into the interstitial fluid. They typically diffuse into capillaries and are transported by the blood (▌Fig. 48-3b). Invertebrate endocrine systems are largely neuroendocrine. In vertebrates, the hypothalamus produces several neurohormones that link the nervous system with the pituitary gland, an endocrine gland that secretes various hormones.

Some local regulators are considered hormones

A **local regulator** is a signaling molecule that diffuses through the interstitial fluid and acts on nearby cells. Some endocrinologists include at least certain local regulators as hormones. Certain chemical compounds that act as classical hormones (transported by the blood) function as local regulators under some conditions.

In **autocrine signaling,** a hormone or other regulator acts on the very cells that produce it (▌Fig. 48-3c). For example, the female hormone estrogen, which functions as a classical hormone, can also exert an autocrine effect that stimulates additional estrogen

ANABOLIC STEROIDS AND OTHER ABUSED HORMONES

An estimated 1 million people in the United States, half of them adolescents, abuse anabolic steroids, a group of synthetic hormones. Anabolic steroids are synthetic androgens (male reproductive hormones) that were developed in the 1930s to prevent muscle atrophy in patients with diseases that prevented them from moving. In the 1950s, anabolic steroids became popular with professional athletes, who used them to increase muscle mass, physical strength, endurance, and aggressiveness. In truth, their athletic performance was probably enhanced, at least in part, by drug-induced euphoria and increased enthusiasm for training.

As with other hormones, the concentration of steroids circulating in the body is precisely regulated, so use of anabolic steroids interferes with normal physiological processes. These drugs remain in the body for a long time. Their metabolites (breakdown products) can be detected in the urine for up to 6 months.

Even during short-term use and at relatively low doses, anabolic steroids have a significant effect on mood and behavior. At higher doses, users experience disturbed thought processes, forgetfulness, and confusion and often find themselves easily distracted. The term "steroid rage" refers to the mood swings, unpredictable anger,

increased aggressiveness, and irrational behavior exhibited by many users.

Anabolic steroids elevate blood pressure, cause liver damage and tumors, and increase low-density lipoprotein (LDL; "bad" cholesterol) concentration, raising the risk of cardiovascular disease. In adolescents, these steroids cause severe acne and stunt growth by prematurely closing the growth plates in bones. Abuse of these hormones reduces sexual function and in males, can shrink the testes, leading to sterility. In females, they cause infertility as well as masculine features such as facial hair and decreased breast size.

When their serious side effects became known in the 1960s, anabolic steroid use became controversial, and in 1973 the Olympic Committee banned their use. They are now prohibited worldwide by amateur and professional sports organizations. However, according to the U.S. Drug Enforcement Administration, a multimillion-dollar black market exists for these synthetic hormones. They are both injected and taken in pill form. Steroid abusers who share needles or use nonsterile techniques when they inject steroids are at risk for contracting hepatitis, HIV, and other serious infections.

The typical anabolic steroid user is a male (95%) athlete (65%), most often a

football player, weight lifter, or wrestler. Surprisingly, though, about 10% of male high school students have used anabolic steroids, and about one third of these students are not even on a high school team. These adolescents use the hormone only to change their physical appearance—to pump up their muscles ("bulk up")—and increase endurance. Many anabolic steroid users have difficulty realistically perceiving their body images. They remain unhappy even after dramatic increases in muscle mass.

Some people also abuse other hormones, including human growth hormone (GH) and erythropoietin. Human growth hormone, like anabolic steroids, helps build muscle mass but also causes acromegaly, a condition in which certain bones thicken abnormally. Growth hormone can also cause diabetes and increase heart size, which can lead to heart failure. Erythropoietin, a hormone produced by the kidney, increases the concentration of red blood cells. Although increased oxygen transport enhances the performance of an endurance athlete by as much as 10%, abnormally high concentrations of red blood cells can cause serious cardiovascular problems. Erythropoietin abuse has caused the deaths of several athletes.

---

secretion. At times, estrogen acts on nearby cells, a type of local regulation known as **paracrine signaling** (❚ Fig. 48-3d).

Local regulators, which use autocrine or paracrine signaling, include local chemical mediators such as growth factors, histamine, nitric oxide, and prostaglandins. More than 50 **growth factors,** typically peptides, stimulate cell division and normal development in specific types of cells. Recall from Chapter 44 that **histamine** is stored in mast cells and is released in response to allergic reactions, injury, or infection. Histamine causes blood vessels to dilate and capillaries to become more permeable. Histamine also stimulates the stomach to release acid. **Nitric oxide (NO),** another local regulator, is a gas produced by many types of cells, including those lining blood vessels. It relaxes nearby smooth muscle fibers; as a result, the blood vessel dilates. (See discussion of nitric oxide in Chapter 6.)

Prostaglandins are modified fatty acids released continuously by the cells of most tissues. These local hormones use paracrine signaling. Although present in very small quantities, prostaglandins affect a wide range of physiological processes. They modify cyclic AMP levels and interact with other hormones to regulate various metabolic activities.

The major prostaglandin target is smooth muscle. Some prostaglandins stimulate smooth muscle to contract, whereas others cause relaxation. Thus, some reduce blood pressure, whereas others raise it. Prostaglandins synthesized in the temperature-regulating center of the hypothalamus cause fever. In fact, nonsteroidal anti-inflammatory drugs (NSAIDs), such as aspirin and ibuprofen, reduce fever and decrease inflammation and pain by inhibiting prostaglandin synthesis.

Because prostaglandins are involved in the regulation of so many metabolic processes, they have great potential for a variety of clinical uses. Physicians use them to induce labor in pregnant women, to induce abortion, and to promote the healing of ulcers in the stomach and duodenum. Prostaglandins may someday be used to treat a variety of illnesses, including asthma, arthritis, kidney disease, certain cardiovascular disorders, and some forms of cancer.

Review

❚ What is neuroendocrine signaling?

❚ What are local regulators?

❚ How are autocrine and paracrine signaling different?

❚ What are two functions of prostaglandins?

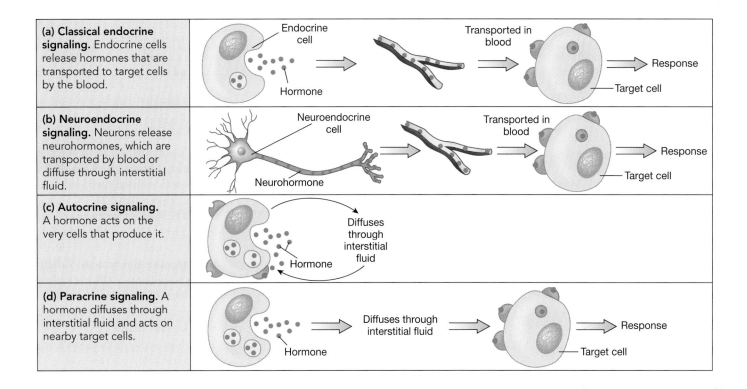

(a) Classical endocrine signaling. Endocrine cells release hormones that are transported to target cells by the blood.	Endocrine cell · Hormone → Transported in blood → Response · Target cell
(b) Neuroendocrine signaling. Neurons release neurohormones, which are transported by blood or diffuse through interstitial fluid.	Neuroendocrine cell · Neurohormone → Transported in blood → Response · Target cell
(c) Autocrine signaling. A hormone acts on the very cells that produce it.	Hormone → Diffuses through interstitial fluid
(d) Paracrine signaling. A hormone diffuses through interstitial fluid and acts on nearby target cells.	Hormone → Diffuses through interstitial fluid → Response · Target cell

Figure 48-3 Some types of endocrine signaling

MECHANISMS OF HORMONE ACTION

Learning Objective

5 Compare the mechanism of action of small, lipid-soluble hormones with that of hydrophilic hormones; include the role of second messengers, such as cyclic AMP.

A hormone present in the interstitial fluid remains "unnoticed" until it reaches its target cells, which have receptors that recognize and bind it. Hormone receptors are large proteins or glycoproteins. The receptor site is somewhat similar to a lock, and the hormones are like different keys. Only the hormone that fits the specific receptor can influence the metabolic machinery of the cell. Receptors are responsible for the specificity of the endocrine system.

Receptors are continuously synthesized and degraded. Their numbers are increased or decreased by **receptor up-regulation** and **receptor down-regulation.** For example, when the concentration of a hormone is too high for an extended period, cells decrease the number of receptors for that hormone. Down-regulation decreases the sensitivity of target cells to the hormone, so it dampens the hormone's ability to carry out its function. Up-regulation occurs when the concentration of a particular hormone is too low. Increasing the number of receptors on the plasma membrane increases the hormone's effect on the cell.

Some hormones enter target cells and activate genes

Steroid hormones and thyroid hormones are relatively small, lipid-soluble (hydrophobic) molecules that easily pass through the plasma membrane of the target cell. Specific protein receptors in the cytoplasm or in the nucleus bind with the hormone to form a hormone–receptor complex (❙ Fig. 48-4). For example, thyroid hormones bind with receptors in the nucleus that are transcription factors. When activated by the hormone, they activate or repress transcription of messenger RNA coding for specific proteins. Proteins that are synthesized produce changes in structure or metabolic activity.

Gene activation and protein synthesis may take several hours. Recent research suggests that steroid hormones may also have more rapid mechanisms of action that do not require protein synthesis. For example, investigators have identified an estrogen receptor in the membrane of the endoplasmic reticulum. When this receptor is activated by estrogen binding, it initiates rapid signaling that leads to changes in the cell that take place within minutes, or even seconds.

Many hormones bind to cell-surface receptors

Because peptide hormones are hydrophilic, they are not soluble in the lipid layer of the plasma membrane and do not enter the target cell. Instead, they bind to a specific cell-surface receptor in

Key Point

Steroid and thyroid hormones are small, lipid-soluble molecules that pass through the plasma membrane and activate or repress specific genes. Proteins synthesized cause the changes that we recognize as the hormone's action.

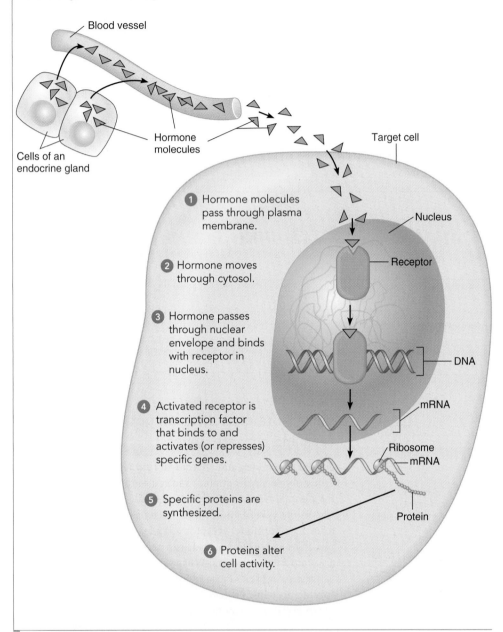

Blood vessel

Cells of an endocrine gland

Hormone molecules

Target cell

① Hormone molecules pass through plasma membrane.

② Hormone moves through cytosol.

③ Hormone passes through nuclear envelope and binds with receptor in nucleus.

④ Activated receptor is transcription factor that binds to and activates (or represses) specific genes.

⑤ Specific proteins are synthesized.

⑥ Proteins alter cell activity.

Nucleus

Receptor

DNA

mRNA

Ribosome
mRNA

Protein

Figure 48-4 *Animated* Mechanism of action of steroid and thyroid hormones

---

the plasma membrane. Two main types of cell-surface receptors that bind hormones are G protein–linked receptors and enzyme-linked receptors. These receptors are discussed in detail in Chapter 6. Here we will review some basic concepts.

G protein–linked receptors initiate signal transduction

G protein–linked receptors are transmembrane proteins that initiate **signal transduction;** they convert an extracellular hormone signal into an intracellular signal that affects some intra-

cellular process. The hormone does not enter the cell. It serves as the first messenger and relays information to a **second messenger,** or intracellular signal. The second messenger then signals effector molecules that carry out the action.

G protein–linked receptors activate **G proteins,** a group of integral regulatory proteins (Fig. 48-5). The G indicates that they bind **guanosine triphosphate (GTP),** which, like ATP, is an important molecule in energy reactions. When the system is inactive, G protein binds to **guanosine diphosphate (GDP),** which is similar to ADP, the hydrolyzed form of ATP. G proteins shuttle a signal between the receptor and a second messenger.

When a hormone-bound receptor binds to it, a stimulatory G protein releases GDP and replaces it with GTP. The G protein undergoes a change in shape that allows it to bind with and activate **adenylyl cyclase,** an enzyme on the cytoplasmic side of the plasma membrane. Once activated, adenylyl cyclase catalyzes the conversion of ATP to **cyclic AMP (cAMP).** By coupling the hormone–receptor complex to an enzyme that generates a signal, G proteins amplify hormone effects, and many second-messenger molecules are rapidly produced. One type of G protein, G_s, stimulates adenylyl cyclase, and another, G_i, inhibits it.

Cyclic AMP activates **protein kinases,** enzymes that phosphorylate (add a phosphate group to) certain specific proteins. When a protein is phosphorylated, its function is altered, and it triggers a chain of reactions leading to some metabolic change. The substrates for protein kinases are different in various cell types, so the effect of the enzyme varies. In skeletal muscle cells, a protein kinase triggers the breakdown of glycogen to glucose, whereas in cells of the hypothalamus, a protein kinase activates the gene that encodes a growth-inhibiting hormone.

Some G proteins use phospholipids as second messengers (see Chapter 6). Certain hormone–receptor complexes activate a G protein that then activates phospholipase C, a membrane-bound enzyme (see Fig. 6-9). This enzyme splits a membrane phospholipid into two products, **inositol trisphosphate (IP$_3$)** and **diacylglycerol (DAG).** Both act as second messengers. DAG (in combination with calcium ions) activates a protein kinase that

Peptide hormones bind with receptors on the plasma membrane. The receptor is a signal transducer that converts the hormone signal to an intracellular signal, which is relayed by a second messenger.

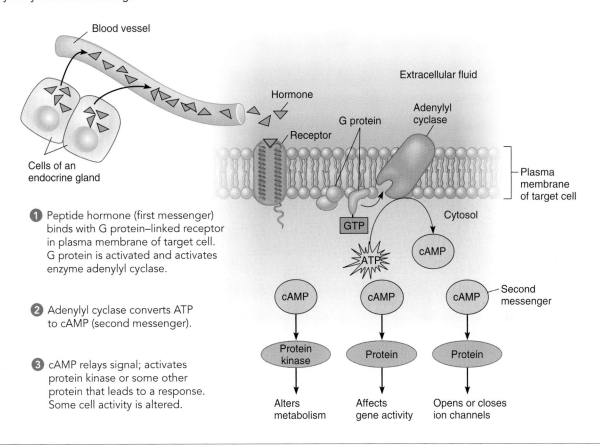

① Peptide hormone (first messenger) binds with G protein–linked receptor in plasma membrane of target cell. G protein is activated and activates enzyme adenylyl cyclase.

② Adenylyl cyclase converts ATP to cAMP (second messenger).

③ cAMP relays signal; activates protein kinase or some other protein that leads to a response. Some cell activity is altered.

Figure 48-5 *Animated* Mechanism of action of peptide hormones

Many peptide hormones signal target cells by way of a G protein–linked receptor. Cyclic AMP, or some other second messenger, relays the signal as part of a cascade of reactions that changes some cell process (the hormone action).

phosphorylates a variety of proteins. IP₃ opens calcium channels in the endoplasmic reticulum, releasing calcium ions into the cytosol. Calcium ions bind to certain proteins, such as **calmodulin.** The activated calmodulin then activates other proteins.

Enzyme-linked receptors function directly

Enzyme-linked receptors are transmembrane proteins with a hormone-binding site outside the cell and an enzyme site inside the cell (see Fig. 6-5c). These receptors are not linked to G proteins. They function directly as enzymes or are directly linked to enzymes. Most enzyme-linked receptors are **receptor tyrosine kinases** that bind growth factors and other signal molecules, including insulin. When the receptor is activated, it phosphorylates the amino acid tyrosine in specific signaling proteins inside the cell.

Review

▍ What are the roles of receptors and second messengers in hormone action?

▍ How do steroid and peptide hormones differ in their mechanisms of action?

▍ What is the mechanism of action of a hormone that uses cyclic AMP as a second messenger?

INVERTEBRATE NEUROENDOCRINE SYSTEMS

Learning Objective

6 Identify four functions of invertebrate hormones, and describe the hormonal regulation of insect development.

Among invertebrates, hormones are secreted mainly by neurons rather than by endocrine glands. These *neurohormones* regulate regeneration in hydras, flatworms, and annelids; color changes in crustaceans; and growth, development, metabolic rate, gamete production, and reproduction, including reproductive behavior, in many other groups. Trends in the evolution of invertebrate endocrine systems include a larger number of both neurohormones and hormones secreted by endocrine glands and a greater role of hormones in physiological processes.

Insects have endocrine glands and neuroendocrine cells. Their various hormones and neurohormones interact with one another to regulate metabolism, growth, and development, including molting and metamorphosis. Hormonal control of development in insects varies among species. Generally, some environmental factor (such as temperature change) activates neuroendocrine cells in the brain. These cells then produce a neurohormone referred to as **brain hormone (BH)**, which is transported down axons and stored in the paired **corpora cardiaca** (sing., *corpus cardiacum*; ▌ Fig. 48-6). When released from the corpora cardiaca, BH stimulates the **prothoracic glands**, endocrine glands in the prothorax, to produce **molting hormone (MH)**, also called **ecdysone** (see Fig. 48-2b). Molting hormone, a steroid hormone, stimulates growth and molting.

In the immature insect, paired endocrine glands called **corpora allata** (sing., *corpus allatum*) secrete **juvenile hormone (JH)**. This hormone suppresses metamorphosis at each larval molt so that the insect increases in size while remaining in its immature state; after the molt, the insect is still in a larval stage. When the concentration of JH decreases below some critical level, metamorphosis occurs, and the insect is transformed into a pupa (see Chapter 30). In the absence of JH, the pupa molts and becomes an adult. The nervous system regulates the secretory activity of the corpora allata, and the amount of JH decreases with successive molts.

Review

▌ What are four actions of hormones in invertebrates?
▌ What is the function of juvenile hormone in insects?

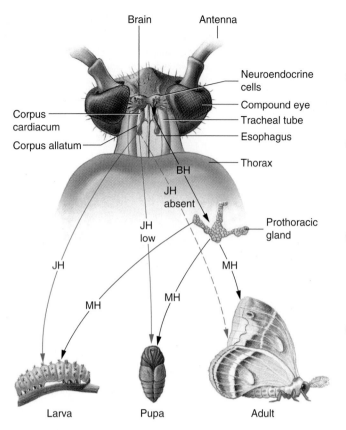

Figure 48-6 Regulation of growth and molting in insects

Hormones regulate insect development. Most insects go through a series of larval stages and then undergo metamorphosis to the adult form. This dissection of an insect head shows the brain, paired corpora allata, and paired corpora cardiaca.

1 Neuroendocrine cells secrete brain hormone (BH), which is stored in corpora cardiaca.

2 When released, BH stimulates prothoracic glands to secrete molting hormone (MH), which stimulates growth and molting.

3 In immature insect, corpora allata secrete juvenile hormone (JH), which suppresses metamorphosis at each larval molt.

4 Metamorphosis to adult form occurs when MH acts in absence of JH.

THE VERTEBRATE ENDOCRINE SYSTEM

Learning Objectives

7 Identify the classical vertebrate endocrine glands, and describe the actions of their hormones.

8 Describe the mechanisms by which the hypothalamus and pituitary gland integrate regulatory functions; describe the actions of the hypothalamic and pituitary hormones.

9 Describe the actions of the thyroid and parathyroid hormones, their regulation, and the effects of malfunction.

10 Contrast the actions of insulin and glucagon, and describe diabetes mellitus and hypoglycemia, including the metabolic effects of these disorders.

11 Describe the actions and regulation of the adrenal hormones, including their role in helping the body respond to stress.

Vertebrate hormones regulate such diverse activities as growth, development, reproduction, metabolic rate, fluid balance, blood homeostasis, and response to stress. Most vertebrates have similar endocrine glands, although the actions of some hormones may be different in various groups. ▌ Table 48-1 gives the sources, target tissues, and principal physiological actions of some of the major vertebrate hormones. The hypothalamus and pituitary gland regulate many of these hormones. The principal human endocrine glands are illustrated in ▌ Figure 48-7.

Homeostasis depends on normal concentrations of hormones

When a disorder or disease process affects an endocrine gland, the rate of secretion may become abnormal. If **hyposecretion** (abnormally reduced output) occurs, target cells are deprived

TABLE 48-1

Some Endocrine Glands and Their Hormones

Gland	Hormone	Target Tissue	Principal Actions
Hypothalamus	Releasing and inhibiting hormones	Anterior lobe of pituitary	Regulate secretion of hormones by the anterior pituitary
Posterior pituitary (storage and release of hormones produced by hypothalamus)	Oxytocin	Uterus	Stimulates contraction
		Mammary glands	Stimulates ejection of milk into ducts
	Antidiuretic hormone (ADH)	Kidneys (collecting ducts)	Stimulates reabsorption of water
Anterior pituitary	Growth hormone (GH)	General	Stimulates growth of skeleton and muscle
	Prolactin	Mammary glands	Stimulates milk production
	Melanocyte-stimulating hormones (MSH)	Pigment cells in skin	Stimulate melanin production
	Thyroid-stimulating hormone (TSH)	Thyroid gland	Stimulates secretion of thyroid hormones; helps regulate bone remodeling
	Adrenocorticotropic hormone (ACTH)	Adrenal cortex	Stimulates secretion of adrenal cortical hormones
	Gonadotropic hormones* (follicle-stimulating hormone [FSH]; luteinizing hormone [LH])	Gonads	Stimulate gonad function and growth
Thyroid gland	Thyroxine (T_4) and triiodothyronine (T_3)	General	Stimulate metabolic rate; regulate energy metabolism
	Calcitonin	Bone	Lowers blood calcium level
Parathyroid glands	Parathyroid hormone	Bone, kidneys, digestive tract	Regulates blood calcium level
Pancreas	Insulin	General	Lowers blood glucose concentration
	Glucagon	Liver, adipose tissue	Raises blood glucose concentration
Adrenal medulla	Epinephrine and norepinephrine	Muscle; blood vessels; liver; adipose tissue	Help body cope with stress; increase metabolic rate; raise blood glucose level; increase heart rate and blood pressure
Adrenal cortex	Mineralocorticoids	Kidney tubules	Maintain sodium and potassium balance
	Glucocorticoids	General	Help body cope with long-term stress; raise blood glucose level
Pineal gland	Melatonin	Hypothalamus	Important in biological rhythms
Ovary*	Estrogens	General; uterus	Develop and maintain sex characteristics in female; stimulate growth of uterine lining
	Progesterone	Uterus; breast	Stimulates development of uterine lining
Testis	Testosterone	General; reproductive structures	Develops and maintains sex characteristics in males; promotes spermatogenesis

*Reproductive hormones are discussed in Chapter 49; see Tables 49-1 and 49-2.

of needed stimulation. If **hypersecretion** (abnormally increased output) occurs, the target cells may be overstimulated. In some endocrine disorders, an appropriate amount of hormone is secreted, but the target cell receptors do not function properly. As a result, the target cells cannot respond to the hormone. Any of these abnormalities leads to loss of homeostasis, resulting in predictable metabolic malfunctions and clinical symptoms (Table 48-2).

The hypothalamus regulates the pituitary gland

Most endocrine activity is controlled directly or indirectly by the **hypothalamus,** which links the nervous and endocrine systems both anatomically and physiologically. The **pituitary gland** is connected to the hypothalamus by the pituitary stalk. In response to input from other areas of the brain and from hormones in the

only about 0.5 g (0.02 oz), the pituitary gland produces and secretes at least seven peptide hormones that exert far-reaching influence over body activities. The human pituitary gland consists of two main parts: the *anterior lobe* and the *posterior lobe*. In some vertebrates, an intermediate lobe is present.

The posterior lobe of the pituitary gland releases hormones produced by the hypothalamus

The **posterior lobe** of the pituitary gland develops from brain tissue. This neuroendocrine organ *secretes* two peptide hormones, **oxytocin** and **antidiuretic hormone,** or **ADH,** also known as *vasopressin.* (See Chapter 46 for a discussion of the function of ADH in regulating water reabsorption in the kidneys.) These hormones are actually *produced* by neuroendocrine cells in two distinct areas of the hypothalamus. Enclosed within vesicles, oxytocin and ADH are transported down the axons of these neuroendocrine cells into the posterior lobe of the pituitary gland (❚ Fig. 48-8). The vesicles are stored in the axon terminals until the neuron is stimulated. Then the hormones are released and diffuse into surrounding capillaries.

Oxytocin concentration in the blood rises toward the end of pregnancy, stimulating the strong contractions of the uterus needed to expel a baby. Oxytocin is sometimes administered clinically to initiate or speed labor. After birth, when an infant sucks at its mother's breast, sensory neurons signal the pituitary to release oxytocin. The hormone stimulates contraction of smooth muscle cells surrounding the milk glands so that milk is let down into the ducts, making it available to the nursing infant. Because oxytocin also stimulates the uterus to contract, breastfeeding promotes recovery of the uterus to nonpregnant size.

The anterior lobe of the pituitary gland regulates growth and other endocrine glands

The **anterior lobe** of the pituitary develops from epithelial cells rather than neural cells. The anterior lobe functions like a classical endocrine gland: it receives signals by way of the blood and releases its hormones into the blood. The *hypothalamus* produces several **releasing hormones** and **inhibiting hormones** that regu-

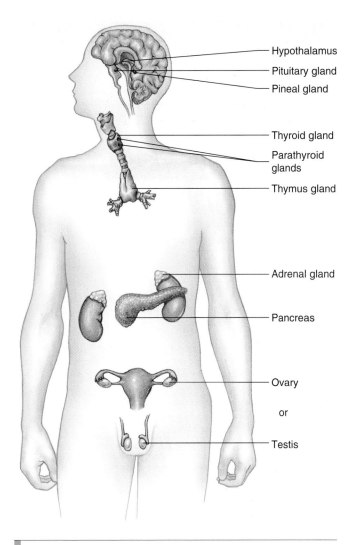

Figure 48-7 *Animated* The classical human endocrine glands

Many of the hormones produced by these glands and their actions are discussed in this chapter.

blood, neurons of the hypothalamus secrete neurohormones that regulate specific physiological processes.

Because secretions of the pituitary gland control the activities of several other endocrine glands, biologists consider it the master gland of the body. Although it is the size of a large pea and weighs

TABLE 48-2

Consequences of Endocrine Malfunction

Hormone	Hyposecretion	Hypersecretion
Growth hormone	Pituitary dwarfism	Gigantism if malfunction occurs in childhood; acromegaly in adult
Thyroid hormones	Cretinism (in children); myxedema, a condition of prounounced adult hypothyroidism; dietary iodine deficiency leads to hyposecretion and goiter	Hyperthyroidism; increased metabolic rate, nervousness, irritability; goiter; Graves' disease
Parathyroid hormone	Spontaneous discharge of nerves; spasms; tetany; death	Weak, brittle bones; kidney stones
Insulin	Diabetes mellitus	Hypoglycemia
Hormones of adrenal cortex	Addison's disease	Cushing's disease

late the production and secretion of specific hormones by the anterior pituitary gland.

Releasing and inhibiting hormones produced by the hypothalamus enter capillaries and pass through special portal veins that connect the hypothalamus with the anterior lobe of the pituitary. (These portal veins, like the hepatic portal vein, do not deliver blood to a larger vein directly; they connect two sets of capillaries.) Within the anterior lobe of the pituitary, the portal veins divide into a second set of capillaries. Releasing and inhibiting hormones pass through the capillary walls into the tissue of the anterior lobe of the pituitary.

The anterior lobe secretes prolactin, melanocyte-stimulating hormones, growth hormone, and several **tropic hormones** that stimulate other endocrine glands (❙ Fig. 48-9). **Prolactin** stimulates the cells of the mammary glands to produce milk in a nursing mother. Recently, investigators have discovered that this hormone has diverse roles in various vertebrate groups. For example, in fishes and amphibians it helps maintain salt balance by decreasing Na^+ and K^+ loss in the urine. Prolactin also inhibits metamorphosis in amphibians and stimulates molting in reptiles. This versatile hormone stimulates lipid metabolism in birds and enhances carbohydrate metabolism in mammals.

Melanocyte-stimulating hormones (MSH) are a group of peptide hormones secreted by the anterior pituitary gland and also by certain neurons in the hypothalamus. In some fishes, amphibians, and reptiles, MSH causes skin darkening, which is important in camouflage in some animals. In humans, MSH also suppresses appetite and is important in regulating energy and body weight.

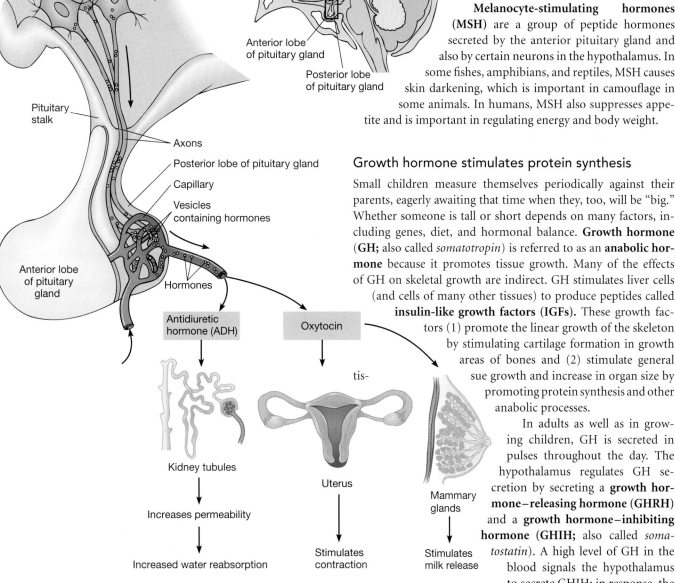

Growth hormone stimulates protein synthesis

Small children measure themselves periodically against their parents, eagerly awaiting that time when they, too, will be "big." Whether someone is tall or short depends on many factors, including genes, diet, and hormonal balance. **Growth hormone (GH;** also called *somatotropin*) is referred to as an **anabolic hormone** because it promotes tissue growth. Many of the effects of GH on skeletal growth are indirect. GH stimulates liver cells (and cells of many other tissues) to produce peptides called **insulin-like growth factors (IGFs).** These growth factors (1) promote the linear growth of the skeleton by stimulating cartilage formation in growth areas of bones and (2) stimulate general sue growth and increase in organ size by promoting protein synthesis and other anabolic processes.

In adults as well as in growing children, GH is secreted in pulses throughout the day. The hypothalamus regulates GH secretion by secreting a **growth hormone–releasing hormone (GHRH)** and a **growth hormone–inhibiting hormone (GHIH;** also called *somatostatin).* A high level of GH in the blood signals the hypothalamus to secrete GHIH; in response, the pituitary slows its release of GH. A low level of GH in the blood stimulates the hypothalamus to secrete GHRH, which stimulates the pituitary gland to release more

Figure 48-8 *Animated* Synthesis and secretion of posterior pituitary hormones

Neuroendocrine cells of the hypothalamus manufacture hormones that are secreted by the posterior lobe of the pituitary gland. The axons of these neurons extend into the posterior lobe. The hormones are packaged in vesicles that are transported through the axons and stored in their ends. When needed, the hormones are secreted, enter the blood, and are transported by the circulatory system.

GH. Many other factors, including blood sugar level, amino acid concentration, and stress, influence GH secretion.

Research supports the age-old notions that to grow, children need plenty of sleep, a proper diet, and regular exercise. Secretion of GH increases during exercise, probably because rapid metabolism by muscle cells lowers the blood glucose level. GH is secreted about 1 hour after the onset of deep sleep and in a series of pulses 2 to 4 hours after a meal.

Emotional support is also necessary for proper growth. Growth is retarded in children who are deprived of cuddling, playing, and other forms of nurture, even when their needs for food and shelter are met. In extreme cases, childhood stress can produce a form of retarded development known as *psychosocial dwarfism*. Some emotionally deprived children exhibit abnormal sleep patterns, which may be the basis for decreased secretion of GH.

Other hormones also influence growth. Thyroid hormones appear necessary for normal GH secretion and function and for normal tissue response to IGFs. Sex hormones must be present for the adolescent growth spurt to occur. However, the presence of sex hormones eventually causes the growth centers within the long bones to ossify so that further increase in height is impossible even when GH is present.

Inappropriate amounts of growth hormone secretion result in abnormal growth

Have you ever wondered why some people fail to grow normally? Deficiency in GH during childhood is called **pituitary dwarfism** because it is usually caused by insufficient production of GH by the pituitary gland. However, dwarfism also results when the liver produces insufficient IGF

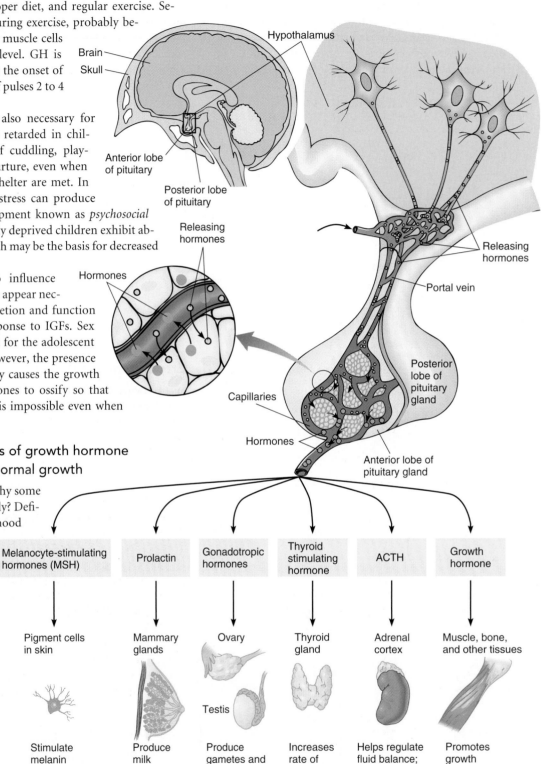

Figure 48-9 *Animated* Synthesis and secretion of anterior pituitary hormones

The hypothalamus secretes several specific releasing and inhibiting hormones that are transported to the anterior lobe of the pituitary gland by way of portal veins. These regulatory hormones stimulate or inhibit the release of specific hormones by cells of the anterior lobe. The anterior lobe secretes hormones that act on a wide variety of target tissues. (ACTH, adrenocorticotropic hormone)

or when target tissues do not respond to IGF. Although miniature, a pituitary dwarf has normal intelligence and is usually well proportioned. If the growth centers in the long bones are still active when this condition is diagnosed, it can often be effectively treated by human GH injection. GH can be synthesized by recombinant DNA technology.

A person grows abnormally tall when the anterior pituitary secretes excessive amounts of GH during childhood. This condition is called **gigantism.** If hypersecretion of GH occurs during adulthood, the individual cannot grow taller. Instead, connective tissue thickens, and bones in the hands, feet, and face may increase in diameter. This condition is known as **acromegaly** (*acromegaly* means "large extremities").

Thyroid hormones increase metabolic rate

The **thyroid gland** is located in the neck region, in front of the trachea and below the larynx. Two of the **thyroid hormones—thyroxine,** also known as T_4, and **triiodothyronine,** or T_3, are synthesized from the amino acid tyrosine and from iodine. Thyroxine has four iodine atoms attached to each molecule; T_3 has three. In most target tissues, T_4 is converted to T_3, the more active form (see Fig. 48-2c). We describe the actions of calcitonin, another hormone secreted by the thyroid gland, when we discuss the parathyroid glands.

In vertebrates, thyroid hormones are essential for normal growth and development. These hormones increase the rate of metabolism in most body tissues. After T_3 binds with its receptor in the nucleus of a target cell, the T_3 receptor complex induces or suppresses synthesis of specific enzymes and other proteins. Thyroid hormones also help regulate the synthesis of proteins necessary for cell differentiation. For example, tadpoles cannot develop into adult frogs without thyroxine.

Thyroid secretion is regulated by negative feedback systems

The regulation of thyroid hormone secretion depends on a negative feedback loop between the anterior pituitary and the thyroid gland (▌ Fig. 48-10). When the concentration of thyroid hormones in the blood rises above normal, the anterior pituitary secretes less **thyroid-stimulating hormone (TSH):**

> Thyroid hormone concentration high ⟶ inhibits anterior pituitary ⟶ secretes less TSH ⟶ thyroid gland secretes less hormone ⟶ homeostasis

When the concentration of thyroid hormones decreases, the pituitary secretes more TSH. TSH acts by way of cAMP to promote synthesis and secretion of thyroid hormones and also to promote increased size of the thyroid gland itself. The effect of TSH can be summarized as follows:

> Thyroid hormone concentration low ⟶ anterior pituitary secretes more TSH ⟶ thyroid gland secretes more hormone ⟶ homeostasis

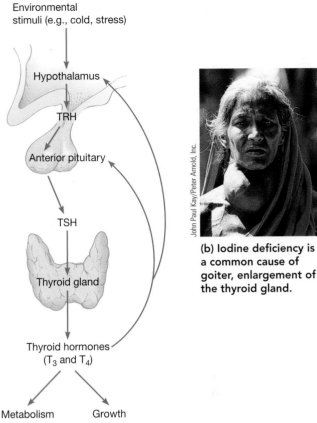

(b) Iodine deficiency is a common cause of goiter, enlargement of the thyroid gland.

(a) **Regulation of thyroid hormone secretion by negative feedback.** An increase in concentration of thyroid hormones above normal levels signals the anterior pituitary to slow thyroid-stimulating hormone (TSH) production. Thus, thyroid hormones limit their own production by negative feedback. Green arrows indicate stimulation; red arrows indicate inhibition. (TRH, TSH-releasing hormone)

Figure 48-10 *Animated* Regulation of thyroid hormone secretion

In some mammals, changes in weather affect the hypothalamus. Exposure to very cold temperature stimulates the hypothalamus to increase secretion of **TSH-releasing hormone** (TRH). This action causes a dramatic increase in thyroid hormone secretion. Metabolic rate increases, which also increases heat production, so the body temperature rises. Increased body temperature has a negative feedback effect, limiting further secretion of thyroid hormones.

Malfunction of the thyroid gland leads to specific disorders

Extreme hypothyroidism during infancy and childhood results in low metabolic rate and can lead to **cretinism,** a condition characterized by retarded mental and physical development. When diagnosed early enough, hypothyroidism can be treated with thyroid hormones, and cretinism can be prevented.

An adult who feels like sleeping all the time, has little energy, and is mentally slow or confused may be suffering from hypo-

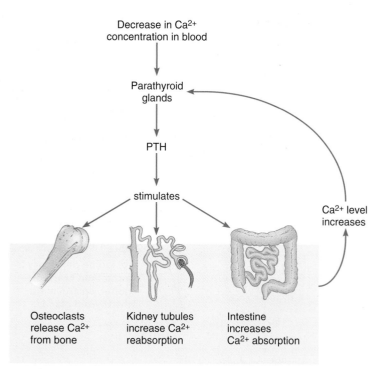

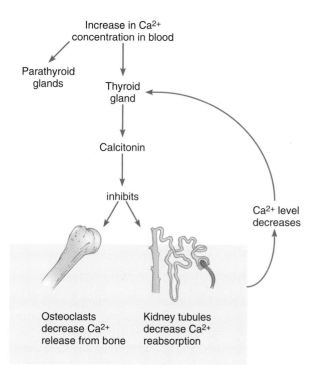

(a) Calcium concentration too low. When calcium concentration decreases below the normal range, the parathyroid glands secrete parathyroid hormone (PTH). PTH stimulates homeostatic mechanisms that restore appropriate calcium concentration.

(b) Calcium concentration too high. When calcium concentration increases above its set point, the parathyroid glands are inhibited. The thyroid gland secretes calcitonin, which inhibits Ca²⁺ release from bone and decreases Ca²⁺ reabsorption in the kidneys.

Figure 48-11 Regulation of calcium homeostasis

Green arrows indicate stimulation; red arrows indicate inhibition.

thyroidism. When there is almost no thyroid function, the basal metabolic rate is reduced by about 40% and the patient develops **myxedema,** characterized by a slowing down of physical and mental activity. Hypothyroidism is treated by oral administration of the missing hormone.

Hyperthyroidism does not cause abnormal growth but does increase metabolic rate by 60% or even more. This increase in metabolism results in the rapid use of nutrients, causing the individual to be hungry and to eat more. But this is not sufficient to meet the demands of the rapidly metabolizing cells, so people with this condition often lose weight. They also tend to be nervous, irritable, and emotionally unstable. The most common form of hyperthyroidism is **Graves' disease,** an autoimmune disease. Abnormal antibodies bind to TSH receptors and activate them. This leads to increased production of thyroid hormones.

An abnormal enlargement of the thyroid gland, or **goiter,** may be associated with either hypersecretion or hyposecretion (see Fig. 48-10b). In Graves' disease, the abnormal antibodies that activate TSH receptors can cause the development of a goiter. Hyposecretion can be caused by dietary iodine deficiency. Without iodine the gland cannot make thyroid hormones, so their concentration in the blood decreases. The anterior pituitary compensates by secreting large amounts of TSH, which stimulates growth of the thyroid gland, sometimes to gigantic proportions. However, enlargement of the gland does not increase production

of the hormones, because the needed ingredient is still missing. Seafood is a rich source of iodine, and iodine is also added to table salt as a nutritional supplement. In fact, thanks to iodized salt, goiter is no longer common in the United States and other highly developed countries. In other parts of the world, however, hundreds of thousands still suffer from this easily preventable disorder.

The parathyroid glands regulate calcium concentration

The **parathyroid glands** are typically embedded in the connective tissue surrounding the thyroid gland. These glands secrete **parathyroid hormone (PTH),** a polypeptide that helps regulate the calcium level of the blood and interstitial fluid (Fig. 48-11a). Parathyroid hormone acts by way of a G protein–linked receptor and cAMP to stimulate calcium release from bones and to increase calcium reabsorption from the kidney tubules. It also activates vitamin D, which then increases the amount of calcium absorbed from the intestine.

Calcitonin, a peptide hormone secreted by the thyroid gland, works antagonistically to parathyroid hormone (Fig. 48-11b). When the concentration of calcium rises above homeostatic levels, calcitonin is released and rapidly inhibits removal of calcium from bone.

The islets of the pancreas regulate glucose concentration

In addition to secreting digestive enzymes (see Chapter 45), the pancreas is an important endocrine gland. Its hormones, insulin and glucagon, are secreted by cells that occur in little clusters, the **islets of Langerhans,** throughout the pancreas (Fig. 48-12). About 1 million islets are present in the human pancreas. They consist mainly of **beta cells,** which secrete insulin, and **alpha cells,** which secrete glucagon.

Insulin lowers the concentration of glucose in the blood

Insulin is an anabolic hormone that promotes storage of fuel molecules. It stimulates cells of many tissues, including liver, muscle, and fat cells, to take up glucose from the blood by facilitated diffusion (see Chapter 5). Once glucose enters muscle cells, it is either used immediately as fuel or stored as glycogen. Insulin also inhibits liver cells from releasing glucose. Thus, insulin activity *lowers* the glucose level in the blood.

Insulin helps regulate fat and protein metabolism. It reduces the use of fatty acids as fuel and instead stimulates their storage in adipose tissue. Insulin has an anabolic effect on protein metabolism, resulting in a net increase in protein. It promotes protein synthesis by increasing the number of amino acid transporters in the plasma membrane, thus stimulating the transport of certain amino acids into cells. Insulin also promotes transcription and translation.

Glucagon raises the concentration of glucose in the blood

The effects of **glucagon** are opposite to those of insulin. Glucagon's main action is to raise blood glucose level. It does this by stimulating liver cells to convert glycogen to glucose, a process known as *glycogenolysis.* Glucagon also stimulates *gluconeogenesis,* the production of glucose from other metabolites. Glucagon mobilizes fatty acids by promoting breakdown of fat and inhibiting triacylglycerol synthesis. This hormone also stimulates the liver to degrade proteins and inhibits protein synthesis.

Glucose concentration regulates insulin and glucagon secretion

The concentration of glucose in the blood directly controls insulin and glucagon secretion (Fig. 48-13). After a meal, when the blood glucose level rises as a result of intestinal absorption, beta cells are stimulated to increase insulin secretion. Then, as cells remove glucose from the blood, decreasing its concentration, insulin secretion decreases.

High blood glucose concentration ⟶ beta cells secrete insulin ⟶ blood glucose concentration decreases ⟶ homeostasis

When you have not eaten for several hours, the concentration of glucose in your blood begins to fall. When it decreases from

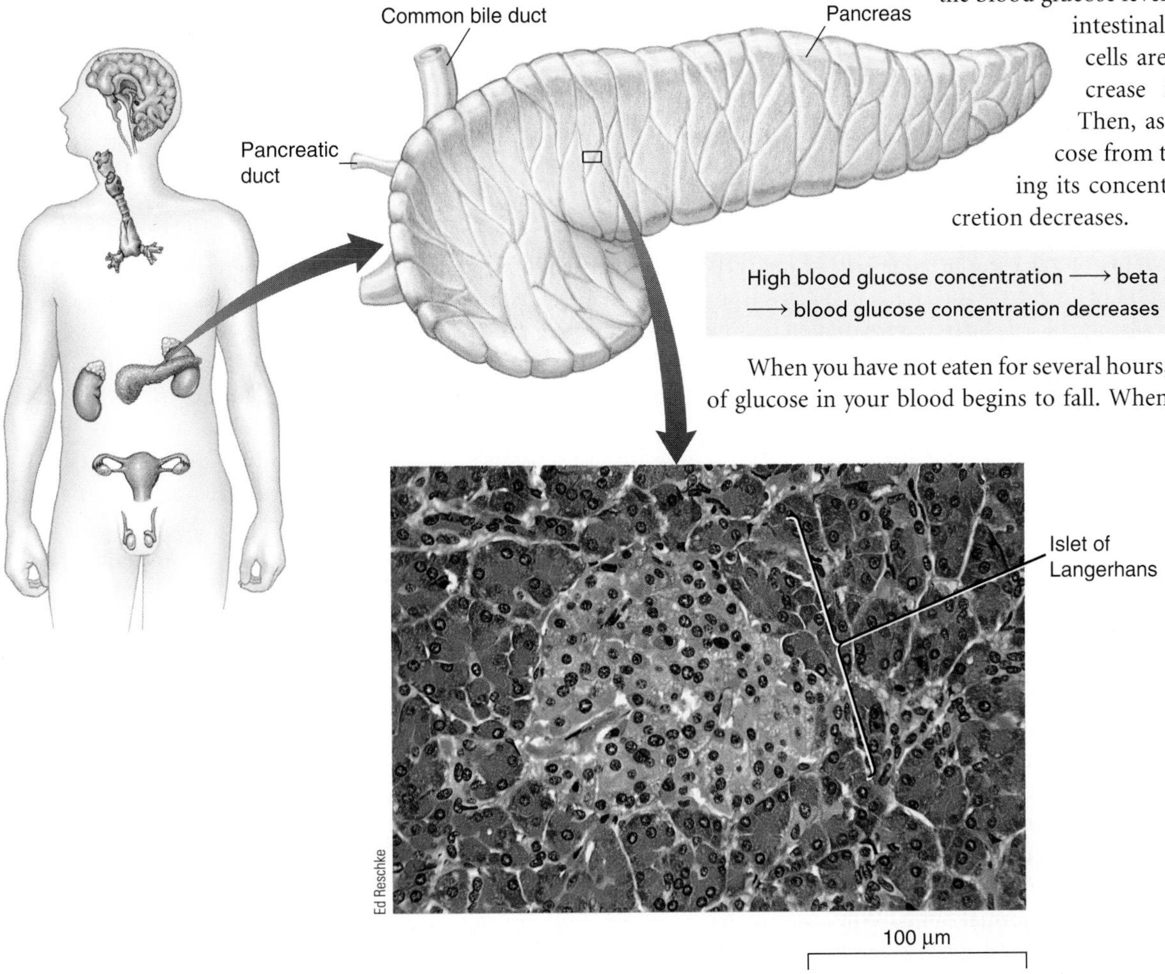

Common bile duct

Pancreatic duct

Pancreas

Islet of Langerhans

Ed Reschke

100 μm

Figure 48-12 Islets of Langerhans

Scattered throughout the pancreas, clusters of endocrine cells (such as the one shown in this LM) secrete insulin and glucagon.

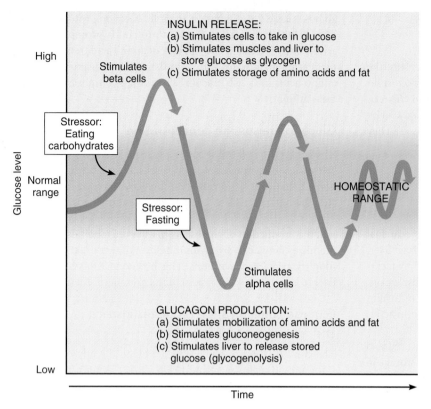

INSULIN RELEASE:
(a) Stimulates cells to take in glucose
(b) Stimulates muscles and liver to store glucose as glycogen
(c) Stimulates storage of amino acids and fat

Stimulates beta cells

Stressor: Eating carbohydrates

Stressor: Fasting

HOMEOSTATIC RANGE

Stimulates alpha cells

GLUCAGON PRODUCTION:
(a) Stimulates mobilization of amino acids and fat
(b) Stimulates gluconeogenesis
(c) Stimulates liver to release stored glucose (glycogenolysis)

High

Normal range

Low

Glucose level

Time

Figure 48-13 *Animated* Regulation of glucose concentration

Insulin and glucagon work antagonistically to regulate the concentration of glucose in the blood.

Diabetes mellitus is a serious disorder of carbohydrate metabolism

Diabetes mellitus, the most common endocrine disorder, is a serious health problem, and according to the World Health Organization, its global prevalence is increasing. More than 150 million people worldwide have this disorder, and the number is rapidly increasing. Diabetes, a leading cause of premature death, is a major cause of cardiovascular disease, blindness, nerve disease, kidney disorders, and gangrene of the limbs. (Note that diabetes mellitus is an entirely different disorder from diabetes insipidus, which is discussed in Chapter 47.) Approximately one half of the people suffering from diabetes are unaware that they have this disorder; consequently, they do not receive treatment and are at high risk of its deadly effects.

Diabetes mellitus is actually a group of related disorders characterized by high blood glucose concentrations. Two main types of diabetes are type 1 and type 2. *Type 1 diabetes* is an autoimmune disease in which antibodies mark the beta cells for destruction. T cells destroy the beta cells, resulting in insulin deficiency. Insulin injections are needed to correct the carbohydrate imbalance that results. This disorder usually develops before age 30, often during childhood. Type 1 diabetes is caused by a combination of genetic predisposition and environmental factors, possibly including infection by a virus.

its normal fasting level of about 90 mg of glucose per 100 mL of blood to about 70 mg of glucose, the alpha cells of the islets increase their secretion of glucagon. Glucose is mobilized from storage in the liver cells, and blood glucose concentration returns to normal:

Low blood glucose concentration ⟶ alpha cells secrete glucagon ⟶ blood glucose concentration increases ⟶ homeostasis

Alpha cells respond to the glucose concentration within their own cytoplasm, which reflects the blood glucose level. When blood glucose level is high, there is generally a high level of glucose within the alpha cells, and glucagon secretion is inhibited.

Note that these are negative feedback systems and that insulin and glucagon work antagonistically to keep blood glucose concentration within normal limits. When the glucose level rises, insulin release brings it back to normal; when the glucose concentration falls, glucagon acts to raise it again. The insulin–glucagon system is a powerful, fast-acting mechanism for keeping blood glucose level within normal limits. Why is it important to maintain a constant blood glucose level? Recall that brain cells ordinarily are unable to use other nutrients as fuel, so they depend on a continuous supply of glucose. As we will discuss, several other hormones also affect blood glucose concentration.

More than 90% of diabetics have *type 2 diabetes*. This disorder develops gradually, usually in overweight people. In type 2 diabetes, normal (or greater than normal) concentrations of insulin are present in the blood, but target cells cannot use it, a condition known as **insulin resistance.** In skeletal muscle cells, the major cause of insulin resistance appears to be that glucose transporters are not synthesized or do not function properly. Insulin resistance has a strong genetic component. Investigators have also reported an association between increased fat tissue and insulin resistance.

Many type 2 diabetics can keep their blood glucose levels within normal range by diet management, weight loss, and regular exercise. When this treatment approach is not effective, they are treated with oral drugs that stimulate insulin secretion and promote its actions. Approximately one third of type 2 diabetics eventually need insulin injections.

More than 20% of obese children have impaired glucose tolerance, an early warning sign of diabetes. Impaired glucose tolerance, which affects more than 200 million people worldwide, is defined as hyperglycemia (a blood glucose concentration that is abnormally high) following ingestion of a large amount of glucose (a glucose load).

Similar metabolic disturbances occur in both types of diabetes mellitus: disruption of carbohydrate, fat, and protein metabolism and electrolyte imbalance.

1. *Glucose use decreases.* In a diabetic, the cells have difficulty taking up glucose from the blood, and its accumulation

causes hyperglycemia. Instead of the normal fasting concentration of about 90 mg of glucose per 100 mL of blood, the level may exceed 140 mg per 100 mL and may reach concentrations of more than 500 mg per 100 mL. When glucose concentration exceeds its **tubular transport maximum (Tm),** the maximum rate at which it can be reabsorbed, glucose is excreted in the urine (see Chapter 47). Despite the large quantities of glucose in the blood, insulin-dependent cells in a diabetic can take in only about 25% of the glucose they require for fuel.

2. *Fat mobilization increases.* Cells turn to fat and protein for energy. The absence of insulin promotes the mobilization of fat stores, providing nutrients for cellular respiration. But unfortunately, the blood lipid level may reach 5 times the normal level, leading to the development of atherosclerosis (see Chapter 43). Increased fat metabolism increases the formation of ketone bodies, which build up in the blood. This accumulation can cause *ketoacidosis,* a condition in which the body fluids and blood become too acidic. When the ketone level in the blood rises, ketones appear in the urine, another clinical indication of diabetes mellitus. If severe, ketoacidosis can lead to coma and death.

3. *Protein use increases.* Lack of insulin also results in increased protein breakdown relative to protein synthesis, so the untreated diabetic becomes thin and emaciated.

4. *Electrolyte imbalance.* When ketone bodies and glucose are excreted in the urine, water follows by osmosis; as a result, urine volume increases. The resulting dehydration causes thirst and challenges electrolyte homeostasis. When ketones are excreted, they also take sodium, potassium, and some other cations with them, contributing to electrolyte imbalance.

The dramatic increase in type 2 diabetes is a public health problem resulting from a sedentary lifestyle and an increasing prevalence of obesity. Researchers are looking for causes and cures of diabetes. Some research groups are focusing on education and lifestyle changes, whereas others are looking for the genes that contribute to diabetes or for signaling molecules involved in glucose metabolism.

Researchers have discovered that adipose tissue produces several hormones and other signaling molecules that play a role in insulin resistance and diabetes. These include leptin, resistin, and adiponectin. Recall from Chapter 46 that *leptin* signals centers in the brain about the status of energy stores in adipose tissue. The brain then adjusts feeding behavior and energy metabolism. Obese people produce large amounts of leptin but are resistant to its effects. *Resistin* antagonizes the action of insulin and may contribute to insulin resistance. *Adiponectin* promotes insulin's effects, but obese people produce low amounts.

In hypoglycemia the blood glucose concentration is too low

Low blood glucose concentration, or *hypoglycemia,* sometimes occurs in people who later develop diabetes. These individuals may have impaired glucose tolerance in which the islets overreact to glucose challenge. When carbohydrates are ingested, there is a delayed insulin response, followed after about 3 hours by too much insulin secretion. This hypersecretion causes glucose levels to fall, and the individual feels drowsy. If this reaction is severe, the person may become uncoordinated or even unconscious.

Serious hypoglycemia can develop if diabetics receive injections of too much insulin or if the islets, because of a tumor, secrete too much insulin. The blood glucose concentration may then fall drastically, depriving brain cells of their needed fuel supply. These events can lead to **insulin shock,** a condition in which the patient may appear drunk or may become unconscious, suffer convulsions, and even die.

The adrenal glands help the body respond to stress

The paired **adrenal glands** are small, yellow masses of tissue that lie in contact with the upper ends of the kidneys (❚ Fig. 48-14). Each gland consists of a central portion, the **adrenal medulla,** and a larger outer section, the **adrenal cortex.** Although joined anatomically, the adrenal medulla and cortex develop from different types of tissue in the embryo and function as distinctly different glands. However, both secrete hormones that help regulate metabolism and help the body respond to stress.

The adrenal medulla initiates an alarm reaction

The adrenal medulla is a neuroendocrine gland that is coupled to the sympathetic nervous system. It develops from neural tissue, and its secretion is controlled by sympathetic nerves. The adrenal

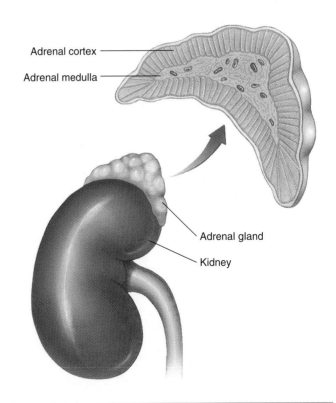

Adrenal cortex

Adrenal medulla

Adrenal gland

Kidney

Figure 48-14 Adrenal gland

The adrenal glands are located above the kidneys. Each gland consists of a central adrenal medulla surrounded by an adrenal cortex.

medulla secretes **epinephrine** and **norepinephrine.** Chemically, these hormones are very similar; they are amino acid derivatives that belong to the chemical group known as **catecholamines** (see Fig. 48-2c). Most of the hormone output of the adrenal medulla is epinephrine. Recall that norepinephrine also serves as a neurotransmitter released by sympathetic neurons and by some neurons in the central nervous system.

Under normal conditions, the adrenal medulla continuously secretes small amounts of both epinephrine and norepinephrine. Their secretion is under neural control. When anxiety is aroused, the hypothalamus sends signals via sympathetic nerves to the adrenal medulla. Sympathetic neurons release acetylcholine, which triggers the adrenal medulla to release larger amounts of epinephrine and norepinephrine.

During a stressful situation, adrenal medullary hormones initiate an *alarm reaction* that enables you to think more quickly, fight harder, or run faster than usual. Metabolic rate increases by as much as 100%. Blood is rerouted to those organs essential for emergency action (▌Fig. 48-15). Blood vessels going to the brain, muscles, and heart are dilated, whereas those to the skin and kidneys are constricted. Constriction of blood vessels serving the skin has the added advantage of decreasing blood loss in case of hemorrhage (and explains the sudden paling that comes with fear or rage). At the same time, the heart beats faster. Strength of muscle contraction increases. Thresholds in the reticular activating system of the brain are lowered, so you become more alert (see Chapter 41). The adrenal medullary hormones also raise fatty acid and glucose levels in the blood, ensuring needed fuel for extra energy.

The adrenal cortex helps the body deal with chronic stress

The adrenal cortex synthesizes steroid hormones from cholesterol (which in turn is made in the liver from acetyl coenzyme A). Although the adrenal cortex produces more than 30 types of steroids, it secretes only three types of hormones in significant amounts: sex hormone precursors, mineralocorticoids, and glucocorticoids.

In both sexes, the adrenal cortex secretes sex hormone precursors, such as androgens (masculinizing hormones). Certain tissues convert sex hormone precursors to **testosterone,** the principal male sex hormone, and **estradiol,** the principal female sex hormone. In males, androgen production by the adrenal cortex is not significant, because the testes produce testosterone. However, in females, the adrenal cortex secretes the androgen used to produce most of the testosterone circulating in the blood.

The principal **mineralocorticoid** is **aldosterone.** Recall from Chapter 47 that aldosterone helps regulate fluid balance by regulating salt balance. In response to aldosterone, the kidneys reabsorb more sodium

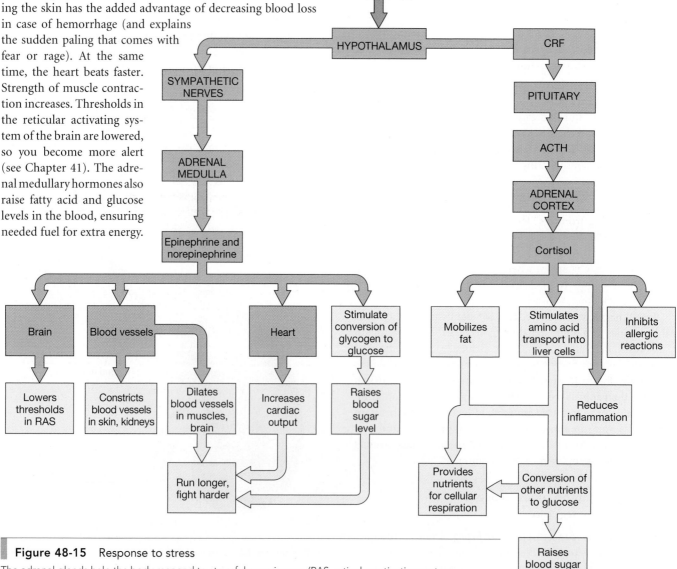

Figure 48-15 Response to stress

The adrenal glands help the body respond to stressful experiences. (RAS, reticular activating system; CRF, corticotropin-releasing factor; ACTH, adrenocorticotropic hormone)

and excrete more potassium. As a result of increased sodium, extracellular fluid volume increases, which results in greater blood volume and elevated blood pressure.

When the adrenal glands do not produce enough aldosterone, large amounts of sodium are excreted in the urine. Water leaves the body with the sodium (because of osmotic pressure), and the blood volume may be so markedly reduced that the patient dies of low blood pressure.

Cortisol, also called *hydrocortisone,* accounts for about 95% of the **glucocorticoid** activity of the human adrenal cortex (see Fig. 48-2b). Cortisol helps ensure adequate fuel supplies for the cells when the body is under stress. Its principal action is to stimulate production of glucose from other nutrients in liver cells. Cortisol helps provide nutrients for glucose production by stimulating amino acid transport into liver cells (see Fig. 48-15). It also promotes mobilization of fats so that glycerol from triacylglycerol molecules is available for conversion to glucose. These actions ensure that glucose and glycogen are produced in the liver, and the concentration of glucose in the blood rises. Thus, the adrenal cortex provides an important backup system for the adrenal medulla, ensuring adequate glucose supplies when the body is under stress and in need of extra energy.

During stress, the brain and the adrenal glands work together to help the body respond effectively (see Fig. 48-15; also see *Focus On: The Neurobiology of Traumatic Experience,* in Chapter 41). Almost any type of stress stimulates the hypothalamus to secrete **corticotropin-releasing factor (CRF),** which stimulates the anterior pituitary to secrete **adrenocorticotropic hormone (ACTH).** ACTH regulates both glucocorticoid and aldosterone secretion. ACTH is so potent that it can result in a 20-fold increase in cortisol secretion within minutes. When the body is not under stress, high levels of cortisol in the blood inhibit both CRF secretion by the hypothalamus and ACTH secretion by the pituitary.

Destruction of the adrenal cortex and the resulting decrease in aldosterone and cortisol secretion cause **Addison's disease.** Reduction in cortisol prevents the body from regulating the blood glucose concentration because the liver cannot synthesize enough glucose. The cortisol-deficient patient also loses the ability to respond to stress. If cortisol levels are significantly depressed, even the stress of mild infections can cause death.

Glucocorticoids are used clinically to reduce inflammation in allergic reactions, infections, arthritis, and certain types of cancer. These hormones inhibit production of prostaglandins (which are mediators of inflammation). Glucocorticoids also reduce inflammation by decreasing capillary permeability, thereby reducing swelling. In addition, they reduce the effects of histamine and so are used to treat allergic symptoms.

When used in large amounts over long periods, glucocorticoids can cause serious side effects. Although they help stabilize lysosome membranes so that they do not destroy tissues with their potent enzymes, the ability of lysosomes to degrade foreign molecules is also decreased. Glucocorticoids decrease interleukin-1 production, blocking cell-mediated immunity and reducing the patient's ability to fight infections. Other side effects include ulcers, hypertension, diabetes mellitus, and atherosclerosis.

Abnormally large amounts of glucocorticoids, whether due to disease or drugs, can result in **Cushing's disease.** In this condition, fat is mobilized from the lower part of the body and deposited about the trunk. Edema gives the patient's face a full-moon appearance. Blood glucose level rises to as much as 50% above normal, causing *adrenal diabetes.* If this condition persists for several months, the beta cells in the pancreas may "burn out" from secreting excessive amounts of insulin, leading to permanent diabetes mellitus. Reduction in protein synthesis causes weakness and decreases immune responses, and a person who is untreated may die of infection.

Many other hormones are known

Many other tissues of the body secrete hormones. The **pineal gland,** located in the brain, produces **melatonin,** which is derived from the amino acid tryptophan, and influences biological rhythms and the onset of sexual maturity. In humans, melatonin facilitates the onset of sleep. Exposure to light suppresses melatonin secretion.

Several hormones secreted by the digestive tract and by adipose tissue regulate digestive processes (see Chapter 46). The **thymus gland** produces **thymosin,** a hormone that plays a role in immune responses. **Atrial natriuretic factor (ANF),** secreted by the atrium of the heart, promotes sodium excretion and lowers blood pressure (see Chapter 47). We discuss reproductive hormones in Chapter 49.

Review

- Why is the hypothalamus considered the link between the nervous and endocrine systems? What is the role of the anterior lobe of the pituitary? Of the posterior lobe?
- How are thyroid hormone secretion and parathyroid hormone secretion regulated?
- What are the antagonistic actions of insulin and glucagon in regulating blood glucose level? What is insulin resistance?
- What are the actions of epinephrine and norepinephrine? How is the adrenal medulla regulated?
- How do the adrenal glands help the body respond to stress?

SUMMARY WITH KEY TERMS

Learning Objectives

1 Compare endocrine with nervous system function, and describe how these systems work together to regulate body processes (page 1029).

   - The **endocrine system** consists of endocrine glands, cells, and tissues that secrete **hormones,** chemical signals that regulate physiological processes. Most endocrine responses are slow, but long-lasting. The endocrine system signals a wide range of target cell types.

   - The nervous system responds rapidly to stimuli by transmitting electrical and chemical signals. Neurons signal other neurons, muscle cells, and gland cells, including endocrine cells. The nervous system helps regulate many endocrine responses.

2 Summarize the regulation of endocrine action by negative feedback systems (page 1029).

■ Hormone secretion is typically regulated by **negative feedback systems.** A hormone is released in response to some change in a steady state and triggers a response that counteracts the changed condition; this process restores homeostasis.

3 Identify four main chemical groups to which hormones are assigned, and give two examples for each group (page 1029).

■ Hormones can be grouped as fatty acid derivatives, steroids, amino acid derivatives, or peptides and proteins. Prostaglandins and the juvenile hormone of insects are **fatty acid derivatives.** Hormones secreted by the adrenal cortex, ovary, and testis, as well as the molting hormone of insects are **steroid hormones.**

■ Thyroid hormones and epinephrine are **amino acid derivatives.** Antidiuretic hormone (ADH) and glucagon are examples of **peptide hormones.** Insulin is a small protein.

ThomsonNOW™ **Watch the mechanisms of action of steroid and peptide hormones by clicking on the figures in ThomsonNOW.**

4 Compare four types of endocrine signaling (page 1031).

■ In **classical endocrine signaling,** discrete **endocrine glands,** glands without ducts, secrete hormones into the interstitial fluid. Hormones are transported by the blood. They bind with receptors on or in specific **target cells.**

■ In **neuroendocrine signaling,** neurons secrete **neurohormones.** These hormones are transported down axons and then secreted and transported by the blood.

■ In **autocrine signaling,** a hormone (or other signal molecule) is secreted into the interstitial fluid and then acts on the very cell that produced it. In **paracrine signaling,** a hormone (or other signal molecule) diffuses through interstitial fluid and acts on nearby target cells.

■ Some **local regulators** are considered hormones; they use autocrine and paracrine signaling. Local regulators include **growth factors,** peptides that stimulate cell division and development, and **prostaglandins,** a group of local hormones that help regulate many metabolic processes.

5 Compare the mechanism of action of small, lipid-soluble hormones with that of hydrophilic hormones; include the role of second messengers, such as cyclic AMP (page 1033).

■ Steroid hormones and thyroid hormones are small, lipid-soluble molecules that pass through the plasma membrane and combine with receptors within the target cell; the hormone–receptor complex may activate or repress transcription of messenger RNA coding for specific proteins.

■ Peptide hormones are hydrophilic and do not enter target cells. They combine with receptors on the plasma membrane of target cells. Many hormones bind to **G protein–linked receptors** that act via **signal transduction.** An extracellular hormone signal is transduced into an intracellular signal by the receptor.

■ Most peptide hormones are first messengers that carry out their actions by way of **second messengers,** such as **cyclic AMP (cAMP).** The G protein–linked receptor activates a **G protein.** The G protein either stimulates or inhibits an enzyme that affects the second messenger. For example, G proteins stimulate or inhibit **adenylyl cyclase,** the enzyme that catalyzes the conversion of ATP to cAMP.

■ Many second messengers stimulate the activity of **protein kinases,** enzymes that phosphorylate specific proteins that affect the activity of the cell.

■ Some G proteins use phospholipid derivatives as second messengers. **Inositol trisphosphate (IP$_3$)** and **diacylglycerol (DAG)** are second messengers that increase calcium concentration and activate enzymes. Calcium ions bind with **calmodulin,** which activates certain enzymes.

■ **Receptor tyrosine kinases** are **enzyme-linked receptors** that bind growth factors, including insulin and nerve growth factor.

6 Identify four functions of invertebrate hormones, and describe the hormonal regulation of insect development (page 1035).

■ Invertebrate hormones and neurohormones help regulate metabolism, growth and development, regeneration, molting, metamorphosis, reproduction, and behavior.

■ Hormones control development in insects. When stimulated by some environmental factor, neuroendocrine cells in the insect brain secrete **brain hormone (BH).** BH stimulates the prothoracic glands to produce **molting hormone (MH),** which stimulates growth and molting.

■ In the immature insect, the **corpora allata** secrete **juvenile hormone (JH),** which suppresses metamorphosis at each larval molt. The amount of JH decreases with successive molts.

7 Identify the classical vertebrate endocrine glands, and describe the actions of their hormones (page 1036).

■ Consult Table 48-1 for a description of the main endocrine glands and their actions. Vertebrate hormones regulate growth and development, reproduction, salt and fluid balance, many aspects of metabolism, and behavior.

■ Endocrine disorders can result from **hyposecretion** (abnormally reduced output) or **hypersecretion** (abnormally increased output) of hormones.

8 Describe the mechanisms by which the hypothalamus and pituitary gland integrate regulatory functions; describe the actions of the hypothalamic and pituitary hormones (page 1036).

■ Nervous and endocrine regulation is integrated in the **hypothalamus,** which regulates the activity of the **pituitary gland.**

■ The neurohormones **oxytocin** and **antidiuretic hormone (ADH)** are produced by the hypothalamus and released by the **posterior lobe** of the pituitary. Oxytocin stimulates contraction of the uterus and stimulates ejection of milk by the mammary glands. ADH stimulates reabsorption of water by the kidney tubules.

■ The hypothalamus secretes **releasing hormones** and **inhibiting hormones** that regulate the hormone output of the **anterior lobe** of the pituitary gland.

■ The anterior lobe of the pituitary secretes prolactin, melanocyte-stimulating hormones, growth hormone, and several **tropic hormones** that stimulate other endocrine glands. **Melanocyte-stimulating hormones (MSH)** suppress appetite and are important in regulating energy and body weight. **Prolactin** stimulates the mammary glands to produce milk.

■ **Growth hormone (GH)** is an **anabolic hormone** that stimulates body growth by promoting protein synthesis. GH stimulates the liver to produce **insulin-like growth factors (IGFs),** which promote skeletal growth and general tissue growth.

9 Describe the actions of the thyroid and parathyroid hormones, their regulation, and the effects of malfunction (page 1036).

■ The **thyroid gland** secretes **thyroid hormones: thyroxine,** or **T$_4$,** and **triiodothyronine,** or **T$_3$.** Thyroid hormones stimulate the rate of metabolism.

- Regulation of thyroid secretion depends mainly on a negative feedback system between the anterior pituitary gland and the thyroid gland.

- Hyposecretion of thyroxine during childhood may lead to **cretinism;** during adulthood it may result in **myxedema. Goiter,** an abnormal enlargement of the thyroid gland, is associated with both hyposecretion and hypersecretion. The most common cause of hyperthyroidism is **Graves' disease,** an autoimmune disease.

- The **parathyroid glands** secrete **parathyroid hormone (PTH),** which regulates the calcium level in the blood. Parathyroid hormone increases calcium concentration by stimulating calcium release from bones, increasing calcium reabsorption by kidney tubules, and increasing calcium reabsorption from the intestine.

- **Calcitonin,** secreted by the thyroid gland, acts antagonistically to parathyroid hormone.

ThomsonNOW Explore the human endocrine glands and hormones by clicking on the figures in ThomsonNOW.

10 Contrast the actions of insulin and glucagon, and describe diabetes mellitus and hypoglycemia, including the metabolic effects of these disorders (page 1036).

- The **islets of Langerhans** in the pancreas secrete **insulin** and **glucagon.** Insulin and glucagon secretion is regulated directly by glucose concentration. Insulin stimulates cells to take up glucose from the blood, so it lowers blood glucose concentration. Glucagon raises blood glucose concentration by stimulating conversion of glycogen to glucose and by stimulating production of glucose from other nutrients.

- In **diabetes mellitus,** either insulin deficiency or **insulin resistance** results in decreased use of glucose, increased fat mobilization, increased protein use, and electrolyte imbalance. In hypoglycemia, impaired glucose tolerance results in a delayed insulin response, followed by hypersecretion of insulin. The glucose concentration falls, causing drowsiness.

11 Describe the actions and regulation of the adrenal hormones, including their role in helping the body respond to stress (page 1036).

- The **adrenal glands** secrete hormones that help the body respond to stress.

- The **adrenal medulla** secretes **epinephrine** and **norepinephrine,** hormones that increase heart rate, metabolic rate, and strength of muscle contraction. These hormones reroute blood to organs needed for fight or flight. The adrenal medulla is regulated by the nervous system.

- The **adrenal cortex** secretes sex hormones; **mineralocorticoids,** such as **aldosterone;** and **glucocorticoids,** such as **cortisol.** Aldosterone increases the rate of sodium reabsorption and potassium excretion by the kidneys.

- During stress, the adrenal cortex ensures adequate fuel supplies for the rapidly metabolizing cells. Cortisol promotes glucose synthesis from other nutrients. During stress, the hypothalamus secretes **corticotropin-releasing factor (CRF),** which stimulates the anterior pituitary gland to secrete **adrenocorticotropic hormone (ACTH).** ACTH regulates both aldosterone and cortisol secretion.

ThomsonNOW Learn more about the hormonal response to stress by clicking on the figure in ThomsonNOW.

TEST YOUR UNDERSTANDING

1. Which of the following is *not* true of endocrine glands? (a) they secrete hormones (b) they have ducts (c) their product is typically transported by the blood (d) they are typically regulated by negative feedback (e) an experimental animal exhibits symptoms of deficiency when they are removed

2. A cell secretes a product that diffuses through the interstitial fluid and acts on nearby cells. This is an example of (a) neuroendocrine secretion (b) autocrine signaling (c) paracrine signaling (d) classical endocrine control (e) peptide hormone function

3. Paracrine regulators that are derived from fatty acids and are found in many different organs are (a) prostaglandins (b) thyroid hormones (c) growth factors (d) anabolic steroids (e) G proteins

4. Which of the following is/are *true* of steroid hormones? (a) hydrophilic (b) secreted by the posterior pituitary (c) typically work through G proteins and cyclic AMP (d) typically bind with receptor in nucleus and affect transcription (e) a and c

5. Which of the following is *not* a correct pair? (a) neurohormone; insect brain hormone (b) calcium; calmodulin (c) posterior lobe of pituitary; releasing hormone (d) anterior lobe of pituitary; growth hormone (e) thyroid hyposecretion; cretinism

6. Which of the following activates a second messenger? (a) hormone–receptor complex (b) calcium ions (c) inositol trisphosphate (IP_3) (d) cyclic AMP (e) diacylglycerol (DAG)

7. Growth hormone (a) is regulated mainly by calcium level (b) stimulates the liver to produce insulin-like growth factors (c) is a catabolic hormone (d) stimulates metabolic rate (e) signals the hypothalamus to produce a releasing hormone

8. Arrange the following events into an appropriate sequence.

   1. high thyroid hormone concentration 2. anterior pituitary inhibited 3. homeostasis 4. lower level of thyroid-stimulating hormone 5. thyroid gland secretes less thyroid hormone

   (a) 1, 2, 4, 5, 3 (b) 5, 4, 3, 2, 1 (c) 1, 2, 5, 4, 3 (d) 4, 5, 2, 3, 1 (e) 1, 4, 2, 5, 3

9. Arrange the following events into an appropriate sequence.

   1. blood glucose concentration increases 2. alpha cells in islets stimulated 3. homeostasis 4. low blood glucose concentration 5. glucagon secretion increases

   (a) 1, 2, 3, 5, 4 (b) 5, 4, 2, 1, 3 (c) 1, 2, 5, 4, 3 (d) 4, 2, 5, 1, 3 (e) 4, 5, 1, 2, 3

10. Parathyroid hormone (a) increases glucose level in blood (b) helps body respond to stress (c) increases permeability of kidney tubules to water (d) promotes uptake of amino acids (e) increases calcium concentration in blood

11. Cortisol (a) decreases glucose level in blood (b) helps body respond to stress (c) increases permeability of kidney tubules to water (d) promotes uptake of amino acids (e) increases calcium concentration in blood

12. Which of the following is *not* a correct pair? (a) thyroid gland; calcitonin (b) islets of Langerhans; glucagon (c) posterior lobe of pituitary; oxytocin (d) anterior lobe of pituitary; cortisol (e) adrenal medulla; epinephrine

13. Which of the following occurs in diabetes mellitus? (a) decreased use of glucose (b) decreased fat metabolism (c) decreased protein use (d) increased concentration of TSH-releasing hormone (e) b and c

14. Insulin resistance is associated with (a) low insulin secretion by the islets of Langerhans (b) type 1 diabetes (c) impaired use of insulin by target cells (d) hypoglycemia (e) hypersecretion of glucagon

15. Aldosterone (a) is released by posterior pituitary (b) is an androgen (c) secretion is stimulated by an increase in thyroid-stimulating hormone (d) is an enzyme that converts epinephrine to norepinephrine (e) increases sodium reabsorption

16. Which of the following is an action of epinephrine? (a) decreases glucose use (b) increases cardiac output (c) constricts blood vessels in brain (d) reduces inflammation (e) mobilizes fat

17. Label the diagram. Use Figure 48-7 to check your answers.

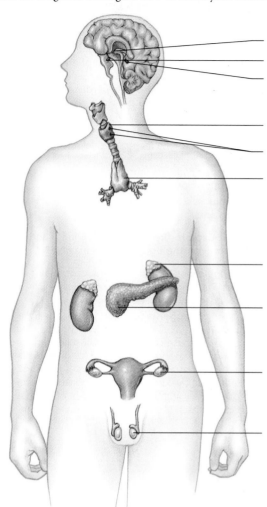

CRITICAL THINKING

1. How do receptors impart specificity within the endocrine system? What might be some advantages of having complex mechanisms for hormone action (such as second messengers)?

2. Why do you think it is important to maintain a constant blood glucose level? Several hormones discussed in this chapter affect carbohydrate metabolism. Why is it important to have more than one? How do they interact?

3. An injection of too much insulin may cause a diabetic to go into insulin shock, in which the person may appear drunk or may become unconscious, suffer convulsions, and even die. From what you know about the actions of insulin, explain the physiological causes of insulin shock.

4. **Evolution Link.** The receptor for aldosterone appears to have evolved much earlier than aldosterone itself. Propose a hypothesis to explain how this might have occurred.

5. **Evolution Link.** You would think that the genes that increase the risk for type 2 diabetes would be disadvantageous, but these genes have been preserved by natural selection. The "thrifty gene" hypothesis holds that these genes allow efficient utilization of food (and fat accumulation) when food is available to increase chances of survival in times of food scarcity. Use this hypothesis to explain the rising rate of diabetes in the United States.

Additional questions are available in ThomsonNOW at www.thomsonedu.com/login

49

Reproduction

Bruce Watkins/Animals Animals

Mating nudibranchs (*Chromodoris*). Photographed in Indonesia.

KEY CONCEPTS

Most animals reproduce sexually by fusion of sperm and egg, but some animals reproduce asexually and some can reproduce either way, depending on conditions.

In vertebrates, gonads (testes or ovaries) produce gametes and reproductive hormones.

In humans and other mammals, reproduction is regulated by hormones produced by the hypothalamus, pituitary gland, and gonads.

In human females, hormones maintain a monthly menstrual cycle that prepares the body for possible pregnancy; ovulation typically occurs midcycle.

When a secondary oocyte is fertilized, development begins and the embryo implants in the wall of the uterus; hormones from the developing embryo, from the corpus luteum, and later from the placenta maintain the pregnancy.

The survival of each species requires that its members produce new individuals to replace those that die. The ability to reproduce and perpetuate its species is a basic characteristic of living things. To paraphrase American sociobiologist E. O. Wilson, animal reproduction is really an animal's way of making more copies of its genes. How this is accomplished depends on the relative benefits (and costs) of the reproductive strategies available in any given situation.

The mating nudibranchs (marine slugs) in the photograph are *hermaphrodites*. Each animal has both male and female organs and produces both sperm and eggs. However, they do not self-fertilize. Two nudibranchs come together to mate and exchange sperm through a tube near the head. Each then goes its own way and lays masses containing millions of eggs.

In this chapter we summarize some major features of animal reproduction and discuss the evolutionary advantages of sexual reproduction. We then focus on the process of human reproduction and on the hormones that regulate reproduction. We conclude the chapter with discussions of contraception and sexually transmitted disease. ■

ASEXUAL AND SEXUAL REPRODUCTION

Learning Objective

1 Compare the benefits of asexual and sexual reproduction, and describe each mode of reproduction, giving specific examples.

Most animals carry on *sexual reproduction,* and some carry on *asexual reproduction.* Some animals reproduce asexually under some conditions and sexually at other times. As we will discuss, many variations of both asexual and sexual reproduction have evolved.

Asexual reproduction is an efficient strategy

In **asexual reproduction,** a single parent gives rise to offspring that are genetically identical to the parent (unless there are mutations). Many invertebrates, including sponges, cnidarians, and some rotifers, flatworms, and annelids, can reproduce asexually. Some vertebrates also reproduce asexually under certain conditions. Asexual reproduction is an adaptation of some sessile animals that cannot move about to search for mates. For animals that do move about, asexual reproduction can be advantageous when the population density is low and mates are not readily available.

In asexual reproduction, a single parent may split, bud, or fragment to give rise to two or more offspring. Sponges and cnidarians are among the animals that can reproduce by **budding.** A small part of the parent's body separates from the rest and develops into a new individual (Fig. 49-1). Sometimes the buds remain attached and become more or less independent members of a colony.

Oyster farmers learned long ago that when they tried to kill sea stars by chopping them in half and throwing the pieces back into the sea, the number of sea stars preying on the oyster bed doubled! In some flatworms, nemerteans, and annelids, this ability to regenerate is part of a method of reproduction known as **fragmentation.** The body of the parent breaks into several pieces; each piece regenerates the missing parts and develops into a whole animal.

Parthenogenesis ("virgin development") is a form of asexual reproduction in which an unfertilized egg develops into an adult animal. The adult is typically haploid. Parthenogenesis is common among insects (especially honeybees and wasps) and crustaceans; it also occurs among some other invertebrate and vertebrate groups, including some species of nematodes, gastropods, fishes, amphibians, and reptiles.

Although a few species appear to reproduce solely by parthenogenesis, in most species episodes of parthenogenesis alternate with periods of sexual reproduction. Parthenogenesis may occur for several generations, followed at some point by sexual reproduction in which males develop, produce sperm, and mate with the females to fertilize their eggs. In some species, parthenogenesis is a means of rapidly producing individuals when conditions are favorable.

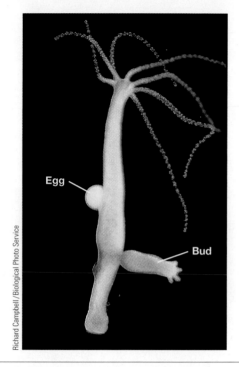

Richard Campbell / Biological Photo Service

Figure 49-1 Asexual reproduction by budding

A part of *Hydra's* body grows outward, then separates and develops into a new individual. The region of the parent body that buds is not specialized exclusively for reproduction. The *Hydra* shown here is also reproducing sexually, as evidenced by the egg (*left*).

Most animals reproduce sexually

Sexual reproduction in animals involves the production and fusion of two types of **gametes**—sperm and eggs. Typically, two different individuals are required. A male parent contributes **sperm,** and a female parent contributes an egg, or **ovum** (pl., *ova*). The sperm provides genes coding for some of the male parent's traits, and the egg contributes genes coding for some of the female parent's traits. The egg is typically large and nonmotile, with a store of nutrients that supports the development of the embryo. The sperm is usually small and motile and adapted to propel itself by beating its long, whiplike flagellum.

When sperm and egg unite, a **zygote,** or fertilized egg, is produced. The zygote develops into a new animal, similar to both parents but not identical to either. Sexual reproduction typically involves remarkably complex structural, functional, and behavioral processes. In vertebrates, hormones secreted by the hypothalamus, pituitary gland, and gonads regulate these processes.

Many aquatic animals practice **external fertilization** in which the gametes meet outside the body (Fig. 49-2a). Mating partners usually release eggs and sperm into the water simultaneously. Gametes live for only a short time, and many are lost in the water; some are eaten by predators. However, so many gametes are released that sufficient numbers of sperm and egg cells meet to perpetuate the species.

In **internal fertilization,** matters are left less to chance. The male generally delivers sperm cells directly into the body of the

(a) External fertilization. Like many aquatic animals, these spawning frogs (*Rana temporaria*) release their gametes into the water. The female lays a mass of eggs, while the male mounts her and simultaneously deposits his sperm in the water.

(b) Internal fertilization. In most terrestrial animals, such as these lions (*Panthera leo*), the male deposits sperm inside the female body. Internal fertilization is also practiced by some fishes and some aquatic reptiles and mammals.

Figure 49-2 *Animated* External and internal fertilization

female. Her moist tissues provide the watery medium required for the movement of sperm, and the gametes fuse inside the body. Most terrestrial animals, sharks, and aquatic reptiles, birds, and mammals practice internal fertilization (❙ Fig. 49-2b).

Hermaphroditism is a form of sexual reproduction in which a single individual produces both eggs and sperm. A few hermaphrodites, such as the tapeworm, are capable of self-fertilization. More typically, two animals come together and fertilize one another's eggs (see photograph of mating nudibranchs at beginning of chapter). The common earthworm is also a hermaphrodite. Two animals copulate, and mutual cross-fertilization occurs, with each inseminating the other. In some hermaphroditic species, self-fertilization is prevented by the development of testes and ovaries at different times.

Sexual reproduction increases genetic variability

Asexual reproduction is actually the fastest and most efficient way to reproduce. Compared to asexual reproduction, sexual reproduction is more expensive in terms of energy, because the animal must produce gametes and find mates. It is also less efficient, because two cells, rather than just one, are required to make a new organism. Why, then, do most animals reproduce sexually?

Many biologists hold that a major benefit of sexual reproduction is that it increases the *fitness* (reproductive success) of off-spring. In contrast to asexual reproduction, in which an animal passes all of its genes to its offspring, sexual reproduction has the biological advantage of promoting genetic variety among the members of a species. Each offspring is the product of a particular combination of genes contributed by both parents rather than a genetic copy of a single individual. By combining inherited traits of two parents, sexual reproduction gives rise to at least some offspring that may be better able to survive than either parent. Also, because the offspring are diploid, they have a backup copy of their genes in case one copy gets damaged by mutation.

Although they generally agree that sexual reproduction has some selective advantage, biologists do not agree on the details. They are exploring several hypotheses, including the following. Sexual reproduction is advantageous because it permits beneficial mutations from each parent to come together in offspring that can reproduce and spread these mutations through the population. For example, certain beneficial mutations may permit animals to protect themselves from predators or resist parasites. Sexual reproduction provides a mechanism for such mutations to spread through the population.

Sexual reproduction also removes harmful mutations from a population. Mutations occur constantly, and most mutations are harmful. When animals reproduce asexually, all the offspring inherit all the harmful mutations. As mutations accumulate in a population, individuals carry a bigger and bigger load of harmful genes. In contrast, when animals with different mutations mate,

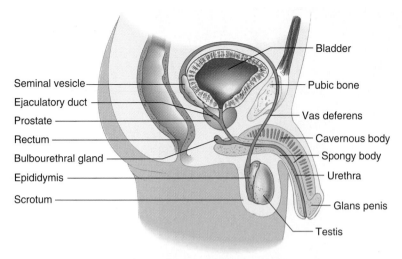

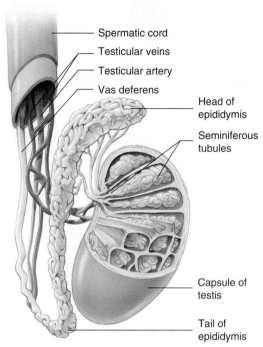

Seminal vesicle

Ejaculatory duct

Prostate

Rectum

Bulbourethral gland

Epididymis

Scrotum

Bladder

Pubic bone

Vas deferens

Cavernous body

Spongy body

Urethra

Glans penis

Testis

(a) The scrotum, penis, and pelvic region of the human male are shown in sagittal section to illustrate their internal structure.

Figure 49-3 *Animated* Male reproductive system

Spermatic cord

Testicular veins

Testicular artery

Vas deferens

Head of epididymis

Seminiferous tubules

Capsule of testis

Tail of epididymis

(b) The testis, epididymis, and spermatic cord are shown partly dissected and exposed. The testis is shown in sagittal section to illustrate the arrangement of the seminiferous tubules.

offspring inherit varying numbers and combinations of mutations. Offspring that inherit too many harmful mutations are selected against. They may not live to reproduce, so their harmful mutations are removed from the population.

Wayne Getz, an applied mathematician at the University of California, Berkeley, developed a mathematical model that predicts whether asexual or sexual reproduction will exist within a population. According to his model, clones of animals produced by asexual reproduction would be favored in an unchanging environment. The model further predicts that sexual reproduction will be adaptive in an unstable, changing environment and that both types of reproduction can coexist under conditions of moderate change.

Competing hypotheses are an expected part of the scientific process because scientific discovery is rarely a straightforward, linear sequence of question–answer, question–answer. In fact, as scientific knowledge expands, many creative hypotheses are discarded as dead ends. Sometimes researchers show that competing hypotheses may each explain part of the problem under scrutiny.

Review

- How would you distinguish among budding, fragmentation, and parthenogenesis?
- What are the advantages and disadvantages of asexual reproduction compared to sexual reproduction?

HUMAN REPRODUCTION: THE MALE

Learning Objectives

2 Describe the structure and function of each organ of the human male reproductive system.

3 Trace the passage of sperm cells through the human male reproductive system from their origin in the seminiferous tubules to their expulsion from the body in the semen. (Include a description of spermatogenesis.)

4 Describe endocrine regulation of reproduction in the human male.

The human male, like other male mammals, has the reproductive role of producing sperm cells and delivering them into the female reproductive tract. When a sperm combines with an egg, the sperm contributes its genes and determines the sex of the offspring.

The testes produce gametes and hormones

In humans and other vertebrates, **spermatogenesis,** the process of sperm cell production, occurs in the paired male gonads, or **testes** (sing., *testis*) (▌Figure 49-3a). Spermatogenesis takes place within a vast tangle of hollow tubules, the **seminiferous tubules,** within each testis (▌Fig. 49-3b). Spermatogenesis begins with undifferentiated cells, the **spermatogonia** in the walls of the tubules (▌Fig. 49-4).

The spermatogonia, which are diploid cells, divide by mitosis and produce more spermatogonia. Some enlarge and become **primary spermatocytes,** which undergo **meiosis** and produce haploid gametes. (You may want to review the discussion of meiosis in Chapter 10.) In many animals, gamete production occurs only in the spring or fall, but humans have no special breeding season. In the human adult male, spermatogenesis proceeds continuously, and millions of sperm are produced each day.

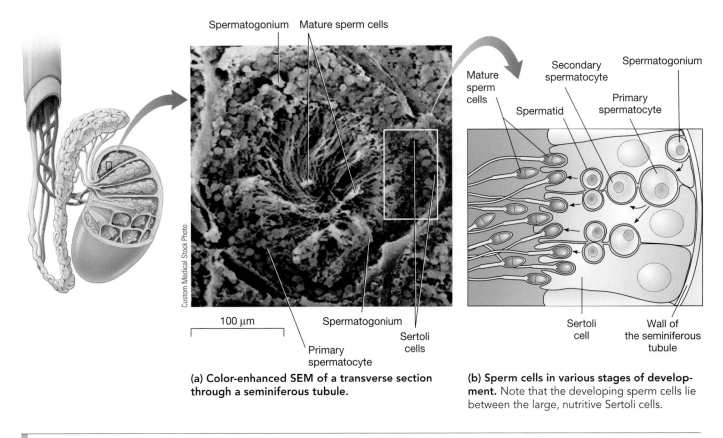

Spermatogonium Mature sperm cells

100 μm

Primary spermatocyte

Spermatogonium

Sertoli cells

Mature sperm cells

Spermatid

Secondary spermatocyte

Primary spermatocyte

Spermatogonium

Sertoli cell

Wall of the seminiferous tubule

Custom Medical Stock Photo

(a) Color-enhanced SEM of a transverse section through a seminiferous tubule.

(b) Sperm cells in various stages of development. Note that the developing sperm cells lie between the large, nutritive Sertoli cells.

Figure 49-4 *Animated* Spermatogenesis in the seminiferous tubules

Each primary spermatocyte undergoes a first meiotic division, which produces two haploid **secondary spermatocytes** (▌Fig. 49-5). During the second meiotic division, each of the two secondary spermatocytes gives rise to two haploid **spermatids.** Four spermatids are produced from the original primary spermatocyte. Each spermatid differentiates into a mature sperm. The sequence is as follows:

Spermatogonium (diploid) ⟶ primary spermatocyte (diploid) ⟶ two secondary spermatocytes (haploid) ⟶ four spermatids (haploid) ⟶ four mature sperm (haploid)

Each mature sperm consists of a head, midpiece, and flagellum (▌Fig. 49-6). The head consists almost entirely of the nucleus. Part of the nucleus is covered by the **acrosome,** a vesicle that differentiates from the Golgi complex. The acrosome contains proteins, including enzymes that help the sperm penetrate the egg.

Mitochondria, located in the midpiece of the sperm, provide the energy for movement of the flagellum. The sperm flagellum has the typical eukaryotic 9 + 2 arrangement of microtubules. During its development, most of the sperm's cytoplasm is discarded and is phagocytosed by the large **Sertoli cells** that ring the fluid-filled lumen of the seminiferous tubule. These cells provide nutrients for the developing sperm cells. Sertoli cells also secrete hormones and other signaling molecules.

Each Sertoli cell extends from the outer membrane of the seminiferous tubule to its lumen. Sertoli cells are joined to one another by *tight junctions* at a place just within the outer membrane of the tubule (see Chapter 5 for discussion of tight junctions). Together, Sertoli cells form a blood–testis barrier that prevents harmful substances from entering the tubule and interfering with spermatogenesis. This barrier also prevents sperm from passing out of the tubule into the blood, where they could stimulate an immune response. The tight junctions between Sertoli cells form compartments that separate sperm cells in various stages of development.

Human sperm cells cannot develop at body temperature. Although the testes develop within the abdominal cavity of the male embryo, about 2 months before birth they descend into the **scrotum,** a skin-covered sac suspended from the groin. The scrotum serves as a cooling unit, maintaining sperm below body temperature. In rare cases, the testes do not descend. If this condition is not corrected by surgery or hormone treatment, the seminiferous tubules eventually degenerate and the male becomes **sterile,** unable to produce offspring.

The scrotum is an outpocketing of the pelvic cavity and is connected to it by the **inguinal canals.** As they descend, the testes pull their blood vessels, nerves, and conducting tubes after them. The inguinal region is a weak place in the abdominal wall. Straining the abdominal muscles by lifting heavy objects sometimes tears the inguinal tissue. A loop of intestine can then bulge into the scrotum through the tear, a condition known as an *inguinal hernia.*

A series of ducts store and transport sperm

Sperm cells leave the seminiferous tubules of each testis and pass into a larger coiled tube, the **epididymis.** There, sperm finish maturing and are stored. During ejaculation, sperm pass from each epididymis into a sperm duct, the **vas deferens** (pl., *vasa*

deferentia). The vas deferens extends from the scrotum through the inguinal canal and into the pelvic cavity.

Each vas deferens empties into a short **ejaculatory duct,** which passes through the prostate gland and then opens into the single **urethra.** The urethra, which at different times conducts urine and semen, passes through the penis to the outside of the body. Thus, the sperm pass through the following structures:

Seminiferous tubules $\longrightarrow$ epididymis $\longrightarrow$ vas deferens $\longrightarrow$
ejaculatory duct $\longrightarrow$ urethra $\longrightarrow$ release from body

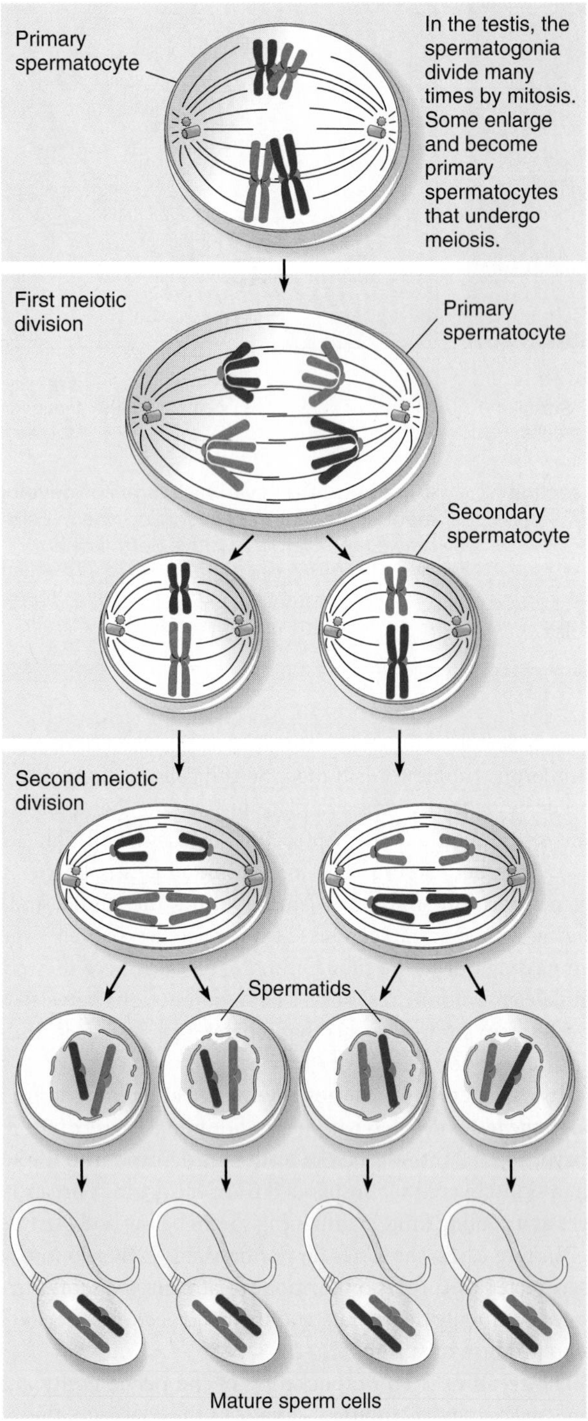

The accessory glands produce the fluid portion of semen

As sperm travel through the conducting tubes, they mix with secretions from three types of accessory glands. Approximately 3 mL of **semen** is ejaculated during sexual climax. Semen consists

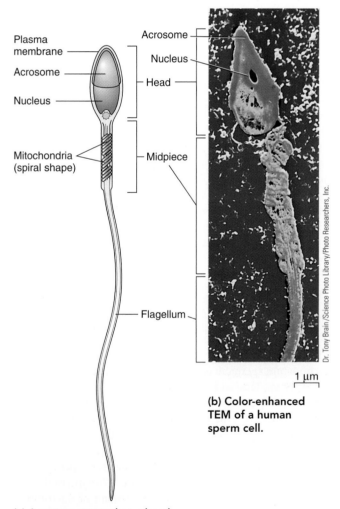

Dr. Tony Brain /Science Photo Library/Photo Researchers, Inc.

(b) Color-enhanced TEM of a human sperm cell.

(a) A mature sperm has a head, midpiece, and flagellum. The acrosome contains enzymes important in penetrating the egg.

Figure 49-5   Spermatogenesis

Note in this example that four chromosomes are present in the primary spermatocyte (2*n*) and that meiosis produces the haploid (*n*) number (2) in the secondary spermatocytes, spermatids, and mature sperm cells.

Figure 49-6   Structure of a mature sperm

of about 200 million sperm cells suspended in the secretions of these glands.

The paired **seminal vesicles** secrete a fluid rich in fructose and prostaglandins into the vasa deferentia (see Fig. 49-3). The fructose provides energy for the sperm after they are ejaculated. Prostaglandins stimulate contractions of the uterus, which help move sperm up the female reproductive tract. The single **prostate gland** secretes an alkaline fluid containing calcium, citric acid, and enzymes. The prostatic fluid may be important in neutralizing the acidic environment of the vagina and in increasing sperm cell motility. During sexual arousal, the paired **bulbourethral glands,** located on each side of the urethra, release a mucous secretion. This fluid lubricates the penis, facilitating its penetration into the vagina.

A major cause of male infertility is insufficient sperm production. When sperm counts drop below 35 million per milliliter of semen, fertility is impaired, and males with a sperm count lower than 20 million per milliliter are usually considered sterile. When a couple's attempts to produce a child are unsuccessful, a sperm count and analysis may be performed in a clinical laboratory. Sometimes semen is found to contain large numbers of abnormal sperm or, occasionally, no sperm at all.

In the United States in the 1970s, an average healthy young man produced about 100 million sperm per milliliter of semen. Today that average has dropped to about 60 million. Although the cause of this decrease is not known, low sperm counts have been linked to a variety of environmental factors, including chronic marijuana use, alcohol abuse, and cigarette smoking. Studies also show that men who smoke tobacco are more likely than nonsmokers to produce abnormal sperm. Exposure to industrial and environmental toxins such as DDT and PCBs (polychlorinated biphenyls) may contribute to low sperm count and sterility. The use of anabolic steroids by athletes to accelerate muscle development can cause sterility in both males and females (see *Focus On: Anabolic Steroids and Other Abused Hormones,* in Chapter 48).

The penis transfers sperm to the female

The **penis** is an erectile copulatory organ that delivers sperm into the female reproductive tract. It is a long shaft that enlarges to form an expanded tip, the **glans.** Part of the loose-fitting skin of the penis folds down and covers the proximal portion of the glans, forming a cuff called the **prepuce,** or foreskin. The foreskin is removed during circumcision (a procedure commonly performed on male babies either for hygienic or religious reasons).

Under the skin, the penis consists of three parallel columns of **erectile tissue:** two **cavernous bodies** and one **spongy body** (Fig. 49-7). The spongy body surrounds the portion of the urethra that passes through the penis. Erectile tissue contains numerous blood vessels. When the male is sexually stimulated, parasympathetic neurons dilate the arteries in the penis. As blood fills the blood vessels of the erectile tissue, the tissue swells. This compresses veins that conduct blood away from the penis, slowing the outflow of blood. Thus, more blood enters the penis than can leave, further engorging the erectile tissue with blood. **Penile erection** occurs; the penis increases in length, diameter, and firmness. Although the human penis contains no bone, penis bones do occur in some other mammals, such as bats, rodents,

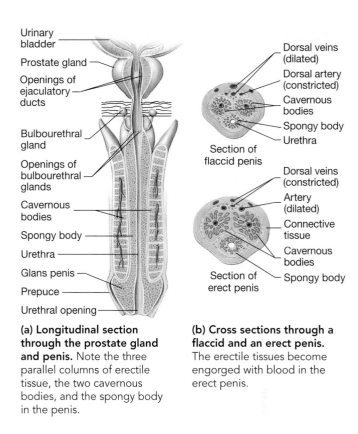

(a) Longitudinal section through the prostate gland and penis. Note the three parallel columns of erectile tissue, the two cavernous bodies, and the spongy body in the penis.

(b) Cross sections through a flaccid and an erect penis. The erectile tissues become engorged with blood in the erect penis.

Figure 49-7 Internal structure of the penis

and some primates. Erectile dysfunction, the chronic inability to sustain an erection, prevents effective sexual intercourse. Associated with a variety of physical causes and psychological issues, this common disorder is now treated with sildenafil (Viagra). This drug blocks the action of an enzyme that breaks down the signaling molecule that sustains erection (see Chapter 6 for a more detailed discussion).

Testosterone has multiple effects

Testosterone is the principal **androgen,** or male sex hormone. It is a steroid produced by the **interstitial cells** between the seminiferous tubules in the testes. Testosterone has many functions, beginning during early embryonic development when it stimulates development of the primary male reproductive organs (Table 49-1). Later in development, testosterone stimulates descent of the testes into the scrotum.

Puberty is the period of sexual maturation during which the secondary sex characteristics begin to develop and the individual becomes capable of reproducing. In males, puberty begins between ages 10 and 12 and continues until ages 16 to 18. Testosterone directly affects muscle and bone and stimulates the adolescent growth spurt in males at puberty. It produces the male's **primary sex characteristics:** growth of the reproductive organs and spermatogenesis. Testosterone also stimulates the development of the **secondary sex characteristics** at puberty, including growth of facial and body hair, muscle development, and the increase in vocal cord length and thickness that causes the voice to deepen. Testosterone is necessary for normal sex drive.

TABLE 49-1

Principal Male Reproductive Hormones

Endocrine Gland and Hormones	Principal Target Tissue	Principal Actions
Hypothalamus Gonadotropin-releasing hormone (GnRH)	Anterior pituitary	Stimulates release of FSH and LH
Anterior Pituitary Follicle-stimulating hormone (FSH)	Testes	Stimulates development of seminiferous tubules; stimulates spermatogenesis
Luteinizing hormone (LH)	Testes	Stimulates interstitial cells to secrete testosterone
Testes Testosterone	General	*Before birth:* stimulates development of primary sex organs and descent of testes into scrotum *At puberty:* responsible for growth spurt; stimulates development of reproductive structures and secondary sex characteristics *In adult:* maintains secondary sex characteristics; stimulates spermatogenesis
Inhibin	Anterior pituitary	Inhibits FSH secretion

In some of its target tissues, testosterone is converted to other steroids. Interestingly, in brain cells testosterone is converted to *estradiol,* the principal female sex hormone. The implications of this transformation are not yet understood.

The hypothalamus, pituitary gland, and testes regulate male reproduction

As you read the following description of male hormone action and regulation, follow the steps in ▌Figure 49-8. Use the numbered circles as a guide. When a boy is about 10 years old the hypothalamus begins to secrete **gonadotropin-releasing hormone (GnRH)** ❶. GnRH stimulates the anterior pituitary to secrete the *gonadotropic hormones* **follicle-stimulating hormone (FSH)** and **luteinizing hormone (LH)** ❷. Both FSH and LH are glycoproteins that use cyclic AMP as a second messenger (see Chapter 48). FSH stimulates Sertoli cells to secrete **androgen-binding protein (ABP)** and other signaling molecules that are necessary for spermatogenesis ❸. LH stimulates the interstitial cells to secrete testosterone.

FSH, LH, and testosterone all directly or indirectly stimulate testosterone secretion and spermatogenesis. A high concentration of testosterone in the testes is required for spermatogenesis. Testosterone and FSH stimulate Sertoli cells to produce ABP, which binds to testosterone and concentrates it in the tubules ❹. Testosterone also maintains the male secondary sex characteristics ❺.

Reproductive hormone concentrations are regulated by negative feedback mechanisms (see Fig. 49-8b). Testosterone acts on the hypothalamus, decreasing its secretion of GnRH, which decreases FSH and LH secretion by the pituitary ❻. Testosterone also directly inhibits the anterior lobe of the pituitary by blocking the normal actions of GnRH on LH synthesis and release.

FSH secretion is inhibited mainly by **inhibin,** a peptide hormone secreted by Sertoli cells ❼. FSH itself stimulates inhibin secretion. The endocrine regulation of reproductive function is complex, and it is likely that other hormones and signaling molecules will be identified.

Insufficient testosterone results in sterility. If a male is **castrated,** that is, the testes are removed, before puberty, he is deprived of testosterone and becomes a eunuch. He retains childlike sexual organs and does not develop secondary sexual characteristics. If castration occurs after puberty, increased secretion of male hormones by the adrenal glands helps maintain masculinity.

Review

- ▌ What is the physiological basis for erection of the penis?
- ▌ What are the functions of the testes?
- ▌ Trace the passage of sperm from a seminiferous tubule through the male reproductive system until it leaves the male body during ejaculation.
- ▌ What are the actions of testosterone? Give an overview of the endocrine regulation of male reproduction.

HUMAN REPRODUCTION: THE FEMALE

Learning Objectives

5 Describe the structure and function of each organ of the human female reproductive system.

6 Trace the development of a human egg and its passage through the female reproductive system until it is fertilized.

7 Describe the endocrine regulation of reproduction in the human female, and identify the important events of the menstrual cycle, such as ovulation and menstruation.

The female reproductive system produces oocytes (immature gametes), receives the penis and sperm released from it during

Hormones from the hypothalamus, anterior pituitary, and testes regulate male reproduction by negative feedback systems.

1. Hypothalamus secretes GnRH, which stimulates anterior pituitary.

2. Pituitary secretes FSH and LH.

3. FSH stimulates Sertoli cells to secrete ABP and other signaling molecules necessary for spermatogenesis. LH stimulates interstitial cells to secrete testosterone.

4. Testosterone stimulates Sertoli cells, leading to stimulation of spermatogenesis.

5. Testosterone maintains secondary sex characteristics.

6. Testosterone inhibits GnRH secretion by hypothalamus and FSH and LH secretion by pituitary.

7. Inhibin inhibits FSH secretion.

(a) Overview of hormone action.

(b) Negative feedback systems regulate hormone concentrations.

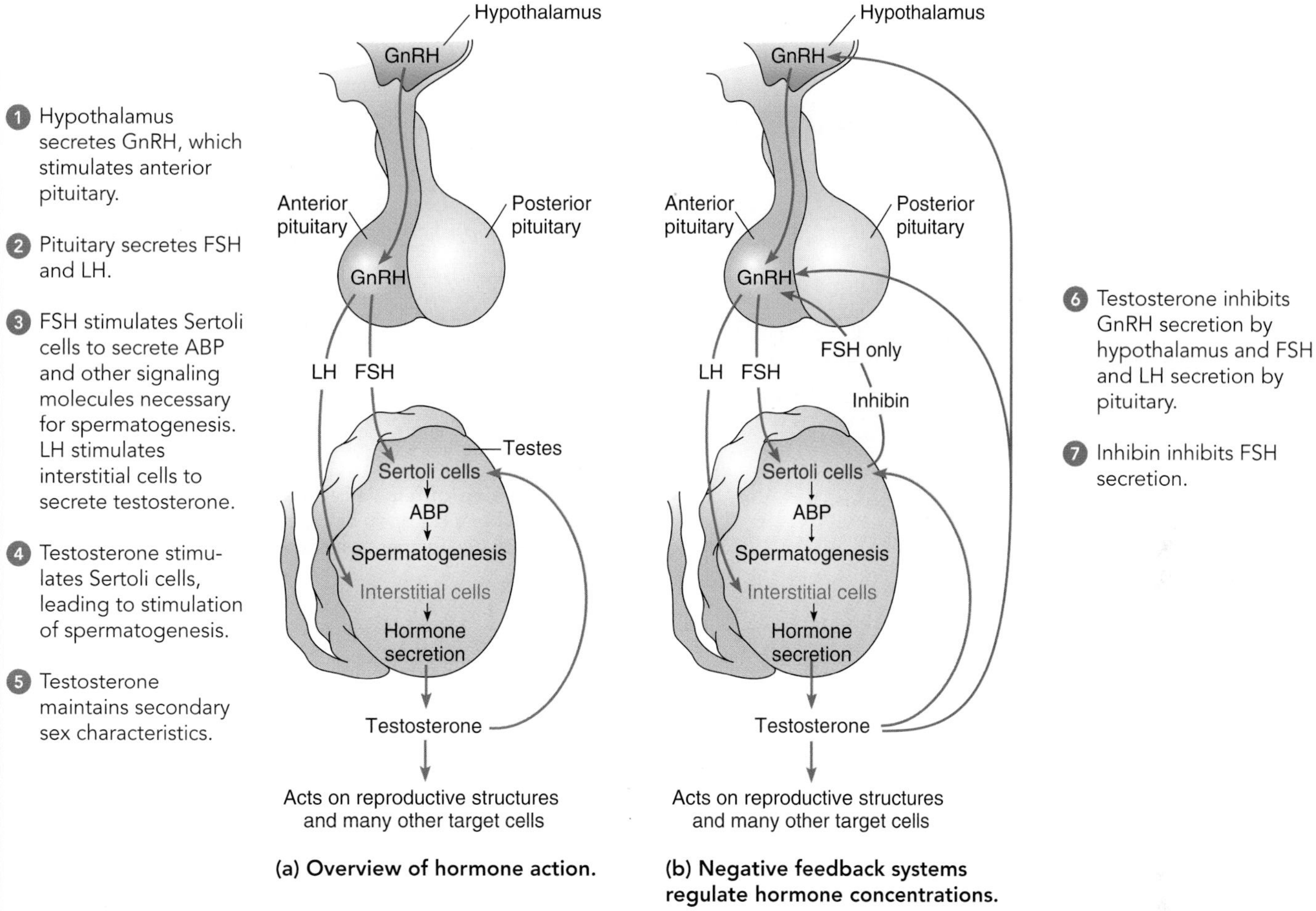

Figure 49-8 *Animated* Regulation of male reproduction

Green arrows indicate stimulation; red arrows indicate inhibition. (GnRH, gonadotropin-releasing hormone; FSH, follicle-stimulating hormone; LH, luteinizing hormone; ABP, androgen-binding protein)

sexual intercourse, houses and nourishes the embryo during prenatal development, gives birth, and produces milk for the young (lactation). Hormones secreted by the hypothalamus, pituitary gland, and ovaries interact to regulate and coordinate these processess.

The ovaries produce gametes and sex hormones

Like the male gonads, the female gonads, or **ovaries,** produce both gametes and sex hormones. About the size and shape of large almonds, the ovaries lie close to the lateral walls of the pelvic cavity and are held in position by several connective tissue ligaments (Fig. 49-9). Internally, the ovary consists mainly of connective tissue containing scattered ova in various stages of maturation.

The process of ovum production, called **oogenesis,** begins in the ovaries. Before birth, hundreds of thousands of **oogonia** are present in the ovaries. All of a female's oogonia form during embryonic development. No new oogonia are formed after birth. During prenatal development, the oogonia increase in size and become **primary oocytes.** By the time of birth, they are in the prophase of the first meiotic division. At this stage, they enter a resting phase that lasts throughout childhood and into adult life.

A primary oocyte and the **granulosa cells** surrounding it together make up a **follicle** (Fig. 49-10). The granulosa cells are connected by tight junctions that form a protective barrier around the oocyte. With the onset of puberty, a few follicles begin

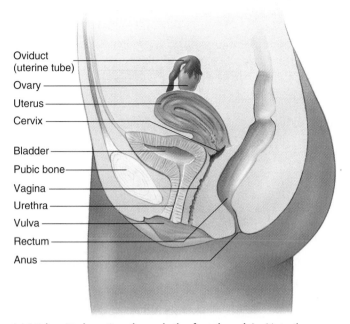

Oviduct (uterine tube)
Ovary
Uterus
Cervix
Bladder
Pubic bone
Vagina
Urethra
Vulva
Rectum
Anus

(a) Midsagittal section through the female pelvis. Note the position of the uterus relative to the vagina.

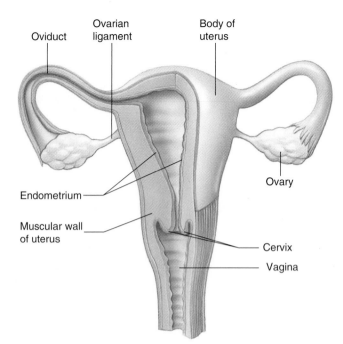

Oviduct
Ovarian ligament
Body of uterus
Endometrium
Muscular wall of uterus
Ovary
Cervix
Vagina

(b) Anterior view of female reproductive system. Some organs have been cut open to expose their internal structure. Connective tissue ligaments anchor the reproductive organs in place.

Figure 49-9 *Animated*   Female reproductive system

to mature each month in response to FSH secreted by the anterior pituitary gland. As a follicle grows, the granulosa cells proliferate, forming several layers. Connective tissue cells surrounding the granulosa cells differentiate to form a layer of **theca cells.**

As the follicle matures, the primary oocyte completes its first meiotic division. The two haploid cells produced differ in size (❚ Fig. 49-11). The smaller one, the first **polar body,** may later divide, forming two polar bodies, but these eventually disintegrate. The larger cell, the **secondary oocyte,** proceeds to the second meiotic division but remains in metaphase II until it is fertilized. When meiosis continues, the second meiotic division gives rise to a single ovum and a second polar body. The polar bodies are small and apparently dispose of unneeded chromosomes with a minimal amount of cytoplasm. The sequence is as follows:

Oogonium (diploid) ⟶ primary oocyte (diploid) ⟶ secondary oocyte + first polar body (both haploid) ⟶ (after fertilization) ovum + second polar body (both haploid)

Recall that in the male, each primary spermatocyte gives rise to four functional sperm cells. In contrast, each primary oocyte generates only one ovum.

As an oocyte develops, it becomes separated from its surrounding follicle cells by a layer of glycoproteins called the **zona pellucida.** As the follicle develops, follicle cells secrete fluid, which collects in the antrum (space) between them (see Fig. 49-10). The follicle cells also secrete **estrogens,** female sex hormones. The principal estrogen is **estradiol** (see Fig. 48-2b). Typically, only one follicle fully matures each month. Several others may develop for a while and then deteriorate by apoptosis.

As a follicle matures, it moves closer to the surface of the ovary, eventually resembling a fluid-filled bulge on the ovarian surface. Follicle cells secrete proteolytic enzymes that break down a small area of the ovary wall. During **ovulation,** the secondary oocyte ejects through the ovary wall and into the pelvic cavity. The portion of the follicle that remains in the ovary develops into the **corpus luteum,** a temporary endocrine gland that secretes estrogen and **progesterone.**

The oviducts transport the secondary oocyte

Almost immediately after ovulation, the secondary oocyte is swept into the funnel-shaped opening of the **oviduct,** also called the **uterine tube** or *fallopian tube.* Action of cilia on the epithelial lining of the oviduct both sweeps the secondary oocyte into the oviduct and moves it along toward the uterus. Fertilization takes place within the oviduct. If fertilization does not occur, the secondary oocyte degenerates there.

Scarring of the oviducts (for example, by *pelvic inflammatory disease,* caused by a sexually transmitted disease) can block the tubes so that the fertilized ovum cannot pass to the uterus. Women with blocked oviducts may be infertile. Sometimes partial constriction of the oviduct results in *tubal pregnancy,* in which the embryo begins to develop in the wall of the oviduct because it cannot progress to the uterus. Oviducts are not adapted to bear the burden of a developing embryo; thus, the oviduct and the embryo it contains must be surgically removed before it ruptures and endangers the life of the mother.

The uterus incubates the embryo

The oviducts open into the upper corners of the pear-shaped **uterus** (see Fig. 49-9b). About the size of a fist, the uterus (or womb) occupies a central position in the pelvic cavity. It has

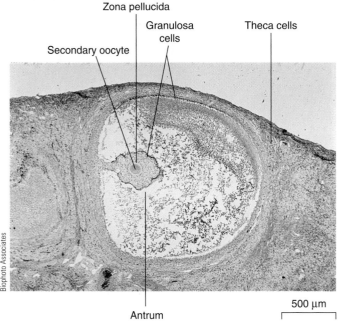

(a) LM of a developing follicle. The secondary oocyte is surrounded by the zona pellucida (a layer of glycoproteins) and by granulosa cells. Connective tissue cells surrounding the granulosa cells form a layer of theca cells.

thick walls of smooth muscle and an epithelial lining, the **endometrium,** that thickens each month in preparation for possible pregnancy. Up to 15% of women (more than 5 million women in the United States alone) are affected by **endometriosis,** a painful disorder in which fragments of the endometrium migrate to other areas, such as the oviducts or ovaries. Like pelvic inflammatory disease, endometriosis causes scarring that can lead to infertility.

If a secondary oocyte is fertilized, the tiny embryo enters the uterus and implants in the endometrium. As it grows and develops, it is sustained by nutrients and oxygen delivered by surrounding maternal blood vessels. If fertilization does not occur during the monthly cycle, the endometrium sloughs off and is discharged in the process known as **menstruation.**

The lower portion of the uterus, called the **cervix,** extends slightly into the vagina. The cervix is a common site of cancer in women. Detection is usually possible by the routine Papanicolaou test (Pap smear), in which a few cells are scraped from the cervix during a regular gynecological examination and studied microscopically. When cervical cancer is detected at very early stages of malignancy, the chances that the patient can be cured are good.

The vagina receives sperm

The **vagina** is an elastic, muscular tube that extends from the uterus to the exterior of the body. The vagina serves as a receptacle for sperm during sexual intercourse and as part of the birth canal (see Fig. 49-9).

The vulva are external genital structures

The female external genitalia, collectively known as the **vulva,** include several structures. Liplike folds, the **labia minora,** surround the vaginal and urethral openings (Fig. 49-12). The area enclosed by the labia minora is the **vestibule** of the vagina. Vestibular glands secrete a lubricating mucus into the vestibule. The **hymen** is a thin ring of tissue that forms a border around the entrance to the vagina.

Anteriorly, the labia minora merge to form the prepuce of the **clitoris,** a small erectile structure comparable to the male glans penis. Like the penis, the clitoris contains erectile tissue that becomes engorged with blood during sexual excitement. Rich in nerve endings, the clitoris is highly sensitive to touch, pressure, and temperature and serves as a center of sexual sensation in the female.

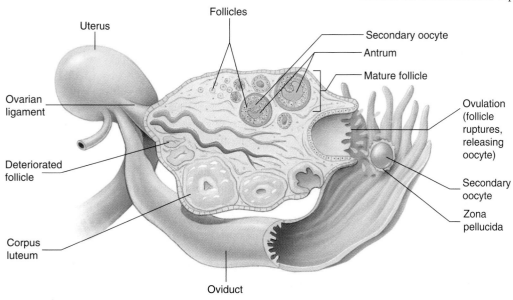

(b) Follicles in the ovary. This is a composite drawing; all of these stages of development would not be present at the same time.

Figure 49-10 *Animated* Follicle development in the ovary

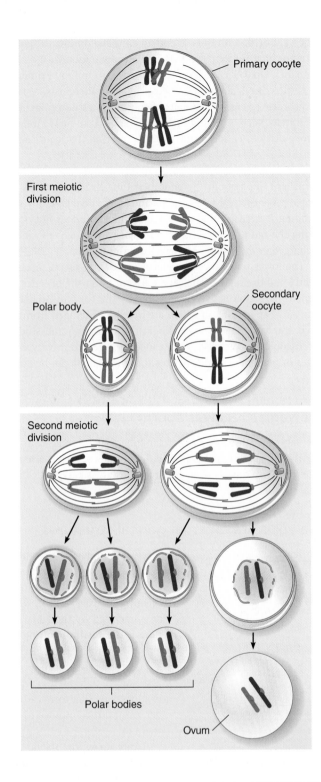

Figure 49-11 Oogenesis

Before birth, oogonia (*not shown*) divide many times by mitosis. Some oogonia differentiate to become primary oocytes that undergo meiosis. Only one functional ovum is produced from each primary oocyte. The other cells produced are polar bodies that degenerate. The first polar body sometimes divides as shown but often just deteriorates. The second meiotic division is completed after fertilization. Note in this example that four chromosomes are present in the primary oocyte (2n) and that meiosis produces the haploid (n) number (2) in the polar bodies, secondary oocyte, and mature ovum.

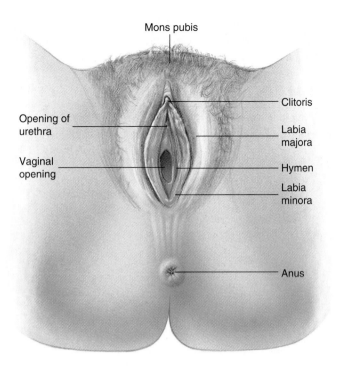

Figure 49-12 External female genital structures

Collectively, the structures shown (excluding the anus) are referred to as the vulva.

External to the delicate labia minora are the thicker **labia majora.** The **mons pubis** is the mound of fatty tissue just above the clitoris at the junction of the thighs and torso. At puberty the mons pubis and labia majora become covered by coarse pubic hair.

The breasts function in lactation

Each breast consists of 15 to 20 lobes of glandular tissue. The amount of adipose tissue around these lobes determines the size of the breasts and accounts for their softness. Gland cells are arranged in grapelike clusters called **alveoli** (Fig. 49-13). Ducts from each cluster join to form a single duct from each lobe, producing 15 to 20 tiny openings on the surface of each nipple. The breasts are a common site of cancer in women (see *Focus On: Breast Cancer*).

Lactation is the production of milk for nourishing the young. During pregnancy, high concentrations of the female reproductive hormones, estrogen and progesterone, stimulate the breasts to increase in size. For the first couple of days after childbirth, the mammary glands produce a fluid called *colostrum,* which contains protein and lactose but little fat. After birth the hormone **prolactin,** secreted by the anterior pituitary, stimulates milk production. When a baby suckles, the posterior pituitary releases **oxytocin,** which stimulates ejection of milk from the alveoli into the ducts.

Breastfeeding promotes recovery of the uterus because oxytocin released during breastfeeding stimulates the uterus to con-

BREAST CANCER

Breast cancer is the most common type of cancer among women, other than skin cancer. It is a leading cause of cancer deaths in women, second only to lung cancer. More than 40,000 women die of breast cancer each year in the United States alone. Risk factors for breast cancer include being overweight (especially after menopause), alcohol use (more than two drinks per day), and lack of exercise. Some studies show that cigarette smoking is a risk factor.

Women with a family history of breast cancer are at higher risk. An estimated 10% of breast cancers are familial, and about half of these patients have mutations in a tumor suppressor gene, either *BRCA1* or *BRCA2*. When the *normal* protein product of the *BRCA1* gene is phosphorylated by a specific protein kinase, it interacts with the normal protein product of the *BRCA2* gene and certain other compounds to repair DNA damage.

About 50% of breast cancers begin in the upper, outer quadrant of the breast. As a malignant tumor grows, it may adhere to the deep tissue of the chest wall. Sometimes it extends to the skin, causing dimpling. Eventually, the cancer spreads to the lymphatic system. About two thirds of breast cancers have metastasized (spread) to the lymph nodes by the time they are first diagnosed. When diagnosis and treatment begin early, 86% of patients survive for 5 years, and 65% survive for 20 years or longer. Untreated patients have a 5-year survival rate of only 20%.

Mastectomy (surgical removal of the breast) and radiation treatment are common methods of treating breast cancer. Lumpectomy (surgical removal of only the affected portion of the breast) in conjunction with radiation treatment appears to be as effective as mastectomy in some cases. Chemotherapy is useful in preventing metastasis, especially in premenopausal patients. A recent development in cancer treatment is the use of *biological response modifiers*, which include substances such as interferons, interleukins, and monoclonal antibodies (see Chapter 44).

About one third of breast cancers are estrogen dependent; that is, their growth depends on circulating estrogens. Removing the ovaries in patients with these tumors relieves the symptoms and may cause remission of the disease for months or even years. An important approach has been to develop pharmacological agents that antagonize the action of estrogen receptors. One challenge has been to inhibit estrogen in the breast while at the same time retaining its beneficial effects on other parts of the body.

According to the American Cancer Society, when breast cancer is confined to the breast, the 5-year survival rate is almost 100%. Because early detection of breast cancer greatly increases the chances of cure and survival, campaigns have been launched to educate women on the importance of self-examination. Mammography, a soft-tissue radiological study

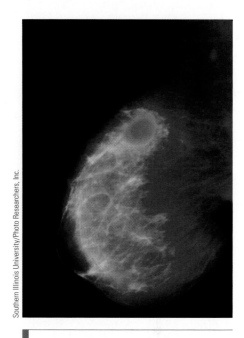

Southern Illinois University/Photo Researchers, Inc.

Mammogram showing area of breast cancer.
Cancer is seen as a dense mass (*red*). The healthy fibrous tissue (*blue*) supports the glandular structures.

of the breast, is helpful in detecting very small lesions that might not be identified by routine examination. In mammography, lesions show on an X-ray plate as areas of increased density (see figure).

---

tract to nonpregnant size. Breastfeeding offers advantages to the baby as well. It promotes a close bond between mother and child and provides milk tailored to the nutritional needs of the human infant. Breast milk contains antibodies, and breast-fed infants have a lower incidence of diarrhea, ear and respiratory infections, and hospital admissions than do formula-fed babies.

The hypothalamus, pituitary gland, and ovaries regulate female reproduction

As in the male, regulation of female reproduction involves many hormones and other signal molecules. In the male, concentration of sex hormones is kept within homeostatic range. In contrast, concentrations of female sex hormones change during a monthly

cycle. ∎ Table 49-2 lists the actions of the principal female reproductive hormones.

Like testosterone in the male, *estrogens* are responsible for growth of the sex organs at puberty, for body growth, and for the development of secondary sexual characteristics. In the female, these include development of the breasts, broadening of the pelvis, and the characteristic development and distribution of muscle and fat responsible for the female body shape. In females, *puberty* typically begins between ages 10 and 12 and continues until ages 14 to 16.

Hormones of the hypothalamus, anterior pituitary, and ovaries regulate the **menstrual cycle,** the monthly sequence of events that prepares the body for possible pregnancy. (The term *menstrual cycle* is sometimes used more narrowly to refer to the changes that occur in the uterus; we use the term here to include

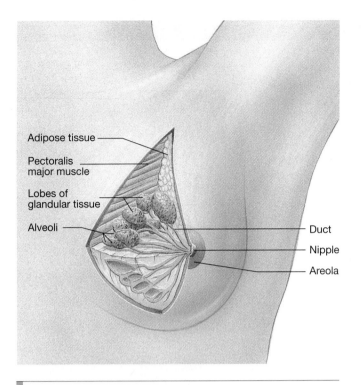

Figure 49-13 *Animated* Structure of the mature female breast

The breast contains lobes of glandular tissue. The lobes consist of alveoli, clusters of gland cells.

Labels on figure:
Adipose tissue
Pectoralis major muscle
Lobes of glandular tissue
Alveoli
Duct
Nipple
Areola

both the ovarian and uterine cycles.) The menstrual cycle runs its course every month from puberty until *menopause,* which occurs at about age 50. Although wide variations exist, a typical menstrual cycle is 28 days long (❙ Fig. 49-14). The first two weeks of the menstrual cycle are the **preovulatory phase.** Ovulation occurs on about the 14th day of the cycle. The third and fourth weeks of the menstrual cycle are the **postovulatory phase.** As you read the following description of the menstrual cycle, follow the steps illustrated in ❙ Figure 49-15. Use the numbered circles as a guide.

The preovulatory phase spans the first 2 weeks of the cycle

The first 5 days or so of the preovulatory phase are known as the menstrual phase. In fact, the first day of the menstrual cycle is marked by the onset of *menstruation,* the monthly discharge through the vagina of blood and tissue from the endometrium. During the first part of the preovulatory phase, the hypothalamus releases *gonadotropin-releasing hormone (GnRH).* This hormone stimulates the anterior pituitary to release gonadotropic hormones: *follicle-stimulating hormone (FSH)* and *luteinizing hormone (LH)* ❶. GnRH, FSH, and LH are the same hormones that help regulate the male reproductive cycle.

The preovulatory phase is also called the *follicular phase* because during this period, FSH stimulates a few follicles to begin to develop ❷. FSH also stimulates the granulosa cells of the follicle to multiply and produce estrogen. Some estrogen diffuses into the blood, but estrogen also has an *autocrine* action on the

granulosa cells that produce it and a *paracrine* effect on nearby granulosa cells (see Chapter 48 for a discussion of autocrine and paracrine regulation). Estrogen stimulates the granulosa cells to multiply, which increases estrogen production. The amount of estrogen secreted by the granulosa cells is enhanced by the action of LH on the theca cells. LH stimulates the theca cells to proliferate and produce androgens that diffuse into the granulosa cells, where they are converted to estrogen ❸. Estrogen stimulates growth of the endometrium, which begins to thicken and develop new blood vessels and glands ❹.

After the first week of the menstrual cycle, usually only one follicle continues to develop. Its granulosa cells become sensitive to LH as well as to FSH. This dominant follicle now secretes enough estrogen to cause a rise in the concentration of estrogen in the blood. *Although still at relatively low concentration,* estrogen inhibits secretion of FSH from the pituitary and may also act

Key Point

Gonadotropic hormones and ovarian hormones regulate the monthly sequence of events that take place within the ovary and uterus.

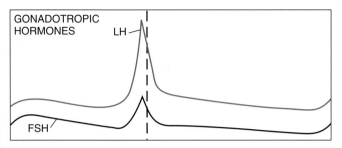

(a) Concentrations of pituitary gonadotropic hormones. Note that the concentrations of both FSH and LH peak just prior to ovulation (*dotted line*).

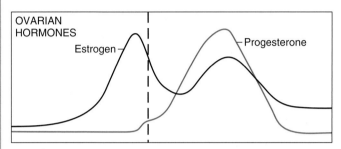

(b) Concentrations of ovarian hormones. Estrogen concentration peaks during the late preovulatory phase. Progesterone, secreted mainly by the corpus luteum, reaches its peak concentration during the postovulatory phase.

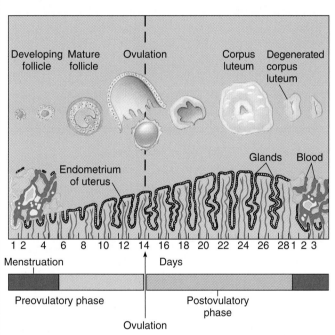

(c) Ovarian and uterine cycles. Hormone concentrations correlate with changes that take place in the ovaries and uterus. If fertilization occurs, the corpus luteum continues to secrete estrogen and progesterone, and menstruation does not occur.

Figure 49-14 *Animated* Endocrine regulation of the menstrual cycle

When fertilization does not occur, the menstrual cycle repeats about every 28 days. Note the changes in hormone concentrations during the menstrual cycle. (FSH, follicle-stimulating hormone; LH, luteinizing hormone)

on the hypothalamus, decreasing secretion of GnRH ⑤. In addition, the granulosa cells secrete *inhibin*, a hormone that inhibits mainly FSH secretion ⑥. As a result of these negative feedback signals, FSH concentration decreases.

As its concentration in the blood peaks during the late preovulatory phase, estrogen signals the anterior pituitary to secrete LH ⑦. This is a positive feedback mechanism. The surge of LH secreted at the middle of the menstrual cycle stimulates the final maturation of the follicle and stimulates *ovulation,* the ejection of the secondary oocyte from the ovary ⑧. The granulosa cells of the follicle decrease their estrogen output, resulting in a temporary decrease in estrogen concentration.

The corpus luteum develops during the postovulatory phase

The postovulatory phase, also called the *luteal phase,* begins after ovulation. LH stimulates development of the corpus luteum, which secretes a large amount of progesterone and estrogen, as well as inhibin ⑨. These hormones stimulate the uterus to continue its preparation for pregnancy. Progesterone stimulates tiny glands in the endometrium to secrete a fluid rich in nutrients.

During the postovulatory phase, the high concentration of progesterone in the blood, along with estrogen, inhibits secretion of GnRH, FSH, and LH ⑩, ⑪. Progesterone is thought to act mainly on the hypothalamus. Inhibin acts on the pituitary

The hypothalamus, anterior pituitary, and ovary secrete hormones that regulate the menstrual cycle through feedback systems.

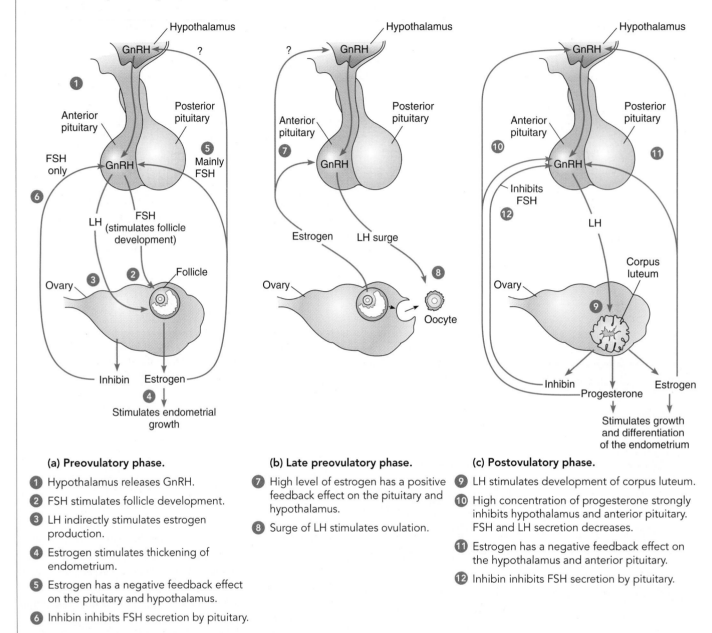

(a) Preovulatory phase.

1 Hypothalamus releases GnRH.

2 FSH stimulates follicle development.

3 LH indirectly stimulates estrogen production.

4 Estrogen stimulates thickening of endometrium.

5 Estrogen has a negative feedback effect on the pituitary and hypothalamus.

6 Inhibin inhibits FSH secretion by pituitary.

(b) Late preovulatory phase.

7 High level of estrogen has a positive feedback effect on the pituitary and hypothalamus.

8 Surge of LH stimulates ovulation.

(c) Postovulatory phase.

9 LH stimulates development of corpus luteum.

10 High concentration of progesterone strongly inhibits hypothalamus and anterior pituitary. FSH and LH secretion decreases.

11 Estrogen has a negative feedback effect on the hypothalamus and anterior pituitary.

12 Inhibin inhibits FSH secretion by pituitary.

Figure 49-15 *Animated* Feedback mechanisms in endocrine regulation of female reproduction

Green arrows indicate stimulation, red arrows indicate inhibition. (GnRH, gonadotropin-releasing hormone; FSH, follicle-stimulating hormone; LH, luteinizing hormone)

to further inhibit FSH secretion 12. As a result of these negative feedback systems, FSH and LH concentrations are low during the postovulatory phase, and no new follicles develop.

If the secondary oocyte is not fertilized, the corpus luteum begins to degenerate after about 8 days. Although the mechanism responsible for corpus luteum degeneration is not completely understood, a decrease in LH may be a factor. In addition, the corpus luteum may become less sensitive to LH.

When the corpus luteum stops secreting progesterone and estrogen, the concentrations of these hormones in the blood fall markedly. As a result, small arteries in the endometrium constrict, reducing the oxygen supply. Menstruation, which marks the beginning of a new cycle, begins as cells die and damaged arteries rupture and bleed. The concentrations of estrogen and progesterone are now too low to inhibit the anterior pituitary, and secretion of FSH and LH increases once again.

Menstrual cycles stop at menopause

At about age 50, a woman enters **menopause**—a period when ova are no longer produced and the woman becomes infertile. Apparently a change in the hypothalamus triggers menopause. Although gonadotropic hormones are produced, the ovaries become less responsive to them, and the oocytes in the ovaries begin to degenerate. The ovaries secrete less estrogen and progesterone, and the menstrual cycle becomes irregular and eventually halts. A sensation of heat ("hot flashes") sometimes occurs, probably because of the effect of decreased estrogen on the temperature-regulating center in the hypothalamus. Despite the physiological changes associated with menopause, a woman's interest in sex and her sexual performance are not usually affected. Although its use is controversial, estrogen replacement therapy relieves many of the symptoms of menopause.

In contrast to humans, females of other species maintain their ability to reproduce throughout the life span. They do not experience menopause. Researchers have suggested that the infertile postmenopausal period allows older women to help their daughters with grandchildren, increasing their survival rate. This practice helps ensure that the older woman's genes will be transmitted to future generations.

Most mammals have estrous cycles

Only humans and some other primates have menstrual cycles. Most other mammals have **estrous cycles.** Although the reproductive organs and hormones are generally similar, there are some important differences. In animals with estrous cycles, the uterus reabsorbs the thickened lining of the endometrium if conception does not occur. Recall that in animals with a menstrual cycle, the endometrium is expelled during menstruation.

In animals with estrous cycles, the female is sexually receptive ("in heat") only during the **estrus** phase of the cycle. The frequency and length of the estrus phase varies. In dogs, the estrus phase typically occurs twice a year and lasts 4 to 13 days. In cats, cycles can occur about every 3 weeks, and the estrus phase averages 7 days. The resting phase of the cycle is prolonged in the fall and early winter. Estrus occurs every 4 to 5 days in rodents. Animals with estrous cycles typically display physiological and behavioral changes that signal sexual readiness to potential mates.

Review

- How is oogenesis different from spermatogenesis? Why are so many sperm produced in the male and so few ova produced in the female?
- What is the function of the corpus luteum? What is the fate of the corpus luteum when the ovum is not fertilized?
- What are specific actions of FSH and LH in the female? Of estrogens? Of progesterone?

Learning Objective

8 Describe the physiological changes that occur during sexual response.

During sexual intercourse, also called **coitus** or copulation, the male deposits semen into the upper end of the vagina. The complex structures of the male and female reproductive systems, and the physiological, endocrine, and psychological processes associated with sexual activity, are adaptations that promote fertilization of the secondary oocyte and development of the resulting embryo.

Sexual stimulation results in two basic physiological responses: (1) vasocongestion, the concentration of blood in reproductive structures, as well as in other areas of the body; and (2) increased muscle tension. Sexual response includes four phases: sexual excitement, plateau, orgasm, and resolution. The *desire* to have sexual activity may be motivated by fantasies or thoughts about sex. This anticipation can lead to (physical) sexual excitement and a sense of sexual pleasure.

Physiologically, the **sexual excitement phase** involves vasocongestion and increased muscle tension. Penile erection is the first male response to sexual excitement. The penis must be erect in order to enter the vagina and function in coitus. In the female, vasocongestion occurs in the vagina, clitoris, and breasts, and the vaginal epithelium secretes a sticky lubricant. Vaginal lubrication is the female's first response to effective sexual stimulation. During the excitement phase, the vagina lengthens and expands in preparation for receiving the penis.

If erotic stimulation continues, sexual excitement heightens to the **plateau phase.** Vasocongestion and muscle tension increase markedly. In both sexes, muscle tension, heart rate, blood pressure, and respiration rate all increase. Sexual intercourse is usually initiated during the plateau phase. The penis creates friction as it is moved inward and outward in the vagina, in actions referred to as pelvic thrusts. Physical sensations resulting from this friction and the psychological experience of emotional intimacy lead to **orgasm,** the intense physical pleasure that is the climax of sexual excitement. In the female, stimulation of the clitoris heightens the sexual excitement that leads to orgasm.

Although it lasts only a few seconds, orgasm is the phase of maximum sexual tension and its release. In both sexes, orgasm is marked by rhythmic contractions of the muscles of the pelvic floor and reproductive structures. These muscular contractions continue at about 0.8-second intervals for several seconds. After the first few contractions, their intensity decreases, and they become less regular and less frequent. Heart rate and respiration more than double, and blood pressure rises markedly, just before and during orgasm. Musculoskeletal contractions occur throughout the body. In the male, orgasm is marked by the *ejaculation* of semen from the penis. No fluid ejaculation accompanies orgasm in the female. Orgasm is followed by the **resolution phase,** a state of well-being during which the body is restored to its unstimulated state.

■ What physiological changes take place during the sexual response cycle?

FERTILIZATION AND EARLY DEVELOPMENT

Learning Objective

9 Describe the processes of human fertilization and early development, and summarize the actions of hormones that regulate pregnancy.

Fertilization is the fusion of sperm and egg. Fertilization and the subsequent establishment of pregnancy together are referred to as **conception.** After ejaculation into the female reproductive tract, sperm remain alive and retain their ability to fertilize an ovum for an estimated 48 to 72 hours. The ovum remains fertile for 12 to 24 hours after ovulation. Therefore, in a very regular 28-day menstrual cycle, sexual intercourse at midcycle is most likely to result in fertilization. However, many women do not have regular menstrual cycles, and many factors can cause an irregular cycle even in women who are generally regular.

When conditions in the vagina and cervix are favorable, sperm begin to arrive at the site of fertilization in the upper oviduct within 5 minutes after ejaculation. At the time of ovulation, when estrogen concentration is high, the cervical mucus has a thin consistency that permits passage of sperm from the vagina into the uterus. After ovulation, when progesterone concentration rises, the cervical mucus becomes thick and sticky, blocking entrance of sperm (as well as bacteria that might harm the developing embryo). Once sperm enter the uterus, contractions of the uterine wall help transport them. These contractions are

induced, in part, by prostaglandins in the semen. The sperm's own motility is important, especially in approaching and fertilizing the ovum.

When a sperm encounters an egg, openings develop in the sperm acrosome, exposing enzymes that digest a path through the zona pellucida surrounding the secondary oocyte. As soon as one sperm enters the secondary oocyte, changes occur that prevent the entrance of other sperm As the fertilizing sperm enters, it usually loses its flagellum (∥ Fig. 49-16). Sperm entry stimulates the secondary oocyte to complete its second meiotic division. The head of the haploid sperm then swells to form the **male pronucleus** and fuses with the **female pronucleus to** form the diploid nucleus of the zygote. The process of fertilization is described in more detail in Chapter 50 (also see *Focus On: Novel Origins*).

If only one sperm is needed to fertilize a secondary oocyte, why are millions ejaculated? Many die as a result of unfavorable pH or phagocytosis by leukocytes and macrophages in the female tract. Only a few hundred succeed in traversing the correct oviduct and reaching the vicinity of the secondary oocyte. If the secondary oocyte is fertilized, development begins as the embryo is slowly moved to the uterus by the cilia lining the oviduct. When it enters the uterus, the embryo consists of a ball of about 32 cells.

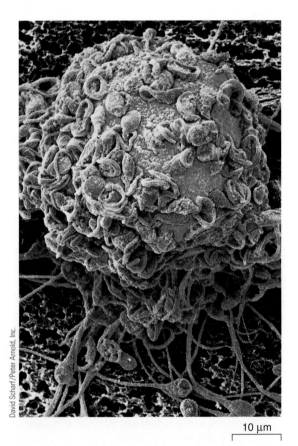

David Scharf/Peter Arnold, Inc.

10 μm

(d) Color-enhanced SEM of human sperm cells surrounding a test ovum. Sperm are being tested for viability.

(a) Sperm release an enzyme that helps disperse the layer of follicle cells (corona radiata) surrounding the secondary oocyte.

Zona pellucida · Corona radiata · Secondary oocyte · Ovum · First polar body · Pronuclei · Polar bodies

(b) After a sperm cell enters, the secondary oocyte completes its second meiotic division, producing an ovum and a polar body.

(c) Pronuclei of sperm and ovum fuse, producing a zygote with the diploid number of chromosomes.

Figure 49-16 *Animated* Fertilization

NOVEL ORIGINS

About 5 million couples in the United States are affected by infertility, the inability to achieve conception after using no contraception for at least 1 year. About 30% of cases involve both male and female factors. Male infertility is often attributed to low sperm count. Among the common causes of female infertility are failure to ovulate, production of infertile eggs (common in older women), and oviduct scarring (often caused by pelvic inflammatory disease). Women with blocked oviducts can usually ovulate and incubate an embryo normally but need clinical assistance in getting the young embryo to the uterus.

In the United States, more than 3 million infertile couples consult health-care professionals each year. Some are helped with conventional treatment, for example, with hormone therapy that regulates ovulation or with fertility drugs. But more than 40,000 couples need more sophisticated clinical help and turn to high-tech, assisted reproductive techniques that have been developed through reproduction and human embryo research. At present these techniques are expensive, and the success rate is low, less than one third.

The most common assisted reproductive procedure is **artificial insemination,** in which a catheter is used to inject sperm directly into the cervix or uterus. Transfer into the uterus is called **intrauterine insemination (IUI).** More than 600,000 IUI procedures are performed annually with a success rate of about 10%. If a male is infertile because of a low sperm count, his sperm can be concentrated, or alternatively, sperm from a donor can be used. Although the sperm donor usually remains anonymous to the prospective birth parents, his genetic qualifications are screened by physicians.

With **in vitro fertilization (IVF),** a woman takes a fertility drug that induces ovulation of several eggs. The eggs are retrieved by needle aspiration through the vagina. The eggs are placed in a dish with sperm and fertilization takes place in vitro (outside the patient's body). Embryos are cultured for 3 to 5 days and then screened for chromosome or gene abnormalities. Healthy ones are transferred into the uterus through the vagina. IVF was first

Newborn bongo with its surrogate mother, an eland.
As a young embryo, the bongo was transplanted into the eland's uterus, where it implanted and developed. Bongos are a rare and elusive species that inhabits dense forests in Africa. The larger and more common elands inhabit open areas.

used in England in 1978 to help a couple who had tried unsuccessfully for several years to have a child. Since that time, thousands of "test-tube babies" have been conceived in this way and born to previously infertile women. The success rate of in vitro fertilization is more than 30%; women less than 35 years old have the highest rate of pregnancy, and the rate declines with age.

In **gamete intrafallopian transfer (GIFT),** a laparoscope (a fiber-optic instrument) is used to guide the transfer of eggs and sperm into a woman's oviduct through a small incision in her abdomen. More than 4000 of these expensive procedures are performed each year with a success rate of about 28%. In **zygote intrafallopian transfer (ZIFT),** ova are fertilized in the laboratory. A laparoscope is used to guide the transfer of the resulting zygotes into the oviduct. In these procedures, the woman's own ova may be used. However, if she does not produce fertile eggs, they can be contributed by a donor (**oocyte donation**).

Another novel procedure is **host mothering.** An embryo is removed from its natu-

ral mother and implanted into a female substitute. The foster mother can support the developing embryo either until birth or temporarily until it is implanted again into the original mother or another host. This technique has proved useful to animal breeders. For example, embryos from prize sheep can be temporarily implanted into rabbits for easy shipping by air and then implanted into a host ewe, perhaps one of inferior quality. Host mothering allows an animal with superior genetic traits to produce more offspring than would be naturally possible. This procedure is also used to increase the populations of certain endangered species (see figure).

Technology is available to freeze the gametes or embryos of many species, including humans, and then transplant them into their donors or into host mothers. Freezing eggs may be an effective strategy for women with cancer who undergo chemotherapy or for young women not yet ready to become parents but who want to preserve young eggs with lower risk for chromosome abnormality. These eggs can be reimplanted at a later time.

After floating free in the uterus for about 3 days, the embryo begins to implant in the thick endometrium on about the seventh day after fertilization (❚ Fig. 49-17). (See Chapter 50 for a discussion of development.)

Membranes that develop around the embryo secrete **human chorionic gonadotropin (hCG),** a peptide hormone that signals the mother's corpus luteum to continue to function. (The presence of hCG in urine or blood is used as an early pregnancy test.) Concentrations of estrogen and progesterone remain high throughout pregnancy. During the first 2 months, the corpus luteum secretes almost all the estrogen and progesterone necessary to maintain the pregnancy. During this time, membranes surrounding the embryo, together with uterine tissue, form the **placenta,** the organ of exchange between the mother and developing embryo. As the corpus luteum slows its secretion and deteriorates after about 3 months, the placenta takes over and secretes large amounts of estrogen and progesterone.

Estrogen and progesterone are necessary to maintain pregnancy. Estrogen stimulates development of the uterine wall, including the muscle needed to expel the fetus during delivery. Progesterone inhibits uterine contractions so that the fetus is not expelled too soon. Because these hormones also inhibit FSH and LH, new follicles do not develop and the menstrual cycle stops during pregnancy.

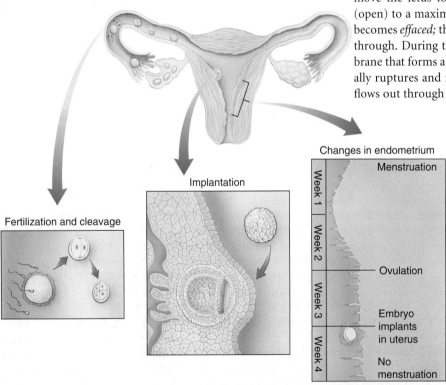

Review

❚ What is the function of hCG?

❚ What are the actions of estrogen and progesterone in maintaining pregnancy?

THE BIRTH PROCESS

Learning Objective

10 Summarize the birth process.

A normal human pregnancy is about 38 weeks long, counting from the day of fertilization, or about 40 weeks from the first day of the last menstrual cycle. The mechanisms that terminate pregnancy and initiate the birth process, called **parturition,** are not fully understood.

At the end of pregnancy, the stretching of the uterine muscle by the growing fetus combined with the effects of increased estrogen concentration and oxytocin produce strong uterine contractions. A long series of involuntary contractions of the uterus are experienced as **labor,** which begins when uterine contractions occur every 10 to 15 minutes.

Labor can be divided into three stages. During the first stage, which typically lasts 5 to 8 hours, the contractions of the uterus move the fetus toward the cervix, causing the cervix to *dilate* (open) to a maximum diameter of 10 cm (4 in). The cervix also becomes *effaced;* that is, it thins out so that the fetal head can pass through. During the first stage of labor, the amnion (the membrane that forms a fluid-filled sac around the embryo/fetus) usually ruptures and releases about a liter of amniotic fluid, which flows out through the vagina.

A positive feedback cycle operates during the birth process. As the baby's head pushes against the cervix, a reflex action causes the uterus to contract. The contraction forces the head against the cervix again, resulting in another contraction, and the positive feedback cycle repeats again and again until the baby descends through the cervix.

During the second stage, which normally lasts between 20 minutes and an hour, the fetus passes through the cervix and the vagina and is born, or "delivered" (❚ Fig. 49-18). With each uterine contraction, the woman bears down so that the fetus is expelled by the combined forces of uterine contractions and contractions of abdominal wall muscles.

At birth, the baby is still connected to the placenta by the umbilical cord. Contractions of the uterus squeeze much of the fetal blood from the placenta into the infant. The cord is tied and cut, separating the child from the mother. (The stump of

Changes in endometrium

Week 1 — Menstruation

Week 2

Week 3 — Ovulation — Embryo implants in uterus

Week 4 — No menstruation

Fertilization and cleavage

Implantation

❚ **Figure 49-17** Events following fertilization

The ovarian and uterine cycles are interrupted when pregnancy occurs. The corpus luteum does not degenerate, and menstruation does not take place. Instead, the wall of the uterus thickens even more, permitting the embryo to implant and develop within it. (Cleavage is the early series of mitotic divisions that converts the zygote to a multicellular embryo.)

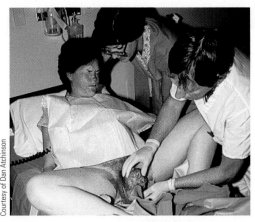

(a) The mother bears down hard with her abdominal muscles, helping to push the baby out. When the head fully appears, the physician or midwife gently grasps it and guides the baby's entrance into the outside world.

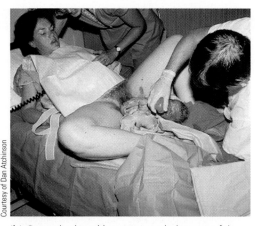

(b) Once the head has emerged, the rest of the body usually follows readily. The physician gently aspirates the mouth and pharynx to clear the upper airway of amniotic fluid, mucus, or blood. At this time the newborn takes its first breath.

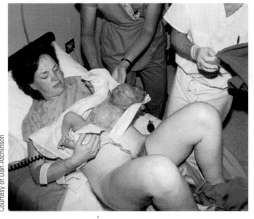

(c) The baby, still attached to the placenta by its umbilical cord, is presented to its mother.

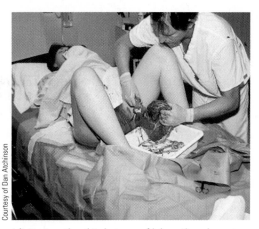

(d) During the third stage of labor, the placenta is delivered.

❚ Figure 49-18 *Animated* The birth process

In about 95% of all human births, the baby descends through the cervix and vagina in the head-down position.

the cord gradually shrivels until nothing remains but the scar, the **navel.**)

During the third stage of labor, which lasts 10 or 15 minutes after the birth, the placenta and the fetal membranes are loosened from the lining of the uterus by another series of contractions and expelled. At this stage they are collectively called the **afterbirth,** because these tissues are expelled from the vagina after the baby has been delivered.

During labor an obstetrician may administer oxytocin to increase the contractions of the uterus or may assist delivery with special forceps or a vacuum device. In some women, the opening between the pelvic bones is too small to permit the passage of the baby vaginally. In this situation, the obstetrician may perform a **cesarean section,** a surgical procedure in which the baby is delivered through an incision made in the abdominal and uterine

walls. A cesarean section may also be performed if the baby's position prevents normal delivery or if the baby shows signs of distress.

Review

❚ What is the role of estrogen in initiating the birth process? Of oxytocin?

❚ BIRTH CONTROL METHODS

Learning Objective

11 Compare the modes of action, effectiveness, advantages, and disadvantages of the methods of birth control discussed; include sterilization and emergency contraception.

When a fertile, heterosexually active woman uses no form of birth control, her chances of becoming pregnant during the course of a year are about 90%. Any method for deliberately separating

TABLE 49-3

Selected Contraceptive Methods

Method	Failure Rate*	Mode of Action	Advantages	Disadvantages
Oral contraceptives	0.3; 5	Inhibit ovulation; may affect endometrium and cervical mucus and prevent implantation	Highly effective; regulate menstrual cycle	Minor discomfort in some women; should not be used by women over age 35 who smoke
Injectable contraceptives (Depo-Provera)	About 1	Inhibit ovulation	Effective; long-lasting	Irregular menstrual bleeding; fertility may not return for 6–12 months after contraceptive is discontinued
Intrauterine device (IUD)	1; 1	Probably stimulates inflammatory response and prevents implantation	Provides continuous protection; highly effective for several years	Cramps; increased menstrual flow; increased risk of pelvic inflammatory disease and infertility; not recommended for women who have not had a child
Spermicides foams, jellies, creams	3; 20	Chemically kill sperm	No known side effects; can be used with a condom or diaphragm to improve efficacy	Messy; must be applied before intercourse
Contraceptive diaphragm (with jelly)†	3; 14	Diaphragm mechanically blocks entrance to cervix; jelly is spermicidal	No side effects	Must be inserted prior to intercourse and left in place for several hours afterward
Condom	2.6; 14	Mechanically prevents sperm from entering vagina	No side effects; some protection against STDs, including HIV	Slightly decreased sensation for male; could break
Rhythm‡ (natural family planning)	13; 19	Abstinence during fertile period	No known side effects	Not very reliable
Withdrawal (coitus interruptus)	9; 22	Male withdraws penis from vagina prior to ejaculation	No side effects	Not reliable; sperm in fluid secreted before ejaculation may be sufficient for conception
Sterilization Tubal sterilization	0.04	Prevents ovum from leaving uterine tube	Most reliable method	Requires surgery; considered permanent
Vasectomy	0.15	Prevents sperm from leaving vas deferens	Most reliable method	Requires surgery; considered permanent
Chance (no contraception)	About 90			

*Lower figure is the failure rate of method; higher figure is rate of method failure plus failure of the user to apply the method correctly. Based on number of failures per 100 women who use method per year in the United States.

†Failure rate is lower when diaphragm is used with spermicides.

‡There are several variations of rhythm method. For those who use the calendar method alone, the failure rate is about 35. However, if body temperature is taken daily and careful records are kept (temperature rises after ovulation), failure rate can be reduced. When women use some method to determine time of ovulation and have intercourse *only* more than 48 hours *after* ovulation, failure rate can be reduced to about 7.

sexual intercourse from reproduction is **contraception** (literally, "against conception"). Although many couples worldwide agree it is best to have babies by choice rather than by chance, most do not have contraceptives available to them. Worldwide, about 133 million births occur each year. Population experts estimate about 33 million (one fourth) of these are unplanned. If we add the estimated 46 million induced abortions performed annually, the total number of unplanned pregnancies is about 79 million each year.

Teen pregnancy is a serious problem. Nearly 1 million teenagers become pregnant each year in the United States, and almost half a million give birth. Thousands of these girls are aged 14 or younger. Many sexually active teenagers do not consistently use birth control. Teens often lack the knowledge and means of protecting themselves from unwanted pregnancy.

Since ancient times, humans have searched for effective methods of preventing fertilization and pregnancy. Of course, fertilization can be avoided by abstaining from sexual intercourse. Some

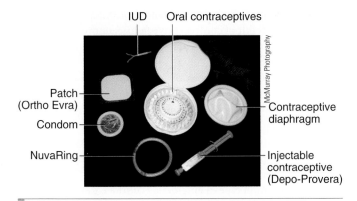

Figure 49-19 Some commonly used contraceptives
The only common male method of contraception is the condom.

couples avoid having intercourse around the time of ovulation. This cyclical temporary abstinence is called the **rhythm method,** or natural family planning. Because the time of ovulation is difficult to determine, the rhythm method has a high failure rate.

In recent years, scientists have developed a variety of contraceptives with a high percentage of reliability. However, the ideal contraceptive has not yet been developed. Among the issues that must be considered in developing contraceptives are safety, effectiveness, cost, convenience, and ease of use. Risks of cancer, birth defects in case the method fails and the woman becomes pregnant, permanent infertility, and side effects such as menstrual abnormalities must be minimized.

Researchers predict that the contraceptive methods of the 21st century will control regulatory peptides and the genes that code for them. For example, the genes that code for pituitary hormones that stimulate the ovaries to release estrogen could be turned off selectively, or other hormone signals could be interrupted. Other approaches being studied are molecular interruption of fertilization and contraceptive vaccines. Researchers are testing a sugar molecule that causes reversible sterility in mice by impairing sperm motility. Studies on male monkeys indicate that immunocontraception may be an effective contraceptive method in the future. Monkeys are immunized with a specific protein from the testis. The antibodies they develop to the protein make them infertile. Some researchers predict that within 15 to 20 years, current methods of contraception will be replaced by more sophisticated molecular methods.

Some of the more common methods of birth control are described in the following paragraphs and in ▌Table 49-3 (see also ▌Fig. 49-19). Oral contraceptives, intrauterine devices (IUDs), and female sterilization account for more than two thirds of all contraception practiced worldwide.

Most hormone contraceptives prevent ovulation

Oral contraceptives, contraceptive patches, and injectable contraceptives are **hormone contraceptives.** More than 80 million women worldwide (more than 8 million in the United States

alone) use oral contraceptives. When taken correctly, oral contraceptives are about 99.7% effective in preventing pregnancy. They are also used to regulate menstrual cycles. The most common preparations are combinations of progestin (synthetic progesterone) and synthetic estrogen. (Natural hormones are destroyed by the liver almost immediately, but synthetic ones are chemically modified during production so that they can be absorbed effectively and metabolized slowly.) In a typical regimen, a woman takes one pill each day for about 3 weeks. Then, for 1 week she takes a sugar pill that allows menstruation to occur because of the withdrawal of the hormones. One oral contraceptive (Seasonale) works on a 91-day regimen, reducing menstruation to four times per year. Other new oral contraceptives are being marketed as methods to manipulate menstrual cycles so that menstruation is shorter.

Oral contraceptives prevent ovulation. When postovulatory levels of ovarian hormones are maintained in the blood, the body is tricked into responding as though conception has occurred. The pituitary gland is inhibited and does not produce the surge of LH that stimulates ovulation.

Studies suggest that women over the age of 35 who smoke or have other risk factors, such as untreated hypertension, should not take oral contraceptives. Women in this category who take oral contraceptives have an increased risk of death from stroke and myocardial infarction. Low-dose oral contraceptive pills appear safe for nonsmokers up to the time of menopause. Oral contraceptives are linked to deaths in about 3 per 100,000 users. This compares favorably with the death rate of about 9 per 100,000 pregnancies.

Trends in contraception are geared toward convenience and improved compliance. The birth control patch (Ortho Evra) delivers estrogen and progestin through the skin. The patch is applied weekly for 3 weeks and removed for the fourth week. During the patch-free week, menstruation usually occurs because the hormone is withdrawn. Another approach is a thin, flexible plastic ring (NuvaRing) that women can flatten like a rubber band and insert into the vagina once a month. The ring releases progestin and estrogen in the amounts present in low-dose birth control pills.

Injectable progestin (Depo-Provera) is injected intramuscularly every 3 months. It prevents ovulation by suppressing anterior pituitary function; it also thickens the cervical mucus (which makes it more difficult for the sperm to reach the egg). Implant systems consisting of one or more plastic progestin rods (each about the size of a cardboard matchstick) are in clinical trials and will be available soon.

Intrauterine devices are widely used

The **intrauterine device (IUD)** is the most widely used method of reversible contraception in the world, used by an estimated 90 million women. However, in the United States only a small percentage of women use them. The IUD is a small device that is inserted into the uterus by a medical professional. IUDs in current use have been shown to be safe and are about 99% effective.

Disadvantages of the IUD include uterine cramping and bleeding in a small percentage of women.

One IUD presently being used in the United States is the Copper T380 (ParaGard), which can be left in place for up to 10 years. A newer IUD (Mirena), also a small, plastic T-shaped device, releases a very small amount of progestin onto the inner wall of the uterus. This IUD can be left in place for up to 5 years and can improve the symptoms of a woman's menstrual period.

The IUD's mode of action is not well understood. The copper or hormone slowly released from an IUD apparently interferes with embryo implantation. In addition, white blood cells mobilized in response to the foreign body in the uterus may produce substances toxic to sperm. Mirena causes thickening of the cervical mucus, preventing passage of sperm into the uterus.

Other common contraceptive methods include the diaphragm and condom

The **contraceptive diaphragm** mechanically blocks the passage of sperm from the vagina into the cervix. It is covered with spermicidal jelly or cream and inserted just prior to sexual intercourse.

The **condom** is also a mechanical method of birth control. The only contraceptive device currently sold for men, the condom provides a barrier that contains the semen so that sperm cannot enter the female tract. The latex condom is the only contraceptive that provides some protection against infection with human immunodeficiency virus (HIV) and other sexually transmitted diseases (STDs). The female condom is a strong, soft, polyurethane sheath inserted in the vagina before sexual intercourse; it provides protection against both pregnancy and sexually transmitted infections.

Emergency contraception is available

Emergency contraception is after-the-fact contraception for rape victims and others who have had unprotected sexual intercourse. If physicians advocated and women used emergency contraception, many of the 3.5 million unwanted pregnancies that occur each year (as a result of contraceptive failure, unprotected sex, or sexual assault) and thousands of abortions could be prevented in the United States alone.

The most common types of emergency contraception are progestin pills (known as Plan B) or combination progestin-and-estrogen pills. Emergency contraception pills are sometimes called morning-after pills, but they can actually be taken up to 5 days after unprotected intercourse. Progestin pills decrease the probability of pregnancy by about 89%. Emergency contraception pills prevent fertilization or ovulation.

A less common type of emergency contraception is insertion of a copper IUD within a week of unprotected intercourse. This method is more than 99% effective in preventing pregnancy.

Sterilization renders an individual incapable of producing offspring

Sterilization is the only method of contraception not affected by noncompliance or failure to use a contraceptive method correctly. Worldwide, female sterilization is the most popular contraceptive method and accounts for one third of all contraceptive use.

Male sterilization is performed by vasectomy

An estimated 1 million **vasectomies** are performed each year in the United States. After using a local anesthetic, the physician makes a small incision on each side of the scrotum. Then each vas deferens is cut and its ends sealed so that they cannot grow back together (❙ Fig. 49-20a). Because testosterone secretion and transport are not affected, vasectomy does not affect masculinity. Sperm continue to be produced, though at a much slower rate, and are destroyed by macrophages in the testes. No change in amount of semen ejaculated is noticed, because sperm account for very little of the semen volume.

By surgically reuniting the ends of the vasa deferentia, surgeons can successfully reverse male sterilization in about 70% of attempts. Apparently, some sterilized men eventually develop antibodies against their own sperm and remain sterile even after their vasectomies have been surgically reversed. Therefore, fertility rates are low (about 30%) for men who undergo vasectomy reversal after 10 or more years.

An alternative to vasectomy reversal is the storage of frozen sperm in sperm banks. If the male should decide to father another child after he has been sterilized, he simply "withdraws" his sperm to artificially inseminate his mate. Sperm banks have been established throughout the United States. Not much is known yet about the effects of long-term sperm storage, and there may be an increased risk of genetic defects.

(a) **Vasectomy.** The vas deferens (sperm duct) on each side is cut and cauterized.

(b) **Tubal sterilization.** Each oviduct is cut and cauterized so that ovum and sperm can no longer meet.

❙ **Figure 49-20** Sterilization

Female sterilization is by tubal sterilization

In the United States **tubal sterilization** is the most common means of birth control. A surgeon performs this procedure while the patient is under general anesthesia by making a small abdominal incision and using a laparoscope to locate and cut the oviducts (▌Fig. 49-20b). The tubes are then sealed (cauterized), or they may be clipped or clamped. Sometimes a small section of the tube is removed. Tubal sterilization can now be accomplished without an incision. A physician inserts two small metal coils through the vagina, cervix, uterus, and up into the oviducts. The coils cause scar tissue to develop, which blocks the oviducts. As in the male, hormone balance and sexual performance are not affected by sterilization.

Abortions can be spontaneous or induced

Abortion is termination of pregnancy that results in the death of the embryo or fetus. **Spontaneous abortions** (commonly referred to as miscarriages) occur without intervention. Embryos that are spontaneously aborted are frequently abnormal. **Induced abortions** are performed deliberately either for therapeutic reasons or as a means of birth control.

Therapeutic abortions are performed when the mother's life is in danger or when there is reason to suspect that the embryo is grossly abnormal. Worldwide, more unplanned than planned pregnancies occur each year, and more than half of the unplanned pregnancies end in abortion. An estimated 46 million induced abortions are performed each year (more than a million in the United States).

Most first-trimester abortions (those performed during the first 3 months of pregnancy) are performed by a suction method or by administering drugs that interrupt the pregnancy and induce labor. In the suction method of abortion, the cervix is dilated, and a suction aspirator is inserted into the uterus. The embryo and other products of conception are evacuated.

Drugs are also used to interrupt pregnancy. The drug *RU-496* (mifepristone) binds competitively with progesterone receptors in the uterus but does not activate them. Because the normal actions of progesterone are blocked, the endometrium breaks down and uterine contractions occur. These conditions prevent implantation of an embryo. The action of methotrexate, a drug used to treat cancer, is similar to that of RU-496. Both of these drugs are used to interrupt pregnancy. Prostaglandins may then be administered to induce the uterine contractions that expel the embryo.

After the first trimester, abortions are performed mainly because of fetal abnormalities or maternal illness. The method most commonly used is dilation and evacuation. The cervix is dilated, forceps are used to remove the fetus, and suction is used to aspirate the remaining contents of the uterus. Drugs such as prostaglandins are being increasingly used to induce labor in second-trimester abortions.

When abortions are performed by skilled medical personnel, the mortality rate in the United States is less than 1 per 100,000.

In countries where abortion is illegal, the death rate is as high as 700 per 100,000. These statistics can be contrasted with the death rate from pregnancy and childbirth of about 9 per 100,000. The World Health Organization estimates that one third of all maternal deaths result from poorly done illegal abortions.

Review

▌ What are three types of hormonal contraception?
▌ What is emergency contraception?

▌ SEXUALLY TRANSMITTED DISEASES

Learning Objective

12 Identify common sexually transmitted diseases, and describe their symptoms, effects, and treatments.

Sexually transmitted diseases (STDs), also called *venereal diseases (VDs),* are, next to the common cold, the most prevalent communicable diseases in the world. The World Health Organization has estimated that more than 250 million people are infected each year with **gonorrhea,** and more than 50 million are infected with **syphilis.**

In the United States, more than 65 million people are currently living with an incurable STD, and more than 12 million people (including 3 million teens) are diagnosed with new STD infections each year. One in three sexually active young people will acquire an STD by age 24. One out of four women and one out of five men are infected with **genital herpes. Chlamydia,** another common STD, causes **pelvic inflammatory disease (PID),** an infection of the female reproductive organs that often leads to sterility.

Human papillomavirus (HPV) is now the most common STD in the United States. An estimated 20 million people in the United States are currently infected with HPV, and about 75% of sexually active men and women have been infected with this virus at some time during their lives. In most infected individuals, the immune system destroys the virus before it causes serious health problems. However, HPV can cause abnormal Pap smears and cervical cancer. Some strains of HPV cause genital warts. Two new vaccines (Gardasil and Cervarix) prevent infection by several harmful HPV strains. Health professionals anticipate that such vaccines will lead to the decline of genital warts and cervical cancer. However, for the vaccine to be most effective, girls must be vaccinated before they become sexually active. For this reason, these vaccines are being offered to girls as young as 9 years old.

Some common STDs are described in ▌Table 49-4. (See Chapter 44 for a discussion of HIV.)

Review

▌ What is the most common STD in the United States?
▌ What is PID?

TABLE 49-4

Some Common Sexually Transmitted Diseases

Disease and Causative Organism	Course of Disease	Treatment
Human papillomavirus (HPV; about 30 strains of HPV can infect reproductive organs)	Some strains cause abnormal Pap smears and cervical cancer; other strains cause genital warts; vaccines now available	Immune modifiers; interferon
Chlamydia (*Chlamydia trachomatis*, a bacterium)	Discharge and burning with urination or asymptomatic; men 15–30 years old with multiple sex partners are most at risk; in women can cause pelvic inflammatory disease (PID), infection of reproductive organs and pelvic cavity that may lead to sterility (> 15% of cases)	Antibiotics
Gonorrhea (*Neisseria gonorrhoeae*, a gonococcus bacterium)	Bacterial toxin may produce redness and swelling at infection site; symptoms in males are painful urination and discharge of pus from penis; in about 60% of infected women no symptoms initially; can spread to epididymis (in males) or oviducts and ovaries (in females), causing sterility; can cause widespread infection; damage to heart valves, meninges (outer coverings of brain and spinal cord), and joints	Antibiotics
Syphilis (*Treponema pallidum*, a spirochete bacterium)	Bacteria enter body through defect in skin; primary chancre (small, painless ulcer) at site of initial infection; highly infectious at this stage; secondary stage, widespread rash and influenza-like symptoms; scaly lesions may occur that are highly infectious; latent stage that follows can last 20 years; eventually, lesions called gummae may form that damage liver, brain, bone, or spleen; death results in 5%–10% of untreated cases	Penicillin
Genital herpes (herpes simplex type 2 virus)	Tiny, painful blisters on genitals; may develop into ulcers; influenza-like symptoms may occur; recurs periodically; threat to fetus or newborn infant	No effective cure; some drugs may shorten outbreaks or reduce severity; medication can be taken to prevent outbreak
Trichomoniasis (a protozoon)	Itching, discharge, soreness; can be contracted from dirty toilet seats and towels; may be asymptomatic in males	Metronidazole (an antibiotic)
"Yeast" infections (genital candidiasis; *Candidiasis albicans*)	Irritation, soreness, discharge; especially common in females; rare in males; can be acquired nonsexually	Antifungal drugs
Acquired immunodeficiency syndrome (AIDS)* (caused by human immunodeficiency virus, HIV)	Influenza-like symptoms: swollen lymph glands, fever, night sweats, weight loss; decreased immunity, leading to pneumonia, rare forms of cancer	No effective cure; a variety of drugs reduce symptoms and prolong life

*AIDS was discussed in Chapter 44.

SUMMARY WITH KEY TERMS

Learning Objectives

1 Compare the benefits of asexual and sexual reproduction, and describe each mode of reproduction, giving specific examples (page 1052).

- In **asexual reproduction,** a single parent endows its offspring with a set of genes identical to its own (unless there are mutations). Asexual reproduction is energy efficient, and it is most successful in a stable environment.

- In **budding,** a part of the parent's body grows and separates from the rest of the body. In **fragmentation,** the parent's body may break into several pieces; each piece can develop into a new animal. In **parthenogenesis,** an unfertilized egg develops into an adult.

- In **sexual reproduction,** offspring are produced by the fusion of two types of gametes—**egg** and **sperm.** When sperm and egg fuse, a fertilized egg, or **zygote,** forms. Sexual reproduction promotes genetic variety and is especially adaptive in an unstable, changing environment.

- In **external fertilization,** mating partners typically release eggs and sperm into the water simultaneously. In **internal fertilization,** the male delivers sperm into the female's body.

- In **hermaphroditism,** a single individual produces both eggs and sperm.

2 Describe the structure and function of each organ of the human male reproductive system (page 1054).

- The human male reproductive system includes the testes, which produce sperm and testosterone; a series of conducting ducts; accessory glands; and the **penis.**

- The **testes,** housed in the **scrotum,** contain the **seminiferous tubules,** where **spermatogenesis** (sperm production) takes place. The **interstitial cells** in the testes secrete testosterone. **Sertoli cells** produce signaling molecules and a fluid that nourishes sperm cells.

- Sperm complete their maturation and are stored in the **epididymis** and **vas deferens.** During ejaculation, sperm pass from the vas deferens to the **ejaculatory duct** and then into the **urethra,** which passes through the penis.

- Each ejaculate of semen contains about 200 million sperm suspended in the secretions of the **seminal vesicles** and **prostate gland.** The **bulbourethral glands** release a mucous secretion.

- The penis consists of three columns of **erectile tissue,** two **cavernous bodies** and one **spongy body** that surrounds the urethra. When its erectile tissue becomes engorged with blood, the penis becomes erect.

ThomsonNOW **Learn more about the male reproductive system by clicking on the figure in ThomsonNOW.**

3   Trace the passage of sperm cells through the human male reproductive system from their origin in the seminiferous tubules to their expulsion from the body in the semen. (Include a description of spermatogenesis.) (page 1054)

- Spermatogenesis takes place in the seminiferous tubules of the testes. **Spermatogonia** divide by mitosis; some differentiate and become **primary spermatocytes,** which undergo **meiosis.** The first meiotic division produces two **secondary spermatocytes.** In the second meiotic division, each secondary spermatocyte produces two **spermatids.** Each spermatid differentiates to form a mature sperm. The head of a sperm consists of the nucleus and a cap, or **acrosome,** containing enzymes that help the sperm penetrate the egg.

- Sperm pass in sequence through the seminiferous tubules of the testis, epididymis, vas deferens, ejaculatory duct, and urethra.

ThomsonNOW **Learn more about spermatogenesis by clicking on the figure in ThomsonNOW.**

4   Describe endocrine regulation of reproduction in the human male (page 1054).

- Testosterone establishes and maintains male **primary sex characteristics** and **secondary sex characteristics.**

- Endocrine regulation of male reproduction involves the hypothalamus, pituitary gland, and testes. The hypothalamus secretes **gonadotropin-releasing hormone (GnRH),** which stimulates the anterior pituitary gland to secrete the gonadotropic hormones: **follicle-stimulating hormone (FSH)** and **luteinizing hormone (LH).**

- FSH, LH, and testosterone directly or indirectly stimulate sperm production. LH stimulates the interstitial cells of the testes to produce testosterone. FSH stimulates the Sertoli cells to produce (1) **androgen-binding protein (ABP),** which binds to testosterone and concentrates it; and (2) **inhibin,** a hormone that inhibits FSH secretion.

ThomsonNOW **Learn more about the hormonal control of sperm production by clicking on the figure in ThomsonNOW.**

5   Describe the structure and function of each organ of the human female reproductive system (page 1058).

- The **ovaries** produce gametes and the steroid hormones **estrogen** and **progesterone.** Fertilization takes place in the **oviducts** (uterine tubes).

- The **uterus** serves as an incubator for the developing embryo. The epithelial lining of the uterus, the **endometrium,** thickens each month in preparation for possible pregnancy. The lower part of the uterus, the **cervix,** extends into the vagina.

- The **vagina** receives the penis during sexual intercourse and is the lower part of the birth canal. The **vulva** include

the **labia majora, labia minora, vestibule** of the vagina, **clitoris,** and **mons pubis.**

- The breasts function in **lactation,** production of milk for the young. Each breast consists of 15 to 20 lobes of glandular tissue. Gland cells are arranged in **alveoli.** The hormone **prolactin** stimulates milk production; **oxytocin** stimulates ejection of milk from the alveoli into the ducts, making it available to the nursing infant.

ThomsonNOW **Learn more about the female reproductive system by clicking on the figure in ThomsonNOW.**

6   Trace the development of a human egg and its passage through the female reproductive system until it is fertilized (page 1058).

- **Oogenesis** takes place in the ovaries. **Oogonia** differentiate into **primary oocytes.** A primary oocyte and the **granulosa cells** surrounding it make up a **follicle.**

- As the follicle grows, connective tissue cells surrounding the granulosa cells form a layer of **theca cells.** As the follicle matures, the primary oocyte undergoes the first meiotic division, giving rise to a **secondary oocyte** and a **polar body.**

- During **ovulation,** the secondary oocyte is ejected from the ovary and enters one of the paired oviducts, where it may be fertilized. The part of the follicle remaining in the ovary develops into a **corpus luteum,** a temporary endocrine gland.

ThomsonNOW **Learn more about oogenesis by clicking on the figure in ThomsonNOW.**

7   Describe the endocrine regulation of reproduction in the human female, and identify the important events of the menstrual cycle, such as ovulation and menstruation (page 1058).

- Endocrine regulation of female reproduction involves the hypothalamus, pituitary gland, and ovaries. The first day of the **menstrual cycle** is marked by the beginning of menstrual bleeding. Ovulation occurs at about day 14 in a typical 28-day menstrual cycle.

- During the **preovulatory phase,** gonadotropin-releasing hormone (GnRH) from the hypothalamus stimulates the anterior lobe of the pituitary to secrete follicle-stimulating hormone (FSH) and luteinizing hormone (LH). FSH stimulates follicle development and stimulates the granulosa cells to produce estrogen. LH stimulates theca cells to multiply and produce androgens, which are converted to estrogen.

- Estrogen is responsible for primary and secondary female sex characteristics and stimulates development of the endometrium.

- After the first week, only one follicle continues to develop. Although at relatively low concentration, estrogen inhibits FSH secretion by negative feedback. Granulosa cells produce inhibin, which also inhibits FSH secretion.

- During the late preovulatory phase, estrogen concentration peaks and through positive feedback signals the anterior pituitary to secrete LH. LH stimulates final maturation of the follicle and stimulates ovulation.

- During the **postovulatory phase,** LH promotes development of the corpus luteum. The corpus luteum secretes progesterone and estrogen, which stimulate final preparation of the uterus for pregnancy. During the postovulatory phase, progesterone, along with estrogen, inhibits secretion of GnRH, FSH, and LH.

- If fertilization does not occur, the corpus luteum degenerates, concentrations of estrogen and progesterone in the blood fall, and menstruation occurs.

ThomsonNOW Explore hormonal control of the menstrual cycle by clicking on the figures in ThomsonNOW.

8 Describe the physiological changes that occur during sexual response (page 1067).
- Vasocongestion and increased muscle tension are physiological responses to sexual stimulation. The phases of sexual response include **sexual excitement, plateau, orgasm,** and **resolution.**

9 Describe the processes of human fertilization and early development, and summarize the actions of hormones that regulate pregnancy (page 1068).
- Human **fertilization** is the fusion of secondary oocyte and sperm to form a zygote. Fertilization and establishment of pregnancy together are called **conception.**
- If the secondary oocyte is fertilized, development begins and the embryo implants in the uterus. Membranes that develop around the embryo secrete **human chorionic gonadotropin (hCG),** a hormone that maintains the corpus luteum. During the first 2 to 3 months of pregnancy, the corpus luteum secretes the large amounts of estrogen and progesterone needed to maintain pregnancy. After 3 months, the **placenta,** the organ of exchange between mother and embryo, assumes this function.

ThomsonNOW Learn more about fertilization by clicking on the figure in ThomsonNOW.

10 Summarize the birth process (page 1070).
- Several hormones, including estrogen, oxytocin, and prostaglandins, regulate **parturition,** the birth process. Labor can be divided into three stages; the baby is delivered during the second stage.

11 Compare the modes of action, effectiveness, advantages, and disadvantages of the methods of birth control discussed; include sterilization and emergency contraception (page 1071).
- Effective methods of **contraception** include **hormone contraceptives,** such as oral contraceptives and injectable progestin; **intrauterine devices (IUDs); condoms; contraceptive diaphragms;** and **sterilization (vasectomy** in the male and **tubal sterilization** in the female); see Table 49-3. Emergency contraception can be used to prevent unwanted pregnancy after rape or unprotected intercourse.
- **Spontaneous abortions** (miscarriages) occur without intervention. **Induced abortions** include **therapeutic abortions,** performed when the mother's health is in danger or when the embryo is thought to be grossly abnormal. Abortions are also induced as a means of birth control.

12 Identify common sexually transmitted diseases, and describe their symptoms, effects, and treatments (page 1075).
- Among the common types of **sexually transmitted diseases (STDs)** are **human papillomavirus (HPV), chlamydia** (the most common cause of **pelvic inflammatory disease**), genital herpes, gonorrhea, syphilis, and **HIV.** See Table 49-4.

TEST YOUR UNDERSTANDING

1. Which of the following is *not* an example of asexual reproduction? (a) budding (b) external fertilization (c) fragmentation (d) parthenogenesis (e) regeneration to form a new individual from a fragment

2. Hermaphroditism (a) is a form of asexual reproduction (b) occurs when an unfertilized egg develops into an adult animal (c) is a form of sexual reproduction in which an animal produces both eggs and sperm (d) typically involves self-fertilization (e) typically requires only one animal

3. The seminiferous tubules (a) are the site of spermatogenesis (b) produce most of the seminal fluid (c) empty directly into the vas deferens (d) are located within the cavernous body (e) receive fluid from the bulbourethral glands

4. Arrange the following stages into the correct sequence.

    1. spermatogonium 2. spermatid 3. primary spermatocyte
    4. secondary spermatocyte 5. sperm

    (a) 2, 1, 3, 4, 5 (b) 5, 1, 3, 4, 2 (c) 4, 3, 1, 2, 5 (d) 1, 4, 3, 2, 5
    (e) 1, 3, 4, 2, 5

5. Which sequence best describes the passage of sperm?

    1. seminiferous tubules 2. vas deferens 3. epididymis
    4. ejaculatory duct 5. urethra

    (a) 3, 1, 2, 4, 5 (b) 1, 3, 2, 4, 5 (c) 5, 4, 2, 3, 1 (d) 1, 3, 4, 2, 5
    (e) 3, 1, 4, 2, 5

6. Which of the following characteristics is *not* associated with testosterone? (a) maintenance of secondary sex characteristics (b) responsible for primary sex characteristics (c) principal androgen (d) protein hormone (e) necessary for spermatogenesis

7. Androgen-binding protein (a) is secreted by Sertoli cells (b) stimulates estrogen production (c) inhibits secretion of FSH (d) inhibits spermatogenesis (e) two of the preceding answers are correct

8. Which of the following cells is haploid? (a) primary oocyte (b) oogonium (c) secondary oocyte (d) corpus luteum (e) follicle cell

9. The corpus luteum (a) is surrounded by ova (b) degenerates if fertilization occurs (c) develops in the preovulatory phase (d) is maintained by prostaglandins (e) serves as a temporary endocrine gland

10. After ovulation, the secondary oocyte enters the (a) ovary (b) corpus luteum (c) cervix (d) oviduct (e) vagina

11. The endometrium (a) is the muscle layer of the uterus (b) is thickest during the preovulatory phase (c) is the site of embryo implantation (d) lines the vagina (e) is directly affected by FSH

12. The hormone that reaches its highest level during the post-ovulatory phase is (a) progesterone (b) estrogen (c) FSH (d) LH (e) testosterone

13. Uterine contraction is strongly stimulated by (a) progesterone (b) FSH (c) LH (d) inhibin (e) oxytocin

14. Which of the following is *not* a hormonal type of contraception? (a) Depo-Provera (b) contraceptive diaphragm (c) injectable progestin (d) birth control patch (e) oral contraceptives

15. Pelvic inflammatory disease is commonly caused by (a) syphilis (b) gonorrhea (c) genital herpes (d) HIV (e) chlamydia

16. Label the diagram. Use Figure 49-9b to check your answers.

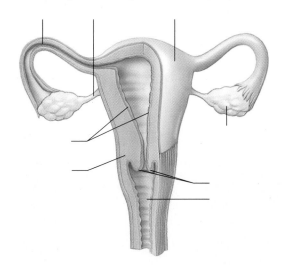

CRITICAL THINKING

1. Contrast the biological advantages of hermaphroditism that involves cross-fertilization with hermaphroditism that involves self-fertilization.

2. What would happen if ovulation occurred but no corpus luteum developed?

3. Why do you think a variety of effective methods of male contraception have not yet been developed? If you were a researcher, what are some approaches you might take to development of a method of male contraception?

4. **Evolution Link.** Asexual reproduction is more common in habitats with few parasites. Sexual reproduction is more common in habitats with larger numbers of parasites. Develop a hypothesis to explain why.

5. **Analyzing Data.** Examine Figure 49-14. Which hormones are present in largest concentration just before ovulation? During which part of the menstrual cycle is the endometrium most prepared to receive an embryo? Which hormones stimulate thickening of the endometrium?

Additional questions are available in ThomsonNOW at www.thomsonedu.com/login

50

Animal Development

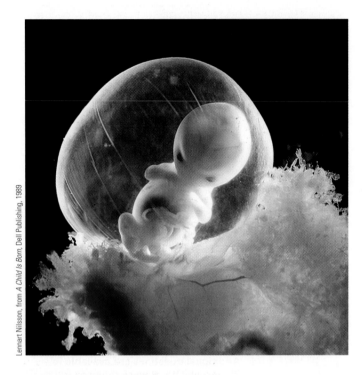

Lennart Nilsson, from *A Child Is Born*, Dell Publishing, 1989

Human embryo in its seventh week of development. The embryo is 2 cm (0.8 in) long.

KEY CONCEPTS

The development of form requires not only cell division and growth but also cell determination and cell differentiation, and pattern formation and morphogenesis.

Fertilization includes contact and recognition between egg and sperm, regulated sperm entry, activation of the fertilized egg, and fusion of egg and sperm pronuclei.

Cleavage, a series of rapid cell divisions without growth, provides cellular building blocks for development.

Gastrulation establishes ectoderm, mesoderm, and endoderm; each gives rise to specific types of tissues.

Neurulation, the origin of the central nervous system, is one of the earliest events in organogenesis (organ development).

Extraembryonic membranes (amnion, chorion, allantois, and yolk sac) have evolved in terrestrial vertebrates as adaptations to reproduction on land.

Development includes all the changes that take place during the entire life of an individual. In this chapter, however, we focus mainly on the fertilization of the egg to form a zygote and the subsequent development of the young animal before birth or hatching.

Just how does a microscopic, unicellular zygote give rise to the bones, muscles, brain, and other structures of a complex animal? The zygote divides by mitosis, forming an **embryo** that subsequently undergoes an orderly sequence of cell divisions. In animals, growth (that is, increase in mass) occurs primarily by an increase in the number of cells; but it also occurs by an increase in cell size, as in fat cells. Although the very early embryo usually does not grow, later cell divisions typically contribute to growth. However, by itself cell division, which is governed by a genetic program that interacts with environmental signals (see Chapter 10), would produce only a formless heap of similar cells. We therefore begin this chapter with a discussion of the fundamental processes that contribute to the development of form: cell determination and cell differentiation, and pattern formation and morphogenesis.

The photograph shows a human embryo, as well as part of the placenta (large, fluffy mass in the lower right-hand corner). In this chapter you have an opportunity to compare and contrast the sequences of developmental events in several different animals in addition to humans. Researchers have chosen these model organisms because they have certain desirable characteristics. For example, the embryos of certain types of animals, such as echinoderms and amphibians, can be obtained in large quantities, which facilitates studies in which biochemical changes are correlated with developmental landmarks. The embryos of echinoderms are particularly easy to observe because they are transparent. The embryos of echinoderms, amphibians, and birds are easily grown in the laboratory. As you read, note the intimate interrelationships among the sequential developmental processes, as well as the fundamental similarities among the early developmental events in the different animal groups featured. ∎

DEVELOPMENT OF FORM

Learning Objectives

1   Analyze the relationship between cell determination and cell differentiation, and between pattern formation and morphogenesis.
2   Relate differential gene expression to nuclear equivalence.

The development of the form of an animal is a consequence of a balanced combination of several fundamental processes that go beyond cell division and growth: cell determination and cell differentiation, and pattern formation and morphogenesis.

As embryonic development proceeds, cell division gives rise to increasing numbers of cells, which serve as the building blocks of development. At various times certain cells become biochemically and structurally specialized to carry out specific functions through a process known as **cell differentiation.** In our discussion of the genetic control of development in Chapter 17, you learned how cell differentiation occurs through **cell determination,** a series of molecular events in which the activities of certain genes are altered in ways that cause a cell to progressively commit to a particular differentiation pathway. This process proceeds even though there may not be any immediate changes in the cell's morphology. You also evaluated evidence supporting the principle of **nuclear equivalence,** which states that in most cases neither cell determination nor cell differentiation entails a loss of genetic information from the cell nucleus. That is, the nuclei of virtually all differentiated cells of an animal contain the same genetic information present in the zygote, but each cell type *expresses* a different subset of that information. In fact, nuclear equivalence is the principle that underlies the cloning of organisms.

Cell differentiation is therefore an expression of changes in the activity of specific genes, which in turn is influenced by a variety of factors inside and outside the cell. This **differential gene expression** is responsible for variations in chemistry, behavior, and structure among cells. Through this process, an embryo can develop into an organism of more than 200 types of cells, each specialized to perform specific functions. However, not all cells differentiate. Some, known as **stem cells,** remain in a relatively undifferentiated state and retain the ability to give rise to various cell types.

Cell differentiation by itself does not explain development. The differentiated cells must become progressively organized, shaping the intricate pattern of tissues and organs that characterizes a multicellular animal. This development of form, known as **morphogenesis,** proceeds through the process of **pattern formation.** Pattern formation is a series of steps requiring signaling between cells, changes in the shapes of certain cells, precise cell migrations, interactions with the extracellular matrix, and even **apoptosis** (programmed cell death; see Chapter 4) of some cells.

Review

∎   Which typically comes first, cell differentiation or cell determination?
∎   What are some kinds of events that contribute to pattern formation? How does pattern formation lead to morphogenesis?

FERTILIZATION

Learning Objectives

3   Describe the four processes involved in fertilization.
4   Describe fertilization in echinoderms, and point out some ways in which mammalian fertilization differs.

In Chapter 49 you studied **spermatogenesis** and **oogenesis,** the processes by which meiosis leads to the formation of haploid cells, which differentiate as sperm and eggs, respectively. In **fertilization** a usually flagellated, motile sperm fuses with a much larger, immotile ovum to produce a **zygote,** or fertilized egg.

Fertilization has two important genetic consequences: restoration of the diploid chromosome number and, in mammals and many other animals, determination of the sex of the offspring (see Chapter 11). Fertilization also has profound physiological effects, because it activates the egg, thus initiating reactions that permit development.

Fertilization involves four events, some of which may occur simultaneously: (1) The sperm contacts the egg and recognition occurs; (2) the sperm or sperm nucleus enters the egg; (3) the egg becomes activated, and certain developmental changes begin; and (4) the sperm and egg nuclei fuse. These events do not necessarily follow the same temporal sequence in all animals. Unless otherwise stated, our discussion applies to sea urchins and other echinoderms such as sea stars, which have been studied intensively because they produce large numbers of gametes and because fertilization is external.

The first step in fertilization involves contact and recognition

Although eggs are immotile, they are active participants in fertilization. An egg is surrounded by a plasma membrane and by one or more external coverings that are important in fertilization. For example, a mammalian egg is enclosed by a thick, noncellular, **zona pellucida,** which is surrounded by a layer of granulosa cells derived from the follicle in which the egg developed.

The egg coverings not only facilitate fertilization by sperm of the same species but also bar interspecific fertilization, a particularly important function in species that practice external fertilization. External to the plasma membrane of a sea urchin egg are two acellular layers that interact with sperm: a very thin **vitelline envelope** and, outside this, a thick glycoprotein layer called the **jelly coat.** Sea urchin sperm become motile when released into sea water, and their motility increases when they reach the vicinity of a sea urchin egg. Motility improves the probability that sperm will encounter the egg, but motility alone may not be sufficient to ensure actual contact with the egg. In sea urchins, as well as certain fishes and some cnidarians, the egg or one of its coverings secretes a chemotactic substance that attracts sperm from the same species. However, for many species, researchers have not found a specific chemical that attracts the sperm to the egg.

When a sea urchin sperm contacts the jelly coat, the sperm undergoes an **acrosome reaction,** in which the membranes surrounding the acrosome (the cap at the head of the sperm; see Fig. 49-6) fuse, and pores in the membrane enlarge. Calcium ions from the sea water move into the acrosome, which swells and begins to disorganize. The acrosome then releases proteolytic enzymes that digest a path through the jelly coat to the vitelline envelope of the egg. If the sperm and egg are of the same species, **bindin,** a species-specific binding protein located on the acrosome, adheres to species-specific bindin receptors located on the egg's vitelline envelope.

Before a mammalian sperm participates in fertilization, it first undergoes **capacitation,** a maturation process in the female reproductive tract. During capacitation, which in humans can take several hours, sperm become increasingly motile and capable of undergoing an acrosome reaction when they encounter an egg. Evidence indicates that mammalian fertilization requires interaction between sperm and species-specific glycoproteins in the egg's zona pellucida.

Sperm entry is regulated

In sea urchins, once contact occurs between acrosome and vitelline envelope, enzymes dissolve a bit of the vitelline envelope in the area of the sperm head. The plasma membrane of the egg is covered with microvilli, several of which elongate to surround the head of the sperm. As they do so, the plasma membranes of sperm and egg fuse and a **fertilization cone** is formed that contracts to draw the sperm into the egg (▍Fig. 50-1).

Fertilization of the egg by more than one sperm, a condition known as **polyspermy,** results in an offspring with extra sets of chromosomes, which is usually lethal. Two reactions, known as the *fast* and *slow blocks* to polyspermy, prevent this occurrence. In the fast block to polyspermy, the egg plasma membrane depolarizes, which prevents its fusion with additional sperm. An unfertilized egg is polarized, that is, the cytoplasm is negatively charged relative to the outside. However, within a few seconds after sperm fusion, ion channels in the plasma membrane open,

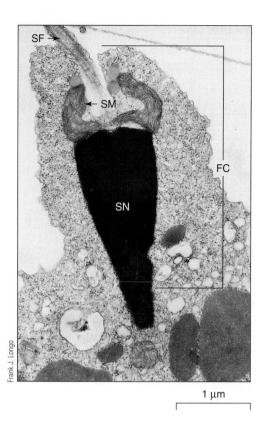

1 µm

Figure 50-1   Fertilization

In this TEM, a fertilization cone (FC) forms as a sperm enters a sea urchin egg. (SN, sperm nucleus; SM, sperm mitochondrion; SF, sperm flagellum)

permitting positively charged calcium ions to diffuse across the membrane and depolarize the egg.

The slow block to polyspermy, the **cortical reaction,** requires up to several minutes to complete, but it is a complete block. Binding of the sperm to receptors on the vitelline envelope activates one or more *signal transduction* pathways (see Chapter 6) in the egg. These events cause calcium ions stored in the egg endoplasmic reticulum to be released into the cytosol. This rise in the level of intracellular calcium causes thousands of cortical granules, membrane-enclosed vesicles in the egg cortex (region close to the plasma membrane), to release enzymes, various proteins, and other substances by exocytosis into the space between the plasma membrane and the vitelline envelope (▍Fig. 50-2). Some of the enzymes dissolve the protein linking the two membranes; thus, the space enlarges as additional substances released by the cortical granules raise the osmotic pressure, which causes an influx of water from the surroundings. Thus, the vitelline envelope becomes elevated away from the plasma membrane and forms the fertilization envelope, a hardened covering that prevents entry of additional sperm.

In mammals, a fertilization envelope does not form, but the enzymes released during exocytosis of the cortical granules alter the sperm receptors on the egg's zona pellucida so that no additional sperm bind to them.

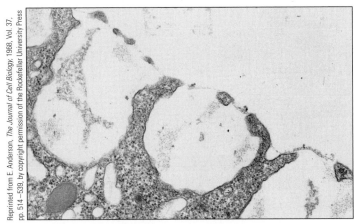

Reprinted from E. Anderson, *The Journal of Cell Biology*, 1968, Vol. 37, pp. 514–539, by copyright permission of the Rockefeller University Press

1 μm

Figure 50-2 The cortical reaction

Three cortical granules are undergoing exocytosis in this TEM of the plasma membrane of a sea urchin egg shortly after fertilization. This release of the contents of the cortical granules initiates the slow block to polyspermy. (From E. Anderson, Reprinted from *The Journal of Cell Biology*, 1968, Vol. 37, pgs 514–539, by copyright permission of the Rockefeller University Press.)

Not all species have these types of blocks to polyspermy. In some, such as salamanders, several sperm usually enter the egg, but only one sperm nucleus fuses with the egg nucleus, and the rest degenerate.

Fertilization activates the egg

Release of calcium ions into the egg cytoplasm does more than stimulate the cortical reaction; it also triggers the *activation program,* a series of metabolic changes within the egg. Aerobic respiration increases; certain maternal enzymes and other proteins become active; and within a few minutes after sperm entry, a burst of protein synthesis occurs. In addition, the egg nucleus is stimulated to complete meiosis. (Recall from Chapter 49 that in most animal species, including mammals, at the time of fertilization an egg is actually a secondary oocyte, arrested early in the second meiotic division.)

In some species, an egg can be artificially activated without sperm penetration by swabbing it with blood and pricking the plasma membrane with a needle, by calcium injection, or by certain other treatments. These haploid eggs go through some developmental stages *parthenogenetically,* that is, without fertilization (see Chapter 49), but they cannot complete development. Special mechanisms of egg activation have evolved in species that are naturally parthenogenetic.

Sperm and egg pronuclei fuse, restoring the diploid state

After the sperm nucleus enters the egg, researchers think it is guided toward the egg nucleus by a system of microtubules that forms within the egg. The sperm nucleus swells and forms the

male pronucleus. The nucleus formed during completion of meiosis in the egg becomes the **female pronucleus.** (Recall from Figure 49-11 that the other nucleus formed during meiosis II in the egg becomes the second polar body.) The haploid male and female pronuclei then fuse to form the diploid nucleus of the zygote, and DNA synthesis occurs in preparation for the first cell division.

Review

- What mechanisms ensure fertilization by only one sperm of the same species?
- How do the mechanisms of fertilization ensure control of both quality (fertilization by a sperm of the same species) and quantity (fertilization by only one sperm)?

CLEAVAGE

Learning Objectives

5 Trace the generalized pattern of early development of the embryo from zygote through early cleavage and formation of the morula and blastula.

6 Contrast early development, including cleavage in the echinoderm (or in amphioxus), the amphibian, and the bird, paying particular attention to the importance of the amount and distribution of yolk.

Despite its simple appearance, the zygote is **totipotent,** that is, it gives rise to all the cell types of the new individual. Because the ovum is very large compared with the sperm, the bulk of the zygote cytoplasm and organelles comes from the ovum. However, the sperm and ovum usually contribute equal numbers of chromosomes.

Shortly after fertilization, the zygote undergoes **cleavage,** a series of rapid mitotic divisions with no period of growth during each cell cycle. For this reason, although the cell number increases, the embryo does not increase in size. The zygote initially divides to form a 2-celled embryo. Then each of these cells undergoes mitosis and divides, bringing the number of cells to 4. Repeated divisions increase the number of cells, called **blastomeres,** that make up the embryo. At about the 32-cell stage the embryo is a solid ball of blastomeres called a **morula.** Eventually, anywhere from 64 to several hundred blastomeres form the **blastula,** which is usually a hollow ball with a fluid-filled cavity, the **blastocoel.**

The pattern of cleavage is affected by yolk

Many animal eggs contain **yolk,** a mixture of proteins, phospholipids, and fats that serves as food for the developing embryo. The amount and distribution of yolk vary among different animal groups, depending on the needs of the embryo. Mammalian eggs have very little yolk, because the embryo obtains maternal nutritional support throughout most of its development; whereas the eggs of birds and reptiles must contain sufficient yolk to sustain the embryo until hatching. Echinoderm eggs typically need only enough yolk to nourish the embryo until it becomes a tiny larva capable of obtaining its own food.

Key Point

Cleavage in sea stars is radial and holoblastic.

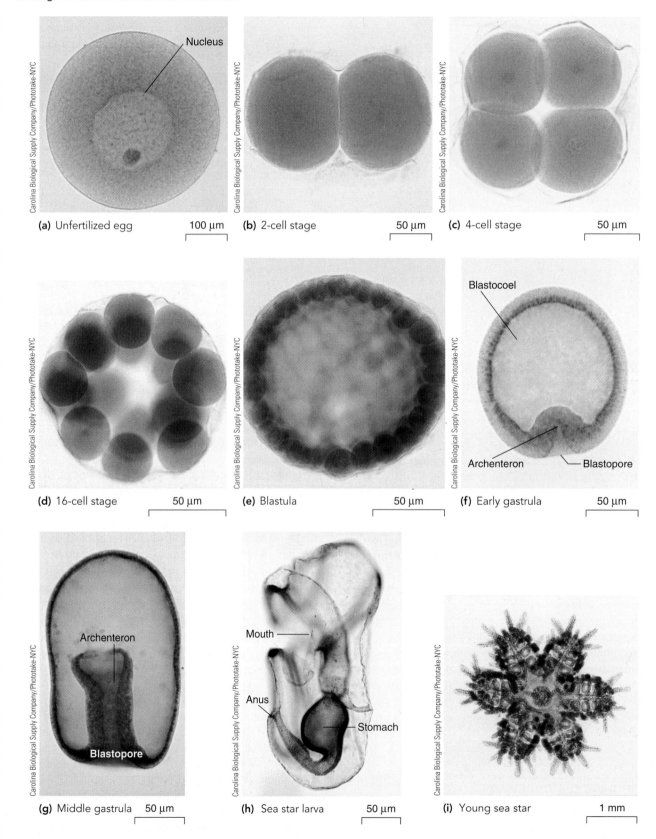

(a) Unfertilized egg — Nucleus — 100 µm

(b) 2-cell stage — 50 µm

(c) 4-cell stage — 50 µm

(d) 16-cell stage — 50 µm

(e) Blastula — 50 µm

(f) Early gastrula — Blastocoel — Archenteron — Blastopore — 50 µm

(g) Middle gastrula — Archenteron — Blastopore — 50 µm

(h) Sea star larva — Mouth — Anus — Stomach — 50 µm

(i) Young sea star — 1 mm

Carolina Biological Supply Company/Phototake-NYC

Figure 50-3 LMs showing sea star development (See caption on facing page.)

Most invertebrates and simple chordates have **isolecithal eggs** with relatively small amounts of yolk uniformly distributed through the cytoplasm. Isolecithal eggs divide completely; this is called **holoblastic cleavage.** Cleavage of these eggs is radial or spiral. Radial cleavage is characteristic of *deuterostomes,* such as chordates and echinoderms; spiral cleavage is common in the embryos of *protostomes,* such as annelids and mollusks (see Fig. 29-5).

In **radial cleavage,** the first division is vertical and splits the egg into two equal cells. The second cleavage division, also vertical, is at right angles to the first division and separates the two cells into four equal cells. The third division is horizontal, at right angles to the other two, and separates the four cells into eight cells: four above and four below the third line of cleavage. This pattern of radial cleavage occurs in echinoderms and in the cephalochordate amphioxus (▌Figs. 50-3 and 50-4).

In **spiral cleavage,** after the first two divisions, the plane of cytokinesis is diagonal to the polar axis (▌Fig. 50-5). This results in a spiral arrangement of cells, with each cell located above and between the two underlying cells. This pattern is typical of annelids and mollusks.

Many vertebrate eggs are **telolecithal,** meaning they have large amounts of yolk concentrated at one end of the cell, known as the **vegetal pole.** The opposite, more metabolically active, pole is the **animal pole.** The eggs of amphibians are moderately telolecithal (mesolecithal). Although cleavage is radial and holoblastic, the divisions in the vegetal hemisphere are slowed by the presence of the inert yolk. As a result, the blastula consists of many small cells in the animal hemisphere and fewer but larger cells in the vegetal hemisphere (▌Fig. 50-6). The blastocoel is displaced toward the animal pole.

The telolecithal eggs of reptiles and birds have very large amounts of yolk at the vegetal pole and only a small amount of cytoplasm concentrated at the animal pole. The yolk of such eggs never cleaves. Cell division is restricted to the **blastodisc,** the small disc of cytoplasm at the animal pole (▌Fig. 50-7); this type of cleavage is termed **meroblastic cleavage.** In birds and some reptiles, the blastomeres form two layers separated by the blastocoel cavity: an upper *epiblast* and a lower, thin layer of flat cells, the *hypoblast.*

Cleavage may distribute developmental determinants

The pattern of cleavage in a particular species depends on both yolk and other factors. Recall from Chapter 17 that some animals have relatively rigid developmental patterns, known as **mosaic**

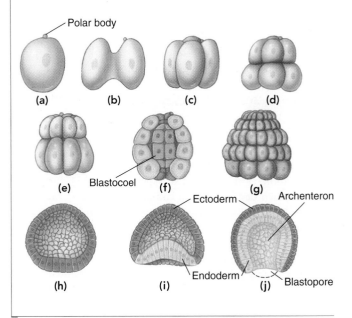

Key Point

Cleavage in amphioxus is radial and holoblastic.

Figure 50-4 Cleavage and gastrulation in amphioxus

As in the sea star, cleavage is holoblastic and radial. The embryos are shown from the side. **(a)** Mature egg with polar body. **(b–e)** The 2-, 4-, 8-, and 16-cell stages. **(f)** Embryo cut open to show the blastocoel. **(g)** Blastula. **(h)** Blastula cut open. **(i)** Early gastrula showing beginning of invagination at vegetal pole. **(j)** Late gastrula. Invagination is completed, and the blastopore has formed.

development. This type of development is largely a consequence of the unequal distribution of important materials in the cytoplasm of the zygote. Because the zygote cytoplasm is not homogeneous, cytoplasmic developmental determinants portioned out to each new cell during cleavage may be different. Such differences help determine the course of development. At the other extreme are mammals, which have zygotes with very homogeneous cytoplasm. They exhibit highly **regulative development,** in which individual cells produced by the cleavage divisions are equivalent, thus allowing the embryo to develop as a self-regulating whole. Developmental patterns of most animals fall somewhere on a continuum between these two extremes.

In some species the distribution of developmental determinants in the unfertilized egg is maintained in the zygote. In others sperm penetration initiates rearrangement of the cytoplasm. For example, fertilization in the amphibian egg causes a shift of some of the cortical cytoplasm. These cytoplasmic movements can be easily followed because the egg cortex contains dark pigment granules. As a result of this rearrangement, a crescent-shaped region of underlying lighter-colored (gray) cytoplasm becomes evident directly opposite the point on the cell where the sperm penetrated the egg. This **gray crescent** region is thought to contain growth factors

Figure 50-3 Continued

(a) The isolecithal egg has a small amount of uniformly distributed yolk. **(b–e)** The cleavage pattern is radial and holoblastic (the entire egg becomes partitioned into cells). **(f, g)** The three germ layers form during gastrulation. The blastopore is the opening into the developing gut cavity, the archenteron. The rudiments of organs are evident in the sea star larva **(h)** and the young sea star **(i).** All views are side views with the animal pole at the top, except **(c)** and **(i),** which are top views. Note that the sea star larva is bilaterally symmetrical, but differential growth produces a radially symmetrical young sea star.

and other developmental determinants. The first cleavage bisects the gray crescent, distributing half to each of the first two blastomeres; in this way the position of the gray crescent establishes the future right and left halves of the embryo. As cleavage continues, the gray crescent material becomes partitioned into certain blastomeres. Those that contain parts of the gray crescent eventually develop into the dorsal region of the embryo.

Experiments have confirmed the importance of determinants in the gray crescent to development. If the first two frog blastomeres are separated experimentally, each develops into a complete tadpole (Fig. 50-8a). When the plane of the first division is altered so that the gray crescent is completely absent from one of the cells, that cell does not develop normally (Fig. 50-8b).

Key Point

Cleavage in annelids is spiral and holoblastic.

(a) Zygote **(b)** 2-cell stage **(c)** 4-cell stage

(d) 8-cell stage **(e)** 16-cell stage **(f)** 32-cell stage **(g)** Trochophore larva

Apical organ
Brain rudiment
Stomach
Intestine
Mouth
Anus
Mesoderm band
Protonephridium
Esophagus

Figure 50-5   Spiral cleavage in an annelid embryo

(a–f) Top views of the animal pole. The successive cleavage divisions occur in a spiral pattern as illustrated. **(g)** A typical trochophore larva. The upper half of the trochophore develops into the extreme anterior end of the adult worm; all the rest of the adult body develops from the lower half.

Cleavage provides building blocks for development

Cleavage partitions the zygote into many small cells that serve as the basic building blocks for subsequent development. At the end of cleavage, the small cells that make up the blastula begin to move about with relative ease, arranging themselves into the patterns necessary for further development. Surface proteins are important in helping cells "recognize" one another and therefore in determining which ones adhere to form tissues.

Key Point

Cleavage in the moderately telolecithal frog egg is radial and holoblastic but is retarded by yolk in the vegetal hemisphere.

Animal pole

Vegetal pole

(a)   **(b)**   **(c)**   **(d)**

Figure 50-6   *Animated*   Cleavage pattern in a frog egg

(a–d) Although cleavage is holoblastic, the large amount of yolk concentrated in the vegetal hemisphere slows cleavage. As a result, fewer cells develop at the vegetal hemisphere than at the animal hemisphere. These embryos are shown from the side.

Key Point

Cleavage in bird embryos is restricted to the blastodisc (meroblastic cleavage).

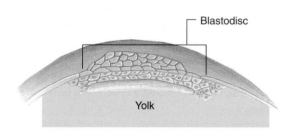

Blastodisc

Yolk

(a) Early blastodisc formation. The blastodisc is a small disc of cytoplasm on the upper surface of the egg yolk. This cutaway view shows cells on the blastodisc surface, as well as in the interior.

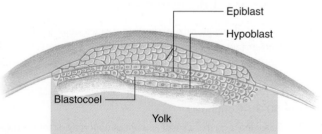

Epiblast
Hypoblast
Blastocoel
Yolk

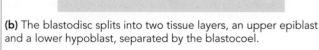

(b) The blastodisc splits into two tissue layers, an upper epiblast and a lower hypoblast, separated by the blastocoel.

Figure 50-7   Cleavage in a bird embryo

QUESTION: Is the development of a frog egg influenced by cytoplasmic developmental determinants?

HYPOTHESIS: The gray crescent, which is normally bisected by the first cleavage division, thus establishing right and left halves of the embryo, is the location of crucial developmental determinants.

EXPERIMENT: The two blastomeres resulting from the first cleavage division are separated by the researcher and allowed to develop independently.

First cleavage (control)　　First cleavage (experimental)

(a) The first cleavage is allowed to occur normally, and each separated blastomere includes half of the gray crescent.

(b) The plane of cleavage is altered by the experimenter such that only one blastomere contains the gray crescent.

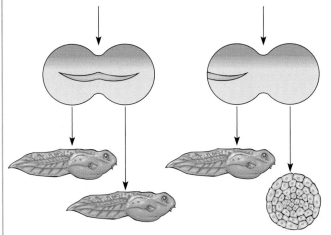

Each of the separated blastomeres contained half of the gray crescent, and each developed into a normal tadpole.

The separated blastomere that received the entire gray crescent developed normally, but the blastomere lacking the gray crescent gave rise to a disorganized ball of cells.

RESULTS AND CONCLUSION: Because only blastomeres containing all or part of the gray crescent developed into tadpoles, the gray crescent marks the location of essential developmental determinants in the fertilized egg.

Figure 50-8 *Animated* Cytoplasmic determinants in frog development

Review

- What is radial cleavage? Spiral cleavage?
- What type of cleavage is characteristic of telolecithal eggs of reptiles and birds?
- How does the cytoplasm influence early development?

GASTRULATION

Learning Objective

7 Identify the significance of gastrulation in the developmental process, and compare gastrulation in the echinoderm (or in amphioxus), the amphibian, and the bird.

The process by which the blastula becomes a three-layered embryo, or **gastrula,** is called **gastrulation.** Thus, early development proceeds through the following stages:

Zygote ⟶ early cleavage stages ⟶ morula ⟶ blastula ⟶ gastrula

During gastrulation, the embryo begins to approximate its body plan as cells arrange themselves into three distinct **germ layers,** or embryonic tissue layers: the outermost layer, the **ectoderm;** the innermost, the **endoderm;** and the **mesoderm,** which develops between them. Additional cell divisions take place during gastrulation, and the germ layers become established through a combination of processes. Many cells lose their old cell-to-cell contacts and establish new ones through cell recognition and adhesion processes involving interactions among the integrins and other plasma membrane proteins and the extracellular matrix (see Chapters 4 and 5). Many cells undergo cytoskeletal changes, particularly alterations in the distribution of actin microfilaments; these changes in the internal architecture of the cells allow them to change shape and/or undergo specific, directional, amoeboid movements. As a result of these movements, many cells ultimately take up new positions in the interior of the embryo.

Each of the germ layers develops into specific parts of the embryo. As you study the illustrations, keep in mind that a conventional color code accepted by developmental biologists is used to depict the germ layers: endoderm, yellow; mesoderm, red or pink; and ectoderm, blue.

The amount of yolk affects the pattern of gastrulation

The simple type of gastrulation that occurs in echinoderms and in amphioxus is illustrated in Figures 50-3f and g and 50-4i and j, respectively. Gastrulation begins when a group of cells at the vegetal pole undergoes a series of changes in shape that cause that part of the blastula wall to first flatten and then bend inward (invaginate). The invaginated wall eventually meets the opposite wall, obliterating the blastocoel.

You can roughly demonstrate this type of gastrulation by pushing inward on the wall of a partly deflated rubber ball until it rests against the opposite wall. In a similar way the embryo is converted into a double-walled, cup-shaped structure. The new internal wall lines the **archenteron,** the newly formed cavity of the developing gut. The opening of the archenteron to the exterior, the **blastopore,** is the site of the future anus in deuterostomes.

This simple type of gastrulation cannot occur in the amphibian embryo, because the large yolk-laden cells in the vegetal half of the blastula obstruct any inward movement at the vegetal pole.

Key Point

Gastrulation forms the three germ layers: endoderm, ectoderm, and mesoderm.

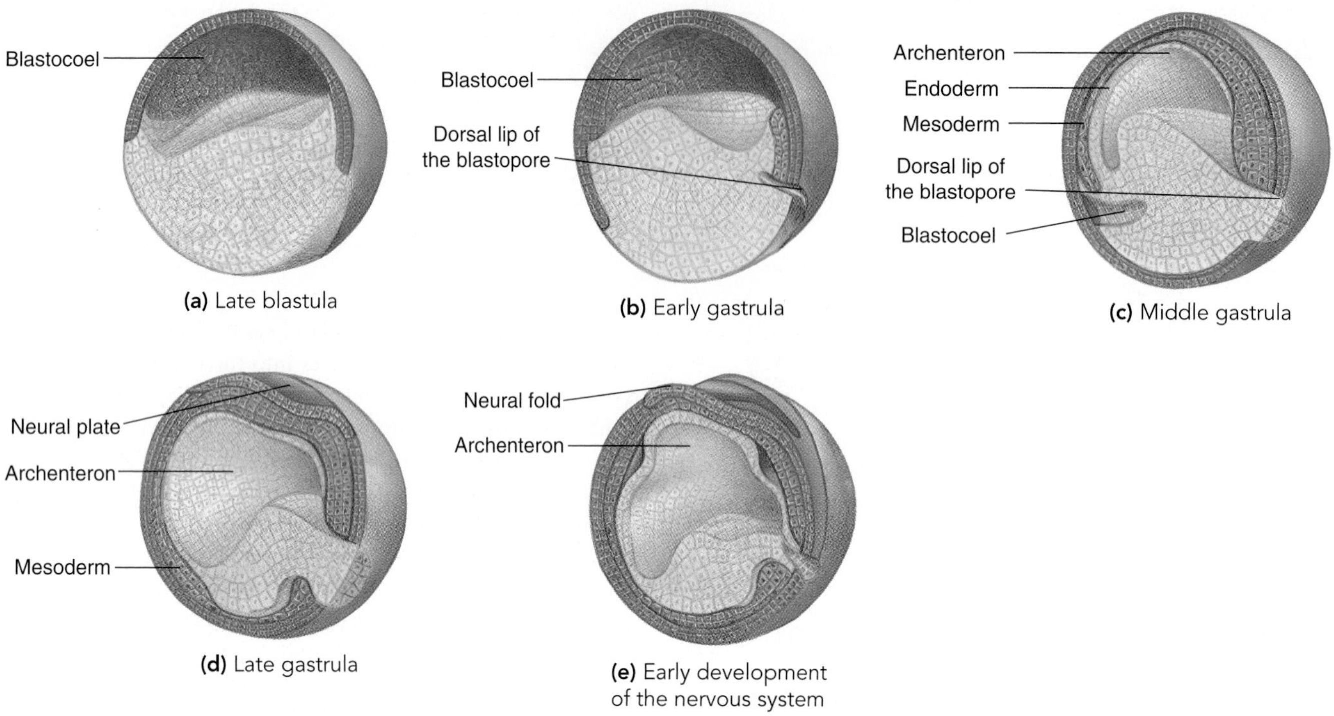

(a) Late blastula

Blastocoel

(b) Early gastrula

Blastocoel

Dorsal lip of the blastopore

(c) Middle gastrula

Archenteron

Endoderm

Mesoderm

Dorsal lip of the blastopore

Blastocoel

(d) Late gastrula

Neural plate

Archenteron

Mesoderm

(e) Early development of the nervous system

Neural fold

Archenteron

Figure 50-9 *Animated* Gastrulation in a frog embryo

The embryos in the diagrams are cut in half longitudinally to show the formation of the germ layers.

Instead, cells from the animal pole move down the embryo surface; when they reach the region derived from the gray crescent, they move into the interior. This inward movement is accomplished as the cells change shape, first becoming flask- or bottle-shaped (so that most of their mass is actually below the surface) and then sinking into the interior as they lose their remaining connections with other cells on the surface. This spot on the embryo's surface, referred to as the **dorsal lip of the blastopore,** is marked by a dimple, shaped like a C lying on its side (❚ Fig. 50-9). As the process continues, the blastopore becomes ring-shaped as cells lateral, and then ventral, to its dorsal lip become involved in similar movements. The yolk-filled cells fill the space enclosed by the lips of the blastopore, forming the yolk plug.

The archenteron forms and is lined on all sides by cells that have moved in from the surface. At first, the archenteron is a narrow slit, but it gradually expands at the anterior end of the embryo. As a result, the blastocoel progressively shrinks and eventually disappears.

Although the details differ somewhat, gastrulation in birds is basically similar to amphibian gastrulation. Cells that make up the upper layer (the epiblast) migrate toward the midline to form a thickened region known as the **primitive streak,** which elongates and narrows as it develops. At its center is a narrow furrow, the **primitive groove.** The primitive streak is a dynamic structure. Its cells constantly change as they migrate in from the epiblast, sink inward at the primitive groove, then move out laterally

and anteriorly in the interior (❚ Fig. 50-10). The primitive groove is the functional equivalent of the blastopore of the echinoderm, amphioxus, and amphibian embryos. However, a bird embryo contains no cavity homologous to the archenteron.

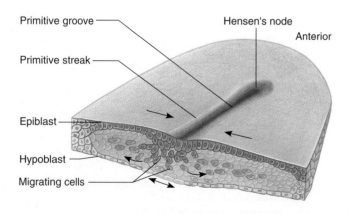

Primitive groove

Primitive streak

Epiblast

Hypoblast

Migrating cells

Hensen's node

Anterior

Figure 50-10 Gastrulation in birds

A three-layered embryo forms as cells move toward the primitive streak, move inward at the primitive groove, and migrate laterally and anteriorly in the interior. Cells shown just above the double-headed arrow have migrated into the hypoblast; they are moving laterally and will eventually become endoderm.

TABLE 50-1

Fate of the Germ Layers Formed at Gastrulation

Ectoderm

Nervous system and sense organs

Outer layer of skin (epidermis) and its associated structures (nails, hair, etc.)

Pituitary gland

Mesoderm

Notochord

Skeleton (bone and cartilage)

Muscles

Circulatory system

Excretory system

Reproductive system

Inner layer of skin (dermis)

Outer layers of digestive tube and of structures that develop from it, such as part of respiratory system

Endoderm

Lining of digestive tube and of structures that develop from it, such as lining of the respiratory system

---

outgrowths of the digestive tract (including the liver, pancreas, and lungs), are all of endodermal origin. Skeletal tissue, muscle, and the circulatory, excretory, and reproductive systems all are derived from mesoderm (Table 50-1; see Fig. 17-1).

The notochord, brain, and spinal cord are among the first organs to develop in the early vertebrate embryo (Fig. 50-11; see also Fig. 50-9d and e). First, the notochord, which is mesodermal tissue, grows forward along the length of the embryo as a cylindrical rod of cells. The notochord eventually is replaced by the vertebral column, although notochord remnants will remain in the cartilage discs between the vertebrae.

The developing notochord has yet another crucial function. Numerous experiments in which researchers have transplanted portions of the notochord mesoderm to other locations in the embryo show that the notochord causes the overlying ectoderm to thicken and differentiate to form the precursor of the central nervous system, the **neural plate.** Such phenomena, in which certain cells stimulate or otherwise influence the differentiation of their neighbors, are examples of **induction** (see Chapter 17).

Repeated testing supports the hypothesis that induction of the neural plate cells does not require direct cell-to-cell contact

At the anterior end of the primitive streak, cells destined to form the **notochord** (supporting rod of mesodermal, cartilage-like cells that serves as the flexible skeletal axis in all chordate embryos) concentrate in a thickened knot known as **Hensen's node.** These cells sink into the interior and then move anteriorly just beneath the epiblast, forming a narrow extension from the node. Cells that will form other types of mesoderm move laterally and anteriorly from the primitive streak, between the epiblast (which becomes ectoderm) and the hypoblast. Other cells that form the endoderm displace the hypoblast cells and cause them to move laterally. These displaced cells will form part of the extraembryonic membranes, discussed in a later section.

Review

- What is the archenteron? Is an archenteron formed in sea stars? In amphibians? In birds?
- How does the amount and distribution of yolk influence gastrulation?

ORGANOGENESIS

Learning Objectives

8 Define *organogenesis*, and summarize the fate of each of the germ layers.

9 Trace the early development of the vertebrate nervous system.

Gastrulation leads to **organogenesis,** or organ formation, in which the cells of the three germ layers continue the processes of pattern formation that lead to the formation of specific structures. The ectoderm eventually forms the outer layer of the skin and gives rise to the nervous system and sense organs. Tissues that eventually line the digestive tract, and organs that develop as

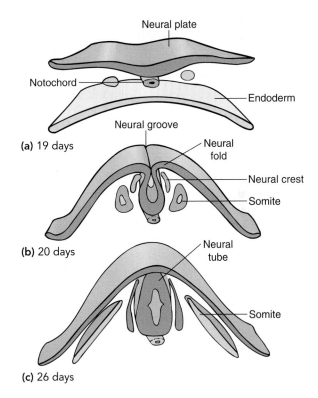

Figure 50-11 *Animated* **Development of the human nervous system**

(a) Approximately 19 days. The neural plate has indented to form a shallow groove flanked by neural folds. **(b)** Approximately 20 days. The neural folds approach one another. The neural crest cells, derived from ectoderm, will migrate in the embryo and give rise to sensory neurons; the somites are blocks of mesoderm that become the vertebrae and other segmented body parts. **(c)** Approximately 26 days. The neural folds have joined to form the hollow neural tube, which gives rise to the brain at the anterior end of the embryo and the spinal cord posteriorly. The endoderm is not shown in **(b)** and **(c)**.

with the developing notochord cells. Researchers therefore think that this induction involves the diffusion of some type of signal molecule, although they have not yet identified the chemical signal itself. The induction of the neural plate by the notochord is a good example of the importance of a cell's position in relation to other cells. That position is often critical in determining the fate of a cell because it determines the cell's exposure to substances released from other cells.

The cells of the neural plate undergo changes in shape that cause the central cells of the neural plate to move downward and form a depression called the **neural groove;** the cells flanking the groove on each side form **neural folds.** Continued changes in cell shape bring the folds closer together until they meet and fuse to form the **neural tube.** In this process, the hollow neural tube comes to lie beneath the surface. The ectoderm overlying it will form the outer layer of skin. The anterior portion of the neural tube grows and differentiates into the brain; the rest of the tube develops into the spinal cord. The brain and the spinal cord remain hollow, even in the adult; the ventricles of the brain and central canal of the spinal cord are persistent reminders of their embryonic origins.

Various motor nerves grow out of the developing brain and spinal cord, but sensory nerves have a separate origin. When the neural folds fuse to form the neural tube, bits of nervous tissue known as the **neural crest** arise from ectoderm on each side of the tube. Neural crest cells migrate downward from their original position and form the dorsal root ganglia of the spinal nerves and the postganglionic sympathetic neurons. From sensory cells in the dorsal root ganglia, dendrites grow out to the sense organs, and axons grow into the spinal cord. Other neural crest cells migrate to various locations in the embryo. They give rise to parts of certain sense organs, nearly all pigment-forming cells in the body, and parts of the head. Some head derivatives of the neural crest are the cartilage and bone of the skull, face, and jaw; connective tissue in the eye and face dermis; sensory nerves and other parts of the peripheral nervous system; and ganglia of the autonomic nervous system.

As the nervous system develops, other organs also begin to take shape. Blocks of mesoderm known as **somites** form on either side of the neural tube. These will give rise to the vertebrae, muscles, and other components of the body axis. In addition to the skeleton and muscles, mesodermal structures include the kidneys, reproductive structures, and the circulatory organs. In fact, the heart and blood vessels are among the first structures to form, and they must function while still developing.

The development of the four-chambered heart of birds and mammals is a good example of organogenesis (Fig. 50-12; see Chapter 43). The heart originates through the fusion of paired blood vessels. Venous blood enters the single atrium; passes into the single ventricle, which is located above the atrium; and is then pumped into the embryo. During subsequent development, the atrium undergoes a process of torsion that brings it to a position above the ventricle, and partitions form that divide the atrium and the ventricle into right and left chambers.

The digestive tract is first formed as a separate foregut and hindgut as the body wall grows and folds; this process cuts them off as two simple tubes. As the embryo grows, these tubes, which

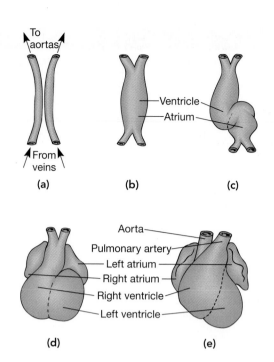

Figure 50-12 Formation of the heart in birds and mammals

These ventral views of successive stages show that the heart forms from the fusion of two blood vessels and that at first the end that receives blood from the veins is at the bottom. The developing heart twists to carry the atrium to a position above the ventricle, and the chambers then divide to form the four-chambered heart.

are lined with endoderm, grow and become greatly elongated. The liver, pancreas, and trachea originate as hollow, tubular outgrowths from the gut. As the trachea grows downward, it gives rise to the paired lung buds, which develop into lungs. The most anterior part of the foregut becomes the pharynx. A series of small outpocketings of the pharynx, the **pharyngeal pouches,** bud out laterally. These pouches meet the **branchial grooves,** a corresponding set of inpocketings from the overlying ectoderm. The arches of tissue formed between the grooves are called **branchial arches.** They contain the rudimentary skeletal, neural, and vascular elements of the face, jaws, and neck.

In fishes and some amphibians, the pharyngeal pouches and branchial grooves meet and form a continuous passage from the pharynx to the outside; these gill slits function as respiratory organs. In terrestrial vertebrates, each branchial groove remains separated from the corresponding pharyngeal pouch by a thin membrane of tissue, and these structures give rise to organs more appropriate for life on land. For example, the middle-ear cavity and its connection to the pharynx, the eustachian tube, derive from a pharyngeal pouch.

Review

- How do cell division, growth, cell differentiation, and morphogenesis interact in the formation of the neural tube?
- What is the developmental significance of the neural crest cells?
- What adult structures develop from each germ layer?

EXTRAEMBRYONIC MEMBRANES

Learning Objective

10 Give the origins and functions of the chorion, amnion, allantois, and yolk sac.

In terrestrial vertebrates, the three germ layers also give rise to four **extraembryonic membranes:** the chorion, amnion, allantois, and yolk sac (▌ Fig. 50-13). Although they develop from the germ layers, these membranes are not part of the embryo itself and are discarded at hatching or birth. The extraembryonic membranes are adaptations to the challenges of embryonic development on land. They protect the embryo, prevent it from drying out, and help in obtaining food and oxygen and eliminating wastes.

The outermost membrane, the **chorion,** encloses the entire embryo. Lying underneath the eggshell in birds and reptiles, it functions as the major organ of gas exchange. As you will see in the next section, its functions have been extended in humans and most other mammals.

Like the chorion, the **amnion** encloses the entire embryo. The amniotic cavity, the space between the embryo and the amnion, becomes filled with amniotic fluid secreted by the membrane. Recall from Chapter 31 that terrestrial vertebrates are known as *amniotes,* because their embryos develop within this pool of fluid. The amniotic fluid prevents the embryo from drying out and permits it a certain freedom of motion. The fluid also serves as a protective cushion that absorbs shocks and prevents the amniotic membrane from sticking to the embryo. In current medical practice a sample of amniotic fluid is obtained through a procedure known as *amniocentesis* and analyzed for biochemical or chromosomal abnormalities that indicate a possible birth defect (see Fig. 16-11).

The **allantois** is an outgrowth of the developing digestive tract. In reptiles and birds, it stores nitrogenous wastes. In humans, the allantois is small and nonfunctional, except that its blood vessels contribute to the formation of umbilical vessels joining the embryo to the placenta. In vertebrates with yolk-rich eggs, the **yolk sac** encloses the yolk, slowly digests it, and makes it available to the embryo. A yolk sac, connected to the embryo by a yolk stalk, develops even in mammalian embryos that have little or no yolk. Its walls serve as temporary centers for the formation of blood cells.

Review

- What are the roles of the chorion and amnion in terrestrial vertebrate embryos?
- How do the functions of the allantois and yolk sac differ between birds and mammals?

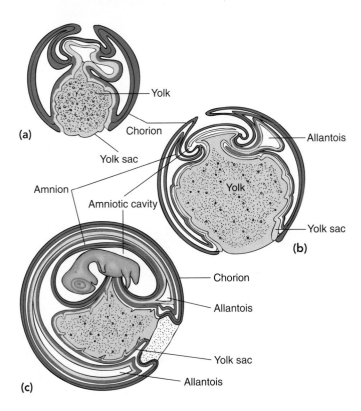

▌ Figure 50-13 Extraembryonic membranes

The formation of the extraembryonic membranes of the chick is illustrated at **(a)** 4 days, **(b)** 5 days, and **(c)** 9 days of development. Each of the membranes develops from a combination of two germ layers. The chorion and amnion form from lateral folds of the ectoderm and mesoderm that extend over the embryo and fuse. The allantois and the yolk sac develop from endoderm and mesoderm.

HUMAN DEVELOPMENT

Learning Objectives

11 Describe the general course of early human development, including fertilization, the fates of the trophoblast and inner cell mass, implantation, and the role of the placenta.

12 Contrast postnatal with prenatal life, describing several changes that occur at or shortly after birth that allow the neonate to live independently.

13 Describe some anatomical and physiological changes that occur with aging.

The human **gestation period,** the duration of pregnancy, averages 266 days (38 weeks, or about 9 months) from the time of fertilization to the birth of the baby (▌ Table 50-2). We will use this system of timing when discussing the events of early development. Because the time of fertilization is not easily marked, obstetricians usually time the pregnancy by counting from the date of onset of the woman's last menstrual period; by this calculation, an average pregnancy is 280 days (40 weeks).

Fertilization occurs in the oviduct, and within 24 hours the human zygote has divided to become a two-celled embryo (▌ Fig. 50-14). Cleavage continues as the embryo is propelled along the oviduct by ciliary action.

When the embryo enters the uterus on about the fifth day of development, the *zona pellucida* (its surrounding coat) is dissolved. During the next few days, the embryo floats free in the uterine cavity and is nourished by a nutritive fluid secreted by the glands of the uterus. Its cells arrange themselves, forming a blastula, which in mammals is called a **blastocyst.** The outer layer of cells, the **trophoblast,** eventually forms the chorion and amnion that surround the embryo. A little cluster of cells, the **inner**

TABLE 50-2

Some Important Developmental Events in the Human Embryo

Time from Fertilization	Event
24 hours	Embryo reaches two-cell stage
3 days	Morula reaches uterus
7 days	Blastocyst begins to implant
2.5 weeks	Notochord and neural plate are formed; tissue that will give rise to heart is differentiating; blood cells are forming in yolk sac and chorion
3.5 weeks	Neural tube forming; primordial eye and ear visible; pharyngeal pouches forming; liver bud differentiating; respiratory system and thyroid gland just beginning to develop; heart tubes fuse, bend, and begin to beat; blood vessels are laid down
4 weeks	Limb buds appear; three primary divisions of brain forming
2 months	Muscles differentiating; embryo capable of movement; gonad distinguishable as testis or ovary; bones begin to ossify; cerebral cortex differentiating; principal blood vessels assume final positions
3 months	Sex can be determined by external inspection; notochord degenerates; lymph glands develop
4 months	Face begins to look human; lobes of cerebrum differentiate; eyes, ears, and nose look more "normal"
Third trimester	Downy hair covers the fetus, then later is shed; neuron myelination begins; tremendous growth of body
266 days	Birth

cell mass, projects into the cavity of the blastocyst. The inner cell mass gives rise to the embryo proper.

On about the seventh day of development, the embryo begins the process of **implantation,** in which it embeds in the endometrium of the uterus (▮ Fig. 50-15). The trophoblast cells in contact with the endometrium secrete enzymes that erode an area just large enough to accommodate the tiny embryo. As the embryo slowly works its way into the underlying connective and vascular tissues, the opening in the endometrium repairs itself. All further development of the embryo takes place *within* the endometrium.

The placenta is an organ of exchange

In placental mammals, the **placenta** is the organ of exchange between mother and embryo (see Fig. 50-15). The placenta provides nutrients and oxygen for the fetus and removes wastes, which the mother then excretes. In addition, the placenta is an endocrine organ that secretes estrogens and progesterone to maintain pregnancy. The placenta develops from both the chorion of the embryo and the uterine tissue of the mother. In early development, the chorion grows rapidly, invading the endometrium and forming fingerlike projections known as *chorionic villi.* The

chorionic villus sampling technique allows prenatal detection of certain genetic disorders (see Fig. 16-12). The villi become vascularized (infiltrated with blood vessels) as the embryonic circulation develops.

As the human embryo grows, the **umbilical cord** develops and connects the embryo to the placenta (see Fig. 50-15). The umbilical cord contains the two umbilical arteries and the umbilical vein. The umbilical arteries connect the embryo to a vast network of capillaries developing within the chorionic villi. Blood from the villi returns to the embryo through the umbilical vein.

The placenta consists of the portion of the chorion that develops villi, together with the underlying uterine tissue that contains maternal capillaries and small pools of maternal blood. The fetal blood in the capillaries of the chorionic villi comes in close contact with the mother's blood in the tissues between the villi. The two circulations are always separated by a membrane through which substances may diffuse or be actively transported. Maternal and fetal blood do not normally mix in the placenta or any other place.

The placenta produces several hormones. From the time the embryo first begins to implant itself, its trophoblastic cells release **human chorionic gonadotropin (hCG),** which signals the

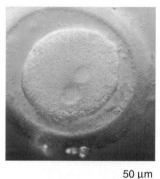

50 µm

(a) Male and female pronuclei prior to fusion

50 µm

(b) 2-cell stage

50 µm

(c) 8-cell stage

50 µm

(d) Cleavage continues, giving rise to a morula

▮ **Figure 50-14** *Animated* Cleavage in a human embryo
(From Lennart Nilsson, *Being Born,* pp. 14, 15, 17. Putnam Publishing Group, 1992.)

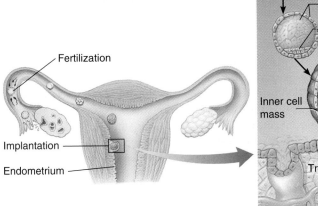

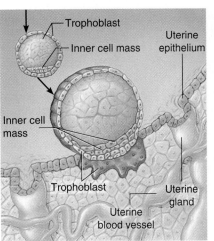

(a) 7 days. About 7 days after fertilization, the blastocyst drifts to an appropriate site along the uterine wall and begins to implant. The cells of the trophoblast proliferate and invade the endometrium.

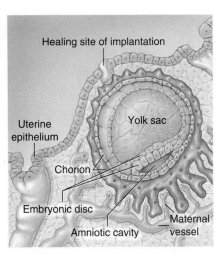

(b) 10 days. About 10 days after fertilization, the chorion has formed from the trophoblast.

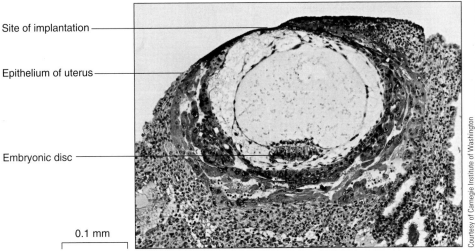

Courtesy of Carnegie Institute of Washington

(c) 12 days. This LM shows an implanted blastocyst at about 12 days after fertilization.

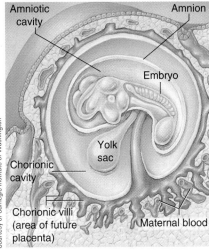

(d) 25 days. After 25 days, the maternal blood vessels provide the embryo with oxygen and nutrients. Note the specialized region of the chorion that will become the placenta.

(e) 45 days. At about 45 days, the embryo and its membranes together are about the size of a Ping-Pong ball, and the mother still may be unaware of her pregnancy. The amnion filled with amniotic fluid surrounds and cushions the embryo. The yolk sac has been incorporated into the umbilical cord. Blood circulation has been established through the umbilical cord to the placenta.

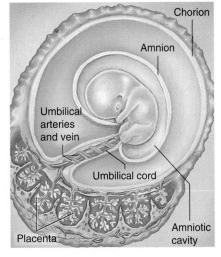

Figure 50-15 *Animated* Implantation and early development in the uterus

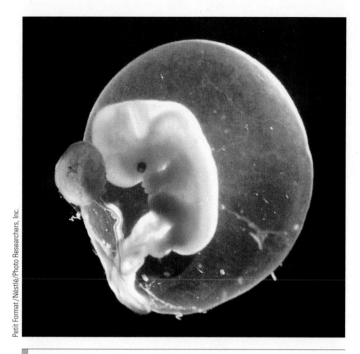

Figure 50-16 *Animated* The second month of development

The amnion is prominent as a transparent fluid-filled sac surrounding this 5.5-week-old human embryo, which is 1 cm (0.4 in) long. Note the limb buds and the eyes. The balloonlike object at the left, connected to the embryo by a stalk, is the yolk sac. This embryo is in a period of very rapid development, as you can see by a comparison with the chapter opening photograph of a human embryo in its seventh week.

corpus luteum (see Chapter 49) that pregnancy has begun. In response, the corpus luteum increases in size and releases large amounts of progesterone and estrogens. These hormones stimulate continued development of the endometrium and placenta.

Without hCG, the corpus luteum would degenerate and the embryo would abort and be flushed out with the menstrual flow. The woman would probably not even know she had been pregnant. If the corpus luteum is removed before about the 11th week after fertilization, the embryo spontaneously aborts. After that time, the placenta itself produces enough progesterone and estrogens to maintain pregnancy.

Organ development begins during the first trimester

Gastrulation occurs during the second and third weeks of development. Then the notochord begins to form and induces formation of the neural plate. The neural tube develops, and the forebrain, midbrain, and hindbrain are evident by the fifth week of development. A week or so later the forebrain begins to grow outward, forming the rudiments of the cerebral hemispheres.

The heart begins to develop and after 3.5 weeks begins to beat spontaneously. Pharyngeal pouches, branchial grooves, and branchial arches form in the region of the developing pharynx. In the floor of the pharynx, a tube of cells grows downward to form the primordial trachea, which gives rise to the lung buds. The di-

gestive system also gives rise to outgrowths that develop into the liver, gallbladder, and pancreas. A thin tail becomes evident but does not grow as rapidly as the rest of the body, and so becomes inconspicuous by the end of the second month. Near the end of the fourth week, the limb buds begin to differentiate; these eventually give rise to arms and legs.

All the organs continue to develop during the second month (▌ Fig. 50-16). Muscles develop, and the embryo becomes capable of movement. The brain begins to send impulses that regulate the functions of some organs, and a few simple reflexes are evident. After the first two months of development, the embryo is referred to as a **fetus** (▌ Fig. 50-17).

By the end of the **first trimester** (the first three months of development), the fetus is about 56 mm (2.2 in) long and weighs about 14 g (0.5 oz). Although small, it looks human. The external genital structures have differentiated, indicating the sex of the fetus. Ears and eyes approach their final positions. Some of the skeleton becomes distinct, and the developing vertebral column has replaced the notochord. The fetus performs breathing movements, pumping amniotic fluid into and out of its lungs, and even makes sucking motions.

Development continues during the second and third trimesters

During the second trimester (months 4 through 6), the fetal heart can be heard with a stethoscope. The fetus moves freely within the amniotic cavity, and during the fifth month the mother usually becomes aware of relatively weak fetal movements ("quickening"). The fetus grows rapidly during the final trimester (months

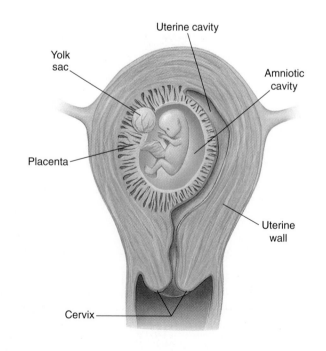

Figure 50-17 *Animated* Human fetus at 10 weeks

Note the position of the fetus within the uterine wall.

7 through 9), and final differentiation of tissues and organs occurs. If born at 24 weeks (out of 40 weeks), the fetus has only about a 50% chance of surviving, even with the best of medical care. Its brain is not yet sufficiently developed to sustain vital functions such as rhythmic breathing, and the kidneys and lungs are immature.

During the seventh month the cerebrum grows rapidly and develops convolutions. The grasping and sucking reflexes are evident, and the fetus may suck its thumb. Any infant born before 37 weeks (out of 40) is considered premature. Premature birth is an increasing problem in the United States, its incidence having risen from 9.6% in 1983 to 12.3% in 2003 (an almost 30% increase). Almost half a million premature babies are born each year. Babies born at 28 weeks or less are at greatest risk of death and severe defects. Babies born closer to 37 weeks have a greater chance of survival with competent medical care but may still face multiple long-term health problems.

At birth the average full-term baby weighs about 3000 g (6.6 lb) and measures about 52 cm (20 in) in total length. A baby that weighs less than 2500 g (5.5 lb) is considered low birth weight and is at risk even if not premature.

More than one mechanism can lead to a multiple birth

Occasionally the cells of the two-cell embryo separate, and each cell develops into a complete organism. Or sometimes the inner cell mass subdivides, forming two groups of cells, each of which develops independently. Because these cells have identical sets of genes, the individuals formed are exactly alike—**monozygotic,** or **identical, twins.** Very rarely, the two inner cell masses do not completely separate and so give rise to **conjoined twins,** who are physically attached and usually share one or more body parts.

Dizygotic twins, also called **fraternal twins,** develop when two eggs are ovulated and each is fertilized by a different sperm. Each zygote has its own distinctive genetic endowment, so the individuals produced are not identical. They may not even be of the same sex. Similarly, triplets (and other multiple births) may be either identical or fraternal.

Before fertility-inducing agents became available in the United States, twins were born once in about 80 births (about 30% of twins are monozygotic), triplets once in 80^2 (or 1 in 6400), and quadruplets once in 80^3 (or 1 in 512,000). However, the March of Dimes reports that between 1980 and 2000, the rate of twin births increased by about 74% and the rate of triplet births increased fivefold. This increase is attributed to widespread use of drugs to improve fertility, which increases the chance of dizygotic twinning.

A family history of twinning increases the probability of having dizygotic twins. However, giving birth to monozygotic twins is probably not influenced by heredity, age of the mother, or other known factors. An estimated two thirds of multiple pregnancies end in the birth of a single baby; the other embryo(s) may be absorbed within the first 10 weeks of pregnancy, or spontaneously aborted. Ultrasound imaging techniques are used to produce a type of image known as a **sonogram,** which can provide valuable data regarding the presence of multiple embryos. This is medically important, because multiple births are associated with higher infant mortality, which is largely a consequence of an increased risk of prematurity and low birth weight.

Environmental factors affect the embryo

We all know that the growth and development of babies are influenced by the food they eat, the air they breathe, the disease organisms that infect them, and the chemicals or drugs to which they are exposed. **Prenatal** development is also affected by these environmental influences. Life before birth is even more sensitive to environmental changes than it is for the fully formed baby. Although there is no direct mixing of maternal and fetal blood, diffusion and various other mechanisms allow many substances—nutrients, drugs, pathogens, and gases—to travel across the placenta.

▌ Table 50-3 describes some environmental influences on development. Some of these are **teratogens,** drugs or other substances that interfere with morphogenesis and so cause malformations. Many factors, such as smoking, alcohol use, and poor nutrition, contribute to low birth weight, a condition responsible for many infant deaths.

About 5% of newborns (more than 150,000 babies per year) in the United States have a defect of clinical significance. Such birth defects account for about 22% of deaths among newborns. Birth defects may be caused by genetic or environmental factors, or a combination of the two. (Genetic factors were discussed in Chapter 16.) In this section we examine some environmental conditions that affect the well-being of the embryo.

Timing is important. Each developing structure has a critical period during which it is most susceptible to unfavorable conditions. Generally, this critical period occurs early in the structure's development, when interference with cell movements or divisions may prevent formation of normal shape or size, resulting in permanent malformation. Because most structures form during the first 3 months of embryonic life, the embryo is most susceptible to environmental factors during this early period. During part of this time, the woman may not even realize she is pregnant and so may not take special precautions to minimize potential dangers.

Physicians diagnose some defects while the embryo is in the uterus. In some cases, treatment is possible before birth. Amniocentesis and chorionic villus sampling, discussed in Chapter 16, are techniques used to detect certain defects. Ultrasound imaging techniques discussed previously help diagnose defects and determine the position of the fetus. The available evidence indicates that unlike imaging techniques that use radiation, ultrasound may be harmless to the fetus. New methods currently under development include special types of MRI and high-resolution, three-dimensional ultrasound imaging (▌ Fig. 50-18).

The neonate must adapt to its new environment

Important changes take place after birth. During prenatal life, the fetus received both food and oxygen from the mother through the placenta. Now the newborn's own digestive and respiratory

TABLE 50-3

Environmental Influences on the Embryo

Factor	Example and Effect	Comment
Nutrition	Severe protein malnutrition doubles number of defects, fewer brain cells are produced, and learning ability may be permanently affected; deficiency of folic acid (a vitamin) linked to CNS defects such as spina bifida (open spine), low birth weight.	Growth rate mainly determined by rate of net protein synthesis by embryo's cells.
Medications	Many medications, even aspirin, affect development of the fetus.	
Excessive vitamins	Vitamin D essential, but excessive amounts may result in a form of mental retardation; an excess of vitamins A and K may also be harmful.	Vitamin supplements are normally prescribed for pregnant women, but only the recommended dosage should be taken.
Thalidomide	Thalidomide, marketed in Europe as a mild sedative, was responsible for serious malformations in about 10,000 babies born in the late 1950s in 20 countries; principal defect was phocomelia, a condition in which babies are born with extremely short limbs, often with no fingers or toes.	This teratogenic drug interferes with the development of blood vessels and nerves; most hazardous when taken during fourth to fifth weeks, when limbs are developing; thalidomide has been approved for the treatment of leprosy and certain cancers (e.g., myeloma), and clinical trials are being conducted to test its usefulness in the treatment of other cancers and some of the complications of AIDS.*
Accutane (isotretinoin)	Accutane, a synthetic derivitive of vitamin A, is used for the treatment of cystic acne. A woman who takes Accutane during pregnancy has a 1 in 5 chance of having a child with serious malformations of the brain (causing mental retardation), head and face, thymus gland (interfering with immune function) or heart.	Although Accutane is marketed with severe restrictions to prevent pregnant women from taking this teratogenic drug, several children are born each year with birth defects attributable to Accutane exposure.
Pathogens		
Rubella	Rubella (German measles) virus crosses placenta and infects embryo; interferes with normal metabolism and cell movements; causes syndrome that involves blinding cataracts, deafness, heart malformations, and mental retardation; risk is greatest (about 50%) when rubella is contracted during the first month of pregnancy; risk declines with each succeeding month.	Rubella epidemic in the United States in 1963–1965 resulted in about 20,000 fetal deaths and 30,000 infants born with serious defects; immunization is available but must be administered several months before conception.
HIV*	HIV can be transmitted from mother to baby before birth, during birth, or by breastfeeding.	See discussion of AIDS in Chapter 44.
Syphilis	Syphilis is transmitted to fetus in about 40% of infected women; fetus may die or be born with defects and congenital syphilis.	Pregnant women are routinely tested for syphilis during prenatal examinations; most cases can be safely treated with antibiotics.
Ionizing radiation	When a pregnant woman is subjected to X-rays or other forms of radiation, infant has a higher risk of birth defects and leukemia.	Radiation was one of the earliest causes of birth defects to be recognized.
Recreational and/or abused substances		
Alcohol	When a woman drinks heavily during pregnancy, the baby may be born with fetal alcohol syndrome (FAS), which includes certain physical deformities and mental and physical retardation; low birth weight and structural abnormalities have been associated with as little as two drinks a day; some cases of hyperactivity and learning disabilities have occurred.	Fetal alcohol syndrome is the leading cause of preventable mental retardation in the United States.
Cigarette smoking	Cigarette smoking reduces the amount of oxygen available to the fetus because some of the maternal hemoglobin is combined with carbon monoxide; may slow growth and can cause subtle forms of damage.	Mothers who smoke deliver babies with lower-than-average birth weights and have a higher incidence of spontaneous abortions, stillbirths, and neonatal deaths, especially from SIDS;* studies also indicate a possible relationship between maternal smoking and slower intellectual development in offspring.
Cocaine	Constricts fetal arteries, resulting in retarded development and low birth weight; in severe cases, there may be learning and behavioral problems, heart defects, and other medical problems.	Cocaine users frequently abuse alcohol as well, so it is difficult to separate the effects.
Heroin	High rates of mortality and prematurity; low birth weight.	Infants that survive are born addicted and must be treated for weeks or months; risk of SIDS is increased 10-fold.

*AIDS, acquired immunodeficiency syndrome; HIV, human immunodeficiency virus; SIDS, sudden infant death syndrome

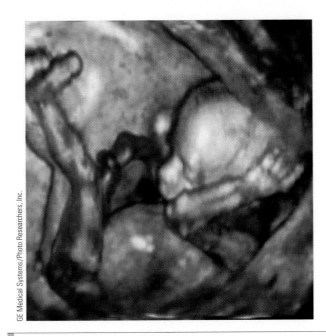

Figure 50-18 Three-dimensional ultrasound image of a human fetus at 12 to 20 weeks gestation

Note the enhanced soft tissue detail.

systems must function. Correlated with these changes are several major changes in the circulatory system.

Normally, the **neonate** (newborn infant) begins to breathe within a few seconds of birth and cries within half a minute. If anesthetics have been given to the mother, however, the fetus may also be anesthetized, and its breathing and other activities may be depressed. Some infants may not begin breathing until several minutes have passed. This is one reason why many women request childbirth methods that minimize the use of medication.

Researchers think that the neonate's first breath is initiated by the accumulation of carbon dioxide in the blood after the umbilical cord is cut. Carbon dioxide stimulates the respiratory centers in the medulla. The resulting expansion of the lungs enlarges its blood vessels (which in the uterus were partially collapsed). Blood from the right ventricle flows in increasing amounts through these larger pulmonary vessels. (During fetal life, blood bypasses the lungs in two ways: by flowing through an opening, the *foramen ovale,* which shunts blood from the right atrium to the left atrium; and by flowing through an arterial duct connecting the pulmonary artery and aorta. Both of these routes close off after birth.)

Aging is not a uniform process

We have examined briefly the development of the embryo and fetus, the birth process, and the adjustments required of the neonate. The human life cycle then proceeds through the stages of infant, child, adolescent, young adult, middle-aged adult, and elderly adult.

Development encompasses any biological change within an organism over time, including the changes commonly called **ag-** **ing.** Changes during the aging process decrease function in the older individual. The declining capacities of the various systems in the human body, although most apparent in the elderly, may begin much earlier in life.

The aging process is far from uniform among different individuals or in various parts of the body. The body systems generally decline at different times and rates. On average, between the ages of 30 and 75 a man loses 64% of his taste buds, 44% of the glomeruli in his kidneys, and 37% of the axons in his spinal nerves. His nerve impulses are propagated at a 10% slower rate, the blood supply to his brain is 20% less, his glomerular filtration rate has decreased 31%, and the vital capacity of his lungs has declined 44%. However, the human body has considerable functional capacity in reserve, so bodily functions are usually adequate. Furthermore, there is evidence that some of these declines can be significantly lessened by modifying the lifestyle (such as diet and exercise). Women also undergo body system declines in function, although, on average, they live about 8 years longer than men.

Although marked improvements in medicine and public health have led to longer life expectancies, there has been no corresponding increase in the *maximum* life expectancy. Relatively little is known about the aging process itself; this is now an active field of investigation, and researchers are gaining considerable insight through genetic studies on model organisms such as the nematode worm, *Caenorhabditis elegans;* the fruit fly, *Drosophila melanogaster;* and the laboratory mouse, *Mus musculus* (see Chapter 17).

Homeostatic response to stress decreases during aging

Research findings support the idea that most aging occurs because a combination of inheritance and environmental stress makes the individual less able to respond to additional stressors. One major question is whether a genetic program has evolved to cause aging or whether genetic involvement in the process is more circumstantial. Most available evidence favors the latter view.

However, some genetically programmed developmental events do seem related to the aging process. Cells that normally stop dividing when they differentiate appear more subject to the changes of aging than cells that continue to divide throughout life. Furthermore, researchers have hypothesized that certain parts of the body begin to malfunction because a genetic program stops key cells from dividing and replenishing themselves. In one model known as **cell aging,** when grown in culture, normal human cells eventually lose their ability to divide. Furthermore, cells taken from an older person divide fewer times than those from a younger person. Cell aging appears related to the fact that most normal human somatic cells are genetically programmed to lose the ability to produce active telomerase, an enzyme that replicates the DNA of the end caps (telomeres) of the chromosomes (see Chapter 12).

Genetically programmed cell death, *apoptosis,* is an essential developmental mechanism that may play a role in aging. However, researchers think that most aging is not a direct consequence of the genetic program leading to apoptosis; instead, mistakes in the *control* of apoptosis may lead to certain degenerative conditions, such as Alzheimer's disease, or to some of the cell deaths that occur following a heart attack or stroke.

Researchers are investigating genes that affect how the body is maintained and repaired in the face of various stressors. Like other life processes, aging may be accelerated by certain environmental influences and may vary because of inherited differences among individuals. Evidence also indicates that premature aging can be precipitated by hormonal changes; by various malfunctions of the immune system, including autoimmune responses; by accumulation of specific waste products within the cells; by changes in the molecular structure of macromolecules such as collagen; and by damage to DNA from continued exposure to cosmic radiation and X-rays.

Of course, researchers tend to be less interested in mechanisms that accelerate aging and more interested in those that might increase the maximum life span. In nematode worms, fruit flies, and mice, decreasing the activity of a gene encoding

IGF-1 (insulin-like growth factor 1), a cell signaling molecule, can increase the life span by 100%. Severe caloric restriction in rats, mice, and some other mammals can also delay aging. Some researchers have obtained experimental evidence that calorie restriction is a stressor that activates genes responsible for regulating an organism's adaptive response to stress. They are in the process of identifying those genes and investigating interactions among them, with the goal of developing treatments capable of activating these adaptive pathways.

Review

- How does a human blastocyst form and implant?
- How does the placenta form, and what is its function?
- What adaptations must the neonate make immediately after birth?
- During which trimester does each of the following occur: cerebrum develops convolutions; limb buds develop; mother feels fetal movements?
- What changes take place during the aging process?
- How are model organisms contributing to the study of aging?

SUMMARY WITH KEY TERMS

Learning Objectives

1 Analyze the relationship between cell determination and cell differentiation, and between pattern formation and morphogenesis (page 1081).

- **Cell determination** is a series of molecular events that lead to **cell differentiation,** in which a cell becomes specialized and carries out specific functions. Relatively undifferentiated cells are known as **stem cells.**
- **Morphogenesis,** the development of form, occurs through **pattern formation,** a series of events in which cells may communicate by signaling, migrate, undergo changes in shape, and even undergo **apoptosis** (programmed cell death).

2 Relate differential gene expression to nuclear equivalence (page 1081).

- Usually no genetic changes occur in cell determination and cell differentiation (the principle of **nuclear equivalence**). Instead, various types of differentiated cells express different subsets of their genes (**differential gene expression**).

3 Describe the four processes involved in fertilization (page 1081).

- Contact and recognition occur between noncellular egg coverings and sperm.
- Sperm entry is regulated to prevent interspecific fertilization and **polyspermy,** which is fertilization of the egg by more than one sperm.
- Fertilization activates the egg, triggering the events of early development.
- Sperm and egg pronuclei fuse and initiate DNA synthesis.

4 Describe fertilization in echinoderms, and point out some ways in which mammalian fertilization differs (page 1081).

- The coverings of echinoderm eggs are the **vitelline envelope** and the **jelly coat;** a **zona pellucida** encloses the mammalian egg.

- On contact, a sperm undergoes an **acrosome reaction,** which facilitates penetration of the egg coverings; in mammals the acrosome reaction is preceded by **capacitation,** a maturation process that results in the ability of a sperm to fertilize an egg.
- Sea urchin fertilization is followed by a fast block to polyspermy (depolarization of the plasma membrane) and a slow block to polyspermy (the **cortical reaction**); changes in the zona pellucida prevent polyspermy in mammals.

5 Trace the generalized pattern of early development of the embryo from zygote through early cleavage and formation of the morula and blastula (page 1083).

- The zygote undergoes **cleavage,** a series of rapid cell divisions without a growth phase. The main effect of cleavage is to partition the zygote into many small **blastomeres.**
- Cleavage forms a solid ball of cells called the **morula** and then usually a hollow ball of cells called the **blastula.**

6 Contrast early development, including cleavage in the echinoderm (or in amphioxus), the amphibian, and the bird, paying particular attention to the importance of the amount and distribution of yolk (page 1083).

- The **isolecithal** eggs of most invertebrates and simple chordates have evenly distributed **yolk.** They undergo **holoblastic cleavage,** which involves division of the entire egg.
- In the moderately **telolecithal** eggs of amphibians, a concentration of yolk at the **vegetal pole** slows cleavage so that only a few large cells form there, compared to a large number of smaller cells at the **animal pole.**
- The highly telolecithal eggs of reptiles and birds, with a large concentration of yolk at one end, undergo **meroblastic cleavage,** which is restricted to the **blastodisc.**
- Animals whose zygotes have relatively homogeneous cytoplasm exhibit **regulative development,** in which the

embryo develops as a self-regulating whole. The relatively rigid developmental patterns of some animals (**mosaic development**) is a consequence of the unequal distribution of cytoplasmic components.

■ The **gray crescent** of an amphibian zygote determines the body axis of the embryo.

7 Identify the significance of gastrulation in the developmental process, and compare gastrulation in the echinoderm (or in amphioxus), the amphibian, and the bird (page 1087).

■ In **gastrulation,** the basic body plan is laid down as three **germ layers** form: the outer **ectoderm,** the inner **endoderm,** and the **mesoderm** between them.

■ The **archenteron,** the forerunner of the digestive tube, forms in some groups; its opening to the exterior is the **blastopore.**

■ In gastrulation in the sea star and in amphioxus, cells from the blastula wall invaginate and eventually meet the opposite wall, forming the archenteron.

■ In the amphibian, invagination at the vegetal pole is obstructed by large, yolk-laden cells; instead, cells from the animal pole move down over the yolk-rich cells and invaginate, forming the **dorsal lip of the blastopore.**

■ In the bird, invagination occurs at the **primitive streak,** and no archenteron forms.

ThomsonNOW™ **Explore the frog life cycle by clicking on the figures in ThomsonNOW.**

8 Define *organogenesis*, and summarize the fate of each of the germ layers (page 1089).

■ **Organogenesis** is the process of organ formation.

■ Ectoderm becomes the nervous system, sense organs, and outer layer of skin (epidermis).

■ Mesoderm becomes the notochord, skeleton, muscles, circulatory system, and inner layer of skin (dermis).

■ Endoderm becomes the lining of the digestive tube.

9 Trace the early development of the vertebrate nervous system (page 1089).

■ The developing **notochord** is responsible for **induction,** which causes the ectoderm to differentiate and form the central nervous system.

■ The brain and spinal cord develop from the **neural tube.**

ThomsonNOW™ **Explore nervous system development by clicking on the figure in ThomsonNOW.**

10 Give the origins and functions of the chorion, amnion, allantois, and yolk sac (page 1091).

■ The **chorion** and **amnion** derive from ectoderm and mesoderm; the **allantois** and **yolk sac,** from endoderm and mesoderm.

■ The chorion is used in gas exchange. The amnion is a fluid-filled sac that surrounds the embryo and keeps it moist; it also acts as a shock absorber. The allantois stores nitrogenous wastes. The yolk sac makes food available to the embryo.

11 Describe the general course of early human development, including fertilization, the fates of the trophoblast and inner cell mass, implantation, and the role of the placenta (page 1091).

■ Fertilization occurs in the oviduct.

■ Cleavage takes place as the embryo is moved down the oviduct.

■ In the uterus, the embryo develops into a **blastocyst** consisting of an outer **trophoblast,** which gives rise to the chorion and amnion, and an **inner cell mass,** which becomes the embryo proper. The blastocyst undergoes **implantation** in the endometrium.

■ The **umbilical cord** connects the embryo to the **placenta,** the organ of exchange between the maternal and fetal circulation. The placenta derives from the embryonic chorion and maternal tissue.

ThomsonNOW™ **Learn more about cleavage, implantation, and the development of the human embryo by clicking on the figures in ThomsonNOW.**

12 Contrast postnatal with prenatal life, describing several changes that occur at or shortly after birth that allow the neonate to live independently (page 1091).

■ Human **prenatal** development requires 266 days from the time of fertilization; organogenesis begins during the **first trimester.** After the first 2 months of development, the embryo is referred to as a **fetus.** Growth and refinement of the organs continue in the second and third trimesters.

■ The **neonate** (newborn) must undergo rapid adaptations, especially changes in the respiratory and digestive systems.

13 Describe some anatomical and physiological changes that occur with aging (page 1091).

■ The **aging** process is marked by a decrease in homeostatic response to stress.

■ All body systems decline with age, but not at the same rate.

TEST YOUR UNDERSTANDING

1. The mechanism that leads directly to morphogenesis is (a) differential gene expression (b) pattern formation (c) nuclear equivalence (d) cell differentiation (e) cell determination

2. The main function of the acrosome reaction is to (a) activate the egg (b) improve sperm motility (c) prevent interspecific fertilization (d) facilitate penetration of the egg coverings by the sperm (e) cause fusion of the sperm and egg pronuclei

3. The fast block to polyspermy in sea urchins (a) is a depolarization of the egg plasma membrane (b) requires exocytosis of the cortical granules (c) includes the elevation of the fertiliza-

tion envelope (d) involves the hardening of the jelly coat (e) is a complete block

4. Place the following events of sea urchin fertilization in the proper sequence.

1. fusion of egg and sperm pronuclei 2. DNA synthesis
3. increased protein synthesis 4. release of calcium ions into the egg cytoplasm

(a) 4, 3, 1, 2 (b) 3, 2, 4, 1 (c) 2, 3, 1, 4 (d) 1, 2, 3, 4
(e) 4, 1, 2, 3

5. The cleavage divisions of a sea urchin embryo (a) occur in a spiral pattern (b) do not include DNA synthesis (c) do not include cytokinesis (d) are holoblastic (e) do not occur at the vegetal pole

6. Meroblastic cleavage is typical of embryos formed from _____ eggs. (a) moderately telolecithal (b) highly telolecithal (c) isolecithal (d) b and c (e) a, b, and c

7. The primitive groove of the bird embryo is the functional equivalent of the _____ in the amphibian embryo. (a) yolk plug (b) archenteron (c) blastocoel (d) gray crescent (e) blastopore

8. Which of the following are mismatched? (a) endoderm; lining of the digestive tube (b) ectoderm; circulatory system (c) mesoderm; notochord (d) mesoderm; reproductive system (e) ectoderm; sense organs

9. Which of the following has three germ layers? (a) morula (b) gastrula (c) blastula (d) blastocyst (e) trophoblast

10. An unidentified substance (or substances) released from the developing notochord causes the overlying ectoderm to form the neural plate. This phenomenon is known as (a) activation (b) determination (c) induction (d) implantation (e) mosaic development

11. Which of the following consists of both fetal and maternal tissues? (a) umbilical cord (b) placenta (c) amnion (d) allantois (e) yolk sac

12. The embryo proper of a mammal develops from the (a) trophoblast (b) umbilical cord (c) inner cell mass (d) entire blastocyst (e) yolk sac

13. Which of the following statements about vertebrate organogenesis is *not* true? (a) the notochord, brain, and spinal cord are among the first organs to develop in the early embryo (b) the developing notochord causes the overlying ectoderm to differentiate into the neural plate (c) the neural folds meet and fuse, forming the four-chambered heart (d) some neural crest cells differentiate into neurons (e) blocks of mesoderm known as somites form on either side of the neural tube

14. On about the seventh day of development, the human embryo (a) implants in the wall of the uterus (b) has a fully developed placenta for obtaining nutrients and oxygen (c) releases human chorionic gonadotropin (d) a and c (e) a, b, and c

CRITICAL THINKING

1. For almost 200 years, scientists debated whether an egg or sperm cell contains a completely formed, miniature human (preformation) or if structures develop gradually from a formless zygote (epigenesis). Relate these views to current concepts of development.

2. Not all teratogenic medications are banned by the U.S. Food and Drug Administration. Why?

3. **Evolution Link.** Consider the chordate characteristics described in Chapter 31 (notochord, dorsal hollow nerve cord, pharynx with slits, and postanal tail). How are they evident in a human embryo? Do they have the same functions in humans as they do in a less specialized chordate, such as amphioxus? Why have they persisted in humans?

4. **Evolution Link.** What is the adaptive value of developing a placenta?

5. **Evolution Link.** Can you think of a rationale for natural selection for increased life span in humans, even though most individuals live past reproductive age?

 Additional questions are available in ThomsonNOW at www.thomsonedu.com/login

51

Animal Behavior

E. S. Ross

The sand wasp (*Philanthus triangulum*) digging a burrow. How does the wasp know how to carry out the complex set of behaviors that ensure her reproductive success?

KEY CONCEPTS

Proximate causes of behavior, which include genetic, developmental, and physiological processes, explain *how* an animal carries out a specific behavior. Ultimate causes are the evolutionary explanations for *why* a certain behavior occurs.

The capacity for behavior is inherited, but behavior is modified in response to environmental experience; this is learned behavior.

A society is a group of individuals of the same species that may work together in an adaptive manner.

In sexual selection, individuals with reproductive advantages are selected over others of the same sex and species.

Natural selection favors altruism among relatives; this is an indirect way to perpetuate some of the helper's own alleles.

Kin selection is a type of natural selection that increases inclusive fitness through successful reproduction of close relatives.

Suppose your professor gave you a hypodermic syringe full of poison and told you to find a particular type of insect, one that you had never seen before and that was armed with active defenses. You then had to inject the ganglia of your prey's nervous system (about which you had been taught nothing) with just enough poison to paralyze but not kill it. You would have difficulty accomplishing these tasks—but a solitary wasp no larger than the first joint of your thumb does it all with elegance and surgical precision, without instruction.

Philanthus, the sand wasp, captures a bee (or beetle), stings it, and places the paralyzed insect in a burrow excavated in the sand (see photograph). She then lays an egg on her prey, which is devoured alive by the larva that hatches from that egg. From time to time, *Philanthus* returns to her hidden burrow to reprovision it, until autumn, when the larva becomes a hibernating pupa. Her offspring will repeat this *behavior*, precisely executing each step without ever having seen it done.

An animal's **behavior** is what it *does* and how it does it, usually in response to stimuli in its environment. A dog may wag its tail, a bird may sing, a butterfly may release a volatile sex attractant. Behavior is just as diverse as biological structure and just as characteristic of a given species as its anatomy or physiology. Like its morphology and physiology, an animal's behavior is the product of natural selection on phenotypes and indirectly on the genotypes that code for those phenotypes. Thus, an animal's repertoire of behavior is a set of adaptations that equip it for survival in a particular environment.

Whether biologists study behavior in an animal's natural environment or in the laboratory, they must consider that what an animal does cannot be isolated from the way in which it lives. **Behavioral ecology** is the study of behavior in natural environments from an evolutionary perspective. For more than two decades, behavioral ecology has been the main approach of biologists who study animal behavior. Before this approach emerged, the study of animal behavior was called *ethology*, and this term is sometimes still used to refer to the overall study of animal behavior.

In this chapter, we consider how an animal's behavior contributes to its reproductive success and to the survival of its species. We discuss types of learning and then focus on the *hows* and *whys* of some key types of behavior, including migration, foraging behavior, and social behavior. We explore sexual selection and helping behavior. ∎

BEHAVIOR AND ADAPTATION

Learning Objective

1 Distinguish between proximate and ultimate causes of behavior, and apply the concepts of ultimate cause and cost–benefit analysis to decide whether a particular behavior is adaptive.

In considering the reproductive behavior of *Philanthus*, we might wonder *how* she accomplishes her task, and *why* she behaves as she does. Early investigators of animal behavior focused on *how* questions. These questions address **proximate causes,** immediate causes such as the genetic, developmental, and physiological processes that permit the animal to carry out the particular behavior.

Today, biologists also ask *why* questions that address the **ultimate causes** of behavior. These questions, which have evolutionary explanations, ask *why* a particular behavior has evolved. Ultimate considerations address costs and benefits of behavior patterns. When studying ultimate causes, we may ask what the adaptive value of a particular behavior might be. An understanding of behavior requires consideration of both proximate and ultimate causes.

Behavioral ecologists use **cost–benefit analysis** to understand specific behaviors. A behavior may help an animal obtain food or water, protect itself, reproduce, or acquire and maintain territory in which to live. The benefits typically contribute to **direct fitness,** which is an individual's reproductive success, measured by the number of viable offspring it produces. Reproduction is, of course, the key to evolutionary success.

Behaviors also involve costs. For example, while a parent is off hunting for food for its offspring, the young it has left alone may be killed by predators. If the benefits are greater than the costs, a behavior is adaptive.

The ultimate cause of a behavior is therefore to increase the probability that the genes of the individual animal will be passed to future generations. Certain responses may lead to the death of the individual while increasing the chance that copies of its genes will survive through the enhanced production or survival of its offspring or other relatives.

Review

∎ In what ways are the behaviors of *Philanthus,* the sand wasp, adaptive?

INTERACTION OF GENES AND ENVIRONMENT

Learning Objective

2 Describe the interactions of heredity, environment, and maturation in animal behavior.

Early biologists debated nature versus nurture, that is, the relative importance of genes compared with environmental experience. They defined **innate behavior** (inborn behavior, popularly referred to as *instinct*) as genetically programmed and **learned behavior** as behavior that has been modified in response to environmental experience. More recently, behavioral ecologists have recognized that no true dichotomy exists. All behavior has a genetic basis. Even the *capacity* for learned behavior is inherited. However, behavior is modified by the environment in which an animal lives; behavior is a product of the interaction between genetic capacity and environmental influences. Thus, behavior begins with an inherited framework that experience can modify.

We can think of a range of behaviors from the more rigidly genetically programmed types through those that, although they have a genetic component, are extensively developed through experience. The sand wasp *Philanthus,* discussed in the chapter introduction, efficiently carries out a largely genetically programmed sequence of behaviors. How to dig the burrow, how to cover it, how to kill the bees—these behaviors are genetically determined. Yet some of her behavior is learned. There is no way her ability to locate the burrow could be genetically programmed. Because a burrow can be dug only in a suitable spot, its location must be learned *after* it is dug. When *Philanthus* covers her burrow with sand, she takes precise bearings of its location by circling the area a few times before flying off again to hunt.

The Dutch ethologist Niko Tinbergen studied this behavior of *Philanthus.* He surrounded the wasp's burrow with a circle of pine cones as potential landmarks (∎ Fig. 51-1). Before the wasp returned with another bee, Tinbergen rearranged or removed the pine cones. Without them, the wasp could not find her burrow. When Tinbergen moved the pine cones to an area where there was no burrow, the female wasp responded as though the burrow were there. When the investigator completely removed them, the female seemed very confused. Only when Tinbergen restored the cones to their original location could the wasp find her burrow. When he substituted a ring of stones for the cones, the wasp responded as though the burrow were in the center of the stones. Thus, *Philanthus* responds to the *arrangement* of the cones rather than to the cones themselves. Tinbergen's findings demonstrate that for *Philanthus*, landmark learning is critical for burrow locating.

Studies of fruit fly courtship and mating have provided interesting examples of interaction between genes and behavior. A ritual consisting of a complex sequence of steps, almost like a dance, must occur before mating takes place. This courtship ritual involves an exchange of visual, auditory, tactile, and chemical signals between the male and female. J. B. Hall, of Brandeis University, and his colleagues have identified more than a dozen genes controlling these actions, suggesting that courtship behavior is largely inherited and pre-programmed. Nevertheless, the fruit fly has the capacity to learn from experience, a capacity that, of course, is also inherited.

The interaction between genes and environment has been studied in many vertebrates. Several species of the love-bird (*Agapornis*) differ not only in appearance but in behavior. One species uses its bill to transport small pieces of bark for building a nest. Another species tucks nest-building materials under its rump feathers. Hybrid birds attempt to tuck material under their feathers, then try to carry it in their beaks, repeating the pattern several times. Eventually, most of the birds carry the material in their bills, but it takes them up to 3 years to perfect this behavior, and most continue to make futile attempts to tuck material into their feathers. Thus, the method of transporting materials is inherited but somewhat flexible. (In nature, hybrids would likely experience dramatically reduced fitness because of the long delay in breeding onset.)

Behavior depends on physiological readiness

Although behavior involves all body systems, it is influenced mainly by the nervous and endocrine systems. The capacity for behavior depends on the genetic characteristics that govern the development and functions of these systems. Before an animal can exhibit any

Key Experiment

QUESTION: How does *Philanthus* locate her burrow when she returns to provision it with food?

HYPOTHESIS: *Philanthus* uses visual cues to locate her burrow.

EXPERIMENT: ❶ While *Philanthus* was in the burrow, Tinbergen placed a circle of pine cones around its entrance.

❷ While the wasp was out foraging, Tinbergen moved the cones to a new position.

❸ In a second experiment, Tinbergen left the pine cones around the burrow but rearranged them in a triangle shape in the wasp's absence. Nearby, he made a circle of stones.

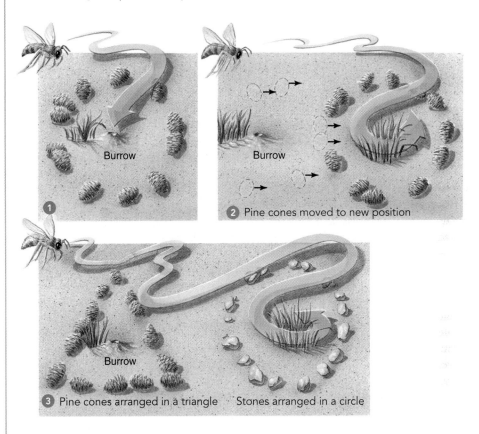

RESULTS: ❶ When *Philanthus* emerged from her burrow, she responded to the circle of pine cones by circling the site, an orientation flight. When she returned with prey, she flew to the middle of the circle of pine cones.

❷ When Tinbergen relocated the circle of pine cones, *Philanthus* flew to the middle of the new circle because she had learned the position of the burrow in relation to the cones.

❸ When Tinbergen replaced the cones with stones, *Philanthus* located her burrow in the middle of the ring of stones.

CONCLUSION: *Philanthus* used visual cues to locate her burrow. She responded to the *arrangement* of cones rather than to the cones themselves. Landmark learning is critical for locating the burrow.

Figure 51-1   Niko Tinbergen's experiments determining how the wasp *Philanthus* locates her burrow

pattern of behavior, it must be physiologically ready to produce the behavior. For example, breeding behavior does not ordinarily occur among birds or most mammals unless sufficient concen-

Figure 51-2 Egg-rolling behavior in the European graylag goose

In this rather rigid behavioral pattern, the goose reaches out by extending her neck and uses her bill to pull the egg back into the nest. If the investigator quickly removes the egg while the goose is in the process of reaching for it or pulling it back, she continues the behavior to completion, as though pulling the now absent egg back to the nest.

trations of sex hormones are present in their blood. A human baby cannot walk until its muscles and neurons are sufficiently developed. These states of physiological readiness are themselves produced by a continuous interaction with the environment. The level of sex hormones in a bird's blood may be determined by seasonal variations in day length. The baby's muscles develop with experience, as well as age.

Several factors influence the development of song in male white-crowned sparrows (generally only male songbirds sing a complex song). These birds show considerable regional variation in their song. During early development, young sparrows normally hear adult males sing the distinctive song of their population. Days 10 to 52 after hatching are a critical period for learning the song. When he is several months old, a young male sparrow "practices" the song over several weeks until he eventually sings in the local "dialect."

In laboratory experiments, birds kept in isolation and deprived of the acoustic experience of hearing the song of mature males eventually sing a poorly developed but recognizable white-crowned sparrow song. When a young white-crowned sparrow is permitted to interact socially with a strawberry finch (which belongs to a different genus of birds), it learns the song of the finch rather than its own species-specific song. This occurs even if the white-crowned sparrow can hear the song of other sparrows but does not interact with them socially. From these experiments, investigators have concluded that the white-crowned sparrow is hatched equipped with a rough genetic pattern of its song, but social and acoustic stimuli are important in developing its ability to sing its specific song.

Many behavior patterns depend on motor programs

Many behaviors that we think of as automatic depend on coordinated sequences of muscle actions called **motor programs.** Some motor programs—for example, walking in newborn gazelles—seem mainly innate. Others, such as walking in human infants, have a greater learned component.

A classic example of a motor program in vertebrates is egg rolling in the European graylag goose (❚ Fig. 51-2). Once activated by a simple sensory stimulus, egg-rolling behavior contin-

ues to completion regardless of sensory feedback. There is little flexibility. Ethologists called this behavior a *fixed action pattern (FAP),* but behavioral ecologists now use the term **behavioral pattern** instead. They have shown that behavior is not really fixed; it can be modified by experience. Certain behavioral patterns can be elicited by a **sign stimulus,** or **releaser,** a simple signal that triggers a specific behavioral response. A wooden egg is a sign stimulus that elicits egg-rolling behavior in the graylag goose.

In a set of classic experiments performed in 1951, Tinbergen showed that the red stripe on the ventral surface (belly) of a male stickleback fish is a sign stimulus. The ventral surface of the male stickleback is normally silver but turns red at the beginning of the breeding season. The male stickleback selects a territory, builds a nest, and defends his territory against rival males. After a female lays her eggs, she leaves and the male guards the eggs and, later, the newly hatched fish. Tinbergen showed that the red stripe triggers aggressive behavior by a male whose territory is being invaded. Tinbergen found that crude models painted with a red belly were more likely to be attacked than more realistic models lacking the red belly (❚ Fig. 51-3). Aggressive behavior is triggered by the red belly rather than by recognition based on a combination of features.

Review

■ Give an example showing that behavior capacity is inherited and is modified by learning.

■ How does physiological readiness affect innate behavior? How does it affect learned behavior?

LEARNING FROM EXPERIENCE

Learning Objective

3 Discuss the adaptive significance of habituation, imprinting, classical conditioning, operant conditioning, and cognition.

Based on our discussion thus far, it should be clear that much inherited behavior can be modified by experience. **Learning** involves persistent changes in behavior that result from experience. The capacity to learn new responses to new situations is adaptive, enabling animals to survive as their environment changes.

QUESTION: What triggers aggressive behavior in the male stickleback?

HYPOTHESIS: The red ventral surface (belly) of a male stickleback is a sign stimulus that triggers aggressive behavior in other male sticklebacks.

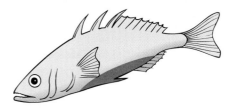

EXPERIMENT: Niko Tinbergen presented male sticklebacks with two series of models. The models in the neutral series were the neutral color of the male in the nonbreeding season. The models in the red series were red on the ventral surface.

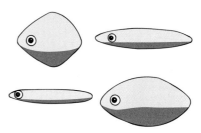

RESULTS: The male fish did not attack realistic models of another male stickleback when they lacked a red belly but did attack other models, however unrealistic, that had red ventral surfaces.

CONCLUSION: Aggressive behavior in the male stickleback is triggered by the red belly of another male fish. The red belly is a sign stimulus, or releaser, for aggressive behavior.

Figure 51-3 Niko Tinbergen's experiments on aggressive behavior in the male stickleback fish

Learning abilities are biased; information most important to survival appears most easily learned. The same rat that may have taken a dozen trials to learn the artificial task of pushing a lever to get a reward learns from one experience to avoid a food that has made it ill. People who poison rats to get rid of them can readily appreciate the adaptive value of this learning ability. Such quick learning in response to an unpleasant experience forms the basis of **warning coloration,** or **aposematic coloration,** which is found in many poisonous insects and brilliantly colored, but distasteful, bird eggs. Once made ill by such a meal, predators quickly learn to avoid them. In the following sections, we consider several types of learning, including habituation, imprinting, classical conditioning, operant conditioning, and cognition.

An animal habituates to irrelevant stimuli

Habituation is a type of learning in which an animal learns to ignore a repeated, irrelevant stimulus, that is, one that neither rewards nor punishes. Pigeons gathered in a city park learn by repeated harmless encounters that humans are not dangerous to them and behave accordingly. This behavior benefits them. A pigeon intolerant of people might waste energy by flying away each time a human approached and might not get enough to eat. Many African animals habituate to humans on photo safari and to the vans that transport them (Fig. 51-4). Urban humans habituate to the noise of traffic. In fact, many urban dwellers report that they do not sleep well when they visit a quiet rural area.

Imprinting occurs during an early critical period

Anyone who has watched a mother goose with her goslings must have wondered how she can "keep track" of such a horde of almost identical little creatures tumbling about in the grass, let alone distinguish them from those of another goose (Fig. 51-5). Although she can recognize her offspring to some extent, basically *they* have the responsibility of keeping track of *her*. The survival of a gosling requires that it quickly learn to discriminate its mother (care provider) from other geese.

Imprinting, a type of social learning based on early experience, has been studied in some mammals as well as birds. It occurs during a **critical period,** usually within a few hours or days after birth (or hatching). Konrad Lorenz, an Austrian physician and early ethologist, discovered that a newly hatched bird imprints on the first moving object it sees—even a human or an inanimate object such as a colored sphere or light. Although the process of imprinting is genetically determined, the bird *learns* to respond to a particular animal or object.

McMurray Photography

Figure 51-4 Habituation

After repeated safe encounters with vans transporting humans on photo safari, many animals, including giraffes, zebras, and lions in the Serengeti, learn to ignore them. Elephants typically ignore the vans unless the driver provokes them by moving too close. In that event an elephant may challenge and even charge the van.

J. H. Dick / VIREO

Figure 51-5 Imprinting

Parent–offspring bonds generally form during an early critical period. These goslings have imprinted on their mother, an upland goose (*Chloephaga picta*).

Among many types of birds, especially ducks and geese, older embryos can exchange calls with their nest mates and parents through the porous eggshell. When they hatch, at least one parent is normally on hand, emitting the characteristic sounds with which the hatchlings are already familiar. If the parent moves, the chicks follow. This movement and the sounds produce imprinting. During a brief critical period after hatching, the chicks learn the appearance of the parent.

Imprinting in some mammals depends on scent. Baby shrews, for example, become imprinted on the odor of their mother (or any female nursing them). In many species, the mother also learns to distinguish her offspring during a critical period. The mother in some species of hooved mammals, such as sheep, will accept her offspring for only a few hours after its birth. If they are kept apart past that time, the young are rejected. Normally, the mother learns to distinguish her own offspring from those of others by olfactory cues.

In classical conditioning, a reflex becomes associated with a new stimulus

In **classical conditioning,** an association forms between some normal body function and a new stimulus. If you have observed dog or cat behavior, you know that the sound of a can opener at dinner time can captivate a pet's attention. Early in the 20th century Ivan Pavlov, a Russian physiologist, discovered that if he rang a bell just before he fed a dog, the dog learned to associate the sound of the bell with food. Eventually, the dog salivated even when the bell was rung in the absence of food (❙ Fig. 51-6).

Pavlov called the physiologically meaningful stimulus (food, in this case) the *unconditioned stimulus.* The normally irrelevant stimulus (the bell) that became a substitute for it was the *conditioned stimulus.* Salivating to the food was an *unconditioned response,* whereas salivating to the bell was a *conditioned response.* Because a dog does not normally salivate at the sound of a bell, the association was clearly learned. It could also be forgotten. If

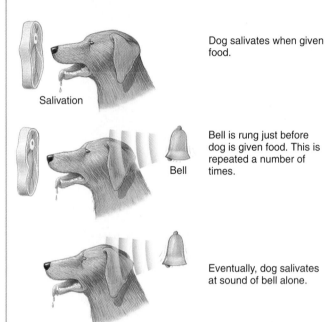
Figure 51-6 Classical conditioning: Pavlov's dog

the bell no longer signaled food, the dog eventually stopped responding to it. Pavlov called this latter process **extinction.** Classical conditioning can be adaptive. For example, if an animal is threatened by a predator in a particular area, it associates the place with danger and avoids that area in the future. In this case the animal learns to *avoid* a negative stimulus rather than to expect a reward.

In operant conditioning, spontaneous behavior is reinforced

In **operant conditioning,** the animal must do something to gain a reward (positive reinforcement) or avoid punishment. Operant conditioning has been studied in many animals, including flat-

Figure 51-7 Operant conditioning

When the pelican chick pecks its parent's beak, the chick is positively reinforced with food. After many tries, the chick learns to be more accurate in begging for food from a parent.

worms, insects, spiders, birds, and mammals. In a typical laboratory experiment, a rat is placed in a cage containing a movable bar. When random actions of the rat result in pressing the bar, a pellet of food rolls down a chute to the rat. Thus, the rat is positively reinforced for pressing the bar. Eventually, the rat learns the association and presses the bar to obtain food.

In negative reinforcement, removal of a stimulus increases the probability that a behavior will occur. For example, a rat may be subjected to an unpleasant stimulus, such as a low-level electric shock. When the animal presses a bar, this negative reinforcer is removed.

Many variations of these techniques have been developed. A pigeon may be trained to peck at a lighted circle to obtain food, a chimpanzee may learn to perform some task to get tokens that can be exchanged for food, or children may learn to stay quietly in their seats at school to obtain the teacher's praise.

Operant conditioning plays a role in the development of some behaviors that appear genetically programmed. An example is the feeding behavior of gull chicks. Herring gull chicks peck the beaks of the parents, which stimulates the parents to regurgitate partially digested food for them. The chicks are attracted by two stimuli: the general appearance of the parent's beak with its elongated shape and distinctive red spot, and its downward movement as the parent lowers its head. Like the rat's chance pressing of the bar, the chicks' initial exploratory pecking behavior is sufficiently functional for them to get their first meal, but they waste a lot of energy in pecking. Some pecks are off target and

are therefore not rewarded. However, pecking behavior becomes more efficient with practice. Thus, a behavior that might appear to be entirely genetic is improved by learning (❚ Fig. 51-7).

Animal cognition is controversial

Cognition is the process of gaining knowledge; it includes thinking, processing information, learning, reasoning, and awareness of thoughts, perceptions, and self. Cognition allows humans to carry out high-level mental functions. For example, we can make difficult decisions; solve problems, including mathematical problems; plan; and develop and comprehend languages. How closely animal cognition approximates human thought processes has been hotly debated and is the focus of a great deal of research. One of the most controversial issues is whether any animals have a sense of self-awareness.

Some animals have a form of cognition called **insight learning,** the ability to adapt past experiences that may involve different stimuli to solve a new problem. A dog can be placed in a blind alley that it must circumvent to reach a reward. The difficulty lies in the fact that the animal must move *away* from the reward to get to it. Typically, the dog flings itself at the barrier nearest the food. Eventually, by trial and error, the frustrated dog may find its way around the barrier and reach the reward.

In contrast to the dog, a chimpanzee placed in a new situation is likely to make new associations between tasks it has learned previously to solve the problem (❚ Fig. 51-8). Primates are especially skilled at insight learning, but some other mammals and a few birds also seem to have this ability to some degree.

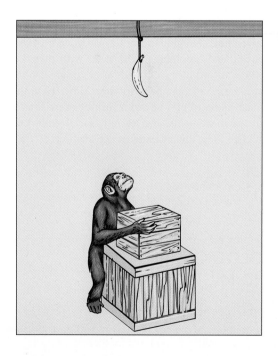

Figure 51-8 Cognition

Confronted with the problem of reaching food hanging from the ceiling, a chimpanzee may draw on past experience to solve the problem. When provided with boxes, many chimps stack them until they can climb and reach the food. What former experience might the chimp be applying to this new situation?

Figure 51-9 Cheetah cubs playing

Play may serve as a means of practicing behavior that will be useful later in life, possibly in hunting, fighting for territory, or competing for mates.

Play may be practice behavior

Perhaps you have watched a kitten pouncing on a dead leaf or practicing a carnivore neck bite or a hind-claw disemboweling stroke on a littermate without causing injury. Many animals, especially young birds and mammals, appear to practice adult patterns of behavior while they play (❚ Fig. 51-9). They may improve their ability to escape, kill prey, or perform sexual behavior. Play may be an example of operant conditioning in action.

Some investigators have suggested that young animals play just to have "fun." Dolphins seem to play just for pleasure. For example, bottlenose dolphins swirl water with their fins, then blow bubbles to produce rings and helices of air. Dolphins develop more complex play behavior over time. Hypotheses for the ultimate causes of play behavior include exercising, learning to coordinate movements, and learning social skills. Of course, there could be some other unknown explanation.

Review

❚ How is imprinting adaptive?
❚ How is operant conditioning adaptive?

BIOLOGICAL RHYTHMS AND MIGRATION

Learning Objectives

4 Give examples of biological rhythms, and describe some of the mechanisms responsible for them.
5 Analyze costs and benefits of migrations, and distinguish between directional orientation and navigation.

Biologists have identified many types of biological rhythms, including daily, monthly, and annual rhythms. Among many animals, physiological cycles such as body-temperature fluctuations and hormone secretion are rhythmic. Human body temperature, for example, follows a typical daily curve. Biological rhythms control various behaviors, including activity, sleep, feeding, drinking, and migration.

Biological rhythms affect behavior

The behavior of many animals, like the activities of many plants, is organized around **circadian** (meaning "approximately 1 day") **rhythms,** which are daily, 24-hour cycles of activity. **Diurnal** animals, such as honeybees and pigeons, are most active during the day. Most bats, moths, and cats are **nocturnal** animals, most active during the hours of darkness. **Crepuscular** animals, like many mosquitoes and fiddler crabs, are busiest at dawn or dusk, or both. Generally, there are ecological reasons for these patterns. If an animal's food is most plentiful in the early morning, for example, its cycle of activity must be regulated so that it becomes active shortly before dawn.

Some biological rhythms of animals reflect the **lunar** (moon) **cycle.** The most striking rhythms are those in marine organisms that are attuned to changes in tides and phases of the moon. For instance, a combination of tidal, lunar, and annual rhythms governs the reproductive behavior of the grunion, a small fish that lives off the Pacific coast of North America. Grunions swarm from April through June on those three or four nights when the highest tides of the year occur. At precisely the high point of the tide, the fish squirm onto the beach and deposit eggs and sperm in the sand. They return to the sea in the next wave. By the time the next tide reaches that portion of the beach 15 days later, the young fish have hatched in the damp sand and are ready to enter the sea. This synchronization may help protect fish eggs from aquatic predators.

An animal's metabolic processes and behavior are typically synchronized with the cyclic changes in its external environment. Its behavior anticipates these regular changes. The little fiddler crabs of marine beaches often emerge from their burrows at low tide to engage in social activities such as territorial disputes. To avoid being washed away, they must return to their burrows before the tide returns. How do the crabs "know" that high tide is about to occur? One might guess that the crabs recognize clues present in the seashore. However, when the crabs are isolated in the laboratory away from any known stimulus that could relate to time and tide, their characteristic behavioral rhythms persist.

Many biological rhythms are regulated by *internal* timing mechanisms that serve as **biological clocks.** As illustrated by the fiddler crabs, these timing mechanisms do not simply respond to environmental cues but are capable of sustaining biological rhythms independently. Such internal clocks have been identified in almost every eukaryote, as well as some bacteria. Molecular biologists have demonstrated that genes control biological clocks. In *Drosophila,* seven genes produce clock proteins that seem to interact in feedback loops in many cell types both inside and outside the nervous system. Researchers have identified similar genes and proteins in many animal groups, including mammals. The principal clock is located in specific areas of the central nervous system.

In mammals, the master clock is located in the **suprachiasmatic nucleus (SCN)** in the hypothalamus. This SCN clock generates approximately 24-hour cycles even without input from the environment. However, the SCN-generated cycles are normally adjusted every day based on visual input received from the retina. Signals from the retina reflecting changes in light intensity adjust internal circadian rhythms to light–dark cycles in the environment. The SCN sends rhythmic signals, in the form of neuropeptides, to the **pineal gland,** an endocrine gland located in the brain. In response, the pineal gland secretes *melatonin,* a hormone that promotes sleep in humans. Clock genes in the SCN are turned on and off by the very proteins they encode, setting up feedback loops that have a 24-hour cycle.

In addition to the master clock, most cells have timing mechanisms that use many of the same clock proteins. These peripheral clocks may function independently of the SCN.

Migration involves interactions among biological rhythms, physiology, and environment

Ruby-throated hummingbirds cross the vast distance of the Gulf of Mexico twice each year, and the sooty tern travels across the entire South Atlantic from Africa to reach its tiny island breeding grounds south of Florida. Birds, butterflies, fishes, sea turtles, wildebeest, zebras, and whales are among the many animals that travel long distances. Behavioral ecologists define **migration** as a periodic long-distance travel from one location to another. Many migrations involve astonishing feats of endurance and navigations.

Why do animals migrate? Ultimate causes of migration apparently involve the advantages of moving from an area that seasonally becomes less hospitable to a region more likely to support reproduction or survival. Seasonal changes include shifts in climate, availability of food resources, and safe nesting sites. For example, as winter approaches, many birds migrate to a warmer region.

An interesting example of migration is the annual journey of millions of monarch butterflies (*Danaus plexippus*) from Canada and the continental United States to Mexico, a journey of 2520 km (about 1520 mi) for some. Their dramatic annual migration appears related to the availability of milkweed plants on which females lay their eggs. On hatching, the larvae (caterpillars) feed on the leaves of the milkweed plant. In cold regions these plants die in late autumn and grow again with the warmer weather of spring. The wintering destinations of monarchs also may be determined by temperature and humidity. Possible benefits include the opportunity to winter in an area that offers an abundant food supply, nonfreezing temperatures, and moist air. Thus, the investment made by migrating monarch butterflies in their long journey increases their chances of surviving the winter.

The benefits of migration are not without costs in time, energy, and even survival. Many weeks may be spent each year on energy-demanding journeys. Some animals may become lost or die along the way from fatigue or predation. When in unfamiliar areas, migrating individuals are often at greater risk from predators. In recent years, human activities have interfered with migrations of many kinds of animals. For example, after millions of years of biological success, survival of sea turtles is threatened by fishing and shrimping nets in which they become entangled.

As a result, thousands of migrating turtles drown each year. Sea turtles also die as a result of ingesting floating plastic bags they mistake for jellyfish.

Proximate causes of migration include signals from the environment that trigger physiological responses leading to migration. In migratory birds, for example, the pineal gland senses changes in day length and then releases hormones that cause restless behavior. The birds show an increased readiness to fly and to fly for longer periods.

How do migrating animals find their way? The term **directional orientation** refers to travel in a specific direction. To travel in a straight line toward a destination requires a sense of direction, or **compass sense.** Many animals use the sun to orient themselves. Because the sun appears to move across the sky each day, an animal must have a sense of time. Biological clocks seem to sense time and regulate circadian rhythms.

Navigation is more complex, requiring both compass sense and **map sense,** which is an "awareness" of location. Navigation involves the use of cues to change direction when necessary to reach a specific destination. When navigating, an animal must integrate information about distance and time, as well as direction.

DNA tests confirm that loggerhead sea turtles that hatch on Florida beaches along the Atlantic swim hundreds of miles across the ocean to the Mediterranean Sea, an area rich in food. Several years later, those that survive to become adults mate, and the females navigate back, often to the same beach, to lay their eggs. Their journey requires both compass sense and map sense.

Marine biologist Kenneth Lohmann, of the University of North Carolina, fitted hatchling turtles with harnesses connected to a swivel arm in the center of a large tank (Fig. 51-10). A computer connected to the swivel arm recorded the turtles' swimming movements. Manipulating light and magnetic fields, Lohmann demonstrated that both these environmental cues are important in turtle migration. The turtles swam toward the east until Lohmann reversed the magnetic field. The turtles reversed their direction to swim toward the new "magnetic east," which was now actually west. Recent work suggests that young turtles use wave direction to help set magnetic direction preference.

Figure 51-10 Navigation by light and magnetic field

To study turtle navigation, researcher Kenneth Lohmann harnessed leatherback turtles (*Dermochelys coriacea*), such as this hatchling, and wired them to a computer that recorded their swimming direction.

Some biologists suspect that turtles also use their sense of smell to guide them, particularly to guide the females to the very same beach to lay their eggs. Similarly, adult salmon use the unique odors of different streams to help find their way back to the same stream from which they hatched.

Birds and some other animals that navigate by day rely on the position of the sun (our local star); those that travel at night use the stars to guide them. When birds see the star patterns of the night sky, they seek to fly in the appropriate migration direction for their species, even when they have had no opportunity to learn this feat from other birds.

Studies by Stephen Emlen, of Cornell University, showed that young indigo buntings learn the constellations, using the position of the North Star as their reference point. (Other stars in the Northern Hemisphere appear to rotate about the North Star.) When Emlen rearranged the constellations in a planetarium sky, birds learned the altered patterns of stars and later attempted to fly in a direction consistent with the artificial pattern. These and other studies suggested that birds have a genetic ability to learn constellation patterns and use them to orient during migration.

Investigators have long observed that some species of birds navigate even when the sky is overcast and they cannot see the stars. They hypothesized that these birds use Earth's magnetic field to navigate. Studies of garden warblers indicate that as these birds make their way from central Europe to Africa each winter, they navigate both by the stars and by Earth's magnetic field. The warblers use stars to determine the general direction of travel, then use magnetic information to refine and correct their course. In addition to birds and sea turtles, honeybees, some fishes, and some other animals are sensitive to Earth's magnetic field and use it as a guide.

Review

▌ Why is it adaptive for some species to be diurnal but for others to be nocturnal or crepuscular?

▌ What is the difference between directional orientation and navigation?

FORAGING BEHAVIOR

Learning Objective

6 Discuss the hypothesis that optimal foraging behavior is adaptive.

Feeding behavior, or **foraging,** involves locating and selecting food, as well as gathering and capturing food. Some behavioral ecologists study the costs and benefits of searching for and selecting certain types of food, as well as the mechanisms used to locate prey. For example, many camouflage strategies have evolved that make potential prey difficult to detect. As predators experience repeated success in locating a particular prey species, they are thought to develop a *search image,* a constellation of cues that help them identify hidden prey.

Why do grizzly bears spend hours digging arctic ground squirrels out of burrows while ignoring larger prey such as caribou? It is more energy efficient to dig for squirrels because the

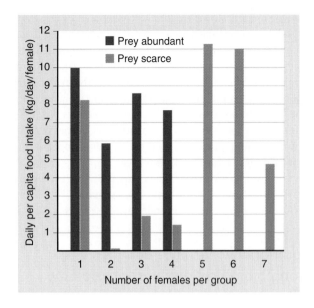

Figure 51-11 Optimal foraging and group size in lions

During times of prey scarcity, optimal foraging is related to group size: Lions hunt most successfully alone or in large groups of five to seven. Observations of group size suggest that the benefits of optimal defense often outweigh the benefits of optimal foraging. (Based on data from C. Packer, D. Scheel, and A. E. Pusey, "Why Lions Form Groups: Food Is Not Enough," *American Naturalist,* Vol. 136, July 1990.)

bears' efforts most probably will be rewarded, whereas caribou are more likely to escape, leaving the bears hungry. This is an example of **optimal foraging,** the most efficient way for an animal to obtain food. When animals maximize energy obtained per unit of foraging time, they may maximize their reproductive success. Many factors, such as avoiding predators while foraging, must be considered in determining efficient or optimal strategies.

In habitats where the most preferred food items are abundant and an animal does not have to travel far to obtain them, animals can afford to be very selective. In contrast, the optimal strategy in poor habitats, where it takes longer to find the best food items, is to select less preferred items. Animals may learn to forage efficiently through operant conditioning, that is, by randomly trying various strategies and selecting the one associated with the most rewards (and the fewest hunger pangs at the end of the day).

Foraging is also affected by the forager's risk of predation. British biologist Guy Cowlishaw studied foraging behavior in a population of baboons in Namibia. He found that the baboons spent more time foraging in a habitat in which food was relatively scarce than in an area with more abundant food that had a high risk of predation by lions and leopards.

Lions live, and often forage, in social units called *prides.* A pride typically consists of a group of related adult females, their cubs, and unrelated males. Biologist Craig Packer, of the University of Minnesota, and his research team have studied lion behavior in Serengeti National Park, Tanzania. Packer has radio-collared at least one female in each of 21 prides and has tracked them for several years. During the season when prey is abundant, lions hunt wildebeest, gazelle, and zebra that have migrated into the area. When these herds migrate out of the area during the dry season, these prey become scarce, and lions feed mainly on warthog and Cape buffalo.

Packer has reported that when prey is abundant, the size of the foraging group has little effect on daily food intake. Whether a lion hunts individually or in small (two to four females) or large (five to seven) groups, food is plentiful and is captured by individuals or groups of any size. During the season when prey is scarce, however, lions are more successful if they hunt alone or in large groups (■ Fig. 51-11). Nevertheless, Packer's data from radio-collared lions indicated that females typically forage in as large a group as they can, even if the group consists of three or four females, an approach that decreases foraging efficiency. Thus, selection pressure seems to be stronger to protect cubs and defend territory against larger prides than to forage maximally (alone).

Review

■ What is optimal foraging?
■ How is optimal foraging adaptive?

Figure 51-12 Social behavior in zebras

The social unit consists of a fairly stable group of females with young and a dominant stallion. The stripes of a group of zebras confuse predators.

COMMUNICATION AND LIVING IN GROUPS

Learning Objectives

7 Analyze social behavior in terms of costs and benefits, and describe modes of animal communication.

8 Describe the adaptive significance of social organization, using establishment of a dominance hierarchy as an example.

9 Distinguish between home range and territory, and describe costs and benefits of territoriality.

The mere presence of more than one individual does not mean a behavior is social. Many factors of the physical environment bring animals together in **aggregations,** but whatever interaction they experience may be circumstantial. A light shining in the dark is a stimulus that draws large numbers of moths, and the high humidity under a log attracts wood lice. Although these aggregations may have adaptive value, they are not truly social, because the animals are not communicating with one another.

We can define **social behavior** as the interaction of two or more animals, usually of the same species. Many animals benefit from living in groups. Schools of fishes tend to confuse predators, so individuals within the school may be less vulnerable to predators than a solitary fish would be (see Fig. 52-1b). Zebra stripes appear to be a visual antipredator adaptation. When viewed from a distance, the stripes tend to visually break up the form of the animal so that individuals cannot be distinguished. A herd of zebras confuses predators, whereas a solitary animal is easy prey for lions, cheetahs, or spotted hyenas (■ Fig. 51-12).

Social foraging is an adaptive strategy used routinely by many animal species. A pack of wolves has greater success in hunting than an individual wolf would have. Among some birds of prey, such as certain hawks, group hunting locates prey more quickly.

Social behavior offers fitness benefits and increases chances of perpetuating the genes that produce such behavior. However, social behavior also has certain costs. Living together means increased competition for food and habitats, increased risk of attracting predators, and increased risk of transmitting disease.

An organized **society** is an actively cooperating group of individuals belonging to the same species and often closely related. Many insects, such as tent caterpillars, which spin communal nests, cooperate socially. A school of fish, a flock of birds, a pack of wolves, and a hive of bees are other examples of societies. Some societies are loosely organized, whereas others are elaborately structured.

Communication is necessary for social behavior

One animal can influence the behavior of another only if the two of them can exchange mutually recognizable signals (■ Fig. 51-13). **Communication** is most evident when one animal performs an act that changes the behavior of another. Animals may communicate to hold a group together, warn of danger, signal social status, indicate willingness to accept or provide care, identify members of the same species, or indicate sexual maturity or readiness. Communication may also be important in finding food, as in the elaborate dances of honeybees. Animals use electrical, tactile, visual, auditory, and chemical signals to transmit information to one another.

Certain fishes (gymnotids) emit electric pulses that generate electric fields in the water. These electric pulses can be used for navigation and communication, including territorial threat, in a fashion similar to bird vocalization. As Harvard sociobiologist Edward O. Wilson has said, "The fish, in effect, sing electrical songs."

Examples of animal touch include grooming and sharing food. Social insects, such as ants, groom and stroke one another

(a) A male spring peeper frog (*Hyla crucifer*) calling to locate a mate.

(b) Communicating with language. Chimpanzee Tatu (*top*) is signing "food" to Washoe (*below*), the first chimp to learn American Sign Language from a human. Washoe then taught other chimps how to sign.

Figure 51-13 Animal communication

with their antennae and mouthparts. Many types of bird chicks peck at their parent's beak to stimulate feeding.

Visual signals can be transmitted quickly

Visual signals are fast, communicate a great deal of information, and indicate the position of the animal sending the signal. An animal's own body, such as its coloration, may serve as the visual message. Numerous species communicate with visual displays that serve as social signals. For example, many birds present visual displays along with their songs.

Auditory signals can be conveyed in the dark

Most animals cannot communicate effectively with visual signals when it is dark. However, animals can hear at night as well as during the day. Auditory signals can also be transmitted over longer distances than visual signals. Sound can move in all directions at once, and an auditory signal can be rapidly terminated. Most animals that communicate with auditory signals use **calls,** short, simple sounds. Birds and mammals use more complex auditory signals. In many bird species, males announce their presence and willingness to interact socially by singing. Some birds respond to a particular sound by matching it.

Among nonhuman mammals, only bottlenose dolphins are known to match sounds in communicating. These dolphins respond to the whistle of a member of its own species by imitating and emitting the same sound. Researchers suggest that vocal matching was important in the evolution of human language.

Orcas (formerly known as killer whales) communicate by sounds and songs. Members of a given pod (social group) have an average of 12 different calls, which vary in pitch and duration and appear to reflect their "emotional" state.

Some animals communicate by scent

Dogs and wolves mark territory by frequent urination. Antelopes, deer, and cats rub facial gland secretions on conspicuous objects in their vicinity and urinate on the ground. **Pheromones** are chemical signals secreted into the environment that convey information between members of a species. These small, volatile molecules provide a simple, widespread means of communication. Animals use pheromones to communicate danger, ownership of territory, and availability for mating. For example, female moths release pheromones that attract males. Ants mark their trails with pheromones.

Most pheromones elicit a very specific, immediate, but transitory type of behavior. Others trigger hormonal activities that result in slow but long-lasting responses. Some pheromones may act in both ways. An advantage to pheromone communication is that relatively little energy is needed to synthesize the simple, but distinctive, organic compounds involved. Members of the same species have receptors that fit the molecular configuration of the pheromone; other species usually ignore it or do not detect it at all. Pheromones are effective in the dark, can pass around obstacles, and last for several hours or longer. Major disadvantages of pheromone communication are slow transmission and limited information content. Some animals secrete several pheromones with different meanings.

Pheromones are important in attracting the opposite sex and in sex recognition in many species. Many social insects use these chemical signals to regulate reproduction. Humans have taken advantage of some sex-attractant pheromones to help control pests such as gypsy moths by luring the males to traps baited with synthetic versions of female gypsy moth pheromones. In honeybees, certain fatty acids are mixed with hydrocarbons from wax glands, and this mixture is transferred onto worker bees when they touch the comb. This mixture serves as a pheromone that identifies all the bees that belong to a particular hive. Should a bee from another hive attempt to enter, guard bees sting and may even kill the foreigner.

In vertebrates, pheromones affect sexual cycles and reproductive behavior, including choice of a mate, and may play a role in defending territory. Among some mammals, an ovulating female (one physiologically ready to mate) releases pheromones as part of her vaginal secretion. When males detect these chemical odors, their sexual interest increases. When the odor of a male mouse is introduced among a group of females, the reproductive cycles of the female mice synchronize. In some species of mice, the odor of a strange male, a sign of high population density, causes a newly impregnated female to abort.

The extent to which humans respond to pheromones is the focus of research. One interesting finding suggests that an unconsciously perceived body odor can synchronize the menstrual cycles of women who associate closely (for instance, college roommates or cell mates in prison). A recent study of the effects of human steroids used in commercial fragrances, such as perfumes, suggests that such compounds may act as subtle signals that modulate mood and behavior.

As discussed in Chapter 42, mammals detect pheromones with specialized chemoreceptor cells that make up the **vomeronasal organ** in the epithelium of the nose. The vomeronasal sensory neurons signal the amygdala and hypothalamus, brain structures that regulate emotional responses and certain endocrine processes. When neurons of the vomeronasal system are damaged in virgin mice, they do not mate. Biologists have identified about 100 genes that code for pheromone receptors in the mouse and rat. These receptors initiate **signal transduction** processes that involve G proteins.

Animals benefit from social organization

In the spring, female paper wasps awaken from hibernation and begin to build a nest together. During the early course of construction, a series of squabbles among the females takes place in which the combatants bite one another's bodies or legs. Finally, one of the wasps emerges as dominant. After that, she is rarely challenged. This queen wasp spends more and more time tending the nest and less and less time out foraging for herself. She takes the food she needs from the others as they return.

The queen then begins to take an interest in raising a family—her family. Because she is almost always in the nest, she can prevent other wasps from laying eggs in the brood cells by rushing at them, jaws agape. Because of her supreme dominance, the queen can bite any other wasp without serious risk of retaliation.

The other wasps of the nest are further organized into a **dominance hierarchy,** a ranking of social status in which each wasp has more status than the wasps that are lower in rank. Wasps lower in the hierarchy are subordinate to those above them, as follows:

Queen ⟶ wasp A ⟶ wasp B ⟶ . . . wasp M ⟶ wasp N

Once a dominance hierarchy is established, little or no time is wasted in fighting. When challenged, subordinate wasps communicate with submissive poses that inhibit the queen's aggressive behavior. Consequently, few or no colony members are lost through wounds sustained in fighting one another. This social organization ensures greater reproductive success for the colony.

In many species, males and females have separate dominance systems. For example, both female and male chimpanzees establish dominance hierarchies. Some females apparently become dominant through aggressive behavior, whereas others achieve high rank by virtue of their mother's status. Dominant females are more successful reproductively than those with lower social status. In many monogamous animals, especially birds, the female acquires the dominance status of her mate by virtue of their relationship and the male's willingness to defend his mate.

Like many fishes and some invertebrates, certain coral reef fishes (labrids) are capable of sex reversal. The largest, most dominant individual of a group is always male, and the remaining fish within his territory are all female. If the male dies or is removed, the most dominant female becomes the new male. Should any harm come to him, the next-ranking female undergoes sex reversal, takes charge, and protects the group territory. Other fishes exhibit the reverse behavior, in which the most dominant fish is always female. In such species, size is less important for aggressive defense than for maximal egg production.

In establishing dominance, males may expend energy in posturing, roaring, leaping about, or sometimes fighting fiercely. These behaviors appear to be a test of male quality. The male with the greatest endurance will likely gain dominance and will have the greatest opportunity to mate and perpetuate his genes. In many animals, social dominance is a function of aggressiveness (❙ Fig. 51-14). Tremendous energy is often needed for the

Gerald Lacz/Peter Arnold, Inc.

❙ **Figure 51-14** Communicating dominance

This baboon (*Papio*) bares his teeth and screams in an unmistakable show of aggression, a signal that allows him to establish and maintain dominance. In many social groups, animals signal other members to establish and maintain dominance.

Figure 51-15 Home range

The Cape buffalo (*Syncerus caffer*) lives in herds of several hundred animals. Each herd, such as this one photographed in the Serengeti, has a fairly constant home range. Buffalo protect herd members, especially calves.

fights engaged in by some birds and many mammals, including baboons, rams, and elephant seals. Establishing dominance is often strenuous and dangerous.

Sex hormones, particularly the male reproductive hormone testosterone, increase aggressiveness. The female hormone estrogen sometimes reduces dominance. Among chickens, the rooster is the most dominant. If a hen receives testosterone injections, her place in the dominance hierarchy shifts upward. When male rhesus monkeys are dominant, their testosterone levels are much higher than when they have been defeated. Not only can testosterone increase dominance but dominance may even increase testosterone production. It is not always easy to determine cause and effect. (Studies suggest that the effects of testosterone on human mood and behavior may be different from those in other vertebrates.)

Social status affects many organs and systems, including brain structure. A recent study showed that dominant male rats had more new neurons in the hippocampus than subordinate rats or controls did.

Many animals defend a territory

Most animals have a **home range,** a geographic area they seldom leave (❚ Fig. 51-15). Because the animal has the opportunity to become familiar with everything in that range, it has an advantage over its competitors, predators, and prey in negotiating the terrain and finding food. Some, but not all, animals exhibit **territoriality.** They defend a **territory,** a portion of the home range, often against other individuals of their species and sometimes against individuals of other species. Many species are territorial for only part of the year, often during the breeding season, but others are territorial throughout the year. Territoriality has been positively correlated with the availability of needed resources that occur in small areas that can be defended.

Territoriality is easily studied in birds. Typically, the male chooses a territory at the beginning of the breeding season. This behavior results from high sex hormone concentrations in his blood. The males of adjacent territories fight until territorial boundaries become established. Generally, the dominance of a male is directly associated with how close he is to the center of his territory. Close to home he acts like a lion, but when invading some other bird's territory, he may behave more like a lamb. Sometimes males respect a neutral line, an area between their territories in which neither is dominant.

Bird songs announce the territory and often substitute for fighting. Songs also announce to eligible females that a proper-tied male resides in the territory. Typically, male birds take up a conspicuous station, sing, and sometimes display striking patterns of coloration or aerial acrobatics to their neighbors, their rivals, and sometimes their mates.

The costs of territoriality include the time and energy expended in staking out and defending a territory and the risks involved in fighting for it. Benefits often include exclusive rights to food within the territory and greater reproductive success. The males of a lion pride father all the cubs born within that pride. Their ability to defend their territory from invasion by strange males ensures their reproductive success. Among many species, animals that fail to establish territories fail to reproduce. Territoriality also tends to reduce conflict among members of the same species and ensures efficient use of environmental resources by encouraging individuals to spread throughout a habitat.

Usually, territorial behavior is related to the specific lifestyle of the animal and to whatever aspect of its environment is most critical to its reproductive success. For instance, seabirds may range over hundreds of square kilometers of open water but exhibit territorial behavior only at crowded nesting sites on an island. The nesting sites are their scarcest resource and the one for which competition is keenest.

Review

- What distinguishes an organized society from a loose aggregation of organisms?
- How does an animal establish dominance? What costs and benefits accrue to dominant individuals in a dominance hierarchy? To subordinate individuals?
- What is territoriality? What are some of its benefits?

SEXUAL SELECTION

Learning Objective

10 Define *sexual selection*, and describe different types of mating systems and approaches to parental care.

Sexual selection is a type of natural selection for successful mating. Animals vary in their ability to compete for mates. Individuals with some inherited advantage, such as males with large size, have a greater chance to mate and pass on their genes to the next generation, and the beneficial trait becomes more common in the population over time. Thus, sexual selection results in the reproductive advantage that some individuals have over others of the same species and sex.

For a male, reproductive success depends on how many females he can impregnate. For a female, reproductive success depends on how many eggs she can produce during her reproductive lifetime, on the quality of the sperm that fertilize them, and on the survival of her offspring to reproductive age. Recall that an animal's reproductive success is a measure of its *direct fitness.* We can distinguish between *intrasexual selection* and *intersexual selection.*

Animals of the same sex compete for mates

In **intrasexual selection,** individuals of the same sex actively compete for mates. Typically in such populations abundant males compete for a limited number of receptive females. Competition may take the form of physical combat. Among many species, baboons and elephant seals, for example, the largest male has the advantage.

Often, males establish dominance in more subtle ways. Many males use elaborate ornamental displays to establish dominance. For example, the male deer with the largest antlers may discourage his competitors. The male bird with the most dramatic plumage and brightest colors may have a psychological advantage over competing males.

Several studies confirm that males ranking higher in a dominance hierarchy mate more frequently than males that rank lower. However, exceptions demonstrate the complexity of reproductive behavior among some species. For example, among baboons with a definite dominance hierarchy, males lower in the hierarchy copulated with females as frequently as did males

of higher status. Investigators observed, though, that dominant males copulated more frequently with females who were in *estrus,* their fertile period.

Animals choose quality mates

Females frequently have the opportunity to select a sexual partner from among several males. In **intersexual selection,** females select their mates on the basis of some physical trait or some resource offered by the winning suitors. The physical trait typically indicates genetic quality or good health. In many species, success of a male in dominance encounters with other males indicates his quality to the female, and she allows the victorious male to court her. Some lower-ranking males develop alternative strategies for attracting females. For instance, a male may win a female's interest by protecting her baby, even though it is not his own.

Females of some species select their mates based on ornamental displays. A female fish may select the most brightly colored male. Female birds often select males with the showiest plumage and brightest colors, and female deer prefer males that display elaborate antlers. Expression of ornamental traits may give the female important information about the male's physical condition and ability to fight. The size of the antlers of red deer, for example, may indicate combat effectiveness, as well as proper nutrition and good health. Female lions are attracted to males with thick, dark manes, an indicator of good nutrition and plenty of testosterone (which regulates growth of hair and melanin production.) Thus, in some species ornamentation signals good health and good genes.

Among crickets and many other insect species, a courting male offers a gift of food to a prospective mate. Studies show that the larger or higher quality the food offering is, the better the chances the male will be accepted.

Among some species of insects, birds, and bats, males gather in a small display area called a **lek,** where they compete for females. When a receptive female appears, males may excitedly display themselves and compete for her attention. Among some species, the female selects a male based on his location in the lek rather than on his appearance. The dominant male may occupy a central position and be chosen by most of the females. The female selects a male, mates with him, and then leaves the lek. The male remains to woo other females.

Courtship rituals ensure that the male is indeed a male and is a member of the same species. Rituals provide the female further opportunity to evaluate him. In some species, courtship may also be necessary as a signal to trigger nest building or ovulation. Courtship rituals can last seconds, hours, or days and often involve a series of behavioral patterns (❚ Fig. 51-16). The first display by the male releases a counterbehavior by the female. This, in turn, releases additional male behavior, and so on until the pair is physiologically ready for copulation. Specific cues enable courtship rituals to function as reproductive isolating mechanisms among species (see Fig. 20-3).

An extreme courtship ritual has been described for redback spiders, which are closely related to the black widow spider. During copulation, the small male spider positions himself above his larger mate's jaws. During 65% of matings, the finale is that the

(a) The male great frigatebird (*Frigata minor*) inflates his red throat sac in display as part of a courtship ritual. Photographed on Christmas Island in the Pacific.

(b) Egrets (*Egretta rufescens*) performing a mating dance.

Figure 51-16 Courtship rituals

Elaborate courtship rituals help animals determine that a potential mate is of the same species and gives each animal a chance to evaluate the other as a potential mate.

female eats her suitor. The apparent explanation for this behavior is that the risk-taking male is able to copulate for a longer period and thus fertilize more eggs than noncannibalized males and that the female is more likely to reject additional mates.

Sexual selection favors polygynous mating systems

In most species, males make little parental investment in their offspring, apart from providing sperm. Males ensure reproductive success by impregnating many females, increasing the probability that their genes will be propagated in multiple offspring. Thus, sexual selection often favors **polygyny,** a mating system in which males fertilize the eggs of many females during a breeding season.

In the mating system known as **polyandry,** one female mates with several males. Benefits may include receiving gifts from several males or enlisting several males' help in caring for the young. Data collected at Jane Goodall's research center at Gombe National Park in Tanzania indicate that female chimpanzees practice polyandry. Females mate with males of their own group, but when they are most fertile, many females slip off to neighboring communities and copulate with less familiar males. Perhaps these sexual rendezvous protect against inbreeding.

Investigators also hypothesize that mating with many males provides insurance against infanticide. Males do not aggress against infants of mothers with whom they have copulated. Recent studies indicate that females with multiple mates are more fertile and produce more offspring.

Polyandry and polygyny sometimes occur in the same species. After mating, a female giant water bug attaches a clutch of eggs to her mate's back (Fig. 51-17a). He cares for the eggs until

they hatch. She then may mate with a different male and glue the new clutch of eggs to his back. However, if the male has more space for eggs, he may mate with another female.

Apparently, in most species it is not certain who fathered the offspring. However, raising some other male's offspring is a genetic disadvantage, so males in some species may compromise mate chasing in favor of **mate guarding.** The male guards his partner after copulation to ensure that she does not copulate with another male. Mate-guarding behavior is likely to occur when the female is receptive and has eggs that might be fertilized by another male. For example, dominant male African elephants guard a female only during the phase of estrus when she is most likely to have a fertile egg. Before that time, or later in estrus after mating has already occurred, younger male elephants of lesser social status can copulate with the female. A high cost of mate guarding is the loss of opportunity for a dominant male to mate with other females.

Sexual selection favors males that inseminate many females and produce many offspring. Perhaps for this reason, **monogamy,** a mating system in which a male mates with only one female during a breeding season, is not common. Less than 10% of mammals are monogamous.

Researchers long thought monogamy was common among birds, because many species form **pair bonds,** stable relationships that ensure cooperative behavior in mating and the rearing of the young. However, genetic evidence shows that some offspring are fathered by males other than the one caring for them. For example, DNA tests show that among eastern bluebirds, 15% to 20% of the chicks are fathered by other males. The extra-pair male contributes new genes, and the female produces offspring with greater genetic variability, increasing their chances for survival. Some researchers distinguish between *social monogamy,* in which animals form a pair bond, and *sexual monogamy.*

(a) Giant water bug (Belostomatidae) male carrying eggs on his back.

(b) A female crocodile in South Africa transports hatchlings in her mouth from their nest site to Lake St. Lucia.

(c) Adult robins invest energy in feeding their young.

(d) Female cheetah (*Acinonyx jubatus*) stands watch as her cubs eat the Thompson's gazelle she has hunted and killed for them.

Figure 51-17 Caring for the young

Parental investment in offspring increases the probability that the young will survive.

Monogamy does occur in some species, typically when males are needed to protect and feed the young. For example, the California mouse is sexually, as well as socially, monogamous. The offspring need their parents' body heat to survive, and the parents take turns keeping them warm.

Some animals care for their young

Parental investment is the contribution that each parent makes in producing and rearing offspring. Most animals do not invest time and energy in caring for their offspring, because the costs of parenting are high. For example, an animal busy feeding and protecting offspring might produce fewer additional offspring. There are also risks in protecting the young from predators. The benefit of parental investment is the greater probability that each offspring will survive.

Natural selection has favored parental care in species in which the female produces few young or reproduces only once during a breeding season. Parental care is an important part of successful reproduction in some invertebrates, including some cnidarians (jellyfish), rotifers, mollusks, and arthropods (crustaceans, insects, spiders, and scorpions), and is a common practice among vertebrates (Fig. 51-17b–d).

Females of many vertebrate species produce relatively few, large eggs. Because of the time and energy invested in producing eggs and carrying the developing embryo, the female has more to lose than the male if the young do not develop. Thus, females are more likely than males to brood eggs and young, and usually females invest more in parental care. Parental care is especially skewed toward the female in mammals because females provide milk to nourish their young. Investing time and effort in care of the young is usually less advantageous to a male (assuming the

Courtesy Dr. Bryan D. Neff

Figure 51-18 Deception in mating

A large parental (adult male) bluegill sunfish (*Lepomis macrochirus*) courts a female (*center*). Another male, impersonating a female, swims close to the female (*right*). The parental is not aware that the sneaker is a rival male. In about 20% of spawnings, a sneaker fertilizes some of the eggs and then swims away.

female can handle the job by herself), because time spent in parenting is time lost from inseminating other females.

In some situations, a male benefits by helping rear his own young or even those of a genetic relative. Receptive females may be scarce, breeding territories may be difficult to establish or guard, and gathering sufficient food may require more effort than one parent can provide. In some habitats, the young need protection against predators or cannibalistic males of the same species. Among wolves and many other carnivores, for example, males defend the young and the food supply.

Among many species of fish, the male cares for the young. Apparently, parenting costs to the males are less than they would be for the females. A male fish must have a territory to attract a female. While guarding his territory, the male guards the fertilized eggs as well. Having eggs also appears to make the male more attractive to potential mates. In contrast, when a female is caring for her eggs or young, she is not able to feed optimally. As a result, her body size is smaller and her fertility is reduced.

Behavioral ecologist Bryan D. Neff, of the University of Western Ontario, recently showed that male bluegill sunfish (*Lepomis macrochirus*) adjust their level of caregiving according to their degree of certainty that the offspring are indeed their own. Adult males, called parentals, defend a nest, attract females, and care for the eggs and newly hatched offspring. Some males, called cuckolders or sneaker males, sneak into nests and fertilize eggs during spawning. Other cuckolders mimic females. The parental male is tricked into regarding the cuckold as another female, allowing it access to the eggs (▌Fig. 51-18). Neff placed sneaker males in transparent containers so they could not fertilize eggs, but the parentals could see them. In response, parentals reduced their level of caring for the eggs (▌Fig. 51-19). However, after the eggs hatched, the parentals increased their level of care. Apparently, olfactory cues indicated that these young fish were indeed the offspring of the parental caring for them (that is, they smelled right).

In a second experiment, Neff exchanged about one third of the eggs between two nests. Parentals continued to care for the eggs, but after the eggs hatched, parentals decreased their level of care. Neff showed that parentals responded to olfactory cues. At least some of their offspring did not smell right. (The odor may be produced by the offspring's urine.) Neff suggests that an animal learns its own odor (or appearance) and forms a template of what its relatives should smell (or look) like. If they don't match up closely enough, they may be rejected.

Review

- What is sexual selection? What are some factors that influence mate choice?
- Under what conditions does sexual selection favor polygynous (or polyandrous) mating systems? Why is monogamy uncommon?
- What are some advantages of courtship rituals?
- Why do more females than males invest in caring for their offspring? Explain in terms of costs, benefits, and fitness.

HELPING BEHAVIOR

Learning Objective

11 Relate the concepts of inclusive fitness and kin selection to altruism.

Among wild turkeys and many other species, subordinate males help dominant males attract a mate, but the subordinate males do not reproduce. If an animal's principal evolutionary mandate is to perpetuate its genes, we must wonder why some animals spend time and energy in helping others.

Altruistic behavior can be explained by inclusive fitness

In a type of cooperative behavior known as **altruism,** one individual behaves in a way that seems to benefit others rather than itself, with no potential payoff. In fact, the helper experiences a fitness cost, while the recipient of the altruistic behavior benefits.

Key Experiment

QUESTION: Does a male bluegill sunfish adjust his level of caregiving if he perceives that he is not the genetic parent of the eggs or offspring he is guarding?

HYPOTHESIS: Behavioral ecologist Bryan D. Neff hypothesized that male bluegill sunfish adjust the level of their parental care according to their level of perceived paternity.

EXPERIMENT: Dr. Neff manipulated visual and olfactory cues. In experiment 1, Dr. Neff manipulated sneaker males. He placed two transparent containers with sneaker males near the nests of 34 parentals (adult males) and left them there during spawning. Empty containers were placed around 20 control nests. After the fish spawned, a potential egg predator (a pumpkinseed sunfish in a clear plastic container) was presented, and the parental's defensive behavior was observed and quantified for both experimental and control groups. In experiment 2, Dr. Neff removed one third of the eggs from the nests of 20 parentals and replaced them with eggs from other nests. Sham egg swaps were carried out on eggs of 15 control parentals. Parental defensive behavior was observed before and after the eggs hatched.

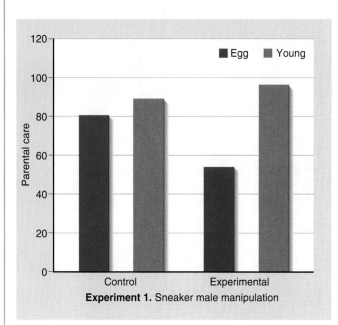

Experiment 1. Sneaker male manipulation

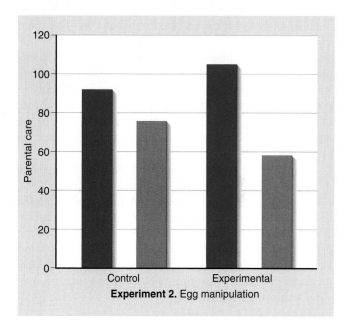

Experiment 2. Egg manipulation

RESULTS: Experiment 1: Parentals reduced their level of caring for eggs. Eight of the males in the experimental group abandoned their nests, and egg defense was significantly lower in this group compared to that in the control group. However, after eggs hatched, there was little difference in parental care between the two groups. Experiment 2: During the egg phase, there was little difference in level of parental care between experimental and control groups. However, after eggs hatched, the experimental parentals significantly decreased their level of care.

CONCLUSION: Male bluegill sunfish adjust their level of parental care according to their level of perceived paternity. In Experiment 1, parentals provided less care when they perceived that the eggs may have been fertilized by sneaker males. After the eggs hatched, olfactory cues indicated that the offspring were indeed their own, and their level of care increased. In Experiment 2, parentals cared for the eggs even though some had been swapped. However, after they hatched, olfactory cues from the offspring indicated that they were not the parental's own offspring, and parentals provided less care.

Figure 51-19 Decisions about parental care

How could true altruistic behavior be adaptive if the animal's response decreases its reproductive success relative to individuals that do not exhibit the altruistic behavior? In 1964, the British evolutionary biologist William D. Hamilton offered a plausible explanation. He suggested that evolution does not distinguish between genes transmitted directly from parent to offspring and those transmitted indirectly through close relatives.

According to Hamilton, natural selection favors animals that help a relative, because the relative's offspring carries some of the helper's alleles. **Inclusive fitness** is the sum of direct fitness (which can be measured by the number of alleles an animal perpetuates in its offspring) plus indirect fitness (the alleles it helps perpetuate in the offspring of its kin).

We can measure the potential genetic gain from helping relatives by the **coefficient of relatedness,** the probability that two individuals inherit the same uncommon allele from a recent common ancestor. The coefficient of relatedness for two animals that are not related is 0, whereas between a parent and its offspring this coefficient is 0.5, indicating that half of their genes are the same. The coefficient of relatedness for siblings is also 0.5; for first cousins it is 0.125. The higher the coefficient of relatedness, the more likely an animal will help a relative.

Hamilton developed a mathematical model to show how an allele that promotes altruistic behavior can increase in frequency in a population. According to **Hamilton's rule,** an altruistic act is adaptive if its indirect fitness benefits are high for the animals that profit, if the recipients are close relatives of the altruist, and if the direct fitness cost to the altruisitic animal is low.

Behavioral ecologists have DNA-fingerprinted lions in the Serengeti National Park and in the Ngorongoro Crater in Tanzania. Their findings confirm that lion brothers that cooperate in prides have a greater probability of perpetuating their genes than those that go off on their own. This is true even if the lion perpetuates his genes only by proxy, through his brother. **Kin selection,** also called *indirect selection,* is a form of natural selection that increases inclusive fitness through the breeding success of close relatives.

The bee society provides an interesting example of kin selection. Recall that worker bees do not reproduce, but they nourish larval bees. The worker bees have up to three quarters of their genes in common with the larval bees, so they are more likely to pass on copies of their genes to the next generation by raising these individuals than they would by producing their own offspring.

Among some birds (such as Florida jays), nonreproducing individuals aid in the rearing of the young. Nests tended by these additional helpers as well as parents produce more young than do nests with the same number of eggs overseen only by parents. The nonreproducing helpers are close relatives who increase their own biological success by ensuring the successful, though indirect, perpetuation of their genes. Helpers may be prevented from producing their own offspring by limiting factors such as a shortage of mates or territories.

Helping behavior may have alternative explanations

Some behavioral ecologists are questioning the role of kin selection in helping behavior. One of the time-honored examples of kin selection has been sentinel behavior, keeping watch for predators and warning other members of the group when danger threatens. Until recently, biologists thought that sentinels were at greater risk for predation, yet selflessly engaged in risky behavior that benefited their relatives. This explanation appears valid for some species, including prairie dogs and ground squirrels (❚ Fig. 51-20). However, recent studies suggest an opposing hypothesis for guarding behavior in other species.

U.S. biologist Peter A. Bednekoff developed a model to explain how guarding could result from selfish behavior. Bednekoff observed that for some species sentinels are not at higher risk than others in their group. In fact, the opposite may be true, and sentinels may actually have an advantage. By detecting predators first, the sentinels can more easily escape.

Studies of sentinel behavior in the African mongoose, commonly known as the meerkat, support Bednekoff's model. During 2000 hours of researcher observation, no sentinel was killed. In fact, investigators concluded that guarding is a selfish behavior in meerkat populations, because sentinels were able to escape quickly into their nearby burrows. This research is an important

Wu Wai Ping/Bruce Coleman, Inc.

Figure 51-20 Kin selection in prairie dogs

Low-ranking members of this social rodent group act as sentries, risking their own lives by exposing themselves outside their burrows. However, by protecting their siblings, they ensure that the genes shared in common will be perpetuated in the population.

reminder that it is risky for biologists to generalize the meanings of complex social behavior from one species to another. Many explanations are possible. Recent findings suggest that helpers often derive direct benefits in safety and eventual reproductive success. Among some species, for example, younger animals who help are permitted to remain in the social group and derive the benefits of group protection. These helpers are also positioned to take over territories and mates when older members of the group die.

Some animals help nonrelatives

Occasionally, an animal who is not related to another helps a nonrelative in a fight, grooms it, or even shares food. Evolutionary biologist Robert L. Trivers, of Rutgers University, explained this behavior in 1971 with his hypothesis of **reciprocal altruism.** According to this hypothesis, one animal helps a nonrelative with no immediate benefit, but at some later time the animal that was helped repays the debt. In this way, the original helper experiences a net fitness benefit. A vampire bat that has just fed regurgitates blood for hungry bats that share the roost. If a bat does not share blood with a neighbor who has fed it in the past, the "cheater" will probably not be helped in the future.

Review

- How does kin selection explain the evolution of altruistic behavior?
- What other hypotheses may account for cooperative behavior?

HIGHLY ORGANIZED SOCIETIES

Learning Objective

12 Contrast a society of social insects with a vertebrate society, and give examples of cultural variation in vertebrate populations.

Characteristics of a highly organized society include cooperation and division of labor among animals of different sexes, age groups, or castes. An intricate system of communication reinforces the organization of the society. The members tend to remain together and to resist attempts by outsiders to enter the group.

Social insects form elaborate societies

By cooperation and division of labor, some insects construct elaborate nests and raise young by mass-production methods. Bees, ants, wasps, and termites form the most organized insect societies. (The first three of these all belong, not coincidentally, to the same order, Hymenoptera.) Insect societies are held together by a complex system of sign stimuli that are keyed to social interaction, and their behaviors tend to be quite rigid.

The social organization of honeybees has been studied more extensively than that of any other social insect. Instructions for the honeybee society are inherited and preprogrammed, and the size and structure of the bee's nervous system permit only a limited range of behavioral variation. Yet these insects are not automatons. Within those limits, the complex bee society can respond with some flexibility to food and other stimuli in the environment.

A honeybee society generally consists of a single adult queen, up to 80,000 worker bees (all female), and at certain times, a few males called drones that fertilize newly developed queens. The queen, the only female in the hive capable of reproduction, deposits about 1000 fertilized eggs per day in the wax cells of a comb.

The composition of a bee society is generally controlled by a pheromone secreted by the queen. It inhibits the workers from raising a new queen and prevents development of ovaries in the workers (▌Fig. 51-21). If the queen dies or if the colony becomes so large that the inhibiting effect of the pheromone dissipates, the workers begin to feed some larvae special food that promotes their development into new queens.

The queen bee stores sperm cells from previous matings in a seminal receptacle. If she releases sperm to fertilize an egg as it is

Treat Davidson/Photo Researchers, Inc.

▌ **Figure 51-21** Maintaining a complex social structure

Numerous worker honeybees surround their queen (*center*). Workers constantly lick secretions from the queen bee. These secretions, transmitted throughout the hive, suppress the activity of the workers' ovaries.

laid, the resulting offspring is female; otherwise, it is male. Thus, males develop by parthenogenesis from unfertilized eggs and are haploid. Because a drone is haploid, each of his sperm cells has *all* his chromosomes; that is, meiosis does not occur during sperm production. The queen bee stores this sperm throughout her lifetime and uses it to produce worker bees.

The worker bees of a hive are more closely related to one another than would be sisters born of a diploid father. Indeed, they have up to three quarters of their genes in common. (Assuming the queen bred with a single drone, they share half of the queen's chromosomes and all of the drone's.) As a consequence, they are more closely related to one another than they would be to their own offspring, if they could have any. (A worker bee's female offspring would have only half of its genes in common with its worker mother.) New queens are also their sisters. Worker bees are therefore more likely to pass on copies of their genes to the next generation by raising these individuals than they would if they were to produce their own offspring.

Division of labor among worker bees is mostly determined by age. The youngest worker bees serve as nurses that nourish larval bees. After about a week as nurse bees, workers begin to produce wax and build and maintain the wax cells. Older workers are foragers, bringing home nectar and pollen. Most worker bees die at the ripe old age of 42 days.

Bees have the most sophisticated mode of communication known among nonmammals. They perform a stereotyped series of body movements called a *dance*. The dance re-enacts, in miniature, the bee's flight to the food. When a honeybee scout locates a rich source of nectar, it apparently observes the angles among the food, hive, and sun. The forager bee then communicates the direction and distance of the food source relative to the hive (▌Fig. 51-22). If the food supply is within about 52 m (about 55 yd), the scout performs a **round dance,** which generally excites the other bees and prompts them to fly short distances in all directions from the hive until they find the nectar. If the food is distant, however, the scout performs a **waggle dance,** which follows a figure-eight pattern.

In the 1940s, the German zoologist Karl von Frisch pioneered studies in bee communication. He found that the waggle dance conveys information about both distance and direction. During the straight run part of the figure eight, the number and frequency of the waggles indicate the distance. The orientation of the movements indicates the direction of the food source in reference to the position of the sun. For example, if the bee dances straight up, the nectar is located directly toward the sun. If the bee dances 40 degrees to the left of the vertical surface of the comb, the nectar is located 40 degrees to the left of a line between the hive and the sun.

The forager bee typically performs the dance in the dark hive on the vertical surface of the comb. How then does its audience "see" the dance? Investigators have shown that the bee produces sounds as it dances. The auditory signals apparently communicate the position of the dancing bee.

Vertebrate societies tend to be relatively flexible

Vertebrate societies usually have nothing comparable to the physically and behaviorally specialized castes of honeybees or ants. An exception is the naked mole rat of southern Africa, a rodent that has a social structure closely resembling that of the social insects. Although most vertebrate societies seem simpler than insect societies, they are also more flexible. Vertebrate societies share a great range and plasticity of potential behaviors and can effectively modify behavior to meet environmental challenges.

The behavioral plasticity of vertebrates makes possible the transmission of culture in some bird and mammal species. **Culture** is behavior common to a population, learned from other members of the group and transmitted from one generation to another. Culture is not inherited. It is maintained by *social learning*—for example, through imitation or teaching.

About 17 behaviors that may be considered cultural have been described among cetaceans (whales, dolphins, porpoises). For example, female orcas teach their offspring to hunt seals according to the custom of their particular group. Orcas also have local dialects that have been documented for at least six generations, suggesting that they are passed on from parent to offspring. Separate dialects are maintained even when various populations of orcas interact socially. Bottlenose dolphins are well known for their ability to learn by imitation. Behavioral ecologists have documented foraging practices that appear to be cultural—transmitted by social learning.

Teaching is an efficient form of social learning. If adult animals can teach younger ones, critical information can be rapidly transferred from one generation to the next. Researchers Alex Thornton, of the University of Cambridge, and Katherine McAulliffe recently reported that helper meerkats teach pups skills for capturing prey. Meerkats feed on a variety of animals, many difficult to subdue. Some, such as scorpions, are potentially dangerous. Young pups cannot find and capture their own prey, and they emit begging calls that induce helpers to feed them. Helpers monitor and nudge the pups to handle the prey efficiently. As they grow older, the pups' begging calls change. In response,

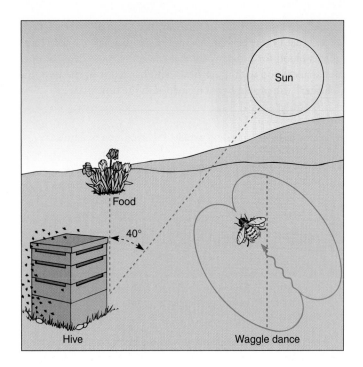

Figure 51-22 *Animated* The honeybee waggle dance

The scout is waggling upward, indicating that the food is toward the sun, and inclined 40 degrees to the left, which reveals the angle of the food source relative to the sun. The dance takes place inside the hive.

helpers modify their teaching, giving the pups the opportunity to practice a greater range of skills. The helpers gradually introduce the pups to live prey.

Recent studies of nonhuman primates demonstrate that some groups develop local customs and teach them to their offspring. Using data provided by researchers at seven chimpanzee research sites, animal behaviorists identified 39 chimpanzee behaviors that they considered cultural variations. These behaviors included various ways of using tools, courtship rituals, and grooming techniques. Each of these local customs was learned from other chimpanzees in the population, and customs varied among different populations. For example, chimpanzees on the western side of the Sassandra-N'Zo River use stone hammers to crack coula nuts. Researchers have photographed mother chimpanzees teaching this behavior to their offspring. Just a few miles away on the eastern side of the river, chimpanzees do not crack nuts, even though they are available. When a chimp migrates and joins a new population, it can transmit knowledge learned in its previous culture. In this way, customs spread from one population to another.

As researchers continue to study cultural variation among chimpanzee populations, learning ever more about our closest relatives, it is important to remember that the capacity to learn such behaviors is the product of natural selection. The behaviors that are maintained are adaptive for these animals in their environment. Studying culture in nonhuman primates may provide insights into the evolutionary origins of human culture. Human society is based to a great extent on the symbolic transmission of culture through spoken and written language.

Sociobiology explains human social behavior in terms of adaptation

Sociobiology focuses on the evolution of social behavior through natural selection. In his landmark book, *Sociobiology: The New Synthesis,* published in 1975, Edward O. Wilson combined principles of population genetics, evolution, and animal behavior to present a comprehensive view of the evolution of social behavior. Wilson's work influenced the development of behavioral ecology, and many of the concepts discussed in this chapter, such as paternal investment in care of the young, are based on contributions made by sociobiologists. Like Darwin and many other biologists of the past, Wilson and other sociobiologists suggest that human behavior can be studied in evolutionary terms.

Sociobiology is controversial, at least in part because of its possible ethical implications. This approach is sometimes viewed as denying that human behavior is flexible enough to permit substantial improvements in the quality of our social lives. Yet sociobiologists agree with their critics that human behavior is flexible, and at least some of the debate focuses on the degree to which human behavior is genetically determined and the extent to which it can be modified.

As sociobiologists acknowledge, people can, through culture, change their way of life far more profoundly in a few years than could a hive of bees or a troop of baboons in hundreds of generations of genetic evolution. This capacity to make changes is indeed genetically determined, and that is a great gift. How we use it and what we accomplish with it are not a gift but a responsibility on which our own well-being and the well-being of other species depend.

Review

- How does the dance of bees compare with human language?
- How are vertebrate societies different from those of social insects?

SUMMARY WITH KEY TERMS

Learning Objectives

1   Distinguish between proximate and ultimate causes of behavior, and apply the concepts of ultimate cause and cost–benefit analysis to decide whether a particular behavior is adaptive (page 1102).

- **Behavior** is what an animal does and how it does it, usually in response to stimuli in its environment. **Proximate causes** of behavior are immediate causes, such as genetic, developmental, and physiological processes that permit the animal to carry out a specific behavior. Proximate causes answer *how* questions. **Ultimate causes** are the evolutionary explanations for *why* a certain behavior occurs.

- We can use **cost–benefit analysis** to determine whether a behavior is adaptive. Benefits contribute to **direct fitness,** the animal's reproductive success measured by the number of viable offspring. If benefits outweigh costs, the behavior is adaptive.

2   Describe the interactions of heredity, environment, and maturation in animal behavior (page 1102).

- Behavior results from the interaction of genes (**innate behavior**) and environmental factors. The capacity for behavior is inherited, but behavior is modified in response to environmental experience.

- An organism must be mature—physiologically ready to produce a given behavior—before it can perform that pattern of behavior. Walking and many other behaviors that we view as automatic are **motor programs,** coordinated sequences of muscle actions.

- A **behavioral pattern** is an automatic behavior that once activated by a simple sensory stimulus, continues to completion regardless of sensory feedback. A behavioral pattern can be triggered by a specific unlearned **sign stimulus,** or **releaser.**

3   Discuss the adaptive significance of habituation, imprinting, classical conditioning, operant conditioning, and cognition (page 1104).

- **Learning** is a change of behavior that results from experience. **Habituation** is a type of learning in which an animal learns to ignore a repeated, irrelevant stimulus so the animal can focus on finding food and carrying out other life activities.

- **Imprinting** establishes a parent–offspring bond during a **critical period** early in development, ensuring that the offspring recognizes the mother.

- In **classical conditioning,** an association is formed between some normal body function and a new stimulus. This type of learning allows an animal to make an association between two stimuli.

- In **operant conditioning,** an animal learns a behavior to receive positive reinforcement or to avoid punishment. This type of learning is important in many natural situations, including young herring gulls perfecting their pecking behavior to obtain food.

- **Cognition** is the process of gaining knowledge; it includes thinking, processing information, learning, reasoning, and awareness of thoughts, perceptions, and self. **Insight learning** is the ability to adapt past experiences that may involve different stimuli to solve a new problem.

4   Give examples of biological rhythms, and describe some of the mechanisms responsible for them (page 1108).

- It is adaptive for an organism's metabolic processes and behavior to be synchronized with cyclical changes in the environment. In many species, physiological processes and activity follow **circadian rhythms,** which are daily cycles of activity.

- **Diurnal** animals are most active during the day; **nocturnal** animals, at night; and **crepuscular** animals, at dawn and/or dusk. Some biological rhythms reflect the **lunar cycle,** or the changes in tides due to changes in the phase of the moon.

- Many biological rhythms are regulated by internal timing mechanisms that serve as **biological clocks.** In mammals the master clock is located in the **suprachiasmatic nucleus (SCN)** in the hypothalamus.

5   Analyze costs and benefits of migrations, and distinguish between directional orientation and navigation (page 1108).

- **Migration** is periodic long-distance travel from one location to another. Proximate causes of migration include physiological responses that are triggered by environmental changes. Ultimate causes of migration include the benefit of moving away from an area that seasonally becomes too cold, dry, or depleted of food to a more hospitable area. Costs of migration include time, energy, and greater risk of predation.

- **Directional orientation** is travel in a specific direction and requires **compass sense,** a sense of direction. Many migrating animals use the sun to orient themselves.
- **Navigation** requires both compass sense and **map sense,** an awareness of location. Birds that navigate at night use the stars as guides. Birds and many other animals also use Earth's magnetic field to navigate.

6 Discuss the hypothesis that optimal foraging behavior is adaptive (page 1110).

- **Optimal foraging,** the most efficient strategy for an animal to get food, can enhance reproductive success.

7 Analyze social behavior in terms of costs and benefits, and describe modes of animal communication (page 1111).

- **Social behavior** is adaptive interaction, usually among members of the same species. Many animal societies are characterized by a means of communication, cooperation, division of labor, and a tendency to stay together. Benefits include cooperative foraging or hunting and defense from predators. Costs include increased competition for food and habitats and increased risks of attracting predators and transmitting disease.
- Animal **communication** involves the exchange of mutually recognizable signals, which can be electrical, tactile, visual, auditory, or chemical. **Pheromones** are chemical signals that convey information between members of a species.

8 Describe the adaptive significance of social organization, using establishment of a dominance hierarchy as an example (page 1111).

- Social organization ensures greater reproductive success for the members of a society. A **dominance hierarchy** is a ranking of status within a group in which more dominant members are accorded benefits (such as food or mates) by subordinates, often without overt aggressive behavior. Social ranking reduces costly physical fights.

9 Distinguish between home range and territory, and describe costs and benefits of territoriality (page 1111).

- Animals often inhabit a **home range,** a geographic area that they seldom leave but do not necessarily defend. A defended area within a home range is called a **territory,** and the defensive behavior is **territoriality.** Some animals, such as lions, engage in group territoriality. Costs of territoriality include time and energy expended in staking out and defending a territory; and risks, in fighting for it. Benefits include rights to food in the territory and reduction in conflict among members of a population.

10 Define *sexual selection,* and describe different types of mating systems and approaches to parental care (page 1115).

- **Sexual selection,** a type of natural selection, occurs when individuals vary in their ability to compete for mates.

Individuals with reproductive advantages are selected over others of the same sex and species.

- Mate choice may be influenced by dominance, gifts, ornaments, and courtship displays. Males of some species gather in a **lek,** a small display area where they compete for females. **Courtship rituals** ensure that the male is a member of the same species and permit the female to assess the quality of the male.
- Sexual selection often favors **polygyny,** a mating system in which a male mates with many females. In **polyandry,** a female mates with several males. **Monogamy,** mating with a single partner during a breeding season, is less common. Among many species, males engage in **mate guarding,** especially when the female is most fertile, to prevent other males from fertilizing her eggs.
- A **pair bond** is a stable relationship between a male and a female that may involve cooperative behavior in mating and in rearing the young. **Parental investment** in care of eggs and offspring increases the probability that offspring will survive. A high investment in parenting is typically less advantageous to the male than to the female.

11 Relate the concepts of inclusive fitness and kin selection to altruism (page 1118).

- **Altruism** is a type of cooperative behavior in which one individual appears to behave in a way that benefits others rather than itself. **Inclusive fitness** is the sum of an individual's direct fitness and indirect fitness (number of offspring of kin). This concept suggests that natural selection favors animals that help a relative because this is an indirect way to perpetuate some of the helper's own alleles.
- According to **Hamilton's rule,** an altruistic act is adaptive if its indirect fitness benefits are high for the animals that are helped, if the recipients are close relatives of the altruist, and if the direct fitness cost to the altruist is low.
- **Kin selection** is a type of natural selection that increases inclusive fitness through successful reproduction of close relatives. Some types of cooperative behavior may increase the direct fitness of the helper.
- In a type of cooperative behavior known as **reciprocal altruism,** the helper does not immediately benefit but is helped later by the animal it helped.

12 Contrast a society of social insects with a vertebrate society, and give examples of cultural variation in vertebrate populations (page 1121).

- A **society** is a group of individuals of the same species that work together in an adaptive manner. Insect societies tend to be rigid, with the role of the individual narrowly defined. Insect division of labor is mainly determined by age.

ThomsonNOW™ **See honeybee dances in action by clicking on the figure in ThomsonNOW.**

- Vertebrate societies are more flexible than insect societies. Some bird and mammal species develop **culture,** learned behavior common to a population that is transmitted from one generation to the next. Symbolic transmission of culture is important in human societies.

TEST YOUR UNDERSTANDING

1. The contemporary study of behavior in natural environments from the point of view of adaptation is (a) ecology (b) ethology (c) behavioral ecology (d) evolutionary ecology (e) behavioral science

2. To understand *why* the graylag goose "rolls" a nonexistent egg toward her nest, you would explore (a) proximate causes (b) ultimate causes (c) pheromones (d) hormonal factors (e) its social learning

3. The responses of an organism to signals from its environment are its (a) behavior (b) culture (c) ultimate behavior (d) releasers (e) motor programs

4. A behavioral pattern is elicited by a (a) motor program (b) sign stimulus (c) releaser (d) a, b, and c (e) b and c

5. A form of learning in which a young animal forms a strong attachment to a moving object (usually its parent) within a few hours of birth is (a) classical conditioning (b) operant conditioning (c) imprinting (d) insight learning (e) parental investment

6. An animal learns to ignore a repeated, irrelevant stimulus. This behavior is (a) classical conditioning (b) operant conditioning (c) imprinting (d) insight learning (e) habituation

7. Salivation by a student when the noon bell rings is an example of (a) classical conditioning (b) operant conditioning (c) imprinting (d) insight learning (e) habituation

8. Both compass sense and map sense are necessary for (a) navigation (b) migration (c) directional orientation (d) both navigation and migration (e) the action of biological clocks

9. In optimal foraging, (a) animals always hunt in social groups (b) an animal obtains food in the most efficient way (c) animals can rarely afford to be selective (d) groups of five to seven animals are most successful (e) animals are typically most successful when they hunt alone

10. Chemical signals that convey information between members of a species are (a) pheromones (b) hormones (c) neurotransmitters (d) leks (e) neuropeptides

11. The benefits of territoriality include (a) rights to defend a home range (b) increased reproductive success (c) energy investment in staking out and defending the area (d) monogamy (e) pair bonding

12. Sexual selection (a) is a form of natural selection (b) occurs when animals are very similar in their ability to compete for mates (c) results in animals that have lower direct fitness (d) occurs mainly among animals that practice polygyny (e) occurs mainly among animals that practice polyandry

13. Mate choice is least likely influenced by (a) ornamental displays such as antlers (b) dominance (c) competition in a lek (d) courtship behavior (e) fidelity of the male

14. The round dance of the bee (a) indicates that food is close to the hive (b) communicates direction (c) results in bees flying long distances in all directions (d) indicates that food is distant from the hive (e) indicates both direction and height of food

15. Behavior that appears to have no payoff, that is, an individual appears to act to benefit others rather than itself, is known as (a) mutualism (b) altruism (c) reciprocal altruism (d) inclusive fitness (e) helping behavior

16. Inclusive fitness (a) considers only the genes an animal transmits to its own offspring (b) suggests that natural selection favors animals that help a relative (c) explains territoriality (d) is a form of social behavior (e) is a form of reciprocal altruism

17. Kin selection (a) increases inclusive fitness (b) is a way of perpetuating genes of nonrelatives (c) accounts for some forms of migration (d) typically involves mate guarding (e) involves ornamental displays and use of a lek

18. Culture (a) is common among invertebrate groups (b) does not vary within the same species (c) is a shared derived character of humans (d) has a strong genetic component in primates (e) is behavior learned from other members of the group and shared by members of a population

CRITICAL THINKING

1. Behavioral ecologists have demonstrated that young tiger salamanders can become cannibals and devour other salamanders. Investigators observed that the cannibal salamanders preferred eating unrelated salamanders over their own cousins and preferred eating cousins over their siblings. Explain this behavior based on what you have learned in this chapter.

2. How is the society of a social insect different from human society? What are some similarities between the transmission of information by heredity and by culture? What are some differences?

3. **Evolution Link.** What might be the adaptive value of sea turtle migration? Consider how this adaptive value has changed if, as a result of human activities, migration now puts sea turtles at greater risk than if they restricted their habitat to a single location. Discuss the possible evolutionary mechanisms by which the behavior of these species may (or may not)

adapt to these environmental pressures. What conservation efforts should we take to increase the probability of successful migration?

4. **Analyzing Data.** Look at Figure 51-11, and identify the groups most likely to be hungry at the end of the day's hunt during times of prey scarcity. Develop a hypothesis to explain why these groups hunt as they do.

5. **Analyzing Data.** Look at the two graphs in Figure 51-19. In which experiment did the parentals in the experimental group guard the eggs more closely? In which experiment did the experimental parentals guard the young more closely? Account for these differences.

Additional questions are available in ThomsonNOW at www.thomsonedu.com/login

Introduction to Ecology: Population Ecology

Jim Brandenburg/Minden Pictures

A population of Mexican poppies. Mexican poppies (*Eschsolzia mexicana*), which thrive on gravelly desert slopes, bloom in the desert after the winter rains.

KEY CONCEPTS

A population can be described in terms of its density, dispersion, birth and death rates, growth rate, survivorship, and age structure.

Changes in population size are caused by natality, mortality, immigration, and emigration.

Population size may be influenced by density-dependent factors and density-independent factors.

Life history traits of a population are adaptations that affect the ability of individuals to survive and reproduce.

A metapopulation consists of two or more local populations with dispersal occurring among them.

Human population structure differs among countries in ways that are primarily related to differences in level of development.

The science of **ecology** is the study of how living organisms and the physical environment interact in an immense and complicated web of relationships. Biologists call the interactions among organisms **biotic factors,** and those between organisms and their nonliving, physical environment, **abiotic factors.** Abiotic factors include precipitation, temperature, pH, wind, and chemical nutrients. Ecologists formulate hypotheses to explain such phenomena as the distribution and abundance of life, the ecological role of specific species, the interactions among species in communities, and the importance of ecosystems in maintaining the health of the biosphere. They then test these hypotheses.

The focus of ecology can be local or global, specific or generalized, depending on what questions the scientist is asking and trying to answer. Ecology is the broadest field in biology, with explicit links to evolution and every other biological discipline. Its universality also encompasses subjects that are not traditionally part of biology. Earth science, geology, chemistry, oceanography, climatology, and meteorology are extremely important to ecology, especially when ecologists examine the abiotic environment of planet Earth. Because humans are part of Earth's web of life, all our activities, including economics and politics, have profound ecological implications. **Environmental science,** a scientific discipline with ties to ecology, focuses on how humans interact with the environment.

As you learned in Chapter 1, most ecologists are interested in the levels of biological organization including and above the level of the individual organism: population, community, ecosystem, landscape, and biosphere. Each level has its own characteristic composition, structure, and functioning.

An individual belongs to a **population,** a group consisting of members of the same species that live together in a prescribed area at the same time. The boundaries of the area are defined by the ecologist performing a particular study. A population ecologist might study a population of microorganisms, animals, or plants, like the Mexican poppies in the photograph, to see how individuals within it live and interact with one another, with other species in their community, and with their physical environment.

In this chapter we begin our study of ecological principles by focusing on the study of populations as functioning systems and end with a discussion of the human population. Subsequent chapters examine the interactions among different populations within communities (Chapter 53), the dynamic exchanges between communities and their physical environments (Chapter 54), the characteristics of Earth's major biological ecosystems (Chapter 55), and some of the impacts of humans on the biosphere (Chapter 56). ■

FEATURES OF POPULATIONS

Learning Objective

1   Define *population density* and *dispersion,* and describe the main types of population dispersion.

Populations exhibit characteristics distinctive from those of the individuals of which they are composed. Some features discussed in this chapter that characterize populations are population density, population dispersion, birth and death rates, growth rates, survivorship, and age structure.

Although communities consist of all the populations of all the different species that live together within an area, populations have properties that communities lack. Populations, for example, share a common gene pool (see Chapter 19). As a result, allele frequency changes resulting from natural selection occur in populations. Natural selection, therefore, acts directly to produce adaptive changes in populations and only indirectly affects the community level.

Population ecology considers both the number of individuals of a particular species that are found in an area and the dynamics of the population. **Population dynamics** is the study of changes in populations—how and why those numbers increase or decrease over time. Population ecologists try to determine the processes common to all populations. They study how a population interacts with its environment, such as how individuals in a population compete for food or other resources, and how predation, disease, and other environmental pressures affect the population. Population growth, whether of bacteria, maples, or giraffes, cannot increase indefinitely because of such environmental pressures.

Additional aspects of populations that interest biologists are their reproductive success or failure (extinction), their evolution, their genetics, and the way they affect the normal functioning of communities and ecosystems. Biologists in applied disciplines, such as forestry, agronomy (crop science), and wildlife management, must understand population ecology to manage populations of economic importance, for example, forests, field crops, game animals, or fishes. Understanding the population dynamics of endangered and threatened species plays a key role in efforts to prevent their slide to extinction. Knowledge of population ecology helps in efforts to prevent the increase of pest populations to levels that cause significant economic or health impacts.

Density and dispersion are important features of populations

The concept of population size is meaningful only when the boundaries of that population are defined. Consider, for example, the difference between 1000 mice in 100 hectares (250 acres) and 1000 mice in 1 hectare (2.5 acres). Often a population is too large to study in its entirety. Researchers examine such a population by sampling a part of it and then expressing the population in terms of density. Examples include the number of dandelions per square meter of lawn, the number of water fleas per liter of pond water, or the number of cabbage aphids per square centimeter of cabbage leaf. **Population density,** then, is the number of individuals of a species per unit of area or volume at a given time.

Different environments vary in the population density of any species they can support. This density may also vary in a single habitat from season to season or year to year. For example, red grouse are ground-dwelling game birds whose populations are managed for hunting. Consider two red grouse populations in the treeless moors of northwestern Scotland, at locations only 2.5 km (1.5 mi) apart. At one location the population density remained stationary during a 3-year period, but at the other site it almost doubled in the first 2 years and then declined to its initial density in the third year. The reason was likely a difference in habitat. Researchers had experimentally burned the area where the population density increased initially and then decreased. Young heather shoots (*Calluna vulgaris*) produced after the burn provided nutritious food for the red grouse. So population density may be determined in large part by biotic or abiotic factors in the environment that are external to the individuals in the population.

The individuals in a population often exhibit characteristic patterns of **dispersion,** or spacing, relative to one another. Individuals may be spaced in a random, clumped, or uniform dispersion. **Random dispersion** occurs when individuals in a population are spaced throughout an area in a manner that is unrelated to the presence of others (❙ Fig. 52-1a). Of the three major types of dispersion, random dispersion is least common and hardest to observe in nature, leading some ecologists to question its existence. Trees of the same species, for example, sometimes appear to be distributed randomly in a tropical rain forest. However, an international team of 13 ecologists studied six tropical forest plots that were 25 to 52 hectares (62 to 130 acres) in area and reported that most of the 1000 tree species observed were clumped and not randomly dispersed. (Ecologists determine both clumped and uniform dispersion by statistically testing for differences from an assumed random distribution.) Random dispersion may occur infrequently, because important environmental factors affecting dispersion usually do not occur at random. Flour beetle larvae in

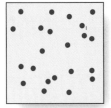

(a) Random dispersion, illustrated for comparison, rarely, if ever, occurs in nature.

Andrew G. Wood/Photo Researchers, Inc.

(b) Clumped dispersion is evident in the schooling behavior of certain fish species. Shown are bluestripe snappers (*Lutjanus kasmira*), photographed in Hawaii. This introduced fish, which grows to 30 cm (12 in), may be displacing native fish species in Hawaiian waters.

Robert Hernandez/Photo Researchers, Inc.

(c) Uniform dispersion is characteristic of these nesting Cape gannets (*Morus capensis*) on the coast of South Africa. The birds space their nests more or less evenly.

Figure 52-1 *Animated* Dispersion of individuals within a population

a container of flour are randomly dispersed, but their environment (flour) is unusually homogeneous.

Perhaps the most common spacing is **clumped dispersion,** also called **aggregated distribution** or **patchiness,** which occurs when individuals are concentrated in specific parts of the habitat. Clumped dispersion often results from the patchy distribution of resources in the environment. It also occurs among animals because of the presence of family groups and pairs, and among plants because of limited seed dispersal or asexual reproduction. An entire grove of aspen trees, for example, may originate asexually from a single plant. Clumped dispersion may sometimes be advantageous, because social animals derive many benefits from their association. Many fish species, for example, associate in dense schools for at least part of their life cycle, possibly because schooling may reduce the risk of predation for any particular individual (Fig. 52-1b). The many pairs of eyes of schooling fish tend to detect predators more effectively than a single pair of eyes of a single fish. When threatened, schooling fish clump together more closely, so it is difficult for a predator to single out an individual.

Uniform dispersion occurs when individuals are more evenly spaced than would be expected from a random occupation of a given habitat. A nesting colony of seabirds, in which the birds are nesting in a relatively homogeneous environment and place their nests at a more or less equal distance from each other, is an example of uniform dispersion (Fig. 52-1c). What might this spacing pattern tell us? In this case, uniform dispersion may occur as a result of nesting territoriality. Aggressive interactions among the nesting birds as they peck at one another from their nests cause each pair to place its nest just beyond the reach of nearby nesting birds. Uniform dispersion also occurs when competition among individuals is severe, when plant roots or leaves that have been shed produce toxic substances that inhibit the growth of nearby plants, or when animals establish feeding or mating territories.

Some populations have different spacing patterns at different ages. Competition for sunlight among same-aged sand pine in a Florida scrub community resulted in a change over time from either random or clumped dispersion when the plants were young, to uniform dispersion when the plants were old. Sand pine is a fire-adapted plant with cones that do not release their seeds until they have been exposed to high temperatures (45°C to 50°C or higher). As a result of seed dispersal and soil conditions following a fire, the seedlings grow back in dense stands that exhibit random or slightly clumped dispersion. Over time, however, many of the more crowded trees tend to die from shading or competition, resulting in uniform dispersion of the surviving trees (Table 52-1).

TABLE 52-1		
Dispersion in a Sand Pine Population in Florida		
Tree Trunks Examined	Density (per m²)	Dispersion
All (alive and dead)	0.16	Random
Alive only	0.08	Uniform

Source: Adapted from A. M. Laessle, "Spacing and Competition in Natural Stands of Sand Pine," *Ecology* 46:65–72, 1965. Data were collected 51 years after a fire.

CHANGES IN POPULATION SIZE

Learning Objectives

2 Explain the four factors (natality, mortality, immigration, and emigration) that produce changes in population size, and solve simple problems involving these changes.

3 Define *intrinsic rate of increase* and *carrying capacity,* and explain the differences between J-shaped and S-shaped growth curves.

One goal of science is to discover common patterns among separate observations. As mentioned previously, population ecologists wish to understand general processes that are shared by many different populations, so they develop mathematical models based on equations that describe the dynamics of a single population. Population models are not perfect representations of a population, but models help illuminate complex processes. Moreover, mathematical modeling enhances the scientific process by providing a framework with which experimental population studies can be compared. We can test a model and see how it fits or does not fit with existing data. Data that are inconsistent with the model are particularly useful, because they demand that we ask how the natural system differs from the mathematical model that we developed to explain it. As more knowledge accumulates from observations and experiments, the model is refined and made more precise.

Population size, whether of sunflowers, elephants, or humans, changes over time. On a global scale, this change is ultimately caused by two factors, expressed on a per capita (that is, per individual) basis: **natality,** the average per capita birth rate, and **mortality,** the average per capita death rate. In humans the birth rate is usually expressed as the number of births per 1000 people per year and the death rate as the number of deaths per 1000 people per year.

To determine the rate of change in population size, we must also take into account the time interval involved, that is, the change in time. To express change in equations, we employ the Greek letter delta (Δ). In equation (1), ΔN is the change in the number of individuals in the population, Δt the change in time, N the number of individuals in the existing population, b the natality, and d the mortality.

$$(1)\ \Delta N/\Delta t = N(b - d)$$

The **growth rate** (r), or rate of change (increase or decrease) of a population on a per capita basis is the birth rate minus the death rate:

$$(2)\ r = b - d$$

As an example, consider a hypothetical human population of 10,000 in which there are 200 births per year (that is, by convention, 20 births per 1000 people) and 100 deaths per year (10 deaths per 1000 people):

$$r = 20/1000 - 10/1000 = 0.02 - 0.01 = 0.01,\ \text{or 1% per year}$$

A modification of equation (1) tells us the rate at which the population is growing at a particular instant in time, that is, its instantaneous growth rate (dN/dt). (The symbols dN and dt are the mathematical differentials of N and t, respectively; they are not products, nor should the d in dN or dt be confused with the death rate, d.) Using differential calculus, this growth rate can be expressed as follows:

$$(3)\ dN/dt = rN$$

where N is the number of individuals in the existing population, t the time, and r the per capita growth rate.

Because $r = b - d$, if individuals in the population are born faster than they die, r is a positive value and population size increases. If individuals in the population die faster than they are born, r is a negative value and population size decreases. If r is equal to zero, births and deaths match and population size is stationary despite continued reproduction and death.

Dispersal affects the growth rate in some populations

In addition to birth and death rates, **dispersal,** which is movement of individuals among populations, must be considered when examining changes in populations on a *local* scale. There are two types of dispersal: immigration and emigration. **Immigration** occurs when individuals enter a population and thus increase its size. **Emigration** occurs when individuals leave a population and thus decrease its size. The growth rate of a local population must take into account birth rate (b), death rate (d), immigration rate (i), and emigration rate (e) on a per capita basis. The per capita growth rate equals the birth rate minus the death rate, plus the immigration rate minus the emigration rate:

$$(4)\ r = (b - d) + (i - e)$$

For example, the growth rate of a human population of 10,000 that has 200 births (by convention, 20 per 1000), 100 deaths (10 per 1000), 10 immigrants (1 per 1000), and 100 emigrants (10 per 1000) in a given year would be calculated as follows:

$$r = (20/1000 - 10/1000) + (1/1000 - 10/1000)$$

$$r = 0.001,\ \text{or 0.1% per year}$$

Each population has a characteristic intrinsic rate of increase

The maximum rate at which a population of a given species could increase under ideal conditions, when resources are abundant and its population density is low, is known as its **intrinsic rate of increase** (r_{max}). Different species have different intrinsic rates of increase. A particular species' intrinsic rate of increase is influenced by several factors. These include the age at which

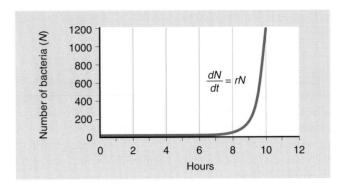

Figure 52-2 *Animated* Exponential population growth

When bacteria divide every 20 minutes, their numbers (expressed in millions) increase exponentially. The curve of exponential population growth has a characteristic J shape. The ideal conditions under which bacteria or other organisms reproduce exponentially rarely occur in nature, and when these conditions do occur, they are of short duration.

reproduction begins, the fraction of the **life span** (duration of the individual's life) during which the individual is capable of reproducing, the number of reproductive periods per lifetime, and the number of offspring the individual is capable of producing during each period of reproduction. These factors, which we discuss in greater detail later in the chapter, determine whether a particular species has a large or small intrinsic rate of increase.

Generally, large species such as blue whales and elephants have the smallest intrinsic rates of increase, whereas microorganisms have the greatest intrinsic rates of increase. Under ideal conditions (an environment with unlimited resources), certain bacteria can reproduce by binary fission every 20 minutes. At this rate of growth, a single bacterium would increase to a population of more than 1 billion in just 10 hours!

If we plot the population size versus time, under optimal conditions, the graph has a J shape that is characteristic of **exponential population growth,** which is the accelerating population growth rate that occurs when optimal conditions allow a constant per capita growth rate (■ Fig. 52-2). When a population grows exponentially, the larger that population gets, the faster it grows.

Regardless of which species we are considering, whenever a population is growing at its intrinsic rate of increase, population size plotted versus time gives a curve of the same shape. The only variable is time. It may take longer for an elephant population than for a bacterial population to reach a certain size (because elephants do not reproduce as rapidly as bacteria), but both populations will always increase exponentially as long as their per capita growth rates remain constant.

No population can increase exponentially indefinitely

Certain populations may grow exponentially for brief periods. Exponential growth has been experimentally demonstrated in certain insects and in bacterial and protistan cultures (by contin-

ually supplying nutrients and removing waste products). However, organisms cannot reproduce indefinitely at their intrinsic rate of increase, because the environment sets limits. These limits include such unfavorable environmental conditions as the limited availability of food, water, shelter, and other essential resources (resulting in increased competition) as well as limits imposed by disease and predation.

In the earlier example, bacteria in nature would never be able to reproduce unchecked for an indefinite period, because they would run out of food and living space, and poisonous wastes would accumulate in their vicinity. With crowding, bacteria would also become more susceptible to parasites (high population densities facilitate the spread of infectious organisms such as viruses among individuals) and predators (high population densities increase the likelihood of a predator catching an individual). As the environment deteriorated, their birth rate (*b*) would decline and their death rate (*d*) would increase. Conditions might worsen to a point where *d* would exceed *b,* and the population would decrease. The number of individuals in a population, then, is controlled by the ability of the environment to support it. As the number of individuals in a population (*N*) increases, environmental limits act to control population growth.

Over longer periods, the rate of population growth may decrease to nearly zero. This leveling out occurs at or near the limits of the environment to support the population. The **carrying capacity (*K*)** represents the largest population that can be maintained for an indefinite period by a particular environment, assuming there are no changes in that environment. In nature, the carrying capacity is dynamic and changes in response to environmental changes. An extended drought, for example, could decrease the amount of vegetation growing in an area, and this change, in turn, would lower the carrying capacity for deer and other herbivores in that environment.

When a population regulated by environmental limits is graphed over longer periods, the curve has a characteristic S shape. The curve shows the population's initial exponential increase (note the curve's J shape at the start, when environmental limits are few), followed by a leveling out as the carrying capacity of the environment is approached (■ Fig. 52-3). The S-shaped growth curve, also called **logistic population growth,** can be modeled by a modified growth equation called a *logistic equation.* The logistic model of population growth was developed to explain population growth in continually breeding populations. Similar models exist for populations that have specific breeding seasons.

The logistic model describes a population increasing from a small number of individuals to a larger number of individuals that are ultimately limited by the environment. The logistic equation takes into account the carrying capacity of the environment:

$$(5)\ dN/dt = rN[(K - N)/K]$$

Note that part of the equation is the same as equation (3). The added element $[(K - N)/K]$ reflects a decline in growth as a population size approaches its carrying capacity. When the number of organisms (*N*) is small, the rate of population growth is unchecked by the environment, because the expression $[(K - N)/K]$ has a value of almost 1. But as the population (*N*) begins to ap-

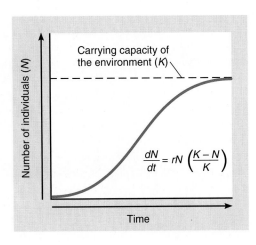

Figure 52-3 *Animated* Carrying capacity and logistic population growth

In many laboratory studies, exponential population growth slows as the carrying capacity (K) of the environment is approached. The logistic model of population growth, when graphed, has a characteristic S-shaped curve.

proach the carrying capacity (K), the growth rate declines because the value of $[(K - N)/K]$ approaches zero.

Although the S curve is an oversimplification of how most populations change over time, it does appear to fit some populations that have been studied in the laboratory, as well as a few that have been studied in nature. For example, Georgyi F. Gause, a Russian ecologist who conducted experiments during the 1930s, grew a population of a single species, *Paramecium caudatum,* in a test tube. He supplied a limited amount of food (bacteria) daily and replenished the growth medium occasionally to eliminate the accumulation of metabolic wastes. Under these conditions, the population of *P. caudatum* increased exponentially at first, but then its growth rate declined to zero, and the population size leveled off (see Fig. 53-4, middle graph).

A population rarely stabilizes at K (carrying capacity) but may temporarily rise higher than K. It will then drop back to, or below, the carrying capacity. Sometimes a population that overshoots K will experience a *population crash,* an abrupt decline from high to low population density. Such an abrupt change is commonly observed in bacterial cultures, zooplankton, and other populations whose resources have been exhausted.

The carrying capacity for reindeer, which live in cold northern habitats, is determined largely by the availability of winter forage. In 1910, humans introduced a small herd of 26 reindeer onto one of the Pribilof Islands of Alaska. The herd's population increased exponentially for about 25 years until there were approximately 2000 reindeer, many more than the island could support, particularly in winter. The reindeer overgrazed the vegetation until the plant life was almost wiped out. Then, in slightly over a decade, as reindeer died from starvation, the number of reindeer plunged to 8, one third the size of the original introduced population. Recovery of subarctic and arctic vegetation after overgrazing by reindeer can take 15 to 20 years, during which time the carrying capacity for reindeer is greatly reduced.

Review

- What effect does each of the following have on population size: (1) natality, (2) mortality, (3) immigration, and (4) emigration?
- How does a J-shaped population growth curve differ from an S-shaped curve in terms of intrinsic rate of increase and carrying capacity?
- What would be the main difference between graphs representing the long-term growth of two populations of bacteria cultured in test tubes, one in which the nutrient medium is replenished, and the other in which it is not replenished?

FACTORS INFLUENCING POPULATION SIZE

Learning Objective

4 Contrast the influences of density-dependent and density-independent factors on population size, and give examples of each.

Certain natural mechanisms influence population size. Factors that affect population size fall into two categories: density-dependent factors and density-independent factors. These two sets of factors vary in importance from one species to another and, in most cases, probably interact simultaneously to determine the size of a population.

Density-dependent factors regulate population size

Sometimes the influence of an environmental factor on the individuals in a population varies with the density or crowding of that population. If a change in population density alters how an environmental factor affects that population, then the environmental factor is said to be a **density-dependent factor.**

As population density increases, density-dependent factors tend to slow population growth by causing an increase in death rate and/or a decrease in birth rate. The effect of these density-dependent factors on population growth increases as the population density increases; that is, density-dependent factors affect a larger proportion, not just a larger number, of the population. Density-dependent factors can also affect population growth when population density declines, by decreasing the death rate and/or increasing the birth rate. Thus, density-dependent factors tend to regulate a population at a relatively constant size that is near the carrying capacity of the environment. (Keep in mind, however, that the carrying capacity of the environment frequently changes.) Density-dependent factors are an excellent example of a **negative feedback system** (Fig. 52-4).

Predation, disease, and competition are examples of density-dependent factors. As the density of a population increases, predators are more likely to find an individual of a given prey species. When population density is high, the members of a population encounter one another more frequently, and the chance of their transmitting parasites and infectious disease organisms

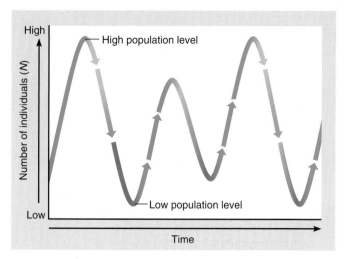

Density-dependent factors are increasingly severe: Population peaks and begins to decline.

Density-dependent factors are increasingly relaxed: Population bottoms out and begins to increase.

Figure 52-4 Density-dependent factors and negative feedback

When the number of individuals in a population increases, density-dependent factors cause a decline in the population. When the number of individuals in a population decreases, a relaxation of density-dependent factors allows the population to increase.

Figure 52-5 Lemming

The brown lemming (*Lemmus trimucronatus*) lives in the arctic tundra. Although lemming populations have been studied for decades, much about the cyclic nature of lemming population oscillations and their effects on the rest of the tundra ecosystem is still not well understood. This species ranges from Alaska eastward to the Hudson Bay.

increases. As population density increases, so does competition for resources such as living space, food, cover, water, minerals, and sunlight; eventually, the point may be reached at which many members of a population fail to obtain the minimum amount of whatever resource is in shortest supply. At higher population densities, density-dependent factors raise the death rate and/or lower the birth rate, inhibiting further population growth. The opposite effect occurs when the density of a population decreases. Predators are less likely to encounter individual prey, parasites and infectious diseases are less likely to be transmitted from one host to another, and competition among members of the population for resources such as living space and food declines.

Density-dependent factors may explain what makes certain populations fluctuate cyclically over time

Lemmings are small, stumpy-tailed rodents that are found in colder regions of the Northern Hemisphere (❙ Fig. 52-5). They are herbivores that feed on sedges and grasses in the arctic tundra. It has long been known that lemming populations have a 3- to 4-year cyclical oscillation that is often described as "boom or bust." That is, the population increases dramatically and then crashes; the population peaks may be 100 times as high as the low points in the population cycle. Many other populations, such as snowshoe hares and red grouse, also exhibit cyclic fluctuations.

What is the driving force behind these fluctuations? Several hypotheses have been proposed to explain the cyclical periodicity of lemming and other boom-or-bust populations; many

hypotheses involve density-dependent factors. One possibility is that as a prey population becomes more dense, it overwhelms its food supply; as a result, the population declines. In the lemming example, researchers have studied the shape of several lemming population curves during their oscillations; the data suggest that lemming populations crash because they overgraze the plants, not because predators eat them.

Another explanation is that the population density of predators, such as long-tailed jaegers (birds that eat lemmings), increases in response to the increasing density of prey. Few jaegers breed when the lemming population is low. However, when the lemming population is high, most jaegers breed, and the number of eggs per clutch is greater than usual. As more predators consume the abundant prey, the prey population declines. Later, with fewer prey in an area, the population of predators declines (some disperse out of the area, and fewer offspring are produced).

In 2003, researchers at the University of Helsinki, Finland, reported the results of a long-term study of collared lemmings in Greenland. They designed a model based on their data indicating that the 4-year fluctuations in lemming populations are not affected by availability of food or living space. Instead, predation by the stoat, a member of the weasel family that preys almost exclusively on lemmings, drives the population cycle. Three other lemming predators—the snowy owl, the arctic fox, and the long-tailed skua—stabilize the population cycle that the stoat establishes.

Parasites may also interact with their hosts to cause regular cyclic fluctuations. Detailed studies of red grouse have shown that even managed populations in wildlife preserves may have significant cyclic oscillations. Reproduction in red grouse is related to the density of parasitic nematodes (roundworms) living in adult intestines. Fewer birds breed successfully when adults are infected with worms; thus, a high density of worms leads to a population crash. Hypothesizing that red grouse populations fluctuate in response to the parasites, ecologists at the University of Stirling in Scotland recently reduced or eliminated population fluctuations in several red grouse populations. They accomplished this by catching and orally treating the birds with a chemical that causes worms to be ejected from their bodies.

Competition is an important density-dependent factor

Competition is an interaction among two or more individuals that attempt to use the same essential resource, such as food, water, sunlight, or living space, that is in limited supply. The use of the resource by one of the individuals reduces the availability of that resource for other individuals. Competition occurs both within a given population (**intraspecific competition**) and among populations of different species (**interspecific competition**). We consider the effects of intraspecific competition here; interspecific competition is discussed in Chapter 53.

Individuals of the same species compete for a resource in limited supply by interference competition or by exploitation competition. In **interference competition,** also called **contest competition,** certain dominant individuals obtain an adequate supply of the limited resource at the expense of other individuals in the population; that is, the dominant individuals actively interfere with other individuals' access to resources. In **exploitation competition,** also called **scramble competition,** all the individuals in a population "share" the limited resource more or less equally so that at high population densities none of them obtains an adequate amount. The populations of species in which exploitation competition operates often oscillate over time, and there is always a risk that the population size will drop to zero. In contrast, those species in which interference competition operates experience a relatively small drop in population size, caused by the death of individuals that are unable to compete successfully.

Intraspecific competition among red grouse involves interference competition. When red grouse populations are small, the birds are less aggressive, and most young birds establish a feeding territory (an area defended against other members of the same species). However, when the population is large, establishing a territory is difficult because there are more birds than there are territories, and the birds are much more aggressive. Those birds without territories often die from predation or starvation. Thus, birds with territories use a larger share of the limited resource (the territory with its associated food and cover), whereas birds without territories cannot compete successfully.

The moose population on Isle Royale, Michigan, the largest island in Lake Superior, provides a vivid example of exploitation competition that is similar to that of the reindeer population on the Pribilof Islands (discussed earlier). Isle Royale differs from most islands in that large mammals can walk to it when the lake freezes over in winter. The minimum distance to be walked is 24 km (15 mi), however, so this movement has happened infrequently. Around 1900, a small herd of moose wandered across the ice of frozen Lake Superior and reached the island for the first time. By 1934, the moose population on the island had increased to about 3000 and had consumed almost all the edible vegetation. In the absence of this food resource, there was massive starva-

tion in 1934. More than 60 years later, in 1996, a similar die-off claimed 80% of the moose after they had again increased to a high density. Thus, exploitation competition for scarce resources can result in dramatic population oscillations.

The effects of density-dependent factors are difficult to assess in nature

Most studies of density dependence have been conducted in laboratory settings where all density-dependent (and density-independent) factors except one are controlled experimentally. But populations in natural settings are exposed to a complex set of variables that continually change. As a result, in natural communities it is difficult to evaluate the relative effects of different density-dependent factors.

Ecologists from the University of California at Davis noted that few spiders occur on tropical islands inhabited by lizards, whereas more spiders and more species of spiders are found on lizard-free islands. Deciding to study these observations experimentally, David Spiller and Thomas Schoener staked out plots of vegetation (mainly seagrape shrubs) and enclosed some of them with lizard-proof screens (Fig. 52-6). Some of the plots were emptied of all lizards; each control enclosure had approximately nine lizards. Nine web-building spider species were observed in the enclosures. Spiders were counted approximately 30 times from 1989 to 1994. During the 4.5 years of observations reported here, spider population densities were higher in the lizard-free enclosures than in enclosures with lizards. Moreover, the enclosures without lizards had more species of spiders. Therefore, we might conclude that lizards control spider populations.

But even this relatively simple experiment may be explained by a combination of two density-dependent factors, predation (lizards eat spiders) and interspecific competition (lizards compete with spiders for insect prey—that is, both spiders and lizards eat insects). In this experiment, the effects of the two density-dependent factors in determining spider population size cannot be evaluated separately. Additional lines of evidence support the actions of both competition and predation at these sites.

Density-independent factors are generally abiotic

Any environmental factor that affects the size of a population but is not influenced by changes in population density is called a **density-independent factor.** Such factors are typically abiotic. Random weather events that reduce population size serve as density-independent factors. These often affect population density in unpredictable ways. A killing frost, severe blizzard, or hurricane, for example, may cause extreme and irregular reductions in a vulnerable population, regardless of its size, and thus may be considered largely density-independent.

Consider a density-independent factor that influences mosquito populations in arctic environments. These insects produce several generations per summer and achieve high population densities by the end of the season. A shortage of food does not seem to be a limiting factor for mosquitoes, nor is there any shortage

Key Experiment

QUESTION: What is the influence of density-dependent factors on population size in a natural habitat?

HYPOTHESIS: Lizards reduce the size of spider populations.

EXPERIMENT: Researchers conducted a field experiment in which they constructed enclosed plots (see photograph); spiders and insects could pass freely into and out of the enclosures, but the lizards could not. Some plots included lizards; the other plots were lizard-free. The spiders were periodically counted over a period of 4.5 years. An earlier experiment indicated that the enclosures themselves have no effect on the number of web-building spiders.

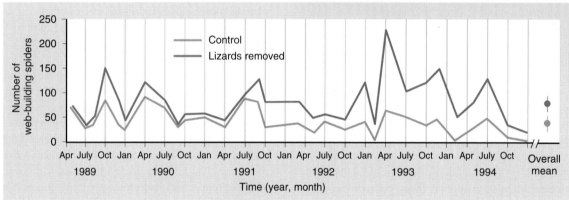

RESULTS AND CONCLUSION: The graph of the mean number of web-building spiders at each count per enclosure with lizards (*control, blue*) and per enclosure with lizards removed (*red*) demonstrates that the number of web-building spiders was consistently higher in the absence of lizards. Lizards may control spider populations by preying on them, by competing with them for insect prey, or by a combination of these factors.

Figure 52-6 Interaction of density-dependent factors

(Graph from D. A. Spiller and T. W. Schoener, "Lizards Reduce Spider Species Richness by Excluding Rare Species," *Ecology*, Vol. 79, No. 2, 1998. Copyright © 1998 Ecological Society of America. Reprinted with permission.)

of ponds in which to breed. What puts a stop to the skyrocketing mosquito population is winter. Not a single adult mosquito survives winter, and the entire population must grow afresh the next summer from the few eggs and hibernating larvae that survive. Thus, severe winter weather is a density-independent factor that affects arctic mosquito populations.

Density-independent and density-dependent factors are often interrelated. Social animals, for example, often resist dangerous weather conditions by collective behavior, as in the case of sheep huddling together in a snowstorm. In this case it appears that the greater the population density of the sheep, the

better their ability to resist the environmental stress of a density-independent event (such as a snowstorm).

Review

■ What are three examples of density-dependent factors that affect population growth?

■ What are three density-independent factors?

LIFE HISTORY TRAITS

Learning Objectives

5   Contrast semelparous and iteroparous reproduction.

6   Distinguish among species exhibiting an *r* strategy, those with a *K* strategy, and those that do not easily fit either category.

7   Describe Type I, Type II, and Type III survivorship curves, and explain how life tables and survivorship curves indicate mortality and survival.

Each species is uniquely suited to its lifestyle. Many years pass before a young magnolia tree flowers and produces seeds, whereas a poppy plant grows from seed, flowers, and dies in a single sea-

Figure 52-7 Semelparity

Agaves flower once and then die. The agave is a succulent with sword-shaped leaves arranged as a rosette around a short stem. Shown is *Agave shawii*, whose leaves grow to 61 cm (2 ft). Note the floral stalk, which can grow more than 3 m (10 ft) tall.

R. Gustafson/Visuals Unlimited

son. A mating pair of black-browed albatrosses produces a single chick every year, but a mating pair of gray-headed albatrosses produces a single chick biennially (every other year).

Species that expend their energy in a single, immense reproductive effort are said to be **semelparous.** Most insects and invertebrates, many plants, and some species of fish exhibit semelparity. Pacific salmon, for example, hatch in fresh water and swim to the ocean, where they live until they mature. Adult salmon swim from the ocean back into the same rivers or streams in which they hatched to spawn (reproduce). After they spawn, the salmon die.

Agaves are semelparous plants that are common in arid tropical and semitropical areas. The thick, fleshy leaves of the agave plant are crowded into a rosette at the base of the stem. Commonly called the century plant because it was mistakenly thought to flower only once in a century, agaves can flower after they are 10 years old or so, after which the entire plant dies (Fig. 52-7).

Many species are **iteroparous** and exhibit repeated reproductive cycles—that is, reproduction during several breeding seasons—throughout their lifetimes. Iteroparity is common in most vertebrates, perennial herbaceous plants, shrubs, and trees. The timing of reproduction, earlier or later in life, is a crucial aspect of iteroparity and involves trade-offs. On the one hand, reproducing earlier in life may mean a reduced likelihood of survival (because an individual is expending energy toward reproduction instead of its own growth), which reduces the potential for later reproduction. On the other hand, reproducing later in life means the individual has less time for additional reproductive events.

Ecologists try to understand the adaptive consequences of various **life history traits,** such as semelparity and iteroparity. Adaptations such as reproductive rate, age at maturity, and **fecundity** (potential capacity to produce offspring), all of which are a part of a species' life history traits, influence an organism's survival and reproduction. The ability of an individual to reproduce successfully, thereby making a genetic contribution to future generations of a population, is called its **fitness** (recall the discussion of fitness in Chapter 19).

Although many different life histories exist, some ecologists recognize two extremes: *r*-selected species and *K*-selected species. Keep in mind, as you read the following descriptions of *r* selection and *K* selection, that these concepts, although useful, oversimplify most life histories. Species tend to possess a combination of *r*-selected and *K*-selected traits, as well as traits that cannot be classified as either *r*-selected or *K*-selected. In addition, some populations within a species may exhibit the characteristics of *r* selection, whereas other populations in different environments may assume *K*-selected traits.

Populations described by the concept of *r* **selection** have traits that contribute to a high population growth rate. Recall that *r* designates the per capita growth rate. Because such organisms have a high *r*, biologists call them **r strategists** or **r-selected species.** Small body size, early maturity, short life span, large broods, and little or no parental care are typical of many *r* strategists, which are usually opportunists found in variable, temporary, or unpredictable environments where the probability of long-term survival is low. Some of the best examples of *r* strategists are insects, such as mosquitoes, and common weeds, such as the dandelion.

In populations described by the concept of *K* **selection,** traits maximize the chance of surviving in an environment where the number of individuals is near the carrying capacity (*K*) of the environment. These organisms, called **K strategists** or **K-selected species,** do not produce large numbers of offspring. They characteristically have long life spans with slow development, late reproduction, large body size, and a low reproductive rate. *K* strategists tend to be found in relatively constant or stable environments, where they have a high competitive ability. Redwood trees are classified as *K* strategists. Animals that are *K* strategists typically invest in parental care of their young. Tawny owls (*Strix aluco*), for example, are *K* strategists that pair-bond for life, with both members of a pair living and hunting in adjacent, well-defined territories. Their reproduction is regulated in accordance with the resources, especially the food supply, present in their territories. In an average year, 30% of the birds do not breed at all. If food supplies are more limited than initially indicated, many of

TABLE 52-2

Life Table for a Cohort of 530 Gray Squirrels (*Sciurus carolinensis*)

Age Interval (years)	Number Alive at Beginning of Age Interval	Proportion Alive at Beginning of Age Interval	Proportion Dying during Age Interval	Death Rate for Age Interval
0–1	530	1.000	0.747	0.747
1–2	134	0.253	0.147	0.581
2–3	56	0.106	0.032	0.302
3–4	39	0.074	0.031	0.418
4–5	23	0.043	0.021	0.488
5–6	12	0.022	0.013	0.591
6–7	5	0.009	0.006	0.666
7–8	2	0.003	0.003	1.000
8–9	0	0.000	0.000	—

Source: Adapted from R. L. Smith and T. M. Smith, *Elements of Ecology,* 4th ed., table 13.1, p. 150, Benjamin/Cummings Science Publishing, San Francisco, 1998.

those that do breed fail to incubate their eggs. Rarely do the owls lay the maximum number of eggs that they are physiologically capable of laying, and breeding is often delayed until late in the season, when the rodent populations on which they depend have become large. Thus, the behavior of tawny owls ensures better reproductive success of the individual and leads to a stable population at or near the carrying capacity of the environment. Starvation, an indication that the tawny owl population has exceeded the carrying capacity, rarely occurs.

Life tables and survivorship curves indicate mortality and survival

A **life table** can be constructed to show the mortality and survival data of a population or **cohort,** a group of individuals of the same age, at different times during their life span. Insurance companies were the first to use life tables, to calculate the relationship between a client's age and the likelihood of the client surviving to pay enough insurance premiums to cover the cost of the policy. Ecologists construct such tables for animals and plants, based on data that rely on a variety of population sampling methods and age determination techniques.

❚ Table 52-2 shows a life table for a cohort of 530 gray squirrels. The first two columns show the units of age (years) and the number of individuals in the cohort that were alive at the beginning of each age interval (the actual data collected in the field by the ecologist). The values for the third column (the proportion alive at the beginning of each age interval) are calculated by dividing each number in column 2 by 530, the number of squirrels in the original cohort. The values in the fourth column (the proportion dying during each age interval) are calculated using the values in the third column—by subtracting the number of survivors at the beginning of the next interval from those alive at the beginning of the current interval. For example, the proportion dying during interval 0–1 years = 1.000 − 0.253 = 0.747. The last column, the death rate for each age interval, is calculated by dividing the pro-

portion dying during the age interval (column 4) by the proportion alive at the beginning of the age interval (column 3). For example, the death rate for the age interval 1–2 years is 0.147 ÷ 0.253 = 0.581.

Survivorship is the probability that a given individual in a population or cohort will survive to a particular age. Plotting the logarithm (base 10) of the number of surviving individuals against age, from birth to the maximum age reached by any individual, produces a **survivorship curve.** ❚ Figure 52-8 shows the three main survivorship curves that ecologists recognize.

In Type I survivorship, as exemplified by bison and humans, the young and those at reproductive age have a high probability of surviving. The probability of survival decreases more rapidly with increasing age; mortality is concentrated later in life. ❚ Figure 52-9 shows a survivorship curve for a natural population of Drummond phlox, an annual native to East Texas that became widely distributed in the southeastern United States after it escaped from cultivation. Because most Drummond phlox seedlings survive to reproduce, after germination the plant exhibits a Type I survivorship that is typical of annuals.

In Type III survivorship, the probability of mortality is greatest early in life, and those individuals that avoid early death subsequently have a high probability of survival, that is, the probability of survival increases with increasing age. Type III survivorship is characteristic of oysters; young oysters have three free-swimming larval stages before settling down and secreting a shell. These larvae are vulnerable to predation, and few survive to adulthood.

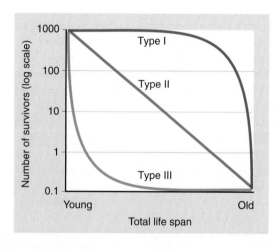

❚ **Figure 52-8** *Animated* Survivorship curves

These curves represent the ideal survivorships of species in which mortality is greatest in old age (Type I), spread evenly across all age groups (Type II), and greatest among the young (Type III). The survivorship of most organisms can be compared to these curves.

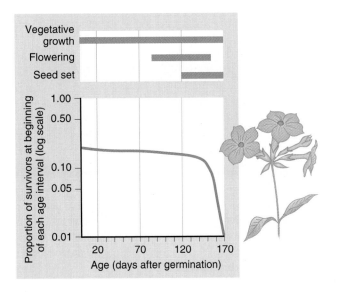

Figure 52-9 Survivorship curve for a Drummond phlox population

Drummond phlox has a Type I survivorship after germination of the seeds. Bars above the graph indicate the various stages in the Drummond phlox life history. Data were collected at Nixon, Texas, in 1974 and 1975. Survivorship on the *y*-axis begins at 0.296 instead of 1.00 because the study took into account death during the seed dormancy period prior to germination (*not shown*). (From W. J. Leverich and D. A. Levin, "Age-Specific Survivorship and Reproduction in *Phlox drummondii*," *American Naturalist* 113(6):1148, 1979. Copyright © 1979 University of Chicago Press. Reprinted with permission.)

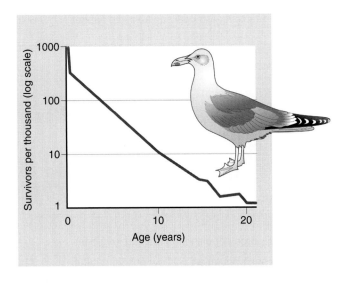

Figure 52-10 Survivorship curve for a herring gull population

Herring gulls (*Larus argentatus*) have Type III survivorship as chicks and Type II survivorship as adults. Data were collected from Kent Island, Maine, during a 5-year period in the 1930s; baby gulls were banded to establish identity. The very slight increase just prior to 20 years of age is due to sampling error. (R. A. Paynter Jr., "A New Attempt to Construct Lifetables for Kent Island Herring Gulls," *Bull. Mus. Comp. Zool.* 133(11):489–528, 1966.)

In Type II survivorship, which is intermediate between Types I and III, the probability of survival does not change with age. The probability of death is equally likely across all age groups, resulting in a linear decline in survivorship. This constancy probably results from essentially random events that cause death with little age bias. Although this relationship between age and survivorship is rare, some lizards have Type II survivorship.

The three survivorship curves are generalizations, and few populations exactly fit one of the three. Some species have one type of survivorship curve early in life and another type as adults. Herring gulls, for example, start out with a Type III survivorship curve but develop a Type II curve as adults (❚ Fig. 52-10). The survivorship curve shown in this figure is characteristic of birds in general. Note that most death occurs almost immediately after hatching, despite the protection and care given to the chicks by the parent bird. Herring gull chicks die from predation or attack by other herring gulls, inclement weather, infectious diseases, or starvation following death of the parent. Once the chicks become independent, their survivorship increases dramatically, and death occurs at about the same rate throughout their remaining lives. Few or no herring gulls die from the degenerative diseases of "old age" that cause death in most humans.

Review

- What are the advantages of semelparity? Of iteroparity? Are there disadvantages?
- Why is parental care of young a common characteristic of *K* strategists?
- Do all survivorship curves neatly fit the Type I, II, or III models? Explain.

METAPOPULATIONS

Learning Objective

8   Define *metapopulation,* and distinguish between source habitats and sink habitats.

The natural environment is a heterogeneous **landscape** consisting of interacting ecosystems that provide a variety of habitat patches. Landscapes, which are typically several to many square kilometers in area, cover larger land areas than individual ecosystems. Consider a forest, for example. The forest landscape is a mosaic of different elevations, temperatures, levels of precipitation, soil moisture, soil types, and other properties. Because each species has its own habitat requirements, this heterogeneity in physical properties is reflected in the different organisms that occupy the various patches in the landscape (❚ Fig. 52-11). Some species occur in very narrow habitat ranges, whereas others have wider habitat distributions.

Population ecologists have discovered that many species are not distributed as one large population across the landscape. Instead, many species exist as a series of local populations distributed in distinct habitat patches. Each local population has its own characteristic demographic features, such as birth, death, emigration, and immigration rates. A population that is divided into

(a) This early spring view of Deep Creek Valley in the Great Smoky Mountains National Park gives some idea of the heterogeneity of the landscape. During the summer, when all the vegetation is a deep green, the landscape appears homogeneous.

Figure 52-11 The mosaic nature of landscapes

(Adapted from R. H. Whittaker, "Vegetation of the Great Smoky Mountains," *Ecological Monographs*, Vol. 26, 1956.)

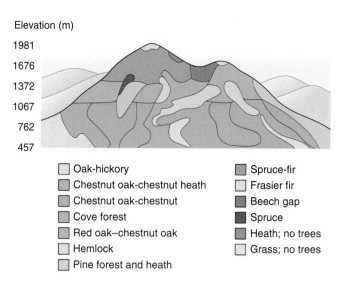

Elevation (m)

1981
1676
1372
1067
762
457

☐ Oak-hickory
☐ Chestnut oak-chestnut heath
☐ Chestnut oak-chestnut
☐ Cove forest
☐ Red oak–chestnut oak
☐ Hemlock
☐ Pine forest and heath

☐ Spruce-fir
☐ Frasier fir
☐ Beech gap
☐ Spruce
☐ Heath; no trees
☐ Grass; no trees

(b) An evaluation of the distribution of vegetation on a typical west-facing slope in the Great Smoky Mountains National Park reveals that the landscape consists of patches. Chestnut oak is a species of oak (*Quercus prinus*); chestnut heath is an area within chestnut oak forest where the trees are widely scattered and the slopes underneath are covered by a thick growth of laurel (*Kalmia*) shrubs; cove forest is a mixed stand of deciduous trees.

several local populations among which individuals occasionally disperse (emigrate and immigrate) is known as a **metapopulation.** For example, note the various local populations of red oak on the mountain slope in Figure 52-11b.

The spatial distribution of a species occurs because different habitats vary in suitability, from unacceptable to preferred. The preferred sites are more productive habitats that increase the likelihood of survival and reproductive success for the individuals living there. Good habitats, called **source habitats,** are areas where local reproductive success is greater than local mortality. **Source populations** generally have greater population densities than populations at less suitable sites, and surplus individuals in the source habitat disperse and find another habitat in which to settle and reproduce.

Individuals living in lower-quality habitats may suffer death or, if they survive, poor reproductive success. Lower-quality habitats, called **sink habitats,** are areas where local reproductive success is less than local mortality. Without immigration from other areas, a **sink population** declines until extinction occurs. If a local population becomes extinct, individuals from a source habitat may recolonize the vacant habitat at a later time. Source and sink habitats, then, are linked to one another by dispersal (❚ Fig. 52-12).

Metapopulations are becoming more common as humans alter the landscape by fragmenting existing habitats to accommodate homes and factories, agricultural fields, and logging. As a result, the concept of metapopulations, particularly as it relates to endangered and threatened species, has become an important area of study in conservation biology.

Review

❚ What is a metapopulation?

❚ How do you distinguish between a source habitat and a sink habitat?

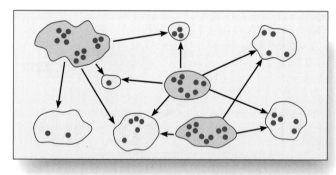

⬤ Source population in a suitable habitat
◯ Sink population in a low-quality habitat
• Individual within a local population
→ Dispersal event

Figure 52-12 Source and sink populations in a hypothetical metapopulation

The local populations in the habitat patches shown here collectively make up the metapopulation. Source habitats provide individuals that emigrate and colonize sink habitats. Studied over time, the metapopulation is shown to exist as a shifting pattern of occupied and vacant habitat patches.

HUMAN POPULATIONS

Learning Objectives

9 Summarize the history of human population growth.

10 Explain how highly developed and developing countries differ in population characteristics such as infant mortality rate, total fertility rate, replacement-level fertility, and age structure.

11 Distinguish between people overpopulation and consumption overpopulation.

Now that we have examined some of the basic concepts of population ecology, we can apply those concepts to the human population. Examine ▌Figure 52-13, which shows the world increase in the human population since the development of agriculture approximately 10,000 years ago. Now look back at Figure 52-2, and compare the two curves. The characteristic J curve of exponential population growth shown in Figure 52-13 reflects the decreasing amount of time it has taken to add each additional billion people to our numbers. It took thousands of years for the human population to reach 1 billion, a milestone that took place around 1800. It took 130 years to reach 2 billion (in 1930), 30 years to reach 3 billion (in 1960), 15 years to reach 4 billion (in 1975), 12 years to reach 5 billion (in 1987), and 12 years to reach 6 billion (in 1999). The United Nations projects that the human population will reach about 7 billion by 2013.

Thomas Malthus (1766–1834), a British clergyman and economist, was one of the first to recognize that the human population cannot continue to increase indefinitely (see Chapter 18). He pointed out that human population growth is not always desirable (a view contrary to the beliefs of his day and to those of many people even today) and that the human population is capable of increasing faster than the food supply. He maintained that the inevitable consequences of population growth are famine, disease, and war.

The world population is currently increasing by about 80 million people per year. It reached 6.5 billion in mid-2006. This change was not caused by an increase in the birth rate (*b*). In fact, the world birth rate has actually declined during the past 200 years. The increase in population is due instead to a dramatic decrease in the death rate (*d*), which has occurred primarily because greater food production, better medical care, and improved sanitation practices have increased the life expectancies of a great majority of the global population. For example, from 1920 to 2000, the death rate in Mexico fell from approximately 40 per 1000 individuals to 4 per 1000, whereas the birth rate dropped from approximately 40 per 1000 individuals to 24 per 1000 (▌Fig. 52-14).

The human population has reached a turning point. Although our numbers continue to increase, the world per capita growth rate (*r*) has declined over the past several years, from a peak of 2.2% per year in the mid-1960s to 1.2% per year in 2006. Population experts at the United Nations and the World Bank have projected that the growth rate will continue to slowly decrease until zero population growth is attained. Thus, exponential growth of the human population may end, and if it does, the S curve will replace the J curve. Demographers project that **zero population**

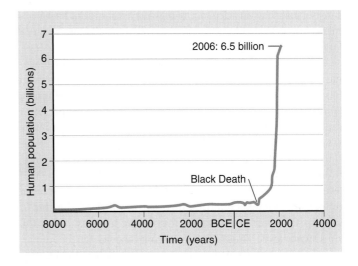

Figure 52-13 Human population growth

During the last 1000 years, the human population has been increasing nearly exponentially. Population experts predict that the population will level out during the 21st century, forming the S curve observed in other species. (Black Death refers to a devastating disease, probably bubonic plague, that decimated Europe and Asia in the 14th century.)

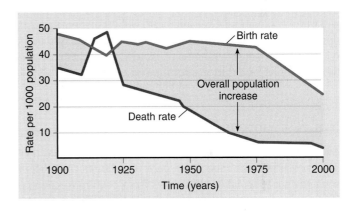

Figure 52-14 *Animated* Birth and death rates in Mexico, 1900 to 2000

Both birth and death rates declined during the 20th century, but because the death rate declined much more than the birth rate, Mexico has experienced a high growth rate. (The high death rate prior to 1920 was caused by the Mexican Revolution.) (Population Reference Bureau)

growth, the point at which the birth rate equals the death rate (*r* = 0), will occur toward the end of the 21st century.

The latest (2006) projections available through the Population Reference Bureau, based on data from the United Nations, forecast that the human population will be almost 9.3 billion in the year 2050. This is a "medium" projection; such population projections are "what if" exercises. Given certain assumptions about future tendencies in natality, mortality, and dispersal, population experts can calculate an area's population for a given number of years into the future. Population projections indicate the changes that may occur, but they must be interpreted with care because they vary depending on what assumptions have been made. Small differences in fertility, as well as death rates, produce large differences in population forecasts.

The main unknown factor in any population growth scenario is Earth's carrying capacity. According to Joel Cohen, who has a joint professorship at Rockefeller University and Columbia University, most published estimates of how many people Earth can support range from 4 billion to 16 billion. These estimates vary so widely because of the assumptions that are made about standard of living, resource consumption, technological innovations, and waste generation. If we want all people to have a high level of material well-being equivalent to the lifestyles common in highly developed countries, then Earth will clearly be able to support far fewer humans than if everyone lives just above the subsistence level. Thus, Earth's carrying capacity for the human population is not decided simply by natural environmental constraints. Human choices and values have to be factored into the assessment.

It is also not clear what will happen to the human population if or when the carrying capacity is approached. Optimists suggest that the human population will stabilize because of a decrease in the birth rate. Some experts take a more pessimistic view and predict that the widespread degradation of our environment caused by our ever-expanding numbers will make Earth uninhabitable for humans and other species. These experts contend that a massive wave of human suffering and death will occur. Some experts think that the human population has already exceeded the carrying capacity of the environment, a potentially dangerous situation that threatens our long-term survival as a species.

Not all countries have the same growth rate

Although world population figures illustrate overall trends, they do not describe other important aspects of the human population story, such as population differences from country to country. **Human demographics,** the science that deals with human population statistics such as size, density, and distribution, provides information on the populations of various countries. As you probably know, not all parts of the world have the same rates of population increase. Countries can be classified into two groups, highly developed and developing, based on their rates of population growth, degrees of industrialization, and relative prosperity (❚ Table 52-3).

Highly developed countries, such as the United States, Canada, France, Germany, Sweden, Australia, and Japan, have low rates of population growth and are highly industrialized relative to the rest of the world. Highly developed countries have the lowest birth rates in the world. Indeed, some highly developed countries such as Germany have birth rates just below that needed to sustain the population and are thus declining slightly in numbers. Highly developed countries also have low **infant mortality rates** (the number of infant deaths per 1000 live births). The infant mortality rate of the United States was 6.7 in 2006, for example, compared with a world infant mortality rate of 52.

Highly developed countries also have longer life expectancies (78 years at birth in the United States versus 67 years worldwide) and higher average GNI PPP per capita ($41,950 in the United States versus $9190 worldwide). GNI PPP per capita is gross national income (GNI) in purchasing power parity (PPP) divided by midyear population. It indicates the amount of goods and services an average citizen of that particular region or country could buy in the United States.

Developing countries fall into two subcategories: moderately developed and less developed. Mexico, Turkey, Thailand, and most countries of South America are examples of *moderately developed countries.* Their birth rates and infant mortality rates are higher than those of highly developed countries, but the rates are declining. Moderately developed countries have a medium level of industrialization, and their average GNI PPP per capita is lower than those of highly developed countries.

Bangladesh, Niger, Ethiopia, Laos, and Cambodia are examples of *less developed countries.* These countries have the highest birth rates, the highest infant mortality rates, the lowest life expectancies, and the lowest average GNI PPP per capita in the world.

One way to represent the population growth of a country is to determine the **doubling time,** the amount of time it would take for its population to double in size, assuming its current growth rate did not change. A simplified formula for doubling time (t_d) is

---

TABLE 52-3

Comparison of 2006 Population Data in Developed and Developing Countries

	Developed	Developing	
	(Highly Developed) United States	(Moderately Developed) Brazil	(Less Developed) Ethiopia
Fertility rate	2.0	2.3	5.4
Doubling time	116 years	50 years	29 years
Infant mortality rate	6.7 per 1000	27 per 1000	77 per 1000
Life expectancy at birth	78 years	72 years	49 years
GNI PPP per capita (U.S. $)	$41,950	$8,230	$1,000
Women using modern contraception	68%	70%	14%

Source: Population Reference Bureau.

TABLE 52-4

Fertility Changes in Selected Developing Countries

Country	Total Fertility Rate	
	1960–1965	2006
Bangladesh	6.7	3.0
Brazil	6.2	2.3
China	5.9	1.6
Egypt	7.1	3.1
Guatemala	6.9	4.4
India	5.8	2.9
Kenya	8.1	4.9
Mexico	6.8	2.4
Nepal	5.9	3.7
Nigeria	6.9	5.9
Thailand	6.4	1.7

Source: Population Reference Bureau.

$t_d = 70 \div r$. (The actual formula involves calculus and is beyond the scope of this text.)

A look at a country's doubling time identifies it as a highly, moderately, or less developed country: the shorter the doubling time, the less developed the country. At rates of growth in 2006, the doubling time is 29 years for Ethiopia, 30 years for Laos, 54 years for Turkey, 100 years for Thailand, 117 years for the United States, and 175 years for France.

It is also instructive to examine **replacement-level fertility,** the number of children a couple must produce to "replace" themselves. Replacement-level fertility is usually given as 2.1 children in highly developed countries and 2.7 children in developing countries. The number is always greater than 2.0 because some children die before they reach reproductive age. Higher infant mortality rates are the main reason that replacement levels in developing countries are greater than in highly developed countries. The **total fertility rate**—the average number of children born to a woman during her lifetime—is 2.7 worldwide, which is well above replacement levels. However, it has fallen below replacement levels, to 1.6, in the highly developed countries. The total fertility rate in the United States is higher than in other industrialized countries—2.0.

The population in many developing countries is beginning to approach stabilization. The fertility rate must decline for the population to stabilize (**Table 52-4;** note the general decline in total fertility rate from the 1960s to 2006 in selected developing countries). The total fertility rate in develop-

ing countries has decreased from an average of 6.1 children per woman in 1970 to 3.4 in 2006 (excluding China), or 2.9 if China is included. Fertility is highly variable among developing countries, and although the fertility rates in most of these countries have declined, it should be remembered that most still exceed replacement-level fertility. Consequently, the populations in these countries are still increasing. Also, even when fertility rates equal replacement-level fertility, population growth will still continue for some time. To understand why this is so, we now examine the age structure of various countries.

The age structure of a country helps predict future population growth

To predict the future growth of a population, it is important to know its **age structure,** which is the number and proportion of people at each age in a population. The number of males and number of females at each age, from birth to death, are represented in an **age structure diagram.**

The overall shape of an age structure diagram indicates whether the population is increasing, stationary, or shrinking (**Fig. 52-15**). The age structure diagram for less developed countries is shaped like a pyramid. Because the largest percentage of the population is in the prereproductive age group (that is, 0 to 14 years of age), the probability of future population growth is great. A strong **population growth momentum** exists because

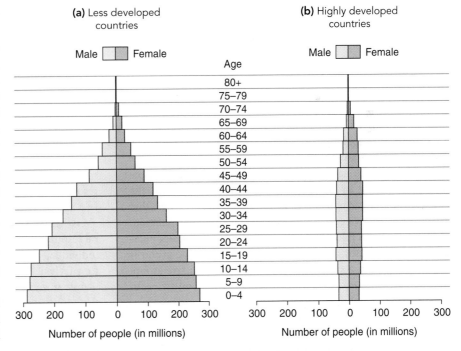

Figure 52-15 *Animated* Age structure diagrams

These age structure diagrams for **(a)** less developed countries and **(b)** highly developed countries indicate that less developed regions have a greater percentage of young people than do highly developed countries. As a result, less developed countries are projected to have greater population growth than are highly developed countries. (Adapted from Population Reference Bureau using data from United Nations, World Population Prospects: The 2002 Revision [*medium scenario*], 2003.)

(a) The rapidly increasing number of people in developing countries overwhelms their natural resources, even though individual resource requirements may be low. Shown is a typical Indian family, from Ahraura Village, with all their possessions.

(b) People in highly developed countries consume a disproportionate share of natural resources. Shown is a typical American family from Pearland, Texas, with all their possessions.

Figure 52-16   People and natural resources

when all these children mature, they will become the parents of the next generation, and this group of parents will be larger than the previous group. Thus, even if the fertility rates in these countries decline to replacement level (that is, couples have smaller families than their parents did), the population will continue to grow for some time. Population growth momentum can have either a positive value (that is, the population will grow) or a negative value (that is, the population will decline). However, it is usually discussed in a positive context, to explain how the future growth of a population is affected by its present age distribution.

In contrast, the more tapered bases of the age structure diagrams of highly developed countries with slowly growing, stable, or declining populations indicate that a smaller proportion of the population will become the parents of the next generation. The age structure diagram of a stable population, one that is neither growing nor shrinking, demonstrates that the number of people at prereproductive and reproductive ages are approximately the same. Also, a larger percentage of the population is older, that is, postreproductive, than in a rapidly increasing population. Many countries in Europe have stable populations. In a population that is shrinking in size, the prereproductive age group is smaller than either the reproductive or postreproductive group. Germany, Russia, and Bulgaria are examples of countries with slowly shrinking populations.

Worldwide, 29% of the human population is younger than age 15. When these people enter their reproductive years, they have the potential to cause a large increase in the growth rate. Even if the birth rate does not increase, the growth rate will increase simply because more people are reproducing.

Most of the world population increase since 1950 has taken place in developing countries as a result of the younger age structure and the higher-than-replacement-level fertility rates of their populations. In 1950, 66.8% of the world's population was in the developing countries in Africa, Asia (minus Japan), and Latin America. Between 1950 and 2005, the world's population more than doubled in size, but most of that growth occurred in developing countries. As a reflection of this, in 2006 the people in developing countries had increased to about 81% of the world's population. Most of the population increase that will occur during the 21st century will also take place in developing countries, largely the result of their young age structures. These countries, most of which are poor, are least able to support such growth.

Environmental degradation is related to population growth and resource consumption

The relationships among population growth, use of natural resources, and environmental degradation are complex, but we can make two useful generalizations. First, although the amount of resources essential to an individual's survival may be small, a rapidly increasing population (as in developing countries) tends to overwhelm and deplete a country's soils, forests, and other natural resources (❙ Fig. 52-16a). Second, in highly developed nations individual resource demands are large, far above requirements for survival. To satisfy their desires rather than their basic needs, people in more affluent nations exhaust resources and degrade the global environment through excessive consumption and "throwaway" lifestyles (❙ Fig 52-16b).

Rapid population growth can cause natural resources to be overexploited. For example, large numbers of poor people must grow crops on land that is inappropriate for farming, such as mountain slopes or some tropical rain forests. Although this practice may provide a short-term solution to the need for food, it does not work in the long run, because when these lands are cleared for farming, their agricultural productivity declines rapidly and severe environmental deterioration occurs.

The effects of population growth on natural resources are particularly critical in developing countries. The economic growth of developing countries is often tied to the exploitation of natural resources, often for export to highly developed countries. Developing countries are faced with the difficult choice of exploiting natural resources to provide for their expanding populations in the short term (to pay for food or to cover debts) or conserving those resources for future generations. It is instructive to note that the economic growth and development of the United States and of other highly developed nations came about through the exploitation and, in some cases, the destruction of their natural resources.

Although resource issues are clearly related to population size (more people use more resources), an equally, if not more, important factor is a population's resource consumption. People in highly developed countries are extravagant consumers—their use of resources is greatly out of proportion to their numbers. A single child born in a highly developed country such as the United States causes a greater impact on the environment and on resource depletion than a dozen or more children born in a developing country. Many natural resources are needed to provide the automobiles, air conditioners, disposable diapers, cell phones, DVDs, computers, clothes, athletic shoes, furniture, boats, and other "comforts" of life in highly developed nations. Thus, the disproportionately large consumption of resources by the United States and other highly developed countries affects natural resources and the environment as much as or more than the population explosion in the developing world does.

A country is overpopulated if the level of demand on its resource base results in damage to the environment. If we compare human impact on the environment in developing and highly developed countries, we see that a country can be overpopulated in two ways. **People overpopulation** occurs when the environment is worsening from too many people, even if those people consume few resources per person. People overpopulation is the current problem in many developing nations.

In contrast, **consumption overpopulation** occurs when each individual in a population consumes too large a share of resources. The effect of consumption overpopulation on the environment is the same as that of people overpopulation—pollution and degradation of the environment. Most affluent highly developed nations, including the United States, suffer from consumption overpopulation. Highly developed nations represent less than 20% of the world's population, yet they consume significantly more than half its resources. According to the Worldwatch Institute, a private research institution in Washington, D.C., highly developed nations account for the lion's share of total resources consumed, as follows: 86% of aluminum used, 76% of timber harvested, 68% of energy produced, 61% of meat eaten, and 42% of fresh water consumed. These nations also generate 75% of the world's pollution and waste.

Review

- What is replacement-level fertility? Why is it greater in developing countries than in highly developed countries?
- How can a single child born in the United States have a greater effect on the environment and natural resources than a dozen children born in Kenya?

SUMMARY WITH KEY TERMS

Learning Objectives

1 Define *population density* and *dispersion,* and describe the main types of population dispersion (page 1127).

- **Population density** is the number of individuals of a species per unit of area or volume at a given time.
- Population **dispersion** (spacing) may be **random dispersion** (unpredictably spaced), **clumped dispersion** (clustered in specific parts of the habitat), or **uniform dispersion** (evenly spaced).

2 Explain the four factors (natality, mortality, immigration, and emigration) that produce changes in population size, and solve simple problems involving these changes (page 1129).

- Population size is affected by the average per capita birth rate (*b*), average per capita death rate (*d*), and two measures of dispersal: average per capita **immigration** rate (*i*), and average per capita **emigration** rate (*e*).
- The **growth rate (r)** is the rate of change (increase or decrease) of a population on a per capita basis. On a global scale (when dispersal is not a factor), $r = b - d$. Populations increase in size as long as the average per capita birth rate (**natality**) is greater than the average per capita death rate (**mortality**).
- For a local population (where dispersal is a factor), $r = (b - d) + (i - e)$.

3 Define *intrinsic rate of increase* and *carrying capacity,* and explain the differences between J-shaped and S-shaped growth curves (page 1129).

- **Intrinsic rate of increase (r_{max})** is the maximum rate at which a species or population could increase in number under ideal conditions.
- Although certain populations exhibit an accelerated pattern of growth known as **exponential population growth** for limited periods of time (the J-shaped curve), eventually the growth rate decreases to around zero or becomes negative.
- Population size is modified by limits set by the environment. The **carrying capacity (K)** of the environment is the largest population that can be maintained for an indefinite time by a particular environment. **Logistic population growth,** when graphed, shows a characteristic S-shaped curve. Seldom do natural populations follow the logistic growth curve very closely.

ThomsonNOW™ **Learn more about population growth curves by clicking on the figures in ThomsonNOW.**

4 Contrast the influences of density-dependent and density-independent factors on population size, and give examples of each (page 1131).

- **Density-dependent factors** regulate population growth by affecting a larger proportion of the population as population density rises. Predation, disease, and competition are examples.

- **Density-independent factors** limit population growth but are not influenced by changes in population density. Hurricanes and blizzards are examples.

5 Contrast semelparous and iteroparous reproduction (page 1134).

- **Semelparous** species expend their energy in a single, immense reproductive effort. **Iteroparous** species exhibit repeated reproductive cycles throughout their lifetimes.

6 Distinguish among species exhibiting an *r* strategy, those with a *K* strategy, and those that do not easily fit either category (page 1134).

- Although many different combinations of **life history traits** exist, some ecologists recognize two extremes: *r* strategy and *K* strategy.

- An *r* strategy emphasizes a high growth rate. Organisms characterized as **r strategists** often have small body sizes, high reproductive rates, and short **life spans,** and they typically inhabit variable environments.

- A *K* strategy maintains a population near the carrying capacity of the environment. Organisms characterized as **K strategists** often have large body sizes, low reproductive rates, and long life spans, and they typically inhabit stable environments.

- The two strategies oversimplify most life histories. Many species combine *r*-selected and *K*-selected traits, as well as traits that cannot be classified as either *r*-selected or *K*-selected.

7 Describe Type I, Type II, and Type III survivorship curves, and explain how life tables and survivorship curves indicate mortality and survival (page 1134).

- A **life table** shows the mortality and survival data of a population or **cohort**—a group of individuals of the same age—at different times during their life span.

- **Survivorship** is the probability that a given individual in a population or cohort will survive to a particular age. There are three general **survivorship curves:** Type I survivorship, in which mortality is greatest in old age; Type II survivorship, in which mortality is spread evenly across all age groups; and Type III survivorship, in which mortality is greatest among the young.

ThomsonNOW™ **Explore survivorship curves by clicking on the figure in ThomsonNOW.**

8 Define *metapopulation,* and distinguish between source and sink habitats (page 1137).

- Many species exist as a **metapopulation,** a set of local populations among which individuals are distributed in distinct habitat patches across a **landscape,** which is a large area of terrain (several to many square kilometers) composed of interacting ecosystems.

- Within a metapopulation, individuals occasionally disperse from one local habitat to another, by emigration and

immigration. **Source habitats** are preferred sites where local reproductive success is greater than local mortality. Surplus individuals disperse from source habitats. **Sink habitats** are lower-quality habitats where individuals may suffer death or, if they survive, poor reproductive success. If extinction of a local **sink population** occurs, individuals from a **source population** may recolonize the vacant habitat at a later time.

9 Summarize the history of human population growth (page 1139).

- The world population increases by about 80 million people per year and reached 6.5 billion in mid-2006.

- Although our numbers continue to increase, the per capita growth rate (*r*) has declined over the past several years, from a peak in 1965 of about 2% per year to a 2006 growth rate of 1.2% per year.

- Scientists who study **human demographics** (human population statistics) project that the world population will become stationary (*r* = 0, or **zero population growth**) by the end of the 21st century.

ThomsonNOW™ **Explore human population and the future by clicking on the figures in ThomsonNOW.**

10 Explain how highly developed and developing countries differ in population characteristics such as infant mortality rate, total fertility rate, replacement-level fertility, and age structure (page 1139).

- **Highly developed countries** have the lowest birth rates, lowest **infant mortality rates,** lowest **total fertility rates,** longest life expectancies, and highest GNI PPP per capita (a measure of the amount of goods and services the average citizen could purchase in the United States). **Developing countries** have the highest birth rates, highest infant mortality rates, highest total fertility rates, shortest life expectancies, and lowest GNI PPP per capita.

- The **age structure** of a population greatly influences population dynamics. It is possible for a country to have **replacement-level fertility** and still experience population growth if the largest percentage of the population is in the prereproductive years. A young age structure causes a positive **population growth momentum** as the very large prereproductive age group matures and becomes parents.

11 Distinguish between people overpopulation and consumption overpopulation (page 1139).

- Developing countries tend to have **people overpopulation,** in which population increase degrades the environment even though each individual uses few resources.

- Highly developed countries tend to have **consumption overpopulation,** in which each individual in a slow-growing or stationary population consumes a large share of resources, which results in environmental degradation.

TEST YOUR UNDERSTANDING

1. Population _____ is the number of individuals of a species per unit of habitat area or volume at a given time. (a) dispersion (b) density (c) survivorship (d) age structure (e) demographics

2. Which of the following patterns of cars parked along a street is an example of uniform dispersion? (a) five cars parked next to one another in the middle, leaving two empty spaces at one end and three empty spaces at the other end (b) five cars

parked in this pattern: car, empty space, car, empty space, and so on (c) five cars parked in no discernible pattern, sometimes having empty spaces on each side and sometimes parked next to another car

3. This type of dispersion occurs when individuals are concentrated in certain parts of the habitat. (a) clumped (b) uniform (c) random (d) a and b (e) b and c

4. The average per capita birth rate, or _____, increases population size, whereas the average per capita death rate, or _____, decreases population size. (a) natality; demography (b) exploitation competition; interference competition (c) mortality; natality (d) total fertility rate; mortality (e) natality; mortality

5. The per capita growth rate of a population where dispersal is not a factor is expressed as (a) $i + e$ (b) $b - d$ (c) dN/dt (d) $rN(K - N)$ (e) $(K - N) \div K$

6. The maximum rate at which a population could increase under ideal conditions is known as its (a) total fertility rate (b) survivorship (c) intrinsic rate of increase (d) doubling time (e) age structure

7. When r is a positive number, the population size is (a) stable (b) increasing (c) decreasing (d) either increasing or decreasing, depending on interference competition (e) either increasing or stable, depending on whether the species is semelparous

8. In a graph of population size versus time, a J-shaped curve is characteristic of (a) exponential population growth (b) logistic population growth (c) zero population growth (d) replacement-level fertility (e) population growth momentum

9. The largest population that can be maintained by a particular environment for an indefinite period is known as a (a) semelparous population (b) population undergoing exponential growth (c) metapopulation (d) population's carrying capacity (e) source population

10. Giant bamboos live many years without reproducing, then send up a huge flowering stalk and die shortly thereafter. Giant bamboo is therefore an example of (a) iteroparity (b) a source population (c) a metapopulation (d) an r strategist (e) semelparity

11. Predation, disease, and competition are examples of _____ factors. (a) density-dependent (b) density-independent (c) survivorship (d) dispersal (e) semelparous

12. _____ competition occurs within a population, and _____ competition occurs among populations of different species. (a) interspecific; intraspecific (b) intraspecific; interspecific (c) Type I survivorship; Type II survivorship (d) interference; exploitation (e) exploitation; interference

13. Population experts project that during the 21st century, the human population will (a) increase the most in highly developed countries (b) increase the most in developing countries (c) increase at similar rates in all countries (d) decrease in developing countries but stabilize in highly developed countries (e) increase and then decrease dramatically in all countries

14. A highly developed country has a (a) long doubling time (b) low infant mortality rate (c) high GNI PPP per capita (d) a and b (e) a, b, and c

15. The continued growth of a population with a young age structure, even after its fertility rate has declined, is known as (a) population doubling (b) iteroparity (c) population growth momentum (d) r selection (e) density dependence

CRITICAL THINKING

1. How might pigs at a trough be an example of exploitation competition? Of interference competition?

2. Explain why the population size of a species that competes by interference competition is often near the carrying capacity, whereas the population size of a species that competes by exploitation competition is often greater than or below the carrying capacity.

3. A female elephant bears a single offspring every 2 to 4 years. Based on this information, which survivorship curve do you think is representative of elephants? Explain your answer.

4. **Evolution Link.** In developing his theory of evolution by natural selection, Charles Darwin considered four main observations: heritable variation among individuals in a population, overproduction of offspring, limits on population growth, and differential reproductive success. Compare how these observations may apply to populations that tend to fit the concepts of r selection and K selection.

5. **Analyzing Data.** In Bolivia, 39% of the population is younger than age 15, and 4% is older than 65. In Austria, 16% of the population is younger than 15, and 16% is older than 65. Which country will have the highest growth rate over the next two decades? Why?

6. **Analyzing Data.** The 2006 population of the Netherlands was 16.4 million, and its land area is 15,768 square miles. The 2006 population of the United States was 299.1 million, and its land area is 3,717,796 square miles. Which country has the greater population density?

7. **Analyzing Data.** The population of India in 2006 was 1121.8 million, and its growth rate was 1.7% per year. Calculate the 2007 population of India.

8. **Analyzing Data.** The world population in 2006 was 6.5 billion, and its annual growth rate was 1.2%. If the birth rate was 21 per 1000 people in the year 2006, what was the death rate, expressed as number per 1000 people?

Additional questions are available in ThomsonNOW at www.thomsonedu.com/login

Community Ecology

Michael P. Gadomski/Photo Researchers, Inc.

Community in a rotting log. Fallen logs, called "nurse logs," shelter plants and other organisms and enrich the soil as they decay. Photographed in Pennsylvania.

In Chapter 52 we examined the dynamics of populations. In the natural world, most populations are part of a **community,** which consists of an association of populations of different species that live and interact in the same place at the same time. The definition for *community* is deliberately broad because it refers to ecological categories that vary greatly in size, lack precise boundaries, and are rarely completely isolated.

Smaller communities nest within larger communities. A forest is a community, but so is a rotting log in that forest (see photograph). Insects, plants, and fungi invade a fallen tree as it undergoes decay. Termites and other wood-boring insects forge tunnels through the bark and wood. Other insects, plant roots, and fungi follow and enlarge these openings. Mosses and lichens on the log's surface trap rainwater and extract minerals (inorganic nutrients), and fungi and bacteria speed decay, which provides nutrients for other inhabitants. As decay progresses, small mammals burrow into the wood and eat the fungi, insects, and plants.

Organisms exist in an abiotic (nonliving) environment that is as essential to their lives as their interactions with one another. Minerals, air, water, and sunlight are just as much a part of a honeybee's environment, for example, as the flowers it pollinates and from which it takes nectar. A biological community and its abiotic environment together compose an **ecosystem.** Like communities, ecosystems are broad entities that refer to ecological units of various sizes.

We now present some of the challenges in trying to find common patterns and processes that govern communities. The living community is emphasized in this chapter, and the abiotic components of ecosystems, including energy flow and trophic structure (feeding relationships), nutrient cycling, and climate, are considered in Chapter 54. ■

KEY CONCEPTS

A community consists of populations that live in the same place at the same time.

Community ecologists focus on questions concerning the number of species, the relative abundance of each species, interactions among species, and resistance of the community to environmental disturbances.

Most ecologists think that communities are loose associations of organisms that have similar environmental requirements and therefore live together in the same environment.

Following a disturbance, communities undergo succession from pioneer communities to mature communities.

Field experiments with plants beginning in the 1990s suggest that more diverse communities have a greater stability, or resistance to change.

COMMUNITY STRUCTURE AND FUNCTIONING

Learning Objectives

1. Define *ecological niche,* and distinguish between an organism's fundamental niche and its realized niche.
2. Define *competition,* and distinguish between interspecific and intraspecific competition.
3. Summarize the concepts of the competitive exclusion principle, resource partitioning, and character displacement.
4. Define *predation,* and describe the effects of natural selection on predator–prey relationships.
5. Distinguish among mutualism, commensalism, and parasitism, and give examples of each.
6. Distinguish between keystone species and dominant species.

Communities exhibit characteristic properties that populations lack. These properties, known collectively as *community structure* and *community functioning,* include the number and types of species present, the relative abundance of each species, the interactions among different species, community resilience to disturbances, energy and nutrient flow throughout the community, and productivity. **Community ecology** is the description and analysis of patterns and processes within the community. Finding common patterns and processes in a wide variety of communities—for example, a pond community, a pine forest community, and a sagebrush desert community—helps ecologists understand community structure and functioning.

Communities are exceedingly difficult to study, because a large number of individuals of many different species interact with one another and are interdependent in a variety of ways. Species compete with one another for food, water, living space, and other resources. (Used in this context, a *resource* is anything from the environment that meets needs of a particular species.) Some organisms kill and eat other organisms. Some species form intimate associations with one another, whereas other species seem only distantly connected. Certain species interact in positive ways, in a process known as **facilitation,** which modifies and enhances the local environment for other species. For example, alpine plants in harsh mountain environments grow faster and larger and reproduce more successfully when other plants are growing nearby. Also, each organism plays one of three main roles in community life: producer, consumer, or decomposer. Unraveling the many positive and negative, direct and indirect interactions of organisms living together as a community is one of the goals of community ecologists.

Community interactions are often complex and not readily apparent

During the late 1990s, an intricate and fascinating relationship emerged among acorn production, white-footed mice, gypsy moth population growth, and the potential occurrence of Lyme disease in humans (Fig. 53-1). Large-scale experiments conducted in oak forests of the northeastern United States linked bumper acorn crops, which occur every 3 to 4 years, to booming mouse populations (the mice eat the acorns). Because the mice also eat gypsy moth pupae, high acorn conditions also lead to low populations of gypsy moths. This outcome helps the oaks,

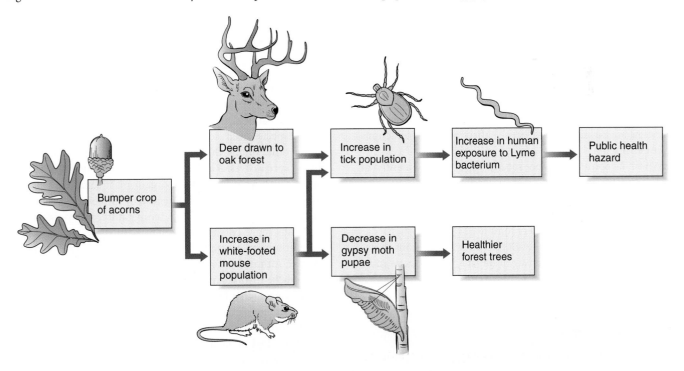

Figure 53-1 Connections to the size of an acorn crop

When there is a bumper crop of acorns, more mice survive and breed in winter, and more deer are attracted to oak forests. Both mice and deer are hosts of ticks that may carry the Lyme disease bacterium to humans. Abundant mice also reduce gypsy moth populations, thereby improving the health of forest trees.

because gypsy moths cause serious defoliation. However, abundant acorns also attract tick-bearing deer. The ticks' hungry offspring feed on the mice, which often carry the Lyme disease–causing bacterium. The bacterium infects the maturing ticks and, in turn, spreads to humans who are bitten by affected ticks. Thus, it may be possible to predict which years pose the greatest potential threat of Lyme disease to humans, based on when oaks are most productive.

An additional complication emerged in the Lyme disease relationships in the early 2000s. Scientists reported that ticks in areas with high biological diversity seldom transmit Lyme disease to humans. In contrast, ticks in areas with less biological diversity, such as where forests are cleared or otherwise degraded by human activities, are more likely to pass Lyme disease to humans. Because the white-footed mouse is one of the last species to disappear in low-diversity areas, it becomes the main food source for ticks, which pick up the disease and transmit it to humans. In high-diversity areas, ticks feed on many species of mammals, birds, and reptiles that may be infected with Lyme disease but

(a) Green anole

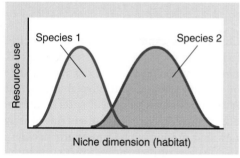

(b) Brown anole

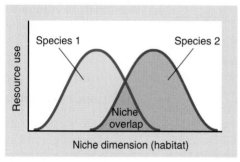

(c) Fundamental niches

(d) Realized niches

Figure 53-2 Effect of competition on an organism's realized niche

(a) The green anole (*Anolis carolinensis*) is the only anole species native to North America. Males are about 12.5 cm (5 in) long; females are slightly smaller. **(b)** The 15.2-cm (6-in) long brown anole (*A. sagrei*) was introduced to Florida. **(c, d)** Positions of the two species along a single niche dimension (in this case, habitat). Species 1 represents the green anole, and Species 2 represents the brown anole. **(c)** The fundamental niches of the two lizards overlap. **(d)** The brown anole outcompetes the green anole in the area where their niches overlap, restricting the niche of the green anole.

do not readily transmit it to ticks. Thus, many of the ticks remain disease-free. Lyme disease is an excellent example of how human activities—in this case, transforming the land and reducing the number of species living in an area—affect natural interactions to our detriment.

Within a community, no species exists independently of other species. As the preceding example shows, the populations of a community interact with and affect one another in complex ways that are not always obvious. Three main types of interactions occur among species in a community: competition, predation, and symbiosis. Before we address these interactions, however, we need to examine the way of life of a given species in its community.

The niche is a species' ecological role in the community

Every species is thought to have its own ecological role within the structure and function of a community; we call this role its **ecological niche.** Although the concept of ecological niche has been in use in ecology since early in the 20th century, Yale ecologist G. E. Hutchinson first described in 1957 the multidimensional

nature of the niche that is accepted today. An ecological niche takes into account all biotic and abiotic aspects of the species' existence, that is, all the physical, chemical, and biological factors that the species needs to survive, remain healthy, and reproduce. Among other things, the niche includes the local environment in which a species lives—its **habitat.** A niche also encompasses what a species eats, what eats it, what organisms it competes with, and how it interacts with and is influenced by the abiotic components of its environment, such as light, temperature, and moisture. The niche thus represents the totality of adaptations by a species to its environment, its use of resources, and the lifestyle to which it is suited. Although a complete description of an organism's ecological niche involves many dimensions and is difficult to define precisely, ecologists usually confine their studies to one or a few niche variables, such as feeding behaviors or ability to tolerate temperature extremes.

The ecological niche of a species is far broader in theory than in actuality. A species is usually capable of using much more of its environment's resources or of living in a wider assortment of habitats than it actually does. The potential ecological niche of a species is its **fundamental niche,** but various factors, such as competition with other species, may exclude it from part of this fundamental niche. Thus, the lifestyle a species actually pursues and the resources it actually uses make up its **realized niche.**

An example may help make this distinction clear. The green anole, a lizard native to Florida and other southeastern states, perches on trees, shrubs, walls, or fences during the day, waiting for insect and spider prey (❚ Fig. 53-2a). In the past these little lizards were widespread in Florida. Several years ago, however, a related species, the brown anole, was introduced from Cuba into southern Florida and quickly became common (❚ Fig. 53-2b). Suddenly the green anoles became rare, apparently driven out of their habitat by competition from the slightly larger brown anoles. Careful investigation disclosed, however, that green anoles were still around. They were now confined largely to the vegetation in wetlands and to the foliated crowns of trees, where they were less easily seen.

The habitat portion of the green anole's fundamental niche includes the trunks and crowns of trees, exterior house walls, and many other locations. Once the brown anoles became established in the green anole habitat, they drove the green anoles from all but wetlands and tree crowns; competition between the two species shrank the green anoles' realized niche (❚ Fig. 53-2c and d). Because communities consist of numerous species, many of which compete to some extent, the interactions among them produce each species' realized niche.

Limiting resources restrict the ecological niche of a species

A species' structural, physiological, and behavioral adaptations determine its tolerance for environmental extremes. If any feature of an environment lies outside the bounds of its tolerance, then the species cannot live there. Just as you would not expect to find a cactus living in a pond, you would not expect water lilies in a desert.

The environmental factors that actually determine a species' ecological niche are difficult to identify. For this reason, the concept of ecological niche is largely abstract, although some of its dimensions can be experimentally determined. Any environmental resource that, because it is scarce or unfavorable, tends to restrict the ecological niche of a species is called a **limiting resource.**

Most of the limiting resources that have been studied are simple variables, such as the soil's mineral content, temperature extremes, and precipitation amounts. Such investigations have disclosed that any factor exceeding the tolerance of a species, or present in quantities smaller than the required minimum, limits the presence of that species in a community. By their interaction, such factors help define the ecological niche for a species.

Limiting resources may affect only part of an organism's life cycle. For example, although adult blue crabs live in fresh or slightly brackish water, they do not become permanently established in such areas because their larvae (immature forms) require salt water. Similarly, the ring-necked pheasant, a popular game bird native to Eurasia, has been widely introduced in North America but has not become established in the southern United States. The adult birds do well there, but the eggs do not develop properly in the warmer southern temperatures.

Biotic and abiotic factors influence a species' ecological niche

In the 1960s, U.S. ecologist Joseph H. Connell investigated biotic and abiotic factors that affect the distribution of two barnacle species in the rocky *intertidal zone* along the coast of Scotland. The intertidal zone is a challenging environment, and organisms in the intertidal zone must tolerate exposure to the drying air during low tides. Barnacles are sessile crustaceans whose bodies are covered by a shell of calcium carbonate (see Fig. 30-22a). When the shell is open, feathery appendages extend to filter food from the water.

Along the coast of Scotland, more adults of one barnacle species, *Balanus balanoides,* are attached on lower rocks in the intertidal zone than are adults of the other species, *Chthamalus stellatus* (❚ Fig. 53-3). The distributions of the two species do not overlap, although immature larvae of both species live together in the intertidal zone. Connell manipulated the two populations to determine what factors were affecting their distribution. When he removed *Chthamalus* from the upper rocks, *Balanus* barnacles did not expand into the vacant area. Connell's experiments showed that *Chthamalus* is more resistant than *Balanus* to desiccation when the tide retreats. However, when Connell removed *Balanus* from the lower rocks, *Chthamalus* expanded into the lower parts of the intertidal zone. The two species compete for space, and *Balanus,* which is larger and grows faster, outcompetes the smaller *Chthamalus* barnacles.

Competition among barnacle species for the limiting resource of living space was one of the processes that Connell's research demonstrated. We now examine other aspects of competition that various ecologists have revealed in both laboratory and field studies.

Competition is intraspecific or interspecific

Competition occurs when two or more individuals attempt to use the same essential resource, such as food, water, shelter, living space, or sunlight. Because resources are often in limited supply in the environment, their use by one individual decreases the amount available to others (❚ Table 53-1). If a tree in a dense forest grows taller than surrounding trees, for example, it absorbs more of the incoming sunlight. Less sunlight is therefore available for nearby trees that are shaded by the taller tree. Competition occurs among individuals within a population (**intraspecific competition**) or between different species (**interspecific competition**). (Intraspecific competition was discussed in Chapter 52.)

Ecologists traditionally assumed that competition is the most important determinant of both the number of species found in a community and the size of each population. Today ecologists recognize that competition is only one of many interacting biotic and abiotic factors that affect community structure. Furthermore, competition is not always a straightforward, direct interaction. A variety of flowering plants, for example, live in a young pine forest and presumably compete with the conifers for such resources as soil moisture and soil minerals. Their relationship, however, is more complex than simple competition. The flowers produce nectar that is consumed by some insect species that

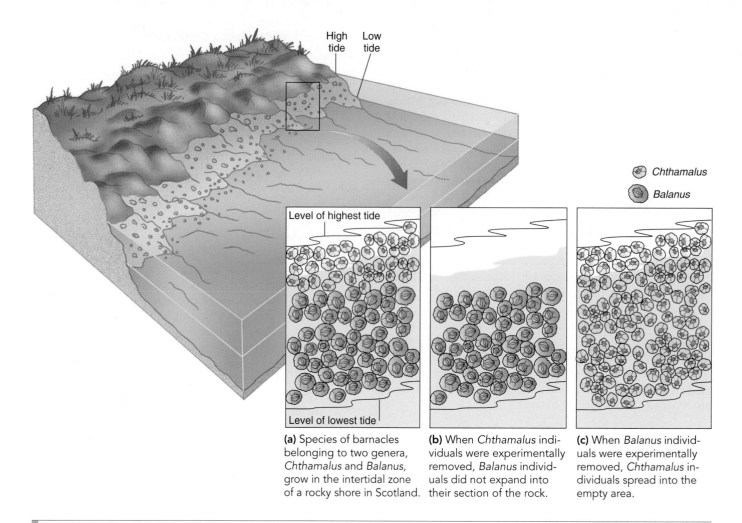

Chthamalus
Balanus

Level of highest tide

Level of lowest tide

(a) Species of barnacles belonging to two genera, *Chthamalus* and *Balanus,* grow in the intertidal zone of a rocky shore in Scotland.

(b) When *Chthamalus* individuals were experimentally removed, *Balanus* individuals did not expand into their section of the rock.

(c) When *Balanus* individuals were experimentally removed, *Chthamalus* individuals spread into the empty area.

Figure 53-3 Abiotic and biotic factors that affect barnacle distribution

(After J. H. Connell, "The Influence of Interspecific Competition and Other Factors on the Distribution of the Barnacle *Chthamalus stellatus,*" *Ecology,* Vol. 42, 1961.)

also prey on needle-eating insects, thereby reducing the number of insects feeding on pines. It is therefore difficult to assess the overall effect of flowering plants on pines. If the flowering plants were removed from the community, would the pines grow faster because they were no longer competing for necessary resources? Or would the increased presence of needle-eating insects (caused by fewer predatory insects) inhibit pine growth?

Short-term experiments in which one competing plant species is removed from a forest community often have demonstrated an improved growth for the remaining species. However, very few studies have tested the long-term effects on forest species of the removal of a single competing species. These long-term effects may be subtle, indirect, and difficult to ascertain; they may lessen or negate the short-term effects of competition for resources.

Competition between species with overlapping niches may lead to competitive exclusion

When two species are similar, as are the green and brown anoles or the two species of barnacles, their fundamental niches may overlap. However, based on experimental and modeling work, many ecologists think that no two species indefinitely occupy the same niche in the same community. According to the **competitive exclusion principle,** it is hypothesized that one species excludes

TABLE 53-1

Ecological Interactions among Species

Interaction	Effect on Species 1	Effect on Species 2
Competition between species 1 and species 2	Harmful	Harmful
Predation of species 2 (prey) by species 1 (predator)	Beneficial	Harmful
Symbiosis		
Mutualism of species 1 and species 2	Beneficial	Beneficial
Commensalism of species 1 with species 2	Beneficial	No effect
Parasitism by species 1 (parasite) on species 2 (host)	Beneficial	Harmful

another from its niche as a result of interspecific competition. Although it is possible for different species to compete for some necessary resource without being total competitors, two species with absolutely identical ecological niches cannot coexist. Coexistence occurs, however, if the overlap between the two niches is reduced. In the lizard example, direct competition between the two species was reduced as the brown anole excluded the green anole from most of its former habitat.

The initial evidence that interspecific competition contributes to a species' realized niche came from a series of laboratory experiments by the Russian biologist Georgyi F. Gause in the 1930s (see descriptions of other experiments by Gause in Chapter 52). In one study Gause grew populations of two species of protozoa, *Paramecium aurelia* and the larger *P. caudatum,* in controlled conditions (❚ Fig. 53-4). When grown in separate test tubes, that is, in the absence of the second species, the population of each species quickly increased to a level imposed by the resources and remained there for some time thereafter. When grown together, however, only *P. aurelia* thrived, whereas *P. caudatum* dwindled and eventually died out. Under different sets of culture conditions, *P. caudatum* prevailed over *P. aurelia.* Gause interpreted these results to mean that although one set of conditions favored one species, a different set favored the other. Nonetheless, because both species were so similar, in time one or the other would eventually triumph. Similar experiments with competing species of fruit flies, mice, beetles, and annual plants have supported Gause's results: one species thrives, and the other eventually dies out.

Thus, competition has adverse effects on species that use a limited resource and may result in competitive exclusion of one or more species. It therefore follows that over time natural selection should favor individuals of each species that avoid or reduce competition for environmental resources. Reduced competition among coexisting species as a result of each species' niche differing from the others in one or more ways is called **resource partitioning.** Resource partitioning is well documented in animals; studies in tropical forests of Central and South America demonstrate little overlap in the diets of fruit-eating birds, primates, and bats that coexist in the same habitat. Although fruits are the primary food for several hundred bird, primate, and bat species, the wide variety of fruits available has allowed fruit eaters to specialize, which reduces competition.

Resource partitioning may also include timing of feeding, location of feeding, nest sites, and other aspects of a species' ecological niche. Princeton ecologist Robert MacArthur's study of five North American warbler species is a classic example of resource partitioning (❚ Fig. 53-5). Although initially the warblers' niches seemed nearly identical, MacArthur found that individuals of each species spend most of their feeding time in different portions of the spruces and other conifer trees they frequent. They also move in different directions through the canopy, consume different combinations of insects, and nest at slightly different times.

Difference in root depth is an example of resource partitioning in plants. For example, three common annuals found in certain abandoned fields are smartweed, Indian mallow, and bristly foxtail. Smartweed roots extend deep into the soil, Indian mal-

QUESTION: Can competitive exclusion be demonstrated in the laboratory under controlled conditions?

HYPOTHESIS: When two species of paramecia (ciliated protozoa) are grown together in a mixed culture, one species outcompetes the other.

EXPERIMENT: Gause grew two species of paramecia, both separately and together.

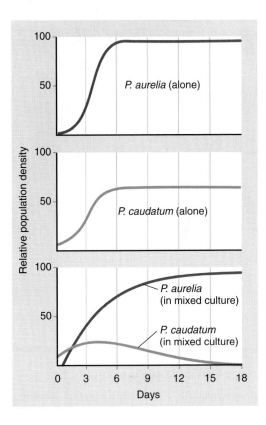

RESULTS AND CONCLUSION: The top and middle graphs show how each species of *Paramecium* flourishes when grown alone. The bottom graph shows how they grow together, in competition with each other. In a mixed culture, *P. aurelia* outcompetes *P. caudatum,* resulting in competitive exclusion.

Figure 53-4 *Animated* G. F. Gause's classic experiment on interspecific competition

(Adapted from G. F. Gause, *The Struggle for Existence,* Williams & Wilkins, Baltimore, 1934.)

low roots grow to a medium depth, and bristly foxtail roots are shallow. This difference reduces competition for the same soil resources—water and minerals—by allowing the plants to exploit different portions of the resource. Soil depth is not the only example of resource partitioning of soil resources in plants. In 2002, biologists reported in the journal *Nature* that the type of nitrogen absorbed from the soil and the timing of nitrogen use vary among different plants in an arctic plant community.

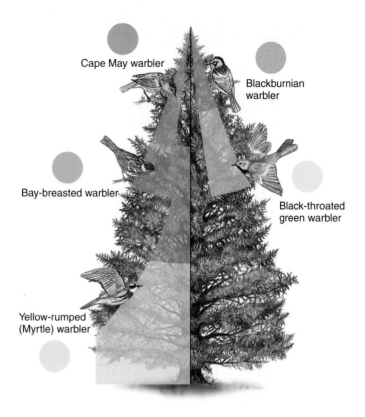

Figure 53-5 *Animated* Resource partitioning

Each warbler species spends at least half its foraging time in its designated area of a spruce tree, thereby reducing the competition among warbler species. (After R. H. MacArthur, "Population Ecology of Some Warblers of Northeastern Coniferous Forests," *Ecology*, Vol. 39, 1958.)

Character displacement is an adaptive consequence of interspecific competition

Sometimes populations of two similar species occur together in some locations and separately in others. Where their geographic distributions overlap, the two species tend to differ more in their structural, ecological, and behavioral characteristics than they do where each occurs in separate geographic areas. Such divergence in traits in two similar species living in the same geographic area is known as **character displacement.** Biologists think that character displacement reduces competition between two species because their differences give them somewhat different ecological niches in the same environment.

There are several well-documented examples of character displacement between two closely related species. The flowers of two *Solanum* species in Mexico are quite similar in areas where either one or the other occurs. However, where their distributions overlap, the two species differ significantly in flower size and are pollinated by different kinds of bees. In other words, character displacement reduces interspecific competition, in this case for the same animal pollinator.

The bill sizes of Darwin's finches provide another example of character displacement (■ Fig. 53-6). On large islands in the Galápagos where the medium ground finch (*Geospiza fortis*) and the small ground finch (*G. fuliginosa*) occur together, their bill depths are distinctive. *Geospiza fuliginosa* has a smaller bill depth

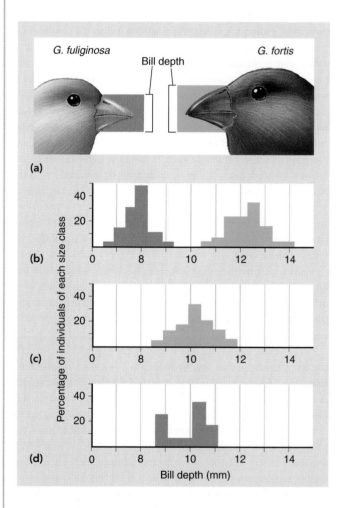
Figure 53-6 Character displacement

(After D. Lack, *Darwin's Finches*, Cambridge University Press, Cambridge, 1947.)

that enables it to crack small seeds, whereas *G. fortis* has a larger bill depth that enables it to crack medium-sized seeds. However, *G. fortis* and *G. fuliginosa* also live on separate islands. Where the two species live separately, bill depths are about the same intermediate size, perhaps because there is no competition from the other species.

Although these examples of the coexistence of similar species are explained in terms of character displacement, the character displacement hypothesis has been demonstrated in field experiments in only a few instances.

Natural selection shapes the body forms and behaviors of both predator and prey

Predation is the consumption of one species, the **prey,** by another, the **predator** (see Table 53-1). Predation includes animals eating other animals, as well as animals eating plants (*herbivory*). Predation has resulted in an evolutionary "arms race," with the evolution of predator strategies (more efficient ways to catch prey) and prey strategies (better ways to escape the predator). A predator that is more efficient at catching prey exerts a strong selective force on its prey, which over time may evolve some sort of countermeasure that reduces the probability of being captured. The countermeasure acquired by the prey, in turn, acts as a strong selective force on the predator. This type of interdependent evolution of two interacting species is known as **coevolution** (see Chapter 36).

We now consider several adaptations related to predator–prey interactions. These include predator strategies (pursuit and ambush) and prey strategies (plant defenses and animal defenses). Keep in mind as you read these descriptions that such strategies are not "chosen" by the respective predators or prey. New traits arise randomly in a population as a result of genetic changes. Some new traits are beneficial, some are harmful, and others have no effect. Beneficial strategies, or traits, persist in a population because such characteristics make the individuals that have them well suited to thrive and reproduce. In contrast, characteristics that make the individuals that have them poorly suited to their environment tend to disappear in a population.

Pursuit and ambush are two predator strategies

A brown pelican sights its prey—a fish—while in flight. Less than 2 seconds after diving into the water at a speed as great as 72 km/h (45 mph), it has its catch. Orcas (formerly known as killer whales) hunt in packs and often herd salmon or tuna into a cove so they are easier to catch. Any trait that increases hunting efficiency, such as the speed of brown pelicans or the intelligence of orcas, favors predators that pursue their prey. Because these carnivores must process information quickly during the pursuit of prey, their brains are generally larger, relative to body size, than those of the prey they pursue.

Ambush is another effective way to catch prey. The yellow crab spider, for example, is the same color as the white or yellow flowers on which it hides (Fig. 53-7). This camouflage keeps unwary insects that visit the flower for nectar from noticing the spider until it is too late. It also fools birds that prey on the crab spider.

Predators that *attract* prey are particularly effective at ambushing. For example, a diverse group of deep-sea fishes called

Figure 53-7 Ambush

A yellow crab spider (*Misumena vatia*) blends into its surroundings, waiting for an unwary insect to visit the flower. An effective predator strategy, ambush relies on surprising the prey.

anglerfish possess rodlike luminescent lures close to their mouths to attract prey.

Chemical protection is an effective plant defense against herbivores

Plants cannot escape predators by fleeing, but they have several physical and chemical adaptations that protect them from being eaten. The presence of spines, thorns, tough leathery leaves, or even thick wax on leaves discourages foraging herbivores from grazing. Other plants produce an array of protective chemicals that are unpalatable or even toxic to herbivores. The active ingredients in such plants as marijuana and tobacco affect hormone activity and nerve, muscle, liver, and kidney functions and may discourage foraging by herbivores. Interestingly, many of the chemical defenses in plants are useful to humans. India's neem tree, for example, contains valuable chemicals effective against more than 100 species of herbivorous insects, mites, and nematodes. Nicotine from tobacco, pyrethrum from chrysanthemum, and rotenone from the derris plant are other examples of chemicals extracted and used as insecticides.

Milkweeds are an excellent example of the biochemical coevolution between plants and herbivores (Fig. 53-8). Milkweeds produce alkaloids and cardiac glycosides, chemicals that are poisonous to all animals except a small group of insects. The ability to either tolerate or metabolize the milkweed toxins has evolved in these insects. They eat milkweeds and avoid competition from other herbivorous insects because few others tolerate milkweed toxins. Predators also learn to avoid these insects, which accumulate the toxins in their tissues and therefore become toxic themselves. The black, white, and yellow coloration of the monarch caterpillar, a milkweed feeder, clearly announces its toxicity to predators that have learned to associate bright colors with illness. Conspicuous colors or patterns, which advertise a species' unpalatability to potential predators, are known as **aposematic coloration** (pronounced "ap″uh-suh-mat′ik"; from the Greek *apo*, "away," and *semat*, "a mark or sign"), or **warning coloration.**

Figure 53-8 Plant chemical defenses

Toxic chemicals protect the common milkweed (*Asclepias syriaca*). Its leaves are poisonous to most herbivores except monarch caterpillars (*Danaus plexippus*) and a few other insects. Monarch caterpillars have bright aposematic coloration. Photographed in Michigan.

Figure 53-9 Cryptic coloration

The mossy leaf-tailed gecko (*Uroplatus sikorae*) hunts for insects by night and sleeps pressed against a tree branch by day. It is virtually invisible when it sleeps (look closely to see its head and front foot). Photographed in a rain forest in southern Madagascar.

Animal prey possess various defensive adaptations to avoid predators

Many animals, such as prairie voles and woodchucks, flee from predators by rapidly running to their underground burrows. Others have mechanical defenses, such as the barbed quills of a porcupine and the shell of a pond turtle. To discourage predators, the porcupine fish inflates itself to 3 times its normal size by pumping water into its stomach (see Fig. 31-14a). Some animals live in groups—for example, a herd of antelope, colony of honeybees, school of anchovies, or flock of pigeons. Because a group has so many eyes, ears, and noses watching, listening, and smelling for predators, this social behavior decreases the likelihood of a predator catching any one of them unaware.

Chemical defenses are also common among animal prey. The South American poison arrow frog (*Dendrobates*) stores poison in its skin. (These frogs obtain the toxins from ants and other insects in their diet.) The frog's bright aposematic coloration prompts avoidance by experienced predators (see Fig. 31-18a). Snakes and other animals that have tried once to eat a poisonous frog do not repeat their mistake! Other examples of aposematic coloration occur in the striped skunk, which sprays acrid chemicals from its anal glands, and the bombardier beetle, which spews harsh chemicals at potential predators (see Fig. 7-8).

Some animals have **cryptic coloration,** colors or markings that help them hide from predators by blending into their physical surroundings. Certain caterpillars resemble twigs so closely that you would never guess they were animals unless they moved. Pipefish are slender green fish that are almost perfectly camouflaged in green eelgrass. Leaf-tailed geckos in southern Madagascar resemble dead leaves or mossy bark, depending on the species (Fig. 53-9). Such cryptic coloration has been preserved and accentuated by means of natural selection. Predators are less likely to capture animals with cryptic coloration. Such animals are therefore more likely to live to maturity and produce offspring that also carry the genes for cryptic coloration.

Sometimes a defenseless species (a *mimic*) is protected from predation by its resemblance to a species that is dangerous in some way (a *model*). Such a strategy is known as **Batesian mimicry.** Many examples of this phenomenon exist. For example, a harmless scarlet king snake looks so much like a venomous coral snake that predators may avoid it (Fig. 53-10).

In **Müllerian mimicry,** different species (*co-models*), all of which are poisonous, harmful, or distasteful, resemble one another. Although their harmfulness protects them as individual species, their similarity in appearance works as an added advantage, because potential predators more easily learn a single common aposematic coloration. Scientists hypothesize that viceroy and monarch butterflies are an example of Müllerian mimicry (see *Focus On: Batesian Butterflies Disproved*).

BATESIAN BUTTERFLIES DISPROVED

The monarch butterfly (*Danaus plexippus*) is an attractive insect found throughout much of North America (see figure, *left side*). As a caterpillar, it feeds exclusively on milkweed leaves. The milky white liquid produced by the milkweed plant contains poisons that the insect tolerates but that remain in its tissues for life. When a young bird encounters and tries to eat its first monarch butterfly, the bird becomes sick and vomits. Thereafter, the bird avoids eating the distinctively marked insect.

Many people confuse the viceroy butterfly (*Limenitis archippus*; see figure, *right side*) with the monarch. The viceroy, which is found throughout most of North America, is approximately the same size, and the color and markings of its wings are almost identical to those of the monarch. As caterpillars, viceroys eat willow and poplar leaves, which apparently do not contain poisonous substances.

During the past century, it was thought that the viceroy butterfly was a tasty food for birds but that its close resemblance to monarchs gave it some protection against being eaten. In other words, birds that had learned to associate the distinctive markings and coloration of the monarch butterfly with its bad taste tended to avoid viceroys because they were similarly marked. The viceroy butterfly was there-

fore considered a classic example of Batesian mimicry.

In 1991, ecologists David Ritland and Lincoln Brower of the University of Florida reported the results of an experiment that tested the long-held notion that birds like the taste of viceroys but avoid eating them because of their resemblance to monarchs. They removed the wings of different kinds of butterflies—monarchs, viceroys, and several tasty species—and fed the seemingly identical wingless bodies to red-winged blackbirds. The results were surprising: monarchs and viceroys were equally distasteful to the birds.

As a result of this work, ecologists are re-evaluating the evolutionary significance of different types of mimicry. Rather than an example of Batesian mimicry, monarchs and viceroys may be an example of Müllerian mimicry, in which two or more different species that are distasteful or poisonous have come to resemble one another during the course of evolution. This likeness

Müllerian mimicry. Evidence suggests that monarch (*left*) and viceroy (*right*) butterflies are an example of Müllerian mimicry, in which two or more poisonous, harmful, or distasteful organisms resemble each other.

Thomas C. Emmel

provides an adaptive advantage because predators learn quickly to avoid all butterflies with the coloration and markings of monarchs and viceroys. As a result, fewer butterflies of either species die, and more individuals survive to reproduce.

The butterfly study provides us with a useful reminder about the process of science. Expansion of knowledge in science is an ongoing enterprise, and newly acquired evidence helps scientists re-evaluate current models or ideas. Thus, scientific knowledge and understanding are continually changing.

Suzanne L. and Joseph T. Collins/Photo Researchers, Inc.

(a) Scarlet king snake

Suzanne L. Collins/Photo Researchers, Inc.

(b) Eastern coral snake

Figure 53-10 *Animated* Batesian mimicry

In this example, **(a)** the scarlet king snake (*Lampropeltis triangulum elapsoides*) is the mimic, and **(b)** the eastern coral snake (*Micrurus ful-*vius fulvius*) is the model. Note that the red and yellow warning colors touch on the coral snake but do not touch on the harmless mimic.

Symbiosis involves a close association between species

Symbiosis is any intimate relationship or association between members of two or more species. Usually symbiosis involves one species living on or in another species. The partners of a symbiotic relationship, called **symbionts,** may benefit from, be unaffected by, or be harmed by the relationship (see Table 53-1). Examples of symbiosis occur across all of the domains and kingdoms of life. Most of the thousands, or perhaps even millions, of symbiotic associations are products of coevolution. Symbiosis takes three forms: mutualism, commensalism, and parasitism.

Benefits are shared in mutualism

Mutualism is a symbiotic relationship in which both partners benefit. Mutualism is either obligate (essential for the survival of both species) or facultative (either partner can live alone under certain conditions).

The association between **nitrogen-fixing bacteria** of the genus *Rhizobium* and legumes (plants such as peas, beans, and clover) is an example of mutualism (see Fig. 54-9). Nitrogen-fixing bacteria, which live inside nodules on the roots of legumes, supply the plants with most of the nitrogen they require to manufacture such nitrogen-containing compounds as chlorophylls, proteins, and nucleic acids. The legumes supply sugars and other energy-rich organic molecules to their bacterial symbionts.

Another example of mutualism is the association between reef-building animals and dinoflagellates called **zooxanthellae.** These symbiotic algae live inside cells of the coral polyp (the coral forms a vacuole around the algal cell), where they photosynthesize and provide the animal with carbon and nitrogen compounds as well as oxygen (Fig. 53-11). Zooxanthellae stimulate the growth of corals, and calcium carbonate skeletons form around their bodies much faster when the algae are present. The corals, in turn, supply the zooxanthellae with waste products such as

ammonia, which the algae use to make nitrogen compounds for both partners.

Mycorrhizae are mutualistic associations between fungi and the roots of plants. The association is common; biologists think that about 80% of all plant species have mycorrhizae. The fungus, which grows around and into the root as well as into the surrounding soil, absorbs essential minerals, especially phosphorus, from the soil and provides them to the plant. In return, the plant provides the fungus with organic molecules produced by photosynthesis. Plants grow more vigorously in the presence of mycorrhizal fungi (see Figs. 26-19 and 35-10), and they better tolerate environmental stressors such as drought and high soil temperatures. Indeed, some plants cannot maintain themselves under natural conditions if the fungi with which they normally form mycorrhizae are not present.

Commensalism is taking without harming

Commensalism is a type of symbiosis in which one species benefits and the other one is neither harmed nor helped. One example of commensalism is the relationship between social insects and scavengers, such as mites, beetles, or millipedes, that live in the social insects' nests. Certain types of silverfish, for example, move along in permanent association with marching columns of army ants and share the abundant food caught in the ant raids. The army ants derive no apparent benefit or harm from the silverfish.

Another example of commensalism is the relationship between a host tree and its epiphytes, which are smaller plants, such as orchids, ferns, or Spanish moss, attached to the host's branches (Fig. 53-12). The epiphyte anchors itself to the tree but does not obtain nutrients or water directly from it. Living on the tree enables it to obtain adequate light, water (as rainfall dripping down the branches), and required minerals (washed out of the tree's leaves by rainfall). Thus, the epiphyte benefits from the association, whereas the tree is apparently unaffected. (Epiphytes harm their host, however, if they are present in a large enough number to block sunlight from the host's leaves; in this instance, the relationship is no longer commensalism.)

Parasitism is taking at another's expense

Parasitism is a symbiotic relationship in which one member, the **parasite,** benefits, whereas the other, the **host,** is adversely affected. The parasite obtains nourishment from its host. A parasite rarely kills its host directly but may weaken it, rendering it more vulnerable to predators, competitors, or abiotic stressors. When a parasite causes disease and sometimes the death of a host, it is called a **pathogen.**

Ticks and other parasites that live outside the host's body are called **ectoparasites;** parasites such as tapeworms that live within the host are called **endoparasites.** Parasitism is a successful lifestyle; by one estimate, more than two thirds of all species are parasites, and more than 100 species of parasites live in or on the human species alone! Examples of human parasites include *Entamoeba histolytica,* an amoeba that causes amoebic dysentery; *Plasmodium,* an apicomplexan that causes malaria; and a variety of parasitic worms, such as blood flukes, tapeworms, pinworms, and hookworms.

Figure 53-11 Mutualism

The greenish brown specks in these polyps of stony coral (*Pocillopora*) are zooxanthellae, algae that live symbiotically within the coral's translucent cells and supply the coral with carbon and nitrogen compounds. In return, the coral provides the zooxanthellae with nitrogen in the form of ammonia.

Jeff Lepore/Photo Researchers, Inc.

Figure 53-12 Commensalism

Spanish moss (*Tillandsia usneoides*) is a gray-colored epiphyte that hangs suspended from larger plants in the southeastern United States. Spanish moss is not a moss but a flowering plant in the pineapple family. Although it is often 6 m (20 ft) or longer, it is nonparasitic and does not harm the host tree. Photographed in autumn in North Carolina, hanging from a bald cypress whose foliage has turned color.

Since the 1980s, wild and domestic honeybees in the United States have been dying off; for example, beekeepers lost roughly half of their bees in 2005, according to the U.S. Department of Agriculture. Although habitat loss, severe weather, and pesticide use have contributed to the problem, varroa mites and smaller tracheal mites have been a major reason for the honeybee decline (❚ Fig. 53-13). Honeybees pollinate up to $15 billion worth of apples, almonds, and other crops each year and produce about $250 million worth of honey, so their decline is a major threat to U.S. agriculture. Entomologists are searching for mite-resistant bees among both North American and European honeybees; the entomologists plan to breed the mite-resistant bees with local honeybees to incorporate genetic resistance into the local bees. Farmers are also experimenting with mite-resistant bee species other than honeybees to pollinate crops.

Keystone species and dominant species affect the character of a community

Certain species, called **keystone species,** are crucial in determining the nature of the entire community, that is, its species composition and ecosystem functioning. Other species of a community

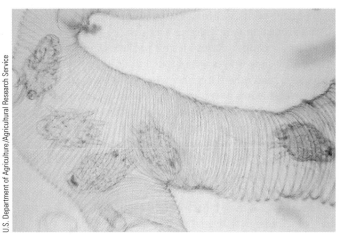

U.S. Department of Agriculture/Agricultural Research Service

100 μm

Figure 53-13 Parasitism

Microscopic tracheal mites (*Acarapis woodi*) are endoparasites that live in the tracheal tubes of honeybees, clogging their airways so they cannot breathe efficiently. Tracheal mites also suck the bees' circulatory fluid, weakening and eventually killing them. Entomologists think the larger varroa mites (*not shown*), which are ectoparasites that also feed on the circulatory fluid, are more devastating to honeybee populations than tracheal mites. Mites may also transmit viruses to their honeybee hosts.

depend on or are greatly affected by the keystone species. Keystone species are usually not the most abundant species in the community. Although present in relatively small numbers, the individuals of a keystone species profoundly influence the entire community because they often affect the amount of available food, water, or some other resource. Thus, the impact of keystone species is greatly disproportionate to their abundance. Identifying and protecting keystone species are crucial goals of conservation biology (see Chapter 56). If a keystone species disappears from a community, many other species in that community may become more common, more rare, or even disappear.

One problem with the concept of keystone species is that it is difficult to measure all the direct and indirect impacts of a keystone species on an ecosystem. Consequently, most evidence for the existence of keystone species is based on indirect observations rather than on experimental manipulations. For example, consider the fig tree. Because fig trees produce a continuous crop of fruits, they may be keystone species in tropical rain forests of Central and South America. Fruit-eating monkeys, birds, bats, and other fruit-eating vertebrates of the forest do not normally consume large quantities of figs in their diets. During that time of the year when other fruits are less plentiful, however, fig trees become important in sustaining fruit-eating vertebrates. It is therefore assumed that should the fig trees disappear, most of the fruit-eating vertebrates would also disappear. In turn, should the fruit eaters disappear, the spatial distribution of other fruit-bearing plants would become more limited because the fruit eaters help disperse their seeds. Thus, protecting fig trees in such tropical rainforest ecosystems probably increases the likelihood that monkeys, birds, bats, and many other tree species will survive. The question is whether this anecdotal evidence of fig trees as keystone species is strong enough for policymakers to grant special protection to fig trees.

Many keystone species are top predators, for example, the gray wolf. Where wolves were hunted to extinction, populations of elk, deer, and other larger herbivores increased exponentially. As these herbivores overgrazed the vegetation, many plant species that could not tolerate such grazing pressure disappeared. Smaller animals such as rodents, rabbits, and insects declined in number because the plants they depended on for food were now less abundant. The number of foxes, hawks, owls, and badgers that prey on these small animals decreased, as did the number of ravens, eagles, and other scavengers that eat wolf kill. Thus, the disappearance of the wolf resulted in communities with considerably less biological diversity.

The reintroduction of wolves to Yellowstone National Park in 1995 has given ecologists a unique opportunity to study the impacts of a keystone species. The wolf's return has already caused substantial changes for other residents in the park. The top predator's effects have ranged from altering relationships among predator and prey species to transforming vegetation profiles. Coyotes are potential prey for wolves, and wolf packs have decimated some coyote populations. A reduction in coyotes has allowed populations of the coyotes' prey, such as ground squirrels and chipmunks, to increase. Scavengers such as ravens, bald eagles, and grizzly bears have benefited from dining on scraps from wolf kills.

The pruning effect of wolves on prey populations such as elk should cause broader, long-term impacts. Yellowstone's booming elk population of 35,000 eventually may decline by as much as 20%, which would relieve heavy grazing pressure on plants and encourage a more lush and varied plant composition. Richer vegetation should support more herbivores such as beavers and snow hares, which in turn will support small predators such as foxes, badgers, and martens.

Dominant species influence a community as a result of their greater size or abundance

In contrast to keystone species, which have a large impact out of proportion to their abundance, **dominant species** greatly affect the community because they are very common. Trees, the dominant species of forests, change the local environment. Trees provide shade, which changes both the light and moisture availability on the forest floor. Trees provide numerous habitats and *microhabitats* (such as a hole in a tree trunk) for other species. Forest trees also provide food for many organisms and therefore play a large role in energy flow through the forest ecosystem. Similarly, cordgrass (*Spartina*) is the dominant species in salt marshes, prairie grass in grasslands, and kelp in kelp beds. Animals are also dominant species. For example, corals are dominant species in coral reefs, and cattle are dominant species in overgrazed rangelands. Typically, a community has one or a few dominant species, and most other species are relatively rare.

Review

- How are acorns, gypsy moths, and Lyme disease related?
- Why is an organism's realized niche usually narrower, or more restricted, than its fundamental niche?
- What is the principle of competitive exclusion?
- What are the three kinds of symbiosis?

COMMUNITY BIODIVERSITY

Learning Objectives

7   Summarize the main determinants of species richness in a community, and relate species richness to community stability.
8   State the results of the experiment on the distance effect in South Pacific bird species.

Species richness and species diversity vary greatly from one community to another and are influenced by many biotic and abiotic factors. **Species richness** is the number of species in a community. Tropical rain forests and coral reefs are examples of communities with extremely high species richness. In contrast, geographically isolated islands and mountaintops exhibit low species richness.

The term *species diversity* is sometimes used interchangeably with species richness, but more precisely, **species diversity** is a measure of the relative importance of each species within a community, based on its abundance, productivity, or size (❚ Fig. 53-14). Ecologists have developed various mathematical expressions to represent species diversity. These *diversity indices* enable ecologists to compare species diversity in different communities. Conservation biologists use diversity indices as part of a comprehensive approach to saving biodiversity.

Ecologists seek to explain why some communities have more species than others

What determines the number of species in a community? No single conclusive answer exists, but several explanations appear

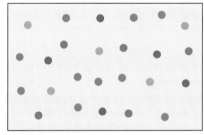

(a) High species diversity

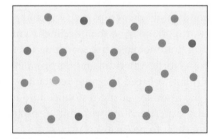

(b) Low species diversity

❚ **Figure 53-14   Species diversity**

Shown are two hypothetical communities, each with five species. Each circle represents one individual; each color represents a different species. Both communities have the same species richness, but the community in **(a)** has a greater species diversity than the community in **(b)**.

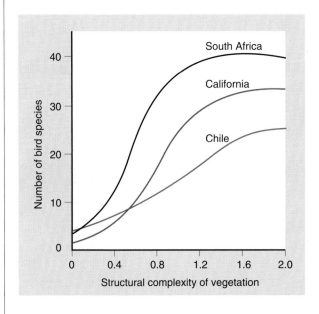

Figure 53-15 Effect of community complexity on species richness

Data were compiled in comparable chaparral habitats (shrubby and woody areas) in Chile, South Africa, and California. The structural complexity of vegetation (*x-axis*) is a numerically assigned gradient of habitats, based on height and density of vegetation, from low complexity (very dry scrub) to high complexity (woodland). (After M. L. Cody and J. M. Diamond, eds., *Ecology and Evolution of Communities,* Harvard University, Cambridge, MA, 1975.)

QUESTION: Does the geographic isolation of a community affect its species richness?

HYPOTHESIS: Islands close to the mainland or to a large island have a higher species richness than the islands that are farther from the mainland.

EXPERIMENT: The number of bird species was catalogued on South Pacific islands that are various distances from New Guinea, a source of colonizing species for these smaller islands. Each point on the graph represents a different island.

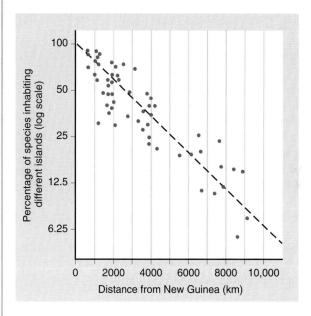

RESULTS AND CONCLUSION: The percentage of South Pacific bird species found on each island in the South Pacific is related to its distance from New Guinea. (New Guinea has 100% of the bird species living in that region.) Thus, species richness declines as the distance from New Guinea increases.

Figure 53-16 *Animated* The distance effect
(After J. M. Diamond, "Biogeographic Kinetics: Estimation of Relaxation Times for Avifaunas of Southwest Pacific Islands," *Proceedings of the National Academy of Sciences,* Vol. 69, 1972.)

plausible. These include the structural complexity of habitats, geographic isolation, habitat stress, closeness to the margins of adjacent communities, dominance of one species over others, and geologic history. Although these and other environmental factors have positive or negative effects on species richness, there are exceptions and variations in every explanation. Some explanations vary at different spatial scales: an explanation that seems to work at a large geographic scale (such as a continent) may not work at a smaller local scale (such as a meadow).

In many habitats, species richness is related to the structural complexity of habitats. In terrestrial environments, the types of plants growing in an area typically determine structural complexity. A structurally complex community, such as a forest, offers a greater variety of potential ecological niches than does a simple community, such as an arid desert or semiarid grassland (Fig. 53-15). An already complex habitat, such as a coral reef, may become even more complex if species potentially capable of filling vacant ecological niches evolve or migrate into the community, because these species create "opportunities" for additional species. Thus, it appears that species richness is self-perpetuating to some degree.

Species richness is inversely related to the geographic isolation of a community. Isolated island communities are generally much less diverse than are communities in similar environments found on continents. This is due partly to the *distance effect,* the difficulty encountered by many species in reaching and successfully colonizing the island (Fig. 53-16). Also, sometimes species become locally extinct as a result of random events. In isolated habitats, such as islands or mountaintops, locally extinct species are not readily replaced. Isolated areas are usually small and have fewer potential ecological niches.

Generally, species richness is inversely related to the environmental stress of a habitat. Only those species capable of tolerating extreme conditions live in an environmentally stressed community. Thus, the species richness of a highly polluted stream is low compared with that of a nearby pristine stream. Similarly, the species richness of high-latitude (farther from the equator) communities exposed to harsh climates is lower than that of lower-latitude (closer to the equator) communities with milder climates (▌Fig. 53-17). This observation, known as the *species richness–energy hypothesis,* suggests that different latitudes affect species richness because of variations in solar energy. Greater energy may permit more species to coexist in a given region. Although the equatorial countries of Colombia, Ecuador, and Peru occupy only 2% of Earth's land, they contain a remarkable 45,000 native plant species. The continental United States and Canada, with a significantly larger land area, host a total of 19,000 native plant species. Ecuador alone contains more than 1300 native species of birds—twice as many as the United States and Canada combined.

Species richness is usually greater at the margins of distinct communities than in their centers. The reason is that an **ecotone,** a transitional zone where two or more communities meet, contains all or most of the ecological niches of the adjacent communities as well as some that are unique to the ecotone (see Fig. 55-25). This change in species composition produced at ecotones is known as the **edge effect.**

Species richness is reduced when any one species enjoys a position of dominance within a community; a dominant species may appropriate a disproportionate share of available resources, thus crowding out, or outcompeting, other species. Ecologist James H. Brown of the University of New Mexico has addressed species composition and richness in experiments conducted since 1977 in the Chihuahuan Desert of southeastern Arizona. In one experiment, the removal of three dominant species, all kangaroo rats, from several plots resulted in an increased diversity of other rodent species. This increase was ascribed both to lowered competition for food and to an altered habitat, because the abundance of grass species increased dramatically after the removal of the kangaroo rats.

Species richness is greatly affected by geologic history. Many scientists think that tropical rain forests are old, stable communities that have undergone relatively few widespread disturbances through Earth's history. (In ecology, a **disturbance** is any event in time that disrupts community or population structure.) During this time, myriad species evolved in tropical rain forests. In con-

trast, glaciers have repeatedly altered temperate and arctic regions during Earth's history. An area recently vacated by glaciers will have low species richness because few species have had a chance to enter it and become established. The idea that older, more stable habitats have greater species richness than habitats subjected to frequent, widespread disturbances is known as the *time hypothesis.*

Key Point

Species richness increases along a polar ⟶ equatorial gradient. This correlation has been observed for many organisms, from plants to primates, in both terrestrial and marine environments.

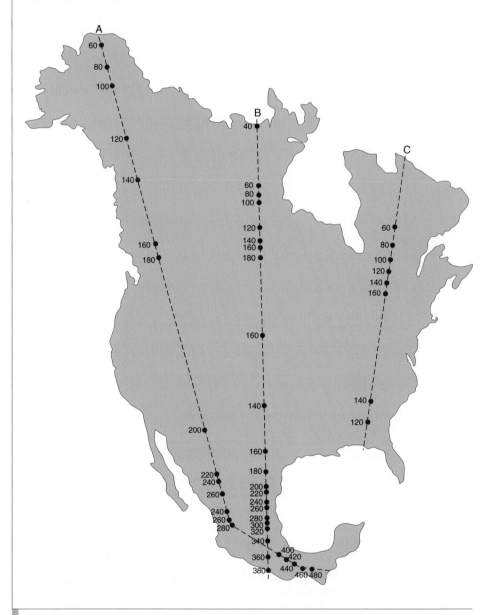

Figure 53-17 *Animated* Effect of latitude on species richness of breeding birds in North America

The species richness for three north–south transects are shown. Note that the overall number of breeding bird species is greater at lower latitudes toward the equator than at higher latitudes. However, this pattern is strongly modified by other factors, such as precipitation and surface features (such as mountains). (After R. E. Cook, "Variation in Species Density of North American Birds," *Systematic Zoology,* Vol. 18, 1969.)

Species richness may promote community stability

Traditionally, most ecologists assumed that community stability—the ability of a community to withstand disturbances—is a consequence of community complexity. That is, ecologists hypothesized that a community with considerable species richness is more stable than a community with less species richness. According to this view, the greater the species richness, the less critically important any single species should be. With many possible interactions within the community, it appears unlikely that any single disturbance could affect enough components of the system to make a significant difference in its functioning.

Supporting evidence for this hypothesis is found in the fact that destructive outbreaks of pests are more common in cultivated fields, which have a low species richness, than in natural communities with a greater species richness. As another example, the almost complete loss of the American chestnut tree to the chestnut blight fungus had little ecological impact on the moderately diverse Appalachian woodlands of which it was formerly a part.

Ongoing studies by David Tilman of the University of Minnesota and John Downing of the University of Iowa have strengthened the link between species richness and community stability. In their initial study, they established and monitored 207 plots of various grassland species in Minnesota for 7 years. During the study period, Minnesota's worst drought in 50 years occurred (1987–1988). The ecologists found that those plots with the greatest number of plant species lost less ground cover, as measured by dry weight, and recovered faster than species-poor plots. Later studies by Tilman and his colleagues supported these conclusions and showed a similar effect of species richness on community stability during nondrought years. Similar work by almost three dozen ecologists at eight grassland sites in Europe also supports the link between species richness and community stability.

Some scientists do not agree with the conclusions of Tilman and other research groups that the presence of more species confers greater community stability. These critics argue that it is difficult to separate species number from other factors that could affect productivity. They suggest it would be better to start with established ecosystems and study what happens to their productivity when plants are removed.

Another observation that adds a layer of complexity to the species richness–community stability debate is that populations of individual species within a species-rich community often vary significantly from year to year. It may seem paradoxical that variation within populations of individual species relates to the stability of the entire community. When you consider all the interactions among the organisms in a community, however, it is obvious that some species benefit at the expense of others. If one species declines in a given year, other species that compete with it may flourish. Thus, if an ecosystem contains more species, it is likely that at least some will be resistant to any given disturbance.

Review

- How is the species richness of a community related to geographic isolation?

- How is the species richness of a community related to the structural complexity of habitats?
- How is the species richness of a community related to the environmental stress of a habitat?

COMMUNITY DEVELOPMENT

Learning Objectives

9 Define *succession*, and distinguish between primary and secondary succession.
10 Describe the intermediate disturbance hypothesis.
11 Discuss the two traditional views of the nature of communities, Clements's organismic model and Gleason's individualistic model.

A community does not spring into existence full blown but develops gradually through a series of stages, each dominated by different organisms. The process of community development over time, which involves species in one stage being replaced by different species, is called **succession.** An area is initially colonized by certain early-successional species that give way over time to others, which in turn give way much later to late-successional species.

Succession is usually described in terms of the changes in the species composition of an area's vegetation, although each successional stage also has its own characteristic kinds of animals and other species. The time involved in succession is on the order of tens, hundreds, or thousands of years, not the millions of years involved in the evolutionary time scale.

Ecologists distinguish between two types of succession, primary and secondary. **Primary succession** is the change in species composition over time in a habitat that was not previously inhabited by organisms. No soil exists when primary succession begins. Bare rock surfaces, such as recently formed volcanic lava and rock scraped clean by glaciers, are examples of sites where primary succession might take place.

The Indonesian island of Krakatoa has provided scientists with a perfect long-term study of primary succession in a tropical

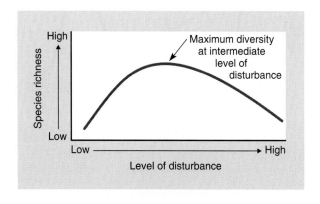

Figure 53-18 Intermediate disturbance hypothesis
Species richness is greatest at an intermediate level of disturbance. (After J. H. Connell, "Diversity in Tropical Rain Forests and Coral Reefs," *Science,* Vol. 199, 1978.)

Most communities are individualistic associations of species rather than distinct units that act like "superorganisms."

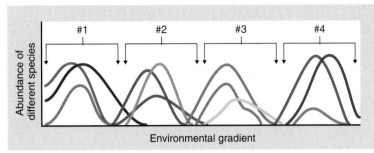

(a) Organismic model. According to this hypothesis, communities are organized as distinct units. Each of the four communities shown (*numbered brackets*) consists of an assemblage of distinct species (*each colored curve represents a single species*). The arrows indicate ecotones, regions of transition along community boundaries.

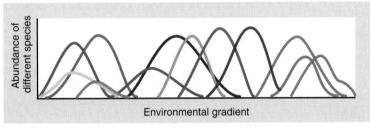

(b) Individualistic model. This model predicts a more random assemblage of species along a gradient of environmental conditions.

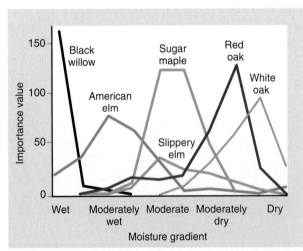

(c) Tree species in Wisconsin forests, which are distributed along a moisture gradient, more closely resemble the individualistic model. The "importance value" (*y-axis*) of each species in a given location combines three aspects (population density, frequency, and size).

Figure 53-19 Nature of communities

(**c,** Adapted from J. T. Curtis, *The Vegetation of Wisconsin,* University of Wisconsin Press, Madison, 1959.)

rain forest. In 1883, a volcanic eruption destroyed virtually all life on the island. Ecologists have surveyed the ecosystem during the more than 100 years since the devastation to document the return of life-forms. As of the 1990s, ecologists had found that the progress of primary succession was extremely slow, in part because of Krakatoa's isolation (recall the distance effect discussed earlier in the chapter). Many species are limited in their ability to disperse over water. Krakatoa's forest, for example, may have only one tenth the tree species richness of undisturbed tropical rain forest of nearby islands. The lack of plant diversity has, in turn, limited the number of colonizing animal species. In a forested area of Krakatoa where zoologists would expect more than 100 butterfly species, for example, there are only 2 species.

Secondary succession is the change in species composition that takes place after some disturbance removes the existing vegetation; soil is already present at these sites. Abandoned agricultural fields or open areas produced by forest fires are common examples of sites where secondary succession occurs. During the summer of 1988, wildfires burned approximately one third of Yellowstone National Park. This natural disaster provided a valuable chance for ecologists to study secondary succession in areas that had been forests. After the conflagration, gray ash covered the forest floor, and most of the trees, although standing, were charred and dead. Secondary succession in Yellowstone has occurred rapidly since 1988. Less than a year later, in the spring of 1989, trout lily and other herbs sprouted and covered much of the ground. By 1998, a young forest of knee-high to shoulder-high lodgepole pines dominated the area. Douglas fir seedlings also began appearing in 1998. Ecologists continue to monitor the changes in Yellowstone as secondary succession unfolds.

Disturbance influences succession and species richness

Early studies suggested that succession inevitably progressed to a stable and persistent community, known as a *climax community,* which was determined solely by climate. Periodic disturbances, such as fires or floods, were not thought to exert much influence on climax communities. If the climax community was disturbed in any way, it would return to a self-sustaining, stable equilibrium in time.

This traditional view of stability has fallen out of favor. The apparent end-point stability of species composition in a "climax" forest is probably the result of how long trees live relative to the human life span. It is now recognized that forest communities never reach a state of permanent equilibrium but instead exist in a state of continual disturbance. The species composition and relative abundance of each species vary in a mature community over a range of environmental gradients, although the community retains a relatively uniform appearance overall.

Because all communities are exposed to periodic disturbances, both natural and human induced, ecologists have long tried to understand the effects of disturbance on species richness.

A significant advance was the development of the **intermediate disturbance hypothesis** by Joseph H. Connell. When he examined species richness in tropical rain forests and coral reefs, he proposed that species richness is greatest at moderate levels of disturbance (▌ Fig. 53-18 on page 1161). At a moderate level of disturbance, the community is a mosaic of habitat patches at different stages of succession; a range of sites exists, from those that were recently disturbed to those that have not been disturbed for many years. When disturbances are frequent or intense, only those species best adapted to earlier stages of succession persist, whereas low levels of disturbance allow late-successional species to dominate to such a degree that other species disappear. For example, when periodic wildfires are suppressed in forests, reducing disturbance to a low level, some of the "typical" forest herbs decline in number or even disappear.

One of the difficulties with the intermediate disturbance hypothesis is defining precisely what constitutes an "intermediate" level of disturbance. Despite this problem, the intermediate disturbance hypothesis has important ramifications for conservation biology because it tells us that we cannot maintain a particular community simply by creating a reserve around it. Both natural and human-induced disturbances will cause changes in species composition, and humans may have to intentionally intervene to maintain the species richness of the original community. What kind of and how much human intervention is necessary to maintain species richness is controversial.

Ecologists continue to study community structure

One of the major issues in community ecology, from the early 1900s to the present, is the nature of communities. Are communities highly organized systems of predictable species, or are they abstractions produced by the minds of ecologists?

U.S. botanist Frederick E. Clements (1874–1945) was struck by the worldwide uniformity of large tracts of vegetation, for example, tropical rain forests in South America, Africa, and Southeast Asia. He also noted that even though the species composition of a community in a particular habitat may be different from that of a community in a habitat with a similar climate elsewhere in the world, overall the components of the two communities are usually similar. He viewed communities as something like "super-organisms," whose member species cooperated with one another

in a manner that resembled the cooperation of the parts of an individual organism's body. Clements's view was that a community went through certain stages of development, like the embryonic stages of an organism, and eventually reached an adult state; the developmental process was succession, and the adult state was the climax community. This cooperative view of the community, called the **organismic model,** stresses the interaction of the members, which tend to cluster in tightly knit groups within discrete community boundaries (▌ Fig. 53-19a).

Opponents of the organismic model, particularly U.S. ecologist Henry A. Gleason (1882–1975), held that biological interactions are less important in the production of communities than are environmental gradients (such as climate and soil) or even chance. Indeed, the concept of a community is questionable. It may be a classification category with no reality, reflecting little more than the tendency of organisms with similar environmental requirements to live in similar places. This school of thought, called the **individualistic model,** emphasizes species individuality, with each species having its own particular abiotic living requirements. It holds that communities are therefore not interdependent associations of organisms. Rather, each species is independently distributed across a continuum of areas that meets its own individual requirements (▌ Fig. 53-19b).

Debates such as this one over the nature of communities are an integral part of the scientific process because they fuel discussion and research that lead to a better understanding of broad scientific principles. Studies testing the organismic and individualistic hypotheses of communities do not support Clements's interactive concept of communities as discrete units. Instead, most studies favor the individualistic model. As shown in ▌ Figure 53-19c, tree species in Wisconsin forests are distributed in a gradient from wet to dry environments. Also, studies of plant and animal movements during the past 14,000 years support the individualistic model, because it appears that individual species, not entire communities, became redistributed in response to changes in climate.

Review

- What is succession?
- How does primary succession differ from secondary succession?
- What is Connell's intermediate disturbance hypothesis?
- How do the organismic and individualistic models of the nature of communities differ?

SUMMARY WITH KEY TERMS

Learning Objectives

1 Define *ecological niche,* and distinguish between an organism's fundamental niche and its realized niche (page 1147).

- The distinctive lifestyle and role of an organism in a community is its **ecological niche.** An organism's ecological niche takes into account all abiotic and biotic aspects of the organism's existence. An organism's **habitat** (where it lives) is one of the parameters used to describe the niche.

- Organisms potentially exploit more resources and play a broader role in the life of their community than they

actually do. The potential ecological niche for an organism is its **fundamental niche,** whereas the niche it actually occupies is its **realized niche.**

2 Define *competition,* and distinguish between interspecific and intraspecific competition (page 1147).

- **Competition** occurs when two or more individuals attempt to use the same essential resource, such as food, water, shelter, living space, or sunlight.

- Competition occurs among individuals within a population (**intraspecific competition**) or between different species (**interspecific competition**).

3 Summarize the concepts of the competitive exclusion principle, resource partitioning, and character displacement (page 1147).
- According to the **competitive exclusion principle,** two species cannot occupy the same niche in the same community for an indefinite period; one species is excluded by another as a result of competition for a limiting resource. In a classic experiment demonstrating competitive exclusion, Gause grew two species of paramecia. In separate test tubes, the population of each species increased and stabilized. When grown together, one species thrived and the other died out. Gause concluded that because both species were so similar, in time one or the other would eventually triumph at the expense of the other.
- The evolution of differences in resource use, known as **resource partitioning,** reduces competition between similar species.
- Competition among some species is reduced by **character displacement,** in which their structural, ecological, and behavioral characteristics diverge where their ranges overlap. The bill sizes of Darwin's finches are example of character displacement. Where the medium ground finch and the small ground finch occur together, their bill depths are distinctive. However, where the two species live separately, bill depths are about the same intermediate size.

4 Define *predation*, and describe the effects of natural selection on predator–prey relationships (page 1147).
- **Predation** is the consumption of one species (the **prey**) by another (the **predator**). During **coevolution** between predator and prey, more efficient ways to catch prey evolve in the predator species, and better ways to escape the predator evolve in the prey species.

5 Distinguish among mutualism, commensalism, and parasitism, and give examples of each (page 1147).
- **Symbiosis** is any intimate or long-term association between two or more species. The three types of symbiosis are mutualism, commensalism, and parasitism.
- In **mutualism,** both partners benefit. Three examples of mutualism are **nitrogen-fixing bacteria** and legumes, **zooxanthellae** and corals, and **mycorrhizae** (fungi and plant roots).
- In **commensalism,** one organism benefits and the other is unaffected. Two examples of commensalism are silverfish and army ants, and epiphytes and larger plants.
- In **parasitism,** one organism (the **parasite**) benefits while the other (the **host**) is harmed. One example of parasitism is mites that grow in or on honeybees. Some parasites are **pathogens** that cause disease.

6 Distinguish between keystone species and dominant species (page 1147).
- **Keystone species** are present in relatively small numbers but are crucial in determining the species composition and ecosystem functioning of the entire community.
- In contrast to keystone species, which have an impact that is out of proportion to their abundance, **dominant spe-**

cies greatly affect the community of which they are a part because they are very common.

7 Summarize the main determinants of species richness in a community, and relate species richness to community stability (page 1158).
- Community complexity is expressed in terms of **species richness,** the number of species within a community, and **species diversity,** a measure of the relative importance of each species within a community based on abundance, productivity, or size.
- Species richness is often great in a habitat that has structural complexity; in a community that is not isolated (the distance effect) or severely stressed; in a community where more energy is available (the species richness–energy hypothesis); in **ecotones** (transition zones between communities); and in communities with long histories without major **disturbances,** events that disrupt community or population structure.
- Several studies suggest that species richness may promote community stability.

8 State the results of the experiment on the distance effect in South Pacific bird species (page 1158).
- The number of bird species found on each island in the South Pacific is inversely related to its geographic isolation—that is, to its distance from New Guinea. Species richness declines as the distance from New Guinea increases.

9 Define *succession*, and distinguish between primary and secondary succession (page 1161).
- **Succession** is the orderly replacement of one community by another. **Primary succession** occurs in an area that has not previously been inhabited (such as bare rock). **Secondary succession** begins in an area where there was a pre-existing community and well-formed soil (such as abandoned farmland).

10 Describe the intermediate disturbance hypothesis (page 1161).
- Disturbance affects succession and species richness. According to the **intermediate disturbance hypothesis,** species richness is greatest at moderate levels of disturbance, which create a mosaic of habitat patches at different stages of succession.

11 Discuss the two traditional views of the nature of communities, Clements's organismic model and Gleason's individualistic model (page 1161).
- The **organismic model** views a community as a "superorganism" that goes through certain stages of development (succession) toward adulthood (climax). In this view, biological interactions are primarily responsible for species composition, and organisms are highly interdependent.
- Most ecologists support the **individualistic model,** which challenges the concept of a highly interdependent community. According to this model, abiotic environmental factors are the primary determinants of species composition in a community, and organisms are somewhat independent of each other.

ThomsonNOW™ **Explore interspecific competition, resource partitioning, mimicry, and the distance effect by clicking on the figures in ThomsonNOW.**

TEST YOUR UNDERSTANDING

1. An association of populations of different species living together in one area is a(an) (a) organismic model (b) ecological niche (c) ecotone (d) community (e) habitat

2. Community structure and community functioning include all of the following *except* (a) community resilience to disturbances (b) the reproductive success of an individual organism

(c) the interactions among different species (d) the number and types of species present and the relative abundance of each species (e) energy flow throughout the community

3. A limiting resource does all of the following *except* (a) tends to restrict the ecological niche of a species (b) is in short supply relative to a species' need for it (c) limits the presence of a species in a given community (d) results in an intermediate disturbance (e) may be limiting for only part of an organism's life cycle

4. Monarch and viceroy butterflies are probably an example of (a) Batesian mimicry (b) character displacement (c) resource partitioning (d) Müllerian mimicry (e) cryptic coloration

5. A symbiotic association in which organisms are beneficial to one another is known as (a) predation (b) interspecific competition (c) intraspecific competition (d) commensalism (e) mutualism

6. A species' _____ is the totality of its adaptations, its use of resources, and its lifestyle. (a) habitat (b) ecotone (c) ecological niche (d) competitive exclusion (e) coevolution

7. Competition with other species helps determine an organism's (a) ecotone (b) fundamental niche (c) realized niche (d) limiting resource (e) ecosystem

8. "Complete competitors cannot coexist" is a statement of the principle of (a) primary succession (b) limiting resources (c) Müllerian mimicry (d) competitive exclusion (e) character displacement

9. The _____ signifies that species richness is greater where two communities meet than at the center of

either community. (a) edge effect (b) fundamental niche (c) character displacement (d) realized niche (e) limiting resource

10. Primary succession occurs on (a) bare rock (b) newly cooled lava (c) abandoned farmland (d) a and b (e) a, b, and c

11. The tendency for two similar species to differ from one another more markedly in areas where they occur together is known as (a) Müllerian mimicry (b) Batesian mimicry (c) resource partitioning (d) competitive exclusion (e) character displacement

12. An unpalatable species demonstrates its threat to potential predators by displaying (a) character displacement (b) limiting resources (c) cryptic coloration (d) aposematic coloration (e) competitive exclusion

13. An ecologist studying several forest-dwelling, insect-eating bird species does not find any evidence of interspecific competition. The most likely explanation is (a) lack of a keystone species (b) low species richness (c) pronounced intraspecific competition (d) coevolution of predator–prey strategies (e) resource partitioning

14. Support for the individualistic model of community structure includes (a) the decline of honeybees because of two species of parasitic mites (b) the identification of fig trees as a keystone species in tropical forests (c) the competitive exclusion of one *Paramecium* species by another (d) the distribution of trees along a moisture gradient in Wisconsin forests (e) the effects of the removal of a dominant rodent species from an Arizona desert

CRITICAL THINKING

1. In what symbiotic relationships are humans involved?

2. Describe the ecological niche of humans. Do you think our realized niche has changed during the past 1000 years? Why or why not?

3. How is the vertical distribution of barnacles in Scotland an example of competitive exclusion?

4. In your opinion, are humans a dominant species or a keystone species? Explain your answer.

5. Many plants that produce nodules for nitrogen-fixing bacteria are common on disturbed sites. Explain how these plants might simultaneously compete with and facilitate other plant species.

6. **Evolution Link.** The rough-skinned newt, which lives in western North America, stores a poison in its skin and is avoided by predators. However, several populations of garter snakes have undergone one or a few mutations that enable them to tolerate the toxin, and these snakes eat the newts with no ill effects. How has natural selection affected this predator–prey relationship? Based on what you have learned about evolutionary arms races, predict what may happen to the newts and the poison-resistant garter snakes over time.

7. **Evolution Link.** Competition is an important part of Darwin's theory of evolution by natural selection, and the evolution of features that reduce competition increases a population's overall fitness. Relate this idea to character displacement and resource partitioning in Darwin's finches.

8. **Analyzing Data.** Examine the top and middle graphs in Figure 53-4. Are these examples of exponential or logistic population growth? Where is *K* in each graph? (You may need to refer to Chapter 52 to answer these questions.)

9. **Analyzing Data.** Examine Figure 53-16. Approximately what percentage of species have colonized islands 2000 km from New Guinea? 6000 km from New Guinea? Based on your answers, explain the relationship between species richness and geographic isolation.

Additional questions are available in ThomsonNOW at www.thomsonedu.com/login

Ecosystems and the Biosphere

Michael Stubben / Visuals Unlimited

Beaver pond. Beavers constructed a dam out of branches and mud (*at far end of pond*), converting a small creek into a pond. Photographed on Shingle Creek, Utah.

KEY CONCEPTS

Ecosystems have two sets of interacting parts: the living, or biotic, component and the nonliving, or abiotic, component.

The abiotic environment—including solar radiation, the atmosphere, the ocean, climate, and fire—helps shape the biotic portion of ecosystems.

Studying the energy content of the different trophic levels provides insight into how energy flows through the biotic and abiotic components of ecosystems.

Carbon, nitrogen, water, and other materials cycle through both biotic and abiotic parts of ecosystems.

Ecosystems are regulated from the top down (for example, predators whose effects move down through the food web) or from the bottom up (for example, the availability of food, which exerts an influence from the base of the food web).

Almost completely isolated from everything in the universe but sunlight, planet Earth has often been compared to a vast spaceship whose life-support system consists of the communities of organisms that inhabit it, plus energy from the sun. These organisms produce oxygen, transfer energy, and recycle water and minerals (inorganic nutrients) with great efficiency. Yet none of these ecological processes would be possible without the abiotic (nonliving) environment of spaceship Earth. As the sun warms the planet, it powers the hydrologic cycle (causes precipitation), drives ocean currents and atmospheric circulation patterns, and produces much of the climate to which organisms have adapted. The sun also supplies the energy that almost all organisms use to carry on life processes.

Individual communities and their abiotic environments are **ecosystems,** which are the basic units of ecology. An ecosystem encompasses all the interactions among organisms living together in a particular place, and among those organisms and their abiotic environment.

Ecosystem interactions are complex because each organism responds not only to other organisms but to conditions in the atmosphere, soil, and water. In turn, organisms exert an effect on the abiotic environment, as when a beaver dam creates a pond in a formerly forested area (see photograph). The pond is formed as the beaver builds an island lodge that will be safe from predators. However, the beaver dam also regulates the flow of water in the stream or river: it holds back water during rainy periods and releases a controlled amount of water throughout the year, even during periods of drought.

Like communities, ecosystems vary in size, lack precise boundaries, and are nested within larger communities. Earth's largest ecosystem is the biosphere, which consists of all of Earth's communities and their interactions and connections with the planet's abiotic environment—its water, soil, rock, and atmosphere. ∎

ENERGY FLOW THROUGH ECOSYSTEMS

Learning Objectives

1   Summarize the concept of energy flow through a food web.
2   Explain typical pyramids of numbers, biomass, and energy.
3   Distinguish between gross primary productivity and net primary productivity.

The passage of energy in a one-way direction through an ecosystem is known as **energy flow.** Energy enters an ecosystem as radiant energy (sunlight), a tiny portion (less than 1%) of which *producers* trap and use during photosynthesis. The energy, now in chemical form, is stored in the bonds of organic (carbon-containing) molecules such as glucose. When cellular respiration breaks these molecules apart, energy becomes available (in the form of ATP) to do work, such as repairing tissues, producing body heat, moving about, or reproducing. As the work is accomplished, energy escapes the organisms and dissipates into the environment as heat. Ultimately, this heat energy radiates into space. Thus, once an organism has used energy, the energy is unavailable for reuse (∎ Fig. 54-1; see also the discussion of the second law of thermodynamics in Chapter 7).

In an ecosystem, energy flow occurs in **food chains,** in which energy from food passes from one organism to the next in a sequence. **Primary producers,** also called *autotrophs,* or simply, *producers,* form the beginning of the food chain by capturing the sun's energy through photosynthesis. Producers, by incorporating the chemicals they manufacture into their own *biomass* (living material), become potential food resources for other organisms. Plants are the most significant producers on land, whereas algae and cyanobacteria are important producers in aquatic environments.

All other organisms in a community are **consumers,** also called *heterotrophs,* that extract energy from organic molecules produced by other organisms. **Herbivores** are consumers that eat plants, from which they obtain the chemical energy of the producers' molecules and the building materials used to construct their own tissues. Herbivores are, in turn, consumed by **carnivores,** consumers that reap the energy stored in the herbivores' molecules. Other consumers, called **omnivores,** eat a variety of organisms, both plant and animal.

Some consumers, called **detritus feeders,** or **detritivores,** eat **detritus,** which is dead organic matter that includes animal carcasses, leaf litter, and feces. Detritus feeders and microbial decomposers destroy dead organisms and waste products. **Decomposers,** also called **saprotrophs,** include microbial heterotrophs that supply themselves with energy by breaking down organic molecules in the remains (carcasses and body wastes) of all members of the

---

Key Point

Ecologists gain insights into how ecosystems function by examining energy flow and the energy content of each trophic level.

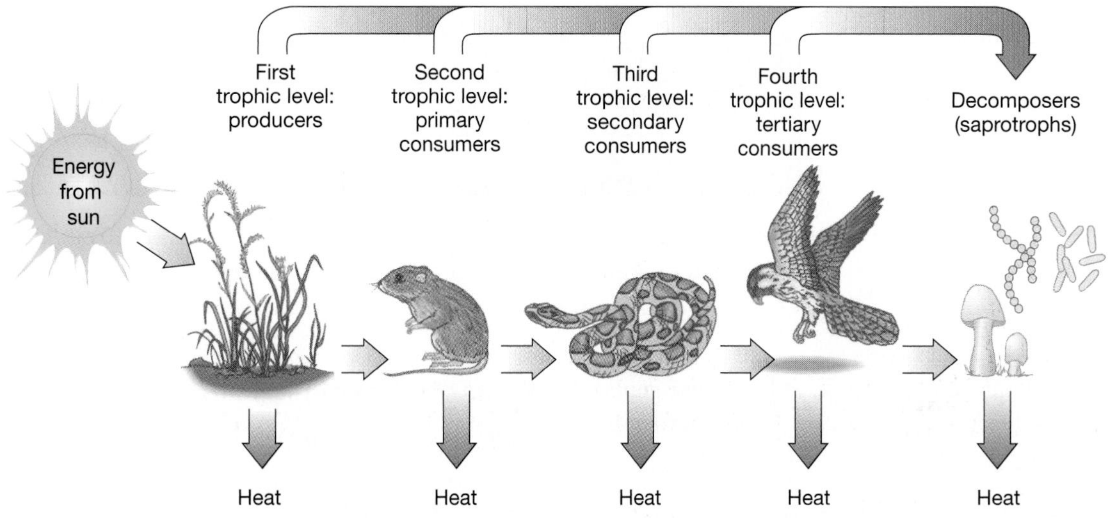

Figure 54-1   *Animated*   Energy flow through ecosystems

Energy enters ecosystems from an external source (the sun) and exits as heat loss. As stipulated by the second law of thermodynamics, most of the energy acquired by a given trophic level is released into the environment as heat and is therefore unavailable to the next trophic level.

food chain. They typically release simple inorganic molecules, such as carbon dioxide and mineral salts, that may be reused by producers. Most bacteria and fungi are important decomposers.

Simple food chains such as the one just described rarely occur in nature, because few organisms eat just one kind, or are eaten by just one other kind, of organism. More typically, the flow of energy and materials through ecosystems takes place in accordance with a range of food choices for each organism. In an ecosystem of average complexity, hundreds of alternative pathways are possible. Thus, a **food web,** which is a complex of interconnected food chains in an ecosystem, is a more realistic model of the flow of energy and materials through ecosystems (▮ Fig. 54-2).

Food webs are divided into **trophic levels** (from the Greek *tropho,* which means "nourishment"). Producers (organisms that photosynthesize) occupy the first trophic level, primary consumers (herbivores) occupy the second, secondary consumers (carnivores and omnivores) the third, and so on (see Fig. 54-1).

Because food chains and food webs are descriptions of "who eats whom," they indicate the negative effects predators have on their prey. For example, consider a simple food chain: grass ⟶ field mouse ⟶ owl. The owl, which kills and eats mice, obviously exerts a negative effect on the mouse population; in like manner, field mice, which eat grass seeds, reduce the grass population.

A trophic level in a food web also influences other trophic levels to which it is not directly linked. Producers and top carnivores do not exert direct effects on one another, yet each indirectly affects the other. In our example, the owls help the grasses by keeping the population of seed-eating mice under control. Likewise, the grasses benefit owls by supporting a population of mice on which the owl population feeds. These indirect interactions may be as important in food-web dynamics as direct predator–prey interactions.

The most important thing to remember about energy flow in ecosystems is that it is linear, or one-way. That is, as long as it is not used to do biological work, energy moves along a food web from one trophic level to the next trophic level. Once an organism has used energy, however, it is lost as heat and is unavailable to any other organism in the ecosystem.

Food webs in all but the simplest ecosystems are too complex to depict all the species and links actually present. Rarely do food-web diagrams take into account that some links are strong and others weak. Moreover, food webs change over time, with additions and deletions of links.

Figure 54-2 *Animated* A food web at the edge of an Eastern deciduous forest
This diagram is greatly simplified compared with what actually happens in nature.

Ecological pyramids illustrate how ecosystems work

Ecologists sometimes compare trophic levels by determining the number of organisms, the biomass, or the relative energy found at each level. This information is presented graphically as **ecological pyramids.** The base of each ecological pyramid represents the producers, the next level is the primary consumers (herbivores), the level above that is the secondary consumers (carnivores), and

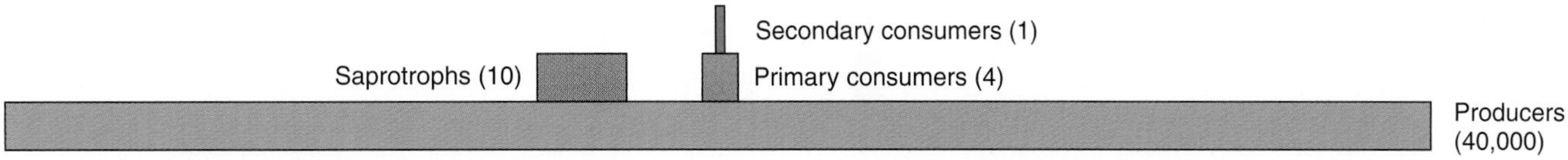

Secondary consumers (1)
Saprotrophs (10)   Primary consumers (4)

Producers
(40,000)

(a) A pyramid of biomass for a tropical forest in Panama.

Primary consumers (21)
Producers (4)

(b) An inverted biomass pyramid, such as that for plankton in the English Channel, occurs when a highly productive lower trophic level experiences high rates of turnover. Plankton are free-floating, mainly microscopic algae and animals.

Figure 54-3   Pyramids of biomass

These pyramids are based on the biomass at each trophic level and generally have a pyramid shape with a large base and progressively smaller areas for each succeeding trophic level. Biomass values are in grams of dry weight per square meter. (**a, b,** Adapted from E. P. Odum, *Fundamentals of Ecology*, 3rd ed., W. B. Saunders Company, Philadelphia, 1971, and based on studies by F.B. Golley and G. I. Child [a] and H. W. Harvey [b].)

so on. The relative area of each bar of the pyramid is proportional to what is being demonstrated.

A **pyramid of numbers** shows the number of organisms at each trophic level in a given ecosystem, with a larger area illustrating greater numbers for that section of the pyramid. In most pyramids of numbers, fewer organisms occupy each successive trophic level. Thus, in African grasslands the number of herbivores, such as zebras and wildebeests, is greater than the number of carnivores, such as lions. Inverted pyramids of numbers, in which higher trophic levels have more organisms than lower trophic levels, are often observed among decomposers, parasites, and herbivorous insects. One tree provides food for thousands of leaf-eating insects, for example. Pyramids of numbers are of limited usefulness, because they do not indicate the biomass of the organisms at each level and they do not indicate the amount of energy transferred from one level to another.

A **pyramid of biomass** illustrates the total biomass at each successive trophic level. **Biomass** is a quantitative estimate of the total mass, or amount, of living material; it indicates the amount of fixed energy at a particular time. Biomass units of measure vary: biomass may be represented as total volume, dry weight, or live weight. Typically, these pyramids illustrate a progressive reduction of biomass in succeeding trophic levels (Fig. 54-3a). Assuming an average biomass reduction of about 90% for each trophic level,[1] 10,000 kg of grass should support 1000 kg of grasshoppers, which in turn support 100 kg of frogs. If we use this logic, the biomass of frog eaters (such as snakes) could weigh, at most, only about 10 kg. From this brief exercise, you see that although carnivores do not eat producers, a large producer biomass is required to support carnivores in a food web.

Occasionally we find an inverted pyramid of biomass in which the primary consumers outweigh the producers (Fig. 54-3b). In these instances, herbivores such as fishes and zooplankton (protozoa, tiny crustaceans, and immature stages of many aquatic animals) consume large numbers of producers, which are usually unicellular algae that are short-lived and reproduce quickly. Thus, although at any point in time relatively few algae are present, the rate of biomass production of the primary consumers is much less than that of the producers.

A **pyramid of energy** indicates the energy content, often expressed as kilocalories per square meter per year, of the biomass of each trophic level. A common method ecologists use to measure energy content is to burn a sample of tissue in a calorimeter; the heat released during combustion is measured to determine the energy content of the organic material in the sample. Energy pyramids, which always have large bases and get progressively smaller through succeeding trophic levels, show that most energy dissipates into the environment when there is a transition from one trophic level to the next. Less energy reaches each successive trophic level from the level beneath it because those organisms at the lower level use some of the energy to perform work, and some of it is lost (Fig. 54-4). (Remember that no biological process is 100% efficient.) The second law of thermodynamics explains why there are few trophic levels: Food webs are short because of the dramatic reduction in energy content that occurs at each successive trophic level.

Ecosystems vary in productivity

The **gross primary productivity (GPP)** of an ecosystem is the rate at which energy is captured during photosynthesis.[2] Thus, GPP is the total amount of photosynthetic energy captured in a given period. Of course, plants and other producers must respire to provide energy for their life processes, and cellular respiration acts as a drain on photosynthetic output. Energy that remains in plant tissues after cellular respiration has occurred is called **net primary productivity (NPP).** That is, NPP is the amount of biomass (the energy stored in plant tissues) found in excess of that broken down by a plant's cellular respiration for normal daily activities. NPP represents the rate at which this organic matter is actually incorporated into plant tissues to produce growth.

Net primary productivity	=	gross primary productivity	−	plant respiration
(plant growth per unit area per unit time)		(total photosynthesis per unit area per unit time)		(per unit area per unit time)

[1] The 90% reduction in biomass is an approximation; actual biomass reduction from one trophic level to the next varies widely in nature.

[2] Gross and net primary productivities are referred to as primary because plants and other producers occupy the first position in food webs.

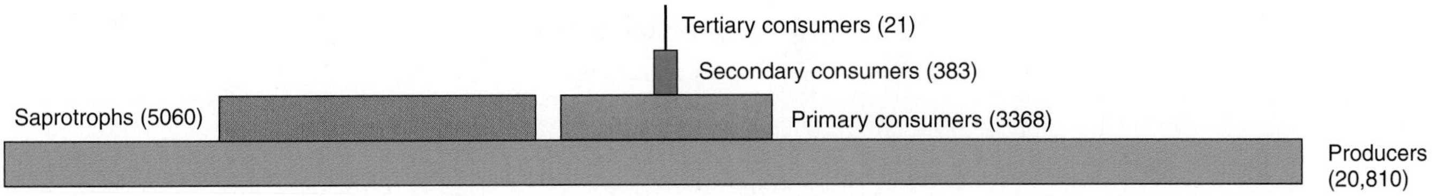

Figure 54-4 Pyramid of energy

A pyramid of energy for Silver Springs, Florida, represents energy flow, the functional basis of ecosystem structure. Energy values are in kilocalories per square meter per year. Note the substantial loss of usable energy from one trophic level to the next. The Silver Springs ecosystem is complex, but tape grass (producers), spiral-shelled snails (primary consumers), young river turtles (secondary consumers), gar (fish; tertiary consumers), and bacteria and fungi (saprotrophs) are representative organisms. When they are young, river turtles are carnivores and consume snails, aquatic insects, and worms; as adults, river turtles are herbivores. (Based on H. T. Odum, "Trophic Structure and Productivity of Silver Springs, Florida," *Ecological Monographs*, Vol. 27, 1957.)

Only the energy represented by net primary productivity is available for consumers, and of this energy only a portion is actually used by them. Both GPP and NPP are expressed as energy per unit area per unit time (for example, kilocalories of energy fixed by photosynthesis per square meter per year) or as dry weight (that is, grams of carbon incorporated into tissue per square meter per year).

Herbivores and other consumers eventually consume all of a plant's net primary production. What happens to this energy? Consider the transfer of net primary production from a plant to a deer that eats the plant. Much of the energy stored in the plant material that the deer consumes—about 25%—is not digested and is lost in its feces. (This energy is not lost from the ecosystem, because detritivores and decomposers will make use of it; it is lost from the deer's point of view, however.) Perhaps 55% of the energy that the deer takes in as food is released during cellular respiration and used to do work such as muscle contraction and to maintain and repair the deer's body. The remaining energy—less than 20%—is available to produce new biomass, that is, new tissues. This net energy available for biomass production by consumer organisms is called **secondary productivity.** An ecosystem's secondary productivity is based on its primary productivity.

Many factors may interact to determine primary productivity. Some plants are more efficient than others in fixing carbon. Environmental factors are also important. These include the availability of solar energy, minerals, and water; other climate factors; the degree of maturity of the community; and the severity of human modification of the environment. Many of these factors are difficult to assess, particularly on a large scale. The summer of 2003, which was extremely hot and dry in Europe, provided ecologists there with a chance to measure primary productivity changes in response to these unusual conditions. Based on their measurements, they estimate that GPP throughout Europe was reduced by 30% as a result of the heat and drought.

Ecosystems differ strikingly in their primary productivities (▌Fig. 54-5 and ▌Table 54-1). On land, tropical rain forests have the highest productivity, probably as a result of the abundant rainfall, warm temperatures, and intense sunlight. As you might expect, tundra, because of its short, cool growing season, and deserts, because of their lack of precipitation, are the least productive terrestrial ecosystems. In ecosystems with comparable annual temperatures (such as temperate deciduous forest, temperate grassland, and temperate desert), water availability affects NPP. Availability of essential minerals such as nitrogen and phosphorus also affects NPP.

Wetlands (swamps and marshes) connect terrestrial and aquatic environments and are extremely productive. The most productive aquatic ecosystems are algal beds, coral reefs, and estuaries. The lack of available minerals in the sunlit region of the open ocean makes this area extremely unproductive, equivalent to an aquatic desert. Earth's major aquatic and terrestrial ecosystems are discussed in Chapter 55.

TABLE 54-1

Net Primary Productivities (NPP) for Selected Ecosystems

Ecosystem	Average NPP (g dry matter/m²/year)
Algal beds and reefs	2500
Tropical rain forest	2200
Swamp and marsh	2000
Estuaries	1500
Temperate evergreen forest	1300
Temperate deciduous forest	1200
Savanna	900
Boreal (northern) forest	800
Woodland and shrubland	700
Agricultural land	650
Temperate grassland	600
Upwelling zones in ocean	500
Lake and stream	250
Arctic and alpine tundra	140
Open ocean	125
Desert and semidesert scrub	90
Extreme desert (rock, sand, ice)	3

Source: Based on R. H. Whittaker, *Communities and Ecosystems*, 2nd ed., Macmillan, New York, 1975.

As primary productivity increases, species richness declines

Ecologists have observed that an ecosystem's species richness declines with increasing productivity. For example, the resource-poor depths of the Atlantic Ocean's abyssal plain have more species richness than productive shallow waters near the coasts. Ecologists are designing experiments to help explain the pattern, which has been documented with rodents in Israel, birds in South America, and large mammals in Africa. Mathematical ecosystem models suggest that a less productive environment has a *patchy distribution of resources* that reduces competition and allows a greater variety of organisms to co-exist (see Chapter 53).

The bad news for global biodiversity is that humans are constantly enriching the environment, for example, with nitrogen and phosphorus inputs from fossil fuels, fertilizers, and livestock. This continual fertilization may make the Earth's ecosystems more and more productive, a shift that some ecologists think could cost the world a substantial loss of species richness. (Other factors that affect species richness were discussed in Chapter 53.)

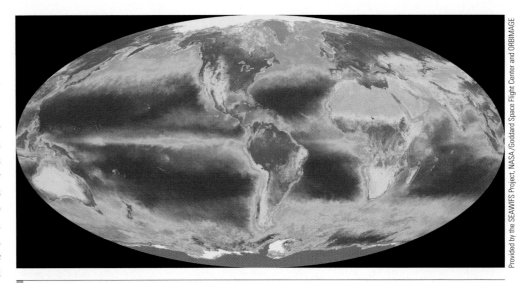

Figure 54-5   A measure of Earth's primary productivity

The data are from a satellite launched as part of NASA's Mission to Planet Earth in 1997. The satellite measured the amount of plant life on land as well as the concentration of phytoplankton (algae) in the ocean. On land, the most productive areas, such as tropical rain forests, are dark green, and the least productive ecosystems (deserts) are orange. In the ocean and other aquatic ecosystems, the most productive regions are red, followed by orange, yellow, green, and blue (the least productive). Data are not available for the gray area.

Humans consume an increasingly greater percentage of global primary productivity

Humans consume far more of Earth's resources than do any of the other millions of animal species. Peter Vitousek and coworkers at Stanford University calculated in 1986 how much of the global NPP is appropriated for the human economy. When both direct and indirect human impacts are accounted for, Vitousek estimated that humans use 32% to 40% of land-based annual NPP. Since 1986, scientists have done much research on global ecology, resulting in improved data sets. In 2001, Stuart Rojstaczer and coworkers at Duke University used current satellite-based data to determine a conservative estimate of land-based annual NPP appropriation by humans at 32%.

The take-home message from Vitousek's and Rojstaczer's research is simple. Essentially, human use of global productivity is competing with other species' needs for energy. Our use of so much of the world's productivity may contribute to the loss, through extinction or genetic impoverishment, of many species that have unique roles in maintaining functional ecosystems. Clearly, at these levels of consumption and exploitation of the Earth's resources, human population growth threatens the planet's ability to support all of its occupants.

Food chains and poisons in the environment

You have seen how energy flows through food chains in ecosystems. Before leaving the discussion of food chains, let us consider how certain toxins, including some pesticides, radioactive isotopes, heavy metals such as mercury, and industrial chemicals such as PCBs, enter and pass through food chains. The effects of the pesticide DDT on some bird species first drew attention to the problem. Falcons, pelicans, bald eagles, ospreys, and many other birds are very sensitive to traces of DDT in their tissues. A substantial body of scientific evidence indicates that one effect of DDT on these birds is that their eggs have extremely thin, fragile shells that usually break during incubation, causing the chicks' deaths. In 1962, U.S. biologist Rachel Carson published *Silent Spring,* which heightened public awareness about the dangers of DDT and other pesticides. After 1972, the year DDT was banned in the United States, the reproductive success of many birds gradually improved.

The impact of DDT on birds is the result of three characteristics of DDT (and other toxins that cause problems in food webs): its persistence, bioaccumulation, and biological magnification. Some toxins are extremely stable and may take many years to break down into less toxic forms. The **persistence** of synthetic pesticides and industrial chemicals is a result of their novel chemical structures. These toxins accumulate in the environment because ways to degrade them have not evolved in natural decomposers such as bacteria.

When an organism does not metabolize (break down) or excrete a persistent toxin, the toxin simply gets stored, usually in fatty tissues. Over time, the organism may accumulate high concentrations of the toxin. The buildup of such a toxin in an organism's body is known as **bioaccumulation.**

QUESTION: Does DDT exhibit biological magnification as it moves through a food chain?

HYPOTHESIS: DDT, a persistent insecticide, increases in concentration at each level of a food chain.

EXPERIMENT: Biologists sampled the concentration of DDT in various organisms of a Long Island salt marsh.

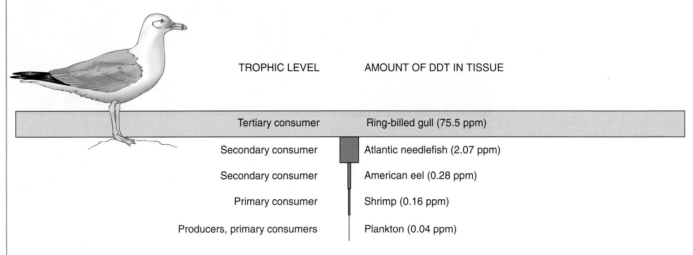

TROPHIC LEVEL	AMOUNT OF DDT IN TISSUE
Tertiary consumer	Ring-billed gull (75.5 ppm)
Secondary consumer	Atlantic needlefish (2.07 ppm)
Secondary consumer	American eel (0.28 ppm)
Primary consumer	Shrimp (0.16 ppm)
Producers, primary consumers	Plankton (0.04 ppm)

RESULTS AND CONCLUSION: The level of DDT increased in the tissues of various organisms as DDT moved through the food chain from producers to consumers. Ring-billed gulls at the top of the food chain had approximately 1 million times as much DDT in their tissues as the concentration of DDT in the water (0.00005 ppm). (Plankton consisted of a mixture of phytoplankton and zooplankton.)

Figure 54-6   Biological magnification of DDT (expressed as parts per million) in a Long Island salt marsh

(Based on data from G. M. Woodwell, C. F. Worster Jr., and P. A. Isaacson, "DDT Residues in an East Coast Estuary: A Case of Biological Concentration of a Persistent Insecticide," *Science,* Vol. 156, May 12, 1967.)

Organisms at higher trophic levels in food webs tend to store greater concentrations of bioaccumulated toxins in their bodies than do those at lower levels. The increase in concentration as the toxin passes through successive levels of the food web is known as **biological magnification.**

As an example of the concentrating characteristic of persistent toxins, consider a food chain studied in a Long Island salt marsh that was sprayed with DDT over a period of years for mosquito control (▌ Figure 54-6). Although this example involved a bird at the top of the food chain, it is important to recognize that *all* top carnivores, from fishes to humans, are at risk from biological magnification of persistent toxins. Because of this risk, currently approved pesticides have been tested to ensure they do not persist and accumulate in the environment.

Review

▌ How does energy flow through a food web such as a deciduous forest?

▌ What are trophic levels, and how are they related to ecological pyramids?

▌ How do gross primary productivity (GPP) and net primary productivity (NPP) differ?

CYCLES OF MATTER IN ECOSYSTEMS

Learning Objective

4   Describe the main steps in each of these biogeochemical cycles: the carbon, nitrogen, phosphorus, and hydrologic cycles.

Matter moves in numerous cycles from one part of an ecosystem to another—that is, from one organism to another (in food chains) and from living organisms to the abiotic environment and back again. We call these cycles of matter **biogeochemical cycles** because they involve biological, geologic, and chemical interactions. For all practical purposes, matter cannot escape from Earth's boundaries. The materials organisms use cannot be "lost," although this matter can end up in locations outside the reach of organisms for a long period. Usually materials are reused and often recycled both within and among ecosystems.

We discuss four different biogeochemical cycles of matter—carbon, nitrogen, phosphorus, and water—as representative of all biogeochemical cycles. These four cycles are particularly

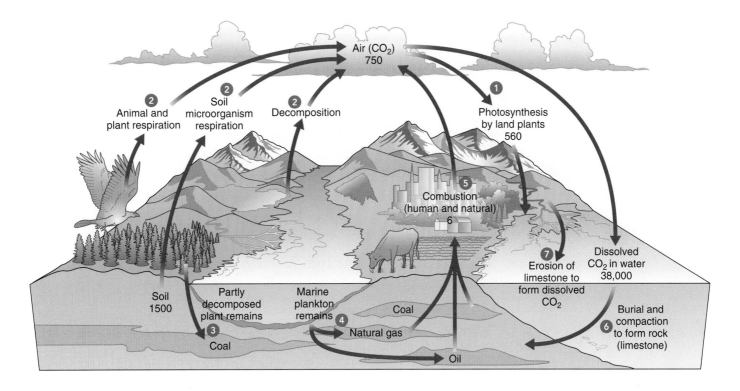

Figure 54-7 *Animated* A simplified diagram of the carbon cycle

All but a tiny fraction of Earth's estimated 10^{23} g of carbon is buried in sedimentary rocks and fossil fuel deposits. The values shown for some of the active pools in the global carbon budget are expressed as 10^{15} g of carbon. For example, the soil contains an estimated 1500×10^{15} g of carbon. (Values from W. H. Schlesinger, *Biogeochemistry: An Analysis of Global Change*, 2nd ed., Academic Press, San Diego, 1997, and several other sources.)

important to organisms because they involve materials used to make the chemical components of cells. Carbon, nitrogen, and water have gaseous components and so cycle over large distances of the atmosphere with relative ease. Phosphorus does not have a gaseous phase, and as a result, only local cycling of phosphorus occurs easily.

Carbon dioxide is the pivotal molecule in the carbon cycle

Proteins, nucleic acids, lipids, carbohydrates, and other molecules essential to life contain carbon. Carbon is present in the atmosphere as the gas carbon dioxide (CO_2), which makes up approximately 0.04% of the atmosphere. It is also present in the ocean and fresh water as dissolved carbon dioxide, that is, carbonate (CO_3^{2-}) and bicarbonate (HCO_3^-); other forms of dissolved inorganic carbon; and dissolved organic carbon from decay processes. Carbon is also present in rocks such as limestone ($CaCO_3$). The global movement of carbon between the abiotic environment, including the atmosphere and ocean, and organisms is known as the **carbon cycle** (❚ Fig. 54-7). Refer to the figure as you read the following description.

As shown in ❶, during photosynthesis, plants, algae, and cyanobacteria remove carbon dioxide from the air and fix, or incorporate, it into complex organic compounds such as glucose. Plants use much of the glucose to make cellulose, starch, amino acids, nucleic acids, and other compounds. Thus, photosynthesis

incorporates carbon from the abiotic environment into the biological compounds of producers.

❷ Many of these compounds are used as fuel for cellular respiration by the producer that made them, by a consumer that eats the producer, or by a decomposer that breaks down the remains of the producer or consumer. The process of cellular respiration returns carbon dioxide to the atmosphere. A similar carbon cycle occurs in aquatic ecosystems between aquatic organisms and dissolved carbon dioxide in the water (*not shown in the figure*).

Sometimes the carbon in biological molecules is not recycled back to the abiotic environment for some time. A large amount of carbon is stored in the wood of trees, where it may stay for several hundred years or even longer. ❸ Also, millions of years ago vast coal beds formed from the bodies of ancient trees that were buried and subjected to anaerobic conditions before they had fully decayed. ❹ Similarly, the oils of unicellular marine organisms probably gave rise to the underground deposits of oil and natural gas that accumulated in the geologic past. Coal, oil, and natural gas, called **fossil fuels** because they formed from the remains of ancient organisms, are vast deposits of carbon compounds, the end products of photosynthesis that occurred millions of years ago. Fossil fuels are nonrenewable resources; that is, Earth has a finite, or limited, supply of them. Although natural forces still form fossil fuels today, they are forming too slowly to replace the fossil fuel reserves we are using.

❺ The process of burning, or combustion, may return the carbon in coal, oil, natural gas, and wood to the atmosphere. In combustion, organic molecules are rapidly oxidized (combined

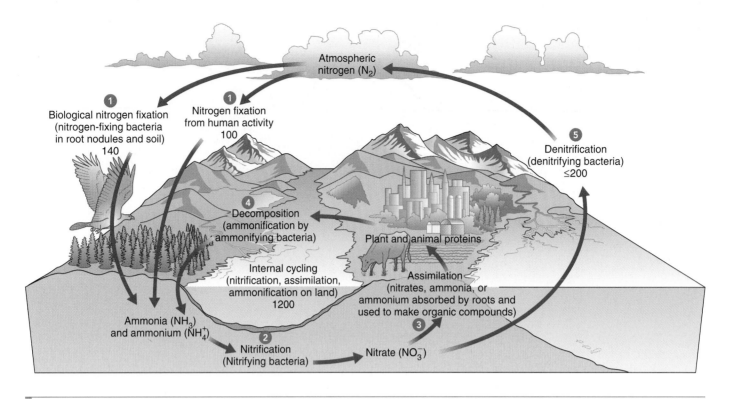

Figure 54-8 *Animated* A simplified diagram of the nitrogen cycle

The largest pool of nitrogen, estimated at 3.9×10^{21} g, is in the atmosphere. The values shown for selected nitrogen fluxes in the global nitrogen budget are expressed as 10^{12} g of nitrogen per year and represent terrestrial values. For example, each year humans fix an estimated 100×10^{12} g of nitrogen. (Values from W. H. Schlesinger, *Biogeochemistry: An Analysis of Global Change*, 2nd ed., Academic Press, San Diego, 1997, and several other sources.)

with oxygen) and converted to carbon dioxide and water with an accompanying release of light and heat.

6 An even greater amount of carbon that is stored for millions of years is incorporated into the shells of marine organisms. When these organisms die, their shells sink to the ocean floor, and sediments cover them, forming seabed deposits thousands of meters thick. The deposits are eventually cemented together to form limestone, a sedimentary rock. Earth's crust is dynamically active, and over millions of years, sedimentary rock on the bottom of the seafloor may lift to form land surfaces. The summit of Mount Everest, for example, consists of sedimentary rock. **7** When the process of geologic uplift exposes limestone, chemical and physical weathering processes slowly erode it away. This returns carbon to the water and atmosphere, where it is available to participate in the carbon cycle once again.

Thus, photosynthesis removes carbon from the abiotic environment and incorporates it into biological molecules. Cellular respiration, combustion, and erosion of limestone return carbon to the water and atmosphere of the abiotic environment.

Human activities have disturbed the global carbon budget

Before the Industrial Revolution, around 1850, the global carbon cycle was in a steady state. Enormous amounts of carbon moved to and from the atmosphere, ocean, and terrestrial ecosystems, but these movements within the global carbon cycle just about canceled out one another. Since 1850, our industrial society has required a lot of energy, and we have burned increasing amounts of fossil fuels—coal, oil, and natural gas—to obtain this energy. This trend, along with a greater combustion of wood as a fuel and the burning of large sections of tropical forest, has released CO_2 into the atmosphere at a rate greater than the natural carbon cycle can handle.

The level of CO_2 increased dramatically beginning in the last half of the 20th century (see Fig. 56-14), and this rise of CO_2 in the atmosphere may cause human-induced changes in climate called *global warming*. Global warming could result in a rise in sea level, changes in precipitation patterns, death of forests, extinction of organisms, and problems for agriculture. It could force the displacement of thousands or even millions of people, particularly from coastal areas. A more thorough discussion of increasing atmospheric CO_2 and the potential impact of global warming is found in Chapter 56.

Bacteria are essential to the nitrogen cycle

Nitrogen is crucial for all organisms because it is an essential part of proteins, nucleic acids, and chlorophyll. Because Earth's atmosphere is about 78% nitrogen gas (N_2), it would appear there could be no possible shortage of nitrogen for organisms. However, molecular nitrogen is so stable that it does not readily combine with other elements. Therefore, the N_2 molecule must be broken apart before the nitrogen atoms combine with other elements to form proteins, nucleic acids, and chlorophyll. Chemical reactions that break up N_2 and combine nitrogen with such elements as oxygen and hydrogen require a great deal of energy.

The **nitrogen cycle,** in which nitrogen cycles between the abiotic environment and organisms, has five steps: nitrogen fixation, nitrification, assimilation, ammonification, and denitrification (❙ Fig. 54-8). Bacteria are exclusively involved in all these steps except assimilation. Refer to the figure as you read the following paragraphs, which describe the nitrogen cycle that occurs on land; a similar cycle occurs in aquatic ecosystems.

In ❶, the first step in the nitrogen cycle, biological **nitrogen fixation** involves conversion of gaseous nitrogen (N_2) to ammonia (NH_3). This process is called *nitrogen fixation* because nitrogen is fixed into a form that organisms can use. Combustion, volcanic action, lightning discharges, and industrial processes also fix nitrogen as nitrate (NO_3^-); each of these supplies enough energy to break apart molecular nitrogen. *Nitrogen-fixing bacteria,* including cyanobacteria and certain other free-living and symbiotic bacteria, carry on biological nitrogen fixation in soil and aquatic environments. Nitrogen-fixing bacteria employ an enzyme called **nitrogenase** to break up molecular nitrogen and combine the resulting nitrogen atoms with hydrogen.

Because nitrogenase functions only in the absence of oxygen, the bacteria that fix nitrogen insulate the enzyme from oxygen in some way. Some nitrogen-fixing bacteria live beneath layers of oxygen-excluding slime on the roots of several plant species. Other important nitrogen-fixing bacteria, in the genus *Rhizobium,* live in oxygen-excluding swellings, or **nodules,** on the roots of legumes such as beans and peas and some woody plants (❙ Fig. 54-9). The nodules contain **leghemoglobin,** an oxygen-binding protein similar in structure to hemoglobin. Even though the nodules exclude much of the oxygen, the small amount that is present binds to leghemoglobin. Leghemoglobin supplies oxygen that the respiring bacteria need.

In aquatic environments, cyanobacteria perform most nitrogen fixation. Filamentous cyanobacteria have special oxygen-excluding cells called **heterocysts** that function to fix nitrogen (see Fig. 24-18). Some water ferns have cavities in which cyanobacteria live, in a manner comparable to the way *Rhizobium* lives in root nodules of legumes. Other cyanobacteria fix nitrogen in symbiotic association with cycads and other terrestrial plants or as the photosynthetic partner of certain lichens.

The reduction of nitrogen gas to ammonia by nitrogenase is a remarkable accomplishment by living organisms, achieved without the tremendous heat, pressure, and energy required during the manufacture of nitrogen fertilizers. Even so, nitrogen-fixing bacteria must expend the energy in 12 g of glucose (or the equivalent) to fix a single gram of nitrogen biologically.

The second step of the nitrogen cycle is **nitrification** (see Fig. 54-8 ❷), the conversion of ammonia (NH_3) or ammonium (NH_4^+, formed when water reacts with ammonia) to nitrate (NO_3^-). Soil bacteria are responsible for the two-phase process of nitrification, which furnishes these bacteria, called *nitrifying bacteria,* with energy.

❸ In the third step, called **assimilation,** roots absorb ammonia (NH_3), ammonium (NH_4^+), or nitrate (NO_3^-) that nitrogen fixation and nitrification formed and incorporate the nitrogen into proteins, nucleic acids, and chlorophyll. When animals con-

Hugh Spencer/Photo Researchers, Inc.

Figure 54-9 Root nodules and nitrogen fixation

Root nodules on the roots of a pea plant provide an oxygen-free environment for nitrogen-fixing *Rhizobium* bacteria that live in them.

sume plant tissues, they assimilate nitrogen by taking in plant nitrogen compounds and converting them to animal nitrogen compounds.

❹ The fourth step is **ammonification,** which is the conversion of organic nitrogen compounds into ammonia (NH_3) and ammonium ions (NH_4^+). Ammonification begins when organisms produce nitrogen-containing wastes such as urea in urine and uric acid in the wastes of birds (see Fig. 47-1). As these substances, along with the nitrogen compounds in dead organisms, decompose, nitrogen is released into the abiotic environment as ammonia (NH_3). The bacteria that perform ammonification in both the soil and aquatic environments are called *ammonifying bacteria.* The ammonia produced by ammonification is available for the processes of nitrification and assimilation. Most available nitrogen in the soil derives from the recycling of organic nitrogen by ammonification.

❺ The fifth step of the nitrogen cycle is **denitrification,** which is the reduction of nitrate (NO_3^-) to gaseous nitrogen (N_2). Denitrifying bacteria reverse the action of nitrogen-fixing and nitrifying bacteria by returning nitrogen to the atmosphere as nitrogen gas. Denitrifying bacteria are anaerobic and therefore live and grow best where there is little or no free oxygen. For example, they are found deep in the soil near the water table, an environment that is nearly oxygen-free.

Human activities have changed the global nitrogen budget

Human activities have disturbed the balance of the global nitrogen cycle. During the 20th century, humans more than doubled the amount of fixed nitrogen (nitrogen that has been chemically

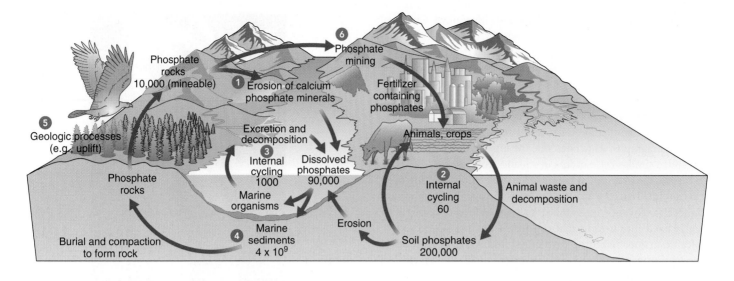

Figure 54-10 *Animated* A simplified diagram of the phosphorus cycle

Some values of the global phosphorus budget are given, in units of 10^{12} g phosphorus per year. For example, each year an estimated 60×10^{12} g of phosphorus cycles from the soil to terrestrial organisms and back to the soil. (Values from W. H. Schlesinger, *Biogeochemistry: An Analysis of Global Change,* 2nd ed., Academic Press, San Diego, 1997, and several other sources.)

combined with hydrogen, oxygen, or carbon) entering the global nitrogen cycle. The excess nitrogen is seriously altering many terrestrial and aquatic ecosystems.

Large quantities of nitrogen fertilizer are produced from nitrogen gas for agriculture. The increasing use of fertilizer has resulted in higher crop yields, but there are negative environmental impacts from human-produced nitrogen. Nitrogen fertilizer is extremely mobile and is easily transferred from the land to rivers to estuaries to the ocean. Thus, the overuse of commercial fertilizer on the land causes water-quality problems that may help explain long-term declines in many coastal fisheries. The amount of nitrate or ammonium in most aquatic ecosystems is in limited supply and therefore limits the growth of algae. Rain washes fertilizer into rivers and lakes, where it stimulates the growth of algae, some of which are toxic. As these algae die, their decomposition by bacteria robs the water of dissolved oxygen, which in turn causes other aquatic organisms, including many fishes, to die of suffocation.

Nitrates from fertilizer also leach (dissolve and wash down) through the soil and contaminate groundwater. Many people who live in rural areas drink groundwater, and groundwater contaminated by nitrates is dangerous, particularly for infants and small children.

Another human activity that affects the nitrogen cycle is the combustion of fossil fuels. When fossil fuels are burned, the nitrogen locked in organic compounds in the fuel is chemically altered and transferred to the atmosphere. In addition, the high temperature of combustion converts some atmospheric nitrogen to **nitrogen oxides.** Automobile exhaust is one of the main sources of nitrogen oxides.

Nitrogen oxides are a necessary ingredient in the production of **photochemical smog,** a mixture of several air pollutants that injure plant tissues, irritate eyes, and cause respiratory problems in humans. Nitrogen oxides also react with water in the atmosphere to form nitric acid (HNO_3) and nitrous acid (HNO_2). When these and other acids leave the atmosphere as **acid deposition,** they decrease the pH of surface waters (lakes and streams) and soils. Acid deposition has been linked to declining animal populations in aquatic ecosystems. On land, acid deposition alters soil chemistry so that certain essential minerals, such as calcium and potassium, wash out of the soil and are therefore unavailable for plants. Nitrous oxide (N_2O), one of the nitrogen oxides, retains heat in the atmosphere (like CO_2) and so promotes global warming. Nitrous oxide also contributes to the depletion of ozone in the stratosphere. (See Chapter 2 for a discussion of acids and pH, and Chapter 56 for a discussion of global warming and stratospheric ozone depletion.)

The phosphorus cycle lacks a gaseous component

Phosphorus does not exist in a gaseous state and therefore does not enter the atmosphere. In the **phosphorus cycle,** phosphorus cycles from the land to sediments in the ocean and back to the land (Fig. 54-10). Refer to the figure as you read the following description.

1 As water runs over rocks containing phosphorus, it gradually erodes the surface and carries off inorganic phosphate (PO_4^{3-}). **2** The erosion of phosphorus rocks releases phosphate into the soil, where it is taken up by roots in the form of inorganic phosphates. Once in cells, phosphates are incorporated into a variety of biological molecules, including nucleic acids, ATP, and the phospholipids that make up cell membranes. Animals obtain most of their required phosphorus from the food they eat, although in some places drinking water may contain a substantial amount of inorganic phosphate. Phosphate released by decomposers becomes part of the pool of inorganic phosphate in

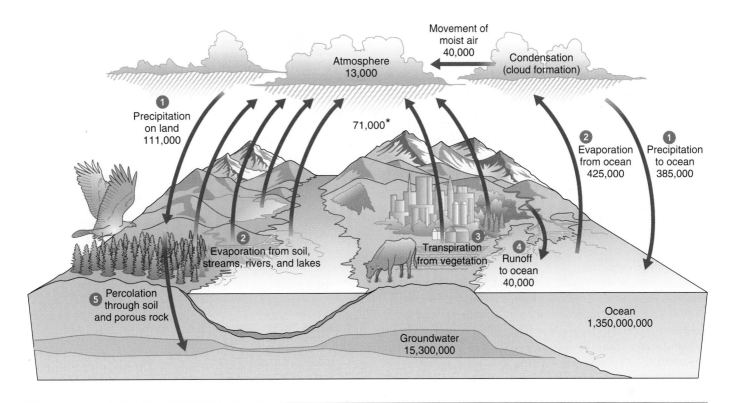

Figure 54-11 *Animated* A simplified diagram of the hydrologic cycle

The global water budget values shown for pools are expressed as cubic kilometers; values for fluxes (movements associated with arrows) are in cubic kilometers per year. The starred value (71,000 km³/yr) is the sum of both transpiration and evaporation from soil, streams, rivers, and lakes. (Values from W. H. Schlesinger, *Biogeochemistry: An Analysis of Global Change,* 2nd ed., Academic Press, San Diego, 1997, and several other sources.)

the soil that plants reuse. Thus, like carbon and nitrogen, phosphorus moves through the food web as one organism consumes another.

❸ Phosphorus cycles through aquatic ecosystems in much the same way as through terrestrial ecosystems. Dissolved phosphate enters aquatic ecosystems through absorption by algae and aquatic plants, which zooplankton and larger organisms consume. In turn, a variety of fishes and mollusks eat the zooplankton. Ultimately, decomposers break down wastes and dead organisms to release inorganic phosphate into the water, making it available for use again by aquatic producers.

❹ Phosphate can be lost for varying time periods from biological cycles. Streams and rivers carry some phosphate to the ocean, where it is deposited on the seafloor and remains for millions of years. ❺ The geologic process of uplift may someday expose these seafloor sediments as new land surfaces, from which phosphate will be once again eroded. ❻ Phosphate deposits are also mined for agricultural use in phosphate fertilizers.

Humans affect the natural cycling of phosphorus

In natural terrestrial communities, very little phosphorus is lost from the cycle, but few communities today are in a natural state, that is, unaltered in some way by humans. Land-denuding practices, such as the clear-cutting of timber, and erosion of agricultural and residential lands accelerate phosphorus loss from the

soil. For practical purposes, phosphorus that washes from the land into the ocean is permanently lost from the terrestrial phosphorus cycle (and from further human use), because it remains in the ocean for millions of years.

Water moves among the ocean, land, and atmosphere in the hydrologic cycle

Life would be impossible without water, which makes up a substantial part of the mass of most organisms. All species, from bacteria to plants and animals, use water as a medium for chemical reactions as well as for the transport of materials within and among cells. (Recall from Chapter 2 that water has many unique properties that help shape the continents, moderate climate, and allow organisms to survive.)

In the **hydrologic cycle** water continuously circulates from the ocean to the atmosphere to the land and back to the ocean (❙ Fig. 54-11). Refer to the figure as you read the following description. ❶ Water moves from the atmosphere to the land and ocean in the form of precipitation (rain, sleet, snow, or hail). ❷ Water that evaporates from the ocean surface and from soil, streams, rivers, and lakes eventually condenses and forms clouds in the atmosphere. ❸ In addition, **transpiration,** the loss of water vapor from land plants, adds a considerable amount of water vapor to the atmosphere. Roughly 97% of the water a plant absorbs from the soil is transported to the leaves, where it is lost by transpiration.

Water may evaporate from land and re-enter the atmosphere directly. ❹ Alternatively, it may flow in rivers and streams to coastal **estuaries,** where fresh water meets the ocean. The movement of surface water from land to ocean is called *runoff,* and the area of land drained by runoff is called a *watershed.* ❺ Water also percolates (seeps) downward in the soil to become *groundwater,* where it is trapped and held for a time. The underground caverns and porous layers of rock in which groundwater is stored are called *aquifers.* Groundwater may reside in the ground for hundreds to many thousands of years, but eventually it supplies water to the soil, streams and rivers, plants, and the ocean. The human removal of more groundwater than precipitation or melting snow recharges, called *aquifer depletion,* eliminates groundwater as a water resource.

Regardless of its physical form (solid, liquid, or vapor) or location, every molecule of water eventually moves through the hydrologic cycle. Tremendous amounts of water cycle annually between Earth and its atmosphere. The volume of water entering the atmosphere from the ocean each year is estimated at about 425,000 km³. Approximately 90% of this water re-enters the ocean directly as precipitation over water; the remainder falls on land. As is true of the other cycles, water (in the form of glaciers, polar ice caps, and certain groundwater) can be lost from the cycle for thousands of years.

Review

■ What are the roles of the following processes in the carbon cycle: photosynthesis, cellular respiration, combustion, and erosion?

■ What is accomplished in each of the five steps in the nitrogen cycle?

■ How does phosphorus cycle without a gaseous component?

ECOSYSTEM REGULATION FROM THE BOTTOM UP AND THE TOP DOWN

Learning Objective

5   Distinguish between bottom-up and top-down processes in ecosystem regulation.

One question that ecologists have recently considered is which regulatory process—bottom-up or top-down—is more significant in the regulation of various ecosystems. Both energy flow and cycles of matter are involved in bottom-up and top-down processes. **Bottom-up processes** are based on food webs that, as you know, always have producers at the first (lowest) trophic level (❚ Fig. 54-12a). In a sense, the biogeochemical cycles that regenerate nutrients such as nitrates and phosphates for producers to assimilate are located "under" the first trophic level. Thus, bottom-up processes regulate ecosystem function by nutrient cycling and by availability of other resources. If bottom-up processes dominate an ecosystem, the availability of resources such as water or soil minerals controls the number of producers, which in turn

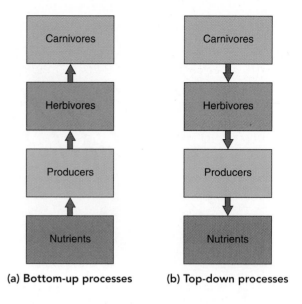

(a) Bottom-up processes   **(b) Top-down processes**

Figure 54-12   Bottom-up and top-down processes

controls the number of herbivores, which controls the number of carnivores. (Recall from Chapter 53 that *limiting resources* tend to restrict the ecological niches of organisms, thereby affecting population size.)

Bottom-up processes apparently predominate in certain aquatic ecosystems in which nitrogen or phosphorus is limiting. An experiment in which phosphorus was added to a phosphorus-deficient river (the Kaparuk River in Alaska) resulted in an increase in algae, followed over time by increased populations of aquatic insects, other invertebrates, and fishes.

In contrast, **top-down processes** regulate ecosystem function by trophic interactions, particularly from the highest trophic level (❚ Fig. 54-12b). Ecosystem regulation by top-down processes occurs because carnivores eat herbivores, which in turn eat producers, which affects levels of nutrients. If top-down processes dominate an ecosystem, the effects of an increase in the population of top predators cascade down the food web through the herbivores and producers. In fact, top-down processes are also known as a *trophic cascade.* A change in the feeding preferences of orcas, formerly known as killer whales, off the coast of Alaska provides an excellent example of a trophic cascade. A few decades ago, when orcas began preying on sea otters, the sea otter population sharply declined. As sea otters have declined, the number of sea urchins, which sea otters eat, has increased. Sea urchins eat kelp, the producers at the base of the food web; the increase in the number of sea urchins has caused a decline in kelp populations.

Top-down regulation appears to predominate in ecosystems with few trophic levels and low species richness. Such ecosystems may have only one or a few species of dominant herbivores, but those species have a strong impact on the producer populations. An excellent example of top-down regulation was discussed in Chapter 52: reindeer introduced to the Pribilof Islands of Alaska overgrazed the vegetation until the plants were almost wiped out.

It may be that top-down and bottom-up regulatory processes are not mutually exclusive. In a 13-year study of a thorn-scrub community in north-central Chile, ecologists demonstrated that top-down regulation predominates in certain small desert mammals and plant species. However, during periodic El Niño events (discussed later in the chapter), the increase in precipitation resulted in bottom-up increases in producers and consumers. In this ecosystem, it appears that both top-down and bottom-up regulatory processes are important in trophic dynamics over an extended period.

Review

■ Biologists think the reintroduction of wolves to Yellowstone National Park, which began in 1995, will ultimately result in a more varied and lush plant composition. Is this an example of bottom-up or top-down processes? Why?

ABIOTIC FACTORS IN ECOSYSTEMS

Learning Objectives

6 Summarize the effects of solar energy on Earth's temperatures.
7 Discuss the roles of solar energy and the Coriolis effect in the production of global air and water flow patterns.
8 Give two causes of regional precipitation differences.
9 Discuss the effects of fire on certain ecosystems.

We have seen how ecosystems depend on the abiotic environment to supply energy and essential materials (in biogeochemical cycles). Other abiotic factors such as solar radiation, the atmosphere, the ocean, climate, and fire also affect ecosystems. For a given abiotic factor, each organism living in an ecosystem has an optimal range in which it survives and reproduces. Water and temperature are probably the two abiotic factors that most affect organisms in ecosystems.

The sun warms Earth

The sun makes life on Earth possible. Without the sun's energy, the temperature on planet Earth would approach absolute zero (0 K or −273°C), and all water would be frozen, even in the ocean. The sun powers the hydrologic cycle, carbon cycle, and other biogeochemical cycles and is the primary determinant of climate. Photosynthetic organisms capture the sun's energy and use it to

make organic compounds that almost all forms of life require. Most of our fuels, such as wood, oil, coal, and natural gas, represent solar energy captured by photosynthetic organisms. Without the sun, almost all life on Earth would cease (see *Focus On: Life without the Sun* for an interesting exception).

The sun's energy, which is the product of a massive nuclear fusion reaction, is emitted into space in the form of electromagnetic radiation, especially ultraviolet, visible, and infrared radiation. About one billionth of the total energy the sun releases strikes the atmosphere, and of this tiny trickle of energy, a minute part operates the biosphere. On average, clouds and surfaces—especially snow, ice, and the ocean—immediately reflect away 30% of the solar radiation that falls on Earth (■ Fig. 54-13). Earth's surface and atmosphere absorb the remaining 70%, which runs the water cycle, drives winds and ocean currents, powers photosynthesis, and warms the planet. Ultimately, the continual radiation of long-wave infrared (heat) energy returns all this energy to space. If heat gains did not exactly balance losses, Earth would heat up or cool down.

Temperature changes with latitude

The most significant local variations in Earth's temperature are produced because the sun's energy does not uniformly reach all places. Our planet's roughly spherical shape and the tilted angle of its axis produce significant variation in the exposure of the

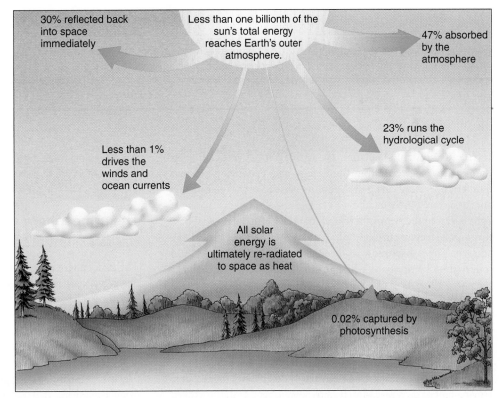

30% reflected back into space immediately

Less than one billionth of the sun's total energy reaches Earth's outer atmosphere.

47% absorbed by the atmosphere

23% runs the hydrological cycle

Less than 1% drives the winds and ocean currents

All solar energy is ultimately re-radiated to space as heat

0.02% captured by photosynthesis

Figure 54-13 The fate of solar radiation that reaches Earth

Most of the energy released by the sun never reaches Earth. The solar energy that does reach Earth warms the planet's surface, drives the hydrologic and other biogeochemical cycles, produces the climate, and powers almost all life through the process of photosynthesis.

LIFE WITHOUT THE SUN

In 1977, an oceanographic expedition aboard the research submersible *Alvin* studied the Galápagos Rift, a deep—greater than 2500 m (8200 ft)—cleft in the ocean floor off the coast of Ecuador. The expedition revealed a series of hydrothermal vents where sea water apparently had penetrated and been heated by the hot rocks below. During its time within Earth, the water had also become charged with mineral compounds, including hydrogen sulfide (H_2S), which is toxic to most species.

However, hydrothermal vents support a rich variety of life-forms (see figure), in contrast with the surrounding seafloor. Many of the species are not found in other habitats. For example, giant, blood-red tube worms almost 3 m (10 ft) in length cluster in great numbers around the vents. Other animals around the hydrothermal vents include unique species of clams, crabs, barnacles, and mussels. Since 1977, hydrothermal vent communities have been identified at many other oceanic sites, such as mid-ocean ridges.

Scientists initially wondered what energy source sustains these organisms. Most deep-sea communities depend on the organic matter that drifts down from the surface waters; in other words, they rely on energy ultimately derived from photosynthesis. Hydrothermal vent communities, however, are too densely clustered and too productive to be dependent on chance encounters with organic material from surface waters. Instead, chemoautotrophic prokaryotes occupy the base of the food web in these aquatic oases. These prokaryotes have enzymes that catalyze the oxidation of hydrogen sulfide to water and sulfur or sulfate. Such chemical reactions are exergonic and provide the energy required to fix CO_2, which is dissolved in the water, into organic compounds. Many of the animals consume the prokaryotes directly by filter-feeding, but others, such as the giant tube worms, get their energy from prokaryotes that live in their tissues.

Scientists continue to generate questions about hydrothermal vent communities.

How do the organisms find and colonize vents, which are ephemeral and widely scattered on the ocean floor? How have the inhabitants of these communities adapted to survive the harsh living conditions, including high pressure, high temperatures, and toxic chemicals? As vent community research continues, scientists hope to discover answers to these and other questions.

A hydrothermal vent community. Chemoautotrophic prokaryotes living in the tissues of these tube worms (*Riftia pachyptila*) extract energy from hydrogen sulfide to manufacture organic compounds. Because these worms lack digestive systems, they depend on the organic compounds provided by the endosymbiotic prokaryotes, along with materials filtered from the surrounding water and digested extracellularly. Also visible in the photograph are some filter-feeding mollusks (*yellow*) and a crab (*white*).

surface to sunlight. The sun's rays strike almost vertically near the equator, concentrating the energy and producing warmer temperatures. Near the poles the sun's rays strike more obliquely and, as a result, are spread over a larger surface area. Also, rays of light entering the atmosphere obliquely near the poles must pass through a deeper envelope of air than those entering near the equator. This causes more of the sun's energy to be scattered and reflected back into space, which further lowers temperatures near the poles. Thus, the solar energy that reaches polar regions is less concentrated and produces lower temperatures.

Temperature changes with season

Earth's inclination on its axis (23.5 degrees from a line drawn perpendicular to the orbital plane) primarily determines the seasons. During half of the year (March 21 to September 22) the Northern Hemisphere tilts toward the sun, concentrating the sunlight and making the days longer (❚ Fig. 54-14). During the other half of the year (September 22 to March 21) the Northern Hemisphere tilts away from the sun, giving it a lower concentration of sunlight and shorter days. The orientation of the Southern Hemisphere is just the opposite at these times. Summer in the Northern Hemisphere corresponds to winter in the Southern Hemisphere.

The atmosphere contains several gases essential to organisms

The atmosphere is an invisible layer of gases that envelops Earth. Oxygen (21%) and nitrogen (78%) are the predominant gases in the atmosphere, accounting for about 99% of dry air; other gases, including argon, carbon dioxide, neon, and helium, make up the remaining 1%. In addition, water vapor and trace amounts of various air pollutants, such as methane, ozone, dust particles, pollen, microorganisms, and chlorofluorocarbons (CFCs) are present. Atmospheric oxygen is essential to plants, animals, and other organisms that respire aerobically; and plants and other photosynthetic organisms also require carbon dioxide.

The atmosphere performs several essential ecological functions. It protects Earth's surface from most of the sun's ultraviolet radiation and X-rays and from lethal amounts of cosmic rays from space. Without this atmospheric shielding, life as we know it would cease. Although the atmosphere protects Earth from high-energy radiation, visible light and some infrared radiation can penetrate, and they warm the surface and the lower atmosphere. This interaction between the atmosphere and solar energy is responsible for weather and climate.

Organisms depend on the atmosphere, but they also help maintain and, in certain instances, modify its composition. For example, atmospheric oxygen increased to its present level as a result of billions of years of photosynthesis. Today, an approximate balance between oxygen-producing photosynthesis and oxygen-using aerobic respiration helps maintain the level of atmospheric oxygen.

The sun drives global atmospheric circulation

In large measure, differences in temperature that are due to variations in the amount of solar energy at different locations on Earth drive the circulation of the atmosphere. The warm surface near the equator heats the air with which it comes into contact, causing this air to expand and rise. As the warm air rises, it flows away from the equator, cools, and sinks again (❙ Fig. 54-15). Much of it recirculates to the same areas it left, but the remainder splits and flows in two directions, toward the poles. The air chills enough to sink to the surface at about 30 degrees north and south latitudes. Similar upward movements of warm air and its subsequent flow toward the poles occur at higher latitudes, farther from the equator. At the poles, the cold polar air sinks and flows toward the lower latitudes, generally beneath the warm air that simultaneously flows toward the poles. The constant motion of air transfers heat from the equator toward the poles, and as the air returns, it cools the land over which it passes. This continuous turnover moderates temperatures over Earth's surface.

The atmosphere exhibits complex horizontal movements

In addition to global circulation patterns, the atmosphere exhibits complex horizontal movements called **winds.** The nature of wind, with its turbulent gusts, eddies, and lulls, is difficult to understand or predict. It results in part from differences in atmospheric pressure and from Earth's rotation.

The gases that constitute the atmosphere have weight and exert a pressure that is, at sea level, about 1013 millibars (14.7 lb/in^2). Air pressure is variable, however, and changes with altitude, temperature, and humidity. Winds tend to blow from areas of high atmospheric pressure to areas of low pressure; the greater the difference between the high and low pressure areas, the stronger the wind.

Earth's rotation influences the direction that wind blows. Because Earth rotates from west to east, wind swerves to the right in the Northern Hemisphere and to the left in the Southern Hemi-

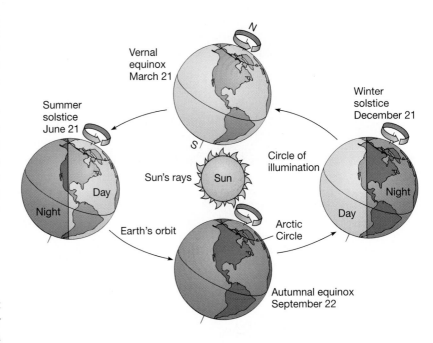

Figure 54-14 Seasonal changes in temperature

Earth's inclination on its axis remains the same as Earth travels around the sun. Thus, the sun's rays hit the Northern Hemisphere obliquely during winter months and more directly during summer months. In the Southern Hemisphere, the sun's rays are oblique during the winter, which corresponds to the Northern Hemisphere's summer. At the equator, the sun's rays are approximately vertical on March 21 and September 22.

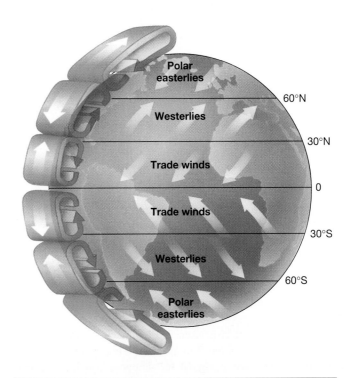

Figure 54-15 *Animated* Atmospheric circulation

The greatest solar energy input occurs at the equator and heats air most strongly in that area. The air rises and travels poleward (*left*) but is cooled in the process so that much of it descends again around 30 degrees latitude in both hemispheres. At higher latitudes the patterns of air movement are more complex.

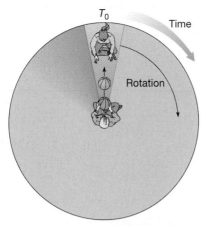

(a) Imagine that while sitting on the merry-go-round, you throw a ball to a friend (also on the merry-go-round) at time T_0, when the merry-go-round is rotating clockwise.

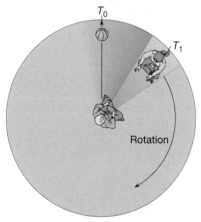

(b) At time T_1, the ball appears to curve to your left instead of going straight. Because of Earth's rotation, winds and ocean currents curve to the left (counterclockwise) in the Southern Hemisphere and to the right (clockwise) in the Northern Hemisphere.

Figure 54-16 The Coriolis effect as demonstrated by a merry-go-round

The center of the merry-go-round (*shown from above*) corresponds to the South Pole; and the outer edge, to the equator.

sphere. This tendency of moving air to be deflected from its path by Earth's rotation is known as the **Coriolis effect.**

The Coriolis effect is visualized by imagining that you and a friend are sitting about 3 m apart on a merry-go-round that is turning clockwise (▌Fig. 54-16). Suppose you throw a ball directly to your friend. By the time the ball reaches the place where your friend was, he or she is no longer in that spot. From where you are sitting, the ball will have swerved far to the left of your friend. This is how the Coriolis effect works in the Southern Hemisphere. To visualize how the Coriolis effect works in the Northern Hemisphere, imagine that you and your friend are sitting on the same merry-go-round, only this time it is moving

counterclockwise. Now, when you throw the ball, it will appear to you to swerve far to the right of your friend.

The global ocean covers most of Earth's surface

The global ocean is a huge body of salt water that surrounds the continents and covers almost three fourths of Earth's surface. It is a single, continuous body of water, but geographers divide it into four sections separated by the continents: the Pacific, Atlantic, Indian, and Arctic Oceans. The Pacific Ocean, which covers one third of Earth's surface and contains more than half of Earth's water, is the largest by far.

Winds drive surface ocean currents

The persistent prevailing winds blowing over the ocean produce mass movements of surface ocean water known as **ocean currents** (▌Fig. 54-17). The prevailing winds generate circular ocean currents called *gyres.* For example, in the North Atlantic, the tropical trade winds tend to blow toward the west, whereas the westerlies in the midlatitudes blow toward the east (see Fig. 54-15). This helps establish a clockwise gyre in the North Atlantic. Thus, surface ocean currents and winds tend to move in the same direction, although there are many variations on this general rule.

The Coriolis effect is partly responsible for the paths that surface ocean currents travel. Earth's rotation from west to east causes surface ocean currents to swerve to the right in the Northern Hemisphere, producing a clockwise gyre of water currents. In the Southern Hemisphere, ocean currents swerve to the left, producing a counterclockwise gyre.

The ocean interacts with the atmosphere

The ocean and the atmosphere are strongly linked. Wind from the atmosphere affects the ocean currents, and heat from the ocean affects atmospheric circulation. One of the best examples of the interaction between ocean and atmosphere is the **El Niño–Southern Oscillation (ENSO)** event. ENSO is a periodic warming of surface waters of the tropical eastern Pacific that alters both oceanic and atmospheric circulation patterns and results in unusual weather in areas far from the tropical Pacific. Normally, westward-blowing trade winds restrict the warmest waters to the western Pacific (near Australia). Every 3 to 7 years, however, the trade winds weaken and the warm water mass expands eastward to South America, raising surface temperatures in the eastern Pacific. Ocean currents, which normally flow westward in this area, slow down, stop altogether, or even reverse and go eastward. The phenomenon is called El Niño (Spanish for "the child") because the warming usually reaches the fishing grounds off Peru just before Christmas. Most ENSOs last from 1 to 2 years.

An ENSO event changes biological productivity in parts of the ocean. The warmer sea-surface temperatures and accompanying changes in ocean circulation patterns off the west coast of South America prevent colder, nutrient-laden deeper waters from **upwelling** (coming to the surface) (▌Fig. 54-18). The lack of nutrients in the water results in a severe decrease in the populations of anchovies and many other marine fishes. For example, during the

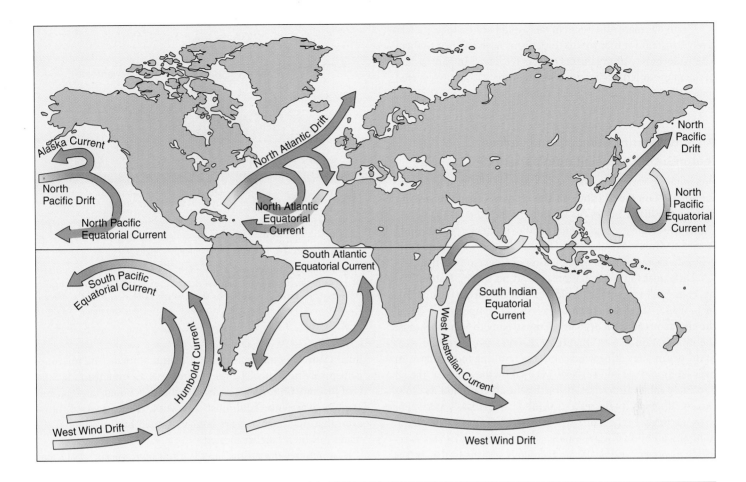

Figure 54-17 *Animated* Major surface ocean currents

1982 to 1983 ENSO, one of the worst ever recorded, the anchovy population decreased by 99%. Other species, such as shrimp and scallops, thrive during an ENSO event. Along the Pacific coast of North America, ENSO shifts the distribution of tropical fishes northward and even affects the salmon run in Alaska.

Climate profoundly affects organisms

Climate is the average weather conditions, plus extremes (records), that occur in a given place over a period of years. The two most important factors that determine an area's climate are temperature (both average temperature and temperature extremes) and precipitation (both average precipitation and seasonal distribution). Other climate factors include wind, humidity, fog, cloud cover, and lightning-caused wildfires. Unlike weather, which changes rapidly, climate generally changes slowly, over hundreds or thousands of years.

Day-to-day variations, day-to-night variations, and seasonal variations are also important dimensions of climate that affect organisms. Latitude, elevation, topography, vegetation, distance from the ocean or other large bodies of water, and location on a continent or other landmass all influence temperature, precipitation, and other aspects of climate.

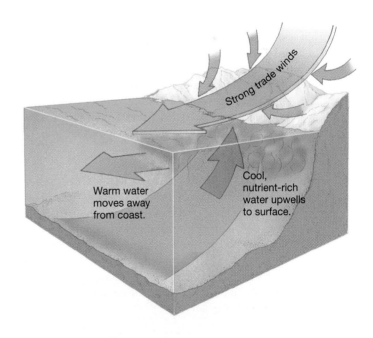

Figure 54-18 *Animated* Upwelling
Coastal upwelling, where deeper waters come to the surface, occurs in the Pacific Ocean along the South American coast. Upwelling provides nutrients for large numbers of microscopic algae, which in turn support a complex food web. Coastal upwelling weakens considerably during years with ENSO events, temporarily reducing fish populations.

Ecosystems and the Biosphere   **1183**

Earth has many different climates, and because they are relatively constant for many years, organisms have adapted to them. The wide variety of organisms on Earth evolved in part because of the many different climates, ranging from cold, snow-covered, polar climates to hot, tropical climates where it rains almost every day.

Air and water movements and surface features affect precipitation patterns

Precipitation varies from one location to another and has a profound effect on the distribution and kinds of organisms present. One of the driest places on Earth is in the Atacama Desert in Chile, where the average annual rainfall is 0.05 cm (0.02 in). In contrast, Mount Waialeale in Hawaii, Earth's wettest spot, receives an average annual precipitation of 1200 cm (472 in).

Differences in precipitation depend on several factors. The heavy-rainfall areas of the tropics result mainly from the uplifting of moisture-laden air. High surface-water temperatures (recall the enormous amount of solar energy striking the equator) cause the evaporation of vast quantities of water from tropical parts of the ocean. Prevailing winds blow the resulting moist, warm air over landmasses. Land surfaces warmed by the sun heat the air and cause moist air to rise. As it rises, the air cools and its moisture-holding ability decreases. (Cool air holds less water vapor than warm air.) When air reaches its saturation point, it cannot hold any additional water vapor; clouds form, and water is released as precipitation. The air eventually returns to the surface on both sides of the equator near the Tropics of Cancer and Capricorn (latitudes 23.5 degrees north and south, respectively). By then, most of the moisture has precipitated so that dry air returns to the equator. This dry air makes little biological difference over the ocean, but its lack of moisture produces some of the great subtropical deserts, such as the Sahara.

Air is also dried during long journeys over landmasses. Near the windward (side from which the prevailing wind blows) coasts of continents, rainfall may be heavy. However, in the temperate zones—the areas between the tropical and polar zones—continental interiors are usually dry because they are far from the ocean that replenishes water vapor in the air passing over it.

Mountains force air to rise, removing moisture from humid air. As it gains altitude, the air cools, clouds form, and precipitation typically occurs, primarily on the windward slopes of the mountains. As the air mass moves down on the other side of the mountain, it is warmed and clouds evaporate, thereby lessening the chance of precipitation of any remaining moisture. This situation exists on the West Coast of North America, where precipitation falls on the western slopes of mountains that are close to the coast. The dry lands on the sides of the mountains away from the prevailing wind (in this case, east of the mountain range) are called **rain shadows** (❚ Fig. 54-19).

Microclimates are local variations in climate

Differences in elevation, in the steepness and direction of slopes, and therefore in exposure to sunlight and prevailing winds may produce local variations in climate known as **microclimates,** which are sometimes quite different from their overall surroundings. Patches of sun and shade on a forest floor, for example, pro-

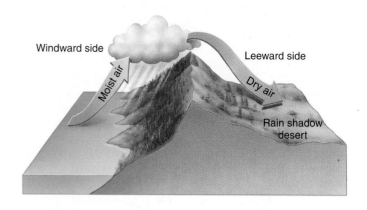

Figure 54-19 *Animated* Rain shadow

A rain shadow is arid or semiarid land that occurs on the leeward side of a mountain. Such a rain shadow occurs east of the Cascade Range in Washington State. The western side of Olympic National Park receives more than 500 cm of precipitation annually, whereas the eastern side receives 40 to 50 cm.

duce a variety of microclimates for plants, animals, and microorganisms living there. The microclimate of an organism's habitat is of primary importance because that is the climate an organism actually experiences and must cope with. (Keep in mind, however, that microclimates are largely affected by the regional climates in which they are located.)

Sometimes organisms modify their own microclimate. For example, trees modify the local climate within a forest so that in summer the temperature is usually lower, and the relative humidity greater, than outside the forest. The temperature and humidity beneath the litter of the forest floor differ still more; in the summer the microclimate of this area is considerably cooler and moister than the surrounding forest. As another example, many desert-dwelling animals burrow to avoid surface climate conditions that would kill them in minutes. The cooler daytime microclimate in their burrows permits them to survive until night, when the surface cools off and they come out to forage or hunt.

Fires are a common disturbance in some ecosystems

Wildfires, which are fires started by lightning, are an important ecological force in many geographic areas. Those areas most prone to wildfires have wet seasons followed by dry seasons. Vegetation that grows and accumulates during the wet season dries out enough during the dry season to burn easily. When lightning hits vegetation or ground litter, it ignites the dry organic material and a fire spreads through the area.

Fires have several effects on organisms. First, combustion frees the minerals that were locked in dry organic matter. The ashes remaining after a fire are rich in potassium, phosphorus, calcium, and other minerals essential for plant growth. With the arrival of precipitation, vegetation flourishes following a fire. Second, fire removes plant cover and exposes the soil. This change stimulates the germination and establishment of seeds requiring bare soil, as well as encourages the growth of shade-intolerant

Figure 54-20 Fire as a tool of ecological management

Here, a controlled burn helps maintain a ponderosa pine (*Pinus ponderosa*) stand in Oregon.

plants. Third, fire causes increased soil erosion because it removes plant cover, leaving the soil more vulnerable to wind and water.

African savanna, California chaparral, North American grasslands, and ponderosa pine forests of the western United States are some fire-adapted ecosystems (see Chapter 55). Fire helps maintain grasses as the dominant vegetation in grasslands by removing fire-sensitive hardwood trees.

Humans try to prevent fires, and sometimes this effort has disastrous consequences. When fire is excluded from a fire-adapted ecosystem, deadwood and other plant litter accumulate. As a result, when a fire does occur, it is much more destructive. The sometimes deadly wildfires in Colorado and other western states and provinces are blamed in part on decades of suppressing fires in the region. Prevention of fire also converts grassland to woody vegetation and facilitates the invasion of fire-sensitive trees into fire-adapted forests.

Controlled burning is a tool of ecological management in which the undergrowth and plant litter are deliberately burned under controlled conditions before they have accumulated to dangerous levels (▌ Fig. 54-20). Controlled burns are also used to suppress fire-sensitive trees, thereby maintaining the natural fire-adapted ecosystem. However, little scientific evidence exists on when controlled burning is appropriate or how often it should be carried out. Moreover, there are major practical and political problems in implementing controlled burns.

Review

- ▌ What basic forces determine the circulation of the atmosphere?
- ▌ What basic forces produce the main ocean currents?
- ▌ What are some of the factors that produce regional differences in precipitation?

STUDYING ECOSYSTEM PROCESSES

Learning Objective

10 Briefly describe some of the long-term ecological research conducted at Hubbard Brook Experimental Forest.

Ecologists conduct detailed ecosystem studies in laboratory simulations and in the field to measure such processes as energy flow, the cycling of nutrients, and the effects of natural and human-induced disturbances (such as air pollution, tree harvesting, and land-use changes). Some ecosystem studies, such as those performed at the Hubbard Brook Experimental Forest (HBEF), a 3100-hectare (7750-acre) reserve in the White Mountain National Forest in New Hampshire, are long term. Beginning in the early 1960s and continuing to the present, HBEF has been the site of numerous studies that address the hydrology (for example, precipitation, surface runoff, and groundwater flow), biology, geology, and chemistry of forests and associated aquatic ecosystems.

Figure 54-21 *Animated* Catchment at Hubbard Brook Experimental Forest

Catchments measure the quantity, timing, and quality of water flowing from a forested watershed.

The National Science Foundation (NSF) has designated HBEF as one of its 24 long-term ecological research sites.

Many of the experiments at HBEF are based on field observations. For example, salamander populations were originally surveyed in forest communities in 1970 and have been resurveyed in recent years. Other studies involve manipulative experiments. In 1978, scientists added dilute sulfuric acid to a small stream in HBEF to study the chemical and biological effects of acidification. This experiment was of practical value, because *acid deposition,* a form of air pollution, has acidified numerous lakes and streams in industrialized countries.

Several researchers have studied the effects on HBEF stream ecosystems of **deforestation,** the clearance of large expanses of forest for agriculture or other uses. When a forest is removed, the total amount of water and minerals that flow into streams increases drastically. ▮ Figure 54-21 shows a concrete dam called a *catchment,* constructed across a stream to measure the flow of water and chemical components out of the ecosystem. Typically, outflow is measured in two separate ecosystems, one of which serves as a control and one of which is experimentally manipulated. These studies demonstrate that deforestation causes soil erosion and leaching of essential minerals that result in decreased soil fertility. The summer temperatures in streams run-

ning through deforested areas are higher than in shady streams running through uncut forests. Many stream organisms do not fare well in deforested areas, in part because they are adapted to cooler temperatures.

Detailed studies such as those at HBEF provide ecologists with insights into how ecological processes function in individual ecosystems. Ecologists compare these data with similar information from other ecosystem studies to develop generalized insights into how ecosystems are structured and how they function. Long-term ecosystem experiments enable ecologists to evaluate and predict the effects of environmental change, including human-induced change.

Ecosystem experiments also contribute to our practical knowledge about how to maintain water quality, wildlife habitat, and productive forests. **Ecosystem management,** a conservation approach that emphasizes restoring and maintaining the quality of an entire ecosystem rather than the conservation of individual species, makes use of such knowledge.

Review

▮ What are some of the environmental effects observed in the deforestation study at Hubbard Brook Experimental Forest?

SUMMARY WITH KEY TERMS

Learning Objectives

1 Summarize the concept of energy flow through a food web (page 1167).

▪ **Energy flow** through an ecosystem is linear, from the sun to producer to consumer to decomposer. Much of this energy is converted to heat as it moves from one organism to another, so organisms occupying the next **trophic level** cannot use it.

▪ Trophic relationships may be expressed as **food chains** or, more realistically, as **food webs,** which show the many alternative pathways that energy may take among the producers, consumers, and decomposers of an ecosystem.

ThomsonNOW™ **Learn more about how trophic levels interact by clicking on the figures in ThomsonNOW.**

2 Explain typical pyramids of numbers, biomass, and energy (page 1167).

▪ **Ecological pyramids** typically express the progressive reduction in numbers of organisms, biomass, and energy found in successive trophic levels. A **pyramid of numbers** shows the number of organisms at each trophic level in a given ecosystem. A **pyramid of biomass** shows the total biomass at each successive trophic level. A **pyramid of energy** indicates the energy content of the biomass of each trophic level.

3 Distinguish between gross primary productivity and net primary productivity (page 1167).

▪ **Gross primary productivity (GPP)** of an ecosystem is the rate at which photosynthesis captures energy. **Net primary productivity (NPP)** is the energy that remains (as biomass) after plants and other producers carry out cellular respiration.

4 Describe the main steps in each of these biogeochemical cycles: the carbon, nitrogen, phosphorus, and hydrologic cycles (page 1172).

▪ Carbon dioxide is the important gas of the **carbon cycle.** Carbon enters plants, algae, and cyanobacteria as CO_2, which photosynthesis incorporates into organic molecules. Cellular respiration, combustion, and erosion of limestone return CO_2 to the water and atmosphere, where it is again available to producers.

▪ The **nitrogen cycle** has five steps. **Nitrogen fixation** is the conversion of nitrogen gas to ammonia. **Nitrification** is the conversion of ammonia or ammonium to nitrate. **Assimilation** is the conversion of nitrates, ammonia, or ammonium to proteins, chlorophyll, and other nitrogen-containing compounds by plants; the conversion of plant proteins into animal proteins is also assimilation. **Ammonification** is the conversion of organic nitrogen to ammonia and ammonium ions. **Denitrification** is the conversion of nitrate to nitrogen gas.

▪ The **phosphorus cycle** has no biologically important gaseous compounds. Phosphorus erodes from rock as inorganic phosphate, which the roots of plants absorb from the soil. Animals obtain the phosphorus they need from their diets. Decomposers release inorganic phosphate into the environment. When phosphorus washes into the ocean and is subsequently deposited in seabeds, it is lost from biological cycles for millions of years.

▪ The **hydrologic cycle** involves an exchange of water between the land, ocean, atmosphere, and organisms. Water enters the atmosphere by evaporation and **transpiration** and leaves the atmosphere as precipitation. On land, water filters through the ground or runs off to lakes, rivers, and the ocean. Aquifers are underground caverns and porous layers of rock in which groundwater is stored. Runoff is the movement of surface water from land to ocean.

ThomsonNOW Explore the carbon, nitrogen, phosphorus, and hydrologic cycles by clicking on the figures in ThomsonNOW.

5 Distinguish between bottom-up and top-down processes in ecosystem regulation (page 1178).

■ If **bottom-up processes** dominate an ecosystem, the availability of resources such as minerals controls the number of producers (that is, the lowest trophic level), which in turn controls the number of herbivores, which in turn controls the number of carnivores.

■ **Top-down processes** regulate ecosystems from the highest trophic level—by consumers eating producers. If top-down processes dominate an ecosystem, an increase in the number of top predators cascades down the food web through the herbivores and producers.

6 Summarize the effects of solar energy on Earth's temperatures (page 1179).

■ Of the solar energy that reaches Earth, 30% is immediately reflected away; the atmosphere and surface absorb the remaining 70%. Ultimately, all absorbed solar energy is reradiated into space as infrared (heat) radiation.

■ A combination of Earth's roughly spherical shape and the tilted angle of its axis concentrates solar energy at the equator and dilutes it at the poles; the tropics are hotter and less variable in climate than are temperate and polar areas.

7 Discuss the roles of solar energy and the Coriolis effect in the production of global air and water flow patterns (page 1179).

■ Visible light and some infrared radiation warm the surface and the lower part of the atmosphere. Atmospheric heat transferred from the equator to the poles produces movement both of warm air toward the poles and of cool air toward the equator, thus moderating the extremes of global climate.

■ **Winds** result in part from differences in atmospheric pressure and from the **Coriolis effect,** the tendency of moving air or water, because of Earth's rotation, to be deflected to the right in the Northern Hemisphere and to the left in the Southern hemisphere.

■ Surface **ocean currents** result in part from prevailing winds and the Coriolis effect.

ThomsonNOW Learn more about atmospheric circulation, ocean currents, and upwelling by clicking on the figures in ThomsonNOW.

8 Give two causes of regional precipitation differences (page 1179).

■ Latitude, elevation, topography, vegetation, distance from the ocean or other large bodies of water, and location on a continent or other landmass influence precipitation.

■ Precipitation is greatest where warm air passes over the ocean, absorbs moisture, and then cools, such as in areas where mountains force humid air upward. Deserts develop in the **rain shadows** of mountain ranges or in continental interiors.

ThomsonNOW Learn more about rain shadows by clicking on the figure in ThomsonNOW.

9 Discuss the effects of fire on certain ecosystems (page 1179).

■ Fire frees the minerals locked in dry organic matter, removes plant cover and exposes the soil, and increases soil erosion. Many ecosystems, such as savanna, chaparral, grasslands, and certain forests, contain fire-adapted organisms.

10 Briefly describe some of the long-term ecological research conducted at Hubbard Brook Experimental Forest (page 1185).

■ Hubbard Brook Experimental Forest (HBEF) is the site of numerous studies that address the hydrology (for example, precipitation, surface runoff, and groundwater flow), biology (for example, effects of deforestation and changes in salamander populations), geology, and chemistry (for example, acid precipitation) of forests and associated aquatic ecosystems.

ThomsonNOW Learn more about the Hubbard Brook Experimental Forest by clicking on the figure in ThomsonNOW.

TEST YOUR UNDERSTANDING

1. A community and its abiotic environment best define a(an) (a) biogeochemical cycle (b) biosphere (c) ecosystem (d) food web (e) trophic level

2. The movement of matter is _____ in ecosystems, and the movement of energy is _____. (a) linear; linear (b) linear; cyclic (c) cyclic; cyclic (d) cyclic; linear (e) cyclic; linear or cyclic

3. A complex of interconnected food chains in an ecosystem is called a(an) (a) ecosystem (b) pyramid of numbers (c) pyramid of biomass (d) biosphere (e) food web

4. Which of the following shows the correct flow of energy through ecosystems? (a) sun ⟶ primary consumer ⟶ secondary consumer ⟶ producer (b) sun ⟶ producer ⟶ secondary consumer ⟶ primary consumer (c) sun ⟶ secondary consumer ⟶ primary consumer ⟶ producer (d) sun ⟶ producer ⟶ primary consumer ⟶ secondary consumer (e) sun ⟶ primary consumer ⟶ producer ⟶ secondary consumer

5. The quantitative estimate of the total amount of living material is called (a) biomass (b) energy flow (c) gross primary productivity (d) plant respiration (e) net primary productivity

6. Which of the following equations shows the relationship between gross primary productivity (GPP) and net primary productivity (NPP)? (a) GPP = NPP − photosynthesis (b) NPP = GPP − photosynthesis (c) GPP = NPP − plant respiration (d) NPP = GPP − plant respiration (e) NPP = GPP − animal respiration

7. Which of the following processes increase(s) the amount of atmospheric carbon in the carbon cycle? (a) photosynthesis (b) cellular respiration (c) combustion (d) a and c (e) b and c

8. In the carbon cycle, carbon is found in (a) limestone rock (b) oil, coal, and natural gas (c) living organisms (d) the atmosphere (e) all of the preceding

9. In the nitrogen cycle, gaseous nitrogen is converted to ammonia during (a) nitrogen fixation (b) nitrification (c) assimilation (d) ammonification (e) denitrification

10. The conversion of ammonia to nitrate, known as _____, is a two-step process performed by soil bacteria. (a) nitrogen fixation (b) nitrification (c) assimilation (d) ammonification (e) denitrification

11. This biogeochemical cycle does not have a gaseous component but cycles from the land to sediments in the ocean and back to the land. (a) carbon cycle (b) nitrogen cycle (c) phosphorus cycle (d) hydrologic cycle (e) neither a nor c has a gaseous component

12. Which of the following processes is *not* directly involved in the hydrologic cycle? (a) transpiration (b) evaporation (c) precipitation (d) nitrification (e) condensation

13. The _____, which results from the rotation of Earth, displaces the paths of atmospheric and ocean currents to the right in the Northern Hemisphere and to the left in the Southern Hemisphere. (a) upwelling (b) prevailing wind (c) microclimate (d) El Niño–Southern Oscillation (e) Coriolis effect

14. The periodic warming of surface waters of the tropical eastern Pacific that alters both oceanic and atmospheric circulation patterns is known as (a) upwelling (b) prevailing wind (c) ocean current (d) El Niño–Southern Oscillation (e) Coriolis effect

15. A mountain range may produce a downwind arid (a) upwelling (b) rain shadow (c) ocean current (d) microclimate (e) ecological pyramid

16. Which of the following is an example of top-down processes? (a) orcas in the Pacific prey on sea otters, causing an increase in the sea urchin population and a decline in kelp populations (b) experiments at Hubbard Brook address the hydrology of forests and associated aquatic ecosystems (c) one tree provides food for thousands of insects (d) in African grasslands, the number of herbivores is greater than the number of carnivores (e) there is bioaccumulation of DDT in a Long Island salt marsh

CRITICAL THINKING

1. Describe the simplest stable ecosystem you can imagine.

2. How might a food web change if all decomposers were eliminated from it?

3. What trophic level(s) is/are you eating if you eat the following:

   a. steamed broccoli

   b. chicken

   c. tuna

   d. mushrooms

4. Why is the cycling of matter essential to the long-term continuance of life?

5. What would happen to the nitrogen cycle if all bacteria were absent? Explain your answer.

6. What is the energy source that powers the wind?

7. How do ocean currents affect climate on land?

8. Would the microclimate of an ant be the same as that of an elephant living in the same area? Why or why not?

9. **Evolution Link.** Ecologist Charles Krebs said that "evolution operates through natural selection, which is ecology in action." Explain what he meant, and relate his idea to one example of ecosystem regulation from the bottom up or the top down.

10. **Analyzing Data.** Examine Figure 54-3b. Explain why this pyramid of biomass is inverted. In other words, how can 4 g of producers support 21 g of primary consumers?

Additional questions are available in ThomsonNOW at www.thomsonedu.com/login

55

Ecology and the Geography of Life

Black-tailed prairie dog. Prairie dogs (*Cynomys ludovicianus*) never wander far from their burrows, which they use to escape from predators.

Raymond Gehman /Corbis

KEY CONCEPTS

Climate, particularly temperature and precipitation, affects the distribution of Earth's major biomes, such as tropical rain forests and tundra.

Abiotic factors—such as water salinity, amount of dissolved oxygen, availability of essential minerals, and water depth—influence the distribution of organisms in aquatic ecosystems.

Species richness varies widely among different biomes and aquatic ecosystems.

Ecotones—areas of transition where two communities meet and intergrade—provide diverse conditions that encourage species richness.

Earth has six biogeographic realms, each consisting of a major landmass separated by the ocean, mountains, or a desert.

Earth has many different environments. *Natural selection* affects an organism's ability to survive and reproduce in a given environment. In natural selection, both **abiotic** (nonliving) and **biotic** (living) factors eliminate the least-fit individuals in a population. Over time, succeeding generations of organisms that live in each biome or major aquatic ecosystem become better adapted to local environmental conditions.

Black-tailed prairie dogs are superbly adapted to their environment. Their teeth and digestive tracts are modified to eat and easily digest the seeds and leaves of grasses that grow in great profusion on the Great Plains of western North America.

Prairie dogs live in large colonies of about 500 individuals. The eyes of every individual in the colony watch for potential danger, and when they see it, prairie dogs call out to warn the rest of the colony (see photograph). When danger approaches, each prairie dog dives into its underground home. Each burrow has at least two openings and consists of an elaborate network of long tunnels with several chambers: a nursery for the young, a sleeping chamber, a toilet chamber, and a listening chamber (close to an entrance). Piles of excavated soil surround the burrow entrances and help prevent flooding during rainstorms.

To survive winter, black-tailed prairie dogs hibernate in their burrows. Their metabolism slows, and they subsist on the stored fat in their bodies. They do not hibernate deeply, however, and may leave their burrows to look for food when the weather warms.

As we have seen for prairie dogs, structural, behavioral, and physiological adaptations for its own particular environment and lifestyle have evolved in each species. As we examine Earth's major terrestrial and aquatic ecosystems, including the species characteristic of each, think about the variety of adaptations that natural selection has produced in organisms in response to their particular environments. ∎

BIOMES

Learning Objectives

1 Define *biome,* and briefly describe the nine major terrestrial biomes, giving attention to the climate, soil, and characteristic plants and animals of each.

2 Describe at least one human effect on each of the biomes discussed.

A **biome** is a large, relatively distinct terrestrial region that has similar climate, soil, plants, and animals regardless of where it occurs. Because it covers such a large geographic area, a biome encompasses many interacting landscapes. Recall from Chapter 52 that a **landscape** is a large land area (several to many square kilometers) composed of interacting ecosystems.

Biomes largely correspond to major climate zones, with temperature and precipitation being most important (∎ Fig. 55-1). Near the poles, temperature is generally the overriding climate factor; whereas in tropical and temperate regions, precipitation becomes more significant than temperature (∎ Fig. 55-2). Other abiotic factors to which biomes are sensitive include temperature extremes, rapid temperature changes, floods, droughts, strong winds, and fires (see section on fires in Chapter 54).

We discuss nine major biomes in this chapter: tundra, boreal forest, temperate rain forest, temperate deciduous forest, temperate grassland, chaparral, desert, savanna, and tropical rain forest (∎ Fig. 55-3). Although we discuss each biome as a distinct entity, biomes intergrade into one another at their boundaries.

Tundra is the cold, boggy plains of the far north

Tundra (also called **arctic tundra**) occurs in extreme northern latitudes wherever snow melts seasonally (∎ Fig. 55-4). The Southern Hemisphere has no equivalent of the arctic tundra because it has no land in the proper latitudes. A similar ecosystem located in the higher elevations of mountains, above the tree line, is called **alpine tundra** to distinguish it from arctic tundra (see *Focus On: The Distribution of Vegetation on Mountains* on page 1195).

Arctic tundra has long, harsh winters and extremely short summers. Although the growing season, with its warmer temperatures, is as short as 50 days, the days are long. Above the Arctic Circle the sun does not set at all for many days in midsummer, although the amount of light at midnight is only one tenth that at noon. There is little precipitation (10 to 25 cm, or 4 to

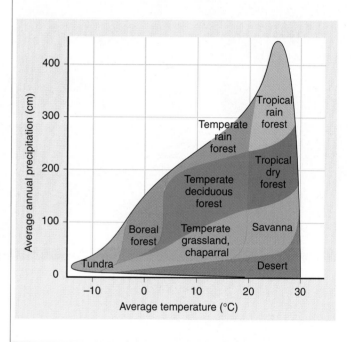

Figure 55-1 Using precipitation and temperature to identify biomes

Factors such as soil type, fire, and seasonality of climate affect whether temperate grassland or chaparral develops. (Adapted from R. H. Whittaker, *Communities and Ecosystems,* 2nd ed., Macmillan, New York, 1975.)

10 in, per year) over much of the tundra, and most of it falls during summer months.

Tundra soils tend to be geologically young, because most were formed only after the last Ice Age.[1] These soils are usually nutrient poor and have little organic litter (dead leaves and stems, animal

---

[1]Glacier ice, which occupied about 30% of Earth's land during the last Ice Age, began retreating about 17,000 years ago. Today, glacier ice occupies about 10% of the land surface.

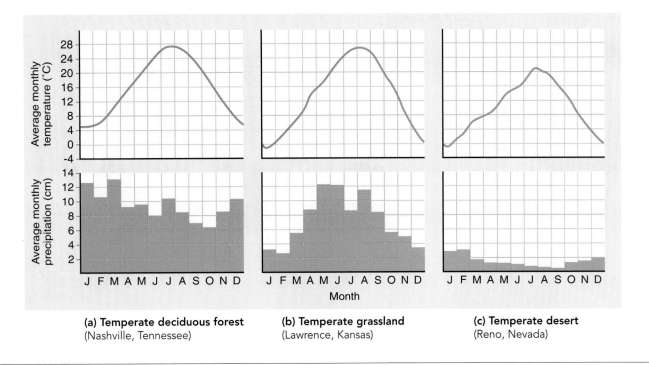

(a) Temperate deciduous forest
(Nashville, Tennessee)

(b) Temperate grassland
(Lawrence, Kansas)

(c) Temperate desert
(Reno, Nevada)

Figure 55-2 Significance of precipitation in temperate biomes

Average monthly temperature (*top*) is approximately the same in each location in North America. However, precipitation (*bottom*) varies a great deal, resulting in **(a)** deciduous forest, where precipitation is plentiful; **(b)** grassland, where it is less plentiful and more seasonal; and **(c)** desert, where it is quite low.

droppings, and remains of organisms) in the uppermost layer of soil. Although the soil surface melts during the summer, tundra has a layer of permanently frozen ground called **permafrost** that varies in depth and thickness. Because permafrost interferes with drainage, the thawed upper zone of soil is usually waterlogged during the summer. Limited precipitation, combined with low temperatures, flat topography (surface features), and permafrost, produces a landscape of broad, shallow lakes, sluggish streams, and bogs.

Low species richness and low primary productivity characterize tundra. Mosses, lichens (such as reindeer moss), grasses, and grasslike sedges dominate tundra vegetation; most of these short plants are herbaceous perennials that live 20 to 100 years. No readily recognizable trees or shrubs grow except in sheltered locations, although dwarf willows, dwarf birches, and other dwarf trees are common.

Year-round animal life of the tundra includes voles, weasels, arctic foxes, gray wolves, snowshoe hares, ptarmigan, snowy owls, musk oxen, and lemmings (see the discussion of lemming population cycles in Chapter 52). In the summer, caribou migrate north to the tundra to graze on sedges, grasses, and dwarf willow. Dozens of bird species also migrate north in summer to nest and feed on abundant insects. Mosquitoes, blackflies, and deerflies survive the winter as eggs or pupae and occur in great numbers during summer weeks.

Tundra regenerates quite slowly after it has been disturbed. Even casual use by hikers causes damage. Long-lasting injury, likely to persist for hundreds of years, was done to large portions of the arctic tundra as a result of oil exploration and military use.

Boreal forest is the evergreen forest of the north

Just south of the tundra is the **boreal forest,** or **taiga,** which stretches across both North America and Eurasia. Boreal forest is the world's largest biome, covering approximately 11% of Earth's land (Fig. 55-5). A biome comparable to the boreal forest is not found in the Southern Hemisphere, because it has no land at the corresponding latitudes. Winters are extremely cold and severe, although not as harsh as in the tundra. Boreal forest receives little precipitation, perhaps 50 cm (20 in) per year, and its soil is typically acidic, is low in minerals (inorganic nutrients), and has a deep layer of partly decomposed conifer needles at the surface. (Conifers are cone-bearing evergreens.) Boreal forest contains numerous ponds and lakes in water-filled depressions that grinding ice sheets dug during the last Ice Age.

Black and white spruces, balsam fir, eastern larch, and other conifers dominate the boreal forest, but **deciduous** trees such as aspen or birch, which shed their leaves in autumn, form striking stands. Conifers have many drought-resistant adaptations, such as needlelike leaves with a minimal surface area to reduce water loss (see Chapter 33). Such an adaptation enables conifers to withstand the "drought" of the northern winter months, when

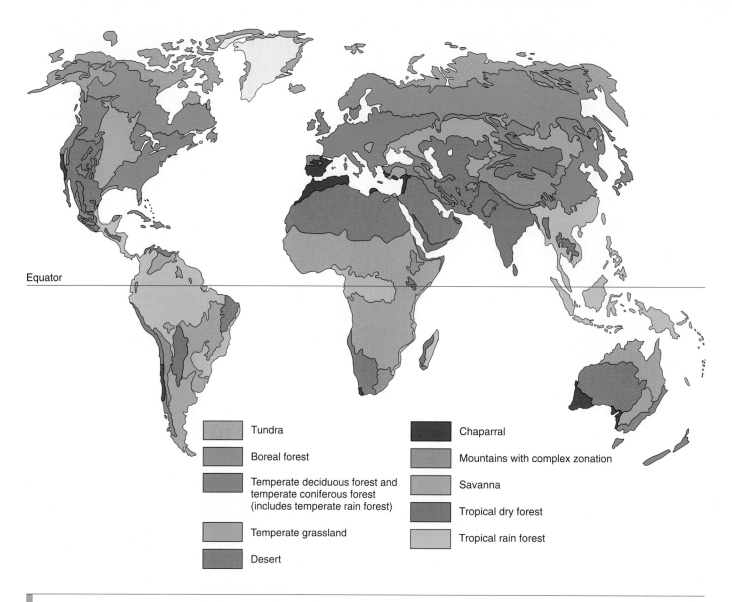

	Tundra		Chaparral
	Boreal forest		Mountains with complex zonation
	Temperate deciduous forest and temperate coniferous forest (includes temperate rain forest)		Savanna
			Tropical dry forest
	Temperate grassland		Tropical rain forest
	Desert		

Equator

Figure 55-3 *Animated* The world's major biomes

This simplified diagram shows sharp boundaries between biomes. Biomes actually intergrade at their boundaries, sometimes over large areas. Note that forested mountains such as the Appalachian Moun-tains are keyed according to their predominant vegetation type, whereas mountains with variable vegetation are keyed as "Mountains with complex zonation." (Based on data from the World Wildlife Fund.)

roots do not absorb water because the ground is frozen. Natural selection also favors conifers in the boreal forest because, being evergreen, they resume photosynthesis as soon as warmer temperatures return.

Animal life of the boreal forest includes some larger species, such as caribou (which migrate from the tundra to the boreal forest for winter), wolves, bears, and moose. However, most animal life is medium-sized to small and includes rodents, rabbits, and fur-bearing predators such as lynx, sable, and mink. Most species of birds are seasonally abundant but migrate to warmer climates for winter. Insects are numerous, but there are few amphibians and reptiles except in the southern boreal forest.

Most of the boreal forest is not suitable for agriculture be-cause of its short growing season and mineral-poor soil. The boreal forest, which is harvested primarily by clear-cutting, is currently the primary source of the world's industrial wood and wood fiber. The annual loss of boreal forests in Canada, Siberia, and Alaska is estimated to encompass an area twice as large as the Amazonian rain forests of Brazil.

Temperate rain forest has cool weather, dense fog, and high precipitation

Coniferous **temperate rain forest** grows on the northwestern coast of North America. Similar vegetation exists in southeastern Australia and in southwestern South America. Annual precipita-tion in this biome is high, from 200 to 380 cm (80 to 150 in);

Figure 55-4 Arctic tundra

Because of the short growing season and permafrost in arctic tundra, only small, hardy plants grow in the northernmost biome that encircles the Arctic Ocean. Photographed during autumn in Northwest Territories, Canada.

Figure 55-5 Boreal forest

Boreal forest is coniferous forest that occurs in cold regions of the Northern Hemisphere adjacent to the tundra. Photographed in Yukon, Canada.

condensation of water from dense coastal fog augments the annual precipitation. The proximity of temperate rain forest to the coastline moderates the temperature so that seasonal fluctuation is narrow; winters are mild, and summers are cool. Temperate rain forest has a relatively nutrient-poor soil, although its organic content may be high. Cool temperatures slow the activity of bacterial and fungal decomposers. Thus, needles and large fallen branches and trunks accumulate on the ground as litter that takes many years to decay and release inorganic minerals to the soil.

The dominant vegetation type in the North American temperate rain forest is large evergreen trees, such as western hemlock, Douglas fir, Sitka spruce, and western red cedar. Temperate rain forest is rich in epiphytic vegetation, which consists of smaller plants that grow nonparasitically on the trunks and branches of large trees (▌Fig. 55-6). Epiphytes in this biome are mainly mosses, lichens, and ferns, all of which also carpet the ground. Squirrels, wood rats, mule deer, elk, numerous bird species (such as jays, nuthatches, and chickadees), several species of reptiles (such as painted turtles and western terrestrial garter snakes), and amphibians (such as Pacific giant salamanders and Pacific treefrogs) are common temperate rainforest animals.

Temperate rain forest, one of the richest wood producers in the world, supplies us with lumber and pulpwood. It is also one of the most complex ecosystems in terms of species richness. Care must be taken to avoid overharvesting original old-growth forest, because such an ecosystem takes hundreds of years to develop. When the logging industry harvests old-growth forest, it typically replants the area with a monoculture (a single species) of trees that it harvests in 40- to 100-year cycles. Thus, the old-

Figure 55-6 Temperate rain forest

Large amounts of precipitation characterize temperate rain forest. Note the epiphytes hanging from the branches of coniferous trees. Photographed in Olympic National Park in Washington State.

Barbara Miller/Biological Photo Service

Figure 55-7 Temperate deciduous forest

The broad-leaf trees that dominate the temperate deciduous forest shed their leaves before winter. Photographed during autumn in Pennsylvania.

growth forest ecosystem, once harvested, never has a chance to redevelop. A small fraction of the original, old-growth temperate rain forest in Washington, Oregon, and northern California remains untouched. These stable forest ecosystems provide biological habitat for many organisms, including 40 endangered and threatened species.

Temperate deciduous forest has a canopy of broad-leaf trees

Seasonality (hot summers and cold winters) is characteristic of **temperate deciduous forest,** which occurs in temperate areas where precipitation ranges from about 75 to 126 cm (30 to 50 in) annually. Typically, the soil of a temperate deciduous forest consists of both a topsoil rich in organic material and a deep, clay-rich lower layer. As organic materials decay, mineral ions are released. If roots of living trees do not absorb these ions, they leach into the clay, where they may be retained.

Broad-leaf hardwood trees, such as oak, hickory, maple, and beech, that lose their foliage annually dominate temperate deciduous forests of the northeastern and Mid-Atlantic United States (▌Fig. 55-7). The trees of the temperate deciduous forest form a dense canopy that overlies saplings and shrubs.

Temperate deciduous forests originally contained a variety of large mammals such as mountain lions, wolves, bison, and other

species now regionally extinct, plus deer, bears, and many small mammals and birds (such as wild turkeys, blue jays, and scarlet tanagers). Both reptiles (such as box turtles and rat snakes) and amphibians (such as spotted salamanders and wood frogs) abounded, together with a denser and more varied insect life than exists today.

In Europe and North America, logging and land clearing for farms, tree plantations, and cities have removed much of the original temperate deciduous forest. Where it has regenerated, temperate deciduous forest is often in a seminatural state—that is, highly modified by humans for recreation, livestock foraging, timber harvest, and other uses. Although these returning forests do not have the biological diversity of virgin stands, many forest organisms have successfully become re-established.

Worldwide, temperate deciduous forest was among the first biomes to be converted to agricultural use. In Europe and Asia, many soils that originally supported temperate deciduous forest have been cultivated by traditional agricultural methods for thousands of years without substantial loss in fertility. During the 20th century, however, intensive agricultural practices were adopted; these, along with overgrazing and deforestation, contributed to the degradation of some agricultural lands.

Temperate grasslands occur in areas of moderate precipitation

Summers are hot, winters are cold, fires help shape the landscape, and rainfall is often uncertain in **temperate grasslands.** Annual precipitation averages 25 to 75 cm (10 to 30 in). In grasslands with less precipitation, minerals tend to accumulate in a marked layer just below the topsoil. These minerals tend to leach out of the soil in areas with more precipitation. Grassland soil contains considerable organic material because surface parts of many grasses die off each winter and contribute to the organic content of the soil (the roots and rhizomes survive underground). Many grasses are sod formers: their roots and rhizomes form a thick, continuous underground mat.

Moist temperate grasslands, also known as *tallgrass prairies,* occur in the United States in Iowa, western Minnesota, eastern Nebraska, and parts of other Midwestern states and across Canada's prairie provinces. Although few trees grow except near rivers and streams, grasses, some as tall as 2 m (6.5 ft), grow in great profusion in the deep, rich soil (▌Fig. 55-8). Before most of this area was converted to arable land, it was covered with herds of grazing animals, particularly bison. The principal predators were wolves, although in sparser, drier areas coyotes took their place. Smaller fauna included prairie dogs and their predators (foxes, black-footed ferrets, and birds of prey such as prairie falcons), western meadowlarks, bobolinks, reptiles (such as gopher snakes and short-horned lizards), and great numbers of insects.

Shortgrass prairies, in which the dominant grasses are less than 0.5 m (1.6 ft) tall, are temperate grasslands that receive less

THE DISTRIBUTION OF VEGETATION ON MOUNTAINS

Hiking up a mountain is similar to traveling toward the North Pole with respect to the major life zones encountered (see figure). This elevation–latitude similarity occurs because the temperature drops as one climbs a mountain, just as it does when one travels north; the temperature drops about 6°C (11°F) with each 1000-m increase in elevation. The types of species growing on the mountain change as the temperature changes.

Deciduous trees, which shed their leaves every autumn, may cover the base of a mountain in Colorado, for example. At higher elevations, where the climate is colder and more severe, a coniferous *subalpine forest* resembling boreal forest grows. Spruces and firs are the dominant trees here. Higher still, the forest thins, and the trees become smaller, gnarled, and shrublike. These twisted, shrublike trees, called *krummholz* (a German word meaning "crooked wood"), are found at their elevational limit (the *tree line*). The exact elevation at which the tree line occurs depends on the latitude and distance from the ocean. In the Rocky Mountains, between 35° and 50° north latitude, the tree line drops 100 m with each 1 degree latitude northward.

Above the tree line, where the climate is quite cold, a kind of tundra occurs, with vegetation composed of grasses, sedges, and small tufted plants, most of which are hardy perennials. Some alpine plants (for example, buttercups) are lowland species that have adapted to the alpine environment, whereas other plants (for example, mountain douglasia) live exclusively in the mountains. This tundra is called *alpine tundra* to distinguish it from arctic tundra. At the top of the mountain, a permanent ice cap or snowcap might be

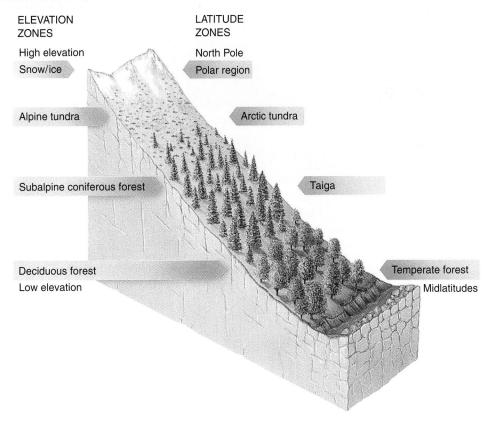

Comparison of elevation and latitude zones.

The cooler temperatures at higher elevations of a mountain produce a series of ecosystems similar to those encountered when going toward the North Pole.

found, similar to the nearly lifeless polar land areas.

Important environmental differences exist between high elevations and high latitudes that affect the types of organisms found in each place. Alpine tundra typically lacks permafrost and receives more precipitation than does arctic tundra. High elevations of temperate mountains do not

have the great extremes of day length that are associated with the changing seasons in high-latitude biomes. The intensity of solar radiation is greater at high elevations than at high latitudes. At high elevations, the sun's rays pass through less atmosphere, which results in greater exposure to ultraviolet radiation (less is filtered out by the atmosphere) than occurs at high latitudes.

precipitation than the moister grasslands just described but more precipitation than deserts. In the United States, shortgrass prairies occur in the eastern half of Montana, the western half of South Dakota, and parts of other Midwestern states, as well as western Alberta in Canada. The plants grow in less abundance than in the moister grasslands, and occasionally some bare soil is exposed.

The North American grassland, particularly the tallgrass prairie, was so well suited to agriculture that little of it remains. More than 90% has vanished under the plow, and the remainder is so fragmented that almost nowhere can we see even an approximation of what European settlers saw when they settled in the Midwest. Today, the tallgrass prairie is considered North America's rarest biome.

Figure 55-8 Temperate grassland

The Nature Conservancy owns this tallgrass prairie preserve in Oklahoma. Like other moist temperate grasslands, it is mostly treeless but contains a profusion of grasses and other herbaceous flowering plants. As bison graze on the plants, they affect community structure and diversity.

Chaparral is a thicket of evergreen shrubs and small trees

Some hilly temperate environments have mild winters with abundant rainfall, combined with extremely dry summers. Such Mediterranean climates, as they are called, occur not only in the area around the Mediterranean Sea but also in California, Western Australia, portions of Chile, and South Africa. In southern California this environment is called **chaparral**. This vegetation type is also known as maquis in the Mediterranean region, mallee scrub in Australia, matorral in Chile, and Cape scrub in Africa. Chaparral soil is thin and infertile. Frequent fires occur naturally in this environment, particularly in late summer and autumn.

Chaparral vegetation looks strikingly similar in different areas of the world, even though the individual species are quite different. A dense growth of evergreen shrubs, often of drought-resistant pine or scrub oak trees, dominates chaparral (▌Fig. 55-9). During the rainy winter season the landscape may be lush and green, but during the hot, dry summer the plants lie dormant. Trees and shrubs often have hard, small, leathery leaves that resist water loss. Many plants are also fire-adapted and grow best in the months following a fire. Such growth is possible because fire releases minerals that were tied up in the plants that burned. With the new availability of essential minerals, plants sprout vigorously during winter rains. Mule deer, wood rats, brush rabbits, skinks and other lizards, and many species of birds (such as Anna's hummingbird, scrub jay, and bushtit) are common animals of the chaparral.

Fires, which occur at irregular intervals in California chaparral vegetation, are often quite costly because they consume expensive homes built on the hilly chaparral landscape. Unfortunately, efforts to control naturally occurring fires sometimes backfire.

Denser, thicker vegetation tends to accumulate when periodic fires are prevented; then, when a fire does occur, it is much more severe. Removing the chaparral vegetation, whose roots hold the soil in place, also causes problems—witness the mud slides that sometimes occur during winter rains in these areas.

Deserts are arid ecosystems

Deserts are dry areas found in temperate (*cold deserts*) and subtropical or tropical regions (*warm deserts*). North America has four distinct deserts. The Great Basin Desert in Nevada, Utah, and neighboring states is a cold desert dominated by sagebrush. The Mojave Desert in Nevada and California is a warm desert known for its Joshua trees. The Chihuahuan Desert, home of century plants (agaves; see Fig. 52-7), is a warm desert found in Texas, New Mexico, and Mexico. The warm Sonoran Desert, with its many species of cacti, is found in Arizona, California, and Mexico (▌Fig. 55-10).

The low water-vapor content of the desert atmosphere leads to daily temperature extremes of heat and cold, so a major change in temperature occurs in each 24-hour period. Deserts vary greatly depending on the amount of precipitation they receive, which is generally less than 25 cm (10 in) per year. A few deserts are so dry that virtually no plant life occurs in them. As a result of sparse vegetation, desert soil is low in organic material but often high in mineral content, particularly the salts NaCl, $CaCO_3$, and $CaSO_4$.

Desert vegetation includes both perennials (cacti, yuccas, Joshua trees, and sagebrushes) and, after a rain, flowering annuals. Desert plants tend to have reduced leaves or no leaves, an adaptation that conserves water. In cacti such as the giant saguaro, for example, the stem carries out photosynthesis and also expands accordion-style to store water; the leaves are modi-

Figure 55-9 Chaparral

Chaparral, which consists primarily of drought-resistant evergreen shrubs and small trees, develops where hot, dry summers alternate with mild, rainy winters. Photographed in the Santa Lucia Mountains, California.

fied into spines, which discourage herbivores. Other desert plants shed their leaves for most of the year, growing only during the brief moist season.

Desert animals tend to be small. During the heat of the day, they remain under cover or return to shelter periodically, whereas at night they come out to forage or hunt. In addition to desert-adapted insects, there are many specialized desert reptiles (such as desert iguanas, desert tortoises, and rattlesnakes) and a few desert-adapted amphibians (such as western spadefoot toads). Mammals include such rodents as the American kangaroo rat, which does not need to drink water but subsists solely on the water content of its food (primarily seeds and insects). American deserts are also home to jackrabbits, and kangaroos live in Australian deserts. Carnivores such as the African fennec fox and some birds of prey, especially owls, live on rodents and rabbits. During the driest months of the year, many desert insects, amphibians, reptiles, and mammals tunnel underground, where they remain inactive; this period of dormancy is known as *estivation*.

Humans have altered North American deserts in several ways. Off-road vehicles damage desert vegetation, which sometimes takes years to recover. When the top layer of desert soil is disturbed, erosion occurs more readily, and less vegetation grows to support native animals. Another problem is that certain cacti and desert tortoises have become rare as a result of poaching. Houses, factories, and farms built in desert areas require vast quantities of water, which must be imported from distant areas. Irrigation of desert soils often causes them to become salty and unfit for crops or native vegetation. Increased groundwater consumption by many desert cities has caused groundwater levels to drop. Aquifer depletion in U.S. deserts is particularly critical in southern Arizona and southwestern New Mexico.

Figure 55-10 Desert

Summer rainfall characterizes the warmer deserts of North America, such as the Sonoran Desert shown here. The Sonoran Desert contains many species of cacti, including the large, treelike saguaro, which grows 15 to 18 m (50 to 60 ft) tall. Photographed in Arizona.

Savanna is a tropical grassland with scattered trees

The **savanna** biome is a tropical grassland with widely scattered clumps of low trees (▌Fig. 55-11). Savanna is found in areas of relatively low or seasonal rainfall with prolonged dry periods. Temperatures in savannas vary little throughout the year, and precipitation, not temperature as in temperate grasslands, regulates seasons. Annual precipitation is 85 to 150 cm (34 to 60 in). Savanna soil is low in essential minerals, in part because it is strongly leached. Savanna soil is often rich in aluminum, which resists leaching, and in places the

Figure 55-11 Savanna

In this photograph of African savanna in Tanzania, the smaller animals in the foreground are Thomson's gazelles (*Gazella thomsonii*), and the larger ones near the trees are wildebeests (*Connochaetes taurinus*).

(a) A broad view of tropical rain forest vegetation along a riverbank in Southeast Asia. Except at river banks, tropical rain forest has a closed canopy that admits little light to the forest floor.

(b) Thick, epiphyte-covered lianas grow into the canopy using a tree trunk for support. Photographed in Costa Rica.

Figure 55-12 Tropical rain forest

aluminum reaches levels that are toxic to many plants. Although the African savanna is best known, savanna also occurs in South America and northern Australia.

Wide expanses of grasses interrupted by occasional trees characterize savanna. Trees such as *Acacia* bristle with thorns that provide protection against herbivores. Both trees and grasses have fire-adapted features, such as extensive underground root systems, that enable them to survive seasonal droughts as well as periodic fires that sweep through the savanna.

The world's greatest assemblage of hooved mammals occurs in the African savanna. Here live great herds of herbivores, including wildebeests, antelopes, giraffes, zebras, and elephants. Large predators, such as lions and hyenas, kill and scavenge the herds. In areas of seasonally varying rainfall, the herds and their predators may migrate annually.

Savanna is rapidly being converted to rangeland for cattle and other domesticated animals, which are replacing the big herds of wild animals. The problem is particularly acute in Africa, which has the most rapidly growing human population of any continent. In some places severe overgrazing by domestic animals has contributed to the conversion of marginal savanna into desert, a process known as **desertification.** In desertification, the reduced grass cover caused by overgrazing allows wind and water to erode the soil; erosion removes the topsoil and decreases the soil's ability to support crops or livestock. (Desertification is not restricted

to savanna. Temperate grasslands and tropical dry forests can also be degraded to desert.)

There are two basic types of tropical forests

There are many kinds of tropical forests, but ecologists generally classify them as one of two types: tropical dry forests or tropical rain forests. **Tropical dry forests** occur in regions with a wet season and a dry season (usually 2 to 3 months each year). Annual precipitation is 150 to 200 cm (60 to 80 in). During the dry season, many tropical trees shed their leaves and remain dormant, much as temperate trees do during the winter. India, Brazil, Thailand, and Mexico are some of the countries that have tropical dry forests. Tropical dry forests intergrade with savanna on their dry edges and with tropical rain forests on their wet edges. Logging and overgrazing by domestic animals have fragmented and degraded many tropical dry forests.

The annual precipitation of **tropical rain forests** is 200 to 450 cm (80 to 180 in). Much of this precipitation, which occurs almost daily, comes from locally recycled water that enters the atmosphere by transpiration from the forest's own trees. Tropical rain forests are often located in areas with ancient, highly weathered, mineral-poor soil. Little organic matter accumulates in such soils. Because temperatures are high and soil moisture is abundant year-round, decay organisms and detritus-feeding ants and termites decompose organic litter quite rapidly. Vast networks of roots and mycorrhizae quickly absorb minerals from decomposing materials. Thus, minerals of tropical rain forests are tied up in the vegetation rather than in the soil.

Tropical rain forests are found in Central and South America, Africa, and Southeast Asia. Tropical rain forest is very productive despite the scarcity of minerals in the soil. Its plants, stimulated by abundant solar energy and precipitation, capture considerable energy by photosynthesis. Of all the biomes, the tropical rain forest is unrivaled in species richness.

Most trees of tropical rain forests are evergreen flowering plants (▌ Fig. 55-12a). A fully developed rain forest has three or more distinct stories of vegetation. The topmost story consists of the crowns of the oldest, tallest trees, some 50 m (164 ft) or more in height; these trees are exposed to direct sunlight and are subject to the warmest temperatures, lowest humidities, and strongest winds. The middle story, which reaches a height of 30 to 40 m (100 to 130 ft), forms a continuous canopy of leaves overhead that lets in little sunlight for the support of the sparse understory. Only 2% to 3% of the light bathing the forest canopy reaches the forest understory, which consists of both smaller plants specialized for life in shade and seedlings of taller trees. Vegetation of tropical rain forests is not dense at ground level except near stream banks or where a fallen tree has opened the canopy.

Tropical rainforest trees support extensive epiphytic communities of smaller plants such as orchids and bromeliads. Although epiphytes grow in crotches of branches, on bark, or even on the leaves of their hosts, most use their host trees only for physical support, not for nourishment.

Because little light penetrates to the understory, many plants living there are adapted to climb already-established host trees. Lianas (woody tropical vines), some as thick as a human thigh, twist up through the branches of tropical rainforest trees (▌ Fig. 55-12b). Once in the canopy, lianas grow from the upper branches of one forest tree to another, connecting the tops of the trees and providing a walkway for many of the canopy's residents.

Not counting bacteria and other soil-dwelling organisms, about 90% of tropical rainforest organisms live in the middle and upper canopies. Rainforest animals include the most abundant and varied insect, reptile, and amphibian fauna on Earth. Birds, too, are diverse, with some specialized to consume fruits (such as parrots), some to consume nectar (such as hummingbirds and sunbirds), and others to consume insects. Most rainforest mammals, such as sloths and monkeys, live only in the trees and never climb down to the ground surface. Some large ground-dwelling mammals, including elephants, are also found in tropical rain forests.

Unless strong conservation measures are initiated soon, human population growth and agricultural and industrial expansion in tropical countries will spell the end of tropical rain forests by the middle of the 22nd century. Many rainforest species may become extinct before they are even identified and scientifically described. (Tropical rainforest destruction is discussed in detail in Chapter 56.)

Review

▌ What climate and soil factors produce the major biomes?

▌ What representative organisms are found in each of these forest biomes: (1) boreal forest, (2) temperate deciduous forest, (3) temperate rain forest, (4) tropical rain forest?

▌ In which biome do you live? Does it match the description given in this text? If not, explain the discrepancy.

▌ How does tundra compare to desert? How does temperate grassland compare to savanna?

▌ AQUATIC ECOSYSTEMS

Learning Objectives

3 Explain the important environmental factors that affect aquatic ecosystems.

4 Distinguish among plankton, nekton, and benthos.

5 Briefly describe the various freshwater, estuarine, and marine ecosystems, giving attention to the environmental characteristics and representative organisms of each.

6 Describe at least one human effect on each of the aquatic ecosystems discussed.

Aquatic "biomes" do not exist, in the sense that aquatic ecologists do not distinguish aquatic ecosystems based on the dominant form of vegetation. Aquatic ecosystems are classified primarily on abiotic factors, such as salinity, that help determine an aquatic life zone's boundaries. **Salinity,** the concentration of dissolved salts in a body of water, affects the kinds of organisms present in aquatic ecosystems, as does the amount of dissolved oxygen. Water greatly interferes with the penetration of light, so floating aquatic organisms that photosynthesize remain near the water's surface, and vegetation attached to the bottom grows only in shallow water. In addition, low levels of essential minerals often limit the number and distribution of organisms in certain aquatic environments. Other abiotic determinants of species composition in aquatic ecosystems include water depth, temperature, pH, and presence or absence of waves and currents.

Aquatic ecosystems contain three main ecological categories of organisms: free-floating plankton, strongly swimming nekton, and bottom-dwelling benthos. **Plankton** are usually small or microscopic organisms that are relatively feeble swimmers. For the most part, they are carried about at the mercy of currents and waves. They are unable to swim far horizontally, but some species are capable of large vertical migrations and are found at different depths of water at different times of the day or at different seasons. Plankton are generally subdivided into two major categories: phytoplankton and zooplankton. **Phytoplankton** (photosynthetic cyanobacteria and free-floating algae) are producers that form the base of most aquatic food webs. **Zooplankton** are nonphotosynthetic organisms that include protozoa, tiny crustaceans, and the larval stages of many animals. **Nekton** are larger, more strongly swimming organisms such as fishes, turtles, and whales. **Benthos** are bottom-dwelling organisms that fix themselves to one spot (sponges, oysters, and barnacles), burrow into the sand (many worms and echinoderms), or simply walk or swim about on the bottom (crayfish, aquatic insect larvae, and brittle stars).

Freshwater ecosystems are closely linked to land and marine ecosystems

Freshwater ecosystems include streams and rivers (flowing-water ecosystems), ponds and lakes (standing-water ecosystems), and marshes and swamps (freshwater wetlands). Each type of freshwater ecosystem has its own specific abiotic conditions and characteristic organisms. Although freshwater ecosystems occupy a relatively small portion—about 2%—of Earth's surface, they are important in the hydrologic cycle: they assist in recycling precipitation that flows as surface runoff to the ocean (see discussion of hydrologic cycle in Chapter 54). Freshwater habitats also provide homes for large numbers of species.

Streams and rivers are flowing-water ecosystems

Many different conditions exist along the length of a stream or river (▌Fig. 55-13). The nature of a **flowing-water ecosystem** changes greatly from its source (where it begins) to its mouth (where it empties into another body of water). Headwater streams (small streams that are the sources of a river) are usually shallow, clear, cold, swiftly flowing, and highly oxygenated. In contrast, rivers downstream from the headwaters are wider and deeper, cloudy (that is, they contain suspended particulates), not as cold, slower flowing, and less oxygenated. Surrounding forest may shade certain parts of the stream or river, whereas other parts may be exposed to direct sunlight. Along parts of a stream or river, groundwater wells up through sediments on the bottom; this local input of water moderates the water temperature so that summer temperatures are cooler and winter temperatures are warmer than in adjacent parts of the flowing-water ecosystem.

The kinds of organisms in flowing-water ecosystems vary greatly from one stream to another, depending primarily on the strength of the current. In streams with fast currents, inhabitants have adaptations such as suckers to attach themselves to rocks so they are not swept away. The larvae of blackflies, for example, attach themselves with suction discs located on the ends of their abdomens. Some stream inhabitants, such as immature water-penny beetles, have flattened bodies that enable them to slip under or between rocks. The water-penny beetle larva gets its common name from its flattened, nearly circular shape. Alternatively, inhabitants such as the brown trout are streamlined and muscular enough to swim in the current.

Streams and rivers depend on land for much of their energy. In headwater streams, up to 99% of the energy input comes from detritus (dead organic material such as leaves) carried from the land into streams and rivers by wind or surface runoff. Downstream, rivers contain more producers and therefore depend slightly less on detritus as a source of energy than do the headwaters.

Human activities have several adverse impacts on rivers and streams, including water pollution and the effects of dams built to contain the water of rivers or streams. Pollution alters the physical environment of a flowing-water ecosystem and changes the biotic component downstream from the pollution source. Dams change the nature of flowing-water ecosystems, both upstream and downstream from the dam location. A dam causes water to back up, resulting in flooding of large areas of land and formation of a reservoir, which destroys terrestrial habitat. Below the

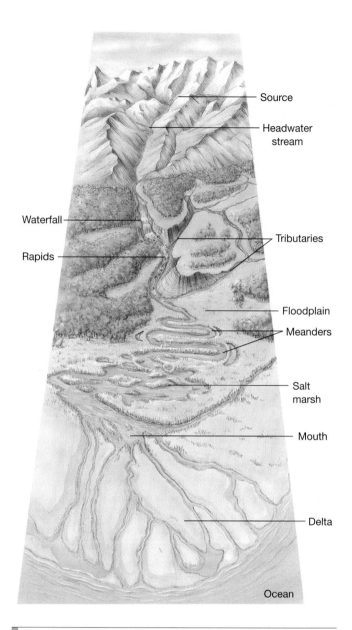

Figure 55-13   Features of a typical river

The river begins at a source, often high in mountains and fed by melting snows or glaciers. Headwater streams flow downstream rapidly, often over rocks (as rapids) or bluffs (as waterfalls). Along the way, tributaries feed into the river. As the river's course levels, the river flows more slowly and winds from side to side, forming meanders. The floodplain is the area on either side of the river that is subject to flooding. Near the ocean, the river may form a salt marsh where fresh water from the river and salt water from the ocean mix. Sediments deposited by the river as it empties into the ocean form the delta, a fertile, low-lying plain at the river's mouth.

dam, the once-powerful river is reduced to a relative trickle, so the flowing-water ecosystem is altered.

Ponds and lakes are standing-water ecosystems

Zonation characterizes **standing-water ecosystems.** A large lake has three basic zones: the littoral, limnetic, and profundal zones (▌Fig. 55-14). Smaller lakes and ponds typically lack a profundal zone.

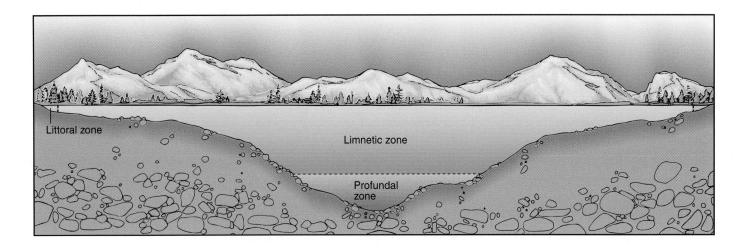

Figure 55-14 *Animated* Zonation in a large lake

A lake is a standing-water ecosystem surrounded by land. Here, vegetation around the lake is not drawn to scale.

The **littoral zone** is a shallow-water area along the shore of a lake or pond. It includes rooted, emergent vegetation, such as cattails and burreeds, plus several deeper-dwelling aquatic plants and algae. The littoral zone is the most productive zone of the lake. Photosynthesis is greatest in the littoral zone, in part because light is abundant and because the littoral zone receives nutrient inputs from surrounding land that stimulate the growth of plants and algae. In addition, ponds and lakes—like streams and rivers—depend on detritus carried from the land for much of their energy. Animals of the littoral zone include frogs and their tadpoles, turtles, worms, crayfish and other crustaceans, insect larvae, and many fishes such as perch, carp, and bass. Surface dwellers, such as water striders and whirligig beetles, are found in the quieter areas.

The **limnetic zone** is the open water beyond the littoral zone, that is, away from the shore; it extends down as far as sunlight penetrates to permit photosynthesis. The main organisms of the limnetic zone are microscopic phytoplankton and zooplankton. Larger fishes also spend some of their time in the limnetic zone, although they may visit the littoral zone to feed and reproduce. Because of the depth of this zone, less vegetation grows in the limnetic zone than in the littoral zone.

Beneath the limnetic zone of a large lake is the **profundal zone.** Because light does not penetrate effectively to this depth, plants and algae do not live in this zone. Food drifts into the profundal zone from the littoral and limnetic zones. Bacteria decompose dead plants and animals that reach the profundal zone, thus liberating minerals. These minerals are not effectively recycled, because no photosynthetic organisms are present to absorb them and incorporate them into the food web. As a result, the profundal zone tends to be both mineral rich and anaerobic (oxygen deficient), with few organisms other than anaerobic bacteria occupying it.

Thermal stratification in temperate lakes. The marked layering of large temperate lakes caused by light penetration is accentuated by **thermal stratification,** in which the temperature changes sharply with depth (Fig. 55-15a). Thermal stratifica-

tion occurs because the summer sunlight penetrates and warms surface water, making it less dense. (Recall from Chapter 2 that the density of water is greatest at 4°C; both above and below this temperature, water is less dense.) In summer, cool (and therefore more dense) water remains at the lake bottom and is separated from warm (and therefore less dense) water above by an abrupt temperature transition called the **thermocline.** Seasonal distribution of temperature and oxygen (more oxygen dissolves in water at cooler temperatures) affects the distribution of fish in the lake.

In temperate lakes, falling temperatures in autumn cause a mixing of the lake waters called the **fall turnover** (Fig. 55-15b ❶). As surface water cools, its density increases, and it sinks and displaces the less dense, warmer, mineral-rich water beneath. Warmer water then rises to the surface where it, in turn, cools and sinks. This process of cooling and sinking continues until the lake reaches a uniform temperature throughout. In winter ❷, surface water cools to below 4°C, its temperature of greatest density. Ice, which forms at 0°C, is less dense than cold water. Thus, ice forms on the surface, and the water on the lake bottom is warmer than the ice on the surface. In the spring ❸, a **spring turnover** occurs as ice melts and surface water reaches 4°C. Surface water again sinks to the bottom, and bottom water returns to the surface. As summer arrives ❹, thermal stratification occurs once again.

The mixing of deeper, nutrient-rich water with surface, nutrient-poor water during the fall and spring turnovers brings essential minerals to the surface and oxygenated water to the bottom. The sudden presence of large amounts of essential minerals in surface waters encourages the development of large algal and cyanobacterial populations, which may form temporary **blooms** in the fall and spring.

Increased nutrients and algal growth. The presence of high levels of plant and algal nutrients such as nitrogen and phosphorus cause *enrichment,* the fertilization of a body of water. Excess amounts of these nutrients enter waterways from sewage and from fertilizer runoff from lawns and fields. The water in an enriched pond or lake is cloudy because of the vast numbers of

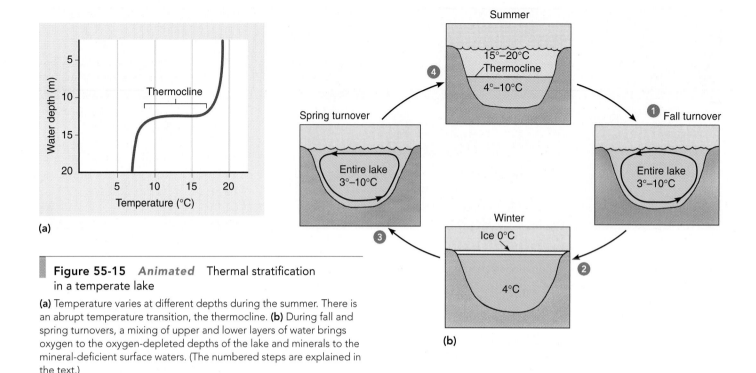

Figure 55-15 *Animated* Thermal stratification in a temperate lake

(a) Temperature varies at different depths during the summer. There is an abrupt temperature transition, the thermocline. **(b)** During fall and spring turnovers, a mixing of upper and lower layers of water brings oxygen to the oxygen-depleted depths of the lake and minerals to the mineral-deficient surface waters. (The numbered steps are explained in the text.)

algae and cyanobacteria that the nutrients support. The species composition is different in enriched and unenriched lakes. For example, an unenriched lake in the northeastern United States may contain pike, sturgeon, and whitefish in the deeper, colder part of the lake where there is a higher concentration of dissolved oxygen. In contrast, the deeper, colder levels of water in enriched lakes are depleted of dissolved oxygen because of the greater amount of decomposition on the lake floor. Fishes such as pike, sturgeon, and whitefish die out, and fishes such as catfish and carp, which tolerate lower concentrations of dissolved oxygen, replace them.

Enrichment is reversible and has declined in North America since the 1970s because the passage of legislation limits the phosphate content of detergents and because better sewage treatment plants have been constructed. Agriculture is the leading source of water-quality impairment of U.S. surface waters today. Fertilizer runoff, as well as animal wastes and plant residues in waterways, still cause enrichment problems.

Freshwater wetlands are transitional between aquatic and terrestrial ecosystems

Freshwater wetlands, which are usually covered by shallow water for at least part of the year, have characteristic soils and water-tolerant vegetation. They include marshes, dominated by grasslike plants, and swamps, in which woody trees or shrubs dominate (❚ Fig. 55-16). Freshwater wetlands also include hardwood bottomland forests (lowlands along streams and rivers that are periodically flooded), prairie potholes (small, shallow ponds that formed when glacial ice melted at the end of the last Ice Age), and peat moss bogs (peat-accumulating wetlands where sphagnum moss dominates).

Wetlands are valued as a wildlife habitat for migratory waterfowl and many other bird species, beavers, otters, muskrats, and game fishes. Wetlands are holding areas for excess water when rivers flood their banks. The floodwater stored in wetlands then drains slowly back into the rivers, providing a steady flow of water throughout the year. Wetlands also serve as groundwater recharging areas. One of their most important roles is to trap and hold pollutants in the flooded soil, thereby cleansing and purify-

Figure 55-16 Freshwater swamp

Trees, such as bald cypress (*shown*), dominate freshwater swamps. In this wetland, photographed in northeastern Texas, a floating carpet of tiny aquatic plants covers the water's surface.

Figure 55-17 Salt marsh

Cordgrass (*Spartina alterniflora*) is the dominant vegetation in this salt marsh along a creek in North Carolina. This photograph was taken at low tide, showing an exposed oyster bed and an immature little blue heron (*Egretta caerulea*).

ing the water. Such important environmental functions as these are known as **ecosystem services.**

At one time wetlands were considered wastelands, areas to be filled in or drained so that farms, housing developments, and industrial plants could be built on them. Wetlands are also breeding places for mosquitoes and therefore were viewed as a menace to public health. The crucial ecosystem services that wetlands provide are widely recognized today, and wetlands have some legal protection. Agriculture, pollution, engineering (dams), and urban and suburban development still threaten wetlands, however.

Estuaries occur where fresh water and salt water meet

Where the ocean meets the land, there may be one of several kinds of ecosystems: a rocky shore, a sandy beach, an intertidal mudflat, or a tidal estuary. An **estuary** is a coastal body of water, partly surrounded by land, with access to the open ocean and a large supply of fresh water from rivers. Water levels in an estuary rise and fall with the tides, and salinity fluctuates with tidal cycles, the time of year, and precipitation. Salinity also changes gradually within the estuary, from unsalty fresh water at the river entrance to salty ocean water at the mouth of the estuary. Because estuaries undergo marked daily, seasonal, and annual variations in temperature, salinity, and other physical properties, estuarine organisms have a wide tolerance to such changes.

Estuaries are among the most fertile ecosystems in the world, often having much greater productivities than the adjacent ocean or freshwater river (see Table 54-1). This high productivity is the result of four factors: (1) The action of tides promotes a rapid circulation of nutrients and helps remove waste products. (2) Minerals are transported from land into streams and rivers that empty into the estuary. (3) A high level of light penetrates the shallow water. (4) The presence of many plants provides an

extensive photosynthetic carpet and also mechanically traps detritus, forming the basis of detritus food webs. Most commercially important fishes and shellfish spend their larval stages in estuaries among the protective tangle of decaying stems.

Temperate estuaries usually contain **salt marshes,** shallow wetlands in which salt-tolerant grasses dominate (Fig. 55-17). Uninformed people have often seen salt marshes as worthless, empty stretches of land. As a result, people have used them as dumps, severely polluting them, or filled them with dredged bottom material to form artificial land for residential and industrial development. A large part of the estuarine environment is lost in this way, along with many of its ecosystem services, for example, biological habitats, sediment and pollution trapping, groundwater supply, and storm buffering (salt marshes absorb much of the energy of a storm surge and therefore prevent flood damage elsewhere).

Mangrove forests, the tropical equivalent of salt marshes, cover perhaps 70% of tropical and subtropical coastal mudflats where tides and waves fluctuate (Fig. 55-18). Like salt marshes, mangrove forests provide valuable ecosystem services. Mangrove roots stabilize the sediments, preventing coastal erosion and providing a barrier against the ocean during storms. Their interlacing roots are breeding grounds and nurseries for commercially important fish and shellfish species, such as blue crabs, shrimp, mullet, and spotted sea trout. Mangrove branches are nesting sites for many species of birds, such as pelicans, herons, egrets, and roseate spoonbills. Mangroves are under assault from coastal development, including aquaculture facilities, and unsustainable

Figure 55-18 Mangroves

Red mangroves (*Rhizophora mangle*) have stiltlike roots that support the tree. Many animals live in the complex root systems of mangrove forests. Photographed at low tide along the coast of Florida, near Miami.

Ecology and the Geography of Life   **1203**

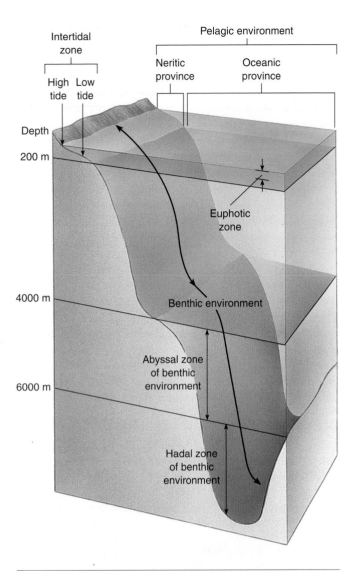

Intertidal zone

High tide Low tide

Pelagic environment

Neritic province Oceanic province

Depth

200 m

Euphotic zone

4000 m

Benthic environment

Abyssal zone of benthic environment

6000 m

Hadal zone of benthic environment

Figure 55-19 *Animated* Zonation in the ocean

The ocean has three main life zones: the intertidal zone, the benthic environment, and the pelagic environment. The pelagic environment consists of the neritic and oceanic provinces. (The slopes of the ocean floor are not as steep as shown; they are exaggerated to save space.)

logging. Some countries, such as the Philippines, Bangladesh, and Guinea-Bissau, have cut down more than two thirds of their mangrove forests.

Marine ecosystems dominate Earth's surface

Although lakes and the ocean are comparable in many ways, they have many differences. Depths of even the deepest lakes do not approach those of the ocean, which has areas that extend more than 6 km (3.6 mi) below the sunlit surface. Tides and currents profoundly influence the ocean. Gravitational pulls of both sun and moon produce two tides a day throughout the ocean, but the height of those tides varies with season, local topography, and phases of the moon (full moons and new moons cause the highest tides).

The immense, complex marine environment is subdivided into several zones: the intertidal zone, the benthic (ocean floor) environment, and the pelagic (ocean water) environment (▌ Fig. 55-19). The pelagic environment is, in turn, divided into two provinces—the neritic province and the oceanic province.

The intertidal zone is transitional between land and ocean

The **intertidal zone** is the shoreline area between low and high tide. Although high levels of light and nutrients, together with an abundance of oxygen, make the intertidal zone a biologically productive environment, it is also a stressful one. If an intertidal beach is sandy, inhabitants must contend with a constantly shifting environment that threatens to engulf them and gives them scant protection against wave action. Consequently, most sand-dwelling organisms, such as mole crabs, are continual and active burrowers. Because they follow the tides up and down the beach, most do not have any notable adaptations to survive drying or exposure.

A rocky shore provides a fine anchorage for seaweeds and invertebrate animals. However, it is exposed to constant wave action when immersed during high tides and to drying and temperature changes when exposed to the air during low tides. A typical rocky-shore inhabitant has some way of sealing in moisture, perhaps by closing its shell, if it has one, plus a powerful means of anchoring itself to rocks. Mussels, for example, have horny, threadlike anchors, and barnacles have special cement glands. Rocky-shore intertidal algae (seaweeds) usually have thick, gummy polysaccharide coats, which dry out slowly when exposed, and flexible bodies not easily broken by wave action (▌ Fig. 55-20). Some rocky-shore community inhabitants hide in burrows or crevices at low tide.

Seagrass beds, kelp forests, and coral reefs are part of the benthic environment

The **benthic environment** is the ocean floor. It is divided into zones based on distance from land, light availability, and depth. The benthic environment consists of sediments (mostly sand and mud) in which many marine animals, such as worms and clams, burrow. Bacteria are common in marine sediments, and living bacteria have been found in ocean sediments more than 500 m (1625 ft) below the ocean floor at several different sites in the Pacific Ocean. The **abyssal zone** is that part of the benthic environment that extends from a depth of 4000 to 6000 m (2.5 to 3.7 mi.). (In Chapter 54, *Focus On: Life without the Sun* describes some of the unusual organisms in hydrothermal vents in the abyssal zone.) The **hadal zone** is that part of the benthic environment deeper than 6000 m.

Here we describe benthic communities in shallow ocean waters—seagrass beds, kelp forests, and coral reefs. **Sea grasses** are flowering plants that have adapted to complete submersion in ocean water (▌ Fig. 55-21). They are not true grasses. Sea grasses live in shallow water, to depths of 10 m (33 ft), where they receive enough light to photosynthesize efficiently. Extensive beds of sea grasses occur in quiet temperate, subtropical, and tropical waters;

Figure 55-20 Seaweeds in a rocky intertidal zone

Sea palms (*Postelsia*), which are 50 to 75 cm (20 to 30 in) tall, are common on the rocky Pacific coast from Vancouver Island to California. The bases of these brown algae are firmly attached to the rocky substrate, enabling them to withstand heavy surf action. Photographed at low tide.

no sea grasses live in polar waters. Sea grasses have a high primary productivity and are therefore ecologically important in shallow marine areas. Their roots and rhizomes stabilize the sediments, reducing surface erosion. Sea grasses provide food and habitat for many marine organisms. In temperate waters, ducks and geese eat sea grasses; in tropical waters, manatees, green turtles, parrot fish, sturgeon fish, and sea urchins eat them. These herbivores consume only about 5% of the sea grasses. The remaining 95% eventually enter the detritus food web when the sea grasses die and bacteria decompose them. In turn, a variety of animals such as mud shrimp, lug worms, and mullet (a type of fish) consume the bacteria.

Kelps, which may reach lengths of 60 m (200 ft), are the largest brown algae (see Fig. 25-14b). Kelps are common in cooler temperate marine waters of both the Northern and Southern Hemispheres. They are especially abundant in relatively shallow waters (depths of about 25 m, or 82 ft) along rocky coastlines. Kelps are photosynthetic and are therefore the primary food producers for the kelp forest ecosystem. Kelp forests also provide habitats for many marine animals. Tube worms, sponges, sea cucumbers, clams, crabs, fishes (such as tuna), and mammals (such as sea otters) find refuge in the algal blades. Some animals eat the blades, but kelps are consumed mainly in the detritus food web. Bacteria that decompose dead kelp provide food for sponges, tunicates, worms, clams, and snails. Kelp beds support a diversity of life that almost rivals that found in coral reefs.

Coral reefs, which are built from accumulated layers of calcium carbonate ($CaCO_3$), are found in warm (temperature usu-

ally greater than 21°C), shallow sea water. The living portions of coral reefs grow in shallow waters where light penetrates. Many coral reefs are composed principally of red coralline algae that require light for photosynthesis. Coral animals also require light for the large number of symbiotic dinoflagellates, known as **zooxanthellae,** that live and photosynthesize in their tissues (see Fig. 53-11). Although species of coral without zooxanthellae exist, only those with zooxanthellae build reefs. In addition to obtaining food from the zooxanthellae living inside them, coral animals capture food at night; they use their stinging tentacles to paralyze small animals that drift nearby. Coral reefs grow slowly in warm, shallow water, as coral organisms build on the calcareous remains of countless organisms before them. The waters in which coral reefs are found are often poor in nutrients. Other factors favor high productivity, however, including the presence of symbiotic zooxanthellae, warm temperatures, and plenty of sunlight.

Coral reef ecosystems are the most diverse of all marine environments and contain hundreds or even thousands of species of fishes and invertebrates, such as giant clams, sea urchins, sea stars, sponges, brittle stars, sea fans, and shrimp (Fig. 55-22). The Great Barrier Reef, along the northeastern coast of Australia, occupies only 0.1% of the ocean's surface, but 8% of the world's fish species live there. The multitude of relationships and interactions that occur at coral reefs is comparable only to those in tropical rain forests among terrestrial ecosystems. As in the rain forest, competition is intense, particularly for light and space to grow.

Coral reefs are ecologically important because they both provide habitat for a wide variety of marine organisms and protect coastlines from shoreline erosion. They also provide humans with seafood, pharmaceuticals, and income from tourism and recreation. Although coral formations are important ecosystems, they are being degraded and destroyed. According to the UN Environment Program, 27% of the world's coral reefs are at high

Figure 55-21 Seagrass bed

Turtle grasses (*Thalassia*) have numerous invertebrates and algal epiphytes attached to their leaves. These shallow underwater meadows of marine flowering plants are ecologically important for shelter and food for many organisms. Photographed off the coast of Mexico.

risk. Coral reefs of southeastern Asia, which contain the most species of all coral reefs, are the most threatened of any region.

In some areas, silt washing downstream from clear-cut inland forests has smothered reefs under a layer of sediment. Some scientists hypothesize that high salinity resulting from the diversion of fresh water to supply the growing human population is killing Florida reefs. Overfishing, pollution from sewage discharge and agricultural runoff, oil spills, boat groundings, fishing with dynamite or cyanide, hurricane damage, disease, coral bleaching, land reclamation, tourism, and the mining of corals for building material are also taking a heavy toll. (Coral bleaching is discussed in Chapter 29.)

Figure 55-22 Coral reef organisms

A panoramic view of a coral reef in the Indian Ocean off the coast of the Maldives shows the many animals that live on and around coral reefs.

The neritic province consists of shallow waters close to shore

The **neritic province** is open ocean that overlies the continental shelves, that is, the ocean floor from the shoreline to a depth of 200 m (650 ft). Organisms that live in the neritic province are floaters or swimmers. The upper reaches of the neritic province make up the **euphotic zone,** which extends from the surface to a depth of approximately 100 m (325 ft). Enough light penetrates the euphotic zone to support photosynthesis.

Large numbers of phytoplankton, particularly diatoms in cooler waters and dinoflagellates in warmer waters, produce food by photosynthesis and are thus the base of food webs. Zooplankton (including tiny crustaceans; jellyfish; comb jellies; protists such as foraminiferans; and larvae of barnacles, sea urchins, worms, and crabs) feed on phytoplankton.

Plankton-eating nekton, such as herring, sardines, squid, manta rays, and baleen whales, consume zooplankton (Fig. 55-23). These, in turn, become prey for carnivorous nekton such as sharks, tuna, dolphins, and toothed whales. Nekton are thought to be mostly confined to the shallower neritic waters (less than 60 m, or 195 ft, deep) because that is where their food is. However, not much is known about the behavior and migration patterns of marine nekton. Many fishes appear to be wide ranging. For example, an individual fish tagged on one side of an ocean may be recaptured on the other side a few months later. Whether the fish traveled alone or in a school is not known at this time.

The oceanic province makes up most of the ocean

The average depth of the world's ocean is 4000 m (2.4 mi). The **oceanic province** is that part of the open ocean that covers the deep-ocean basin, that is, the ocean floor at depths more than 200 m. It is the largest marine environment and contains about

75% of the ocean's water. Cold temperatures, high hydrostatic pressure, and an absence of sunlight characterize the oceanic province; these environmental conditions are uniform throughout the year.

Most organisms of the oceanic province depend on **marine snow,** organic debris that drifts down into the **aphotic** ("without light") **region** from the upper, lighted regions. Organisms of this little-known realm are filter-feeders, scavengers,

Figure 55-23 Fish in the open ocean

Pacific sardines (*Sardinops sagax*) are one of the world's important commercial fishes and provide an important source of protein for the human diet. Sardines are also an important food source for other marine animals.

or predators. Many are invertebrates, some of which attain great sizes. The giant squid, for example, measures up to 18 m (59 ft) in length, including its tentacles. Fishes of the oceanic province are strikingly adapted to darkness and food scarcity. For example, the gulper eel's huge jaws enable it to swallow large prey (Fig. 55-24). (An organism that encounters food infrequently needs to eat as much as possible when it has the chance.) Many animals of the oceanic province have illuminated organs that enable them to see one another to mate or to capture food. Adapted to drifting or slow swimming, these fishes often have reduced bone and muscle mass.

Figure 55-24   Gulper eel from the oceanic province

The gulper eel (*Saccopharynx lavenbergi*) uses its "trap-door" jaws to swallow prey as large as itself. The tail, of which only a small portion is shown, makes up most of the length of a gulper eel's body. Gulper eels grow to 1.8 m (6 ft). Shown is a live specimen, photographed in an aquarium aboard ship after being captured at a depth of 1500 m (4921 ft) off Southern California.

Human activities are harming the ocean

Because the ocean is so vast, it is hard to visualize that human activities could affect, much less harm, it. They do, however. Development of resorts, cities, industries, and agriculture along coasts alters or destroys many coastal ecosystems, including mangrove forests, salt marshes, seagrass beds, and coral reefs. Coastal and marine ecosystems receive pollution from land, from rivers emptying into the ocean, and from atmospheric contaminants that enter the ocean via precipitation. Disease-causing viruses and bacteria from human sewage contaminate shellfish and other seafood and sometimes threaten public health. Millions of tons of trash, including plastic, fishing nets, and packaging materials, end up in coastal and marine ecosystems; some of this trash entangles and kills marine organisms. Less visible contaminants of the ocean include fertilizers, pesticides, heavy metals, and synthetic chemicals from agriculture and industry.

Offshore mining and oil drilling pollute the neritic province with oil and other contaminants. Millions of ships dump oily ballast and other wastes overboard in the neritic and oceanic provinces. Fishing is highly mechanized, and new technologies can remove every single fish in a targeted area of the ocean. Scallop dredges and shrimp trawls are dragged across the benthic environment, destroying entire communities with a single swipe. Most countries recognize the importance of the ocean to life on this planet, but few have the resources or programs to protect and manage the ocean effectively.

Review

- What are plankton, nekton, and benthos?
- What environmental factors are most important in determining the adaptations of organisms that live in aquatic environments?

- How do you distinguish between freshwater wetlands and estuaries? Between flowing-water and standing-water ecosystems?
- What are the four main marine environments?
- Which aquatic ecosystem is often compared to tropical rain forests? Why?

ECOTONES

Learning Objective

7   Define and describe some of the features of an ecotone.

We have discussed the various terrestrial biomes and aquatic ecosystems as if they were distinct and separate entities, but within landscapes, ecosystems intergrade with one another at their boundaries. You learned in Chapter 53 that the transition zone where two communities or biomes meet and intergrade is called an **ecotone.** Ecotones range in size from quite small, such as the area where an agricultural field meets a woodland or where a stream flows through a forest, to continental in scope. For example, at the border between tundra and boreal forest an extensive ecotone exists that consists of tundra vegetation interspersed with small, scattered conifers. Such ecotones provide habitat diversity. In fact, often a greater variety and density of organisms populate ecotones than live in either adjacent ecosystem (Fig. 55-25).

QUESTION: Does species richness vary among ecotones and their adjoining communities?

HYPOTHESIS: Ecotones have greater species richness than the communities they connect.

EXPERIMENT: Plant species were sampled for two communities in southwestern Oregon and for the ecotone between them. The communities, one with nonserpentine soil and one with serpentine soil, are defined largely by soil conditions. Nonserpentine soils (samples 1 to 10) are "normal" soils. Serpentine soils (samples 18 to 28) contain high levels of elements such as chromium, nickel, and magnesium that are toxic to many plants.

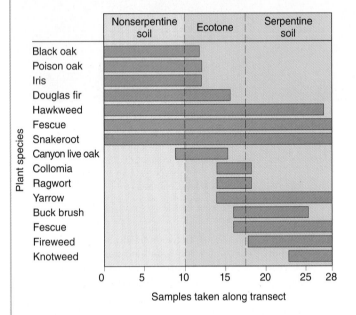

RESULTS AND CONCLUSION: Shown are the various plant species found in the two communities (yellow and blue) and in the ecotone between them (green). The ecotone had a greater richness than either adjoining community.

Figure 55-25 Ecotones and species richness

(Modified from C. D. White, "Vegetation-Soil Chemistry Correlations in Serpentine Ecosystems," Ph.D. dissertation, University of Oregon, Eugene, 1971. Reprinted with permission of Dr. Charles D. White.)

Ecologists who study ecotones look for adaptations that enable organisms to survive there. They also examine the relationship between species richness and ecotones and how ecotones change over time. Long-term studies of ecotones have revealed they are far from static. The ecotone boundary between desert and semiarid grassland in southern New Mexico, for example, has moved during the past 50 years as the desert ecosystem has expanded into the grassland.

Review

▪ What is an ecotone?

▪ Where do ecotones occur?

BIOGEOGRAPHY

Learning Objective

8   Define *biogeography,* and briefly describe Wallace's biogeographic realms.

The study of the geographic distribution of plants and animals is called **biogeography** (see Chapter 18). Biogeographers search for patterns in geographic distribution and try to explain how such patterns arose, including where populations originated, how they spread, and when. Biogeographers recognize that geologic and climate changes such as mountain building, continental drift, and periods of extensive glaciation influence the distribution of species. Biogeography is linked to evolutionary history and provides insights into how organisms may have interacted in ancient ecosystems. Studying biogeography helps us relate ancient ecosystems to modern ones, because ancient ecosystems form a continuum with modern ecosystems.

One of the basic tenets of biogeography is that each species originated only once. The particular place where this occurred is known as the species' **center of origin.** The center of origin is not a single point but the distribution of the population when the new species originated. From its center of origin, each species spreads until a barrier of some kind halts it; examples of barriers include the ocean, a desert, or a mountain range; unfavorable climate; or the presence of organisms that compete successfully for food or shelter.

Most plant and animal species have characteristic geographic distributions. The **range** of a particular species is that portion of Earth in which it is found. The range of some species may be a relatively small area. For example, wombats, the marsupial equivalent of groundhogs, are found only in drier parts of southeastern Australia and nearby islands. Such localized, native species are said to be **endemic,** that is, they are not found anywhere else in the world. In contrast, some species have a nearly worldwide distribution and occur on more than one continent or throughout much of the ocean. Such species are said to be **cosmopolitan.**

One of the early observations of biogeographers is that the ranges of different species do not include everywhere that they *could* survive. Central Africa has elephants, gorillas, chimpanzees, lions, antelopes, umbrella trees, and guapiruvu trees, whereas areas in South America with a similar climate have none of these. These animals and plants originated in Africa after continental drift had already separated the supercontinent Pangaea into several landmasses. The organisms could not expand their range into South America because the Atlantic Ocean was an impassable barrier. Likewise, the ocean was a barrier to South American monkeys, sloths, tapirs, balsa trees, and snakewood trees, none of which are found in Africa.

Land areas are divided into six biogeographic realms

As the various continents were explored and their organisms studied, biologists observed that the world could be divided into major blocks of vegetation, such as forests, grasslands, and deserts, and that these vegetation types corresponded to specific climates. The relationship between animal distribution, geography, and climate was not deduced until 1876. At that time, Alfred Wallace, who discovered the same theory of evolution by natural selection as Charles Darwin, divided Earth's land areas into six major biogeographic realms: the Palearctic, Nearctic, Neotropical, Ethiopian, Oriental, and Australian (**Fig. 55-26**). A major barrier separates each of the six biogeographic realms from the others and helps maintain each region's biological distinctiveness.

Many biologists quickly embraced Wallace's classification, which is still considered valid today. However, human activities, such as the intentional and unintentional introduction of foreign species, are contributing to a homogenization of the biogeographic realms (see Chapter 56).

Refer to Figure 55-26 as we briefly consider some of the characteristic animals in each realm. The *Nearctic* and *Palearctic realms* are more closely related than the other regions, especially in their northern parts, where they share many animals such as wolves, hares, and caribou. This similarity may be due to a land bridge that has periodically connected Siberia and Alaska. This bridge was present late in the Pleistocene epoch, about 10,000 years ago. Animals adapted to cold environments may have dispersed between Asia and North America along this bridge.

The *Neotropical realm* was almost completely isolated from the Nearctic realm and other landmasses for most of the past 70 million years. During this time, many marsupial species evolved. The isthmus of Panama, which formed a dry-land connection about 3 million years ago, linked North and South America and provided a route for dispersal. Only three species, the opossum, armadillo, and porcupine, are descendants of animals that survived the northward dispersal from South America, but many species, such as the tapir and llama, are descendants of animals that survived the southward dispersal from North America.

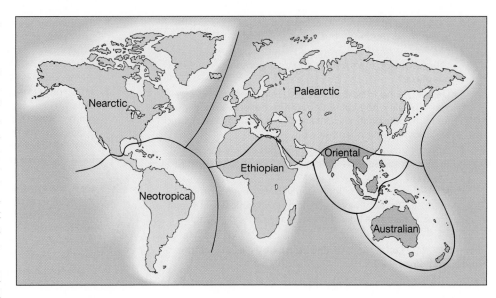

Figure 55-26 Wallace's biogeographic realms
Certain unique species characterize each of the six biogeographic realms.

Competition from these species caused many of South America's marsupial species to go extinct.

The Sahara Desert separates the *Ethiopian realm,* which contains the most varied vertebrates of all six realms, from other landmasses. Some overlap exists between the Ethiopian realm and *Oriental realm* because a land bridge with a moist climate linked Africa to Asia during the Miocene and Pliocene epochs. The Oriental realm has the fewest endemic species of all the tropical realms.

The *Australian realm* has not had a land connection with other regions for more than 85 million years. It has no native placental mammals, and marsupials and monotremes, including the duck-billed platypus and the spiny anteater, dominate it. Adaptive radiation of the marsupials during their long period of isolation led to species with ecological niches similar to those of placental mammals of other realms (see Fig. 31-28).

Review

- What is biogeography?
- Which biogeographic realm has been separated from the other biogeographic realms for the longest period? What animals characterize this biogeographic realm?

SUMMARY WITH KEY TERMS

Learning Objectives

1 Define *biome,* and briefly describe the nine major terrestrial biomes, giving attention to the climate, soil, and characteristic plants and animals of each (page 1190).

- A **biome** is a large, relatively distinct terrestrial region with characteristic climate, soil, plants, and animals.
- A frozen layer of subsoil (**permafrost**) and low-growing vegetation that is adapted to extreme cold and a short growing season characterize **tundra,** the northernmost biome.

- Coniferous trees adapted to cold winters; a short growing season; and acidic, mineral-poor soil dominate the **boreal forest,** or **taiga.**
- Large conifers dominate **temperate rain forest,** which receives high precipitation.
- **Temperate deciduous forest** occurs where precipitation is relatively high and soils are rich in organic matter. Broad-leaf trees that lose their leaves seasonally dominate temperate deciduous forest.

- **Temperate grassland** typically has a deep, mineral-rich soil and has moderate but uncertain precipitation.
- Thickets of small-leaf evergreen shrubs and trees and a climate of wet, mild winters and dry summers characterize **chaparral.**
- **Desert,** found in both temperate (cold deserts) and subtropical or tropical regions (warm deserts) with low levels of precipitation, contains organisms with specialized water-conserving adaptations.
- Tropical grassland, called **savanna,** has widely scattered trees interspersed with grassy areas. Savanna occurs in tropical areas with low or seasonal rainfall.
- Mineral-poor soil and high rainfall that is evenly distributed throughout the year characterize **tropical rain forest.** Tropical rain forest has high species richness and high productivity.

ThomsonNOW **Learn more about biomes by clicking on the figure in ThomsonNOW.**

2 Describe at least one human effect on each of the biomes discussed (page 1190).
- Oil exploration and military exercises result in long-lasting damage to tundra. Clear-cut logging destroys boreal and temperate rain forests. Temperate deciduous forests are removed for logging and for use as farms, tree plantations, and land development.
- Human population growth and its accompanying agricultural and industrial expansion threaten most of the world's tropical rain forests.
- Farmland has replaced most temperate grasslands; savannas are increasingly converted to rangeland for cattle. Development of hilly chaparral results in mud slides and costly fires. Land development in deserts reduces wildlife habitat.

3 Explain the important environmental factors that affect aquatic ecosystems (page 1199).
- In aquatic ecosystems, important environmental factors include **salinity** (concentration of dissolved salts), amount of dissolved oxygen, availability of light, levels of essential minerals, water depth, temperature, pH, and presence or absence of waves and currents.

4 Distinguish among plankton, nekton, and benthos (page 1199).
- Aquatic life is ecologically divided into **plankton** (free-floating organisms), **nekton** (strongly swimming organisms), and **benthos** (bottom-dwelling organisms).
- **Phytoplankton** are photosynthetic algae and cyanobacteria that form the base of the food web in most aquatic communities. **Zooplankton** are nonphotosynthetic organisms that include protozoa, tiny crustaceans, and the larval stages of many animals.

5 Briefly describe the various freshwater, estuarine, and marine ecosystems, giving attention to the environmental characteristics and representative organisms of each (page 1199).
- Freshwater ecosystems include flowing-water ecosystems (streams and rivers), standing water ecosystems (ponds and lakes), and freshwater wetlands. In **flowing-water ecosystems** the water flows in a current. Flowing-water ecosystems have few phytoplankton and depend on detritus from the land for much of their energy.
- Large **standing-water ecosystems** (freshwater lakes) are divided into zones on the basis of water depth. The marginal **littoral zone** contains emergent vegetation and algae and is very productive. The **limnetic zone,** open water away from the shore that extends as far down as sunlight penetrates, contains phytoplankton, zooplankton, and larger fishes. The deep, dark **profundal zone** holds little life other than bacterial decomposers.

- **Freshwater wetlands,** lands that are transitional between freshwater and terrestrial ecosystems, are usually covered at least part of the year by shallow water and have characteristic soils and vegetation. Freshwater wetlands perform many valuable **ecosystem services.**
- An **estuary** is a coastal body of water, partly surrounded by land, with access to both the ocean and a large supply of fresh water from rivers. Salinity fluctuates with tidal cycles, the time of year, and precipitation. Temperate estuaries usually contain **salt marshes,** whereas **mangrove forests** dominate tropical coastlines.
- Four important marine environments are the intertidal zone, the benthic environment, the neritic province, and the oceanic province. The **intertidal zone** is the shoreline area between low and high tides. Organisms of the intertidal zone have adaptations to resist wave action and the extremes of being covered by water (high tide) and exposed to air (low tide).
- The **benthic environment** is the ocean floor. **Sea grasses, kelps,** and **coral reefs** are important benthic communities in shallow ocean waters.
- The **neritic province** is open ocean from the shoreline to a depth of 200 m. Organisms that live in the neritic province are all floaters or swimmers. Phytoplankton are the base of the food web in the **euphotic zone,** where enough light penetrates to support photosynthesis.
- The **oceanic province** is that part of the open ocean that is deeper than 200 m. The uniform environment is one of darkness, cold temperature, and high pressure. Animal inhabitants of the oceanic province are either predators or scavengers that subsist on **marine snow,** detritus that drifts down from other areas of the ocean.

ThomsonNOW **Learn more about lake and ocean zonation by clicking on the figures in ThomsonNOW.**

6 Describe at least one human effect on each of the aquatic ecosystems discussed (page 1199).
- Water pollution and dams adversely affect flowing-water ecosystems. Increased nutrients, supplied by human activities, stimulate algal growth, resulting in enriched ponds and lakes. Agriculture, pollution, and land development threaten wetlands and estuaries. Pollution, coastal development, offshore mining and oil drilling, and overfishing threaten marine ecosystems.

7 Define and describe some of the features of an ecotone (page 1207).
- An **ecotone** is the transition zone where two communities or biomes meet and intergrade. Ecotones provide habitat diversity and are often populated by a greater variety of organisms than lives in either adjacent ecosystem.

8 Define *biogeography,* and briefly describe Wallace's biogeographic realms (page 1208).
- **Biogeography** is the study of the geographic distribution of plants and animals, including where populations came from, how they got there, and when. Each species originated only once, at its **center of origin.** From its center of origin, each species spreads until a physical, environmental, or biological barrier halts it. The **range** of a particular species is that portion of the Earth in which it is found.
- Alfred Wallace divided Earth's land areas into six major biogeographic realms: the Palearctic, Nearctic, Neotropical, Ethiopian, Oriental, and Australian. Each realm is biologically distinctive because a mountain range, desert, ocean, or other barrier separates it from the others. Today, human activities are contributing to a homogenization of the biogeographic realms.

TEST YOUR UNDERSTANDING

1. The northernmost biome, known as _____, typically has little precipitation, a short growing season, and permafrost. (a) chaparral (b) boreal forest (c) tundra (d) northern deciduous forest (e) savanna

2. South of tundra is the _____, which consists of coniferous forests with many lakes. (a) chaparral (b) boreal forest (c) alpine tundra (d) northern deciduous forest (e) permafrost

3. Forests of the northeastern and Mid-Atlantic United States, which have broad-leaf hardwood trees that lose their foliage annually, are called (a) temperate deciduous forests (b) tropical dry forests (c) boreal forests (d) temperate rain forests (e) tropical rain forests

4. The deepest, richest soil in the world occurs in (a) temperate rain forest (b) tropical rain forest (c) savanna (d) temperate grassland (e) chaparral

5. This biome, with its thicket of evergreen shrubs and small trees, is found in areas with Mediterranean climates. (a) temperate rain forest (b) tropical rain forest (c) savanna (d) temperate grassland (e) chaparral

6. This biome is a tropical grassland interspersed with widely spaced trees. (a) temperate rain forest (b) tropical rain forest (c) savanna (d) temperate grassland (e) chaparral

7. This biome has the greatest species richness. (a) temperate rain forest (b) tropical rain forest (c) savanna (d) temperate grassland (e) chaparral

8. Organisms in aquatic environments fall into three categories: free-floating _____, strongly swimming _____, and bottom-dwelling _____. (a) nekton; benthos; plankton (b) nekton; plankton; benthos (c) plankton; benthos; nekton (d) plankton; nekton; benthos (e) benthos; nekton; plankton

9. Temperate-zone lakes are thermally stratified, with warm and cold layers separated by a transitional (a) aphotic region (b) thermocline (c) barrier reef (d) ecotone (e) littoral zone

10. Emergent vegetation grows in the _____ zone of freshwater lakes. (a) littoral (b) limnetic (c) profundal (d) neritic (e) intertidal

11. This coastal body of water has access to the open ocean and a large supply of fresh water from rivers. (a) intertidal zone (b) estuary (c) freshwater wetland (d) neritic province (e) standing-water ecosystem

12. The _____ is open ocean from the shoreline to a depth of 200 m. (a) benthic environment (b) intertidal zone (c) neritic province (d) oceanic province (e) aphotic region

13. Sea grasses (a) occur in shallow water of the ocean's benthic environment (b) may reach lengths of 60 m (c) contain symbiotic algae known as zooxanthellae (d) are common on rocky shores in the intertidal zone (e) are the main producers in freshwater wetlands

14. The transition zone where two ecosystems or biomes meet and intergrade is called a(an) (a) biosphere (b) aphotic region (c) thermocline (d) biogeographic realm (e) ecotone

15. Which biogeographic realm has been separated from the other landmasses for more than 85 million years? (a) Ethiopian (b) Palearctic (c) Nearctic (d) Oriental (e) Australian

CRITICAL THINKING

1. In which biomes would migration be most common? Hibernation? Estivation? Explain your answers.

2. Develop a hypothesis to explain why animals adapted to the desert are usually small. How would you test your hypothesis?

3. Why do most of the animals of the tropical rain forest live in trees?

4. What would happen to the organisms in a river with a fast current if a dam were built? Would there be any differences in habitat if the dam were upstream or downstream of the organisms in question? Explain your answers.

5. **Evolution Link.** When a black-tailed prairie dog or other small animal dies, other prairie dogs bury it. Develop a hypothesis to explain how this behavior may be adaptive. How would you test this hypothesis?

6. **Analyzing Data.** Examine Figure 55-1. What is the lowest average annual precipitation characteristic of tropical rain forests? The highest? What is the range of average temperature in tropical rain forests?

7. **Analyzing Data.** Examine Figure 55-25. How many of the sampled species are found in the nonserpentine soil? In the serpentine soil? In the ecotone? What generalization about ecotones do these data support?

Additional questions are available in ThomsonNOW at www.thomsonedu.com/login

Global Environmental Issues

Steve Nesbitt, Florida Fish and Wildlife Conservation Commission

Lucky, the whooping crane (*Grus americana*). Photographed with one of his parents in Florida.

KEY CONCEPTS

Conservation biology is a multidisciplinary science that studies human effects on biodiversity and devises practical solutions to address these effects.

Today's mass extinction event is unique, because never before in Earth's history have so many species been threatened with extinction in such a compressed time period.

Globally, removal of large expanses of forest for timber, agriculture, and other uses is contributing to the loss of biological diversity and valuable ecosystem services.

There is strong evidence of global climate change, which scientists predict will adversely affect biodiversity, human health, and agriculture.

Although the stratospheric ozone layer is slowly recovering, harmful effects on biodiversity and humans will continue for several decades.

The whooping crane is a magnificent North American bird named for its loud, penetrating voice. It stands about 1.4 m (4.5 ft) tall and lives for about 50 years. During the 20th century, the whooping crane population dropped to only 16 in the wild, and their breeding range shrank from much of central North America to one remote location in Canada. Whooping crane numbers declined primarily due to the loss of their wetland habitat, which humans fill in for farmland, housing developments, and industrial plants.

Thanks to intensive efforts by conservation organizations and government agencies in Canada and the United States, the wild flock of whooping cranes still exists and now contains about 220 birds. The wild population spends summer in Canada and migrates to the Texas coast to winter.

Concerned that bad weather, disease, or some other natural or human-caused disaster could wipe out the single wild population, in the 1990s biologists started reintroducing a second population of whooping cranes to the eastern part of North America. Captive-bred whooping cranes are released in central Florida to establish this population, where a population of related sandhill cranes with similar habitat requirements resides.

A breeding pair of whooping cranes made history in 2002 when they successfully reproduced. An eagle killed one of the offspring, but the other chick, named Lucky, survived (see photograph). Lucky is the first whooping crane to be produced by captive-reared, released parents and the first to hatch in the wild in the United States since 1939.

Whooping cranes are an excellent example of the kinds of injuries humans inflict on biological diversity. They also exemplify human efforts to maintain the world's declining biological diversity. We conclude this text by focusing our attention on four serious environmental issues that affect the biosphere: declining biological diversity, deforestation, global warming, and ozone depletion in the stratosphere. ∎

THE BIODIVERSITY CRISIS

Learning Objectives

1 Identify the various levels of biodiversity: genetic diversity, species richness, and ecosystem diversity.
2 Distinguish among threatened species, endangered species, and extinct species.
3 Discuss at least four causes of declining biological diversity, and identify the most important cause.

The human species (*Homo sapiens*) has been present on Earth for about 800,000 years (see Chapter 22), which is a brief span of time compared with the age of our planet, some 4.6 billion years. Despite our relatively short tenure on Earth, our biological impact on other species is unparalleled. Our numbers have increased dramatically—the human population reached 6.5 billion in mid-2006—and we have expanded our biological range, moving into almost every habitat on Earth.

Wherever we have gone, we have altered the environment and shaped it to meet our needs. In only a few generations we have transformed the face of Earth, placed a great strain on Earth's resources and resilience, and profoundly affected other species. As a result of these changes, many people are concerned with **environmental sustainability,** the ability to meet humanity's current needs without compromising the ability of future generations to meet their needs. The impact of humans on the environment merits special study in biology, not merely because we ourselves are humans but also because our impact on the rest of the biosphere is so extensive.

Extinction, the death of a species, occurs when the last individual member of a species dies (see Chapter 20). Although extinction is a natural biological process, human activities have greatly accelerated it so that it is happening at a rate as much as 1000 times the normal background rate of extinction, according to the 2005 Millennium Ecosystem Assessment report. The burgeoning human population has forced us to spread into almost all areas of Earth. Whenever humans invade an area, the habitats of many plants and animals are disrupted or destroyed, which can contribute to their extinction.

Biological diversity, also called **biodiversity,** is the variation among organisms (❙ Fig. 56-1). Biological diversity includes much more than simply the number of species, called **species richness,** of archaea, bacteria, protists, plants, fungi, and animals. Biological diversity occurs at all levels of biological organization, from populations to ecosystems. It takes into account **genetic diversity,** the genetic variety within a species, both among individuals within a given population and among geographically separate populations. (An individual species may have hundreds of genetically distinct populations.) Biological diversity also includes **ecosystem diversity,** the variety of ecosystems found on Earth—the forests, prairies, deserts, lakes, coastal estuaries, coral reefs, and other ecosystems.

Biologists must consider all three levels of biological diversity—species richness, genetic diversity, and ecosystem diversity—as they address the human impact on biodiversity. For example, the disappearance of populations (that is, a decline in genetic diversity) indicates an increased risk that a species will become extinct (a decline in species richness). Biologists perform detailed analyses to quantify how large a population must be to ensure a given species' long-term survival. The smallest population that has a high chance of enduring into the future is known as the **minimum viable population (MVP).**

Biological diversity is currently decreasing at an unprecedented rate (see *Focus On: Declining Amphibian Populations*). According to the UN Global Biodiversity Assessment, which is based on the work of about 1500 scientists from around the world, more than 31,000 plant and animal species are currently threatened with extinction. More recent surveys indicate that this number is probably too conservative. For example, the World Conservation Union Red List of Threatened Plants, based on 20 years of data collection and analysis around the world, lists about 34,000 species of plants currently threatened with extinction. Because Red Lists are not yet available for many tropical countries, it is difficult to know the true scale of the biodiversity crisis.

We may have entered the greatest period of mass extinction in Earth's history, but the current situation differs from previous periods of mass extinction in several respects. First, its cause is directly attributable to human activities. Second, it is occurring in a tremendously compressed period (just a few decades, as opposed to hundreds of thousands of years), much faster than rates of speciation (or replacement). Perhaps even more sobering, larger numbers of plant species are becoming extinct today than in previous mass extinctions. Because plants are the base of terrestrial food webs, extinction of animals that depend on plants is not far behind. It is crucial that we determine how the loss of biodiversity affects the stability and functioning of ecosystems, which make up our life-support system.

The legal definition of an **endangered species,** as stipulated by the U.S. Endangered Species Act, is a species in imminent danger of extinction throughout all or a significant part of its range. (The area in which a particular species is found is its **range.**) Unless humans intervene, an endangered species will probably become extinct.

When extinction is less imminent but the population of a particular species is quite small, the species is classified as **threatened.** The legal definition of a threatened species is a species likely to become endangered in the foreseeable future, throughout all or a significant part of its range.

Genetic diversity in an American oystercatcher population

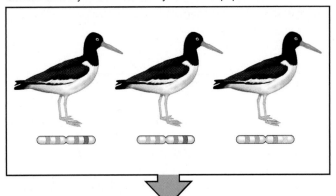

Species richness in an Eastern sandy shore ecosystem

Ecosystem diversity across an entire region

Figure 56-1 Levels of biodiversity

The single unduplicated chromosome below each oystercatcher represents the genetic variation among individuals within a population. Conservation at all three ecological levels must occur to protect biological diversity.

Endangered or threatened species represent a decline in biological diversity, because as their numbers decrease, their genetic diversity is severely diminished. Endangered and threatened species are at greater risk of extinction than species with greater genetic variability, because long-term survival and evolution depend on genetic diversity (see the section on genetic drift in Chapter 19).

Endangered species have certain characteristics in common

Many threatened and endangered species share certain characteristics that may make them more vulnerable to extinction. Some of these characteristics include having an extremely small (localized) range, requiring a large territory, living on islands, having a low reproductive success, needing specialized breeding areas, and possessing specialized feeding habits.

Many endangered species have a small natural range, which makes them particularly prone to extinction if their habitat is altered. The Tiburon mariposa lily, for example, is found nowhere in nature except on a single hilltop near San Francisco. Development of that hilltop would almost certainly cause the extinction of this species.

Species that need extremely large territories may be threatened with extinction when all or part of their territory is modified by human activity. The California condor, for example, is a scavenger bird that lives off carrion and requires hundreds of square kilometers of undisturbed territory to find adequate food. The condor is slowly recovering from the brink of extinction. From 1987 to 1992, it existed only in zoos and was not found in nature. A program to reintroduce zoo-bred California condors into the wild began in 1992. By 2006, there were about 150 condors living in captivity and about 125 condors flying free in California and adjacent areas in Mexico and Arizona. The condor story is not an unqualified success, because the condors have yet to become a self-sustaining population that replaces its numbers without augmentation by captive breeding.

Many **endemic** island species are endangered. (*Endemic* means they are not found anywhere else in the world.) These species often have small populations that cannot be replaced by immigration if their numbers decline. Because they evolved in isolation from competitors, predators, and disease organisms, island species have few defenses when such organisms are introduced, usually by humans. It is not surprising that of the 171 bird species that have become extinct in the past few centuries, 155 of them lived on islands.

For a species to survive, its members must be present within their range in large enough numbers for males and females to mate. The minimum viable population that ensures reproductive success varies from one type of organism to another. For all species, when the population size falls below the MVP, the population goes into a decline and becomes susceptible to extinction.

Endangered species often have low reproductive rates. The female blue whale produces a single calf every other year, whereas no more than 6% of swamp pinks, an endangered species of small flowering plant, produce flowers in a given year (Fig. 56-2). Some endangered species breed only in specialized areas; for example, the green sea turtle lays its eggs on just a few beaches.

Highly specialized feeding habits endanger a species. In nature, the giant panda eats only bamboo. Periodically all the bam-

DECLINING AMPHIBIAN POPULATIONS

Since the 1970s, many of the world's frog populations have dwindled or disappeared. According to the World Conservation Union Global Amphibian Assessment (GAA), a 3-year study published in 2004, 43% of the world's 5743 described species of amphibians have decreasing populations. Since 1980, 129 species have disappeared and are presumed extinct.

The declines are not limited to areas with obvious habitat destruction (such as drainage of wetlands where frogs live), degradation from pollutants, or overharvesting. Some remote, pristine locations also show dramatic declines in amphibians. No single factor seems responsible for these declines. Potential factors for which there is strong evidence include pollutants, increased UV radiation, infectious diseases, and climate warming.

Agricultural chemicals are implicated in amphibian declines in California's Sierra Nevada mountains. Frog populations on the eastern slopes are relatively healthy, but about eight species are declining on the western slopes, where prevailing winds carry residues of 15 different pesticides from the Central Valley's agricultural region. Agricultural chemicals are also implicated in amphibian declines on the Eastern Shore of Maryland and in Ontario, Canada.

Increased UV radiation caused by ozone thinning is another possible culprit. Most amphibians possess an enzyme that helps repair DNA damage from natural UV radiation. Species suffering declines may have a limited ability to repair DNA damage caused by natural UV radiation.

Infectious diseases are strongly implicated in some of the declines. In 1998, a fungus in the chytrid taxon was first connected to massive declines in more than 12 species in Australia. Currently, the GAA reports that worldwide this fungus threatens about 200 amphibian species with extinction.

Golden toads and about 20 other frog species that have disappeared in mountain areas of Costa Rica may have succumbed to climate warming. In recent years, increasing global temperatures have reduced moisture levels in the cloud forests of Costa Rica's central highlands. The organisms in these tropical mountain forests depend on the mist, particularly during the dry winter season. Scientists have correlated less frequent dry-season mist with declines in the abundance of many animal species, including the frogs that disappeared.

Amphibian Deformities

In 1995, while on a field trip to a local pond, schoolchildren in Minnesota discovered that almost half of the leopard frogs they caught were deformed. The children found frogs with extra legs, extra toes, eyes located on the shoulder or back, deformed jaws, missing legs, missing toes, and missing eyes (see photograph). Deformed frogs usually die early, before they reproduce. Predators easily catch frogs with extra or missing legs. Since 1995, 46 states have reported abnormally large numbers of deformities (as high as 60% in some local populations) in more than 60 amphibian species. Canada and other countries have also reported frog deformities.

Biologists have investigated many possible causes that produce amphibian deformities during development, and no single factor explains the deformities found in all locations. Several pesticides affect

Frans Lanting/Minden Pictures

Frog deformities.
Pollution, ultraviolet radiation, and parasites are implicated in the recent widespread increase in developmental abnormalities in amphibians.

normal development in frog embryos: Lab experiments using traces of the pesticide S-methoprene and its breakdown products have caused frog deformities similar to those in Vermont ponds close to where S-methoprene is used to control mosquitoes and fleas. Ultraviolet radiation causes mutations in DNA. Laboratory tests have also demonstrated that infecting tadpoles with a parasitic flatworm causes them to develop deformities like the ones observed in field sites in Santa Clara County, California.

---

boo plants in a given area flower and die together; when this occurs, panda populations face starvation. Like many other endangered species, giant pandas are also endangered because their habitat has been fragmented, and there are few intact habitats where they can survive. China's 1100 wild giant pandas live in 24 isolated habitats that occupy a small fraction of their historic range.

Human activities contribute to declining biological diversity

Species become endangered and extinct for a variety of reasons, including the destruction or modification of habitats and the production of pollution. Humans also upset the delicate balance of organisms in a given area by introducing foreign species or

Jeffrey Lepore/Photo Researchers, Inc.

Figure 56-2 The swamp pink, an endangered species

The swamp pink (*Helonias bullata*) lives in boggy areas of the eastern United States. Photographed in Killens Pond State Park, Delaware.

by controlling native pests or predators. Illegal hunting and uncontrolled commercial harvesting are also factors. ▌Table 56-1 summarizes an extensive study of the human threats to biological diversity in the United States. These data illustrate the overall importance of habitat loss and degradation. Some interesting variations among the groups of organisms are also apparent. Pollution, for example, is a very serious threat to reptiles, fishes, and invertebrates but a relatively minor threat to plants.

To save species, we must protect their habitats

Most species facing extinction today are endangered by destruction, fragmentation, or degradation of natural habitats. Building roads, parking lots, bridges, and buildings; clearing forests to grow crops or graze domestic animals; and logging forests for timber all take their toll on natural habitats. Draining marshes converts aquatic habitats to terrestrial ones, whereas building dams and canals floods terrestrial habitats (▌Fig. 56-3). Because most organisms require a particular type of environment, habitat destruction reduces their biological range and ability to survive.

Humans often leave small, isolated patches of natural landscape that roads, fences, fields, and buildings completely surround. In ecological terms, *island* refers not only to any landmass surrounded by water but also to any isolated habitat surrounded by an expanse of unsuitable territory. Accordingly, a small patch of forest surrounded by agricultural and suburban lands is considered an island.

Habitat fragmentation, the breakup of large areas of habitat into small, isolated segments (that is, islands), is a major threat to the long-term survival of many populations and species (▌Fig. 56-4). Species from the surrounding "developed" landscape may intrude into the isolated habitat (the *edge effect,* discussed in Chapter 53), whereas rare and wide-ranging species that require a large patch of undisturbed habitat may disappear altogether. Habitat fragmentation encourages the spread of *invasive species,* disease organisms, and other "weedy" species and makes it difficult for organisms to migrate. Generally, habitat fragments support only a fraction of the species found in the original, unaltered environment. However, much remains to be learned about precisely how habitat fragmentation affects specific populations, species, and ecosystems.

Human activities that produce acid precipitation and other forms of pollution indirectly modify habitats left undisturbed and in their natural state. Acid precipitation has contributed to the decline of large stands of forest trees and to the biological death of many freshwater lakes, for example, in the Adirondack Mountains and in Nova Scotia. Other types of pollutants, such as

---

TABLE 56-1

Percentages of Imperiled U.S. Species That Are Threatened by Various Human Activities*

Activity	All Species (1880)†	Plants (1055)	Mammals (85)	Birds (98)	Reptiles (38)	Fishes (213)	Invertebrates (331)
Habitat loss/ degradation	85%‡	81%	89%	90%	97%	94%	87%
Exotic species	49%	57%	27%	69%	37%	53%	27%
Pollution	24%	7%	19%	22%	53%	66%	45%
Overexploitation	17%	10%	45%	33%	66%	13%	23%

Source: Adapted from D. S. Wilcove et al., "Quantifying Threats to Imperiled Species in the United States," *BioScience*, Vol. 48 (8), Table 2, Aug. 1998.

*This includes species classified as imperiled by The Nature Conservancy and all species listed as endangered or threatened under the Endangered Species Act or formally proposed for listing.

†Numbers in parentheses are the total number of species evaluated. The "All Species" category represents about 75% of imperiled species in the United States.

‡Because many of the species are affected by more than one human activity, the percentages in each column do not total 100.

Figure 56-3 Habitat destruction

This tiny island lies in the Panama Canal. It was once a hilltop in a forest that was flooded when the canal was built.

chemicals from industry and agriculture, organic pollutants from sewage, acid wastes seeping from mines, and thermal pollution from the heated wastewater of industrial plants, also adversely affect organisms.

To save native species, we must control intrusions of invasive species

Biotic pollution, the introduction of a foreign species into an area where it is not native, often upsets the balance among the organisms living in that area and interferes with the ecosystem's normal functioning. Unlike other forms of pollution, which may be cleaned up, biotic pollution is usually permanent. The foreign species may prey on native species or compete with them for food or habitat. If the foreign species causes economic or environmental harm, it is known as an **invasive species.** Generally, a foreign competitor or predator harms local organisms more than do native competitors or predators. Most invasive species lack natural agents (such as parasites, predators, and competitors) that would otherwise control them. Also, without a shared evolutionary history, most native species typically are less equipped to cope with invasive species. Although foreign species sometimes spread into new areas on their own, humans are usually responsible for such introductions, either knowingly or accidentally.

One of North America's greatest biological threats is the zebra mussel, a native of the Caspian Sea. It was probably introduced by a foreign ship that flushed ballast water into the Great Lakes in 1985 or 1986. Since then, the tiny freshwater mussel, which clusters in extraordinary densities, has massed on hulls of boats, piers, buoys, water intake systems and, most damaging of all, on native clam and mussel shells. The zebra mussel's strong appetite for algae, phytoplankton, and zooplankton is also cutting into the food supply of native fishes, mussels, and clams, threatening

Key Experiment

QUESTION: How does habitat fragmentation affect predation of juvenile bay scallops in the marine environment?

HYPOTHESIS: Predation of juvenile bay scallops is greater in patchy areas of sea grass than in undisturbed (continuous) seagrass meadows.

EXPERIMENT: Predation rates of juvenile bay scallops were measured over a 4-week period in three seagrass environments: continuous, patchy, and very patchy.

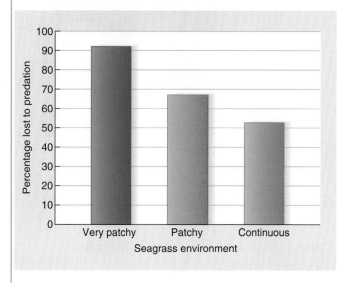

RESULTS AND CONCLUSION: The lowest rate of predation (52%) occurred in the continuous seagrass meadow, and the highest predation rate (92%) occurred in the very patchy environment. The patchy (fragmented) areas provided the predators with easier access to the young scallops. It was more difficult for the predators to intrude from surrounding areas into the continuous seagrass meadow.

Figure 56-4 Effects of habitat fragmentation

(E. A. Irlandi, W. G. Ambrose, and B. A. Orando, "Landscape Ecology and the Marine Environment: How Spatial Configuration of Seagrass Habitat Influences Growth and Survival of the Bay Scallop," *Oikos,* Vol. 72, 1995.)

their survival. The zebra mussel is currently found throughout most of the Mississippi River and its tributary rivers. According to the U.S. Coast Guard, the United States spends about $5 billion each year to control the spread of the zebra mussel and to repair damage such as clogged pipes (❙ Fig. 56-5).

Islands are particularly susceptible to the introduction of invasive species. In Hawaii, the introduction of sheep has imperiled both the mamane tree (because the sheep eat it) and a species of honeycreeper, an endemic bird that relies on the tree for food. Hawaii's plants evolved in the absence of herbivorous mammals and therefore have no defenses against introduced sheep, pigs, goats, and deer.

Figure 56-5 Zebra mussels clog a pipe

Thumbnail-sized zebra mussels (*Dreissena polymorpha*) have caused billions of dollars in damage in addition to displacing native clams and mussels.

Figure 56-6 Illegal commercial harvesting

These hyacinth macaws (*Anodorhynchus hyacinthus*) were seized in French Guiana in South America as part of the illegal animal trade there. The hyacinth macaw population is much reduced in South America.

Other human activities affect biodiversity directly or indirectly

Sometimes species become endangered or extinct as a result of deliberate efforts to eradicate or control their numbers, often because they prey on game animals or livestock. In the past, ranchers, hunters, and government agents decimated populations of large predators such as the wolf, mountain lion, and grizzly bear. Some animals are killed because their lifestyles cause problems for humans. The Carolina parakeet, a beautiful green, red, and yellow bird endemic to the southeastern United States, was extinct by 1920, exterminated by farmers because it ate fruit from their trees.

Ranchers and farmers poisoned and trapped prairie dogs and pocket gophers so extensively that they disappeared from most of their original range. Because numbers of prairie dogs sharply decreased, the black-footed ferret, a natural predator of these animals, became endangered. By the winter of 1985 to 1986, only 10 ferrets were known to exist. A successful captive-breeding program enabled biologists to release black-footed ferrets, beginning in 1991. Black-footed ferrets have successfully reproduced since being reintroduced, but an outbreak of disease (plague) in black-tailed prairie dogs spread to the ferrets in Wyoming, killing the colony there. Recovery efforts for these charismatic animals continue in Colorado, Montana, and New Mexico, although ferrets remain vulnerable to human development of the prairie habitat.

Unregulated hunting, or overhunting, has caused the extinction of certain species in the past but is now strictly controlled in most countries. The passenger pigeon was one of the most common birds in North America in the early 1800s, but a century of overhunting resulted in its extinction in the early 1900s.

Illegal commercial hunting, or *poaching*, endangers larger animals such as the tiger, cheetah, and snow leopard, whose beautiful furs are quite valuable. Rhinoceroses are slaughtered for their horns (used for ceremonial dagger handles in the Middle East and for purported medicinal purposes in Asian medicine). Bears are killed for their gallbladders (which Asian doctors use to treat ailments ranging from indigestion to hemorrhoids). Bush meat—meat from rare primates, elephants, anteaters, and such—is sold to urban restaurants. Although laws protect these animals, demand for their products on the black market has promoted illegal hunting, particularly in impoverished countries where a sale of contraband products can support a family for months.

Commercial harvest is the collection of live organisms from nature. Most commercially harvested organisms end up in zoos, aquaria, biomedical research labs, circuses, and pet stores. For example, several million birds are commercially harvested each year for the pet trade, but many die in transit, and many more die from improper treatment in their owners' homes. At least 40 parrot species are now threatened or endangered, in part because of commercial harvest. Although it is illegal to capture endangered animals from the wild, a thriving black market exists, mainly because collectors in the United States, Europe, and Japan pay extremely large sums for rare tropical birds (❙ Figure 56-6).

Commercial harvest also threatens plants. Many unique or rare plants have been so extensively collected from the wild that they are now classified as endangered. These include certain carnivorous plants, wildflower bulbs, cacti, and orchids. In contrast, carefully monitored and regulated commercial use of animal and plant resources creates an economic incentive to ensure that these resources do not disappear.

Review

- What are the three levels of biological diversity?
- Which organism is more likely to become extinct, an endangered species or a threatened species? Explain your answer.
- How does habitat fragmentation contribute to declining biological diversity?
- How do invasive species contribute to the biodiversity crisis?

CONSERVATION BIOLOGY

Learning Objectives

4 Define *conservation biology,* and compare in situ and ex situ conservation measures.

5 Describe the benefits and shortcomings of the U.S. Endangered Species Act and the Convention on International Trade in Endangered Species of Wild Flora and Fauna.

Conservation biology is the scientific study of how humans impact organisms and of the development of ways to protect biological diversity. Conservation biologists develop models, design experiments, and perform fieldwork to address a wide range of questions. For example, what are the processes that influence a decline in biological diversity? How do we protect and restore populations of endangered species? If we are to preserve entire ecosystems and landscapes, which ones are the most important to save?

Conservation biologists have determined that a single large area of habitat capable of supporting several populations is more effective at safeguarding an endangered species than several habitat fragments, each capable of supporting a single population. Also, a large area of habitat typically supports greater species richness than several habitat fragments.

Conservation is more successful when habitat areas for a given species are in close proximity rather than far apart. If an area of habitat is isolated from other areas, individuals may not effectively disperse from one habitat to another. Because the presence of humans adversely affects many species, habitat areas that lack roads or are inaccessible to humans are better than human-accessible areas.

According to conservation biologists, it is more effective and, ultimately, more economical to preserve intact ecosystems in which many species live than to try to preserve individual species. Conservation biologists generally consider it a higher priority to preserve areas with greater biological diversity.

Conservation biology includes two problem-solving approaches that save organisms from extinction: in situ and ex situ conservation. **In situ conservation,** which includes the establishment of parks and reserves, concentrates on preserving biological diversity in nature. A high priority of in situ conservation is identifying and protecting sites that harbor a great deal of diversity. With increasing demands on land, however, in situ conservation cannot preserve all types of biological diversity. Sometimes only ex situ conservation can save a species.

Ex situ conservation conserves individual species in human-controlled settings. Breeding captive species in zoos and storing seeds of genetically diverse plant crops are examples of ex situ conservation.

In situ conservation is the best way to preserve biological diversity

Protecting animal and plant habitats—that is, conserving and managing ecosystems as a whole—is the single best way to protect biological diversity. Many nations have set aside areas for wildlife habitats. Such natural ecosystems offer the best strategy for the long-term protection and preservation of biological diversity. Currently more than 100,000 national parks, marine sanctuaries, wildlife refuges, forests, and other areas are protected throughout the world. These encompass a total area almost as large as Canada. However, many protected areas have multiple uses that sometimes conflict with the goal of preserving species. National parks provide for recreational needs, for example, whereas national forests are used for logging, grazing, and mineral extraction. The mineral rights to many wildlife refuges are privately owned, and some wildlife refuges have had oil, gas, and other mineral development.

Protected areas are not always effective in preserving biological diversity, particularly in developing countries where biological diversity is greatest, because there is little money or expertise to manage them. Another shortcoming of the world's protected areas is that many are in lightly populated mountain areas, tundra, and the driest deserts, places that often have spectacular scenery but relatively few kinds of species. In reality, such remote areas are often designated reserves because they are unsuitable for commercial development. In contrast, ecosystems in which biological diversity is greatest often receive little attention. Protected areas are urgently needed in tropical rain forests, the tropical grasslands and savannas of Brazil and Australia, and dry forests that are widely scattered around the world. Desert organisms are underprotected in northern Africa and Argentina, and the species of many islands and temperate river basins also need protection.

These areas are part of what some biologists have identified as the world's 25 **biodiversity hotspots** (❚ Fig. 56-7). The hotspots collectively make up 1.4% of Earth's land but contain as many as 44% of all vascular plant species, 29% of the world's endemic bird species, 27% of endemic mammal species, 38% of endemic reptile species, and 53% of endemic amphibian species. Nearly 20% of the world's human population lives in the hotspots. Fifteen of the 25 hotspots are tropical, and 9 are mostly or solely islands. The 25 hotspots are somewhat arbitrary, as additional hotspots are recognized when different criteria are used.

Not all conservation biologists support focusing limited conservation resources on protecting biodiversity hotspots, because doing so ignores the many **ecosystem services** provided by entire ecosystems that are rapidly vanishing due to habitat destruction. For example, the vast boreal forests in Canada and Russia are critically important to the proper functioning of the global carbon and nitrogen cycles. Yet conservation organizations do not

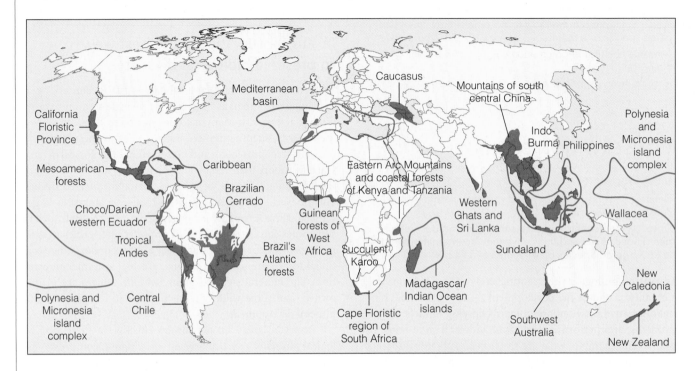

Figure 56-7   Biodiversity hotspots

Rich in endemic species, these hotspots are under pressure from the number of humans living in them. (Data from Conservation International.)

recognize these forests, which are being logged at unprecedented rates, as hotspots. Landscape ecology takes a larger view of ecosystem conservation.

Landscape ecology considers ecosystem types on a regional scale

The subdiscipline of ecology that studies the connections in a heterogeneous **landscape** consisting of multiple interacting ecosystems is known as **landscape ecology**. Increasingly, biologists are focusing efforts on preserving biodiversity in landscapes. But what is the minimum ecosystem/landscape size needed to preserve species numbers and distributions?

One long-term study addressing this question in tropical rain forests is the Minimum Critical Size of Ecosystems Project, which is studying a series of Amazonian rainforest fragments varying in size from 1 to 100 hectares (▌Fig. 56-8). Preliminary data from this study indicate that smaller forest fragments do not maintain their ecological integrity. For example, large trees near forest edges often die or are damaged from exposure to wind, desiccation (from lateral exposure of the forest fragment to sunlight), and invasion by parasitic woody vines. In addition, biologists have documented that various species adapted to forest interiors

Figure 56-8   Minimum Critical Size of Ecosystems Project

Shown are 1-hectare and 10-hectare plots of a long-term (more than two decades) study under way in Brazil on the effects of fragmentation on Amazonian rain forest. Plots with an area of 100 hectares are also under study, along with identically sized sections of intact forest, which are controls.

do not prosper in the smaller fragments and eventually die out, whereas species adapted to forest edges invade the smaller fragments and thrive.

To remedy widespread habitat fragmentation and its associated loss of biodiversity, conservation biologists have proposed linking isolated fragments with **habitat corridors,** strips of habitat connecting isolated habitat patches. Habitat corridors allow wildlife to move about so they can feed, mate, and recolonize habitats after local extinctions take place. Habitat corridors range from a local scale (such as an overpass that allows wildlife to cross a road safely) to a landscape scale that connects separate reserves (▌Fig. 56-9). The minimum corridor width for landscape-scale corridors varies depending on the size of home ranges for various species. For example, to link two reserves and protect wolves in Minnesota, the corridor must have a minimum width of 12 km (7.2 mi), whereas the minimum corridor width for bobcats in South Carolina must be 2.5 km (1.5 mi).

Restoring damaged or destroyed habitats is the goal of restoration ecology

Although preserving habitats is an important part of conservation biology, the realities of our world, including the fact that the land-hungry human population continues to increase, dictate a variety of other conservation measures. Sometimes scientists reclaim disturbed lands and convert them into areas with high biological diversity. **Restoration ecology,** in which the principles of ecology are used to return a degraded environment to one that is more functional and sustainable, is an important part of in situ conservation.

Since 1934, the University of Wisconsin–Madison Arboretum has carried out one of the most famous examples of ecological restoration (▌Fig. 56-10). Several distinct natural communities have been carefully developed on damaged agricultural land. These communities include a tallgrass prairie, a xeric (dry) prairie, and several types of pine and maple forests native to Wisconsin.

Restoration of disturbed lands not only creates biological habitats but also has additional benefits, such as the regeneration of soil that agriculture or mining damaged. The disadvantages of restoration ecology include the time and expense required to restore an area. Nonetheless, restoration ecology is an important aspect of conservation biology, as it is thought that restoration will reduce extinctions.

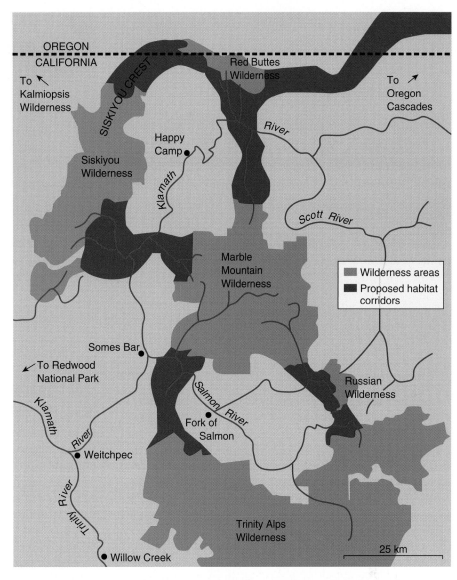

Figure 56-9 Proposed habitat corridors

Five wilderness areas are already established in northern California and southern Oregon. These areas would more effectively preserve biodiversity if they were joined by habitat corridors to form a single landscape-scale conservation unit. To date, the U.S. Forest Service has not adopted this habitat corridor proposal. (*Arrows show the direction of other nearby wilderness areas.*) (From F. Pace, "The Klamath Corridors: Preserving Biodiversity in the Klamath National Forest," in *Landscape Linkages and Biodiversity,* edited by W. E. Hudson, pp. 105–116, Island Press, Washington, D.C., 1991. Reprinted by permission of Island Press.)

Ex situ conservation attempts to save species on the brink of extinction

Zoos, aquaria, and botanical gardens often attempt to save certain endangered species from extinction. Eggs are collected from nature, or the remaining few animals are captured and bred in zoos and other research facilities.

Special techniques, such as artificial insemination and host mothering, increase the number of offspring. In **artificial insemination,** sperm is collected from a suitable male of a rare species and used to impregnate a female, perhaps located in another zoo in a different city or even in another country. In **host**

(a) The restoration of the prairie by the University of Wisconsin–Madison Arboretum was at an early stage in 1935. The men are digging holes to plant prairie grass sod.

(b) The prairie as it looks today. This picture was taken at about the same location as the 1935 photograph.

Figure 56-10 Restoring damaged lands

mothering, a female of a rare species is treated with fertility drugs, which cause her to produce multiple eggs. Some of these eggs are collected, fertilized with sperm, and surgically implanted into females of a related but less rare species, which later give birth to offspring of the rare species (see *Focus On: Novel Origins,* in Chapter 49). Plans are under way to clone endangered species, such as the giant panda, that do not reproduce well in captivity. Another technique involves hormone patches, which are being developed to stimulate reproduction in endangered birds; the patch is attached under the female bird's wing.

A few spectacular successes have occurred in captive-breeding programs, in which large enough numbers of a species have been produced to re-establish small populations in the wild. Conservation efforts, for example, let the bald eagle make a remarkable comeback in the lower 48 states, from 417 nesting pairs in 1963 to more than 7000 pairs in 2006. In 1994, the bald eagle was removed from the endangered list and transferred to the less critical threatened list, and in 2006, the U.S. Fish and Wildlife Service (FWS) removed it from the threatened list.

Attempting to save a species on the brink of extinction is usually extremely expensive; therefore, only a small proportion of endangered species can be saved. Moreover, zoos, aquaria, and botanical gardens do not have the space to try to save all endangered species. This means that conservation biologists must set priorities on which species to attempt to save. Zoos have traditionally focused on large, charismatic animals, such as pandas, bald eagles, and whooping cranes, because the public is more interested in them. Such conservation efforts ignore millions of less glamorous but ecologically important species. However, zoos, aquaria, and botanical gardens serve a useful purpose because they educate the public about the value of biodiversity. Clearly, controlling unchecked human development so that species do not become endangered in the first place is a more effective way to protect and maintain natural habitat.

The Endangered Species Act provides some legal protection for species and habitats

In 1973, the *Endangered Species Act (ESA)* was passed in the United States, authorizing the FWS to protect endangered and threatened species in the United States and abroad. Many other countries now have similar legislation. The FWS conducts a detailed study of a species to determine if it should be listed as endangered or threatened. Since the passage of the ESA, more than 1260 U.S. species have been listed as endangered or threatened (∎ Table 56-2). The ESA provides legal protection to listed species so that their danger of extinction is reduced. For example, the act makes it illegal to sell or buy any product made from an endangered or threatened species.

The ESA requires officials of the FWS to select critical habitats and design a recovery plan for each species listed. The recovery plan includes an estimate of the current population size, an analysis of what factors contributed to its endangerment, and a list of activities that may help the species recover.

The Endangered Species Act is considered one of the strongest pieces of environmental legislation in the United States, in part because species are designated as endangered or threatened entirely on biological grounds. Currently, economic considerations cannot influence the designation of endangered or threatened species. Biologists generally agree that as a result of passage of the ESA in 1973, fewer species became extinct than would have had the law never been passed.

The ESA is also one of the most controversial pieces of environmental legislation. For example, the ESA does not provide compensation for private property owners who suffer financial losses because they cannot develop their land if a threatened or endangered species lives there. The ESA has also interfered with some federally funded development projects.

The ESA was scheduled for congressional reauthorization in 1992 but has been entangled since then in disagreements between

TABLE 56-2

Organisms Listed as Endangered or Threatened in the United States, 2005

Type of Organism	Number of Endangered Species	Number of Threatened Species
Mammals	69	9
Birds	77	13
Reptiles	14	22
Amphibians	11	10
Fishes	71	43
Snails	21	11
Clams	62	8
Crustaceans	18	3
Insects	35	9
Spiders	12	0
Flowering plants	571	144
Conifers and cycads	2	1
Ferns and other plants	24	2

Source: U.S. Fish and Wildlife Service.

conservation advocates and those who support private property rights. Conservation advocates think the ESA does not do enough to save endangered species, whereas those who own land on which rare species live think the law goes too far and infringes on property rights. Some critics—notably business interests and private property owners—view the ESA as an impediment to economic progress. Those who defend the ESA point out that of 34,000 past cases of endangered species versus development, only 21 cases were not resolved by some sort of compromise. When the black-footed ferret was reintroduced on the Wyoming prairie, for example, it was classified as an "experimental, nonessential species" so that its reintroduction would not block ranching and mining in the area. Thus, the ferret release program obtained the support of local landowners, support that was deemed crucial to ferret survival in nature.

This type of compromise is essential to the success of saving endangered species because, according to the U.S. General Accounting Office, more than 90% of endangered species live on at least some privately owned lands. Critics of the ESA think the law should be changed so that private landowners are given economic incentives to protect endangered species living on their lands. For example, tax cuts for property owners who are good land stewards could make the presence of endangered species on their properties an asset instead of a liability.

Defenders of the ESA agree that the law is not perfect. About 10 U.S. species have recovered enough to be delisted—that is, no longer classified as endangered or threatened. However, the FWS says that hundreds of listed species are stable or improving; they expect as many as several dozen additional species to be delisted over the next few decades.

Another problem with the ESA is that it currently takes, on average, about 17 years for a candidate species with declining numbers to be listed as threatened or endangered. This backlog is due to both the high cost of evaluating each species and the limited funds allocated to the FWS for these studies. Meanwhile, 27 species on the ESA's "candidate list" have become extinct since 1973, without ever being classified as threatened or endangered so that they could have some measure of protection.

The ESA is geared more to saving a few popular or unique endangered species than to saving the much larger number of less glamorous species that perform valuable ecosystem services. About one third of the annual funding for the ESA is used to help just 10 species. Yet it is the less glamorous organisms, such as plants, fungi, bacteria, and insects, that play central roles in ecosystems and contribute most to their functioning. Bacteria and fungi, for example, provide the critically important ecosystem service of breaking down dead organic materials into simple substances (CO_2, water, and minerals) that are subsequently recycled to plants and other autotrophs.

Conservationist biologists would like to see the ESA strengthened in such a way as to manage whole ecosystems and maintain complete biological diversity rather than attempt to save endangered species as isolated entities. This approach offers collective protection to many declining species rather than to single species.

International agreements provide some protection of species and habitats

At the international level, 169 countries participate in the *Convention on International Trade in Endangered Species of Wild Flora and Fauna (CITES)*, which went into effect in 1975. Originally drawn up to protect endangered animals and plants considered valuable in the highly lucrative international wildlife trade, CITES bans hunting, capturing, and selling of endangered or threatened species and regulates trade of organisms listed as potentially threatened. Unfortunately, enforcement of this treaty varies from country to country, and even where enforcement exists, the penalties are not very severe. As a result, illegal trade in rare, commercially valuable species continues.

The goals of CITES often stir up controversy over such issues as who actually owns the world's wildlife and whether global conservation concerns take precedence over competing local interests. These conflicts often highlight socioeconomic differences between wealthy consumers of CITES products and poor people who trade the endangered organisms.

The case of the African elephant illustrates these controversies. Listed as an endangered species since 1989 to halt the slaughter of elephants for the ivory trade, the species has recovered in southern Africa (Namibia, Botswana, and Zimbabwe). When elephant populations grow too large for their habitat, they root out and knock over so many smaller trees that the forest habitat can support fewer other species. The African people living near the elephants want to cull the herd periodically to sell elephant meat, hides, and ivory for profit. In the late 1990s and early 2000s, CITES transferred elephant populations in Namibia, Botswana, and Zimbabwe to a less restrictive listing to allow a one-time trade of legally obtained stockpiled ivory (from animals that died of natural causes). The money earned from the sale of the ivory is funding conservation programs and community development projects for people living near the elephants.

Another international treaty, the *Convention on Biological Diversity*, requires that each signatory nation inventory its own biodiversity and develop a **national conservation strategy,** a detailed plan for managing and preserving the biological diversity of that specific country. Currently, 188 nations participate in the Convention on Biological Diversity.

Review

- What is conservation biology?
- Are landscape biology and restoration ecology examples of in situ or ex situ conservation?
- Which type of conservation measure, in situ or ex situ, helps the greatest number of species? Why?
- Why is the U.S. Endangered Species Act so controversial?

DEFORESTATION

Learning Objectives

6   Discuss the ecosystem services of forests, and describe the consequences of deforestation.

7   State at least three reasons why forests (tropical rain forests and boreal forests) are disappearing today.

The most serious problem facing the world's forests is **deforestation,** which is the temporary or permanent clearance of large expanses of forest for agriculture or other uses (▌ Fig. 56-11). According to the UN Food and Agriculture Organization (FAO), which provides world statistics on forest cover, forests are currently shrinking about 9 million hectares (22.2 million acres) each year.

When forests are destroyed, they no longer make valuable contributions to the environment or to the people who depend on them. Deforestation increases soil erosion and thus decreases soil fertility. Soil erosion causes increased sedimentation of waterways, which harms downstream aquatic ecosystems by reducing light penetration, covering aquatic organisms, and filling in waterways. Uncontrolled soil erosion, particularly on steep deforested slopes, causes mudflows that endanger human lives and property and reduces production of hydroelectric power as silt builds up behind dams. In drier areas, deforestation can lead to the formation of deserts.

Deforestation contributes to the loss of biological diversity. Many species have limited ranges within a forest, particularly in the tropics, so these species are especially vulnerable to habitat destruction or modification. Migratory species, such as birds and butterflies, also suffer from tropical deforestation.

By trapping and absorbing precipitation, forests on hillsides and mountains help protect nearby lowlands from floods. When a forest is cut down, the watershed cannot absorb and hold water as well, and the total amount of surface runoff flowing into rivers and streams increases. This not only causes soil erosion but puts lowland areas at extreme risk of flooding.

Deforestation may affect regional and global climate changes. Transpiring trees release substantial amounts of moisture into the air. This moisture falls back to the surface in the hydrologic cycle. When a large forest is removed, rainfall may decline, and

Figure 56-11 *Animated* Deforestation

Aerial view of clear-cut areas in the Gifford Pinchot National Forest in southwestern Washington State. The lines are roads built at taxpayer expense to haul away logs.

droughts may become common in that region. Studies suggest that the local climate has become drier in parts of Brazil where tracts of the rain forest have been burned. Temperatures may also rise slightly in a deforested area, because there is less evaporative cooling from the trees.

Deforestation may increase global temperature by releasing carbon stored in the trees into the atmosphere as carbon dioxide, which enables the air to retain heat. The carbon in forests is released immediately if the trees are burned or more slowly when unburned parts decay. If trees are harvested and logs are removed, roughly one half of the forest carbon remains as dead materials (branches, twigs, roots, and leaves) that decompose, releasing carbon dioxide. When an old-growth forest is harvested, it may take 200 years for the replacement forest to accumulate the amount of carbon that was stored in the original forest.

Why are tropical rain forests disappearing?

Most of the remaining undisturbed tropical rain forests, in the Amazon and Congo River basins of South America and Africa, are being cleared and burned at a rate unprecedented in human history. Tropical rain forests are also being destroyed at an extremely rapid rate in southern Asia, Indonesia, Central America, and the Philippines.

Several studies show a strong statistical correlation between population growth and deforestation. More people need more food, so they clear forests for agricultural expansion. However, tropical deforestation cannot be attributed simply to population pressures. The main causes of deforestation vary from place to place, and a variety of economic, social, and governmental factors interact to cause deforestation. Government policies sometimes provide incentives that favor the removal of forests. For example, in the late 1950s the Brazilian government constructed the Belem–Brasilia Highway, which cut through the Amazon Basin and opened the Amazonian frontier for settlement. Sometimes economic conditions encourage deforestation. The farmer who converts more forest to pasture can maintain a larger herd of cattle, which is a good hedge against inflation.

If we keep in mind that tropical deforestation is a complex problem, three agents are probably the most immediate causes of deforestation in tropical rain forests: subsistence agriculture, commercial logging, and cattle ranching. Other reasons for the destruction of tropical forests include the development of hydroelectric power, which inundates large areas of forest; mining, particularly when ore smelters burn charcoal produced from rainforest trees; and plantation-style agriculture of crops such as citrus fruits and bananas.

Subsistence agriculture, in which a family produces enough food to feed itself, accounts for perhaps 60% of tropical deforestation. In many developing countries where tropical rain forests are located, the majority of people do not own the land on which they live and work. Most subsistence farmers have no place to go except into the forest, which they clear to grow food. Land reform in Brazil, Madagascar, Mexico, the Philippines, Thailand, and many other countries would make the land owned by a few available to everyone, thereby easing the pressure of subsistence farmers on tropical forests. This scenario is unlikely, because wealthy landowners have more economic and political clout than impoverished peasants.

Subsistence farmers often follow loggers' access roads until they find a suitable spot. They first cut down the trees and allow them to dry; then they burn the area and plant crops immediately after burning (❚ Fig. 56-12). This is known as **slash-and-burn agriculture.** Yields from the first crop are often quite high, because the nutrients that were in the trees are now available in the soil. However, soil productivity declines rapidly, and subsequent crops are poor. In a few years the farmer must move to a new part of the forest and repeat the process. Cattle ranchers often claim the abandoned land for grazing, because land that is not fertile enough to support crops can still support livestock.

Slash-and-burn agriculture done on a small scale, with periods of 20 to 100 years between cycles, is sustainable. The forest regrows rapidly after a few years of farming. But when millions of people try to obtain a living in this way, the land is not allowed to lie uncultivated long enough to recover. Globally, at least 200 million subsistence farmers obtain a living from slash-and-burn agriculture, and the number is growing. To compound the problem, there is only half as much forest available today as there was 50 years ago.

About 20% of tropical deforestation is the result of commercial logging, and vast tracts of tropical rain forests, particularly in Southeast Asia, are harvested for export abroad. Most tropical countries allow commercial logging to proceed much faster than is sustainable, because it supplies them with much-needed revenues. In the final analysis, uncontrolled tropical deforestation does not contribute to economic development; rather, it reduces or destroys the value of an important natural resource.

Approximately 12% of tropical rainforest destruction is carried out to provide open rangeland for cattle. Cattle ranching is particularly important in Central America. Much of the beef raised on these ranches, which foreign companies often own, is exported to restaurant chains in North America and Europe. After the forests are cleared, cattle graze on the land for as long as

Figure 56-12 Clearing the forest
Tropical rain forest in Brazil is burned to provide agricultural land. This type of cultivation is known as slash-and burn agriculture.

20 years, after which time the soil fertility is depleted. When this occurs, shrubby plants, or *scrub savanna*, take over the range.

Why are boreal forests disappearing?

Tropical rain forests are not the only forests at risk from deforestation. Extensive logging of certain boreal forests began in the late 1980s and continues to the present. Coniferous evergreen trees such as spruce, fir, cedar, and hemlock dominate these northern forests of Alaska, Canada, Scandinavia, and northern Russia.

Boreal forests, harvested primarily by clear-cut logging, are currently the primary source of the world's industrial wood and wood fiber. The annual loss of boreal forests is estimated to encompass an area twice as large as the Amazonian rain forests of Brazil. About 1 million hectares (2.5 million acres) of Canadian forests are logged annually, and most of Canada's forests are under logging tenures. (*Tenures* are agreements between provinces and companies that give companies the right to cut timber.) Canada is the world's biggest timber exporter, and most of its forest products are exported to the United States. On the basis of current harvest quotas, logging in Canada is unsustainable. Extensive tracts of boreal forests in Russia are also harvested, although exact estimates are unavailable. Alaska's boreal forests are also at risk because the U.S. government may increase logging on public lands there in the future.

Review

❚ What are three ecosystem services that forests provide?

❚ What are two reasons for deforestation in tropical rain forests? What is the main reason for deforestation of boreal forests?

GLOBAL WARMING

Learning Objectives

8 Name at least three greenhouse gases, and explain how greenhouse gases contribute to global warming.

9 Describe how global warming may affect sea level, precipitation patterns, organisms (including humans), and food production.

Earth's average temperature is based on daily measurements from several thousand land-based meteorological stations around the world, as well as data from weather balloons, orbiting satellites, transoceanic ships, and hundreds of sea-surface buoys with temperature sensors. Earth's average surface temperature increased 0.6°C (1.1°F) during the 20th century, and the 1990s was the warmest decade of the century (❚ Fig. 56-13). The early 2000s continued the warming trend. Since the mid-19th century, the warmest year on record was 2005, followed by 1998 and 2002.

Scientists around the world have studied **global warming** for the past 50 years. As the evidence has accumulated, those most qualified to address the issue have reached a strong consensus that the 21st century will experience significant climate change and that human activities will be responsible for much of this change.

In response to this consensus, governments around the world organized the UN Intergovernmental Panel on Climate Change (IPCC). With input from hundreds of climate experts, the IPCC provides the most definitive scientific statement about global warming. The 2001 IPCC Third Assessment Report concluded that human-produced air pollutants continue to change the atmosphere, causing most of the warming observed in the past 50

years. Scientists can identify the human influence on climate change despite questions about *how much* of the recent warming stems from natural variations. The IPCC report projects up to a 5.8°C (10.4°F) increase in global temperature by the year 2100, although the warming will probably not occur uniformly from region to region. Thus, Earth may become warmer during the 21st century than it has been for thousands of years, based on paleoclimate data.

Almost all climate experts agree with the IPCC's assessment that the warming trend has already begun and will continue throughout the 21st century. However, scientists are uncertain over how rapidly the warming will proceed, how severe it will be, and where it will be most pronounced. (Recall that uncertainty and debate are part of the scientific process and that scientists can never claim to know a "final answer.") As a result of these uncertainties, many people, including policymakers, are confused about what we should do. Yet the stakes are quite high because human-induced global warming has the potential to disrupt Earth's climate for a very long time.

Greenhouse gases cause global warming

Carbon dioxide (CO_2) and certain other trace gases, including methane (CH_4), surface ozone (O_3),[1] nitrous oxide (N_2O), and chlorofluorocarbons (CFCs), are accumulating in the atmosphere as a result of human activities (❚ Table 56-3). The concentration of atmospheric CO_2 has increased from about 280 parts per million (ppm) approximately 250 years ago (before the Industrial Revolution began) to 377 ppm in 2004 (❚ Fig. 56-14). Burning carbon-containing fossil fuels—coal, oil, and natural gas—accounts for

TABLE 56-3

Changes in Selected Atmospheric Greenhouse Gases, Preindustrial to Present

Gas	Estimated Pre–1750 Concentration	Present Concentration
Carbon dioxide	280 ppm*	377 ppm[§]
Methane	730 ppb[†]	1847 ppb
Nitrous oxide	270 ppb	319 ppb
Tropospheric ozone	25 ppb	34 ppb
CFC-12	0 ppt[‡]	545 ppt
CFC-11	0 ppt	253 ppt

Source: Carbon Dioxide Information Analysis Center, Environmental Sciences Division, Oak Ridge National Laboratory.

*ppm = parts per million.

[†]ppb = parts per billion.

[‡]ppt = parts per trillion.

[§]Derived from in situ sampling at Mauna Loa, Hawaii. All other data from Mace Head, Ireland, monitoring site.

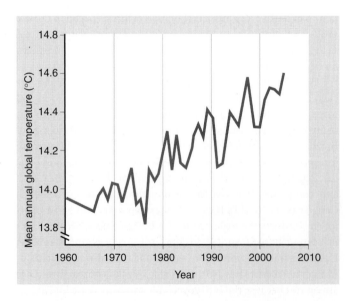

❚ **Figure 56-13** Mean annual global temperature, 1960 to 2005

Data are presented as surface temperatures (°C) for 1960, 1965, and every year thereafter. The measurements, which naturally fluctuate, clearly show the warming trend of the last several decades. (Surface Air Temperature Analysis, Goddard Institute for Space Studies, NASA.)

---

[1]Surface ozone (more precisely, ozone in the troposphere) is a greenhouse gas as well as a component of photochemical smog. Ozone in the upper atmosphere, the stratosphere, provides an important planetary service that is discussed later in this chapter.

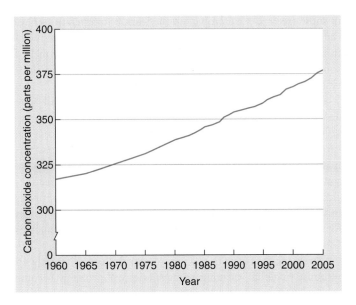

Figure 56-14 *Animated* Carbon dioxide in the atmosphere, 1960 to 2005

Note the steady increase in the concentration of atmospheric carbon dioxide. Measurements are taken at the Mauna Loa Observatory, Hawaii, far from urban areas where factories, power plants, and motor vehicles emit carbon dioxide. (Scripps Institution of Oceanography, University of California, La Jolla, California.)

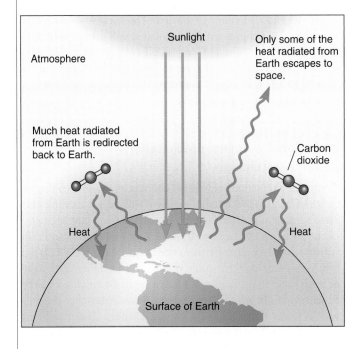

Figure 56-15 *Animated* Enhanced greenhouse effect

Visible light passes through the atmosphere to Earth's surface. However, carbon dioxide and other greenhouse gases direct the longer wavelength heat (infrared radiation), which would normally be radiated to space, back to the surface. Therefore, the buildup of greenhouse gases in the atmosphere results in global warming.

about three fourths of human-made carbon dioxide emissions to the atmosphere. Land conversion, such as when forests are logged or burned, also releases carbon dioxide. (Burning releases carbon dioxide into the atmosphere; also, because trees normally remove carbon dioxide from the atmosphere during photosynthesis, tree removal prevents this process.) The levels of the other trace gases associated with global warming are also rising.

Global warming occurs because these gases absorb infrared radiation, that is, heat, in the atmosphere. This absorption slows the natural heat flow into space, warming the lower atmosphere. Some of the heat from the lower atmosphere is transferred to the ocean and raises its temperature as well. This retention of heat in the atmosphere is a natural phenomenon that has made Earth habitable for its millions of species. However, as human activities increase the atmospheric concentration of these gases, the atmosphere and ocean continue to warm, and the overall global temperature rises.

Because carbon dioxide and other gases trap the sun's radiation somewhat like glass does in a greenhouse, the natural trapping of heat in the atmosphere is called the **greenhouse effect,** and the gases that absorb infrared radiation are known as **greenhouse gases.** The additional warming produced when increased levels of gases absorb additional infrared radiation is called the **enhanced greenhouse effect** (❚ Fig. 56-15).

Although current rates of fossil fuel combustion and deforestation are high, causing the carbon dioxide level in the atmosphere to increase markedly, scientists think the warming trend is slower than the increasing level of carbon dioxide might indicate. The reason is that water requires more heat to raise its temperature than gases in the atmosphere do (recall the high specific heat

of water discussed in Chapter 2). As a result, the ocean takes longer to warm than the atmosphere. Most climate scientists think that warming will be more pronounced in the second half of the 21st century than in the first half.

What are the probable effects of global warming?

We now consider some of the probable effects of global warming, including changes in sea level; changes in precipitation patterns; effects on organisms, including humans; and effects on agriculture. These changes will persist for centuries because many greenhouse gases remain in the atmosphere for hundreds of years. Furthermore, even after greenhouse gas concentrations have stabilized, scientists think that Earth's mean surface temperature will continue to rise, because the ocean adjusts to climate change on a delayed time scale.

With global warming, the global average sea level is rising

As Earth's overall temperature increases by just a few degrees, there could be a major thawing of glaciers and the polar ice caps. In addition to sea-level rise caused by the retreat of glaciers and thawing of polar ice, the sea level will probably rise because of thermal expansion of the warming ocean. Water, like other substances, expands as it warms.

The IPCC estimates that sea level will rise an additional 0.5 m (20 in) by 2100. Such an increase will flood low-lying coastal areas, such as parts of southern Louisiana and South Florida. Coastal areas that are not inundated will more likely suffer erosion and other damage from more frequent and more intense weather events such as hurricanes. Countries particularly at risk include Bangladesh, Egypt, Vietnam, Mozambique, and many island nations such as the Maldives.

With global warming, precipitation patterns will change

Computer simulations of weather changes as global warming occurs indicate that precipitation patterns will change, causing some areas such as midlatitude continental interiors to have more frequent droughts. At the same time, heavier snowstorms and rainstorms may cause more frequent flooding in other areas. Changes in precipitation patterns could affect the availability and quality of fresh water in many places. Arid or semiarid areas, such as the Sahel region just south of the Sahara Desert, may have the most serious water shortages as the climate changes. Closer to home, water experts predict water shortages in the American West, because warmer winter temperatures will cause more precipitation to fall as rain rather than snow; melting snow currently provides 70% of stream flows in the West during summer months.

The frequency and intensity of storms over warm surface waters may also increase. Scientists at the National Oceanic and Atmospheric Administration developed a computer model that examines how global warming may affect hurricanes. When the model was run with a sea-surface temperature 2.2°C warmer than today, more intense hurricanes resulted. (The question of whether hurricanes will occur *more frequently* in a warmer climate remains uncertain.) Changes in storm frequency and intensity are expected because as the atmosphere warms, more water evaporates, which in turn releases more energy into the atmosphere (recall the discussion of water's heat of vaporization in Chapter 2). This energy generates more powerful storms.

With global warming, the ranges of organisms are changing

Dozens of studies report on the effects of global warming on organisms. For example, researchers determined that populations of zooplankton in the California Current have declined 80% since 1951, apparently because the current has warmed slightly. (The California Current flows from Oregon southward along the California coast.) The decline in zooplankton has affected the entire ecosystem's food web, and populations of seabirds and plankton-eating fishes have also declined.

Warmer temperatures in Antarctica—during the past 50 years the average annual temperature on the Antarctic Peninsula has increased 2.6°C (5°F)—have contributed to reproductive failure in Adélie penguins. The birds normally lay their eggs

in snow-free rocky outcrops, but the warmer temperatures have caused more snowfall (recall that warmer air holds more moisture), which melts when the birds incubate the eggs. The melted snow forms cold pools of slush that kill the developing chick embryos.

Biologists generally agree that global warming will have an especially severe impact on plants, which cannot migrate as quickly as animals when environmental conditions change. (The speed of seed dispersal has definite limitations.) During past climate warmings, such as during the glacial retreat that took place some 12,000 years ago, the upper limit of dispersal for tree species was probably 200 km (124 mi) per century. If Earth warms as much during the 21st century as projections indicate, the ranges for some temperate tree species may shift northward as much as 480 km (300 mi) (❚ Fig. 56-16).

Each species reacts to changes in temperature differently. In response to global warming, some species will probably become extinct, particularly those with narrow temperature requirements, those confined to small reserves or parks, and those living in fragile ecosystems. Other species may survive in greatly reduced numbers and ranges. Ecosystems considered most vulnerable to species loss in the short term are polar seas, coral reefs and atolls, prairie wetlands, coastal wetlands, tundra, taiga, tropical forests, and mountains, particularly alpine tundra.

In response to global warming, some species may disperse into new environments or adapt to the changing conditions in their present habitats. Global warming may not affect certain species, whereas other species may emerge as winners, with greatly expanded numbers and ranges. Those considered most likely to prosper include weeds, pests, and disease-carrying organisms, all of which are generalists that are already common in many different environments.

Global warming will have a more pronounced effect on human health in developing countries

Data linking climate warming and human health problems are accumulating. Since 1950, the United States has experienced an increased frequency of extreme heat-stress events, which are extremely hot, humid days during summer months. Medical records show that heat-related deaths among elderly and other vulnerable people increase during these events.

Climate warming may also affect human health indirectly. Mosquitoes and other disease carriers could expand their range into the newly warm areas and spread malaria, dengue fever, yellow fever, Rift Valley fever, and viral encephalitis. As many as 50 million to 80 million additional cases of malaria could occur annually in tropical, subtropical, and temperate areas. According to the World Health Organization, during 1998, the second-warmest year on record, the incidence of malaria, Rift Valley fever, and cholera surged in developing countries.

Highly developed countries are less vulnerable to such disease outbreaks because of better housing (which keeps mosquitoes outside), medical care, pest control, and public health measures such as water treatment plants. Texas reported a few cases of dengue fever in the late 1990s, for example, whereas nearby Mexico had thousands of cases during that period.

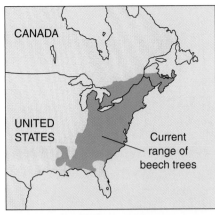

(a) Present range of American beech trees.

(b) One projected range of beech trees after global warming occurs.

Figure 56-16 Climate change and beech trees in North America

(Adapted from M. B. Davis and C. Zabinski, "Changes in Geographical Range Resulting from Greenhouse Warming: Effects on Biodiversity in Forests," in *Global Warming and Biological Diversity*, edited by R. L. Peters and T. E. Lovejoy, Yale University Press, New Haven, Connecticut, 1992.)

Global warming may increase problems for agriculture

Several studies show that the rising sea level will inundate river deltas, which are some of the world's best agricultural lands. The Nile River (Egypt), Mississippi River (United States), and Yangtze River (China) are examples of river deltas that have been studied. Certain agricultural pests and disease-causing organisms will probably proliferate. As mentioned earlier, global warming may also increase the frequency and duration of droughts and, in some areas, crop-damaging floods.

On a regional scale, current global-warming models forecast that agricultural productivity will increase in some areas and decline in others. Models suggest that Canada and Russia will increase their agricultural productivity in a warmer climate, whereas tropical and subtropical regions, where many of the world's poorest people live, will decline in agricultural productivity. Central America and Southeast Asia may experience some of the greatest declines in agricultural productivity.

Review

- What is the enhanced greenhouse effect?
- How do greenhouse gases cause the enhanced greenhouse effect?
- What are some of the significant problems that global warming may cause during the 21st century?

DECLINING STRATOSPHERIC OZONE

Learning Objectives

10 Distinguish between surface ozone and stratospheric ozone.

11 Cite the causes and potential effects of ozone destruction in the stratosphere.

Ozone (O_3) is a form of oxygen that is a human-made pollutant in the lower atmosphere but a naturally produced, essential part of the stratosphere (see Fig. 21-6). The **stratosphere,** which encircles the planet some 10 to 45 km (6 to 28 mi) above the surface, contains a layer of ozone that shields the surface from much of the ultraviolet radiation from the sun (❚ Fig. 56-17). If ozone disappeared from the stratosphere, Earth would become unlivable for most forms of life. Ozone in the lower atmosphere is converted back to oxygen in a few days and so does not replenish the ozone depleted in the stratosphere.

A slight thinning in the ozone layer over Antarctica forms naturally for a few months each year. In 1985, however, scientists observed a greater thinning than usual. This increased thinning, which begins each September, is commonly referred to as the "ozone hole" (❚ Fig. 56-18). There, ozone levels decrease as much as 67% each year. During the 1990s, the ozone-thinned area continued to grow, and by 2000 it had reached the record size of 28.3 million km² (11.3 million mi²), larger than the North American continent.

In addition, worldwide levels of stratospheric ozone have been falling for several decades. According to the National Center for Atmospheric Research, since the 1970s ozone levels over Europe and North America have dropped almost 10%.

Certain chemicals destroy stratospheric ozone

Both chlorine- and bromine-containing substances catalyze ozone destruction. The primary chemicals responsible for ozone loss in the stratosphere are a group of chlorine compounds called *chlorofluorocarbons (CFCs).* Chlorofluorocarbons have been used as propellants in aerosol cans, coolants in air conditioners and refrigerators, foam-blowing agents for insulation and packaging, and solvents and cleaners for the electronics industry. Additional compounds that also attack ozone include halons (found in many fire extinguishers), methyl bromide (a pesticide), methyl chloroform (an industrial solvent), and carbon tetrachloride (used in many industrial processes, including the manufacture of pesticides and dyes).

After release into the lower troposphere, CFCs and similar compounds slowly drift up to the stratosphere, where ultraviolet radiation breaks them down, releasing chlorine. Similarly, the breakdown of halons and methyl bromide releases bromine. The thinning in the ozone layer over Antarctica occurs annually between September and November (spring in the Southern Hemisphere). At this time, two important conditions occur: Sunlight returns to the polar region, and the *circumpolar vortex,* a mass of cold air that circulates around the southern polar region and isolates it from the warmer air in the rest of the planet, is well developed.

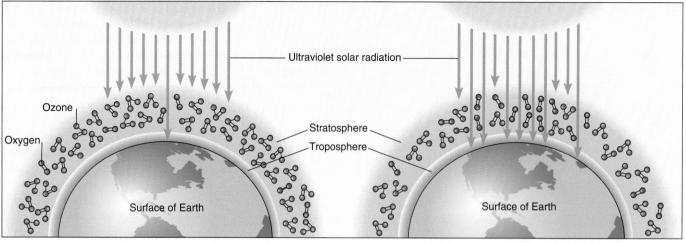

(a) Stratospheric ozone absorbs 99% of incoming ultraviolet radiation, effectively shielding Earth's surface.

(b) When stratospheric ozone is reduced, more high-energy ultraviolet radiation penetrates the atmosphere to the surface, where it harms organisms.

Figure 56-17 *Animated* Ultraviolet radiation and the ozone layer

The cold air causes polar stratospheric clouds to form; these clouds contain ice crystals to which chlorine and bromine adhere, making them available to destroy ozone. The sunlight promotes the chemical reaction in which chlorine or bromine breaks ozone molecules apart, converting them into oxygen molecules. The chemical reaction in which ozone is destroyed does not alter the chlorine or bromine, and thus a single chlorine or bromine atom breaks down many thousands of ozone molecules. The chlorine and bromine remain in the stratosphere for many years. When the circumpolar vortex breaks up, the ozone-depleted air spreads northward, diluting ozone levels in the stratosphere over South America, New Zealand, and Australia.

Ozone depletion harms organisms

With depletion of the ozone layer, more ultraviolet radiation reaches the surface. Excessive exposure to UV radiation is linked to human health problems, such as cataracts, skin cancer, and a weakened immune system. The lens of the eye contains transparent proteins that are replaced at a very slow rate. Exposure to excessive UV radiation damages these proteins, and over time, the damage accumulates; a cataract forms as the lens becomes cloudy. Cataracts can be surgically treated, but millions of people in developing nations cannot afford the operation so remain partially or totally blind.

Scientists are concerned that increased levels of UV radiation may disrupt ecosystems. For example, the productivity of Antarctic phytoplankton, the microscopic drifting algae that are the base of the Antarctic food web, has declined from increased exposure to UVB. (UVB, one of three types of UV radiation, is particularly damaging to DNA; see the section on vertebrate skin in Chapter 39.) Research shows that surface UV radiation inhibits photosynthesis in these phytoplankton. Biologists have also documented direct damage to natural populations of Antarctic fish. Increased DNA mutations in icefish eggs and larvae (young fish) were matched to increased levels of UV radiation; research-

ers are currently studying if these mutations lessen the animals' ability to survive.

There is also concern that high levels of UV radiation may damage crops and forests, but the effects of UVB radiation on plants are very complex and have not been adequately studied. Plants interact with many other species in both natural ecosystems and agricultural ecosystems, and the effects of UV radiation on each of these organisms, in turn, affect plants indirectly. For example, exposure to higher levels of UVB radiation may increase wheat yields by inhibiting fungi that cause disease in wheat. In contrast, exposure to higher levels of UV radiation may decrease cucumber yields because it increases the incidence of disease.

International cooperation is helping repair the ozone layer

In 1987, representatives from many countries met in Montreal to sign the **Montreal Protocol,** an agreement that originally stipulated a 50% reduction of CFC production by 1998. After scientists reported that decreases in stratospheric ozone occurred over the heavily populated midlatitudes of the Northern Hemisphere in all seasons, the Montreal Protocol was modified to include stricter measures to limit CFC production. Industrial companies that manufacture CFCs quickly developed substitutes.

Production of CFCs, carbon tetrachloride, and methyl chloroform was completely phased out in the United States and other highly developed countries in 1996, with the exception of a relatively small amount exported to developing countries. Existing stockpiles could be used after the deadline, however. Developing countries phased out CFC use in 2005. Methyl bromide was supposed to be phased out by 2005 in highly developed countries, which are responsible for 80% of the global use of that chemical. However, these countries have been given more time to find effective substitutes. Hydrochlorofluorocarbons (HCFCs) will be phased out in 2030.

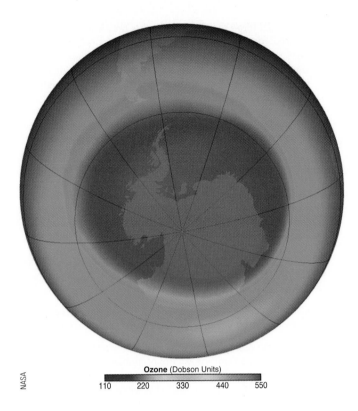

NASA

Figure 56-18 Ozone thinning

A computer-generated image of part of the Southern Hemisphere, taken in September 2005, reveals the ozone thinning (*purple and blue areas*). The ozone-thinned area is not stationary but is moved about by air currents. (Dobson units measure stratospheric ozone. They are named after British scientist Gordon Dobson, who studied ozone levels over Antarctica during the late 1950s.)

Satellite measurements taken in 1997 provided the first evidence that the levels of ozone-depleting chemicals were starting to decline in the stratosphere. In the early 2000s, the first signs of recovery of the ozone layer were evident: the *rate* of stratospheric ozone depletion was declining.

However, atmospheric levels of two chemicals (CFC-12 and halon-1211) still produced by developing countries may have increased. CFCs are extremely stable, and those being used today probably will continue to deplete stratospheric ozone for at least 50 years. Assuming that countries continue to adhere to the Montreal Protocol, however, scientists expect that human-exacerbated ozone thinning will gradually decline over time.

Review

▮ What environmental service does the stratospheric ozone layer provide?

▮ How does ozone depletion occur?

▮ What are some of the consequences of the thinning ozone layer?

CONNECTIONS AMONG ENVIRONMENTAL PROBLEMS

Learning Objective

12 Describe how an environmental problem such as the biodiversity crisis is related to human population growth.

We have pointed out several connections among the four environmental problems discussed in this chapter. For example, deforestation, global warming, and ozone depletion will likely cause future reductions in biological diversity. Similarly, any environmental problem that you can think of, even if not discussed in this chapter, is related to other environmental concerns. Most important, however, is that all environmental problems are connected to overpopulation. Stated simply, human population growth is outstripping Earth's resources and straining our ecological support systems.

We have seen that the rate of human population growth is greatest in developing countries but that highly developed nations are also significantly overpopulated, in the sense that they have a high per capita consumption of resources (see Chapter 52). **Consumption overpopulation,** the disproportionately large consumption of resources by people in highly developed countries, affects the environment *as much as or more than* the population explosion in the developing world.

Humans share much in common with the fate of other organisms on the planet. We are not immune to the environmental damage we have produced. We differ from other organisms, however, in our capacity to reflect on the consequences of our actions and to alter our behavior accordingly. Humans, both individually and collectively, can bring about change. The widespread substitution of constructive change for destructive change is the key to ensuring the biosphere's survival.

Review

▮ How is human population growth related to destruction of natural habitats?

SUMMARY WITH KEY TERMS

Learning Objectives

1 Identify the various levels of biodiversity: genetic diversity, species richness, and ecosystem diversity (page 1213).

▮ **Biological diversity** is the variety of organisms considered at three levels: populations, species, and ecosystems. **Genetic diversity** is the genetic variety within a species, both within a given population and among geographically separate populations. **Species richness** is the number of species of archaea, bacteria, protists, plants, fungi, and animals. **Ecosystem diversity** is the variety of Earth's ecosystems, such as forests, prairies, deserts, lakes, coastal estuaries, and coral reefs.

2 Distinguish among threatened species, endangered species, and extinct species (page 1213).

▮ A species becomes **extinct** when its last individual member dies. A species whose severely reduced numbers throughout all or a significant part of its **range** put it in imminent danger of extinction is classified as an **endangered species.** When extinction is less imminent but the population is quite small, a species is classified as a **threatened species.**

3 Discuss at least four causes of declining biological diversity, and identify the most important cause (page 1213).

  ▮ Human activities that reduce biological diversity include habitat loss and **habitat fragmentation,** pollution, introduction of **invasive species,** pest and predator control, illegal commercial hunting, and **commercial harvest.** Of these, habitat loss and fragmentation are the most significant.

4 Define *conservation biology,* and compare in situ and ex situ conservation measures (page 1219).

  ▮ **Conservation biology** is the study of how humans affect organisms and of the development of ways to protect biological diversity.

  ▮ Efforts to preserve biological diversity in the wild, known as **in situ conservation,** are urgently needed in the world's **biodiversity hotspots.** The study of the connections in a heterogeneous **landscape** consisting of multiple interacting ecosystems is known as **landscape ecology.** Increasingly, biologists are focusing efforts on preserving biodiversity in entire ecosystems and landscapes.

  ▮ **Ex situ conservation** involves conserving individual species in human-controlled settings. Breeding captive species in zoos and storing seeds of genetically diverse plant crops are examples.

5 Describe the benefits and shortcomings of the U.S. Endangered Species Act and the Convention on International Trade in Endangered Species of Wild Flora and Fauna (page 1219).

  ▮ The Endangered Species Act (ESA) authorizes the U.S. Fish and Wildlife Service to protect endangered and threatened species, both in the United States and abroad. The ESA is controversial; it does not compensate private property owners for financial losses because they cannot develop their land if a threatened or endangered species lives there. Conservationists would like to strengthen the ESA to manage whole ecosystems and maintain complete biological diversity rather than attempt to save endangered species as isolated entities.

  ▮ At the international level, the Convention on International Trade in Endangered Species of Wild Flora and Fauna (CITES) protects endangered animals and plants considered valuable in the highly lucrative international wildlife trade. Enforcement of this treaty varies from country to country; where enforcement exists, penalties are not very severe, so illegal trade in rare, commercially valuable species continues.

6 Discuss the ecosystem services of forests, and describe the consequences of deforestation (page 1224).

  ▮ Forests provide many **ecosystem services,** including wildlife habitat, protection of watersheds, prevention of soil erosion, moderation of climate, and protection from flooding.

  ▮ The greatest problem facing forests today is **deforestation,** the temporary or permanent clearance of forests for agriculture or other uses. Deforestation increases soil erosion and thus decreases soil fertility. Deforestation contributes to loss of biological diversity. When a forest is cut down, the watershed cannot absorb and hold water as well, and the total amount of surface runoff flowing into rivers and streams increases. Deforestation may affect regional and global climate changes and may contribute to an increase in global temperature.

ThomsonNOW™ **Learn more about deforestation by clicking on the figure in ThomsonNOW.**

7 State at least three reasons why forests (tropical rain forests and boreal forests) are disappearing today (page 1224).

  ▮ Forests are destroyed to provide subsistence farmers with agricultural land, to produce timber, to provide open rangeland for cattle, and to supply fuel wood. **Subsistence agriculture,** in which a family produces enough food to feed itself, accounts for perhaps 60% of tropical rainforest deforestation. Extensive clear-cut logging of certain boreal forests of Alaska, Canada, and Russia is currently the primary source of the world's industrial wood and wood fiber.

8 Name at least three greenhouse gases, and explain how greenhouse gases contribute to global warming (page 1226).

  ▮ **Greenhouse** gases—carbon dioxide, methane, surface ozone, nitrous oxide, and chlorofluorocarbons—cause the **greenhouse effect,** in which the atmosphere retains heat and warms Earth's surface. Increased levels of CO_2 and other greenhouse gases in the atmosphere are causing concerns about an **enhanced greenhouse effect,** which is additional warming produced by increased levels of gases that absorb infrared radiation.

ThomsonNOW™ **Watch the development of the enhanced greenhouse effect by clicking on the figure in ThomsonNOW.**

9 Describe how global warming may affect sea level, precipitation patterns, organisms (including humans), and food production (page 1226).

  ▮ During the 21st century, **global warming** may cause a rise in sea level. Precipitation patterns may change, resulting in more frequent droughts in some areas and more frequent flooding in other areas.

  ▮ Global warming is causing some species to shift their ranges; biologists think that global warming will cause some species to go extinct, some to be unaffected, and others to expand their numbers and ranges. Data linking climate warming and human health problems (particularly in developing countries) are accumulating. Problems for agriculture include increased flooding, increased droughts, and declining agricultural productivity in tropical and subtropical areas.

10 Distinguish between surface ozone and stratospheric ozone (page 1229).

  ▮ **Ozone (O_3)** is a form of oxygen that is a human-made pollutant in the lower atmosphere but a naturally produced, essential part of the stratosphere. The **stratosphere,** which encircles Earth some 10 to 45 km above the surface, contains a layer of ozone that shields the surface from much of the damaging ultraviolet radiation from the sun.

11 Cite the causes and potential effects of ozone destruction in the stratosphere (page 1229).

  ▮ The total amount of ozone in the stratosphere is declining, and large areas of ozone thinning develop over Antarctica each year. Chlorofluorocarbons and similar chlorine- and bromine-containing compounds attack the ozone layer.

  ▮ Excessive exposure to UV radiation is linked to human health problems, including cataracts, skin cancer, and a weakened immune system. Increased levels of UV radiation may disrupt ecosystems, such as the Antarctic food web. There is also concern that high levels of UV radiation may damage crops and forests.

ThomsonNOW™ **Learn more about ultraviolet radiation and the ozone layer by clicking on the figure in ThomsonNOW.**

12 Describe how an environmental problem such as the biodiversity crisis is related to human population growth (page 1231).

■ Overpopulation increases destruction of natural habitats, as people convert natural environments to agricultural land and log forests for needed lumber. **Consumption overpopulation**—the disproportionately large consumption of resources by people in highly developed countries—harms the environment as much as the population explosion in the developing world.

TEST YOUR UNDERSTANDING

1. Which of the following statements about extinction is *not* correct? (a) extinction is the permanent loss of a species (b) extinction is a natural biological process (c) once a species is extinct, it never reappears (d) human activities have little impact on extinctions (e) thousands of plant and animal species are currently threatened with extinction

2. An endangered species (a) is severely reduced in number (b) is in imminent danger of becoming extinct throughout all or a significant part of its range (c) usually does not have reduced genetic variability (d) is not in danger of extinction in the foreseeable future (e) a and b

3. The most important reason for declining biological diversity is (a) air pollution (b) introduction of foreign (invasive) species (c) habitat destruction and fragmentation (d) illegal commercial hunting (e) commercial harvesting

4. Habitat corridors (a) surround a given habitat (b) cut through a continuous habitat, producing an edge effect (c) vary in width depending on the species they are designed to protect (d) are an important strategy of ex situ conservation (e) have been widely adopted by restoration ecologists

5. In situ conservation (a) includes breeding captive species in zoos (b) includes seed storage of genetically diverse crops (c) concentrates on preserving biological diversity in the wild (d) focuses exclusively on large, charismatic animals (e) a and b

6. Restoration ecology (a) is the study of how humans impact organisms (b) returns a degraded environment as close as possible to its former state (c) is an example of ex situ conservation (d) has been used to successfully reverse the decline in amphibian populations (e) is an important provision of the Endangered Species Act

7. Which of the following is linked to the decline of amphibian populations? (a) agricultural chemicals (b) increased UV radiation (c) infectious diseases (d) global climate warming (e) all of the preceding

8. When forests are destroyed, (a) soil fertility increases (b) soil erosion decreases (c) some forest organisms decline in number (d) the danger of flooding in nearby lowlands declines (e) carbon dioxide is removed from the atmosphere and stored in soil

9. About 60% of tropical rainforest deforestation is the result of (a) commercial logging (b) cattle ranching (c) hydroelectric dams (d) mining (e) subsistence agriculture

10. Deforestation of boreal forests is primarily the result of (a) commercial logging (b) cattle ranching (c) hydroelectric dams (d) mining (e) subsistence agriculture

11. Which of the following gases contributes to both global warming and thinning of the ozone layer? (a) CO_2 (b) CH_4 (c) surface O_3 (d) CFCs (e) N_2O

12. Global warming occurs because (a) carbon dioxide and other greenhouse gases react chemically to produce excess heat (b) Earth has too many greenhouses and other glassed buildings (c) volcanic eruptions produce large quantities of sulfur and other greenhouse gases (d) carbon dioxide and other greenhouse gases trap infrared radiation in the atmosphere (e) carbon dioxide and other greenhouse gases allow excess heat to pass out of the atmosphere

13. What gas is a human-made pollutant in the lower (surface) atmosphere but a natural and beneficial gas in the stratosphere? (a) CO_2 (b) CH_4 (c) O_3 (d) CFCs (e) N_2O

14. Where is stratospheric ozone depletion most pronounced? (a) over Antarctica (b) over the equator (c) over South America (d) over North America and Europe (e) over Alaska and Siberia

15. The Montreal Protocol is an agreement to (a) phase out greenhouse gas emissions (b) curtail CFC production (c) design a recovery plan for each endangered species (d) curtail deforestation (e) prevent the selling of products made from endangered or threatened species

CRITICAL THINKING

1. If half of the world's biological diversity were to disappear, how would it affect your life?

2. Why might captive-breeding programs that reintroduce species into natural environments fail?

3. Conservation biologists often say that their discipline is less about biology than it is about economics and human decision making. What do you think they mean?

4. Explain why the hotspot approach to protecting biodiversity may be less important than some other approaches.

5. If you were given the task of developing a policy for the United States to deal with global climate change during the next 50 years, what would you propose? Explain your answer.

6. A more descriptive name for *Homo sapiens* is "*Homo dangerous.*" Explain this specific epithet, given what you have learned in this chapter.

7. **Evolution Link.** Because new species will eventually evolve to replace those that humans are driving to extinction, why is declining biological diversity such a threat to us?

8. **Evolution Link.** Biologists have wondered how introduced species that would probably have limited genetic variation (due to the founder effect) survive and adapt so successfully that they become invasive. Part of the answer may be that invasive species are the result of multiple introductions instead of a single one. Explain how multiple introductions from a species' native area to an introduced area could increase that species' invasion success.

9. **Evolution Link.** Conservation biologists have altered the evolution of salmon populations in captive-breeding programs. Wild female salmon tend to produce fewer large eggs, because the large eggs contain more nutrients for the offspring, giving each individual a greater chance to survive. After just a few generations, however, captive-bred females now lay greater numbers of small eggs. Suggest a possible adaptive advantage for many, small eggs in the captive-bred environment. What would you predict regarding the probably reproductive success of captive-bred females released in the wild?

Additional questions are available in ThomsonNOW at www.thomsonedu.com/login

Periodic Table of the Elements

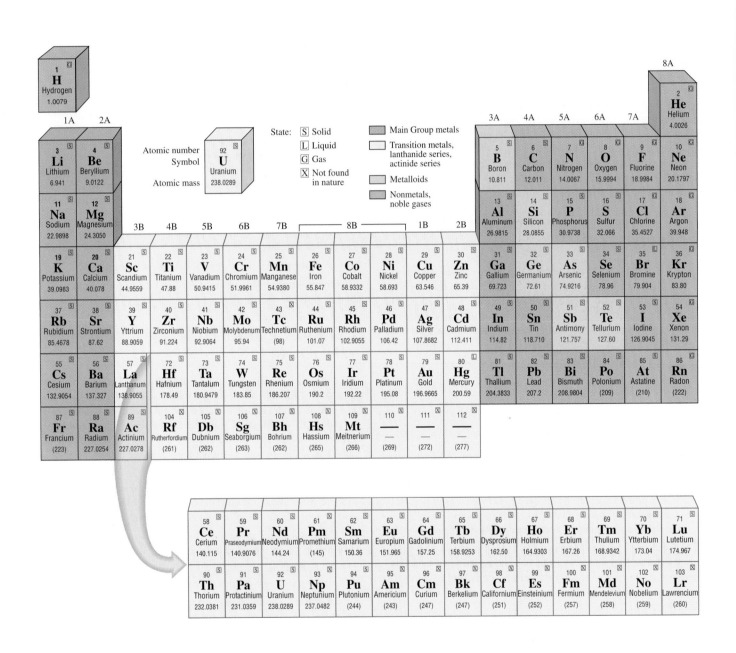

Appendix B

Classification of Organisms

The system of cataloguing organisms used in this book is described in Chapter 1 and in Chapters 23 through 31. In this eighth edition of *Biology*, we use the three-domain and six-kingdom classification. The three domains are **Bacteria, Archaea,** and **Eukarya** (eukaryotes). The six kingdoms are **Bacteria** (which corresponds to domain Bacteria); **Archaea** (which corresponds to domain Archaea); **Protista, Plantae, Fungi,** and **Animalia** (these four kingdoms are assigned to domain Eukarya). In this classification overview, we have included select groups. (We have also omitted many groups, especially extinct ones.) We have omitted viruses from this survey, because they do not fit into the three domains or six kingdoms.

PROKARYOTES

Domains Bacteria and Archaea are made up of prokaryotic organisms. They are distinguished from eukaryotic organisms by their smaller ribosomes and absence of membranous organelles, including the absence of a discrete nucleus surrounded by nuclear envelope. Prokaryotes reproduce mainly asexually by binary fission. When present, flagella are simple and solid; they do not have the 9 + 2 microfilament structure typical of eukaryotes.

DOMAIN BACTERIA, KINGDOM BACTERIA

Very large, diverse group of prokaryotic organisms. Typically unicellular, but some form colonies or filaments. Mainly heterotrophic, but some groups are photosynthetic or chemosynthetic. Bacteria are nonmotile or move by rotating flagella. Typically have peptidoglycan in their cell walls. Estimated 100,000 to 200,000 species.

Bacterial nomenclature and taxonomic practices are controversial and changing. See Table 24-4.

Proteobacteria. Large, diverse group of gram-negative bacteria. Five subgroups are designated alpha, beta, gamma, delta, and epsilon. Proteobacteria include rickettsias, enterobacteria, purple sulfur bacteria, myxobacteria.

Cyanobacteria. Gram-negative, photosynthetic.

Gram-positive bacteria. Diverse group; includes actinomycetes, lactic acid bacteria, mycobacteria, streptococci, staphylococci, clostridia. Thick cell wall of peptidoglycan; many produce spores.

Mycoplasmas. Lack cell walls. Extremely small bacteria bounded by plasma membrane.

Chlamydias. Lack peptidoglycan in their cell walls. Energy parasites dependent on host for ATP.

Spirochetes. Spiral-shaped bacteria with flexible cell walls.

DOMAIN ARCHAEA, KINGDOM ARCHAEA

Prokaryotes with unique cell membrane structure and cell walls lacking peptidoglycan. Also distinguished by their ribosomal RNA, lipid structure, and specific enzymes. Archaea are found in extreme environments—hot springs, sea vents, dry and salty seashores, boiling mud, and near ash-ejecting volcanoes. Three main types based on their metabolism and ecology are *methanogens, extreme halophiles,* and *extreme thermophiles.* Fewer than 225 named species. See Table 24-3 for a classification based on molecular data (as outlined in *Bergey's Manual*).

Methanogens. Anaerobes that produce methane gas from simple carbon compounds.

Extreme halophiles. Inhabit saturated salt solutions.

Extreme thermophiles. Grow at 70°C or higher; some thrive above boiling point.

DOMAIN EUKARYA, KINGDOM PROTISTA

Primarily unicellular or simple multicellular eukaryotic organisms that do not form tissues and that exhibit relatively little division of labor. Most modes of nutrition occur in this kingdom. Life cycles may include both sexually and asexually reproducing phases and may be extremely complex, especially in parasitic forms. Locomotion is by cilia, flagella, amoeboid movement, or by other means. Flagella and cilia have 9 + 2 structure.

Excavates. Anaerobic zooflagellates that have deep, or excavated, oral groove. Have atypical mitochondria or lack them. Most are endosymbionts. Include diplomonads and trichonymphs.

    Diplomonads. Excavates with one or two nuclei, no mitochondria, no Golgi complex, and up to eight flagella.

    Trichonymphs. Excavates with hundreds of flagella; live in guts of termites and wood-eating cockroaches.

Discicristates. Zooflagellates with disc-shaped cristae in their mitochondria. Include euglenoids and trypanosomes.

    Euglenoids and trypanosomes. Unicellular flagellates; some free-living and some pathogenic. Many are heterotrophic, but some (about one third) are photosynthetic. At least 900 species.

Alveolates. Unicellular protists with alveoli, flattened vesicles located inside plasma membrane. Include ciliates, dinoflagellates, and apicomplexans.

    Ciliates. Unicellular protists that move by means of cilia. Reproduction is asexual by binary fission or sexual by conjugation. About 7200 species.

Dinoflagellates. Unicellular (some colonial), photosynthetic, biflagellate. Cell walls composed of overlapping cell plates; contain cellulose. Contain chlorophylls *a* and *c* and carotenoids, including fucoxanthin. About 2000 to 4000 species.

Apicomplexans. Parasitic unicellular protists that lack specific structures for locomotion. At some stage in life cycle, they develop spores (small infective agents). Some pathogenic. About 3900 species.

Heterokonts. Diverse group; all have motile cells with two flagella, one with tiny, hairlike projections off shaft. Include water molds, diatoms, golden algae, and brown algae.

Water molds. Consist of branched, coenocytic mycelia. Cellulose and/or chitin in cell walls. Produce biflagellate asexual spores. Sexual stage involves production of oospores. Some parasitic. About 700 species.

Diatoms. Unicellular (some colonial), photosynthetic. Most nonmotile, but some move by gliding. Cell walls composed of silica rather than cellulose. Contain chlorophylls *a* and *c* and carotenoids, including fucoxanthin. At least 100,000 species estimated.

Golden algae. Unicellular (some colonial), photosynthetic, biflagellate (some lack flagella). Cells covered by tiny scales of either silica or calcium carbonate. Contain chlorophylls *a* and *c* and carotenoids, including fucoxanthin. About 1000 species.

Brown algae. Multicellular, often quite large (kelps). Photosynthetic; contain chlorophylls *a* and *c* and carotenoids, including fucoxanthin. Biflagellate reproductive cells. About 1500 species.

"Plants." Photosynthetic organisms with chloroplasts bounded by outer and inner membranes. This monophyletic group includes organisms currently placed in two kingdoms: Plantae (land plants) and Protista (red algae and green algae).

Red algae. Most multicellular (some unicellular), mainly marine. Some (coralline algae) have bodies impregnated with calcium carbonate. No motile cells. Photosynthetic; contain chlorophyll *a*, carotenoids, phycocyanin, and phycoerythrin. About 5000 species.

Green algae. Unicellular, colonial, siphonous, and multicellular forms. Some motile and flagellate. Photosynthetic; contain chlorophylls *a* and *b* and carotenoids. About 17,000 species.

Cercozoa. Amoeboid cells, often with hard outer shells through which cytoplasmic projections extend. Include foraminiferans and actinopods.

Foraminiferans. Unicellular protists that produce calcareous tests (shells) with pores through which cytoplasmic projections extend, forming sticky net to entangle prey.

Actinopods. Unicellular protists that produce axopods (long, filamentous cytoplasmic projections) that protrude through pores in their siliceous shells.

Amoebozoa. Amoeboid protists that lack tests and move by means of lobose pseudopodia. Include amoebas, plasmodial slime molds, and cellular slime molds.

Amoebas. Free-living and parasitic unicellular protists whose movement and capture of food are associated with pseudopodia.

Plasmodial slime molds. Spend part of life cycle as thin, streaming, multinucleate plasmodium that creeps along on decaying leaves or wood. Flagellate or amoeboid reproductive cells; form spores in sporangia. About 700 species.

Cellular slime molds. Vegetative (nonreproductive) form is unicellular; move by pseudopods. Amoeba-like cells aggregate to form multicellular pseudoplasmodium that eventually develops into fruiting body that bears spores. About 50 species.

Opisthokonts. Monophyletic group that includes members of three kingdoms: Fungi, Animalia, and Protista (choanoflagellates). Opisthokonts (Greek *opistho*, "rear," and *kontos*, "pole") have single posterior flagellum in flagellate cells.

Choanoflagellates. Zooflagellates with single flagellum surrounded at base by collar of microvilli. May be related to common ancestor of animals.

DOMAIN EUKARYA, KINGDOM PLANTAE

Multicellular eukaryotic organisms with differentiated tissues and organs. Cell walls contain cellulose. Cells frequently contain large vacuoles; photosynthetic pigments in plastids. Photosynthetic pigments are chlorophylls *a* and *b* and carotenoids. Nonmotile. Reproduce both asexually and sexually, with alternation of gametophyte (*n*) and sporophyte (2*n*) generations.

Phylum Bryophyta. *Mosses.* Nonvascular plants that lack xylem and phloem. Marked alternation of generations with dominant gametophyte generation. Motile sperm. Gametophytes generally form dense green mat consisting of individual plants. At least 9900 species.

Phylum Hepatophyta. *Liverworts.* Nonvascular plants that lack xylem and phloem. Marked alternation of generations with dominant gametophyte generation. Motile sperm. Gametophytes of certain species have flat, liverlike thallus; other species more mosslike in appearance. About 6000 species.

Phylum Anthocerophyta. *Hornworts.* Nonvascular plants that lack xylem and phloem. Marked alternation of generations with dominant gametophyte generation. Motile sperm. Gametophyte is small, flat, green thallus with scalloped edges. Spores produced on erect, hornlike stalk. About 100 species.

Phylum Pteridophyta. Vascular plants with dominant sporophyte generation. Reproduce by spores. Motile sperm.

Ferns. Generally homosporous. Gametophyte is free-living and photosynthetic. About 11,000 species.

Whisk ferns. Homosporous. Sporophyte stem branches dichotomously; lacks true roots and leaves. Gametophyte is subterranean and nonphotosynthetic and forms mycorrhizal relationship with fungus. About 12 species.

Horsetails. Homosporous. Sporophyte has hollow, jointed stems and reduced, scalelike leaves. Gametophyte is tiny photosynthetic plant. About 15 species.

Phylum Lycopodiophyta. *Club mosses.* Sporophyte plants are vascular with branching rhizomes and upright stems that bear microphylls. Although modern representatives are small, some extinct species were treelike. Some homosporous; others heterosporous. Motile sperm. About 1200 species.

Phylum Coniferophyta. *Conifers.* Heterosporous vascular plants with woody tissues (trees and shrubs) and needle-shaped or scalelike leaves. Most are evergreen. Seeds usually borne naked on surface of cone scales. Nutritive tissue in seed is haploid female gametophyte tissue. Nonmotile sperm. About 630 species.

Phylum Cycadophyta. *Cycads.* Heterosporous, vascular, dioecious plants that are small and shrubby or larger and palmlike. Produce naked seeds in conspicuous cones. Flagellate sperm. About 140 species.

Phylum Ginkgophyta. *Ginkgo.* Broadleaf deciduous trees that bear naked seeds directly on branches. Dioecious. Contain vascular tissues. Flagellate sperm. Ginkgo tree is only living representative. One species.

Phylum Gnetophyta. *Gnetophytes.* Woody shrubs, vines, or small trees that bear naked seeds in cones. Contain vascular tissues. Possess many features similar to flowering plants. About 70 species.

Phylum Anthophyta. *Flowering plants* or *angiosperms.* Largest, most successful group of plants. Heterosporous; dominant sporophytes with extremely reduced gametophytes. Contain vascular tissues. Bear flowers, fruits, and seeds (enclosed in fruit; seeds contain endosperm as nutritive tissue). Double fertilization. More than 300,000 species.

DOMAIN EUKARYA, KINGDOM FUNGI

Eukaryotic, mainly multicellular organisms with cell walls containing chitin. Body form is often mycelium. Cells usually haploid or dikaryotic, with brief diploid period following fertilization. Heterotrophs that secrete digestive enzymes onto food source and then absorb predigested food. Most decomposers, but some parasites. Reproduce by means of spores, which may be produced sexually or asexually. No flagellate stages except in chytrids. Classified as opisthokonts (along with choanoflagellates and animals) because flagellate cells, where present, have a single posterior flagellum (Greek *opistho,* "rear," and *kontos,* "pole").

Phylum Chytridiomycota. *Chytridiomycetes* or *chytrids.* Parasites and decomposers found mainly in fresh water. Motile cells (gametes and zoospores) contain single, posterior flagellum. Reproduce both sexually and asexually. About 790 species.

Phylum Zygomycota. *Zygomycetes (molds).* Important decomposers; some are insect parasites. Produce sexual resting spores called *zygospores;* and nonmotile, haploid, asexual spores in sporangium. Hyphae are coenocytic. Many are heterothallic (two mating types). About 1060 species.

Phylum Glomeromycota. *Glomeromycetes.* Symbionts that form intracellular mycorrhizal associations within roots of most trees and herbaceous plants. Reproduce asexually with large, multinucleate spores called *blastospores.* About 160 species.

Phylum Ascomycota. *Ascomycetes* or *sac fungi (yeasts, powdery mildews, molds, morels, truffles).* Important symbionts; 98% of lichen-forming fungi are ascomycetes; some form mycorrhizae. Sexual reproduction: form ascospores in sacs called *asci.* Asexual reproduction: produce spores called *conidia,* which pinch off from conidiophores. Hyphae usually have perforated septa. Dikaryotic stage. About 32,300 species.

Phylum Basidiomycota. *Basidiomycetes* or *club fungi (mushrooms, bracket fungi, puffballs).* Many form mycorrhizae with tree roots. Sexual reproduction: form basidiospores on basidium. Asexual reproduction uncommon. Heterothallic. Hyphae usually have perforated septa. Dikaryotic stage. About 22,300 species.

DOMAIN EUKARYA, KINGDOM ANIMALIA

Eukaryotic, multicellular heterotrophs with differentiated cells. In most animals, cells are organized to form tissues, tissues form organs, and tissues and organs form specialized organ systems that carry on specific functions. Most have well-developed nervous system and respond adaptively to changes in their environment. Most are capable of locomotion during some time in their life cycle. Most diploid and reproduce sexually; flagellate haploid sperm unites with large, non-motile, haploid egg, forming diploid zygote that undergoes cleavage. Classified as opisthokonts (along with choanoflagellates and fungi) because flagellate cells, where present, have a single posterior flagellum (Greek *opistho,* "rear," and *kontos,* "pole").

Phylum Porifera. *Sponges.* Mainly marine; solitary or colonial. Body bears many pores through which water circulates. Food is filtered from water by collar cells (choanocytes). Asexual reproduction by budding; external sexual reproduction in which sperm are released and swim to internal egg. Larva is motile. About 10,000 species.

Phylum Cnidaria. *Hydras, jellyfish, sea anemones, corals.* Marine, with a few freshwater species; solitary or colonial; polyp and medusa forms. Radial symmetry. Tentacles surrounding mouth. Stinging cells (cnidocytes) contain stinging structures called *nematocysts.* Planula larva. About 10,000 species.

Phylum Ctenophora. *Comb jellies.* Marine; free-swimming. Biradial symmetry. Two tentacles and eight longitudinal rows of cilia resembling combs; animal moves by means of these bands of cilia. About 100 species.

Protostomes

Coelomates (characterized by true coelom, that is, body cavity completely lined with mesoderm). Spiral, determinate cleavage, and mouth typically develops from blastopore. Two branches of protostomes are Lophotrochozoa and Ecdysozoa.

Lophotrochozoa

Include the platyhelminths, nemerteans (ribbon worms), mollusks, annelids, the lophophorate phyla, and the rotifers. The name *Lophotrochozoa* comes from the names of the two major animal groups included: the Lophophorata and the Trochozoa. The Lophophorata include the lophophorate phyla, characterized by a lophophore, a ciliated ring of tentacles surrounding the mouth. The name *Trochozoa* is derived from the trochophore larva that characterizes its two major groups—the mollusks and annelids.

Phylum Platyhelminthes. *Flatworms.* Acoelomate (no body cavity); region between body wall and internal organs filled with tissue. Planarians are free-living; flukes and tapeworms are parasitic. Body dorsoventrally flattened; cephalization; three tissue layers. Simple nervous system with ganglia in head region. Excretory organs are protonephridia with flame cells. About 20,000 species.

Phylum Nemertea. *Proboscis worms* (also called *ribbon worms*). Long, dorsoventrally flattened body with complex proboscis used for defense and for capturing prey. Functionally acoelomate but have small true coelom in proboscis. Definite organ systems. Complete digestive tract. Circulatory system with blood. About 1000 species.

Phylum Mollusca. *Snails, clams, squids, octopods.* Unsegmented, soft-bodied animals usually covered by dorsal shell. Have ventral, muscular foot. Most organs located above foot in visceral mass. Shell-secreting mantle covers visceral mass and forms mantle cavity, which contains gills. Trochophore and/or veliger larva. About 50,000 species.

Phylum Annelida. *Segmented worms: polychaetes, earthworms, leeches.* Both body wall and internal organs are segmented. Body segments separated by septa. Some have nonjointed appendages. Setae used in locomotion. Closed circulatory system; metanephridia; specialized regions of digestive tract. Trochophore larva. About 15,000 species.

Phylum Brachiopoda. *Lamp shells.* One of lophophorate phyla. Marine; body enclosed between two shells. About 350 species.

Phylum Phoronida. One of lophophorate phyla. Tube-dwelling marine worms. About 12 species.

Phylum Bryozoa. One of lophophorate phyla. Mainly marine; sessile colonies produced by asexual budding. About 5000 species.

Phylum Rotifera. *Wheel animals.* Aquatic, microscopic, wormlike animals. Anterior end has ciliated crown that looks like wheel when cilia beat. Posterior end tapers to foot. Characterized by pseudocoelom (body cavity not completely lined with mesoderm). Constant number of cells in adult. About 1800 species.

Ecdysozoa

Animals in this group characterized by ecdysis (molting).

Phylum Nematoda. *Roundworms: ascaris, hookworms, pinworms.* Slender, elongated, cylindrical worms; covered with cuticle. Characterized by pseudocoelom (body cavity not completely lined with mesoderm). Free-living and parasitic forms. About 15,000 species.

Phylum Arthropoda. *Arachnids (spiders, mites, ticks), crustaceans (lobsters, crabs, shrimp), insects, centipedes, millipedes.* Segmented animals with paired, jointed appendages and hard exoskeleton made of chitin. Open circulatory system with dorsal heart. Hemocoel occupies most of body cavity, and coelom is reduced. More than 1 million species.

Deuterostomes

Coelomates with radial, indeterminate cleavage. Blastopore develops into anus, and mouth forms from second opening.

Phylum Echinodermata. *Sea stars, sea urchins, sand dollars, sea cucumbers.* Marine animals. Pentaradial symmetry as adults; bilateral symmetry as larvae. Endoskeleton of small, calcareous plates. Water vascular system; tube feet for locomotion. About 7000 species.

Phylum Hemichordata. *Acorn worms.* Marine animals with ring of cilia around mouth. Anterior muscular proboscis is connected by collar region to long, wormlike body. Larval form resembles echinoderm larva. About 85 species.

Phylum Chordata. *Subphylum Urochordata (tunicates), subphylum Cephalochordata (lancelets), subphylum Vertebrata (fishes, amphibians, reptiles, birds, mammals).* Notochord; pharyngeal slits; dorsal, tubular nerve cord; and postanal tail present at some time in life cycle. About 48,000 species.

Appendix C

Understanding Biological Terms

Your task of mastering new terms will be greatly simplified if you learn to dissect each new word. Many terms can be divided into a prefix, the part of the word that precedes the main root; the word root itself; and often a suffix, a word ending that may add to or modify the meaning of the root. As you progress in your study of biology, you will learn to recognize the more common prefixes, word roots, and suffixes. Such recognition will help you analyze new terms so that you can more readily determine their meaning and will help you remember them.

Prefixes

a-, ab- from, away, apart (*abduct,* move away from the midline of the body)

a-, an-, un- less, lack, not (*asymmetrical,* not symmetrical)

ad- (also **af-, ag-, an-, ap-**) to, toward (*adduct,* move toward the midline of the body)

allo- different (*allometric growth,* different rates of growth for different parts of the body during development)

ambi- both sides (*ambidextrous,* able to use either hand)

andro- a man (*androecium,* the male portion of a flower)

anis- unequal (*anisogamy,* sexual reproduction in which the gametes are of unequal sizes)

ante- forward, before (*anteflexion,* bending forward)

anti- against (*antibody,* proteins that have the capacity to react against foreign substances in the body)

auto- self (*autotroph,* organism that manufactures its own food)

bi- two (*biennial,* a plant that takes two years to complete its life cycle)

bio- life (*biology,* the study of life)

circum-, circ- around (*circumcision,* a cutting around)

co-, con- with, together (*congenital,* existing with or before birth)

contra- against (*contraception,* against conception)

cyt- cell (*cytology,* the study of cells)

di- two (*disaccharide,* a compound made of two sugar molecules chemically combined)

dis- apart (*dissect,* cut apart)

ecto- outside (*ectoderm,* outer layer of cells)

end-, endo- within, inner (*endoplasmic reticulum,* a network of membranes found within the cytoplasm)

epi- on, upon (*epidermis,* upon the dermis)

ex-, e-, ef- out from, out of (*extension,* a straightening out)

extra- outside, beyond (*extraembryonic membrane,* a membrane that encircles and protects the embryo)

gravi- heavy (*gravitropism,* growth of a plant in response to gravity)

hemi- half (*cerebral hemisphere,* lateral half of the cerebrum)

hetero- other, different (*heterozygous,* having unlike members of a gene pair)

homeo- unchanging, steady (*homeostasis,* reaching a steady state)

homo-, hom- same (*homologous,* corresponding in structure; *homozygous,* having identical members of a gene pair)

hyper- excessive, above normal (*hypersecretion,* excessive secretion)

hypo- under, below, deficient (*hypotonic,* a solution whose osmotic pressure is less than that of a solution with which it is compared)

in-, im- not (*incomplete flower,* a flower that does not have one or more of the four main parts)

inter- between, among (*interstitial,* situated between parts)

intra- within (*intracellular,* within the cell)

iso- equal, like (*isotonic,* equal osmotic pressure)

macro- large (*macronucleus,* a large, polyploid nucleus found in ciliates)

mal- bad, abnormal (*malnutrition,* poor nutrition)

mega- large, great (*megakaryocyte,* giant cell of bone marrow)

meso- middle (*mesoderm,* middle tissue layer of the animal embryo)

meta- after, beyond (*metaphase,* the stage of mitosis after prophase)

micro- small (*microscope,* instrument for viewing small objects)

mono- one (*monocot,* a group of flowering plants with one cotyledon, or seed leaf, in the seed)

oligo- small, few, scant (*oligotrophic lake,* a lake deficient in nutrients and organisms)

oo- egg (*oocyte,* cell that gives rise to an egg cell)

paedo- a child (*paedomorphosis,* the preservation of a juvenile characteristic in an adult)

para- near, beside, beyond (*paracentral,* near the center)

peri- around (*pericardial membrane,* membrane that surrounds the heart)

photo- light (*phototropism,* growth of a plant in response to the direction of light)

poly- many, much, multiple, complex (*polysaccharide,* a carbohydrate composed of many simple sugars)

post- after, behind (*postnatal,* after birth)

pre- before (*prenatal,* before birth)

pseudo- false (*pseudopod,* a temporary protrusion of a cell, i.e., "false foot")

retro- backward (*retroperitoneal,* located behind the peritoneum)

semi- half (*semilunar,* half-moon)

sub- under (*subcutaneous tissue*, tissue immediately under the skin)

super, supra- above (*suprarenal*, above the kidney)

sym- with, together (*sympatric speciation*, evolution of a new species within the same geographic region as the parent species)

syn- with, together (*syndrome*, a group of symptoms that occur together and characterize a disease)

trans- across, beyond (*transport*, carry across)

Suffixes

-able, -ible able (*viable*, able to live)

-ad used in anatomy to form adverbs of direction (*cephalad*, toward the head)

-asis, -asia, -esis condition or state of (*euthanasia*, state of "good death")

-cide kill, destroy (*biocide*, substance that kills living things)

-emia condition of blood (*anemia*, a blood condition in which there is a lack of red blood cells)

-gen something produced or generated or something that produces or generates (*pathogen*, an organism that produces disease)

-gram record, write (*electrocardiogram*, a record of the electrical activity of the heart)

-graph record, write (*electrocardiograph*, an instrument for recording the electrical activity of the heart)

-ic adjective-forming suffix that means *of* or *pertaining to* (*ophthalmic*, of or pertaining to the eye)

-itis inflammation of (*appendicitis*, inflammation of the appendix)

-logy study or science of (*cytology*, study of cells)

-oid like, in the form of (*thyroid*, in the form of a shield, referring to the shape of the thyroid gland)

-oma tumor (*carcinoma*, a malignant tumor)

-osis indicates disease (*psychosis*, a mental disease)

-pathy disease (*dermopathy*, disease of the skin)

-phyll leaf (*mesophyll*, the middle tissue of the leaf)

-scope instrument for viewing or observing (*microscope*, instrument for viewing small objects)

Some Common Word Roots

abscis cut off (*abscission*, the falling off of leaves or other plant parts)

angi, angio vessel (*angiosperm*, a plant that produces seeds enclosed within a fruit or "vessel")

apic tip, apex (*apical meristem*, area of cell division located at the tips of plant stems and roots)

arthr joint (*arthropods*, invertebrate animals with jointed legs and segmented bodies)

aux grow, enlarge (*auxin*, a plant hormone involved in growth and development)

blast a formative cell, germ layer (*osteoblast*, cell that gives rise to bone cells)

brachi arm (*brachial artery*, blood vessel that supplies the arm)

bry grow, swell (*embryo*, an organism in the early stages of development)

cardi heart (*cardiac*, pertaining to the heart)

carot carrot (*carotene*, a yellow, orange, or red pigment in plants)

cephal head (*cephalad*, toward the head)

cerebr brain (*cerebral*, pertaining to the brain)

cervic, cervix neck (*cervical*, pertaining to the neck)

chlor green (*chlorophyll*, a green pigment found in plants)

chondr cartilage (*chondrocyte*, a cartilage cell)

chrom color (*chromosome*, deeply staining body in nucleus)

cili small hair (*cilium*, a short, fine cytoplasmic hair projecting from the surface of a cell)

coleo a sheath (*coleoptile*, a protective sheath that encircles the stem in grass seedlings)

conjug joined (*conjugation*, a sexual phenomenon in certain protists)

cran skull (*cranial*, pertaining to the skull)

decid falling off (*deciduous*, a plant that sheds its leaves at the end of the growing season)

dehis split (*dehiscent fruit*, a fruit that splits open at maturity)

derm skin (*dermatology*, study of the skin)

ecol dwelling, house (*ecology*, the study of organisms in relation to their environment, i.e., "their house")

enter intestine (*enterobacteria*, a group of bacteria that includes species that inhabit the intestines of humans and other animals)

evol to unroll (*evolution*, descent with modification, gradual directional change)

fil a thread (*filament*, the thin stalk of the stamen in flowers)

gamet a wife or husband (*gametangium*, the part of a plant, protist, or fungus that produces reproductive cells)

gastr stomach (*gastrointestinal tract*, the digestive tract)

glyc, glyco sweet, sugar (*glycogen*, storage form of glucose)

gon seed (*gonad*, an organ that produces gametes)

gutt a drop (*guttation*, loss of water as liquid "drops" from plants)

gymn naked (*gymnosperm*, a plant that produces seeds that are not enclosed with a fruit, i.e., "naked")

hem blood (*hemoglobin*, the pigment of red blood cells)

hepat liver (*hepatic*, of or pertaining to the liver)

hist tissue (*histology*, study of tissues)

hydr water (*hydrolysis*, a breakdown reaction involving water)

leuk white (*leukocyte*, white blood cell)

menin membrane (*meninges*, the three membranes that envelop the brain and spinal cord)

morph form (*morphogenesis*, development of body form)

my, myo muscle (*myocardium*, muscle layer of the heart)

myc a fungus (*mycelium*, the vegetative body of a fungus)

nephr kidney (*nephron*, microscopic unit of the kidney)

neur, nerv nerve (*neuromuscular*, involving both the nerves and muscles)

occiput back part of the head (*occipital*, back region of the head)

ost bone (*osteology*, study of bones)

path disease (*pathologist*, one who studies disease processes)

ped, pod foot (*bipedal*, walking on two feet)

pell skin (*pellicle*, a flexible covering over the body of certain protists)

phag eat (*phagocytosis*, process by which certain cells ingest particles and foreign matter)

phil love (*hydrophilic*, a substance that attracts, i.e., "loves," water)

phloe bark of a tree (*phloem*, food-conducting tissue in plants that corresponds to bark in woody plants)

phyt plant (*xerophyte*, a plant adapted to xeric, or dry, conditions)

plankt wandering (*plankton*, microscopic aquatic protists that float or drift passively)

rhiz root (*rhizome*, a horizontal, underground stem that superficially resembles a root)

scler hard (*sclerenchyma*, cells that provide strength and support in the plant body)

sipho a tube (*siphonous*, a type of tubular body form found in certain algae)

som body (*chromosome*, deeply staining body in the nucleus)

sor heap (*sorus*, a cluster or "heap" of sporangia in a fern)

spor seed (*spore*, a reproductive cell that gives rise to individual offspring in plants, protists, and fungi)

stom a mouth (*stoma*, a small pore, i.e., "mouth," in the epidermis of plants)

thigm a touch (*thigmotropism*, plant growth in response to touch)

thromb clot (*thrombus*, a clot within a blood vessel)

troph nourishment (*heterotroph*, an organism that must depend on other organisms for its nourishment)

tropi turn (*thigmotropism*, growth of a plant in response to contact with a solid object, such as a tendril "turning" or wrapping around a wire fence)

visc pertaining to an internal organ or body cavity (*viscera*, internal organs)

xanth yellow (*xanthophyll*, a yellowish pigment found in plants)

xyl wood (*xylem*, water-conducting tissue in plant, the "wood" of woody plants)

zoo an animal (*zoology*, the science of animals)

Appendix D

Abbreviations

The biological sciences use a great many abbreviations and with good reason. Many technical terms in biology and biological chemistry are both long and difficult to pronounce. Yet it can be difficult for beginners, when confronted with something like NADPH or EPSP, to understand the reference. Here, for your ready reference, are some of the common abbreviations used in biology.

A adenine
ABA abscisic acid
ABC transporters ATP-binding cassette transporters
ACTH adrenocorticotropic hormone
AD Alzheimer's disease
ADA adenosine deaminase
ADH antidiuretic hormone
ADP adenosine diphosphate
AIDS acquired immunodeficiency syndrome
AMP adenosine monophosphate
amu atomic mass unit (dalton)
APC anaphase-promoting complex *or* antigen-presenting cell
ATP adenosine triphosphate
AV node or valve atrioventricular node or valve (of heart)
B lymphocyte or B cell lymphocyte responsible for antibody-mediated immunity
BAC bacterial artificial chromosome
BH brain hormone (of insects)
BMR basal metabolic rate
bya billion years ago
C cytosine
C_3 three-carbon pathway for carbon fixation (Calvin cycle)
C_4 four-carbon pathway for carbon fixation
CAM crassulacean acid metabolism
cAMP cyclic adenosine monophosphate
CAP catabolite activator protein
CD4 T cell T helper cell (TH); has a surface marker designated CD4
CD8 T cell T cell with a surface marker designated CD8; includes T cytotoxic cells
Cdk cyclin-dependent protein kinase
cDNA complementary deoxyribonucleic acid
CFCs chlorofluorocarbons
CFTR cystic fibrosis transmembrane conductance regulator
CITES Convention on International Trade in Endangered Species of Wild Flora and Fauna
CNS central nervous system
CO cardiac output
CoA coenzyme A
COPD chronic obstructive pulmonary disease
CP creatine phosphate
CPR cardiopulmonary resuscitation
CREB cyclic AMP response element binding protein

CSF cerebrospinal fluid
CVS cardiovascular system
DAG diacylglycerol
DNA deoxyribonucleic acid
DOC dissolved organic carbon
E_A activation energy (of an enzyme)
ECG electrocardiogram
ECM extracellular matrix
EEG electroencephalogram
EKG electrocardiogram
EM electron microscope or micrograph
ENSO El Niño–Southern Oscillation
EPSP excitatory postsynaptic potential (of a neuron)
ER endoplasmic reticulum
ES cells embryonic stem cells
EST expressed sequence tag
F_1 first filial generation
F_2 second filial generation
Fab portion the part of an antibody that binds to an antigen
Factor VIII blood-clotting factor (absent in hemophiliacs)
FAD/FADH$_2$ flavin adenine dinucleotide (oxidized and reduced forms, respectively)
FAP fixed action pattern
Fc portion the part of an antibody that interacts with cells of the immune system
FISH fluorescent in situ hybridization
FSH follicle-stimulating hormone
G guanine
G_1 phase first gap phase (of the cell cycle)
G_2 phase second gap phase (of the cell cycle)
G3P glyceraldehyde-3-phosphate
G protein cell-signaling molecule that requires GTP
GA_3 gibberellin
GAA Global Amphibian Assessment (by World Conservation Union)
GABA gamma-aminobutyric acid
GH growth hormone (somatotropin)
GnRH gonadotropin-releasing hormone
GTP guanosine triphosphate
HB hemoglobin
HBEF Hubbard Brook Experimental Forest
HBO$_2$ oxyhemoglobin
HCFCs hydrochlorofluorocarbons
hCG human chorionic gonadotropin
HD Huntington disease
HDL high-density lipoprotein
HFCs hydrofluorocarbons
hGH human growth hormone
HIV human immunodeficiency virus

HLA human leukocyte antigen
HUGO Human Genome Organization
IAA indole acetic acid (natural auxin)
Ig immunoglobulin, as in IgA, IgG, etc.
IGF insulin-like growth factor
IP$_3$ inositol trisphosphate
IPCC United Nations Intergovernmental Panel on Climate Change
IPSP inhibitory postsynaptic potential (of a neuron)
IUCN World Conservation Union
IUD intrauterine device
JH juvenile hormone (of insects)
kb kilobase
LDH lactate dehydrogenase enzyme
LDL low-density lipoprotein
LH luteinizing hormone
LM light microscope or micrograph
LSD lysergic acid diethylamide
LTP long-term potentiation
MAO monoamine oxidase
MAPs microtubule-associated proteins
MHC major histocompatibility complex
MI myocardial infarction
miRNA microribonucleic acid
MPF mitosis-promoting factor
MRI magnetic resonance imaging
mRNA messenger RNA
MSAFP maternal serum α-fetoprotein
MSH melanocyte-stimulating hormone
mtDNA mitochondrial DNA
MTOC microtubule organizing center
MVP minimum viable population
mya million years ago
9 + 2 structure cilium or flagellum (of a eukaryote)
9 × 3 structure centriole or basal body (of a eukaryote)
n, 2n the chromosome number of a gamete and of a zygote, respectively
NAD$^+$/NADH nicotinamide adenine dinucleotide (oxidized and reduced forms, respectively)
NADP$^+$/NADPH nicotinamide adenine dinucleotide phosphate (oxidized and reduced forms, respectively)
NAG *N*-acetyl glucosamine
NK cell natural killer cell
NMDA *N*-methyl-D aspartate (an artificial ligand)
NO nitric oxide
NSF National Science Foundation
P generation parental generation
P53 a tumor suppressor gene
P680 reaction center of photosystem II
P700 reaction center of photosystem I
PABA para-aminobenzoic acid

PAMPs pathogen-associated molecular patterns
PCR polymerase chain reaction
PEP phosphoenolpyruvate
Pfr phytochrome (form that absorbs far-red light)
PGA phosphoglycerate
PID pelvic inflammatory disease
PIF3 phytochrome-interacting factor-3
PKU phenylketonuria
PNS peripheral nervous system
pre-mRNA precursor messenger RNA (in eukaryotes)
Pr phytochrome (form that absorbs red light)
PTH parathyroid hormone
RAS reticular activating system
RBC red blood cell (erythrocyte)
REM sleep rapid-eye-movement sleep
RFLP restriction fragment length polymorphism
RNA ribonucleic acid
RNAi ribonucleic acid interference
rRNA ribosomal RNA
rubisco ribulose bisphosphate carboxylase/oxygenase
RuBP ribulose bisphosphate
S phase DNA synthetic phase (of the cell cycle)
SA node sinoatrial node (of heart)
SCID severe combined immunodeficiency
SEM scanning electron microscope or micrograph
siRNA small interfering ribonucleic acid
snRNA small nuclear ribonucleic acid
snoRNA small nucleolar ribonucleic acid
snRNP small nuclear ribonucleoprotein complex
SRP RNA signal recognition particle ribonucleic acid
SSB protein single-strand binding protein
ssp subspecies
STD sexually transmitted disease
STR short tandem repeat
T thymine
T lymphocyte or T cell lymphocyte responsible for cell-mediated immunity
T$_C$ lymphocyte T cytotoxic cell
T$_H$ lymphocyte T helper cell (CD4 T cell)
TATA box base sequence in eukaryotic promoter
TCA cycle tricarboxylic acid cycle (synonym for citric acid cycle)
TCR T-cell receptor
TEM transmission electron microscope or micrograph
Tm tubular transport maximum
TNF tumor necrosis factor
tRNA transfer RNA
U uracil
UPE upstream promoter element
UV light ultraviolet light
WBC white blood cell (leukocyte)

Appendix E

Answers to Test Your Understanding Questions

Chapter 1: 1. a 2. a 3. e 4. c 5. a 6. b 7. a 8. c
9. c 10. b 11. a 12. b 13. b 14. d 15. c 16. b 17. b
18. e

Chapter 2: 1. c 2. d 3. e 4. a 5. b 6. b 7. c 8. d
9. a 10. b 11. a 12. a 13. d 14. e 15. e 16. d 17. e

Chapter 3: 1. a 2. d 3. b 4. e 5. e 6. d 7. c 8. c
9. a 10. c 11. d 12. a 13. b 14. e 15. c 16. e 17. b

Chapter 4: 1. b 2. a 3. d 4. d 5. d 6. a 7. e 8. d
9. a 10. a 11. b 12. a 13. e 14. a 15. d

Chapter 5: 1. d 2. c 3. c 4. c 5. a 6. b 7. a 8. c
9. b 10. b 11. b 12. e 13. d 14. a 15. c 16. e 17. a

Chapter 6: 1. b 2. d 3. a 4. e 5. e 6. c 7. a 8. b
9. d 10. c 11. a 12. e 13. b 14. d 15. c

Chapter 7: 1. a 2. c 3. c 4. d 5. b 6. d 7. b 8. d
9. e 10. e 11. a 12. e 13. c 14. b 15. a

Chapter 8: 1. c 2. a 3. d 4. e 5. a 6. a 7. b 8. c
9. c 10. a 11. c 12. d 13. e 14. b 15. c

Chapter 9: 1. a 2. d 3. c 4. a 5. a 6. c 7. a 8. e
9. e 10. a 11. c 12. b 13. c 14. a 15. c 16 b

Chapter 10: 1. d 2. c 3. a 4. d 5. e 6. d 7. c 8. a
9. a 10. d 11. c 12. e 13. d 14. b 15. b

Chapter 11: 1. c 2. d 3. c 4. c 5. d 6. b 7. e 8. c
9. e 10. c 11. b 12. a

13. a. all yellow   b. $\frac{1}{2}$ yellow $\frac{1}{2}$ green   c. all yellow
d. $\frac{3}{4}$ yellow $\frac{1}{4}$ green

14. 150

15. The short-winged condition is recessive. Both parents are
heterozygous.

16. Repeated matings of the roan bull and the white cow will yield
an approximate 1:1 ratio of roan-to-white offspring. Repeated matings among roan offspring will yield red, roan, and white offspring in
an approximate 1:2:1 ratio. The mating of two red individuals will
yield only red offspring.

17. There are 36 possible outcomes when a pair of dice is rolled.
There are six ways of obtaining a 7: 1,6; 6,1; 2,5; 5,2; 3,4; and 4,3.
There are five ways of rolling a 6: 1,5; 5,1; 2,4; 4,2; and 3,3. There are
also five ways of rolling an 8: 2,6; 6,2; 3,5; 5,3; and 4,4.

18. The genotype of the brown, spotted rabbit is *bbSS*. The genotype
of the black, solid rabbit is *BBss*. An F_1 rabbit would be black, spot-
ted (*BbSs*). The F_2 is expected to be $\frac{9}{16}$ black, spotted (*B_S_*), $\frac{3}{16}$ black,
solid (*B_ss*), $\frac{3}{16}$ brown, spotted (*bbS_*), and $\frac{1}{16}$ brown, solid (*bbss*).

19. The F_1 offspring are expected to be black, short-haired. There
is a $\frac{1}{16}$ chance of brown and tan, long-haired offspring in the F_2
generation.

20. yes.

21. pleiotropic

22. The rooster is *PpRr;* hen A is *PpRR;* hen B is *Pprr;* and hen C is
PPRR.

23. a. Both parents are heterozygous for a single locus.   b. One parent is heterozygous, and the other is homozygous recessive (for a single locus).   c. Both parents are heterozygous (for two loci).   d. One
parent is heterozygous, and the other is homozygous recessive (for
two loci).

24. These are the expected types of F_2 plants: $\frac{1}{16}$ of the plants produce
2-kg fruits; $\frac{1}{4}$ produce 1. 75-kg fruits; $\frac{3}{8}$ produce 1. 5-kg fruits; $\frac{1}{4}$ produce 1. 25-kg fruits; and $\frac{1}{16}$ produce 1-kg fruits.

25. All male offspring of this cross are barred, and all females are
nonbarred, thus allowing the sex of the chicks to be determined by
their phenotypes.

26. This is a two-point test cross involving linked loci. The parental
class of offspring are *Aabb* and *aaBb;* the recombinant classes are
AaBb and *aabb.* There is 4. 6% recombination, which corresponds
to 4. 6 map units between the loci.

27. If genes B and C are 10 map units apart, gene A is in the middle.
If genes B and C are 2 map units apart, gene C is in the middle.

Chapter 12: 1. c 2. d 3. d 4. e 5. b 6. d 7. b 8. e
9. a 10. a 11. d 12. c 13. e 14. b 15. a 16. d 17. b

Chapter 13: 1. e 2. a 3. b 4. d 5. c 6. b 7. a 8. e
9. c 10. a 11. a 12. b 13. d 14. e 15. d 16. c 17. b
18. e

Chapter 14: 1. a 2. c 3. a 4. e 5. c 6. a 7. c 8. a
9. b 10. c 11. a 12. b 13. e 14. e 15. a 16. a 17. c

Chapter 15: 1. a 2. b 3. e 4. d 5. e 6. a 7. b 8. a
9. c 10. a 11. b 12. c 13. b 14. a

Chapter 16: 1. e 2. a 3. a 4. c 5. c 6. a 7. e 8. d
9. a 10. a 11. d 12. c 13. e 14. e 15. d 16. d 17. e
18. a 19. a 20. b 21. b 22. d 23. b 24. a 25. c 26. a
27. a 28. a

Chapter 17: 1. c 2. d 3. a 4. b 5. a 6. c 7. d 8. b
9. d 10. e 11. b 12. c 13. b 14. e 15. d 16. c

Chapter 18: 1. d 2. b 3. c 4. b 5. c 6. b 7. a 8. c
9. b 10. c 11. e 12. b 13. a 14. b 15. b 16. c

Chapter 19: 1. b 2. b 3. b 4. d 5. b 6. b 7. c 8. d
9. a 10. e 11. e 12. d 13. e 14. c 15. b

Chapter 20: 1. e 2. b 3. a 4. a 5. c 6. b 7. d 8. b
9. b 10. c 11. b 12. e 13. c 14. e 15. d

Chapter 21: 1. a 2. c 3. b 4. d 5. c 6. e 7. c 8. a
9. d 10. a 11. b 12. c 13. b 14. c 15. c

Chapter 22: 1. c 2. a 3. d 4. d 5. a 6. b 7. e 8. c
9. a 10. b 11. b 12. d 13. c 14. c 15. e

Chapter 23: 1. b 2. c 3. a 4. d 5. b 6. b 7. e 8. c
9. a 10. a 11. a 12. e 13. d 14. a 15. b 16. e 17. a

Chapter 24: 1. e 2. a 3. c 4. b 5. c 6. a 7. d 8. b
9. b 10. a 11. b 12. a 13. d 14. c 15. e 16. c 17. b
18. a 19. d 20. a

Chapter 25: 1. c 2. a 3. c 4. d 5. b 6. d 7. d 8. e
9. b 10. b 11. e 12. d 13. a 14. e 15. c

Chapter 26: 1. a 2. b 3. e 4. d 5. d 6. a 7. b 8. c
9. a 10. c 11. e 12. c 13. a 14. b 15. d 16. d 17. c 18. a

Chapter 27: 1. d 2. a 3. b 4. c 5. d 6. b 7. a 8. e
9. c 10. d 11. c 12. c 13. d 14. b 15. d

Chapter 28: 1. e 2. b 3. c 4. d 5. a 6. e 7. d 8. d
9. c 10. b 11. d 12. a 13. c 14. a 15. c

Chapter 29: 1. d 2. e 3. d 4. a 5. b 6. d 7. e 8. a
9. b 10. c 11. a 12. c 13. c 14. e 15. b 16. d

Chapter 30: 1. b 2. a 3. c 4. d 5. c 6. a 7. c 8. a
9. b 10. e 11. c 12. d 13. e 14. b 15. d 16. a 17. d
18. b 19. b 20. e

Chapter 31: 1. d 2. b 3. a 4. a 5. b 6. e 7. c 8. b
9. a 10. d 11. c 12. c 13. a 14. d 15. c 16. e

Chapter 32: 1. c 2. a 3. a 4. c 5. d 6. e 7. b 8. c
9. c 10. b 11. e 12. c 13. b 14. a 15. c 16. b 17. d

Chapter 33: 1. c 2. b 3. d 4. d 5. b 6. e 7. c 8. a
9. d 10. d 11. a 12. e 13. c 14. c 15. a 16. c 17. a

Chapter 34: 1. e 2. b 3. d 4. d 5. e 6. b 7. b 8. a
9. c 10. d 11. b 12. e 13. c 14. b 15. a 16. b

Chapter 35: 1. c 2. b 3. b 4. b 5. e 6. a 7. c 8. c
9. d 10. d 11. a 12. d 13. b 14. e 15. c 16. a

Chapter 36: 1. e 2. c 3. d 4. e 5. c 6. a 7. c 8. b
9. c 10. b 11. d 12. c 13. a 14. d 15. a

Chapter 37: 1. c 2. a 3. a 4. b 5. c 6. c 7. e 8. e
9. b 10. d 11. d 12. b 13. e 14. c 15. e 16. d

Chapter 38: 1. b 2. b 3. d 4. b 5. a 6. c 7. b 8. b
9. a 10. e 11. c 12. e 13. a 14. d 15. b 16. d 17. a
18. a 19. e

Chapter 39: 1. d 2. a 3. c 4. c 5. a 6. b 7. d 8. b
9. a 10. e 11. b 12. a 13. c 14. e 15. b 16. d

Chapter 40: 1. c 2. e 3. a 4. b 5. e 6. d 7. e 8. e
9. a 10. c 11. c 12. c 13. c 14. b 15. d 16. e

Chapter 41: 1. b 2. d 3. c 4. b 5. d 6. a 7. a 8. d
9. a 10. c 11. e 12. d 13. b 14. b 15. d 16. a

Chapter 42: 1. e 2. c 3. a 4. d 5. e 6. c 7. b 8. a
9. e 10. d 11. b 12. a 13. c 14. b 15. c 16. b

Chapter 43: 1. b 2. d 3. b 4. a 5. b 6. c 7. a 8. e
9. a 10. c 11. e 12. d 13. a 14. c 15. e 16. d 17. d

Chapter 44: 1. b 2. a 3. d 4. c 5. e 6. c 7. a 8. b
9. c 10. a 11. d 12. e 13. c 14. e 15. d 16. b 17. a

Chapter 45: 1. c 2. a 3. b 4. a 5. c 6. b 7. b 8. a
9. d 10. e 11. b 12. a 13. e 14. c 15. d 16. e

Chapter 46: 1. a 2. a 3. c 4. d 5. d 6. e 7. e 8. e
9. b 10. d 11. a 12. b 13. c 14. c 15. a 16. e

Chapter 47: 1. c 2. b 3. a 4. c 5. d 6. e 7. b 8. e
9. d 10. d 11. c 12. a 13. b 14. d 15. e

Chapter 48: 1. b 2. c 3. a 4. d 5. c 6. a 7. b 8. a
9. d 10. e 11. b 12. d 13. a 14. c 15. e 16. b

Chapter 49: 1. b 2. c 3. a 4. e 5. b 6. d 7. a 8. c
9. e 10. d 11. c 12. a 13. e 14. b 15. e

Chapter 50: 1. b 2. d 3. a 4. a 5. d 6. b 7. e 8. b
9. b 10. c 11. b 12. c 13. c 14. d

Chapter 51: 1. c 2. b 3. a 4. e 5. c 6. e 7. a 8. a
9. b 10. a 11. b 12. a 13. e 14. a 15. b 16. b 17. a 18. e

Chapter 52: 1. b 2. b 3. a 4. e 5. b 6. c 7. b 8. a
9. d 10. e 11. a 12. b 13. b 14. e 15. c

Chapter 53: 1. d 2. b 3. d 4. d 5. e 6. c 7. c 8. d
9. a 10. d 11. e 12. d 13. e 14. d

Chapter 54: 1. c 2. d 3. e 4. d 5. a 6. d 7. e 8. e
9. a 10. b 11. c 12. d 13. e 14. d 15. b 16. a

Chapter 55: 1. c 2. b 3. a 4. d 5. e 6. c 7. b 8. d
9. b 10. a 11. b 12. c 13. a 14. e 15. e

Chapter 56: 1. d 2. e 3. c 4. c 5. c 6. b 7. e 8. c
9. e 10. a 11. d 12. d 13. c 14. a 15. b

Glossary

abiotic factors Elements of the nonliving, physical environment that affect a particular organism. Compare with *biotic factors*.

abscisic acid (ab-sis′ik) A plant hormone involved in dormancy and responses to stress.

abscission (ab-sizh′en) The normal (usually seasonal) fall of leaves or other plant parts, such as fruits or flowers.

abscission layer The area at the base of the petiole where the leaf will break away from the stem. Also known as *abscission zone*.

absorption (ab-sorp′shun) (1) The movement of nutrients and other substances through the wall of the digestive tract and into the blood or lymph. (2) The process by which chlorophyll takes up light for photosynthesis.

absorption spectrum A graph of the amount of light at specific wavelengths that is absorbed as light passes through a substance. Each type of molecule has a characteristic absorption spectrum. Compare with *action spectrum*.

accessory fruit A fruit consisting primarily of tissue other than ovary tissue, e.g., apple, pear. Compare with *aggregate, simple,* and *multiple fruits*.

acclimatization Adjustment to seasonal changes.

acetyl coenzyme A (acetyl CoA) (as′uh-teel) A key intermediate compound in metabolism; consists of a two-carbon acetyl group covalently bonded to coenzyme A.

acetyl group A two-carbon group derived from acetic acid (acetate).

acetylcholine (ah′see-til-koh′leen) A common neurotransmitter released by cholinergic neurons, including motor neurons.

achene (a-keen′) A simple, dry fruit with one seed in which the fruit wall is separate from the seed coat, e.g., sunflower fruit.

acid A substance that is a hydrogen ion (proton) donor; acids unite with bases to form salts. Compare with *base*.

acid precipitation Precipitation that is acidic as a result of both sulfur and nitrogen oxides forming acids when they react with water in the atmosphere.

acidic solution A solution in which the concentration of hydrogen ions [H^+] exceeds the concentration of hydroxide ions [OH^-]. An acidic solution has a pH less than 7. Compare with *basic solution* and *neutral solution*.

acoelomate (a-seel′oh-mate) An animal lacking a body cavity (coelom). Compare with *coelomate* and *pseudocoelomate*.

acquired immune responses See *specific immune responses*.

acquired immunodeficiency syndrome (AIDS) A serious, potentially fatal disease caused by the human immunodeficiency virus (HIV).

acromegaly (ak″roh-meg′ah-lee) A condition characterized by overgrowth of the extremities of the skeleton, fingers, toes, jaws, and nose. It may be produced by excessive secretion of growth hormone by the anterior pituitary gland.

acrosome reaction (ak′roh-sohm) A series of events in which the acrosome, a caplike structure covering the head of a sperm cell, releases proteolytic (protein-digesting) enzymes and undergoes other changes that permit the sperm to penetrate the outer covering of the egg.

actin (ak′tin) The protein of which microfilaments consist. Actin, together with the protein myosin, is responsible for muscle contraction.

actin filaments Thin filaments consisting mainly of the protein actin; actin and myosin filaments make up the myofibrils of muscle fibers.

actinopods (ak-tin′o-podz) Protozoa characterized by axopods that protrude through pores in their shells. See *radiolarians*.

action potential An electrical signal resulting from depolarization of the plasma membrane in a neuron or muscle cell. Compare with *resting potential*.

action spectrum A graph of the effectiveness of light at specific wavelengths in promoting a light-requiring reaction. Compare with *absorption spectrum*.

activation energy (E_A) The kinetic energy required to initiate a chemical reaction.

activator protein A positive regulatory protein that stimulates transcription when bound to DNA. Compare with *repressor protein*.

active immunity Immunity that develops as a result of exposure to antigens; it can occur naturally after recovery from a disease or can be artificially induced by immunization with a vaccine. Compare with *passive immunity*.

active site A specific region of an enzyme (generally near the surface) that accepts one or more substrates and catalyzes a chemical reaction. Compare with *allosteric site*.

active transport Transport of a substance across a membrane that does not rely on the potential energy of a concentration gradient for the substance being transported and therefore requires an additional energy source (often ATP); includes carrier-mediated active transport, endocytosis, and exocytosis. Compare with *diffusion* and *facilitated diffusion*.

adaptation (1) An evolutionary modification that improves an organism's chances of survival and reproductive success. (2) A decline in the response of a receptor subjected to repeated or prolonged stimulation.

adaptive immune responses See *specific immune responses*.

adaptive radiation The evolution of a large number of related species from an unspecialized ancestral organism.

adaptive zone A new ecological opportunity that was not exploited by an ancestral organism; used by evolutionary biologists to explain the ecological paths along which different taxa evolve.

addiction Physical dependence on a drug, generally based on physiological changes that take place in response to the drug; when the drug is withheld, the addict may suffer characteristic withdrawal symptoms.

adenine (ad′eh-neen) A nitrogenous purine base that is a component of nucleic acids and ATP.

adenosine triphosphate (ATP) (a-den′oh-seen) An organic compound containing adenine, ribose, and three phosphate groups; of prime importance for energy transfers in cells.

adhering junction A type of anchoring junction between cells; connects epithelial cells.

adhesion The property of sticking to some other substance. Compare with *cohesion.*

adipose tissue (ad´i-pohs) Tissue in which fat is stored.

adrenal cortex (ah-dree´nul kor´teks) The outer region of each adrenal gland; secretes steroid hormones, including mineralocorticoids and glucocorticoids.

adrenal glands (ah-dree´nul) Paired endocrine glands, one located just superior to each kidney; secrete hormones that help regulate metabolism and help the body cope with stress.

adrenal medulla (ah-dree´nul meh-dull´uh) The inner region of each adrenal gland; secretes epinephrine and norepinephrine.

adrenergic neuron (ad-ren-er´jik) A neuron that releases norepinephrine or epinephrine as a neurotransmitter. Compare with *cholinergic neuron.*

adventitious (ad˝ven-tish´us) Of plant organs, such as roots or buds, that arise in an unusual position on a plant.

aerobe Organism that grows or metabolizes only in the presence of molecular oxygen. Compare with *anaerobe.*

aerobic (air-oh´bik) Growing or metabolizing only in the presence of molecular oxygen. Compare with *anaerobic.*

aerobic respiration See *respiration.*

afferent (af´fer-ent) Leading toward some point of reference. Compare with *efferent.*

afferent neurons Neurons that transmit action potentials from sensory receptors to the brain or spinal cord. Compare with *efferent neurons.*

age structure The number and proportion of people at each age in a population. Age structure diagrams represent the number of males and females at each age, from birth to death, in the population.

aggregate fruit A fruit that develops from a single flower with many separate carpels, e.g., raspberry. Compare with *simple, accessory,* and *multiple fruits.*

aggregated distribution See *clumped dispersion.*

aging Progressive changes in development in an adult organism.

agnathans (ag-na´thanz) Jawless fishes; historical class of vertebrates, including lampreys, hagfishes, and many extinct forms.

AIDS See *acquired immunodeficiency syndrome.*

albinism (al´bih-niz-em) A hereditary inability to form melanin pigment, resulting in light coloration.

albumin (al-bew´min) A class of protein found in most animal tissues; a fraction of plasma proteins.

aldehyde An organic molecule containing a carbonyl group bonded to at least one hydrogen atom. Compare with *ketone.*

aldosterone (al-dos´tur-ohn) A steroid hormone produced by the vertebrate adrenal cortex; stimulates sodium reabsorption. See *mineralocorticoids.*

algae (al´gee) (sing., *alga*) An informal group of unicellular, or simple multicellular, photosynthetic protists that are important producers in aquatic ecosystems.

allantois (a-lan´toe-iss) An extraembryonic membrane of reptiles, birds, and mammals that stores the embryo's nitrogenous wastes; most of the allantois is detached at hatching or birth.

allele frequency The proportion of a specific allele in the population.

alleles (al-leelz´) Genes governing variation of the same character that occupy corresponding positions (loci) on homologous chromosomes; alternative forms of a gene.

allelopathy (uh-leel´uh-path˝ee) An adaptation in which toxic substances secreted by roots or shed leaves inhibit the establishment of competing plants nearby.

allergen A substance that stimulates an allergic reaction.

allergy A hypersensitivity to some substance in the environment, manifested as hay fever, skin rash, asthma, food allergies, etc.

allometric growth Variation in the relative rates of growth for different parts of the body during development.

allopatric speciation (al-oh-pa´trik) Speciation that occurs when one population becomes geographically separated from the rest of the species and subsequently evolves. Compare with *sympatric speciation.*

allopolyploid (al˝oh-pol´ee-ployd) A polyploid whose chromosomes are derived from two species. Compare with *autopolyploid.*

all-or-none law The principle that neurons transmit an impulse in a similar way no matter how weak or strong the stimulus; the neuron either transmits an action potential (all) or does not (none).

allosteric regulators Substances that affect protein function by binding to allosteric sites.

allosteric site (al-oh-steer´ik) A site on an enzyme other than the active site, to which a specific substance binds, thereby changing the shape and activity of the enzyme. Compare with *active site.*

alpha (α) helix A regular, coiled type of secondary structure of a polypeptide chain, maintained by hydrogen bonds. Compare with *beta (β)-pleated sheet.*

alpine tundra An ecosystem located in the higher elevations of mountains, above the tree line and below the snow line. Compare with *tundra.*

alternation of generations A type of life cycle characteristic of plants and a few algae and fungi in which they spend part of their life in a multicellular *n* gametophyte stage and part in a multicellular 2*n* sporophyte stage.

altruistic behavior Behavior in which one individual helps another, seemingly at its own risk or expense.

alveolus (al-vee´o-lus) (pl., *alveoli*) (1) An air sac of the lung through which gas exchange with the blood takes place. (2) Saclike unit of some glands, e.g., mammary glands. (3) One of several flattened vesicles located just inside the plasma membrane in certain protists.

Alzheimer's disease (AD) A progressive, degenerative brain disorder characterized by amyloid plaques and neurofibrillary tangles.

amino acid (uh-mee´no) An organic compound containing an amino group (—NH$_2$) and a carboxyl group (—COOH); may be joined by peptide bonds to form a polypeptide chain.

amino group A weakly basic functional group; abbreviated —NH$_2$.

aminoacyl-tRNA (uh-mee˝no-ace´seel) Molecule consisting of an amino acid covalently linked to a transfer RNA.

aminoacyl-tRNA synthetase One of a family of enzymes, each responsible for covalently linking an amino acid to its specific transfer RNA.

ammonification (uh-moe˝nuh-fah-kay´shun) The conversion of nitrogen-containing organic compounds to ammonia (NH$_3$) by certain soil bacteria (ammonifying bacteria); part of the nitrogen cycle.

amniocentesis (am˝nee-oh-sen-tee´sis) Sampling of the amniotic fluid surrounding a fetus to obtain information about its development and genetic makeup. Compare with *chorionic villus sampling.*

amnion (am´nee-on) In terrestrial vertebrates, an extraembryonic membrane that forms a fluid-filled sac for the protection of the developing embryo.

amniotes Terrestrial vertebrates: reptiles, birds, and mammals; animals whose embryos are enclosed by an amnion.

amoeba (a-mee´ba) (pl., *amoebas*) A unicellular protozoon that moves by means of pseudopodia.

amphibians Members of vertebrate class that includes salamanders, frogs, and caecilians.

amphipathic molecule (am˝fih-pa´thik) A molecule containing both hydrophobic and hydrophilic regions.

ampulla Any small, saclike extension, e.g., the expanded structure at the end of each semicircular canal of the ear.

amylase (am´-uh-laze) Starch-digesting enzyme, e.g., human salivary amylase or pancreatic amylase.

amyloplasts See *leukoplasts*.

anabolic steroids Synthetic androgens that increase muscle mass, physical strength, endurance, and aggressiveness but cause serious side effects; these drugs are often abused.

anabolism (an-ab´oh-lizm) The aspect of metabolism in which simpler substances are combined to form more complex substances, resulting in the storage of energy, the production of new cell materials, and growth. Compare with *catabolism*.

anaerobe Organism that grows or metabolizes only in the absence of molecular oxygen. See *facultative anaerobe* and *obligate anaerobe*. Compare with *aerobe*.

anaerobic (an″air-oh´bik) Growing or metabolizing only in the absence of molecular oxygen. Compare with *aerobic*.

anaerobic respiration See *respiration*.

anaphase (an´uh-faze) Stage of mitosis in which the chromosomes move to opposite poles of the cell; anaphase occurs after metaphase and before telophase.

anaphylaxis (an″uh-fih-lak´sis) An acute allergic reaction following sensitization to a foreign substance or other substance.

ancestral characters See *shared ancestral characters*.

androgen (an´dro-jen) Any substance that has masculinizing properties, such as a sex hormone. See *testosterone*.

androgen-binding protein (ABP) A protein produced by Sertoli cells in the testes; binds and concentrates testosterone.

anemia (uh-nee´mee-uh) A deficiency of hemoglobin or red blood cells.

aneuploidy (an´you-ploy-dee) Any chromosomal aberration in which there are either extra or missing copies of certain chromosomes.

angiosperms (an´jee-oh-spermz″) The traditional name for flowering plants, a very large (more than 300,000 species), diverse phylum of plants that form flowers for sexual reproduction and produce seeds enclosed in fruits; include monocots and eudicots.

angiotensin I (an-jee-o-ten´sin) A polypeptide produced by the action of renin on a plasma protein (angiotensinogen).

angiotensin II A peptide hormone formed by the action of angiotensin-converting enzyme on angiotensin I; stimulates aldosterone secretion by the adrenal cortex.

animal pole The nonyolky, metabolically active pole of a vertebrate or echinoderm egg. Compare with *vegetal pole*.

anion (an´eye-on) A particle with one or more units of negative charge, such as a chloride ion (Cl^-) or hydroxide ion (OH^-). Compare with *cation*.

anisogamy (an″eye-sog´uh-me) Sexual reproduction involving motile gametes of similar form but dissimilar size. Compare with *isogamy* and *oogamy*.

annelid (an´eh-lid) A member of phylum Annelida; segmented worm such as earthworm.

annual plant A plant that completes its entire life cycle in 1 year or less. Compare with *perennial* and *biennial*.

antenna complex The arrangement of chlorophyll, accessory pigments, and pigment-binding proteins into light-gathering units in the thylakoid membranes of photoautotrophic eukaryotes. See *reaction center* and *photosystem*.

antennae (sing., *antenna*) Sensory structures characteristic of some arthropod groups.

anterior Toward the head end of a bilaterally symmetrical animal. Compare with *posterior*.

anther (an´thur) The part of the stamen in flowers that produces microspores and, ultimately, pollen grains.

antheridium (an″thur-id´ee-im) (pl., *antheridia*) In plants, the multicellular male gametangium (sex organ) that produces sperm cells. Compare with *archegonium*.

anthropoid (an´thra-poid) A member of a suborder of primates that includes monkeys, apes, and humans.

antibody (an´tee-bod″ee) A specific protein (immunoglobulin) that recognizes and binds to specific antigens; produced by plasma cells.

antibody-mediated immunity A type of specific immune response in which B cells differentiate into plasma cells and produce antibodies that bind with foreign antigens, leading to the destruction of pathogens. Compare with *cell-mediated immunity*.

anticodon (an″tee-koh´don) A sequence of three nucleotides in transfer RNA that is complementary to, and combines with, the three-nucleotide codon.

antidiuretic hormone (ADH) (an″ty-dy-uh-ret´ik) A hormone secreted by the posterior lobe of the pituitary that controls the rate of water reabsorption by the kidney.

antigen (an´tih-jen) Any molecule, usually a protein or large carbohydrate, that is specifically recognized as foreign by cells of the immune system.

antigen–antibody complex The combination of antigen and antibody molecules.

antigen-presenting cell (APC) A cell that displays foreign antigens as well as its own surface proteins. Dendritic cells, macrophages, and B cells are APCs.

antimicrobial peptides Soluble molecules that destroy pathogens.

anti-oncogene See *tumor suppressor gene*.

antioxidants Certain enzymes (e.g., catalase and peroxidase), vitamins, and other substances that destroy free radicals and other reactive molecules. Compare with *oxidants*.

antiparallel Said of a double-stranded nucleic acid in which the 5′ to 3′ direction of the sugar–phosphate backbone of one strand is reversed in the other strand.

anus (ay´nus) The distal end and outlet of the digestive tract.

aorta (ay-or´tah) The largest and main systemic artery of the vertebrate body; arises from the left ventricle and branches to distribute blood to all parts of the body except the lungs.

aphotic region (ay-fote´ik) The lower layer of the ocean (deeper than 100 m or so) where light does not penetrate.

apical dominance (ape´ih-kl) The inhibition of lateral buds by a shoot tip.

apical meristem (mehr´ih-stem) An area of dividing tissue, located at the tip of a shoot or root, that gives rise to primary tissues; apical meristems cause an increase in the length of the plant body. Compare with *lateral meristems*.

apicomplexans A group of parasitic protozoa that lack structures for locomotion and that produce sporozoites as infective agents; malaria is caused by an apicomplexan. Also called *sporozoa*.

apoenzyme (ap″oh-en´zime) Protein portion of an enzyme; requires the presence of a specific coenzyme to become a complete functional enzyme.

apomixis (ap″uh-mix´us) A type of reproduction in which fruits and seeds are formed asexually.

apoplast A continuum consisting of the interconnected, porous plant cell walls, along which water moves freely. Compare with *symplast*.

apoptosis (ap-uh-toe´sis) Programmed cell death; apoptosis is a normal part of an organism's development and maintenance. Compare with *necrosis*.

aposematic coloration The conspicuous coloring of a poisonous or distasteful organism that enables potential predators to easily see and recognize it. Also called *warning coloration*. Compare with *cryptic coloration*.

aquaporin One of a family of transport proteins located in the plasma membrane that facilitate the rapid movement of water molecules into or out of cells.

arachnids (ah-rack′nidz) Eight-legged arthropods, such as spiders, scorpions, ticks, and mites.

arachnoid The middle of the three meningeal layers that cover and protect the brain and spinal cord; see *pia mater* and *dura mater*.

archaea (ar′key-ah) Prokaryotic organisms with a number of features, such as the absence of peptidoglycan in their cell walls, that set them apart from the bacteria. Archaea is the name of one of the two prokaryotic domains. Compare with *bacteria*.

Archaean eon The period of Earth's history from its beginnings approximately 4.6 billion up to 2.5 billion years ago; life originated during the Archaean.

archaic *Homo sapiens* Regionally diverse descendants of *H. erectus* or *H. ergaster* that lived in Africa, Asia, and Europe from about 400,000 to 200,000 years ago.

archegonium (ar′ke-go′nee-um) (pl., *archegonia*) In plants, the multicellular female gametangium (sex organ) that contains an egg. Compare with *antheridium*.

archenteron (ark-en′ter-on) The central cavity of the gastrula stage of embryonic development that is lined with endoderm; primitive digestive system.

arctic tundra See *tundra*.

arterial pulse See *pulse, arterial*.

arteriole (ar-teer′ee-ole) A very small artery. Vasoconstriction and vasodilation of arterioles help regulate blood pressure.

artery A thick-walled blood vessel that carries blood away from a heart chamber and toward the body organs. Compare with *vein*.

arthropod (ar′throh-pod) An invertebrate that belongs to phylum Arthropoda; characterized by a hard exoskeleton; a segmented body; and paired, jointed appendages.

artificial insemination The impregnation of a female by artificially introducing sperm from a male.

artificial selection The selection by humans of traits that are desirable in plants or animals and breeding only those individuals that have the desired traits.

ascocarp (ass′koh-karp) The fruiting body of an ascomycete.

ascomycete (ass″koh-my′seat) Member of a phylum of fungi characterized by the production of nonmotile asexual conidia and sexual ascospores.

ascospore (ass′koh-spor) One of a set of sexual spores, usually eight, contained in a special spore case (an ascus) of an ascomycete.

ascus (ass′kus) A saclike spore case in ascomycetes that contains sexual spores called *ascospores*.

asexual reproduction Reproduction in which there is no fusion of gametes and in which the genetic makeup of parent and of offspring is usually identical. Compare with *sexual reproduction*.

assimilation (of nitrogen) The conversion of inorganic nitrogen (nitrate, NO_3^-, or ammonia, NH_3) to the organic molecules of living things; part of the nitrogen cycle.

association areas Areas of the brain that link sensory and motor areas; responsible for thought, learning, memory, language abilities, judgment, and personality.

association neuron See *interneuron*.

assortative mating Sexual reproduction in which individuals pair nonrandomly, i.e., select mates on the basis of phenotype.

asters Clusters of microtubules radiating out from the poles in dividing cells that have centrioles.

astrocyte A type of glial cell; some are phagocytic; others regulate the composition of the extracellular fluid in the central nervous system.

atherosclerosis (ath″ur-oh-skle-row′sis) A progressive disease in which lipid deposits accumulate in the inner lining of arteries, leading eventually to impaired circulation and heart disease.

atom The smallest quantity of an element that retains the chemical properties of that element.

atomic mass The total number of protons and neutrons in an atom; expressed in atomic mass units or daltons.

atomic mass unit (amu) The approximate mass of a proton or neutron; also called a *dalton*.

atomic number The number of protons in the atomic nucleus of an atom, which uniquely identifies the element to which the atom corresponds.

ATP See *adenosine triphosphate*.

ATP synthase Large enzyme complex that catalyzes the formation of ATP from ADP and inorganic phosphate by chemiosmosis; located in the inner mitochondrial membrane, the thylakoid membrane of chloroplasts, and the plasma membrane of bacteria.

atrial natriuretic peptide (ANP) A hormone released by the atrium of the heart; helps regulate sodium excretion and lowers blood pressure.

atrioventricular (AV) node (ay″tree-oh-ven-trik′you-lur) Mass of specialized cardiac tissue that receives an impulse from the sinoatrial node (pacemaker) and conducts it to the ventricles.

atrioventricular (AV) valve (of the heart) A valve between each atrium and its ventricle that prevents backflow of blood. The right AV valve is the tricuspid valve; the left AV valve is the mitral valve.

atrium (of the heart) (ay′tree-um) A heart chamber that receives blood from the veins.

australopithecines Early hominids that lived between about 4 mya and 1 mya, based on fossil evidence. Includes several species in two genera, *Ardipithecus* and *Australopithecus*.

autocrine regulation A type of regulation in which a signaling molecule (e.g., a hormone) is secreted into interstitial fluid and then acts on the cells that produce it. Compare with *paracrine regulation*.

autoimmune disease (aw″toh-ih-mune′) A disease in which the body produces antibodies against its own cells or tissues. Also called *autoimmunity*.

autonomic nervous system (aw-tuh-nom′ik) The portion of the peripheral nervous system that controls the visceral functions of the body, e.g., regulates smooth muscle, cardiac muscle, and glands. Its divisions are the sympathetic and parasympathetic nervous systems. Compare with *somatic nervous system*.

autopolyploid A polyploid whose chromosomes are derived from a single species. Compare with *allopolyploid*.

autoradiography Method for detecting radioactive decay; radiation causes the appearance of dark silver grains in special X-ray film.

autosome (aw′toh-sohm) A chromosome other than the sex (X and Y) chromosomes.

autotroph (aw′toh-trof) An organism that synthesizes complex organic compounds from simple inorganic raw materials; also called *producer* or *primary producer*. Compare with *heterotroph*. See *chemoautotroph* and *photoautotroph*.

auxin (awk′sin) A plant hormone involved in various aspects of growth and development, such as stem elongation, apical dominance, and root formation on cuttings, e.g., indole acetic acid (IAA).

avirulence Properties that render an infectious agent nonlethal, i.e., unable to cause disease in its host. Compare with *virulence.*

Avogadro's number The number of units (6.02×10^{23}) present in one mole of any substance.

axillary bud A bud in the axil of a leaf. Compare with *terminal bud.*

axon (aks'on) The long extension of the neuron that transmits nerve impulses away from the cell body. Compare with *dendrite.*

axopods (aks'o-podz) Long, filamentous, cytoplasmic projections characteristic of actinopods.

B cell (B lymphocyte) The type of white blood cell responsible for antibody-mediated immunity. When stimulated, B cells differentiate to become plasma cells that produce antibodies. Compare with *T cell.*

bacillus (bah-sill'us) (pl., *bacilli*) A rod-shaped bacterium. Compare with *coccus, spirillum, vibrio,* and *spirochete.*

background extinction The continuous, low-level extinction of species that has occurred throughout much of the history of life. Compare with *mass extinction.*

bacteria (bak-teer'ee-ah) Prokaryotic organisms that have peptidoglycan in their cell walls; most are decomposers, but some are parasites and others are autotrophs. Bacteria is the name of one of the two prokaryotic domains. Compare with *archaea.*

bacterial artificial chromosome (BAC) Genetically engineered segments of chromosomal DNA with the ability to carry large segments of foreign DNA.

bacteriophage (bak-teer'ee-oh-fayj) A virus that infects a bacterium (literally, "bacteria eater"). Also called *phage.*

balanced polymorphism (pol"ee-mor'fizm) The presence in a population of two or more genetic variants that are maintained in a stable frequency over several generations.

bark The outermost covering over woody stems and roots; consists of all plant tissues located outside the vascular cambium.

baroreceptors (bare"oh-ree-sep'torz) Receptors within certain blood vessels that are stimulated by changes in blood pressure.

Barr body A condensed and inactivated X chromosome appearing as a distinctive dense spot in the interphase nucleus of certain cells of female mammals.

basal angiosperms (bay'sl) Three clades of angiosperms that are thought to be ancestral to all other flowering plants. Compare with *core angiosperms.*

basal body Structure involved in the organization and anchorage of a cilium or flagellum. Structurally similar to a centriole; each is in the form of a cylinder composed of nine triplets of microtubules (9×3 structure).

basal metabolic rate (BMR) The amount of energy expended by the body at resting conditions, when no food is being digested and no voluntary muscular work is being performed.

base (1) A substance that is a hydrogen ion (proton) acceptor; bases unite with acids to form salts. Compare with *acid.* (2) A nitrogenous base in a nucleotide or nucleic acid. See *purines* and *pyrimidines.*

basement membrane The thin, noncell layer of an epithelial membrane that attaches to the underlying tissues; consists of tiny fibers and polysaccharides produced by the epithelial cells.

base-substitution mutation A change in one base pair in DNA. See *missense mutation* and *nonsense mutation.*

basic solution A solution in which the concentration of hydroxide ions [OH⁻] exceeds the concentration of hydrogen ions [H⁺]. A basic solution has pH greater than 7. Compare with *acidic solution* and *neutral solution.*

basidiocarp (ba-sid'ee-o-karp) The fruiting body of a basidiomycete, e.g., a mushroom.

basidiomycete (ba-sid"ee-o-my'seet) Member of a phylum of fungi characterized by the production of sexual basidiospores.

basidiospore (ba-sid'ee-o-spor) One of a set of sexual spores, usually four, borne on a basidium of a basidiomycete.

basidium (ba-sid'ee-um) The clublike spore-producing organ of basidiomycetes that bears sexual spores called *basidiospores.*

basilar membrane The multicellular tissue in the inner ear that separates the cochlear duct from the tympanic canal; the sensory cells of the organ of Corti rest on this membrane.

Batesian mimicry (bate'see-un mim'ih-kree) The resemblance of a harmless or palatable species to one that is dangerous, unpalatable, or poisonous so that predators are more likely to avoid them. Compare with *Müllerian mimicry.*

behavioral ecology The scientific study of behavior in natural environments from the evolutionary perspective.

behavioral isolation A prezygotic reproductive isolating mechanism in which reproduction between similar species is prevented because each group exhibits its own characteristic courtship behavior; also called *sexual isolation.*

bellwether species An organism that provides an early warning of environmental damage. Examples include lichens, which are very sensitive to air pollution, and amphibians, which are sensitive to a wide variety of environmental stressors.

benthos (ben'thos) Bottom-dwelling sea organisms that fix themselves to one spot, burrow into the sediment, or simply walk about on the ocean floor.

berry A simple, fleshy fruit in which the fruit wall is soft throughout, e.g., tomato, banana, grape.

beta (β) oxidation Process by which fatty acids are converted to acetyl CoA before entry into the citric acid cycle.

beta (β)-pleated sheet A regular, folded, sheetlike type of protein secondary structure, resulting from hydrogen bonding between two different polypeptide chains or two regions of the same polypeptide chain. Compare with *alpha (α) helix.*

biennial plant (by-en'ee-ul) A plant that takes 2 years to complete its life cycle. Compare with *annual* and *perennial.*

bilateral symmetry A body shape with right and left halves that are approximately mirror images of each other. Compare with *radial symmetry.*

bile The fluid secreted by the liver; emulsifies fats.

binary fission (by'nare-ee fish'un) Equal division of a prokaryotic cell into two; a type of asexual reproduction.

binomial system of nomenclature (by-nome'ee-ul) System of naming a species by the combination of the genus name and a specific epithet.

bioaccumulation The buildup of a persistent toxic substance, such as certain pesticides, in an organism's body.

biodiversity See *biological diversity.*

biofilm An irregular layer of microorganisms embedded in the slime they secrete and concentrated at a solid or liquid surface.

biogenic amines A class of neurotransmitters that includes norepinephrine, serotonin, and dopamine.

biogeochemical cycle (bye"o-jee"o-kem'ee-kl) Process by which matter cycles from the living world to the nonliving, physical environment and back again, e.g., the carbon cycle, the nitrogen cycle, and the phosphorus cycle.

biogeography The study of the past and present geographic distributions of organisms.

bioinformatics The storage, retrieval, and comparison of biological information, particularly DNA or protein sequences within a given species and among different species.

biological clocks Mechanisms by which activities of organisms are adapted to regularly recurring changes in the environment. See *circadian rhythm*.

biological diversity The variety of living organisms considered at three levels: genetic diversity, species diversity, and ecosystem diversity. Also called *biodiversity*.

biological magnification The increased concentration of toxic chemicals, such as PCBs, heavy metals, and certain pesticides, in the tissues of organisms at higher trophic levels in food webs.

biological species concept See *species*.

biomass (bye′o-mas) A quantitative estimate of the total mass, or amount, of living material in a particular ecosystem.

biome (by′ohm) A large, relatively distinct terrestrial region characterized by a similar climate, soil, plants, and animals, regardless of where it occurs on Earth.

bioremediation A method to clean up a hazardous waste site that uses microorganisms to break down toxic pollutants, or plants to selectively accumulate toxins.

biosphere All of Earth's living organisms, collectively.

biotic factors Elements of the living world that affect a particular organism, i.e., its relationships with other organisms. Compare with *abiotic factors*.

biotic potential See *intrinsic rate of increase*.

bipedal Walking on two feet.

bipolar cell A type of neuron in the retina of the eye; receives input from the photoreceptors (rods and cones) and synapses on ganglion cells.

biramous appendages Appendages with two jointed branches at their ends; characteristic of crustaceans.

bivalent (by-vale′ent or biv′ah-lent) See *tetrad*.

blade (1) The thin, expanded part of a leaf. (2) The flat, leaflike structure of certain multicellular algae.

blastocoel (blas′toh-seel) The fluid-filled cavity of a blastula.

blastocyst The mammalian blastula. See *blastula*.

blastodisc A small disc of cytoplasm at the animal pole of a reptile or bird egg; cleavage is restricted to the blastodisc (meroblastic cleavage).

blastomere A cell of an early embryo.

blastopore (blas′toh-pore) The primitive opening into the body cavity of an early embryo that may become the mouth (in protostomes) or anus (in deuterostomes) of the adult organism.

blastula (blas′tew-lah) In animal development, a hollow ball of cells produced by cleavage of a fertilized ovum. In mammalian development, known as a *blastocyst*.

blood A fluid, circulating connective tissue that transports nutrients and other materials through the bodies of many types of animals.

blood pressure The force exerted by blood against the inner walls of the blood vessels.

bloom The sporadic occurrence of huge numbers of algae in freshwater and marine ecosystems.

body mass index (BMI) An index of weight in relation to height; calculated by dividing the square of the weight (square kilograms) by height (meters).

Bohr effect Increased oxyhemoglobin dissociation due to lowered pH; occurs as carbon dioxide concentration increases.

bolting The production of a tall flower stalk by a plant that grows vegetatively as a rosette (growth habit with a short stem and a circular cluster of leaves).

bond energy The energy required to break a particular chemical bond.

bone tissue Principal vertebrate skeletal tissue; a type of connective tissue.

boreal forest (bor′ee-uhl) The northern coniferous forest biome found primarily in Canada, northern Europe, and Siberia; also called *taiga*.

bottleneck A sudden decrease in a population size caused by adverse environmental factors; may result in genetic drift; also called *genetic bottleneck* or *population bottleneck*.

bottom-up processes Control of ecosystem function by nutrient cycles and other parts of the abiotic environment. Compare with *top-down processes*.

Bowman's capsule A double-walled sac of cells that surrounds the glomerulus of each nephron.

brachiopods (bray′kee-oh-podz) The phylum of solitary marine invertebrates having a pair of shells and, internally, a pair of coiled arms with ciliated tentacles; one of the lophophorate phyla.

brain A concentration of nervous tissue that controls neural function; in vertebrates, the anterior, enlarged portion of the central nervous system.

brain stem The part of the vertebrate brain that includes the medulla, pons, and midbrain.

branchial Pertaining to the gills or gill region.

brassinosteroid (BR) One of a group of steroids that function as plant hormones and are involved in several aspects of growth and development.

bronchiole (bronk′ee-ole) Air duct in the lung that branches from a bronchus; divides to form air sacs (alveoli).

bronchus (bronk′us) (pl., *bronchi*) One branch of the trachea and its immediate branches within the lung.

brown alga One of a phylum of predominantly marine algae that are multicellular and contain the pigments chlorophyll *a* and *c*, and carotenoids, including fucoxanthin.

bryophytes (bry′oh-fites) Nonvascular plants including mosses, liverworts, and hornworts.

bryozoans Animals belonging to phylum Bryozoa, one of the three lophophorate phyla; form sessile colonies by asexual budding.

bud An undeveloped shoot that develops into flowers, stems, or leaves. Buds are enclosed in bud scales.

bud scale A modified leaf that covers and protects a dormant bud.

bud scale scar Scar on a twig left when a bud scale abscises from the terminal bud.

budding Asexual reproduction in which a small part of the parent's body separates from the rest and develops into a new individual; characteristic of yeasts and certain other organisms.

buffer A substance in a solution that tends to lessen the change in hydrogen ion concentration (pH) that otherwise would be produced by adding an acid or base.

bulb A globose, fleshy, underground bud that consists of a short stem with fleshy leaves, e.g., onion.

bundle scar Marks on a leaf scar left when vascular bundles of the petiole break during leaf abscission.

bundle sheath cells Tightly packed cells that form a sheath around the veins of a leaf.

bundle sheath extension Support cells that extend from the bundle sheath of a leaf vein toward the upper and/or lower epidermis.

buttress root A bracelike root at the base of certain trees that provides upright support.

C₃ plant Plant that carries out carbon fixation solely by the Calvin cycle. Compare with *C₄ plant* and *CAM plant*.

C₄ plant Plant that fixes carbon initially by a pathway, in which the reaction of CO_2 with phosphoenolpyruvate is catalyzed by PEP carboxylase in leaf mesophyll cells; the products are transferred to the bundle sheath cells, where the Calvin cycle takes place. Compare with *C₃ plant* and *CAM plant*.

calcitonin (kal-sih-toh′nin) A hormone secreted by the thyroid gland that rapidly lowers the calcium content in the blood.

callus (kal′us) Undifferentiated tissue formed on an explant (excised tissue or organ) in plant tissue culture.

calmodulin A calcium-binding protein; when bound, it alters the activity of certain enzymes or transport proteins.

calorie The amount of heat energy required to raise the temperature of 1g of water 1°C; equivalent to 4.184 joules. Compare with *kilocalorie.*

Calvin cycle Cyclic series of reactions in the chloroplast stroma in photosynthesis; fixes carbon dioxide and produces carbohydrate. See C_3 *plant.*

calyx (kay′liks) The collective term for the sepals of a flower.

CAM plant Plant that carries out crassulacean acid metabolism; carbon is initially fixed into organic acids at night in the reaction of CO_2 and phosphoenolpyruvate, catalyzed by PEP carboxylase; during the day the acids break down to yield CO_2, which enters the Calvin cycle. Compare with C_3 *plant* and C_4 *plant.*

cambium See *lateral meristems.*

Cambrian Radiation A span of 40 million years, beginning about 542 mya, during which many new animal groups appeared in the fossil record.

cAMP See *cyclic AMP.*

cancer cells See *malignant cells.*

CAP See *catabolite activator protein.*

capillaries (kap′i-lare-eez) Microscopic blood vessels in the tissues that permit exchange of materials between cells and blood.

capillary action The ability of water to move in small-diameter tubes as a consequence of its cohesive and adhesive properties.

capping See *mRNA cap.*

capsid Protein coat surrounding the nucleic acid of a virus.

capsule (1) The portion of the moss sporophyte that contains spores. (2) A simple, dry, dehiscent fruit that develops from two or more fused carpels and opens along many sutures or pores to release seeds. (3) A gelatinous coat that surrounds some bacteria.

carbohydrate Compound containing carbon, hydrogen, and oxygen, in the approximate ratio of C:2H:O, e.g., sugars, starch, and cellulose.

carbon cycle The worldwide circulation of carbon from the abiotic environment into living things and back into the abiotic environment.

carbon fixation reactions Reduction reactions of photosynthesis in which carbon from carbon dioxide becomes incorporated into organic molecules, leading to the production of carbohydrate. See *Calvin cycle.*

carbonyl group A polar functional group consisting of a carbon attached to an oxygen by a double bond; found in aldehydes and ketones.

carboxyl group A weakly acidic functional group; abbreviated —COOH.

carcinogen (kar-sin′oh-jen) An agent that causes cancer or accelerates its development.

cardiac cycle One complete heartbeat.

cardiac muscle Involuntary, striated type of muscle found in the vertebrate heart. Compare with *smooth muscle* and *skeletal muscle.*

cardiac output The volume of blood pumped by the left ventricle into the aorta in 1 minute.

cardiovascular disease Disease of the heart or blood vessels; the leading cause of death in most industrial societies.

carnivore (kar′ni-vor) An animal that feeds on other animals; flesh eater; also called *secondary* or *tertiary consumer.* Secondary consumers eat primary consumers (herbivores), whereas tertiary consumers eat secondary consumers.

carotenoids (ka-rot′n-oidz) A group of yellow to orange plant pigments synthesized from isoprene subunits; include carotenes and xanthophylls.

carpel (kar′pul) The female reproductive unit of a flower; carpels bear ovules. Compare with *pistil.*

carrier-mediated active transport Transport across a membrane of a substance from a region of low concentration to a region of high concentration; requires both a transport protein with a binding site for the specific substance and an energy source (often ATP).

carrier-mediated transport Any form of transport across a membrane that uses a membrane-bound transport protein with a binding site for a specific substance; includes both facilitated diffusion and carrier-mediated active transport.

carrying capacity The largest population that a particular habitat can support and sustain for an indefinite period, assuming there are no changes in the environment.

cartilage A flexible skeletal tissue of vertebrates; a type of connective tissue.

caryopsis See *grain.*

Casparian strip (kas-pare′ee-un) A band of waterproof material around the radial and transverse walls of endodermal root cells.

caspase Any of a group of proteolytic enzymes that are active in the early stages of apoptosis.

catabolism The aspect of metabolism in which complex substances are broken down to form simpler substances; catabolic reactions are particularly important in releasing chemical energy stored by the cell. Compare with *anabolism.*

catabolite activator protein (CAP) A positively acting regulator that becomes active when bound to cAMP; active CAP stimulates transcription of the *lac* operon and other operons that code for enzymes used in catabolic pathways.

catalyst (kat′ah-list) A substance that increases the speed at which a chemical reaction occurs without being used up in the reaction. Enzymes are biological catalysts.

catecholamine (cat″eh-kole′-ah-meen) A class of compounds including dopamine, epinephrine, and norepinephrine; these compounds serve as neurotransmitters and hormones.

cation A particle with one or more units of positive charge, such as a hydrogen ion (H^+) or calcium ion (Ca^{2+}). Compare with *anion.*

CD4 T cell See *T helper cell.*

CD8 T cell See *T cytotoxic cell.*

cDNA library A collection of recombinant plasmids that contain complementary DNA (cDNA) copies of mRNA templates. The cDNA, which lacks introns, is synthesized by reverse transcriptase. Compare with *genomic DNA library.*

cell The basic structural and functional unit of life, which consists of living material enclosed by a membrane.

cell cycle Cyclic series of events in the life of a dividing eukaryotic cell; consists of mitosis, cytokinesis, and the stages of interphase.

cell-cycle control system Regulatory molecules that control key events in the cell cycle; common to all eukaryotes.

cell determination See *determination.*

cell differentiation See *differentiation.*

cell fractionation The technique used to separate the components of cells by subjecting them to centrifugal force. See *differential centrifugation* and *density gradient centrifugation.*

cell-mediated immunity A type of specific immune response carried out by T cells. Compare with *antibody-mediated immunity.*

cell plate The structure that forms during cytokinesis in plants, separating the two daughter cells produced by mitosis.

cell signaling Mechanisms of communication between cells. Cells signal one another with secreted signaling molecules, or a signaling molecule on one cell combines with a receptor on another cell. See *signal transduction*.

cell theory The theory that the cell is the basic unit of life, of which all living things are composed, and that all cells are derived from pre-existing cells.

cell wall The structure outside the plasma membrane of certain cells; may contain cellulose (plant cells), chitin (most fungal cells), peptidoglycan and/or lipopolysaccharide (most bacterial cells), or other material.

cellular respiration See *respiration.*

cellular slime mold A phylum of funguslike protists whose feeding stage consists of unicellular, amoeboid organisms that aggregate to form a pseudoplasmodium during reproduction.

cellulose (sel′yoo-lohs) A structural polysaccharide consisting of beta glucose subunits; the main constituent of plant primary cell walls.

Cenozoic era A geologic era that began about 66 million years ago and extends to the present.

center of origin The geographic area where a given species originated.

central nervous system (CNS) In vertebrates, the brain and spinal cord. Compare with *peripheral nervous system.*

centrifuge A device used to separate cells or their components by subjecting them to centrifugal force.

centriole (sen′tree-ohl) One of a pair of small, cylindrical organelles lying at right angles to each other near the nucleus in the cytoplasm of animal cells and certain protist and plant cells; each centriole is in the form of a cylinder composed of nine triplets of microtubules (9 × 3 structure).

centromere (sen′tro-meer) A specialized constricted region of a chromatid; contains the kinetochore. In cells at prophase and metaphase, sister chromatids are joined in the vicinity of their centromeres.

centrosome (sen′tro-sowm) An organelle in animal cells that is the main microtubule-organizing center; typically contains a pair of centrioles and is important in cell division.

cephalization The evolution of a head; the concentration of nervous tissue and sense organs at the front end of the animal.

cephalochordates Members of the chordate subphylum that includes the lancelets.

cerebellum (ser-eh-bel′um) A convoluted subdivision of the vertebrate brain concerned with coordination of muscular movements, muscle tone, and balance.

cerebral cortex (ser-ee′brul kor′teks) The outer layer of the cerebrum composed of gray matter and consisting mainly of nerve cell bodies.

cerebrospinal fluid (CSF) The fluid that bathes the central nervous system of vertebrates.

cerebrum (ser-ee′brum) A large, convoluted subdivision of the vertebrate brain; in humans, it functions as the center for learning, voluntary movement, and interpretation of sensation.

chaos The tendency of a simple system to exhibit complex, erratic dynamics; used by some ecologists to model the state of flux displayed by some populations.

chaparral (shap″uh-ral′) A biome with a Mediterranean climate (mild, moist winters and hot, dry summers). Chaparral vegetation is characterized by drought-resistant, small-leaved evergreen shrubs and small trees.

chaperones See *molecular chaperones.*

character displacement The tendency for two similar species to diverge (become more different) in areas where their ranges overlap; reduces interspecific competition.

Chargaff's rules A relationship in DNA molecules based on nucleotide composition data; the number of adenines equals the number of thymines, and the number of guanines equals the number of cytosines.

chelicerae (keh-lis′er-ee) The first pair of appendages in certain arthropods; clawlike appendages located immediately anterior to the mouth and used to manipulate food into the mouth.

chemical bond A force of attraction between atoms in a compound. See *covalent bond, hydrogen bond,* and *ionic bond.*

chemical compound Two or more elements combined in a fixed ratio.

chemical evolution The origin of life from nonliving matter.

chemical formula A representation of the composition of a compound; the elements are indicated by chemical symbols with subscripts to indicate their ratios. See *molecular formula, structural formula,* and *simplest formula.*

chemical symbol The abbreviation for an element; usually the first letter (or first and second letters) of the English or Latin name.

chemiosmosis Process by which phosphorylation of ADP to form ATP is coupled to the transfer of electrons down an electron transport chain; the electron transport chain powers proton pumps that produce a proton gradient across the membrane; ATP is formed as protons diffuse through transmembrane channels in ATP synthase.

chemoautotroph (kee″moh-aw′toh-trof) Organism that obtains energy from inorganic compounds and synthesizes organic compounds from inorganic raw materials; includes some bacteria. Compare with *photoautotroph, photoheterotroph,* and *chemoheterotroph.*

chemoheterotroph (kee″moh-het′ur-oh-trof) Organism that uses organic compounds as a source of energy and carbon; includes animals, fungi, and many bacteria. Compare with *photoautotroph, photoheterotroph,* and *chemoautotroph.*

chemoreceptor (kee″moh-ree-sep′tor) A sensory receptor that responds to chemical stimuli.

chemotroph (kee′moh-trof) Organism that uses organic compounds or inorganic substances, such as iron, nitrate, ammonia, or sulfur, as sources of energy. Compare with *phototroph.* See *chemoautotroph* and *chemoheterotroph.*

chiasma (ky-az′muh) (pl., *chiasmata*) An X-shaped site in a tetrad (bivalent) usually marking the location where homologous (nonsister) chromatids previously crossed over.

chimera (ky-meer′uh) An organism consisting of two or more kinds of genetically dissimilar cells.

chitin (ky′tin) A nitrogen-containing structural polysaccharide that forms the exoskeleton of insects and the cell walls of many fungi.

chlorophyll (klor′oh-fil) A group of light-trapping green pigments found in most photosynthetic organisms.

chlorophyll-binding proteins About 15 different proteins associated with chlorophyll molecules in the thylakoid membrane.

chloroplasts (klor′oh-plastz) Membranous organelles that are the sites of photosynthesis in eukaryotes; occur in some plant and algal cells.

cholinergic neuron (kohl″in-air′jik) A neuron that releases acetylcholine as a neurotransmitter. Compare with *adrenergic neuron.*

chondrichthyes (kon-drik′-thees) The class of cartilaginous fishes that includes the sharks, rays, and skates.

chondrocytes Cartilage cells.

chordates (kor′dates) Deuterostome animals that, at some time in their lives, have a cartilaginous, dorsal skeletal structure called a *notochord;* a dorsal, tubular, nerve cord; pharyngeal gill grooves; and a postanal tail.

chorion (kor′ee-on) An extraembryonic membrane in reptiles, birds, and mammals that forms an outer cover around the embryo, and in mammals contributes to the formation of the placenta.

chorionic villus sampling (CVS) (kor″ee-on′ik) Study of extra-embryonic cells that are genetically identical to the cells of an embryo, making it possible to assess its genetic makeup. Compare with *amniocentesis.*

choroid layer A layer of cells filled with black pigment that absorbs light and prevents reflected light from blurring the image that falls on the retina; the layer of the eyeball outside the retina.

chromatid (kroh′mah-tid) One of the two identical halves of a duplicated chromosome; the two chromatids that make up a chromosome are referred to as *sister chromatids.*

chromatin (kro′mah-tin) The complex of DNA and protein that makes up eukaryotic chromosomes.

chromoplasts Pigment-containing plastids; usually found in flowers and fruits.

chromosome theory of inheritance A basic principle in biology that states that inheritance can be explained by assuming that genes are linearly arranged in specific locations along the chromosomes.

chromosomes Structures in the cell nucleus that consist of chromatin and contain the genes. The chromosomes become visible under the microscope as distinct structures during cell division.

chylomicrons (kie-low-my′kronz) Protein-covered fat droplets produced in the intestinal cells; they enter the lymphatic system and are transported to the blood.

chytrid See *chytridiomycete.*

chytridiomycete (ki-trid″ee-o-my′seat) A member of a phylum of fungi characterized by the production of flagellate cells at some stage in their life history. Also called *chytrid.*

ciliate (sil′e-ate) A unicellular protozoon covered by many short cilia.

cilium (sil′ee-um) (pl., *cilia*) One of many short, hairlike structures that project from the surface of some eukaryotic cells and are used for locomotion or movement of materials across the cell surface.

circadian rhythm (sir-kay′dee-un) An internal rhythm that approximates the 24-hour day. See *biological clocks.*

circulatory system The body system that functions in internal transport and protects the body from disease.

cisternae (sing., *cisterna*) Stacks of flattened membranous sacs that make up the Golgi complex.

citrate (citric acid) A six-carbon organic acid.

citric acid cycle Series of chemical reactions in aerobic respiration in which acetyl coenzyme A is completely degraded to carbon dioxide and water with the release of metabolic energy that is used to produce ATP; also known as the *Krebs cycle* and the *tricarboxylic acid (TCA) cycle.*

clade A group of organisms containing a common ancestor and all its descendants; a monophyletic group.

cladistics An approach to classification based on recency of common ancestry rather than degree of structural similarity. Also called *phylogenetic systematics.* Compare with *phenetics* and *evolutionary systematics.*

cladogram A branching diagram that illustrates taxonomic relationships based on the principles of cladistics.

class A taxonomic category made up of related orders.

classical conditioning A type of learning in which an association is formed between some normal response to a stimulus and a new stimulus, after which the new stimulus elicits the response.

cleavage Series of mitotic cell divisions, without growth, that converts the zygote to a multicellular blastula.

cleavage furrow A constricted region of the cytoplasm that forms and progressively deepens during cytokinesis of animal cells, thereby separating the two daughter cells.

cline Gradual change in phenotype and genotype frequencies among contiguous populations that is the result of an environmental gradient.

clitoris (klit′o-ris) A small, erectile structure at the anterior part of the vulva in female mammals; homologous to the male penis.

cloaca (klow-a′ka) An exit chamber in some animals that receives digestive wastes and urine; may also serve as an exit for gametes.

clonal selection Lymphocyte activation in which a specific antigen causes activation, cell division, and differentiation only in cells that express receptors with which the antigen binds.

clone (1) A population of cells descended by mitotic division from a single ancestral cell. (2) A population of genetically identical organisms asexually propagated from a single individual. Also see *DNA cloning.*

cloning The process of forming a clone.

closed circulatory system A type of circulatory system in which the blood flows through a continuous circuit of blood vessels; characteristic of annelids, cephalopods, and vertebrates. Compare with *open circulatory system.*

closed system An entity that does not exchange energy with its surroundings. Compare with *open system.*

club mosses A phylum of seedless vascular plants with a life cycle similar to that of ferns.

clumped dispersion The spatial distribution pattern of a population in which individuals are more concentrated in specific parts of the habitat. Also called *aggregated distribution* and *patchiness.* Compare with *random dispersion* and *uniform dispersion.*

cnidarians (ni-dah′ree-anz) Phylum of animals that have stinging cells called *cnidocytes,* two tissue layers, and radial symmetry; include hydras and jellyfish.

cnidocytes Stinging cells characteristic of cnidarians.

coated pit A depression in the plasma membrane, the cytosolic side of which is coated with the protein clathrin; important in receptor-mediated endocytosis.

cochlea (koke′lee-ah) The structure of the inner ear of mammals that contains the auditory receptors (organ of Corti).

coccus (kok′us) (pl., *cocci*) A bacterium with a spherical shape. Compare with *bacillus, spirillum, vibrio,* and *spirochete.*

codominance (koh″dom′in-ants) Condition in which two alleles of a locus are expressed in a heterozygote.

codon (koh′don) A triplet of mRNA nucleotides. The 64 possible codons collectively constitute a universal genetic code in which each codon specifies an amino acid in a polypeptide, or a signal to either start or terminate polypeptide synthesis.

coelacanths A genus of lobe-finned fishes that have survived to the present day.

coelom (see′lum) The main body cavity of most animals; a true coelom is lined with mesoderm. Compare with *pseudocoelom.*

coelomate (seel′oh-mate) Animal that has a true coelom. Compare with *acoelomate* and *pseudocoelomate.*

coenocyte (see′no-site) An organism consisting of a multinucleate cell, i.e., the nuclei are not separated from one another by septa.

coenzyme (koh-en′zime) An organic cofactor for an enzyme; generally participates in the reaction by transferring some component, such as electrons or part of a substrate molecule.

coenzyme A (CoA) Organic cofactor responsible for transferring groups derived from organic acids.

coevolution The reciprocal adaptation of two or more species that occurs as a result of their close interactions over a long period.

cofactor A nonprotein substance needed by an enzyme for normal activity; some cofactors are inorganic (usually metal ions); others are organic (coenzymes).

cohesion The property of sticking together. Compare with *adhesion*.

cohort A group of individuals of the same age.

colchicine A drug that blocks the division of eukaryotic cells by binding to tubulin subunits, which make up the microtubules, the major component of the mitotic spindle.

coleoptile (kol-ee-op′tile) A protective sheath that encloses the young stem in certain monocots.

collagens (kol′ah-gen) Proteins found in the collagen fibers of connective tissues.

collecting duct A tube in the kidney that receives filtrate from several nephrons and conducts it to the renal pelvis.

collenchyma (kol-en′kih-mah) Living cells with moderately but unevenly thickened primary cell walls; collenchyma cells help support the herbaceous plant body.

colony An association of loosely connected cells or individuals of the same species.

commensalism (kuh-men′sul-izm) A type of symbiosis in which one organism benefits and the other one is neither harmed nor helped. Compare with *mutualism* and *parasitism*.

commercial harvest The collection of commercially important organisms from the wild. Examples include the commercial harvest of parrots (for the pet trade) and cacti (for houseplants).

community An association of populations of different species living together in a defined habitat with some degree of interdependence. Compare with *ecosystem*.

community ecology The description and analysis of patterns and processes within the community.

compact bone Dense, hard bone tissue found mainly near the surfaces of a bone.

companion cell A cell in the phloem of flowering plants that governs loading and unloading sugar into the sieve tube element for translocation.

compass sense The sense of direction an animal requires to travel in a straight line toward a destination.

competition The interaction among two or more individuals that attempt to use the same essential resource, such as food, water, sunlight, or living space. See *interspecific* and *intraspecific competition*. See *interference* and *exploitation competition*.

competitive exclusion principle The concept that no two species with identical living requirements can occupy the same ecological niche indefinitely.

competitive inhibitor A substance that binds to the active site of an enzyme, thus lowering the rate of the reaction catalyzed by the enzyme. Compare with *noncompetitive inhibitor*.

complement A group of proteins in blood and other body fluids that are activated by an antigen–antibody complex and then destroy pathogens.

complementary DNA (cDNA) DNA synthesized by reverse transcriptase, using RNA as a template.

complete flower A flower that has all four parts: sepals, petals, stamens, and carpels. Compare with *incomplete flower*.

compound eye An eye, such as that of an insect, consisting of many light-sensitive units called *ommatidia*.

concentration gradient A difference in the concentration of a substance from one point to another, as for example, across a cell membrane.

condensation synthesis A reaction in which two monomers are combined covalently through the removal of the equivalent of a water molecule. Compare with *hydrolysis*.

cone (1) In botany, a reproductive structure in many gymnosperms that produces either microspores or megaspores. (2) In zoology, one of the conical photoreceptive cells of the retina that is particularly sensitive to bright light and, by distinguishing light of various wavelengths, mediates color vision. Compare with *rod*.

conidiophore (kah-nid′e-o-for″) A specialized hypha that bears conidia.

conidium (kah-nid′e-um) (pl., *conidia*) An asexual spore that is usually formed at the tip of a specialized hypha called a *conidiophore*.

conifer (kon′ih-fur) Any of a large phylum of gymnosperms that are woody trees and shrubs with needlelike, mostly evergreen leaves and with seeds in cones.

conjugation (kon″jew-gay′shun) (1) A sexual process in certain protists that involves exchange or fusion of a cell with another cell. (2) A mechanism for DNA exchange in bacteria that involves cell-to-cell contact.

connective tissue Animal tissue consisting mostly of intercellular substance (fibers scattered through a matrix) in which the cells are embedded, e.g., bone.

conservation biology A multidisciplinary science that focuses on the study of how humans impact organisms and on the development of ways to protect biological diversity.

constitutive gene A gene that is constantly transcribed.

consumer See *heterotroph*.

consumption overpopulation A situation in which each individual in a human population consumes too large a share of resources; results in pollution, environmental degradation, and resource depletion. Compare with *people overpopulation*.

contest competition See *interference competition*.

continental drift The theory that continents were once joined and later split and drifted apart.

contraception Any method used to intentionally prevent pregnancy.

contractile root (kun-trak′til) A specialized type of root that contracts and pulls a bulb or corm deeper into the soil.

contractile vacuole A membrane-enclosed organelle found in certain freshwater protists, such as *Paramecium;* appears to have an osmoregulatory function.

control group In a scientific experiment, a group in which the experimental variable is kept constant. The control provides a standard of comparison used to verify the results of the experiment.

controlled mating A mating in which the genotypes of the parents are known.

convergent circuit (kun-vur′jent) A neural pathway in which a postsynaptic neuron is controlled by signals coming from two or more presynaptic neurons. Compare with *divergent circuit*.

convergent evolution (kun-vur′jent) The independent evolution of structural or functional similarity in two or more distantly related species, usually as a result of adaptations to similar environments.

core angiosperms The clade to which most angiosperm species belong. Core angiosperms are divided into three subclades: magnoliids, monocots, and eudicots. Compare with *basal angiosperms*.

corepressor Substance that binds to a repressor protein, converting it to its active form, which is capable of preventing transcription.

Coriolis effect (kor″e-o′lis) The tendency of moving air or water to be deflected from its path to the right in the Northern Hemisphere and to the left in the Southern Hemisphere. Caused by the direction of Earth's rotation.

cork cambium (kam′bee-um) A lateral meristem that produces cork cells and cork parenchyma; cork cambium and the tissues it produces make up the outer bark of a woody plant. Compare with *vascular cambium.*

cork cell A cell in the bark that is produced outwardly by the cork cambium; cork cells are dead at maturity and function for protection and reduction of water loss.

cork parenchyma (par-en′kih-mah) One or more layers of parenchyma cells produced inwardly by the cork cambium.

corm A short, thickened underground stem specialized for food storage and asexual reproduction, e.g., crocus, gladiolus.

cornea (kor′nee-ah) The transparent covering of an eye.

corolla (kor-ohl′ah) Collectively, the petals of a flower.

corpus callosum (kah-loh′sum) In mammals, a large bundle of nerve fibers interconnecting the two cerebral hemispheres.

corpus luteum (loo′tee″um) The temporary endocrine tissue in the ovary that develops from the ruptured follicle after ovulation; secretes progesterone and estrogen.

cortex (kor′teks) (1) The outer part of an organ, such as the cortex of the kidney. Compare with *medulla.* (2) The tissue between the epidermis and vascular tissue in the stems and roots of many herbaceous plants.

cortical reaction Process occurring after fertilization that prevents additional sperm from entering the egg; also known as the "slow block to polyspermy."

cortisol A steroid hormone, secreted by the adrenal cortex, that helps the body adjust to long-term stress; stimulates conversion of other nutrients to glucose in the liver, resulting in increased blood glucose concentration.

cosmid cloning vector A cloning vector with features of both bacteriophages and plasmids and with the ability to carry large segments of foreign DNA.

cosmopolitan species Species that have a nearly worldwide distribution and occur on more than one continent or throughout much of the ocean. Compare with *endemic species.*

cotransport The active transport of a substance from a region of low concentration to a region of high concentration by coupling its transport to the transport of a substance down its concentration gradient.

cotyledon (kot″uh-lee′dun) The seed leaf of a plant embryo, which may contain food stored for germination.

cotylosaurs The first reptiles; also known as *stem reptiles.*

countercurrent exchange system A biological mechanism that enables maximum exchange between two fluids. The two fluids must be flowing in opposite directions and have a concentration gradient between them.

coupled reactions A set of reactions in which an exergonic reaction provides the free energy required to drive an endergonic reaction; energy coupling generally occurs through a common intermediate.

covalent bond The chemical bond involving shared pairs of electrons; may be single, double, or triple (with one, two, or three shared pairs of electrons, respectively). Compare with *ionic bond* and *hydrogen bond.*

covalent compound A compound in which atoms are held together by covalent bonds; covalent compounds consist of molecules. Compare with *ionic compound.*

cranial nerves The 10 to 12 pairs of nerves in vertebrates that emerge directly from the brain.

cranium The bony framework that protects the brain in vertebrates.

crassulacean acid metabolism See *CAM plant.*

creatine phosphate An energy-storing compound in muscle cells.

cretinism (kree′tin-izm) A chronic condition caused by lack of thyroid secretion during fetal development and early childhood; results in retarded physical and mental development if untreated.

cri du chat syndrome A human genetic disease caused by losing part of the short arm of chromosome 5 and characterized by mental retardation, a cry that sounds like a kitten mewing, and death in infancy or childhood.

cristae (kris′tee) (sing., *crista*) Shelflike or fingerlike inward projections of the inner membrane of a mitochondrion.

Cro-Magnons Prehistoric humans (*Homo sapiens*) with modern features (tall, erect, lacking a heavy brow) who lived in Europe some 30,000 years ago.

cross bridges The connections between myosin and actin filaments in muscle fibers; formed by the binding of myosin heads to active sites on actin filaments.

crossing-over A process in which genetic material (DNA) is exchanged between paired, homologous chromosomes.

cryptic coloration Colors or markings that help some organisms hide from predators by blending into their physical surroundings. Compare with *aposematic coloration.*

cryptochrome A proteinaceous pigment that strongly absorbs blue light; implicated in resetting the biological clock in plants, fruit flies, and mice.

ctenophores (ten′oh-forz) Phylum of marine animals (comb jellies) whose bodies consist of two layers of cells enclosing a gelatinous mass. The outer surface is covered with comblike rows of cilia, by which the animal moves.

cuticle (kew′tih-kl) (1) A noncell, waxy covering over the epidermis of the aerial parts of plants that reduces water loss. (2) The outer covering of some animals, such as roundworms.

cyanobacteria (sy-an″oh-bak-teer′ee-uh) Prokaryotic photosynthetic microorganisms that possess chlorophyll and produce oxygen during photosynthesis. Formerly known as blue-green algae.

cycad (sih′kad) Any of a phylum of gymnosperms that live mainly in tropical and semitropical regions and have stout stems (to 20 m in height) and fernlike leaves.

cyclic AMP (cAMP) A form of adenosine monophosphate in which the phosphate is part of a ring-shaped structure; acts as a regulatory molecule and second messenger in organisms ranging from bacteria to humans.

cyclic electron transport In photosynthesis, the cyclic flow of electrons through photosystem I; ATP is formed by chemiosmosis, but no photolysis of water occurs, and O_2 and NADPH are not produced. Compare with *noncyclic electron transport.*

cyclin-dependent kinases (Cdks) Protein kinases involved in controlling the cell cycle.

cyclins Regulatory proteins whose levels oscillate during the cell cycle; activate cyclin-dependent kinases.

cystic fibrosis A genetic disease with an autosomal recessive inheritance pattern; characterized by secretion of abnormally thick mucus, particularly in the respiratory and digestive systems.

cytochromes (sy′toh-krohmz) Iron-containing heme proteins of an electron transport system.

cytokines Signaling proteins that regulate interactions between cells in the immune system. Important groups include interferons, interleukins, tumor necrosis factors, and chemokines.

cytokinesis (sy″toh-kih-nee′sis) Stage of cell division in which the cytoplasm divides to form two daughter cells.

cytokinin (sy″toh-ky′nin) A plant hormone involved in various aspects of plant growth and development, such as cell division and delay of senescence.

cytoplasm The plasma membrane and cell contents with the exception of the nucleus.

cytosine A nitrogenous pyrimidine base that is a component of nucleic acids.

cytoskeleton The dynamic internal network of protein fibers that includes microfilaments, intermediate filaments, and microtubules.

cytosol The fluid component of the cytoplasm in which the organelles are suspended.

cytotoxic T cell See *T cytotoxic cell.*

dalton See *atomic mass unit (amu).*

day-neutral plant A plant whose flowering is not controlled by variations in day length that occur with changing seasons. Compare with *long-day, short-day,* and *intermediate-day plants.*

deamination (dee-am-ih-nay′shun) The removal of an amino group (—NH₂) from an amino acid or other organic compound.

decarboxylation A reaction in which a molecule of CO₂ is removed from a carboxyl group of an organic acid.

deciduous A term describing a plant that sheds leaves or other structures at regular intervals, e.g., during autumn. Compare with *evergreen.*

decomposers Microbial heterotrophs that break down dead organic material and use the decomposition products as a source of energy. Also called *saprotrophs* or *saprobes.*

deductive reasoning The reasoning that operates from generalities to specifics and can make relationships among data more apparent. Compare with *inductive reasoning.* See *hypothetico-deductive approach.*

deforestation The temporary or permanent removal of forest for agriculture or other uses.

dehydrogenation (dee-hy″dro-jen-ay′shun) A form of oxidation in which hydrogen atoms are removed from a molecule.

deletion (1) A chromosome abnormality in which part of a chromosome is missing, e.g., cri du chat syndrome. (2) The loss of one or more base pairs from DNA, which can result in a frameshift mutation.

demographics The science that deals with human population statistics, such as size, density, and distribution.

denature (dee-nay′ture) To alter the physical properties and three-dimensional structure of a protein, nucleic acid, or other macromolecule by treating it with excess heat, strong acids, or strong bases.

dendrite (den′drite) A branch of a neuron that receives and conducts nerve impulses toward the cell body. Compare with *axon.*

dendritic cells A set of immune cells present in many tissues that capture antigens and present them to T cells.

dendrochronology (den″dro-kruh-naal′uh-gee) A method of dating using the annual rings of trees.

denitrification (dee-nie″tra-fuh-kay′shun) The conversion of nitrate (NO₃⁻) to nitrogen gas (N₂) by certain bacteria (denitrifying bacteria) in the soil; part of the nitrogen cycle.

dense connective tissue A type of tissue that may be irregular, as in the dermis of the skin, or regular, as in tendons.

density-dependent factor An environmental factor whose effects on a population change as population density changes; tends to retard population growth as population density increases and enhance population growth as population density decreases. Compare with *density-independent factor.*

density gradient centrifugation Procedure in which cell components are placed in a layer on top of a density gradient, usually a sucrose solution and water. Cell structures migrate during centrifugation, forming a band at the position in the gradient where their own density equals that of the sucrose solution.

density-independent factor An environmental factor that affects the size of a population but is not influenced by changes in population density. Compare with *density-dependent factor.*

deoxyribonucleic acid (DNA) Double-stranded nucleic acid; contains genetic information coded in specific sequences of its constituent nucleotides.

deoxyribose Pentose sugar lacking a hydroxyl (—OH) group on carbon-2′; a constituent of DNA.

depolarization (dee-pol″ar-ih-zay′shun) A decrease in the charge difference across a plasma membrane; may result in an action potential in a neuron or muscle cell.

derived characters See *shared derived characters.*

dermal tissue system The tissue that forms the outer covering over a plant; the epidermis or periderm.

dermis (dur′mis) The layer of dense connective tissue beneath the epidermis in the skin of vertebrates.

desert A temperate or tropical biome in which lack of precipitation limits plant growth.

desertification The degradation of once-fertile land into nonproductive desert; caused partly by soil erosion, deforestation, and overgrazing by domestic animals.

desmosomes (dez′moh-sohmz) Buttonlike plaques, present on two opposing cell surfaces, that hold the cells together by means of protein filaments that span the intercellular space.

determinate growth Growth of limited duration, as for example, in flowers and leaves. Compare with *indeterminate growth.*

determination The developmental process by which one or more cells become progressively committed to a particular fate. Determination is a series of molecular events usually leading to differentiation. Also called *cell determination.*

detritivore (duh-try′tuh-vore) An organism, such as an earthworm or crab, that consumes fragments of freshly dead or decomposing organisms; also called *detritus feeder.*

detritus (duh-try′tus) Organic debris from decomposing organisms.

detritus feeder See *detritivore.*

deuteromycetes (doo″ter-o-my′seats) An artificial grouping of fungi characterized by the absence of sexual reproduction but usually having other traits similar to ascomycetes; also called *imperfect fungi.*

deuterostome (doo′ter-oh-stome) Major division of the animal kingdom in which the anus develops from the blastopore; includes the echinoderms and chordates. Compare with *protostome.*

development All the progressive changes that take place throughout the life of an organism.

diabetes mellitus (mel′i-tus) The most common endocrine disorder. In type 1 diabetes, there is a marked decrease in the number of beta cells in the pancreas, resulting in insulin deficiency. In the more common type 2 diabetes, insulin receptors on target cells do not bind with insulin (insulin resistance).

diacylglycerol (DAG) (di″as-il-glis′er-ol) A lipid consisting of glycerol combined chemically with two fatty acids; also called *diglyceride.* Compare with *monoacylglycerol* and *triacylglycerol.*

dialysis The diffusion of certain solutes across a selectively permeable membrane.

diaphragm In mammals, the muscular floor of the chest cavity; contracts during inhalation, expanding the chest cavity.

diastole (di-ass′toh-lee) Phase of the cardiac cycle in which the heart is relaxed. Compare with *systole.*

diatom (die′eh-tom″) A usually unicellular alga that is covered by an ornate, siliceous shell consisting of two overlapping halves; an important component of plankton in both marine and fresh waters.

dichotomous branching (di-kaut′uh-mus) In botany, a type of branching in which one part always divides into two more or less equal parts.

diencephalon See *forebrain*.

differential centrifugation Separation of cell particles according to their mass, size, or density. In differential centrifugation, the supernatant is spun at successively higher revolutions per minute.

differential gene expression The expression of different subsets of genes at different times and in different cells during development.

differentiated cell A specialized cell; carries out unique activities, expresses a specific set of proteins, and usually has a recognizable appearance.

differentiation (dif″ah-ren-she-ay′shun) Development toward a more mature state; a process changing a young, relatively unspecialized cell to a more specialized cell. Also called *cell differentiation*.

diffusion The net movement of particles (atoms, molecules, or ions) from a region of higher concentration to a region of lower concentration (i.e., down a concentration gradient), resulting from random motion. Compare with *facilitated diffusion* and *active transport*.

digestion The breakdown of food to smaller molecules.

diglyceride See *diacylglycerol*.

dihybrid cross (dy-hy′brid) A genetic cross that takes into account the behavior of alleles of two loci. Compare with *monohybrid cross*.

dikaryotic (dy-kare-ee-ot′ik) Condition of having two nuclei per cell (i.e., $n + n$), characteristic of certain fungal hyphae. Compare with *monokaryotic*.

dimer An association of two monomers (e.g., a disaccharide or a dipeptide).

dinoflagellate (dy″noh-flaj′eh-late) A unicellular, biflagellate, typically marine alga that is an important component of plankton; usually photosynthetic.

dioecious (dy-ee′shus) Having male and female reproductive structures on separate plants; compare with *monoecious*.

dipeptide See *peptide*.

diploid (dip′loyd) The condition of having two sets of chromosomes per nucleus. Compare with *haploid* and *polyploid*.

diplomonads Small, mostly parasitic zooflagellates with one or two nuclei, no mitochondria, and one to four flagella.

direct fitness An individual's reproductive success, measured by the number of viable offspring it produces. Compare with *inclusive fitness*.

directed evolution See *in vitro evolution*.

directional selection The gradual replacement of one phenotype with another because of environmental change that favors phenotypes at one of the extremes of the normal distribution. Compare with *stabilizing selection* and *disruptive selection*.

disaccharide (dy-sak′ah-ride) A sugar produced by covalently linking two monosaccharides (e.g., maltose or sucrose).

disomy The normal condition in which both members of a chromosome pair are present in a diploid cell or organism. Compare with *monosomy* and *trisomy*.

dispersal The movement of individuals among populations. See *immigration* and *emigration*.

dispersion The pattern of distribution in space of the individuals of a population relative to their neighbors; may be clumped, random, or uniform.

disruptive selection A special type of directional selection in which changes in the environment favor two or more variant phenotypes at the expense of the mean. Compare with *stabilizing selection* and *directional selection*.

distal Remote; farther from the point of reference. Compare with *proximal*.

distal convoluted tubule The part of the renal tubule that extends from the loop of Henle to the collecting duct. Compare with *proximal convoluted tubule*.

disturbance In ecology, any event that disrupts community or population structure.

divergent circuit A neural pathway in which a presynaptic neuron stimulates many postsynaptic neurons. Compare with *convergent circuit*.

diving reflex A group of physiological mechanisms, such as decrease in metabolic rate, that are activated when a mammal dives to its limit.

dizygotic twins Twins that arise from the separate fertilization of two eggs; commonly known as *fraternal twins*. Compare with *monozygotic twins*.

DNA See *deoxyribonucleic acid*.

DNA cloning The process of selectively amplifying DNA sequences so their structure and function can be studied.

DNA fingerprinting The analysis of DNA extracted from an individual, which is unique to that individual; also called *DNA typing*.

DNA ligase Enzyme that catalyzes the joining of the 5′ and 3′ ends of two DNA fragments; essential in DNA replication and used in recombinant DNA technology.

DNA methylation A process in which gene inactivation is perpetuated by enzymes that add methyl groups to DNA.

DNA microarray A diagnostic test involving thousands of DNA molecules placed on a glass slide or chip.

DNA polymerases Family of enzymes that catalyze the synthesis of DNA from a DNA template by adding nucleotides to a growing 3′ end.

DNA probe See *genetic probe*.

DNA provirus Double-stranded DNA molecule that is an intermediate in the life cycle of an RNA tumor virus (retrovirus).

DNA replication The process by which DNA is duplicated; ordinarily a semiconservative process in which a double helix gives rise to two double helices, each with an "old" strand and a newly synthesized strand.

DNA sequencing Procedure by which the sequence of nucleotides in DNA is determined.

DNA typing See *DNA fingerprinting*.

domain (1) A structural and functional region of a protein. (2) The broadest taxonomic category; each domain includes one or more kingdoms.

dominance hierarchy A linear "pecking order" into which animals in a population may organize according to status; regulates aggressive behavior within the population.

dominant allele (al-leel′) An allele that is always expressed when it is present, regardless of whether it is homozygous or heterozygous. Compare with *recessive allele*.

dominant species In a community, a species that as a result of its large biomass or abundance exerts a major influence on the distribution of populations of other species.

dopamine A neurotransmitter of the biogenic amine group.

dormancy A temporary period of arrested growth in plants or plant parts such as spores, seeds, bulbs, and buds.

dorsal (dor′sl) Toward the uppermost surface or back of an animal. Compare with *ventral*.

dosage compensation Genetic mechanism by which the expression of X-linked genes in mammals is made equivalent in XX females and XY males by rendering all but one X chromosome inactive.

double fertilization A process in the flowering plant life cycle in which there are two fertilizations; one fertilization results in formation of a zygote, whereas the second results in formation of endosperm

double helix The structure of DNA, which consists of two anti-parallel polynucleotide chains twisted around each other.

doubling time The amount of time it takes for a population to double in size, assuming that its current rate of increase does not change.

Down syndrome An inherited condition in which individuals have abnormalities of the face, eyelids, tongue, and other parts of the body and are physically and mentally retarded; usually results from trisomy of chromosome 21.

drupe (droop) A simple, fleshy fruit in which the inner wall of the fruit is a hard stone, e.g., peach, cherry.

duodenum (doo″o-dee′num) The portion of the small intestine into which the contents of the stomach first enter.

duplication An abnormality in which a set of chromosomes contains more than one copy of a particular chromosomal segment; the translocation form of Down syndrome is an example.

dura mater The tough, outer meningeal layer that covers and protects the brain and spinal cord. Also see *arachnoid* and *pia mater*.

dynamic equilibrium The condition of a chemical reaction when the rate of change in one direction is exactly the same as the rate of change in the opposite direction, i.e., the concentrations of the reactants and products are not changing, and the difference in free energy between reactants and products is zero.

ecdysis Molting; shedding outer skin; common process in insects, crustaceans, snakes.

ecdysone (ek′dih-sone) See *molting hormone*.

Ecdysozoa A branch of the protostomes that includes animals that molt, such as the rotifers, nematodes, and arthropods.

echinoderms (eh-kine′oh-derms) Phylum of spiny-skinned marine deuterostome invertebrates characterized by a water vascular system and tube feet; include sea stars, sea urchins, and sea cucumbers.

echolocation Determination of the position of objects by detecting echoes of high-pitched sounds emitted by an animal; a type of sensory system used by bats and dolphins.

ecological niche See *niche*.

ecological pyramid A graphical representation of the relative energy value at each trophic level. See *pyramid of biomass* and *pyramid of energy*.

ecological succession See *succession*.

ecology (ee-kol′uh-jee) A discipline of biology that studies the interrelations among living things and their environments.

ecosystem (ee′koh-sis-tem) The interacting system that encompasses a community and its nonliving, physical environment. Compare with *community*.

ecosystem management A conservation focus that emphasizes restoring and maintaining ecosystem quality rather than the conservation of individual species.

ecosystem services Important environmental services, such as clean air to breathe, clean water to drink, and fertile soil in which to grow crops, that ecosystems provide.

ecotone The transition zone where two communities meet and intergrade.

ectoderm (ek′toh-derm) The outer germ layer of the early embryo; gives rise to the skin and nervous system. Compare with *mesoderm* and *endoderm*.

ectoparasite A tick or other parasite that lives outside its host's body. Compare with *endoparasite*.

ectotherm An animal whose temperature fluctuates with that of the environment; may use behavioral adaptations to regulate tem-perature; sometimes referred to as *cold-blooded*. Compare with *endotherm*.

edge effect The ecological phenomenon in which ecotones between adjacent communities often contain a greater number of species or greater population densities of certain species than either adjacent community.

Ediacaran period (ee-dee-ack′uh-ran″) The last (most recent) period of the Proterozoic eon, from 600 million to 542 million years ago; named for early animal fossils found in the Ediacara Hills in South Australia.

effector A muscle or gland that contracts or secretes in direct response to nerve impulses.

efferent (ef′fur-ent) Leading away from some point of reference. Compare with *afferent*.

efferent neurons Neurons that transmit action potentials from the brain or spinal cord to muscles or glands. Compare with *afferent neurons*.

ejaculation (ee-jak″yoo-lay′shun) A sudden expulsion, as in the ejection of semen from the penis.

electrolyte A substance that dissociates into ions when dissolved in water; the resulting solution can conduct an electric current.

electron A particle with one unit of negative charge and negligible mass, located outside the atomic nucleus. Compare with *neutron* and *proton*.

electron configuration The arrangement of electrons around the atom. In a Bohr model, the electron configuration is depicted as a series of concentric circles.

electron microscope A microscope capable of producing high-resolution, highly magnified images through the use of an electron beam (rather than light). Transmission electron microscopes (TEMs) produce images of thin sections; scanning electron microscopes (SEMs) produce images of surfaces.

electron shell Group of orbitals of electrons with similar energies.

electron transport system A series of chemical reactions during which hydrogens or their electrons are passed along an electron transport chain from one acceptor molecule to another, with the release of energy.

electronegativity A measure of an atom's attraction for electrons.

electrophoresis, gel See *gel electrophoresis*.

electroreceptor A receptor that responds to electrical stimuli.

element A substance that cannot be changed to a simpler substance by a normal chemical reaction.

elimination Ejection of undigested food from the body. Compare with *excretion*.

El Niño–Southern Oscillation (ENSO) (el nee′nyo) A recurring climatic phenomenon that involves a surge of warm water in the Pacific Ocean and unusual weather patterns elsewhere in the world.

elongation (in protein synthesis) Cyclic process by which amino acids are added one by one to a growing polypeptide chain. See *initiation* and *termination*.

embryo (em′bree-oh) (1) A young organism before it emerges from the egg, seed, or body of its mother. (2) Developing human until the end of the second month, after which it is referred to as a fetus. (3) In plants, the young sporophyte produced following fertilization and subsequent development of the zygote.

embryo sac The female gametophyte generation in flowering plants.

embryo transfer See *host mothering*.

emergent properties Characteristics of an object, process, or behavior that could not be predicted from its component parts; emergent properties can be identified at each level as we move up the hierarchy of biological organization.

emigration The movement of individuals out of a population. Compare with *immigration*.

enantiomers (en-an'tee-oh-merz) Two isomeric chemical compounds that are mirror images.

end product inhibition See *feedback inhibition*.

endangered species A species whose numbers are so severely reduced that it is in imminent danger of extinction throughout all or part of its range. Compare with *threatened species*.

endemic species Localized, native species that are not found anywhere else in the world. Compare with *cosmopolitan species*.

endergonic reaction (end'er-gon"ik) A nonspontaneous reaction; a reaction requiring a net input of free energy. Compare with *exergonic reaction*.

endocrine gland (en'doh-crin) A gland that secretes hormones directly into the blood or tissue fluid instead of into ducts. Compare with *exocrine gland*.

endocrine system The body system that helps regulate metabolic activities; consists of ductless glands and tissues that secrete hormones.

endocytosis (en"doh-sy-toh'sis) The active transport of substances into the cell by the formation of invaginated regions of the plasma membrane that pinch off and become cytoplasmic vesicles. Compare with *exocytosis*.

endoderm (en'doh-derm) The inner germ layer of the early embryo; becomes the lining of the digestive tract and the structures that develop from the digestive tract—liver, lungs, and pancreas. Compare with *ectoderm* and *mesoderm*.

endodermis (en"doh-der'mis) The innermost layer of the plant root cortex. Endodermal cells have a waterproof Casparian strip around their radial and transverse walls that ensures that water and minerals enter the xylem only by passing through the endoderm cells.

endolymph (en'doh-limf) The fluid of the membranous labyrinth and cochlear duct of the ear.

endomembrane system See *internal membrane system*.

endometrium (en"doh-mee'tree-um) The uterine lining.

endoparasite A parasite such as a tapeworm that lives within the host. Compare with *ectoparasite*.

endoplasmic reticulum (ER) (en'doh-plaz"mik reh-tik'yoo-lum) An interconnected network of internal membranes in eukaryotic cells enclosing a compartment, the ER lumen. Rough ER has ribosomes attached to the cytosolic surface; smooth ER, a site of lipid biosynthesis, lacks ribosomes.

endorphins (en-dor'finz) Neuropeptides released by certain brain neurons; block pain signals.

endoskeleton (en"doh-skel'eh-ton) Bony and/or cartilaginous structures within the body that provide support. Compare with *exoskeleton*.

endosperm (en'doh-sperm) The 3n nutritive tissue that is formed at some point in the development of all angiosperm seeds.

endospore A resting cell formed by certain bacteria; highly resistant to heat, radiation, and disinfectants.

endostyle In nonvertebrate chordates, a groove in the floor of the pharynx that secretes mucus and traps food particles in sea water passing through the pharynx. In vertebrates, the thyroid gland is derived from the endostyle.

endosymbiont (en"doe-sim'bee-ont) An organism that lives inside the body of another kind of organism. Endosymbionts may benefit their host (mutualism) or harm their host (parasitism).

endothelium (en-doh-theel'ee-um) The tissue that lines the cavities of the heart, blood vessels, and lymph vessels.

endotherm (en'doh-therm) An animal that uses metabolic energy to maintain a constant body temperature despite variations in environmental temperature; e.g., birds and mammals. Compare with *ectotherm*.

endotoxin A poisonous substance in the cell walls of gram-negative bacteria. Compare with *exotoxin*.

energy The capacity to do work; expressed in kilojoules or kilocalories.

energy of activation See *activation energy*.

enhanced greenhouse effect See *greenhouse effect*.

enhancers Regulatory DNA sequences that can be located long distances away from the actual coding regions of a gene.

enkephalins (en-kef'ah-linz) Neuropeptides released by certain brain neurons that block pain signals.

enterocoely (en'ter-oh-seely) The process by which the coelom forms as a cavity within mesoderm produced by outpocketings of the primitive gut (archenteron); characteristic of many deuterostomes. Compare with *schizocoely*.

enthalpy The total potential energy of a system; sometimes referred to as the "heat content of the system."

entropy (en'trop-ee) Disorderliness; a quantitative measure of the amount of the random, disordered energy that is unavailable to do work.

environmental resistance Unfavorable environmental conditions, such as crowding, that prevent organisms from reproducing indefinitely at their intrinsic rate of increase.

environmental sustainability The ability to meet humanity's current needs without compromising the ability of future generations to meet their needs.

enzyme (en'zime) An organic catalyst (usually a protein) that accelerates a specific chemical reaction by lowering the activation energy required for that reaction.

enzyme–substrate complex The temporary association between enzyme and substrate that forms during the course of a catalyzed reaction; also called *ES complex*.

eon The largest division of the geologic time scale; eons are divided into eras.

eosinophil (ee-oh-sin'oh-fil) A type of white blood cell whose cytoplasmic granules absorb acidic stains; functions in parasitic infestations and allergic reactions.

epidermis (ep-ih-dur'mis) (1) An outer layer of cells that covers the body of plants and functions primarily for protection. (2) The outer layer of vertebrate skin.

epididymis (ep-ih-did'ih-mis) (pl., *epididymides*) A coiled tube that receives sperm from the testis and conveys it to the vas deferens.

epigenetic inheritance Inheritance that involves changes in how a gene is expressed without any change in that gene's nucleotide sequence.

epiglottis A thin, flexible structure that guards the entrance to the larynx, preventing food from entering the airway during swallowing.

epinephrine (ep-ih-nef'rin) Hormone produced by the adrenal medulla; stimulates the sympathetic nervous system.

epistasis (ep-ih-sta'-sis) Condition in which certain alleles of one locus alter the expression of alleles of a different locus.

epithelial tissue (ep-ih-theel'ee-al) The type of animal tissue that covers body surfaces, lines body cavities, and forms glands; also called *epithelium*.

epoch The smallest unit of geologic time; a subdivision of a period.

equilibrium See *dynamic equilibrium, genetic equilibrium,* and *punctuated equilibrium*.

era An interval of geologic time that is a subdivision of an eon; eras are divided into periods.

erythroblastosis fetalis (eh-rith″row-blas-toe′sis fi-tal′is) Serious condition in which Rh^+ red blood cells (which bear antigen D) of a fetus are destroyed by maternal anti-D antibodies.

erythrocyte (eh-rith′row-site) A vertebrate red blood cell; contains hemoglobin, which transports oxygen.

erythropoietin (eh-rith″row-poy′ih-tin) A peptide hormone secreted mainly by kidney cells; stimulates red blood cell production.

ES complex See *enzyme–substrate complex.*

esophagus (e-sof′ah-gus) The part of the digestive tract that conducts food from the pharynx to the stomach.

essential nutrient A nutrient that must be provided in the diet because the body cannot make it or cannot make it in sufficient quantities to meet nutritional needs, e.g., essential amino acids and essential fatty acids.

ester linkage Covalent linkage formed by the reaction of a carboxyl group and a hydroxyl group, with the removal of the equivalent of a water molecule; the linkage includes an oxygen atom bonded to a carbonyl group.

estivation A state of torpor caused by lack of food or water during periods of high temperature. Compare with *hibernation.*

estrogens (es′troh-jenz) Female sex hormones produced by the ovary; promote the development and maintenance of female reproductive structures and of secondary sex characteristics.

estuary (es′choo-wear-ee) A coastal body of water that connects to an ocean, in which fresh water from the land mixes with salt water.

ethology (ee-thol′oh-jee) The study of animal behavior under natural conditions from the point of view of adaptation.

ethyl alcohol A two-carbon alcohol.

ethylene (eth′ih-leen) A gaseous plant hormone involved in various aspects of plant growth and development, such as leaf abscission and fruit ripening.

euchromatin (yoo-croh′mah-tin) A loosely coiled chromatin that is generally capable of transcription. Compare with *heterochromatin.*

eudicot (yoo-dy′kot) One of the two clades of flowering plants; eudicot seeds contain two cotyledons, or seed leaves. Compare with *monocot.*

euglenoids (yoo-glee′noidz) A group of mostly freshwater, flagellate, unicellular algae that move by means of an anterior flagellum and are usually photosynthetic.

eukaryote (yoo″kar′ee-ote) An organism whose cells have nuclei and other membrane-enclosed organelles. Includes protists, fungi, plants, and animals. Compare with *prokaryote.*

euphotic zone The upper reaches of the ocean, in which enough light penetrates to support photosynthesis.

eustachian tube (yoo-stay′shee-un) The auditory tube passing between the middle-ear cavity and the pharynx in vertebrates; permits the equalization of pressure on the tympanic membrane.

eutrophic lake A lake enriched with nutrients such as nitrate and phosphate and consequently overgrown with plants or algae.

evergreen A plant that sheds leaves over a long period, so some leaves are always present. Compare with *deciduous.*

Evo Devo The study of the evolution of the genetic control of development.

evolution Any cumulative genetic changes in a population from generation to generation. Evolution leads to differences in populations and explains the origin of all the organisms that exist today or have ever existed.

evolutionary species concept An alternative to the biological species concept in which for a population to be declared a separate species, it must have undergone evolution long enough for statistically significant differences to emerge. Compare with *species.*

evolutionary systematics An approach to classification that considers both evolutionary relationships and the extent of divergence that has occurred since a group branched from an ancestral group. Compare with *cladistics* and *phenetics.*

excitatory postsynaptic potential (EPSP) A change in membrane potential that brings a neuron closer to the firing level. Compare with *inhibitory postsynaptic potential (IPSP).*

excretion (ek-skree′shun) The discharge from the body of a waste product of metabolism (not to be confused with the elimination of undigested food materials). Compare with *elimination.*

excretory system The body system in animals that functions in osmoregulation and in the discharge of metabolic wastes.

exergonic reaction (ex′er-gon″ik) A reaction characterized by a release of free energy. Also called *spontaneous reaction.* Compare with *endergonic reaction.*

exocrine gland (ex′oh-crin) A gland that excretes its products through a duct that opens onto a free surface, such as the skin (e.g., sweat glands). Compare with *endocrine gland.*

exocytosis (ex″oh-sy-toh′sis) The active transport of materials out of the cell by fusion of cytoplasmic vesicles with the plasma membrane. Compare with *endocytosis.*

exon (1) A protein-coding region of a eukaryotic gene. (2) The mRNA transcribed from such a region. Compare with *intron.*

exoskeleton (ex″oh-skel′eh-ton) An external skeleton, such as the shell of mollusks or outer covering of arthropods; provides protection and sites of attachment for muscles. Compare with *endoskeleton.*

exotoxin A poisonous substance released by certain bacteria. Compare with *endotoxin.*

explicit memory Factual knowledge of people, places, or objects; requires conscious recall of the information.

exploitation competition An intraspecific competition in which all the individuals in a population "share" the limited resource equally, so at high population densities, none of them obtains an adequate amount. Also called *scramble competition.* Compare with *interference competition.*

exponential population growth The accelerating population growth rate that occurs when optimal conditions allow a constant per capita growth rate. Compare with *logistic population growth.*

ex situ conservation Conservation efforts that involve conserving individual species in human-controlled settings, such as zoos. Compare with *in situ conservation.*

exteroceptor (ex′tur-oh-sep″tor) One of the sense organs that receives sensory stimuli from the outside world, such as the eyes or touch receptors. Compare with *interoceptor.*

extinction The elimination of a species; occurs when the last individual member of a species dies.

extracellular matrix (ECM) A network of proteins and carbohydrates that surrounds many animal cells.

extraembryonic membranes Multicellular membranous structures that develop from the germ layers of a terrestrial vertebrate embryo but are not part of the embryo itself. See *chorion, amnion, allantois,* and *yolk sac.*

F_1 generation (first filial generation) The first generation of hybrid offspring resulting from a cross between parents from two different true-breeding lines.

F_2 generation (second filial generation) The offspring of the F_1 generation.

facilitated diffusion The passive transport of ions or molecules by a specific carrier protein in a membrane. As in simple diffusion, net transport is down a concentration gradient, and no additional energy has to be supplied. Compare with *diffusion* and *active transport.*

facilitation (1) In neurology, a process in which a neuron is brought close to its threshold level by stimulation from various presynaptic neurons. (2) In ecology, a situation in which one species has a positive effect on other species in the community, for example, by enhancing the local environment.

facultative anaerobe An organism capable of carrying out aerobic respiration but able to switch to fermentation when oxygen is unavailable, e.g., yeast. Compare with *obligate anaerobe*.

FAD/FADH$_2$ Oxidized and reduced forms, respectively, of flavin adenine dinucleotide, a coenzyme that transfers electrons (as hydrogen) in metabolism, including cellular respiration.

fallopian tube See *oviduct*.

family A taxonomic category made up of related genera.

fatty acid A lipid that is an organic acid containing a long hydrocarbon chain, with no double bonds (saturated fatty acid), one double bond (monounsaturated fatty acid), or two or more double bonds (polyunsaturated fatty acid); components of triacylglycerols and phospholipids, as well as monoacylglycerols and diacylglycerols.

fecundity The potential capacity of an individual to produce offspring.

feedback inhibition A type of enzyme regulation in which the accumulation of the product of a reaction inhibits an earlier reaction in the sequence; also known as *end product inhibition*.

fermentation An anaerobic process by which ATP is produced by a series of redox reactions in which organic compounds serve both as electron donors and terminal electron acceptors.

fern One of a phylum of seedless vascular plants that reproduce by spores produced in sporangia; ferns undergo an alternation of generations between the dominant sporophyte and the gametophyte (prothallus).

fertilization The fusion of two *n* gametes; results in the formation of a 2*n* zygote. Compare with *double fertilization*.

fetus The unborn human offspring from the third month of pregnancy to birth.

fiber (1) In plants, a type of sclerenchyma cell; fibers are long, tapered cells with thick walls. Compare with *sclereid*. (2) In animals, an elongated cell such as a muscle or nerve cell. (3) In animals, the microscopic, threadlike protein and carbohydrate complexes scattered through the matrix of connective tissues.

fibrin An insoluble protein formed from the plasma protein fibrinogen during blood clotting.

fibroblasts Connective tissue cells that produce the fibers and the protein and carbohydrate complexes of the matrix of connective tissues.

fibronectins Glycoproteins of the extracellular matrix that bind to integrins (receptor proteins in the plasma membrane).

fibrous root system A root system consisting of several adventitious roots of approximately equal size that arise from the base of the stem. Compare with *taproot system*.

Fick's law of diffusion A physical law governing rates of gas exchange in animal respiratory systems; states that the rate of diffusion of a substance across a membrane is directly proportional to the surface area and to the difference in pressure between the two sides.

filament In flowering plants, the thin stalk of a stamen; the filament bears an anther at its tip.

first law of thermodynamics The law of conservation of energy, which states that the total energy of any closed system (any object plus its surroundings, i.e., the universe) remains constant. Compare with *second law of thermodynamics*.

fitness See *direct fitness*.

fixed action pattern (FAP) An innate behavior triggered by a sign stimulus.

flagellum (flah-jel′um) (pl., *flagella*) A long, whiplike structure extending from certain cells and used in locomotion. (1) Eukaryote flagella consist of two central, single microtubules surrounded by nine double microtubules (9 + 2 structure), all covered by a plasma membrane. (2) Prokaryote flagella are filaments rotated by special structures located in the plasma membrane and cell wall.

flame cells Collecting cells that have cilia; part of the osmoregulatory system of flatworms.

flavin adenine dinucleotide See *FAD/FADH$_2$*.

flowering plants See *angiosperms*.

flowing-water ecosystem A river or stream ecosystem.

fluid-mosaic model The currently accepted model of the plasma membrane and other cell membranes, in which protein molecules "float" in a fluid phospholipid bilayer.

fluorescence The emission of light of a longer wavelength (lower energy) than the light originally absorbed.

fluorescent in situ hybridization (FISH) A technique to detect specific DNA segments by hybridization directly to chromosomes; visualized microscopically by using a fluorescent dye.

follicle (fol′i-kl) (1) A simple, dry, dehiscent fruit that develops from a single carpel and splits open at maturity along one suture to liberate the seeds. (2) A small sac of cells in the mammalian ovary that contains a maturing egg. (3) The pocket in the skin from which a hair grows.

follicle-stimulating hormone (FSH) A gonadotropic hormone secreted by the anterior lobe of the pituitary gland; stimulates follicle development in the ovaries of females and sperm production in the testes of males.

food chain The series of organisms through which energy flows in an ecosystem. Each organism in the series eats or decomposes the preceding organism in the chain. See *food web*.

food web A complex interconnection of all the food chains in an ecosystem.

foramen magnum The opening in the vertebrate skull through which the spinal cord passes.

foraminiferan (for″am-in-if′er-an) A marine protozoon that produces a shell, or test, that encloses an amoeboid body.

forebrain In the early embryo, one of the three divisions of the developing vertebrate brain; subdivides to form the telencephalon, which gives rise to the cerebrum, and the diencephalon, which gives rise to the thalamus and hypothalamus. Compare with *midbrain* and *hindbrain*.

forest decline A gradual deterioration (and often death) of many trees in a forest; can be caused by a combination of factors, such as acid precipitation, toxic heavy metals, and surface-level ozone.

fossil Parts or traces of an ancient organism usually preserved in rock.

fossil fuel Combustible deposits in Earth's crust that are composed of the remnants of prehistoric organisms that existed millions of years ago, e.g., oil, natural gas, and coal.

founder cell A cell from which a particular cell lineage is derived.

founder effect Genetic drift that results from a small population colonizing a new area.

fovea (foe′vee-ah) The area of sharpest vision in the retina; cone cells are concentrated here.

fragile site A weak point at a specific location on a chromosome where part of a chromatid appears attached to the rest of the chromosome by a thin thread of DNA.

fragile X syndrome A human genetic disorder caused by a fragile site that occurs near the tip on the X chromosome; effects range from mild learning disabilities to severe mental retardation and hyperactivity.

frameshift mutation A mutation that results when one or two nucleotide pairs are inserted into or deleted from the DNA. The change causes the mRNA transcribed from the mutated DNA to have an altered reading frame such that all codons downstream from the mutation are changed.

fraternal twins See *dizygotic twins.*

free energy The maximum amount of energy available to do work under the conditions of a biochemical reaction.

free radicals Toxic, highly reactive compounds with unpaired electrons that bond with other compounds in the cell and interfere with normal function.

frequency-dependent selection Selection in which the relative fitness of different genotypes is related to how frequently they occur in the population.

freshwater wetlands Land that is transitional between freshwater and terrestrial ecosystems and is covered with water for at least part of the year, e.g., marshes and swamps.

frontal lobes In mammals, the anterior part of the cerebrum.

fruit In flowering plants, a mature, ripened ovary. Fruits contain seeds and usually provide seed protection and dispersal.

fruiting body A multicellular structure that contains the sexual spores of certain fungi; refers to the ascocarp of an ascomycete and the basidiocarp of a basidiomycete.

fucoxanthin (few″koh-zan′thin) The brown carotenoid pigment found in brown algae, golden algae, diatoms, and dinoflagellates.

functional genomics The study of the roles of genes in cells.

functional group A group of atoms that confers distinctive properties on an organic molecule (or region of a molecule) to which it is attached, e.g., hydroxyl, carbonyl, carboxyl, amino, phosphate, and sulfhydryl groups.

fundamental niche The potential ecological niche that an organism could occupy if there were no competition from other species. Compare with *realized niche.*

fungus (pl., *fungi*) A heterotrophic eukaryote with chitinous cell walls and a body usually in the form of a mycelium of branched, threadlike hyphae. Most fungi are decomposers; some are parasitic.

G protein One of a group of proteins that bind GTP and are involved in the transfer of signals across the plasma membrane.

G₁ phase The first gap phase within the interphase stage of the cell cycle; G_1 occurs before DNA synthesis (S phase) begins. Compare with *S* and G_2 *phases.*

G₂ phase Second gap phase within the interphase stage of the cell cycle; G_2 occurs after DNA synthesis (S phase) and before mitosis. Compare with *S* and G_1 *phases.*

gallbladder A small sac that stores bile.

gametangium (gam″uh-tan′gee-um) Special multicellular or unicellular structure of plants, protists, and fungi in which gametes are formed.

gamete (gam′eet) A sex cell; in plants and animals, an egg or sperm. In sexual reproduction, the union of gametes results in the formation of a zygote. The chromosome number of a gamete is designated *n*.

gametic isolation (gam-ee′tik) A prezygotic reproductive isolating mechanism in which sexual reproduction between two closely related species cannot occur because of chemical differences in the gametes.

gametogenesis The process of gamete formation. See *spermatogenesis* and *oogenesis.*

gametophyte generation (gam-ee′toh-fite) The *n*, gamete-producing stage in the life cycle of a plant. Compare with *sporophyte generation.*

gamma-aminobutyric acid (GABA) A neurotransmitter that has an inhibitory effect.

ganglion (gang′glee-on) (pl., *ganglia*) A mass of neuron cell bodies.

ganglion cell A type of neuron in the retina of the eye; receives input from bipolar cells.

gap junction Structure consisting of specialized regions of the plasma membrane of two adjacent cells; contains numerous pores that allow the passage of certain small molecules and ions between them.

gastrin (gas′trin) A hormone released by the stomach mucosa; stimulates the gastric glands to secrete pepsinogen.

gastrovascular cavity A central digestive cavity with a single opening that functions as both mouth and anus; characteristic of cnidarians and flatworms.

gastrula (gas′troo-lah) A three-layered embryo formed by the process of gastrulation.

gastrulation (gas-troo-lay′shun) Process in embryonic development during which the three germ layers (ectoderm, mesoderm, and endoderm) form.

gel electrophoresis Procedure by which proteins or nucleic acids are separated on the basis of size and charge as they migrate through a gel in an electric field.

gene A segment of DNA that serves as a unit of hereditary information; includes a transcribable DNA sequence (plus associated sequences regulating its transcription) that yields a protein or RNA product with a specific function.

gene amplification The developmental process in which certain cells produce multiple copies of a gene by selective replication, thus allowing for increased synthesis of the gene product. Compare with *nuclear equivalence* and *genomic rearrangement.*

gene flow The movement of alleles between local populations due to the migration of individuals; can have significant evolutionary consequences.

gene locus See *locus.*

gene pool All the alleles of all the genes present in a freely interbreeding population.

gene therapy Any of a variety of methods designed to correct a disease or alleviate its symptoms through the introduction of genes into the affected person's cells.

genetic bottleneck See *bottleneck.*

genetic code See *codon.*

genetic counseling Medical and genetic information provided to couples who are concerned about the risk of abnormality in their children.

genetic drift A random change in allele frequency in a small breeding population.

genetic engineering Manipulation of genes, often through recombinant DNA technology.

genetic equilibrium The condition of a population that is not undergoing evolutionary change, i.e., in which allele and genotype frequencies do not change from one generation to the next. See *Hardy–Weinberg principle.*

genetic polymorphism (pol″ee-mor′fizm) The presence in a population of two or more alleles for a given gene locus.

genetic probe A single-stranded nucleic acid (either DNA or RNA) used to identify a complementary sequence by hydrogen-bonding to it.

genetic recombination See *recombination, genetic.*

genetic screening A systematic search through a population for individuals with a genotype or karyotype that might cause a serious genetic disease in them or their offspring.

genetics The science of heredity; includes genetic similarities and genetic variation between parents and offspring or among individuals of a population.

genome (jee′nome) Originally, all the genetic material in a cell or individual organism. The term is used in more than one way depending on context, e.g., an organism's haploid genome is all the DNA contained in one haploid set of its chromosomes, and its mitochondrial genome is all the DNA in a mitochondrion. See *human genome*.

genomic DNA library A collection of recombinant plasmids in which all the DNA in the genome is represented. Compare with *cDNA library*.

genomic imprinting See *imprinting*, first definition.

genomic rearrangement A physical change in the structure of one or more genes that occurs during the development of an organism and leads to an alteration in gene expression; compare with *nuclear equivalence* and *gene amplification*.

genomics The emerging field of biology that studies the entire DNA sequence of an organism's genome to identify all the genes, determine their RNA or protein products, and ascertain how the genes are regulated.

genotype (jeen′oh-type) The genetic makeup of an individual. Compare with *phenotype*.

genotype frequency The proportion of a particular genotype in the population.

genus (jee′nus) A taxonomic category made up of related species.

geometric isomer One of two or more chemical compounds having the same arrangement of covalent bonds but differing in the spatial arrangement of their atoms or groups of atoms.

germ layers In animals, three embryonic tissue layers: endoderm, mesoderm, and ectoderm.

germ line cell In animals, a cell that is part of the line of cells that will ultimately undergo meiosis to form gametes. Compare with *somatic cell*.

germination Resumption of growth of an embryo or spore; occurs when a seed or spore sprouts.

germplasm Any plant or animal material that may be used in breeding; includes seeds, plants, and plant tissues of traditional crop varieties and the sperm and eggs of traditional livestock breeds.

gibberellin (jib″ur-el′lin) A plant hormone involved in many aspects of plant growth and development, such as stem elongation, flowering, and seed germination.

gills (1) The respiratory organs characteristic of many aquatic animals, usually thin-walled projections from the body surface or from some part of the digestive tract. (2) The spore-bearing, platelike structures under the caps of mushrooms.

ginkgo (ging′ko) A member of an ancient gymnosperm group that consists of a single living representative (*Ginkgo biloba*), a hardy, deciduous tree with broad, fan-shaped leaves and naked, fleshy seeds (on female trees).

gland See *endocrine gland* and *exocrine gland*.

glial cells (glee′ul) In nervous tissue, cells that support and nourish neurons; they also communicate with neurons and have several other functions; also see *astrocyte, microglia,* and *oligodendrocyte*.

globulin (glob′yoo-lin) One of a class of proteins in blood plasma, some of which (gamma globulins) function as antibodies.

glomeromycetes A group of fungal symbionts that form arbuscular mycorrhizae with the roots of many plants; belong to phylum Glomeromycota.

glomerulus (glom-air′yoo-lus) The cluster of capillaries at the proximal end of a nephron; the glomerulus is surrounded by Bowman's capsule.

glucagon (gloo′kah-gahn) A hormone secreted by the pancreas that stimulates glycogen breakdown, thereby increasing the concentration of glucose in the blood. Compare with *insulin*.

glucose A hexose aldehyde sugar that is central to many metabolic processes.

glutamate An amino acid that functions as the major excitatory neurotransmitter in the vertebrate brain.

glyceraldehyde-3-phosphate (G3P) Phosphorylated three-carbon compound that is an important intermediate in glycolysis and in the Calvin cycle.

glycerol A three-carbon alcohol with a hydroxyl group on each carbon; a component of triacylglycerols and phospholipids, as well as monoacylglycerols and diacylglycerols.

glycocalyx (gly″koh-kay′lix) A coating on the outside of an animal cell, formed by the polysaccharide portions of glycoproteins and glycolipids associated with the plasma membrane.

glycogen (gly′koh-jen) The principal storage polysaccharide in animal cells; formed from glucose and stored primarily in the liver and, to a lesser extent, in muscle cells.

glycolipid A lipid with covalently attached carbohydrates.

glycolysis (gly-kol′ih-sis) The first stage of cellular respiration, literally the "splitting of sugar." The metabolic conversion of glucose into pyruvate, accompanied by the production of ATP.

glycoprotein (gly′koh-pro-teen) A protein with covalently attached carbohydrates.

glycosidic linkage Covalent linkage joining two sugars; includes an oxygen atom bonded to a carbon of each sugar.

glyoxysomes (gly-ox′ih-sohmz) Membrane-enclosed structures in cells of certain plant seeds; contain a large array of enzymes that convert stored fat to sugar.

gnetophyte (nee′toe-fite) One of a small phylum of unusual gymnosperms that have some features similar to those of flowering plants.

goblet cells Unicellular glands that secrete mucus.

goiter (goy′ter) An enlargement of the thyroid gland.

golden alga A member of a phylum of algae, most of which are biflagellate, are unicellular, and contain pigments, including chlorophylls *a* and *c* and carotenoids, including fucoxanthin.

Golgi complex (goal′jee) Organelle composed of stacks of flattened, membranous sacs. Mainly responsible for modifying, packaging, and sorting proteins that will be secreted or targeted to other organelles of the internal membrane system or to the plasma membrane; also called *Golgi body* or *Golgi apparatus*.

gonad (goh′nad) A gamete-producing gland; an ovary or a testis.

gonadotropic hormones (go-nad-oh-troh′pic) Hormones produced by the anterior pituitary gland that stimulate the testes and ovaries; include follicle-stimulating hormone (FSH) and luteinizing hormone (LH).

gonadotropin-releasing hormone (GnRH) A hormone secreted by the hypothalamus that stimulates the anterior pituitary to secrete the gonadotropic hormones: follicle-stimulating hormone (FSH) and luteinizing hormone (LH).

graded potential A local change in electrical potential that varies in magnitude depending on the strength of the applied stimulus.

gradualism The idea that evolution occurs by a slow, steady accumulation of genetic changes over time. Compare with *punctuated equilibrium*.

graft rejection An immune response directed against a transplanted tissue or organ.

grain A simple, dry, one-seeded fruit in which the fruit wall is fused to the seed coat, e.g., corn and wheat kernels. Also called *caryopsis*.

granulosa cells In mammals, cells that surround the developing oocyte and are part of the follicle; produce estrogens and inhibin.

granum (pl., *grana*) A stack of thylakoids within a chloroplast.

gravitropism (grav"ih-troh'pizm) Growth of a plant in response to gravity.

gray crescent The grayish area of cytoplasm that marks the region where gastrulation begins in an amphibian embryo.

gray matter Nervous tissue in the brain and spinal cord that contains cell bodies, dendrites, and unmyelinated axons. Compare with *white matter*.

green alga A member of a diverse phylum of algae that contain the same pigments as plants (chlorophylls *a* and *b* and carotenoids).

greenhouse effect The natural global warming of Earth's atmosphere caused by the presence of carbon dioxide and other gases that trap the sun's radiation. The additional warming produced when increased levels of greenhouse gases absorb infrared radiation is known as the *enhanced greenhouse effect*.

greenhouse gases Trace gases in the atmosphere that allow the sun's energy to penetrate to Earth's surface but do not allow as much of it to escape as heat.

gross primary productivity The rate at which energy accumulates (is assimilated) in an ecosystem during photosynthesis. Compare with *net primary productivity*.

ground state The lowest energy state of an atom.

ground tissue system All tissues in the plant body other than the dermal tissue system and vascular tissue system; consists of parenchyma, collenchyma, and sclerenchyma.

growth factors A group of more than 50 extracellular peptides that signal certain cells to grow and divide.

growth hormone (GH) A hormone secreted by the anterior lobe of the pituitary gland; stimulates growth of body tissues; also called *somatotropin*.

growth rate The rate of change of a population's size on a per capita basis.

guanine (gwan'een) A nitrogenous purine base that is a component of nucleic acids and GTP.

guanosine triphosphate (GTP) An energy transfer molecule similar to ATP that releases free energy with the hydrolysis of its terminal phosphate group.

guard cell One of a pair of epidermal cells that adjust their shape to form a stomatal pore for gas exchange.

guttation (gut-tay'shun) The appearance of water droplets on leaves, forced out through leaf pores by root pressure.

gymnosperm (jim'noh-sperm) Any of a group of seed plants in which the seeds are not enclosed in an ovary; gymnosperms frequently bear their seeds in cones. Includes four phyla: conifers, cycads, ginkgoes, and gnetophytes.

habitat The natural environment or place where an organism, population, or species lives.

habitat fragmentation The division of habitats that formerly occupied large, unbroken areas into smaller pieces by roads, fields, cities, and other human land-transforming activities.

habitat isolation A prezygotic reproductive isolating mechanism in which reproduction between similar species is prevented because they live and breed in different habitats.

habituation (hab-it"yoo-ay'shun) A type of learning in which an animal becomes accustomed to a repeated, irrelevant stimulus and no longer responds to it.

hair cell A vertebrate mechanoreceptor found in the lateral line of fishes, the vestibular apparatus, semicircular canals, and cochlea.

half-life The period of time required for a radioisotope to change into a different material.

haploid (hap'loyd) The condition of having one set of chromosomes per nucleus. Compare with *diploid* and *polyploid*.

"hard-wiring" Refers to how neurons signal one another, how they connect, and how they carry out basic functions such as regulating heart rate, blood pressure, and sleep–wake cycles.

Hardy–Weinberg principle The mathematical prediction that allele frequencies do not change from generation to generation in a large population in the absence of microevolutionary processes (mutation, genetic drift, gene flow, natural selection).

haustorium (hah-stor'ee-um) (pl., *haustoria*) In parasitic fungi, a specialized hypha that penetrates a host cell and obtains nourishment from the cytoplasm.

Haversian canals (ha-vur'zee-un) Channels extending through the matrix of bone; contain blood vessels and nerves.

heat The total amount of kinetic energy in a sample of a substance.

heat energy The thermal energy that flows from an object with a higher temperature to an object with a lower temperature.

heat of vaporization The amount of heat energy that must be supplied to change one gram of a substance from the liquid phase to the vapor phase.

helicases Enzymes that unwind the two strands of a DNA double helix.

helper T cell See *T helper cell*.

hemichordates A phylum of sedentary, wormlike deuterostomes.

hemizygous (hem"ih-zy'gus) Possessing only one allele for a particular locus; a human male is hemizygous for all X-linked genes. Compare with *homozygous* and *heterozygous*.

hemocoel Blood cavity characteristic of animals with an open circulatory system.

hemocyanin A hemolymph pigment that transports oxygen in some mollusks and arthropods.

hemoglobin (hee'moh-gloh"bin) The red, iron-containing protein pigment in blood that transports oxygen and carbon dioxide and aids in regulation of pH.

hemolymph (hee'moh-limf) The fluid that bathes the tissues in animals with an open circulatory system, e.g., arthropods and most mollusks.

hemophilia (hee"moh-feel'ee-ah) A hereditary disease in which blood does not clot properly; the form known as *hemophilia A* has an X-linked, recessive inheritance pattern.

Hensen's node See *primitive streak*.

hepatic (heh-pat'ik) Pertaining to the liver.

hepatic portal system The portion of the circulatory system that carries blood from the intestine through the liver.

herbivore (erb'uh-vore) An animal that feeds on plants or algae. Also called *primary consumer*.

heredity The transmission of genetic information from parent to offspring.

hermaphrodite (her-maf'roh-dite) An organism that has both male and female sex organs.

heterochromatin (het"ur-oh-kroh'mah-tin) Highly coiled and compacted chromatin in an inactive state. Compare with *euchromatin*.

heterocyst (het'ur-oh-sist") An oxygen-excluding cell of cyanobacteria that is the site of nitrogen fixation.

heterogametic A term describing an individual that produces two classes of gametes with respect to their sex chromosome constitutions. Human males (XY) are heterogametic, producing X and Y sperm. Compare with *homogametic*.

heterospory (het"ur-os'pur-ee) Production of two types of *n* spores, microspores (male) and megaspores (female). Compare with *homospory*.

heterothallic (het″ur-oh-thal′ik) Pertaining to certain algae and fungi that have two mating types; only by combining a plus strain and a minus strain can sexual reproduction occur. Compare with *homothallic*.

heterotroph (het′ur-oh-trof) An organism that cannot synthesize its own food from inorganic raw materials and therefore must obtain energy and body-building materials from other organisms. Also called *consumer*. Compare with *autotroph*. See *chemoheterotroph* and *photoheterotroph*.

heterozygote advantage A phenomenon in which the heterozygous condition confers some special advantage on an individual that either homozygous condition does not (i.e., *Aa* has a higher degree of fitness than does *AA* or *aa*).

heterozygous (het-ur″oh-zye′gus) Having a pair of unlike alleles for a particular locus. Compare with *homozygous*.

hexose A monosaccharide containing six carbon atoms.

hibernation Long-term torpor in response to winter cold and scarcity of food. Compare with *estivation*.

high-density lipoprotein (HDL) See *lipoprotein*.

hindbrain In the early embryo, one of the three divisions of the developing vertebrate brain; subdivides to form the metencephalon, which gives rise to the cerebellum and pons, and the myelencephalon, which gives rise to the medulla. Compare with *forebrain* and *midbrain*.

histamine (his′tah-meen) Substance released from mast cells that is involved in allergic and inflammatory reactions.

histones (his′tohnz) Small, positively charged (basic) proteins in the cell nucleus that bind to the negatively charged DNA. See *nucleosomes*.

holdfast The basal structure for attachment to solid surfaces found in multicellular algae.

holoblastic cleavage A cleavage pattern in which the entire embryo cleaves; characteristic of eggs with little or moderate yolk (isolecithal or moderately telolecithal), e.g., the eggs of echinoderms, amphioxus, and mammals. Compare with *meroblastic cleavage*.

home range A geographic area that an individual animal seldom or never leaves. Compare with *range*.

homeobox A short (180-nucleotide) DNA sequence that characterizes many homeotic genes as well as some other genes that play a role in development.

homeodomain A functional region of certain transcription factors; consists of approximately 60 amino acids specified by a homeobox DNA sequence and includes a recognition alpha helix, which binds to specific DNA sequences and affects their transcription.

homeostasis (home″ee-oh-stay′sis) The balanced internal environment of the body; the automatic tendency of an organism to maintain such a steady state.

homeotic gene (home″ee-ah′tik) A gene that controls the formation of specific structures during development. Such genes were originally identified through insect mutants in which one body part is substituted for another.

hominid (hah′min-id) Any of a group of extinct and living humans.

hominoid (hah′min-oid) The apes and hominids.

homogametic Term describing an individual that produces gametes with identical sex chromosome constitutions. Human females (XX) are homogametic, producing all X eggs. Compare with *heterogametic*.

homologous chromosomes (hom-ol′ah-gus) Chromosomes that are similar in morphology and genetic constitution. In humans there are 23 pairs of homologous chromosomes; one member of each pair is inherited from the mother, and the other from the father.

homologous features See *homology*.

homology Similarity in different species that results from their derivation from a common ancestor. The features that exhibit such similarity are called *homologous features*. Compare with *homoplasy*.

homoplastic features See *homoplasy*.

homoplasy Similarity in the characters in different species that is due to convergent evolution, not common descent. Characters that exhibit such similarity are called *homoplastic features*. Compare with *homology*.

homospory (hoh″mos′pur-ee) Production of one type of *n* spore that gives rise to a bisexual gametophyte. Compare with *heterospory*.

homothallic (hoh″moh-thal′ik) Pertaining to certain algae and fungi that are self-fertile. Compare with *heterothallic*.

homozygous (hoh″moh-zy′gous) Having a pair of identical alleles for a particular locus. Compare with *heterozygous*.

hormone An organic chemical messenger in multicellular organisms that is produced in one part of the body and often transported to another part where it signals cells to alter some aspect of metabolism.

hornwort A phylum of spore-producing, nonvascular, thallose plants with a life cycle similar to that of mosses.

host mothering The introduction of an embryo from one species into the uterus of another species, where it implants and develops; the host mother subsequently gives birth and may raise the offspring as her own.

***Hox* genes** Clusters of homeobox-containing genes that specify the anterior-posterior axis of various animals during development.

human chorionic gonadotropin (hCG) A hormone secreted by cells surrounding the early embryo; signals the mother's corpus luteum to continue to function.

human genetics The science of inherited variation in humans.

human genome The totality of genetic information in human cells; includes the DNA content of both the nucleus and mitochondria. See *genome*.

human immunodeficiency virus (HIV) The retrovirus that causes AIDS (acquired immunodeficiency syndrome).

human leukocyte antigen (HLA) See *major histocompatibility complex*.

humus (hew′mus) Organic matter in various stages of decomposition in the soil; gives soil a dark brown or black color.

Huntington disease A genetic disease that has an autosomal dominant inheritance pattern and causes mental and physical deterioration.

hybrid The offspring of two genetically dissimilar parents.

hybrid breakdown A postzygotic reproductive isolating mechanism in which, although an interspecific hybrid is fertile and produces a second (F_2) generation, the F_2 has defects that prevent it from successfully reproducing.

hybrid inviability A postzygotic reproductive isolating mechanism in which the embryonic development of an interspecific hybrid is aborted.

hybrid sterility A postzygotic reproductive isolating mechanism in which an interspecific hybrid cannot reproduce successfully.

hybrid vigor The genetic superiority of an F_1 hybrid over either parent, caused by the presence of heterozygosity for a number of different loci.

hybrid zone An area of overlap between two closely related populations, subspecies, or species in which interbreeding occurs.

hybridization (1) Interbreeding between members of two different taxa. (2) Interbreeding between genetically dissimilar parents. (3) In molecular biology, complementary base pairing between nucleic acid (DNA or RNA) strands from different sources.

hydration Process of association of a substance with the partial positive and/or negative charges of water molecules.

hydrocarbon An organic compound composed solely of hydrogen and carbon atoms.

hydrogen bond A weak attractive force existing between a hydrogen atom with a partial positive charge and an electronegative atom (usually oxygen or nitrogen) with a partial negative charge. Compare with *covalent bond* and *ionic bond*.

hydrologic cycle The water cycle, which includes evaporation, precipitation, and flow to the ocean; supplies terrestrial organisms with a continual supply of fresh water.

hydrolysis Reaction in which a covalent bond between two subunits is broken through the addition of the equivalent of a water molecule; a hydrogen atom is added to one subunit and a hydroxyl group to the other. Compare with *condensation synthesis*.

hydrophilic Interacting readily with water; having a greater affinity for water molecules than they have for each other. Compare with *hydrophobic*.

hydrophobic Not readily interacting with water; having less affinity for water molecules than they have for each other. Compare with *hydrophilic*.

hydroponics (hy″dra-paun′iks) Growing plants in an aerated solution of dissolved inorganic minerals, i.e., without soil.

hydrostatic skeleton A type of skeleton found in some invertebrates in which contracting muscles push against a tube of fluid.

hydroxide ion An anion (negatively charged particle) consisting of oxygen and hydrogen; usually written OH^-.

hydroxyl group (hy-drok′sil) Polar functional group; abbreviated —OH.

hyperpolarize To change the membrane potential so that the inside of the cell becomes more negative than its resting potential.

hypertonic A term referring to a solution having an osmotic pressure (or solute concentration) greater than that of the solution with which it is compared. Compare with *hypotonic* and *isotonic*.

hypha (hy′fah) (pl., *hyphae*) One of the threadlike filaments composing the mycelium of a water mold or fungus.

hypocotyl (hy′poh-kah″tl) The part of the axis of a plant embryo or seedling below the point of attachment of the cotyledons.

hypothalamus (hy-poh-thal′uh-mus) Part of the vertebrate brain; in mammals it regulates the pituitary gland, the autonomic system, emotional responses, body temperature, water balance, and appetite; located below the thalamus.

hypothesis A testable statement about the nature of an observation or relationship. Compare with *theory*.

hypothetico-deductive approach Emphasizes the use of deductive reasoning to test hypotheses. Compare with *hypothetico-inductive approach*. See *deductive reasoning*.

hypothetico-inductive approach Emphasizes the use of inductive reasoning to discover new general principles. Compare with *hypothetico-deductive approach*. See *inductive reasoning*.

hypotonic A term referring to a solution having an osmotic pressure (or solute concentration) less than that of the solution with which it is compared. Compare with *hypertonic* and *isotonic*.

hypotrichs A group of dorsoventrally flattened ciliates that exhibit an unusual creeping–darting locomotion.

identical twins See *monozygotic twins*.

illuviation The deposition of material leached from the upper layers of soil into the lower layers.

imaginal discs Paired structures in an insect larva that develop into specific adult structures during complete metamorphosis.

imago (ih-may′go) The adult form of an insect.

imbibition (im″bi-bish′en) The absorption of water by a seed prior to germination.

immigration The movement of individuals into a population. Compare with *emigration*.

immune response Process of recognizing foreign macromolecules and mounting a response aimed at eliminating them. See *specific* and *nonspecific immune responses; primary* and *secondary immune responses*.

immunoglobulin (im-yoon″oh-glob′yoo-lin) See *antibody*.

imperfect flower A flower that lacks either stamens or carpels. Compare with *perfect flower*.

imperfect fungi See *deuteromycetes*.

implantation The embedding of a developing embryo in the inner lining (endometrium) of the uterus.

implicit memory The unconscious memory for perceptual and motor skills, e.g., riding a bicycle.

imprinting (1) The expression of a gene based on its parental origin; also called *genomic imprinting*. (2) A type of learning by which a young bird or mammal forms a strong social attachment to an individual (usually a parent) or object within a few hours after hatching or birth.

in situ conservation Conservation efforts that concentrate on preserving biological diversity in the wild. Compare with *ex situ conservation*.

in vitro Occurring outside a living organism (literally "in glass"). Compare with *in vivo*.

in vitro evolution Test tube experiments that demonstrate that RNA molecules in the RNA world could have catalyzed the many different chemical reactions needed for life. Also called *directed evolution*.

in vitro fertilization The fertilization of eggs in the laboratory prior to implantation in the uterus for development..

in vivo Occurring in a living organism. Compare with *in vitro*.

inborn error of metabolism A metabolic disorder caused by the mutation of a gene that codes for an enzyme needed for a biochemical pathway.

inbreeding The mating of genetically similar individuals. Homozygosity increases with each successive generation of inbreeding. Compare with *outbreeding*.

inbreeding depression The phenomenon in which inbred offspring of genetically similar individuals have lower fitness (e.g., decline in fertility and high juvenile mortality) than do noninbred individuals.

inclusive fitness The total of an individual's direct and indirect fitness; includes the genes contributed directly to offspring and those contributed indirectly by kin selection. Compare with *direct fitness*. See *kin selection*.

incomplete dominance A condition in which neither member of a pair of contrasting alleles is completely expressed when the other is present.

incomplete flower A flower that lacks one or more of the four parts: sepals, petals, stamens, and/or carpels. Compare with *complete flower*.

independent assortment, principle of The genetic principle, first noted by Gregor Mendel, that states that the alleles of unlinked loci are randomly distributed to gametes.

indeterminate growth Unrestricted growth, as for example, in stems and roots. Compare with *determinate growth*.

index fossils Fossils restricted to a narrow period of geologic time and found in the same sedimentary layers in different geographic areas.

indoleacetic acid See *auxin*.

induced fit Conformational change in the active site of an enzyme that occurs when it binds to its substrate.

inducer A molecule that binds to a repressor protein, converting it to its inactive form, which is unable to prevent transcription.

inducible operon An operon that is normally inactive because a repressor molecule is attached to its operator; transcription is activated when an inducer binds to the repressor, making it incapable of binding to the operator. Compare with *repressible operon.*

induction The process by which the differentiation of a cell or group of cells is influenced by interactions with neighboring cells.

inductive reasoning The reasoning that uses specific examples to draw a general conclusion or discover a general principle. Compare with *deductive reasoning.* See *hypothetico-inductive approach.*

infant mortality rate The number of infant deaths per 1000 live births. (A child is an infant during its first 2 years of life.)

inflammatory response The response of body tissues to injury or infection, characterized clinically by heat, swelling, redness, and pain, and physiologically by increased dilation of blood vessels and increased phagocytosis.

inflorescence A cluster of flowers on a common floral stalk.

ingestion The process of taking food (or other material) into the body.

ingroup See *outgroup.*

inhibin A hormone that inhibits FSH secretion; produced by Sertoli cells in the testes and by granulosa cells in the ovaries.

inhibitory postsynaptic potential (IPSP) A change in membrane potential that takes a neuron farther from the firing level. Compare with *excitatory postsynaptic potential (EPSP).*

initiation (of protein synthesis) The first steps of protein synthesis, in which the large and small ribosomal subunits and other components of the translation machinery bind to the 5′ end of mRNA. See *elongation* and *termination.*

initiation codon See *start codon.*

innate behavior Behavior that is inherited and typical of the species; also called *instinct.*

innate immune responses See *nonspecific immune responses.*

inner cell mass The cluster of cells in the early mammalian embryo that gives rise to the embryo proper.

inorganic compound A simple substance that does not contain a carbon backbone. Compare with *organic compound.*

inositol trisphosphate (IP$_3$) A second messenger that increases intracellular calcium concentration and activates enzymes.

insight learning A complex learning process in which an animal adapts past experience to solve a new problem that may involve different stimuli.

instinct See *innate behavior.*

insulin (in′suh-lin) A hormone secreted by the pancreas that lowers blood glucose concentration. Compare with *glucagon.*

insulin-like growth factors (IGFs) Somatomedins; proteins that mediate responses to growth hormone.

insulin resistance See *diabetes mellitus.*

insulin shock A condition in which the blood glucose concentration is so low that the individual may appear intoxicated or may become unconscious and even die; caused by the injection of too much insulin or by certain metabolic malfunctions.

integral membrane protein A protein that is tightly associated with the lipid bilayer of a biological membrane; a transmembrane integral protein spans the bilayer. Compare with *peripheral membrane protein.*

integration The process of summing (adding and subtracting) incoming neural signals.

integrins Receptor proteins that bind to specific proteins in the extracellular matrix and to membrane proteins on adjacent cells; transmit signals into the cell from the extracellular matrix.

integumentary system (in-teg″yoo-men′tar-ee) The body's covering, including the skin and its nails, glands, hair, and other associated structures.

integuments The outer cell layers that surround the megasporangium of an ovule; develop into the seed coat.

intercellular substance In connective tissues, the combination of matrix and fibers in which the cells are embedded.

interference competition Intraspecific competition in which certain dominant individuals obtain an adequate supply of the limited resource at the expense of other individuals in the population. Also called *contest competition.* Compare with *exploitation competition.*

interferons (in″tur-feer′onz) Cytokines produced by animal cells when challenged by a virus; prevent viral reproduction and enable cells to resist a variety of viruses.

interkinesis The stage between meiosis I and meiosis II. Interkinesis is usually brief; the chromosomes may decondense, reverting at least partially to an interphase-like state, but DNA synthesis and chromosome duplication do not occur.

interleukins A diverse group of cytokines produced mainly by macrophages and lymphocytes.

intermediate-day plant A plant that flowers when it is exposed to days and nights of intermediate length but does not flower when the day length is too long or too short. Compare with *long-day, short-day,* and *day-neutral plants.*

intermediate disturbance hypothesis In community ecology, the idea that species richness is greatest at moderate levels of disturbance, which create a mosaic of habitat patches at different stages of succession.

intermediate filaments Cytoplasmic fibers that are part of the cytoskeletal network and are intermediate in size between microtubules and microfilaments.

internal membrane system The group of membranous structures in eukaryotic cells that interact through direct connections by vesicles; includes the endoplasmic reticulum, outer membrane of the nuclear envelope, Golgi complex, lysosomes, and the plasma membrane; also called *endomembrane system.*

interneuron (in″tur-noor′on) A nerve cell that carries impulses from one nerve cell to another and is not directly associated with either an effector or a sensory receptor. Also known as an *association neuron.*

internode The region on a stem between two successive nodes. Compare with *node.*

interoceptor (in′tur-oh-sep″tor) A sense organ within a body organ that transmits information regarding chemical composition, pH, osmotic pressure, or temperature. Compare with *exteroceptor.*

interphase The stage of the cell cycle between successive mitotic divisions; its subdivisions are the G$_1$ (first gap), S (DNA synthesis), and G$_2$ (second gap) phases.

interspecific competition The interaction between members of different species that vie for the same resource in an ecosystem (e.g., food or living space). Compare with *intraspecific competition.*

interstitial cells (of testis) The cells between the seminiferous tubules that secrete testosterone.

interstitial fluid The fluid that bathes the tissues of the body; also called *tissue fluid.*

intertidal zone The marine shoreline area between the high-tide mark and the low-tide mark.

intraspecific competition The interaction between members of the same species that vie for the same resource in an ecosystem (e.g., food or living space). Compare with *interspecific competition.*

intrinsic rate of increase The theoretical maximum rate of increase in population size occurring under optimal environmental conditions. Also called *biotic potential.*

intron A non-protein-coding region of a eukaryotic gene and also of the pre-mRNA transcribed from such a region. Introns do not appear in mRNA. Compare with *exon*.

invasive species A foreign species that, when introduced into an area where it is not native, upsets the balance among the organisms living there and causes economic or environmental harm.

inversion A chromosome abnormality in which the breakage and rejoining of chromosome parts results in a chromosome segment that is oriented in the opposite (reverse) direction.

invertebrate An animal without a backbone (vertebral column); invertebrates account for about 95% of animal species.

ion An atom or group of atoms bearing one or more units of electric charge, either positive (cation) or negative (anion).

ion channels Channels for the passage of ions through a membrane; formed by specific membrane proteins.

ionic bond The chemical attraction between a cation and an anion. Compare with *covalent bond* and *hydrogen bond*.

ionic compound A substance consisting of cations and anions, which are attracted by their opposite charges; ionic compounds do not consist of molecules. Compare with *covalent compound*.

ionization The dissociation of a substance to yield ions, e.g., the ionization of water yields H^+ and OH^-.

iris The pigmented portion of the vertebrate eye.

iron–sulfur world hypothesis The hypothesis that simple organic molecules that are the precursors of life originated at hydrothermal vents in the deep-ocean floor. Compare with *prebiotic soup hypothesis*.

irreversible inhibitor A substance that permanently inactivates an enzyme. Compare with *reversible inhibitor*.

islets of Langerhans (eye′lets of lahng′er-hanz) The endocrine portion of the pancreas that secretes glucagon and insulin, hormones that regulate the concentration of glucose in the blood.

isogamy (eye-sog′uh-me) Sexual reproduction involving motile gametes of similar form and size. Compare with *anisogamy* and *oogamy*.

isolecithal egg An egg containing a relatively small amount of uniformly distributed yolk. Compare with *telolecithal egg*.

isomer (eye′soh-mer) One of two or more chemical compounds having the same chemical formula but different structural formulas, e.g., structural and geometrical isomers and enantiomers.

isoprene units Five-carbon hydrocarbon monomers that make up certain lipids such as carotenoids and steroids.

isotonic (eye″soh-ton′ik) A term applied to solutions that have identical concentrations of solute molecules and hence the same osmotic pressure. Compare with *hypertonic* and *hypotonic*.

isotope (eye′suh-tope) An alternative form of an element with a different number of neutrons but the same number of protons and electrons. See *radioisotopes*.

iteroparity The condition of having repeated reproductive cycles throughout a lifetime. Compare with *semelparity*.

jasmonate One of a group of lipid-derived plant hormones that affect several processes, such as pollen development, root growth, fruit ripening, and senescence; also involved in defense against insect pests and disease-causing organisms.

jelly coat One of the acellular coverings of the eggs of certain animals, such as echinoderms.

joint The junction between two or more bones of the skeleton.

joule A unit of energy, equivalent to 0.239 calorie.

juvenile hormone (JH) An arthropod hormone that preserves juvenile structure during a molt. Without it, metamorphosis toward the adult form takes place.

juxtaglomerular apparatus (juks″tah-glo-mer′yoo-lar) A structure in the kidney that secretes renin in response to a decrease in blood pressure.

K selection A reproductive strategy recognized by some ecologists in which a species typically has a large body size, slow development, and long life span and does not devote a large proportion of its metabolic energy to the production of offspring. Compare with *r selection*.

karyogamy (kar-e-og′uh-me) The fusion of two haploid nuclei; follows fusion (plasmogamy) of cells from two sexually compatible mating types.

karyotype (kare′ee-oh-type) The chromosomal composition of an individual.

keratin (kare′ah-tin) A horny, water-insoluble protein found in the epidermis of vertebrates and in nails, feathers, hair, and horns.

ketone An organic molecule containing a carbonyl group bonded to two carbon atoms. Compare with *aldehyde*.

keystone species A species whose presence in an ecosystem largely determines the species composition and functioning of that ecosystem.

kidney The paired vertebrate organ important in excretion of metabolic wastes and in osmoregulation.

killer T cell See *T cytotoxic cell*.

kilobase (kb) 1000 bases or base pairs of a nucleic acid.

kilocalorie The amount of heat required to raise the temperature of 1 kg of water 1°C; also called *Calorie*, which is equivalent to 1000 calories.

kilojoule 1000 joules. See *joule*.

kinases Enzymes that catalyze the transfer of phosphate groups from ATP to acceptor molecules. See *protein kinases*.

kinetic energy Energy of motion. Compare with *potential energy*.

kinetochore (kin-eh′toh-kore) The portion of the chromosome centromere to which the mitotic spindle fibers attach.

kingdom A broad taxonomic category made up of related phyla; many biologists currently recognize six kingdoms of living organisms.

kin selection A type of natural selection that favors altruistic behavior toward relatives (kin), thereby ensuring that although the chances of an individual's survival are lessened, some of its genes will survive through successful reproduction of close relatives; increases inclusive fitness.

Klinefelter syndrome Inherited condition in which the affected individual is a sterile male with an XXY karyotype.

Koch's postulates A set of guidelines used to demonstrate that a specific pathogen causes specific disease symptoms.

Krebs cycle See *citric acid cycle*.

krummholz The gnarled, shrublike growth habit found in trees at high elevations, near their upper limit of distribution.

labyrinth The system of interconnecting canals of the inner ear of vertebrates.

labyrinthodonts The first successful group of tetrapods.

lactate (lactic acid) A three-carbon organic acid.

lactation (lak-tay′shun) The production or release of milk from the breast.

lacteal (lak′tee-al) One of the many lymphatic vessels in the intestinal villi that absorb fat.

lagging strand A strand of DNA that is synthesized as a series of short segments, called *Okazaki fragments*, which are then covalently joined by DNA ligase. Compare with *leading strand*.

lamins Polypeptides attached to the inner surface of the nuclear envelope that provide a type of skeletal framework.

landscape A large land area (several to many square kilometers) composed of interacting ecosystems.

landscape ecology The subdiscipline in ecology that studies the connections in a heterogeneous landscape.

large intestine The portion of the digestive tract of humans (and other vertebrates) consisting of the cecum, colon, rectum, and anus.

larva (pl., *larvae*) An immature form in the life history of some animals; may be unlike the parent.

larynx (lare′inks) The organ at the upper end of the trachea that contains the vocal cords.

lateral meristems Areas of localized cell division on the side of a plant that give rise to secondary tissues. Lateral meristems, including the vascular cambium and the cork cambium, cause an increase in the girth of the plant body. Compare with *apical meristem.*

leaching The process by which dissolved materials are washed away or carried with water down through the various layers of the soil.

leader sequence Noncoding sequence of nucleotides in mRNA that is transcribed from the region that precedes (is upstream to) the coding region.

leading strand Strand of DNA that is synthesized continuously. Compare with *lagging strand.*

learning A change in the behavior of an animal that results from experience.

legume (leg′yoom) (1) A simple, dry fruit that develops from a single carpel and splits open at maturity along two sutures to release seeds. (2) Any member of the pea family, e.g., pea, bean, peanut, alfalfa.

lek A small territory in which males compete for females.

lens The oval, transparent structure located behind the iris of the vertebrate eye; bends incoming light rays and brings them to a focus on the retina.

lenticels (len′tih-sels) Porous swellings of cork cells in the stems of woody plants; facilitate the exchange of gases.

leptin A hormone produced by adipose tissue that signals brain centers about the status of energy stores.

leukocytes (loo′koh-sites) White blood cells; colorless amoeboid cells that defend the body against disease-causing organisms.

leukoplasts Colorless plastids; include amyloplasts, which are used for starch storage in cells of roots and tubers.

lichen (ly′ken) A compound organism consisting of a symbiotic fungus and an alga or cyanobacterium.

life history traits Significant features of a species' life cycle, particularly traits that influence survival and reproduction.

life span The maximum duration of life for an individual of a species.

life table A table showing mortality and survival data by age of a population or cohort.

ligament (lig′uh-ment) A connective tissue cable or strap that connects bones to each other or holds other organs in place.

ligand A molecule that binds to a specific site in a receptor or other protein.

light-dependent reactions Reactions of photosynthesis in which light energy absorbed by chlorophyll is used to synthesize ATP and usually NADPH. Include *cyclic electron transport* and *noncyclic electron transport.*

lignin (lig′nin) A substance found in many plant cell walls that confers rigidity and strength, particularly in woody tissues.

limbic system In vertebrates, an action system of the brain. In humans, plays a role in emotional responses, motivation, autonomic function, and sexual response.

limiting resource An environmental resource that because it is scarce or unfavorable tends to restrict the ecological niche of an organism.

limnetic zone (lim-net′ik) The open water away from the shore of a lake or pond extending down as far as sunlight penetrates. Compare with *littoral zone* and *profundal zone.*

linkage The tendency for a group of genes located on the same chromosome to be inherited together in successive generations.

lipase (lip′ase) A fat-digesting enzyme.

lipid Any of a group of organic compounds that are insoluble in water but soluble in nonpolar solvents; lipids serve as energy storage and are important components of cell membranes.

lipoprotein (lip-oh-proh′teen) A large molecular complex consisting of lipids and protein; transports lipids in the blood. High-density lipoproteins (HDLs) transport cholesterol to the liver; low-density lipoproteins (LDLs) deliver cholesterol to many cells of the body.

littoral zone (lit′or-ul) The region of shallow water along the shore of a lake or pond. Compare with *limnetic zone* and *profundal zone.*

liver A large, complex organ that secretes bile, helps maintain homeostasis by removing or adding nutrients to the blood, and performs many other metabolic functions.

liverworts A phylum of spore-producing, nonvascular, thallose or leafy plants with a life cycle similar to that of mosses.

local hormones See *local regulators.*

local regulators Prostaglandins (a group of local hormones), growth factors, cytokines, and other soluble molecules that act on nearby cells by paracrine regulation or act on the cells that produce them (autocrine regulation).

locus The place on the chromosome at which the gene for a given trait occurs, i.e., a segment of the chromosomal DNA containing information that controls some feature of the organism; also called *gene locus.*

logistic population growth Population growth that initially occurs at a constant rate of increase over time (i.e., exponential) but then levels out as the carrying capacity of the environment is approached. Compare with *exponential population growth.*

long-day plant A plant that flowers in response to shortening nights; also called *short-night plant.* Compare with *short-day, intermediate-day,* and *day-neutral plants.*

long-night plant See *short-day plant.*

long-term potentiation (LTP) Long-lasting increase in the strength of synaptic connections that occurs in response to a series of high-frequency electrical stimuli. Compare with *long-term synaptic depression (LTD).*

long-term synaptic depression (LTD) Long-lasting decrease in the strength of synaptic connections that occurs in response to low-frequency stimulation of neurons. Compare with *long-term potentiation (LTP).*

loop of Henle (hen′lee) The U-shaped loop of a mammalian kidney tubule that extends down into the renal medulla.

loose connective tissue A type of connective tissue that is widely distributed in the body; consists of fibers strewn through a semifluid matrix.

lophophorate phyla Three related invertebrate protostome phyla, characterized by a ciliated ring of tentacles that surrounds the mouth.

Lophotrochozoa A branch of the protostomes that includes the flatworms, nemerteans (proboscis worms), mollusks, annelids, and the lophophorate phyla.

low-density lipoprotein (LDL) See *lipoprotein.*

lumen (loo′men) (1) The space enclosed by a membrane, such as the lumen of the endoplasmic reticulum or the thylakoid lumen. (2) The cavity or channel within a tube or tubular organ, such as a blood vessel or the digestive tract. (3) The space left within a plant cell after the cell's living material dies, as in tracheids.

lung An internal respiratory organ that functions in gas exchange; enables an animal to breathe air.

luteinizing hormone (LH) (loot′eh-ny-zing) Gonadotropic hormone secreted by the anterior pituitary; stimulates ovulation and maintains the corpus luteum in the ovaries of females; stimulates testosterone production in the testes of males.

lymph (limf) The colorless fluid within the lymphatic vessels that is derived from blood plasma; contains white blood cells; ultimately lymph is returned to the blood.

lymph node A mass of lymph tissue surrounded by a connective tissue capsule; manufactures lymphocytes and filters lymph.

lymphatic system A subsystem of the cardiovascular system; returns excess interstitial fluid (lymph) to the circulation; defends the body against disease organisms.

lymphocyte (lim′foh-site) White blood cell with nongranular cytoplasm that governs immune responses. See *B cell* and *T cell*.

lysis (ly′sis) The process of disintegration of a cell or some other structure.

lysogenic conversion The change in properties of bacteria that results from the presence of a prophage.

lysosomes (ly′soh-sohmz) Intracellular organelles present in many animal cells; contain a variety of hydrolytic enzymes.

lysozyme An enzyme found in many tissues and in tears and other body fluids; attacks the cell wall of many gram-positive bacteria.

macroevolution Large-scale evolutionary events over long time spans. Macroevolution results in phenotypic changes in populations that are significant enough to warrant their placement in taxonomic groups at the species level and higher. Compare with *microevolution*.

macromolecule A very large organic molecule, such as a protein or nucleic acid.

macronucleus A large nucleus found, along with one or several micronuclei, in ciliates. The macronucleus regulates metabolism and growth. Compare with *micronucleus*.

macronutrient An essential element required in fairly large amounts for normal growth. Compare with *micronutrient*.

macrophage (mak′roh-faje) A large phagocytic cell capable of ingesting and digesting bacteria and cell debris. Macrophages are also antigen-presenting cells.

magnoliid One of the clades of flowering plants; magnoliids are core angiosperms that were traditionally classified as "dicots," but molecular evidence indicates they are neither eudicots or monocots.

major histocompatibility complex (MHC) A group of membrane proteins, present on the surface of most cells, that are slightly different in each individual. In humans, the MHC is called the *HLA (human leukocyte antigen) group*.

malignant cells Cancer cells; tumor cells that are able to invade tissue and metastasize.

malignant transformation See *transformation*.

malnutrition Poor nutritional status; results from dietary intake that is either below or above required needs.

Malpighian tubules (mal-pig′ee-an) The excretory organs of many arthropods.

mammals The class of vertebrates characterized by hair, mammary glands, a diaphragm, and differentiation of teeth.

mandible (man′dih-bl) (1) The lower jaw of vertebrates. (2) Jaw-like, external mouthparts of insects.

mangrove forest A tidal wetland dominated by mangrove trees in which the salinity fluctuates between that of sea water and fresh water.

mantle In the mollusk, a fold of tissue that covers the visceral mass and that usually produces a shell.

marine snow The organic debris (plankton, dead organisms, fecal material, etc.) that "rains" into the dark area of the oceanic province from the lighted region above; the primary food of most organisms that live in the ocean's depths.

marsupials (mar-soo′pee-ulz) A subclass of mammals, characterized by the presence of an abdominal pouch in which the young, which are born in a very undeveloped condition, are carried for some time after birth.

mass extinction The extinction of numerous species during a relatively short period of geologic time. Compare with *background extinction*.

mast cell A type of cell found in connective tissue; contains histamine and is important in an inflammatory response and in allergic reactions.

maternal effect genes Genes of the mother that are transcribed during oogenesis and subsequently affect the development of the embryo. Compare with *zygotic genes*.

matrix (may′triks) (1) In cell biology, the interior of the compartment enclosed by the inner mitochondrial membrane. (2) In zoology, nonliving material secreted by and surrounding connective tissue cells; contains a network of microscopic fibers.

matter Anything that has mass and takes up space.

maxillae Appendages used for manipulating food; characteristic of crustaceans.

mechanical isolation A prezygotic reproductive isolating mechanism in which fusion of the gametes of two species is prevented by morphological or anatomical differences.

mechanoreceptor (meh-kan′oh-ree-sep″tor) A sensory cell or organ that perceives mechanical stimuli, e.g., touch, pressure, gravity, stretching, or movement.

medulla (meh-dul′uh) (1) The inner part of an organ, such as the medulla of the kidney. Compare with *cortex*. (2) The most posterior part of the vertebrate brain, lying next to the spinal cord.

medusa A jellyfish-like animal; a free-swimming, umbrella-shaped stage in the life cycle of certain cnidarians. Compare with *polyp*.

megaphyll (meg′uh-fil) Type of leaf found in horsetails, ferns, gymnosperms, and angiosperms; contains multiple vascular strands (i.e., complex venation). Compare with *microphyll*.

megaspore (meg′uh-spor) The *n* spore in heterosporous plants that gives rise to a female gametophyte. Compare with *microspore*.

meiosis (my-oh′sis) Process in which a 2*n* cell undergoes two successive nuclear divisions (meiosis I and meiosis II), potentially producing four *n* nuclei; leads to the formation of gametes in animals and spores in plants.

melanin A dark pigment present in many animals; contributes to the color of the skin.

melanocortins A group of peptides that appear to decrease appetite in response to increased fat stores.

melatonin (mel-ah-toh′nin) A hormone secreted by the pineal gland that plays a role in setting circadian rhythms.

memory cell B or T cell (lymphocyte) that permits rapid mobilization of immune response on second or subsequent exposure to a particular antigen. Memory B cells continue to produce antibodies after the immune system overcomes an infection.

meninges (meh-nin′jeez) (sing., *meninx*) The three membranes that protect the brain and spinal cord: the dura mater, arachnoid, and pia mater.

menopause The period (usually occurring between 45 and 55 years of age) in women when the recurring menstrual cycle ceases.

menstrual cycle (men′stroo-ul) In the human female, the monthly sequence of events that prepares the body for pregnancy.

menstruation (men-stroo-ay′shun) The monthly discharge of blood and degenerated uterine lining in the human female; marks the beginning of each menstrual cycle.

meristem (mer′ih-stem) A localized area of mitotic cell division in the plant body. See *apical meristem* and *lateral meristems*.

meroblastic cleavage Cleavage pattern observed in the telolecithal eggs of reptiles and birds, in which cleavage is restricted to a small disc of cytoplasm at the animal pole. Compare with *holoblastic cleavage*.

mesencephalon See *midbrain*.

mesenchyme (mes′en-kime) A loose, often jellylike connective tissue containing undifferentiated cells; found in the embryos of vertebrates and the adults of some invertebrates.

mesoderm (mez′oh-derm) The middle germ layer of the early embryo; gives rise to connective tissue, muscle, bone, blood vessels, kidneys, and many other structures. Compare with *ectoderm* and *endoderm*.

mesophyll (mez′oh-fil) Photosynthetic tissue in the interior of a leaf; sometimes differentiated into palisade mesophyll and spongy mesophyll.

Mesozoic era That part of geologic time extending from roughly 251 million to 66 million years ago.

messenger RNA (mRNA) RNA that specifies the amino acid sequence of a protein; transcribed from DNA.

metabolic pathway A series of chemical reactions in which the product of one reaction becomes the substrate of the next reaction.

metabolic rate Energy use by an organism per unit time. See *basal metabolic rate (BMR)*.

metabolism The sum of all the chemical processes that occur within a cell or organism; the transformations by which energy and matter are made available for use by the organism. See *anabolism* and *catabolism*.

metamorphosis (met″ah-mor′fuh-sis) Transition from one developmental stage to another, such as from a larva to an adult.

metanephridia (sing., *metanephridium*) The excretory organs of annelids and mollusks; each consists of a tubule open at both ends; at one end a ciliated funnel opens into the coelom, and the other end opens to the outside of the body.

metaphase (met′ah-faze) The stage of mitosis in which the chromosomes line up on the midplane of the cell. Occurs after prometaphase and before anaphase.

metapopulation A population that is divided into several local populations among which individuals occasionally disperse.

metastasis (met-tas′tuh-sis) The spreading of cancer cells from one organ or part of the body to another.

metencephalon See *hindbrain*.

methyl group A nonpolar functional group; abbreviated —CH₃.

micro RNAs (miRNAs) Single-stranded RNA molecules about 21 to 22 nucleotides long that inhibit the translation of mRNAs involved in growth and development.

microclimate Local variations in climate produced by differences in elevation, in the steepness and direction of slopes, and in exposure to prevailing winds.

microevolution Small-scale evolutionary change caused by changes in allele or genotype frequencies that occur within a population over a few generations. Compare with *macroevolution*.

microfilaments Thin fibers consisting of actin protein subunits; form part of the cytoskeleton.

microfossils Ancient traces (fossils) of microscopic life.

microglia Phagocytic glial cells found in the CNS.

micronucleus One or more smaller nuclei found, along with the macronucleus, in ciliates. The micronucleus is involved in sexual reproduction. Compare with *macronucleus*.

micronutrient An essential element that is required in trace amounts for normal growth. Compare with *macronutrient*.

microphyll (mi′kro-fil) Type of leaf found in club mosses; contains one vascular strand (i.e., simple venation). Compare with *megaphyll*.

microsphere A protobiont produced by adding water to abiotically formed polypeptides.

microspore (mi′kro-spor) The *n* spore in heterosporous plants that gives rise to a male gametophyte. Compare with *megaspore*.

microsporidia Small, unicellular, fungal parasites that infect eukaryotic cells; classified with the zygomycetes.

microtubule-associated proteins (MAPs) Include structural proteins that help regulate microtubule assembly and cross-link microtubules to other cytoskeletal polymers; and motors, such as kinesin and dynein, that use ATP to produce movement.

microtubule-organizing center (MTOC) The region of the cell from which microtubules are anchored and possibly assembled. The MTOCs of many organisms (including animals, but not flowering plants or most gymnosperms) contain a pair of centrioles.

microtubules (my-kroh-too′bewls) Hollow, cylindrical fibers consisting of tubulin protein subunits; major components of the cytoskeleton and found in mitotic spindles, cilia, flagella, centrioles, and basal bodies.

microvilli (sing., *microvillus*) Minute projections of the plasma membrane that increase the surface area of the cell; found mainly in cells concerned with absorption or secretion, such as those lining the intestine or the kidney tubules.

midbrain In vertebrate embryos, one of the three divisions of the developing brain. Also called *mesencephalon*. Compare with *forebrain* and *hindbrain*.

middle lamella The layer composed of pectin polysaccharides that serves to cement together the primary cell walls of adjacent plant cells.

midvein The main, or central, vein of a leaf.

migration (1) The periodic or seasonal movement of an organism (individual or population) from one place to another, usually over a long distance. See *dispersal*. (2) In evolutionary biology, a movement of individuals that results in a transfer of alleles from one population to another. See *gene flow*.

mineralocorticoids (min″ur-al-oh-kor′tih-koidz) Hormones produced by the adrenal cortex that regulate mineral metabolism and, indirectly, fluid balance. The principal mineralocorticoid is aldosterone.

minerals Inorganic nutrients ingested as salts dissolved in food and water.

minimum viable population (MVP) The smallest population size at which a species has a high chance of sustaining its numbers and surviving into the future.

mismatch repair A DNA repair mechanism in which special enzymes recognize the incorrectly paired nucleotides and remove them. DNA polymerases then fill in the missing nucleotides.

missense mutation A type of base-substitution mutation that causes one amino acid to be substituted for another in the resulting protein product. Compare with *nonsense mutation*.

mitochondria (my″toh-kon′dree-ah) (sing., *mitochondrion*) Intracellular organelles that are the sites of oxidative phosphorylation in eukaryotes; include an outer membrane and an inner membrane.

mitochondrial DNA (mtDNA) DNA present in mitochondria that is transmitted maternally, from mothers to their offspring. Mitochondrial DNA mutates more rapidly than nuclear DNA.

mitosis (my-toh′sis) The division of the cell nucleus resulting in two daughter nuclei, each with the same number of chromosomes as the parent nucleus; mitosis consists of prophase, prometaphase, metaphase, anaphase, and telophase. Cytokinesis usually overlaps the telophase stage.

mitotic spindle Structure consisting mainly of microtubules that provides the framework for chromosome movement during cell division.

mitral valve See *atrioventricular valve.*

mobile genetic element See *transposon.*

model organism A species chosen for biological studies because it has characteristics that allow for the efficient analysis of biological processes. Most model organisms are small, have short generation times, and are easy to grow and study under controlled conditions.

modern synthesis A comprehensive, unified explanation of evolution based on combining previous theories, especially of Mendelian genetics, with Darwin's theory of evolution by natural selection; also called the *synthetic theory of evolution.*

mole The atomic mass of an element or the molecular mass of a compound, expressed in grams; one mole of any substance has 6.02×10^{23} units (Avogadro's number).

molecular anthropology The branch of science that compares genetic material from individuals of regional human populations to help unravel the origin and migrations of modern humans.

molecular chaperones Proteins that help other proteins fold properly. Although not dictating the folding pattern, chaperones make the process more efficient.

molecular clock analysis A comparison of the DNA nucleotide sequences of related organisms to estimate when they diverged from one another during the course of evolution.

molecular formula The type of chemical formula that gives the actual numbers of each type of atom in a molecule. Compare with *simplest formula* and *structural formula.*

molecular mass The sum of the atomic masses of the atoms that make up a single molecule of a compound; expressed in atomic mass units (amu) or daltons.

molecule The smallest particle of a covalently bonded element or compound; two or more atoms joined by covalent bonds.

mollusks A phylum of coelomate protostome animals characterized by a soft body, visceral mass, mantle, and foot.

molting The shedding and replacement of an outer covering such as an exoskeleton.

molting hormone A steroid hormone that stimulates growth and molting in insects. Also called *ecdysone.*

monoacylglycerol (mon″o-as″il-glis′er-ol) Lipid consisting of glycerol combined chemically with a single fatty acid. Also called *monoglyceride.* Compare with *diacylglycerol* and *triacylglycerol.*

monocot (mon′oh-kot) One of two classes of flowering plants; monocot seeds contain a single cotyledon, or seed leaf. Compare with *eudicot.*

monoclonal antibodies Identical antibody molecules produced by cells cloned from a single cell.

monocyte (mon′oh-site) A type of white blood cell; a large, phagocytic, nongranular leukocyte that enters the tissues and differentiates into a macrophage.

monoecious (mon-ee′shus) Having male and female reproductive parts in separate flowers or cones on the same plant; compare with *dioecious.*

monogamy A mating system in which a male animal mates with a single female during a breeding season.

monoglyceride See *monoacylglycerol.*

monohybrid cross A genetic cross that takes into account the behavior of alleles of a single locus. Compare with *dihybrid cross.*

monokaryotic (mon″o-kare-ee-ot′ik) The condition of having a single *n* nucleus per cell, characteristic of certain fungal hyphae. Compare with *dikaryotic.*

monomer (mon′oh-mer) A molecule that can link with other similar molecules; two monomers join to form a dimer, whereas many form a polymer. Monomers are small (e.g., sugars or amino acids) or large (e.g., tubulin or actin proteins).

monophyletic group (mon″oh-fye-let′ik) A group of organisms that evolved from a common ancestor. Compare with *polyphyletic group* and *paraphyletic group.*

monosaccharide (mon-oh-sak′ah-ride) A sugar that cannot be degraded by hydrolysis to a simpler sugar (e.g., glucose or fructose).

monosomy The condition in which only one member of a chromosome pair is present and the other is missing. Compare with *trisomy* and *disomy.*

monotremes (mon′oh-treemz) Egg-laying mammals such as the duck-billed platypus of Australia.

monounsaturated fatty acid See *fatty acid.*

monozygotic twins Genetically identical twins that arise from the division of a single fertilized egg; commonly known as *identical twins.* Compare with *dizygotic twins.*

morphogen Any chemical agent thought to govern the processes of cell differentiation and pattern formation that lead to morphogenesis.

morphogenesis (mor-foh-jen′eh-sis) The development of the form and structures of an organism and its parts; proceeds by a series of steps known as *pattern formation.*

mortality The rate at which individuals die; the average per capita death rate.

morula (mor′yoo-lah) An early embryo consisting of a solid ball of cells.

mosaic development A rigid developmental pattern in which the fates of cells become restricted early in development. Compare with *regulative development.*

mosses A phylum of spore-producing nonvascular plants with an alternation of generations in which the dominant *n* gametophyte alternates with a 2*n* sporophyte that remains attached to the gametophyte.

motor neuron An efferent neuron that transmits impulses away from the central nervous system to skeletal muscle.

motor program A coordinated sequences of muscle actions responsible for many behaviors we think of as automatic.

motor unit All the skeletal muscle fibers that are stimulated by a single motor neuron.

mRNA cap An unusual nucleotide, 7-methylguanylate, that is added to the 5′ end of a eukaryotic messenger RNA. Capping enables eukaryotic ribosomes to bind to mRNA.

mucosa (mew-koh′suh) See *mucous membrane.*

mucous membrane A type of epithelial membrane that lines a body cavity that opens to the outside of the body, e.g., the digestive and respiratory tracts; also called *mucosa.*

mucus (mew′cus) A sticky secretion composed of covalently linked protein and carbohydrate; serves to lubricate body parts and trap particles of dirt and other contaminants. (The adjectival form is spelled *mucous.*)

Müllerian mimicry (mul-ler′ee-un mim′ih-kree) The resemblance of dangerous, unpalatable, or poisonous species to one another so that potential predators recognize them more easily. Compare with *Batesian mimicry*.

multiple alleles (al-leelz′) Three or more alleles of a single locus (in a population), such as the alleles governing the ABO series of blood types.

multiple fruit A fruit that develops from many ovaries of many separate flowers, e.g., pineapple. Compare with *simple, aggregate,* and *accessory fruits*.

muscle (1) A tissue specialized for contraction. (2) An organ that produces movement by contraction.

mutagen (mew′tah-jen) Any agent capable of entering the cell and producing mutations.

mutation Any change in DNA; may include a change in the nucleotide base pairs of a gene, a rearrangement of genes within the chromosomes so that their interactions produce different effects, or a change in the chromosomes themselves.

mutualism (1) In ecology, a symbiotic relationship in which both partners benefit from the association. Compare with *parasitism* and *commensalism*. (2) In animal behavior, cooperative behavior in which each animal in the group benefits.

mycelium (my-seel′ee-um) (pl., *mycelia*) The vegetative body of most fungi and certain protists (water molds); consists of a branched network of hyphae.

mycorrhizae (my″kor-rye′zee) Mutualistic associations of fungi and plant roots that aid in the plant's absorption of essential minerals from the soil.

mycotoxins Poisonous chemical compounds produced by fungi, e.g., aflatoxins that harm the liver and are known carcinogens.

myelencephalon See *hindbrain*.

myelin sheath (my′eh-lin) The white, fatty material that forms a sheath around the axons of certain nerve cells, which are then called *myelinated fibers*.

myocardial infarction (MI) Heart attack; serious consequence occurring when the heart muscle receives insufficient oxygen.

myofibrils (my-oh-fy′brilz) Tiny threadlike structures in the cytoplasm of striated and cardiac muscle that are composed of myosin filaments and actin filaments; these filaments are responsible for muscle contraction; see *myosin filaments* and *actin filaments*.

myoglobin (my′oh-glo″bin) A hemoglobin-like, oxygen-transferring protein found in muscle.

myosin (my′oh-sin) A protein that together with actin is responsible for muscle contraction.

myosin filaments Thick filaments consisting mainly of the protein myosin; actin and myosin filaments make up the myofibrils of muscle fibers.

n The chromosome number of a gamete. The chromosome number of a zygote is 2*n*. If an organism is not polyploid, the *n* gametes are haploid and the 2*n* zygotes are diploid.

NAD⁺/NADH Oxidized and reduced forms, respectively, of nicotinamide adenine dinucleotide, a coenzyme that transfers electrons (as hydrogen), particularly in catabolic pathways, including cellular respiration.

NADP⁺/NADPH Oxidized and reduced forms, respectively, of nicotinamide adenine dinucleotide phosphate, a coenzyme that acts as an electron (hydrogen) transfer agent, particularly in anabolic pathways, including photosynthesis.

nanoplankton Extremely minute (<20 μm in length) algae that are major producers in the ocean because of their great abundance; part of phytoplankton.

natality The rate at which individuals produce offspring; the average per capita birth rate.

natural killer cell (NK cell) A large, granular lymphocyte that functions in both nonspecific and specific immune responses; releases cytokines and proteolytic enzymes that target tumor cells and cells infected with viruses and other pathogens.

natural selection The mechanism of evolution proposed by Charles Darwin; the tendency of organisms that have favorable adaptations to their environment to survive and become the parents of the next generation. Evolution occurs when natural selection results in changes in allele frequencies in a population.

necrosis Uncontrolled cell death that causes inflammation and damages other cells. Compare with *apoptosis*.

nectary (nek′ter-ee) In plants, a gland or other structure that secretes nectar.

negative feedback mechanism A homeostatic mechanism in which a change in some condition triggers a response that counteracts, or reverses, the changed condition, restoring homeostasis, e.g., how mammals maintain body temperature. Compare with *positive feedback mechanism*.

nekton (nek′ton) Free-swimming aquatic organisms such as fish and turtles. Compare with *plankton*.

nematocyst (nem-at′oh-sist) A stinging structure found within cnidocytes (stinging cells) in cnidarians; used for anchorage, defense, and capturing prey.

nematodes The phylum of animals commonly known as *roundworms*.

nemerteans The phylum of animals commonly known as *ribbon worms;* each has a proboscis (tubular feeding organ) for capturing prey.

neonate Newborn individual.

neoplasm See *tumor*.

nephridial organ (neh-frid′ee-al) The excretory organ of many invertebrates; consists of simple or branching tubes that usually open to the outside of the body through pores; also called *nephridium*.

nephron (nef′ron) The functional, microscopic unit of the vertebrate kidney.

neritic province (ner-ih′tik) Ocean water that extends from the shoreline to where the bottom reaches a depth of 200 m. Compare with *oceanic province*.

nerve A bundle of axons (or dendrites) wrapped in connective tissue that conveys impulses between the central nervous system and some other part of the body.

nerve net A system of interconnecting nerve cells found in cnidarians and echinoderms.

nervous tissue A type of animal tissue specialized for transmitting electrical and chemical signals.

net primary productivity The energy that remains in an ecosystem (as biomass) after cellular respiration has occurred; net primary productivity equals gross primary productivity minus respiration. Compare with *gross primary productivity*.

neural crest (noor′ul) A group of cells along the neural tube that migrate and form various parts of the embryo, including parts of the peripheral nervous system.

neural plasticity The ability of the nervous system to change in response to experience.

neural plate See *neural tube*.

neural transmission See *transmission, neural*.

neural tube The hollow, longitudinal structure in the early vertebrate embryo that gives rise to the brain and spinal cord. The neural tube forms from the neural plate, a flattened, thickened region of the ectoderm that rolls up and sinks below the surface.

neuroendocrine cells Neurons that produce neurohormones.

neurohormones Hormones produced by neuroendocrine cells; transported down axons and released into interstitial fluid; common in invertebrates; in vertebrates, hypothalamus produces neurohormones.

neuron (noor'on) A nerve cell; a conducting cell of the nervous system that typically consists of a cell body, dendrites, and an axon.

neuropeptide One of a group of peptides produced in neural tissue that function as signaling molecules; many are neurotransmitters.

neuropeptide Y A signaling molecule produced by the hypothalamus that increases appetite and slows metabolism; helps restore energy homeostasis when leptin levels and food intake are low.

neurotransmitter A chemical signal used by neurons to transmit impulses across a synapse.

neutral solution A solution of pH 7; there are equal concentrations of hydrogen ions [H^+] and hydroxide ions [OH^-]. Compare with *acidic solution* and *basic solution*.

neutral variation Variation that does not appear to confer any selective advantage or disadvantage to the organism.

neutron (noo'tron) An electrically neutral particle with a mass of 1 atomic mass unit (amu) found in the atomic nucleus. Compare with *proton* and *electron*.

neutrophil (new'truh-fil) A type of granular leukocyte important in immune responses; a type of phagocyte that engulfs and destroys bacteria and foreign matter.

niche (nich) The totality of an organism's adaptations, its use of resources, and the lifestyle to which it is fitted in its community; how an organism uses materials in its environment as well as how it interacts with other organisms; also called *ecological niche*. See *fundamental niche* and *realized niche*.

nicotinamide adenine dinucleotide See *NAD+/NADH*.

nicotinamide adenine dinucleotide phosphate See *NADP+/ NADPH*.

nitric oxide (NO) A gaseous signaling molecule; a neurotransmitter.

nitrification (nie"tra-fuh-kay'shun) The conversion of ammonia (NH_3) to nitrate (NO_3^-) by certain bacteria (nitrifying bacteria) in the soil; part of the nitrogen cycle.

nitrogen cycle The worldwide circulation of nitrogen from the abiotic environment into living things and back into the abiotic environment.

nitrogen fixation The conversion of atmospheric nitrogen (N_2) to ammonia (NH_3) by certain bacteria; part of the nitrogen cycle.

nitrogenase (nie-traa'jen-ase) The enzyme responsible for nitrogen fixation under anaerobic conditions.

nociceptors (no'sih-sep-torz) Pain receptors; free endings of certain sensory neurons whose stimulation is perceived as pain.

node The area on a stem where each leaf is attached. Compare with *internode*.

nodules Swellings on the roots of plants, such as legumes, in which symbiotic nitrogen-fixing bacteria (*Rhizobium*) live.

noncompetitive inhibitor A substance that lowers the rate at which an enzyme catalyzes a reaction but does not bind to the active site. Compare with *competitive inhibitor*.

noncyclic electron transport In photosynthesis, the linear flow of electrons, produced by photolysis of water, through photosystems II and I; results in the formation of ATP (by chemiosmosis), NADPH, and O_2. Compare with *cyclic electron transport*.

nondisjunction Abnormal separation of sister chromatids or of homologous chromosomes caused by their failure to disjoin (move apart) properly during mitosis or meiosis.

nonpolar covalent bond Chemical bond formed by the equal sharing of electrons between atoms of approximately equal electronegativity. Compare with *polar covalent bond*.

nonpolar molecule Molecule that does not have a positively charged end and a negatively charged end; nonpolar molecules are generally insoluble in water. Compare with *polar molecule*.

nonsense mutation A base-substitution mutation that results in an amino acid–specifying codon being changed to a termination (stop) codon; when the abnormal mRNA is translated, the resulting protein is usually truncated and nonfunctional. Compare with *missense mutation*.

nonspecific immune responses Mechanisms such as physical barriers (e.g., the skin) and phagocytosis that provide immediate and general protection against pathogens. Also called *innate immunity*. Compare with *specific immune responses*.

norepinephrine (nor-ep-ih-nef'rin) A neurotransmitter that is also a hormone secreted by the adrenal medulla.

Northern blot A technique in which RNA fragments, previously separated by gel electrophoresis, are transferred to a nitrocellulose membrane and detected by autoradiography or chemical luminescence. Compare with *Southern blot* and *Western blot*.

notochord (no'toe-kord) The flexible, longitudinal rod in the anterior-posterior axis that serves as an internal skeleton in the embryos of all chordates and in the adults of some.

nuclear area Region of a prokaryotic cell that contains DNA; not enclosed by a membrane. Also called *nucleoid*.

nuclear envelope The double membrane system that encloses the cell nucleus of eukaryotes.

nuclear equivalence The concept that the nuclei of all differentiated cells of an adult organism are genetically identical to one another and to the nucleus of the zygote from which they were derived. Compare with *genomic rearrangement* and *gene amplification*.

nuclear pores Structures in the nuclear envelope that allow passage of certain materials between the cell nucleus and the cytoplasm.

nucleoid See *nuclear area*.

nucleolus (new-klee'oh-lus) (pl., *nucleoli*) Specialized structure in the cell nucleus formed from regions of several chromosomes; site of assembly of the ribosomal subunits.

nucleoplasm The contents of the cell nucleus.

nucleoside triphosphate Molecule consisting of a nitrogenous base, a pentose sugar, and three phosphate groups, e.g., adenosine triphosphate (ATP).

nucleosomes (new'klee-oh-sohmz) Repeating units of chromatin structure, each consisting of a length of DNA wound around a complex of eight histone molecules. Adjacent nucleosomes are connected by a DNA linker region associated with another histone protein.

nucleotide (noo'klee-oh-tide) A molecule consisting of one or more phosphate groups, a five-carbon sugar (ribose or deoxyribose), and a nitrogenous base (purine or pyrimidine).

nucleotide excision repair A DNA repair mechanism commonly used to repair a damaged segment of DNA caused by the sun's ultraviolet radiation or by harmful chemicals.

nucleus (new'klee-us) (pl., *nuclei*) (1) The central region of an atom that contains the protons and neutrons. (2) A cell organelle in eukaryotes that contains the DNA and serves as the control center of the cell. (3) A mass of nerve cell bodies in the central nervous system. Compare with *ganglion*.

nut A simple, dry fruit that contains a single seed and is surrounded by a hard fruit wall.

nutrients The chemical substances in food that are used as components for synthesizing needed materials and/or as energy sources.

nutrition The process of taking in and using food (nutrients).

obesity Excess accumulation of body fat; a person is considered obese if the body mass index (BMI) is 30 or higher.

obligate anaerobe An organism that grows only in the absence of oxygen. Compare with *facultative anaerobe.*

occipital lobes Posterior areas of the mammalian cerebrum; interpret visual stimuli from the retina of the eye.

oceanic province That part of the open ocean that overlies an ocean bottom deeper than 200 m. Compare with *neritic province.*

Okazaki fragment One of many short segments of DNA, each 100 to 1000 nucleotides long, that must be joined by DNA ligase to form the lagging strand in DNA replication.

olfactory epithelium Tissue containing odor-sensing neurons.

oligodendrocyte A type of glial cell that forms myelin sheaths around neurons in the CNS.

oligosaccharin One of several signaling molecules in plants that trigger the production of phytoalexins and affect aspects of growth and development.

ommatidium (om″ah-tid′ee-um) (pl., *ommatidia*) One of the light-detecting units of a compound eye, consisting of a lens and a crystalline cone that focus light onto photoreceptors called *retinular cells.*

omnivore (om′nih-vore) An animal that eats a variety of plant and animal materials.

oncogene (on′koh-jeen) An abnormally functioning gene implicated in causing cancer. Compare with *proto-oncogene* and *tumor suppressor gene.*

oocytes (oh′oh-sites) Meiotic cells that give rise to egg cells (ova).

oogamy (oh-og′uh-me) The fertilization of a large, nonmotile female gamete by a small, motile male gamete. Compare with *isogamy* and *anisogamy.*

oogenesis (oh″oh-jen′eh-sis) Production of female gametes (eggs) by meiosis. Compare with *spermatogenesis.*

oospore A thick-walled, resistant spore formed from a zygote during sexual reproduction in water molds.

open circulatory system A type of circulatory system in which the blood bathes the tissues directly; characteristic of arthropods and many mollusks. Compare with *closed circulatory system.*

open system An entity that exchanges energy with its surroundings. Compare with *closed system.*

operant conditioning A type of learning in which an animal is rewarded or punished for performing a behavior it discovers by chance.

operator site One of the control regions of an operon; the DNA segment to which a repressor binds, thereby inhibiting the transcription of the adjacent structural genes of the operon.

operculum In bony fishes, a protective flap of the body wall that covers the gills.

operon (op′er-on) In prokaryotes, a group of structural genes that are coordinately controlled and transcribed as a single message, plus their adjacent regulatory elements.

opisthokont A member of a clade of eukaryotes, including certain protists (choanoflagellates), fungi, and animals; flagellate cells in this group have a single, posterior flagellum.

opposable thumb The arrangement of the fingers so that they are positioned opposite the thumb, enabling the organism to grasp objects.

optimal foraging The process of obtaining food in a manner that maximizes benefits and/or minimizes costs.

orbital Region in which electrons occur in an atom or molecule

order A taxonomic category made up of related families.

organ A specialized structure, such as the heart or liver, made up of tissues and adapted to perform a specific function or group of functions.

organ of Corti The structure within the inner ear of vertebrates that contains receptor cells that sense sound vibrations.

organ system An organized group of tissues and organs that work together to perform a specialized set of functions, e.g., the digestive system or circulatory system.

organelle One of the specialized structures within the cell, such as the mitochondria, Golgi complex, ribosomes, or contractile vacuole; many organelles are membrane-enclosed.

organic compound A compound consisting of a backbone made up of carbon atoms. Compare with *inorganic compound.*

organism Any living system consisting of one or more cells.

organismic respiration See *respiration.*

organogenesis The process of organ formation.

orgasm (or′gazm) The climax of sexual excitement.

origin of replication A specific site on the DNA where replication begins.

osmoconformer An animal in which the salt concentration of body fluids varies along with changes in the surrounding sea water so that it stays in osmotic equilibrium with its surroundings. Compare with *osmoregulator.*

osmoregulation (oz″moh-reg-yoo-lay′shun) The active regulation of the osmotic pressure of body fluids so that they do not become excessively dilute or excessively concentrated.

osmoregulator An animal that maintains an optimal salt concentration in its body fluids despite changes in salinity of its surroundings. Compare with *osmoconformer.*

osmosis (oz-moh′sis) The net movement of water (the principal solvent in biological systems) by diffusion through a selectively permeable membrane from a region of higher concentration of water (a hypotonic solution) to a region of lower concentration of water (a hypertonic solution).

osmotic pressure The pressure that must be exerted on the hypertonic side of a selectively permeable membrane to prevent diffusion of water (by osmosis) from the side containing pure water.

osteichthyes (os″tee-ick′thees) Historically, the vertebrate class of bony fishes. Biologists now divide bony fishes into three classes: Actinopterygii, the ray-finned fishes; Actinistia, the lobe-finned fishes; and Dipnoi, the lungfishes.

osteoblast (os′tee-oh-blast) A type of bone cell that secretes the protein matrix of bone. Also see *osteocyte.*

osteoclast (os′tee-oh-clast) Large, multinucleate cell that helps sculpt and remodel bones by dissolving and removing part of the bony substance.

osteocyte (os′tee-oh-site) A mature bone cell; an osteoblast that has become embedded within the bone matrix and occupies a lacuna.

osteon (os′tee-on) The spindle-shaped unit of bone composed of concentric layers of osteocytes organized around a central Haversian canal containing blood vessels and nerves.

otoliths (oh′toe-liths) Small calcium carbonate crystals in the saccule and utricle of the inner ear; sense gravity and are important in static equilibrium.

outbreeding The mating of individuals of unrelated strains, also called *outcrossing.* Compare with *inbreeding.*

outcrossing See *outbreeding.*

outgroup In cladistics, a taxon that represents an approximation of the ancestral condition; the outgroup is related to the *ingroup* (the members of the group under study) but separated from the ingroup lineage before they diversified.

ovary (oh′var-ee) (1) In animals, one of the paired female gonads responsible for producing eggs and sex hormones. (2) In flowering plants, the base of the carpel that contains ovules; ovaries develop into fruits after fertilization.

oviduct (oh'vih-dukt) The tube that carries ova from the ovary to the uterus, cloaca, or body exterior. Also called *fallopian tube* or *uterine tube.*

oviparous (oh-vip'ur-us) Bearing young in the egg stage of development; egg laying. Compare with *viviparous* and *ovoviviparous.*

ovoviviparous (oh"voh-vih-vip'ur-us) A type of development in which the young hatch from eggs incubated inside the mother's body. Compare with *viviparous* and *oviparous.*

ovulation (ov-u-lay'shun) The release of an egg from the ovary.

ovule (ov'yool) The structure (i.e., megasporangium) in the plant ovary that develops into the seed following fertilization.

ovum (pl., *ova*) Female gamete of an animal.

oxaloacetate Four-carbon compound; important intermediate in the citric acid cycle and in the C_4 and CAM pathways of carbon fixation in photosynthesis.

oxidants Highly reactive molecules such as free radicals, peroxides, and superoxides that are produced during normal cell processes that require oxygen; can damage DNA and other molecules by snatching electrons. Compare with *antioxidants.*

oxidation The loss of one or more electrons (or hydrogen atoms) by an atom, ion, or molecule. Compare with *reduction.*

oxidative phosphorylation (fos"for-ih-lay'shun) The production of ATP using energy derived from the transfer of electrons in the electron transport system of mitochondria; occurs by chemiosmosis.

oxygen-carrying capacity The maximum amount of oxygen transported by hemoglobin.

oxygen debt The oxygen necessary to metabolize the lactic acid produced during strenuous exercise.

oxygen–hemoglobin dissociation curve A curve depicting the percentage saturation of hemoglobin with oxygen, as a function of certain variables such as oxygen concentration, carbon dioxide concentration, or pH.

oxyhemoglobin Hemoglobin that has combined with oxygen.

oxytocin (ok"see-tow'sin) Hormone secreted by the hypothalamus and released by the posterior lobe of the pituitary gland; stimulates contraction of the pregnant uterus and the ducts of mammary glands.

ozone A blue gas, O_3, with a distinctive odor that is a human-made pollutant near Earth's surface (in the troposphere) but a natural and essential component of the stratosphere.

P generation (parental generation) Members of two different true-breeding lines that are crossed to produce the F_1 generation.

P680 Chlorophyll *a* molecules that serve as the reaction center of photosystem II, transferring photoexcited electrons to a primary acceptor; named by their absorption peak at 680 nm.

P700 Chlorophyll *a* molecules that serve as the reaction center of photosystem I, transferring photoexcited electrons to a primary acceptor; named by their absorption peak at 700 nm.

pacemaker (of the heart) See *sinoatrial (SA) node.*

Pacinian corpuscle (pah-sin'ee-an kor'pus-el) A receptor located in the dermis of the skin that responds to pressure.

paedomorphosis Retention of juvenile or larval features in a sexually mature animal.

pair bond A stable relationship between animals of opposite sex that ensures cooperative behavior in mating and rearing the young.

paleoanthropology (pay"lee-o-an-thro-pol'uh-gee) The study of human evolution.

Paleozoic era That part of geologic time extending from roughly 542 million to 251 million years ago.

palindromic Reading the same forward and backward; DNA sequences are palindromic when the base sequence of one strand reads the same as its complement when both are read in the 5′ to 3′ direction.

palisade mesophyll (mez'oh-fil) The vertically stacked, columnar mesophyll cells near the upper epidermis in certain leaves. Compare with *spongy mesophyll.*

pancreas (pan'kree-us) Large gland located in the vertebrate abdominal cavity. The pancreas produces pancreatic juice containing digestive enzymes; also serves as an endocrine gland, secreting the hormones insulin and glucagon.

panspermia The idea that life did not originate on Earth but began elsewhere in the galaxy and drifted through space to Earth.

parabronchi (sing., *parabronchus*) Thin-walled ducts in the lungs of birds; gases are exchanged across their walls.

paracrine regulation A type of regulation in which a signal molecule (e.g., certain hormones) diffuses through interstitial fluid and acts on nearby target cells. Compare with *autocrine regulation.*

paraphyletic group A group of organisms made up of a common ancestor and some, but not all, of its descendants. Compare with *monophyletic group* and *polyphyletic group*

parapodia (par"uh-poh'dee-ah) (sing., *parapodium*) Paired, thickly bristled paddlelike appendages extending laterally from each segment of polychaete worms.

parasite A heterotrophic organism that obtains nourishment from the living tissue of another organism (the host).

parasitism (par'uh-si-tiz"m) A symbiotic relationship in which one member (the parasite) benefits and the other (the host) is adversely affected. Compare with *commensalism* and *mutualism.*

parasympathetic nervous system A division of the autonomic nervous system concerned with the control of the internal organs; functions to conserve or restore energy. Compare with *sympathetic nervous system.*

parathyroid glands Small, pea-sized glands closely adjacent to the thyroid gland; they secrete parathyroid hormone, which regulates calcium and phosphate metabolism.

parathyroid hormone (PTH) A hormone secreted by the parathyroid glands; regulates calcium and phosphate metabolism.

parenchyma (par-en'kih-mah) Highly variable living plant cells that have thin primary walls; function in photosynthesis, the storage of nutrients, and/or secretion.

parsimony The principle based on the experience that the simplest explanation is most probably the correct one.

parthenogenesis (par"theh-noh-jen'eh-sis) The development of an unfertilized egg into an adult organism; common among honeybees, wasps, and certain other arthropods.

partial pressure (of a gas) The pressure exerted by a gas in a mixture, which is the same pressure it would exert if alone. For example, the partial pressure of atmospheric oxygen (P_{O_2}) is 160 mm Hg at sea level.

parturition (par"to-rish'un) The birth process.

passive immunity Temporary immunity that depends on the presence of immunoglobulins produced by another organism. Compare with *active immunity.*

passive ion channel A channel in the plasma membrane that permits the passage of specific ions such as Na^+, K^+, or Cl^-.

patch–clamp technique A method that allows researchers to study the ion channels of a tiny patch of membrane by tightly sealing a micropipette to the patch and measuring the flow of ions through the channels.

patchiness See *clumped dispersion.*

pathogen (path'oh-gen) An organism, usually a microorganism, capable of producing disease.

pathogen-associated molecular patterns (PAMPs) Molecules on bacteria and other pathogens that combine with Toll-like receptors on macrophages, stimulating them to produce cytokines.

pattern formation See *morphogenesis.*

pedigree A chart constructed to show an inheritance pattern within a family through multiple generations.

peduncle The stalk of a flower or inflorescence.

pellicle A flexible outer covering consisting of protein; characteristic of certain protists, e.g., ciliates and euglenoids.

penis The male sexual organ of copulation in reptiles, mammals, and a few birds.

pentose A sugar molecule containing five carbons.

people overpopulation A situation in which there are too many people in a given geographic area; results in pollution, environmental degradation, and resource depletion. Compare with *consumption overpopulation.*

pepsin (pep′sin) An enzyme produced in the stomach that initiates digestion of protein.

pepsinogen The precursor of pepsin; secreted by chief cells in the gastric glands of the stomach.

peptide (pep′tide) A compound consisting of a chain of amino acid groups linked by peptide bonds. A dipeptide consists of two amino acids, a polypeptide of many.

peptide bond A distinctive covalent carbon-to-nitrogen bond that links amino acids in peptides and proteins.

peptidoglycan (pep″tid-oh-gly′kan) A modified protein or peptide having an attached carbohydrate; component of the bacterial cell wall.

peptidyl transferase The ribosomal enzyme that catalyzes the formation of a peptide bond.

perennial plant (purr-en′ee-ul) A woody or herbaceous plant that grows year after year, i.e., lives more than 2 years. Compare with *annual* and *biennial.*

perfect flower A flower that has both stamens and carpels. Compare with *imperfect flower.*

pericentriolar material Fibrils surrounding the centrioles in the microtubule-organizing centers in cells of animals and other organisms having centrioles.

pericycle (pehr′eh-sy″kl) A layer of meristematic cells typically found between the endodermis and phloem in roots.

periderm (pehr′ih-durm) The outer bark of woody stems and roots; composed of cork cells, cork cambium, and cork parenchyma, along with traces of primary tissues.

period An interval of geologic time that is a subdivision of an era; each period is divided into epochs.

peripheral membrane protein A protein associated with one of the surfaces of a biological membrane. Compare with *integral membrane protein.*

peripheral nervous system (PNS) In vertebrates, the nerves and receptors that lie outside the central nervous system. Compare with *central nervous system (CNS).*

peristalsis (pehr″ih-stal′sis) Rhythmic waves of muscular contraction and relaxation in the walls of hollow tubular organs, such as the ureter or parts of the digestive tract, that serve to move the contents through the tube.

permafrost Permanently frozen subsoil characteristic of frigid areas such as the tundra.

peroxisomes (pehr-ox′ih-sohmz) In eukaryotic cells, membrane-enclosed organelles containing enzymes that produce or degrade hydrogen peroxide.

persistence A characteristic of certain chemicals that are extremely stable and may take many years to be broken down into simpler forms by natural processes.

petal One of the parts of the flower attached inside the whorl of sepals; petals are usually colored.

petiole (pet′ee-ohl) The part of a leaf that attaches to a stem.

pH The negative logarithm of the hydrogen ion concentration of a solution (expressed as moles per liter). Neutral pH is 7, values less than 7 are acidic, and those greater than 7 are basic.

phage See *bacteriophage.*

phagocytosis (fag″oh-sy-toh′sis) Literally, "cell eating"; a type of endocytosis by which certain cells engulf food particles, microorganisms, foreign matter, or other cells.

pharmacogenetics A new field of gene-based medicine in which drugs are personalized to match a patient's genetic makeup.

pharyngeal slits (fair-in′jel) Openings that lead from the pharyngeal cavity to the outside; evolved as part of a filter-feeding system in chordates and later became modified for other functions, including gill slits in many aquatic vertebrates.

pharynx (fair′inks) Part of the digestive tract. In complex vertebrates, it is bounded anteriorly by the mouth and nasal cavities and posteriorly by the esophagus and larynx; the throat region in humans.

phenetics (feh-neh′tiks) An approach to classification based on measurable similarities in phenotypic characters without consideration of homology or other evolutionary relationships. Compare with *cladistics* and *evolutionary systematics.*

phenotype (fee′noh-type) The physical or chemical expression of an organism's genes. Compare with *genotype.*

phenotype frequency The proportion of a particular phenotype in the population.

phenylketonuria (PKU) (fee″nl-kee″toh-noor′ee-ah) An inherited disease in which there is a deficiency of the enzyme that normally converts phenylalanine to tyrosine; results in mental retardation if untreated.

pheromone (fer′oh-mone) A substance secreted by an organism to the external environment that influences the development or behavior of other members of the same species.

phloem (flo′em) The vascular tissue that conducts dissolved sugar and other organic compounds in plants.

phosphate group A weakly acidic functional group that can release one or two hydrogen ions.

phosphodiester linkage Covalent linkage between two nucleotides in a strand of DNA or RNA; includes a phosphate group bonded to the sugars of two adjacent nucleotides.

phosphoenolpyruvate (PEP) Three-carbon phosphorylated compound that is an important intermediate in glycolysis and is a reactant in the initial carbon fixation step in C_4 and CAM photosynthesis.

phosphoglycerate (PGA) Phosphorylated three-carbon compound that is an important metabolic intermediate.

phospholipids (fos″foh-lip′idz) Lipids in which two fatty acids and a phosphorus-containing group are attached to glycerol; major components of cell membranes.

phosphorus cycle The worldwide circulation of phosphorus from the abiotic environment into living things and back into the abiotic environment.

phosphorylation (fos″for-ih-lay′shun) The introduction of a phosphate group into an organic molecule. See *kinases.*

photoautotroph An organism that obtains energy from light and synthesizes organic compounds from inorganic raw materials; includes plants, algae, and some bacteria. Compare with *photoheterotroph, chemoautotroph,* and *chemoheterotroph.*

photoheterotroph An organism that can carry out photosynthesis to obtain energy but cannot fix carbon dioxide and therefore requires organic compounds as a carbon source; includes some bacteria. Compare with *photoautotroph, chemoautotroph,* and *chemoheterotroph.*

photolysis (foh-tol′uh-sis) The photochemical splitting of water in the light-dependent reactions of photosynthesis; a specific enzyme is needed to catalyze this reaction.

photon (foh′ton) A particle of electromagnetic radiation; one quantum of radiant energy.

photoperiodism (foh″teh-peer′ee-o-dizm) The physiological response (such as flowering) of plants to variations in the length of daylight and darkness.

photophosphorylation (foh″toh-fos-for-ih-lay′shun) The production of ATP in photosynthesis.

photoreceptor (foh″toh-ree-sep′tor) (1) A sense organ specialized to detect light. (2) A pigment that absorbs light before triggering a physiological response.

photorespiration (foh″toh-res-pur-ay′shun) The process that reduces the efficiency of photosynthesis in C_3 plants during hot spells in summer; consumes oxygen and produces carbon dioxide through the degradation of Calvin cycle intermediates.

photosynthesis The biological process that captures light energy and transforms it into the chemical energy of organic molecules (e.g., carbohydrates), which are manufactured from carbon dioxide and water.

photosystem One of two photosynthetic units responsible for capturing light energy and transferring excited electrons; photosystem I strongly absorbs light of about 700 nm, whereas photosystem II strongly absorbs light of about 680 nm. See *antenna complex* and *reaction center.*

phototroph (foh′toh-trof) Organism that uses light as a source of energy. Compare with *chemotroph.* See *photoautotroph* and *photoheterotroph.*

phototropism (foh″toh-troh′pizm) The growth of a plant in response to the direction of light.

phycocyanin (fy″koh-sy-ah′nin) A blue pigment found in cyanobacteria and red algae.

phycoerythrin (fy″koh-ee-rih′thrin) A red pigment found in cyanobacteria and red algae.

phylogenetic systematics See *cladistics.*

phylogenetic tree A branching diagram that shows lines of descent among a group of related species.

phylogeny (fy-loj′en-ee) The complete evolutionary history of a group of organisms.

phylum (fy′lum) A taxonomic grouping of related, similar classes; a category beneath the kingdom and above the class.

phytoalexins Antimicrobial compounds produced by plants that limit the spread of pathogens such as fungi.

phytochemicals Compounds found in plants that play important roles in preventing certain diseases; some function as antioxidants.

phytochrome (fy′toh-krome) A blue-green, proteinaceous pigment involved in a wide variety of physiological responses to light; occurs in two interchangeable forms depending on the ratio of red to far-red light.

phytoplankton (fy″toh-plank′tun) Microscopic floating algae and cyanobacteria that are the base of most aquatic food webs. Compare with *zooplankton.* See *plankton* and *nanoplankton.*

pia mater (pee′a may′ter) The inner membrane covering the brain and spinal cord; the innermost of the meninges; also see *dura mater* and *arachnoid.*

pigment A substance that selectively absorbs light of specific wavelengths.

pili (pie′lie) (sing., *pilus*) Hairlike structures on the surface of many bacteria; function in conjugation or attachment.

pineal gland (pie-nee′al) Endocrine gland located in the brain.

pinocytosis (pin″oh-sy-toh′sis) Cell drinking; a type of endocytosis by which cells engulf and absorb droplets of liquids.

pioneer The first organism to colonize an area and begin the first stage of succession.

pistil The female reproductive organ of a flower; consists of either a single carpel or two or more fused carpels. See *carpel.*

pith The innermost tissue in the stems and roots of many herbaceous plants; primarily a storage tissue.

pituitary gland (pi-too′ih-tehr″ee) An endocrine gland located below the hypothalamus; secretes several hormones that influence a wide range of physiological processes.

placenta (plah-sen′tah) The partly fetal and partly maternal organ whereby materials are exchanged between fetus and mother in the uterus of placental mammals.

placoderms (plak′oh-durmz) A group of extinct jawed fishes.

plankton Free-floating, mainly microscopic aquatic organisms found in the upper layers of the water; consisting of phytoplankton and zooplankton. Compare with *nekton.*

planula larva (plan′yoo-lah) A ciliated larval form found in cnidarians.

plasma The fluid portion of blood in which red blood cells, white blood cells, and platelets are suspended.

plasma cell Cell that secretes antibodies; a differentiated B lymphocyte (B cell).

plasma membrane The selectively permeable surface membrane that encloses the cell contents and through which all materials entering or leaving the cell must pass.

plasma proteins Proteins such as albumins, globulins, and fibrinogen that circulate in the blood plasma.

plasmid (plaz′mid) Small, circular, double-stranded DNA molecule that carries genes separate from those in the main DNA of a cell.

plasmodesmata (sing., *plasmodesma*) Cytoplasmic channels connecting adjacent plant cells and allowing for the movement of molecules and ions between cells.

plasmodial slime mold (plaz-moh′dee-uhl) A funguslike protist whose feeding stage consists of a plasmodium.

plasmodium (plaz-moh′dee-um) A multinucleate mass of living matter that moves and feeds in an amoeboid fashion.

plasmogamy (1) Fusion of the cytoplasm of two cells without fusion of nuclei. (2) A stage in the asexual reproduction of some fungi; hyphae of two compatible mating types come together, and their cytoplasm fuses.

plasmolysis (plaz-mol′ih-sis) The shrinkage of cytoplasm and the pulling away of the plasma membrane from the cell wall when a plant cell (or other walled cell) loses water, usually in a hypertonic environment.

plastids (plas′tidz) A family of membrane-enclosed organelles occurring in photosynthetic eukaryotic cells; include chloroplasts, chromoplasts, and amyloplasts and other leukoplasts.

platelets (playt′lets) Cell fragments in vertebrate blood that function in clotting; also called *thrombocytes.*

platyhelminths The phylum of acoelomate animals commonly known as *flatworms.*

pleiotropy The ability of a single gene to have multiple effects.

plesiomorphic characters See *shared ancestral characters.*

pleural membrane (ploor′ul) The membrane that lines the thoracic cavity and envelops each lung.

ploidy The number of chromosome sets in a nucleus or cell. See *haploid, diploid,* and *polyploid.*

plumule (ploom′yool) The embryonic shoot apex, or terminal bud, located above the point of attachment of the cotyledon(s).

pluripotent (ploor-i-poh′tent) A term describing a stem cell that can divide to give rise to many, but not all, types of cells in an organism. Compare with *totipotent*.

pneumatophore (noo-mat′uh-for″) Roots that extend up out of the water in swampy areas and are thought to provide aeration between the atmosphere and submerged roots.

polar body A small *n* cell produced during oogenesis in female animals that does not develop into a functional ovum.

polar covalent bond Chemical bond formed by the sharing of electrons between atoms that differ in electronegativity; the end of the bond near the more electronegative atom has a partial negative charge, and the other end has a partial positive charge. Compare with *nonpolar covalent bond*.

polar molecule Molecule that has one end with a partial positive charge and the other with a partial negative charge; polar molecules are generally soluble in water. Compare with *nonpolar molecule*.

polar nucleus In flowering plants, one of two *n* cells in the embryo sac that fuse with a sperm during double fertilization to form the 3*n* endosperm.

pollen grain The immature male gametophyte of seed plants (gymnosperms and angiosperms) that produces sperm capable of fertilization.

pollen tube In gymnosperms and flowering plants, a tube or extension that forms after germination of the pollen grain and through which male gametes (sperm cells) pass into the ovule.

pollination (pol″uh-nay′shen) In seed plants, the transfer of pollen from the male to the female part of the plant.

poly-A tail See *polyadenylation*.

polyadenylation (pol″ee-a-den-uh-lay′shun) That part of eukaryotic mRNA processing in which multiple adenine-containing nucleotides (a poly-A tail) are added to the 3′ end of the molecule.

polyandry A mating system in which a female mates with several males during a breeding season. Compare with *polygyny*.

polygenic inheritance (pol″ee-jen′ik) Inheritance in which several independently assorting or loosely linked nonallelic genes modify the intensity of a trait or contribute to the phenotype in additive fashion.

polygyny A mating system in which a male animal mates with many females during a breeding season. Compare with *polyandry*.

polymer (pol′ih-mer) A molecule built up from repeating subunits of the same general type (monomers); examples include proteins, nucleic acids, or polysaccharides.

polymerase chain reaction (PCR) A method by which a targeted DNA fragment is amplified in vitro to produce millions of copies.

polymorphism (pol″ee-mor′fizm) (1) The existence of two or more phenotypically different individuals within a population. (2) The presence of detectable variation in the genomes of different individuals in a population.

polyp (pol′ip) A hydralike animal; the sessile stage of the life cycle of certain cnidarians. Compare with *medusa*.

polypeptide See *peptide*.

polyphyletic group (pol″ee-fye-let′ik) A group made up of organisms that evolved from two or more different ancestors. Compare with *monophyletic group* and *paraphyletic group*.

polyploid (pol′ee-ployd) The condition of having more than two sets of chromosomes per nucleus. Compare with *diploid* and *haploid*.

polyribosome A complex consisting of a number of ribosomes attached to an mRNA during translation; also known as a *polysome*.

polysaccharide (pol-ee-sak′ah-ride) A carbohydrate consisting of many monosaccharide subunits (e.g., starch, glycogen, and cellulose).

polysome See *polyribosome*.

polyspermy The fertilization of an egg by more than one sperm.

polytene A term describing a giant chromosome consisting of many (usually >1000) parallel DNA double helices. Polytene chromosomes are typically found in cells of the salivary glands and some other tissues of certain insects, such as the fruit fly, *Drosophila*.

polyunsaturated fatty acid See *fatty acid*.

pons (ponz) The white bulge that is the part of the brain stem between the medulla and the midbrain; connects various parts of the brain.

population A group of organisms of the same species that live in a defined geographic area at the same time.

population bottleneck See *bottleneck*.

population crash An abrupt decline in the size of a population.

population density The number of individuals of a species per unit of area or volume at a given time.

population dynamics The study of changes in populations, such as how and why population numbers change over time.

population ecology That branch of biology that deals with the numbers of a particular species that are found in an area and how and why those numbers change (or remain fixed) over time.

population genetics The study of genetic variability in a population and of the forces that act on it.

population growth momentum The continued growth of a population after fertility rates have declined, as a result of a population's young age structure.

poriferans Sponges; members of phylum Porifera.

positive feedback mechanism A homeostatic mechanism in which a change in some condition triggers a response that intensifies the changing condition. Compare with *negative feedback mechanism*.

posterior Toward the tail end of a bilaterally symmetrical animal. Compare with *anterior*.

postsynaptic neuron A neuron that transmits an impulse away from a synapse. Compare with *presynaptic neuron*.

postzygotic barrier One of several reproductive isolating mechanisms that prevent gene flow between species after fertilization has taken place, e.g., hybrid inviability, hybrid sterility, and hybrid breakdown. Compare with *prezygotic barrier*.

potential energy Stored energy; energy that can do work as a consequence of its position or state. Compare with *kinetic energy*.

potentiation A form of synaptic enhancement (increase in neurotransmitter release) that can last for several minutes; occurs when a presynaptic neuron continues to transmit action potentials at a high rate for a minute or longer.

preadaptation A novel evolutionary change in a pre-existing biological structure that enables it to have a different function; feathers, which evolved from reptilian scales, represent a preadaptation for flight.

prebiotic soup hypothesis The hypothesis that simple organic molecules that are the precursors of life originated and accumulated at Earth's surface, in shallow seas or on rock or clay surfaces. Compare with *iron–sulfur world hypothesis*.

predation Relationship in which one organism (the predator) kills and devours another organism (the prey).

pre-mRNA RNA precursor to mRNA in eukaryotes; contains both introns and exons.

prenatal Pertaining to the time before birth.

pressure-flow hypothesis The mechanism by which dissolved sugar is thought to be transported in phloem; caused by a pressure gradient between the source (where sugar is loaded into the phloem) and the sink (where sugar is removed from phloem).

presynaptic neuron A neuron that transmits an impulse to a synapse. Compare with *postsynaptic neuron.*

prezygotic barrier One of several reproductive isolating mechanisms that interfere with fertilization between male and female gametes of different species, e.g., temporal isolation, habitat isolation, behavioral isolation, mechanical isolation, and gametic isolation. Compare with *postzygotic barrier.*

primary consumer See *herbivore.*

primary growth An increase in the length of a plant that occurs at the tips of the shoots and roots due to the activity of apical meristems. Compare with *secondary growth.*

primary immune response The response of the immune system to first exposure to an antigen. Compare with *secondary immune response.*

primary mycelium A mycelium in which the cells are monokaryotic and haploid; a mycelium that grows from either an ascospore or a basidiospore. Compare with *secondary mycelium.*

primary producer See *autotroph.*

primary productivity The amount of light energy converted to organic compounds by autotrophs in an ecosystem over a given period. Compare with *secondary productivity.* See *gross primary productivity* and *net primary productivity.*

primary structure (of a protein) The complete sequence of amino acids in a polypeptide chain, beginning at the amino end and ending at the carboxyl end. Compare with *secondary, tertiary,* and *quaternary protein structure.*

primary succession An ecological succession that occurs on land that has not previously been inhabited by plants; no soil is present initially. See *succession.* Compare with *secondary succession.*

primates Mammals that share such traits as flexible hands and feet with five digits; a strong social organization; and front-facing eyes; includes lemurs, tarsiers, monkeys, apes, and humans.

primer See *RNA primer.*

primitive streak Dynamic, constantly changing structure that forms at the midline of the blastodisc in birds, mammals, and some other vertebrates and is active in gastrulation. The anterior end of the primitive streak is Hensen's node.

primosome A complex of proteins responsible for synthesizing the RNA primers required in DNA synthesis.

principle In science, a statement of a rule that explains how something works. A scientific principle has withstood repeated testing and has the highest level of scientific confidence.

prion (pri'on) An infectious agent that consists only of protein.

producer See *autotroph.*

product Substance formed by a chemical reaction. Compare with *reactant.*

product rule The rule for combining the probabilities of independent events by multiplying their individual probabilities. Compare with *sum rule.*

profundal zone (pro-fun'dl) The deepest zone of a large lake, located below the level of penetration by sunlight. Compare with *littoral zone* and *limnetic zone.*

progesterone (pro-jes'ter-own) A steroid hormone secreted by the ovary (mainly by the corpus luteum) and placenta; stimulates the uterus (to prepare the endometrium for implantation) and breasts (for milk secretion).

progymnosperm (pro-jim'noh-sperm) An extinct group of plants that may have been the ancestors of gymnosperms.

prokaryote (pro-kar'ee-ote) A cell that lacks a nucleus and other membrane-enclosed organelles; includes the bacteria and archaea (kingdoms Eubacteria and Archaea). Compare with *eukaryote.*

prometaphase Stage of mitosis during which spindle microtubules attach to kinetochores of chromosomes, which begin to move toward the cell's midplane; occurs after prophase and before metaphase.

promoter The nucleotide sequence in DNA to which RNA polymerase attaches to begin transcription.

prop root An adventitious root that arises from the stem and provides additional support for a plant such as corn.

prophage (pro'faj) Bacteriophage nucleic acid that is inserted into the bacterial DNA.

prophase The first stage of mitosis. During prophase the chromosomes become visible as distinct structures, the nuclear envelope breaks down, and a spindle forms.

proplastids Organelles that are plastid precursors; may mature into various specialized plastids, including chloroplasts, chromoplasts, or leukoplasts.

proprioceptors (pro"pree-oh-sep'torz) Receptors in muscles, tendons, and joints that respond to changes in movement, tension, and position; enable an animal to perceive the position of its body.

prostaglandins (pros"tah-glan'dinz) A group of local regulators derived from fatty acids; synthesized by most cells of the body and produce a wide variety of effects; sometimes called local *hormones.*

prostate gland A gland in male animals that produces an alkaline secretion that is part of the semen.

proteasome A large multiprotein structure that recognizes and degrades protein molecules tagged with ubiquitin into short, nonfunctional peptide fragments.

protein A large, complex organic compound composed of covalently linked amino acid subunits; contains carbon, hydrogen, oxygen, nitrogen, and sulfur.

protein kinase One of a group of enzymes that activate or inactivate other proteins by phosphorylating (adding phosphate groups to) them.

proteomics The study of all the proteins encoded by the human genome and produced in a person's cells and tissues.

Proterozoic eon The period of Earth's history that began approximately 2.5 billion years ago and ended 542 million years ago; marked by the accumulation of oxygen and the appearance of the first multicellular eukaryotic life-forms.

prothallus (pro-thal'us) (pl., *prothalli*) The free-living, *n* gametophyte in ferns and other seedless vascular plants.

protist (pro'tist) One of a vast kingdom of eukaryotic organisms, primarily unicellular or simple multicellular; mostly aquatic.

protobionts (pro"toh-by'ontz) Assemblages of organic polymers that spontaneously form under certain conditions. Protobionts may have been involved in chemical evolution.

proton A particle present in the nuclei of all atoms that has one unit of positive charge and a mass of 1 atomic mass unit (amu). Compare with *electron* and *neutron.*

protonema (pro"toh-nee'mah) (pl., *protonemata*) In mosses, a filament of *n* cells that grows from a spore and develops into leafy moss gametophytes.

protonephridia (pro"toh-nef-rid'ee-ah) (sing., *protonephridium*) The flame-cell excretory organs of flatworms and some other simple invertebrates.

proto-oncogene A gene that normally promotes cell division in response to the presence of certain growth factors; when mutated, it may become an oncogene, possibly leading to the formation of a cancer cell. Compare with *oncogene.*

protostome (pro'toh-stome) A major division of the animal kingdom in which the blastopore develops into the mouth, and the anus forms secondarily; includes the annelids, arthropods, and mollusks. Compare with *deuterostome.*

protozoa (proh″toh-zoh′a) (sing., *protozoon*) An informal group of unicellular, animal-like protists, including amoebas, foraminiferans, actinopods, ciliates, flagellates, and apicomplexans. (The adjectival form is *protozoan*.)

provirus (pro-vy′rus) A part of a virus, consisting of nucleic acid only, that was inserted into a host genome. See *DNA provirus*.

proximal Closer to the point of reference. Compare with *distal*.

proximal convoluted tubule The part of the renal tubule that extends from Bowman's capsule to the loop of Henle. Compare with *distal convoluted tubule*.

proximate causes (of behavior) The immediate causes of behavior, such as genetic, developmental, and physiological processes that permit the animal to carry out a specific behavior. Compare with *ultimate causes of behavior*.

pseudocoelom (sue″doh-see′lom) A body cavity between the mesoderm and endoderm; derived from the blastocoel. Compare with *coelom*.

pseudocoelomate (sue″doh-seel′oh-mate) An animal having a pseudocoelom. Compare with *coelomate* and *acoelomate*.

pseudoplasmodium (sue″doe-plaz-moh′dee-um) In cellular slime molds, an aggregation of amoeboid cells that forms a spore-producing fruiting body during reproduction.

pseudopodium (sue″doe-poe′dee-um) (pl., *pseudopodia*) A temporary extension of an amoeboid cell that is used for feeding and locomotion.

puff In a polytene chromosome, a decondensed region that is a site of intense RNA synthesis.

pulmonary circulation The part of the circulatory system that delivers blood to and from the lungs for oxygenation. Compare with *systemic circulation*.

pulse, arterial The alternate expansion and recoil of an artery.

punctuated equilibrium The idea that evolution proceeds with periods of little or no genetic change, followed by very active phases, so that major adaptations or clusters of adaptations appear suddenly in the fossil record. Compare with *gradualism*.

Punnett square The grid structure, first developed by Reginald Punnett, that allows direct calculation of the probabilities of occurrence of all possible offspring of a genetic cross.

pupa (pew′pah) (pl., *pupae*) A stage in the development of an insect, between the larva and the imago (adult); a form that neither moves nor feeds and may be in a cocoon.

purines (pure′eenz) Nitrogenous bases with carbon and nitrogen atoms in two attached rings, e.g., adenine and guanine; components of nucleic acids, ATP, GTP, NAD^+, and certain other biologically active substances. Compare with *pyrimidines*.

pyramid of biomass An ecological pyramid that illustrates the total biomass, as, for example, the total dry weight, of all organisms at each trophic level in an ecosystem.

pyramid of energy An ecological pyramid that shows the energy flow through each trophic level of an ecosystem.

pyrimidines (pyr-im′ih-deenz) Nitrogenous bases, each composed of a single ring of carbon and nitrogen atoms, e.g., thymine, cytosine, and uracil; components of nucleic acids. Compare with *purines*.

pyruvate (pyruvic acid) A three-carbon compound; the end product of glycolysis.

quadrupedal (kwad′roo-ped″ul) Walking on all fours.

quantitative trait A trait that shows continuous variation in a population (e.g., human height) and typically has a polygenic inheritance pattern.

quaternary structure (of a protein) The overall conformation of a protein produced by the interaction of two or more polypeptide chains. Compare with *primary, secondary,* and *tertiary protein structure*.

r selection A reproductive strategy recognized by some ecologists, in which a species typically has a small body size, rapid development, and short life span and devotes a large proportion of its metabolic energy to the production of offspring. Compare with *K selection*.

radial cleavage The pattern of blastomere production in which the cells are located directly above or below one another; characteristic of early deuterostome embryos. Compare with *spiral cleavage*.

radial symmetry A body plan in which any section through the mouth and down the length of the body divides the body into similar halves. Jellyfish and other cnidarians have radial symmetry. Compare with *bilateral symmetry*.

radicle (rad′ih-kl) The embryonic root of a seed plant.

radioactive decay The process in which a radioactive element emits radiation, and as a result, its nucleus changes into the nucleus of a different element.

radioisotopes Unstable isotopes that spontaneously emit radiation; also called *radioactive isotopes*.

radiolarians Those actinopods that secrete elaborate shells of silica (glass).

radula (rad′yoo-lah) A rasplike structure in the digestive tract of chitons, snails, squids, and certain other mollusks.

rain shadow An area that has very little precipitation, found on the downwind side of a mountain range. Deserts often occur in rain shadows.

random dispersion The spatial distribution pattern of a population in which the presence of one individual has no effect on the distribution of other individuals. Compare with *clumped dispersion* and *uniform dispersion*.

range The area where a particular species occurs. Compare with *home range*.

ray A chain of parenchyma cells (one to many cells thick) that functions for lateral transport in stems and roots of woody plants.

ray-finned fishes A class (Actinopterygii) of modern bony fishes; contains about 95% of living fish species.

reabsorption The selective removal of certain substances from the glomerular filtrate by the renal tubules and collecting ducts of the kidney, and their return into the blood.

reactant Substance that participates in a chemical reaction. Compare with *product*.

reaction center The portion of a photosystem that includes chlorophyll *a* molecules capable of transferring electrons to a primary electron acceptor, which is the first of several electron acceptors in a series. See *antenna complex* and *photosystem*.

realized niche The lifestyle that an organism actually pursues, including the resources that it actually uses. An organism's realized niche is narrower than its fundamental niche because of interspecific competition. Compare with *fundamental niche*.

receptacle The end of a flower stalk where the flower parts (sepals, petals, stamens, and carpels) are attached.

reception Process of detecting a stimulus.

receptor (1) In cell biology, a molecule on the surface of a cell, or inside a cell, that serves as a recognition or binding site for signaling molecules such as hormones, antibodies, or neurotransmitters. (2) A sensory receptor. See *sensory receptor*.

receptor down-regulation The process by which some hormone receptors decrease in number, thereby suppressing the sensitivity of target cells to the hormone. Compare with *receptor up-regulation*.

receptor up-regulation The process by which some hormone receptors increase in number, thereby increasing the sensitivity of the target cells to the hormone. Compare with *receptor down-regulation*.

receptor-mediated endocytosis A type of endocytosis in which extracellular molecules become bound to specific receptors on the cell surface and then enter the cytoplasm enclosed in vesicles.

recessive allele (al-leel′) An allele that is not expressed in the heterozygous state. Compare with *dominant allele.*

recombinant DNA Any DNA molecule made by combining genes from different organisms.

recombination, genetic The appearance of new gene combinations. Recombination in eukaryotes generally results from meiotic events, either crossing-over or shuffling of chromosomes.

red alga A member of a diverse phylum of algae that contain the pigments chlorophyll *a,* carotenoids, phycocyanin, and phycoerythrin.

red blood cell (RBC) See *erythrocyte.*

red tide A red or brown coloration of ocean water caused by a population explosion, or bloom, of dinoflagellates.

redox reaction (ree′dox) The chemical reaction in which one or more electrons are transferred from one substance (the substance that becomes oxidized) to another (the substance that becomes reduced). See *oxidation* and *reduction.*

reduction The gain of one or more electrons (or hydrogen atoms) by an atom, ion, or molecule. Compare with *oxidation.*

reflex action An automatic, involuntary response to a given stimulus that generally functions to restore homeostasis.

refractory period The brief period that elapses after the response of a neuron or muscle fiber, during which it cannot respond to another stimulus.

regulative development The very plastic developmental pattern in which each individual cell of an early embryo retains totipotency. Compare with *mosaic development.*

regulatory gene Gene that turns the transcription of other genes on or off.

releasing hormone A hormone secreted by the hypothalamus that stimulates secretion of a specific hormone by the anterior lobe of the pituitary gland.

renal (ree′nl) Pertaining to the kidney.

renal pelvis The funnel-shaped chamber of the kidney that receives urine from the collecting ducts; urine then moves into the ureters.

renin (reh′nin) An enzyme released by the kidney in response to a decrease in blood pressure; activates a pathway leading to production of angiotensin II, a hormone that increases aldosterone release; aldosterone increases blood pressure.

replacement-level fertility The number of children a couple must produce to "replace" themselves. The average number is greater than two, because some children die before reaching reproductive age.

replication See *DNA replication.*

replication fork Y-shaped structure produced during the semiconservative replication of DNA.

repolarization The process of returning membrane potential to its resting level.

repressible operon An operon that is normally active, but can be controlled by a repressor protein, which becomes active when it binds to a corepressor; the active repressor binds to the operator, making the operon transcriptionally inactive. Compare with *inducible operon.*

repressor protein A negative regulatory protein that inhibits transcription when bound to DNA; some repressors require a corepressor to be active; some other repressors become inactive when bound to an inducer molecule. Compare with *activator protein.*

reproduction The process by which new individuals are produced. See *asexual reproduction* and *sexual reproduction.*

reproductive isolating mechanisms The reproductive barriers that prevent a species from interbreeding with another species; as a result, each species' gene pool is isolated from those of other species. See *prezygotic barrier* and *postzygotic barrier.*

reptiles A class of vertebrates characterized by dry skin with horny scales and adaptations for terrestrial reproduction; include turtles, snakes, and alligators; reptiles are a paraphyletic group.

residual capacity The volume of air that remains in the lungs at the end of a normal exhalation.

resin A viscous organic material that certain plants produce and secrete into specialized ducts; may play a role in deterring disease organisms or plant-eating insects.

resolution See *resolving power.*

resolving power The ability of a microscope to show fine detail, defined as the minimum distance between two points at which they are seen as separate images; also called *resolution.*

resource partitioning The reduction of competition for environmental resources such as food that occurs among coexisting species as a result of each species' niche differing from the others in one or more ways.

respiration (1) Cellular respiration is the process by which cells generate ATP through a series of redox reactions. In aerobic respiration the terminal electron acceptor is molecular oxygen; in anaerobic respiration the terminal acceptor is an inorganic molecule other than oxygen. (2) Organismic respiration is the process of gas exchange between a complex animal and its environment, generally through a specialized respiratory surface, such as a lung or gill.

respiratory centers Centers in the medulla and pons that regulate breathing.

resting potential The membrane potential (difference in electric charge between the two sides of the plasma membrane) of a neuron in which no action potential is occurring. The typical resting potential is about -70 millivolts. Compare with *action potential.*

restoration ecology The scientific field that uses the principles of ecology to help return a degraded environment as closely as possible to its former undisturbed state.

restriction enzyme One of a class of enzymes that cleave DNA at specific base sequences; produced by bacteria to degrade foreign DNA; used in recombinant DNA technology.

restriction map A physical map of DNA in which sites cut by specific restriction enzymes serve as landmarks.

reticular activating system (RAS) (reh-tik′yoo-lur) A diffuse network of neurons in the brain stem; responsible for maintaining consciousness.

retina (ret′ih-nah) The innermost of the three layers (retina, choroid layer, and sclera) of the eyeball, which is continuous with the optic nerve and contains the light-sensitive rod and cone cells.

retinular cells See *ommatidium.*

retrovirus (ret′roh-vy″rus) An RNA virus that uses reverse transcriptase to produce a DNA intermediate, known as a *DNA provirus,* in the host cell. See *DNA provirus.*

reverse transcriptase An enzyme produced by retroviruses that catalyzes the production of DNA using RNA as a template.

reversible inhibitor A substance that forms weak bonds with an enzyme, temporarily interfering with its function; a reversible inhibitor is either competitive or noncompetitive. Compare with *irreversible inhibitor.*

Rh factors Red blood cell antigens, known as *D* antigens, first identified in *Rhesus* monkeys. People who have these antigens are Rh$^+$; people lacking them are Rh$^-$. See *erythroblastosis fetalis.*

rhizome (ry′zome) A horizontal underground stem that bears leaves and buds and often serves as a storage organ and a means of asexual reproduction, e.g., iris.

rhodopsin (rho-dop′sin) Visual purple; a light-sensitive pigment found in the rod cells of the vertebrate eye; a similar molecule is employed by certain bacteria in the capture of light energy to make ATP.

ribonucleic acid (RNA) A family of single-stranded nucleic acids that function mainly in protein synthesis

ribose The five-carbon sugar present in RNA and in important nucleoside triphosphates such as ATP.

ribosomal RNA (rRNA) See *ribosomes*.

ribosomes (ry′boh-sohmz) Organelles that are part of the protein synthesis machinery of both prokaryotic and eukaryotic cells; consist of a larger and smaller subunit, each composed of ribosomal RNA (rRNA) and ribosomal proteins.

ribozyme (ry′boh-zime) A molecule of RNA that has catalytic properties.

ribulose bisphosphate (RuBP) A five-carbon phosphorylated compound with a high energy potential that reacts with carbon dioxide in the initial step of the Calvin cycle.

ribulose bisphosphate carboxylase/oxygenase See *rubisco*.

RNA interference (RNAi) Phenomenon in which certain small RNA molecules interfere with the expression of genes or their RNA transcripts; RNA interference involves small interfering RNAs, microRNAs, and a few other kinds of short RNA molecules.

RNA polymerase An enzyme that catalyzes the synthesis of RNA from a DNA template.

RNA primer The sequence of about five RNA nucleotides that are synthesized during DNA replication to provide a 3′ end to which DNA polymerase adds nucleotides. The RNA primer is later degraded and replaced with DNA.

RNA world A model that proposes that during the evolution of cells, RNA was the first informational molecule to evolve, followed at a later time by proteins and DNA.

rod One of the rod-shaped, light-sensitive cells of the retina that are particularly sensitive to dim light and mediate black-and-white vision. Compare with *cone*.

root cap A covering of cells over the root tip that protects the delicate meristematic tissue directly behind it.

root graft The process of roots from two different plants growing together and becoming permanently attached to each other.

root hair An extension, or outgrowth, of a root epidermal cell. Root hairs increase the absorptive capacity of roots.

root pressure The pressure in xylem sap that occurs as a result of the active absorption of mineral ions followed by the osmotic uptake of water into roots from the soil.

root system The underground portion of a plant that anchors it in the soil and absorbs water and dissolved minerals.

rough ER See *endoplasmic reticulum*.

rubisco The common name of ribulose bisphosphate carboxylase/oxygenase, the enzyme that catalyzes the fixation of carbon dioxide in the Calvin cycle.

rugae (roo′jee) Folds, such as those in the lining of the stomach.

runner See *stolon*.

S phase Stage in interphase of the cell cycle during which DNA and other chromosomal constituents are synthesized. Compare with G_1 and G_2 *phases*.

saccule The structure within the vestibule of the inner vertebrate ear that along with the utricle houses the receptors of static equilibrium.

salicylic acid A signaling molecule that helps plants defend against insect pests and pathogens such as viruses by helping activate systemic acquired resistance.

salinity The concentration of dissolved salts (e.g., sodium chloride) in a body of water.

salivary glands Accessory digestive glands found in vertebrates and some invertebrates; in humans there are three pairs.

salt An ionic compound consisting of an anion other than a hydroxide ion and a cation other than a hydrogen ion. A salt is formed by the reaction between an acid and a base.

salt marsh A wetland dominated by grasses in which the salinity fluctuates between that of sea water and fresh water; salt marshes are usually located in estuaries.

saltatory conduction The transmission of a neural impulse along a myelinated neuron; ion activity at one node depolarizes the next node along the axon.

saprobe See *decomposer*.

saprotroph (sap′roh-trof) See *decomposer*.

sarcolemma (sar″koh-lem′mah) The muscle cell plasma membrane.

sarcomere (sar′koh-meer) A segment of a striated muscle cell located between adjacent Z lines that serves as a unit of contraction.

sarcoplasmic reticulum The system of vesicles in a muscle cell that surrounds the myofibrils and releases calcium in muscle contraction; a modified endoplasmic reticulum.

saturated fatty acid See *fatty acid*.

savanna (suh-van′uh) A tropical grassland containing scattered trees; found in areas of low rainfall or seasonal rainfall with prolonged dry periods.

scaffolding proteins (1) Proteins that organize groups of intracellular signaling molecules into signaling complexes. (2) Nonhistone proteins that help maintain the structure of a chromosome.

schizocoely (skiz′oh-seely) The process of coelom formation in which the mesoderm splits into two layers, forming a cavity between them; characteristic of protostomes. Compare with *enterocoely*.

Schwann cells Supporting cells found in nervous tissue outside the central nervous system; produce the myelin sheath around peripheral neurons.

sclera (skler′ah) The outer coat of the eyeball; a tough, opaque sheet of connective tissue that protects the inner structures and helps maintain the rigidity of the eyeball.

sclereid (skler′id) In plants, a sclerenchyma cell that is variable in shape but typically not long and tapered. Compare with *fiber*.

sclerenchyma (skler-en′kim-uh) Cells that provide strength and support in the plant body, are often dead at maturity, and have extremely thick walls; includes fibers and sclereids.

scramble competition See *exploitation competition*.

scrotum (skroh′tum) The external sac of skin found in most male mammals that contains the testes and their accessory organs.

second law of thermodynamics The physical law stating that the total amount of entropy in the universe continually increases. Compare with *first law of thermodynamics*.

second messenger A substance, e.g., cyclic AMP or calcium ions, that relays a message from a hormone bound to a cell-surface receptor; leads to some change in the cell.

secondary consumer See *carnivore*.

secondary growth An increase in the girth of a plant due to the activity of the vascular cambium and cork cambium; secondary growth results in the production of secondary tissues, i.e., wood and bark. Compare with *primary growth*.

secondary immune response The rapid production of antibodies induced by a second exposure to an antigen several days, weeks, or even months after the initial exposure. Compare with *primary immune response*.

secondary mycelium A dikaryotic mycelium formed by the fusion of two primary hyphae. Compare with *primary mycelium*.

secondary productivity The amount of food molecules converted to biomass by consumers in an ecosystem over a given period. Compare with *primary productivity*.

secondary structure (of a protein) A regular geometric shape produced by hydrogen bonding between the atoms of the uniform polypeptide backbone; includes the alpha helix and the beta-pleated sheet. Compare with *primary, tertiary,* and *quaternary protein structure.*

secondary succession An ecological succession that takes place after some disturbance destroys the existing vegetation; soil is already present. See *succession.* Compare with *primary succession.*

secretory vesicles Small cytoplasmic vesicles that move substances from an internal membrane system to the plasma membrane.

seed A plant reproductive body consisting of a young, multicellular plant and nutritive tissue (food reserves), enclosed by a seed coat.

seed coat The outer protective covering of a seed.

seed fern An extinct group of seed-bearing woody plants with fern-like leaves; seed ferns probably descended from progymnosperms and gave rise to cycads and possibly ginkgoes.

segmentation genes In *Drosophila,* genes transcribed in the embryo that are responsible for generating a repeating pattern of body segments within the embryo and adult fly.

segregation, principle of The genetic principle, first noted by Gregor Mendel, that states that two alleles of a locus become separated into different gametes.

selectively permeable membrane A membrane that allows some substances to cross it more easily than others. Biological membranes are generally permeable to water but restrict the passage of many solutes.

self-incompatibility A genetic condition in which the pollen cannot fertilize the same flower or flowers on the same plant.

semelparity The condition of having a single reproductive effort in a lifetime. Compare with *iteroparity.*

semen The fluid consisting of sperm suspended in various glandular secretions that is ejaculated from the penis during orgasm.

semicircular canals The passages in the vertebrate inner ear containing structures that control the sense of equilibrium (balance).

semiconservative replication See *DNA replication.*

semilunar valves Valves between the ventricles of the heart and the arteries that carry blood away from the heart; aortic and pulmonary valves.

seminal vesicles (1) In mammals, glandular sacs that secrete a component of semen (seminal fluid). (2) In some invertebrates, structures that store sperm.

seminiferous tubules (sem-ih-nif′er-ous) Coiled tubules in the testes in which spermatogenesis takes place in male vertebrates.

senescence (se-nes′cents) The aging process.

sensory neuron A neuron that transmits an impulse from a receptor to the central nervous system.

sensory receptor A cell (or part of a cell) specialized to detect specific energy stimuli in the environment.

sepal (see′pul) One of the outermost parts of a flower, usually leaf-like in appearance, that protect the flower as a bud.

septum (pl., *septa*) A cross wall or partition, e.g., the walls that divide a hypha into cells.

sequencing See *DNA sequencing.*

serial endosymbiosis The hypothesis that certain organelles such as mitochondria and chloroplasts originated as symbiotic prokaryotes that lived inside other, free-living prokaryotic cells.

serotonin A neurotransmitter of the biogenic amine group.

Sertoli cells (sur-tole′ee) Supporting cells of the tubules of the testis.

sessile (ses′sile) Permanently attached to one location, e.g., coral animals.

setae (sing., *seta*) Bristlelike structures that aid in annelid locomotion.

set point A normal condition maintained by homeostatic mechanisms.

sex chromosome Chromosome that plays a role in sex determination.

sex-influenced trait A genetic trait that is expressed differently in males and females.

sex-linked gene A gene carried on a sex chromosome. In mammals almost all sex-linked genes are borne on the X chromosome, i.e., are X-linked.

sexual dimorphism Marked phenotypic differences between the two sexes of the same species.

sexual isolation See *behavioral isolation.*

sexual reproduction A type of reproduction in which two gametes (usually, but not necessarily, contributed by two different parents) fuse to form a zygote. Compare with *asexual reproduction.*

sexual selection A type of natural selection that occurs when individuals of a species vary in their ability to compete for mates; individuals with reproductive advantages are selected over others of the same sex.

shade avoidance The tendency of plants that are adapted to high light intensities to grow taller when they are closely surrounded by other plants.

shared ancestral characters Traits that were present in an ancestral species that have remained essentially unchanged; suggest a distant common ancestor. Also called *plesiomorphic characters.* Compare with *shared derived characters.*

shared derived characters Homologous traits found in two or more taxa that are present in their most recent common ancestor but not in earlier common ancestors. Also called *synapomorphic characters.* Compare with *shared ancestral characters.*

shoot system The aboveground portion of a plant, such as the stem and leaves.

short tandem repeats (STRs) Molecular markers that are short sequences of repetitive DNA; because STRs vary in length from one individual to another, they are useful in identifying individuals with a high degree of certainty.

short-day plant A plant that flowers in response to lengthening nights; also called *long-night plant.* Compare with *long-day, intermediate-day,* and *day-neutral plants.*

short-night plant See *long-day plant.*

sickle cell anemia An inherited form of anemia in which there is abnormality in the hemoglobin beta chains; the inheritance pattern is autosomal recessive.

sieve tube elements Cells that conduct dissolved sugar in the phloem of flowering plants.

sign stimulus Any stimulus that elicits a fixed action pattern in an animal.

signal amplification The process by which a few signaling molecules can effect major responses in the cell; the strength of each signaling molecule is magnified.

signal transduction A process in which a cell converts and amplifies an extracellular signal into an intracellular signal that affects some function in the cell. Also see *cell signaling.*

signaling molecule A molecule such as a hormone, local regulator, or neurotransmitter that transmits information when it binds to a receptor on the cell surface or within the cell.

signal-recognition particle (SRP) A protein–RNA complex that directs the ribosome–mRNA–polypeptide complex to the surface of the endoplasmic reticulum.

simple fruit A fruit that develops from a single ovary. Compare with *aggregate, accessory,* and *multiple fruits.*

simplest formula A type of chemical formula that gives the smallest whole-number ratio of the component atoms. Compare with *molecular formula* and *structural formula.*

single-strand binding proteins (SSBs) Proteins involved in DNA replication that bind to single DNA strands and prevent the double helix from re-forming until the strands are copied.

sink habitat A lower-quality habitat in which local reproductive success is less than local mortality. Compare with *source habitat.*

sinoatrial (SA) node The mass of specialized cardiac muscle in which the impulse triggering the heartbeat originates; the pacemaker of the heart.

sister chromatids See *chromatid.*

sister taxa Groups of organisms that share a more recent common ancestor with one another than either taxon does with any other group shown on a cladogram.

skeletal muscle The voluntary striated muscle of vertebrates, so called because it usually is directly or indirectly attached to some part of the skeleton. Compare with *cardiac muscle* and *smooth muscle.*

slash-and-burn agriculture A type of agriculture in which tropical rain forest is cut down, allowed to dry, and burned. The crops that are planted immediately afterward thrive because the ashes provide nutrients; in a few years, however, the soil is depleted and the land must be abandoned.

slow block to polyspermy See *cortical reaction.*

small interfering RNAs (siRNAs) Double-stranded RNA molecules about 23 nucleotides in length that silence genes at the post-transcriptional level by selectively cleaving mRNA molecules with base sequences complementary to the siRNA.

small intestine Portion of the vertebrate digestive tract that extends from the stomach to the large intestine.

small nuclear ribonucleoprotein complexes (snRNP) Aggregations of RNA and protein responsible for binding to pre-mRNA in eukaryotes; catalyze the excision of introns and the splicing of exons.

smooth ER See *endoplasmic reticulum.*

smooth muscle Involuntary muscle tissue that lacks transverse striations; found mainly in sheets surrounding hollow organs, such as the intestine. Compare with *cardiac muscle* and *skeletal muscle.*

social behavior Interaction of two or more animals, usually of the same species.

sociobiology The branch of biology that focuses on the evolution of social behavior through natural selection.

sodium–potassium pump Active transport system that transports sodium ions out of, and potassium ions into, cells.

soil erosion The wearing away or removal of soil from the land; although soil erosion occurs naturally from precipitation and run-off, human activities (such as clearing the land) accelerate it.

solute A dissolved substance. Compare with *solvent.*

solvent Substance capable of dissolving other substances. Compare with *solute.*

somatic cell In animals, a cell of the body not involved in formation of gametes. Compare with *germ line cell.*

somatic nervous system That part of the vertebrate peripheral nervous system that keeps the body in adjustment with the external environment; includes sensory receptors on the body surface and within the muscles, and the nerves that link them with the central nervous system. Compare with *autonomic nervous system.*

somatomedins See *insulin-like growth factors.*

somatotropin See *growth hormone.*

somites A series of paired blocks of mesoderm that develop on each side of the notochord in cephalochordates and vertebrates. Somites define the segmentation of the embryo, and in vertebrates, they give rise to the vertebrae, ribs, and certain skeletal muscles.

sonogram See *ultrasound imaging.*

soredium (sor-id′e-um) (pl., *soredia*) In lichens, a type of asexual reproductive structure that consists of a cluster of algal cells surrounded by fungal hyphae.

sorus (soh′rus) (pl., *sori*) In ferns, a cluster of spore-producing sporangia.

source habitat A good habitat in which local reproductive success is greater than local mortality. Surplus individuals in a source habitat may disperse to other habitats. Compare with *sink habitat.*

Southern blot A technique in which DNA fragments, previously separated by gel electrophoresis, are transferred to a nitrocellulose or nylon membrane and detected by autoradiography or chemical luminescence. Compare with *Northern blot* and *Western blot.*

speciation Evolution of a new species.

species According to the biological species concept, one or more populations whose members are capable of interbreeding in nature to produce fertile offspring and do not interbreed with members of other species. Compare with *evolutionary species concept.*

species diversity A measure of the relative importance of each species within a community; represents a combination of species richness and species evenness.

species richness The number of species in a community.

specific epithet The second part of the name of a species; designates a specific species belonging to that genus.

specific heat The amount of heat energy that must be supplied to raise the temperature of 1 g of a substance 1°C.

specific immune responses Defense mechanisms that target specific macromolecules associated with a pathogen. Includes cell-mediated immunity and antibody-mediated immunity. Also known as *acquired* or *adaptive immune responses.*

sperm The motile male gamete of animals and some plants and protists; also called a *spermatozoan.*

spermatid (spur′ma-tid) An immature sperm cell.

spermatocyte (spur-mah′toh-site) A meiotic cell that gives rise to spermatids and ultimately to mature sperm cells.

spermatogenesis (spur″mah-toh-jen′eh-sis) The production of male gametes (sperm) by meiosis and subsequent cell differentiation. Compare with *oogenesis.*

spermatozoan (spur-mah-toh-zoh′un) See *sperm.*

sphincter (sfink′tur) A group of circularly arranged muscle fibers, the contractions of which close an opening, e.g., the pyloric sphincter at the exit of the stomach.

spinal cord In vertebrates, the dorsal, tubular nerve cord.

spinal nerves In vertebrates, the nerves that emerge from the spinal cord.

spindle See *mitotic spindle.*

spine A leaf that is modified for protection, such as a cactus spine.

spiracle (speer′ih-kl) An opening for gas exchange, such as the opening of a trachea on the body surface of an insect.

spiral cleavage A distinctive spiral pattern of blastomere production in an early protostome embryo. Compare with *radial cleavage.*

spirillum (pl., *spirilla*) A long, rigid, helical bacterium. Compare with *spirochete, vibrio, bacillus,* and *coccus.*

spirochete A long, flexible, helical bacterium. Compare with *spirillum, vibrio, bacillus,* and *coccus.*

spleen An abdominal organ located just below the diaphragm that removes worn-out blood cells and bacteria from the blood and plays a role in immunity.

spliceosome A large nucleoprotein particle that catalyzes the reactions that remove introns from pre-mRNA.

spongy mesophyll (mez'oh-fil) The loosely arranged mesophyll cells near the lower epidermis in certain leaves. Compare with *palisade mesophyll*.

spontaneous reaction See *exergonic reaction*.

sporangium (spor-an'jee-um) (pl., *sporangia*) A spore case, found in plants, certain protists, and fungi.

spore A reproductive cell that gives rise to individual offspring in plants, fungi, and certain algae and protozoa.

sporophyll (spor'oh-fil) A leaflike structure that bears spores.

sporophyte generation (spor'oh-fite) The 2*n*, spore-producing stage in the life cycle of a plant. Compare with *gametophyte generation*.

sporozoa See *apicomplexans*.

sporozoite The infective sporelike state in apicomplexans.

stabilizing selection Natural selection that acts against extreme phenotypes and favors intermediate variants; associated with a population well adapted to its environment. Compare with *directional selection* and *disruptive selection*.

stamen (stay'men) The male part of a flower; consists of a filament and anther.

standing-water ecosystem A lake or pond ecosystem.

starch A polysaccharide composed of alpha glucose subunits; made by plants for energy storage.

start codon The codon AUG, which signals the beginning of translation of messenger RNA. Compare with *stop codon*.

stasis Long periods in the fossil record in which there is little or no evolutionary change.

statocyst (stat'oh-sist) An invertebrate sense organ containing one or more granules (statoliths); senses gravity and motion.

statoliths (stat'uh-liths) Granules of loose sand or calcium carbonate found in statocysts.

stele The cylinder in the center of roots and stems that contains the vascular tissue.

stem cell A relatively undifferentiated cell capable of repeated cell division. At each division at least one of the daughter cells usually remains a stem cell, whereas the other may differentiate as a specific cell type.

stereocilia Hairlike projections of hair cells; microvilli that contain actin filaments.

sterilization A procedure that renders an individual incapable of producing offspring; the most common surgical procedures are vasectomy in the male and tubal ligation in the female.

steroids (steer'oids) Complex molecules containing carbon atoms arranged in four attached rings, three of which contain six carbon atoms each and the fourth of which contains five, e.g., cholesterol and certain hormones, including the male and female sex hormones of vertebrates.

stigma The portion of the carpel where pollen grains land during pollination (and before fertilization).

stipe A short stalk or stemlike structure that is a part of the body of certain multicellular algae.

stipule (stip'yule) One of a pair of scalelike or leaflike structures found at the base of certain leaves.

stolon (stow'lon) An aboveground, horizontal stem with long internodes; stolons often form buds that develop into separate plants, e.g., strawberry; also called *runner*.

stomach Muscular region of the vertebrate digestive tract, extending from the esophagus to the small intestine.

stomata (sing., *stoma*) Small pores located in the epidermis of plants that provide for gas exchange for photosynthesis; each stoma is flanked by two guard cells, which are responsible for its opening and closing.

stop codon Any of the three codons in mRNA that do not code for an amino acid (UAA, UAG, or UGA) but signal the termination of translation. Compare with *start codon*.

stratosphere The layer of the atmosphere between the troposphere and the mesosphere. It contains a thin ozone layer that protects life by filtering out much of the sun's ultraviolet radiation.

stratum basale (strat'um bah-say'lee) The deepest sublayer of the human epidermis, consisting of cells that continuously divide. Compare with *stratum corneum*.

stratum corneum The most superficial sublayer of the human epidermis. Compare with *stratum basale*.

strobilus (stroh'bil-us) (pl., *strobili*) In certain plants, a conelike structure that bears spore-producing sporangia.

stroke volume The volume of blood pumped by one ventricle during one contraction.

stroma A fluid space of the chloroplast, enclosed by the chloroplast inner membrane and surrounding the thylakoids; site of the reactions of the Calvin cycle.

stromatolite (stroh-mat'oh-lite) A columnlike rock that consists of many minute layers of prokaryotic cells, usually cyanobacteria.

structural formula A type of chemical formula that shows the spatial arrangement of the atoms in a molecule. Compare with *simplest formula* and *molecular formula*.

structural isomer One of two or more chemical compounds having the same chemical formula but differing in the covalent arrangement of their atoms, e.g., glucose and fructose.

style The neck connecting the stigma to the ovary of a carpel.

subsidiary cell In plants, a structurally distinct epidermal cell associated with a guard cell.

substance P A peptide neurotransmitter released by certain sensory neurons in pain pathways; signals the brain regarding painful stimuli; also stimulates other structures, including smooth muscle in the digestive tract.

substrate A substance on which an enzyme acts; a reactant in an enzymatically catalyzed reaction.

succession The sequence of changes in the species composition of a community over time. See *primary succession* and *secondary succession*.

sucker A shoot that develops adventitiously from a root; a type of asexual reproduction.

sulcus (sul'kus) (pl., *sulci*) A groove, trench, or depression, especially one occurring on the surface of the brain, separating the convolutions.

sulfhydryl group Functional group abbreviated —SH; found in organic compounds called thiols.

sum rule The rule for combining the probabilities of mutually exclusive events by adding their individual probabilities. Compare with *product rule*.

summation The process of adding together excitatory postsynaptic potentials (EPSPs).

suppressor T cell T lymphocyte that suppresses the immune response.

supraorbital ridge (soop"rah-or'bit-ul) The prominent bony ridge above the eye socket; ape skulls have prominent supraorbital ridges.

surface tension The attraction that the molecules at the surface of a liquid may have for one another.

survivorship The probability that a given individual in a population or cohort will survive to a particular age; usually presented as a survivorship curve.

survivorship curve A graph of the number of surviving individuals of a cohort, from birth to the maximum age attained by any individual.

suspensor (suh-spen′sur) In plant embryo development, a multicellular structure that anchors the embryo and aids in nutrient absorption from the endosperm.

sustainability See *environmental sustainability.*

swim bladder The hydrostatic organ in bony fishes that permits the fish to hover at a given depth.

symbiosis (sim-bee-oh′sis) An intimate relationship between two or more organisms of different species. See *commensalism, mutualism,* and *parasitism.*

sympathetic nervous system A division of the autonomic nervous system; its general effect is to mobilize energy, especially during stress situations; prepares the body for fight-or-flight response. Compare with *parasympathetic nervous system.*

sympatric speciation (sim-pa′trik) The evolution of a new species within the same geographic region as the parental species. Compare with *allopatric speciation.*

symplast A continuum consisting of the cytoplasm of many plant cells, connected from one cell to the next by plasmodesmata. Compare with *apoplast.*

synapomorphic characters See *shared derived characters.*

synapse (sin′aps) The junction between two neurons or between a neuron and an effector (muscle or gland).

synapsis (sin-ap′sis) The process of physical association of homologous chromosomes during prophase I of meiosis.

synaptic enhancement An increase in neurotransmitter release thought to occur as a result of calcium ion accumulation inside the presynaptic neuron.

synaptic plasticity The ability of synapses to change in response to certain types of stimuli. Synaptic changes occur during learning and memory storage.

synaptonemal complex The structure, visible with the electron microscope, produced when homologous chromosomes undergo synapsis.

synthetic theory of evolution See *modern synthesis.*

systematics The scientific study of the diversity of organisms and their evolutionary relationships. Taxonomy is an aspect of systematics. See *taxonomy.*

systemic acquired resistance A defensive response in infected plants that helps fight infection and promote wound healing.

systemic anaphylaxis A rapid, widespread allergic reaction that can lead to death.

systemic circulation The part of the circulatory system that delivers blood to and from the tissues and organs of the body. Compare with *pulmonary circulation.*

systems biology A field of biology that synthesizes knowledge of many small parts to understand the whole. Also referred to as *integrative biology* or *integrative systems biology.*

systole (sis′tuh-lee) The phase of the cardiac cycle when the heart is contracting. Compare with *diastole.*

T cell (T lymphocyte) The type of white blood cell responsible for a wide variety of immune functions, particularly cell-mediated immunity. T cells are processed in the thymus. Compare with *B cell.*

T cytotoxic cell (T$_C$) T lymphocyte that destroys cancer cells and other pathogenic cells on contact. Also known as *CD8 T cell* and *killer T cell.*

T helper cell (T$_H$) T lymphocyte that activates B cells (B lymphocytes) and stimulates T cytotoxic cell production. Also known as *CD4 T cell.*

T tubules Transverse tubules; system of inward extensions of the muscle fiber plasma membrane.

taiga (tie′gah) See *boreal forest.*

taproot system A root system consisting of a prominent main root with smaller lateral roots branching off it; a taproot develops directly from the embryonic radicle. Compare with *fibrous root system.*

target cell or tissue A cell or tissue with receptors that bind a hormone.

TATA box A component of a eukaryotic promoter region; consists of a sequence of bases located about 30 base pairs upstream from the transcription initiation site.

taxon A formal taxonomic group at any level, e.g., phylum or genus.

taxonomy (tax-on′ah-mee) The science of naming, describing, and classifying organisms; see *systematics.*

Tay–Sachs disease A serious genetic disease in which abnormal lipid metabolism in the brain causes mental deterioration in affected infants and young children; inheritance pattern is autosomal recessive.

tectorial membrane (tek-tor′ee-ul) The roof membrane of the organ of Corti in the cochlea of the ear.

telencephalon See *forebrain.*

telolecithal egg An egg with a large amount of yolk, concentrated at the vegetal pole. Compare with *isolecithal egg.*

telomerase A special DNA replication enzyme that can lengthen telomeric DNA by adding repetitive nucleotide sequences to the ends of eukaryotic chromosomes; typically present in cells that divide an unlimited number of times.

telomeres The protective end caps of chromosomes that consist of short, simple, noncoding DNA sequences that repeat many times.

telophase (teel′oh-faze or tel′oh-faze) The last stage of mitosis and of meiosis I and II when, having reached the poles, chromosomes become decondensed, and a nuclear envelope forms around each group.

temperate deciduous forest A forest biome that occurs in temperate areas where annual precipitation ranges from about 75 cm to 125 cm.

temperate grassland A grassland characterized by hot summers, cold winters, and less rainfall than is found in a temperate deciduous forest biome.

temperate rain forest A coniferous biome characterized by cool weather, dense fog, and high precipitation, e.g., the north Pacific coast of North America.

temperate virus A virus that integrates into the host DNA as a prophage.

temperature The average kinetic energy of the particles in a sample of a substance.

temporal isolation A prezygotic reproductive isolating mechanism in which genetic exchange is prevented between similar species because they reproduce at different times of the day, season, or year.

tendon A connective tissue structure that joins a muscle to another muscle, or a muscle to a bone. Tendons transmit the force generated by a muscle.

tendril A leaf or stem that is modified for holding or attaching onto objects.

tension–cohesion model The mechanism by which water and dissolved inorganic minerals are thought to be transported in xylem; water is pulled upward under tension because of transpiration while maintaining an unbroken column in xylem because of cohesion; also called *transpiration–cohesion model.*

teratogen Any agent capable of interfering with normal morphogenesis in an embryo, thereby causing malformations; examples include radiation, certain chemicals, and certain infectious agents.

terminal bud A bud at the tip of a stem. Compare with *axillary bud.*

termination (of protein synthesis) The final stage of protein synthesis, which occurs when a termination (stop) codon is reached, causing the completed polypeptide chain to be released from the ribosome. See *initiation* and *elongation.*

termination codon See *stop codon.*

territoriality Behavior pattern in which one organism (usually a male) stakes out a territory of its own and defends it against intrusion by other members of the same species and sex.

tertiary consumer See *carnivore.*

tertiary structure (of a protein) (tur′she-air″ee) The overall three-dimensional shape of a polypeptide that is determined by interactions involving the amino acid side chains. Compare with *primary, secondary,* and *quaternary protein structure.*

test A shell.

test cross The genetic cross in which either an F_1 individual, or an individual of unknown genotype, is mated to a homozygous recessive individual.

testis (tes′tis) (pl., *testes*) The male gonad that produces sperm and the male hormone testosterone; in humans and certain other mammals; the testes are located in the scrotum.

testosterone (tes-tos′ter-own) The principal male sex hormone (androgen); a steroid hormone produced by the interstitial cells of the testes; stimulates spermatogenesis and is responsible for primary and secondary sex characteristics in the male.

tetrad The chromosome complex formed by the synapsis of a pair of homologous chromosomes (i.e., four chromatids) during meiotic prophase I; also known as a *bivalent.*

tetrapods (tet′rah-podz) Four-limbed vertebrates: the amphibians, reptiles, birds, and mammals.

thalamus (thal′uh-mus) The part of the vertebrate brain that serves as a main relay center, transmitting information between the spinal cord and the cerebrum.

thallus (thal′us) (pl., *thalli*) The simple body of an alga, fungus, or nonvascular plant that lacks root, stems, or leaves, e.g., a liverwort thallus or a lichen thallus.

theca cells The layer of connective tissue cells that surrounds the granulosa cells in an ovarian follicle; stimulated by luteinizing hormone (LH) to produce androgens, which are converted to estrogen in the granulosa cells.

theory A widely accepted explanation supported by a large body of observations and experiments. A good theory relates facts that appear unrelated; it predicts new facts and suggests new relationships. Compare with *hypothesis.*

therapsids (ther-ap′sidz) A group of mammal-like reptiles of the Permian period; gave rise to the mammals.

thermal stratification The marked layering (separation into warm and cold layers) of temperate lakes during the summer. See *thermocline.*

thermocline (thur′moh-kline) A marked and abrupt temperature transition in temperate lakes between warm surface water and cold deeper water. See *thermal stratification.*

thermodynamics Principles governing energy transfer (often expressed in terms of heat transfer). See *first law of thermodynamics* and *second law of thermodynamics.*

thermoreceptor A sensory receptor that responds to heat.

thigmomorphogenesis (thig″moh-mor-foh-jen′uh-sis) An alteration of plant growth in response to mechanical stimuli, such as wind, rain, hail, and contact with passing animals.

thigmotropism (thig′moh-troh′pizm) Plant growth in response to contact with a solid object, such as the twining of plant tendrils.

threatened species A species in which the population is small enough for it to be at risk of becoming extinct throughout all or part of its range but not so small that it is in imminent danger of extinction. Compare with *endangered species.*

threshold level The potential that a neuron or other excitable cell must reach for an action potential to be initiated.

thrombocytes See *platelets.*

thylakoid lumen See *thylakoids.*

thylakoids (thy′lah-koidz) An interconnected system of flattened, saclike, membranous structures inside the chloroplast.

thymine (thy′meen) A nitrogenous pyrimidine base found in DNA.

thymus gland (thy′mus) An endocrine gland that functions as part of the lymphatic system; processes T cells; important in cell-mediated immunity.

thyroid gland An endocrine gland that lies anterior to the trachea and releases hormones that regulate the rate of metabolism.

thyroid hormones Hormones, including thyroxin, secreted by the thyroid gland; stimulate rate of metabolism.

tidal volume The volume of air moved into and out of the lungs with each normal resting breath.

tight junctions Specialized structures that form between some animal cells, producing a tight seal that prevents materials from passing through the spaces between the cells.

tissue A group of closely associated, similar cells that work together to carry out specific functions.

tissue culture The growth of tissue or cells in a synthetic growth medium under sterile conditions.

tissue engineering A developing technology that is striving to grow human tissues and organs (for transplantation) in cell cultures.

tissue fluid See *interstitial fluid.*

tolerance A decreased response to a drug over time.

Toll-like receptors Cell-surface receptors on phagocytes that recognize certain common features of classes of pathogens. See also *pathogen-associated molecular patterns (PAMPs).*

tonoplast The membrane surrounding a vacuole.

top-down processes Control of ecosystem function by trophic interactions, particularly from the highest trophic level. Compare *bottom-up processes.*

topoisomerases (toe-poe-eye-sahm′er-ases) Enzymes that relieve twists and kinks in a DNA molecule by breaking and rejoining the strands.

torpor An energy-conserving state of low metabolic rate and inactivity. See *estivation* and *hibernation.*

torsion The twisting of the visceral mass characteristic of gastropod mollusks.

total fertility rate The average number of children born to a woman during her lifetime.

totipotent (toh-ti-poh′tent) A term describing a cell or nucleus that contains the complete set of genetic instructions required to direct the normal development of an entire organism. Compare with *pluripotent.*

trace element An element required by an organism in very small amounts.

trachea (tray′kee-uh) (pl., *tracheae*) (1) Principal thoracic air duct of terrestrial vertebrates; windpipe. (2) One of the microscopic air ducts (or tracheal tubes) branching throughout the body of most terrestrial arthropods and some terrestrial mollusks.

tracheal tubes See *trachea.*

tracheid (tray′kee-id) A type of water-conducting and supporting cell in the xylem of vascular plants.

tract A bundle of nerve fibers within the central nervous system.

transcription The synthesis of RNA from a DNA template.

transcription factors DNA-binding proteins that regulate transcription in eukaryotes; include positively acting activators and negatively acting repressors.

transduction (1) The transfer of a genetic fragment from one cell to another, e.g., from one bacterium to another, by a virus. (2) In the nervous system, the conversion of energy of a stimulus to electrical signals.

transfer RNA (tRNA) RNA molecules that bind to specific amino acids and serve as adapter molecules in protein synthesis. The tRNA anticodons bind to complementary mRNA codons.

transformation (1) The incorporation of genetic material into a cell, thereby changing its phenotype. (2) The conversion of a normal cell to a cancer cell (called a malignant transformation).

transgenic organism A plant or animal that has foreign DNA incorporated into its genome.

translation The conversion of information provided by mRNA into a specific sequence of amino acids in a polypeptide chain; process also requires transfer RNA and ribosomes.

translocation (1) The movement of organic materials (dissolved food) in the phloem of a plant. (2) Chromosome abnormality in which part of one chromosome has become attached to another. (3) Part of the elongation cycle of protein synthesis in which a transfer RNA attached to the growing polypeptide chain is transferred from the A site to the P site.

transmembrane protein An integral membrane protein that spans the lipid bilayer.

transmission, neural The conduction of a neural impulse along a neuron or from one neuron to another.

transpiration The loss of water vapor from the aerial surfaces of a plant (i.e., leaves and stems).

transpiration–cohesion model See *tension–cohesion model.*

transport vesicles Small cytoplasmic vesicles that move substances from one membrane system to another.

transposon (tranz-poze′on) A DNA segment that is capable of moving from one chromosome to another or to different sites within the same chromosome; also called a *mobile genetic element.*

transverse tubules See *T tubules.*

triacylglycerol (try-ace″il-glis′er-ol) The main storage lipid of organisms, consisting of a glycerol combined chemically with three fatty acids; also called *triglyceride.* Compare with *monoacylglycerol* and *diacylglycerol.*

tricarboxylic acid (TCA) cycle See *citric acid cycle.*

trichome (try′kohm) A hair or other appendage growing out from the epidermis of a plant.

tricuspid valve See *atrioventricular valve.*

triglyceride See *triacylglycerol.*

triose A sugar molecule containing three carbons.

triplet A sequence of three nucleotides that serves as the basic unit of genetic information.

triplet code The sequences of three nucleotides that compose the codons, the units of genetic information in mRNA that specify the order of amino acids in a polypeptide chain.

trisomy (try′sohm-ee) The condition in which each chromosome has two copies, except one, which is present in triplicate; the cell contains one more chromosome than the diploid number. Compare with *monosomy* and *disomy.*

trochophore larva (troh′koh-for) A larval form found in mollusks and many polychaetes.

trophic level (troh′fik) Each sequential step of matter and energy in a food web, from producers to primary, secondary, or tertiary consumers; each organism is assigned to a trophic level based on its primary source of nourishment.

trophoblast (troh′foh-blast) The outer cell layer of a late blastocyst, which in placental mammals gives rise to the chorion and to the fetal contribution to the placenta.

tropic hormone (trow′pic) A hormone, produced by one endocrine gland, that targets another endocrine gland.

tropical dry forest A tropical forest where enough precipitation falls to support trees but not enough to support the lush vegetation of a tropical rain forest; often occurs in areas with pronounced rainy and dry seasons.

tropical rain forest A lush, species-rich forest biome that occurs in tropical areas where the climate is very moist throughout the year. Tropical rain forests are also characterized by old, infertile soils.

tropism (troh′pizm) In plants, a directional growth response that is elicited by an environmental stimulus.

tropomyosin (troh-poh-my′oh-sin) A muscle protein involved in regulation of contraction.

true-breeding strain A genetic strain of an organism in which all individuals are homozygous at the loci under consideration.

tube feet Structures characteristic of echinoderms; function in locomotion and feeding.

tuber A thickened end of a rhizome that is fleshy and enlarged for food storage, e.g., white potato.

tubular transport maximum (Tm) The maximum rate at which a substance is reabsorbed from the renal tubules of the kidney.

tumor A mass of tissue that grows in an uncontrolled manner; a neoplasm.

tumor necrosis factors (TNFs) Cytokines that kill tumor cells and stimulate immune cells to initiate an inflammatory response.

tumor suppressor gene A gene (also known as an *anti-oncogene*) whose normal role is to block cell division in response to certain growth-inhibiting factors; when mutated, may contribute to the formation of a cancer cell. Compare with *oncogene.*

tundra (tun′dra) A treeless biome between the boreal forest in the south and the polar ice cap in the north that consists of boggy plains covered by lichens and small plants. Also called *arctic tundra.* Compare with *alpine tundra.*

tunicates Chordates belonging to subphylum Urochordata; sea squirts.

turgor pressure (tur′gor) Hydrostatic pressure that develops within a walled cell, such as a plant cell, when the osmotic pressure of the cell's contents is greater than the osmotic pressure of the surrounding fluid.

Turner syndrome An inherited condition in which only one sex chromosome (an X chromosome) is present in cells; karyotype is designated XO; affected individuals are sterile females.

tyrosine kinase An enzyme that phosphorylates the tyrosine part of proteins.

tyrosine kinase receptor A plasma membrane receptor that phosphorylates the tyrosine part of proteins; when a ligand binds to the receptor, the conformation of the receptor changes and it may phosphorylate itself as well as other molecules; important in immune function and serves as a receptor for insulin.

ultimate causes (of behavior) Evolutionary explanations for why a certain behavior occurs. Compare with *proximate causes of behavior.*

ultrasound imaging A technique in which high-frequency sound waves (ultrasound) are used to provide an image (sonogram) of an internal structure.

ultrastructure The fine detail of a cell, generally only observable by use of an electron microscope.

umbilical cord In placental mammals, the organ that connects the embryo to the placenta.

uniform dispersion The spatial distribution pattern of a population in which individuals are regularly spaced. Compare with *random dispersion* and *clumped dispersion.*

unsaturated fatty acid See *fatty acid.*

upstream promoter elements (UPEs) Components of a eukaryotic promoter, found upstream of the RNA polymerase-binding site; the strength of a promoter is affected by the number and type of UPEs present.

upwelling An upward movement of water that brings nutrients from the ocean depths to the surface. Where upwelling occurs, the ocean is very productive.

uracil (yur´ah-sil) A nitrogenous pyrimidine base found in RNA.

urea (yur-ee´ah) The principal nitrogenous excretory product of mammals; one of the water-soluble end products of protein metabolism.

ureter (yur´ih-tur) One of the paired tubular structures that conducts urine from the kidney to the bladder.

urethra (yoo-ree´thruh) The tube that conducts urine from the bladder to the outside of the body.

uric acid (yoor´ik) The principal nitrogenous excretory product of insects, birds, and reptiles; a relatively insoluble end product of protein metabolism; also occurs in mammals as an end product of purine metabolism.

urinary bladder An organ that receives urine from the ureters and temporarily stores it.

urinary system The body system in vertebrates that consists of kidneys, urinary bladder, and associated ducts.

urochordates A subphylum of chordates; includes the tunicates.

uterine tube (yoo´tur-in) See *oviduct.*

uterus (yoo´tur-us) The hollow, muscular organ of the female reproductive tract in which the fetus undergoes development.

utricle The structure within the vestibule of the vertebrate inner ear that, along with the saccule, houses the receptors of static equilibrium.

vaccine (vak-seen´) A commercially produced, weakened or killed antigen associated with a particular disease that stimulates the body to make antibodies.

vacuole (vak´yoo-ole) A fluid-filled, membrane-enclosed sac found within the cytoplasm; may function in storage, digestion, or water elimination.

vagina The elastic, muscular tube, extending from the cervix to its external opening, that receives the penis during sexual intercourse and serves as the birth canal.

valence electrons The electrons in the outer electron shell, known as the *valence shell,* of an atom; in the formation of a chemical bond, an atom can accept electrons into its valence shell or donate or share valence electrons.

van der Waals interactions Weak attractive forces between atoms; caused by interactions among fluctuating charges.

vas deferens (vas def´ur-enz) (pl., *vasa deferentia*) One of the paired sperm ducts that connects the epididymis of the testis to the ejaculatory duct.

vascular cambium A lateral meristem that produces secondary xylem (wood) and secondary phloem (inner bark). Compare with *cork cambium.*

vascular tissue system The tissues specialized for translocation of materials throughout the plant body, i.e., the xylem and phloem.

vasoconstriction Narrowing of the diameter of blood vessels.

vasodilation Expansion of the diameter of blood vessels.

vector (1) Any carrier or means of transfer. (2) Agent, e.g., a plasmid or virus, that transfers genetic information. (3) Agent that transfers a parasite from one host to another.

vegetal pole The yolky pole of a vertebrate or echinoderm egg. Compare with *animal pole.*

vein (1) A blood vessel that carries blood from the tissues toward a chamber of the heart (compare with *artery*). (2) A strand of vascular tissue that is part of the network of conducting tissue in a leaf.

veliger larva The larval stage of many marine gastropods (snails) and bivalves (e.g., clams); often is a second larval stage that develops after the trochophore larva.

ventilation The process of actively moving air or water over a respiratory surface.

ventral Toward the lowermost surface or belly of an animal. Compare with *dorsal.*

ventricle (1) A cavity in an organ. (2) One of the several cavities of the brain. (3) One of the chambers of the heart that receives blood from an atrium.

vernalization (vur″nul-uh-zay´shun) The induction of flowering by a low-temperature treatment.

vertebrates A subphylum of chordates that possess a bony vertebral column; includes fishes, amphibians, reptiles, birds, and mammals.

vesicle (ves´ih-kl) Any small sac, especially a small, spherical, membrane-enclosed compartment, within the cytoplasm.

vessel element A type of water-conducting cell in the xylem of vascular plants.

vestibular apparatus Collectively, the saccule, utricle, and semicircular canals of the inner ear.

vestigial (ves-tij´ee-ul) Rudimentary; an evolutionary remnant of a formerly functional structure.

vibrio A spirillum (spiral-shaped bacterium) that is shaped like a comma. Compare with *spirillum, spirochete, bacillus,* and *coccus.*

villus (pl., *villi*) A multicellular, minute, elongated projection from the surface of an epithelial membrane, e.g., villi of the mucosa of the small intestine.

viroid (vy´roid) A tiny, naked, infectious particle consisting only of nucleic acid.

virulence Properties that render an infectious agent pathogenic (and often lethal) to its host. Compare with *avirulence.*

virus A tiny pathogen consisting of a core of nucleic acid usually encased in protein and capable of infecting living cells; a virus is characterized by total dependence on a living host.

viscera (vis´ur-uh) The internal body organs, especially those located in the abdominal or thoracic cavities.

visceral mass The concentration of body organs (viscera) located above the foot in mollusks.

vital capacity The maximum volume of air a person exhales after filling the lungs to the maximum extent.

vitamin A complex organic molecule required in very small amounts for normal metabolic functioning.

vitelline envelope An acellular covering of the eggs of certain animals (e.g., echinoderms), located just outside the plasma membrane.

viviparous (vih-vip´er-us) Bearing living young that develop within the body of the mother. Compare with *oviparous* and *ovoviviparous.*

voltage-activated ion channels Ion channels in the plasma membrane of neurons that are regulated by changes in voltage. Also called *voltage-gated channels.*

vomeronasal organ In mammals, an organ in the epithelium of the nose, made up of specialized chemoreceptor cells that detect pheromones.

vulva The external genital structures of the female.

warning coloration See *aposematic coloration.*

water mold A funguslike protist with a body consisting of a coenocytic mycelium that reproduces asexually by forming motile zoospores and sexually by forming oospores.

water potential Free energy of water; the water potential of pure water is zero and that of solutions is a negative value. Differences in water potential are used to predict the direction of water movement (always from a region of less negative water potential to a region of more negative water potential).

water vascular system Unique hydraulic system of echinoderms; functions in locomotion and feeding.

wavelength The distance from one wave peak to the next; the energy of electromagnetic radiation is inversely proportional to its wavelength.

weathering processes Chemical or physical processes that help form soil from rock; during weathering processes, the rock is gradually broken into smaller and smaller pieces.

Western blot A technique in which proteins, previously separated by gel electrophoresis, are transferred to paper. A specific labeled antibody is generally used to mark the location of a particular protein. Compare with *Southern blot* and *Northern blot.*

white matter Nervous tissue in the brain and spinal cord that contains myelinated axons. Compare with *gray matter.*

wild type The phenotypically normal (naturally occurring) form of a gene or organism.

wobble The ability of some tRNA anticodons to associate with more than one mRNA codon; in these cases the 5′ base of the anticodon is capable of forming hydrogen bonds with more than one kind of base in the 3′ position of the codon.

work Any change in the state or motion of matter.

X-linked gene A gene carried on an X chromosome.

X-ray diffraction A technique for determining the spatial arrangement of the components of a crystal.

xylem (zy′lem) The vascular tissue that conducts water and dissolved minerals in plants.

XYY karyotype Chromosome constitution that causes affected individuals (who are fertile males) to be unusually tall, with severe acne.

yeast A unicellular fungus (ascomycete) that reproduces asexually by budding or fission and sexually by ascospores.

yolk sac One of the extraembryonic membranes; a pouchlike outgrowth of the digestive tract of embryos of certain vertebrates (e.g., birds) that grows around the yolk and digests it. Embryonic blood cells are formed in the mammalian yolk sac, which lacks yolk.

zero population growth Point at which the birth rate equals the death rate. A population with zero population growth does not change in size.

zona pellucida (pel-loo′sih-duh) The thick, transparent covering that surrounds the plasma membrane of a mammalian ovum.

zooflagellate A unicellular, nonphotosynthetic protozoon that has one or more long, whiplike flagella.

zooplankton (zoh″oh-plank′tun) The nonphotosynthetic organisms present in plankton, e.g., protozoa, tiny crustaceans, and the larval stages of many animals. See *plankton.* Compare with *phytoplankton.*

zoospore (zoh′oh-spore) A flagellated motile spore produced asexually by certain algae, chytrids, and water molds and other protists.

zooxanthellae (zoh″oh-zan-thel′ee) (sing., *zooxanthella*) Endosymbiotic, photosynthetic dinoflagellates found in certain marine invertebrates; their mutualistic relationship with corals enhances the corals' reef-building ability.

zygomycetes (zy″gah-my′seats) Fungi characterized by the production of nonmotile asexual spores and sexual zygospores.

zygosporangium (zy″gah-spor-an′gee-um) A thick-walled sporangium containing a zygospore.

zygospore (zy′gah-spor) A sexual spore produced by a zygomycete.

zygote The 2*n* cell that results from the union of *n* gametes in sexual reproduction. Species that are not polyploid have haploid gametes and diploid zygotes.

zygotic genes Genes that are transcribed after fertilization, either in the zygote or in the embryo. Compare with *maternal effect genes.*

Index

ostracoderm fossils, 677
Camellias, 613
Camera eye, *911*
Camouflage
  ambush as predator strategy and, 1153, *1153*
  of amphibians, 683
  as defense (cryptic coloration), 664, 1154, *1154*
  of leafy sea dragon, *681*
  of sponge crab, *660*
cAMP. *See* Cyclic adenosine monophosphate
Campis radicans (trumpet vine), *772*
Canaliculi, in bone, 815, 832
Canary grass coleoptiles, 792–793, *793*
Cancer, 816
  alcohol use/abuse and, 887
  anti-cancer agent, 575
  breast, 958, 1062, 1063, *1063*
  cervical, 505, *505*, 816, 1075, *1076*
  colorectal, 816, 1000, *1000*
  immune system and, 950, 960–961, *961*
  leukemia, 925
  skin, 816, 829
  virus causing, *505*
Cancer Genome Project, 387
Cancer vaccines, 960–961
Candida, 576
Candidiasis albicans (yeast infection), *1076*
Canidae (family), *491*, 492
Canine distemper, 505, *505*
Canines (canine teeth), 993–994, *994*
Canis familiaris (domestic dog), 491
Canis latrans (coyote), *491*, *994*, 1158
Canis lupus familiaris (dog), *491*
Cannon, Walter, 817
Cap, of mushrooms, 568
Capacitation, 1082
Cape buffalo (*Syncerus caffer*), *1111*, 1114
Cape gannets (*Morus capensis*), *1128*
Cape ground squirrel, 823
Cape scrub, in Africa, 1196
Capillaries, *979*
  blood pressure in, *936*
  glomerular, 1018–1020, *1018–1021*
  in invertebrates, 921
  lymph, 939, *939*
  peritubular, 1019, *1019*
  pulmonary, 937
  in vertebrates, 922, 925–926, *926*, *927*
Capillary bed, *939*
Capillary network, 925, 927
Capillary permeability, in inflammation, 949
Capping, *See also* 5' cap
Capsella bursa-pastoris (shepherd's purse), *776*
Capsid, 501–502
Capsomeres, 501
Capsule
  bacterial, 513
  of moss sporophyte, 586, *587*
  simple fruit, *778*, 779
Captive-breeding programs, 1212, 1218, 1222
Carageenan, 545
Carapace, of lobsters, 660
Carbohydrate metabolism, 1044–1045
Carbohydrates, 1001
  complex, 1001
  diabetes and, 1044–1045
  digestion of, 997–998, *998*
  refined, 1001
Carbon,
  in early biosphere, 453
  in soil for plant growth, 761, *762*
Carbon cycle, *1173*, 1173–1174
Carbon dioxide,
  breathing regulation and, 981
  in carbon cycle, *1173*, 1173–1174
  concentration in lungs, 978–981, 982
  deforestation and, 1224
  diffusion of, 972–973
  fixation, 1180
  fixation by cyanobacteria, 523
  global warming and, 1226–1227, *1227*
  human activities affecting levels of, 1174
  hyperventilation and, 982
  initiation of breathing in neonate, 1097
  partial pressure of, *979*, *979*

plants and, 582, 589
  in soil, 757, 760
  stomatal opening and, 724
  transport of, 980–981, *981*
Carbon monoxide, *856*, 859
Carbon tetrachloride, 1230
Carbon-14, in fossil dating, 400
Carbonate, 1173
Carbonic acid, 757–758, 980
Carbonic anhydrase, 980, *981*
Carboniferous period,
  amniote radiation, 685
  coal formation and, 590, 591, *591*
  hexapod fossils from, 661
Carboxypeptidase, 998, *998*
Carcharodon carcharias (great white shark), *678*
Carcinogens (cancer-causing agents),
  576, 816
Carcinomas, 816
Cardamine pratensis (bitter cress), *781*, *781*
Cardiac arrest, 935
Cardiac center, of brain, 932, *932*, 936
Cardiac cycle, 931, *931*
Cardiac murmur, 932
Cardiac muscle, *815*, 816–817, 842
  action potential of, 930, *930*
  neurotransmitter action on, 932, *932*, 933
Cardiac output (CO), 932, *932*, 934
Cardiac valves, 929, *929*
Cardiopulmonary resuscitation (CPR), 982
Cardiovascular disease, 919–920, 934
  diabetes and, 935
  fat/cholesterol and, 920, 934, 935, 1001–1002,
    1003
  smoking and, 935, 984
Cardiovascular system, 817, *818*, *820*, 919–920
  evolution of, 927–928, *928*
Cardon cactus (*Pachycereus*), *772*
Carmona, Richard H., 984
Carnivores (order Carnivora), 485, 694, 990,
    991, *994*, 1167
  pursuit of prey and, 1153
Carnivorous plants, 728, *728*
Carnivorous reptiles, 686
Carolina parakeet, 1218
Carotenoids, 1004
  in brown algae, 544
  in diatoms, 541
  in dinoflagellates, 539
  in euglenoids, 536
  in golden algae, 543
  in green algae, 545
  in leaf abscission, 726
  in red algae, 545
Carotid arteries, 937, *937*, *938*
Carotid bodies, 981
Carpels, 608, *609*, 610, *768*, 769
  evolution of, 611, 613, *613*
Carrier-mediated transport,
  and mineral transport in plant roots, 752
Carrion plant, 772
Carrots (*Daucus carota*), *748*, 748
Carrying capacity (K), 1130–1131, *1131*, 1135,
    1140
Carson, Rachel, 1171
Cartilage, *813*, 814–815
  articular, 833
  bone compared to, 680
Cartilaginous fishes, *675*, *676*, 678, 678–679, *679*,
    1015, 1016
  skeletons of, 831
Caryopses (grains), *778*, 779
Cassowarys, 690
Castings, earthworm, 760
Castration, 1058
Catabolism/catabolic pathways, 1001
Catalase, 1003
Catalogue of Life, 483
Catalpa bignonioides (southern catalpa), *717*
Catalysts,
  See also Enzymes
Catchment, *1185*, 1186
Catecholamines, 857, 1046. *See also* Epinephrine;
    Norepinephrine

Caterpillars, 798
  tent, 1111
Cation exchange, 759, *759*
Cations,
  clay particles binding, 759, *759*
Catkins, of oaks, 698
Cats, 692, *694*
  communication by scent, 1112
  domestic (*Felis catus*), *485*
Cattle, *694*
  as dominant species, 1158
  ranching, 1198, 1225
  See also Cows
Caudal (posterior) end of animal, *623*, 624
Caudipteryx fossils, *688*, 689
Cavalier-Smith, Thomas, 563
Cavernous bodies, penile, *1054*, 1057, *1057*
CD4 (T helper/T_H) cells, 950, *951*, 952, 953, *953*,
    962
CD4 receptors, on T helper cells, 509
CD8 (T cytotoxic/T_C) cells, 950, *951*, 952, *953*,
    960, 961
Cecum, *993*, 1000
Cedar, Western red (*Thuja plicata*), and
    mycorrhizae, 570, *572*
Cell(s),
  aging of, 1097
  destruction by virus lytic cycle, 503, *504*
  evolution of viruses from, 502
  neural crest, 674
  prokaryotic, *500*, 512–518
  size of, 807
  specialized, 620
  of sponges, 631
  See also Animal cells; Eukaryotes/eukaryotic
    cells; Plant cells; Prokaryotes/prokaryotic
    cells
Cell biologists, 532
Cell body, of neuron, 817, 847, *847*
Cell constant animals, 653
Cell death
  programmed (apoptosis),
    948, 967, 1081
    in aging, 1097–1098
    natural killer cells causing, 948
    T cell, 950
Cell determination, 1081
Cell differentiation,
  animal cells, 1081
  plant cells, 703, 710
Cell division
  in animal development, 1080, 1081, 1083–
    1086, *1084–1087*
  in eukaryotic cells (*see also* Cell cycle)
  in plant cells, 710
    in apical meristems (primary growth), 710,
      711, 711–712
    cytokinins promoting, 795–797, *796*, *797*
    in lateral meristems (secondary growth),
      711, 712, 712
Cell elongation, in plant growth, 710, *710*, 711
  auxins in, 792–794, *793–795*, 795
  ethylene in, 797
  gibberellins in, 795
Cell junctions,
  See also specific type
Cell membrane. *See* Cell(s); Plasma membrane
  (cell)
Cell signaling,
  in *Dictyostelium*, 551
  endocrine, 1031–1032, *1033*
  in immune response, 944–945
  neurons in, 845 (*see also* Neural signaling)
  in plants
    hormones in, 791, 791–792, *792*, 798–799
    pest control and, 798–799
  in rhizobia–plant relationship, 756–757, *758*
Cell structure,
  See also Cell walls
Cell turgor, 762
Cell typing, in tissue transplants, 964

Cell walls,
  in fungi, 556, 558, 560
  plant, 702–706, *704*, *705*
  prokaryotic/bacterial, 512–513, *513*,
    514
  protists, 540, 541, 545
Cell-mediated immunity, 950, *951*, 952, *953*
Cell-surface receptors, 1033–1035, *1034*
  immune response and, 952
Cellular respiration, 978
  aerobic,
    in mitochondria, 970
  See also ATP
Cellular slime molds, *533–535*, *550*, 551, A-3
Cellulose, 705–706, 750
  decomposed by fungi, 570
  protists and, 536
Cellulose microfibril, 705
Cenozoic era,
  mammal evolution in, 691
Center of origin, 1208
Centipedes (class Chilopoda), 657, *657*, *658*
Central canal, spinal cord, 871, *872*
Central nervous system (CNS), 861, 866,
    867, *868*, 871–880, 894
  drugs affecting, 886–887, 888–889
  glial cells in, *845*, 845–846, 847, 848–849,
    849
  in humans, 871, 871–880
  neuron production, 847–848
  structures of, 846, *868*, 868–883
  See also Brain; Spinal cord
Central sulcus, 870, 872, *875*
Century plant, 1135, *1135*
Cephalaspidomorphi (class), *675*, 676, *676*, 677
Cephalic (anterior) end of animal, *623*, 624
Cephalization, 623, 642, 663, 674, 866
Cephalochordata (subphylum), 672, 673, *673*,
    675. *See also* Lancelets
Cephalopods (class Cephalopoda), 646, *646*,
    647, 649
  circulatory system of, 921
  nervous system of, 866–867
  photoreceptors of, 909
Cephalosporin antibiotic, 575
Cephalothorax
  of chelicerates, 657, *658*
  of lobsters, *660*, 661
Ceratitis capitata (Mediterranean fruit fly), *910*
Ceratium, *539*
Cercozoa, 534, *535*, 546–547, *549*, A-3
Cereal crops, 608
Cerebellum, 868–869, *870*, 874, 875
Cerebral cortex, 690, 870, 872, 873, 874, *874*, 875.
    See also Cerebrum
Cerebral ganglia
  in annelids, 866, *867*
  in arthropods, 655, 866
  in earthworms, 651, *651*, 866, *867*
  evolution of, 866
  in flatworms, 866, *867*
  in human autonomic nervous system, 884
  in human central nervous system, *885*
  in mollusks, 866–867
Cerebral hemispheres, 870, 872, 873
Cerebrospinal fluid (CSF), 849, 871, *871*
Cerebrovascular accident (CVA/stroke), 934
Cerebrum, 690, 869–870, *870*, 872–873, *874*–
    877
Cervarix, 1075
Cervical cancer, 505, *505*, 816, 1075, *1076*
Cervical mucus, and fertilization, 1068
Cervical region, of vertebral column, 831
Cervix, uterine, *1060*, 1061, 1068, 1070
Cesarean section, 1071
Cetaceans (order Cetacea), *694*,
    1122. *See also specific type*
Chalk formations, foraminiferan tests forming,
    546, *549*
Chambered nautilus, 649
Channels. *See* Ion channels
Chaos carolinense, *549*
Chaparral, 1196, *1196*
Chara (stonewort), *546*, 583